Haematology

OXFORD
TRANSFUSION & TRANSPLANTATION SCIENCE
EDITED BY Robin Knight

OXFORD
BIOMEDICAL SCIENCE PRACTICE
EXPERIMENTAL & PROFESSIONAL SKILLS

OXFORD
CYTOPATHOLOGY
EDITED BY Behdad Shambayati

OXFORD
CLINICAL BIOCHEMISTRY
EDITED BY Nessar Ahmed

OXFORD
DATA HANDLING AND ANALYSIS
Andrew Blann

OXFORD
CLINICAL IMMUNOLOGY
EDITED BY Angela Hall, Chris Scott & Matthew Buckland

OXFORD
MEDICAL MICROBIOLOGY
EDITED BY Michael Ford

OXFORD
CELL STRUCTURE & FUNCTION
EDITED BY Guy Orchard & Brian Nation

OXFORD
HISTOPATHOLOGY
EDITED BY Guy Orchard & Brian Nation

fundamentals OF
biomedical science

Fundamentals of Biomedical Science

Haematology

Third edition

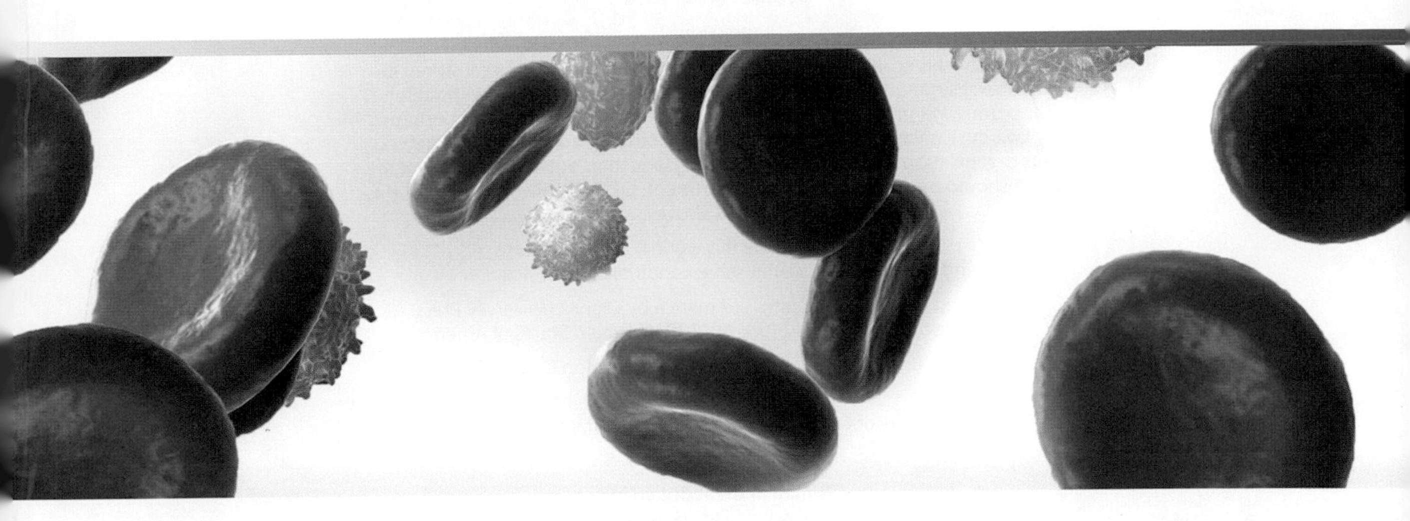

Dr Gary W. Moore
BSc DBMS CSci FIBMS CBiol FRSB CertMHS
Director of Clinical Diagnostic Research, Technoclone GmbH, Vienna
Honorary Consultant Biomedical Scientist, Specialist Haemostasis Unit, Addenbrooke's Hospital, Cambridge
Visiting Professor, Faculty of Science and Technology, Middlesex University

Gavin Knight
BSc (Hons) MSc SFHEA FIBMS CSci
Principal Lecturer, School of Pharmacy and Biomedical Science
Programme Lead, Biomedical Sciences
University of Portsmouth

Dr Andrew D. Blann
PhD FRCPath FRCPE FIBMS CSci
Institute of Biomedical Science
Coldbath Square, London

OXFORD
UNIVERSITY PRESS

OXFORD
UNIVERSITY PRESS
Great Clarendon Street, Oxford, OX2 6DP,
United Kingdom

Oxford University Press is a department of the University of Oxford.
It furthers the University's objective of excellence in research, scholarship,
and education by publishing worldwide. Oxford is a registered trade mark of
Oxford University Press in the UK and in certain other countries

First edition 2010
Second edition 2016

Impression: 1

Published in the United States of America by Oxford University Press
198 Madison Avenue, New York, NY 10016, United States of America

British Library Cataloguing in Publication Data
Data available

Library of Congress Control Number: 2020948776

ISBN 978-0-19-882609-5

Printed in Great Britain by
Bell & Bain Ltd., Glasgow

Contents

An introduction to the Fundamentals of Biomedical Science series

Laboratory scientists form the foundation of modern health care, from cancer screening to diagnosing HIV, from blood transfusion for surgery to infection control. Without laboratory scientists, the diagnosis of disease, the evaluation of the effectiveness of treatments, and research into the causes and cures of disease would not be possible. However, the path to becoming a laboratory scientist is a challenging one: trainees must not only assimilate knowledge from a range of disciplines, but must understand—and demonstrate—how to apply this knowledge in a practical, hands-on environment.

The *Fundamentals of Biomedical Science* series is written to reflect the challenges of laboratory science education and training today. It blends essential basic science with insights into laboratory practice to show you how an understanding of the biology of disease is coupled with the analytical approaches that lead to diagnosis. Produced in collaboration with the Institute of Biomedical Science, the series provides coverage of the full range of disciplines to which a laboratory scientist might be exposed.

Learning from the series

The *Fundamentals of Biomedical Science* series draws on a range of learning features to help readers master both biomedical science theory and biomedical science practice.

Case studies illustrate how the biomedical science theory and practice presented throughout the series relate to situations and experiences that are likely to be encountered routinely in the laboratory. Answers to questions posed by some case studies are available in the book's online resources.

Patient history

- A 32-year-old woman suffered a deep vein thrombosis during her first pregnancy.
- She reported a family history of thrombosis.
- Coagulation screening and hereditary thrombophilia screening were performed one

Additional information to augment the main text appears in **boxes**.

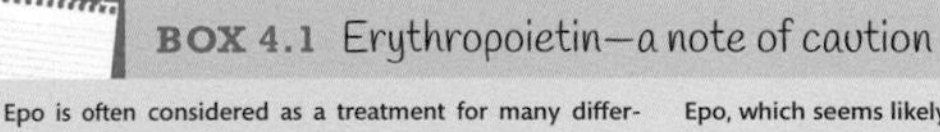

Epo is often considered as a treatment for many different types of anaemia. However, this may be unwise as the indiscriminate use of Epo may not be advisable, especially in patients with cancer and/or end-stage renal disease (ESRD). One of the characteristics of cancer is that it is often associated with anaemia, so that it has been believed (quite reasonably) that such patients would benefit from treatment with Epo. However, an additional characteristic of tumours

Epo, which seems likely to explai adverse effect of this growth facto due mainly to tumour progressio tion that the anaemia often asso beneficially treated with (recomb lenged by a study which reported in those using this hormone. An although Epo did indeed increas

Method boxes walk through the key protocols that the reader is likely to encounter in the laboratory on a regular basis.

METHOD Generating banding patterns in cytoge investigations

- A variety of different methods can be used to produce banding patterns on chromosomes of interest. It is important when considering these techniques to remember that metaphase chromosomes are required. Using these techniques, cells are encouraged to progress through the

morphology because of its r staining characteristics.

- *Q-banding* uses a fluorescent s to produce a distinct banding p tern is not visible to the nake

Key Points

Erythropoiesis is the process of the development of the red blood cell. Beginning in the bone marrow, immature stages include the erythroblast, the nucleated red blood cell, and the reticulocyte. The key growth factor is Epo. Under-production of red cells leads to anaemia, while over-production leads to erythrocytosis and polycythaemia.

Key points reinforce the key concepts that the reader should master from having read the material presented, while **Chapter summary** points act as end-of-chapter checklists for readers to verify that they have remembered correctly the principal themes and ideas presented within each chapter.

proliferation and differentiation.

polycythaemia vera
A malignant disease of increased numbers of red blood cells.

myelofibrosis
A malignant disease of the bone marrow.

essential

other pathways, such as that of stem cell factor, can also part-re
proteins can negatively regulate Epo mRNA via binding in the E

Once in the circulation, Epo interacts with its target cells of er
structure in the cell membrane—the Epo receptor (EpoR). Lo
a 56kDa protein, the EpoR being a dimer that spans the cell m
at the cell surface of erythroid cells is tightly regulated—arou
progenitor and 300 on a late-stage erythroblast. Upon bin
enzyme, Janus kinase 2 (**JAK2**) attached to the intracellular tai
this in turn phosphorylates a tyrosine residue. This phosphor

Key terms provide on-the-page explanations of terms with which the reader may not be familiar; in addition, each title in the series features a **glossary**, in which the key terms featured in that title are collated.

SELF-CHECK 4.1

How does the red blood cell differ from almost all other cells of the body, and how does this relate to function?

SELF-CHECK 4.2

What are the principal differences between the reticulocyte and the mature red blood cell?

Self-check questions throughout each chapter provide the reader with a ready means of checking that they have understood the material they have just encountered; answers to self-check questions are available in the book's online resources.

Discussion questions

4.1 How is the red blood cell adapted for its purpose?

4.2 What are the different types of Hb within the cell and why do they change as the embryo develops into a foetus, an infant, and an adult?

4.3 Explain the different features of the morphology of the red cell.

Discussion questions are provided at the end of each chapter to encourage the reader to analyse and reflect on the material they have just read.

Cross references

We discuss anaemia in Chapters 5 and 6, leukaemia in Chapters 10–12, and wider aspects of erythrocytosis and polycythaemia vera in Chapter 11.

hormones (thyroxine and triiodothyronine) stimulate red blo
part responsible for the anaemia of hypothyroidism.

Fully functioning erythropoiesis also requires a host of mine
vitamin C, copper, vitamin E, vitamins B_6 and B_{12}, thiamin
to promote stress erythroid cell expansion and differentiati
monocytes, inducing their differentiation to erythroblasti
may result from a deficiency of any of these micronutrients
Chapters 5 and 6.

Cross references help the reader to see biomedical science as a unified discipline, making the connections between topics presented within each volume, and across all volumes in the series.

Online learning materials

Each title in the *Fundamentals of Biomedical Science* series is supported by online resources, which features additional materials for students, trainees, and lecturers.

www.oup.com/he/moore-fbms3e

Guides to key experimental skills and methods

Video walk-throughs of key experimental skills are provided to help you master the essential skills that are the foundation of biomedical science practice.

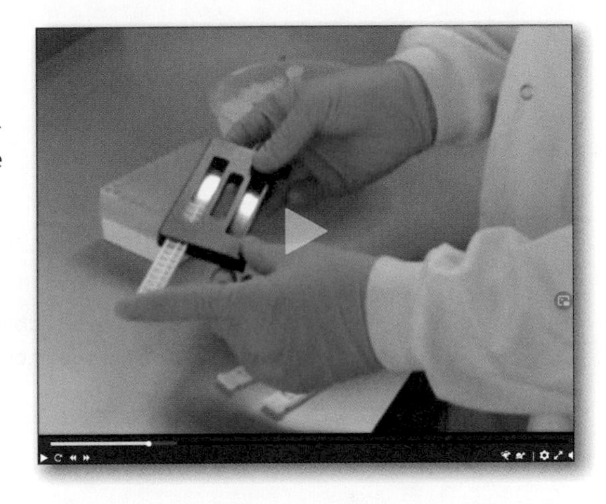

Biomedical science in practice

Interviews with practising laboratory scientists working in a range of disciplines give a valuable insight into the reality of work in a biomedical science laboratory.

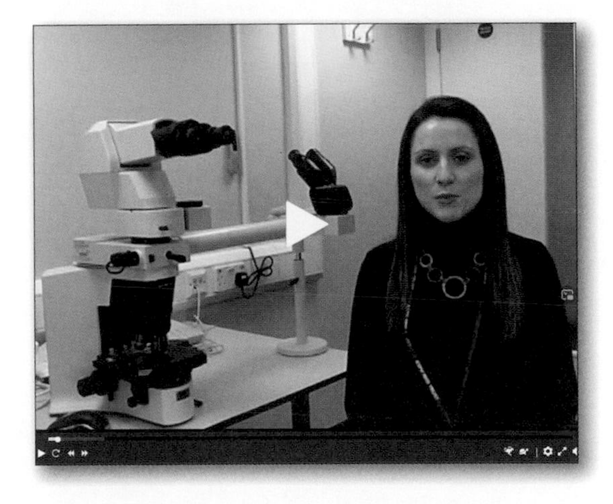

Lecturer support materials

The online resources for each title in the series also features figures from the book in electronic format, for registered adopters to download for use in lecture presentations, and other educational resources.

To register as an adopter visit **www.oup.com/he/moore-fbms3e** and follow the on-screen instructions.

Any comments?

We welcome comments and feedback about any aspect of this series.
Just visit **www.oxfordtextbooks.co.uk/orc/feedback** and share your views.

Contributors

Dr Andrew D. Blann PhD, FRCPath, FRCPE, FIBMS, CSci
Institute of Biomedical Science, Coldbath Square, London

Dr David A. Gurney MSc, DBMS, CSci, FIBMS
Lead Scientist/Operational Manager, Coagulation Laboratory, Department of Haematology, University Hospitals NHS Trust Birmingham, UK

Ms Pam Holtom
Senior Biomedical Scientist, Heartlands Hospital, Birmingham, UK

Dr Ian Jennings PhD, CSci, FIBMS
Scientific Programme Manager, UKNEQAS (Blood Coagulation), Sheffield, UK

Gavin Knight BSc (Hons) MSc SFHEA FIBMS CSci
Principal Lecturer, School of Pharmacy and Biomedical Science, Programme Lead, Biomedical Sciences, University of Portsmouth

Dr Sukhjinder Marwah PhD, CSci, FIBMS
Clinical Scientist, Department of Haematology and Blood Transfusions, City Hospital, Birmingham, UK

Dr Gary W. Moore BSc, DBMS, CSci, FIBMS, CBiol, FRSB, CertMHS
Director of Clinical Diagnostic Research, Technoclone GmbH, Vienna
Honorary Consultant Biomedical Scientist, Specialist Haemostasis Unit, Addenbrooke's Hospital, Cambridge
Visiting Professor, Faculty of Science and Technology, Middlesex University

Dr Jane M. Needham PhD, CSci, FRCPath, FIBMS
Consultant Biomedical Scientist, Haemophilia, Haemostasis and Thrombosis Centre, Basingstoke and North Hampshire NHS Foundation Trust, UK

Online materials developed by

Sheelagh Heugh, Principal Lecturer in Clinical Biochemistry, Haematology and Transfusion Science, Faculty of Human Sciences, London Metropolitan University

Dr Ken Hudson, Lecturer in Biomedical Science, Faculty of Human Sciences, London Metropolitan University

William Armour, Senior Lecturer in Biomedical Science, Faculty of Human Sciences, London Metropolitan University

Table 1: Reference ranges

	Reference range
Red cells	
Haemoglobin (Hb)	
Male	133–167 g/L
Female	118–148 g/L
Red blood cell count (RBCC)	
Male	4.3–5.7 × 10^{12}/L
Female	3.9–5.0 × 10^{12}/L
Haematocrit (Hct)	
Male	0.35–0.53 L/L
Female	0.33–0.47 L/L
Reticulocytes	25–125 × 10^{9}/L (0.5–3.5% of total red cell count)
Red cell distribution width (RDW)	
As coefficient of variation	10.3–15.3%
As standard deviation	34.5–50.5 fL
Red cell indices	
Mean cell volume (MCV)	77–98 fL
Mean cell haemoglobin (MCH)	26–33 pg
Mean cell haemoglobin concentration (MCHC)	330–370 pg/L
Erythrocyte sedimentation rate (ESR)	<10 mm/hour
Micronutrients and plasma proteins	
Iron	10–37 μmol/L
Total iron binding capacity	
Men	54–72 μmol/L
Women	55–81 μmol/L
Transferrin	2–4 g/L
Transferrin saturation	
Men	18–40%
Women	13–37%
Ferritin	
Male	25–380 μg/L
Post-menopausal female	28–365 μg/L
Pre-menopausal female	7.5–224 μg/L
Vitamin B_{12}	160–925 ng/L 120–680 pmol/L
Vitamin B_6*	17–279 nmol/L 2.9–47.0 μg/L
Folate	
Serum	>7 nmol/L >3 μg/L
Red blood cell	>340 nmol/L >150 μg/L

(Continued)

TABLE 1 *(Continued)*

Methylmalonic acid	0.08–0.28 μg/L
Plasma viscosity	1.5–1.72 mPa.s at 25°C 1.16–1.33 mPa.s at 37°C
White blood cells	
White cell count (WCC)	4.0–10.0 × 10^9/L
The differential:	Absolute count
Neutrophils	2.0–7.0 × 10^9/L
Lymphocytes	1.0–3.0 × 10^9/L
Monocytes	0.2–1.0 × 10^9/L
Eosinophils	0.02–0.5 × 10^9/L
Basophils	0.02–0.1 × 10^9/L
Blasts/atypical cells	~0.01 × 10^9/L
Haemostasis	
Platelets	143–400 × 10^9/L
Fibrinogen	1.5–4.0 g/L
Prothrombin time (PT)	11–14 seconds
Partial thromboplastin time (PTT)	24–34 seconds
International normalized ratio (INR)	2.0–3.0 or 3.0–4.0 (therapeutic ranges for monitoring vitamin K antagonist anticoagulation)
D-dimers	<500 units/mL (strongly dependent on method)

*Vitamin B_6 has several isoforms; pyridoxal 5′-phosphate is most commonly reported.

Note

These reference ranges are pooled from recent authoritative textbooks and guidelines. The figures vary with time, with location, and according to the method of analysis, and must reflect the nature of the population that the hospital serves. Accordingly, the practitioner *must* work to their local reference range.

Abbreviations

A23187	calcium ionophere
α_2AP	alpha 2-antiplasmin
α_2M	alpha 2-macroglobulin
aβ_2GPI	beta 2-glycoprotein I antibodies
ABL	Abelson
AC	acetylation
ACD	anaemia of chronic disease
aCL	anticardiolipin antibodies
aCML	atypical chronic myeloid leukaemia
ADAMTS-13	*a* zinc- and calcium-dependent *d*isintegrin *a*nd *m*etalloprotease with *t*hrombo*s*pondin type 1 motifs, member *13*
ADCC	antibody directed cellular cytotoxicity
ADP	adenosine diphosphate
AF	atrial fibrillation
Ag	antigen
AHG	antihuman globulin
aHUS	atypical haemolytic uraemic syndrome
AIDS	acquired immunodeficiency syndrome
AIHA	autoimmune haemolytic anaemia
AITL	angioimmunoblastic T-cell lymphoma
ALA	aminolevulinic acid
ALCL	anaplastic large cell lymphoma
ALK	anaplastic lymphoma kinase
ALK-1	activin receptor-like kinase 1
ALL	acute lymphoblastic leukaemia
ALOT	acute leukaemia orientation tube
AMC	absolute monocyte count
AMKL	acute megakaryoblastic leukaemia
AML	acute myeloid leukaemia
AML-MRC	acute myeloid leukaemia with myelodysplasia-related changes
AMML	acute myelomonocytic leukaemia
ANAE	alpha-naphthyl acetate esterase
ANC	absolute neutrophil count
AP	arteriolar-pericyte
APA	antiphospholipid antibodies
APC	activated protein C or antigen-presenting cell
APCR	activated protein C resistance
API2	apoptosis inhibitor 2
APL	acute promyelocytic leukaemia
APML	acute promyelocytic leukaemia
APS	antiphospholipid syndrome
APTT	activated partial thromboplastin time
ARF	alternate reading frame
ARU	aspirin reaction units
ASLA	activated seven lupus anticoagulant assay
AT	antithrombin
ATL	adult T-cell leukaemia
ATM	ataxia telangiectasia mutated
ATP	adenosine triphosphate
ATPase	adenosine triphosphatase
ATRA	all-*trans* retinoic acid
AVWD	acquired von Willebrand disease
β_2GPI	beta 2-glycoprotein I
BC	blast crisis
BCL	B-cell lymphoma
B-CLL	B-cell chronic lymphocytic leukaemia
BCP	B-cell precursor
BCR	breakpoint cluster region *or* B-cell receptor
BCSH	British Committee for Standards in Haematology
BEN	benign ethnic neutropenia
bFGF	basic fibroblast growth factor
BJP	Bence-Jones proteins
BL	Burkitt lymphoma
BM	bone marrow
BMP	bone morphogenic proteins
BNF	British National Formulary
BNLI	British National Lymphoma Investigation
BP	binding protein
BPI	bacterial permeability-inducing factor
B-PLL	B-cell prolymphocytic leukaemia
BSAP	B-cell-specific activator protein
BSH	British Society for Haematology
BSS	Bernard–Soulier syndrome
BTK	Bruton's tyrosine kinase
BU	Bethesda unit
C3	complement component 3

C4bBP C4b-binding protein
CAE chloroacetate esterase
c-ALL common-acute lymphoblastic leukaemia
CATT card agglutination test for trypanosomiasis
CBF core binding factor
Cbl cobalamin
CCR chemokine receptor
CD cluster of differentiation
CDA congenital dyserythropoietic anaemia
CDK cyclin-dependent kinase
CDR complementarity-determining region
CEBPA CCAAT enhancer binding protein α gene
CEL chronic eosinophilic leukaemia
CFU colony-forming unit
CFU-Baso CFU for basophils
CFU-E CFU for erythrocytes
CFU-EMk CFU for erythrocytes and megakaryocytes
CFU-Eo CFU for eosinophils
CFU-GEMM CFU for granulocytes, erythrocytes, monocytes, and megakaryocytes
CFU-GMo CFU for granulocytes and monocytes
CFU-L CFU for lymphocytes
CFU-M CFU for monocytes
CFU-Mk CFU for megakaryocytes
CFU-N CFU for neutrophils
CGH comparative genomic hybridization
CGL chronic granulocytic leukaemia
C_H constant domain of an antibody heavy chain
CHL classical Hodgkin lymphoma
CI coagulation index
Cip/Kip CDK inhibitory proteins
CKI cyclin kinase inhibitor
C_L constant domain of an antibody light chain
CLIA chemiluminescent immunoassay
CLL chronic lymphocytic leukaemia
CLP common lymphocyte progenitors
CLPO chronic lymphoproliferative disorders
CMAF cellular musculoaponeurotic fibrosarcoma
CML chronic myeloid leukaemia
CMML chronic myelomonocytic leukaemia
CMPD chronic myeloproliferative disease
CMPD-u chronic myeloproliferative disease—unclassified
CMV cytomegalovirus
CNL chronic neutrophilic leukaemia
CNS central nervous system
CNS DLBCL central nervous system diffuse large B-cell lymphoma
CoA coenzyme A *or* co-activator(s)
COPD congestive obstructive pulmonary disease
COSHH Control of Substances Hazardous to Health
COX cyclooxygenase
CPD continuing professional development
CPGenIII coproporphyrinogen III
CRAB hypercalcaemia, renal insufficiency, anaemia, bone lesions
CRP C-reactive protein
CSF colony-stimulating factor
CSci Chartered Scientist
CSR class-switching recombination
CYP cytochrome P
2,3-DPG 2,3-diphosphoglycerate
Da Dalton
DAB 3,3'-diaminobenzidine
DAF decay-accelerating factor
DAPTT dilute activated partial thromboplastin time
DARC Duffy antigen receptor for chemokines
DAT direct antiglobulin test
DBA Diamond–Blackfan anaemia
DBL-EBP Duffy binding-like erythrocyte-binding protein
DDT dichlorodiphenyltrichloroethane
del deletion
DFXaI direct activated factor X inhibitor
DIC disseminated intravascular coagulation
DISC death-inducing signalling complex
DLBCL diffuse large B-cell lymphoma
DMT-1 divalent metal transporter-1
DNA deoxyribonucleic acid
DNMT DNA methyltransferases
DOAC direct oral anticoagulant
DPT dilute prothrombin time
DRVVT dilute Russell's viper venom time
DS Down syndrome
DS-AMKL Down syndrome-associated acute megakaryoblastic leukaemia
DSCR Down syndrome critical region
dsDNA double-stranded DNA
DSPA Desmodus salivary plasminogen activator
DTI direct thrombin inhibitor

dTMP	deoxythymidine monophosphate
DTS	dense tubular system
dUMP	deoxyuridine monophosphate
DVT	deep vein thrombosis
EBER	Epstein–Barr virus-encoded RNA
EBV	Epstein–Barr virus
EBNA	Epstein–Barr virus nuclear antigen
ECP	eosinophil cationic protein
ECPR	endothelial protein C receptor
EDN/EPX	eosinophil-derived neurotoxin/eosinophil protein X
EDS	Ehlers–Danlos syndrome
EDTA	ethylene diamine tetra-acetic acid
EGIL	European Group for the Immunological Classification of Leukaemias
EHL	extended half-life
EL	erythroleukaemia
ELF-EMF	extremely low-frequency electromagnetic field
ELIFA	enzyme-linked immunofiltration assay
ELISA	enzyme-linked immunosorbent assay
ELP	electrophoretic strip
Eos	eosinophil
EPCR	endothelial protein C receptor
Epo	erythropoietin
EpoR	erythropoietin receptor
EQA	external quality assessment
ESR	erythrocyte sedimentation rate
ESRD	end-stage renal disease
ET	Ecarin time
ET	essential thrombocythaemia
ETO	eight-twenty-one
EVI1	ecotropic viral integration site-1
EZF-1	elongation factor for Z-1
FA	Fanconi anaemia
FAB	French–American–British
FBC	full blood count
Fc	crystallizable fraction (of an immunoglobulin)
FcγR	gamma receptor for the IgG fragment crystallizable region
FDC	follicular dendritic cell
FDP	fibrin(ogen) degradation products
FFC	fluorescence flow cytometry
FGF	fibroblast growth factor
FGFR	fibroblast growth factor receptor
FII	factor II or prothrombin
FIIa	activated factor II or thrombin
FISH	fluorescence in situ hybridization
FITC	fluorescein isothiocyanate
FIX	factor IX
FIXa	activated factor IX
FL	follicular lymphoma
fL	femtolitre (10-15 litre)
FLIFA	fluorescence-linked immunofiltration assay
FLT3	fms-related tyrosine kinase 3
FPA	fibrinopeptide A
FPB	fibrinopeptide B
FRET	fluorescence resonance energy transfer
FSC	forward scatter
FTCL	follicular T-cell lymphoma
FV	factor V
FVa	activated factor V
FVII	factor VII
FVIIa	activated factor FVII
FVIII	factor VIII
FVIIIa	activated factor VIII
FVL	factor V Leiden
FX	factor X
FXa	activated factor X
FXI	factor XI
FXIa	activated factor XI
FXII	factor XII
FXIIa	activated factor XII
FXIII	factor XIII
FXIIIa	activated factor XIII
G0 phase	quiescent phase (of the cell cycle)
G1 phase	Gap 1 phase
G2 phase	Gap 2 phase
G6PD	glucose-6-phosphate dehydrogenase
Gb3	globotriaosyl ceramide
GCB	germinal centre-like B-cell-like
G-CSF	granulocyte colony-stimulating factor
GEC	glycolytic enzymes complex
GEMM	granulocyte, erythrocyte, monocyte, megakaryocyte
GGT	gamma glutamyl-transferase
GLA	gamma-carboxyglutamic acid

GLC	glycolytic enzyme complex
Glu	glutamic acid
Glut	glutamine
GM-CSF	granulocyte macrophage colony-stimulating factor
GMNPT	geometric mean normal prothrombin time
GP	general practitioner
GP	glycoprotein
GPA	glycophorin A
GPC	glycophorin C
GPI	glycosyl phosphatidylinositol
GPIa	glycoprotein Ia
GPIb	glycoprotein Ib
GPIb-IX-V	glycoprotein Ib-IX-V complex
GPIIbIIIa	glycoprotein IIb-IIIa complex
GPLU	(Ig)G (anti)phospholipid unit
GPVI	glycoprotein VI
GS	glutathione
GSH	reduced glutathione
GST	glutathione S-transferase
GT	Glanzmann's thrombasthenia
H	Histone
H&E	haematoxylin and eosin
Haly-PA	Haly plasminogen activator
HAT	histone acetyltransferase
HAX	HS-1-associated protein X
Hb	haemoglobin
HbF	foetal haemoglobin
HBS	heparin binding site (of antithrombin)
HbS	sickle haemoglobin
HC II	heparin cofactor II
HCL	hairy cell leukaemia
HCPC	Health and Care Professions Council
Hct	haematocrit
HDAC	histone deacetylase
HDFN	haemolytic disease of the foetus and newborn
HELLP	haemolysis, elevated liver enzymes and low platelets
HES	hypereosinophilic syndrome
HH	hereditary haemochromatosis
HHT	hereditary haemorrhagic telangiectasia
HHV8	human herpesvirus 8
HIF	hypoxia-inducible factor
HIT	heparin-induced thrombocytopenia
HIV	human immunodeficiency virus
HJV	haemojuvelin
HL	Hodgkin lymphoma
HLA	human leucocyte antigen
HMW	high molecular weight
HMWK	high-molecular-weight kininogen
hpf	high-power field
HPFH	hereditary persistence of foetal haemoglobin
HPLC	high-performance/pressure liquid chromatography
HPP	hereditary pyropoikilocytosis
HRG	histidine-rich glycoprotein
HRP	horse radish peroxidase
HRP-1	histidine-rich protein 1 (likewise HRP-2 and HRP-3)
HRS	Hodgkin-Reed-Sternberg
HS	hereditary spherocytosis
HSt	hereditary stomatocytosis
HTLV-1	human lymphotrophic virus-1
HUS	haemolytic uraemic syndrome
IAP	inhibitor of apoptosis
IAT	indirect antiglobulin test
IBD	inflammatory bowel disease
IBMS	Institute of Biomedical Science
ICAM	intercellular adhesion molecule
ICSH	International Committee for Standards in Haematology
IEF	isoelectric focusing
IF	intrinsic factor
IFN	interferon
Ig	immunoglobulin
IGH	immunoglobulin heavy-chain gene
IGVH	immunoglobulin variable heavy chain gene
IL	interleukin
IM	infectious mononucleosis
IMF	idiopathic myelofibrosis
INK	inhibitors of CDK4
INR	International Normalized Ratio
inv	inversion
IPSID	immunoproliferative small intestine disease
IPSS	International Prognostic Scoring System

IQC	internal quality control
IRMA	immunoradiometric assay
IRP	International Reference Preparation
IRS	indoor residual spraying
ISCN	International System for Human Cytogenetic Nomenclature
ISI	International Sensitivity Index
ISO	International Organization for Standardization
ITD	internal tandem duplication
ITN	insecticide-treated net
ITP	idiopathic thrombocytopenic purpura
IU	international unit
JAK	Janus-activated kinase
JAK2	Janus kinase 2
JBL	journal-based learning
JMD	juxtamembrane domain
JMML	juvenile myelomonocytic leukaemia
KAHRP	knob-associated histidine-rich protein
KCT	kaolin clotting time
KLF-1	Kruppel-like factor-1
KMS	Kasabach-Merritt syndrome
LA	lupus anticoagulant
LBL	lymphoblastic lymphoma
LDCHL	lymphocyte-depleted classical Hodgkin lymphoma
LDH	lactate dehydrogenase
LDT	lymphocyte doubling time
LGL	large granular lymphocytes
LIA	latex immunoassay
LLIN	long-lasting insecticidal nets
LMAN-1	lectin mannose binding protein 1
LMP	latent membrane protein 1
LMWH	low molecular weight heparin
LN	lymph node
LPL	lymphoplasmacytic lymphoma
LPS	lipopolysaccharide
LRM	leucine-rich motif
LST	lymphoid screening tube
LUC	large unstained/unclassified cell
LV-PA	Lachesis venom plasminogen activator
mAb	monoclonal antibody
MAHA	microangiopathic haemolytic anaemia
MALT	mucosa-associated lymphoid tissue
MAPK	mitogen-activated protein kinase
MBD	methyl-binding domain
MBP	major basic protein
MCCHL	mixed cellularity classical Hodgkin lymphoma
MCFD2	multiple coagulation factor deficiency 2
MCH	mean cell haemoglobin
MCHC	mean cell haemoglobin concentration
MCL	mantle cell lymphoma
MCR	Manchester comparative reagent
M-CSF	macrophage colony-stimulating factor
MCV	mean cell volume
MDS	myelodysplastic syndrome
MDS-U	myelodysplastic syndrome unclassified
MECP2	methyl-CpG-binding protein 2
M/E ratio	myeloid/erythroid ratio
MESA	mature erythrocyte surface antigen
Met-Hb	methaemoglobin
MF	mycosis fungoides
MF	myelofibrosis
MGUS	monoclonal gammopathy of undetermined significance
MHC	Major Histocompatibility Complex
MI	myocardial infarction
MIB-1	molecular immunology Bortsel-1
MIP-1α	macrophage inflammatory protein-1α
miRNA	microRNA
MKL1	megakaryoblastic leukaemia-1
MLL	mixed lineage leukaemia
MM	multiple myeloma
MMA	methylmalonic acid
MMSET	multiple myeloma SET domain
MPAL	mixed-phenotype acute leukaemias
MPD	myeloproliferative disorders
M-phase	mitosis phase
MPLU	(Ig)M (anti)phospholipid unit
MPN	myeloproliferative neoplasm
MPO	myeloperoxidase
M-protein	monoclonal protein
MRC	myelodysplasia-related changes
MRI	magnetic resonance imaging
mRNA	messenger RNA

MRP1 multidrug resistance protein-1
MTHFR methylene tetrahydrofolate reductase
MZL marginal zone lymphoma
NAD nicotinamide adenine dinucleotide
NADH nicotinamide adenine dinucleotide hydride
NADP nicotinamide adenine dinucleotide phosphate
NADPH nicotinamide adenine dinucleotide phosphate hydride
NAIT neonatal alloimmune thrombocytopenia
NE neutrophil elastase
NEQAS National External Quality Assessment Scheme
NET neutrophil extracellular traps
NFκB nuclear factor kappa B
NGCB non-germinal centre B cell
NHL non-Hodgkin lymphoma
NHS National Health Service
NK natural killer (cell)
NPM nucleophosmin
NSAID non-steroidal anti-inflammatory drug
NSCHL nodular sclerosis classical Hodgkin lymphoma
NTBI non-transferrin bound iron
NuMA nuclear matrix associated
NUP214 nucleoporin 214 kDa
NuRD nucleosome remodelling (and histone) deacetylation (complex)
OAF osteoclast activating factors
OD optical density
OPG osteoprotegerin
Pa Pascal
PA pernicious anaemia
PAF platelet activating factor
PAI-1 plasminogen activator inhibitor type 1
PAI-2 plasminogen activator inhibitor type 2
PAI-3 plasminogen activator inhibitor type 3
PAR-1 protease-activated receptor-1
PAS Periodic acid–Schiff
PAU platelet-aggregation units
PB Paul–Bunnell
PBX1 pre-B cell leukaemia homeobox 1
PC protein C
PCD plasma cell disorders
PCFT proton-coupled folate transporter
PCH paroxysmal cold haemoglobinuria
PCI protein C inhibitor
PCL plasma cell leukaemia
PCP pneumocystis pneumonia
PCR polymerase chain reaction
PCST plasma cell screening tube
pDC plasmacytic dendritic cells
PDGFR platelet-derived growth factor receptor
PE pulmonary embolism
PEG polyethylene glycol
PET positron emission tomography
PF4 platelet factor 4
PfEMP 1 *Plasmodium falciparum* erythrocyte membrane protein 1
PfEMP 2 *P. falciparum* erythrocyte membrane protein 2
PfEMP 3 *P. falciparum* erythrocyte membrane protein 3
PfLDH *P. falciparum* lactate dehydrogenase
pg picogram (10^{-12} gram)
PGF platelet growth factor
Ph Philadelphia chromosome
PHD prolyl-hydroxylase domain
PiCT prothrombinase-induced clotting time
PIVKA proteins induced by vitamin K absence/antagonism
PK prekallikrein
PK pyruvate kinase
PKDL post kala-azar dermal leishmaniasis
PLC plasma cell leukaemia
pLDH *Plasmodium* lactate dehydrogenase
PLL prolymphocytic leukaemia
PLZF promyelocytic leukaemia zinc finger
PMBL primary mediastinal large B-cell lymphoma
PML promyelocytic leukaemia gene
PN-2 protease nexin-2
PNH paroxysmal nocturnal haemoglobinuria
PNP platelet neutralization procedure
POCT point-of-care test
POEMS polyneuropathy, organomegaly, endocrinopathy, monoclonal protein, skin changes
PPIX protoporphyrin IX
PPgenIX protoporphyrinogen IX
PPP platelet-poor plasma
pRB retinoblastoma protein
PRCA pure red cell aplasia
PRP platelet-rich plasma

PRU	P2Y12 reaction units
PS	protein S
PT	prothrombin time
PTCL	peripheral T-cell lymphoma
PTLD	post-transplant lymphoproliferative disorders
PTT	partial thromboplastin time
PV	parasitophorous vacuole or polycythaemia vera or plasma viscosity
PVM	parasitophorous vacuole membrane
PZ	protein Z
QBC	quantitative buffy coat
RA	refractory anaemia
RA	rheumatoid arthritis
RAEB	refractory anaemia with excess blasts
RAEB-t	refractory anaemia with excess blasts in transformation
RANK	receptor activator of nuclear factor κB
RANKL	receptor activator of nuclear factor κB ligand
RARA	retinoic acid receptor alpha gene
RARE	retinoic acid response element
RARS	refractory anaemia with ringed sideroblasts
RASA	ring-infected erythrocyte membrane surface antigen
RBC	red blood cell
RBCC	red blood cell count
RBM15	RNA binding motif protein 15
RCC	refractory cytopenia of childhood
RCM	red cell mass
RCMD	refractory cytopenia with multilineage dysplasia
RDW	red cell distribution width
RE	restriction enzyme
REAL	Revised European and American classification of Lymphoid (neoplasms/malignancies)
Rh	Rhesus
RhAG	rhesus-associated glycoprotein
RIA	radio immunoassay
RIPA	ristocetin-induced platelet aggregation
RN	refractory neutropenia
RNA	ribonucleic acid
ROS	reactive oxygen species
RPN1	ribophorin 1
RS	reactive site
RT	refractory thrombocytopenia
RT	reptilase time
RUNX	runt-related transcription factor
RXR	retinoid X receptor
s	second
SAK	staphylokinase
SBB	Sudan black B
SCF	stem cell factor
SCOCS	surface-connected open canalicular system
SCT	silica clotting time
SDS-PAGE	sodium dodecyl sulphate polyacrylamide gel electrophoresis
SEAO	South East Asian ovalocytosis
Serpin	serine protease inhibitor
SFMC	soluble fibrin monomer complexes
SHM	somatic hypermutation
SK	streptokinase
SLE	systemic lupus erythematosus
SLL	small lymphocytic lymphoma
smIg	surface membrane immunoglobulin
SMK	sinusoid-megakaryocyte
SMMHC	smooth muscle myosin heavy chain
SMZL	splenic marginal zone lymphoma
SNP	single-nucleotide polymorphism
SOP	standard operating procedure
S phase	synthesis phase
SSC	side scatter
SST	small sample tube
STAT	signal transducer and activator of transcription
STEAP	six-transmembrane epithelial antigen of the prostate
STEC-HUS	Shiga toxin-producing Escherichia Coli-haemolytic uraemic syndrome
sTfR	soluble transferrin receptor
STI	signal transduction inhibitors
t	translocation
TAFI	thrombin activatable fibrinolysis inhibitor
TAM	transient abnormal myelopoiesis
t-AML	therapy-related acute myeloid leukaemia
TAR	thrombocytopenia with absent radius
TAT	thrombin-antithrombin complexes
TC	cytotoxic T-cell
TC	transcobalamin
TCBIR	transcobalamin-receptor
TCL	T-cell lymphoma
TCR	T-cell receptor

TDT terminal deoxynucleotidyl transferase
TEG thromboelastography
TEL translocation ETS leukaemia
TF tissue factor
TFH T-follicular helper
TFPI tissue factor pathway inhibitor
TfR transferrin receptor
TGF transforming growth factor
TGFβ1 transforming growth factor β1
TH helper T cell
THF tetrahydrofolate
TIBC total iron-binding capacity
T-LGL T-cell large granular lymphocytic leukaemia
TM thrombomodulin
TMD transient myeloproliferative disease
t-MDS therapy-related myelodysplastic syndrome
t-MDS/MPN therapy-related myelodysplastic/myeloproliferative neoplasm
TNF tumour necrosis factor
TNFα tumour necrosis factor α
Topo II topoisomerase II
t-PA tissue plasminogen activator
T-PLL T-prolymphocytic leukaemia
TPM3 tropomyosin 3
Tpo thrombopoietin
TRAP thrombin receptor agonist peptide
TSG tumour suppressor gene
TSV-PA *Trimeresurus stejnegeri* viper plasminogen activator
TSVT Taipan snake venom time
TT thrombin time
TTP thrombotic thrombocytopenic purpura
TXA2 thromboxane A2
u unit
U&E urea and electrolytes
U46619 9,11-dideoxy-9α,11α-methanoepoxy PGF2α
UFH unfractionated heparin
UK NEQAS United Kingdom National External Quality Assurance Scheme
UKAS United Kingdom Accreditation Service
u-PA urinary plasminogen activator (or urokinase)
u-PAR urinary plasminogen activator receptor
V/Q ventilation/perfusion
VCA viral capsid antigen
V_H variable domain of an antibody heavy chain
VKDB vitamin K deficiency bleeding
VKORC1 vitamin K epoxide reductase complex subunit 1
V_L variable domain of an antibody light chain
VSG variant surface glycoproteins
VTE venous thromboembolism
VW Ag II von Willebrand antigen II
VWD von Willebrand disease
VWF von Willebrand factor
VWF:Ag von Willebrand factor: antigen
VWF:CB von Willebrand factor: collagen binding (activity)
VWF:FVIIIB von Willebrand factor: factor VIII binding
VWF:RCo von Willebrand factor: ristocetin co-factor (activity)
WAS Wiskott–Aldrich syndrome
WASp Wiskott–Aldrich syndrome protein
WBC white blood cell
WBCC white blood cell count
WHO World Health Organization
WM Waldenström macroglobulinaemia
ZAP-70 zeta-associated protein 70
ZPI protein Z-dependent protease inhibitor

PART 1

Haematology and Haemopoiesis

1

Introduction to haematology

Gary Moore, Gavin Knight, Andrew Blann, and Alexis Henley

This chapter introduces haematology, not only as the science of the study of blood itself but also about how the subject relates to other disciplines in pathology. You will get a feel for haematology in the wider aspects of health care and for how, as haematologists, you will need to relate to your colleagues working in other disciplines within pathology, in hospitals, and with the professional bodies.

Learning objectives

After studying this chapter, you should confidently be able to:

- explain key aspects of the science of haematology;
- appreciate the role of the scientist in the haematology laboratory;
- describe the role of haematology in the provision of health care;
- outline the overlap of haematology and other pathology disciplines;
- comment on the role of professional and regulatory bodies.

1.1 What is haematology?

Put simply, haematology is the study of blood, an organ unusual in that it interacts with almost all other organs. The haematology laboratory in a healthcare setting is concerned with diagnosing and monitoring the management of diseases of the blood and blood-forming organs. Blood cells are manufactured in **bone marrow** and released into the **peripheral blood** once they are mature. Blood is a dynamic and crucial fluid, providing molecular and cellular transport and many regulatory functions. Blood interfaces with all organs and tissues in the body, carrying essential substances such as oxygen and nutrients to the cells, and waste products

bone marrow
Soft tissue located inside hollow bones responsible for the production and maturation of blood cells.

peripheral blood
The blood that is contained within the circulatory system.

Cross references

Red blood cells and their associated disease states are described in further detail in Chapters 4–6.

White blood cells and their associated disease states are described in further detail in Chapters 8–12.

Platelets and their associated disease states are described in further detail in Chapters 11, 13, 14, 16, and 17.

away from cells to the excretory organs. As such, it has a critical role in ensuring adequate whole-body physiology and **homeostasis**. It follows that adverse changes to the blood will have numerous consequences, many of which can be serious and life-threatening.

Conversely, adverse changes to organs and tissues may translate into changes in the composition of the blood that are secondary to the primary disease. It is in this latter capacity that blood can be used in the clinical laboratory for the detection and monitoring of various diseases and their treatments.

Blood is composed of approximately 45% blood cells, which are classified into three main types:

- **red blood cells** (RBCs) that carry oxygen to the tissues;
- **white blood cells** (WBCs) that function primarily as defence against infection;
- **platelets** that prevent blood loss at sites of injury by combining with specialized proteins to form a clot.

The remaining 55% is **plasma**, which is an aqueous solution that acts as the transport medium for blood cells, dissolved nutrients, and plasma proteins, including those involved in **blood coagulation**.

Scientists working in the haematology laboratory perform and interpret an array of diverse blood tests concerned with the investigation of the number, structure, and function of the cellular and molecular elements of blood and the investigation and control of bleeding and clotting disorders.

Blood tests are performed either on whole blood, plasma, or **serum**, depending on the investigation required. Blood is collected by **venepuncture** into specially designed bottles. Blood tests that require whole blood or plasma are collected into blood tubes containing **anticoagulants** to prevent the blood clotting before it is analysed; blood for tests on serum is collected into plain tubes. Anticoagulants are also used in the clinic as therapeutics to reduce the risk of a clot forming within the body: full details are in Chapter 17.

homeostasis
The maintenance of stable physiological systems.

blood coagulation
The process where specialized proteins interact to form a clot (in conjunction with platelets).

serum
The fluid that remains after the blood has been allowed to clot.

venepuncture
The process of obtaining intravenous access to obtain a sample of blood via a needle.

anticoagulant
A physiological or pharmacological mechanism that retards clotting processes.

SELF-CHECK 1.1

Name the three different types of blood cell.

1.1.1 A classification of haematology

Haematology tends to be considered under the three main areas of RBCs, WBCs, and **haemostasis**, which are then further subclassified.

Red blood cells

Haematologists are interested in the number and function of RBCs, their size, and the amount and quality of the **haemoglobin** that they carry. By far the most common condition concerning RBCs is **anaemia**, which is a reduction in the oxygen-carrying capacity of the blood arising from reduced or abnormal haemoglobin inside the RBCs, and/or reduced numbers of RBCs. At the polar extreme are the diseases **erythrocytosis** and **polycythaemia**, where there are too many RBCs, leading to a high haemoglobin level. Anaemia leads to clinical symptoms of lethargy, weakness, dizziness, and feeling faint. If the anaemia worsens, patients can experience shortness of breath, palpitations, headaches, and sore mouth and gums. The different types of anaemia result from a variety of underlying medical disorders.

haemostasis
The interplay of cellular and molecular processes that maintain blood fluidity and that also generate blood clots at sites of injury, regulate clot formation, and degrade clots.

haemoglobin
Metalloprotein inside RBCs that is responsible for oxygen transport.

erythrocytosis
A condition characterized by a high red blood cell count, generally a response to a factor such as hypoxia or increased levels of erythropoietin. The haematocrit and haemoglobin are also increased. It is described in more detail in Chapter 3.

Common causes of anaemia are:

- iron deficiency, which results in a defect in production of the haem (iron-containing) component of haemoglobin due to the lack of ferrous iron;
- vitamin B_{12} and/or **folate** deficiency, which affect the production of DNA;
- other diseases and conditions such as malignancy, renal disease, liver disease, lead poisoning, and infection;
- hereditary conditions, such as **sickle cell disease** and **thalassaemia,** where there is a defect in the haemoglobin molecule;
- acute and chronic blood loss.

polycythaemia
A malignancy of RBCs that causes too many to be produced. This leads to a high red cell count, haematocrit, and haemoglobin. It is described in more detail in Chapter 11.

sickle cell disease
An inherited disorder of haemoglobin of varying severity. The name arises from the deformed shape the RBCs take when the abnormal haemoglobin inside them polymerizes.

thalassaemia
A family of inherited disorders of haemoglobin where there is an imbalance in globin chain production.

White blood cells

White blood cells, also known as leucocytes, are part of the immune system, and there are five recognizable types found in peripheral blood in health. They help protect the body against **infections** caused by microbes such as viruses and bacteria. The most common serious disease of WBCs is **leukaemia**, a type of **neoplasia**.

The haematology laboratory investigates the number of each type of WBC in the peripheral blood and whether they are mature or immature cells.

- Raised numbers of normally functioning white cells can be seen in bacterial, viral, and fungal infections, where extra cells become available to deal with the microorganisms.
- Raised numbers of immature white cells are commonly found in many leukaemias. These cells represent the uncontrolled proliferation of malignant **clones** in the bone marrow or lymphoid tissue entering the peripheral blood.
- Low numbers of WBCs can be caused by some medications and cytotoxic chemotherapy. Reduced numbers of white cells can also be seen in diseases such as **aplastic anaemia**. The consequence of depleted numbers of WBCs is the body's inability to fight infection effectively, a condition known as immunodeficiency.

Cross references

Anaemia—this major group of diseases of RBCs is described in Chapters 5 and 6.
Leukaemia and related disorders are described in further detail in Chapters 9-12.

infection
The presence of sufficiently high numbers of a microorganism to invoke clinical symptoms and provoke a defensive response.

leukaemia
A haemoproliferative disorder characterized by increased numbers of blood cells in the bone marrow and peripheral blood.

neoplasia
An abnormal proliferation of cells that can be benign or progress to malignancy. The word comes from the Greek *neo* = new, *plasia* = formation.

clone
A cell, group of cells, or organism descended from and genetically identical to a single common ancestor.

Haemostasis

Haemostasis comprises an integrated group of balanced cellular and molecular processes designed to minimize the loss of blood upon damage to blood vessels, i.e. **haemorrhage**, a condition which can be life-threatening. However, excessive and/or inappropriate activity of the coagulation system, generally resulting in **thrombosis**, can also be dangerous. Unimpaired platelet function and certain coagulation proteins are crucial to effective haemostasis.

Haematologists are also involved in the diagnosis and management of patients whose blood has a predisposition to clot and of those people who have an underlying bleeding disorder, for example **haemophilia**.

The haemostasis laboratory plays a key role in monitoring patients who are receiving medication for thrombosis. Anticoagulant medication inhibits the ability of the blood to clot, but too much can cause the patient to haemorrhage, therefore regular monitoring is often required.

In addition to the categories of diseases discussed above, haematologists are also involved in the diagnosis of some parasitic blood infections, such as **malaria**. (See also Box 1.1.)

aplastic anaemia
A serious disease where the bone marrow does not produce enough blood cells.

haemorrhage
Excessive bleeding caused by a breakdown in haemostasis.

thrombosis
The process of the (generally inappropriate) formation of blood clots.

haemophilia
Hereditary bleeding disorder caused by a deficiency in clotting factors.

malaria
An infectious disease found in tropical and subtropical regions. It is caused by protozoan parasites of *Plasmodium* species that are carried by mosquitoes.

SELF-CHECK 1.2

What are the three main areas of haematology?

BOX 1.1 What will this book achieve?

All the diseases you have just been introduced to, and their detection in the laboratory by scientists, are discussed in more detail in the chapters that follow:

- Chapter 2 introduces basic blood tests;
- Chapter 3 describes how the blood cells develop;
- Chapters 4-6 consider the red blood cell in health and disease;
- Chapter 7 introduces you to blood-borne parasites;
- Chapter 8 discusses white blood cells in health and disease;
- Chapters 9-12 introduce biological mechanisms leading to cancer of the blood, and the classification of these cancers;
- Chapters 13-17 focus on haemostasis and the consequences of its failure;
- Chapter 18 consolidates some key issues in haematology in three complex case studies.

1.2 The role of the scientist in the haematology laboratory

The laboratory is populated with a number of different healthcare professionals. Some, such as support staff, do not need to be fully versed in the fine points of the subject; these are the domain of two closely linked sets of scientists who carry out a wide range of laboratory tests to produce and interpret results that assist in the diagnosis and treatment of disease. Clinical scientists often focus on developing new techniques, on troubleshooting, and on the more unusual cases that pass through the laboratory. Biomedical scientists are generally more concerned with the routine day-to-day running of the department and its many and varied methods and procedures. Both sets of scientists are likely to interpret results, and so interact with, and advise, medical staff on the probable implications of a patient's results. It follows that there is often a blurring of the roles of the two scientist groups, and with appropriate training, they may well be interchangeable.

oncology
The area of medicine that deals with the development, diagnosis, treatment, and prevention of tumours.

haemato-oncology
Cancers of the blood.

The scientist specializing in haematology also plays an essential role in supporting many hospital departments, such as accident and emergency, intensive care, operating theatres, special care baby units, and **oncology**. Without the contribution of the scientist, these departments could not function as effectively.

The scientist working in the haematology laboratory is particularly important in supporting medical colleagues in the specialist areas of **haemato-oncology**, haemostasis, and

haemoglobinopathies. Indeed, as your career progresses you may decide to specialize in one of these subdisciplines.

haemoglobinopathy
Disease (such as thalassaemia and sickle cell disease) resulting from mutations in the globin genes and so abnormal haemoglobin synthesis.

Many of the analytical techniques employed in haematology are also common to other pathology disciplines (e.g. immunoassay, genetic techniques (such as polymerase chain reaction), and microscopy), which will equip you with transferable scientific skills, whilst other scientific methods are specific to haematology.

1.2.1 Common haematology techniques

Whilst haematology laboratories employ a wide range of analyses and analytical techniques, there are some that are performed in large numbers on a daily basis throughout the UK. Many are basic screening tests that form a diagnostic springboard for the initiation of follow-up investigations to identify and characterize specific disease states. (See also Box 1.2.)

Full blood count

The **full blood count (FBC)** is the single most commonly performed routine haematological blood test. It provides information on the number and size of RBCs, WBCs, and platelets. It also measures the concentration of haemoglobin in the blood. The FBC is performed on highly specialized automated analysers. This is a first-line test that is important in providing information on the type of follow-up investigation that may be required.

Cross reference
Haemorrhage, thrombosis, and haemostasis are fully explained in Chapters 13–16.

Blood films

Despite the increasing sophistication of autoanalysers and their ability to accurately recognize blood cells, there is still a place for the light microscope. Microscopy has an important role in the haematology laboratory as it enables the size, maturity, shape, content, and other aspects of the physiology and pathology of blood cells to be assessed. A drop of blood is smeared onto a glass slide, dried, fixed, and stained. Different components within particular cells take up different stains so that they can be identified under a microscope. Blood films are used to investigate the various causes of red cell, white cell, and platelet disorders, and parasitic infections of the blood. Haematological disease, and many disorders generating secondary haematological changes, give rise to a vast array of abnormal morphological findings that can be apparent on a stained blood film. Building up the knowledge and skills to be able to identify the myriad **morphological** changes that can be encountered in haematological practice will be one of the biggest and most fascinating challenges of your career. Bone marrow is also examined in a similar way to whole blood.

Cross reference
Details of the FBC are provided in Chapter 2, and the importance of the bone marrow is outlined in Chapter 3, but these are also themes throughout the book.

morphological
The external appearance (of cells).

BOX 1.2 A rewarding career

One fascinating aspect of practice as a scientist in haematology is that the nature of many haematological diseases is evident when marrying FBC data with blood film appearances and other results. You literally see the disease for yourself when examining a blood film. Indeed, it is not unknown for scientists to discover serious disorders such as leukaemia and malaria before they are clinically evident to medical staff, and that can be immensely rewarding.

Erythrocyte sedimentation rate

An **erythrocyte sedimentation rate (ESR)** is performed to empirically assess the inflammatory response to tissue injury and response to treatment, and so has clinical value in monitoring diseases like **rheumatoid arthritis**. However, an abnormal ESR may also be present in non-inflammatory disease such as cancer and anaemia, and many pregnant women have an abnormal ESR that may be entirely due their pregnancy, and not to ill health. The ESR is a physical property not of blood cells, nor of the plasma proteins, but of whole blood, and measures the **rheological** properties of the blood. The ESR is a test that can be performed either manually or on a specific autoanalyser. A test allied to the ESR is **plasma viscosity**, which is determined by water content and macromolecular components. Pathological alterations in levels of certain plasma constituents will alter plasma viscosity and inform diagnostic decisions.

Coagulation screen

This group of tests measures the time it takes for blood plasma to clot in response to specific stimuli and identifies which particular areas of the biochemistry of blood coagulation may be abnormal. These are front-line tests routinely employed in hospitals to monitor patients who are bleeding or undergoing an operation. A test called the **International Normalized Ratio** (INR) is used to monitor patients who are on the therapeutic anticoagulant drug **warfarin**. **Coagulation screening** can be performed manually, but is more commonly performed on automated coagulation analysers.

Haematinic assays

This type of assay directly measures the concentration of factors such as iron, **ferritin**, vitamin B_{12}, and folate to indicate nutritional and other causes of anaemia. Due to the type of random access analyser that performs the analyses, **haematinic** assays may be performed in the biochemistry laboratory. The results of haematinic assays often dictate the type of therapy patients receive when they are anaemic.

Immunophenotyping

Flow cytometry is a highly specialized technique that allows the measurement of multiple physical characteristics of a single cell, such as its size and granularity. The technique can be extended by pre-treating cells with fluorochrome-conjugated monoclonal antibodies, which are then analysed in an instrument called a **fluorescence flow cytometer**. This technique, **immunophenotyping**, can determine the presence of certain molecules on the surface of the cell, and is essential for the accurate identification of a number of types of haematological malignancy.

Haemoglobin-variant detection

Separation methods such as **electrophoresis** and **high-pressure liquid chromatography** (HPLC) are used to identify variants of haemoglobin. These techniques are important in the diagnosis of haemoglobinopathies such as sickle cell anaemia and thalassaemia.

Molecular genetic techniques

Analysis of DNA is an important tool for the investigation of many inherited haematological disorders such as haemophilia and haemoglobinopathies. Genetic analysis is also important

rheumatoid arthritis

An inflammatory disease that mainly affects the joints.

rheology/rheological

The study of the physical nature of blood or plasma. The principal measurements are the ESR and viscosity.

International Normalized Ratio (INR)

A system established by the World Health Organization (WHO) to standardize the prothrombin time (PT) reporting system for patients receiving vitamin K antagonist oral anticoagulants.

warfarin

A common vitamin K antagonist oral anticoagulant drug used to prevent the recurrence of thrombosis.

ferritin

The main storage protein for iron.

fluorescence flow cytometry

Cells treated with fluorescent dyes move in a liquid stream past a laser beam. Analysis is based on the size, granularity, and fluorescence of the individual cell.

electrophoresis

Migration of dispersed particles (perhaps molecules) relative to a fluid under the influence of an electric field.

high-pressure [/performance] liquid chromatography

Column chromatography technique used to separate, identify, and quantify compounds.

in the management of leukaemia because specific gene defects can indicate the severity of disease and the intensity of the treatment required.

All these techniques, plus many others, are described in detail in later chapters within this book.

Point-of-care testing (POCT)

Traditionally, blood tests are performed in a centralized laboratory by specifically educated, trained, and registered practitioners, who are usually biomedical scientists. Technological advances have led to the development of portable analytical devices with limited or single-test repertoires that allow basic pathology tests to be performed away from the centralized laboratory, either at the bedside, in the outpatient clinic, or in the general practice. This has had two major effects on professional practice:

1. Scientists can, on occasions, practise away from the central laboratory and enjoy a degree of patient contact, such as in an anticoagulant clinic.
2. Other healthcare staff, and even patients, can access and operate the devices.

POCT is more commonly available for analytes associated with the biochemistry laboratory (such as blood glucose), one major exception being the INR for warfarin monitoring. Additionally, haematology POCT is available for haemoglobin, other coagulation tests, and malaria screening.

1.2.2 Additional roles of scientists

As well as analytical investigations, biomedical and clinical scientists have responsibility for a number of other important areas that maintain effective laboratory practice.

Quality management

A key role of all scientists is to ensure that the quality of the results produced by each technique and operator is maintained, which is achieved in a variety of ways. The laboratory will participate in **external quality assessment (EQA)** programmes for all the tests they perform. Examples of such a scheme are those run by the National External Quality Assessment Scheme (NEQAS) in the UK. Additionally, **internal quality controls (IQCs)** are performed to monitor the performance of an assay in real time.

Quality is also maintained by assessing the competency of laboratory staff, at initial training and at regular intervals, and by the laboratory operating a quality management system, which encompasses document control and a rolling audit programme. Much of the document control centres around **standard operating procedures (SOP)**, which as the name suggests, are documents that explicitly state how every aspect of a given procedure must be performed. This is a way of ensuring, as far as is possible, that every test is performed the same way—irrespective of the operator—in order to achieve consistent high-quality results. The SOPs are regularly reviewed and updated where necessary.

The quality assessment body for pathology laboratories in the UK is the United Kingdom Accreditation Service (UKAS). Each medical laboratory is expected to meet a number of defined requirements, or clauses, for quality and competence and is assessed at regular intervals for compliance to the International Organization for Standardization (ISO) 15189 accreditation standard for medical laboratories. It is a requirement of the ISO 15189 standard that all laboratories have a quality manager who ensures compliance to the requirements of the accreditation standard. (See also Boxes 1.3 and 1.4.)

Cross reference

There is a chapter in the *Biomedical Science Practice* volume in this series that covers quality assurance in more detail.

risk assessment
The determination and documentation of the quantitative or qualitative value of risk related to a specific situation/procedure and a recognized hazard.

Health and safety

The clinical laboratory can be a dangerous place. The nature of the work means that potentially biohazardous blood specimens are handled and that the analytical work involves the use of chemicals many of which may be harmful if not controlled. With safety in mind, radioisotopes are very rarely used, and if so, are subject to considerable regulation.

As with all workplace environments, laboratories are subject to legislative health and safety requirements, for example Control of Substances Hazardous to Health (COSHH). **Risk assessments** must be performed for each procedure in the laboratory's repertoire. Laboratory personnel must be exposed to minimal risk, which is achieved in a variety of ways, including the use of personal protective equipment such as laboratory coats, gloves, and safety goggles. Automated procedures have reduced the necessity of the individual coming into direct contact with blood and reagents. In hospital laboratories, a senior biomedical scientist is often designated as the health and safety officer.

Cross reference
Health and safety is covered in more detail within the *Biomedical Science Practice* volume in this series.

BOX 1.3 Quality definitions

It is important to recognize the difference between EQA and IQC because they assess the quality of results in different contexts.

External quality assessment (EQA)

A central agency supplies all registered laboratories with blood samples for analysis by locally employed techniques for each test performed. Local results are compared to those from laboratories throughout the UK to identify consensus or performance anomalies. It must be noted that EQA is only a retrospective evaluation of performance.

Internal quality control (IQC)

Samples of known values are analysed simultaneously with patient samples to monitor the performance of an assay.

BOX 1.4 United Kingdom Accreditation Service (UKAS)

UKAS is the national accreditation body for the UK. It assesses organizations against internationally accepted and agreed standards and grants accredited status to those that fulfil the standards relevant to their practice. The standard used to assess pathology laboratories is the ISO 15189 for medical laboratories, which covers areas such as quality management, performance and validation of analytical processes, personnel, training and education, and health and safety. All UK pathology laboratories must be registered with UKAS. Accreditation of a laboratory by UKAS provides evidence and reassurance to service users that a laboratory has been independently assessed against a recognized standard to evidence it is operating at an internationally acceptable level.

Training and education

Many hospital laboratories have approved training status from the Institute of Biomedical Science (IBMS), the UK's principal professional body in biomedical science. Training is overseen by a scientist who is designated as the training officer and is often the link between the laboratory and the university. Laboratories may have B.Sc. students on work placement or have staff attending university to obtain higher degrees. Scientists must prove that they are competent to perform each individual procedure/assay, and competency assessment records are kept for each employee.

In order to be retained on the government's register of approved healthcare practitioners (the **Health and Care Professions Council (HCPC)**, explained in more detail in section 1.4.3), ongoing **continuing professional development (CPD)** must be undertaken. Each laboratory needs to demonstrate how it assists personnel to develop, and this includes ensuring each member of staff has successfully completed their annual CPD return. This may be achieved by organizing journal clubs or ongoing lecture/tutorial programmes.

Health and Care Professions Council (HCPC)
The government body that regulates professionals in a variety of healthcare disciplines to ensure they are fit to practice.

Continuing professional development (CPD)
The process of maintaining currency in one's field via practical and educational activities that maintain up-to-date and informed professional practice.

As additional educational resources, many departments arrange in-house lectures/courses delivered by internal and external speakers. Early in your career you will likely be a delegate, but later you could be asked to contribute presentations of your own. Further into your career, when you have amassed sufficient knowledge and experience, you may be invited to be a visiting lecturer at a university to educate the next generation of biomedical scientists.

SELF-CHECK 1.3

What are the major roles of the biomedical scientist in the haematology laboratory?

1.3 The role of haematology in the provision of health care

The UK spends 4% of its annual healthcare budget on pathology. Haematology laboratories are performing an increasing number and broader range of tests than ever before for the healthcare community.

As healthcare professionals, we aim to provide a framework for the identification (diagnosis) and treatment of human disease. Broadly speaking, this can be envisaged as comprising two stages: initially when the individual self-refers to their general practitioner (GP), which is termed the **primary care** setting; then, any particular case may require additional investigations or treatments not generally available to the GP, so that the patient is referred to and then cared for by a hospital, which is **secondary care**. Some haematology tests have a role in primary care, where the GP may call on the pathology laboratories for help in diagnosis and (if needed) to monitor the effects of treatment. However, the haematology laboratory is important in virtually all aspects of secondary care.

Exceptionally difficult or complex cases may demand additional tests that could be beyond the scope of a routine laboratory. If so, then referral to a specialist centre at a different hospital may be required—this is described as **tertiary care**. Such specialist centres are often linked to a university, i.e. will be part of a university teaching hospital.

There are many ways to classify human disease. As an example to give you a flavour of the contribution of haematology diagnostics to overall health care, one model is to consider three broad areas of pathology—cancer, connective tissue disease (such as rheumatoid arthritis, osteoarthritis, and their allied conditions), and cardiovascular disease (to include its risk factors such as diabetes and dyslipidaemia). Together, these constitute 70–80% of the healthcare burden of the developed world. The remaining conditions include, for example, infectious diseases and psychiatric illness.

For many patients, one of the first presentations to their GP will be for a group of symptoms that could indicate anaemia. As you will see in Chapters 4–6, this may be due to problems with the RBCs, but may also be due to disease of the blood vessels, heart disease, and/or lung disease. This group of symptoms may be caused by cardiovascular disease, by a connective tissue disease, or by cancer—the symptoms of anaemia (whether actual anaemia or not) are common in all three conditions. These symptoms and history are likely to prompt the GP to order an FBC. However, patients with cancer or connective tissue disease will have a separate group of signs and symptoms, as well as abnormal blood results.

The winter months bring their excess burden of colds and influenza, especially to the elderly, and so with it sequelae such as bacterial throat infections and chest infections. A common prescription to deal with these infections is broad-spectrum antibiotics. These infections can have several effects on the blood—notably an increase in the WBC count and changes to the ESR. However, the necessity for a prolonged prescription of antibiotics over several months may suggest something more serious, especially if accompanied by symptoms of anaemia, such as leukaemia, where the malignant cells are made at the expense of normal cells. The late stages of this serious disease include an increased risk of bruising and bleeding, signs that can be investigated by the haematologist.

myocardium
Heart muscle.

Cardiovascular disease can lead to heart failure, irregular heartbeat, or cardiac arrest. The most common reason for cardiac arrest is the presence of clots within the coronary arteries that prevent blood flow to the **myocardium**. Once deprived of this blood, with its life-giving oxygen, the muscle cells of the heart will fail and die, leading to an often terminal heart attack. It follows that the causative process is the development of a clot (thrombus), which may erupt from areas of damage (lesions) within the coronary artery itself, or by thrombus formation elsewhere in the body which travels to the heart and there becomes lodged. Thus, one role of the haematology laboratory in the care of the patient with cardiovascular disease is to provide knowledge of the coagulation system. However, patients with cancer are also at risk of thrombosis, and, conversely, a small proportion of people who find themselves with an unexplained clot in the legs or the lungs have an undiagnosed cancer.

SELF-CHECK 1.4

What percentage of the annual healthcare budget is spent on pathology?

Thus, the haematology laboratory can help with the diagnosis and management of all three of the major human disease groups.

1.3.1 The overlap with other pathology disciplines

Although most scientists will specialize within a particular pathology discipline, it is important to remember that each area contributes information towards a complete understanding of disease.

Blood transfusion

The position of the **blood transfusion** laboratory is clear, as in the vast majority of pathology departments the haematology and blood transfusion laboratory are scientifically and organizationally linked. Most scientists specializing in haematology will also have trained in blood transfusion. However, the many differences between these two arms of laboratory science are emphasized by the requirement to deal with each part in a separate textbook of this series.

blood transfusion
The science of ensuring the safe transfer of blood and other substances from one person to another.

The blood transfusion laboratory is concerned with the preparation of blood and blood products for transfusion, which includes compatibility testing between donor and recipient blood. The findings of anaemia, thrombocytopenia, or abnormal blood coagulation in the haematology laboratory can lead to requests for the blood transfusion laboratory to prepare red cells, platelets, or plasma for infusion into the patient. A considerable degree of quality control is required, as errors made in the blood bank can be fatal.

Cross reference
Blood transfusion is the subject of a separate textbook in this series, namely *Transfusion and Transplantation Science.*

Immunology

Immunology involves the study of the immune system and its disorders. Deficiencies of the immune system are investigated, together with their association to infection, tumour growth, autoimmunity (e.g. rheumatoid arthritis), and allergies. The immunology laboratory is also involved in tissue-matching for organ transplants, and plays an important role in monitoring and treating patients with acquired immune deficiency syndrome (AIDS).

There is a clear link between the immunology and haematology disciplines. Immunology is concerned with white cell function and antibody production, whilst haematology looks at white cell numbers, morphology, and some aspects of function. There is an inevitable crossover; for instance, leukaemias are recognized initially from the FBC, blood film, and bone marrow results and then characterized using immunophenotyping, all of which are performed in the haematology laboratory.

Biochemistry

The biochemistry laboratory is concerned with the study of changes in the chemical composition of blood and other body fluids to assist in the diagnosis and monitoring of disease, for example blood sugar in diabetes and liver function tests in liver disease. The laboratory is also involved in other analyses, such as toxicology investigation, prenatal screening for Down syndrome, and for neural tube defects.

There is overlap with haematology, as conditions such as liver disease may also exhibit abnormal blood cells and deranged blood clotting. Renal function is investigated in the biochemistry laboratory, but if disease in this organ is present, the resultant anaemia is diagnosed and monitored in a haematology laboratory. In this instance, the anaemia occurs due to reduction in levels of a hormone (erythropoietin) produced in the kidneys that stimulates red cell production.

Bacteriology

Bacteriology, a branch of microbiology, is involved in the identification of microorganisms that cause infections such as food poisoning, bacterial meningitis, and septicaemia. Microorganisms are cultured and subjected to tests (such as molecular genetics) to establish their identity. Subsequent tests identify which antibiotics will be effective in treating the specific organisms that have been isolated. Infection is also of interest to the haematologist, as it can result in a high WBC count; in cases of severe septicaemia, very abnormal clotting is seen accompanied by depleted platelet numbers, which have severe consequences for the patient. White cells can adopt altered morphology in response to infection.

Cross reference
You will meet blood-borne parasites and their laboratory detection in Chapter 7.

The bacteriology laboratory is increasingly involved in the monitoring of hospital-acquired infections and also investigates for gut parasites. Malaria is a parasitic infection with a life-cycle phase that occurs partly in the bloodstream, and thus it is detected in the haematology laboratory, as are other blood-borne parasites.

Virology

The virology laboratory diagnoses infectious diseases such as hepatitis, rubella, human immunodeficiency virus (HIV), severe acute respiratory syndrome coronavirus 2 (SARS-CoV-2), and influenza. Various types of hepatitis virus cause liver disease, which will lead to both abnormal biochemistry and haematology results. The treatment for conditions such as infection with HIV can lead to a low WBC count, therefore the haematology laboratory will be co-involved in the monitoring of the patient's treatment. Some viral infections, such as infectious mononucleosis (with the Epstein–Barr virus, causing glandular fever), present with characteristic morphological changes in white cells that are detected both microscopically and serologically in the haematology laboratory.

Histopathology

Cross reference
Infectious mononucleosis and its investigation are discussed in detail in Chapter 8.

The histopathology laboratory processes biopsy, surgical resections, and post-mortem tissue samples. The most closely related area to haematology is examining trephines from bone marrow and biopsies from lymphoma patients. A co-morbidity to many types of cancer is an anaemia, and cancer may also cause a deep vein thrombosis. Consequently, an abnormal haematology profile may be the first indication of a malignancy that may ultimately be diagnosed by histopathologists.

Cytology

In cytology, cell samples are examined for the presence of cancerous and precancerous cells, including cervical screening. As outlined above, cancer may cause both an anaemia and a deep vein thrombosis.

Molecular genetics

Cross reference
Cytogenetic analysis and molecular genetics are discussed in much greater detail in Chapters 9–12.

Cytogeneticists study chromosomes, which is key to understanding certain diseases such as Down syndrome, and abnormalities such as Robertsonian translocation. Moving to looking within whole chromosomes, molecular genetic analysis plays an important role in the diagnosis and management of diseases such as haemoglobinopathy, leukaemia, and lymphomas, and clinicians need to marry the results from both laboratories when diagnosing subtypes of haematological malignancies and making treatment decisions.

Stem cell laboratory

Stem cell laboratories prepare harvested stem cells for transfusion into patients, and are important for the treatment of some patients with haematological malignancies. Cytogenetics and stem cell laboratories are only found in teaching hospitals.

SELF-CHECK 1.5

Which discipline is most closely related to haematology?

1.4 The role of the professional body

A **professional body** is a learned organization that represents a particular profession by maintaining control or oversight of the legitimate practice of that profession. It seeks to further the profession, the interests of individuals engaged in that profession, and to safeguard the public interest.

The major professional body for biomedical and clinical scientists employed in UK pathology laboratories is the IBMS. However, the diversity in the biomedical/clinical sciences and the requirement to specialize often demands additional professional bodies. As far as haematology is concerned, this is the British Society for Haematology, which predominantly represents medical practitioners, but also other healthcare staff whose practice involves/encompasses haematology.

1.4.1 The Institute of Biomedical Science

The Institute of Biomedical Science (IBMS) is the professional body for biomedical and clinical scientists in the United Kingdom. The institute was founded in 1912 and aims to promote biomedical/clinical science and its practitioners. There are approximately 20,000 members. The main roles of the IBMS include setting standards of practice to protect patients, assessing competence for biomedical scientists to practise, accrediting university degrees, providing postgraduate professional qualifications, and organizing a CPD scheme. Additionally, the Institute plays a key role in assessing qualifications for registration with the HCPC.

The IBMS holds a biennial conference and publishes monthly *The Biomedical Scientist*, which contains science articles, news, and job adverts. The Institute's scientific publication is the *British Journal of Biomedical Science* containing peer-reviewed scientific papers. It is not mandatory for biomedical or clinical scientists to be members of the IBMS, but there are many benefits, including the authority to adopt designatory initials that indicate the class of membership, such as MIBMS (member) or FIBMS (fellow). (See also Box 1.5.)

The IBMS website gives full details. http://www.ibms.org.

1.4.2 The British Society for Haematology

The British Society for Haematology (BSH) is the main haematology society in the UK; its main objective is to advance the study and practice of haematology. The BSH provides education, information, and networking to haematologists. The Society holds an annual scientific meeting and publishes the peer-reviewed journal *British Journal of Haematology*. It is likely that senior members of staff will be members of this society, whilst specialists may also be members of groups with more focused appeal (such as coagulation specialists and the British Society for Thrombosis and Haemostasis).

A major role of the BSH is to publish guidelines from the British Committee for Standards in Haematology. For full information about the BSH, see http://www.b-s-h.org.uk.

BOX 1.5 Continuing professional development

The material learnt during your undergraduate degree that qualifies you to enter the profession will not provide you with all the knowledge you need for your entire career. Medicine, biomedical science, and technology progress at a rapid pace and it is a requirement that health professionals constantly strive to maintain and update their knowledge and professional practice. The IBMS has a mature scheme that allocates credits for a range of professional and educational activities. If a practitioner accrues sufficient credits within a specified time period, they are awarded a CPD Diploma.

CPD activities must be varied in nature to update different areas of practice. Suitable CPD activities include attending conferences and lectures, learning new laboratory procedures/techniques, reading up-to-date research articles, attending journal clubs, giving lectures/tutorials, publishing scientific papers, and writing reflective statements on workplace learning. The IBMS publishes monthly journal-based learning (JBL) exercises where practitioners read a relevant article and then answer questions that have been set by IBMS examiners. Pass marks in JBL attract additional CPD points.

1.4.3 Health and Care Professions Council (HCPC)

It is mandatory that all healthcare scientists practising in the NHS in the UK are registered with the HCPC. The HCPC is the regulator and exists to protect the public. Practitioners have to reach specific standards of education and professional competence to gain entry onto the register and attain the status of a registered healthcare professional. All healthcare professionals are required to continue to keep their knowledge and skills up to date while they are registered and practising in their profession. At re-registration a percentage of individuals are selected at random for audit, whereby they are expected to provide a summary of practice for the previous two years, evidence of varied CPD activities, and a statement with evidence that demonstrates how they have met the standards. The HCPC website gives full details: http://www.hpc-uk.org.

1.4.4 Chartered Scientist

A chartered professional is a practitioner who has achieved a specific level of skill or competence in their area of work. Although it is not a requirement to practice, chartered status is a qualification that acts as a form of accreditation by evidencing education and competence at a high level.

Chartered Scientist (CSci) is awarded by the Science Council in the UK to recognize high levels of scientific skill and experience independent of the discipline in which a scientist undertakes their practice. A scientist achieving chartered status obtains prestigious professional acknowledgement of working at the full professional level, which can be recognized on an equal footing with scientists in other disciplines. Additionally, benchmarking professional scientists at the same high level can engender public trust and confidence in science and scientists.

Applicants for CSci are normally expected to be qualified to at least Master's level and must demonstrate high-level competencies in the following five areas:

- application of knowledge and understanding;
- personal responsibility;
- interpersonal skills;
- professionalism;
- professional standards.

Once registered, chartered scientists must maintain their status by meeting CPD requirements each year. Biomedical and clinical scientists in the UK can apply for CSci through the IBMS.

SELF-CHECK 1.6

What are the major roles of the professional body?

SELF-CHECK 1.7

Why is CPD important to effective professional practice as a laboratory scientist?

Chapter summary

- Haematology is the study of diseases of the blood and blood-forming organs.
- The scientist in haematology performs analyses that assist in the differential diagnosis and management of disease.
- Haematology inevitably overlaps with other physiologies and disciplines within pathology.
- The roles of the various professional bodies include setting professional, scientific, educational, and quality-control standards.

Discussion questions

1.1 Why is haematology considered a discrete discipline when the white cell aspects could be part of the immunology discipline and the measurement of certain molecules and elements could be contained within biochemistry?

1.2 What effect would the loss of a haematology service have on healthcare delivery in a hospital?

Answers to self-check questions and case study questions are provided as part of this book's online resources.

Visit www.oup.com/he/moore-fbms3e.

2

Major haematology parameters and basic techniques

Andrew Blann, Gary Moore, and Gavin Knight

In this chapter, we will build on the introduction to certain aspects of haematology made in Chapter 1, and will describe, in depth, principal haematology tests (such as the full blood count) and many of the basic techniques that are used to define particular indices, such as haemoglobin.

Learning objectives

After studying this chapter, you should confidently be able to:

- appreciate the importance of different anticoagulants and glass/plastic tubes for the different blood tests requested;
- describe major haematology techniques: spectrometry, microscopy, light scatter, impedance technology, and calibration;
- be aware of the major components of the full blood count, which are the red blood cell indices, the white blood cell indices, and platelets;
- describe the differences in the values of the erythrocyte sedimentation rate and plasma viscosity;
- describe Köhler illumination and explain how this method improves sample clarity when using light microscopy;
- understand coagulation screening;
- grasp the important aspects of the micronutrients iron, vitamin B_{12}, and folate, as well as plasma proteins such as transferrin and ferritin.

2.1 Obtaining a sample

Almost all of the haematologist's work is on blood or bone marrow.

2.1.1 Obtaining a sample of blood

Venepuncture is the process of obtaining a sample of blood, generally from a vein (usually the median cubital vein) on the inside of the elbow joint (hence venous blood). The skin at this location is very soft, and veins are close to the surface of the skin (that is, they are superficial) (Figure 2.1). Until recently, most blood samples were obtained using a needle that had to be fitted onto a syringe. However, this method has been superseded in many hospitals by the use of **BD Vacutainers**®, glass tubes with an inbuilt vacuum, in which a needle draws blood directly from the vein without the use of a syringe (Figure 2.2). However, blood may also be obtained via a winged infusion set, or a semi-permanently sited cannula, and, rarely, arterial blood may be obtained. Venepuncture itself is a learned skill that requires completion of a training course. These are generally organized by the training section of the hospital. However, small quantities of blood can be obtained from soft tissues of the thumb, the ear lobe, or the heel. This is done by making a small incision in the soft tissue with a sharp needle called a lancet and then allowing the blood to drip directly into a blood collection tube–this method is particularly useful if a full blood sample is not required, or if a sample is required from a child or an infant from whom venepuncture may be difficult.

2.1.2 Obtaining a sample of bone marrow

All blood cells arise from the bone marrow, the anatomical site where blood cells are generated, so that analysis of bone marrow can in many cases be very informative. However, obtaining a sample of the bone marrow can be difficult. Two approaches are common. The first is to push

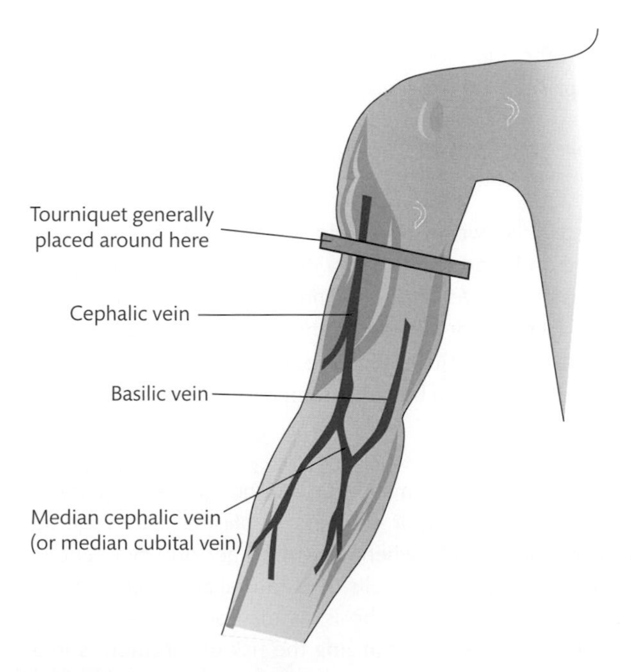

FIGURE 2.1
Anatomy of the arm showing superficial veins suitable for venepuncture.

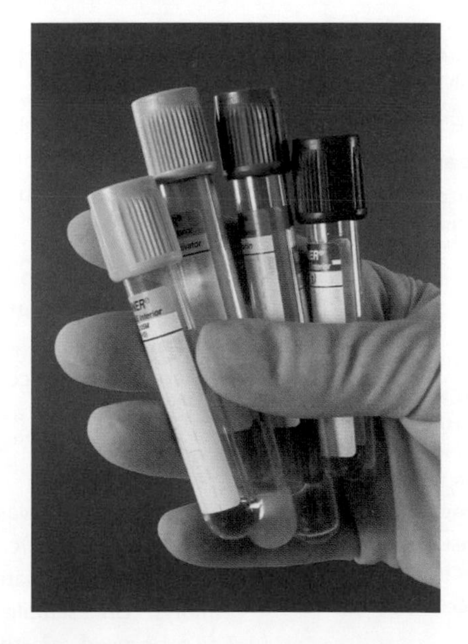

FIGURE 2.2
BD Vacutainers® commonly used in haematology. They have different coloured tops that indicate a particular anticoagulant (or lack thereof) and a label for patient details.© Claire Paxton and Jacqui Farrow/Science Photo Library.

Cross reference

The bone marrow and the process of bone marrow aspiration are discussed in more detail in Chapters 3 and 10.

a heavy-duty needle into a particular bone (such as the breastbone or the hip), and then suck out some of the bone marrow. This procedure is referred to as a **bone marrow aspiration**. The second is to obtain a larger sample of bone that will include bone tissue as well as bone marrow. This is called a **trephine** procedure, and requires an even more substantial needle and should only be used to biopsy tissue from the hip.

2.2 Anticoagulants

micronutrients

Minerals and vitamins absorbed from the diet and required in trace amounts for the correct function of a cell or biochemical process.

fibrinogen

Fibrin precursor and ligand that facilitates platelet-platelet aggregation and forms a mesh that is the basis of the thrombus.

sodium citrate

An anticoagulant chemical added to blood to allow the measurement of coagulation proteins, prothrombin time, and the activated partial thromboplastin time.

heparin

A natural anticoagulant often added to blood to allow additional analyses of white blood cells.

therapeutic anticoagulation

The use of anticoagulants as a treatment for the increased risk of thrombosis.

Once removed from the body, blood will rapidly (in perhaps two-five minutes) form a semi-solid mass of cells (the clot) and a fluid. Separation of the clot from the fluid in a **centrifuge** provides a clear, often yellowish liquid called serum, required for the estimation of **micronutrients**, such as iron and vitamin B_{12}, proteins, and for other tests. A large number of blood cells and coagulation proteins (such as **fibrin**, formed from **fibrinogen**) will make up the clot, so it cannot be used for most tests of red blood cells, white blood cells, platelets, and the proteins involved in coagulation. So in order to be able to analyse these cells and proteins, we must ensure that the blood does not clot, and do so with anticoagulants. In some cases, BD Vacutainers® are supplied with an anticoagulant already present. Other BD Vacutainers® can be obtained with no anticoagulant, and will therefore ultimately provide serum.

Different blood tests require different anticoagulants:

- A **full blood count** (FBC), which provides information about the numbers of red blood cells, white blood cells, and platelets performed on blood that is anticoagulated with a sodium or potassium salt of ethylene diamine tetra-acetic acid (**EDTA**). This blood sample can also be used for more specialized tests, such as for analysing the integrity of the membrane of the red blood cell and for producing a blood film.
- **Coagulation** tests are invariably performed on plasma obtained from whole blood that has been anticoagulated with **sodium citrate** (of course, if blood was allowed to clot, the coagulation proteins would have been used in the clotting process). The plasma itself is obtained after the BD Vacutainer®, or other tube containing the blood, has been centrifuged. The blood cells, lying below the plasma, are generally discarded. However, at least one common coagulation test (the International Normalized Ratio, INR) *can* be performed on whole blood obtained from a finger prick as part of the monitoring of the effect of an oral anticoagulant such as warfarin.
- The erythrocyte sedimentation rate (ESR), which gives a global score of a physical property of the blood, can be assessed on blood that is held within its own dedicated glass tube: blood clotting in this tube is also prevented by sodium citrate. The result itself is determined by a combination of factors that include the nature of the red blood cell and the presence of certain plasma proteins. However, some haematology analysers are designed to be able to provide an ESR result on the same sample of blood as is used for the FBC.

Cross references

Red blood cells, and diseases associated with them, are described in considerable detail in Chapters 4-6, blood-borne parasites in Chapter 7.

White blood cells are discussed in more detail in Chapters 8-12.

We discuss platelets further in Chapters 13-16.

Details of the anticoagulant action of heparin and warfarin are given in Chapter 16.

Like the ESR, the plasma viscosity test provides information on another global property of the plasma; specifically, it describes the 'thickness' or 'thinness' of plasma. Rarely, blood may be anticoagulated with **heparin**—for example, when we wish to perform cytogenetic analysis on a sample or immunophenotype the white cells. This natural anticoagulant can also be used as a therapy to reduce the risk of thrombosis, as may occur after surgery for hip replacement, for example. The process of minimizing the risk of thrombosis in a clinical setting (called **therapeutic anticoagulation**) demands an entire section of its own (Chapter 17).

Key Points

- **Substances such as micronutrients, lipids, and proteins are measured in the non-cellular component of the blood. This is serum (the fluid that remains after the blood has clotted) and plasma. The latter can only be obtained if the blood has been anti-coagulated with an agent such as EDTA or sodium citrate.**
- **EDTA functions as an anticoagulant by irreversibly chelating the calcium ions that are essential in the blood-clotting processes you will meet in Chapter 13. Chelating agents such as EDTA form soluble complex molecules upon binding to certain metal ions which inactivate the metal ions. The word 'chelate' is derived from the Greek for the claw of a lobster because there are two caliper-like groups in the molecule that function as associating units and fasten to the central atom to produce heterocyclic rings.**
- **Trisodium citrate functions as an anticoagulant by forming a loose and reversible ionic complex with calcium ions. This makes it ideal for coagulation tests because many of these tests involve the addition of reagents containing calcium ions to initiate clotting *in vitro*. Additionally, platelets retain functional ability in citrate-anticoagulated blood but not in EDTA, so platelet function testing is performed on citrated samples.**

One problem with some anticoagulants is that they interfere with the blood itself. The best-known artefacts of anticoagulants include a swelling of red blood cells and platelets by EDTA. It follows that the FBC needs to be analysed within a few hours if we are to be sure that the indices of red blood cell and platelet volume remain accurate.

SELF-CHECK 2.1

What are the major differences between serum and plasma?

SELF-CHECK 2.2

What are the two major laboratory anticoagulants used in haematology, and which of these is required for each of the major blood tests?

SELF-CHECK 2.3

Which micronutrients are essential for the production of healthy blood cells?

2.3 An introduction to major techniques

Once a blood sample has been obtained, it must be analysed. Almost all routine haematology results are obtained from tests performed on highly sophisticated machines, or from viewing blood films that have been stained with special dyes (which allows them to be seen with a microscope). Without doubt, the routine haematology laboratory (and, indeed, the biochemistry laboratory) is dominated by one or more autoanalysers that will provide the FBC, and others that provide coagulation test results. Although there are technical variations, these autoanalysers all use a small number of basic techniques. These are generally spectrometry, impedance, cytochemistry, and flow cytometry. Plasma proteins are often quantified by an immunoassay.

Key Points

Relating a particular result (such as a high white blood cell count) to a particular clinical condition (such as septicaemia) is often one of the more rewarding aspects of haematology. However, time and time again an experienced practitioner will be called on to provide details of the basic science underlying an unusual test or to explain an atypical observation.

2.3.1 Spectrometry

This technique is used to provide the haemoglobin result. A small sample of blood is mixed with a non-ionic detergent that destroys the membrane of the red blood cell so that a red/pink 'soup' is created, of which the dominant component is haemoglobin. However, there are several different subtypes of haemoglobin, so an additional chemical (such as a mixture of potassium, ferricyanide, and cyanide ions, known as Drabkin's solution) is required to convert them to a single species, the concentration of which is measured by a spectrometer. This technique converts most types of haemoglobin to one single species (that is, cyanmethaemoglobin) and is recommended by the World Health Organization (WHO) for measuring haemoglobin concentrations. The machine assesses the extent to which the passage of a beam of light is reduced by the density of the solution. Thus, the density of the red/pink colour is proportional to the haemoglobin present in the original blood sample according to the law of Beer and Lambert.

Cross reference

Different types of haemoglobin are described in Chapter 4.

2.3.2 Impedance

This process relies on the ability of an ionic fluid (an electrolyte) to assist the passage of electricity from one electrode to another. However, this passage of electricity can be interrupted by particles. In haematology autoanalysers, this principle is exploited in a chamber with a small pore that allows blood cells to pass through. Each time a blood cell passes through this pore, the passage of electricity is impeded by an amount proportional to the size of that cell by the non-conductive properties of the cells' phospholipid membranes. Hence the higher the number of cells present, the greater is the frequency of disturbances in the flow of electricity. Sophisticated software can convert the number of disturbances not only to the number of cells, but also the size of the cell that is responsible for impeding the current. This method is capable of providing a red blood cell count, white cell count, platelet count, and the mean cell volume.

2.3.3 Cytochemistry

Cytochemistry utilizes the presence of certain enzymes (such as peroxidase and acid phosphatase) and other molecules (such as nucleic acids, fats, and iron) within some normal cells, and also in abnormal cells, such as in certain types of leukaemia. As regards the latter, cytochemistry can be used to identify cells of a particular lineage (myeloid or lymphoid) or stage of the disease. This technique can be used by an autoanalyser, but it can also be used on a film of blood or bone marrow dried onto a glass slide.

Cross reference

Cytochemistry can be used for the investigation of haematological malignancies (Chapter 10) and in iron-deficient anaemia (Chapter 5).

2.3.4 Flow cytometry

This process is allied to both microscopy and chemistry. In its most simple form, it relies on the disruption of a fine beam of light, perhaps provided by a laser. The degree to which this beam of light is interrupted (that is, forward scattered light) is proportional to the size of the cell, whilst the degree to which light is scattered obliquely (side scattered light) tells us of the granularity of the cell (Figure 2.3). The frequency of the interruptions of the beam of light gives the number of cells. In some analysers, chemicals are used to strip the cytoplasm from the cell so that only the nucleus is analysed: in others, certain chemicals are added to a population of cells to determine those that have enzymes that can metabolize the chemical, and so help in identification. These methods are used by the manufacturers of several different types of analyser in routine laboratory use to determine the number of the different types of white blood cells, these being the neutrophil, lymphocyte, monocyte, eosinophil, and basophil (Figure 2.4).

2.3.5 Fluoresence flow cytometry

Flow cytometry can be extended, with the use of antibodies (almost always monoclonal) conjugated to fluorochromes (a chemical that absorbs light at one wavelength, and emits it at a second wavelength), so that different subtypes of cells can be quantified. Such an instrument is called a fluorescence flow cytometer (FFC), the principles of which are explained in Figure 2.5. Probably the most common use of this process in haematology is in the diagnosis and management of white cell malignancies, principally leukaemia. However, in addition, this powerful technique can also be used in the diagnosis and management of a wide range of red cell, white cell, and platelet disorders (Table 2.1). The value of this technique is demonstrated by the frequent referrals to FFC analysis in chapters to come.

Cross reference

Use of FFC analysis in leukaemia classification is outlined in Chapters 10-12.

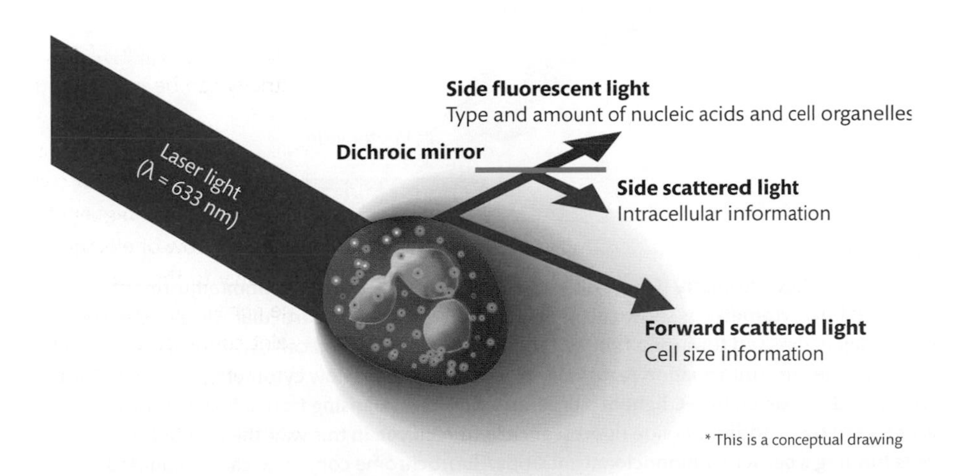

FIGURE 2.3

The principle of the scattering of light in flow cytometry. The cell interrupts the passage of the light (forward scattered light) and so gives information about its size. Light is also deflected by the internal components of the cell, dominated by the nucleus and granules (if present), hence side scattered light. Figure courtesy of Sysmex.

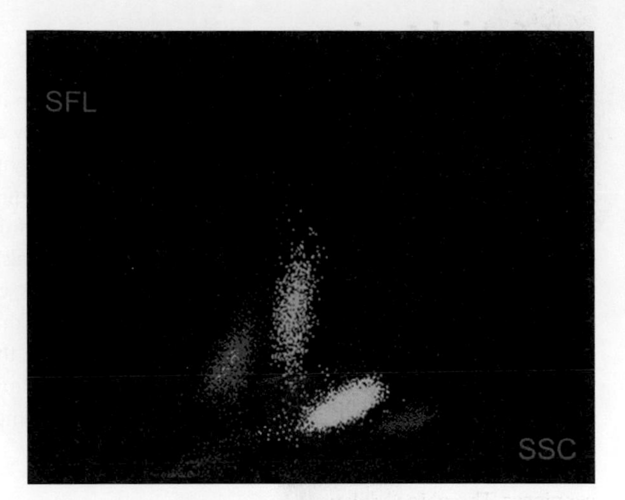

FIGURE 2.4
White blood cell populations identified by flow cytometry. The patterns of side scatter light (SSCL) and forward scatter light (SFL) are reasonably specific for each of the different white blood cells. In this instance, the flow cytometer software has arbitrarily given each population a different colour to help identification. Figure courtesy of Sysmex.

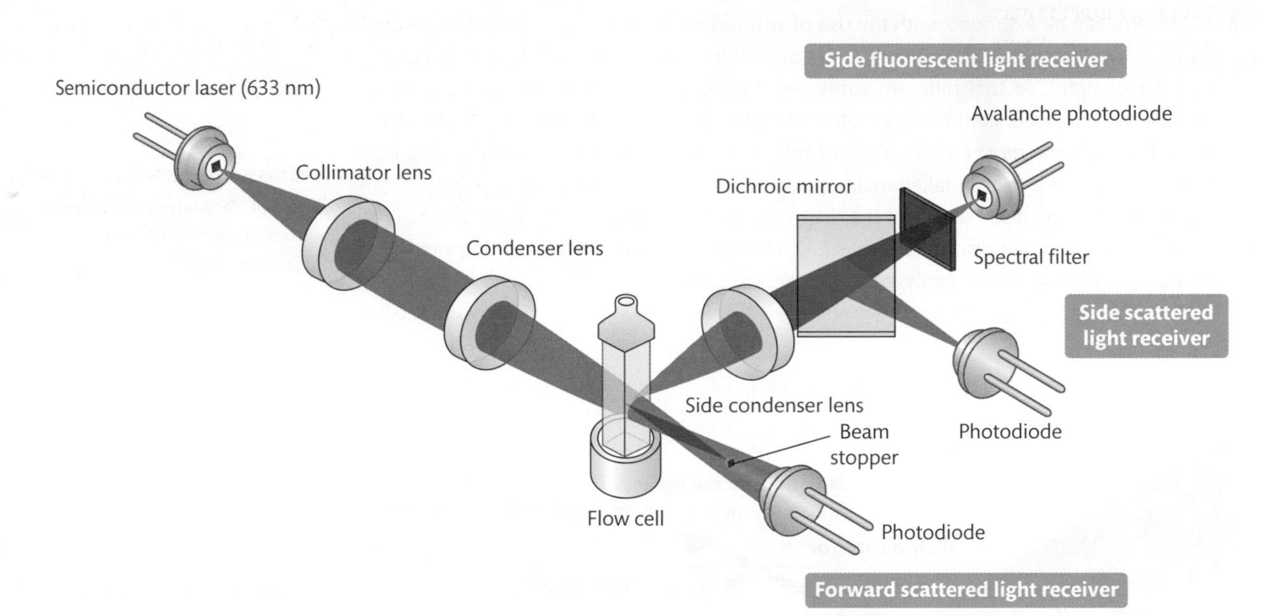

FIGURE 2.5
Fluoresence flow cytometry (FFC). This figure illustrates the relatively minor adjustment of standard flow cytometry as shown in Figure 2.3. In FFC, light of a particular wavelength from a laser is again directed towards a flow cell containing the target red cells, white cells, or platelets. SFL gives the same information regarding cell size as standard flow cytometry, and again SSCL is detected. However, in FFC light of a different wavelength, arising from a fluorochrome, is detected by an additional side fluorescence light receiver. In this way, the number of cells binding a particular monoclonal antibody/fluorochrome conjugate can be counted. Furthermore, the intensity of the collected SSCL is proportional to the amount of fluorochrome that each cell has bound, giving us information about the relative density of expression of the molecule to which the monoclonal antibody/fluorochrome is directed. The process requires the cell population to be incubated with a fluorochrome-linked monoclonal antibody to the cell surface molecule of interest (such as CD3 for T lymphocytes). Typical fluorochromes include fluoroscein isothiocyanate (FITC). Figure courtesy of Sysmex.

2.3.6 Light microscopy

Although modern haematology autoanalysers give a great deal of information on the different types of cells in the blood, light microscopy remains an important aspect of laboratory haematology, enabling the morphology of blood cells to be evaluated and a manual differential count to be performed. Although the FBC can provide a broad range of information about a specimen, the appearance of red cells, white cells, and platelets can provide more information useful to clinicians when considering diagnosis. Light microscopy can also identify parasites, such as those that cause malaria.

Cross references

More information on microscopy can be found in the *Biomedical Science Practice* volume of this series.

Chapter 7 shows several blood-borne parasites.

Microscope preparation

Keeping the light microscope in the best possible condition is essential to obtain the most information from a patient's blood film. The light microscope should be set up to ensure full and consistent illumination across the field of view—the area of the blood film seen through the eyepieces. The objective lens should be free from grease and cleaned using an appropriate lens cloth. Dust should be removed from the light source. If in any doubt about how to clean your microscope, consult the manufacturer's instructions or your laboratory's standard operating procedure. The structure of a modern light microscope is provided in Figure 2.6. The combination of a well-cared-for microscope and a good-quality blood film is necessary for diagnostic purposes.

Köhler illumination

The method used to achieve consistent illumination is called **Köhler illumination**. Used daily, Köhler illumination ensures optimal conditions for examining a specimen microscopically.

The first step in achieving Köhler illumination involves selecting the ×10 objective lens, placing a blood film on the stage, and focusing the microscope. Next, the field diaphragm is reduced

TABLE 2.1 **Examples of the use of fluoresence flow cytometry analysis.**

Red blood cells	• Classification and the determination of the frequencies and polymorphisms of blood group molecules (such as molecules of the Rh system, collectively CD240) • Confirming or refuting the preliminary diagnosis of paroxysmal nocturnal haemoglobinuria (CD55 and CD59)
White blood cells	• Monitoring of hairy cell leukaemia (which can call for combination of up to eight CD molecules) • Classification of a lymphocytic leukaemia as being of B-cell (CD5, CD19, CD23) or T-cell (CD3, CD4, CD8) origin • To improve a scoring system to help determine those patients with the myelodysplastic syndrome who are at risk of progression to acute myeloid leukaemia
Platelets	• Determining the degree of activation of cells (such as expression of CD62P) • Confirming or refuting the preliminary diagnosis of Glanzmann's thrombasthenia by the expression of glycoprotein IIbIIIa

CD = cluster of differentiation, a simple international classification system for cell surface molecules.
For additional details of all points, see respective chapters.

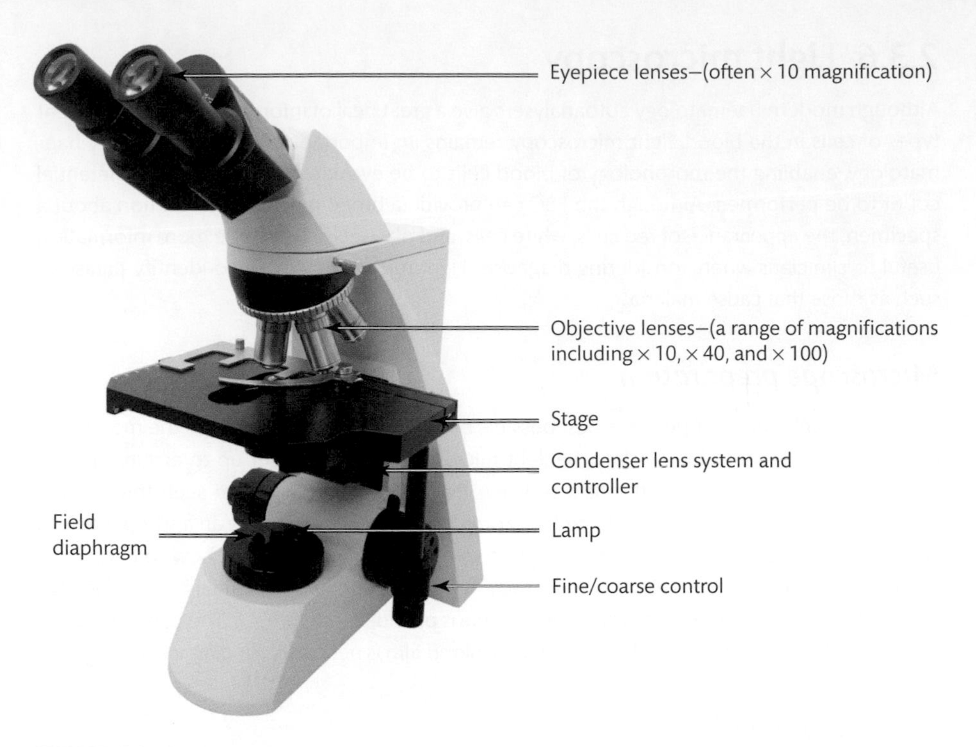

FIGURE 2.6
The basic structure of a modern light microscope. True magnification involves multiplying the objective magnification by the eyepiece magnification: e.g. 40 × 100 = ×400 magnification.
© Dragan Trifunovic/istockphoto.com.

to enable visualization of a small point of light through the eyepiece. The *x* and *y* condenser adjustment screws are used to centre the small point of light within the field of view. At this point, the field diaphragm is centred under the condenser. Finally, the field diaphragm must be opened just sufficiently to ensure that the entire field of view is illuminated. Consistent and reproducible illumination has now been achieved.

SELF-CHECK 2.4

List the steps involved in achieving Köhler illumination. Outline the benefits of setting a microscope up in this manner.

Specimen examination

A blood film should first be examined without the use of the microscope, i.e. macroscopically, to examine the coloration of the film, the thickness of the spread, and distribution of blood on the slide. Abnormal cellular aggregates are usually appreciable macroscopically. Once the requirements of macroscopic examination have been satisfied, microscopic examination can begin.

At low power, for example with a ×10 objective lens, assess the distribution of cells to identify any abnormal white cells, cell clumps, or fibrin strands present. Once the low power assessment has been completed, a ×40 or ×50 objective lens can be used to assess the detail of the cells.

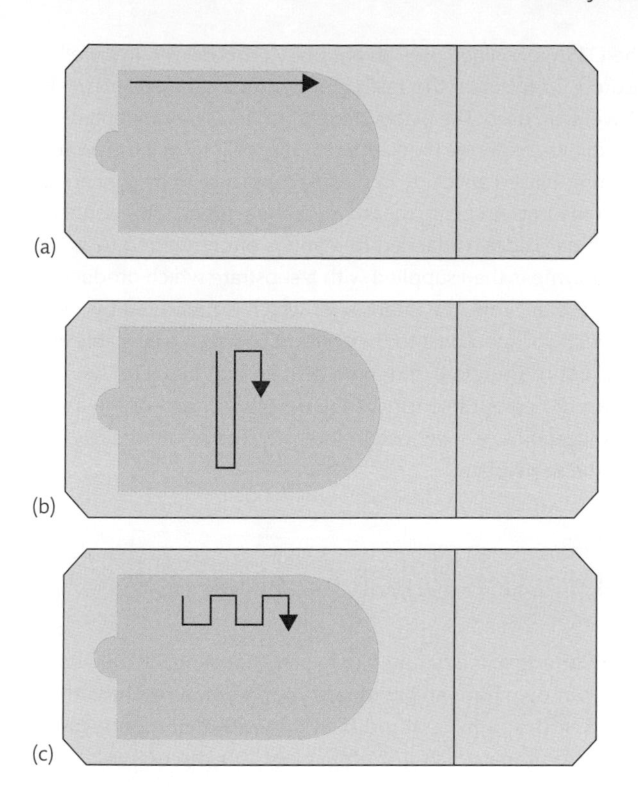

FIGURE 2.7
Diagrams of blood films showing tracking patterns employed in differential white blood cells: (a) tracking along the length of the film; (b) modified battlement method; and (c) battlement method; two fields are counted close to the edge parallel to the edge of the film, then four fields at right angles, then two fields parallel to the edge and so on. From Bain BJ. *Blood Cells: A Practical Guide*, 4th edn. Blackwell, Oxford, 2006.

The ×40 objective is the standard lens to use, with air and oil lenses available. Remember, total magnification equals objective lens magnification multiplied by eyepiece lens magnification. A ×400 magnification in conjunction with the *battlement method* should be used to obtain a good cross section of the blood film. See Figure 2.7 for examples of the different methods employed to examine blood films. At this high magnification, crucial characteristics of blood cells may be determined. (See also Box 2.1.)

Each different type of cell line should be examined, regardless of the reasoning for the request to examine the blood film. Often, having a striking abnormality under the microscope can distract the novice into examining only the most abnormal of cells, although, in practice, it is important to assess all cell lineages for number and appearance.

2.3.7 Immunoassays

Whilst they are not generally part of routine screening, immunoassays are commonly used as follow-up tests to measure the plasma or serum concentrations of various proteins, including those involved in coagulation (in plasma) and others to do with iron and vitamins (in serum). Immunoassays use two antibodies: one immobilized antibody that recognizes a specific protein (an antigen), or a specific section of a protein, to capture the molecule, and a second antibody, which is linked to a variety of detection methods. Immunoassay methods in common use include:

- **enzyme-linked immunosorbent assays (ELISA);**
- latex immunoassays (LIA);
- immunonephelometry.

Enzyme-linked immunosorbent assay (ELISA)
An important technique in many disciplines of biomedical science, generally for detecting molecules within fluids.

You can see in Figure 2.8 that the ELISA technique uses an antibody immobilized in a well to capture the protein being measured. To facilitate the testing of multiple samples, plastic or polystyrene plates containing 96 wells are used. The plates are referred to as the solid phase of the assay. Patient plasmas, often in diluted form, are then added to the wells so that the protein being measured will bind to the immobilized antibody in direct proportion to its concentration. A second antibody specific to the protein being measured is then added. This antibody has an enzyme (such as horseradish peroxidase) attached to it and is often referred to as the 'tag' or 'secondary antibody'. The enzyme is then supplied with a substrate which produces a coloured product when cleaved by the enzyme, the intensity of which is measured by spectrometry. The colour intensity is directly proportional to the amount of enzyme available as a result of binding to the captured protein. Therefore, the more protein that binds to the capture antibody, the greater the measured colour intensity at the detection stage of the assay. A patient deficient in the protein will generate a lower colour intensity. The results are read off a standard curve that is assayed in the same plate.

BOX 2.1 Practice identifying blood cells

For educational purposes, try obtaining some samples and corresponding FBC results from your laboratory. Produce your own films and compare your manual results with those obtained from the analyser. If you have any difficulties with cell identification, you can refer these to your training officer (or equivalent) for discussion and additional help.

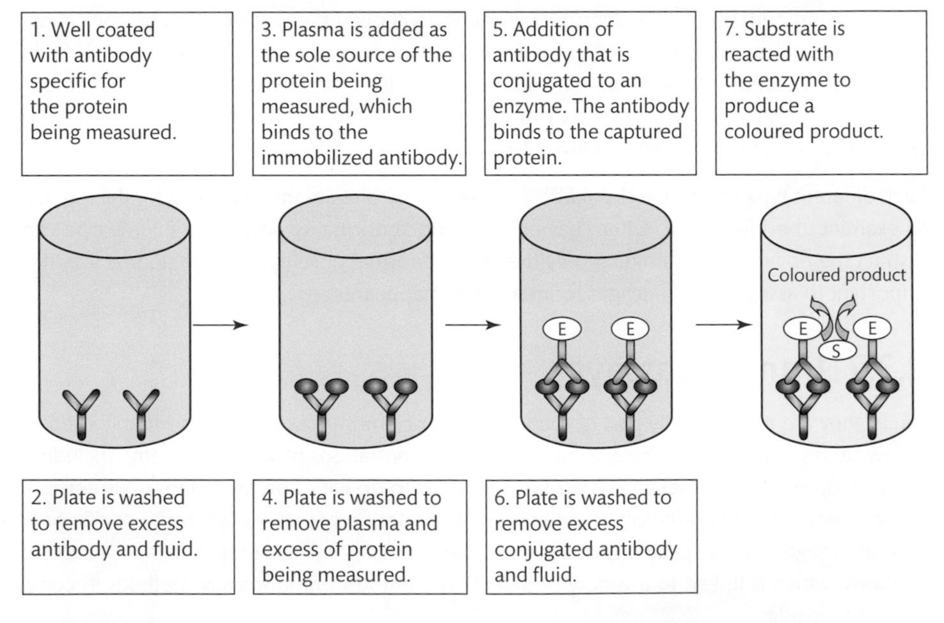

FIGURE 2.8

Principle of ELISA technique. This is referred to as a sandwich ELISA because the protein is indeed sandwiched between the primary capture and the secondary tag antibodies.

METHOD Standard curves

Standard curves are used in the measurement of concentrations of substances such as proteins or DNA. An assay is first performed with various known concentrations of the substance being measured; these measurements are referred to as the standards. Each concentration will generate a different endpoint reading. Examples of endpoints are optical density, luminescence, fluorescence, radioactivity, or clotting times.

You can see in Figure 2.9 that the standard curve is prepared by plotting concentration on the *x*-axis and endpoint value on the *y*-axis. A line or curve is fitted through the points for the standards. Note in Figure 2.9 that the standard curve is in fact a straight line. This is the most accurate, and so scientifically the most acceptable, part of the standard curve. Above and below this straight line section the relationship between the concentration and the endpoint value is difficult to interpret and so is of less rigorous scientific value.

Once the standard curve is prepared, the assay can be performed both on control samples and on patient samples containing unknown concentrations of the analyte. For each patient sample endpoint result, the graph is read horizontally across from its value on the *y*-axis until it intersects with the standard curve. This value is then read off the *x*-axis by reading down vertically. The concentration of the analyte in the patient sample is the value on the *x*-axis. Most analysers will automatically prepare the standard curves and read the values of controls and patient samples. (See also Box 2.2.)

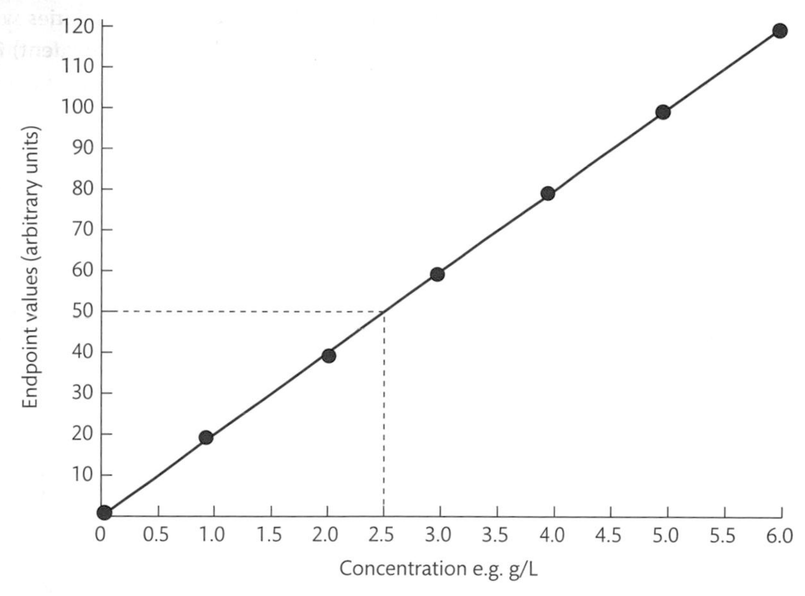

FIGURE 2.9

Standard curve. The result for each standard is plotted against concentration and a line drawn to connect the points. The diagram shows a patient sample with an endpoint result of 50 being read off the standard curve to generate a result of 2.5 g/L.

In LIA techniques, the capture antibody is immobilized on microscopic latex beads. You can see in Figure 2.10 that adding a plasma sample to a solution of these beads causes agglutination of the beads in direct proportion to the amount of protein present. The degree of agglutination is measured by light scatter.

Cross references

Further details of quality control procedures are given in the *Biomedical Science Practice* text in this series.

Further details on spectrometry, automation, immunoassay, and microscopy are present in the *Biomedical Science Practice* text in this series.

The technique of light scatter is also the basis of **nephelometry**, which measures the intensity of light scatter that occurs when light is transmitted through a reaction mixture containing particulate matter. Nephelometry is performed in a nephelometer. Antibodies to the antigen of choice (such as the marker of inflammation, **C-reactive protein (CRP)**) are mixed in an optimum concentration of each so that exceptionally small aggregates are formed. Light (such as may be provided by a laser) will be scattered in proportion to the concentration of the aggregates, and from this the concentration of the molecule being measured can be determined. Because this process uses antibodies, it is more correctly referred to as immunonephelometry. A closely related technique is **turbidimetry**, which measures the loss of intensity

BOX 2.2 What's the difference between a standard and an internal control?

A standard is a preparation with a known, exact concentration (or other value such as INR) for use in the generation of standard curves. A control preparation has no fixed value, but has a range of expected values. Such internal controls are analysed at regular intervals for tests that are performed throughout the day and night, such as FBC or **prothrombin time (PT)**, or each time a batch of samples is performed for tests that may be analysed just once a day, a few times a week, or less frequently. We recognize that operator, reagent, and analyser variability mean that you won't get exactly the same result each time a sample is analysed. As long as the difference in results is within acceptable limits of analytical variability and will not alter clinical decision-making, then a result, or batch of results, can be reported. Controls are used to check that a procedure is generating results within an acceptable range.

prothrombin time (PT)

One of the major coagulation tests: used to assess key aspects of the coagulation pathway.

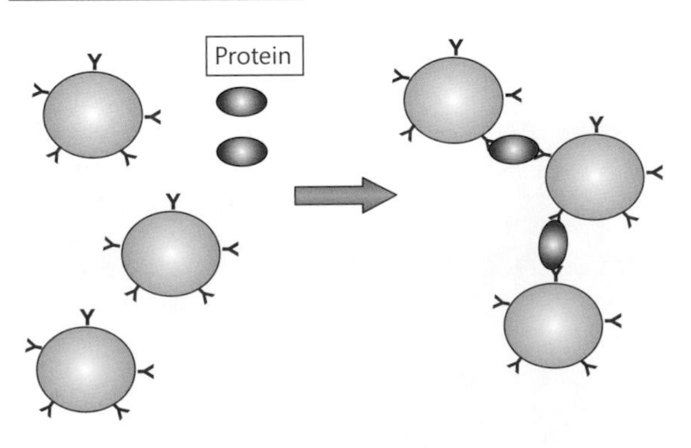

FIGURE 2.10

Latex particles act as the solid phase to anchor the capture antibody. The protein being measured in the patient's plasma acts as a bridge between coated latex particles, leading to a degree of agglutination in direct proportion to the amount of measured protein in the plasma. The amount of agglutination is measured by turbidimetry.

of light transmitted through a reaction mixture containing particulate matter, whereby the unscattered light is measured. Turbidimetry is performed using a turbidimeter.

Immunoassays rarely used nowadays include radioimmunoassay (RIA), immunoradiometric assay (IRMA), and immunodiffusion. Because of the inherent danger of radioisotopes, the former are being replaced by ELISA, so that radioisotope work has been almost completely eliminated from the routine lab (although it may still be used in specialized referral centres). Immunodiffusion suffers from a lack of sensitivity. However, in exceptional circumstances, these methods may be called upon.

Key Points

Immunoassays merely capture the protein and measure its concentration, but not its function. In the context of immunoassays, the protein being measured is referred to as the antigen.

SELF-CHECK 2.5

What are the commonalities and differences between ELISA and LIA techniques?

2.3.8 Genetic/molecular biology

Like fluorescence flow cytometry, techniques involving genes and DNA are common to all aspects of haematology, and indeed to all branches of biomedical science. The common aspect is that we seek differences in the formation and/or expression of genes (that is, mutations), but also changes in entire chromosomes and section of chromosomes. Much research on the gene, its regulation, and interactions with the constituents of the cytoplasm in the preceding decades led to a seminal textbook, *Molecular Biology of the Gene*, the title of which pays homage to Nobel-laureate Francis Crick, who described 'the central dogma of molecular biology' in 1958. Many therefore take 'molecular biology', and perhaps 'molecular techniques', to be referring exclusively to genes and genetics, whereas clearly the biology of molecules broadly addresses other areas, such as the glycoprotein coagulation factors. Other authorities refer more accurately to these techniques as 'molecular genetics', to distinguish them from Mendelian genetics of inheritance and the karyotyping of chromosomes. Nevertheless, regardless of nomenclature, even the most sceptical would be hard pressed to deny the fact that genetic/molecular biology has brought a revolution in the diagnosis and management of hundreds, if not thousands, of different human diseases. This technology is now considered of such importance that many large hospitals have their own laboratory dedicated to this form of analysis. (See also Box 2.3.)

Further developments in molecular genetics include non-coding RNA molecules, which can be small (20–22 nucleotides in length, hence miRNAs), long (around 200 nucleotides in length, hence lncRNA), or circular. They have roles in regulating gene expression in malignancies and in genes involved in certain metabolic pathways. Measurement of non-coding RNAs has yet to enter routine practice, but some (particularly miRNAs) may do in the coming decade. Point-of-care genetic testing can also be expected to grow: one such test determines polymorphisms in a metabolic enzyme that influences the effectiveness of

BOX 2.3 Molecular genetics and haematology

A major triumph of the genetics of biomedical science has been the demonstration that almost all cases of the white blood cell disease of chronic myeloid leukaemia are caused by a specific genetic abnormality called *BCR-ABL*. Similarly, almost all cases of the red blood cell disease polycythaemia are due to a mutation of a gene called *JAK2*, and many people with the platelet disease essential thrombocythaemia (ET) also have this mutation. However, those with ET and a non-mutated *JAK2* may have a mutation in another gene, *MPL*. But despite this, there are still some ET patients with normal *JAK2* and *MPL*, leading to the possibility of other genes with a potential role in this disease. Recent advances have indeed found a third gene, *CALR*, mutations in which may also be linked to ET, and which may differentiate alternate forms of ET. This is an example of how molecular genetics can help define a particular disease so that treatment can begin as soon as possible. Further details are presented in Chapter 11.

Reference: Trifa AP et al. CALR versus JAK2 mutated essential thrombocythaemia—a report on 141 patients. British Journal of Haematology 2015: 168: 151–3.

an antiplatelet drug. The greater part of our use of genetics/molecular biology in haematology is disease of white cells, and accordingly is an integral part of Chapters 9–12, where additional details of methods and their application are presented. Just one example of the relevance of molecular genetics to the study of leukaemia is the observation that parts of chromosome 15 and chromosome 17 come together to form a new gene (*PML-RARA*) that leads to acute promyelocytic leukaemia. Other aspects of use of genetic techniques in haematology are shown in Table 2.2.

TABLE 2.2 Molecular genetics in haematology.

Red blood cells	• Genetic determinants of haemolysis in sickle cell anaemia • Mutations in the receptor for erythropoietin as a cause of polycythaemia • Mutations in the genes for the components of the membrane as a cause of hereditary spherocytosis
White blood cells	• Carriage in almost all cases of hairy cell leukaemia of (and likely to be caused by) a single mutation in *BRAF* • Telomere dysfunction as a factor in outcome in chronic lymphocytic leukaemia • Translocation of BCL and ABL as a cause of chronic myeloid leukaemia • Understanding the genetic basis of the pathogenesis of myelodysplastic syndromes
Platelets	• Mutations in the gene for the thrombopoietin receptor as a cause of thrombocytosis • Genetic basis of different forms of haemorrhagic disease such as haemophilia and von Willebrand disease • Genetic variability in the metabolism of the vitamin K antagonist warfarin

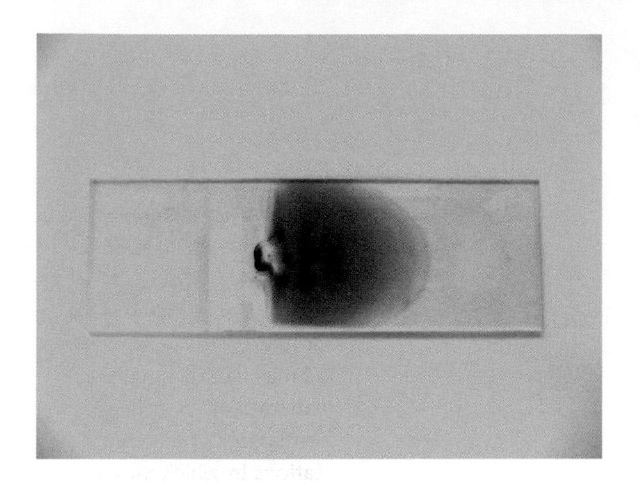

FIGURE 2.11
The appearance of a blood film. As the blood is drawn out from left to right, cells become more spread out and so fine details of individual cells can be noted.© Institute of Biomedical Science (IBMS) and Sysmex Haematology Morphology Training CD-ROM, 2009.

2.4 The blood film

The blood film provides the opportunity to view the components of blood under a microscope. A drop of blood, generally from the same tube that provides the FBC, is smeared onto a glass slide, allowed to dry in air at room temperature, and fixed in methanol. It is then covered with stains such as those originally developed by **Romanowsky**, but subsequently refined by workers such as Jenner, Giemsa, and Leishman. After a set time, the stains are washed off with buffered saline and the slide is once more allowed to air-dry. Figure 2.11 shows a blood film, where a drop of blood has been spread out from left (where it is thickest) to the right (where the film is thin). Laboratory staff working in haematology should be competent to produce high-quality blood films, but this process is now automated in a number of laboratories to improve the speed and consistency of producing blood films.

Romanowsky
A type of stain specially developed to enable the examination of different blood cells. Romanowsky stains contain polychromed methylene blue and eosin.

The slide is now ready to be viewed by light microscopy. But first we need to establish the correct degree of magnification. Multiplying the eyepiece magnification (often ×10) by the magnification of the objective lens (e.g. ×40) provides the total magnification (in this example, ×400). If this is too low (such as ×100), we will be unable to establish fine details of the cells, but a very high magnification (such as ×1000) permits many fine details to be seen, but generally only one or two cells at a time—this is too slow for a busy laboratory. Therefore, the midpoint, perhaps a magnification of ×400, is preferred as the fine details of many cells can be viewed at the same time.

Cross reference
See the chapter on microscopy in the *Biomedical Science Practice* text in this series for more information about magnification.

The second thing we must do is to find the correct place on the film to examine the sample, so we can adequately assess the morphology of the blood cells. The best place for these observations is where the cells are close together—not too far away from each other where the film is too thin, nor where the film is too thick so that individual cells cannot accurately be defined. This is illustrated by Figure 2.12, which shows three parts of the same slide and the differences in the density of the red blood cells. In Figure 2.12(a) the cells are far apart, but in Figure 2.12(c) they are too close together and clearly overlap. Choose an area such as in Figure 2.12(b).

Having established the power of magnification and where on the slide to look, we can begin our examination of blood cells. Figure 2.13 shows a typical blood film at a high power of magnification. The principal feature is a white blood cell (top left), characterized by a three-lobed nucleus that had taken up stains so that it now appears dark purple/black. A second feature is the large number of roughly round cells that are all a single colour—these are red blood cells. A close viewing of these cells reveals some degree of variety in both the size of the cells and the

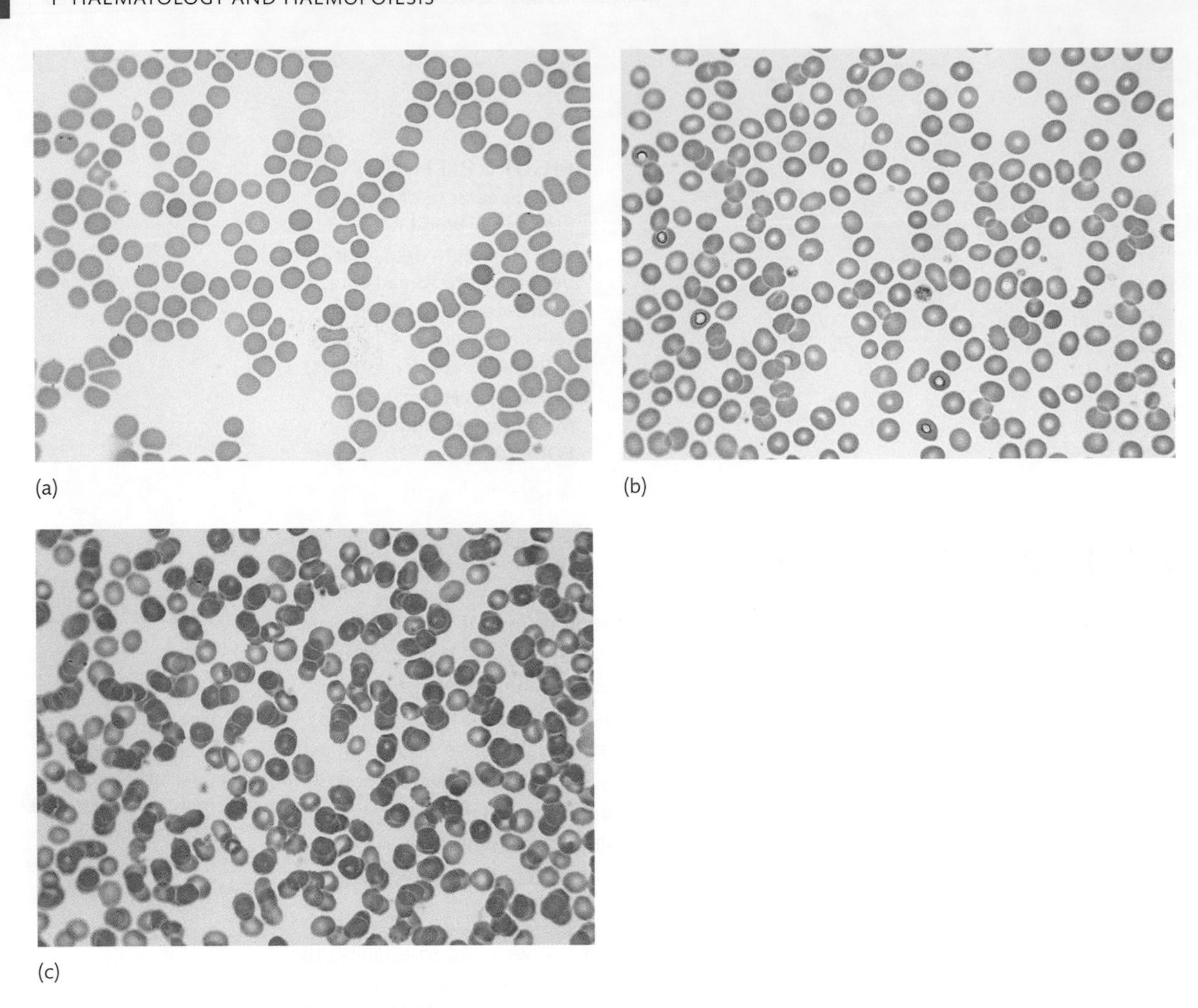

FIGURE 2.12
Peripheral blood film examination. Blood smears are stained with a Romanowsky stain. When viewing morphology, choose an area of the film where red cells are close together, i.e. (b). Incorrect places are where the cell layer is too thin and cells become flattened (a), and where the film is too thick and the cells overlap (c). Note that there are no white blood cells present. (Magnification ×400.)

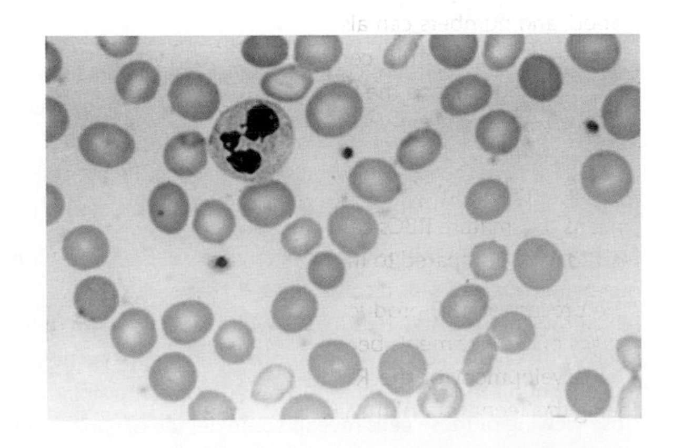

FIGURE 2.13
A typical blood film at a high power of magnification, showing red cells, a white cell, and platelets. Different blood cells can be identified by the different patterns of dye uptake. This figure shows a single white blood cell—a neutrophil (with its purple and irregular nucleus of three lobes), many slightly smaller red blood cells, and a smaller number of platelets (the three small purple dots). (Magnification ×800.)

density of the grey colouring. The importance of this variation in size and degree of coloration has significance, as we will see in Chapters 4–6. The third feature is the presence of three small purple bodies—these are platelets.

With the advent of automated autoanalysers that can provide an excellent breakdown of the different types of blood cells, the blood film has become less important and now is not routinely examined. However, there is still a place for the microscopic examination of a blood film; such examination may be necessary to confirm the autoanalyser's profile or to perform other assessments of the blood, such as for malarial parasites, the presence of bodies inside the red blood cell, for unusual white blood cells, or to look for evidence of partially destroyed or damaged red blood cells.

2.5 The full blood count

The FBC is the most requested and most important blood test in haematology. It provides a package of information on red blood cells, white blood cells, and platelets.

2.5.1 Red blood cells

Red blood cells (RBCs), with a diameter of approximately 7 μm when mature, are unusual among cells of the body as they lack a nucleus. Consequently, they are easy to identify in a blood film (Figures 2.12 and 2.13). The full analysis of RBCs includes several tests.

Haemoglobin (Hb) is undoubtedly the index most frequently referred to in biomedical haematology. It is an iron-containing protein that absorbs oxygen from areas of high-oxygen content (that is, at the lungs), and then releases it in areas where oxygen levels are low (i.e. in the tissues). The reference range for Hb varies between the sexes. Pre-menopausal women lose blood with each menstrual period, but in post-menopausal women levels are still lower than age-matched men, as the latter produce testosterone, which stimulates red cell production. Men who have lost their testes to disease or through trauma generally have an Hb level comparable to that of post-menopausal women.

Cross reference

Reference ranges are included in Table 1 of this book and are further discussed in section 2.10 of this chapter.

The level of Hb in a sample of venous blood is reported by the laboratory in grams per litre (g/L), so that in an adult, a result of 138 g/L would be within the reference range for both adult males and females. Hb is normally carried in the blood inside RBCs, which are also called **erythrocytes** (*erythro* = red, *-cyte* = cell). There are approximately 640 million molecules of Hb in each red cell. The number of RBCs in a sample of blood is reported as the **red blood cell count (RBCC)**. Because these cells lack a nucleus and are biconcave discs, they have the flexibility to pass along the smallest capillaries. They are the most abundant cell in the blood, and numbers can also vary in health between the sexes, and also in various diseases. The function of the red cells is to carry oxygenated Hb (often described as oxyhaemoglobin) from the lungs to the tissues (where oxygen is given up), and then the deoxygenated Hb returns from the tissue to the lungs via the venous circulation. Thus a crucial component of this cell is the ability of oxygen to pass freely into the cell at the lungs and out of the cell at the tissues, and this in turn demands a highly specialized cell membrane. The lack of a nucleus means that mature RBCs cannot repair or divide, so the lifespan of the RBC is relatively short, at 120 days, compared to most nucleated cells.

The process of RBC production starts in the bone marrow, where cells pass through various stages of development, becoming increasingly mature. The **reticulocyte** is the final stage of the development of the RBC before full maturation. Indeed, reticulocytes may be seen as being the 'teenagers' of the life cycle of the RBC. Present in normal blood in very low numbers

Cross reference

The process of the development and physiology of RBCs are described in Chapters 3 and 4, respectively.

(perhaps one for each 100 mature red cells), increased numbers are maybe the product of a pathological process or could be the body's response to pregnancy, therapy with iron, B_{12}, or folate, or to blood loss. However, reticulocytes are not always part of the FBC and so, if required, need to be specially requested.

The **haematocrit** (Hct) expresses as a decimal (e.g. 0.42 L/L) or as a percentage (e.g. 42%) that proportion of whole blood that is taken up by all the blood cells. Since there are approximately 600–900 more RBCs per unit volume than either white blood cells or the comparatively tiny platelets, then red cells make up the major proportion of blood, and so the Hct. Consequently, at the practical level, the Hct provides an idea of the proportion of the mass of the RBCs that make up the whole blood volume.

The RBC indices comprise three measures of different aspects of the RBC, its size, and the amount of Hb it contains. These indices are:

- the **mean cell volume** (MCV, typically about 92 femtolitres (fL); 1 fL = 10^{-15} L), which is the volume of the average (mean) RBC. Note the stress is on 'average', as each index is the mean of thousands of individual cells. As it is measured directly in the blood, it takes no account of the number of RBCs or the Hb result;
- **mean cell haemoglobin** (MCH, typically about 29.5 picograms (pg); 1 pg = 10^{-12} g), which, as the name implies, reports the average amount (mass) of Hb in the average cell. It does not take into account the volume of the cell;
- **mean cell haemoglobin concentration** (MCHC, typically about 330 g/L), which is the average concentration of Hb inside the average cell and does take into account the MCV.

Most laboratory autoanalysers measure Hb, the red cell count, and MCV directly from the blood. Three other indices (Hct, MCH, and MCHC) are calculated from these three root measurements:

- The Hct is obtained by multiplying together the RBCC and the MCV, then dividing the result by 1000, so that:

$$\text{Hct} = (\text{MCV} \times \text{RBCC}) / 1000.$$

For example, if the MCV is 85 fL and the RBCC is 4.2×10^{12}/L, then the Hct = 0.357 L/L. In practice this could be rounded up to 0.36 L/L. However, this result may also be expressed as whole numbers as a percentage, which would be 36% (see Figure 2.16).

- The MCH is obtained by dividing the Hb by the RBCC, so that:

$$\text{MCH} = \text{Hb} / \text{RBCC}.$$

For example, if the Hb is 138 g/L and the RBCC is 4.2×10^{12}/L, then the MCH = 32.8 pg. In practice, this could be rounded up to 33 pg.

- The MCHC is obtained by dividing the Hb result by the product of the MCV and the RBCC, and then multiplying by 100. But note that the RBCC and the MCV also make up the Hct. So that:

$$\text{MCHC} = \{\text{Hb} / \text{MCV} \times \text{RBCC}\} \times 100.$$

For example, if the Hb is 138 g/L, the MCV is 85 fL, and the RBCC is 4.2×10^{12}/L, then the MCHC is 386 g/L.

So because these three latter red cell values (Hct, MCH, and MCHC) are mathematically derived from the three root measurements (Hb, MCV, and RBCC), it is entirely possible for either one or more of the Hct, the MCH, or the MCHC to be outside the reference range whilst the other indices are apparently normal. Indeed, in the third example above, the Hb, MCV, and RBCC are all within the reference range, but their mathematical product, MCHC, is outside the reference range. Consequently, one must consider all six red cell indices (and possibly some other blood tests) together to obtain a full picture.

Similarly, consider an Hct such as 0.42 L/L. Because the Hct is the product of two other indices (the MCV and the RBCC), then the same Hct would be present in one sample whose RBCC is 6×10^{12}/L and whose MCV is 70 fL, and in a different sample where the RBCC is 7×10^{12}/L, but whose MCV is 60 fL. This is because $(6 \times 70)/100$ is the same as $(7 \times 60)/100$, that is 0.42. As we see in Chapters 5 and 6, this may have consequences for the way in which we interpret these results.

Each blood sample has an MCV that is the average size of all red cells tested. Cells that are considered small, below the bottom of the MCV reference range (in our case, 77 fL) are described as **microcytes**, whereas cells considered large (which is an MCV greater than 98 fL) are **macrocytes**. Cells with an MCV within the reference range (77–98 fL) are **normocytes**. However, this classification system of taking an average masks the possibility of considerable variation in the range of possible sizes within that same sample. The variation in the sizes of the red cell population can be assessed by the **red cell distribution width (RDW)**. This is derived by the haematology autoanalyser from all the measured red cell volumes which contribute to the MCV. Suppose our MCV is 90 fL. This average number may be derived from a population whose actual cells vary from 80 fL to 100 fL, or in a different sample whose cells vary from 85 fL to 95fL. The greater the RDW, then the greater the variation in the sizes of the RBCs. A common reason for an increased MCV and RDW is the presence of reticulocytes. This index may be also useful in establishing a diagnosis, and we will return to it in Chapter 6.

SELF-CHECK 2.6

List the red cell indices.

Key Points

A greater overall burden of human disease can be accounted for by abnormalities in RBC functioning than in the pathology of white blood cells or platelets.

2.5.2 White blood cells

White blood cells (WBCs), or **leucocytes** (*leuco-* = white, *-cyte* = cell), are collectively responsible for defending us from attack by microorganisms such as viruses, bacteria, and parasites, when raised levels of WBCs can be expected. These cells are also important in hypersensitivity and allergy. The process by which certain WBCs ingest and destroy pathogens is called **phagocytosis**, and the cells that perform this function are phagocytes. Increased numbers of WBCs (i.e. a **leucocytosis**) may also be present in a number of conditions, such as rheumatoid arthritis, cancer, after surgery, and, as we shall see in detail, in leukaemia. WBCs also help defend us from pathogenic microbes by generating antibodies to help seek out and destroy the invaders. Haematologists recognize five different types of WBCs that can be found in the (normal) blood—the **neutrophil** (the purple-coloured body in Figure 2.13), **lymphocyte**, **monocyte**, **eosinophil**, and **basophil**. Each type of leucocyte can be defined by their function (Table 2.3) and by their particular morphology (Table 2.4). However, in certain diseases, such as leukaemia, other forms of WBCs appear in the blood.

phagocytosis

The process of the ingestion and destruction of foreign and unwanted material, such as bacteria and effete RBCs. Phagocytosis is performed by phagocytes—principally neutrophils and monocytes/macrophages.

leucocytosis

An increase in the total WBC count above the top of the reference range.

Cross reference

Leukaemia is described in detail in Chapters 9–12.

SELF-CHECK 2.7

What are the differences between the two major groups of WBCs?

SELF-CHECK 2.8

What are the morphological characteristics of the two most frequent WBCs?

Additional details of the functions of leucocytes are presented in Chapter 8.

TABLE 2.3 **Functional characteristics of white blood cells.**

Neutrophils	• Phagocytosis of bacteria and yeast • Participation in inflammation • Scavenging and removal of debris
Lymphocytes	• Generation of antibodies (B lymphocytes) • Cooperation in antibody production (T lymphocytes) • Destruction of cells infected with viruses (T lymphocytes)
Monocytes	• Phagocytosis of bacteria and yeast • Participation in inflammation • Scavenging and removal of debris • Cooperation with lymphocytes in generating antibodies (as macrophage antigen-presenting cells) • Release of cytokines (such as interleukins) • Participation in haemostasis (expression of tissue factor)
Eosinophils	• Protection against parasitic infection (such as helminths) • Participation in allergic responses
Basophils	• Participation in hypersensitivity reactions • Release of histamine and heparin

TABLE 2.4 **Key morphological features of white blood cells.**

Mononuclear leucocytes	Lymphocytes: small, the nucleus occupies up to 95% of the cell. Monocytes: large, the nucleus occupies perhaps 70–80% of the cell. Rarely, there may be granules in the cytoplasm.
Polymorphonuclear leucocytes or granulocytes	Neutrophils: the nucleus generally has between three and five lobes. Intracytoplasmic constituents are generally at a neutral pH, resulting in purple granules. Eosinophils: the nucleus generally has two large lobes. Intracytoplasmic constituents react with the acidic component of dyes (such as eosin), resulting in red-brown granules. Basophils: the nucleus generally has two lobes. Intracellular constituents have an affinity for the basic component of dyes, resulting in dense purple/black granules that often obscure the nucleus.

When using a light microscope to examine WBCs, the following should be assessed:

- the nuclear structure, including chromatin appearance and presence of nucleoli;
- the colour of the cytoplasm;
- the presence of granules within the cytoplasm, their colour, intensity, and whether or not granules are expected in that particular cell line;
- abnormal maturation of the nucleus and the cytoplasm;
- inclusion bodies, or vacuolation, and the associated lineage;
- the shape of the cell in comparison to what might be considered normal for each species of cell;
- the presence of any progenitor cells, intermediate stages of maturation, or cells otherwise abnormally localized to the peripheral blood.

By consistently following this pattern of investigation, no WBC abnormalities should be overlooked.

Manual differential count

For every automated FBC completed, a total WBC count will also be provided, along with a breakdown of the absolute numbers of the species of WBC and their relative percentages. The term 'white blood cell differential count' refers to the absolute numbers and percentages of the different species of WBC within the total white cell count. Depending upon the specification of the analyser, a three-population differential—assessing granulocytes, monocytes, and lymphocytes or neutrophils, lymphocytes, and mixed (monocytes, eosinophils, and basophils) populations—might be used. However, a five-population differential is preferable—providing neutrophil, eosinophil, basophil, monocyte, and lymphocyte counts. Additionally, there are usually immature granulocyte parameters that can be measured.

Sometimes it is necessary for biomedical scientists to determine the differential count manually, especially when atypical WBCs, such as WBC progenitors or **dysplastic** cells are present.

dysplastic
This refers to cells that have abnormal maturation characteristics.

A minimum of 100 WBCs should be counted at ×400 magnification using the battlement method (Figure 2.7), with the identity of each species of WBC recorded on a cell counter. A well-prepared blood film, ideally using a sample less than three hours old, stained with a good-quality Romanowsky stain should be used to ensure blood cell morphology is maintained. The initial value obtained from the manual differential count is a percentage of each species of WBC. The absolute count, reported as $x \times 10^9$/L, for each species can be easily obtained using the percentages and the total WBC count obtained from the original FBC.

The various species of WBC have different migration patterns when spread on a glass slide. The better the quality of the blood film, the less pronounced the spreading characteristics of these white cells. Nevertheless, as a scientist it is important to obtain accurate and precise results, and to report these appropriately. Larger cells, such as neutrophils and monocytes, tend to spread to the periphery of the film (i.e. to the tip and the edges), whereas lymphocytes may be found at higher levels in the centre of the film. Because the battlement method (Figure 2.7(b)) is completed across the width of the slide, the edges and the middle of the film should be equally assessed, thereby reducing the risk of obtaining inaccurate results.

METHOD Calculating the absolute cell counts from a manual differential

Now consider a patient who has a total WBC of 8.0 × 10^9/L, as determined by a standard haematology analyser. A biomedical scientist completes a 100-cell manual differential count obtaining the following data:

Neutrophils	72%
Lymphocytes	22%
Monocytes	3%
Eosinophils	2%
Basophils	1%
Total =	100%

These data now need to be converted into absolute values, reported as × 10^9/L.

The first step is to convert the percentages into decimal figures:

Neutrophils	0.72
Lymphocytes	0.22
Monocytes	0.03
Eosinophils	0.02
Basophils	0.01
Total =	1.00

Now multiply the original total white cell count (in this case 8.0 × 10^9/L) by each of the decimals:

Neutrophils	0.72 × 8.0 = 5.76
Lymphocytes	0.22 × 8.0 = 1.76
Monocytes	0.03 × 8.0 = 0.24
Eosinophils	0.02 × 8.0 = 0.16
Basophils	0.01 × 8.0 = 0.08
Total WBC =	8.0 × 10^9/L

Remember to add each of your absolute counts together to ensure the final value equals the total white cell count originally obtained. Note that the white cell differential is reported to two decimal places.

Some scientists recommend using a 'mini' battlement approach along the periphery of the slide, whilst others prefer working along the length of the slide from thick to thin in a region near the periphery of the film. Your training officer will have their own views. For the inexperienced morphologist, focusing on the monolayer and working across the width of the slide allows for full appreciation of the morphology without the risk of misinterpreting seemingly 'crushed' cells in the thick regions of the film.

SELF-CHECK 2.9

A patient is shown to have an abnormal differential count using automated methods. Subsequent examination of the requested blood film demonstrates that this differential count is incorrect. A manual differential cell count is completed providing the following percentages:

Neutrophils	64%
Lymphocytes	21%
Monocytes	12%
Eosinophils	3%
Basophils	0%

The total WBC is 16.0 × 10^9/L.

Based on the figures above:

1. calculate the absolute count for each of the WBC species;
2. indicate which of these species falls outside the reference range for the species absolute counts and whether they are high or low.

Using the correct white cell terminology, describe which population(s) of cells is/are abnormal and how.

2.5.3 Platelets

Platelets are small fragments of the cytoplasm of a larger cell found only in the bone marrow, the **megakaryocyte**. Platelets form a clot, or **thrombus**, when aggregated together with the help of the blood protein fibrin, and so reduce blood loss. However, platelets are not simply inert participants in stemming blood flow, but are in fact extremely dynamic, and have numerous granules containing molecules that promote haemostasis. Platelets circulate in the blood for seven to ten days, and are then destroyed in the spleen and the liver.

thrombus
Technical name for a clot, formed by aggregating platelets within a fibrin mesh. RBCs may also be present in a thrombus, especially if formed in a vein.

A low platelet count (possibly caused by drugs, such as quinine, sulphonamides, and other antibiotics; poor production of platelets, as may be present in disease of the bone marrow; or their excessive consumption) is called **thrombocytopenia**. This condition can lead to an increased risk of bruising and bleeding, and is considered in greater detail in Chapter 14. The converse, a raised count, is **thrombocytosis**, and is often present in many physiological and pathological situations. These include infections, after surgery, some autoimmune diseases (such as inflammatory bowel disease and rheumatoid arthritis), and after short but intense bouts of physical activity platelet counts also rise in patients with iron deficiency anaemia. A high platelet count may lead to thrombosis.

Cross reference
Platelets and coagulation are explained in full detail in Chapters 13 and 14.

Platelets appear in a blood film as light-purple or grey bodies that are very much smaller than RBCs. They are anucleate (that is, they do not have a nucleus) and contain granules. Indeed, to the untrained eye they could be mistaken for debris (Figures 2.13 and 2.14).

SELF-CHECK 2.10

What are the key aspects of the platelet?

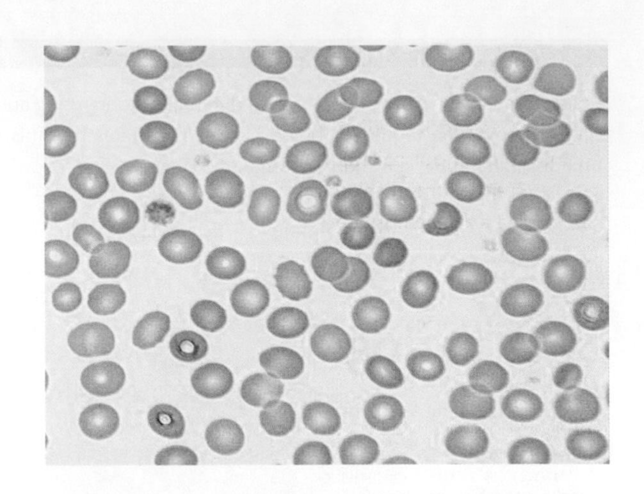

FIGURE 2.14
The platelet. These appear as small pale-blue/purple staining cells that are considerably smaller than RBCs. However, one platelet, near the left margin, is large. (Magnification ×600.) High-power magnification images of platelets are also to be found in Figure 2.13.

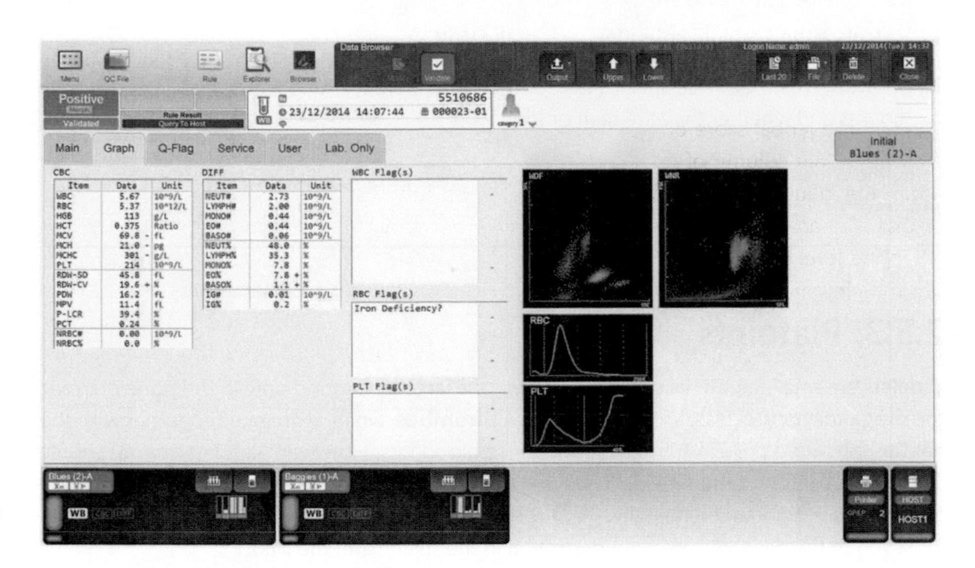

CBC

Item	Data	Unit
WBC	5.67	10^9/L
RBC	5.37	10^12/L
HGB	113	g/L
HCT	0.375	Ratio
MCV	69.8 -	fL
MCH	21.0 -	pg
MCHC	301 -	g/L
PLT	214	10^9/L
RDW-SD	45.8	fL
RDW-CV	19.6 +	%
PDW	16.2	fL
MPV	11.4	fL
P-LCR	39.4	%
PCT	0.24	%
NRBC#	0.00	10^9/L
NRBC%	0.0	%

DIFF

Item	Data	Unit
NEUT#	2.73	10^9/L
LYMPH#	2.00	10^9/L
MONO#	0.44	10^9/L
EO#	0.44	10^9/L
BASO#	0.06	10^9/L
NEUT%	48.0	%
LYMPH%	35.3	%
MONO%	7.8	%
EO%	7.8 +	%
BASO%	1.1 +	%
IG#	0.01	10^9/L
IG%	0.2	%

FIGURE 2.15
Print out from a haematology autoanalyser. The panel on the far left includes the red, white, and platelet indices of the FBC. The panel on its right includes the WBC differential. The figures on the right are cytograms of the red and white cells, and the platelets. Interpretation of the cytograms requires experience.

2.5.4 The haematology analyser

At the centre of every haematology laboratory is an autoanalyser that provides all the indices of the FBC described above, and in many cases, other indices, such as the size of the average platelet (the mean platelet volume) and the number of nucleated RBCs (if any are present), the importance of which will be explained in chapters to come. These other indices are often brought together in a single report, such as that shown in Figure 2.15, although the exact details vary according to the manufacturer of the particular analyser.

Although microscopy is a powerful tool in determining the morphology of different types of WBCs (as in Figures 2.12–2.14), it is not used in a routine setting. Instead, as detailed in sections 2.3.4 and 2.3.5 (where Figures 2.3 and 2.4 apply), all autoanalysers rely on flow cytometry (note not *fluorescence* flow cytometry) to routinely determine the frequency of the different white cell populations. However, the flow cytometry criteria of size and granularity alone are not sufficient to differentiate all cell types, as in addition, the use of cytochemistry is important to identify different species of white cell using the automated analyser.

Key Points

The FBC is undoubtedly the most important set of results that the haematology laboratory can offer our colleagues on the ward, in the operating theatre, and in the clinic. No competent biomedical scientist can practise without a sound knowledge of how it is obtained and the implications of major abnormalities.

2.6 Rheology

Rheology provides information on the physical properties of the blood, and is therefore unlike other tests in that there is no measurement of a defined molecule or cell. The two most common tests are ESR and plasma viscosity testing. However, it is also possible that some services offer whole blood viscosity.

rheology/rheological
The study of the physical nature of blood or plasma—principal measurements are ESR and viscosity.

2.6.1 The ESR

The ESR is a global score of physical aspects of the whole blood. The result is obtained by allowing a thin column of anticoagulated blood in a vertically aligned tube to settle down under the influence of gravity. As it does so, the blood cells will separate from the plasma, so that after an hour, a band of clear plasma will sit on top of the column of blood cells. The fall in the level of the blood cells is then recorded at a rate of millimetres per hour (mm/h). It is therefore unique in requiring no sophisticated machine and few technical skills. In some instances, reference ranges for the ESR may take into account patient age. As patients age, their ESR often rises, but this may well be because of low levels of asymptomatic disease as an ESR can be abnormal in a large number of conditions. These include inflammation, infection, the acute phase response, after surgery, anaemia, leukaemia, and almost all forms of cancer. Indeed, it follows that an abnormal ESR is present in most patients in hospital, and in Chapter 5 there is a section on anaemia of the elderly. The ESR is often influenced by increased levels of molecules (such as fibrinogen), whose concentrations rise as a result of inflammation.

Key Points

In reality, the units for ESR should be reported as millimetres in the *first* hour because the rate of sedimentation falls as time progresses—there is only so far the cells can fall before they form a compact layer at the bottom of the tube. Despite this, most laboratories use mm/h purely for convenience.

2.6.2 Viscosity

Plasma viscosity provides an idea of how thick or thin the plasma has become—whether or not it is thinner, and more like water (tending towards a low result of less than 1.5 mPa/s), or thicker and more like treacle (tending towards a high result, over 1.72 mPa/s). Viscosity records a global property of the plasma, not individual molecules. Indeed, the molecules that make up the major component of plasma viscosity include the clotting proteins fibrinogen and **von Willebrand factor (VWF)**, and also albumin and immunoglobulins. There is also a relationship between plasma viscosity and total plasma protein concentration, and often

von Willebrand factor
A large protein that enables platelets to bind to exposed subendothelium and also stabilizes coagulation factor VIII.

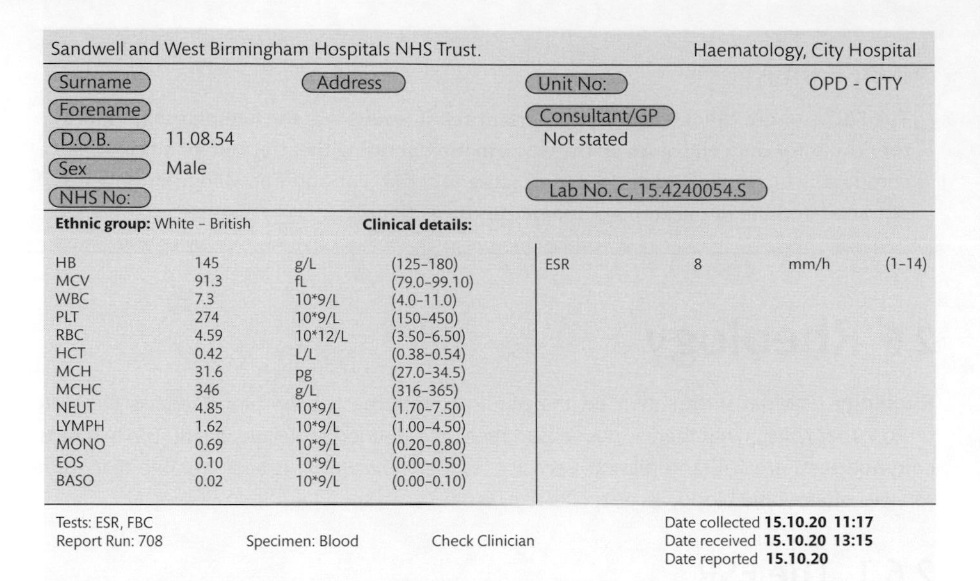

Sandwell and West Birmingham Hospitals NHS Trust. Haematology, City Hospital

Surname | Address | Unit No: | OPD - CITY
Forename | Consultant/GP
D.O.B. 11.08.54 | Not stated
Sex Male
NHS No: | Lab No. C, 15.4240054.S

Ethnic group: White - British **Clinical details:**

Test	Result	Units	Reference range
HB	145	g/L	(125-180)
MCV	91.3	fL	(79.0-99.10)
WBC	7.3	10*9/L	(4.0-11.0)
PLT	274	10*9/L	(150-450)
RBC	4.59	10*12/L	(3.50-6.50)
HCT	0.42	L/L	(0.38-0.54)
MCH	31.6	pg	(27.0-34.5)
MCHC	346	g/L	(316-365)
NEUT	4.85	10*9/L	(1.70-7.50)
LYMPH	1.62	10*9/L	(1.00-4.50)
MONO	0.69	10*9/L	(0.20-0.80)
EOS	0.10	10*9/L	(0.00-0.50)
BASO	0.02	10*9/L	(0.00-0.10)

Test	Result	Units	Reference range
ESR	8	mm/h	(1-14)

Tests: ESR, FBC
Report Run: 708 Specimen: Blood Check Clinician
Date collected **15.10.20 11:17**
Date received **15.10.20 13:15**
Date reported **15.10.20**

FIGURE 2.16
FBC/ESR result. The top section has demographic details of the patient and the hospital. Below it, the left-hand column is the abbreviation of the particular test. To the right are the result, the units, and the reference range (in brackets). The date of the sample is lower right.

with the Hct and the amount of water in the blood. The viscosity of the blood as a whole (i.e. whole-blood viscosity, that is therefore influenced by blood cells) can also be assessed, but is less informative (and is requested far less frequently) than plasma viscosity. Indeed, abnormalities in the RBC are commonly found in several diseases, leading to difficulty in interpretation.

Perhaps the single most important haematological disease where plasma viscosity is grossly abnormal is the WBC malignancy **myeloma**, which is also characterized by anaemia and, generally, a high ESR. Viscosity is high because the abnormal lymphocytes are often generating excessive amounts of antibody-like protein. An increased ESR is inevitably present in the inflammatory autoimmune disease of temporal arteritis, and is a key component of the diagnosis of this condition. Accordingly, a potential diagnosis of temporal arteritis is one of the very few instances where an urgent ESR must be performed, as a delay in diagnosis, and so in the initiation of the immunosuppressive treatment, may lead to blindness.

Plasma viscosity and ESR are often closely related, as they are both influenced by plasma proteins. However, there is debate regarding the clinical value of both tests being offered by the same laboratory, and most offer only one of the two. Indeed, a more precise marker of inflammation is the plasma protein CRP, so much so that some laboratories offer CRP as an alternative to an ESR, partially because, in theory, CRP is not directly influenced by anaemia.

SELF-CHECK 2.11

What are the two major tests of rheology?

2.7 The report form

The most common requests made to the haematology laboratory are the FBC and ESR. Once completed, these results are generally returned to the requesting authority (ward, general practice) as a single report form. As is shown in Figure 2.15, the amount of information provided by the autoanalyser can be overwhelming to the non-haematologist, and a simpler report, summarizing key aspects, is preferred. Figure 2.16 shows just such a simplified report, that of a combined FBC and ESR from a middle-aged male. The importance of the reference range is discussed in section 2.10.

2.8 Haemostasis

Haemostasis is the balanced orchestration of interactions between blood vessels, blood cells, plasma proteins, and some small molecules that maintain blood in a fluid state and also limit and arrest bleeding upon damage to the blood vessel.

2.8.1 The physiology of haemostasis

There are five main components of the system:

1. ***Vascular integrity***–intact blood vessels promote mechanisms that maintain circulating blood in a fluid state. When the vessel is damaged, exposed subendothelial structures in the vessel wall (such as collagen) promote blood clotting processes.

2. ***Primary haemostasis***–VWF binds to structures exposed by vessel damage and then captures platelets at the site of injury. The captured platelets become activated and recruit more platelets, allowing them to aggregate to each other and form the initial physical barrier for preventing blood loss and the entry of microorganisms into the wound.

3. ***Secondary haemostasis***–a clot made entirely of platelets is not robust enough to withstand the blood pressure in many blood vessels, and thus needs a strengthening mechanism. This comes in the form of a series of interlinked enzyme reactions culminating in the generation of the enzyme **thrombin** from its precursor prothrombin. This crucial enzyme then converts fibrinogen, a soluble plasma protein, into an insoluble polymerized form called fibrin. The fibrin forms a mesh around the platelet clot to impart greater structural integrity. This complex series of enzyme reactions is termed **blood coagulation**. The enzymes are referred to as **coagulation factors** and circulate in the plasma as inactive precursors, the majority of which are converted to their active forms by other activated coagulation factors.

4. ***Inhibitors***–although thrombin is a crucial component of coagulation, it is also a potentially lethal enzyme. Consequently, a group of plasma- and membrane-bound proteins exist to regulate and eventually shut down the blood coagulation reactions. Other coagulation factors also have their own inhibitors.

5. ***Fibrinolysis***–blood clots need to be removed once bleeding has stopped and the wound has healed. Platelets are cellular in origin and have autolytic mechanisms. The fibrin clot is digested by the enzyme plasmin, which is generated and controlled in a similar fashion to thrombin by a separate group of activators and inhibitors.

The main events of haemostasis are summarized in Figure 2.17, and the fine detail of the coagulation process is explained in Chapter 13.

Abnormalities in the different areas of haemostasis can lead to disorders that predispose the patient to either **haemorrhagic** or **thrombotic** disease. Table 2.5 outlines the types of disorders that can be associated with abnormalities in each area. Although bleeding or thrombotic disorders are associated with each area, the most common bleeding disorders occur in the areas of primary and secondary haemostasis, whilst the most common thrombotic disorders are a result of abnormalities in the inhibitory mechanisms of blood coagulation. The haemostasis laboratory of a haematology department plays a critical role in the detection and characterization of haemostatic disease in a variety of patients and clinical circumstances, such as:

haemorrhagic disease
Disorders that lead to excessive bleeding.

thrombotic disease
Disorders that lead to thrombosis, which is a partial or complete obstruction of a blood vessel by a blood clot.

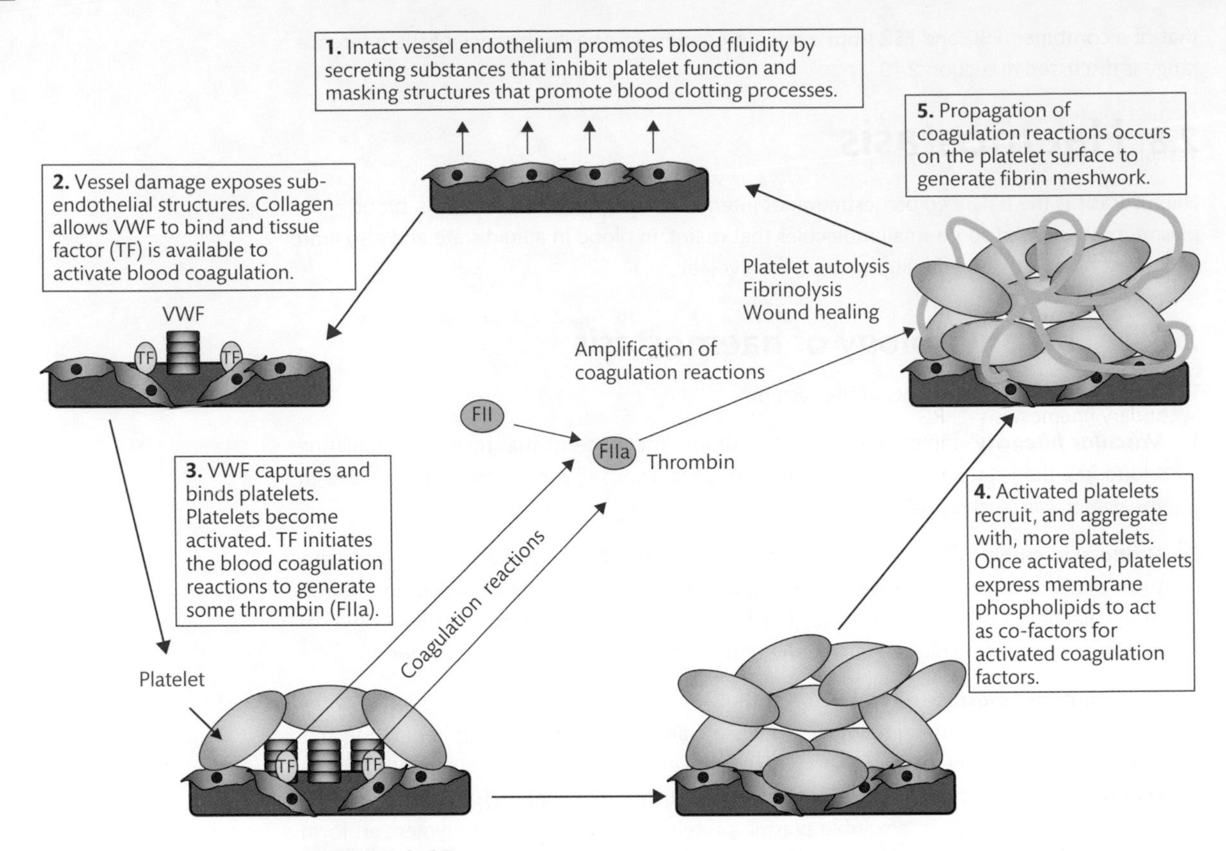

FIGURE 2.17

Main events of haemostasis. 1. Intact blood vessel endothelium is active in dampening down clotting processes. 2. Structures exposed upon damage to the endothelium promote primary and secondary haemostasis. 3. VWF binds to collagen and then tethers platelets, whilst tissue factor initiates the first series of coagulation reactions. 4. Platelets activated at tethering recruit more platelets to the clot and express surface phospholipid. 5. The small amount of thrombin generated at initiation of coagulation promotes further coagulation reactions on the surface of platelets, leading to fibrin formation that strengthens the platelet clot. Upon wound healing, the clot is destroyed (fibrinolysis).

- patients with bleeding symptoms;
- patients with thromboses;
- patients with other disorders that have secondary effects on haemostasis;
- preoperative checks for patients who may have haemostatic disorders that only manifest upon significant challenge;
- monitoring treatment of bleeding disorders;
- monitoring treatment with therapeutic anticoagulants.

Complex specialized laboratory tests are required to characterize the nature of functional platelet abnormalities or to identify specific deficiencies of blood coagulation or inhibitory proteins. It is not possible to perform every test on every patient. Consequently, biomedical scientists perform screening tests that assess overall coagulation function, which, if abnormal, can be further investigated with appropriate specialized tests.

Cross reference

You will meet the design and use of specialized diagnostic haemostasis tests in Chapters 14 and 15.

TABLE 2.5 **Types of disorders associated with the different areas of haemostasis.**

Area of haemostasis	Causes of bleeding disorders	Causes of thrombotic disorders
Vascular integrity	Inability to support a clot as it forms due to structural defects of blood vessels	Secondary vessel structural changes such as those seen in cholesterol deposition
Primary haemostasis	Reduced concentration or impaired function of VWF Reduced numbers or impaired function of platelets	High levels of VWF Elevated platelet numbers or platelets with enhanced function
Secondary haemostasis	Reduced concentration or impaired function of one or more of the coagulation enzymes or cofactors	Elevated levels of one or more of the coagulation enzymes or cofactors
Inhibitors	Enhanced inhibitor function (extremely rare)	Reduced concentration or impaired function of an inhibitor or cofactor
Fibrinolysis	Increased activation of fibrinolysis	Reduced concentration or impaired function of plasminogen Reduced regulation

SELF-CHECK 2.12

What are the five areas of haemostasis and their main roles?

Key Points

Deficiencies can result from:

- **absence of a particular protein;**
- **reduced concentration of a normally functioning molecule;**
- **normal concentration of an abnormally functioning molecule;**
- **reduced concentration of an abnormally functioning molecule.**

2.8.2 Screening tests

The nature of disorders of blood vessels means that abnormalities affecting haemostasis cannot be subjected to laboratory-based investigations in the routine diagnostic setting. Abnormalities of blood vessels, VWF, and platelets can lead to an abnormally long bleeding time. You will also meet analysers in Chapter 14 that assess global VWF and platelet function. They are mainly used for patients whose clinical symptoms suggest a disorder of primary haemostasis.

By far and away the most commonly performed haemostasis screening tests are those for coagulation screening and platelet counts, the latter having been covered in this chapter. Coagulation screening tests artificially segregate groups of coagulation factors into three discrete compartments so that patterns of abnormal results, if present, can indicate which subsequent tests should be chosen to identify single- or multiple-factor deficiencies. The three discrete compartments are:

- the extrinsic pathway;
- the intrinsic pathway;
- the common pathway.

Look at Figure 14.1 (Chapter 14, section 14.2.1) and you will see that the end product of both the extrinsic and intrinsic pathways is activated factor X (FXa), which is the enzyme that begins the common pathway, culminating in fibrin generation. Each screening test is designed to activate the coagulation reactions at a different starting point, but they all have the same endpoint of fibrin generation. The coagulation screening tests in regular use are:

- PT;
- activated partial thromboplastin time;
- thrombin time;
- quantitative measurement of fibrinogen activity.

Prothrombin time (PT)

The PT uses a reagent called thromboplastin that activates coagulation at the start of the extrinsic pathway, which subsequently activates the common pathway. The time taken from the addition of thromboplastin to the patient's plasma to the generation of a fibrin clot is the PT itself, which is recorded in seconds. The coagulation factors of the intrinsic pathway take no part, as there is nothing in the reagent to activate them.

Activated partial thromboplastin time (APTT)

The patient's plasma is incubated with the APTT reagent, which specifically activates the intrinsic pathway, but only up to the point where calcium ions are required. After the incubation period, which is typically three to five minutes, calcium ions are added, allowing the intrinsic pathway to progress to activation of the common pathway and subsequent clot formation. The time taken to clot from the addition of the calcium ions is the APTT itself, which is also recorded in seconds.

Thrombin time (TT)

The TT merely involves adding thrombin to the patient's plasma to bypass all the other coagulation factors and just convert their fibrinogen to fibrin. It is mainly used as a quick method to check the patient's fibrinogen level and is also recorded in seconds.

Fibrinogen

Presence of an adequate level of fibrinogen is crucial if the preceding coagulation factor reactions are to have their desired effect. Measuring fibrinogen activity is performed using a modified version of the TT, where the patient's plasma is diluted and then clotted with a calibrated thrombin reagent. Now that automated analysers are available to perform this test, many laboratories no longer perform TTs.

Cross reference

You can find further detail about coagulation screening tests in Chapter 14 on bleeding disorders.

2.8.3 Analytical platforms

Although the endpoint for all coagulation screening tests is a fibrin clot, there are in fact three main analytical methods for detecting clot formation.

Tilt-tube technique

The biomedical scientist manually pipettes reagents and plasma into test tubes in a 37 °C water bath and estimates clotting times using a stopwatch. Upon addition of the final reagent, the test tube is removed from the water bath, held horizontally, and returned two to three times/second (to maintain temperature) until the reaction mixture is seen to have clotted.

Mechanical clot detection

Reagents and plasma (or whole blood) are manually pipetted into a rotating cuvette, which contains a small metal ball sitting within a magnetic field. The cuvette is contained within a coagulometer that reads the endpoint when formation of the fibrin clot removes the metal ball from the magnetic field. Automated versions are available that use robotics to deliver reagents and plasma and to time the clot formation.

Photo-optical clot detection

Robotics are used to deliver reagents and plasma and endpoints are timed automatically. Reaction cuvettes are placed in the light path of a fixed beam and the increase in turbidity upon fibrin formation alters the light scatter, which is detected photo-optically by either turbidimetry or nephelometry. The analyser plots the change in light transmittance or scatter over time for each test as a coagulation curve, an example of which you can see in Figure 2.18.

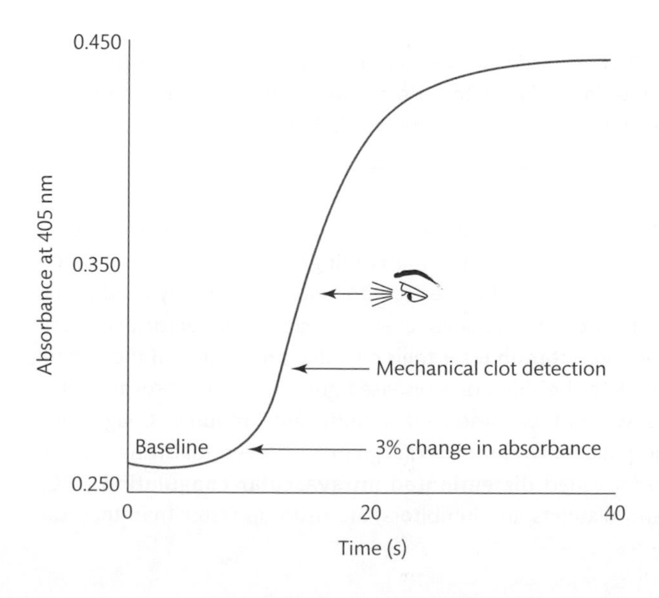

FIGURE 2.18
Coagulation curve for a normal PT. At the addition of the thromboplastin reagent the analyser plots a baseline of light transmittance or scatter, which remains constant until fibrin clot formation begins. Polymerization of fibrin increases the turbidity of the reaction mixture leading to reduced transmission of light through the cuvette, plotted as increase in absorbance, and an increase in light scatter. The analyser can be pre-set to detect a clot at a defined percentage of change in transmittance or increased scattered light intensity as low as a 3% difference from the baseline. You can see from the graph that mechanical clot detection is less sensitive to the early stages of fibrin formation, yet its design means that clots are detected at the point of coagulation of the entire reaction mixture. Photo-optical systems are often pre-set closer to this level. Tilt-tube techniques that rely on the skill and visual acuity of the operator are inevitably less sensitive.

TABLE 2.6 Representative coagulation reference ranges.

Screening test	Reference range (s)	Reference range (ratio)
PT	11–14	0.90–1.10
APTT	24–34	0.85–1.15
TT	12–15	0.80–1.20

Fibrinogen is reported as a concentration, a typical reference range being 1.5–4.0 g/L.

2.8.4 What do the results tell us?

If a patient has reduced function of one or more coagulation factors, it will cause a 'bottleneck' in the coagulation reactions. Consequently, one or more of the screening tests will take longer to clot than normal when compared to a reference range for each test. There is no definitive reference range for each test because reagents vary between manufacturers and an individual reagent may perform differently on alternative analytical platforms. Some laboratories report PT, APTT, and TT results as ratios by dividing the clotting time of the patient's plasma with that of a normal-pool control plasma. Representative reference ranges are given in Table 2.6.

Most hereditary coagulation factor deficiencies affect just one factor, the most common being haemophilia A, which is a deficiency of factor VIII. Look at Figure 14.1 (Chapter 14, section 14.2.1) and you will see that an isolated deficiency in the extrinsic pathway will prolong only the PT, one in the intrinsic pathway will prolong only the APTT, and a common pathway deficiency will prolong both PT and APTT. A fibrinogen deficiency will be evident from its direct measurement and will prolong PT, APTT, and TT. Assessing the screening test results allows biomedical scientists and medical staff to make informed decisions about which specific coagulation factors should be measured with more complex tests to identify the deficiency.

Key Points

Remember that PT, APTT, and TT are artificial ways of causing plasma to clot *in vitro* and the clotting times we generate in the laboratory do not necessarily equate to *in vivo* clotting times, where coagulation interacts with other physiological processes.

Some primary disorders lead to a secondary reduction in production or an increased consumption of coagulation factors and generate abnormal clotting screen results. For instance, most coagulation factors are synthesized in the liver, so if the patient's primary condition is a diseased liver a secondary effect can be reduced levels of coagulation factors. Similarly, some of the clotting factors require **vitamin K** for their effective production. If the patient does not have enough vitamin K in their diet or a diseased gut impairs its absorption, the vitamin K-dependent factors will not be produced in sufficient amounts. Coagulation can be overactivated in some patients with sepsis, malignancy, or even following surgery, leading to a secondary disorder called **disseminated intravascular coagulation** (DIC), where coagulation factors (and platelets and inhibitors) are used up faster than they can be replaced.

Inflammation and pregnancy can increase levels of factor VIII and fibrinogen such that they accelerate *in vitro* coagulation reactions and cause the APTT to fall below the reference range. In some instances, this can mask another abnormality.

Use of screening tests to monitor anticoagulant therapy

Many patients who have had a thrombosis or are at risk of developing a thrombosis are treated with **anticoagulant** drugs that reduce the ability of their coagulation mechanisms to form a clot. The anticoagulant effects of two of the most commonly used drugs, warfarin and **unfractionated heparin** (UFH), can be assessed using standard coagulation screening tests.

unfractionated heparin
An injectable anticoagulant whose effect is monitored by the APTT.

Warfarin achieves therapeutic anticoagulation by interfering with the synthesis of the vitamin K-dependent factors—which are factors II, VII, IX, and X. Patients continue to synthesize these factors but their biological activity is diminished. Warfarin is monitored using the PT because most of the factors that contribute to the clotting time are vitamin K-dependent. Patients are tested at regular intervals to ensure that their dose of warfarin is not too high or too low. If too high, they are at risk of developing dangerous bleeding symptoms because they may not clot at all, even when they need to. If the dose is too low, there will be insufficient protection against developing a thrombosis.

For patients on warfarin, their PT is converted to a parameter called the International Normalized Ratio (INR). The INR is derived from the ratio between the time blood or plasma takes to clot normally compared to the time it takes to clot due to warfarin. So, if a normal PT is 12 seconds, and on warfarin the PT is 24 seconds, then the ratio is 2.0. In view of reagent and analytical platform variability, this ratio is then calculated to the power of a previously calibrated mathematical expression of the sensitivity of the reagent to the effects of warfarin, the International Sensitivity Index (ISI). In our example, if the ISI was 1.0, the ratio would remain as 2.0. If the ISI was 1.3, the resultant INR would be 2.46.

The INR/ISI system ensures that patients get equivalent results irrespective of the reagents and analytical platform used to generate the PT, to prevent unnecessary and potentially dangerous dose alterations. Someone at a relatively low risk of thrombosis receives warfarin doses to maintain their INR between 2.0 and 3.0. However, those at high risk of thrombosis would be prescribed enough warfarin to maintain an INR between 3.0 and 4.0. Approximately 1% of the UK population are receiving warfarin at any one time, and so PT/INR testing forms a large part of the workload in a routine haemostasis laboratory.

low molecular weight heparin (LMWH)
An injectable anticoagulant used principally to treat or reduce the risk of venous thromboembolism.

UFH works by binding to one of the circulating inhibitors of coagulation, **antithrombin**, and amplifying its anticoagulant effect. Antithrombin in its natural state does not interfere with coagulation screening tests, but can markedly prolong the APTT and TT in a patient receiving UFH. The TT is over-sensitive to UFH, so patients are monitored by APTT, the aim being to keep their APTT ratio between 1.5 and 2.5. There is no equivalent to ISI for APTT reagents. In many cases, UFH has been superseded by an improved formulation, **low molecular weight heparin (LMWH)**. The advantage of this preparation over UFH is that it generally does not need to be monitored. However, if its activity is needed, this can be determined by the anti-factor Xa assay.

Cross reference
The importance of DOACs is described in Chapter 17.

A new class of drugs, the **direct oral anticoagulants (DOACs)**, so named because they act directly on certain coagulation molecules (thrombin or Factor Xa), are alternatives to warfarin and LMWH in a small, well-defined number of situations. One advantage of the DOACs is that they do not need to be monitored routinely with a blood test (as does the INR for warfarin), thus providing a considerable laboratory saving and reduced inconvenience to the patients.

Cross reference
You will learn more detail about therapeutic anticoagulation and the role of the laboratory in its monitoring in Chapter 17.

SELF-CHECK 2.13

Name the main coagulation screening tests and the areas of coagulation they assess.

SELF-CHECK 2.14

Which two therapeutic anticoagulants can be monitored with which routine coagulation screening tests?

2.9 Haematinics

Haematinics considers what may be described as a vital supporting cast to the main players of RBCs, WBCs, and platelets. The major micronutrients (iron, vitamin B_{12}, vitamin B_6, and folate) must be provided in the diet. Minor micronutrients include vitamin K, selenium, and zinc. Other micronutrients, essential to other aspects of physiology (but less important in haematology) include vitamins A, C, D, and E, iodine, and copper. The liver and other organs and cells synthesize key plasma proteins crucial to micronutrient metabolism.

2.9.1 Micronutrients

Deficiencies in any one of four micronutrients (iron and vitamins B_6 and B_{12} and folate) may result in impaired production of RBCs by the bone marrow and therefore anaemia.

The oxygen-carrying molecule Hb is a mixture of proteins (globin) and non-proteins (haem). At the centre of the haem molecule is an atom of iron, and this is where the oxygen binds. This iron must be provided in the diet, but we also need healthy intestines to absorb it from our food, and it must be carried from the intestines to the bone marrow. Therefore, lack of iron, either in the diet or because of failure to absorb iron across the intestinal wall, can lead to a particular type of anaemia, called iron-deficiency anaemia, where the RBCs are smaller than usual (as defined by a reduced MCV).

Possibly more complex are the biochemical steps in the synthesis of DNA and haem that rely on vitamins B_6 and B_{12}, and folate. The vitamins are required as cofactors by metabolic enzymes. The B vitamins are needed in the synthesis of haem. Folate is converted by the enzyme folic acid reductase to dihydrofolic acid, itself a root molecule for the production of thymidine, ultimately a constituent of DNA. Failure of the bone marrow to be provided with enough of these micronutrients, either by a poor diet or by inadequate absorption across the gut wall, will lead to a particular type of anaemia, called megaloblastic anaemia, where the RBCs become larger than usual. This complex process is explained in Chapter 4, and the diseases that follow when the process is faulty are described in Chapters 5 and 6.

Cross references

Chapters 5 and 6 describe the many different types of anaemia.

Further details of the involvement of iron, folate, and vitamins B_6 and B_{12} in the synthesis of Hb are presented in Chapter 4.

Insufficient vitamin B_{12} or folate will, unlike insufficient iron, also lead to certain changes in WBCs, and also other cells such as those of the nervous system. One of the principal WBC changes is that the nucleus of the neutrophil condenses into a larger number of lobes than usual (often six or more, compared with the normal three to five lobes). This is called hypersegmentation and the cell that exhibits this is called a **hypersegmented neutrophil**. Most autoanalysers are unable to detect this change, but it can be seen on a blood film; yet another example of where the skill of the scientist comes into its own!

Key Points

The importance of trace amounts of micronutrients can be demonstrated by the marked abnormalities that accompany the deficiencies of iron and the B vitamins.

2.9.2 Plasma proteins

Once iron has been absorbed from the diet by the gastrointestinal tract, its passage into the blood is regulated by the plasma protein **hepcidin**. Once in the blood, it must be collected and moved to the bone marrow by a specialized blood protein, **transferrin**. Without this protein, free iron in the plasma may be excreted by the kidney, so that a lack of transferrin (possibly because of liver disease) may also lead to a deficiency of iron and, hence, anaemia. Iron is a valuable resource, and the body has developed a number of strategies to ensure it has enough stored up to guard against times of possible shortage. **Ferritin** is a specialized plasma protein that can collect and store many atoms of iron in organs such as the liver, spleen, and bone marrow. There are also several other plasma proteins of interest to haematologists, such as those which carry vitamin B_{12} from the stomach to the large intestines (where it is absorbed), and then to the cells where it is needed. These proteins are **intrinsic factor**, **transcobalamin**, and **haptocorrin**. The importance of each of these carrier, regulatory, and storage molecules is explained in Chapters 4 and 5.

2.9.3 The laboratory and haematinics

The study and measurement of haematinics is an area where haematology and biochemistry have shared interests. From a purely practical viewpoint, the measurement of 'haematology' plasma proteins and vitamins demands complex analysers and techniques, some of which are common to other proteins that many may see as being part of the 'biochemistry' profile. Consequently, the exact laboratory where these analyses take place varies from hospital to hospital because of the nature of the particular analyser. Furthermore, analysis of some vitamins is so complex that many smaller labs will find it more convenient to send their samples to a reference laboratory, which may be staffed by haematologists or biochemists. Table 2.7 summarizes haematinics.

SELF-CHECK 2.15

What are the key haematinics?

TABLE 2.7 Key haematinics.

Micronutrients	• Iron: low levels are present in some forms of anaemia; however, high levels may be dangerous and can lead to organ failure • Vitamin B_{12}, Vitamin B_6, and folate: low levels are present in some forms of anaemia
Plasma proteins	• Ferritin: a molecule that stores iron • Transferrin: a molecule that carries iron from the intestines to the bone marrow • Transcobalamin and haptocorrin: molecules that carry vitamin B_{12}

2.10 The reference range

The purpose of the haematology laboratory is to facilitate the diagnosis and then confirm the effect (or lack of effect) of treatment. A key component of this is the need to know what we hope a blood result should be. This is the set of results we refer to, and hence the term, 'reference range'. We prefer this name to alternatives such as 'normal range' or 'range'.

2.10.1 What do we mean by 'reference range'?

The expression 'normal range' is inadequate simply because a result that is normal (that is, is present in a lot of individuals) in a population does not mean it is desirable. A good example of this is a low Hb level that may be endemic in some parts of the world, possibly because of malnutrition, genetics, or parasites—none of which we would consider healthy. In addition, merely because someone appears healthy (i.e. is asymptomatic), it does not automatically follow that their blood result is satisfactory, and vice versa. Similarly, 'target range' is not fully appropriate as it implies a level of a result that we are trying to achieve; this may never be possible in some individuals, resulting in disappointment and a sense of failure. However, there are cases where a target is a useful objective.

It is also worthwhile discussing where 'normal values' come from. Who is normal? Many of us have unsuspected asymptomatic diseases that may well impact on haematology. In the past, results from blood donors were considered to be representative of being 'normal', but we now recognize the short-coming in this definition, as blood donors are in fact highly motivated and health-aware individuals who are therefore, on the whole, 'healthier' than the general population. This is a classic example of selection bias.

It is important to note that reference ranges vary both from hospital to hospital and over time. The former is because different autoanalysers may well give a slightly different result on the same sample of blood. Furthermore, the reference range should serve the local population that the hospital serves, and local populations can vary a great deal. As we improve our knowledge of biomedical science, it becomes clear that some reference ranges need to be changed. In the 1975 edition of a major practical textbook, the middle of the 'normal' range for MCV in the adult was given as 85 fL. In the 2001 edition, in the 'reference range and normal values' table, the mean MCV is given as 92 fL. In 1975, the reference range for neutrophils was 2.0–7.5 $\times 10^9$/L, but in the 2001 edition the range is 2.0–7.0 $\times 10^9$/L. It follows that in 1975 a result of 7.25 was considered to be within the 'normal' range, whereas twenty-six years later the same result is outside this range, and so may be described as a mild neutrophil leucocytosis. Whether or not this is actionable is another question.

A note on units

Not only do the units of blood tests vary around the world (such as total cholesterol being reported in mmol/L in the UK, but as mg/L in the USA and elsewhere) but also in time. Historically, Hb has been described as a percentage, then in units of g/dL. However, the unit (dL, decilitre) is not fully part of the international system, which reports in terms of the litre (L). Hence the Hb unit is now g/L, so that the redundant unit of the result of 13.9 g/dL simply becomes 139 g/L.

2.10.2 Interpretation

All routine haematology (and biochemistry) results sent out from the laboratory are accompanied by a reference range, which is a set of numbers enclosed by brackets (see Figure 2.16). In addition, the laboratory will often draw the reader's attention to those results that are considered to be out of range and therefore worthy of attention. There may be an asterisk or other flag alongside the abnormal result(s). Indeed, for this reason the reference range may also be considered a 'concern range'. This is because a result fractionally outside the reference range does not always carry a serious health hazard. However, the further a particular result is outside the reference range, the more seriously we must address the result, as it may well be the consequence of actual disease, and so should be actioned.

The reference ranges for this volume are presented as Table 1, located immediately before the list of abbreviations in the front of this book.

SELF-CHECK 2.16

In your practice, why should you not use the reference values given in a textbook?

Key Points

Position statement

The authors present a set of reference ranges that they consider appropriate. It does not follow that your particular laboratory is wrong merely because it has a different set of ranges. Each practitioner must work to their own local reference range, not to those presented in this volume (or, indeed, any volume). They are provided here for perspective and to allow for comparison in the various case studies and examples we present.

Chapter summary

- Anticoagulants are used to prevent blood from clotting. These include EDTA, trisodium citrate, and lithium heparin.
- Modern laboratories rely on an autoanalyser to provide the FBC, but the blood film is also an essential tool, especially to examine cell morphology.
- The FBC comprises Hb, the RBCC, Hct, three red cell indices (MCV, MCH, MCHC), the WBC count and differential, and the platelet count.
- ESR and plasma viscosity provide information on physical characteristics of the blood (rheology).
- Köhler illumination is used to 'set up' a light microscope prior to each use.
- Manual differential counts are important for calculating the absolute values for each species of WBC when an automated analyser is unable to provide an accurate differential.
- Haemostasis is the balanced orchestration of vascular integrity, primary and secondary haemostasis, regulatory mechanisms, and fibrinolysis. They maintain blood fluidity and limit and arrest bleeding upon vessel damage.

- Disorders of the different areas of haemostasis can lead to bleeding or thrombotic disorders.
- Blood coagulation disorders are first detected in the laboratory with the coagulation screening tests PT, APTT, TT, and fibrinogen estimation. Particular result patterns inform subsequent diagnostic pathways.
- Essential micronutrients include iron, vitamin B_{12}, and folate. Key plasma proteins include intrinsic factor, transferrin, and ferritin. Together these are often described as haematinics.
- The reference range, generally composed of indices of thousands of presumed healthy individuals, provides an indication of the level of concern about a particular result.

Discussion questions

2.1 Explain the value of each of the six RBC indices.

2.2 Name the essential micronutrients and describe why they are needed.

2.3 What are the clinical significances of the WBC count and the platelet count?

Further reading

- Arroyo AB, de Los Reyes-García AM, Teruel-Montoya R, Vicente V, González-Conejero R, and Martínez C. *MicroRNAs in the haemostatic system: more than witnesses of thromboembolic diseases? Thrombosis Research* 2018: **166**: 1–9.
- Bain BJ, Bates I, and Laffan MS (eds). *Dacie and Lewis: Practical Haematology*, 12th edn. Churchill Livingstone, London, 2016.
- Caldwell I, Ruskova A, Eaddy N, and Bain BJ. *Acute leukemic transformation of myelodysplastic syndrome: Is cytochemistry still relevant? American Journal of Hematology* 2018: **93(1)**: 148–9. https://doi.org/10.1002/ajh.24831.
- Della Starza I, De Novi LA, Nunes V, Del Giudice I, Ilari C, Marinelli M, Negulici AD, Vitale A, Chiaretti S, Foà R, and Guarini A. *Whole-genome amplification for the detection of molecular targets and minimal residual disease monitoring in acute lymphoblastic leukaemia*. *British Journal of Haematology* 2014: **165**: 341–8.
- Falini B, Martelli MP, and Tiacci E. *BRAF V600E mutation in hairy cell leukaemia: From bench to bedside*. *Blood* 2016: **128**: 1918–27.
- Hajizamani S, Shahjahani M, Shahrabi S, and Saki N. *MicroRNAs as prognostic biomarker and relapse indicator in leukemia*. *Clinical and Translational Oncology* 2017: **19(8)**: 951–60.
- Herwald H and Egesten A. *The neutrophil: A beautiful beast or a beastly beauty? Journal of Innate Immunology* 2015: **7**: 555–6.
- Johansson U, Bloxham D, Couzens S, Jesson J, Morilla R, Erber W, and Macey M. *British Committee for Standards in Haematology*. *Guidelines on the use of multicolour flow cytometry in the diagnosis of haematological neoplasms*. *British Journal of Haematology* 2014: **165**: 455–88.

- Jones CH, Pepper C, and Baird DM. ***Telomere dysfunction and its role in haematological cancer***. *British Journal of Haematology* 2012: **156**: 573–87.
- Kohlmann A, Grossmann V, Nadarajah N, and Haferlach T. ***Next-generation sequencing–feasibility and practicality in haematology***. *British Journal of Haematology* 2013: **160**: 736–53.
- Kotaki R, Koyama-Nasu R, Yamakawa N, and Kotani A. ***miRNAs in normal and malignant hematopoiesis***. *International Journal of Molecular Sciences* 2017: **11(July)**: 18(7) pii: E1495.
- So DY, Wells GA, McPherson R, Labinaz M, Le May MR, Glover C, Dick AJ, Froeschl M, Marquis JF, Gollob MH, Tran L, Bernick J, Hibbert B, and Roberts JD. ***A prospective randomized evaluation of a pharmacogenomic approach to antiplatelet therapy among patients with ST-elevation myocardial infarction: the RAPID STEMI study***. *Pharmacogenomics Journal* 2016: **16(1)**: 71–8.
- Waller P and Blann AD. ***Non-coding RNAs–A primer for the laboratory scientist***. *British Journal of Biomedical Science* 2019: **76**: 157–65.
- Zonneveld R, Molema G, and Plotz FB. ***Analyzing neutrophil morphology, mechanics, and motility in sepsis: Options and challenges for novel bedside technologies***. *Critical Care Medicine* 2016: **44(1)**: 218–28.

Answers to self-check questions and case study questions are provided as part of this book's online resources.

Visit www.oup.com/he/moore-fbms3e.

3

Haemopoiesis and the bone marrow

Andrew Blann

This chapter outlines the origin and development of blood cells, the process of haemopoiesis (Greek: *poiesis*—to make). This is the regulated development of blood cells—red blood cells, white blood cells, and platelets—as they progress from being precursor stem cells in the bone marrow to fully functioning mature cells found in the blood. Much of this knowledge has been obtained from the analysis of bone marrow itself. Understanding haemopoiesis is the key to understanding several pathological processes, including anaemia, autoimmunity, and leukaemia, as is developed in the chapters that follow.

Learning objectives

After studying this chapter, you should confidently be able to:

- explain the importance of effective haemopoiesis;
- list the major components of bone marrow;
- describe the mechanisms of haemopoiesis and the importance of growth factors;
- appreciate the complex nature of the molecular genetics of blood cell development;
- explain the value of the analysis of bone marrow;
- comment on the uses of major cytochemical and fluorescence flow cytometry methods in the analysis of bone marrow.

3.1 Overview of the cellular constituents of the blood

Effective **haemopoiesis**, the process of the development of blood cells, is crucial to the health of the individual: it generates mature, functional blood cells that transport oxygen, defend us from infection, and participate in haemostasis. In adult life, haemopoiesis predominantly occurs in the **bone marrow**, the soft tissue within the centre of bones where blood cells develop. Bone marrow, accounting for perhaps 5% of the weight of an adult, consists of two compartments: red marrow (containing the haemopoietic tissue) and yellow marrow (predominantly fat). In the neonate, almost all bone marrow is red, and as we age this is replaced by yellow marrow. In this chapter, when we refer to bone marrow, it is the red component that we are discussing. In exceptional circumstances (such as in certain pathological conditions) haemopoiesis may also occur in other tissues, including the liver, lymph nodes, and spleen. The situation is different in the foetus and the neonate, however, where it is normal for haemopoiesis to occur in the bone marrow, lymph nodes, liver, and spleen, as well as in the yolk sac of the embryo.

In a healthy individual, some 5–10 × 10^{11} blood cells are produced by the bone marrow each day by a process that is highly balanced and regulated by **cytokines**, **growth factors**, hormones, and environmental factors, including the amount of oxygen in the body. Ideally, the number of cells produced each day exactly matches the number of cells that have come to the end of their life cycle, thus maintaining a steady state. Damage and/or disease of the bone marrow (such as may be caused by drugs, radiation, infiltrating cancer, or viruses) can lead to irregularities in the production of blood cells, and this itself can lead to other clinical problems, such as infections. Accordingly, a thorough understanding of the structure and function of the bone marrow and haemopoiesis is necessary in order to grasp the concepts of these diseases, which include aplastic anaemia (where there is a reduction in cell production) and the **haemoproliferative disorders**, some of which are life-threatening. The most common haemoproliferative disorder is **leukaemia**, where an excess of white blood cells is produced.

As indicated in Chapter 1, blood cells are easy to classify into one of three types: red cells, white cells, and platelets. The mature cells that are present in the blood all have their origins in the bone marrow, the location of the **stem cells** that ultimately give rise to the mature cells found in blood. Red cells and white cells mature under the influence of **lineage-specific growth factors**, which act on only one set of cells and on no other (e.g. on cells of the red cell series, but not on cells of the granulocyte series). In doing so, they pass through reasonably well-defined stages of maturity, beginning with blast stages, such as that of the **myeloblast**. Mature platelets are fragments of the cytoplasm of their own dedicated precursor, the megakaryocyte.

cytokines
Small hormone-like intercellular mediators with a diverse range of functions, including the stimulation of the immune system in response to an encounter with a pathogen.

growth factors
Cytokines produced by one type of cell that initiate or promote the growth or differentiation of another cell.

haemoproliferative disorder
A condition characterized by inappropriately increased numbers of circulating blood cells and their precursors.

myeloblast
The least mature, morphologically identifiable cell of the myeloid lineage.

Cross references

We discuss aplastic anaemia in more detail in Chapter 5.

Additional details of the importance of bone marrow and its analysis are presented in Chapter 10.

We present full details of leukaemia in Chapters 11 and 12.

SELF-CHECK 3.1

Why do we need a good knowledge of haemopoiesis?

3.2 Ontogeny of haemopoiesis

Ontogeny effectively means 'where it takes place'. Haemopoiesis begins in the haemogenic endothelium of the aorta-gonad mesonephron and yolk sac of the embryo, with the appearance by the third week after fertilization of primitive stem cells called haemocytoblasts, which have originated from haemogenic endothelial cells, and also primitive blood vessels. However, this is relatively brief, as after about four to six weeks, once the vasculature is developed, stem

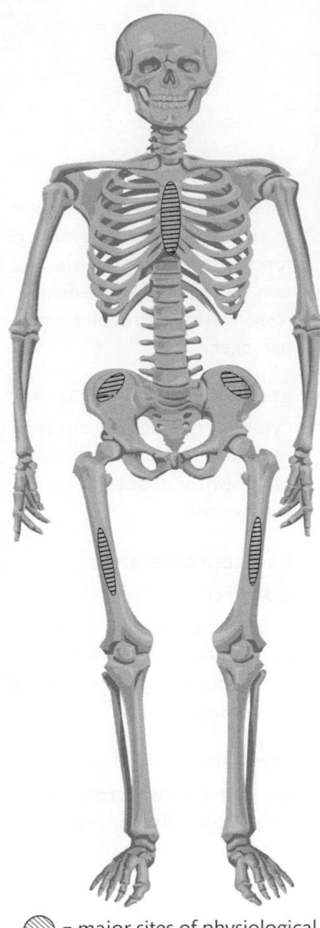

FIGURE 3.2
The skeleton showing major sites of haemopoiesis. The major sites of haemopoiesis in the adult are the sternum, femur, and pelvis. Access to the latter is via the iliac crest. Minor sites include the ribs.

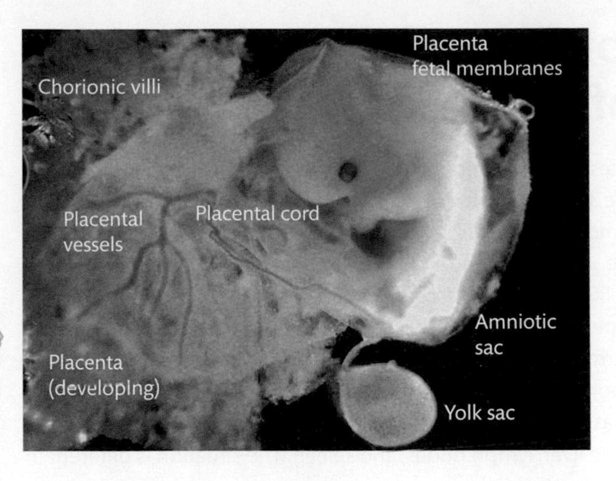

FIGURE 3.1
Embryo: the embryo is situated within the uterus. Attached is a placenta, and below the embryo, a spherical yolk sac. The yolk sac is the primary site for haemopoiesis during embryonic development. © MA Hill, 2004.

cells migrate to the liver and spleen, which slowly take over to become the dominant sites of blood cell production (Figure 3.1). Between the second and seventh months of gestation, the liver becomes the major site of haemopoiesis. Once the bone marrow is sufficiently developed and vascularized, stem cells once more migrate, and begin to take over blood development during the fifth to ninth months of gestation so that, at birth, it is normally the only place where haemopoiesis occurs. Infant haemopoiesis can occur in all bones. As the child develops, however, this falls back into the axial skeleton and proximal ends of the long bones, primarily the femur. In the adult, the major sites of haemopoiesis are the sternum and iliac crests; other sites include the skull, vertebrae, and ribs (Figure 3.2). The importance of these locations will be clear in the study of haemoproliferative disorders, as is explained in Chapters 5, 6, and 9–12.

Haemopoiesis occurring within the bone marrow is termed intramedullary. By contrast, neonate and adult production at other sites (e.g. in the liver, lymph nodes, and spleen) is called **extramedullary haemopoiesis** ('outside the bone marrow'). In the adult, extramedullary haemopoiesis is present only in pathological conditions—for example, when the bone marrow becomes infiltrated with fibrous tissue (**myelofibrosis**), when there is chronic bleeding, in severe haemoglobinopathy, or when there is severe **haemolytic anaemia**. In these cases, the extension of haemopoiesis to sites beyond the bone marrow (the liver and spleen) is simply the body's response to a severe lack of functioning red blood cells and platelets.

SELF-CHECK 3.2

Where does haemopoiesis take place at different times in the development of the individual?

Cross references

Myelofibrosis can also be a cause of anaemia, as discussed in Chapters 5 and 11.

Haemoglobinopathy, and its consequences, and different types of haemolytic anaemia is discussed in Chapter 6.

3.3 Bone marrow architecture and cellularity

The major structural function of bone is to provide support and pivot points for muscles, ligaments, and tendons; protection for delicate tissues (such as the lungs and heart, brain, and spinal cord); and physical support for body organs (liver, spleen, etc.). While many of these physical demands are met by virtue of the strength of bone as a hard and supportive connective tissue, many bones are 'hollow' and are extremely dynamic organs. Spaces within bone that are not there for structural reasons are host to haemopoietic tissues and the blood vessels and other supportive cells that serve them. Fat cells are very common, and a ratio of fat cells (comprising yellow marrow) to haemopoietic cells (comprising red marrow) of 1:1 is often found (although this varies with age). The complex microenvironment that ultimately gives rise to the blood cells can be seen as having three components:

- the haemopoietic tissues, sometimes described as cords, which consist of the stem cells and their progeny, the immature but developing blood cells such as myeloblasts and erythroblasts—more details of these are presented in the sections that follow;
- sinuses—the vascular spaces or pools of blood that are lined with **endothelial cells** to regulate the release of mature and immature cells into the peripheral blood;
- non-haemopoietic cells that support the bone marrow and often produce growth factors. These include:
  - stromal cells (such as **fibroblasts**) that produce the scaffolding (such as collagen) that supports other cells;
  - **macrophages** that produce growth factors (such as transforming growth factor-β and stem cell factor), promote red blood cell production (erythropoiesis), store iron, and perform routine debris removal;
  - adipocytes that store energy in the form of fat.

haemolytic anaemia

The consequences of the premature destruction of red blood cells in the circulation and/or the spleen.

endothelial cell

A specialized cell that forms the internal lining of the blood vessels and sinuses.

A close physical association of various cells is necessary to ensure the correct development of particular mature blood cells via cell–cell contact, with secretion and binding of growth factors. Indeed, there is a growing appreciation that haemopoiesis does not occur throughout the bone marrow, but in well-defined niches. The arteriolar-pericyte (AP) niche consists of endothelial cells and pericytes, both of which regulate haemopoietic stem cells. In support of this concept is the observation that, in the mouse, arterial endothelial cells (but not sinusoidal endothelial cells) secrete nearly all stem cell factors. The sinusoid-megakaryocyte (SMK) niche comprises megakaryocytes—which eventually form platelets—and haemopoietic stem cells: both niches contain chemokine-abundant reticular cells. Platelet and myeloid-biased haemopoietic stem cells, expressing von Willebrand factor (VWF), are abundant in the SMK niches. Depletion of megakaryocytes selectively expands haemopoietic stem cells expressing von Willebrand factor, whereas the depletion of AP niche cells selectively depletes von Willebrand factor lymphoid-biased haematopoietic stem cells. This may help explain altered megakaryopoiesis in type 2B von Willebrand disease—a type of bleeding disorder where platelets cannot adhere adequately to the vascular sub-endothelium. In parallel with these developments is the growing awareness of roles for osteoblasts and osteoclasts, both likely to promote haemopoiesis by secreting growth factors. We revisit the fine structure of the bone marrow in section 3.6.

Cross reference

Additional details of the importance of bone marrow and its analysis are presented in Chapters 9 and 10.

SELF-CHECK 3.3

What are the major types of cell within the bone marrow and what are their particular roles?

Key Points

Good production of blood cells requires not only the progenitor stem cells, but also a healthy bone marrow microenvironment that is composed of several different types of supporting cells. However, in certain circumstances, production of blood cells can occur outside the bone marrow in organs such as the liver and spleen.

3.4 Models of differentiation, stem cells, and growth factors

Blood cells pass through a series of well-defined stages of development and maturation, possibly in different compartments of the bone marrow.

3.4.1 Stem cells

Haemopoiesis begins with a common primitive **pluripotent stem cell** (often also called the haemopoietic stem cell), present at a frequency of some 11,000–22,000 per individual. Each of these cells has the capacity to independently replicate, proliferate, and differentiate into the various **lineage-specific stem cells** that will ultimately give rise to each of the different types of mature blood cells. These relationships are described in Figure 3.3. Pluripotent stem cells have a frequency of perhaps 1 per 1000–2000 marrow cells and are impossible to recognize using normal morphological criteria (resembling, as they do, **lymphocytes**); most are in the resting stage of the cell cycle—stage $\mathbf{G_0}$. However, stem cells (such as the myeloblast) can be recognized by the presence of a certain molecule (**CD34**) on their surface (although CD34 is also present on many non-stem cells). Nevertheless, an important practical aspect of the CD34 molecule is that it can be used to harvest stem cells from the peripheral blood, as is required for stem cell transplantation. Haemopoietic stem cells can be harvested not only from the bone marrow, but also from the blood found within umbilical cords. (See also Box 3.1.)

lymphocyte
A small white blood cell with immunological properties, such as antibody production or the destruction of cells infected with viruses.

Cross reference
The cell cycle is described in greater detail in Chapter 9.

SELF-CHECK 3.4

What is the difference between a pluripotent stem cell and a unipotent stem cell?

3.4.2 Colony-forming units

Much of the current theory of the differentiation and development of blood cells is extrapolated from animal models and from tissue culture work in the laboratory where stem cells, also known as **colony-forming units** (CFUs), can be grown from samples of actual bone marrow cells. The CFU has been, and continues to be, essential in our understanding of the different stages and pathways through which cells develop. These CFUs are presumed to be analogous to stem cells within the bone marrow.

Several different models of haemopoiesis exist, one of which is illustrated in Figure 3.3. This model first recognizes the haemangioblast, a stem cell from which develop the endothelial cells that line

BOX 3.1 Pluripotent stem cells

A cure of diseases such as leukaemia can be undertaken by the transplant of pluripotent stem cells, harvested from the bone marrow, ideally of the patient themselves. However, the process is hampered by several issues, one of which is the small numbers of stem cells harvested, which may be insufficient to effect a cure. One way around this is to be able to reprogramme non-bone marrow cells such as B lymphocytes and skin fibroblasts. Considerable research is being directed at generating induced pluripotent stem cells, which, if successful, will revolutionize the treatment of several diseases.

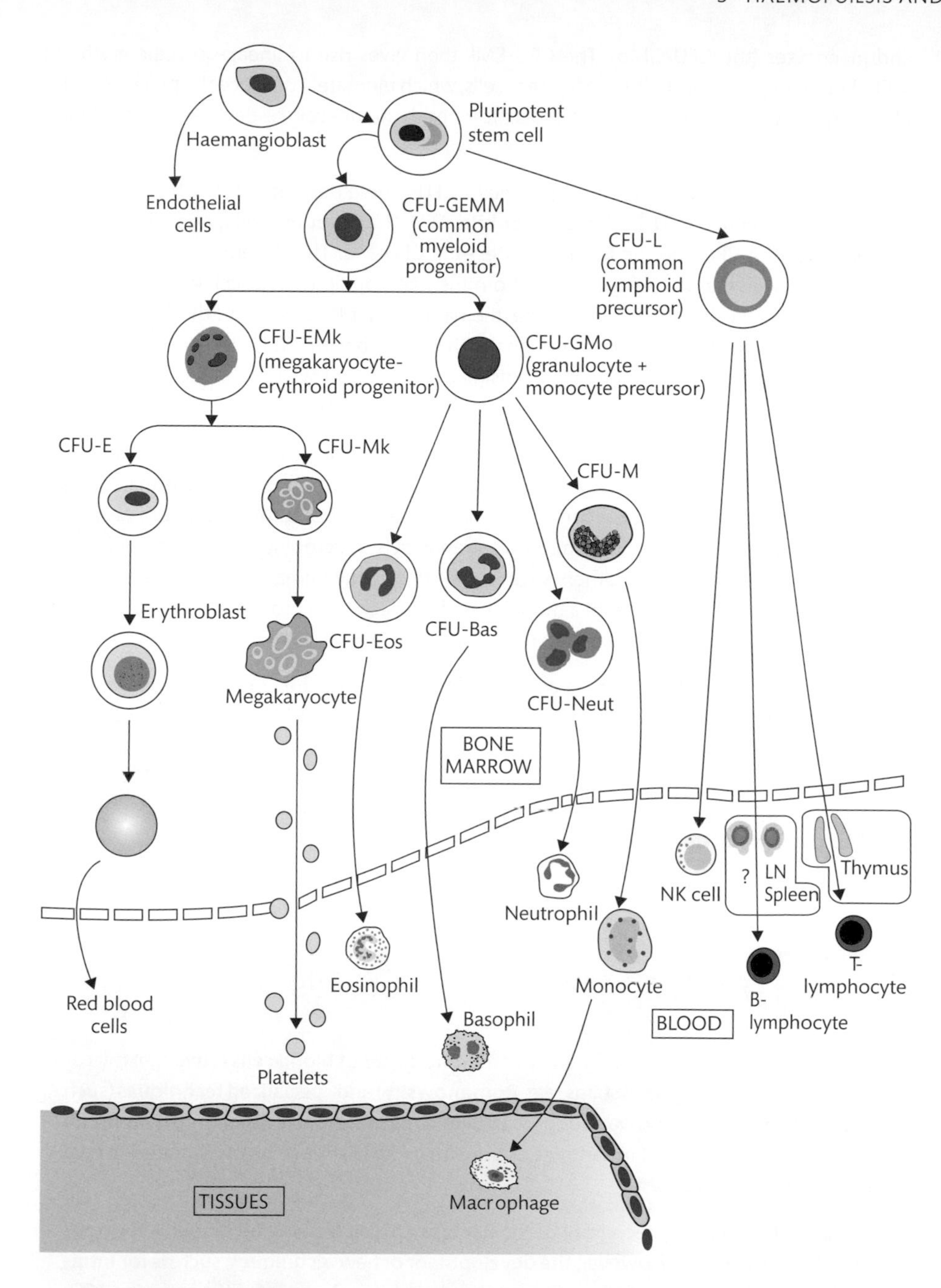

FIGURE 3.3
Cell lineages, haemopoiesis, and cell distribution. The different cell lineages give rise to immature and mature cells within the bone marrow. Mature cells pass from the bone marrow to the blood, and some then pass into the tissues. CFU = colony-forming unit; GEMM = granulocyte, erythrocyte, monocyte, megakaryocyte; E = erythroid; Eos = eosinophil; Bas(o) = basophil; Neut = neutrophil; M = monocyte; NK = natural killer; LN = lymph node.

the blood vessels and all blood cells. The stem cell that gives rise only to blood cells is called the pluripotent stem cell, which in turn gives rise to two stem-cell CFUs, each dedicated to particular cell lineages. One of these CFUs, often referred to as the common myeloid progenitor, will ultimately give rise to **granulocytes**, **erythrocytes**, **monocytes**, and **megakaryocytes**, and so is abbreviated to CFU-GEMM. A second CFU, often described as the common lymphoid precursor, is dedicated to the production of lymphocytes alone, and as such is abbreviated to CFU-L.

The CFU-GEMM then gives rise to two additional stem cells, one progenitor specific for both the erythrocyte and the megakaryocyte (CFU-EMk), and another progenitor for granulocytes and monocytes (the CFU-GMo). The CFU-EMk then gives rise to lineage-specific erythroid (CFU-E) and megakaryocyte (CFU-Mk) stem cells, which ultimately produce the precursors and blasts (erythroblasts, megakaryoblasts) specific for each mature cell, respectively, the red blood cell and the platelet.

The CFU-GMo gives rise to four lineage-specific CFUs, three dedicated to each of the granulocyte lineages (hence CFU-Eo for **eosinophils**, CFU-N for **neutrophils**, and CFU-Baso for **basophils**) and a CFU specific for monocytes—the CFU-Mo. The development pathway for lymphocytes is similar. The CFU-L arising from the haemopoietic stem cell gives rise directly to three lymphoid cells—the B lymphocyte, the natural killer (NK) cell, and the T lymphocyte. However, T lymphocytes must pass through the thymus to become fully functional. In parallel, B lymphocytes may also need to mature outside the bone marrow, but this has not been fully established. (See also Box 3.2.)

Whilst the model we have described is dominant, with cells differentiating and maturing along a defined and fixed lineage, others exist. One such suggests that the granulocyte/macrophage precursor can arise from either a common myeloid precursor or a multilymphoid precursor, and that dendritic cells can arise from both a granulocyte/macrophage precursor and from a common lymphoid precursor. Others suggest that the rigidly demarcated lineage tree (as in Figure 3.3) should be replaced with a model that allows stem and progenitor cells to retain the capacity to step sideways and adopt alternative lineages.

But regardless of the model, there is common agreement that the complex development of blood cells is driven and controlled by local growth factors and hormones, many of which are cytokines. As we have seen, these cytokines are produced by haemopoietic and non-haemopoietic cells both within the bone marrow and in other organs. Some growth factors act specifically on a single lineage of CFUs; others act broadly on CFUs for different classes of blood cell. These points will be reviewed in section 3.4.4.

BOX 3.2 Recognition of cells

The primary tool in the identification of different classes of blood cells is the light microscope. Armed with standard stains (e.g. Romanowsky) and specialized techniques (such as immunocytochemistry), various precursor and mature blood cells of the different cell lineages can be identified in samples of bone marrow that have been aspirated from a bone such as the iliac crest.

The ability to accurately recognize particular types of cell is a skill developed over years of patient observation. However, the development of new techniques, such as for intracellular granules, and for cell-surface molecules of the cluster of differentiation (CD) series, have greatly improved the reliability of cell identification.

The ability of fluorescence flow cytometry and fluorochrome-conjugated antibodies (that is, fluorescence flow cytometry analysis) to accurately identify subpopulations of cells, such as those co-expressing a number of different CD molecules, has revolutionized cell analysis. This process is called immunophenotyping, and a good example is the ability to detect cells that occur at a low frequency (such as residual leukaemic cells), whose detection would not be possible by conventional light microscopy. These issues are expanded upon in Chapter 10.

Romanowsky
This is a family of stains utilized for the visualization of blood cells using light microscopy. These stains contain a mixture of dyes, commonly including azure B and eosin Y.

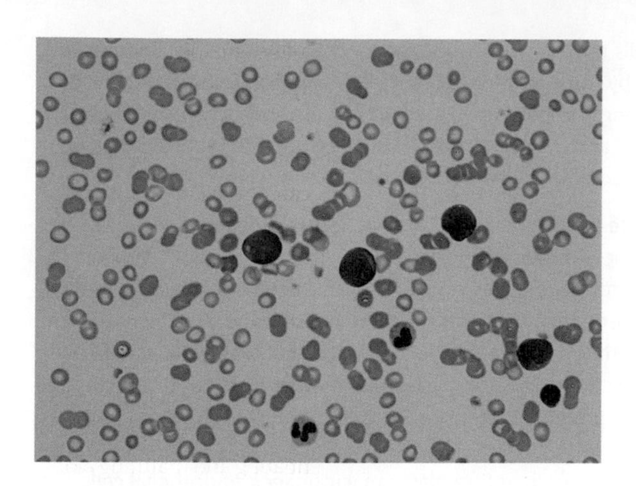

FIGURE 3.4
Blast cells. Four blast cells, the larger of the seven cells shown demonstrating a high nuclear:cytoplasm ratio. The chromatin is open, facilitating transcription and translation. Prominent nucleoli are present. Other white cells in the image include two neutrophils and a lymphocyte (lower right). Courtesy of the Haematology Laboratory, Queen Alexandra Hospital, Cosham.

3.4.3 The blast cell

Whilst the CFU may be seen as a theoretical stage in haemopoiesis, and certainly can be defined in the research laboratory using cell culture, a blast is a stage in development that can be identified in the bone marrow, and, in certain diseases, in the peripheral blood. In the adult, blast cells are almost exclusively found in the bone marrow because they are part of the normal and healthy process of haemopoiesis. In physiology, there are a number of different types of blast cell, and each has an additional qualifying name, which identifies the particular lineage of the mature cell to which it belongs and the type of mature cell it will become upon release into the blood. For example:

- myeloblast—a blast of the myeloid pathway;
- lymphoblast—a blast of the lymphocyte pathway;
- erythroblast—a blast of the erythrocyte (red blood cell) pathway.

However, haemoproliferative diseases such as leukaemia and myeloma are characterized by abnormalities in haemopoiesis. An important aspect of leukaemia is the expansion of blasts (such as myeloblasts) from the bone marrow and the inappropriate appearance of increased numbers of these primitive cells in the peripheral circulation. Blasts can be identified in a peripheral blood film as they are markedly larger than mature white blood cells and have a high nuclear/cytoplasmic ratio. Indeed, it may be that the cytoplasm is reduced to a thin rim around the outside of the cell (Figure 3.4).

Cross references

We discuss granulocytes, eosinophils, basophils, neutrophils, monocytes, and lymphocytes in Chapters 2 and 8.

Lymphocytes are discussed in more detail in the *Clinical Immunology* volume in this series.

We discuss blasts and leukaemia in more detail in Chapters 8–12.

SELF-CHECK 3.5

What are the differences between CFUs, stem cells, and blast cells?

3.4.4 The role of growth factors

The bone marrow microenvironment is crucial for successful haemopoiesis: cells must be able to communicate directly with each other. However, there are also a host of important hormone-like molecules that can indirectly pass messages between cells. These cytokine growth factors—both proteins and glycoproteins—are essential for the developmental control of cells from stage to stage. They are released by one type of cell (such as a macrophage or a stromal cell) to act on different stem cells and blast cells to promote the process of cell development. For example, stem cell factor SCF (also known by some as interleukin (IL)-11) is a ligand for the c-kit receptor, which in turn moves the cell along the cell-cycling pathway, whilst IL-3 acts broadly on cells from the stem cell level down to, and including, the CFUs. These growth factors are listed in Table 3.1, and the role of growth factors in cell development is illustrated in Figure 3.5.

Molecules released by leucocytes that act on other leucocytes and non-leucocytes are **interleukins (ILs)**, and some of these molecules may also have some non-growth factor activity, such as assisting in antibody production. However, in the present setting, the regulation of haemopoiesis is dominated by a series of **colony-stimulating factors** (CSFs, such as SCF mentioned in the previous paragraph) that act on various precursors, either generally or specifically. Growth factors may act on a single cell lineage at a single stage of its development, or on a series of cells at different stages of development. Further, growth factors may be autocrine (acting on cells of the same lineage) or paracrine (acting on cells of a different lineage). Thus, various supporting cells within the bone marrow, such as fibroblasts, macrophages, endothelial cells, etc. and stem cells themselves may both release and be acted on by various growth factors.

An important concept in cell biology is that of programmed cell death, or **apoptosis**. The basis of this process is that cells require the continual receipt of growth factor signals, effectively to keep them alive. Failure to receive these signals (or 'survival factors'), which include **erythropoietin**, triggers physiological changes within the cell. In the case of erythropoietin, the target cell would be a normoblast, and a lack of growth factor sets in motion a train of biochemical events that leads to the activity of certain enzymes (such as those of the caspase family), which in turn lead to the death of the cell. Consequently, apoptosis is sometimes considered to be 'cell suicide'. For many cells in the bone marrow, and elsewhere, growth factors are the crucial signal needed for the cell to remain viable.

erythropoietin
A growth factor, generally derived from the kidney, that promotes the development of red blood cells.

Cross reference
The full implications of apoptosis upon haematological cancer are described in Chapter 9.

TABLE 3.1 Haemopoietic growth factors.

Interleukins (ILs)	• IL-1 stimulates production of granulocyte-macrophage CSF (GM-CSF), granulocyte CSF (G-CSF), monocyte CSF (M-CSF), and IL-6 from a variety of cells within the bone marrow, including stromal cells. • IL-3, IL-4, and IL-6 act on early multipotential cells. • IL-5 is eosinophil CSF.
Colony-stimulating factors (CSFs)	• G-CSF stimulates the differentiation of granulocyte precursors (such as the myeloblast) and also the activity of mature granulocytes. • GM-CSF stimulates the differentiation and maturation of granulocytes and macrophages, and also the function of these cells once mature.
Other factors	• Stem cell factor acts on pluripotent stem cells, causing them to differentiate further, and also has effects on the later maturation of several cell lineages. • Erythropoietin is released mostly from the kidney and travels in the blood to the bone marrow, where it acts on erythroid precursors to stimulate red blood cell production. • Thrombopoietin, produced principally by the liver, stimulates megakaryocyte maturation and thus platelet production. • Tumour necrosis factor (TNF) has actions similar to IL-1.

IL-interleukin; CSF-colony stimulating factor; TNF-tumour necrosis factor.

Advances in biotechnology now allow the large-scale production of many of these growth factors, which have consequently become available as therapeutics for a variety of conditions. For example, patients with renal failure may become anaemic because their kidney fails to produce the erythropoietin required to stimulate the production of red blood cells. This growth factor can now be exogenously supplied as a therapeutic agent in particular cases of anaemia, and can facilitate a rise in the red cell count, which can rectify the anaemia. Similarly, thrombopoietin can be used to stimulate platelet production, and GM-CSF is used to stimulate neutrophil production. However, the most potent use of growth factors is in stimulating stem cell expansion *in vitro* and their subsequent use *in vivo* to aid bone marrow transplantation. Another use of these growth factors is to stimulate cell production in a bone marrow that has been suppressed by cancer chemotherapy or radiotherapy.

A further example of the promiscuous and complex nature of the growth factors is illustrated by the fact that G-CSF, GM-CSF, IL-6, M-CSF, and SCF all act on granulocyte colony formation. Furthermore, IL-6 also acts on non-haemopoietic cells such as hepatocytes (where it influences iron metabolism), neuronal precursor cells, mesangial cells, and osteoclasts. Indeed, IL-6, and probably IL-1, do have an important role in the acute phase response to actual or potential trauma or infection. Those with any interest in immunology will recognize several ILs (such as IL-6) as key cytokines in an ongoing inflammatory response. It is therefore curious to note that IL-1 and IL-3 are involved in the embryonic development of haemopoietic stem cells, an environment unlikely to host a raging inflammatory response. Furthermore, interference with the toll-like receptor 4 (with roles in the innate immune system, recognizing lipopolysaccharide from Gram-negative bacteria) signalling pathway decreases haemopoietic stem cells. These observations underline the complicated nature of haemopoiesis, and how much we have to learn.

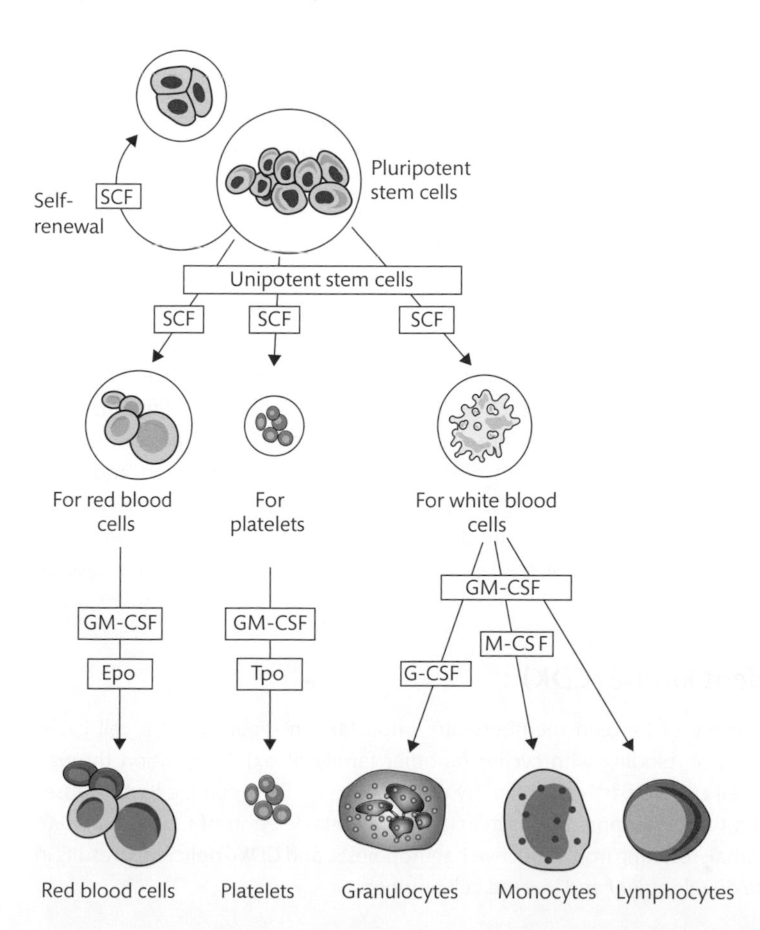

FIGURE 3.5

Simplified role of growth factors in haemopoiesis. Relationships between stem cells and their progeny, and their generation of mature blood cells. Note that GM-CSF can non-specifically stimulate the differentiation of all blood cells, whilst certain growth factors act specifically only on one particular lineage.

SCF = stem cell factor; GM-CSF = granulocyte-monocyte colony-stimulating factor; Epo = erythropoietin; Tpo = thrombopoietin; G-CSF = granulocyte colony-stimulating factor; M-CSF = monocyte colony-stimulating factor.

Much of our knowledge of growth factors, such as erythropoietin, has been obtained from tissue culture experiments. Bone marrow harvested from volunteers can be grown and characterized in tissue culture, and so can be used as a model for the effect of various CSFs on stem cells. For example, incubation of bone marrow with a certain highly purified cocktail of growth factors may well result in the growth or maturation of granulocytes, but not of monocytes. However, in this highly artificial *in vitro* system, we cannot be sure if the events we witness are truly representative of the situation within our own bone marrow *in vivo* in either health or in disease. Nonetheless, the clinical provision of exogenous CSFs such as thrombopoietin as a specialized therapy to those with a failing bone marrow (as in cases of chemotherapy or following bone marrow transplantation) is of undoubted benefit.

SELF-CHECK 3.6

What are the molecular signals that promote haemopoiesis?

3.4.5 The molecular genetics of haemopoiesis

Recent years have seen an explosion in our understanding, at the level of nucleic acids, of the steps involved in haemopoiesis. Reflecting the complexity of these processes, and the wealth of literature, a comprehensive discourse of the subject is beyond the need of even the most dedicated and enthusiastic laboratory scientist. However, some knowledge is required, as errors in molecular genetics lead to malignancy, as will be explained in Chapters 9–12, and elsewhere, and where the clinical implications of mutation in these genes will be discussed. Accordingly, as today's research is tomorrow's routine, it is as well to now introduce a small number of pathways and molecules involved in physiological haemopoiesis. However, by way of caution, certain aspects of what follows is derived from work in mice and/or complex tissue culture experiments; although we may not be able to extrapolate it to our own species, it is nonetheless exceptionally informative.

The GATA transcription family

This family of six nuclear proteins, so named because they bind to combinations of the G, A, and T nucleotides, are coded for by widely dispersed genes at Xp11.23 (*GATA1*), 3q21.3 (*GATA2*), 10p14 (*GATA3*), 8p23.1 (*GATA4*), 20q13.33 (*GATA5*), and 18Q11.2 (*GATA6*). Each protein has specific roles in the regulation of certain developmental and differentiation genes. The first three, GATAs 1, 2, and 3, are essential for normal haemopoiesis. GATA1 drives haemopoietic stem cells towards erythroid/megakaryocyte differentiation, and without it, erythroid progenitors undergo apoptosis, whilst GATA2 expression is essential for the maintenance of the pool of haemopoietic stem cells. GATA3 is expressed in several tissues, including innate lymphoid cells, where it controls the maintenance and proliferation of T lymphocytes.

Cyclin-dependent kinase (CDK)

These enzymes (a family of thirteen members) are important in regulating the cell cycle, becoming activated upon binding with cyclins (another family of six). Progression through the G_1 phase of the cell cycle is part-controlled by the CDK4/6–cyclin D complex, with subsequent regulation of a tumour-suppressor protein. Simultaneous deletion of *CDK4* and *CDK6* are embryonically lethal, resulting from defective haemopoiesis, and *CDK6* deficiency results in defective thymocyte development and low red cell numbers.

Homeobox

This sequence of DNA is widely distributed in all forms of life, and encodes a protein that binds to, and so regulates, other genes. In our species there are some 50 variants, of which HOXB4 (at 17q21.32) is expressed in haemopoietic progenitor cells, where it promotes emergence from the embryonic mesoderm and self-renewal. This may be exploitable in our search for new methods to harvest stem cells for transplantation, and in generating red cells for transfusion.

RUNX1 This gene codes for a transcription factor of the same name that plays a role in guiding the megakaryocyte/erythroid progenitor cell towards megakaryopoiesis by inhibiting erythroid differentiation, itself controlled by further transcription factors named KLF1 and FLI1. Illustrating the seemingly haphazard nature of this area of research, a proto-oncogene implicated in T-cell acute lymphoblastic leukaemia (*TAL-1*) was described in 1990. We now know that the transcription factor coded for by *TAL-1* is in fact an important physiological regulator of haemopoiesis and operates alongside *RUNX1*. Unsurprisingly, *RUNX1* knock-out mice die after some 12–13 days, the embryo livers failing to demonstrate haemopoiesis, and mutations in *RUNX1* are linked to leukaemia.

Notch At the protein level, Notch molecules at the cell surface are activated by their ligands to release an intracellular second messenger that migrates to the nucleus to exert transcriptional activity on various genes, such as those of the NF-κB and GATA2 pathways. These interactions have implications for the generation of embryonic haemopoietic stem cells, in T-cell development, in marginal zone B-cell development, and in myelopoiesis. There is also growing evidence of the involvement of Notch in the bone marrow microenvironment, helping to provide a niche for stem-cell regeneration. Mutations in Notch can cause both oncogene and tumour suppressor activity and have been described in almost all leucocyte malignancies.

miRNAs

miRNAs are a class of small (~22 nucleic acid, hence micro, or mi) non-coding RNAs that have roles in post-transcriptional regulation of protein expression. Target RNAs are repressed by inducing their degradation or blocking, and it has been suggested that more than one-third of all human genes are the targets of miRNAs. Multiple miRNAs, such as miR-17-92, miR-34a, and miR-150 regulate early B lymphocytes, whilst miR-125b guides haemopoiesis towards the myeloid lineage. miRNAs may function as tumour suppressors (deletion of miR-15A and miR-16-1 are linked to chronic lymphocytic leukaemia) and as oncogenes (overexpression of miR-17-92 is linked to B-cell lymphomas). Several miRNAs, such as miR-21 and miR-155, are promising diagnostic markers.

3.5 Specific-lineage haemopoiesis

Each mature blood cell is the end product of the well-regulated differentiation of stem cells and blast cells of a specific lineage, although, as we have noted in Figures 3.3 and 3.5, the initial stages are common to all cell types. Let us now go on to consider the development of the specific cell lines in more detail.

3.5.1 Erythrocyte production (erythropoiesis)

The earliest stem cell in the red cell pathway is the CFU-GEMM, although this cell can also give rise to other stem cells that are independent of red cells, such as granulocytes and monocytes (Figure 3.3). The CFU-GEMM gives rise to the CFU-EMk, which, in turn, produces the earliest recognizable red-cell lineage-specific stem cell, the CFU-E. From the CFU-E arises the earliest

normoblast, or erythroblast
Blast of the erythrocyte (red blood cell) pathway.

reticulocyte
The final stage of the development of a red blood cell before it reaches maturity.

morphologically recognizable erythrocyte precursor, the proerythroblast. Further steps are marked by a steady reduction in the size of the particular blast cell through the **normoblast** (or **erythroblast**) stage, then the nucleated red blood cell. This process of development is 'nursed' by a specialized macrophage in the centre of an erythroblastic island, details of which are presented in Chapter 4. The erythroblast then loses its nucleus entirely to become a **reticulocyte**. This juvenile red blood cell contains remnants of ribosomal ribonucleic acid which, when reacting with certain basic dyes such as new methylene blue, produces a blue or purple precipitate visible by light microscopy. Reticulocytes can also be numerated by a modified flow cytometry technique, as fluorescent dyes such as auramine O and thiazole orange specifically bind to ribonucleic acid. The final step is the transformation of the reticulocyte into the mature red blood cell, a process that takes three to four days, and which may occur within the bone marrow or in the blood.

Cross reference
Erythropoiesis is discussed further in Chapter 4.

Thus, the key steps in the development of the red blood cell from the blast cell to the fully functioning erythrocyte involve a slow and steady reduction in size, with loss of the nucleus. This process occurs in parallel with the development of **haemoglobin**, the oxygen-carrying protein within red blood cells, which becomes more and more prominent as the red cell matures. Several crucial steps in erythropoiesis depend on a red cell specific growth factor, erythropoietin. Additional details regarding erythropoiesis are presented in Chapter 4.

3.5.2 Granulocyte production (granulopoiesis)

granules
Small bodies within the cytoplasm of granulocytes that contain bioactive molecules such as enzymes.

As discussed in Chapter 2, the granulocytes consist of three distinct cells—the neutrophil, the eosinophil, and the basophil, and each have different sets of **granules** in their cytoplasm. The precise make up of granules is important to the correct function of each of the three types of granulocytes, such as in the destruction of bacteria. Small numbers of granules may also be present in other white blood cells, such as monocytes, and occasionally lymphocytes.

The development of granulocytes can be traced from the haemopoietic stem cell to the CFU-GEMM. The CFU-GEMM then gives rise to four CFUs: one for basophils (i.e. CFU-Baso), a second for neutrophils (the CFU-Neut), a third for eosinophils (the CFU-Eos), and a fourth for monocytes (CFU-M) (Figure 3.3). From each of the three granulocyte CFUs (CFU-Neut, CFU-Eos, CFU-Baso) arise neutrophils, eosinophils, and basophils that are found in the blood.

metamyelocyte
The final stage of the development of a granulocyte before it becomes a mature polymorphonuclear leucocyte.

The earliest identifiable neutrophil precursor cell to arise from the CFU-G that can be recognized by conventional light microscopy is the myeloblast (10–20 μm in diameter, and accounting for 4% of all nucleated cells in the bone marrow), which subsequently develops into the slightly larger promyelocyte (15–25 μm). As the latter differentiates into the myelocyte and then the **metamyelocyte** (with its kidney-bean-shaped nucleus, often described as a stab cell), granules appear in larger numbers and become more prominent and dense. Figure 3.6 shows three metamyelocytes. The final step is the shrinking and lobulation of the nucleus, which marks the mature neutrophil, as shown in Figure 2.13 in Chapter 2. (See also Box 3.3.)

The lineage-specific precursors of mature neutrophils, eosinophils, and basophils can be recognized at the myelocyte stage by the pattern and staining properties of their granules. These granules contain enzymes such as myeloperoxidase and collagenase, and other molecules such as heparin, lactoferrin, and histamine.

The final stages of cell development are the condensation of the nucleus into the distinct lobule pattern (generally between three and five small lobes for the neutrophils, and two lobes for eosinophils and basophils) and the presence of specific granules.

In the eosinophil, the granules stain deep-red with Romanowsky stains; by contrast, the granules of basophils are coloured black. Just as the dominant granulocyte in the blood is the

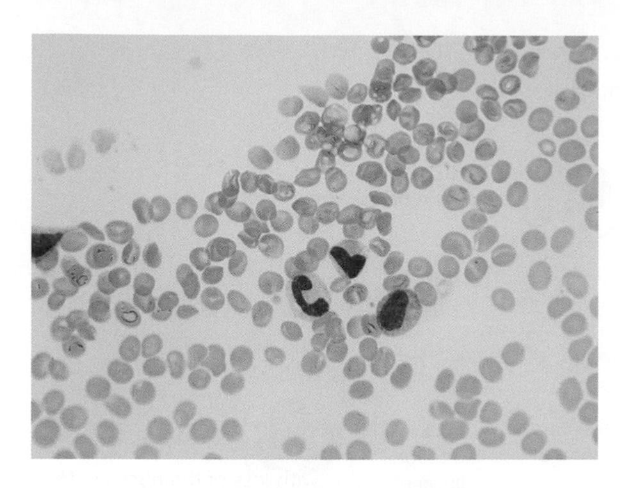

FIGURE 3.6
Metamyelocytes. These are the final 'immature' stage in the development of the neutrophil. The nucleus of the cell on the left has the typical 'stab' or 'band' morphology, the nucleus on the right has a 'kidney bean' shape. (Magnification ×400.)

BOX 3.3 Sex and the neutrophil

The sex of an individual is defined as the presence of two X chromosomes for females, and an X and a Y chromosome for males. An interesting characteristic of neutrophils is that, during the process of nuclear condensation, one of the two X chromosomes in females becomes fully condensed, so that it appears as a small 'dumb-bell' structure. This inactivated X chromosome, or Barr body, was so named following the work of Barr and Bertram in 1949. So although absence of a neutrophil Barr body does not prove male sex, its presence implies female sex. Now see if you can detect a Barr body in Figure 2.13.

neutrophil, the dominant metamyelocyte in the bone marrow is of the neutrophil lineage and it is possible to identify other granulation precursors—both eosinophil metamyelocytes and basophil metamyelocytes—in a sample of bone marrow. The entire process of **granulopoiesis** is shown in Figure 3.7. (See also Box 3.4.)

granulopoiesis
The development of granulocytes.

Cross reference
The full importance of granules and granulocytes is developed in Chapter 8.

BOX 3.4 Granules

The three granulocytes are so named because of the presence of intracellular granules. The precise make-up of these enzymes is specific for each granulocyte.

Neutrophil granules include enzymes such as alkaline phosphatase, elastase, gelatinase, myeloperoxidase, lysozyme, and cathepsins. These granules also include the coagulation protein plasminogen.

Eosinophil granules include the enzyme peroxidase and a molecule called MBP, which has a relative molecular mass of 10 kDa. MBP can also cause the release of histamine from basophils.

Basophil granules include histamine, heparin, hyaluronidate, and serotonin. These molecules may also be chemotactic for neutrophils and eosinophils.

Some of these granules, especially in the neutrophil, are important in the digestion of pathogens by the process of phagocytosis.

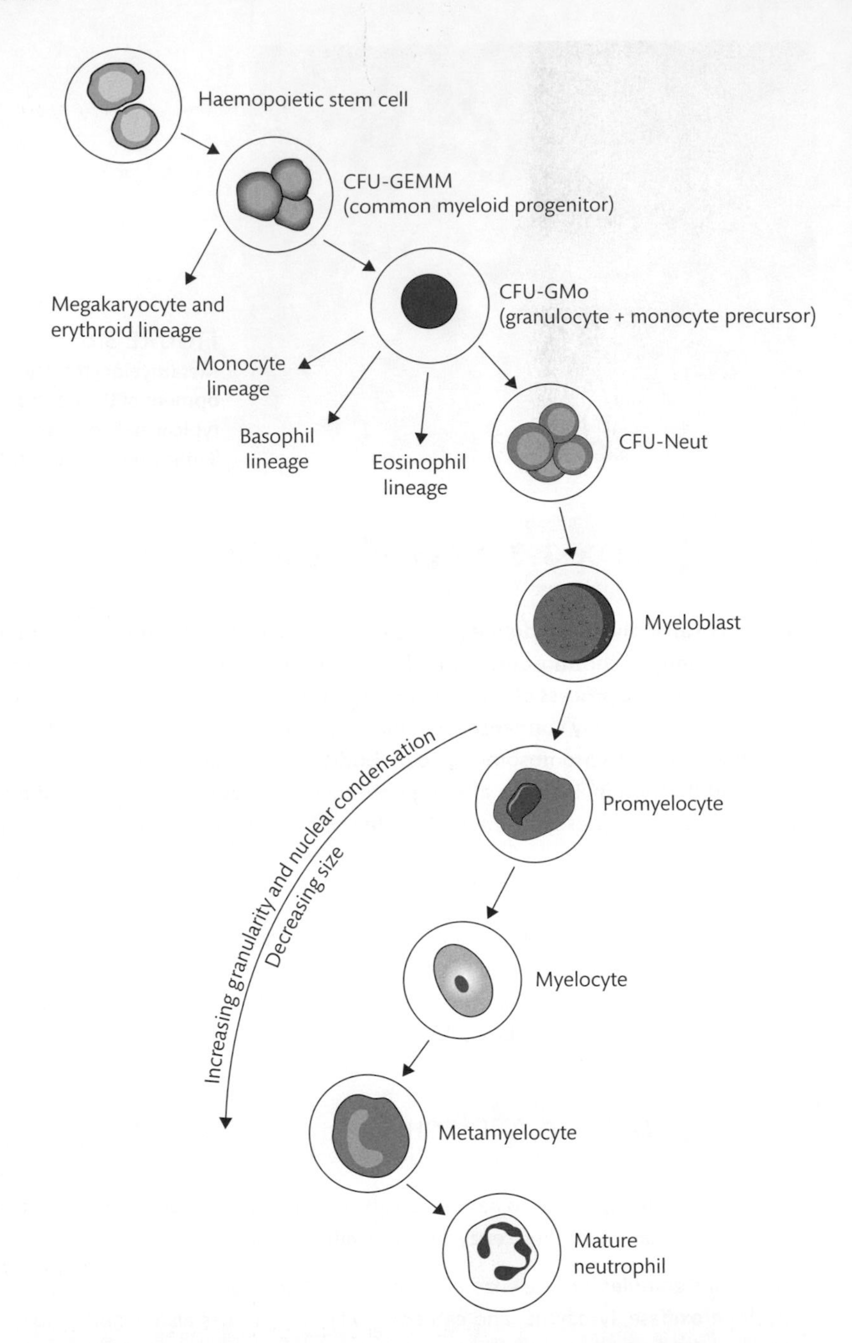

FIGURE 3.7
Neutrophil development. Stages in the development of the neutrophil within the bone marrow. Early stages involve the derivation of the lineage-specific CFU for neutrophils from the multipotent stem cells (CFU-GEMM and the CFU-GMo). Later stages see the development of cells which, as they pass through the blast stage with developing maturity, form a multilobed nucleus and generate intracellular granules, many of which contain bacteriocidal enzymes.

Cross reference

White blood cell granular contents are reviewed in depth in Chapter 8.

SELF-CHECK 3.7

How can you distinguish between the three different types of granulocyte?

3.5.3 Platelet production (thrombopoiesis)

The platelet's lineage-specific progenitor cell, the CFU-Mk, arises from the CFU-EMk, a cell which resembles a small lymphocyte. From the CFU-Mk, which expresses the CD34 membrane antigen, arises the **megakaryoblast**, and then the mature megakaryocyte, which is found in

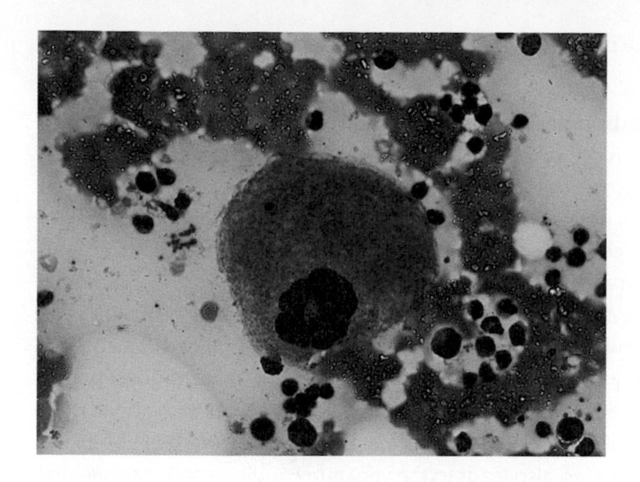

FIGURE 3.8
Megakaryocyte. This is a large bone-marrow-derived cell that produces platelets. The nucleus forms multiple lobes from successive phases of endomitosis, while the cytoplasm becomes increasingly granular. Pro-platelets bud from the cell membrane, each containing a number of granules from the parent megakaryocyte. From these pro-platelets, mature platelets form. Courtesy of the Haematology Laboratory, Queen Alexandra Hospital, Cosham.

normal bone marrow at a frequency of about one in 2000 nucleated cells (Figure 3.8). Notably, some megakaryocytes leave the bone marrow and migrate to the lungs, which may provide an additional site for platelet production.

The megakaryocyte is so named not only because it is very large—the meaning of 'mega'—but because of its unique characteristic of carrying more than the usual numbers of chromosomes. All somatic cells of the body have 46 chromosomes and are described as diploid (denoted 2N). By contrast, the non-somatic cells, the gametes (sperm and ova), have 23 chromosomes, and are described as being haploid (denoted N). However, megakaryocytes often have not merely 46 chromosomes (that is, are 2N), but possibly several multiples of this, such as 3N (with 69 chromosomes) or even higher. Indeed, the **ploidy** number for over half of the megakaryocyte population is 16N. This property of multiple sets of chromosomes is known as **polyploidy**. As a consequence, the megakaryocyte is very large and has a large and irregular nucleus, often with numerous lobes. This large number of chromosomes is achieved through the process of endomitotic replication, essentially mitosis without cytokinesis.

Cross reference

Chromosomal complement and ploids are further discussed in Chapter 9.

Developing megakaryocytes can also be identified by the presence of major membrane glycoproteins on the surface of the cell, many of which are also present on mature platelets (Table 3.2). Megakaryocytes of 2–4N ploidy produce platelet peroxidase and von Willebrand factor but have no **alpha granules**, the intracellular organelles within mature platelets that contain molecules active in coagulation.

Not surprisingly, the size of the megakaryocyte reflects the degree of ploidy—2N cells may have a diameter of 20 μm, whilst 64N cells may be as large as 56 μm. This is in contrast to the

TABLE 3.2 **Identification of megakaryocytes and platelets.**

Membrane glycoproteins	Contents of the cytoplasm
CD41a (the gpIIb/IIIa complex, also known as $\alpha_{IIb}\beta_3$)	Alpha granules, dense granules, lysosomal vesicles
CD69P (P-selectin), CD41b (gpIIb), CD61 (gpIIIa)	von Willebrand factor, platelet factor 4, β-thromboglobulin
CD42a (gpIX), CD42b (gpIb), CD51 (αV)	Fibrinogen, coagulation factors V and VIII

Note: there are different subpopulations of alpha granules: for example, some are rich in von Willebrand factor and poor in fibrinogen, whilst other granules have the opposite composition, being rich in fibrinogen but poor in von Willebrand factor.

'average' nucleated bone marrow cell, which is likely to have a diameter of 14 μm and the red blood cell of 7-8 μm.

thrombopoiesis
The development of platelets (also known as thrombocytes).

thrombopoietin
A growth factor that stimulates megakaryocyte maturation and platelet production.

thrombocytopenia
Low numbers of platelets in the blood, often cited as less than 100,000 per mL, equivalent to less than 100×10^9/L.

The key growth factor for **thrombopoiesis** that mirrors the role of erythropoietin in erythropoiesis is **thrombopoietin**. This growth factor is produced by the liver and other organs (such as the kidney and bone marrow stromal cells) and is encoded for by a gene called *Tpo* on chromosome 3, which produces a protein of relative molecular mass 80-100 kDa. The growth factor itself 'switches on' its target cells—the megakaryocyte and its precursors such as the CFU-Mk—by interacting with a specific receptor coded for by a gene called *c-Mpl*. Thrombopoietin is required by the megakaryocyte to enable cell growth and the production of platelets, but it also acts on platelets themselves to stimulate their participation in coagulation. Indeed, the absence of either *c-Mpl* or *Tpo* results in very low numbers of platelets in the blood (**thrombocytopenia**) in mice, but also in reduced numbers of megakaryocytes and their precursors in the bone marrow. A similar abnormality in *c-Mpl*, which leads to the defective expression or function of the thrombopoietin receptor in our species, also leads to the very rare condition called congenital amegakaryocytic thrombocytopenia.

Thrombopoietin may also act on erythroid precursors, possibly because it shares a high degree of its amino acid sequence (structural homology) with erythropoietin. Plasma levels of thrombopoietin are in the order of 100 pg/L, but this can increase several fold in thrombocytopenia and in acute inflammation.

Platelets, which are generated at a rate of 10^{11} every day, are fragments of the cytoplasm of the megakaryocyte. How do these fragments arise? Pseudopodial projections of the cytoplasm called pro-platelets break off and pass into the circulation. This process eventually consumes all of the cytoplasmic constituents of the megakaryocyte—the cytosol, granules, mitochondria, and other organelles, as well as coagulation molecules, etc. Hence after the production of perhaps 1500-4000 platelets, nothing is left of the remainder of the megakaryocyte except its nucleus. The remnant of the megakaryocyte is eliminated by phagocytosis mediated by bone marrow macrophages, although apoptosis may also be involved.

During the final stage of pro-platelet development, the intracellular contents of what will be the mature platelet are delivered by microtubules. The precise signal for shedding pro-platelets from the megakaryocyte is unknown but does not involve thrombopoietin.

The 'juvenile' platelet, like the 'juvenile' red blood cell (that is, the reticulocyte), has a slightly larger diameter and volume (2-4 μm and 8-11 fL, respectively) than the final mature platelet (diameter 1.5-3.5 μm, volume 6-9 fL). A further feature of juvenile platelets is that they also have intracellular inclusions such as messenger RNA, which can be specifically stained, allowing the 'reticulated platelet' to be detected.

Cross reference
The structure and function of platelets is described in Chapter 2, whilst the role of the platelet in coagulation is described in Chapter 13.

Mature platelets can also bind thrombopoietin as they bear the thrombopoietin receptor, although their ability to detect this does not have a place in the routine laboratory. However, the binding of thrombopoietin enhances the secretion of alpha granules, and also promotes aggregation by agonists such as ADP, collagen, and thrombin. Table 3.2 shows the most common features used for identifying platelets and megakaryocytes: surface CD molecules, organelles, and certain molecules in the cytoplasm.

SELF-CHECK 3.8

What are the similarities between the juvenile form of red blood cells and platelets and their respective mature cells?

3.5.4 Monocyte production

Monocytes share a common stem cell with granulocytes (the CFU-GMo), but production of the mature cell passes to a more focused and dedicated lineage-specific precursor, the CFU-M, which produces only monocytes. At least two species of immature monocytes can be found in the bone marrow—the **monoblast**, which gives rise to the **promonocyte**. The mature monocyte develops from the promonocyte and passes from the bone marrow into the blood. After a period in the circulation (which can vary from hours to days), monocytes migrate into the tissues (such as the skin, the lung, the liver, the lymph nodes, and the spleen).

Once monocytes have arrived in the tissues, they differentiate into macrophages. Some then acquire additional properties, and transform further still into highly specialized cells, such as the Langerhans cell (specific for the spleen), the Kupffer cell (specific for the liver), and antigen-presenting cells (most often found in lymph nodes, occasionally in the skin, and in association with lymphocytes). Granules are often to be found in the cytoplasm of both monocytes and macrophages. This is entirely possible because the monocyte shares a common stem cell with the granulocyte—the CFU-GMo (Figure 3.3).

Cross reference

Additional details of monocytes and lymphocytes are presented in Chapter 8.

3.5.5 Lymphocyte production

The common lymphoid progenitor stem cell (the CFU-L) arises directly from the pluripotent haemopoietic stem cell, and its development is independent of the other blood cells. CFU-L then gives rise to immature **lymphoblasts**, which then differentiate into the mature lymphocytes. Two *major* types of lymphocyte are recognized: B cells and T cells. The former synthesize antibodies and the latter cooperate in antibody production, but also recognize and destroy cells that are infected with viruses. A third lymphocyte, the NK cell (natural killer, also described by some as an NKT cell, acknowledging its link with T cells) is also recognized.

In mammals, the thymus is essential for the development of T lymphocytes, so much so that many infants with the congenital disorder DiGeorge syndrome lack a thymus and also T lymphocytes. This is important because it leads to a condition known as **immunodeficiency**, whereby white blood cells fail to provide an adequate immunological response to a microbial pathogen such as a virus or bacterium. The maturation of the T cell is characterized by the stepwise appearance of certain molecules at the surface of the cell, such as CD3, CD4, and CD8, and by the rearrangement of genes leading to the T-cell receptor (the complex molecule through which the T cell communicates with other cells).

immunodeficiency

A condition wherein the immune system is depleted and so unlikely to be fully able to defend the individual from microbial pathogens.

A similar maturation process for B lymphocytes in birds requires the presence of the Bursa of Fabricius. This organ does not seem to have a mammalian equivalent, and it is assumed that immature B cells gain additional differentiating signals in the bone marrow, liver, lymph nodes, and possibly elsewhere. The maturation of the B cell is accompanied by the stepwise appearance of molecules that include CD19, CD20, and CD24, and by the rearrangement of genes coding for the primary B cell receptor (that is, immunoglobulins).

As is the case for the B lymphocyte, the site of maturation (if it exists) of the NK cell has also not been determined but seems likely to be the thymus. NK cells may be identified by CD30 and CD91, and by whether or not there has been rearrangement of genes for the T-cell receptor (Figure 3.3).

SELF-CHECK 3.9

Where do the different types of lymphocytes complete their maturation?

Key Points

The development of the different blood cells is complex, but certain patterns are present. All mature blood cells are the product of a slow and steady process of maturation that passes through progenitor stages and which relies on the presence of growth factors.

3.6 Bone marrow sampling and analysis

Knowledge of the interrelationships between different cells of the bone marrow is important in numerous diseases, such as aplastic, megaloblastic, and sideroblastic anaemia, and in the proliferative malignancies polycythaemia, essential thrombocythaemia, myeloma, myelofibrosis, and leukaemia. Thus, the contents of the bone marrow provide an important opportunity to confirm or refute a diagnosis made on peripheral blood. In addition, sampling the bone marrow at different stages of a particular disease, such as after a round of chemotherapy, can tell us of the success (or not) of that procedure.

3.6.1 Obtaining bone marrow

In practice, the only method of assessing haemopoiesis is to examine the contents of the bone marrow. This can be achieved in one of two ways. The first is essentially an extension of the method for obtaining a peripheral blood sample from a vein on the inside of the elbow (venepuncture). However, in obtaining bone marrow, a heavy-gauge needle is driven part of the way into the bone, such as the sternum or the iliac crest, to suck out (aspirate) some marrow into a syringe. The sample is called a bone marrow aspirate and is processed as if a sample of blood.

Cross references

Additional details of the importance of bone marrow and its analysis are presented in Chapter 10.

Additional details of the consequences of a vitamin B_{12} deficiency are to be found in Chapter 5.

Aplastic anaemia, myeloma, myelofibrosis, and leukaemia are described fully in Chapters 5 and 9-12.

Aspects of the examination of tissues (cellular pathology, or histopathology) are described in the *Histopathology* volume in this series.

The second method is to deliberately harvest bone tissue itself in the form of a **trephine** biopsy, which also provides a view of the internal architecture of the bone marrow. This process demands a considerably more robust needle. As a trephine biopsy provides actual bone, this tissue must be processed as if it was a piece of normal tissue and will require decalcification before histological examination. An alternative to the 'standard' trephine outlined above is an 'imprint', where the sample is rolled over the surface of a glass slide, in which case it can be processed as if a blood film. The process of obtaining both the aspirate and trephine samples is painful, and so requires anaesthetics.

Once obtained, bone marrow may be analysed by standard techniques, by cytochemistry, and by fluorescence flow cytometry.

3.6.2 Standard analysis of bone marrow

Once aspirated, bone marrow can be spread on to a glass slide, dried, and stained with the same dyes (such as Romanowsky) as is the peripheral blood film. Precursors may be recognized by specific characteristics, such as the presence of prominent nucleoli in a myeloblast, or sheer size of the megakaryocyte (Figure 3.8). Figure 3.9 shows a low-power magnification of a bone marrow aspiration. Compare the greatly increased density of the bone marrow aspiration with Figure 3.10, a low-power figure of peripheral blood, in which only two white blood cells are present. Figure 3.11 shows a trephine bone marrow sample stained with a Romanowsky stain.

The precise composition of the various precursor and mature cells that make up a healthy bone marrow are shown in Table 10.2 of Chapter 10. Further, infiltrating cells, or the overgrowth of non-haemopoietic tissue such as fibroblasts (as may be present in myelofibrosis), and even

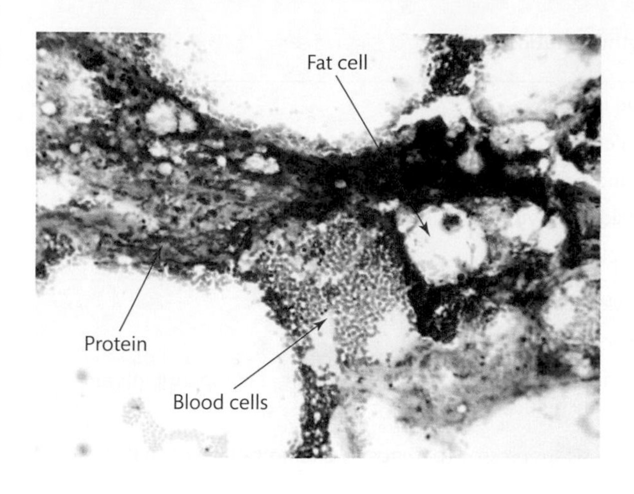

FIGURE 3.9

Bone marrow. This low-power photomicrograph shows a sample of bone marrow aspirate that has been spread on to a glass slide, dried in air, fixed, and then stained with different dyes that are taken up by different cells and tissues. The figure is dominated by a central streak of blue and purple, with several unstained areas. These approximately circular patterns are fat cells that fail to take up any of the dyes. The blue dye has stained acellular protein material such as collagen, which provides physical support to the bone marrow. The other major coloration (purple) results from the uptake of other dyes by developing blood cells. Their approximate circular pattern is clearly visible, but the low-power magnification does not permit other analyses. (Magnification ×100.)

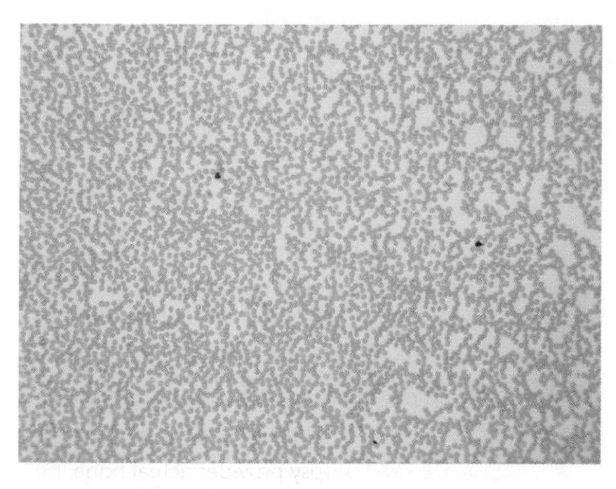

FIGURE 3.10

Peripheral blood. This shows a peripheral blood film at the same low-power magnification as the bone marrow aspiration in Figures 3.9 and Figure 3.11. The two purple cells are possibly a neutrophil on the left and a monocyte on the right. (Magnification ×100.)

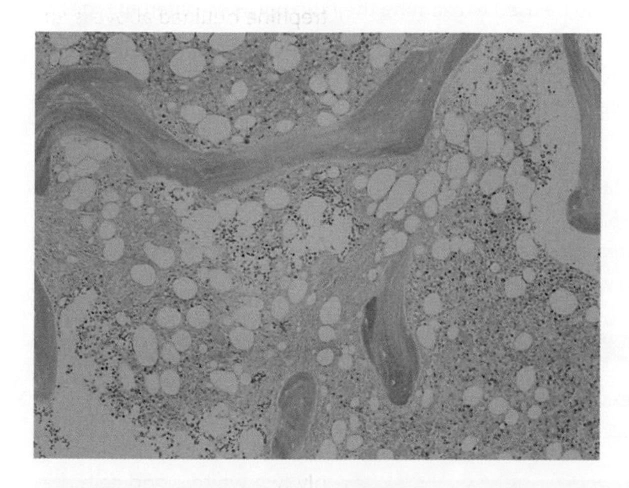

FIGURE 3.11

Bone marrow trephine showing normal bone marrow architecture. The cellular component is haemopoietic (red) marrow, comprising largely blood cell precursors. Clear spaces are adipose tissue or fatty (yellow) bone marrow. Pink swathes running through the marrow are bone trabeculae (haematoxylin and eosin staining, original magnification ×100). Contrast this figure with Figure 3.9.

Courtesy of the Haematology laboratory, Queen Alexandra Hospital, Cosham.

infections (by bacteria or fungi) can also be detected. An aspirate is particularly useful in that it provides information on the number of different cells and the lineages to which these precursors belong, such as myeloid or erythroid.

TABLE 3.3 The myeloid/erythroid ratio.

M/E ratio increased	• Myeloid leukaemia • Most infections • Leukaemoid reactions • Depression of erythropoiesis (as in pure red cell aplasia) • Late myeloma
M/E ratio normal	• Myelosclerosis • Early myeloma • Aplastic anaemia (both myeloid and erythroid compartments are concurrently reduced)
M/E ratio reduced	• Myelopoiesis suppressed (perhaps by chemotherapy or viral infections) • Erythropoiesis enhanced, as may be found: - following severe blood loss - in iron-deficient anaemia - in polycythaemia vera

This ratio provides information on the relative numbers of white blood cells and their precursors compared to red blood cell precursors. In health: at birth 2:1; 0–1 years raised, e.g. 7:1; 1–20 years generally 3:1; adult between 3:1 and 4:1.

In the healthy adult, the number of white blood cell (i.e. myeloid) precursors exceeds the number of red blood cell (i.e. erythroid) precursors by a ratio of about 3–4:1. However, this myeloid/erythroid (M/E) ratio can vary with different diseases, as summarized in Table 3.3. For example, in leukaemia, the marrow will be dominated by leucocyte precursors such as lymphoblasts, myelocytes, and metamyelocytes, so that the M/E ratio will be closer to 2. By the same principle, a bone marrow aspirate can also be used to examine the ratio between different types of white blood precursors. For example, the ratio of granulocyte precursors to lymphocyte precursors may vary between 1:1 and 17:1, especially in neoplastic diseases and in viral infections (when we expect more lymphocyte precursors) or bacterial infections (when we expect more neutrophil precursors).

megaloblasts

Literally 'large blast cells' present in the bone marrow, often as a result of lack of vitamin B_{12} or folate.

In an anaemia related to lack of vitamin B_{12} or folate, we expect the presence of **megaloblasts**, as described in chapter 5, section 5.6.1. Infiltration by metastatic tumours from a distant primary site such as lung, prostate, or breast carcinoma can be studied using a bone marrow aspirate. In both these cases, the M/E ratio will generally be unaltered.

SELF-CHECK 3.10

What are the principal reasons for examining the bone marrow?

SELF-CHECK 3.11

What are the main differences between bone marrow aspiration and the trephine sample?

Key Points

Analysis of the bone marrow is a complex process requiring expert interpretation, and as such provides essential information about the functioning of this organ.

3.6.3 Cytochemical analysis of bone marrow

Cytochemistry is a technique used to probe peripheral blood and bone marrow cells for the presence of different cytoplasmic enzymes and other chemicals and has been developed and refined over decades. The method is very rarely used on a trephine sample. Aspirated cells are spread onto a glass slide, dried, and then subjected to various dyes, chemicals, and buffers. **Cytochemistry** can be used to study both red blood cells and white blood cells, although the focus is on the latter.

Cytochemistry
A technique that uses chemical stains which react with cytoplasmic components and so define different cells in the blood and bone marrow.

Without doubt, the greatest use of these special stains is in the investigation of malignancies such as different leukaemias, which we explore more fully in Chapter 10. For example, the technique can support and refine preliminary diagnoses where some malignancies can be distinguished by different staining patterns (as fully explained in Chapter 11).

The most common staining techniques include:

- myeloperoxidase, an enzyme present most strongly in granulocytes and their precursors, but also sometimes present in red cells and their precursors. The reactants, benzidine dihydrochoride, zinc sulphate, and hydrogen peroxide, produce a black precipitate;
- Sudan black, which stains a component of the granules of granulocytes and monocytes, and so often provides the same information as does myeloperoxidase;
- neutrophil alkaline phosphatase, an enzyme found predominantly in mature neutrophils. Most methods produce a precipitate of bright blue granules;
- acid phosphatase, an enzyme found in granulocytes, but which is generally used to define certain lymphocytic leukaemias in peripheral blood. However, in the bone marrow, macrophages and megakaryocytes are strongly positive.
- Periodic acid-Schiff (PAS), a stain used to stain glycogen and related polysaccharides. Although granulocyte precursors stain weakly, if at all, mature neutrophils show intense magenta staining;
- the esterases, a family of enzymes that are generally used to investigate different leukaemias.

Abnormalities in red cell biology can be investigated using **Perls' stain**, which is also known as Prussian blue. This stain detects the presence of iron-containing molecules inside the cell and so, if found in a red blood cell, defines a **siderocyte**. The same stain when applied to a bone marrow aspirate can identify iron within red blood cell precursors such as **sideroblasts**—these are erythroblasts with inappropriate deposits of iron and are associated with a particular type of anaemia. This stain is therefore important in identifying different types of anaemia, as will be discussed in more detail in Chapters 5 and 6.

siderocyte
A red blood cell containing granules of non-haem iron complexed to other molecules.

sideroblast
An erythroblast—therefore generally found in the bone marrow—containing granules of non-haem iron complexed to other molecules.

SELF-CHECK 3.12

What are the most common cytochemical tests?

3.6.4 Fluorescence flow cytometry

As we have seen in Chapter 2, flow cytometry assesses cell populations according to their size and granularity. It can be modified with fluorochrome-linked antibodies to probe for the presence of different molecules on the surface of the cell, although it can also be modified to look at molecules within the cytoplasm. This technique, fluorescence flow cytometry (FFC), as explained in section 2.3.5, uses monoclonal antibodies conjugated to fluorochromes to defined cell-surface glycoproteins such as CD34, a marker of stem cells. Accordingly, the process is often described as **immunophenotyping**. Tables 3.2 and 3.5 show those CD markers most useful in the identification of other cells. However, the greatest use of FFC as regards the bone marrow is to probe for the presence of malignant cells, as in leukaemia, lymphoma, and myeloma, as is fully explained in Chapter 10.

immunophenotyping
The process of determining the characteristics of a group of cells by probing for the presence of cell-based molecules using monoclonal antibodies conjugated to fluorochromes: synonymous with FFC analysis.

In practice, many laboratories will focus on a panel of only a few selected monoclonal antibodies that can be used to screen for the most common pathological conditions. However, care is required as the expression of a particular CD molecule may well be altered in diseases such as leukaemia. For example, the supposedly specific T-lymphocyte marker CD7 may be found on some granulocyte precursors in an acute myeloid leukaemia (Table 3.4).

SELF-CHECK 3.13

Discuss the value of the common cell-surface markers used in FFC analysis.

Key Points

Special investigations such as flow cytometry and cytochemistry are useful tools for identifying different types of cells in both the bone marrow and in the peripheral blood.

Cross references
Flow cytometry, FFC analysis, immunophenotyping, and cytochemistry are discussed further in Chapters 2 and 10.
Sideroblasts and siderocytes, and their roles in anaemia, are presented in Chapter 5. The importance of CD molecules for the investigation of leukaemia and other haematological neoplasia is presented in Chapters 10–12.

TABLE 3.4 CD molecules and the cell types on which they are found.

CD molecules	Cell type
CD2, CD3, and CD7	T lymphocytes
CD10, CD19, CD20, and CD22	B lymphocytes
CD13, CD33, and CD117	Myeloid cells
CD14	Monocytes
CD34	Haemopoietic precursors (stem cells)
CD45	All white blood cells

This list is certainly not intended to be specific nor exhaustive. CD = cluster of differentiation.

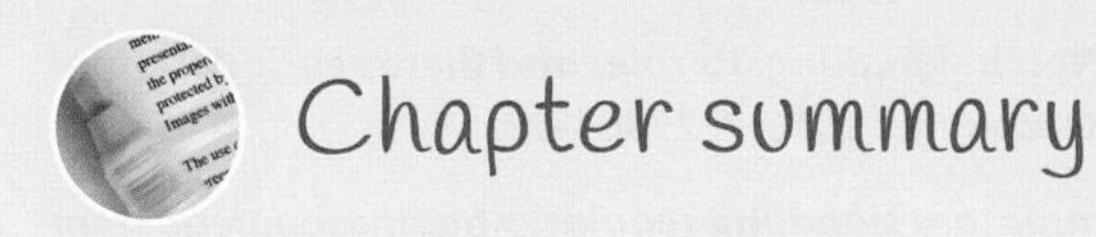

Chapter summary

- The production of blood cells occurs by a process called haemopoiesis.
- In the healthy adult, haemopoiesis occurs only in the bone marrow: in the neonate it may also occur in the liver and spleen.
- Haemopoietic tissue consists of stem cells and supportive tissue such as fibroblasts, macrophages, and endothelial cells.
- Haemopoiesis is driven by growth factors such as erythropoietin, interleukins, and thrombopoietin, and regulated by genes such as those of the GATA and homeobox families.
- Each major group of blood cells has its own specific haemopoietic pathway: erythropoiesis for red blood cells; thrombopoiesis for platelets; and granulopoiesis for granulocytes. Lymphocytes and monocytes also have their own specific pathways.
- Bone marrow aspiration is required to confirm diseases (such as myeloma) suspected of having an origin in this tissue, and to monitor the effect of treatment. In this respect, the M/E ratio is useful.
- Special analyses include cytochemistry and FFC analysis. These are most often used to investigate different types of leukaemia.

Discussion questions

3.1 Explain why a firm knowledge of haemopoiesis is required for the full understanding of diseases such as leukaemia.

3.2 Compare and contrast the two different methods for obtaining a sample of bone marrow.

3.3 What are the merits of cytochemistry and FFC in the analysis of bone marrow?

Further reading

- Calvi LM and Link DC. ***The haematopoietic stem cell niche in homeostasis and disease.*** *Blood* 2015: **126**: 2443–51.
- Ciau-Uitz A, Monteiro R, Kirmizitas A, and Patient R. ***Developmental haematopoiesis: Ontogeny, genetic programming and conservation***. *Experimental Hematology* 2014: **42**: 669–83.
- Crispino JD and Horwitz MS. ***GATA factor mutations in haematological disease***. *Blood* 2017: **129**: 2103–10.
- Deutsch VR and Tomer A. ***Advances in megakaryocytopoiesis and thrombopoiesis: From bench to bedside***. *British Journal of Haematology* 2013: **161**: 778–93.
- Gao X, Xu C, Asada N, and Frenette PS. ***The haemopoietic stem cell niche: From embryo to adult***. *Development* 2018: **145**: dev139691.

- Gu Y, Masiero M, and Banham AH. ***Notch signalling: Its roles and therapeutic potential in haematological malignancies.*** *Oncotarget* 2016: **7**: 29804–23.
- He Q, Zhang C, Wang L et al. ***Inflammatory signaling regulates haematopoietic stem and progenitor cell emergence in vertebrates.*** *Blood* 2015: **125**: 1098-11-6.
- Hitchcock IS and Kaushansky K. ***Thrombopoietin from beginning to end.*** *British Journal of Haematology* 2014: **165**: 259–68.
- Johansson U, Bloxham D, Couzens S, Jesson J, Morilla R, Erber W, and Macey M. ***Guidelines on the use of multicolour flow cytometry in the diagnosis of haematological neoplasms. British Committee for Standards in Haematology.*** *British Journal of Haematology* 2014: **165**: 455–88.
- Kumar V and Delovitch TL. ***Different subsets of natural killer T cells may vary in their roles in health and disease.*** *Immunology* 2014: **142**: 321–36.
- Wiess CN and Ito K. ***A macro view of Micro-RNAs: The discovery of microRNA and their role in haematopoiesis and haematoligic disease.*** *International Reviews Cell Molecular Biology* 2017: **334**: 99–175.

Answers to self-check questions and case study questions are provided in the book's online resources.

Visit www.oup.com/he/moore-fbms3e.

PART 2

Peripheral Blood Cells in Health and Disease

4

The red blood cell in health and disease

Andrew Blann and Pam Holtom

Chapters 1 and 2 introduced some basic aspects of the red blood cell (an erythrocyte, the most abundant cell type in the body [~2.5×10^{13}]) and described how it provides major components of the full blood count (FBC) and the blood film, whilst Chapter 3 briefly described the development of the red blood cell. In the present chapter, we go further to describe the structure and function of the red cell, and how it is adapted for its purpose of carrying oxygen. We also consider the morphology of the cell, and interrelationships between size, shape, and colour. Once we have studied how the red cell operates in health (that is, the normal physiology of the cell), only then can we go on to study it in disease (that is, pathology). The principal disease associated with the red blood cell is anaemia; less common diseases are erythrocytosis and polycythaemia vera. The latter are examined in this chapter, anaemia is addressed in Chapters 5 and 6.

Learning objectives

After studying this chapter, you should confidently be able to:

- explain how the red blood cell is adapted for its purpose;
- appreciate key aspects of the cell biology and molecular genetics of erythropoiesis, and of haemoglobin generation;
- list the different types of haemoglobin and understand the relationship between their structure and function.
- describe the structure of the red blood cell membrane;
- discuss the importance of the red blood cell enzymes and metabolic intermediates;
- identify major variations in the morphology of the red blood cell;
- understand the aetiology and consequences of polycythaemia vera and erythrocytosis.

TABLE 4.1 Relationship between structure and function of the red blood cell.

Feature	Advantage	Disadvantage
Uncomplicated cell membrane	Simple passage of oxygen	Fragile, so relatively susceptible to damage
Lack of a nucleus and most of the enzymes and organelles normally present in a nucleated cell	Flexibility to penetrate fine capillaries	Unable to synthesize essential protective enzymes and other molecules, and lacks the ability to maintain membrane integrity
Membrane lacks HLA molecules	Relatively easy to transplant (that is, as a blood transfusion)	None (?)
All features	Highly specialized	Short lifespan (~120 days)

The red blood cell is one of the body's most highly specialized cells, comprising ~2,800 unique proteins, and dominated by haemoglobin, the ~270 million molecules per cell accounting for ~90% of its dry weight. Specialization is manifested in numerous ways; for example, its membrane is modified for the free and easy passage of oxygen both in and out of the cell. An additional consequence of specialization is that it has a relatively 'uncluttered' cytoplasm, with only those organelles and molecules directly required for its unique function, the carriage of oxygen by haemoglobin. Notably, some 10% of carbon dioxide resultant from tissue respiration is carried by red cells. The degree of specialization is so extreme that it has dispensed with the need for a nucleus and mitochondria. However, this modification comes at a price: the red blood cell is unable to regenerate itself or produce proteins as efficiently as nucleated cells. A consequence of this is a relatively short lifespan—often cited at 120 days for a healthy cell. These specialized features are summarized in Table 4.1. But before examining these specializations in detail, we will review how the red blood cell develops—the process of **erythropoiesis**.

4.1 The development of the red blood cell: erythropoiesis

As introduced in Chapter 3, almost all the development of the red blood cell occurs within the bone marrow. We will look at this process, erythropoiesis, from the perspective of cell biology and molecular genetics, and then consider the key hormone, erythropoietin.

4.1.1 The cell biology of erythropoiesis

The highly regulated process of red cell production (of approximately 2 million erythrocytes per second, over 100 billion every day) involves a series of steps that begin with multipotent stem cells (for instance, the common myeloid precursor (the CFU-GEMM) as explained in Figure 3.3). A key aspect is the divergence of the erythroid lineage from the megakaryocyte lineage, with development via intermediate stages (such as the lineage-specific, colony-forming unit; CFU-E) to the **proerythroblast** (Figure 4.1). The largest of the red cell blasts (14–19 μm in diameter), the proerythroblast steadily shrinks through four to five progressive cell divisions as it passes into the erythroblast (or normoblast) stage (10–15 μm diameter), where haemoglobin synthesis begins. This transition is also accompanied by a reduction in 'blue' staining of the cytoplasm (when Romanowsky stains are used) as the erythroblast passes through basophilic, polychromatic, and orthochromatic stages, reflecting changing protein and nucleic acid constituents,

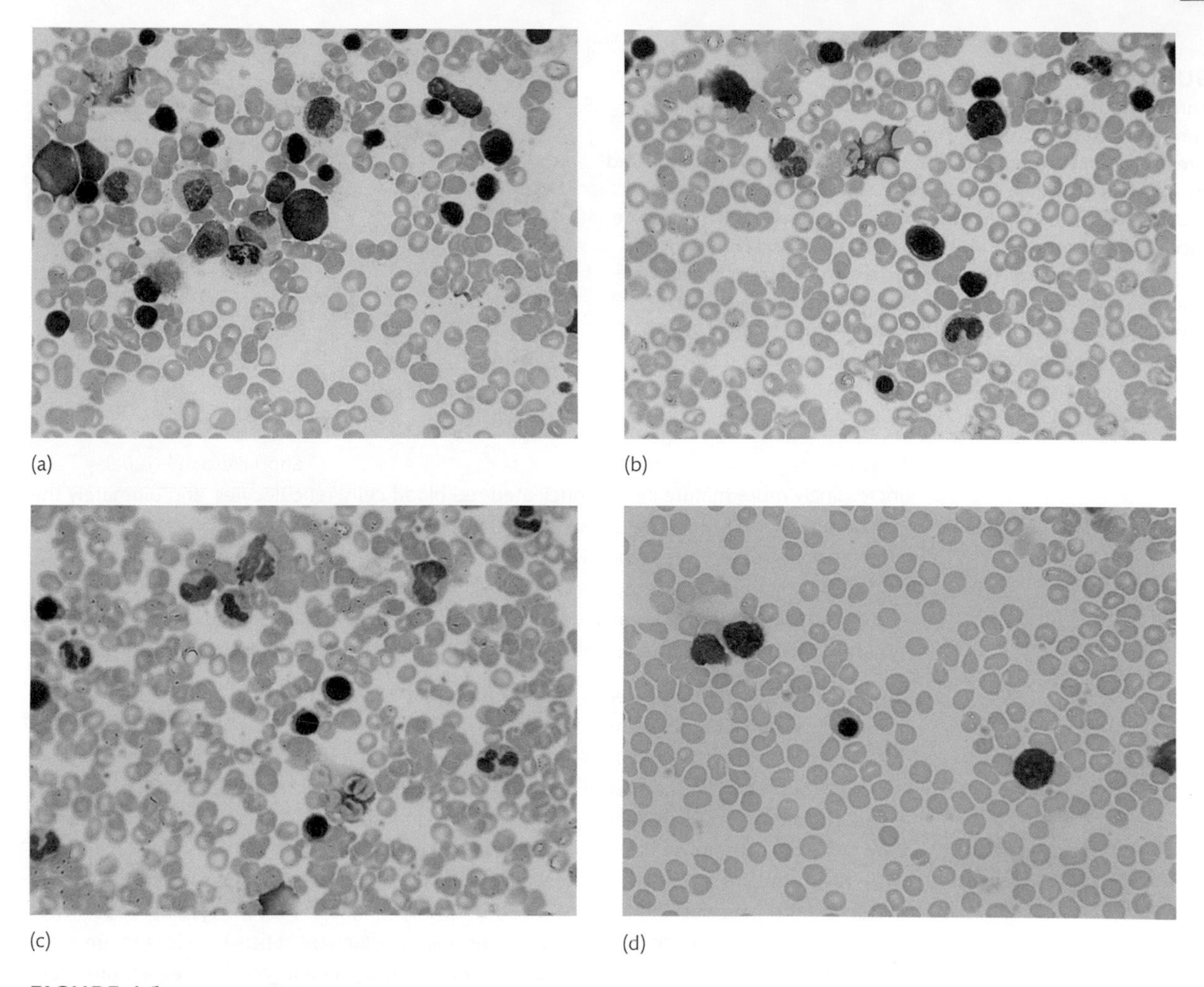

FIGURE 4.1

Erythroid precursors. As erythroid precursors develop in the bone marrow and transform into mature red blood cells, they become progressively smaller and the cytoplasm loses its blue colour. The large cell in the centre of (a) is a proerythroblast, figures (b–d) show erythroblasts in the early, intermediate, and late stages of differentiation, respectively. (Magnification ×400.)

principally ribosomes synthesizing haemoglobin. The stage after the erythroblast/normoblast sees the emergence of the **nucleated red blood cell**, with a diameter of 9–12 μm. These stages are shown in Figure 4.1(a)–(d).

nucleated red blood cell
A stage in the development of the red blood cell when it still retains a nucleus and before it becomes a reticulocyte.

Several of these stages occur in association with specialized bone marrow macrophages, in niches for erythropoiesis called **erythroblastic islands** (Figure 4.2). In the past, it was believed that bone marrow macrophages were present simply to scavenge and phagocytose, and so recycle nuclei extruded from the erythroblast. However, it is now becoming clear that the key role of the macrophage is to provide the cytokines and other signals (such as insulin-like growth factor-1) crucial to the promotion of erythropoiesis. Furthermore, these CD169-bearing 'nurse' macrophages may also directly transfer iron and haem iron to the erythroblast, which, with transferrin receptors, also obtains iron-loaded transferrin from the circulation (details to follow).

The close physical proximity of the macrophage, and its characteristic cytoplasmic protrusions, promote the maturation and proliferation of proerythroblasts and erythroblasts into

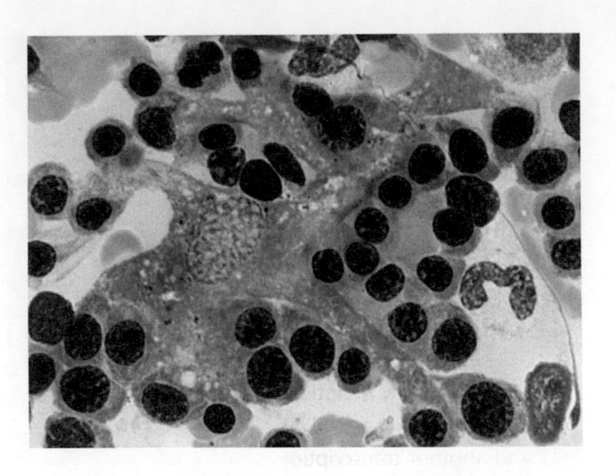

FIGURE 4.2
Erythroblastic islands. Erythropoiesis occurs in a defined microenvironment within the bone marrow–the erythroblastic island. Each centres on a nurse macrophage (stained blue here) which is surrounded by a ring, or crown, of developing erythroid cells. Cytoplasmic protrusions of the macrophage increase the effective cell membrane surface to allow more erythroid cells to be in contact with the macrophage. Within the macrophage, nuclei extruded from nucleated red blood cells are present, and are likely to be recycled. © 2010 Blackwell Publishing Ltd, *British Journal of Haematology*, 150:499. Reproduced with permission from Gerber B, Nair G, and Stuss G. *Erythroblastic islands. British Journal of Haematology* 2010: 150: 499, 499. Doi:10.1111/j.1365-2141.2010.08260.x.

increasingly more mature cells of nucleated red blood cells, reticulocytes, and ultimately the red blood cell itself. The close proximity of these cells is promoted by the interaction of adhesion molecules on both cell types. These adhesion molecules include tetraspanins CD81, CD82, and CD151 on the red cell precursors, and vascular cell adhesion molecule and fibronectin on the nurse macrophages. The early red cell precursors, such as the CFU-E, express integrins α4β1 and α5β1, although the latter is lost as the particular precursor passes to the erythroblast stage.

As erythroid maturation proceeds, the erythroblastic island migrates to regions of the bone marrow (the sinusoids) which are close to blood vessels, probably to facilitate the migration of reticulocytes and erythrocytes into the circulation. Erythroblastic islands are not bone marrow-specific: they have been noted in the embryonic yolk sac, placenta, foetal liver, and in the spleen. There is growing evidence of distinct anatomical niches. The sinusoidal-megakaryocyte niche contains sinusoidal endothelial cells and megakaryocytes, whilst the arteriolar-pericyte niche contains arteriolar endothelial cells, Schwann cells (secreting transforming growth factor-β), and pericytes and are innervated by the sympathetic nervous system. Both niches also include haemopoietic stem cells and chemokine-abundant reticular cells, and are close to osteoblasts (which produce a number of cytokines implicated in stem cell regulation). Stem cell factor synergizes with Epo to promote erythropoiesis, whilst insulin, insulin-like growth factor, activin, and angiotensin II all positively regulate this process. However, TGF-β, TNF-α, and γ-interferon all have a negative effect on erythropoiesis. The presence of certain cluster of differentiation (CD) molecules at the surface of cells at different stages of development can be used to identify those committed to erythroid or megakaryocyte pathways (Table 4.2).

TABLE 4.2 Changes in CD markers in cell development

	MEP	CFU-MK	CFU-E	MK	E
CD34	+	+	–	–	–
CD38	+	–	–		
CD235a/CD71	–	–	–	–	+
CD123	–	+	+	+	–
CD41a/CD42b	–	–	–	+	–

CD = cluster of differentiation; MEP = megakaryocyte/erythroid precursor; CFU-MK = colony-forming unit for megakaryocytes; CFU-E = colony-forming unit for erythroid blasts; MK = megakaryocyte; E = erythrocyte.

4.1.2 The molecular genetics of erythropoiesis

In section 3.4.5 we were briefly introduced to key concepts regarding the genetic regulation of haemopoiesis: we now focus on erythropoiesis. However, we must again be cautious in that much of what follows arises from mouse models, which may not translate. For example, murine erythrocytes are markedly smaller (52 vs 90 fl), have a shorter lifespan (42 v 120 days), a higher oxygen affinity (40 v 25 mm Hg), and lower levels of 2,3-diphosphoglycerate, whilst RHEX, a regulator of erythroid cell expansion, is present in humans, but not in the mouse. Key intracellular regulators of erythropoiesis include protein nuclear transcription factors such as GATA1 (coded for by *GATA1*, deletion of which causes severe anaemia), which directs the CFU-GEMM stem cell down the erythrocyte/megakaryocyte lineage. Without GATA1, erythroid progenitors undergo apoptosis. Other transcription factors drive further specialization: Kruppel-like factor-1 (KLF1) to an erythroid fate, and FLI1 and RUNX1 towards megakaryocytopoiesis. Overexpression of FLI1 and another transcription factor, ERG, converts erythroblasts to megakaryocytes, which subsequently can be induced to produce viable clot-forming platelets. KLF1 is also required for the expulsion of the nucleus from the erythroblast as it matures into a reticulocyte.

However, other genes also have a role in erythropoiesis. Although originally described in an acute T-cell leukaemia, *TAL-1*, which codes for a transcription factor, is highly expressed in all stages of erythroid differentiation: knock-out leads to early embryonic lethal anaemia with complete failure of yolk sac haemopoiesis. Activation of stem cell *HOXB4 in vitro* leads to a modest increase in erythroid cells that express embryonic and foetal haemoglobin (the genetics of haemoglobin are the subject of section 4.3.1). Nevertheless, transcription factors coded for by *GATA1*, *KLF1*, and *TAL1* have been collectively described as the 'core erythroid network', as knock-out mice exhibit severe impairment of erythropoiesis, and in humans loss-of-function mutations lead to anaemia.

Non-coding RNA

It has been estimated that between two-thirds and 90% of the mammalian genome is transcribed, but a much smaller proportion (estimated at <2% to 20%) produces mRNA transcripts with the potential to code for protein, leaving much RNA unaccounted for. Some of these 'missing' transcripts are transfer and ribosomal RNA, others include non-coding RNAs that are long (>200 nucleotides), others that are short. The latter include microRNAs (miRNAs), small (~22 nucleotides) non-coding RNAs with diverse functions in regulating cellular activity, and that are purported to influence more than one-third of all human genes. One of the first reports of these molecules in erythropoiesis showed that miR-221 and miR-222 inhibit normal erythropoiesis and erythroleukaemic cell growth, and observations that followed pointed to roles for miR-103, miR-155, miR-144, and miR-451, and others, in normal erythroid differentiation. Subsequent reports linked some, such as miR-144 and miR-451, to *GATA-1*, whilst others showed that inhibition of these miRNAs lead to a significant decrease in the production of mature erythrocytes. miR-486 is upregulated during erythroid differentiation and forced overexpression leads to an increase in this differentiation. However, at a practical level, miR-486 seems likely to be a mediator of leukaemogenesis, and so may be a novel therapeutic target. A further practical aspect of these molecules is the suggestion that plasma miR-451 may be a relevant biomarker for the pathological erythropoiesis in certain cases of haemoglobinopathy, whilst miR-486 may have a role in the treatment of certain leukaemias.

Thousands of long, non-coding RNAs (lncRNA) have been identified, and the purpose of the vast majority remains obscure. However, many of these are the targets of the key erythroid transcription factors GATA, TAL1, and KLF1. One lncRNA species, alncRNA-EC7, is an enhancer of *SLC4A1*, which codes for the major membrane protein Band 3. Another, lncRNA UCA1, is dynamically regulated during erythroid maturation, with maximal expression in

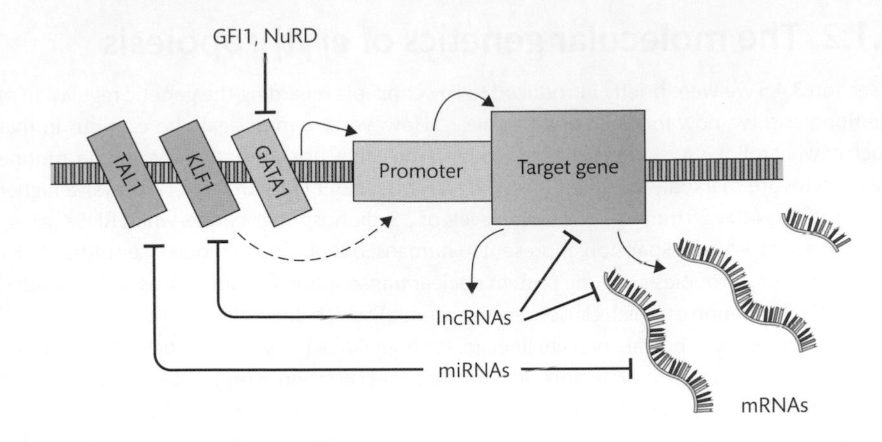

FIGURE 4.3
Gene regulation. The core erythroid network of transcription factors (TFs) such as TAL1, KLF1, and GATA1 regulates the promotor of genes involved in erythropoiesis. TFs may themselves be regulated by other TFs such as GFI1 and the nucleosome remodelling deacetylase (NuRD) complex. Modified from Nandakumar et al. (2016) and elsewhere.

proerythroblasts obtained from CD34^{+} umbilical cord progenitor cells. But once more, we observe caution in interpretation and extrapolation, as only 15% of mouse lncRNAs are expressed in humans, and vice versa.

Figure 4.3 summarizes certain aspects of the regulation of genes.

4.1.3 Erythropoietin (Epo)

Much of the process of cell development described above is driven by growth factors such as stem cell factor and **erythropoietin** (Epo). The latter (coded for by *EPO* at 7q22.1) is an essential erythroid-specific factor for developing proerythroblasts and erythroblasts, also thought to influence CFU-E progenitor cell survival. Indeed, if erythroblasts grown *in vitro* are denied Epo, they rapidly become apoptotic and die. It also has a role in iron metabolism, as erythroblasts respond to Epo by secreting the hormone **erythroferrone** (coded for by *FAM132B* at 2q37.2), which inhibits the action of hepcidin (to be discussed in detail in section 4.3.1). With a relative molecular mass of 30–40 kDa, Epo is produced by several different types of cell, primarily pre-tubal renal fibroblasts, but also neurons, brain pericytes, and glial cells. However, the vast majority, if not all, of plasma Epo arises in the kidney.

The key stimulus for the production of Epo is hypoxia, key intracellular mediators being hypoxia inducible factor 2α and prolyl-4-hydrolase, that move to the nucleus to bind to a hypoxia-responsive element present in the enhancer region of *EPO* to initiate the generation of Epo. However, other pathways, such as that of stem cell factor, can also part-regulate Epo production, and GATA proteins can negatively regulate Epo mRNA via binding in the Epo promotor region.

Once in the circulation, Epo interacts with its target cells of erythroid progenitors via a specific structure in the cell membrane—the Epo receptor (EpoR). Located at 19p13.2, *EPOR* encodes a 56kDa protein, the EpoR being a dimer that spans the cell membrane. The number of EpoRs at the cell surface of erythroid cells is tightly regulated—around 1100 copies on an erythroid progenitor and 300 on a late-stage erythroblast. Upon binding of Epo to its receptor, an enzyme, Janus kinase 2 (**JAK2**) attached to the intracellular tail, becomes phosphorylated, and this in turn phosphorylates a tyrosine residue. This phosphorylated tyrosine then promotes a further phosphorylation in intracellular second messengers (such as STAT and phosphatidylinositol-3-kinase), which move to the nucleus to promote a series of events leading to the transcription of a number of genes, whose downstream effect is to promote cell proliferation and differentiation (Figure 4.4). As we will see in section 4.6, mutations in *JAK2* are the basis of the disease **polycythaemia vera (PV)**, but these are also important in **myelofibrosis** and **essential thrombocythaemia**. This is explained in Chapter 11.

Erythropoietin
A renal-derived hormone that stimulates the production of red blood cells.

Erythroferrone
A hormone produced by erythroblasts that inhibits the production of hepcidin, so promoting iron uptake.

JAK2
An intracellular enzyme activated by the binding of Epo to its receptor that ultimately results in cell proliferation and differentiation.

polycythaemia vera
A malignant disease of increased numbers of red blood cells.

myelofibrosis
A malignant disease of the bone marrow.

essential thrombocythaemia
A malignant disease leading to an increased platelet count.

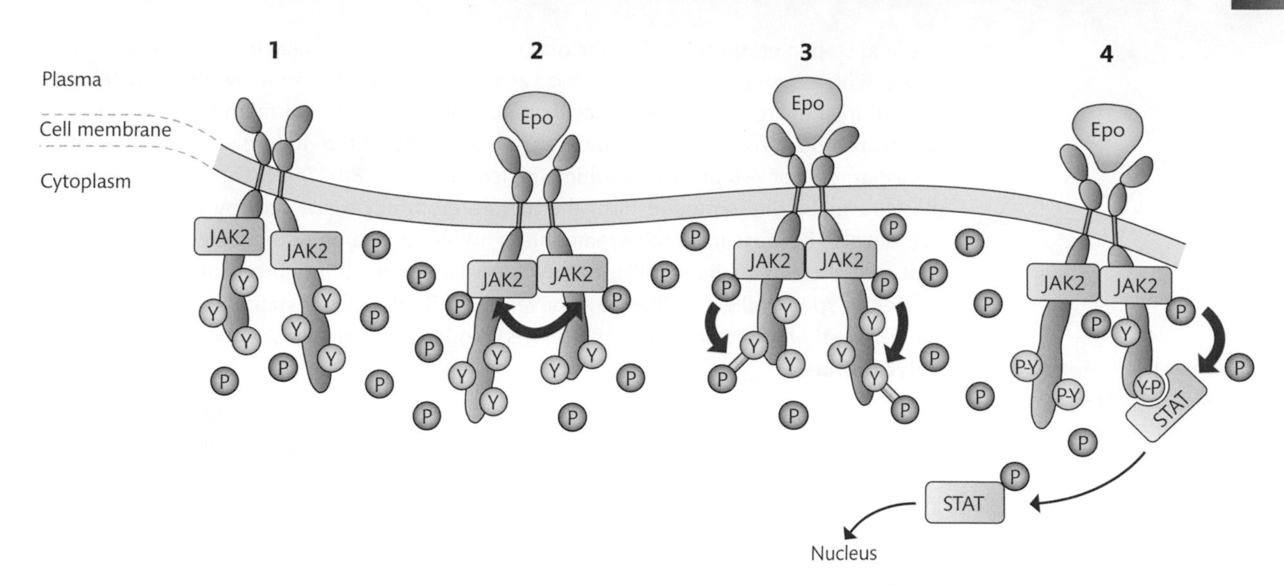

FIGURE 4.4

Activation of JAK2 by the binding of Epo to its receptor. The EpoR is a dimer multi-domain transmembrane glycoprotein. The intracellular portion of each part of the dimer is linked to a JAK2 molecule, and at rest these enzymes are far apart (1). Upon binding of Epo to the extracellular domains of the receptor, the intracellular domains move close together, allowing the two JAK2 molecules to cross-phosphorylate and so activate themselves (2). The phosphorylated JAK2 (JAK2-P) can now phosphorylate tyrosine residues (Y—this letter being the international abbreviation for this amino acid) on other parts of the intracellular domains (3). The phosphorylated tyrosine (Y-P) provides a docking motif for the intracellular second messenger, STAT (4). The JAK2-P phosphorylates the STAT. The phosphorylated STAT (STAT-P) moves off and migrates to the nucleus, where it promotes transcription.

Individuals with renal disease (especially if on dialysis) frequently have low levels of Epo, leading to impaired erythrocyte production, low numbers of circulating red blood cells, and so anaemia (but note Box 4.1). Accordingly, Epo is often an effective therapy. Curiously, there is no sex difference in levels of circulating Epo, despite lower red cell and haemoglobin levels in women. An explanation for this is the speculation that women deliver oxygen to their tissues more efficiently than do men. Epo may well have other roles, such as in iron metabolism, and (in animal models) in protection from diabetes.

4.1.4 The reticulocyte

The next step in the maturation of the red cell involves the loss of its nucleus, marking its transformation into the reticulocyte (diameter 7–10 μm and a volume of ~150 fL). As described above, the loss of the nucleus ('nuclear extrusion') is mediated by the nurse macrophage, possibly under the control of *KLF-1* and miR-30a; the extruded nucleus, a pyrenocyte, encased in plasma membrane. Much of the 40 pg of protein and nucleic acids within the nucleus is then recycled during the formation of new blasts and stem cells. Pyrenocytes express phosphatidylserine on their surface, as do apoptotic cells, which may provide a signal to phagocytic macrophages. Extrusion of the nucleus is followed by the removal of the now redundant intracellular organelles (endoplasmic reticulum, Golgi apparatus, mitochondria) by autophagy and endocytosis. This process is accompanied by a 20% reduction in cell volume and plasma membrane surface area.

The final step in erythropoiesis is the maturation of the reticulocyte into the mature erythrocyte, which then passes into the blood at a rate (in health) of some two million cells per second. In the circulation, the reticulocyte undergoes additional membrane protein and skeletal remodelling to form the biconcave phenotype of the mature erythrocytes, which has a diameter of 7–8 μm and a volume generally around 80–100 fL (Figure 4.5). The reticulocyte can be differentiated from the mature erythrocyte not only by its slightly larger size, but also because it contains remnants of mRNA for haemoglobin, detectable by supravital special stains such as brilliant cresyl blue. However, although reticulocytes generally cannot be specifically identified by conventional Romanowsky staining, a proportion do stain a slightly bluer colour than mature red blood cells. This bluish tinge is referred to as **polychromasia**.

polychromasia
A finding on the blood film that translates as 'many colours'. In practice, there will be red blood cells of a normal colour, but others (reticulocytes) with a blue tinge.

As this final step in red cell maturation can take place in the peripheral blood as well as in the bone marrow, small numbers of reticulocytes (perhaps 0.5–1.5% of the entire red cell population) may be present in healthy blood. However, larger numbers of reticulocytes, perhaps <2.5%, imply an abnormality, such as in certain types of anaemia, as is described in Chapters 5 and 6. Similarly, nucleated red blood cells are seen so rarely in healthy adult blood that their presence is inevitably the result of a pathological process, although nucleated red blood cells may occasionally be seen in healthy neonatal blood. Figure 4.5 shows the different stages through which the red cell passes, and the position of Epo in promoting this pathway.

4.1.5 Further regulation of erythropoiesis

The process of erythropoiesis is generally very tightly regulated to ensure the generation of approximately 5×10^{10} erythrocytes each day, enough to replace those cells destroyed by a particular disease process, or simply by the cell's old age. However, this number may change: many diseases of the bone marrow, such as leukaemia, lead to a reduced production of red blood cells, and, consequently, anaemia. Conversely, increased numbers of red blood cells can also be produced by the bone marrow, leading to **erythrocytosis**. An increased red cell count is also found in polycythaemia.

erythrocytosis
A red cell count above the upper limit of the reference range.

However, the most important regulator, Epo, is not the only growth factor to influence red blood cell production. There are also roles for interleukin-3 and thrombopoietin at the level of the common myeloid progenitor (Figure 4.5). The male sex hormone testosterone also acts as

BOX 4.1 Erythropoietin—a note of caution

Epo is often considered as a treatment for many different types of anaemia. However, this may be unwise as the indiscriminate use of Epo may not be advisable, especially in patients with cancer and/or end-stage renal disease (ESRD). One of the characteristics of cancer is that it is often associated with anaemia, so that it has been believed (quite reasonably) that such patients would benefit from treatment with Epo. However, an additional characteristic of tumours is their reliance on, and extra sensitivity to, certain growth factors. It is now clear that certain cancers can respond to Epo, which seems likely to explain (against expectation) an adverse effect of this growth factor on cancer survival rates, due mainly to tumour progression. Similarly, the presumption that the anaemia often associated with ESRD can be beneficially treated with (recombinant) Epo has been challenged by a study which reported an increased risk of death in those using this hormone. Another study showed that, although Epo did indeed increase levels of haemoglobin (from a mean of 101 g/L to 126 g/L), and several indices of quality of life, there were safety issues connected with its use.

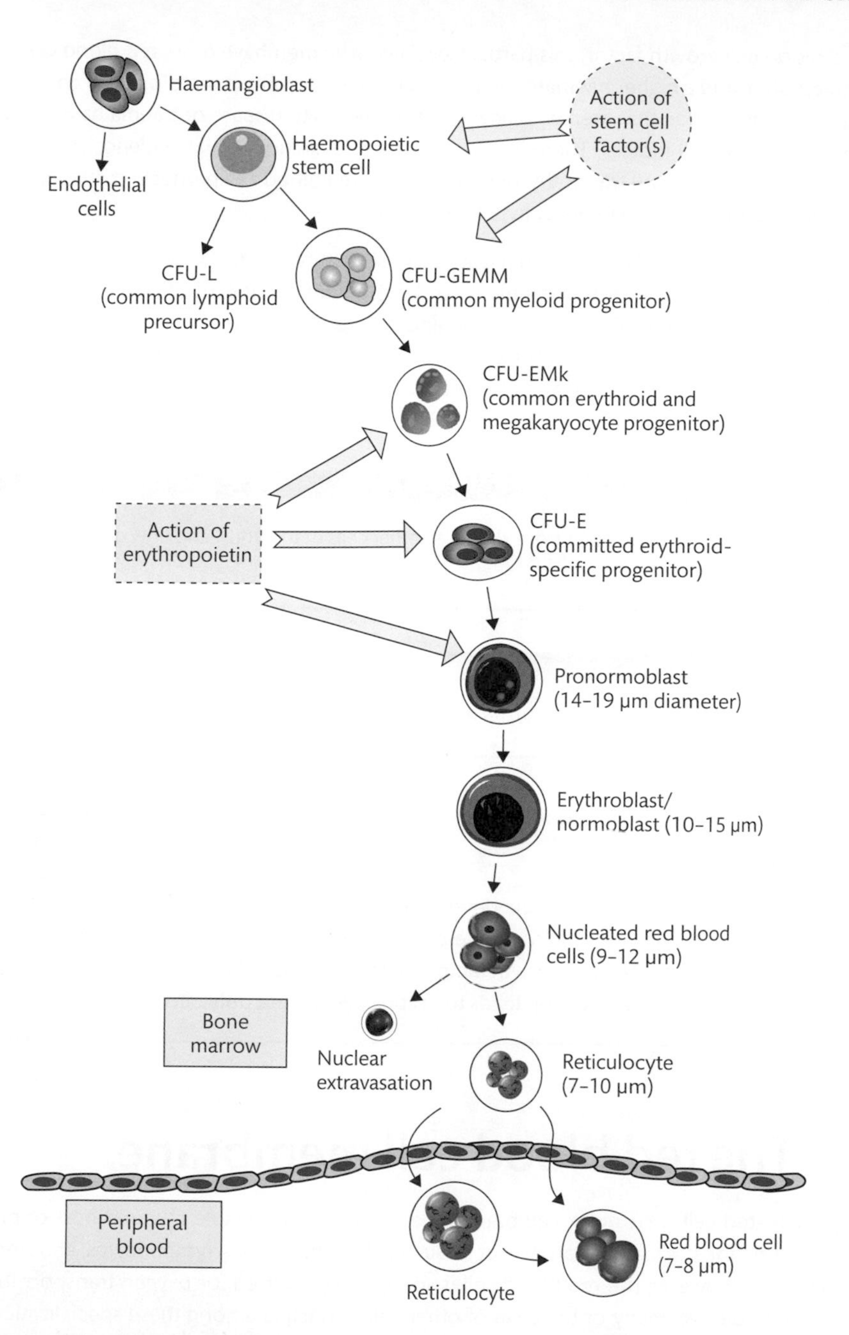

FIGURE 4.5

Erythropoiesis. Stages in the development of the mature red blood cell (erythrocyte) in the bone marrow and peripheral blood. A pathway of stem or precursor cells can be identified, commencing with the haemangioblast, and passing to various colony-forming units (CFUs) and ultimately to the mature cell. The 'early' stem cells such as the CFU-GEMM are stimulated primarily by stem cell growth factors, whereas 'late' stem cells such as the CFU-E are stimulated by Epo. As this pathway proceeds, cells become smaller, and as they do, haemoglobin synthesis becomes more prominent. A crucial stage is the extravasation of the nucleus from the nucleated red blood cell as it becomes a reticulocyte. Some reticulocytes remain in the bone marrow, where they mature into erythrocytes, although other reticulocytes pass from the bone marrow into the peripheral circulation, where their own final maturation takes place.

a red blood cell growth factor; this partially explains why men have more red blood cells and haemoglobin, and a higher haematocrit, than do women. An unfortunate 'proof of concept' of this is the observation that men who have lost their testes to disease or trauma have a reduction in their red cell indices. There is also considerable circumstantial evidence that thyroid hormones (thyroxine and triiodothyronine) stimulate red blood cell production and may be in part responsible for the anaemia of hypothyroidism.

Cross references

We discuss anaemia in Chapters 5 and 6, leukaemia in Chapters 10–12, and wider aspects of erythrocytosis and polycythaemia vera in Chapter 11.

Fully functioning erythropoiesis also requires a host of minerals and vitamins, such as cobalt, vitamin C, copper, vitamin E, vitamins B_6 and B_{12}, thiamine, and riboflavin. Selenium acts to promote stress erythroid cell expansion and differentiation, as well as directly acting on monocytes, inducing their differentiation to erythroblastic island macrophages. Anaemia may result from a deficiency of any of these micronutrients and hormones, as is explored in Chapters 5 and 6.

SELF-CHECK 4.1

How does the red blood cell differ from almost all other cells of the body, and how does this relate to function?

SELF-CHECK 4.2

What are the principal differences between the reticulocyte and the mature red blood cell?

Key Points

Erythropoiesis is the process of the development of the red blood cell. Beginning in the bone marrow, immature stages include the erythroblast, the nucleated red blood cell, and the reticulocyte. The key growth factor is Epo. Under-production of red cells leads to anaemia, while over-production leads to erythrocytosis and polycythaemia.

4.2 The red blood cell membrane

In a nucleated cell, a complex cell membrane is required to enable a whole host of physiological functions, as may be carried out by such cells as fibroblasts, lymphocytes, and smooth muscle cells. However, the red blood cell is so highly specialized for oxygen transport that it need not undertake many of the tasks of other cells. Principal among these specializations is the lack of a nucleus or other organelles such as mitochondria. So, relieved of the burden of the nucleus, the red cell has a physical flexibility or deformability that allows it to pass along the smallest capillaries and so deliver oxygen to the tissues: a nucleus would severely restrict this freedom of movement. Indeed, the flexibility of the red blood cell is so highly developed that it can deform to an extent that its length increases by 250%, whereas an increase in surface area of only 3–4% is likely to lead to cell lysis. Furthermore, a normal 8 μm red blood cell can deform to pass through a 3 μm blood vessel lumen.

However, this high degree of specialization also brings disadvantages, such as a susceptibility to pathological factors and the extremes of physiology that would not normally be a problem for the membrane of a nucleated cell.

The red cell membrane has three components:

- a double layer consisting of equivalent amounts of phospholipids and cholesterol;
- various proteins and glycoproteins embedded within the double phospholipid layer, but which may also have sections inside and/or on the exterior surface of the cell. The molecules which interface with the plasma are collectively called the glycocalyx;
- an internal cytoskeletal scaffold or skeleton that gives the red blood cell its characteristic round but flattened shape—a biconcave disc.

The red cell membrane consists of ~20 major and ~850 minor proteins, making up perhaps 50% of the membrane, the remainder being lipids (40%) and carbohydrates (10%). The general composition of the membrane is illustrated in Figure 4.6.

The lipoprotein bilayer is not symmetrical. Instead, the external layer is rich in phosphatidylcholine and sphingomyelin, whilst the internal layer is dominated by phosphatidylserine and phosphatidylethanolamine. Cholesterol is believed to be equally distributed. However, the phospholipids can move between the inner and outer membrane layers under the direction of different enzymes such as 'flippases', 'floppases', and 'scramblases'. The different proportions of lipids are important because macrophages, primarily in the spleen, are programmed to recognize and destroy (by the process of phagocytosis) those cells that have a high proportion of phosphatidylserine. The fluidity of the lipid component of the membrane is crucial in facilitating the lateral movements of floating 'rafts' made up of complexes of glycoproteins.

4.2.1 Function of red cell membrane components

The lipid bilayer is populated with a large number of glycoproteins at a frequency which varies from a few hundred to over a million copies per cell. The functions of these molecules can be classified in a number of ways: as being important in the transport of ions and molecules across the membrane (Table 4.3), as enzymes (Table 4.4), and with roles in adhesion and/or

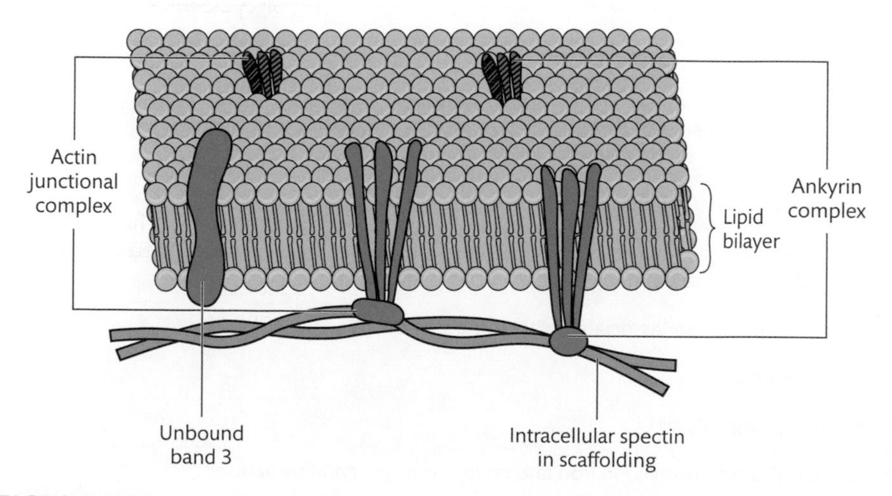

FIGURE 4.6

The general composition of the membrane: a three-dimensional schematic representation of the red cell membrane. Complexes of two sets of molecules (the ankyrin complex and the actin junctional complex) have extracellular and intracellular domains, the latter linking to an internal cytoskeleton of spectrins that give the cell its unique shape. A third complex (composed mostly of Band 3) is not bound to the intracellular scaffolding.

TABLE 4.3 Components of the red blood cell membrane with transport function.

- Band 3 (CD233, the major anion transporter (with the aid of carbonic anhydrase) and central to gas exchange by generating bicarbonate from water and carbon dioxide, and exchanging the bicarbonate ion for the chloride ion, linked to blood group Diego; also links the membrane to the cytoskeleton)
- Aquaporin 1 (transporter of water, oxygen, and carbon dioxide, linked to blood group Colton)
- Aquaporin 3 (transporter of water and glycerol, linked to blood group Gill)
- Glut-1 (glucose and L-dehydroascorbic acid transport)
- Kidd antigen (the JK blood group system, involved in urea transport)
- RhAg (CD241, essential for the expression of rhesus (Rh) molecules, but is also a gas transporter, probably carbon dioxide and/or ammonia/the ammonium ion)
- Rh complex (RhD [= CD240D], RhCcEe [= CD240 CE], possibly involved in ammonia/ammonium ion, and/or oxygen/carbon dioxide transport, and also functions in Band 3/RhAG complex assembly)
- $Na^+/K^+/2Cl^-$ co-transporter
- Na^+/K^+ co-transporter
- Na^+/Cl^- co-transporter
- K^+/Cl^- co-transporter

Note also that several of these molecules have CD designations between 233 and 242 (the 'red cell' area). Furthermore, many molecules may also be present on other cells such as those of the kidney (Band 3), white blood cells and platelets (glycosyl phosphatidylinositol), and the endothelium (DARC).

TABLE 4.4 Enzymes of the red blood cell membrane.

- Na^+/K^+ ATPase
- Ca^{++} ATPase
- Carbonic anhydrase (two forms: type II binds to an intracellular domain on Band 3, type IV to an extracellular domain)
- Kell glycoprotein (CD238, a zinc endopeptidase that cleaves big endothelin-1)
- Peroxiredoxin 2 (an abundant [200,000 copies per cell] antioxidant linked to Band 3 that also activates the calcium ion-dependent potassium channel)
- Acetylcholinesterase (associated with Yt blood group)*
- 5'-nucleotidase*
- Alkaline phosphatase*
- Caeruloplasmin (involved in iron and copper transport and oxidation)*
- ADP-ribosyltransferase (an ectoenzyme linked to blood group DO)
- Glycolytic enzyme complex (GEC)**

*These molecules are anchored to the membrane by glycosyl phosphatidylinositol. **The GEC consists of glyceraldehyde-3-phosphate dehydrogenase, phosphofructokinase, lactic dehydrogenase, pyruvate kinase, aldolase, and enolase, and is linked to ankyrin.

receptor function and/or structural function (Table 4.5). The most abundant of these molecules are **Band 3** (also known as SLC4A1, ~1.2 million copies per cell) and **glycophorin A** (or GPA, ~1 million copies per cell).

Band 3, glycophorin A, glycophorin C, Rh-associated glycoprotein
Components of the membrane of the red cell that provide recognition, transport, and anchorage sites.

Band 3 is worthy of further attention: not only is it the largest molecule in the membrane (~102 kDa), but it is also the most abundant. It forms complexes with several other intra-membrane components, has an anion exchange channel, and is associated with intracellular and extracellular carbonic anhydrase and intracellular glycolytic enzymes with the potential to generate ATP. It is stabilized in the membrane by binding to spectrins via protein 4.2, protein 4.1, and ankyrin, and also binds the antioxidant peroxiredoxin. Finally, Band 3's N-terminus, when the cell is oxygenated, binds and inactivates glycolytic and ATP-generating enzymes. Deoxyhaemoglobin is also able to bind to this site, and when it does so, the glycolytic enzymes are liberated and activated, resulting in the generation of ATP, but also in a weakening of the cytoskeleton. This pathway therefore points to the process of regulation of red cell properties by oxygen.

Band 3 and GPA, alongside other glycophorins and aquaporin 1, provide the support for blood group antigens such as those of the Diego and MNS systems. The glycosylated antigens of the A, B, H, and I systems are carried by Band 3, GLUT-1 (which transports glucose into the cell), RhAG (**Rh-associated glycoprotein**, ~100–200,000 per cell) and aquaporin-1, whilst those of the Rh system (D, C, and E) are non-glycosylated membrane components. The importance of glucose transport is demonstrated by the presence of 200,000–700,000 GLUT-1 molecules per cell, and will be revisited in section 4.3.2.

Enzymes of the membrane (Table 4.4) include those responsible for maintaining the ionic homeostasis of the cytoplasm (sodium, potassium, and calcium), in the metabolism of carbon dioxide and the bicarbonate ion (part-regulated by the enzyme carbonic anhydrase), and in ADP/ATP metabolism. A group of glycolytic enzymes (glyceraldehyde-3-phosphate dehydrogenase, aldolase, phosphofructokinase, and lactate dehydrogenase) are linked to the intracellular portion of the Band 3 molecule.

The third aspect of red cell membrane components includes those of adhesion and as receptors, although these often overlap with structural roles (Table 4.5). For example, the complex of Band 3 and RhAG are involved in anion transport, but are also linked to the intracellular skeletal components, principally the alpha and beta forms of spectrin. Several surface molecules provide defence from attack by complement, and although the presence of a receptor (DARC/Duffy) for cytokines such as interleukin-8 on mature red cells seems curious, it may have other functions (DARC is also present elsewhere, such as on the endothelium, where cytokine activation may be important). Similarly, those molecules that in other settings may have roles in adhesion, in the mature red cell may have other functions, such as promoting the integrity of other components of the membrane. These adhesion molecules may have had roles earlier in the life of the red cell, such as in mediating contact between erythroblasts and macrophages (such as tetraspanins and integrins on the surface of the former).

TABLE 4.5 Membrane components with structural, adhesion, and receptor function.

- Intercellular adhesion molecule-4 (CD242, an integrin-binding protein, associated with blood group LW)
- Lutheran glycoprotein (Lu, CD329, a ligand for laminin 511/512)
- CD47 (part of the Band3/Rh complex, interacts with macrophages)
- Duffy antigen (blood group FY) receptor for chemokines (DARC) (CD234)

(continued)

- Complement component 3b/4b receptor (CD35) (also known as CR1, and also binds immune complexes)
- Tetraspanin (CD151) (might associate with integrins and laminins, linked to blood group RAPH/MER2)
- Glycophorins A, B, C, and D (CD235A, CD235B, CD236C, and CD236D, respectively facilitate the assembly of Band 3 and Band3/RhAG complexes that link the membrane to the internal cytoskeleton and may contribute to the glycocalyx)
- Glycosyl phosphatidylinositol (an anchoring molecule)
- CD59 (membrane inhibitor of reactive lysis–a complement regulatory protein)*
- CD55 (decay-accelerating factor–a complement regulatory protein, also known as C3 convertase inhibitor)*
- CD44 (an adhesion molecule that binds hyaluronic acid, linked to blood group IN)
- CD147 (EMMPRIN, regulates the production of matrix metalloproteinases, linked to blood group OK)
- Semaphorin 7A (adhesion molecule involved in cell migration and linked to blood group JMH)

* These molecules are anchored to the membrane by glycosyl phosphatidylinositol.

4.2.2 Structure of the red blood cell membrane

The skeleton of the red cell membrane is composed principally of spectrins, ankyrin, protein 4.1, and actin and its associated proteins (tropomyosin, tropomodulin, adducin, and dematin). These, with surface and cross-membrane molecules, associate into three different groupings, two of which interact with the major structural molecules inside the cell (**alpha- and beta-spectrins**).

alpha- and beta-spectrin
The major structural components of the matrix that provide 'skeletal' support for the structure of the cell.

The ankyrin complex

This complex is likely to be composed of a Band 3 tetramer (comprising some 40% of all membrane Band 3), ankyrin, two GPA or GBP dimers, two protein 4.2 molecules, CD47, the Landsteiner-Wiener (LW) blood group antigen, proteins involved in glucose metabolism (see Table 4.4), and an Rh complex of RhAG, RhD, and RhCE (Figure 4.7). These are linked to the internal cytoskeleton by molecules that include ankyrin, protein 4.1, and a group of glycolytic enzymes listed in Table 4.3.

The actin junctional complex

This complex also contains Band 3 molecules (but as a dimer, and also comprising some 40% of all membrane Band 3), a GLUT molecule, a stomatin molecule, one each of glycophorins A, C, and D, and the Kell, Duffy, and XK molecules (Figure 4.8). These are linked to the internal cytoskeleton of spectrins via a complex that includes dermatin, the GEC, tropomyosin, tropomodulin, adducin, and protein 4.1R.

The Band 3/GPA complex

This third complex is formed of the remaining 20% of total membrane Band 3, and a molecule of GPA. It lacks the molecules needed to bind to the spectrin scaffolding, and so diffuses freely in the lipid bilayer. Its function is unclear, but it may provide additional transport functions.

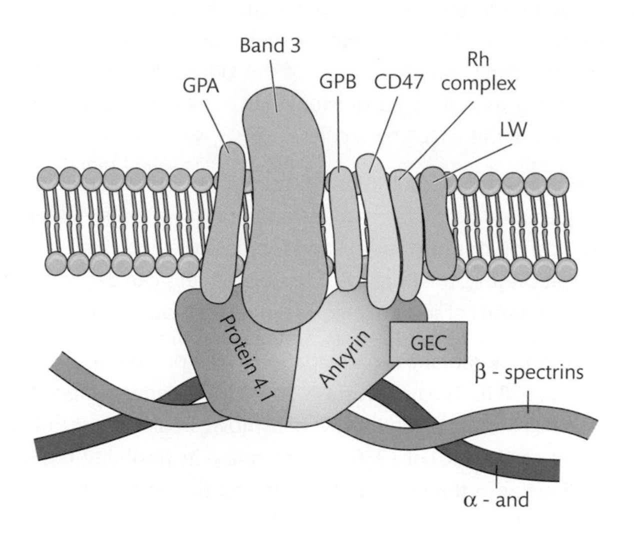

FIGURE 4.7
A representation of the ankyrin complex, so named as an ankyrin molecule links cross-membrane sections with the cytoskeleton. Molecules RhAG, GPA, GPB, LW, CD47, and Band 3 form a heterologous complex within the lipid bilayer. Some molecules span the membrane and interface with molecules of a sub-complex comprising protein 4.2, the glycolytic enzymes complex (GEC) (listed in Table 4.4), and ankyrin itself, which is in contact with the alpha- and beta-spectrins of the cytoskeleton. This figure and Figure 4.8 are simple schematics of the membrane components, and do not imply a precise spatial arrangement: the GPA and B molecules do not necessarily lie between the Band 3 molecules, and CD47 and LW do not necessarily lie on the right of the complex. It is likely that the individual molecules interact, as GPA and GPB facilitate membrane assembly of Band 3, whilst the Rh complex molecules (D and CE) facilitate Band 3 and RhAG complex assembly. Cartoon amended from Lux (2016).

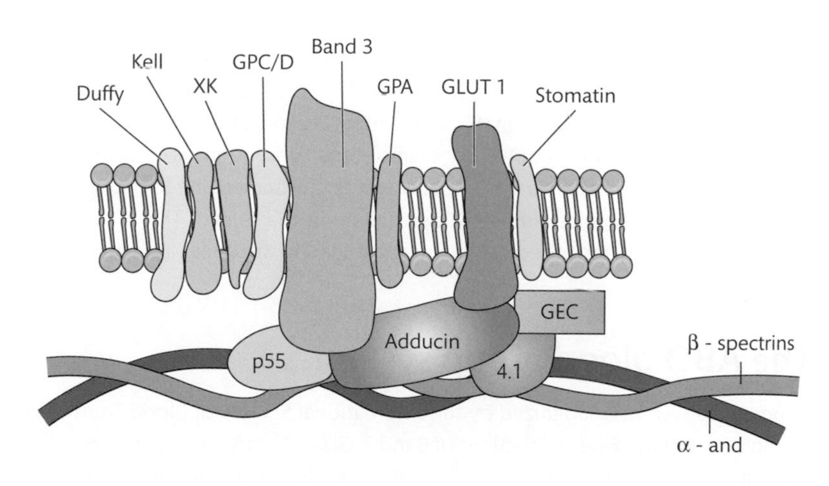

FIGURE 4.8
A representation of the actin junctional complex. This consists of molecules such as Duffy, Kell, XK, Band 3, Glut 1, GPA, GPC, and GPD, which span the membrane and link to a series of other molecules that collectively make up the protein 4.1R complex (adducin, dematin, p55, etc.). This complex in turn links with other sections of the spectrin cytoskeleton that are apart from those complexed to the ankyrin complex. GEC = glycolytic enzymes complex. Cartoon amended from Lux (2016).

The membrane as a functional unit

The interrelationships between the molecules of the cell membrane are crucial in giving the cell its flexibility and its unique double-concave shape. Indeed, the normal biconcave discoid shape maintained by the spectrin skeleton gives the cell a surface area 40% greater than a sphere of similar overall volume. Defects in these structural glycoproteins, some of which are genetic, give rise to changes in the shape and flexibility of the red blood cell.

Many of these changes to the red cell membrane can lead to a particular pathological condition such as anaemia. We revisit the consequences of defects in these membrane components in Chapter 6, one of which is due to problems with glycosyl phosphatidylinositol (GPI). An important 'housekeeping' function of the body is its ability to recognize and destroy abnormal cells using a complex series of plasma proteins called **complement**; red cells with these defects are detected and eliminated, potentially resulting in anaemia.

Notably, an additional feature of the membrane is that it lacks human leucocyte antigen (HLA) molecules. These are highly specialized molecules, which have a number of functions, such as in our defence against viruses—which is a good thing. However, their existence also frustrates our attempts to ensure viable organ transplantation, such as of the kidney. By contrast, red blood cells are far easier to transplant (e.g. as a blood transfusion) than a solid organ such as the heart or kidney because of the absence of these HLA molecules.

A further consequence of the high degree of specialization of the red cell membrane is that it is far more fragile than those of nucleated cells. One manifestation of this is a susceptibility to extremes of certain physiological conditions, such as pH and temperature, which are not a problem for nucleated cells. In the laboratory, the fragility of red blood cells can be assessed by their responses to different concentrations of sodium chloride in the osmotic fragility test.

A final important function of the red cell membrane is participation in the processes of maintaining the mean cell volume (MCV) at ~90 fL. The homeostasis of the MCV depends upon the amount of water within the cell, which in turn involves the laws of osmosis. In order to maintain a mean cell haemoglobin concentration (MCHC) of ~330 g/L and retain its flexibility, the red cell is dependent on the volume of water in the cell, and thus also on the balance between sodium (low levels relative to those of the plasma), potassium (high levels relative to those of the plasma), and calcium cations (which are normally undetectable in the cell). From an electrical perspective, these cations must also be balanced by the sum of intracellular anions, principally chloride and bicarbonate.

The cell achieves a balance between these ions with a number of regulatory co-transporter molecules (also described as pumps, many of which require ATP), which collectively regulate the passage of water and these ions in and out of the cell (Table 4.3). However, if there is a shortage of ATP, then the pumps may be unable to maintain the ionic homeostasis of the cell, leading to changes in cell shape and volume. For example, the consequences of an increase in the intracellular calcium for the cell are a loss of potassium and water leading to shape change, dehydration, less deformability, and, ultimately, destruction in the spleen. This is the pathological basis of several other types of red cell defects that lead to anaemia, as is explained in Chapters 5 and 6.

4.2.3 The ABO blood group system

Of the 33 blood group systems recognized by the International Society of Blood Transfusion, that of ABO is most important. External sections of Band 3, GLUT-1, RhAG, and aquaporin-1 are characterized by the presence of chains of carbohydrates. Variation in genes (alleles, located on chromosome 9q34.1–q34.2) for certain enzymes (glycosyltransferases) in turn lead to variation in the composition of these carbohydrate chains, being a combination of L-fucose, N-acetyl D-galactosamine, and D-galactose. These differences are sufficiently strong to lead to the generation of antibodies, and it is this combination of different carbohydrate antigens, and the antibodies they induce, that gives us the ABO blood group system.

The basic unit of the ABO system is a trimer of acetylgalactosamine-galactose-fucose, and this is the 'O' chain. However, the addition of a further acetylgalactosamine residue to the galactose molecule converts this to an 'A' chain, whilst addition of a second galactose molecule to the existing galactose molecule converts the 'O' trimer to a 'B' chain. These structures are not specific for red cells and are widely expressed on other cells. Those with the 'A' structure on their cells (approximately 42% of the population of the UK) spontaneously develop antibodies to the 'B' structure (i.e. anti-B antibodies) in the neonatal period. In mirror form, those with 'B' on the surface of their cells (9% of the UK) also have anti-A antibodies. Those with the 'O' structure (45% of the UK) develop anti-A and anti-B antibodies. In a small number of people (4% of the

UK), both the 'A' and the 'B' structures are present, so these individuals are of blood group 'AB'. The A and B structures are also known as CD233.

Since we have two copies of chromosome 9, each expressing a different transferase, then several different genotypes are possible. Those expressing two 'A' genes (i.e. AA, being homozygotes) and those with an 'A' gene and an 'O' gene (i.e. AO, being heterozygous), both have the 'A' phenotypes. Similarly, those of blood group 'B' can be homozygotes (BB) or heterozygotes (OB). All those of group 'O' phenotype are genotype OO homozygotes, whilst all those of group 'AB' phenotype are genotype AB heterozygotes. There are several variants of group A, the most common being A_1, A_2, A_3, A_x, and A_{end}, and of group B, the most common being B, B_3, B_x, B_m, and B_{el}.

4.2.4 The RhD blood group system

Thirty-nine years elapsed between the discovery of the ABO system by Landsteiner in 1900 and that of the Rh system. The 'Rh' component is so named because rabbits were immunized with red blood cells from a Rhesus monkey. The resultant antiserum detected an antigen on the surface of approximately 85% of those of white European descent, that antigen being named 'D', related antigens subsequently being named 'C' and 'E', and in some cases 'c' and 'e'. The current view is of two closely related genes, *RHD* and *RHCE*, located at chromosome 1p36-11, that each code for a complex molecule that loops six times on the intra- and extracellular faces of the membrane (Figure 4.9). This structure is clearly far more sophisticated than the relatively simple molecules in Figures 4.6 and 4.7 that have only one transmembrane segment. Mutations in *RHD* and *RHCE* give rise to variations in the extracellular domains of the molecule that may be regarded by some as sufficiently non-self as to induce the formation of the alloantibodies that frustrate our colleagues in blood transfusion. Those lacking the RhD molecules are D- (previously, Rhesus negative), generally due to a homozygous deletion of *RHD*. The RhCE molecule is almost always present. These structures are also known as CD240. (See also Box 4.2.)

Cross reference

Full details of blood groups are to be found in the *Transfusion and Transplantation Science* volume of this series of books.

4.2.5 Minor blood group systems

The ABO and Rh systems are important because they may evoke powerful (and possibly fatal) consequences if there is a mismatch between the patient and the blood they are to receive as a blood transfusion. Differences in the external structure of many other red cell membrane structures give rise to minor blood groups, so named because (in general) they are of less clinical importance than the ABO and Rh systems. However, although often described as minor,

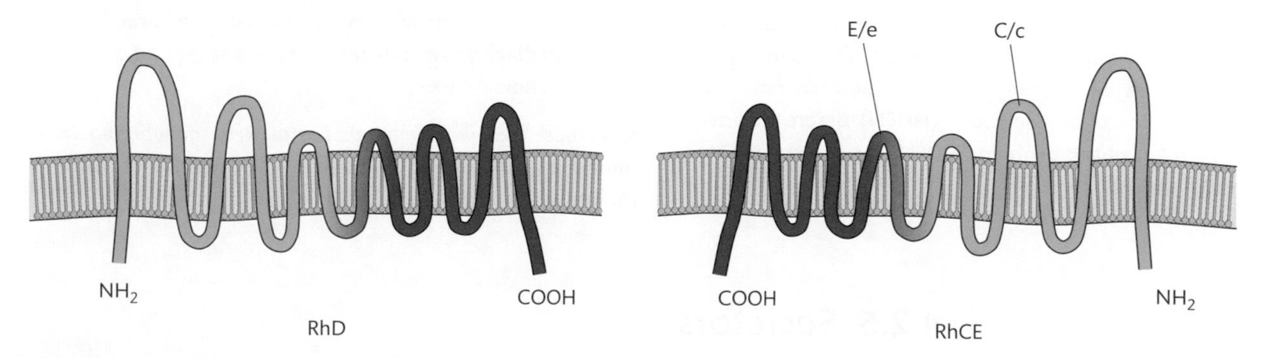

FIGURE 4.9

A representation of the RhD and RhCE molecules. Each consists of a single polypeptide that snakes in and out of the cell membrane. The location of the sites of C/c and E/e specificities are shown on the second and fourth extracellular loops. Cartoon amended from Daniels (2013).

the antigens of the Kell (K, CD238, probably a zinc endopeptidase) system are the most potent immunogens after ABO and Rh, anti-K antibodies having the potential to cause severe haemolytic disease of the foetus and newborn (HDFN) (see Box 4.2). Other minor blood groups include those of MNS (CD235, which functions as a receptor for complement, bacteria and viruses, and is carried by glycophorins A and B), P (a receptor for the bacteria *E. coli*, with three isoforms: P, P_1, and P^k), Lutheran (CD239, probably involved in adhesion and intracellular signalling, the shorthand being Lu), Lewis (Le, possibly a receptor for E-selectin, CD62E), Duffy (Fy, CD234, a receptor for cytokines, with two isotypes giving four possible phenotypes; Fy(a+b-), Fy(a+b+), Fy(a-b+), and Fy(a-b-)), Kidd (Jk, a urea transporter on red cells and renal endothelial cells, with isoforms Jk^a, Jk^b, and Jk3), and Diego (DI, involved in anion transport and providing structural support, being located on Band 3). Alternative versions of these structures (such as Lutheran antigens Lu^a and Lu^b) can give rise to antibodies that can precipitate a transfusion reaction and cross the placenta to cause HDFN. (See also Box 4.3.) The first blood group mapped to the human X chromosome, the XG group system, includes two sialoglycoprotein antigens: Xg^a and CD99. The former are found only on red cells, the latter on all cell types, but their function(s) are unknown, and there does not seem to be a link with any particular clinical syndrome.

BOX 4.2 RhD and haemolytic disease of the foetus and newborn (HDFN)

Approximately 15% of white Europeans lack RhD on the surface of their red cells, and so are described as D negative (D−). Those with RhD are D positive (D+). So, if some D+ cells find their way into someone who is D−, they will be regarded as foreign and will be attacked by the immune system of the D− person, and so destroyed. This is why RhD compatibility is important in blood transfusion. However, a D− woman carrying a D+ foetus (the D antigen being provided by the father of her child) may be 'immunized' by some foetal D+ red cells passing into her circulation at the delivery of her infant. This is not so much of a problem for her as she will recognize and destroy these foreign (D+) cells. The problem is that her immune system will recall this invasion, and should she again carry a D+ foetus, her developing child will be regarded as foreign, and so the subject of attack. This leads to the pathology called 'haemolytic disease of the foetus and newborn' (HDFN). Before the advent of treatment of this problem, HDFN was present in a staggering 1 in 170 births.

Therefore, the ability to determine whether a D− woman is carrying a D+ foetus is important, as any leakage of red cells into her from her child needs to be minimized if HDFN is to be prevented in a subsequent pregnancy. This was previously done by the 'Kleihauer' test, which sought D+ red cells by the presence of foetal haemoglobin. Although this test was superseded by FACS analysis of the mother's blood, both tests were prone to false positives and false negatives. Techniques in molecular genetics now offer the ability to determine, with 100% reliability, the D status of a foetus early in pregnancy. The fine detail is that a polymerase chain reaction can analyse circulating cell-free foetal DNA for the presence of the RhD gene. This is a classic example of how advances in laboratory science lead directly to improved diagnosis and (if necessary) management, and so to better clinical care.

Reference: Akolekar R et al. Foetal RHD genotyping in maternal plasma at 11–13 weeks of gestation. Fetal Diagnostic Therapy 2011: 29: 301–6.

4.2.6 Secretors

The *Se* gene (*FUT2*, mapped to chromosome 19p13.3) controls the ability to secrete Group A and B molecules into the plasma and all body fluids, including faeces, saliva, and semen, leading to a value in forensic science. Obeying Mendelian genetics, the *Se* gene has a silent allele, *se*, giving *Se/Se* homozygotes, *Se/se* heterozygotes (both of which secrete), and the *se/se* silent homozygote (i.e. the absence of an active *FUT2*), which defines non-secretors. Approximately 75% of a Caucasian population are secretors. This is also relevant to the biology of the Lewis antigen system, where *FUT2* and *FUT3* lead to the presence of alternative forms Le^a and Le^b.

The biological purpose of secretion is unclear, but may relate to protection or susceptibility to infections such as by noroviruses and rotaviruses. It may also influence the binding of certain bacteria to mucin glycoprotein on the cell membrane, as salivary carbohydrate structures found in mucins can aggregate some oral bacteria, putting secretors at a theoretical advantage.

SELF-CHECK 4.3

What are the major components of the red blood cell membrane and what are their principal functions?

Key Points

The membrane of the red blood cell is unique. It needs to be relatively fragile to allow the free passage of oxygen and to enable the cell to deform and so flow along the finest capillaries. However, this high degree of specialization brings susceptibility to damaging factors that would not normally be a problem to the membrane of a nucleated cell.

BOX 4.3 Blood groups and disease

Some of the variation in the distribution of certain blood groups can be accounted for by links to disease. Decades ago, astute scientists noted that the request for blood for transfusion in certain types of surgery was linked to ABO blood group. Formal studies subsequently showed that Group O subjects have a 14% reduced risk of squamous cell carcinoma (a malignancy of the skin) and a 4% reduced risk of basal cell carcinoma compared with the other three groups combined. Group O also protects from pancreatic cancer. Group B is linked with ovarian cancer, and Group A carries a greater risk of gastric cancer, whereas Group O is protective. This may be linked to certain strains of *H. pylori*. Following an infection with the bacteria *Vibrio cholerae*, blood group O is linked to an increased risk of severe infection compared to those of groups A, B, and AB. It has been suggested that Darwinian selective pressure explains the high frequency of group B and the low frequency of group O in areas where cholera is endemic. *E. coli* is a common bacterium and in most of us is not pathogenic, but in severe infection it can be fatal. In one such outbreak, 87.5% of those who died were blood group O: if there was no effect of blood group this proportion would be 45%.

The Duffy molecule is the route by which parasites that cause malaria (*Plasmodium vivax* and *P. knowlesi*) enter the red cell. Therefore, people who lack this molecule (that is, are Fy (a-b-)) are at a selective advantage, and indeed it is present at a reduced frequency in populations at risk of this infection. Similarly, MN blood group molecules are a receptor for *Plasmodium falciparum*, so that those who are M-N- are at an advantage. This topic is revisited in Chapter 7.

4.3 The cytoplasm of the red blood cell

The function of the red blood cell is to transport oxygen. This is achieved by the highly specialized protein haemoglobin, but there are several other crucial molecules in this cell. Whilst the transport of oxygen is essential to our survival, it is not without problems: high concentrations of oxygen within the cell can be toxic, as it can form highly reactive chemical species (such as free radicals), which can damage proteins, fats, and carbohydrates. Thus, the red cell must defend itself with antioxidants, such as peroxiredoxin 2, bound to the membrane component Band 3, and the complex amino acid **glutathione** in the cytoplasm.

glutathione
An amino acid that provides protection against toxic reactive oxygen species.

As the red cell has no mitochondria, it cannot use the oxygen it is carrying to generate energy. However, it can generate a limited amount of energy (as ATP) from the anaerobic (without oxygen) conversion of glucose to lactic acid by components of GEC (Table 4.4) that are bound to the cytoplasmic domains of Band 3.

Cross reference
The minor subtypes of haemoglobin are also described in Chapter 2.

The structure of the haemoglobin molecule has two parts—a protein part (**globin**) and a complex non-protein part (**haem**). The haem group contains iron (absorbed from the diet, and to which the oxygen is attached) at the centre of a complex molecule—the porphyrin ring (synthesized in the mitochondrion). The globin part consists of four individual subtypes of globin that come together to form a functional tetramer.

4.3.1 Haemoglobin

Hb is an iron-containing protein synthesized by blast cell precursors in the bone marrow (such as the erythroblasts that ultimately give rise to mature red blood cells). Each red cell contains approximately 640 million haemoglobin molecules. It is designed to absorb oxygen from areas of high oxygen content (that is, at the lungs) where it becomes **oxyhaemoglobin**, and then release it in areas where oxygen levels are low, which is likely to be the case in the body tissues. Emphasizing the importance of oxygen transport, haemoglobin molecules that are not carrying oxygen can often be described as **deoxyhaemoglobin**. Haemoglobin in composed of a protein component (globin) and a non-protein component (haem). Formation of the latter is complex, which we will begin to address by looking at how iron gets to the erythroid precursors in the bone marrow.

Iron

Like folate, vitamins B_6, and B_{12}, iron must also be provided in the diet, the reference nutrient intakes varying with age and sex (Table 4.6).

Around 10% of dietary iron is absorbed in the duodenum and jejunum, which is generally not enough to satisfy erythropoiesis—the balance being made up by recycling. Foods with the highest proportion of iron include liver, meat, eggs, and dried fruit. Iron already in the form of haem—as may be present in liver and muscle haemoglobin and myoglobin, or in the ferrous (Fe^{2+}) state—is absorbed more rapidly than the other form of free inorganic iron found in vegetable matter, which is likely to be in the ferric (Fe^{3+}) state.

divalent metal transporter-1, hephaestin, ferroportin, and hepcidin
Proteins involved in the absorption of iron and its delivery to the circulation.

Free ferrous iron is absorbed by intestinal epithelial cells (enterocytes) via a specific cell membrane molecule (**divalent metal transporter-1**). The intestinal conversion of ferric to ferrous iron is facilitated by ferroreductase enzymes at the apical surface of the enterocytes. Haem appears to enter the cell via a different mechanism, possibly by haem-responsive gene-1 protein, and iron present in haem is released by the intracellular enzyme haemoxgenase-1. This enzyme is also important in iron recycling, as will be explained in section 4.4.2. Once inside the enterocyte, iron

TABLE 4.6 Reference nutrient intake of iron.

Age	Males	Females
0–6 months	1.7–4.3	1.7–4.3
7–12 months	7.8	7.8
1–10 years	6.1–8.7	6.1–8.7
11–18 years	11.3	14.8*
19–50 years	8.7	14.8*
50+ years	8.7	8.7

Units: mg/day. *No increase in pregnancy or lactation.
Source: British Nutrition Foundation www.nutrition.org.

may be transported immediately across the cytoplasm by a metalloprotein chaperone or stored in ferritin (details to follow). It is delivered from the cytoplasm to the circulation by the cell membrane molecule **ferroportin**. This molecule exports only ferrous iron, but it is carried in the blood by transferrin in the ferric form, the conversion being catalysed by hephaestin (a copper-requiring 134 kDa ferroxidase coded for by *HEPH* at Xq12), anchored in the membrane adjacent to ferroportin. Deletion of this gene in mice leads to intestinal iron accumulation. Reflecting its importance in iron metabolism and homeostasis, ferroportin is also abundant on macrophages and hepatocytes, with lower expression on certain other cells, such as cardiomyocytes. Curiously, ferroportin is also a component of erythroid cells: why these cells should seek to export their iron is unclear but may be a route to maintaining non-toxic levels of this cation.

The movement of iron into the plasma by ferroportin is regulated by the liver-derived 25-amino acid peptide **hepcidin**, coded for by *HAMP* on chromosome 19q13.1, and although it is also expressed by pancreatic beta cells, cardiomyocytes, placental syncytiotrophoblasts, various kidney cell types, adipocytes, macrophages, and glial cells, their contribution to plasma is insignificant. The regulation of this gene is of great importance in both the physiology and the pathology of iron metabolism (Box 4.4 and Table 4.7).

Hepcidin controls the expression of ferroportin, which in turn controls the export of iron from intestinal enterocytes, macrophages, Kupffer cells, hepatocytes, and placental cells. At the molecular level, when hepcidin binds to an extracellular aspect of ferroportin, the latter is internalized and degraded in lysosomes. A consequence of this is that iron is not moved out of the cell and into the bloodstream. So, a feedback mechanism exists, which involves the enhanced release of hepcidin by a liver whose iron stores are full. Hepcidin also regulates the release of iron from macrophages

TABLE 4.7 Regulators of hepcidin.

Factors resulting in high hepcidin (so suppressing iron uptake)	Factors leading to low hepcidin (so promoting iron uptake)
Chronic kidney disease, dialysis, genetic factors, infectious/inflammatory disease, intravenous/oral iron administration, red blood cell transfusions, reduced glomerulofiltration rate, replete iron stores	Alcohol abuse, anaemia, chronic hepatitis C virus infection, chronic liver disease, erythropoiesis stimulating agents, Epo, hypoxia, ineffective/expanded erythropoiesis, genetic factors, oestrogens, testosterone

BOX 4.4 Regulation of iron at the gene level

Historically named, reflecting its first isolation (coding for hepatic antimicrobial peptide, as hepcidin was known), hepatic HAMP mRNA is regulated by (at least) three factors. First, IL-6 (and possibly IL-1β) induces HAMP via the STAT3 pathway, leading to increased circulating hepcidin and so (ultimately) reduced passage of iron into the blood. Second, levels of plasma and tissues stores of iron are regulated via a negative feedback pathway involving transferrin receptors (themselves regulated post-transcriptionally by iron-responsive element-binding proteins whose targets are mRNAs). The third route in our species is less well developed, but involves erythropoiesis: when this process is active, hepcidin levels fall as more iron is needed in the generation of red cells. It has been speculated that hypoxia, Epo, and growth differentiation factor 15 (released by maturing erythroblasts) may all, directly or indirectly, suppress HAMP expression and so promote high levels of iron in the plasma (hyperferraemia). HAMP itself is under the control of the intracellular second messengers of the SMAD pathway, that are themselves regulated at the cell surface by haemojuvelin, bone morphogenic proteins BMP2 and BMP6, and the human haemochromatosis protein (HFE [high Fe], coded for by *HFE* on 6p22.2). Loss of control of this pathway leads to haemochromatosis, as explained in Chapter 6. In the macrophage, transcription of ferroportin is suppressed following binding of pathogen-like ligands to toll-like receptors, and ferroportin mRNA is regulated in the cytoplasm by iron regulatory proteins (principally iron-regulatory protein-1) binding to iron-regulatory elements, and miR-485-3p and miR-20a miRNAs.

by its inactivation of ferroportin. This regulatory system explains why the rate of absorption of iron is related to the demands of erythropoiesis (increased absorption when erythropoiesis is active) and of the amount of stored iron (increased absorption when stores are low). A raised hepcidin concentration is key to the development of anaemia of chronic disease (ACD: section 5.5), much of which is inflammatory, and its synthesis is also induced by bacterial infections as a protective mechanism. The most plausible reason for this is that a presumed bacterial infection (which would be associated with increased inflammatory cytokines) can be countered by denying iron to potential pathogens that rely on this micronutrient. Indeed, mice engineered to lack hepcidin (and so have high levels of iron) show markedly less survival when infected with siderophilic bacterial pathogens, supporting the view that hepcidin is part of the innate immune system. Table 4.7 shows regulators of hepcidin. Notably, in the mouse, stimulation of toll-like receptors 2 and 6 (part of the innate immune system) result in decreased ferroportin mRNA, and so a reduction of iron passage into the blood, a pathway that is independent of hepcidin.

Once exported from enterocytes by ferroportin in its ferrous form and converted into its ferric form by hephaestin and caeruloplasmin, the iron is ready to be transported to the bone marrow by a number of possible carriers. The most important of the carrier proteins is transferrin (a 76–80 kDa protein coded for by *TF* at 3q2.1), which can carry two atoms of ferric iron (hence diferric transferrin). When free of iron it is called **apotransferrin**, and one gram of apotransferrin can carry 1.25 mg of iron. The amount of iron being carried by transferrin can be a useful indication of the general iron status of the body, and levels can also be monitored as part of the clinical regulation of iron uptake. This is because the saturation (or binding status) of transferrin with high levels of iron will stimulate hepcidin release, which in turn will reduce the levels of iron passing from the intestines to the blood. Other (minor) carriers of iron include proteins that are synthesized by the liver (such as albumin and lactoferrin), and by anions (such as citrate or acetate), collectively described as non-transferrin-bound iron, which can be imported into hepatocytes by the cation importer ZIP14 and stored there as ferritin. This carrier system is important should transferrin be saturated, as may occur in iron overload (discussed in section 5.4). Figure 4.10 summarizes certain aspects of the absorption, regulation, and carriage of iron, with more details in Figure 4.11, and additional text in section 5.4.1 and Figure 5.3.

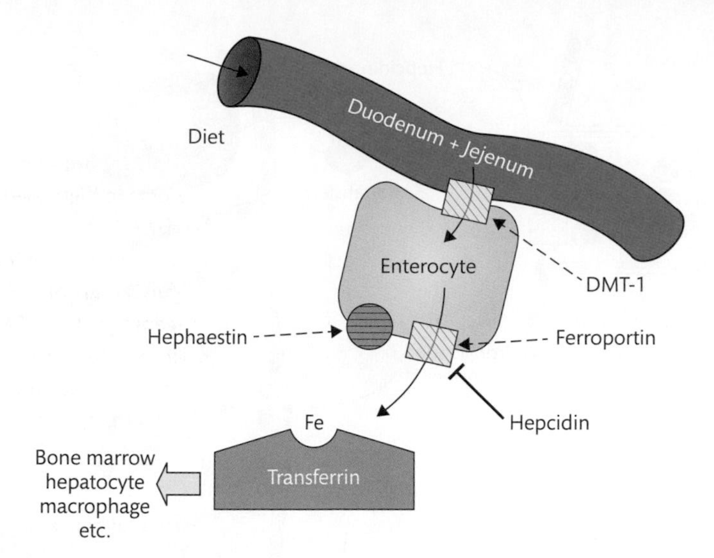

Figure 4.10
Absorption, regulation, and carriage of iron. Intestinal iron is largely absorbed by divalent metal transporter-1 (DMT-1). Some may be stored in the enterocyte, some transported directly to ferroportin, its export vehicle. However, ferroportin itself is regulated by hepcidin. Hephaestin converts ferrous to ferric iron, which is carried to its destinations by transferrin.

The iron-transferrin complex enters the developing red blood cell (such as the erythroblast) by binding to one of the 50,000 transferrin receptors on the surface of the cell, of which there are two forms. **Transferrin receptor 1** (TfR1, CD71, coded for by *TFRC* at 3q29) can be shed from the surface of the erythroblast into the circulation, in which state it can be described as the soluble transferrin receptor (sTfR). Since levels of sTfR reflect levels of membrane-bound transferrin receptor, they may be used as a surrogate laboratory marker for increased activity of the erythroblasts and possibly of erythropoiesis itself. This may be useful in certain types of anaemia as sTfR is not (unlike transferrin and ferritin) an acute phase reactant, and so should not be increased in inflammation. However, the transferrin receptor is also found on macrophages, rapidly dividing cells, and on activated lymphocytes, and so levels may be increased during inflammation.

Once inside the cell, the complex moves into an endosome, where a change in the pH facilitates the release of iron from the transferrin molecule, whilst ferrireductase enzyme reduces the iron to its ferrous form. The free iron is then conducted out of the endosome by DMT-1 (as it was conducted into the enterocyte), and to its destinations by chaperone proteins. The resulting apotransferrin returns to the plasma, from where it can travel back to the intestines to collect more iron. This is not the only pathway that erythroblasts can obtain iron—ferritin released from nurse macrophages can enter the cell by pinocytosis.

The second transferrin receptor, TfR2 (encoded by *TFR2* at 7q22.1), with a 25-fold lower affinity for transferrin, has a minor role, but contributes to hepcidin regulation in the liver, as patients deficient in TfR2 are unable to promote appropriate iron-mediated induction of hepcidin. Accordingly, it may be seen as a mechanism for iron-sensing by the liver. TfR2 is also a component of the EpoR in erythroid precursors, where it may be involved in terminal erythroblast differentiation. In an iron-replete situation, erythroid TfR2 restricts Epo sensitivity to restrict erythropoiesis, but in iron deficiency, downregulation of hepatic TfR2 inhibits hepcidin signalling, so encouraging iron supplies, and enhances Epo sensitivity to stimulate erythropoiesis. The glycolytic enzyme glyceraldehyde-3-phosphate dehydrogenase may also act as a transferrin receptor.

Iron not channelled directly to the bone marrow may instead be stored in organs such as the liver, pancreas, spleen, and bone marrow, in association with proteins **ferritin** and

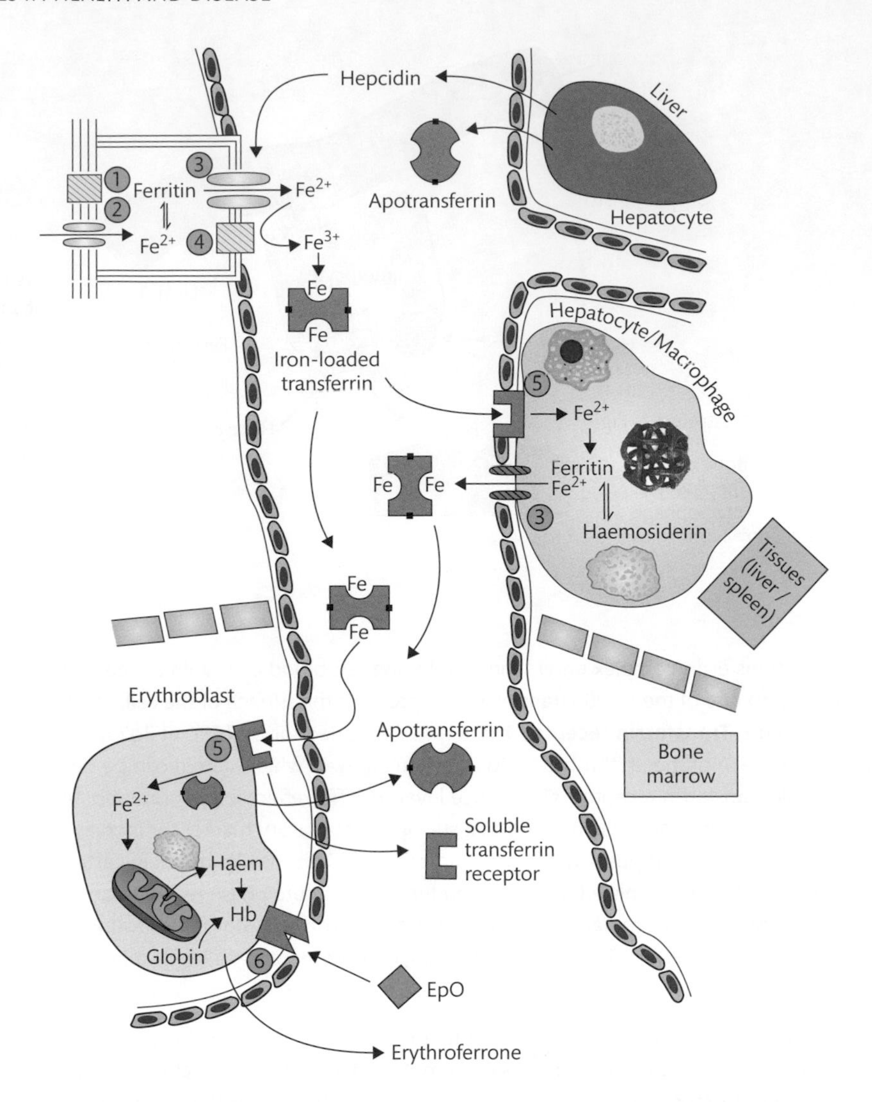

FIGURE 4.11

Absorption, transport, storage, and assimilation of iron into haem. Iron must be provided in the diet and is absorbed across the lumen of the intestines into enterocytes (top left). Some iron may be stored in ferritin, but most passes, via ferroportin, into the plasma. This process is regulated by hepcidin from the hepatocyte (top right). Free ferric iron is collected by apotransferrin (also produced by hepatocytes), and iron-loaded transferrin is carried in the blood to cells such as splenic macrophages and hepatocytes (centre right), where ferrous iron may be stored in ferritin. Alternatively, transferrin may pass to the bone marrow (bottom right), where it is taken up by erythroblasts stimulated by Epo, and which produce erythroferrone. The iron-loaded transferrin passes into cells via the transferrin receptor, which may be shed into the plasma as the sTfR. 1 = Fe-reductase; 2 = divalent metal transporter (DMT)–1; 3 = ferroportin; 4 = hephaestin; 5 = the transferrin receptor; 6 = EpoR; Fe^{2+} = ferrous iron; Fe^{3+} = ferric iron.

haemosiderin. Apoferritin, the form of ferritin that is free of iron, is a very large, spherical molecule composed of heavy and light subunits (coded for at 11q13 and 19q13.3–13.4), with a relative molecular mass of 450–465 kDa. However, it can store up to 5000 atoms of ferric iron in the centre of the sphere, so that the mass of iron-replete ferritin may be increased by 35%. Almost all the iron imported into erythroblasts is directed to the mitochondria, where haem is synthesized.

Haemosiderin is formed by the aggregation of partially digested ferritin and is found in macrophages. Release of iron from haemosiderin is slow, and therefore haemosiderin-complexed iron represents a long-term store that cannot readily be accessed. In this form, it can be detected under light microscopy using a special stain, Perls' Prussian blue. This stain is particularly valuable in assessing iron stores in the bone marrow and liver. Most body iron, generally 65% (the equivalent of maybe 3–4 g) is sequestered within haemoglobin, but a considerable proportion (approximately 30%) is stored in ferritin and haemosiderin. The remainder is in myoglobin (a form of globin present in muscles), in enzymes (such as cytochromes and catalase), and in complexes with transferrin. The passage of iron from the intestines, through the blood, to the cells is illustrated in Figure 4.11.

SELF-CHECK 4.4

Describe the process by which iron gets into the blood.

SELF-CHECK 4.5

Once absorbed, how does the iron get to the bone marrow and then into the developing red blood cell?

Key Points

The synthesis of haem relies on complex metabolic pathways. Key enzymes rely on vitamins as cofactors, so that deficiency of the micronutrients, as well as iron itself, can lead to the failure to produce a mature red blood cell, and so to anaemia.

Haem

Haem is a complex molecule synthesized in the cytoplasm and mitochondria of developing red blood cell precursors (principally the erythroblasts and nucleated red blood cells). Several of the enzymatic reactions in erythropoiesis require micronutrients that must be provided in the diet; in addition to iron, these include **vitamin B_{12}**, **vitamin B_6**, and **folate**. Enzymes involved in DNA synthesis require vitamin B_{12} and folate, whilst vitamins B_6 and B_{12} are required for haem biosynthesis. A shortage of these micronutrients (principally, iron) is a common problem encountered in haematology and will be discussed in the chapters that follow. Isolated vitamin B_6 deficiency is rare in the absence of certain drugs, which include the anti-tuberculosis agent isoniazid, and pencillamine. However, deficiency of vitamin B_{12} and/or folate is relatively common and is discussed in Chapter 5. Reference nutrient intake of these vitamins and folate is shown in Table 4.8.

vitamin B_{12}, vitamin B_6, and folate
Micronutrients essential for erythropoiesis: their absence results in anaemia.

Cross reference
Further details of these vitamins are described in Chapter 5, section 5.4.

The initial major step in the formation of haem is the synthesis of a protoporphyrin ring, an overview of which is given in Figure 4.12. The early steps in the mitochondrion involve the carboxylation of propionyl-coenzyme A (propionyl-CoA, a three-carbon molecule and participant in the Krebs cycle) and so the formation of the four-carbon methylmalonyl-CoA, often described as methylmalonic acid. The enzyme that catalyses this step, methylmalonyl-CoA mutase, contains a derivative of vitamin B_{12} as its coenzyme. The next step is an isomerization which results in the formation of succinyl-CoA.

TABLE 4.8 **Reference nutrient intake of vitamins and folate.**

Age	Vitamin B_6 (males)(mg/d)	Vitamin B_6 (females) (mg/d)	Vitamin B_{12} (μg/d)	Folate (μg/d)
0–6 months	0.2	0.2	0.3	50
7–12 months	0.3–0.4	0.3–0.4	0.4	50
1–10 years	0.7–1.0	0.7–1.0	0.5–1.0	70–150
11–18 years	1.2–1.5	1.0–1.2	1.2–1.5	200**
≥19 years	1.4	1.2*	1.5	200**

*1.7 if lactating; **260 if pregnant.

Source: British Nutrition Foundation www.nutrition.org

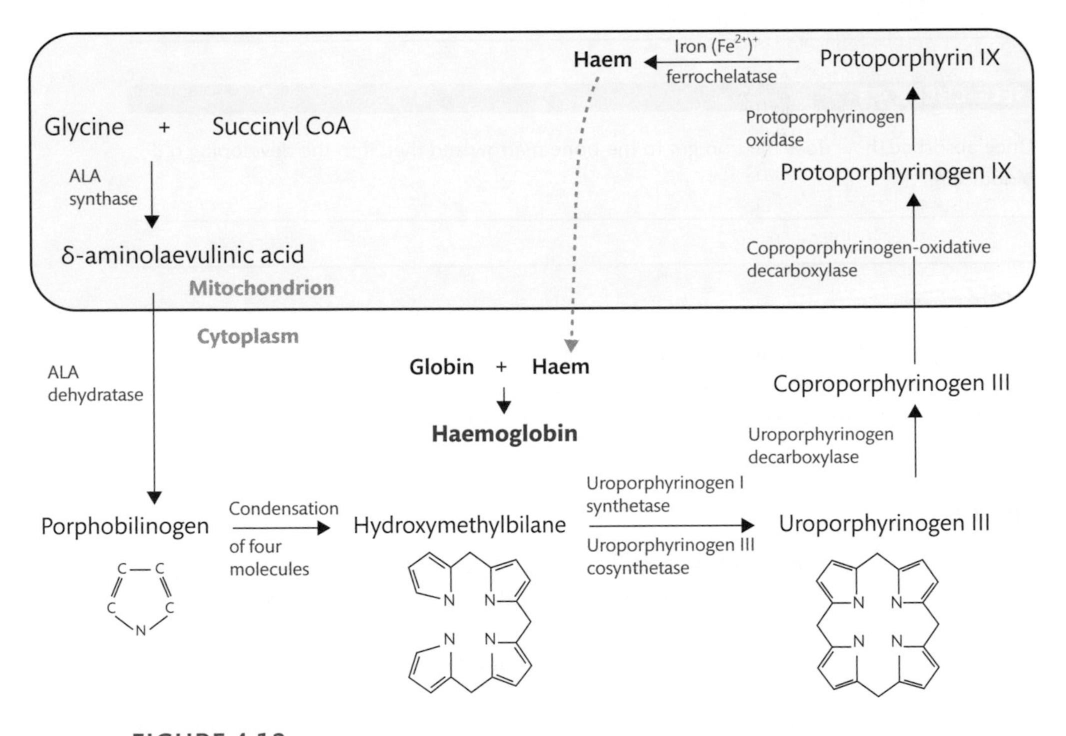

FIGURE 4.12
Biochemical steps in the formation of haem and haemoglobin. This process occurs in cells such as the erythroblast and nucleated red blood cell. Many of these steps occur in mitochondria, although the final assembly of haemoglobin takes place in the cytoplasm.

Succinyl-CoA (an intermediate in the Krebs cycle, and therefore abundant in the mitochondrion) and glycine (transported to the mitochondrial matrix) are then fused into the five-carbon molecule aminolaevulinic acid (ALA) by the enzyme ALA synthase, with a molecule of carbon dioxide as a by-product. Levels of this important and rate-limiting enzyme are regulated at the level of its mRNA by the repressive binding of an iron-regulatory protein. ALA synthase requires a form of vitamin B_6 as a cofactor.

A slightly different form of vitamin B_{12} (cyanocobalamin) is required for another essential metabolic step—that of the conversion of homocysteine and methyl-tetrahydrofolate (itself derived from dietary folate) into tetrahydrofolate, which is required at high demand for DNA synthesis in erythroid precursors. We revisit this metabolism in more detail in Chapter 5, focusing on errors that lead to anaemia.

Later steps involve the generation of a series of complex metabolic intermediates in the cytoplasm. Two molecules of ALA are dehydrated and condensed by the enzyme porphobilinogen synthase to a ring structure of carbon and nitrogen atoms to form the pyrrole structure called porphobilinogen, itself the fundamental building block of haem.

Next, four porphobilinogen molecules are amalgamated to form hydroxymethylbilane, in which the ring structure is almost complete. Closure of this ring is affected with the formation of uroporphyrinogen III. Subsequent steps are the conversion of uroporphyrinogen III to coproporphyrinogen III, which then moves into the mitochondria to become protoporphyrinogen IX, and ultimately protoporphyrin IX.

The final step is the insertion of an atom of ferrous iron into the protoporphyrin ring to create haem. The iron is delivered to the mitochondrion from endosomes by protein chaperones, although it is possible the iron-laded endosomes fuse directly with external mitochondrial membrane in a kiss-and-run transfer. The iron is then transported to the mitochondrial matrix by mitoferrin-1 (a 39kDa protein encoded for by *SLC25A28* at 8p21.2), where it may be stored in a mitochondrial form of ferritin (mitoferritin), formed into iron-sulphur clusters, or inserted into the protoporphyrin ring under the control of an 85kDa enzyme, ferrochelatase, coded for by *FECH* at 18q21.31. The specificity of this enzyme for iron is not 100%, as it may also place an atom of zinc into the ring (zinc, like this form of iron, is a divalent cation—that is, Zn^{2+}) Thus, zinc protoporphyrin is an indirect marker of the levels of iron within the erythroblast: if levels of iron are low, zinc may be inserted instead. The next step is the folding of the globin molecules around the haem to give haemoglobin.

Although the synthesis of this protoporphyrin ring requires two important cofactors (vitamins B_6 and B_{12}) other enzymes require vitamins C, D, and E. Lack of these micronutrients may result in defects in red blood cell maturation, as well as in other areas of metabolism. Furthermore, defects in the various enzymes responsible for the many complicated metabolic pathways in the synthesis of haem will also lead to red cell diseases, principally sideroblastic anaemia, megaloblastic anaemia, and porphyria, as is discussed in Chapter 5. However, haem has important roles in other aspects of metabolism: it is present in myoglobin, in cytochrome, catalase, and peroxidase enzymes, and has a place in tryptophan metabolism. Figure 4.13 shows the molecular structure of haem.

Cross reference

The importance of folate, vitamin B_6, and vitamin B_{12} as micronutrients, as demonstrated by their deficiency, is described in Chapter 5.

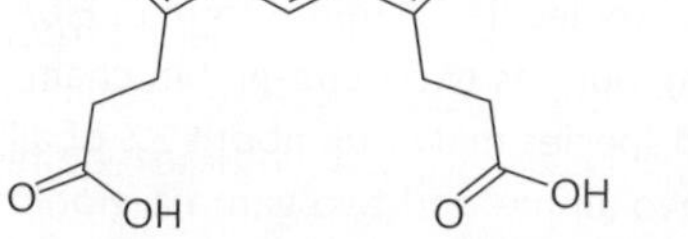

FIGURE 4.13

Molecular structure of haem. The atom of iron (Fe^{2+}) is held in place by four atoms of nitrogen. These atoms are in turn part of the complex porphyrin ring, which itself interacts with globin via two acid groups (-OOH) and two sulphur-rich groups (-S-Cys).

SELF-CHECK 4.6

Why are vitamins needed during erythropoiesis? Where in the red cell are they required?

Globin

Globin is a large globular protein with a relative molecular mass of 16–17 kDa that exists in a number of slightly different forms. There are several subtypes of globin, each coded for by different genes: three functioning genes (two for alpha [*HBA1* and *HBA2*]–, and a single zeta-globin gene [*HBZ*]) and four non-functioning pseudogenes are on chromosomes 16p13.3; whilst five functioning genes (epsilon [*HBE*], beta [*HBB*], delta [*HBD*]–, and gammas 1 and 2 [*HBG1* and *HBG2*])) and one pseudogene are on chromosome 11p15.5. The functioning genes each produce specific messenger RNAs that in turn generate different types of globin protein molecules. For example, the alpha-globin chain contains 141 amino acids, whereas the beta-globin chain has 146 residues. The gene for gamma-globin has two subspecies, each coding for a globin molecule with slightly different amino acid compositions. The gamma-A gene produces a globin molecule with alanine at a certain position, whilst the globin molecule coded for by the gamma-G variant instead has glycine at that position. Synthesis of globin molecules in the cytoplasm for red blood cell precursors, such as the erythroblast and the nucleated red blood cell, occurs in parallel with the synthesis of haem: the two are then combined to form haemoglobin, commonly abbreviated to Hb.

The mature Hb molecule is made up of four individual globin molecules, and so is a **tetramer**–each single unit, or molecule of globin, is a monomer, and possesses one haem ring. Thus the entire molecule has a relative molecular mass of approximately 64–68 kDa. The different globin monomer molecules (alpha-, beta-, delta-, gamma-, etc.) all confer different oxygen-carrying properties on the mature Hb tetramer at different times in the development of the individual.

The precise make-up of the globin tetramer evolves from the embryo and foetus, via the neonate, to the adult, as illustrated in Figure 4.14. Embryonic Hb exists for the first 12 weeks, and consists mostly of alpha-, zeta-, epsilon-, and gamma-globin chains—making up haemoglobins Portland, Gower I, and Gower II (these three Hbs are made up of a 'mix' of the different globins, e.g. Portland Hb has two zeta- and two gamma-globin chains). Zeta chains are considered the embryonic equivalent of alpha chains. By contrast, foetal Hb comprises alpha- and gamma-globin chains.

These different types of Hb are needed because the embryo and foetus are faced with the challenge of obtaining oxygen not from the lungs but from the placenta, and thus ultimately from the maternal circulation. Hence embryonic, and subsequently foetal, Hb must effectively take oxygen from the adult Hb of the mother. It does this by having a considerably greater affinity for oxygen than does the maternal Hb. The foetus and neonate face a similar challenge as they develop, which they partly solve by introducing delta-globin in the Hb molecule in the place of gamma-globin. The major final switch from foetal to adult Hb in the neonate occurs gradually some three to six months after birth as beta-globin slowly replaces delta- and gamma-globin when associating with alpha-globin chains.

The dominant Hb species in the adult, making up perhaps 96–98% of this protein, is HbA. This consists of two alpha-globin molecules and two beta-globin molecules. A minor form of HbA is HbA_2, which has two alpha-globin chains (as does HbA), but has two delta-globin chains in place of the beta-globin chains of the HbA. This second species makes up about 2% of all Hb in the healthy adult. The remainder, HbF, comprises two alpha- and two gamma-globin chains, and makes up the remaining small fraction (less than 1%). Thus the healthy adult is still

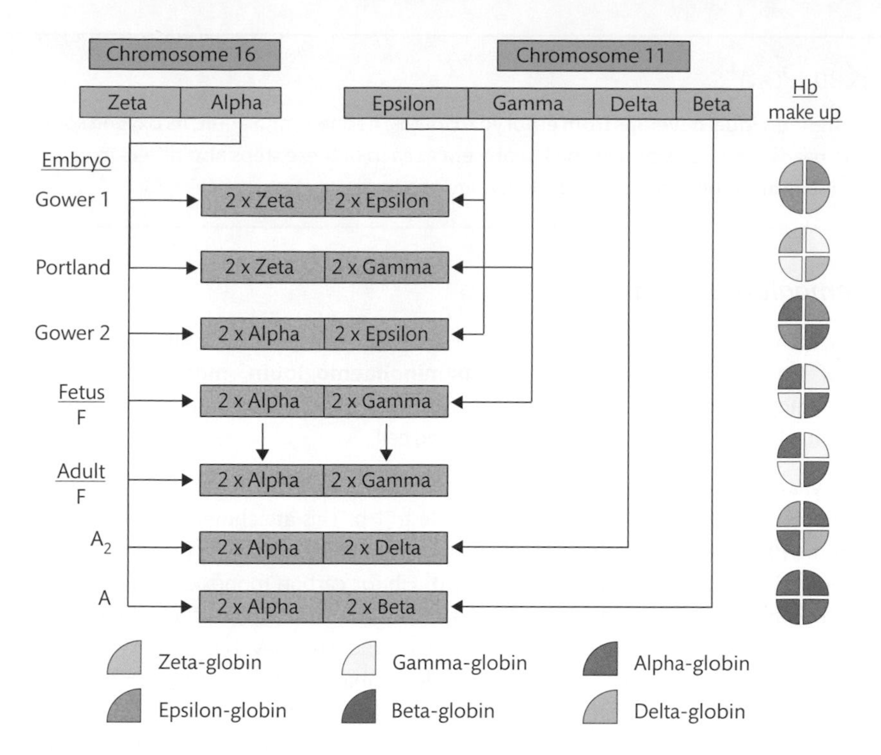

FIGURE 4.14

Ontogeny of Hb. Changes in species of Hb in different stages of development. In the embryo, Gower 1 Hb is composed of two zeta-globin molecules and two epsilon-globin molecules. In Portland Hb, the epsilon-globin is replaced by gamma-globin, whilst Gower 2 Hb is a tetramer of two alpha-globins and two epsilon-globin molecules. In the foetus, foetal Hb consists of two alpha-globins and two gamma-globins. However, a trace of foetal Hb is present in the adult, where two other species are HbA_2 (a tetramer of two alpha-globins and two delta-globins), although the dominant species is HbA, consisting of two alpha-globin molecules and two beta-globin molecules.

expressing genes that were most active in foetal and neonatal life, the precise reason(s) for which is unclear, but which clearly provides some survival advantage. Notably, therefore, all adult Hb consists of two alpha-globin molecules paired up with a second type of molecule, which can be either beta-, delta-, or gamma-globin.

The transition of globin synthesis from embryo to neonate proceeds in an orderly and highly regulated manner. For example, gamma-globin is regulated by transcription factors such as GATA1, TAL1, and KLF1, whilst the developmental switch from gamma- to beta-globin in foetal to neonatal life is regulated by BCL11A, which represses gamma-globin. In human CD34+ stem cells undergoing differentiation, *FOXO3* silencing reduces gamma-globin mRNA, whereas overexpression has the reverse effect. Another potent silencer of HbF is leukaemia/lymphoma-related factor, the knock-out of which in mice leads to an increase in this molecule. In the erythroid precursors of the yolk sac and liver, *HBZ* produces zeta-globin, whilst *HBE* produces epsilon-globin. However, this is relatively short-lived as a 'primitive' switch occurs, where *HBA1* and *HBA2* direct the formation of the two isoforms of alpha-globin. On chromosome 11, *HBG1* and *HBG2* direct the formation of the two isoforms of gamma-globin. Much later, at term and in the initial months of independent life, the 'definitive' switch occurs in the beta locus of erythroblasts in the bone marrow, with the increasing activation of *HBB*, producing beta-globin, and a minor degree of activation of *HBD*, producing delta-globin.

Cross-reference

Abnormalities in the genes for different types of Hb are the basis of an important group of diseases—the haemoglobinopathies—as are explained in Chapter 6.

Key Points

As the individual develops from embryo to foetus, neonate, and adult, its oxygen sources and needs vary. Accordingly, the Hb present at each of these steps also differs in order to maintain an optimum capture and carriage of oxygen.

Haemoglobin variants

Highly sophisticated techniques such as spectroscopy can detect several other minor species of Hb. **Carboxyhaemoglobin**, **carbaminohaemoglobin**, **methaemoglobin**, and **sulphaemoglobin** arise not from different genes for Hb, but are caused by changes to Hb once it has been synthesized and is within the red cell.

Carboxyhaemoglobin, which in normal subjects makes up a maximum of 2% of total Hb, is formed by the attachment of carbon monoxide to Hb. This attachment occurs at the same site of the Hb molecule as oxygen, so that carbon monoxide effectively prevents the carriage of oxygen. This happens because the affinity of Hb for carbon monoxide (that is, the readiness with which carbon monoxide binds) is over 200 times that of the affinity for oxygen, and results in decreased release of oxygen to the tissues. Increased levels of carboxyhaemoglobin are present in habitual tobacco smokers, where levels may rise to 5% of total Hb. High levels of carboxyhaemoglobin can be reversed by hyperventilation with air, although this is quicker if pure oxygen is breathed. However, death occurs when levels of carboxyhaemoglobin exceed 80% of total Hb, as may follow the prolonged inhalation of petrol engine exhaust.

The gaseous product of respiration is carbon dioxide, which must be transported from the tissues to the lung. A tiny fraction is carried in plasma as dissolved gas, but the greater proportion is carried in combination with water as bicarbonic acid, H_2CO_3. However, perhaps 10% of carbon dioxide is carried in a carbamino form, that is, as carbaminohaemoglobin.

Methaemoglobin is a variant of Hb, characterized by the iron being in the oxidized ferric state instead of its usual reduced ferrous state. Methaemoglobin is continuously being formed in the red blood cell by metabolites of oxygen (oxidants), but it can be converted back to 'normal' Hb or deoxyhaemoglobin. Under normal conditions, levels of methaemoglobin (which is essentially inert and does not carry oxygen) do not exceed 2% of the total Hb pool. This is because oxygen does not bind to ferric iron. However, if the red blood cell lacks the ability to return the ferric iron of methaemoglobin to the ferrous state of Hb, then levels of methaemoglobin will rise, and when such levels reach perhaps 10% of total Hb, then clinical signs of headaches, breathing problems, and cyanosis can develop. Fortunately, both methylene blue and ascorbic acid can reverse high levels of methaemoglobin.

Cross reference

Hb measurement is introduced in Chapter 2.

Sulphaemoglobin, an inert non-toxic form of Hb, can be produced by the action of sulphur-containing drugs such as sulphonamides. In this form, oxygen can neither be carried nor delivered, and cyanosis results when levels are 3–5% of that of total Hb. Unlike methaemoglobin, sulphaemoglobin cannot be converted back to Hb.

As outlined in Chapter 2, the World Health Organization recommended method for determining Hb concentration can convert all but sulphaemoglobin to cyanmethaemoglobin.

SELF-CHECK 4.7

What are the major types of globin proteins and how do they come together to form adult Hb?

4.3.2 Red cell enzymes and metabolism

Whilst the major component of the cytoplasm is Hb, there are also numerous enzymes with key roles in the physiology of the cell. These enzymes are also synthesized whilst the developing cell still retains a nucleus, that is, during the erythroblast and nucleated red blood cell stages in the bone marrow. They are mediators in metabolic pathways that generate ATP, provide defence from oxidants, and regulate the carriage of oxygen.

Metabolic pathways

As mature red blood cells lack mitochondria, they are unable to obtain energy by normal aerobic respiration (that is, obtaining energy using oxygen) and so must do so by anaerobic mechanisms. Energy is needed to enable the cell to maintain its shape and deformability, and to resist the toxicity of oxygen. However, the cell can obtain energy by anaerobic respiration via the **Embden–Meyerhof glycolytic pathway**. The substrate for this pathway is glucose, transported into the cell by GLUT1. This molecule is part of the actin junctional complex, the intracellular component of which binds a glycolytic enzyme complex (GEC) of glyceraldehyde-3-phosphate dehydrogenase, phosphofructokinase, lactate dehydrogenase, pyruvate kinase, aldolase, and enolase. This complex is also present on the intracellular part of the ankyrin complex. Consequently, the molecule providing the substrate for the pathway and its constituent enzymes is closely linked.

Embden–Meyerhof glycolytic pathway
A highly complex series of metabolic reactions whereby energy (in the form of ATP, NADH, and NADPH) is generated from glucose.

An early step in this pathway is the generation of glucose-6-phosphate. Most of the glucose-6-phosphate (perhaps 90%) proceeds along the glycolytic pathway to form two molecules of glyceraldehyde-3-phosphate. However, perhaps 10% of the glucose-6-phosphate is diverted to a sub-pathway, the pentose phosphate pathway, where an atom of hydrogen is transferred to NADP, to generate the hydrogen carrier NADPH. The glucose-6 phosphate is therefore transformed to 6-phosphogluconate, which can be further metabolized to phosphoglycerate. A further step involves the transfer of a hydrogen atom from glyceraldehyde-3-phosphate (which is converted to 1,3-diphosphoglycerate) to generate another hydrogen carrier (NADH). We will explain the importance of these hydrogen carriers shortly.

In many cells, such as those of the liver, 1,3-diphosphoglycerate is the substrate for enzymes that generate ATP, the product of which is phosphoenolpyruvate. A further step is the generation of additional ATP from phosphoenolpyruvate by the action of the enzyme **pyruvate kinase**. The final step in this pathway is conversion of pyruvate to lactate. However, the red blood cell has an abundance of an enzyme which can convert 1,3-diphosphoglycerate to **2,3-diphosphoglycerate (2,3-DPG)**, in the Rapoport–Luebering shunt. Although the synthesis of 2,3-DPG sacrifices the production of one molecule of ATP, it can be recycled back into the main metabolic pathway by other enzymes. (A more detailed discussion of the role of this 2,3-DPG follows in section 4.3.3 on the carriage of oxygen by Hb.)

pyruvate kinase
A metabolic enzyme of the glycolytic pathway involved in the generation of ATP.

2,3-diphosphoglycerate (2,3-DPG)
A red blood cell metabolite product of anaerobic respiration but which is also involved in oxygen carriage by Hb.

Defence against oxidation

Apart from the enzymes involved in respiration, the red cell possesses a separate group of enzymes and other molecules designed to help protect it from the cytotoxic effect of oxygen and its metabolites, such as the superoxide radical, which are collectively called **reactive oxygen species (ROS)**. Many ROSs are produced by macrophages and neutrophils as part of our defence against microbial attack, and are therefore beneficial. However, they are non-specific in their action and can also attack healthy tissues. One of the principal protective antioxidant systems is glutathione, a tripeptide of glutamine, cysteine, and glycine; other antioxidants include ascorbic acid, superoxide dismutase, catalase, and peroxiredoxin-2 (bound to

reactive oxygen species (ROS)
Toxic forms of oxygen that can attack and destroy many components of the red cell.

molecules on the inner surface of the cell membrane). In its reduced state, often denoted as GSH, glutathione is a buffer that can limit the potentially damaging effects of these ROSs within the cell by effectively donating hydrogen to the oxygen, so forming water. In this process, glutathione becomes oxidized, a state denoted by GS-, and with another molecule forms the dimer GS-SG. Alternatively, GS- can combine with a sulphur group on a bystanding protein (forming, for instance, GS-S-protein). These oxidized glutathione species can then be converted back to reduced glutathione (GSH) by the enzyme glutathione reductase and the metabolic hydrogen-donating intermediate, NADPH.

An important enzyme in the generation of this NADPH from glucose-6-phosphate and NADP is **glucose-6-phosphate dehydrogenase (G6PD)**. The substrate glucose-6-phosphate is converted into 6-phosphogluconate, which can be fed back into the metabolic pathway, or can move into another metabolic pathway. Therefore, lack of G6PD leads to the impaired generation of NADPH, which in turn leads to a lack of GSH. Low levels of GSH leave the red blood cell open to the toxic effects of oxygen.

The interrelationships between these molecules and their metabolic pathways are summarized in Figure 4.15.

Key Points

The presence of oxygen is essential but is not always beneficial. Instead, it can cause cellular damage. The red blood cell protects itself with several molecules and metabolic pathways to contain these toxic oxygen effects. Loss of this protection leads to premature red cell destruction.

Clinical aspects of metabolism

There are numerous clinical consequences of errors in these pathways. The importance of G6PD is demonstrated by mutations in the gene that gives rise to the intact enzyme. The consequences of these mutations are twofold: the mutated G6PD is not only less efficient than the normal G6PD, but also has a shorter lifespan within the cell, giving rise to what is, in effect, G6PD deficiency. This is in turn manifested as a particular type of anaemia, details of which are presented in Chapter 6.

Other important antioxidant enzymes include **superoxide dismutase** and **catalase**. The former helps protect the red cell by converting oxygen radicals such as superoxide to hydrogen peroxide. Hydrogen peroxide is still toxic to the cell, but can be neutralized by the enzyme catalase, by glutathione, and by other molecules.

Individuals unable to synthesize pyruvate kinase have decreased levels of intracellular ATP, which leads to early destruction of the red cell and so to a different type of anaemia. Because pyruvate kinase effectively removes the metabolites of 2,3-DPG, a deficiency in pyruvate kinase activity may lead to an accumulation of 2,3-DPG in the cell, which could influence the way in which the Hb molecule carries oxygen.

An additional example of the importance of these metabolic enzymes is in the conversion of methaemoglobin back to Hb by NADH. Increased levels of methaemoglobin can be indicative of a deficiency in an enzyme that can use NADH to convert ferric iron to the ferrous iron of Hb. Lack of this enzyme, NADH-linked methaemoglobin reductase, can be congenital and can lead to 10–20% of the total Hb pool being methaemoglobin, a condition associated with mental handicap and cyanosis.

Cross reference

The consequences of the deficiencies of pyruvate kinase and G6PD (that is, anaemia) are explained in Chapter 6.

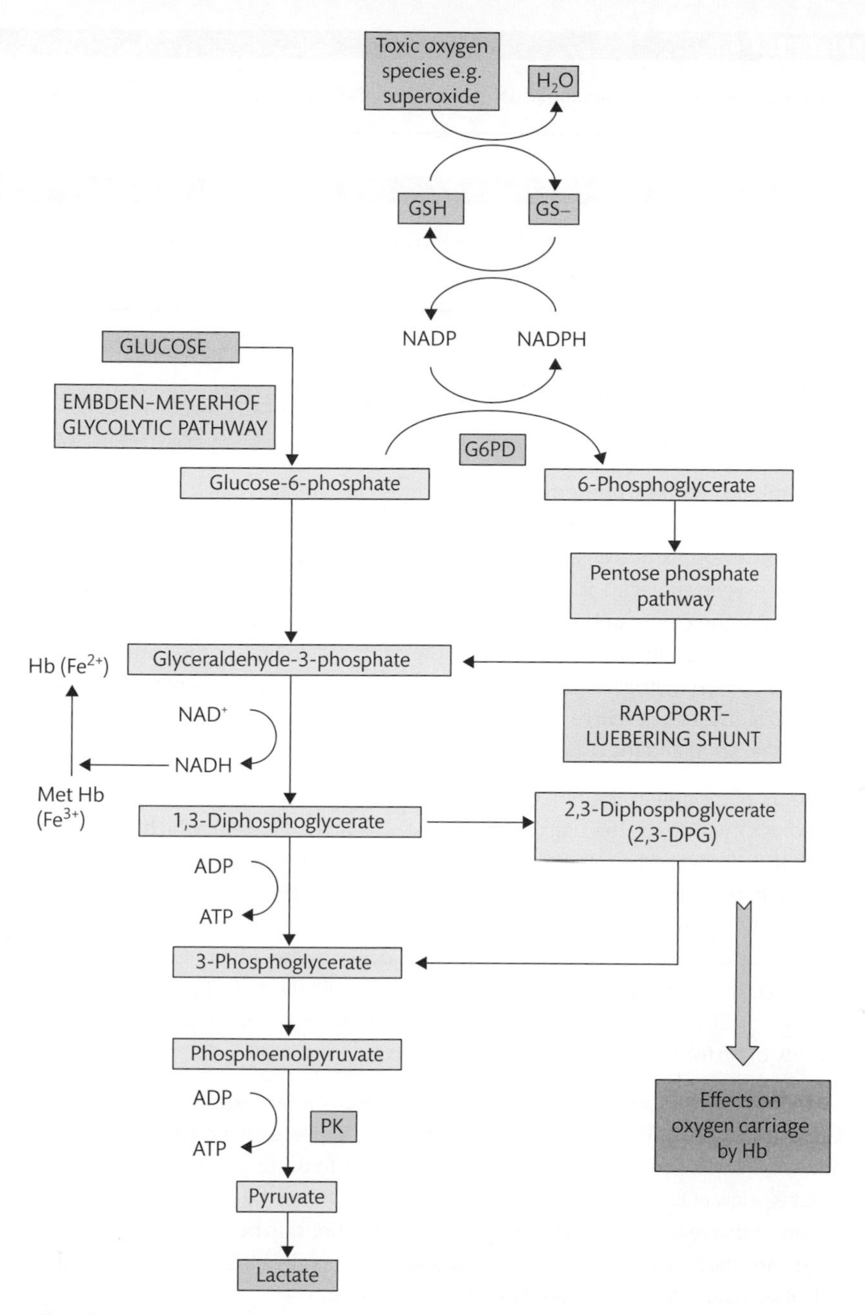

FIGURE 4.15

Biochemical pathways of the red blood cell. The major pathway for the anaerobic generation of energy (in the form of ATP, NADH, and NADPH) is the Embden–Meyerhof glycolytic pathway (central spine). Glucose is first phosphorylated to glucose-6-phosphate (G6P) and is then converted to glyceraldehyde-3-phosphate. However, some G6P is converted via G6PD to 6-phosphoglycerate. This pathway (top right) also generates NADPH from NAD, the former of which ultimately contributes, via glutathione, to the removal of toxic oxygen species. Further down the major pathway (mid-left), NADH is generated from NAD, which has the ability to convert methaemoglobin (Met-Hb) to Hb. A subsequent pathway (mid-right) generates 2,3-DPG, which can influence oxygen transport by Hb. Finally, the enzyme pyruvate kinase (PK) generates ATP from phosphoenolpyruvate, which results in pyruvate and ultimately lactate. At the top, a proton is passed from NADP/NADPH to glutathione (GS-/GSH) to help quench reactive oxygen species.

SELF-CHECK 4.8

What are the purposes of the metabolic pathways within the red blood cell?

SELF-CHECK 4.9

What are the dangers of oxygen within the cell? How do we describe dangerous oxygen-type molecules?

4.3.3 Oxygen transport and the oxygen dissociation curve

The physical and chemical properties of the Hb molecule are such that it exhibits characteristic oxygen-binding properties. Hb (or, more accurately, deoxyhaemoglobin) needs to be able to pick up large amounts of oxygen at the lungs (where it is abundant) to become oxyhaemoglobin. The oxygen molecule is held in place by a complex non-covalent interaction with the atom of iron at the centre of the haem ring. Subsequently, the oxyhaemoglobin must be able to deliver this oxygen in those places where the gas is scarce (e.g. in the tissues), whereupon it reverts to deoxyhaemoglobin. The key to understanding this process of uptake and release is the oxygen dissociation curve, a representation of the relationship between the amount of oxygen in the blood and the tissues and the amount of oxygen carried by Hb.

The amount of oxygen in the blood or tissues can be quantified in terms of partial pressure (denoted pO_2), which has the units of mmHg. At a high pO_2 (12 kPa), which we expect at the lungs, all the Hb should be saturated with oxygen (so oxygen saturation is as close to 100% as is practicable, the reference range being 12–14.6 kPa). Each Hb molecule can carry up to four molecules of oxygen, but the process of uptake is staggered. The uptake and binding of the first molecule of oxygen by deoxyhaemoglobin increases the affinity for the binding of the second molecule, a phenomenon called 'facilitation'. Similarly, once the first two oxygen molecules are bound, the uptake of the third and fourth molecules are increasingly easy. Indeed, the affinity of Hb for the fourth molecule of oxygen is approximately 400 times that of the first.

This enhanced or cooperative uptake is called allosterism—an expression coined largely to explain the changes in the structure of an enzyme to facilitate an increase in its uptake of its ligand. As the oxyhaemoglobin circulates from the lungs to the tissues, where oxygen levels are low (that is, a low pO_2 is present, such as 2 kPa), the Hb is forced by laws of chemistry to give up its oxygen, and so revert to deoxyhaemoglobin. This relationship between the uptake or release of oxygen and the amount of oxygen in the blood or tissues gives the oxygen dissociation curve its typical sigmoid shape, which we can see in Figure 4.16.

A convenient measure of the ability of an individual person to carry oxygen is their P50. This is that particular partial pressure of oxygen at which 50% of their Hb is oxygenated—that is, where there are equal proportions of oxyhaemoglobin and deoxyhaemoglobin. In health, this pO_2 is generally a little under 3.6 kPa. However, this figure can vary with disease or other conditions, and is associated with a shift in the oxygen dissociation curve to the left or right.

- An increased P50 indicates a *shift to the right* and so a decreased affinity of the deoxyhaemoglobin for oxygen. In practice, this means that it is more difficult for Hb to bind oxygen, so that a larger pO_2 is required to maintain the 50% oxygen saturation. However, it also means that it is easier for oxyhaemoglobin to give up its oxygen to the tissues where it is needed, such as in cases of high metabolic activity.

- Conversely, if the P50 is low, we expect a *shift to the left* in the curve, which indicates increased affinity of the Hb for oxygen and so a reduction in the amount of oxygen that is required to carry and maintain a 50% oxygen saturation. The practical consequences of this are that deoxyhaemoglobin can take up oxygen more easily, but that the oxyhaemoglobin is less willing to release it.

Factors influencing oxygen metabolism

Several factors influence the ability of Hb to absorb and/or release oxygen. These can be explained in both left and right shifts in the oxygen dissociation curve.

Metabolic influences on the carriage of oxygen include the hydrogen ion concentration (as pH) inside the red cell. A decrease in pH (as occurs in acidosis) shifts the curve to the right, whilst an increase in pH (as in alkalosis) causes a left shift. However, this effect may be influenced by carbon dioxide and the subsequent formation of the bicarbonate anion and a proton. This is important in, for example, exercising muscles, as the metabolic effects of this exercise, which generate high levels of carbon dioxide and hydrogen ions, and hence low pH, act to shift the oxygen dissociation curve to the right so that for a given pO_2, more oxygen is released to the tissues.

Conversely, in the lungs, levels of carbon dioxide and hydrogen ions are both low, so the pH is high. This shifts the curve to the left, so that more oxygen is absorbed by the deoxyhaemoglobin. An exercising muscle is also generally associated with increased local temperature and the

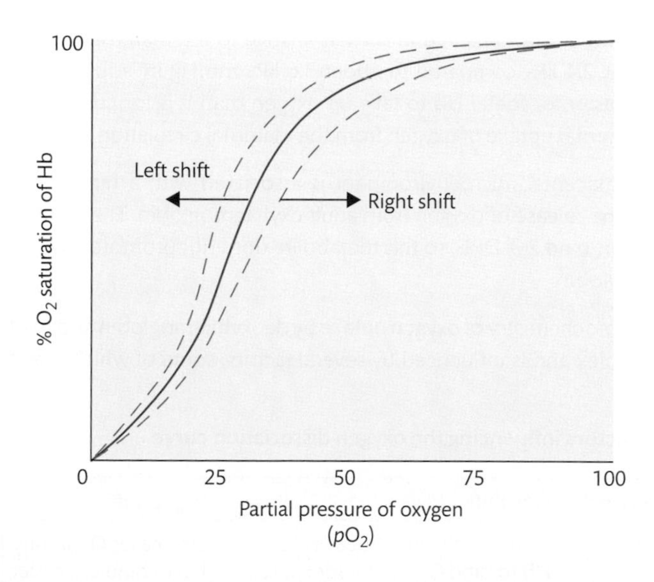

FIGURE 4.16
The oxygen dissociation curve. The normal relationship between the partial pressure of oxygen in the blood and the degree to which Hb is saturated with oxygen is given by the solid line. It is convenient to refer to the degree of oxygen saturation where 50% of the Hb is saturated (that is, the P50, where in theory each molecule of Hb carries two molecules of oxygen from a maximum of four molecules). This equates to a partial pressure of oxygen (pO_2) of approximately 3.6 kPa. This particular partial pressure can be influenced by external factors which force the relationship between the P50 and the pO_2 to vary to the left (as in foetal Hb, where the pO_2 may be, for instance, 2.4 kPa) or to the right (as in acidosis or high levels of 2,3-DPG, giving a pO_2 of maybe 5.3 kPa)—hence the left shift and right shift.

generation of protons, the latter resulting in a low pH. These conditions act to move the curve to the right to enable the release of oxygen to that muscle. At the lungs, where ambient air temperature (which on a cool day may be 10°C) is generally lower than the body temperature of 37°C, a decrease in temperature shifts the curve to the left, enabling oxygen uptake.

In section 4.3.2, we briefly examined 2,3-DPG, a metabolic product of anaerobic respiration in the glycolytic pathway (Figure 4.15). In the red cell, 2,3-DPG is generally present at a similar molar concentration as Hb and can influence the ability of the Hb molecule to carry oxygen. It binds to a specific site between the beta-globin molecules, separate from the oxygen-binding site, when the entire molecule is deoxygenated. In practice, this means that high levels of 2,3-DPG shift the oxygen dissociation curve to the right, and so are associated with a reduced oxygen affinity. Conversely, low levels of 2,3-DPG favour oxygen carriage, as the curve will be shifted to the left. This happens because 2,3-DPG has a greater binding affinity for deoxyhaemoglobin than it does for oxyhaemoglobin. When oxygen is given up to the tissues, which are at low oxygen tension, 2,3-DPG gains access to the sites in the Hb molecule that carry oxygen and prevents re-uptake of oxygen by Hb in these tissues.

As we have already noted, a small amount of carbon dioxide may combine with Hb to form carbaminohaemoglobin, which is associated with a leftward shift in the curve. However, carbon monoxide has a far more powerful effect, as it is bound by Hb far more avidly than is oxygen, and in high concentrations also causes a left shift in the curve. The same leftwards shift is also seen in the presence of high levels of methaemoglobin.

Foetal Hb is a complex of two alpha-globin and two gamma-globin molecules, whereas adult Hb is composed of two alpha-globin and two beta-globin molecules (HbA) and two alpha-globin with two delta-globin molecules (HbA_2). The foetus survives in the uterus because foetal Hb has an oxygen dissociation curve to the left relative to the maternal adult Hb. The P50 of foetal Hb is about 2.4 kPa compared to about 3.6 kPa mmHg in 'adult' Hb. In practice, this means that it is easier for foetal Hb to take up oxygen than it is for adult Hb. This difference enhances the placental uptake of oxygen from the maternal circulation.

In addition, the placental microenvironment is associated with a high concentration of 2, 3-DPG, assisting the release of oxygen from adult oxyhaemoglobin. The gamma-globin chains of foetal Hb do not bind 2,3-DPG, so this metabolite does not promote oxygen release by the foetal oxyhaemoglobin.

It is clear that the biochemistry of oxygen uptake by deoxyhaemoglobin and release by oxyhaemoglobin is complex and is influenced by several factors, some of which may be competing.

TABLE 4.9 **Factors influencing the oxygen dissociation curve.**

Physiological effect	Left shift	Right shift
O_2	Increases O_2 affinity, easier for Hb to bind O_2, harder for Hb to release O_2	Decreases O_2 affinity, harder for Hb to bind O_2, easier for Hb to release O_2
pH	High pH (alkalosis)	Low pH (acidosis)
Temperature	Low	High
Carbon dioxide	Low	High
Carbon monoxide	High	Low
2,3-DPG	Low	High

However, these factors are not present in isolation but act together. In the tissues, the local pH, CO_2, 2,3-DPG, and temperature all act in concert to promote the release of oxygen. At the lungs, the reverse is true as CO_2 is expelled and oxygen taken up. These effects are summarized in Table 4.9.

A second role of the red cell, unrelated to its role as an oxygen carrier, is in the regulation of blood flow. A major vasodilator produced by endothelial cells is nitric oxide (NO). This molecule can be taken up and carried by red cells as nitrosylhaemoglobin and nitrosohaemoglobin, to be given up in cases of low oxygen tension, which then acts to vasodilate and so increase blood flow. In this way, red cells can help regulate blood flow to those areas of hypoxia where oxygen is most needed.

SELF-CHECK 4.10

What is the significance of the left and right shift in the oxygen dissociation curve?

Key Points

The complex mechanics of the oxygen dissociation curve explain how Hb can collect oxygen when local levels are high (as in the lungs) but can then release it when local levels are low (as in the tissues).

4.4 The death of the red cell

At the end of its four-month lifespan (perhaps 120 days), the red cell is destroyed and much of it is recycled. As the red cell has no DNA or ribosomes, it is unable to generate new enzyme molecules, such as G6PD and pyruvate kinase, and an important reason for the death of the cell is that these enzymes are eventually depleted. For example, loss of antioxidant systems such as catalase, superoxide dismutase, and glutathione permit ROSs to damage Hb and the cell membrane, the latter leading to the loss of shape and flexibility. A consequence of this is cell shrinkage and the appearance of phosphatidylserine at the cell surface. This process, known as **eryptosis**, is allied to the programmed cell death, or suicide, of nucleated cells (apoptosis). These changes are noted by phagocytic cells of the **reticuloendothelial system**, such as macrophages in the spleen, which then destroy the aged red cells. As we see in Chapter 6, early and/or inappropriate loss of enzymes such as G6PD can also lead to premature red cell destruction, and may therefore lead to anaemia.

The process that leads to the destruction of red blood cells, or to shortening of their lifespan is called **haemolysis**. When this process occurs in cells of organs such as the spleen and liver it is described as **extravascular haemolysis** (literally, haemolysis outside the blood vessels). When it happens within the circulation it is called **intravascular haemolysis** (haemolysis inside the blood vessels). Most of this red blood cell recycling (90%) is extravascular; the remaining 10% is intravascular.

eryptosis
The process of red blood cell death and haemolysis.

reticuloendothelial system
A collection of white blood cells (such as macrophages) present in organs such as the liver, spleen, and lymph nodes, with roles in immunology and phagocytosis.

haemolysis
The destruction of red blood cells.

4.4.1 The macrophage

When the aged red blood cell, perhaps already undergoing eryptosis, encounters a phagocytic cell, it will be recognized as abnormal and absorbed into phagolysosomes, where it is digested. Alternatively, it may be attacked, and its cell integrity breached, so that it lyses within the plasma (i.e. intravascular haemolysis), its contents being dispersed. Both neutrophils and macrophages may perform these processes, but it is widely believed that they occur in the bone marrow, liver, and spleen, and that the macrophage is the dominant effector. Given the importance of the hepatocyte in iron metabolism, it has been speculated that hepatic macrophages are the leading iron recyclers.

Regardless of anatomical site, once digested by the macrophage, the red cell iron is extracted from the haem and exported via ferroportin to transferrin (as is the case for enterocyte iron) for passage to erythroid precursors in the bone marrow. Alternatively, the iron may be stored in ferritin and haemosiderin. It may also be that iron-loaded macrophages migrate to the bone marrow to unload their cargo, thus bypassing transferrin. The recycling is crucial because, of the daily requirement for iron (only 1–2 mg is provided by intestinal enterocytes), whereas 20–25 mg is provided by iron-recycling macrophages. Between them, these cells must provide 2×10^{15} atoms of iron every second to the bone marrow so that erythropoiesis remains physiological. Hepatic iron deposits in macrophages and hepatocytes range between 300 mg and 1 gram.

4.4.2 Bilirubin

When the Hb molecule is degraded, the proteins, carbohydrates, and lipids are recycled. However, the protoporphyrin ring from the haem molecule is too complex to be recycled and is broken down to iron (recycled as discussed above), carbon monoxide, and **biliverdin** by the enzyme **haemoxygenase**, a 32–36kDa molecule with three isoforms. This enzyme is also responsible for liberating iron from dietary haem absorbed by the enterocytes of the small intestine. This process is not merely about recovering iron—haem is highly cytotoxic (pro-oxidant, pro-inflammatory) and needs to be degraded, a considerable task, since each red cell has 1.2×10^9 haem groups. The biliverdin is then converted to **bilirubin**, the major breakdown product of haem. Haemoxygenase may also be linked to other aspects of iron metabolism, such as ferroportin. Furthermore, a patient deficient in haemoxygenase also has low hepcidin levels and high levels of the sTfR. This is pertinent as in iron deficiency, transferrin receptors are highly expressed by erythroblasts and are shed into the circulation, so that increased sTfR levels reflects increased demand for iron and is likely to represent a more sensitive marker than ferritin or transferrin saturation.

biliverdin
A breakdown product of haem that is converted into bilirubin.

haemoxygenase
A key enzyme in the recycling pathway of the haem molecule that generates iron, carbon monoxide, and biliverdin.

bilirubin
The major breakdown product of haem which is not recycled.

Most of the bilirubin binds to albumin and gives our plasma and serum its mildly lemon-yellow colour. In this form, bilirubin is described as unconjugated, or indirect bilirubin. Upon its passage through the liver, bilirubin may be combined with glucuronic acid, whereupon it is described as conjugated, or direct bilirubin, and is then moved into the gall bladder. Conjugated bilirubin is then excreted via the bile duct and duodenum; it is this form that gives faeces a brown colour. Intestinal bacteria can convert some of this bilirubin to urobilin and urobilinogen. However, some intestinal conjugated bilirubin may be absorbed from the large intestines and may be found in the circulation. In this form, it may be excreted via the kidney, giving the urine its characteristic light-yellow colour.

When healthy, the body is perfectly capable of clearing and recycling millions of senescent red blood cells each day. However, if there is excess red blood cell destruction (that is, pathological haemolysis), then the recycling processes of the liver and other organs can be overwhelmed. One consequence of this is a build-up of bilirubin which, if deposited in tissues, may lead to the yellowing coloration of the skin, the condition we know as **jaundice** (from the French for yellow). This may also manifest itself as more heavily (yellow or even orange) coloured urine and plasma. Hence, clinically evident jaundice may be present in severe haemolytic anaemia and is a clear pathological sign. However, jaundice may also follow severe liver disease, such as obstructive cholestasis, liver cancer, or alcoholic cirrhosis, which does not directly reflect abnormal red blood cell turnover.

jaundice
A yellow coloration reflecting high levels of bilirubin in the plasma.

4.4.3 Scavengers

The processes described above refer mainly to red cell recycling within cells and organs, but some 10% of haemolysis happens in the circulation, and when it does, free Hb and haem within the blood bind to specific scavenger molecules. The majority of free Hb forms a complex with the liver-derived plasma protein **haptoglobin** (coded for by *HP* on 16q22.2). This complex is removed by CD163 receptors on macrophages and seems likely to be a system for a more efficient recycling of iron and the globins. A soluble form of CD163 may bind haptoglobin/Hb complexes, and so add to the scavenging process. Plasma Hb not scavenged by haptoglobin may be degraded by plasma proteases, liberating free haem. This circulating haem can form a complex with another product of the liver, **haemopexin** (a 63 kDa protein coded for by *HPX* on 11p15.4), in which form it can also be cleared by the liver macrophages and hepatocytes via the CD91 receptor and may too be part of the recycling process. CD91 is linked to the low-density lipoprotein receptor. Any free haem not bound to haemopexin may be broken down by plasma haemoxygenase into iron (which may be picked up by apotransferrin), carbon monoxide, and bilirubin, which can be carried by albumin.

haptoglobin and haemopexin
Plasma proteins that complex with, and thus remove, Hb and free haem from the circulation.

These scavenging molecules, being products of the liver, are acute-phase reactants, so that plasma levels rise in inflammation. Conversely, reduced plasma concentrations of haptoglobin and haemopexin may reflect consumption/binding by high levels of plasma Hb and haem, which may, in turn, indicate intravascular haemolysis. Figure 4.17 summarizes the fate of the products of the red blood cell.

SELF-CHECK 4.11

What factors are associated with the destruction of the red blood cell?

SELF-CHECK 4.12

What is the major excretory product of the red blood cell and how is it removed?

4.5 Red cell morphology

Having developed a grasp of the structure and function of the red cell, we now move on to look at the value of the appearance (morphology) of the cell in the practice of clinical haematology. A great deal of important information is obtained from the size, shape, and other features of the red cell. Some of this information can be provided by the haematology autoanalyser, but a key tool is the microscope. The basic morphology of a red blood cell when viewed under a light microscope is of a disc, as with white blood cells. However, this very simple descriptor masks a great variety of different sizes, shapes, and colours of different populations of these cells, as can be seen in Figure 4.18.

4.5.1 Variation in size and shape

Anisocytosis

The average red blood cell is generally taken to have a diameter of ~7 μm and to have a shape that is ideally perfectly round in two dimensions. However, this can vary somewhat according to various pathophysiological conditions. Variation in the size of the cells on the blood film is called **anisocytosis**. The natural distribution of the size of a red cell population demands that

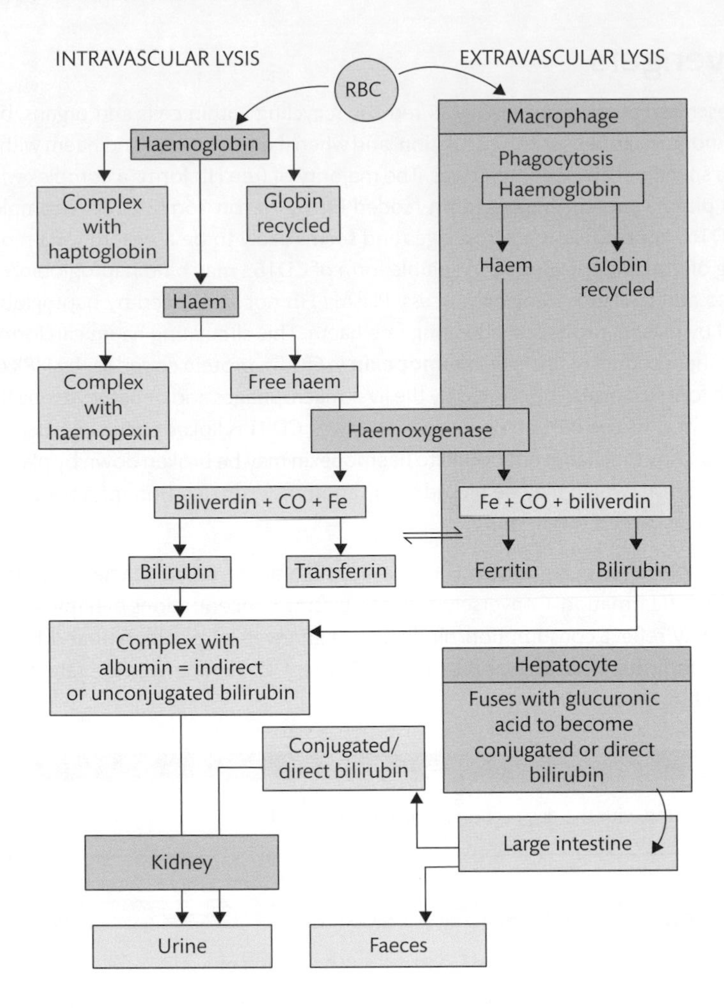

FIGURE 4.17

Destruction of the red blood cell. The fate of the red blood cell is to be lysed either in the circulation (intravascular haemolysis) or in cells of the reticuloendothelial system (extravascular haemolysis, such as in macrophages). Intravascular Hb may bind to haptoglobin or may be broken down to haem and globin (which is recycled). Haem may complex with haemopexin or be broken down further by haemoxygenase. Within the macrophage, extravascular haemolysis sees the globin being recycled and the haem being processed by haemoxygenase. The products of haem are iron, carbon monoxide, biliverdin, and bilirubin. The latter complexes with albumin (and as indirect or unconjugated), but in the liver it combines with glucuronic acid and is described as direct or conjugated bilirubin. The latter then passes into the gall bladder and bile into the intestines, where it may be converted to urobilin. However, some conjugated bilirubin may re-enter the blood and, with unconjugated bilirubin, be excreted by the kidney.

some cells are slightly larger and some are slightly smaller, as shown in Figure 4.18(a). The issues of 'large' and 'small' are described in Chapter 2, and generally depend on the local reference range, which depends on age (e.g. a neonate has larger red blood cells than an adult). When a cell is abnormally large or small, it is referred to as **macrocyte** or as a **microcyte,** respectively**.** A cell with an MCV within the reference range is a **normocyte** (Figures 4.18(a)–(c)). The autoanalyser in the laboratory will automatically give the size of the MCV, but this simple empirical result can mask some important pathology.

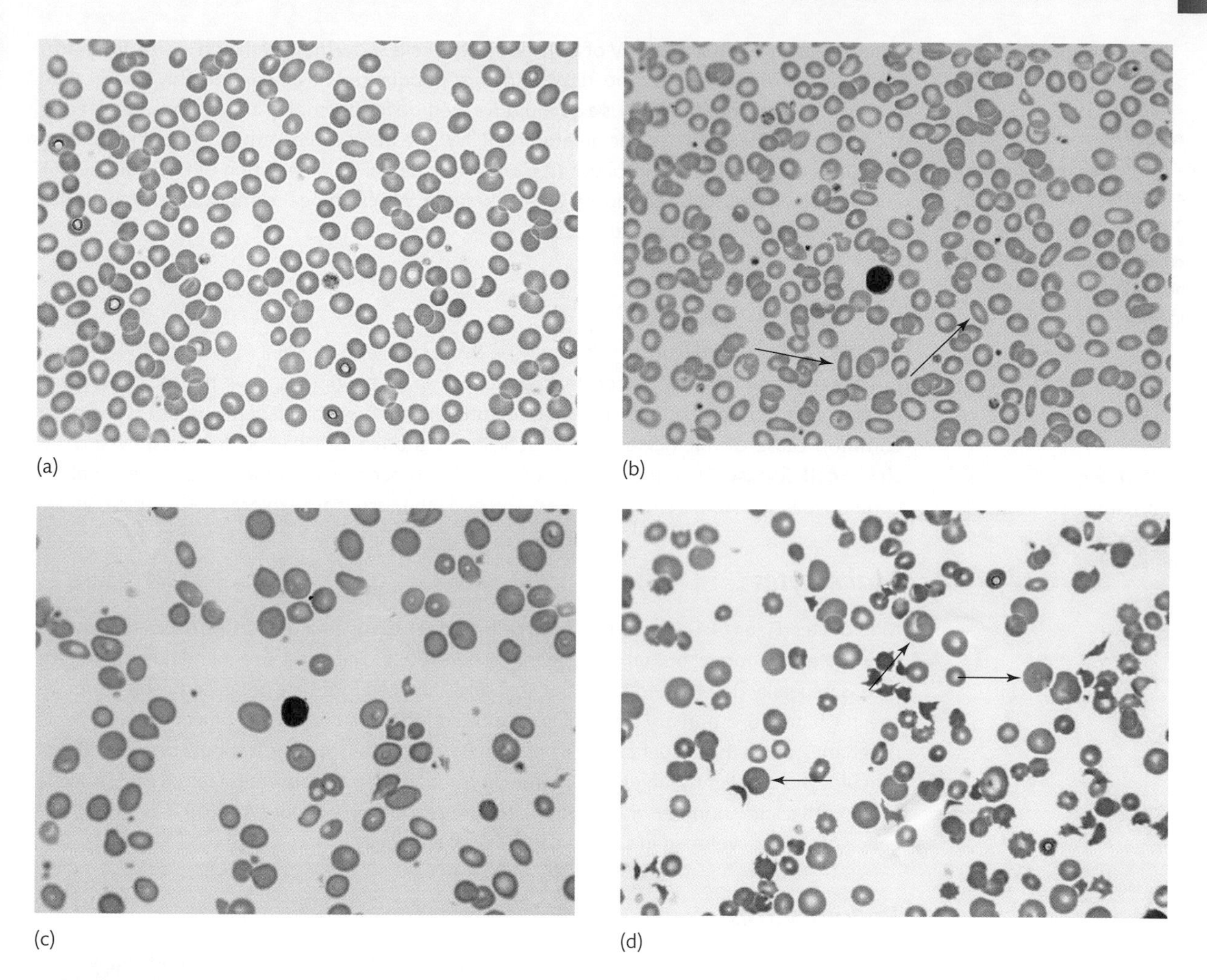

FIGURE 4.18
Red cell morphology. (a) Normal red cells (hence 'normocyte'): biconcave discs, when stained showing a central area of pallor (where there is less Hb). Note the small variation in size and shape. The small purple bodies are platelets. (b) Microcytosis: red cells are smaller than normal; and anisocytosis: there is a variation in size of red cells—some are clearly much larger than others. The dark stained cell is a lymphocyte; the red cell to its right is a microcyte. Some red cells (arrowed) are elongated and appear elliptical—this may be due to a particular pathology described in Chapter 6. (c) Macrocytosis: some red cells are larger than usual; and anisocytosis: there is a variation in size of red cells—some are clearly much larger than others. The dark stained cell is a lymphocyte; a macrocyte is present to its left. Compare this to the difference between the lymphocyte and the microcyte in (b). (d) Polychromasia, anisocytosis, and poikilocytosis: some of the red cells (arrowed) are bluer (and larger) than usual and so are representative of young red cells (reticulocytes). There are also numerous microcytes and several mis-shapen red blood cells (schistocytes, which appear as fragments or debris), the significance of which is discussed in Chapter 6. (Magnification ×400).

Red cell distribution width

This index (the RDW) is the measure of the variation of the MCV as defined by the haematology autoanalyser. Consider a population of red cells with an MCV of 85 fL comprising some slightly smaller (perhaps 81 fL) and others larger (such as 89 fL). Now consider a second population whose cells have the same MCV of 85 fL but whose cells show more variation in size, for example between 76 fL and 95 fL. The RDW of the first population will be considerably smaller than that of the latter. Thus the RDW is a *quantitative* method for describing anisocytosis. The greater the variation in red cell size, the greater the RDW. The RDW can be expressed as the standard deviation of the MCV, in femtolitres (fL), or as the coefficient of variation of RBC volume (%). The RDW increases with age and is higher in women. We examine the importance of this index in greater detail in Chapters 5 and 6.

Cross reference

As explained in Chapter 2, the definition of a normal MCV varies from laboratory to laboratory.

Sickle cell disease and an allied condition, thalassaemia, are discussed in Chapter 6.

Microcytes

Similarly, microcytes (Figure 4.18(b)) may be those cells with an MCV <80, 78, or 75 fL. Again, the exact definition of a microcyte depends on local conditions; thus the local reference range is of importance. Once more, this feature is not merely of academic value, as the most common cause of microcytes is iron deficiency; another is a haemoglobinopathy such as **sickle cell disease**. Therefore the size of the cell can become important as an index to allow certain types of anaemia to be identified. This and other issues are explored in Chapters 5 and 6.

sickle cell disease

A disease that follows from a mutation in a gene for one of the globin molecules.

Macrocytes

Macrocytes (Figure 4.18(c)) are those red cells whose MCV is greater than the top of the laboratory's reference range, but taken by many to be >100 fL. The exact size that defines a macrocyte is generally determined by senior staff and can therefore vary between hospitals. As we see in Chapter 5, a primary cause of an increased MCV is a high intake of alcohol; another is pregnancy; and a third is deficiency in vitamin B_{12} or folate. However, it should be recalled that reticulocytes are slightly larger than mature erythrocytes, so that, overall, a raised MCV may be to do with a large number of these immature red blood cells (Figure 4.18(d)). The full significance of this is developed in Chapters 5 and 6 on anaemia.

Poikilocytosis

The macrocyte, normocyte, and microcyte generally retain their regular, round shape on a blood film, so that the shape of the entire cell population is approximately uniform. However, as we will see in sections to come (e.g. section 6.2.4), numerous pathological situations are characterized by cells with an irregular shape, such as the elliptocyte, the spherocyte (see below) and the poikilocyte (a cell with a tear-drop or pear shape). A blood film where there is a high proportion of cells of an irregular morphology is said to display poikilocytosis. It follows that a film whose cells all retain their round and regular shape, regardless of size, cannot be described in these terms. There is a modest degree of poikilocytosis in Figure 4.18(b), but Figure 4.18(d) exhibits both anisocytosis and poikilocytosis.

4.5.2 Variation in colour: polychromasia

Red blood cells readily take up dyes and so appear under the microscope as if coloured. As with variation in cell size, there is also a variation in this colour. As the stains are taken up by Hb, the density or strength of the stain can help us to estimate the amount of Hb inside the cell and so the ability of that cell to carry oxygen. Typically, red blood cells, because they are not flat (as are coins, for instance) but biconcave, should contain an area of central pallor, which is approximately one-third the size of the red cell. These cells are described as **normochromic**. Cells which fail to take up the dye (because they are deficient in Hb), and so are not particularly well coloured, may be described as **hypochromic**. If there are many such cells present, there will also be hypochromia. Several cells that are both microcytic and hypochromic are shown in Figure 4.18(d). In some cells, there is colour only in a small band around the outside of the cell—the middle of the cell seems to be 'empty'. These hypochromic cells probably have a low mean cell haemoglobin (MCH) and MCHC.

Hyperchromia may be present when a normal amount of Hb (i.e. the MCH) is present with a small cell, so that the MCHC is increased. However, it may also be present in certain cases of megaloblastic anaemia due to vitamin B_{12} or folate deficiency, or if red blood cells lose their central pallor and appear as if **spherocytes**. These cells commonly reflect either an immune process directed against the red cells or an abnormality in the cell membrane (generally a deficiency in structural proteins, such as spectrin or ankyrin) and die prematurely, often in the spleen. This is frequently accompanied by anaemia, additional details of which are presented in Chapter 6.

Red blood cells that are normal in size and colour are referred to with the combined term 'normocytic, normochromic'. However, such red cell appearances can be encountered in pathological situations such as blood loss or the anaemia of chronic diseases, where the disease process does not affect red cell structure or content. As previously described in section 4.1.4, polychromasia refers to a bluish tinge associated with an immature red blood cell population—the reticulocyte. This is illustrated in Figure 4.18(d)—the large reticulocytes are clearly stained blue more strongly. As with variation in size, this characteristic of variation in colour, called anisochromasia, is important in the diagnosis of different types of anaemia, as we see in Chapters 5 and 6.

4.5.3 Other morphological changes

Close examination of the shape of the red cell can provide a wealth of information regarding the likelihood of different diseases. Many of these are discussed in Chapters 5 and 6. For example, sickle cell disease is so named for the curved shape that the red cell tends to form when deprived of oxygen (Figure 6.8). Figure 4.18(d) has many fragments or distorted red blood cells (schistocytes) that indicate an active disease process—the curved cells are *not* sickle cells.

Haematologists also recognize various other irregular shapes that red cells may adopt. These include burr cells (Figure 4.19(a)) and target cells (Figure 4.19(b)). Burr cells are characterized by having a 'spiky' appearance and are also known as acanthocytes (*akanthos* = thorny/spiky; *cyte* = cell). Target cells have a darker area within the central area of pallor ('Mexican hat'). There are also changes that are not due to a particular pathology, but due instead to laboratory artefacts, such as prolonged storage or poor fixation. Figures 4.19(c) and (d) show typical artefactual changes to the morphology of both red blood cells and white blood cells—there are clear differences compared to cells in the other figures.

Key Points

Examination of a blood slide under a light microscope can give useful information about the integrity of the red blood cell, its size, how much Hb it carries, and factors that can cause damage.

4.5.4 Inclusion bodies

Normally, when seen under the light microscope, the red cell should be homogeneous with no unusual internal features. However, in various diseases and conditions, **inclusion bodies** can be seen within the cell that can be of considerable diagnostic value. Inclusion bodies may be parasites, deposits of iron, remnants of DNA, or denatured Hb. Some of these inclusion bodies

inclusion body
An abnormal body within the red blood cell generally associated with a particular pathology.

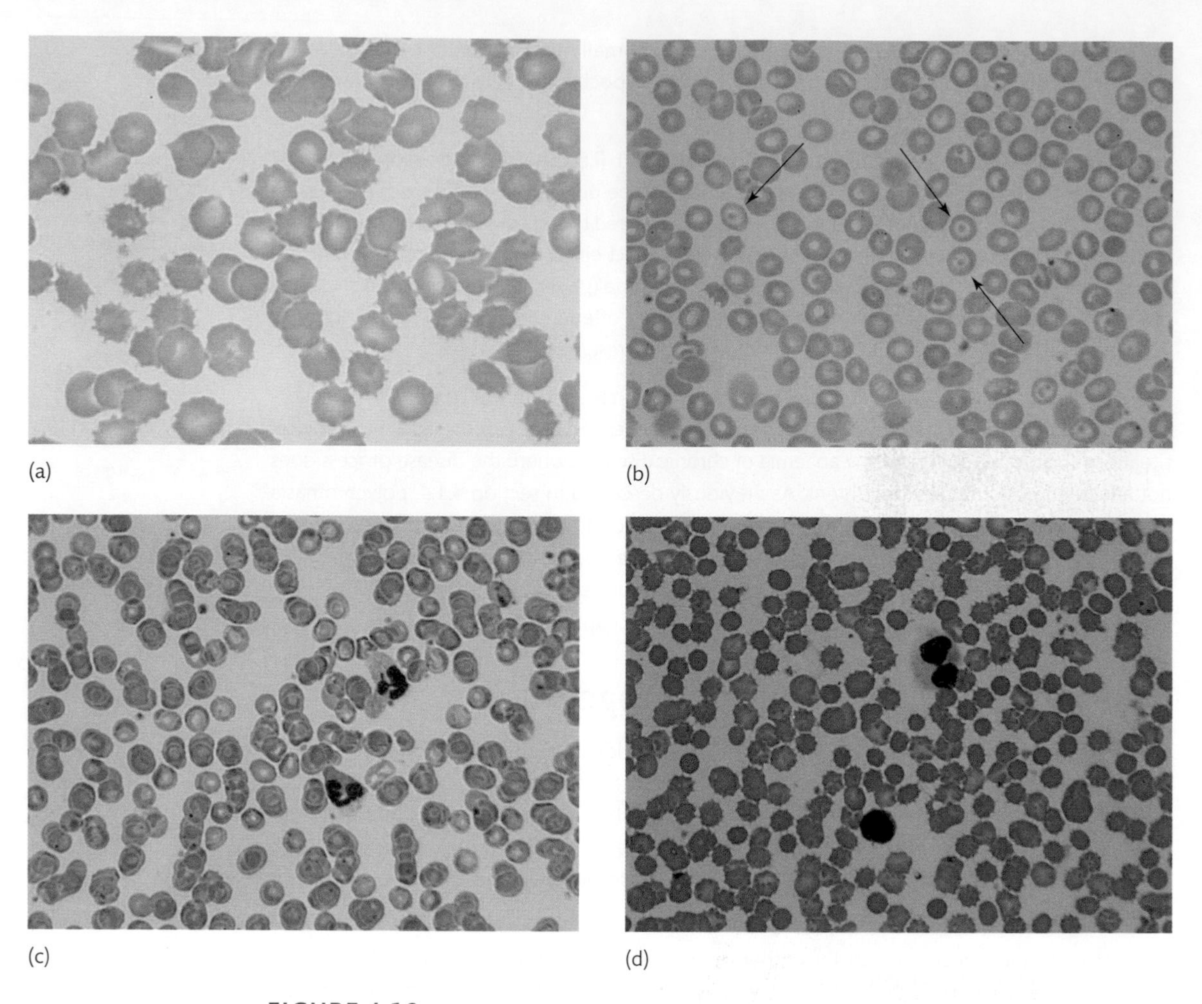

FIGURE 4.19

(a) Burr cells: many of these cells are 'spiky'. (b) Target cells: these have darker areas within the central area of pallor ('Mexican hat'). Three examples are arrowed. (c, d) Storage and fixation changes. Samples taken into EDTA and stored incorrectly or for too long will undergo changes. These may be crenation (irregular, slightly spiky edges) of the red cells and deterioration of the white cells (loss of membrane and nuclear integrity). Poor drying and fixation results in trapped water within the cells and so a poor morphology. One of the white blood cells in (d) (a lymphocyte) has retained a normal morphology whilst the other, a neutrophil, is abnormal. (Magnification varies: ×400 to ×600.)

demand special techniques for visualization as they may not be detectable by conventional staining. More details of particular inclusion bodies follow in Chapter 6 on anaemia.

4.5.5 The red cell cytogram

The haematology analyser gives us six important indices about the red cell—the amount of Hb and number of red cells in a fixed volume of blood, the size of the red cell (MCV), and amount of Hb in a red cell (MCH), the concentration of Hb inside a red cell (MCHC), and the proportion of blood that is made up of red cells (the haematocrit, Hct). However, the analyser can give us another useful tool: the cytogram. This is a two-dimensional graphical representation plotting

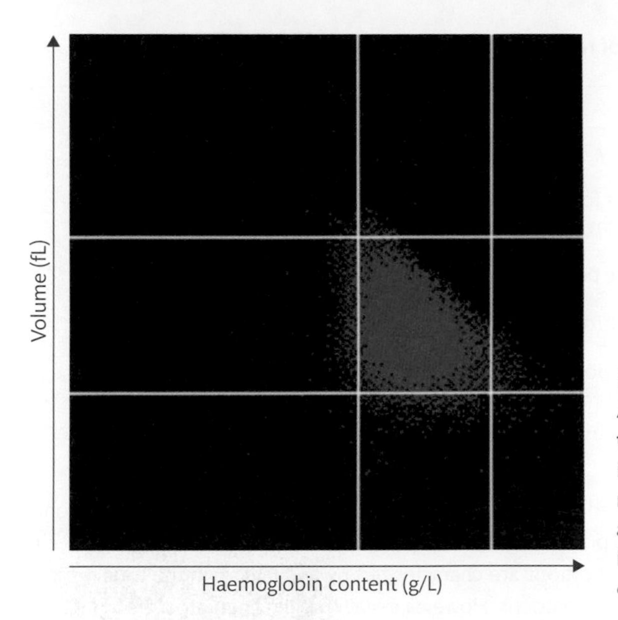

FIGURE 4.20

A red cell cytogram showing the volume of red cells contrasting with their Hb content. From this, the proportion of, for example, microcytic hypochromic cells (those in the bottom left quadrant) can be determined. As we would hope from this normal cytogram, very few cells are present in this location, and there are few cells above the topmost horizontal bar—if present, these would be macrocytes. By permission of, and thanks to, Graham Gibbs MSc CSci FIBMS.

the size of the cell (that is, the MCV) on a vertical (*y*) axis, and the Hb of cell (that is, the MCH) on the horizontal *x*) axis (Figure 4.20). The figure is completed by two horizontal and two vertical lines, representing the upper and lower limit of the reference range of the MCV and the MCH, that intersect to give nine areas. As is explained in Chapter 5, the cytograms of red cells from different conditions adopt certain characteristic shapes, providing an additional tool in the diagnosis of certain diseases.

Table 4.10 lists certain clinical syndromes where changes in morphology may be expected.

SELF-CHECK 4.13

What is the importance of the RDW?

SELF-CHECK 4.14

What are the correct haematological terms for red blood cells that are larger or smaller than normal?

SELF-CHECK 4.15

What is meant when a red cell is described as being hypochromic?

4.6 An introduction to diseases of the red cell

The major disease of red blood cells is undoubtedly anaemia. However, this disease is so common and has so many different facets that it demands not one, but two dedicated chapters (Chapters 5 and 6). In contrast to anaemia, where levels of Hb and the number of red cells are low, there are

TABLE 4.10 Clinical aspects of red cell morphology.

Laboratory	Clinic
Microcytosis	Iron deficiency, haemoglobinopathy, lead poisoning
Macrocytosis	Vitamin B_{12} or folate deficiency, chemotherapy, alcoholism, hypothyroidism
Increased RDW	Inflammation, pulmonary embolism, stroke, reticulocytosis
Schistocytosis	Thrombotic thrombocytopenia purpura, haemolytic uraemic syndrome
Target cells	Liver disease, post-splenectomy, haemoglobinopathy

N.B. this list is not exhaustive: see Chapters 5 and 6.

two conditions where Hb and the red cell count are raised. These are erythrocytosis (in which only the red cell count is increased) and polycythaemia vera (PV) (where red cells and other blood cells are often increased). Both these conditions are characterized by an increase in the total red cell mass of the body, and also a high haematocrit. However, recall that the haematocrit is in fact the mathematical product of the red cell count and the MCV, so that inaccuracies may be present and could confuse the diagnosis. An increased red cell mass brings hyperviscosity, and so a stress on the cardiovascular system, which may precipitate a myocardial infarction or stroke.

Cross reference
Red cell mass is discussed in detail in Chapter 11.

Erythrocytosis and PV, and their consequences, are the product of very dissimilar pathologies, the latter being linked to a mutation in a gene coding for part of the EpoR (the enzyme JAK2). Although the naming of particular processes, diseases, and conditions seems, to many, to be a mystery, there is in fact logic in the system, which can be decoded by breaking down the formal name into its constituent parts, as we saw earlier in this chapter. For example, erythrocytosis describes the number of red blood cells (erythrocytes) being raised in the blood (-osis). Similarly, recall that leucocytosis means white blood cells (leucocyt-) being raised in the blood (-osis). By contrast, polycythaemia means an increase (poly-) in cells (-cyth-) in the blood (-aemia)—it does not specify which cells are increased, but it is taken to be all blood cells. Vera derives from 'true' and is retained for historical reasons.

From a laboratory perspective, Hb and haematocrit are around 180 g/L and 0.54 L/L, respectively in both erythrocytosis and PV. However, on the whole, levels of Epo are higher in erythrocytosis, whilst PV is often characterized by a higher white cell count and platelet count.

4.6.1 Erythrocytosis

At the simplest level, erythrocytosis is defined by a red cell count being above the top of the reference range. It follows that Hb and haematocrit are also likely to be increased. Indeed, one guideline suggests a haematocrit >0.52 in males and >0.48 in females, which must be present for at least two months for erythrocytosis to be confirmed. The World Health Organization focuses on an Hb >185 g/L in males and >165 g/L in females. An additional test is to estimate the total mass of red cells within the body.

Pathology

There are a number of possible reasons for erythrocytosis, and we can focus on four such causes. Type 1, relative erythrocytosis, is due to a reduction in the volume of plasma, as may be present in dehydration. If this is the case, the root pathology lies elsewhere, perhaps as the consequence of renal failure. If so, the condition is not primarily of haematology. Type 2, apparent erythrocytosis, is present in those whose red cell mass is within the reference range but who have an increased haematocrit. Type 3, idiopathic erythrocytosis, describes the situation in which no clear mechanism can be identified. Type 4, absolute erythrocytosis, can be classified as being congenital (type 4A) or acquired (type 4B). A principal cause of type 4A erythrocytosis is an abnormal Hb variant (such as those with a high affinity for oxygen), or in cases of congenitally low levels of 2,3-DPG. Causes of type 4B erythrocytosis are generally reactive to a state of hypoxia. One of the functions of the kidney is to sense levels of oxygen. When such hypoxia occurs, the kidney will increase its release of Epo to stimulate the bone marrow to produce more red cells. Peripheral hypoxia can be assessed by measuring the arterial oxygen saturation, most easily achieved with a pulse oximeter. However, this technique can also give misleading results. Other causes of hypoxia causing type 4B erythrocytosis include lung pathology, such as congestive obstructive pulmonary disease, and cyanotic congenital heart disease.

Although not strictly a pathology, the erythrocytosis of high altitude (as may be found in the Andes and Himalayas) is instructive. The increased red cell count and other indices are a required physiological response to the atmospheric hypoxia and are therefore entirely normal in this setting. When the subject descends to sea level, the erythrocytosis (as it is at 'normal' atmospheric oxygen levels) is an unnecessary burden and slowly resolves. It follows that those unaccustomed to high altitude need weeks of acclimatization to allow their bone marrow to generate a high red cell mass and this is the basis of altitude training by athletes. Type 4C erythrocytosis is most frequently caused by pathological generation of high levels of Epo, as may be present in certain renal neoplasias and gangliomas.

The laboratory in erythrocytosis

In addition to changes in red cell indices (raised Hb, red cell count, Hct), there may also be a mild neutrophilia and a slight thrombocytosis. The reasons are unclear but may be the result of other aspects of the disease process—gross increases in these other cells are found in PV. For example, smokers may have a neutrophil leucocytosis, possibly the result of chronic pulmonary inflammation. Bone marrow examination is rarely required, but may be called on to exclude PV. Estimation of red cell mass requires the use of radioisotopes and is therefore reserved for specialist centres. However, estimation of Epo is probably the most useful diagnostic tool, and although levels are typically raised in erythrocytosis secondary to hypoxia, they may also be increased in polycythaemia veras of various aetiologies.

Management of erythrocytosis

The first step in the management of erythrocytosis is to identify and treat, where possible, contributing factors such as smoking, alcohol, and treatments of hypertension that are based on fluid elimination by diuretics. The standard treatment is venepuncture to reduce the red cell mass and is generally considered if the Hct exceeds 0.54. Patients having had a thrombosis, or who are deemed to be at risk (perhaps by virtue of thrombocytosis and/or hyperviscosity) may require anticoagulation, probably oral. In this high-risk group, venesection may be called for at a lower Hct, such as 0.45, especially in the presence of diabetes or hypertension.

4.6.2 Chuvash erythrocytosis

Originally described in an ethnically homogenous population of a part of Russia, this disease was described as an endemic congenital polycythaemia. However, as no link with JAK2-type PV could be found, it is classified as an erythrocytosis. Regulators of hypoxia inducible factor

(HIF, which acts to promote the generation of Epo) include prolyl-hydroxylase domain (PHD) enzymes, and products of the von Hippel-Lindau (VHL) tumour suppressor gene which prime the HIF for proteolytic degradation in proteasomes. Loss of function mutations in *VHL* (such as *VHL598C-T*, present in the Chuvash population) and PHDs effectively lead to high levels of HIF, a consequence being that the cell continues to behave as if hypoxia is present. This ultimately results in high levels of Epo (up to ten times the normal level) and so a raised Hb of perhaps 180 g/L. Interestingly, the levels of ferritin and sTfR are also elevated, and the frequency of varicose veins (which would seem to be the product of venous congestion) and thrombosis is also increased. The *VHL598C-T* mutation has been described in an Italian population, and other mutations in *VHL* have also been reported, all of which result in phenotypes very similar to Chuvash erythrocytosis.

4.6.3 Polycythaemia vera

The aetiology of this disease is completely different from that of erythrocytosis, in that PV is primarily the consequence of mutations in the genes coding for a part of the EpoR, JAK2. Furthermore, PV is part of a group of diseases called myeloproliferative neoplasms: other members of this group are essential thrombocythaemia and primary myelofibrosis. All are clonal stem cell disorders, and each may progress to acute myeloid leukaemia. The common underlying aetiology of these three diseases is mutation in the gene for JAK2 (which generates an abnormal JAK2 molecule called JAK2-V617F), a component of the intracellular domain of the EpoR (see Figure 4.4). Indeed, this mutation is never found in erythrocytosis. So although PV is quite rightly a red cell disease, it is better placed in Chapter 11, in a section named myeloproliferative neoplasms, where full details are to be found.

Cross reference

Polycythaemia vera is fully discussed in detail in Chapter 11.

SELF-CHECK 4.16

What are the principal differences between erythrocytosis and polycythaemia vera?

Key Points

Despite a commonality in increased numbers of red blood cells, there are fundamental differences between erythrocytosis and polycythaemia. These include aetiology, laboratory findings, and prognosis.

CASE STUDY 1.1 Myocardial infarction in a young woman

A 23-year old woman, a regular general swimmer, is invited to join a competitive swimming club. Relaxing in a public house after an intense training session in the pool and the gym, she complains of chest pain and nausea. Her friends call her general practice for an immediate appointment, which is granted, and 25 minutes later she is having an ECG. The practice nurse is astonished to find a raised S-T segment (evidence of a myocardial infarction), and this is followed up by a near-patient testing device for

troponin-T, which is slightly raised. An ambulance is immediately called, which takes her to the local A&E department, where the ECG is repeated and the abnormality confirmed. Blood tests were as follows:

Index	Result	Reference range (female)
Hb	190 g/L	118–148
Red cell count	5.8×10^{12}/L	3.9–5.0
MCH	32.7 pg	26–33
MCV	86	77–98
MCHC	380	330–370
Hct	0.50	0.33–0.45
Reticulocytes	110×10^9/L	25–125
ESR	18 mm/hour	<10
White blood cell count	16.7×10^9/L	4–10
Neutrophils	12.3×10^9/L	2–7
Lymphocytes	2.2×10^9/L	1–3
Monocytes	1.5×10^9/L	0.2–1.0
Eosinophils	0.5×10^9/L	0.02–0.5
Basophils	0.2×10^9/L	0.02–0.1
Platelets	300×10^9/L	150–400
Sodium	140 mmol/L	135–145
Potassium	4.2 mmol/L	3.8–5.0
Urea	4.2 mmol/L	3.3–6.7
Creatinine	75 μmol/L	71–133
Creatine kinase	625 IU/L	55–170
Creatine kinase MB	49 IU/L	<25
Troponin T	2.5 IU/L	<0.6

The patient's medical history is unremarkable, except that she is taking an oral contraceptive, but her grandfather died of a stroke aged 65, and an uncle died of a heart attack aged 58. The patient clearly has two immediate issues—the heart and the blood cell indices. Are these related? What further investigations should be carried out?

Chapter summary

- The development of the red cell (erythropoiesis) occurs in the bone marrow: precursors such as the proerythroblast and erythroblast give rise to the reticulocyte and the mature erythrocyte.
- The red blood cell is highly specialized for the carriage of oxygen, a consequence of which is the lack of a nucleus and organelles.
- The cell membrane is also highly specialized, with a series of structural proteins providing shape and flexibility.
- The cytoplasm contains Hb, metabolites, and enzymes.
- Hb consists of protein (globin) and non-protein (haem) components.
- There are various types of globin that vary throughout the life of the subject and that have a differing capacity to carry oxygen.
- Haem is a complex molecule, with an atom of iron at its centre.
- Intracellular enzymes are required to provide energy and resist the toxic effects of oxygen.
- Upon the death of the cell, globin proteins, lipids, and iron are recycled. The remaining pyrrole/porphyrin ring is degraded to bilirubin and excreted.
- Anisocytosis describes the variation in size of red blood cells, polychromasia describes variation in their 'colour', and poikilocytosis their shape.
- Red cells may contain one or more different types of inclusion body, which often indicate a particular metabolic problem.
- Erythrocytosis describes an increase in the number of red cells, the Hct, and Hb.
- In polycythaemia vera, there is also a rise in red cell indices, but in addition there are raised platelets and/or white cells, and a risk of leukaemia.

Discussion questions

4.1 How is the red blood cell adapted for its purpose?

4.2 What are the different types of Hb within the cell and why do they change as the embryo develops into a foetus, an infant, and an adult?

4.3 Explain the different features of the morphology of the red cell.

Further reading

- Daniels G. ***Variants of RhD–current testing and clinical consequences***. ***British Journal of Haematology*** **2013: 161: 461–70.**

- Franke K, Gassmann M, and Wielockx B. ***Erythrocytosis: The HIF pathway in control***. *Blood* 2013: **122**: 1122–8.
- Gao X, Xu C, Asada N, and Frenette PS. ***The hematopoietic stem cell niche: From embryo to adult***. *Development* 2018: **145**: dev139691 doi:10.1242/dev
- Girelli D, Nemeth E, and Swinkels DW. ***Hepcidin in the diagnosis of iron disorders***. *Blood* 2016: **127**: 2809–13.
- Kafina MD and Paw BH. ***Intracellular iron and heme trafficking and metabolism in developing erythroblasts***. *Metallomics* 2017: **9**: 1193–203.
- Kim M, Tan YS, Cheng WC, Kingsbury TJ, Heimfeld S, and Civin CI. ***MIR144 and MIR451 regulated human erythropoiesis via RAB14***. *British Journal of Haematology* 2015: **168**: 583–97.
- Lux SE. ***Anatomy of the red cell membrane skeleton: unanswered questions***. *Blood* 2016: **127**: 187–99.
- Mohanty JG, Nagababu E, and Rifkind JM. ***Red blood cell oxidative stress impairs oxygen delivery and induces red blood cell aging***. *Frontiers in Physiology* 2014: **28(5)**: 84. doi:10.3389/fphys.2014.0008.
- Muckenthaler MU, Rivella S, Hentze MW, and Galy B. ***A red carpet for iron metabolism***. *Cell* 2017: **168**: 344–61.
- Nandakumar SK, Ulirsch JC, and Sankaran VG. ***Advances in understanding erythropoiesis: Evolving perspectives***. *British Journal of Haematology* 2016: **173**: 206–18.
- Waye JS and Eng B. ***Diagnostic testing for α-globin gene disorders in a heterogeneous North American population***. *International Journal of Laboratory Hematology* 2013: **35**: 306–13.
- Weiss CN and Ito K. ***A macro view of microRNAs: The discovery of microRNAs and their role in hematopoiesis and hematologic disease***. *International Review of Cell and Molecular Biology* 2017: **334**: 99–175.
- Yeo JH, Lam YW, and Fraser ST. ***Cellular dynamics of mammalian red blood cell production in the erythroblastic island niche***. *Biophysical Reviews* 2019: **11**: 873–94.

Answers to self-check questions and case study questions are provided as part of this book's online resources.

 Visit www.oup.com/he/moore-fbms3e.

5

Anaemia 1: The bone marrow, micronutrients, and disease in other organs

Andrew Blann and Sukhjinder Marwah

Anaemia is undoubtedly the major (clinically significant) disease of red cells, causing a great deal of morbidity and mortality. As the pathology of this disease comprises a range of diverse conditions of varied aetiology, we will split our coverage of anaemia into two chapters. In the present chapter we examine how problems with the bone marrow, micronutrients (iron and vitamins), and disease in other organs can lead to anaemia. In Chapter 6, we will look at the remaining causes of anaemia (the destruction of red cells by factors such as antibodies, and those due to shorter lifespan as consequence of mutations in certain genes) and at other red blood cell disease.

Learning objectives

After studying this chapter, you should confidently be able to:

- appreciate that the principal disease of red blood cells is anaemia;
- explain how anaemia can arise from changes within the bone marrow;
- describe the consequences of low levels of iron;
- understand the pathological consequences of high levels of iron;
- describe the consequences of the poor supply of vitamins B_{12} and B_6 and folate to the bone marrow;

- outline how disease in different body organs can lead to anaemia;
- suggest how the laboratory can diagnose these conditions.

5.1 Introduction to anaemia

Broadly speaking, the pathology of red blood cells considers situations where key indices are either above or below the reference range. However, as we have discussed in Chapter 2, having a result within the reference range is no guarantee of health, just as having a result marginally outside the reference range does not automatically mean ill health. Low numbers of red cells, and/or low levels of haemoglobin generally lead to the clinical condition called anaemia. Considerably less common is the reverse of anaemia, whereby the red blood cell count and haemoglobin are high, which is also likely to lead to illness. This latter situation is present in two closely related conditions—polycythaemia and erythrocytosis, as are discussed in Chapter 4.

5.1.1 Clinical aspects of anaemia

The ultimate biochemical process is respiration: the process whereby energy is extracted from glucose in the mitochondrion and which requires oxygen. The delivery of oxygen to the mitochondrion is the rate-limiting step in the process, a pathway that begins at the lungs and requires an efficient cardiovascular system. Pulmonary and/or cardiovascular disease may lead to insufficient oxygen delivery to the cells of the body, a condition known as hypoxia. Those tissues whose oxygenation is less than optimal are said to be ischaemic. The consequences of hypoxia/ischaemia are therefore that insufficient energy is being generated from glucose and oxygen. This can have many repercussions, such as tiredness and lethargy, but more severely could lead to major disease such as myocardial infarction. However, these same symptoms can arise if the pulmonary and cardiovascular systems are working well, in which case the cause of hypoxia/ischaemia may be those cells responsible for delivering oxygen to the tissues, i.e. red blood cells. When there are too few red blood cells in circulation, or there is a problem with their function, then we consider the major pathology of red blood cells, anaemia.

Recognition and management of anaemia in an individual is also complicated by the variety and severity of signs and symptoms in different people, as set out in Table 5.1. Unfortunately, few of these signs and symptoms are sufficiently sensitive (i.e. are present in a large number of people with the condition) or are specific enough (i.e. are found only in those people with the condition) to be reliable in practice. We must therefore also examine the red blood cell population in addition to signs and symptoms.

Regardless of aetiology, anaemia (as defined by low haemoglobin) is not benign. In both men and women, as the haemoglobin level falls, the likelihood of mortality increases. Accordingly, once diagnosed, anaemia must be treated, not merely because it leads to a poor quality of life, as shown by the range of symptoms in Table 5.1. As we shall see, low haemoglobin and resulting anaemia may be the first signs of more serious and occult disease yet to be discovered, such as cancer. Even 'mild' anaemia brings an increased risk of morbidity and mortality after surgery, and this effect is more evident in those with moderate to severe anaemia before their operation.

Cross reference

Red blood cell function is described in Chapter 4.

TABLE 5.1 Signs and symptoms of anaemia.

Signs	Pallor (especially of the conjunctiva) Tachycardia (pulse rate over 100 beats per minute) Glossitis (swollen and painful tongue—reasonably specific for vitamin B_{12} deficiency) Koilonychia (spoon nails—reasonably specific for iron deficiency) Dark urine (in haemolytic anaemia)
Symptoms	Decreased work capacity, fatigue, lethargy Weakness, dizziness, palpitations Shortness of breath (especially on exertion) 'Tired all the time' In children: decreased IQ, poor concentration, and sleepiness Rarely: headaches, tinnitus, taste disturbance
More severe disease	Jaundice, splenomegaly Hepatomegaly, angina Cardiac failure, fever

5.1.2 Definitions of anaemia

Although anaemia is the primary pathological condition of red blood cells, there is a surprising lack of consensus as to how it should be defined. For example, one textbook defines anaemia (in the adult) as a level of haemoglobin of <140 g/L in adult males, or <120 g/L in adult females; in another textbook a result of <135 g/L in men or <115 g/L in women defines the condition. The World Health Organization, taking a global view, once defined the state of anaemia as when haemoglobin is <130 g/L in men and <120 g/L in women (<110 g/L if pregnant), with other cut-off points in childhood (6-59 months, <100 g/L, 5–14 years, <110 g/L) and adjustment in the presence of smoking and at altitude (https://www.who.int/vmnis/indicators/haemoglobin.pdf).

A more recent definition is that anaemia is a condition in which the number of red blood cells or their oxygen-carrying capacity is insufficient to meet physiologic needs, which vary by age, sex, elevation above sea level, the effect of smoking, and pregnancy status. A United Kingdom guideline recommends that anaemia in pregnancy is defined by haemoglobin <110 g/L in the first trimester, <105 g/L in the second and third trimesters, and <100 g/L in the postpartum period. Notably, this latter definition is purely numerical and does not consider whether the woman is symptomatic. Others characterize the disease as abnormalities in red blood cells that, alongside appropriate symptoms (Table 5.1), call for treatment—a definition that does not specify the level of haemoglobin or the sex of the subject. It follows from this point of view that someone whose haemoglobin is 90 g/L, but is asymptomatic, may not be anaemic. This underlines the important point that anaemia is more about symptoms than defined laboratory numbers, although when the former are sufficiently severe they will inevitably coincide.

Some authorities have proposed that age (in age ranges of 20–49 years and 50+ for women, and of 20–59 years and 60+ for men) and race (lower limit of normal for white men <132 g/L, black men <127 g/L, white women <122 g/L, black women <115 g/L) should also be considered when formulating a reference range. Others suggest degrees of anaemia can be further quantified, such as being mild (haemoglobin >100 g/L) or moderate (haemoglobin 85–100 g/L) (World Health Organization).

Key Points

It is, again, important to emphasize that the practitioner must take note of their local reference range, which varies from laboratory to laboratory.

A further classification of anaemia is based on the size of the red blood cell (i.e. the **mean cell volume, MCV**) as introduced in Chapter 4. In this system, if there is anaemia *and* the red blood cells are small (with a low MCV, perhaps 70 fL), then the anaemia can be described as **microcytic**. Conversely, the anaemia associated with large red cells (where, for instance, the MCV may be 105 fL), is termed **macrocytic**. When the red blood cells are of normal size *and* the subject is anaemic, then the term **normocytic** anaemia is used. As we will see, this classification is very important because it can be used to determine aetiology and, very likely, treatment. (See also Box 5.1.)

mean cell volume (MCV)
The volume of the 'average' red blood cell.

microcytic
Pertaining to a small red blood cell.

macrocytic
Pertaining to a large red blood cell.

normocytic
Pertaining to a normal-sized red blood cell.

Despite the discussion above, the competent practitioner will be aware that the presence of a result outside the reference range does not always confer a label of anaemia. For a particular subject, small red blood cells or large red blood cells (i.e. with many microcytes or macrocytes respectively, so that the MCV is outside the reference range) may be entirely normal, or at least physiologically tolerable, and therefore acceptable. For example, it may be quite normal for a pregnant woman to have red blood cells which are larger than before the pregnancy began—but she would hardly be described as having a disease! Similarly, the red cells of newborns and infants are larger than in the adult, so that in these situations an MCV of 108 fL may not be pathological.

5.1.3 The aetiology of anaemia

Anaemia may be the consequence of several different and distinct pathologies, and which include problems with the bone marrow, with the delivery of micronutrients to the bone marrow, or with other organs. Alternatively, there may be mutations in certain genes (such as those of haemoglobin, components of the cell membrane, or metabolic enzymes) that can lead directly to anaemia. Many of these changes can lead to the early destruction of

BOX 5.1 The definition of anaemia

As has been mentioned in section 5.1.2, the traditional approach to the definition of anaemia is essentially numerical—a level of haemoglobin lower than a particular value, dependent on age and sex, regardless of signs and symptoms. However, an alternative definition positions anaemia as a disease, linked with a set of symptoms and abnormalities in red blood cell biology that needs to be treated. It follows that an elderly woman with an Hb of 117 g/L, who is asymptomatic, may not, on clinical grounds, be treated. In which case, it could be argued that she is not anaemic.

The reference range for haemoglobin for this textbook is 133-167 g/L for men and 118-148 g/L for women. This in no way invalidates or criticizes any other reference range, and vice versa. Each practitioner must act on their local reference range. Other reference ranges for this textbook are given in Table 1 at the front of this book.

haemolytic anaemia
An anaemia where red blood cells are damaged and/or destroyed.

the red blood cell. This destructive process is called haemolysis, leading to the expression '**haemolytic anaemia**'. Naturally, more than one cause may be present in a particular patient at the same time.

The first of these potential causes of anaemia (the bone marrow, problems with micronutrients, and disease in other organs) is discussed in the present chapter. Anaemia due to haemolysis and to gene mutation (causing haemoglobinopathy, membrane defects, and lack of enzymes), will be explained in Chapter 6. Table 5.2 summarizes the aetiology of anaemia.

SELF-CHECK 5.1

What is the basis of the major types of disease of the red blood cell?

SELF-CHECK 5.2

How do we classify anaemia according to the size of the red blood cell?

TABLE 5.2 An aetiological classification of anaemia.

The bone marrow	Suppression of erythropoiesis Infiltration by malignant cells
Problems with iron	Deficiency at the bone marrow Excess absorption Ineffective incorporation into haem
Lack of vitamins	Vitamins B_{12} and B_6 Folic acid
Anaemia associated with disease in other organs	Liver Kidney Reproductive organs Connective tissues Thyroid
Anaemia due to blood loss	Haemorrhage (perhaps due to problems with haemostasis) Trauma (road traffic accident, personal violence, e.g. stab wound)
Haemolysis	Due to antibodies Due to drugs Infections
Anaemia arising from gene mutation	Haemoglobinopathy Membrane defects Enzyme defects

The first four topics are explained in this chapter, the last two topics in Chapter 6. Haemorrhage features in Chapters 14, 16, and 17.

5.2 Anaemia arising from changes in the bone marrow

The bone marrow is an excellent place to start our examination of diseases of the red blood cells as this is the site of their production (and of white cells and platelets). This production process, erythropoiesis, is described in Chapter 4. Interference with this process can lead to a reduction in the numbers of red blood cells being produced, with no change in the production of other blood cells—the white blood cells and platelets. The most common condition that leads to this reduction in red cell numbers alone is described as **pure red cell aplasia (PRCA)**.

Cross reference
Erythropoiesis is discussed in Chapters 3 and 4.

However, in many cases, one or more factors that reduce red cell numbers will also lead to a reduction in white blood cells and platelets. This feature, **pancytopenia**, meaning 'low levels of all types of blood cells', may be caused by a number of factors and often arises when the bone marrow is invaded by cancer originating in other organs. Alternatively, the bone marrow is very sensitive to external factors such as drugs (whether medicinal, such as the antibiotic chloramphenicol; environmental, such as insecticides; or industrial, such as benzene), radiation, and infectious agents (bacteria, such as in tuberculosis; and viruses, such as hepatitis B virus), all of which can lead to pancytopenia. From a historical perspective, the first disease of this nature to be described was aplastic anaemia, which (at the time) focused inappropriately on red cells.

5.2.1 Red cell aplasia

The key to this group of diseases is the major red blood cell precursor—the erythroblast. There are two major variants of this condition. In the first, there is a severe reduction, or even an absence, of erythroblasts. In the second, erythroblast numbers are increased.

PRCA is the most common form of red cell aplasia due to absent or reduced numbers of erythroblasts. It may appear as a congenital condition or it could arise as an acquired syndrome. The most common congenital form of PRCA is **Diamond-Blackfan anaemia** (also called congenital hypoplastic anaemia), having an incidence of only between four and seven per million live births. A family history of anaemia is common and inheritance appears to be autosomal-dominant, whilst the disease is familial in 20%, the majority of cases occurring sporadically. There are often concomitant congenital abnormalities such as of the heart and kidney, and 90% of cases are diagnosed in the first year of life. The dominant feature of the full blood count is a normochromic macrocytic anaemia and reticulocytopenia, whilst the bone marrow is normocellular. Mutations in any one of 15 genes coding for ribosomal proteins are present in 60% of patients, the remainder having no known pathogenic mutation. The ratio of 28S/18S rRNA, coding for the 40S and 60S ribosomal subunits, may be a useful diagnostic tool. The anaemia may respond to glucocorticoids and growth factors (but not erythropoietin, Epo), but the only cure is bone marrow transplantation, perhaps with part-human leucocyte antigen (HLA)-matched nucleated cord blood cells. However, an alternative diagnosis for anaemia between the ages of one and five years is transient erythroblastopenia (that is, lack of erythroblasts) of childhood, which usually develops in the second year of life without a family history of anaemia.

Cross reference
Diamond-Blackfan anaemia is discussed in the context of malignancy in Chapter 9.

Acquired PRCA can be classified as primary (idiopathic—where no clear cause can be identified) or secondary (acquired as the result of exposure to a clear pathogenic agent, such as a drug or a virus). The list of known causes of acquired PRCA is considerable (Table 5.3).

The congenital dyserythropoietic anaemias (CDAs) include several conditions resulting from increased numbers of erythroblasts (that is, erythroid hyperplasia), and which may be large (that is, are megaloblastic). Erythropoiesis is ineffective and the anaemia is generally mild. Red cells from the best-characterized and most common variant (type II) are susceptible to lysis by acidified serum. The molecular basis of this appears to be lack of an enzyme necessary for the correct processing of certain cell membrane proteins, which renders the cell fragile.

5.2.2 Pancytopenia

Unlike PRCA, where only red cell production is suppressed, pancytopenia is characterized by low levels in the blood of all three types of blood cell—red cells, white cells, and platelets. The accompanying anaemia is called aplastic anaemia. The causes of this reduction in cell numbers can be classified as being congenital, acquired, or idiopathic.

Cross reference

Fanconi's anaemia is discussed in greater detail in Chapter 9.

The principal *congenital* cause of pancytopenia (accounting for two-thirds of cases) is **Fanconi's anaemia (FA)**. This inherited bone-marrow failure syndrome is caused by mutations in any one of 17 genes that make up a DNA repair pathway. The protein product of one of these genes, BRCA1, binds to several FA proteins, all of which are essential for normal haemopoiesis. Although the effect of deficiency in BRCA1 in our species in unknown, in the mouse, knock-out of *Brca1* causes Fanconi-like cytopenias. Apart from pancytopenia, there are also likely to be skeletal, renal, and neurological abnormalities alongside short stature and upper limb abnormalities (such as the absence of thumbs). At the nuclear level, the disease is characterized by chromosomal fragility, with breaks in the DNA; a late consequence of this disease is an increased risk of malignancy, including leukaemia, solid tumours, and **myelodysplasia**. Accelerated telomere attrition is characteristic of early myeloid oncogenesis, specifically chromosome 7 loss in myelodysplasia and acute myeloid leukaemia after severe aplastic anaemia. Other rare congenital causes of pancytopenia include the Schwachman–Diamond syndrome and dyskeratosis congenita.

Like PRCA, there are many causes of acquired pancytopenia, and several are common between the two (such as viral hepatitis and systemic lupus erythematosus) (Table 5.4). Although many of the remaining causes of pancytopenia and aplastic anaemia are idiopathic, they often have an immunological basis. Indeed, it has been argued that the aplastic anaemia following pregnancy and viral infections is in fact a direct consequence of an aberrant immune response.

TABLE 5.3 Causes of red cell aplasia.

Congenital	Diamond–Blackfan syndrome, CDA
Infections	Viruses: parvovirus B19, hepatitis A, B, C, and E viruses, Epstein–Barr virus, mumps, cytomegalovirus, human immunodeficiency virus Bacteria: meningococcal and staphylococcal species, tuberculosis
Malignancy	Solid tumours (such as cancer of the thymus, stomach, breast, lung, thyroid, and kidney), haematological tumours (leukaemias, lymphomas, myeloma, myelofibrosis, essential thrombocythaemia, Waldenström macroglobulinaemia), Castleman disease
Autoimmune disease	Systemic lupus erythematosus, rheumatoid arthritis and autoimmune haemolytic anaemia, Sjögren's syndrome, autoantibodies to red cell progenitors, autoantibodies to Epo, T-cell-mediated recognition of red-cell progenitors, inflammatory bowel disease
Other causes	Drugs and chemicals (notably azathioprine, methotrexate, gold, isoniazid, chloramphenicol, recombinant human Epo, and co-trimoxazole), pregnancy, severe renal failure, ABO-incompatible stem cell transplantation, pyoderma gangrenosum, riboflavin deficiency

Note: Many of these also cause pancytopenia, and so aplastic anaemia.

This theory is supported by the observation of immune-mediated, T-lymphocyte destruction of bone marrow cells, and that this can be reversed by immunosuppression with anti-thymocyte globulin and ciclosporin.

SELF-CHECK 5.3

What are the two most common forms of congenital reduction in the number of red blood cells?

5.2.3 The anaemia of haemopoietic cancer

The bone marrow is simply a place of production, perhaps a factory, whose products are blood cells. Any factory whose work space is taken over by external forces or objects, and is thus at a reduced work capacity, will clearly be unable to produce goods. In this respect, the bone marrow is no exception. The major invader is cancer, which can spread from its original site (such as the breast or prostate) to invade other tissues. However, there may also be cancer arising within the bone marrow itself. In both cases, the pathological basis of the disease is that the cancer tissue invariably grows and slowly takes over normal bone marrow tissue. If this is the case, then erythropoiesis will suffer and so anaemia may result.

White blood cell cancer

There are several cancers of the bone marrow itself: the principal conditions are the haemoproliferative diseases of leukaemia and myeloma. Put simply, leukaemia is characterized by a failure of the bone marrow to correctly regulate the number of white blood cells it produces, such that it makes too many. Not only are too many white blood cells produced, but in addition the cells are abnormal in that they are unable to provide support to the body in defending itself from attack by microorganisms. In the early stages of the disease, these cancerous white blood cells remain in the bone marrow, but as the disease develops they pass into the blood. Ultimately, the number of these cancerous white blood cells rises so that the peripheral white blood cell count itself will become elevated.

Myeloma is also a disease of white blood cells, which sees them multiply in an inappropriate manner. In this case, however, the tumour of these abnormal cells generally remains inside the bone marrow until the terminal stages of the disease, when abnormal cells can be found in the blood. A third white blood cell tumour, **lymphoma**, centres on the lymph node. Like myeloma and leukaemia cells, lymphoma cells can also spread from their site of origin (that is, a particular lymph node) and invade other lymph nodes and other organs, notably the spleen, liver, and bone marrow. Thus, when lymphoma cells, which may also be found in the blood, invade the

TABLE 5.4 Causes of pancytopenia.

Congenital	Fanconi's anaemia
Acquired	Chemotherapy (e.g. chloramphenicol, phenylbutazone, gold) Infections (e.g. viral hepatitis, parvovirus B19) Other defined bone marrow disease (e.g. myelofibrosis, myelodysplasia) Other defined disease (e.g. hypersplenism, systemic lupus erythematosus) Severe vitamin B_{12} deficiency
Idiopathic	Frequently shown to be immune mediated

Cross reference

Tumours of the bone marrow and of lymph nodes (leukaemia, myeloma, and lymphoma) are described fully in Chapters 9, 10, 11, and 12.

bone marrow they may displace those stem cells responsible for producing red blood cells and so cause a normocytic anaemia.

Myelodysplasia and myelofibrosis

Despite roughly similar names, these two diseases are profoundly different. Myelodysplasia is not a single disease, and so the myelodysplastic syndrome (often abbreviated to MDS) is a collection of disorders characterized by the clonal proliferation of multipotential haemopoietic stem cells, whereby these cells are disordered and inefficient. Many cells die in the bone marrow, leading to ineffective haemopoiesis and a reduction in the numbers of all circulating blood cells. As perhaps 25% of patients will convert into developing acute myeloid leukaemia, myelodysplasia has been described as 'preleukaemia' or 'smouldering acute leukaemia'.

Unlike myelodysplasia, leukaemia, and myeloma, where the disease arises from haemopoietic precursors, in myelofibrosis the proliferating cell is not haemopoietic (that is, it is not a stem cell). Instead, the bone marrow becomes overgrown with fibroblasts, and, as with leukaemia and lymphoma, there is often splenomegaly. These fibroblasts are not thought to be directly involved in haemopoiesis, but instead are believed to have a supporting role. The fibroblasts are often driven to proliferate and produce collagen by inappropriate responses to growth factors produced by other cells such as monocytes, macrophages, and megakaryocytes. Like cancer, this inappropriate growth of fibroblasts is progressive and eventually leads to deterioration in haemopoiesis and thus to anaemia, treatments for which include blood transfusion and Epo.

Cross reference

Further details of myelodysplasia and myelofibrosis are presented in Chapters 9 and 11. The importance of stem cells is discussed in Chapters 3 and 4.

Chemotherapy

An important component of modern medicine is **chemotherapy** (treatment using drugs). While the common usage of the term 'chemotherapy' is taken to imply the use of drugs in the treatment of cancer, it actually applies to the administration of *any* drug. An important source book and online resource widely available in hospitals in the UK is the British National Formulary (BNF), which lists all drugs licensed by government agencies for the treatment of precise disorders. Each drug has notes regarding its mode of action, dose, and also side effects. Entries include those drugs that are perceived to be supposedly innocuous, with few side effects, such as aspirin (although even aspirin can lead to gastrointestinal upsets in some subjects). Other drugs, such as antibiotics, can cause side effects such as allergy, but in rare cases there may also be suppression of the bone marrow. However, the most dangerous diseases call for the more severe forms of chemotherapy, as are demanded in certain cancers. One of the most frequent side effects of the many different classes of drug used in the treatment of cancer is suppression of the bone marrow; other side effects include jaundice, renal failure, alopecia, and diarrhoea. Fortunately, bone marrow suppression (and other side effects) is almost always reversible upon cessation of use of the particular drug.

A standard treatment for many different types of cancer is the administration of drugs such as methotrexate, cyclophosphamide, busulphan, and vinblastine, which act to destroy the tumour. Most of these drugs are effectively sophisticated poisons, and so are described as cytotoxic. Notably, many of these drugs may be used to treat haematological neoplasia, such as leukaemia. However, a problem is that the cytotoxic drugs themselves can be relatively indiscriminate in terms of the cells they attack, so that effectively any cell in the body can be damaged by chemotherapy. Indeed, the jaundice associated with certain drugs is the consequence of their cytotoxicity on the liver. This adverse effect is particularly true of the bone marrow: it is a highly active tissue, which is very susceptible to cytotoxic drugs of this type. Nevertheless, as we have seen, even supposedly benign and commonly used classes of drugs such as antibiotics (e.g. chloramphenicol) and non-steroidal anti-inflammatory drugs (NSAIDs, e.g. phenylbutazone) can also suppress the bone marrow.

Together, medication drugs cause 15–25% of cases of bone marrow suppression. Indeed, a frequent cause of both PRCA and pancytopenia (and therefore also aplastic anaemia) is chemotherapy (Tables 5.3 and 5.4). Various non-medication drugs and chemicals can also attack the bone marrow; examples include industrial hydrocarbons such as benzene and agricultural drugs such as pesticides.

Standard treatments for chemotherapy-associated anaemia include red blood cell transfusion and erythropoiesis-stimulating agents. If necessary, intravenous iron may also be given, but the effect of this agent is variable, depending on iron stores and inflammation. In such cases, serum hepcidin levels may help predict those patients most likely to benefit from iron therapy.

Cross reference

The role of benzene in the development of malignancies is discussed in Chapter 9.

SELF-CHECK 5.4

What are the principal types of haematological cancer that can lead to anaemia?

5.2.4 Treatment of anaemia resulting from disease of the bone marrow

Treatment of bone-marrow-derived anaemia very much depends on the aetiology and the clinical severity of the particular disease. Where the cause of the anaemia is evident (e.g. where it is being caused by a particular drug), then the disease should be reversible upon withdrawal of the drug. Equally, the aplastic anaemia of pregnancy generally resolves after delivery of the neonate. By contrast, congenital diseases such as Fanconi's anaemia are curable only by bone marrow transplantation, although androgens may help stimulate erythropoiesis.

Some cases of acquired or idiopathic PRCA may be successfully treated by immunosuppression with ciclosporin or corticosteroids (e.g. prednisolone 1 mg/kg per day). Unfortunately, the treatment of most haematological malignancies is with agents that will also further suppress erythropoiesis (including cytotoxic drugs and radiotherapy). However, growth factors such as Epo and granulocyte macrophage colony-stimulating factor (GM-CSF) can be given to help the recovery of the red cell count.

In other circumstances, treatment will be palliative—red cell transfusion for incapacitating or life-threatening anaemia (as described above), platelet transfusion for severe thrombocytopenia, and prophylactic antibacterials and antivirals if the numbers of white blood cells are profoundly low. Intravenous immunoglobulins may be used for active viral infections. If there is iron overload because of hypertransfusion, then iron chelation (the process of the chemical removal of iron) may be necessary.

5.2.5 The role of the laboratory in the investigation of anaemia following bone marrow changes

The initial signs of PRCA are likely to be those of general anaemia—that is, a reduced red cell count and haemoglobin. The haematocrit is unlikely to be below the lower end of the reference range unless the red cell count is considerably reduced. By definition, white cells and platelets will be within the reference range. Consequently, a full blood count is essential. However, key investigations also involve examination of the bone marrow—which in PRCA is normocellular with respect to leucocyte precursors and megakaryocytes, but will reveal grossly reduced or absent erythroblasts.

In Diamond–Blackfan anaemia, the anaemia is often moderate to severe (haemoglobin 20–100 g/L) but the MCV often exceeds the top of the reference range (that is, there is likely to be a macrocytic anaemia). Notably, foetal haemoglobin may be increased, possibly as an adaptation to the anaemia. This is in contrast to the transient erythroblastopenia of childhood, where the MCV is within the reference range and the expression of foetal haemoglobin is normal. In the peripheral circulation there is mild to moderate anaemia (haemoglobin 80–110 g/L) with mild macrocytosis, anisocytosis, and **poikilocytosis**, although in some forms acanthocytes and other bizarre red cells may be present. If a bone marrow examination is thought necessary, PRCA is characterized by low or absent erythroblasts, whereas the CDAs are characterized by increased numbers of erythroblasts.

poikilocytosis
A variation in the shapes of the red cell population. Causes include abnormal erythropoiesis and myelofibrosis.

The anaemia that follows from leukaemia is perhaps the easiest to detect, as a high white blood cell count (and possibly a thrombocytopenia) is often present. The anaemia is most likely to be normocytic and normochromic. A bone marrow aspiration will inevitably be performed and should find all the erythroid precursors (including reticulocytes) to be present but in reduced numbers. The bone marrow in myelodysplasia is usually hypercellular, often with enlarged and abnormal erythroid precursors (such as multinucleate normoblasts), although there may be abnormalities in the precursors of all cell lineages. However, a feature of bone marrow in some types of myelodysplasia is the presence of **ring sideroblasts**. These will be described more fully in section 5.3 on iron.

ring sideroblasts
Abnormal erythroblasts with iron granules arranged in a ring around the nucleus.

The definition of pancytopenia is a reduction in the numbers of red cells, white cells, and platelets. Consequently, it should be relatively simple to diagnose from a full blood count. Generally, the (aplastic) anaemia should be normochromic and normocytic. The bone marrow will be hypoplastic, with the normal haemopoietic tissues being grossly reduced and replaced by fat cells and other non-haemopoietic tissues. However, deviations from this picture occur in Fanconi's anaemia and in acquired aplastic anaemia, where the red cells may be macrocytic. The bone marrow in myelofibrosis will show increased collagen and fibroblasts alongside reduced haemopoietic precursors.

Generally speaking, the reticulocyte count, normally 0.5–2.5% of the red cell count (absolute count 25–125 × 10^9 cells/L), should be raised in most cases of chronic anaemia as a physiological response to the inability of the depleted red cell mass to supply the tissues with sufficient oxygen for their physiological needs. However, in those anaemias where the bone marrow and erythropoiesis is compromised (as in PRCA and aplastic anaemia) reticulocytes should be low, or even absent, from the peripheral blood and grossly reduced in the bone marrow.

Further details of the use of bone marrow analysis in leukaemia and other neoplasia are presented in Chapter 10.

Key Points

The bone marrow is a very active and a very sensitive organ. Generating millions of cells daily, it can be influenced by many conditions not allied directly to haematology, and also by many drugs and other treatments. Consequently, it is not surprising that even the slightest adverse effect could lead directly to changes in blood cell numbers, and so to illness.

5.3 Insufficient iron

Almost all haematological indices have related pathology both above and below their respective reference ranges, and iron exemplifies this double-edged sword. A crucial micronutrient, iron (and its co-micronutrients vitamins B_{12} and B_6 and folic acid) is required for various aspects of human physiology, not merely haemopoiesis. Chapter 4 indicates the steps in the development of red blood cells that require the vitamins and folate as cofactors for key enzymes in the metabolism of haem. The same chapter also explains the need for iron. These micronutrients are obtained from the diet, so that inadequate nutrition (that is, malnutrition) is likely to cause different types of anaemia.

Each molecule of haemoglobin must have four atoms of iron to which oxygen can (transiently) bind and so be carried from the lungs to the tissues. Thus, a lack of iron leads directly to impaired carriage of oxygen. Indeed, iron deficiency, when due to malnutrition and/or malabsorption, is the most common cause of anaemia in the developed and developing world. The World Health Organization estimates that 30% of the world's population (some 2 billion people) is anaemic, and within this group the most common is **iron-deficiency anaemia**, which results from insufficient iron being delivered to the bone marrow. In turn, the most common cause of this deficiency is poor nutrition. However, iron deficiency does not lie behind all anaemia: a numerically minor aspect of iron-related disease is **sideroblastic anaemia**, another being iron overload. We will consider these conditions in turn.

sideroblastic anaemia
An anaemia that results from impaired inclusion of iron into haem, and thus into haemoglobin.

5.3.1 Iron requirements

The total iron load of an average adult is in the region of 3-5 g (towards the lower end of the range in the female) and is distributed in different molecules, cells, and tissues. The majority of iron is found in haemoglobin (generally two-thirds of total body iron) and deposits in stores in the liver, bone marrow, and elsewhere (as ferritin and haemosiderin; 25% of total body iron). The remainder occurs as trace amounts in cells and tissues where, for example, the iron may be needed as a cofactor for metabolic enzymes such as catalase and the cytochromes. Almost all total body ferritin is in the cytoplasm of various cells (such as hepatocytes and macrophages), but a small fraction is present in the plasma at a concentration of 20-300 μg/L. This is important, as it seems likely that the plasma levels of ferritin reflect the total body stores.

Approximately 1-2 mg of iron a day is lost in urine, in shed epithelial cells of the skin and intestines (the latter in faeces), and in sweat; all of these losses must be recovered from the diet. However, daily iron requirements vary with age and sex, as shown in Table 4.6. More is required in menstruation, where 20-25 mg is lost per cycle. This aside, the adult requirement of 9 mg a day is relatively easy to achieve in a Western diet, which provides approximately 15 mg of iron each day, although pregnancy may require a supplement, as 500-1000 mg of extra iron is required by the foetus.

5.3.2 Iron-deficient anaemia

Whilst the biological basis of this anaemia seems clear, the aetiology is diverse, and deficiency can occur at any point in the path of iron from the diet to the erythroblast mitochondria in the bone marrow (see Figure 4.11). The clinico-pathological importance of this micronutrient justifies frequent referral to the physiology of iron, as introduced in section 4.3.1. Figure 5.1 summarizes how malnutrition and specific organ disease can lead to iron deficiency.

FIGURE 5.1
Potential routes to iron-deficient anaemia: there are several alternative aetiologies, such as insufficient iron in the diet (malnutrition), diseases of the stomach (failure of a sufficiently acid environment), upper and lower intestines (malabsorption), and liver disease (failure to produce sufficient transferrin, the iron transport protein).

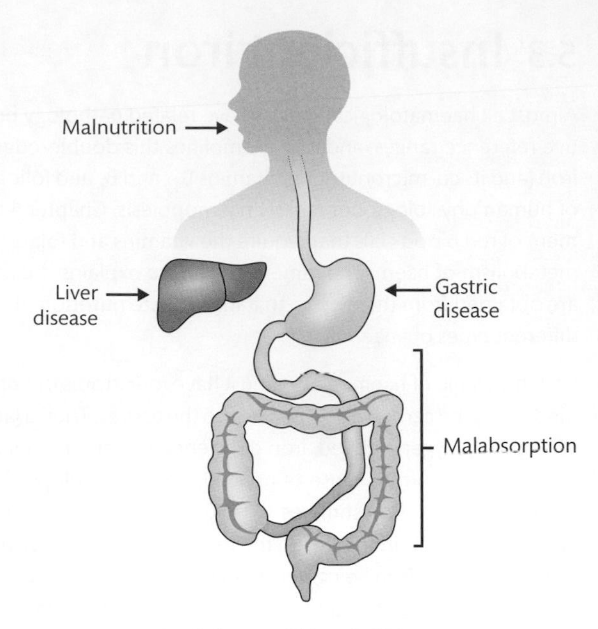

Aetiology of iron deficiency

Absorption of iron

Although, as an element, iron is relatively abundant, it is not present in all foodstuffs. Thus, an iron-poor diet may be unable to supply enough of this micronutrient. As a result, a clinically evident iron deficiency may develop whose aetiology is malnutrition. Indeed, the fortification with iron of common processed foodstuffs (such as bread) has been advocated as one method to alleviate the iron deficiency of malnutrition. However, since iron-recycling macrophages provide 20–25 mg of iron to the erythroblast pool daily, and hepatic iron stores may be 300 mg–1 gm, it may take weeks or perhaps months for any effect of absorption failure to lead to clinically obvious iron-deficient anaemia.

Ferrous iron (Fe^{2+}) is readily absorbed by a transmembrane protein, the divalent metal transporter-1, which is found on the luminal surface of enterocytes of the duodenum and upper jejunum. However, although ferric iron (Fe^{3+}) is poorly absorbed, it can be converted to ferrous iron by the luminal ferrireductase enzyme duodenal cytochrome b. This functions best at low pH, an environment promoted by ascorbic acid and by acid secretions from the stomach. It follows that in the presence of severe gastric pathology, such as cancer, gastritis, or atrophy, iron will fail to be absorbed in sufficient amounts to satisfy daily demand. Thus disease, or surgery to the stomach and intestines, can also lead to the malabsorption of iron (Table 5.5). Indeed, this aspect will be revisited in the section on vitamin B_{12} that follows.

TABLE 5.5 Intestinal factors contributing to reduced iron absorption.

The stomach	Achlorhydria, gastric atrophy, gastritis Alcoholism, gastric carcinoma
Upper and lower digestive tract	Duodenitis, coeliac disease Ulceration, Crohn's disease Other inflammatory bowel diseases Increased hepcidin levels
Surgery	Resection of any of the above tissues

Regulation and transport of iron in the plasma

Once imported by the enterocyte, the movement of iron into the plasma is regulated by the membrane protein ferroportin, levels of which are, in turn, regulated by the liver protein hepcidin. Pathology of this process is discussed in section 5.4. Once exported into the circulation, iron must be picked up by its transport protein, apotransferrin, whereupon the two combine to form transferrin. This protein is produced mainly by the liver, and so plasma levels of this protein may fall in the face of hepatic dysfunction such as that caused by alcoholic liver disease, hepatitis, cirrhosis, primary biliary sclerosis, and hepatoma, and so iron-deficient anaemia.

Upon arrival at the bone marrow, iron-loaded transferrin passes into red cell precursors such as erythroblasts by binding to transferrin receptors on the surface of the cell, which are then internalized into an endosome. Once the endosome is inside the cytoplasm, the iron is uncoupled from the transferrin at low pH, and ferric iron is converted to ferrous iron by a metalloreductase proteins of the STEAP (six-transmembrane epithelial antigen of the prostate) family, principally STEAP3. The iron is then exported via the divalent metal transporter-1: the same molecule involved in moving iron from the lumen of the intestines into the enterocyte (see Figure 4.9). The iron-free apotransferrin leaves the erythroblasts for the plasma, where it picks up more iron, so repeating the process. However, plentiful iron in the diet, a healthy intestine, good carriage to the bone marrow, and presentation of the iron to erythrocyte progenitors may be to no avail if these cells are unable to process the iron into haem. In such cases, the iron and its complexes build up within the mitochrondria of the erythroblast, and histological staining for iron shows it to form in rings.

Cross reference

The movement of iron from the diet to the bone marrow, and its incorporation into haem, is explained in Chapter 4, section 4.3, where Figures 4.10 and 4.11 apply.

Treatment of iron-deficiency anaemia

The dominant treatment of iron-dependent anaemia (once proven with appropriate laboratory investigations) is intuitively simple—oral or intravenous iron to boost the deficiency. The former is generally about 70–80 mg of elemental iron, three times a day. However, given our knowledge of the complex nature of iron metabolism (Figures 5.2 and 5.3), this will may need to be revised in individual cases. All being well with the intestines (which, of course, is not always the case), a single dose of oral iron should in theory be absorbed into the blood. This is likely to stimulate the production of hepcidin, which should then feed back to ferroportin to inhibit the export of more iron into the blood, which somewhat defeats the objective. The same principle is likely to frustrate the effect of intravenous ion. One solution is to reduce the frequency, to perhaps once every three days, to allow the levels of hepcidin to fall, and so release ferroportin to transfer its iron to the plasma.

5.3.3 Sideroblastic anaemia

Sideroblastic anaemia is characterized not by a lack of iron in the diet, in the blood, or in the tissues, but by its failure to be incorporated into haem in red blood cell precursors such as the erythroblast. This results in ineffective erythropoiesis and so anaemia, with low numbers of reticulocytes. Because the delivery of iron to the bone marrow is unimpaired, it builds up and so causes secondary iron overload, with deposits within the bone marrow and in other organs. At the intracellular level, the iron can be shown to localize in random deposits (hence sidero-, from Latin *sideris*, meaning constellation). Iron-rich mitochondria form a ring around the nucleus of the erythroblast, and as such the cells are described as sideroblasts. These ring sideroblasts give the disease, sideroblastic anaemia, its name. Therefore, the condition can only truly be diagnosed by the presence of sideroblasts in a sample of bone marrow. In some cases, red cells can still be produced, but may contain iron granules: such cells are called siderocytes, and the iron inclusion granules are called **Pappenheimer bodies**. Iron deposits in sideroblasts and siderocytes can be detected using Perls' stain (discussed in section 5.5.3). As the leading aetiology of these conditions is genetic, the clinically dominant form is congenital sideroblastic anaemia.

Failure of the synthesis of haem

The metabolic pathways for the generation of haem are complex, as outlined in Figure 4.12 of Chapter 4, and are prone to error. We will now review these pathways and how such errors lead to sideroblastic anaemia. Although certain pathways run in parallel, it is convenient to describe them in an order, with the support of Figure 5.2.

1. An early step is the requirement for thiamine (vitamin B_1), an essential dietary component for the correct functioning of the Krebs cycle, providing not only ATP but also metabolites, such that deficiency leads to numerous clinical conditions. It is transported by SCL19A2 (coded for by *SCL19A2* on 1q23.3): nonsense and missense mutations in *SCL19A2* lead to megaloblastic erythroid maturation and ring sideroblasts.

2. Glycine is transported across the inner mitochondrial membrane by SCL235A38. Missense, nonsense, and splicing errors in *SCL235A38* on 3p22.1 lead to failure to deliver glycine, and so failure to generate a key product in this metabolic pathway.

3. The enzyme regulating the generation of methylmalonyl-coenzyme A requires vitamin B_{12}, and that catalysing the formation of delta-aminolaevulinic acid (ALA) requires vitamin B_6, so that deficiencies of these vitamins will lead to a reduction in the production of haem. A further need is activation by mitochondrial chaperone ClpX (coded for by *CLPX* on 15q22.31). Loss-of-function mutations in *ALAS* (coding for aminolaevulinic acid synthase) lead to impaired haem synthesis: iron cannot be incorporated into the protoporphyrin ring, but instead builds up in the mitochondria of the erythroblast. *ALAS* is found on the X-chromosome, mutations in which lead to a sex-linked variant of sideroblastic anaemia, which predominantly affects males, and comprises ~40% of all cases. Missense and nonsense mutations in *ALAS2* on Xp11.21 cause failure to synthesize ALA from glycine and succinyl-coenzyme A. Glycine transporter SCL235A38 may also have a role in exporting ALA into the cytoplasm.

4. Cytoplasmic ALA is converted to the tetrapyrrole coproporphyrinogen III (CPgenIII), which is then transported through the outer mitochondrial membrane by ABCB6. CPgenIII is converted to protoporphyrinogen IX (PPgenIX) by an oxidative decarboxylation reaction in the inter-membrane space, and this is then transported across the inner mitochondrial membrane into the matrix by TMEM14C. PPgenIX is converted into protoporphyrin IX (PPIX) by protoporphyrin oxidase. It follows that loss-of-function mutations in genes for the enzymes and transporters in this process will lead to failure to generate haem.

5. Abnormalities in the iron-import regulator mitoferrin1 present on the internal mitochondrial membrane lead to low levels of iron within the organelle. Mitoferrin1 may be stabilized on the inner membrane by complexing with ferrochelatase and ATP-binding cassette ABCB10, the latter being under the control of GATA1, providing further insights linking loss-of-function mutations in this pathway to failure of erythropoiesis. In normal erythropoiesis, mitoferritin mRNA levels are low, as are those of the protein, but in hereditary sideroblastic anaemia, mitoferritin overexpression could represent a method of sequestering excess (toxic) iron.

6. Whilst iron may pass directly from mitoferrin to ferrochelatase, it may also be collected by frataxin, a potential iron storage protein and/or transporter. Coded for by *FXN* at 9q21.11, missense mutations causing a lack of frataxin are linked to Friedreich's ataxia.

7. Frataxin can deliver iron (a) to ferrochelatase, where lack of function mutations lead to failure of the insertion of iron into the pyrrole ring to generate haem, and so haemoglobin (b) to iron-sulphur clusters, and (c) to mitochondrial ferritin. Increased levels of this storage molecule may sequester a high proportion of iron to the detriment of ferrochelatase and iron-sulphur clusters. A consequence of the failure to incorporate iron into the protoporphyrin ring is the intra-mitochondrial accumulation of the iron, as noted above, which adds to the deposits in the iron-storage protein mitoferritin and in iron-sulphur clusters.

8. Haem is exported from the mitochondrion via feline leukaemia virus subgroup C cellular receptor, encoded for by *FLVCR1* on 1q32.3. Mutations in this gene produce proteins with folding errors that are unable to export haem, leading to accumulation of toxic haem and failure to generate haemoglobin.

9. Iron-sulphur clusters are formed from iron (delivered by frataxin) and sulphur (delivered by the conversion of cysteine to alanine) in double-dimer (Fe_2S_2) or double-tetramer (Fe_4S_4) isotypes, and may have roles in the electron transfer chain, storage, and/or regulation. One cluster, iron-regulatory protein 1 (IRP1), plays an important role in regulating ALAs—when iron levels are low, IRP1 binds to an iron-regulatory element in the ALAs promotor region. This prevents the accumulation of PPIX intermediates, increased levels of which can promote reactive oxygen species that will be toxic to intracellular enzymes and membranes.

10. Glutaredoxin 5, coded for by *GLRX5* on 14q32.2, is a crucial part of the integrity of iron-sulphur complexes. Splicing defects in *GLRX5* lead to unstable iron-sulphur clusters and mitochondrial iron overload, which in turn lead to repression of *ALAS* and so reduced haem synthesis. Defects in glutaredoxin have been shown to cause sideroblastic anaemia by impairing haem biosynthesis and depleting cytosolic iron.

11. Missense mutations in *ABCB7* on Xq13 lead to failure to export iron-sulphur clusters into the cytoplasm, and so their accumulation in the mitochondria. Markedly underexpressed *ABCB7* is an important mediator of the refractory anaemia with ring sideroblasts, a feature of myelodysplastic syndromes, in which mutations in *SF3B1* (linked to a ribonucleoprotein complex) are also implicated.

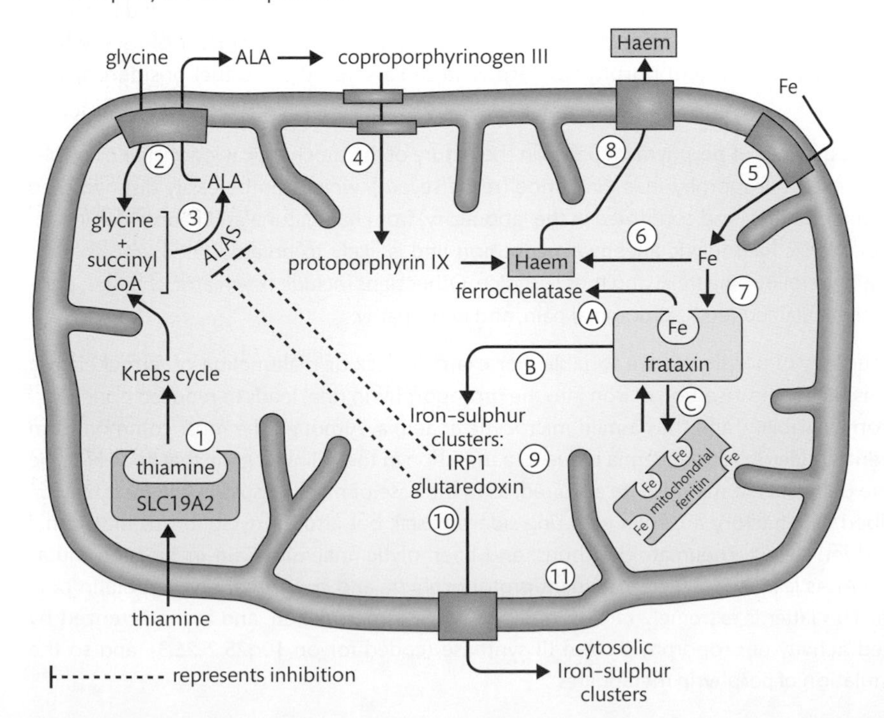

FIGURE 5.2
Mitochondrial pathology causing sideroblastic anaemia. Molecular genetics and other techniques in basic cell biology continue to inform our knowledge of the complex pathways required for the synthesis of haem. See the text for details. Figure modified from Kafina and Paw (2017) and elsewhere.

Other causes of sideroblastic anaemia

Sideroblastic anaemia may also be caused by drugs such as alcohol, isoniazid, and chloramphenicol, by a lack of copper (as it is a cofactor for other enzymes involved in haem synthesis), and by excess ingestion of zinc (as it inhibits the absorption of copper in the intestines). Further, lead poisoning is linked (but not directly related) to sideroblastic anaemia: it interferes with several steps in protoporphyrin synthesis (such as inhibition of the enzyme pyrimidine-5'-nucleotidase) and blocks the placement of iron into the centre of the protoporphyrin ring. However, the anaemia is a relatively late complication of a systemic pathology whose early stages include constipation and peripheral neuropathy. However, there are also other types of sideroblastic anaemia caused by defects in other genes involved in the production of haem, such as for the enzyme pseudouridine synthase.

Treatment of sideroblastic anaemia

In most cases, the treatment of inherited sideroblastic anaemia is with oral vitamin B_6, which should be effective in around one-third of patients. Severe cases may need to be regularly transfused, but this may lead to iron overload, although chelation therapy may be effective in reducing inappropriately high iron stores caused by the failure to utilize iron appropriately (see section 5.4). Treatment of acquired cases may require the removal or replacement of the causative agent, such as a different antibiotic in place of chloramphenicol. A distant hope is to be able to replace the dysfunctional mitochondria.

5.3.4 Porphyria and related conditions

porphyria
Disease caused by abnormalities in the production of haem characterized by anaemia and, in the laboratory, by increased iron.

Porphyria is the collective noun for a number of diseases that follow from defects in the biosynthesis of haem, which lead to the overproduction of porphyrins and/or associated precursors such as ALA and protoporphyrin. The aetiology of these conditions can be genetic (such as loss- or gain-of-function mutations in genes for the numerous biosynthetic enzymes; see Figure 5.2) or (less frequently) can be caused by lead intoxication or drugs. Examples of the latter include barbiturates and certain antibiotics, which may cause an intermittent porphyria. Porphyria can also be the product of sideroblastic anaemia.

The consequences of porphyria depend on the nature of the biochemical lesion. For example, a sign of excess uroporphyrin is pink urine (red if severe), which can be easily distinguished with urine dipsticks (and confirmed in the laboratory) from haematuria and haemoglobinuria. A hypochromic haemolytic anaemia is common and is likely to be associated with splenomegaly, hyperbilirubinaemia, and thus jaundice. Other signs include psychiatric changes, photosensitivity, stained teeth, abdominal pain, and neuropathy.

The aetiology of porphyrias are variable. For example, lack or malfunction of ferrochelatase (responsible for the insertion of iron into the protoporphyrin ring) leads to reduced bone marrow iron availability, and thus a mild microcytic anaemia. Although the most common form of inherited sideroblastic anaemia is due to a mutation in the X-linked gene that codes for the enzyme delta-ALA, it may also be acquired, as in the case of myelodysplasia (where it may be described as refractory anaemia with ring sideroblasts), but also in myelofibrosis, myeloma, myeloid leukaemia, rheumatoid arthritis, and haemolytic anaemia. Gain-in-function mutations in ALAs lead to X-linked-dominant protoporphyria and congenital erythropoietin porphyria. The latter is extremely rare, at one or two cases per million, and is characterized by reduced activity of uroporphyrihogen III synthase (coded for on 10q25.2-26.3), and so the accumulation of porphyrin metabolites.

SELF-CHECK 5.5

What are the principal causes of iron deficiency?

5.4 Iron overload

The presence of too much iron can be just as injurious as iron deficiency. The body has no effective mechanism for actively eliminating iron, so absorption must be carefully regulated. When this mechanism fails, iron stores rise, causing problems for the tissues where it is stored or deposited (Table 5.6). Accordingly, the accurate assessment of the body's iron store is valuable and is most efficiently estimated by determining the amount of iron in the liver by sensitive magnetic resonance imaging. If this is unavailable, serum ferritin is an alternative, as it correlates markedly with the mass of iron in the liver. Iron overload (**haemochromatosis**) can be classified as primary (hereditary) or secondary (acquired), depending on whether it is a consequence of a primary defect in the regulation of iron or is secondary to genetic or acquired disorders, or their treatment.

haemochromatosis
The pathological condition of iron overload.

5.4.1 Iron mis-regulation

We have already noted that the passage of iron from the enterocyte, hepatocyte, and macrophage into the blood is regulated by the action of hepcidin on ferroportin. Lack-of-function mutations in *FPN1* impair the iron-exporting ability of ferroportin, particularly in macrophages, wherein iron levels increase dramatically, leading to the recognized syndrome of Ferroportin disease. A second type of mutation in *FPN1* renders ferroportin resistant to the inhibitory effects of hepcidin, and so the persistent export of the metal and development of haemochromatosis (both ferroportin diseases are discussed below).

In Box 4.4 and section 4.3.1 we noted the importance of molecules such as haemojuvelin (HJV) and bone morphogenic proteins (BMPs), and transferrin receptors 1 and 2 (TfR1, TfR2) in the regulation of hepcidin. To this list we can add the HFE protein (H for High, FE for iron), coded for by *HFE* at chromosome 6p21.3, close to the genes for the HLA molecules. The mature molecule is present on the surface of cells from a variety of tissues—including the duodenum, liver, pancreas, placenta, kidney, and ovary—but principally on the macrophage and hepatocyte. The cartoon of Figure 5.3 illustrates a view of how HFE and other molecules come together to

TABLE 5.6 Pathophysiology of iron overload.

Causes	• Mutations in key genes (e.g. C282Y, causing hereditary haemochromatosis) • Repeated red blood cell transfusions • Increased dietary uptake of iron • Ineffective erythropoiesis (beta-thalassaemia, sideroblastic anaemia) • Decreased hepcidin levels
Consequences: damage to:	• The liver (present in 30% of men and 7% of women with iron overload, causing cirrhosis and possibly carcinoma) • The heart (arrhythmia and heart failure) • Skin (rash, pigmentation, and dermatitis) • Joints (arthralgia, osteoarthritis) • Endocrine organs (diabetes (present in 2–5%) and hypopituitarism)

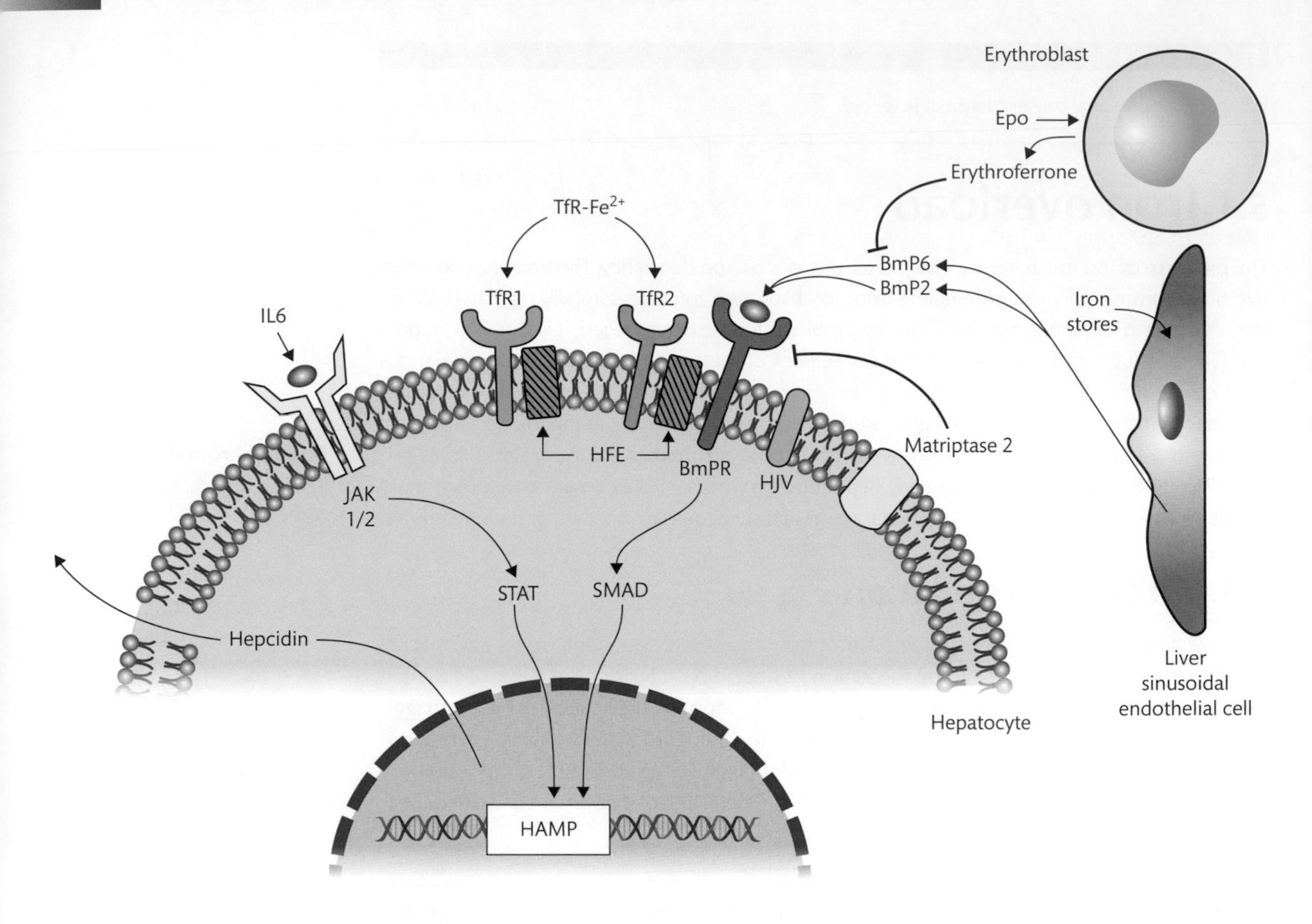

FIGURE 5.3

Regulation of hepcidin. Several pathways interact to regulate the transcription of *HAMP* and so levels of circulating hepcidin. Binding IL-6 by its receptor results in the activation of second messengers of the STAT family. It is also possible the IL-1β, IL22, and IFNα may promote hepcidin in a similar manner. A collection of molecules (TfR2, HFE, BMPR, and HJV) form a complex to respond to BMPs 2 and 6, themselves produced by liver sinusoidal endothelial cells. This complex, and the BMP6, are themselves regulated by matripase2 and erythroferrone, respectively. Cartoon amended from Muckenthaler et al. (2017); Knutson (2017); Pantopoulos (2018) and others.

regulate levels of hepcidin. At the surface of the hepatocyte HFE, HJV, TfR2 (in its role as sensor of iron-loaded transferrin) and a BMP receptor (BMPR) come together to receive BMP2 and BMP6, secreted by sinusoidal endothelial cells, which sense (by an as yet unknown mechanism) iron stores. The precise form of the association between HFE, HJV, and BMPR is also unclear—the former two molecules may be part of the receptor or may support or stabilize its integrity in the membrane. The complex may be part-regulated by the action of TfR1 (normally directed towards cellular iron acquisition) which can bind HFE, and so draw it away from the complex. A further regulation of the complex is the activity of the enzyme matriptase2 (also known as TMPRSS6, a product of a gene of the same name, located at 22q12.3), which can cleave all four components. The expression of *TMPRSS6* is increased by iron deficiency and hypoxia. Docking of BMPs into this complex results in the activation of *HAMP* by SMAD second messengers. A second promoter of *HAMP* transcription is the binding of IL-6 to its receptor, and the activation of the STAT3 second messengers, themselves activated by JAK1/2, part of the intracellular domain of the receptor. This draws striking parallels with the activation of the erythroblast by the docking of Epo to its receptor, as outlined in Figure 4.3. A further regulator is erythroferrone, which blocks the activity of BMP6, and so part-inhibits the BMPR and SMAD second messengers in the activation of *HAMP*. However, given the complexity of these pathways, it will be no surprise to learn that mutations in numerous genes have been linked to abnormal iron processing. For example, loss-of-function variants in *TMPRSS6* are causally associated with high hepcidin levels and reduced serum iron causing iron-refractory iron-deficient anaemia. However, the most common abnormality is in the gene for HFE.

5.4.2 Hereditary haemochromatosis (HH)

Hereditary haemochromatosis (HH) is the most common genetic cause of iron overload, and perhaps 83% of cases of HH are caused by a mutation (C282Y) in *HFE* (thirty other variants in *HFE* have also been described). The genetic cause of C282Y is a G→A transition at nucleotide position 845, which leads to the insertion of tyrosine instead of cysteine at position 282 in the mature HFE protein (hence C282Y). The second most common mutation (C187G) results in another mutation in the mature HFE molecule, His-63→Asp (H63D), and a significant proportion of patients with HH who are heterozygous for C282Y also have the H63D mutation. A third is A193T, resulting in S65C.

C282Y is present in Europeans in its heterozygous form in perhaps 9.2% of people, although this figure varies greatly with region—ranging from approximately 1% in Southern Europe, to 12% in the Netherlands, to 25% in Ireland. Heterozygotes do not have an increased risk of clinically evident HH, although iron levels may be raised in perhaps 25%. African, Middle Eastern, and Australian populations have a prevalence of perhaps 0–0.5%. In its homozygous form, HH is present in 0.4% of Europeans and inevitably has clinical consequences. The mutation is thought to have arisen in north Europe over 6000 years ago, when diets were transitioning from meat-based (hence rich in iron) to cereal-based, its rapid spread presumably being that it conferred a selective advantage in iron acquisition, and possibly in resistance to certain intracellular pathogens such as *M. tuberculosis*.

The consequence of the C282Y mutation is that malfunctioning HFE protein is present at the surface of the cell, and so fails to contribute to the integrity of the HFE/BMPR/HJV complex, the consequence of which is that hepcidin message is not transcribed by *HAMP*. Thus, HFE-related haemochromatosis is characterized by low messenger RNA for the *HAMP* in the liver and low serum hepcidin. Thus, in this setting low levels of hepcidin would be unable to regulate (in this case, inhibit) the movement of iron from the enterocyte into the plasma, resulting in the hyperferraemia (an excess of iron in the blood) which is a characteristic of HH. Although serum hepcidin is low in both those heterozygous and homozygous for the C282Y mutation, levels are strongly related to levels of ferritin in homozygotes for C282Y. Hepcidin levels in urine are negatively correlated with the severity of HH, adding further justification to the measurement of this small peptide.

The dominant form of HH, caused by mutations such as C282Y in *HFE*, may be classified as type 1 HH. Four other types of genetic variants have been described.

- Juvenile HJV-associated HH (type 2A) is due to a mutation, and so inactivation, in the gene for HJV on 1q21. Consequences include the loss of regulation of hepcidin, which leads to increased iron movement from the enterocyte to the blood. Clinical complications develop in the late teens to early twenties.

- Type 2B HH is due to an inactivating mutation in *HAMP*, which also results in no or inactive hepcidin.
- Type 3 HH is associated with a mutation in the gene for the transferrin receptor 2 on 7q22, which also leads to low levels of hepcidin.
- As discussed above, ferroportin disease (type 4A HH) differs from other HH disease as the loss-of-function defect in the ferroportin-encoding *SCL40A1* on 2q32 is autosomal-dominant. With impaired iron export function, there is overexpression of hepcidin with moderate to severe iron overload, mostly in tissue macrophages, and there is low transferrin saturation and serum iron.
- Ferroportin haemochromatosis (type 4B HH) is characterized by a gain-in-function mutation in *SCL40A1*, leading to hepcidin resistance, and is also autosomal-dominant. There is high transferrin saturation, macrophage iron deficiency with high levels of hepcidin.

There are also rare patients with defects in more than one gene responsible for iron metabolism (such as a case with combined TfR2 and HFE inactivation, which presented as juvenile haemochromatosis). Two other syndromes are of note. Acaeruloplasminaemia is a rare recessive disorder caused by a loss-of-function mutation in *CP* on 3q23-q24 and resulting in impaired ferroxidase activity, leading to excessive iron deposits, low transferrin saturation, and mild microcytic anaemia. Atransferrinaemia is a recessive disorder caused by transferrin deficiency due to mutations in *TF* on 3q22.1. Characteristics include very low to undetectable plasma transferrin, leading to impaired erythropoiesis, a microcytic hypochromic anaemia, and iron overload in the liver, heart, and pancreas.

Cross references

The role of hepcidin in the regulation of iron homeostasis is outlined in Chapter 4, Haemoglobinopathy is explained in detail in Chapter 6.

5.4.3 Secondary iron overload

Excess iron can be due to a number of other factors: the indirect effect of excessive iron consumption in the diet, severe liver disease, or as consequence of treatment (Table 5.6). Regarding the latter, patients with severe haemoglobinopathy or myelodysplastic syndrome are likely to require regular blood transfusions and, as a consequence, to develop iron overload. One unit of transfused blood contains approximately 200–250 mg of iron, and, although blood transfusion may ameliorate anaemia, a consequence may well be iron overload. Generally, after the transfusion of 10–20 units, patients become overloaded. Consequently, iron chelation therapy is recommended after approximately 12 months of blood transfusion, but the figures vary markedly between patients. Transfusional iron overload can have serious clinical consequences and, unless body iron is seriously controlled, patients may suffer significant morbidity and mortality.

5.4.4 Consequences of excess iron

In patients with chronic iron overload, cardiac failure is a major, life-threatening complication. After about 100 units of blood transfusion, the deposition of excess iron in the heart muscles leads to myocarditis and cardiac fibrosis, which eventually causes severe arrhythmias or heart failure. Without treatment, these patients survive less than a year. Other complications include the deposition of iron in the liver, which may lead to fibrosis, cirrhosis or cancer, and diabetes mellitus as a result of beta-cell destruction secondary to iron overload in the pancreas. Iron deposits in the pituitary gland may lead to growth hormone failure and infertility (hypogonadism) due to reduced gonadotropin levels, and there may also be arthropathy and osteoporosis. In early-onset juvenile haemochromatosis, the most severe complications are cardiomyopathy and hypogonadism. The relationships between excess iron, biochemistry, and the clinical complications of iron overload are summarized in Figure 5.4. (See also Box 5.2.)

SELF-CHECK 5.6

What are the principal causes of excess body iron?

5.4.5 Treatment of iron overload

Since the introduction of iron-chelating therapy, deaths from iron overload have significantly decreased in patients who receive regular treatment. Iron chelators are usually administered by subcutaneous infusion, and the treatment should be started after 10–15 units of blood transfusion or when ferritin is over 2000 ng/mL. The most widely available iron-chelating agent is desferrioxamine, and the success of treatment is monitored through the assessment of ferritin levels. Desferrioxamine reduces the amount of iron in the liver and also labile plasma iron. In addition, deferasirox shows potential in protecting cells from the damaging effect of intracellular oxidative stress and may (when used in myelodysplasia and aplastic anaemia) have direct anti-inflammatory effects on T lymphocytes.

An alternative to active iron chelation therapy is to simply bleed the patient from a vein in the arm as if they were a blood donor (venesection) and this is the preferred method in hereditary haemochromatosis. Venesection should proceed regularly (perhaps each month) until ferritin levels are at the lower end of the reference range (perhaps 50 μg/L). Furthermore, the response of the full blood count and iron indices to phlebotomy provides an essential assessment of the severity of the iron overload. However, these approaches do not reverse organ damage. Dietary advice is to minimize foodstuffs with a high iron content.

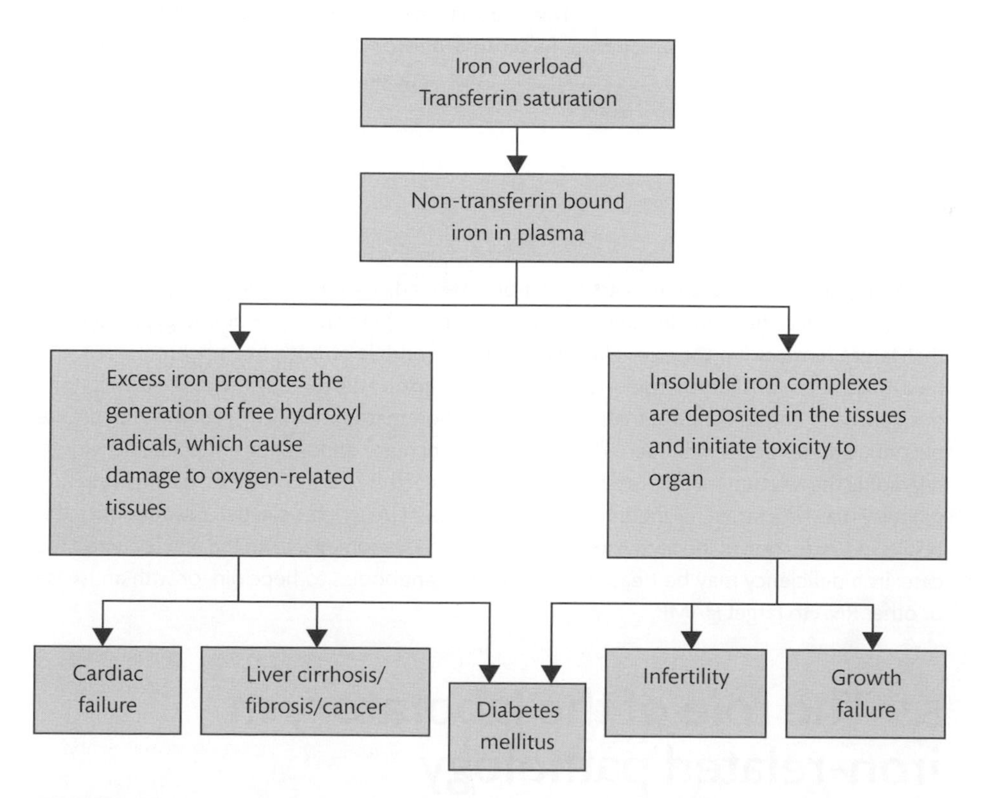

FIGURE 5.4
Clinical consequences of iron overload.

BOX 5.2 Why does excess iron cause disease?

The principal theory linking iron and clinically evident disease involves the generation of reactive oxygen species (ROSs) such as the hydroxyl radical. In normal, healthy individuals, transferrin is only 30% saturated with iron and so has a large reserve capacity to take up further iron. However, in patients with iron overload, transferrin saturation may reach 100% and so the ability of transferrin to absorb further iron is exceeded. This leads to 'free iron' or non-transferrin bound iron (NTBI) which circulates in the plasma. NTBI is not only directly toxic to cells, but is also indirectly toxic in that it has the potential to generate free hydroxyl radicals that propagate oxygen-related tissue damage. NTBI catalyses the Fenton reaction to generate free hydroxyl radicals ($^{\cdot}OH$) in the two following steps:

$$Fe^{3+} + O_2^{\cdot -} \rightarrow Fe^{2+} + O_2$$

$$Fe^{2+} + H_2O_2 \rightarrow Fe^{3+} + OH^- + HO^{\cdot}$$

The hydroxyl radical is generated (i.e. is a product), and so goes on the right-hand side of the second equation. The superoxide radical $O_2^{\cdot -}$ is on the left-hand side of the first equation, and so is a substrate. H_2O_2 is hydrogen peroxide and HO^- is the hydroxide ion. Fortunately, a series of metabolic pathways exists (involving glutathione) to limit the effect of ROSs (see Figure 4.15). The importance of this pathway is illustrated by the consequences of lack of an enzyme (glucose-6-phosphate dehydrogenase) that helps generate glutathione. The consequences of this deficiency often include a haemolytic anaemia, as is discussed in Chapter 6.

There are several opportunities to treat iron overload, based on the pathways shown in Figure 5.3. Synthetic hepcidin analogues (minihepcidins) are short (~9 amino acid) peptides that target ferroportin: they are effective in mouse models of both haemochromatosis and beta-thalassaemia. A further opportunity is to target ferroportin directly with an oral inhibitor—this approach entered clinical trials in 2018. Inhibition matripase2 (TMPRSS6) with antisense oligonucleotides and small interference RNAs can increase endogenous hepcidin activity by degrading the relevant mRNA, resulting in promotion of the BMPR complex. The inflammatory pathway may be a target for inhibition of Janus kinase 2 (JAK2)—in beta-thalassaemic mice this results in a reduction in ineffective erythropoiesis and splenomegaly. But the reverse is also the case: iron deficiency may be treated by therapeutic antibodies to hepcidin, or with antisense or other RNA to target *HAMP*.

5.5 The role of the laboratory in iron-related pathology

The laboratory's role centres on the full blood count, the blood film, various molecules in the plasma, and (rarely) investigation of the bone marrow.

5.5.1 The full blood count and film

The laboratory investigation of anaemia and iron overload demands a full blood count (FBC), which provides six red cell indices. Of these, the MCV is of particular value as it is likely to be reduced in the face of iron deficiency, whereby red cells may be described as microcytic. It follows that an anaemia associated with microcytes is a microcytic anaemia (Figure 5.5). The blood film from an iron-deficient subject is also likely to show microcytic hypochromic cells with pencil cells and occasional target cells. However, not all cases of microcytic anaemia are due to iron deficiency—an alternative cause may be haemoglobinopathy. In addition to a reduced MCV, iron deficiency is generally associated with a reduced mean cell haemoglobin (MCH) and mean cell haemoglobin concentration (MCHC). If this is the case, then the red cells are said to be hypochromic. Furthermore, some autoanalysers can report the proportion of red cells that they consider to be hypochromic—a proportion greater than 2% is generally considered suggestive of iron deficiency.

5.5.2 Plasma markers

Several laboratory markers are important in assessing iron-deficient anaemia. These include serum ferritin (the level of which is generally reduced: male <20 μg/L, female <12 μg/L), decreased iron (male <50 μg/dL, female <40 μg/dL), and reduced transferrin saturation (<20%). However, over-reliance on plasma ferritin may lead to error because, as an acute-phase reactant, levels may be raised in acute or chronic inflammation or in other acute disease or trauma. This can be a great frustration, as anaemia is present in many chronic diseases, as is discussed in section 5.7. Nevertheless, in the absence of illness or inflammation, body stores of iron can be assessed by levels of ferritin. It has been argued that ferritin should not even be in the blood as it stores iron in cells, so that serum ferritin has leaked from cellular stores (or perhaps damaged cells). However, it is certainly the standard test, and each μg in the serum is equivalent to approximately 8–10 mg of stored iron. Equally, if the ferritin concentration is low, regardless of inflammation, this is highly suggestive of iron deficiency.

Whilst serum iron and total iron-binding capacity (TIBC, although sometimes referred to as unsaturated iron-binding capacity) have their place, the percentage of the TIBC that is carrying iron (that is, %Sat) is most widely used. In health, this figure is likely to be perhaps 15–30%, but in iron deficiency it falls to <10%, and in iron overload to up to 100%. However, in functional iron deficiency (defined as a state in which there is insufficient iron incorporation into erythroid precursors despite apparently adequate body iron stores), TIBC may not be useful in monitoring response to treatment, so that other tests such as red cell or reticulocyte markers (such as

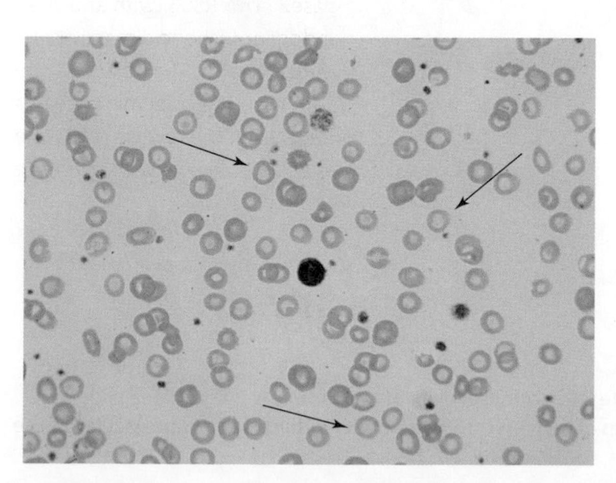

FIGURE 5.5
Iron deficiency, as is characterized by hypochromic microcytes—the cells seem to be 'empty', with a lack of staining in the centre of the cell (examples are arrowed). There is often anisocytosis (variation in cell size). A lymphocyte is also present. The red cells are significantly smaller than the nucleus of the small lymphocyte. (Magnification ×400.)

reticulocyte haemoglobin concentration, CHr), or ferritin, are recommended. The importance of the acid environment of the stomach is underlined by the high rate of iron-deficiency anaemia in patients with autoimmune gastritis, a disease characterized by failure of the stomach wall to produce acid. This condition may also lead to vitamin B_{12} deficiency, as discussed in section 5.6.1. Indeed, deficiency of both micronutrients can co-exist.

Other tests of iron physiology and pathology include levels of the soluble transferrin receptor (sTfR). The increased requirement for iron by the erythroblast leads to high numbers of transferrin receptors at their surface, a proportion of which are shed and can be measured as sTfRs. Thus increased levels of sTFRs reflect increased numbers and activity of erythroblasts (and possibly macrophages), inevitably the result of a pathological process (Figure 5.6). A strength of sTFR measurement is that, unlike transferrin and ferritin, it is not an acute-phase reactant and so is not influenced by inflammation.

5.5.3 Bone marrow

An additional feature of iron-deficient anaemia is the reduction, or even the absence, of stored iron (as ferritin and haemosiderin) in macrophages of the bone marrow, liver, and spleen. Indeed, it is likely that bone marrow stores of iron become depleted before there is evidence of iron deficiency from the red cell indices.

Should initial investigations point to sideroblastic anaemia, a bone marrow aspirate will be required. Sideroblastic anaemia is diagnosed where greater than 15% of erythroblasts are ring sideroblasts. However, a small number of ring sideroblasts may be present in other conditions such as myelodysplasia and copper deficiency, also diagnosed by examination of the bone marrow. The blood film of X-linked sideroblastic anaemia is characterized by a hypochromic microcytic anaemia (although the picture may be dimorphic, leading to an increased red cell distribution width), in the bone marrow by ineffective erythropoiesis, and in the tissues by iron overload. The key laboratory test is **Perls' stain** (also known as Perls' Prussian blue). Iron overload is characterized by increased transferrin saturation and hyperferritinaemia.

Perls' stain
The principal stain of blood, or bone marrow, or tissues (such as of the liver) for deposits of iron.

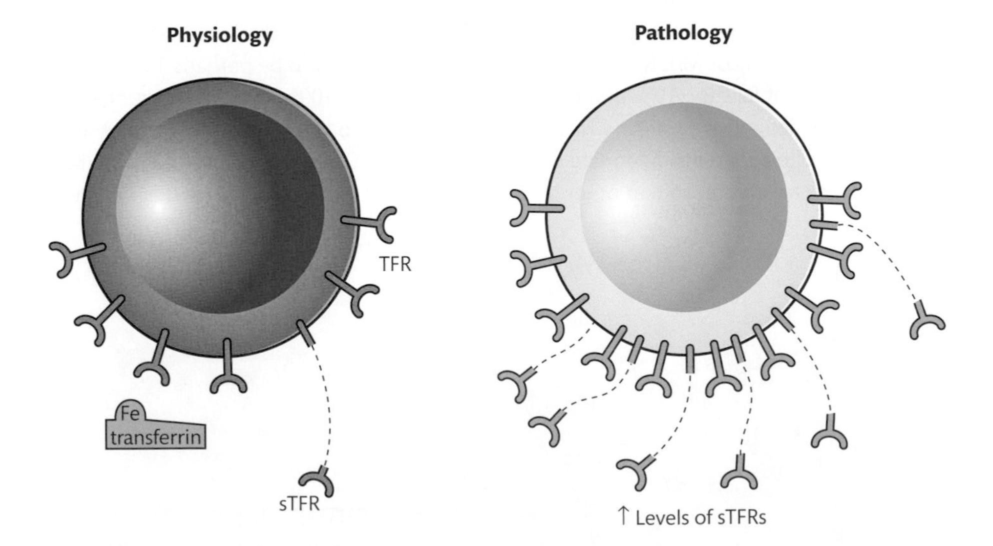

FIGURE 5.6

Soluble transferrin receptors. On the left, normal activity of erythroblasts is associated with low levels of sTFRs. On the right, at times of high erythroblast activity and number, increased levels of sTFRs are detectable.

The sign of lead poisoning in a blood film is the appearance of basophilic stippling of red blood cells. This is due to denatured RNA. There may be ring sideroblasts in the bone marrow and the related anaemia may be hypochromic and/or haemolytic. Occasionally, a dimorphic picture of a hypochromic/microcytic population in conjunction with a normochromic/normocytic population may be present.

5.5.4 Excess iron

In iron overload, the role of the laboratory is different. The liver is the prime site for iron storage, with more than 90% of all excess body iron being deposited as intracellular ferritin and haemosiderin. Liver biopsy is the best method for assessing iron stores and is quantitative, specific, and sensitive. However, it is an invasive and painful procedure that carries the risk of sampling error, especially in a patient with liver damage. A non-invasive alternative is magnetic resonance imaging (MRI) of the liver, as one study reported a respectable correlation coefficient of 0.71 between MRI-defined liver iron concentration and serum ferritin. As mentioned, serum ferritin levels give a good indication of the status of iron stores, but this test is subject to a major confounding influence: inflammation. Consequently, interpretation must be made with caution in the presence of infections and in inflammatory diseases such as rheumatoid arthritis.

If transferrin saturation exceeds 45% and there is raised ferritin (>300 μg/L in men and >200 μg/L in women), then HH may be suspected. Increased liver function tests (such as raised

TABLE 5.7 Major laboratory features of iron-related haematological disease.

Iron deficient anaemia	**Full blood count** Reduced haemoglobin, MCV, MCH, and MCHC. RBCC may be normal but haematocrit more often than not is low (recall that haematocrit is the mathematical product of the RBCC and the MCV). Increased number of hypochromic red blood cells.
	Blood film Increased reticulocytes, following iron replacement therapy. Presence of microcytes (hence reduced MCV), often hypochromic. A population of elliptocytes called pencil cells, is commonly present.
	Bone marrow Examination not usually indicated. However, iron may be reduced and there may be evidence of increased erythropoiesis as the bone marrow attempts to remedy the low red cell mass with compensatory hyperactivity.
	Serology Low serum ferritin (but beware false positive effects of the acute phase response), low transferrin saturation. Increased sTfRs (evidence of increased erythroblast activity).
Iron overload/ sideroblastic anaemia	**Full blood count/film/bone marrow** MCV often raised in acquired cases but low in inherited variants: bone marrow sideroblasts (Perls' stain) and possibly siderocytes on the blood film, presence of Pappenheimer bodies. Dimorphic red cell population of hypochromic/microcytic cells in conjunction with normochromic/normocytic cells resulting in increased RDW.
	Serology High serum iron and ferritin, high transferrin saturation (possibly 100%). Low sTfR levels.
	Other investigations Liver biopsy, magnetic resonance imaging of the liver (for iron stores) and possibly bone marrow aspirates, demonstrate high iron stores (Perls' stain). Refer for genetic screening if HH considered. If treatment by phlebotomy, FBC and iron studies must be performed before and after to monitor the expected fall in iron levels.

transaminase enzymes) imply the involvement of (and damage to) this organ. If this damage is severe, there may be raised serum bilirubin and the classical sign of jaundice.

In the investigation of porphyria, reference laboratories are likely to measure enzymes such as aminolaevulinic acid dehydratase and synthetase, and urobilinogen synthetase in parallel with their particular substrates and products. They may also conduct molecular biology studies to demonstrate mutations in relevant genes. Table 5.7 summarizes the role of the laboratory in the investigation of iron-related disease, whilst Figure 5.7 illustrates the pathways of iron cycling.

The laboratory is not only required for diagnosis, but also to monitor treatment. One of the treatments of iron-deficient anaemia is the provision of iron. This is likely to lead to a burst in erythropoiesis with new, fresh erythrocytes of normal size appearing in the blood. However, this burst of new erythrocyte production may outstrip folate stores, so cells may be folate-deficient. Thus a dimorphic picture may be present on a blood film, consisting of 'old' microcytes alongside 'new' normocytes, as is illustrated in Figure 5.8. Figures 4.18(b) and (d) also illustrate microcytes.

SELF-CHECK 5.7

What are ring sideroblasts and what can they tell us?

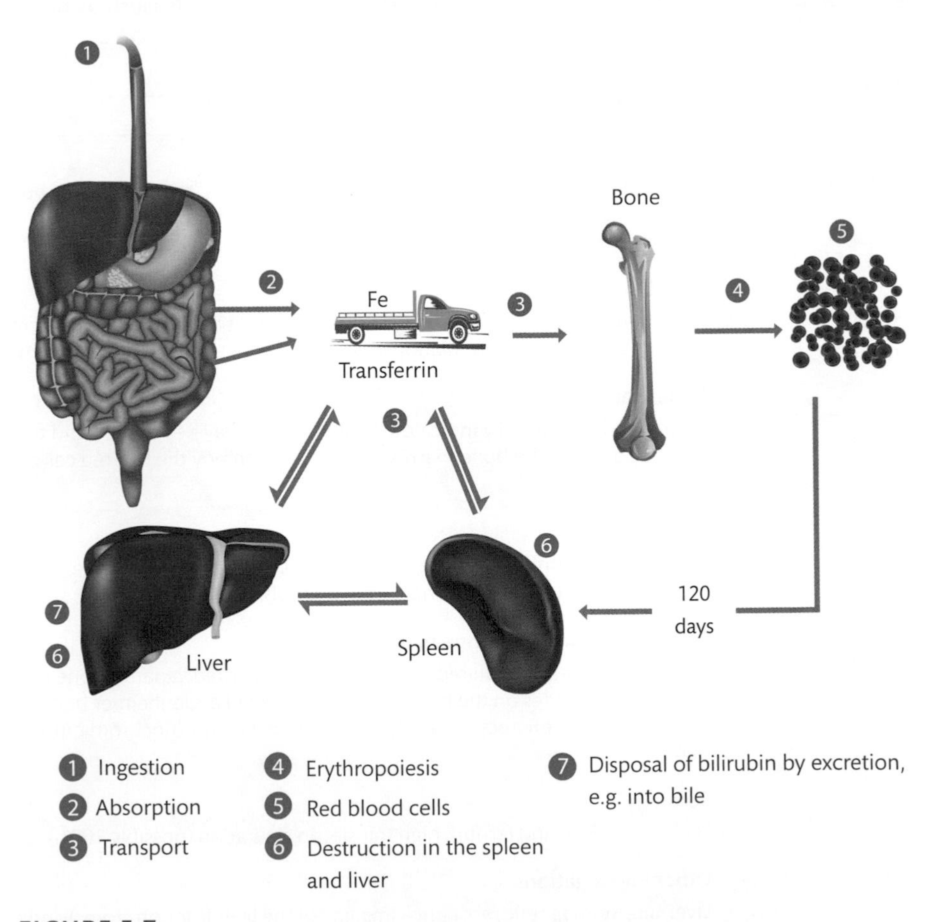

FIGURE 5.7

Pathways of iron cycling. Iron is absorbed by the intestines and is moved around various organs by the carriage molecule transferrin.

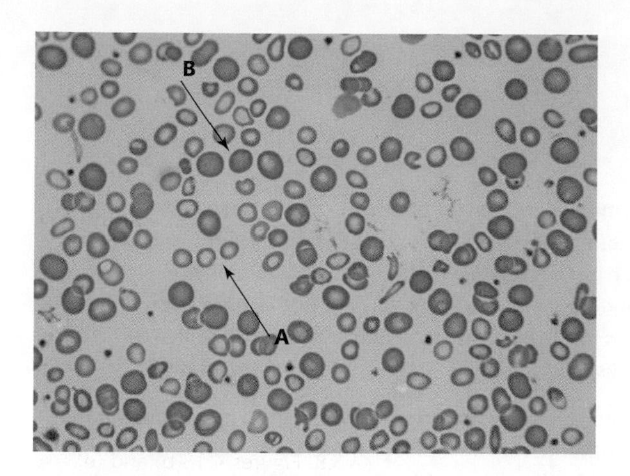

FIGURE 5.8
Following treatment for iron deficiency, or post transfusion, there may be two distinct populations: one of normal cells and another of abnormal cells leading to anisocytosis (variation in size) and anisochromasia (variation in colour). Note in this photograph that the majority of smaller cells (microcytes) are also pale (hypochromic) (examples arrow A), whilst the larger cells (normocytes, new cells having responded to oral iron, or perhaps transfused cells) (examples arrow B) have a good colour. (Magnification ×400.)

Key Points

Iron is clearly a very important micronutrient, required not only for haem, but also for certain enzymes. However, it is also a double-edged sword, because too much iron can be as disabling as too little iron.

5.6 Anaemia arising from lack of vitamins

Section 4.3 indicated the importance of vitamins B_{12} and B_6 and folate in the development of red blood cells, although they have roles in other aspects of metabolism (such as in maintaining the integrity of nerves). Indeed, the haematological presentation of vitamin B_{12} and folate deficiencies are identical, but neuropathy is not present in folate deficiency. Much of the metabolism and pathology of the vitamins have parallels with those of iron: all are micronutrients that must be obtained in the diet, so that deficiency may be the consequence of a poor diet. In addition, since absorption demands a healthy intestine, active intestinal disease or abdominal surgery is a likely contributor to anaemia, as both vitamin B_{12} and folate are absorbed in relatively limited sections of the small intestine: the transporter for folate is found only in the upper part, whereas the receptor for vitamin B_{12} is restricted to the distal part but is also absorbed in the ileum. Like iron, vitamin B_{12} is transported in the blood by specific proteins.

5.6.1 Vitamin B_{12}

Physiology and transport

The crystalline structure of vitamin B_{12}, the largest of all vitamins with a molecular weight of 1355 Da, was elucidated in 1948. It is composed of a ring of four pyrrole units (similar to haem), with an atom of cobalt at its centre. This leads some to describe the vitamin as **cobalamin** (often abbreviated to Cbl), of which there are four isoforms: cyanocobalamin, hydroxycobalamine, methylcobalamin, and deoxyadenosylcobalamin (or simply adenosylcobalamin), the latter two forms being bioactive.

cobalamin
The chemical 'template' for different forms of vitamin B_{12}.

METHOD Staining for iron

The key stain for iron is Perls' stain, developed by the German pathologist, Max Perls. Samples of bone marrow or peripheral blood spread out on glass slides are air-dried and then fixed with methanol for 10–20 min. When dry, they are exposed to acidified potassium ferrocyanide for a further 10 min, washed with tap water and then distilled water, and then exposed for around 10–15 s to a second stain such as aqueous neutral red or eosin as a counterstain. (Iron stain 56°C for 10 min or 30 min at room temperature.) Following a final wash and air-dry, slides may be examined by light microscopy. If present, any iron-containing (siderotic) material or haemosiderin is converted to the blue-coloured product, ferriferrocyanide. Consequently, the stain is also known as Perls' Prussian blue.

A modification of the technique can be used in the histopathology laboratory to stain body tissues or biopsies. Neither ferritin nor apoferritin take up Perls' stain. However, ferritin undergoes degradation to generate haemosiderin, which can be detected in tissue macrophages of the liver (Kupffer cells) and spleen.

When the body is overloaded with iron, it may also be found in other tissues, including cardiac tissue. Siderotic material can also be detected by Romanowsky dyes as basophilic granules, which are a mixture of protein and iron. Such granules are called Pappenheimer bodies.

Figures 5.9 and 5.10 illustrate Perls' staining of a blood and bone marrow sample, respectively. Case study 5.1 is illustrative of the consequences of iron deficiency.

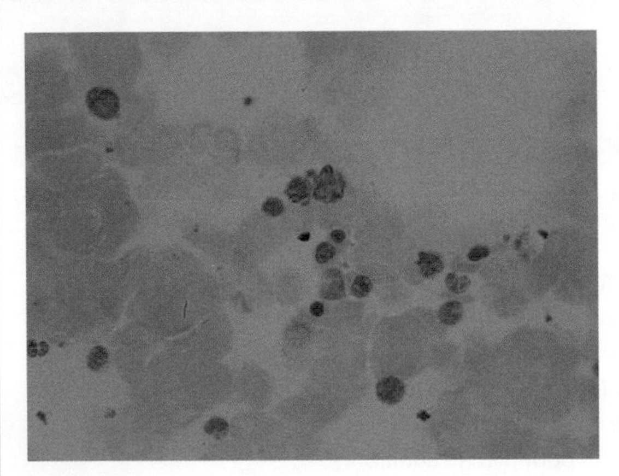

FIGURE 5.9
Perls' stain for iron in a blood film. The blue material is iron, the red-pink materials are counterstained nuclei of white blood cells.

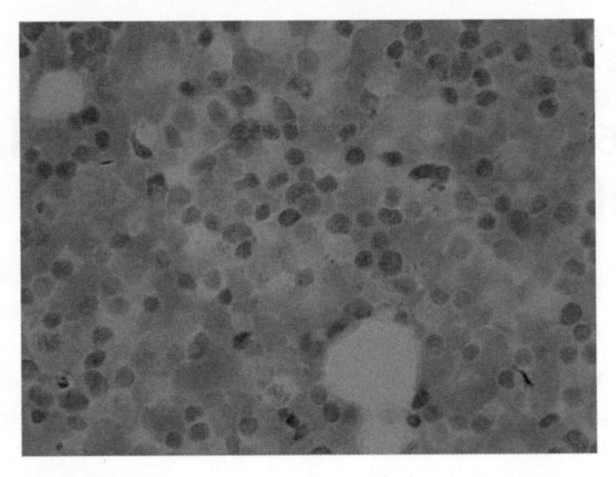

FIGURE 5.10
Perls' stain for iron in a sample of bone marrow. Note the greater density of cells than in the blood film. The blue material is iron, the red-pink materials are counterstained nuclei of white blood cells.

The adult daily requirement of this vitamin is 1.5 μg (2 μg if lactating), which should be easily satisfied by a mixed diet providing >5 μg per day, of which perhaps 2–3 μg are absorbed. Unlike iron, these requirements are independent of age or sex. However, unlike iron, vitamin B_{12} is almost absent from the plant kingdom—it is synthesized only in certain bacteria (such as *Streptomyces* species) and concentrates in the tissues of higher predatory organisms near the top of the food chain. Although bacteria colonizing the intestines produce the vitamin, it is unlikely to be absorbed, as it is produced in the colon.

CASE STUDY 5.1 Consequences of iron deficiency

A 50-year-old woman with long-term Crohn's disease, on various treatments and with a history of surgery to the abdomen, had the following test results following a GP visit, where she complained of tiredness, some lethargy, fevers, and diarrhoea.

Index	Result	Reference range (see front of book)
Haemoglobin	105 g/L	118–148
MCV	70 fL	77–98
WCC	6.1×10^9/L	3.7–9.5
Erythrocyte sedimentation rate (ESR)	26 mm/hour	<10
Platelets	350×10^9/L	143–400
Fibrinogen	3.5 g/L	1.5–4.0

The low haemoglobin was followed up with an assessment of micronutrient status, reporting serum iron 5.4 μmol/L (reference range 10–37) and serum vitamin B_{12} 227 ng/L (reference range 160–925).

Questions

- **Can you make a preliminary diagnosis based on the FBC and ESR? What are the abnormalities?**
- **Are you able to modify this diagnosis in view of the iron and vitamin B_{12} results?**
- **What are your next steps?**

Like iron, the liver (and kidney) can store vitamin B_{12} in milligram quantities. Such quantities are sufficient to meet the needs of the body even if it is deprived of an external source of the vitamin for 6–12 months based on a half-life of about 12 months. Consequently, there may be a considerable period between a failure to consume sufficient quantities of the vitamin, or to absorb it, and the onset of clinical signs of the deficiency.

To be absorbed, the vitamin B_{12} present in a foodstuff must first be liberated and extracted by the combination of the proteolytic enzyme pepsin and the acid environment of the stomach, the latter the result of hydrochloric acid secreted by parietal (oxyntic) cells. Free vitamin B_{12} then binds to **haptocorrin**, a 66 kDa glycoprotein also known as **transcobalamin-1** or the R-protein/factor, coded for by *TCN1* at 11q12.1, secreted into saliva, and possibly into the stomach. Specialist **gastric parietal cells** synthesize and release **intrinsic factor** (IF), which then binds the free vitamin B_{12} (the latter was once known as extrinsic factor) in the upper duodenum, following the release of vitamin B_{12} from the haptocorrin facilitated by the proteolytic action of pancreatic trypsin and other proteases. A 45 kDa molecular weight glycoprotein coded from by *GIF* on 11q12.1, IF is a carrier transport protein that delivers the vitamin B_{12}

gastric parietal cells
Cells of the luminal wall of the stomach which synthesize and secrete the intestinal vitamin B_{12} carrier molecule IF.

to the large intestine, although perhaps 1–5% of free vitamin B_{12} is absorbed along the entire intestine by passive diffusion. Factors such as histamine and gastrin stimulate the release of IF into the gastric juice, whilst pancreatic secretions in the upper small intestines promote the transfer of the vitamin from haptocorrin to IF. In chronic pancreatic disease, release of vitamin B_{12} from salivary haptocorrin is disrupted through lack of bicarbonate and trypsin production.

Absorption of the IF/vitamin B_{12} complex occurs in the enterocytes of the distal/terminal ileum and requires the IF to dock to its receptor on the luminal surface of the enterocyte. This receptor has three components, cubilin, megalin, and amnionless, and may be seen in terms of the receptor divalent metal transporter-1 (DMT-1, see Figure 4.11) that facilitates the absorption of iron. The precise receptor for the vitamin/IF complex is cubilin: the other two molecules are supportive. The vitamin is uncoupled from IF by lysosomal enzymes, is exported with multidrug resistance protein-1 (MRP1), and enters the plasma. There does not seem to be a mechanism for regulating this process, in contrast to the ferroportin/hepcidin mechanism for regulating iron. In the post-enterocyte circulation, vitamin B_{12} binds the specific carrier molecule **transcobalamin-2** (a 45.5 kDa glycoprotein coded for by *TCN2* on chromosome 22) and **haptocorrin/transcobalamin-1**. It is possible that different forms of the vitamin are carried by the two proteins, with haptocorrin acting as a reservoir, as it carries 75–90% of the vitamin. At the surface of the receiving cell, perhaps the erythroblast or the hepatocyte, the complex of vitamin B_{12}/transcobalamin-2 binds to the transcobalamin-receptor (TCBIR, CD320), itself a 114 kD dimer coded for by *CD320*: haptocorrin does not bind the receptor. Once inside the cell, lysosomal degradation of the receptor and the release of vitamin B_{12} from transcobalamin-2 allow the former to participate in metabolism in the cytoplasm and in the mitochondria. Free of vitamin/transcobalamin, the receptor is recycled to the cell surface in a manner similar to that of the transferrin receptor. The presence of vitamin B_{12} in bile is taken as evidence of an enterohepatic circulation, possibly to recycle unbound vitamin, although the contribution of this to physiology is unknown. These steps are summarized in Figure 5.11.

transcobalamin-2
A molecule carrying vitamin B_{12} for delivery to the cell.

haptocorrin/ transcobalamin-1
A molecule carrying vitamin B_{12} in the intestines and in blood.

Functions

Inside the cell, there is evidence that a proportion, if not all, of each of the forms of vitamin B_{12} are reduced to their core cobalamin, whilst other remain unchanged. The vitamin is an essential cofactor for at least two enzymes. The first (in the mitochrondrion) takes methylmalonic acid/methylmalonyl-coenzyme A as a substrate and converts it to succinyl-coenzyme A (succinyl-CoA, needed to form ALA with glycine, and ultimately haem) using one of the forms of vitamin B_{12} (adenosylcobalamin) as a cofactor. This is a key early step in the synthesis of haem (section 4.3.1) (Figure 5.12).

In the second process (in the cytoplasm), homocysteine is converted to methionine by methionine synthase, which requires vitamin B_{12} to support the transfer of a methyl group. This methyl group must be provided by methyltetrahydrofolate (methyl-THF), which then becomes THF. Methyl-THF is then regenerated by MTHF reductase. Therefore, a lack of vitamin B_{12} leads not only to the absence of the products succinyl-CoA and methionine, but also to the build-up of the two substrates for these reactions, which are methylmalonic acid and homocysteine. However, failure of the regeneration of THF also leads to the lack of a further intermediate, 5-10-methylene-THF. This metabolite is required for the synthesis of deoxythymidine monophosphate. So, deprived of vitamin $B_{12,}$ or perhaps with loss-of-function mutations in MTHF reductase, thymidine cannot be generated, and so DNA synthesis is impaired (Figure 5.13).

SELF-CHECK 5.8

Why is the stomach important for the absorption of vitamin B_{12}?

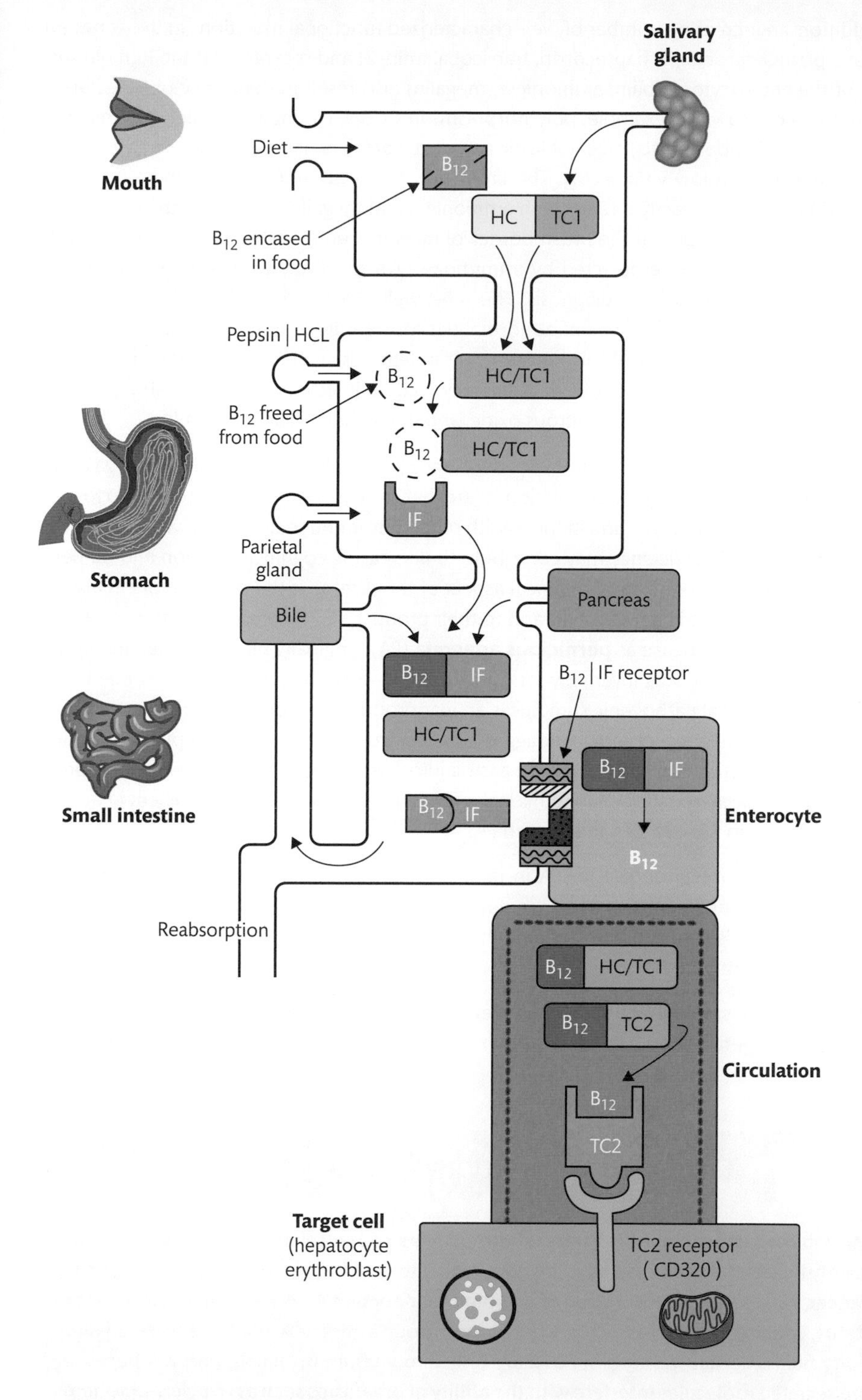

FIGURE 5.11
Absorption and transport of vitamin B_{12}. Clarifying the vitamin from foodstuff begins in the month, where haptocorrin/transcobalamin-1 is produced. Once freed from food in the stomach, it binds haptocorrin and is passed to the upper small intestine, where it is passed to IF. Transported from the upper gastrointestinal tract by IF, the complex docks to a complex receptor on the luminal surface of the enterocyte. It is conducted to its target cell by transcobalamin (TC1), although some is carried by haptocorrin (TC2), and enters the target cells via the transcobalamin-2 receptor. Some unabsorbed complex is recycled via bile. HC = haptocorrin; IF = intrinsic factor; TC1 = transcobalamin-1; TC2 = transcobalamin-2.

Deficiency of vitamin B_{12}

The complex steps in the passage of vitamin B_{12} from the diet to the cell give rise to many possible causes of deficiency. Being widely distributed in animal material (such as fish, meat, eggs, and dairy products), the only form of deficiency arising from malnutrition is in those who follow a strict vegan diet and is relatively rare.

In addition, an increasing number of well-characterized functional mutations in the genes for various plasma carriers (IF, haptocorrin, transcobalamin-2) and receptors (at the luminal surface of the enterocyte—cubilin, amnionless, megalin) also result in deficiency of the vitamin at the bone marrow. For example, polymorphisms in *CUBN*, coding for cubilin, are linked to maternal neural tube defects, megaloblastic anaemia, coronary heart disease, and gastric cancer in patients with low vitamin B_{12}. The latter may be because a mutant cubilin molecule may not be able to correctly interact with amnionless and megalin to facilitate the absorption of the vitamin/IF complex at the brush border of intestinal enterocytes. Low or absent levels of transcobalamins can be detected by immunoassay, a test likely to be offered by a reference laboratory, and useful in diagnosis. The most well-characterized defects in vitamin B_{12} synthetic pathways include those of abnormalities in the mitochondrial synthesis of deoxyadenosylcobalamin. Defects of this nature result in features such as methylmalonic aciduria and homocysteinuria. Other causes of deficiency include infestation with the fish tapeworm *Diphyllobothrium latum*, abuse of nitrous oxide, and the diabetic drug metformin.

However, the leading cause of vitamin B_{12} deficiency (>90% of cases) is autoimmune gastritis, being present in up to 6% of the population, a rate that increases with age (present in up to 20% of those in their sixth decade) and in those with other autoimmune disease. It has been speculated that this subtle deficiency may contribute to decreasing cognitive function and dementia in the elderly. The principal serological aspect of the disease is the presence of circulating autoantibodies to gastric parietal cells and to their product, IF. The presence of these autoantibodies defines the disease as **pernicious anaemia** (PA), originally named in the nineteenth century, as the disease was invariably fatal. How do these antibodies arise? It has been suggested that the initial pathogenic stimulus is an inappropriate response to *Helicobacter pylori*. The autoantibodies direct chronic autoaggression towards the gastric proton pump, resulting in parietal cell destruction and decreased gastric juice secretion (any gastric juice that is present is not necessarily acidic). Notably, this lack of acid may also lead to iron-deficient anaemia, which is present in 20–40% of patients with PA.

pernicious anaemia
The anaemia defined by the autoimmune destruction of gastric parietal cells and/or IF, and which causes vitamin B_{12} deficiency.

However, the pathological basis of PA in patients with acquired hypogammaglobulinaemia cannot be an autoantibody. In such cases, abnormal cell-mediated immunity can generally be demonstrated. This is supported by the high frequency of leucocyte infiltrations in stomach biopsies of patients with PA.

Another frequent serological finding is an elevation in the levels of plasma gastrin (>300 ng/L) and/or pepsinogen. It should be stressed that the vitamin B_{12}-deficient PA should be regarded

BOX 5.3 A multidisciplinary note

Microbiologists will be fully aware of the dangers of *Helicobacter pylori*, which commonly infects the stomach and is linked not only to gastritis, gastric atrophy, and gastric cancer, but also to malabsorption of vitamin B_{12}. In section 4.2.6 we have noted secretor status, under the control of *FUT2*, where blood group antigens A and B are secreted into body fluids. Certain variants in *FUT2* are linked to vitamin B_{12} levels, perhaps because the secreted molecules interfere with the ability of organisms such as *H. pylori* to adhere to gastric tissues. A related gene, *FUT6*, encodes a membrane protein involved in anther blood group, sialyl-Lewis X. These Lewis-associated antigens are also linked to *H. pylori* adherence to the gastric and duodenal mucosa. However, an alternative hypothesis is that variants in *FUT6* influence the glycosylation of vitamin B_{12}-binding proteins and their receptors, and so absorption of the vitamin.

as the long-term consequence of the autoimmune destruction of the parietal cells. A further long-term consequence is an increased (tenfold) rate of gastric tumours. Other forms of non-immune gastric damage (alcoholism, gastric atrophy, non-specific gastritis, surgery, carcinoma) may also lead to a failure to generate acid and/or IF, ultimately leading to vitamin B_{12} deficiency. The increasing therapeutic use of proton-pump inhibitors (designed to reduce gastric acid) and histamine H_2-receptor blockers are an additional possible cause of this disease.

The metabolic pathways reliant on vitamin B_{12} are widely distributed, so that there may be several pathological consequences of vitamin B_{12} deficiency for the body. Non-haematological consequences of vitamin B_{12} deficiency include a swollen and painful tongue (glossitis), peripheral neuropathy/polyneuritis (possibly manifesting as poor motor coordination (ataxia), loss of sensation, tremor, or tingling), and consciousness and personality changes (which may range from simple confusion and irritability, to depression, loss of memory, and even psychosis). Wider aspects of this vitamin are described in Box 5.3.

5.6.2 Vitamin B_6

This vitamin (also known as pyridoxine, as it has a pyridine ring at its core) is an essential cofactor in more than 140 different metabolic reactions, and in haematology is crucial for the conversion of succinyl-CoA and glycine into ALA by the enzyme ALA synthase. Insufficient vitamin B_6 therefore leads to failure to generate ALA, and so haem, and ultimately a sideroblastic anaemia (Figure 5.12).

As it is also required for the generation of cystathione, and then cysteine, from homocysteine via cystathione synthase, low levels of vitamin B_6 lead to the build-up of homocysteine, with hyperhomocysteinaemia and homocysteinuria (Figure 5.13). It is also important in many other metabolic processes, including those of amino acids, sphingolipids, and gluconeogenesis, and can quench ROS.

Vitamin B_6 is common in many foodstuffs (fruit, vegetables (especially potatoes), cereals, and meat) and is easily passively absorbed in the jejunum and ileum, so that the recommended daily requirement of 0.5–2.0 mg (towards the higher end if lactating or pregnant) is easy to achieve. Consequently, dietary deficiency is extremely rare, but may be caused by malabsorption syndromes. Unusually, there seems to be no confirmed intestinal or plasma transporter.

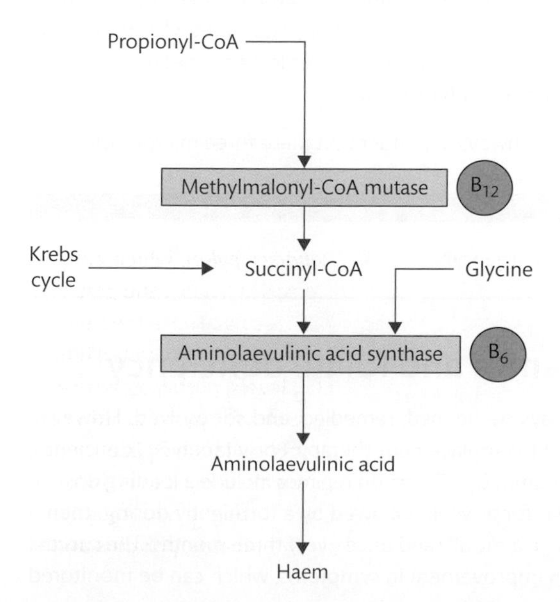

FIGURE 5.12

Role of B vitamins in the synthesis of ALA. B vitamins are essential cofactors to enzymes regulating the production of haem.

5.6.3 Folate

The intestinal absorption of folate (the reduced form of folic acid, or pteroylglutamic acid) is less complex than that of vitamin B_{12} and occurs in the duodenum and upper small intestine. General dietary folates occur largely as polyglutamates, which must be hydrolysed by folate conjugase on the brush borders of enterocytes. Folate may also be generated by intestinal microorganisms. Absorption of the monoglutamate form occurs in part by a proton-coupled folate transporter (PCFT, *SLC46A1*) widely expressed on brush borders. Crossing the enterocyte, folate leaves the basolateral membrane via multidrug resistance protein subtype 3 (MRP3) and is transported to its target organ (such as the liver, where it undergoes first-pass metabolism to a more active form, and bone marrow), becoming tetrahydrofolate (THF), methylene-THF, and then methyl-THF. The daily requirement of folate is approximately 100–200 mg; this should be adequately provided for through a mixed diet, which is able to provide 200–300 mg per day. However, folate body stores are small relative to daily need, so folate deficiency occurs frequently in the absence of supplementation. The importance of folate (and vitamin B_{12}) in other aspects of pathology is demonstrated by the fact that most cases of spina bifida and neural tube defects result from maternal deficiency in these micronutrients during the periconceptual period. This may be related to an increased incidence in pregnant women who are more likely to become folate-deficient, as stores last only a few months, whereas vitamin B_{12} stores are generally sufficient for over a year. Studies in the USA, where foodstuffs are fortified with folate, demonstrate a significant reduction in the rates of cleft palate, pyloric stenosis, and upper limb reduction defects. The metabolic requirement for folate stems from its role as a substrate for thymidine, and thus DNA.

The leading genetic cause of low serum folate is a loss-of-function mutation in the enzyme methyltetrahydrofolate reductase (MTHFR, coded for by *MTHFR* at 1p36.22), which also results in increased homocysteine and low methionine. However, this mutation does not lead to megaloblastic anaemia, and may or may not lead to low folate, but almost always leads to hyperhomocysteinaemia. Hereditary folate malabsorption has been described but is extremely rare. Failure of the folate-dependent conversion of deoxyuridine to deoxythymidine results in impaired DNA synthesis that results in a prolongation of the S phase of the cell cycle, and so maturation arrest. The consequences of this include retardation of DNA replication in erythroblasts, a response to which is increased growth of the cell into a megaloblast, a key feature of the disease.

As a result of the close relationship between the methionine cycle and the folate cycle, should there be a vitamin B_{12} deficiency, folate effectively becomes trapped as methyl-THF because the vitamin is required to convert this metabolite to THF. The clinical and metabolic consequences of this can be alleviated by treatment with folic acid.

Figure 5.13 summaries the biochemical pathways dependent on these three micronutrients.

SELF-CHECK 5.9

What are the primary causes of vitamin B_{12} deficiency?

5.6.4 Treatment of vitamin and folate deficiency

The cause of the deficiency should always be defined, remedied, and so resolved. However, in the short term, the standard treatment is replacement therapy. For vitamin B_{12} deficiency, this is likely to be regular injections of vitamin B_{12}. Common regimes include a loading dose of perhaps 100–1000 μg/day intramuscular for a week, followed by a fortnightly dosing, then a lower maintenance dose of between once a month and once every three months. The success of this regime should be sought with an improvement in symptoms, which can be monitored

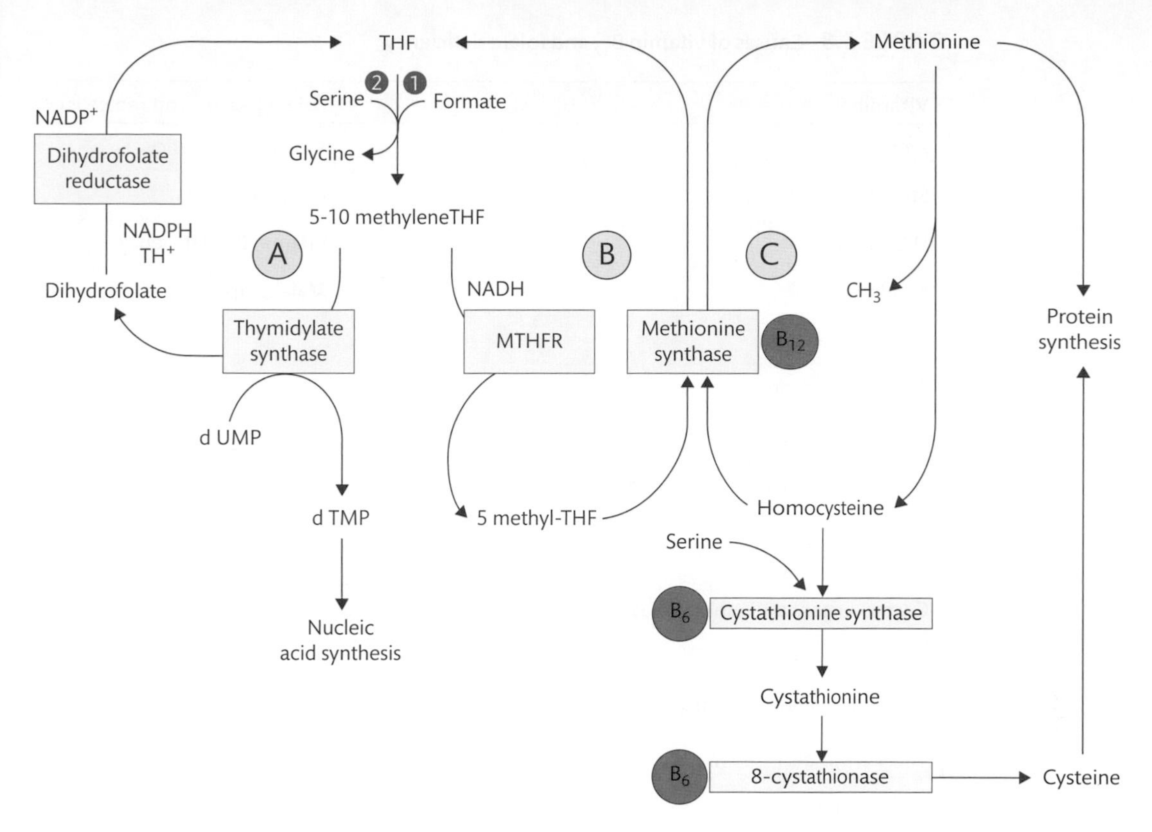

FIGURE 5.13

Simplified representation of the roles of vitamins B_6 and B_{12}, and folate as cofactors. In cycle A on the left, 5-10-methylene-THF, derived from THF, is a substrate for the generation of deoxythymidine monophosphate (dTMP) from deoxyuridine monophosphate (dUMP). THF is regenerated by reduction (top left). In cycle B, 5-methyl-THF is generated by the action of methylene-tetrahydrofolate reductase (MTHFR), which transfers the methyl group to homocysteine, thus forming methionine and regenerating THF. In cycle C, the methyl group can be removed from methionine to regenerate homocysteine, itself a substrate for enzymes ultimately producing cysteine.

with the appropriate blood testing. Many regard this as an improvement over the original treatment of 100–300 g of raw beef liver, to be taken orally, which provided folate and iron in addition to vitamin B_{12}. This regime produced a peak rise in reticulocytes five to seven days after treatment began, alongside acceleration in megaloblast maturation. Notably, a diet of raw beef muscle fails to elicit this response. However, the approach of providing parenteral vitamin B_{12} is not always successful. For example, certain cases are caused by loss-of-function mutations in genes coding for enzymes that convert 'raw' vitamin B_{12} into deoxyadenosylcobalamin (in the mitochondrion) or into methylcobalamin (in the cytoplasm). Treatment in the latter cases is with systemic hydroxocobalamin. The autoimmune form (that is, PA) may be treated with immunosuppression (generally steroids). Folate deficiency is treated with oral supplements.

The successful treatment of vitamin or folate deficiency should be evaluated with FBC and blood film and reticulocyte counts 10–14 days after the start of treatment. This should reveal a rising haemoglobin and falling MCV. Bone marrow examination at this stage is rarely necessary. All indices should have returned to normal after eight weeks. A failure to respond in this way suggests inadequately low levels of supplements or an incorrect diagnosis.

Table 5.8 summarizes the main causes of vitamin B_{12} and folate deficiency.

TABLE 5.8 Causes of vitamin B_{12} and folate deficiency.

Vitamin B_{12}	Folate (serum and red cells)
Pregnancy	Alcohol, pregnancy
Strict veganism	Anticonvulsants
HIV infection	Vitamin B_{12} deficiency
Myeloma	Malabsorption
Folate deficiency	
Sjögren's syndrome	
Use of proton-pump inhibitors or other acid antagonists	
Loss of function mutation in genes for receptors and carrier proteins	
Ileal disease	
Gastric surgery and other disease	
Pernicious anaemia	
Bacterial overgrowth in the intestines	
Pancreatic insufficiency	
Tropical sprue	
Tapeworm infestation	
Unknown causes (present in perhaps 10% of patients)	

5.6.5 The role of the laboratory in vitamin B_{12}-deficient anaemia

As with iron-related pathology, laboratory investigation of vitamin B_{12} deficiency centres on an FBC and blood film. The principal finding in the FBC is that the MCV becomes elevated in both vitamin B_{12} and folate deficiency, hence the red cells may be described as macrocytic, leading (in the presence of a low haemoglobin and symptoms) to a diagnosis of macrocytic anaemia. In addition to increased MCV, vitamin B_{12} and folate deficiency is often associated with elevated MCH, in which case the red cells are said to be hyperchromic. These red cells have increased rigidity and so a 30–50% reduction in red cell lifespan, and there is increased plasma bilirubin and lactate dehydrogenase. Platelet production is reduced to 10% of what might be expected, reflecting ineffective thrombopoiesis, and so that may be thrombocytopenia.

Although the macrocyte is important in the diagnosis of vitamin B_{12} deficiency, two points must be made. The precise definition of this cell depends on local conditions, as some describe the upper limit of the reference range to be, for example, 98 fL, whereas others may put this limit at 100 fL. It follows that in some laboratories, a red blood cell with an MCV of 99 fL may be a macrocyte, but in others the same cell may be a normocyte. Consequently, the practitioner must refer to their own local reference range. The second point refers to the specificity and sensitivity of the macrocyte in defining vitamin B_{12} deficiency: a large number of drugs and conditions are associated with macrocytosis, but not all of these are associated with an anaemia. Consequently, alternative and differential diagnoses must be addressed when considering a diagnosis of macrocytosis and vitamin B_{12} deficiency (Table 5.9). A key physiological cause of macrocytosis is pregnancy. Macrocytes are shown in Figure 5.14.

Similarly, the aware practitioner will avoid the pitfalls in the interpretation of the blood film with a raised MCV, such as 110 fL. The presence of oval macrocytes, anisocytosis, and poikilocytes suggest vitamin B_{12} deficiency, whereas round macrocytes are commonly (but not exclusively) seen in numerous chronic diseases, such as hepatitis, obstructive jaundice, and those conditions listed in Table 5.9. An additional consideration, although rare, is marked reticulocytosis (such as >10% of the total red blood cell count). This should be considered if there is evidence of haemolysis, polychromasia, nucleated red blood cells, spherocytes, or schistocytes.

In the bone marrow, a consequence of the shortage of vitamin B_{12} is that erythoblasts become enlarged and oval shaped, and are referred to as megaloblasts (literally, large blasts). This finding gives the disease an additional and perhaps more scientific name—**megaloblastic anaemia**. Progenitors of other lineages are also abnormal, with the presence of giant metamyelocytes. These abnormalities are caused by defects in DNA synthesis, which retards proliferation and maturation, and the entire marrow is hypercellular. In very severe deficiency this may lead to haemopoietic arrest, and thus pancytopenia.

The phenomenon of increased size continues as the erythroid cells develop within the bone marrow and is the basis of increased numbers of large erythrocytes (macrocytes) in the bone marrow and in the peripheral blood. These DNA defects also explain **hypersegmented neutrophils** present in peripheral blood and are an established feature on a blood film (Figure 5.15).

hypersegmented neutrophil
A principal blood film sign of vitamin B_{12} deficiency: the nucleus of the normal neutrophil may have perhaps three or five lobes; in vitamin B_{12} deficiency, the number of lobes may rise to seven or eight.

A further consequence of ineffective erythropoiesis is a low reticulocyte count and intramedullary haemolysis, which may lead to raised lactate dehydrogenase and bilirubin levels. Because many of these changes are also present in myeloid leukaemia and myelodysplasia, interpretation must be cautious. An additional aspect of megaloblastic anaemia on the blood film is basophilic stippling, although several other conditions (such as lead poisoning) also cause this sign (Figure 5.16).

5.6.6 Clinical investigation and management of low vitamin status

Clinical evaluation of the patient begins with signs and symptoms, and then moves to the laboratory.

TABLE 5.9 Causes of macrocytosis.

Drugs	Chemotherapy: cyclophosphamide, hydroxycarbamide, methotrexate, azathioprine Antimicrobials: pyrimethamine, trimethoprim, zidovudine Anticonvulsants: phenytoin, primidone, valproic acid Others: metformin, sulfasalazine
Pathology	Vitamin B_{12} and or folate deficiency Alcoholism, hypothyroidism, multiple myeloma, liver disease Myelodysplasia, aplastic anaemia, acute leukaemia Reticulocytosis, hyperglycaemia, marked leukocytosis, cold agglutinins

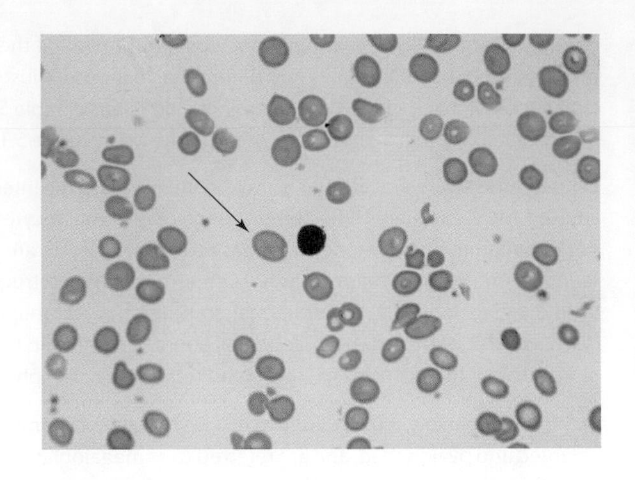

FIGURE 5.14
Macrocytosis may be present in patients deficient in vitamin B_{12}, vitamin B_6, or folate. If so, the macrocytes are often oval and may be accompanied by schistocytes and anisocytosis. Note that a macrocyte is larger than the nearby lymphocyte (arrowed). Contrast this with Figure 5.5, where the microcytes are notably smaller than the lymphocyte. (Magnification ×400.)

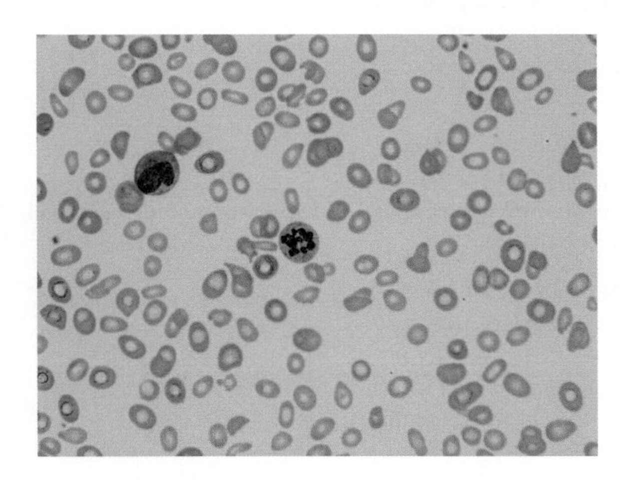

FIGURE 5.15
Hypersegmented neutrophils are a sign of vitamin B_{12} deficiency and are the consequence of abnormal DNA synthesis. Neutrophils normally have between three and five nuclear lobes, but if hypersegmented there may be seven or more lobes, as in the cell in the centre of this figure. This can be compared with Figure 5.16, where the normal neutrophil has few lobes. (Magnification ×400.)

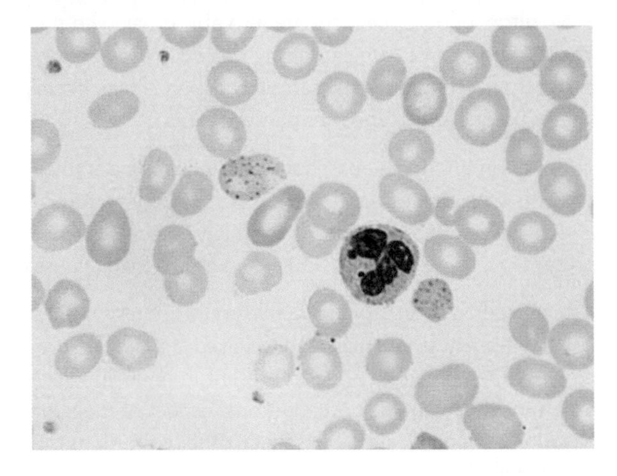

FIGURE 5.16
Basophilic stippling of red cells, also called punctate basophilia, as may be seen in megaloblastic anaemia. However, it is also seen in many other conditions, including lead poisoning, myelodysplasia, haemoglobinopathy, and haemolytic anaemia. This slide shows two excellent examples either side of a neutrophil; note also that two red blood cells are not only oval shaped, but also have a denser blue coloration than others (polychromasia). (Magnification ×600.)

Vitamin B_{12}

The ultimate definition of vitamin B_{12} deficiency is serum levels below the local reference range, and determination of the presence of autoantibodies to IF is crucial. However, the tests for levels of this vitamin may demonstrate poor sensitivity and specificity, poor agreement between laboratories regarding a reference range, and, hence, the definition of a lower limit of a reference range. This can be crucial, as a 'high' lower limit of the reference range will, of course, define fewer cases of deficiency than a 'low' lower limit of the reference range, which will find more cases. The patient will continue with the same signs and symptoms regardless of the laboratory findings, underlining the importance of the symbiosis between the laboratory and the clinic. One guideline (Davalia et al. 2014) recommends a cut-off point for serum vitamin B_{12} of 148 pmol/L (200 ng/L).

Should levels of vitamin B_{12} and/or autoantibodies be borderline or otherwise unclear, a second round of blood tests will be necessary, and include methylmalonic acid (MMA), also known as methylmalonyl-CoA (Figure 5.12). The rationale for this is that lack of the vitamin results in poor activity of the enzyme MMA-mutase, which normally transforms MMA into succinyl-CoA. Thus, lack of activity of this enzyme leads to high (unused) levels of its substrate (that is, MMA). However, MMA may be reduced in acute or chronic renal failure, and very rarely in defects in enzymes involved in the generation of MMA. As such, a cautious approval to interpreting MMA results in patients with renal disease should be adopted.

Rarely, if the combination of vitamin B_{12} and MMA assay is equivocal, then measurement of homocysteine may be necessary. This exploits the requirement of methylcobalamin as an essential cofactor for methionine synthase, which takes homocysteine as a substrate (Figure 5.13). A growing body of opinion suggests that measuring transcobalamin could be valuable in circumstances where vitamin B_{12} could be falsely elevated, such as in cases of renal failure and myeloproliferative disease. In myeloproliferative diseases, transcobalamin is released in large quantities from the abnormal granulocytes. These compete with 'native' transcobalamin when binding vitamin B_{12}, resulting in a failure to provide tissues with bioavailable B_{12}. Notably, the assays of MMA and homocysteine may also be useful in a different setting. High levels of the former will be present in deficiency of the deoxyadenosylcobalamin variant of vitamin B_{12}, whereas hyperhomocysteinaemia will be a consequence of the failure to generate the methylcobalamin variant of the vitamin.

To summarize, investigation and treatment focuses on the presence or absence of objective clinical parameters such as anaemia, glossitis, and neurological symptoms. In the former, the patient presents with symptoms, and a raised MCV leads directly to the measurement of vitamin B_{12} and folate, as low deficiency mimics B_{12} deficiency. Figure 5.17 summarizes these

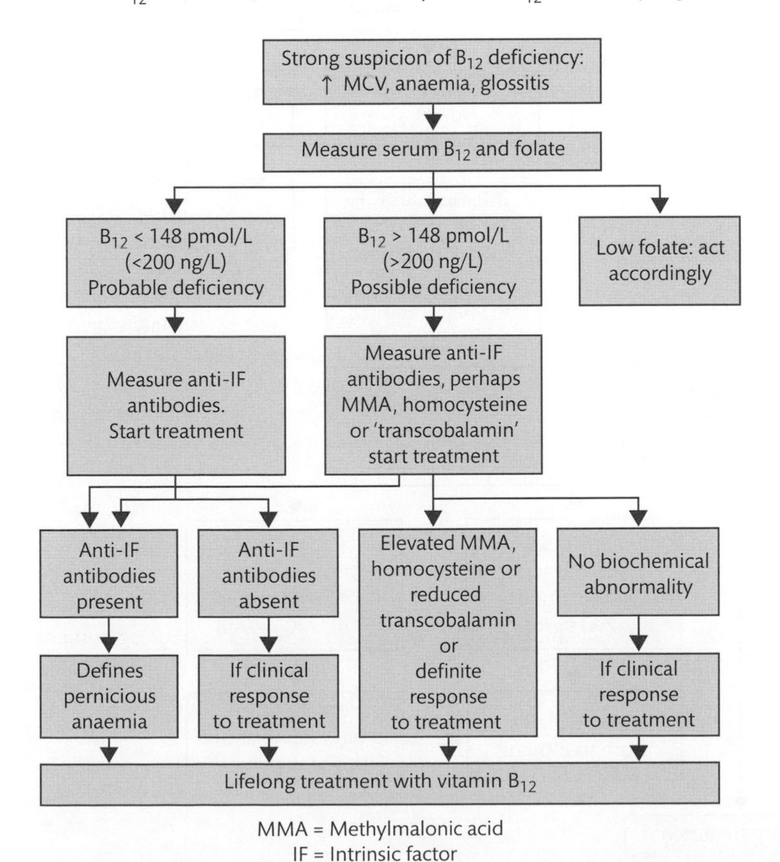

FIGURE 5.17
Investigation and management of possible vitamin B_{12} deficiency in the presence of a strong suspicion of disease. Regardless of the levels of the vitamin, autoantibodies to IF are sought and treatment begins. Measurement of MMA, homocysteine, and transcobalamin may be needed. Modified from Davalia et al. (2014).

aspects in patients with a strong suspicion of deficiency in vitamin B_{12}, Figure 5.18 when serum vitamin B_{12} has been tested in the absence of symptoms.

Once proven to be present, the aetiology of vitamin B_{12} deficiency, if caused by autoantibodies to gastric parietal cells and/or IF, is defined in the laboratory as PA. The presence of these autoantibodies is relatively easy to detect in the immunology laboratory by standard serological techniques such as indirect immunofluorescence, although an enzyme-linked immunofiltration assay (ELIFA) may also be used. In addition, anti-IF antibodies may be found in gastric juice. The laboratory may also offer a test for levels of plasma gastrin or pentagastrin, which are likely to be high in PA.

The (auto)immunological basis of PA is supported by relationships with other autoimmune diseases. In one study, 9% of patients with autoimmune thyroid disease also had PA, compared with 0.1% in the general population. Conversely, thyroid autoantibodies are present in

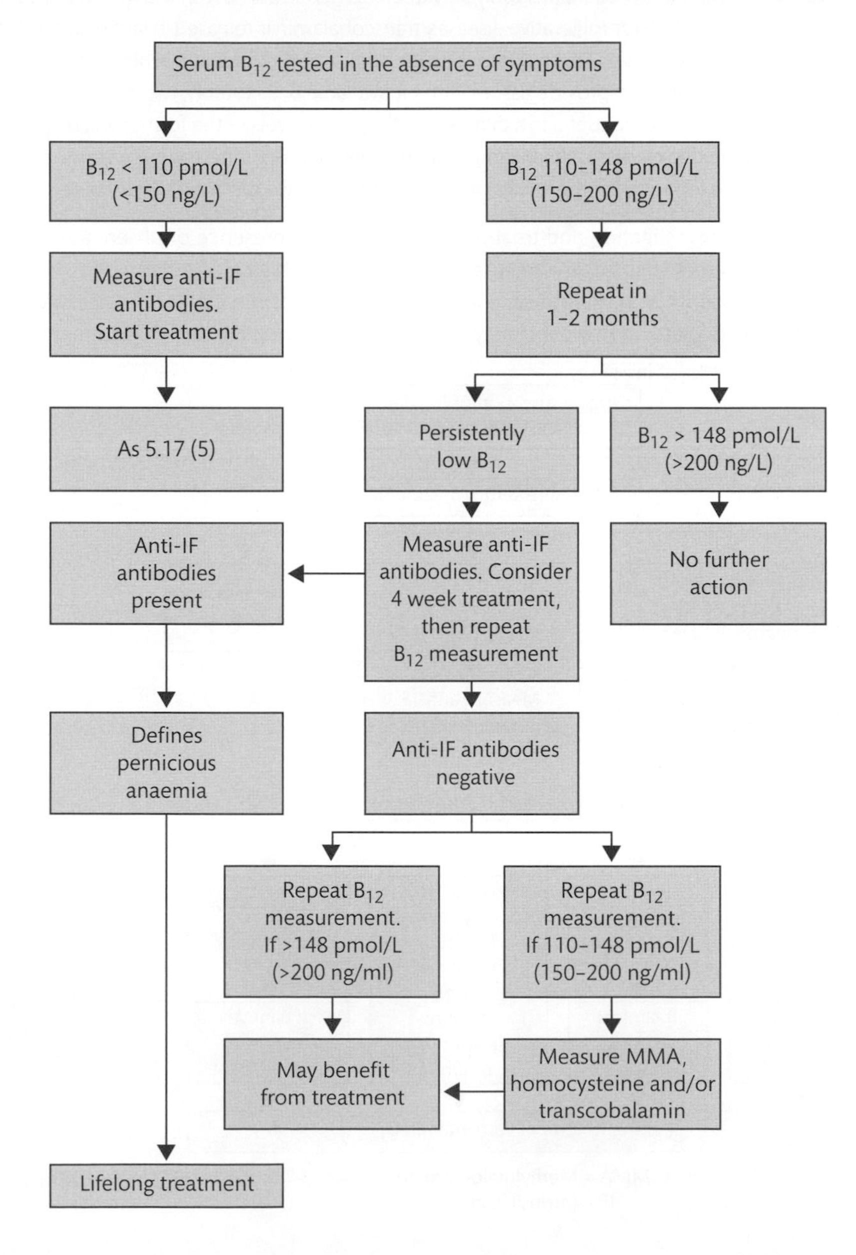

FIGURE 5.18

It is possible that levels of the vitamin are determined in the absence of clear symptoms, or are measured in another setting, and if so, a different path is followed. Once more, levels of the vitamin are determined, and if very low (<110 pmol/L, <150 ng/L), then the pathway in Figure 5.17 applies. For 'standard' low levels (110–48 pg/L, 150–200 ng/L), bloods are repeated and if still low, autoantibodies are sought. Modified from Davalia et al. (2014).

the serum of 55% of patients with PA. The disease is ten times more common in type-1 diabetes than in type-2 disease. Indeed, it is regarded as prudent to prospectively test such patients for autoantibodies perhaps annually, whilst endoscopic surveillance for tumours at five-yearly intervals has been recommended.

An alternative test to help define the nature of the deficiency is of the absorption of vitamin B_{12}. In this test, developed by Schilling, the patient is presented with an oral dose of radiolabelled vitamin B_{12} and a large excess of injected unlabelled cyanocobalamin. The urine and/or blood can then be probed for labelled and unlabelled vitamin to determine the proportion of each type of the vitamin which has been absorbed. A variant of this test is to measure the absorbance of the vitamin in the presence and absence of oral IF. However, the Schilling test has fallen out of favour because of the inherent safety issue associated with radioisotopes, its complex nature, difficulties with its interpretation (such as in renal disease), and the improved performance of alternative assays. Table 5.10 summarizes major laboratory features of vitamin B_{12}-deficient anaemia. Once diagnosis has been confirmed, treatment is with intramuscular injection of the vitamin. Case study 5.2 is illustrative of the consequences of vitamin B_{12} deficiency.

In the UK, the National Institute for Health and Care Excellence (NICE: www.nice.org) offers advice and guidance on behalf of the government's Department of Health. In September 2015, they published Medtech Innovation Briefing MIB40 regarding the Abbott ARCHITECT active-B12 assay for detecting levels of holotranscobalamin (that is, transcobalamin that is carrying the vitamin). The report states that it can be used as a marker of vitamin B_{12} deficiency. Four diagnostic test accuracy studies, using different reference standards, reported greater diagnostic accuracy for the active-B_{12} assay compared with assays measuring other markers of vitamin B_{12} deficiency.

TABLE 5.10 Major laboratory features of vitamin B_{12} deficient anaemia.

Full blood count	Reduced haemoglobin; increased MCV, possibly also in the MCH and MCHC; possible leucopenia and thrombocytopenia
Blood film	Presence of macrocytic red cells, anisocytosis, occasional nucleated red cells; increased number of hyperchromic red blood cells; hypersegmented neutrophils
Bone marrow	Presence of megaloblasts (thereby defining megaloblastic anaemia)
Serology	Presence of autoantibodies to gastric parietal cells and to IF (if present, these define PA); low levels of serum vitamin B_{12}, often with reduced serum and red cell folate

SELF-CHECK 5.10

What are the primary FBC, blood film, and bone marrow abnormalities to be expected in vitamin B_{12} deficiency?

Vitamin B_6

Because vitamin B_6 is abundant and easily absorbed, a deficiency in this vitamin is extremely rare. However, if present, deficiency (probably defined by serum levels <10 nmol/L of pyridoxal phosphate) can be treated with oral vitamin B_6 50–200 mg/day. The elderly, those on dialysis, and alcoholics are at risk. Inadequate levels of vitamin B_6 may cause reduced haem synthesis and thus a haemolytic, sideroblastic, and megaloblastic anaemia, a picture similar to that of vitamin B_{12} deficiency.

CASE STUDY 5.2 Consequences of vitamin B_{12} deficiency

A 75-year-old woman complains of slowly developing fatigue so that she has difficulty getting out of the house. On examination, she has a swollen and painful tongue, and she complains of becoming forgetful, with occasional 'tingling' of her fingertips, and numbness in her toes. Routine bloods were:

Index	Result	Reference range (see front of book)
Haemoglobin	96 g/L	118–148
White cell count	5.7×10^9/L	3.7–9.5
Platelets	198×10^9/L	143–400
MCV	108 fL	77–98
ESR	15 mm/hour	<10
Haematocrit	0.45	0.33–0.47

In view of her signs, symptoms, and history, additional bloods were requested. The results of these were serum iron 11.6 μmol/L (reference range 10–37), serum vitamin B_{12} 78pmol/L (reference range >148 pmol/L).

Questions

- **Can you make a preliminary diagnosis based on the FBC and ESR? What are the abnormalities?**
- **Are you able to modify this diagnosis in view of the iron and vitamin B_{12} results?**
- **What are your next steps?**

Folate

Deficiency of folate is also rare. However, measuring serum folate is problematic because of its liberation from stores and transiently increased levels after a meal, leading to fears of false positives (as may be found in anorexia, acute alcohol consumption, use of anticonvulsants, and pregnancy) and false negatives. Consequently, red blood cell folate has been regarded as a more reliable determinant, as levels are constant throughout the life of the cell. Low red cell folate levels are very common in PA (>50%), possibly because vitamin B_{12} is required for the transfer of methyltetrahydrofolate from the plasma into the red cell. However, a 2014 guideline recommended that routine red cell folate testing is not necessary but may be used if symptoms are strong and serum folate is normal. Plasma homocysteine may be called upon in special circumstances: a result >15 μmol/L may indicate folate deficiency. Other causes of low folate are listed in Table 5.8.

CASE STUDY 5.3 Anaemia after renal transplantation

A 45-year-old man recovers well from his liver transplantation, and tolerates the immunosuppression (azathioprine 50 mg od, tacrolimus 3 mg od) well (blood results 3 February–26 April). However, by June he complains of increasing tiredness and weakness, which by 10 June brings a diagnosis of anaemia, prompting the cessation of the azathioprine. Despite this, although the patient feels a little better, blood results do not respond, and with the result of 24 June, the practitioner recommends oral iron. This has little effect the following week but results later in July improve and are back in the reference range. The patient remained on oral iron for an additional three months.

	Haemoglobin	Red blood cell count	MCV
Reference range	133–167 g/L	4.3–5.7 × 10^{12}/L	77–98 fL
3 February	145	5.3	85
1 March	149	5.5	83
28 March	143	5.3	84
26 April	140	5.0	84
20 May	135	4.9	85
10 June	130	5.0	80
17 June	128	4.8	77
24 June	131	4.9	74
1 July	130	5.1	75
8 July	133	5.0	77
15 July	135	5.2	80
29 July	140	5.2	85
27 August	146	5.3	86

Questions

1. Why was the cessation of azathioprine considered?
2. How would you further describe the anaemia of 10 June?
3. How would you further describe the anaemia of 24 June?
4. What prompted the recommendation of oral iron?
5. Why did it take so long for an anaemia to develop? The patient was doing well in February and March.
6. Could the azathioprine have played any other part?
7. Why has there been no confirmation of iron deficiency with tests such as transferrin, serum iron, etc.?
8. The patient is clearly cured of his anaemia at the end of July and August–why the need to continue iron for another three months?

Key Points

The aetiology and pathology of PA is well understood, and its many links make it a classic of disease unravelling. However, one of its major laboratory signs—macrocytosis—is also found in several other pathological and healthy conditions.

5.7 Anaemia of chronic disease (ACD)

In many instances, the aetiology of an anaemia can be clear-cut, perhaps the best example being dietary and/or iron deficiency. However, iron deficiency at the bone marrow in the light of a good diet must have another cause, such as gastric disease, a genetic mutation, and inflammation, and in some cases, all three (and maybe more). A subtype of ACD is functional iron deficiency, which is defined as inadequate incorporation of iron into erythroid precursors despite adequate body stores of the micronutrient; another is the anaemia of inflammation. The latter is inevitably linked to inflammatory disease in another organ/organ system, and so demands clear evidence of an inflammatory process, such as a raised white blood cell count, although the more specific markers of interleukin-6 (IL-6) and C-reactive protein (CRP) are preferred. The precise definition of the aetiology is important, as immunosuppression may help an anaemia of inflammation (although it may also suppress the bone marrow), whereas it will have on effect on anaemia whose aetiology is not inflammatory.

Cross references

Full details of the role of Epo and erythropoiesis are presented in Chapter 4.

Liver function tests are fully explained in the *Clinical Biochemistry* volume of this series.

The heterogenous condition of the anaemia of chronic disease brings together numerous different aetiological aspects, several of which we have already addressed, and which often overlap. In contrast, acute anaemia is often associated with a clear event such as massive and traumatic blood loss, perhaps after surgery or a road traffic accident, or perhaps an infection-induced haemolysis. This final section will attempt to bring together the theory and laboratory practice of anaemia and look at how these apply to clinically oriented situations.

5.7.1 The liver

The functions of this complex organ include the storage of iron and the synthesis of iron-related molecules. Therefore, both these factors impact upon anaemia. Part of the body's iron stores are carried in hepatocyte and liver macrophages (Kupffer cells) as ferritin and haemosiderin. A failing liver (as may be present in hepatoma, cirrhosis, or hepatocellular carcinoma) will therefore be unable to synthesize these proteins and thus to adequately store and transport iron. This situation can lead to (or at least contribute to) a microcytic anaemia. Severe damage to hepatocytes is associated with acanthocytosis (cells with membrane protrusions like spikes, sometimes described as spur cells, although they are present in several other conditions such as hypothyroidism). Alcoholism may not simply lead directly to liver disease and atrophy, but may also develop into malnutrition and malabsorption, with deficiencies in vitamin B_{12} and folate, which in turn may promote megaloblastic anaemia. However, alcohol abuse can cause macrocytosis in the absence of vitamin B_{12} or folic acid deficiency.

Actual damage to the liver (as may be caused by cytotoxic chemotherapy or toxins such as alcohol) may liberate iron-rich ferritin from intracellular stores. Consequently, levels may appear normal or even high, erroneously suggesting plentiful iron stores. Hypercholesterolaemia, whether the result of obstructive jaundice or genetic-type familial hypercholesterolaemia, can lead to macrocytosis and target cells; in contrast, hepatocellular damage, as may be caused by toxins (such as heavy metals and industrial solvents) is associated with acanthocytosis. Macrocytes are also present in other forms of liver disease.

If there is severe liver disease, it is likely that liver function tests such as alkaline phosphatase, bilirubin, and gamma glutamyl-transferase (GGT) will also be abnormal, whilst alcoholism is associated with increased levels of GGT. This illustrates an additional reason why the experienced practitioner will be aware of issues outside of their immediate discipline—in this case, in biochemistry. Indeed, this underlines the developing field of blood science.

Inflammation, in the form of the acute-phase response, may induce increased levels of ferritin and transferrin. Such inflammation may be the secondary response to a distant infection, to an autoimmune disease (such as rheumatoid arthritis), or to active inflammation of the liver itself (that is, hepatitis) (see section 5.7.6 below). In addition, chronic hepatitis (as may be caused by hepatitis viruses A, B, and C) will also cause blood disease by depleting iron stores and reducing the ability of the liver to maintain haemostasis.

5.7.2 The kidney

The kidney produces the red blood cell growth factor Epo, which is required for effective erythropoieisis (section 4.1.3). Consequently, patients with chronic renal failure, and possibly on dialysis, are at risk of becoming anaemic, and if present, it is generally normocytic. The normal inverse relationship between Epo and haemoglobin is lost when serum creatinine rises above 135 μmol/L, and the degree of hypoxia required to stimulate release of the hormone rises considerably. Fortunately, recombinant Epo can be prescribed to treat anaemia of this aetiology, although this is not always as beneficial in the long term (Box 4.1). Rarely, in frank and serious renal damage, as may be present in trauma or disease such as glomerulonephritis, there may be blood loss into the urine (haematuria). If this is heavy and persistent, a normocytic anaemia may follow, ultimately resulting in microcytic anaemia once iron stores have become depleted.

Measurement of serum Epo may be useful in assessing significant anaemia. However, the level of Epo cannot be taken as a measure of oxygenation because many conditions (such as hyperviscosity, cancer, and pregnancy) suppress Epo production independently of the effects of hypoxia. In parallel to severe liver disease and liver function tests, severe renal disease may be assessed by the biochemistry blood tests of urea and electrolytes (U&E), and also by the estimated glomerular filtration rate (eGFR). Case study 5.3 describes anaemia and renal transplantation.

Cross reference

Urea and electrolytes, and the eGFR, are fully explained in the *Clinical Biochemistry* volume of this series.

5.7.3 Gastrointestinal disease

Sections 5.3–5.6 discuss how disease of the digestive tract can lead to microcytic and macrocytic anaemia because of the malabsorption and consequent deficiency of iron and vitamin B_{12}, respectively. However, anaemia may also arise from other diseases of the intestines, such as the physical loss of blood from bleeding lesions. Such anaemia is generally (and initially) normocytic. These diseases may be present in the upper section of the intestinal tract and include oesophagitis and oesophageal varices.

In the lower section of the intestines, blood may be lost from haemorrhagic ulcers, or as a consequence of colorectal cancer, inflammatory bowel disease (IBD, such as duodenitis, colitis, and Crohn's disease—see section 5.7.6), or haemorrhoids. Indeed, occult bleeding is an early sign of gastrointestinal cancer and may lead to normocytic anaemia. However, if chronic blood loss persists, of whatever aetiology, and iron stores become depleted, then the anaemia may become microcytic.

5.7.4 Reproductive organs

Increased haemoglobin, haematocrit, and red cell numbers in men compared to women is the result of testosterone acting on the bone marrow as a growth factor. Loss of the testes to trauma or disease is associated with a reduction in the red cell indices, often to those of healthy women. Healthy women lose blood approximately each month as part of their menstrual cycle, and this also contributes to their lower haematology indices, as these indices improve after hysterectomy. However, excessive blood loss through menstruation (menorrhagia) can lead to a normocytic anaemia, and as with chronic blood loss for intestinal disease, may eventually deplete iron stores so that the anaemia becomes microcytic.

5.7.5 Endocrine disease

Anaemia is often present in patients with anterior pituitary failure, chronic adrenal insufficiency, and primary hyperparathyroidism, but these are encountered relatively infrequently. However, haematological abnormalities may well be present in both an overactive and underactive thyroid. The anaemia of hypothyroidism may be masked by a reduced blood volume, but acanthocytosis and macrocytosis may be present—the latter generally responding to replacement therapy with thyroxine, which is believed to stimulate erythropoiesis. Conversely, microcytosis is common in hyperthyroidism, even if there is no anaemia; moreover, an increase in plasma volume can mask an increased red cell mass possibly driven by high levels of thyroxine, which has the capacity to overstimulate erythroblasts. Recall also that an abnormal profile may be caused not by the disease process but by its drug treatment: macrocytosis in diabetes is as likely to be caused by the hypoglycaemic therapy metformin.

5.7.6 The anaemia of inflammation

We now know that inflammatory cytokines are present in numerous forms of anaemia, leading to the concept of the anaemia of inflammation, an example of this aetiology being the suppression of bone marrow by IFN-γ. Rheumatologists have long known that patients with chronic inflammatory disease, typified by rheumatoid arthritis (RA) and systemic lupus erythematosus (SLE), are at risk of a normocytic anaemia, and that this correlates with inflammatory markers such as ESR and CRP. Indeed, anaemia is the most common non-connective tissue pathology associated with SLE (renal disease is a close second). However, many diverse inflammatory diseases are associated with anaemia and we have already noted several (such as thyroiditis, glomerulonephritis, Crohn's disease, and viral or parasitic hepatitis), and accordingly these may also be included in the anaemia of inflammation group, and, in addition to a normocytic anaemia, are associated with low serum iron (hypoferraemia), low serum iron-binding capacity, and normal to elevated ferritin concentrations.

As discussed at the beginning of section 5.7, a key driver of this process may be the pro-inflammatory cytokine interleukin-6 (IL-6). Increased levels of this cytokine induce the synthesis of hepcidin, high levels of which may inhibit the entry of iron into the blood. In a study of patients in intensive care, hepcidin was six times higher in those whose anaemia had an inflammatory component compared to those whose anaemia was the consequence of iron deficiency. In another focusing on inflammatory anaemia, urinary hepcidin correlated with serum ferritin, both likely to reflect and be driven by the acute-phase response. Hepcidin is also regulated by bacterial lipopolysaccharide (LPS) and bone morphogenetic proteins. An inflammatory response to a bacterial infection, and a resultant anaemia, may be a defence mechanism in attempting to deny the microbe iron. However, such as response would be accompanied by an increased white blood cell count, implying a red cell specific effect.

A rare and severe disease, haemophagocytic syndrome, may develop as a by-pathology of strong immunological activation, itself potentially developing from a lymphoid malignancy or a connective tissue disease. Haemophagocytic syndrome is characterized in the laboratory by a low haemoglobin (typically 80–90 g/L) and platelets (typically 30–80 × 10^9/L), with a markedly raised lymphocyte count (often 6–16 × 10^9/L), increased lactate dehydrogenase, and grossly raised ferritin (often >1000μg/L, reference range 25–380). On the blood film, red cells can be discerned within what appear to be macrophages but what may actually be histiocytes that have migrated into the blood. In the clinic, the disease is associated with fever and enlargement of the liver and spleen.

5.7.7 Cancer

Several aspects of anaemia in cancer must be considered, as we have observed in section 5.2. One is a direct effect of the tumour on erythropoiesis, which may be present in many malignancies: either in suppressing renal Epo production or by having a direct effect on the bone marrow that is likely to produce a normocytic anaemia. An alternative aspect is the secondary effect of the tumour. Examples of this include the chronic loss of blood from a bleeding gastrointestinal cancer (which can be detected by testing for faecal occult blood), and the effects of liver cancer. A third aspect is treatment. As we have seen in the opening section of this chapter, a side effect of anti-tumour chemotherapy is toxic suppression of the bone marrow, and so PRCA, pancytopenia, and aplastic anaemia.

The presence of leucocytes within tumours implies an inflammatory response, and in many cases, increased levels of inflammatory cytokines can be found. The anaemia of myeloma may relate not only to direct suppression of the bone marrow, but also by increased IL-6 levels, which can promote increased hepcidin and so reduce plasma iron. However, an active inflammatory response is often absent in many malignancies. Increased numbers of platelets are also often present in cancer, and this may be due directly to the malignancy or indirectly, via the response of the body to the malignancy.

5.7.8 Anaemia of the elderly

Of course, being elderly is not a disease, but it is becoming recognized that the high degree of anaemia in the elderly cannot simply be a consequence of old age and is increasingly being cited as a syndrome in its own right. It is estimated that 10% of the over-65s are anaemic, increasing to 20–30% in those aged 85 or over. The elderly tend to accrue disease such as rheumatoid

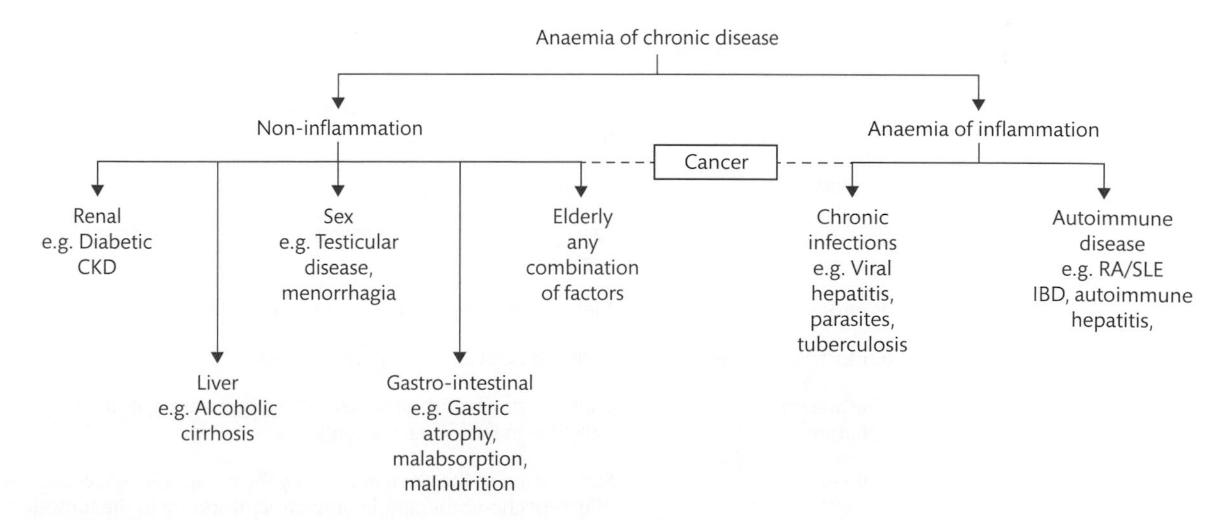

FIGURE 5.19

The anaemia of chronic disease. The anaemia of chronic diseases is a heterologous grouping with diverse aetiologies, the leading being inflammatory or non-inflammatory, which may be caused by several different factors. Different cancers may appear in either major grouping.

arthritis and cancer, and it is well established that renal function deteriorates with age. This anaemia brings reduction in quality of life and cognitive function, but also an increased risk of mortality as, for example, it may be a surrogate of as-yet undiagnosed neoplasia. Each of the factors described above may independently contribute to a minor degree of mild anaemia, but together may precipitate frank disease. Other potential causes of 'unexplained' anaemia of the elderly include low levels of Epo and/or a reduction in Epo signalling, and inflammation. A study of 65–89-year-olds with unexplained normocytic anaemia found a lower eGFR (implying renal dysfunction) and iron alongside raised inflammatory markers (IL-6, IL-8, interferon-γ, and neopterin) and ferritin (but not raised neutrophils) compared to age- and sex-matched controls.

Table 5.11 and Figure 5.19 summarizes the various routes by which chronic disease can cause anaemia.

SELF-CHECK 5.11

What is the pathological basis of the anaemia of chronic disease?

Key Points

The proficient haematologist not only needs a deep understanding of the mechanics of their subject, but also an awareness of the whole body and how other organs influence the blood, both in health and disease.

5.7.9 The laboratory in the anaemia of chronic disease

The key initial laboratory tests are MCV and MCH, and the percentage of hypochromic red cells (%HRC; that is, whose MCH is below the bottom of the reference range, such as <26 pg) is also useful, although not all autoanalysers provide this index. Additional useful tests are the

TABLE 5.11 The anaemia of chronic disease.

The liver	Failure to store iron and synthesize carrier proteins (e.g. transferrin)
The kidney	Failure to produce Epo
Gastrointestinal disease	Failure to absorb micronutrients: occult blood loss from bleeding tumours
Reproductive organs	Menorrhagia: failure of testes to produce testosterone
Endocrine disease	Anaemia associated with hypothyroidism
The anaemia of inflammation	Suppression of erythropoiesis and hypoferraemia, both potentially due to inflammatory cytokines
Cancer	Suppression of the bone marrow by the metastatic process: effects of chemotherapy, inflammatory response to the tumour
Anaemia of the elderly	Several possible overlapping disease processes: renal disease, endocrine dysfunction, stem cell aging, medications acting on the bone marrow, nutrient deficiency, chronic inflammatory disease

reticulocyte count and the reticulocyte haemoglobin content (CHr, effectively the reticulocyte MCH, a result of <25 pg being suggestive of classical iron deficiency). In one study, the CHr was superior to the MCV, serum ferritin, and iron saturation in predicting iron depletion. An alternative to %HRC or CHr in the assessment of functional iron deficiency is red cell zinc protoporphyrin (ZPP) which, if raised, reflects the insertion of zinc into haem instead of iron. Levels of serum ferritin <12 μg/L indicate absent iron stores.

Many (and, crucially, not all) cases of the anaemia of chronic disease are inflammatory, and accordingly many tests (such as ferritin and transferrin) become less reliable. In this respect, a key regulator of iron, hepcidin, is itself regulated by iron availability, hypoxia, Epo, inflammatory cytokine IL-6, and LPS, but also by the hormone growth-differentiation factor 15 (GDF15) a member of the TGF-β superfamily and product of erythroid precursors. Therefore hepcidin (like ferritin and transferrin) is also raised in chronic disease such as rheumatoid arthritis, inflammatory bowel disease, cancer, and chronic kidney disease. GDF15 is strongly expressed in beta thalassaemia, and blocks hepcidin. Although GDF15 (measurable by ELISA) has been linked to many diseases, including severe aplastic anaemia, where it correlated inversely with hepcidin and positively with Epo, it has yet to enter routine practice. Both the anaemia of chronic disease and iron-deficiency anaemia are characterized by low haemoglobin and MCV, but other serum levels of these molecules may be useful in differentiating those cases of the anaemia of chronic disease that have an inflammatory component (such as a chronic infection or an autoimmune disease) from cases of (non-inflammatory) iron-deficiency anaemia (Table 5.12).

It is possible that, like sTfR, measurement of hepcidin and GDF 15 may enter clinical service. However, before this happens there must be an established national quality assurance scheme, and agreement on reference levels across different methodological aspects (e.g. ELISA, mass spectrometry).

Case study 5.4 illustrates an anaemia of chronic disease.

TABLE 5.12 The laboratory in anaemia of chronic disease (ACD) and in iron-deficiency anaemia (IDA).

	IDA	ACD
Haemoglobin	↓	↓
MCV	↓	N
Iron	↓	↓
Ferritin	↓	↑↑
Transferrin	↑	↓
Transferrin saturation	↓↓	↓
sTfR	↑↑	↑
IL-6	N	↑↑ (therefore the anaemia of inflammation)
CRP	↑	↑↑(therefore the anaemia of inflammation)
Epo	↑↑	↑
Hepcidin	↓	↑
Growth-differentiation factor 15	N	↑

Key: ↑ = increased relative to health; ↑↑ = markedly increased relative to health; ↓ = reduced relative to health; ↓↓ = markedly reduced relative to health; N = equivalent to health; MCV = mean cell volume; IL-6 = interleukin-6; CRP = C-reactive protein; sTfR = soluble transferrin receptor. Table modified by permission from Theurl et al. 2010.

CASE STUDY 5.4 An anaemia of chronic disease

An 82-year-old man complains of being tired and lethargic. On examination, he has a distended and painful abdomen. He also complains of periodic diarrhoea and constipation and recalls that he has lost perhaps a stone in weight in the past year. Routine bloods were as follows.

Index	Result	Reference range (see front of book)
Haemoglobin	106 g/L	133–167
WCC	6.3×10^9/L	3.7–9.5
Platelets	268×10^9/L	143–400
MCV	88 fL	77–98
ESR	15 mm/hour	<10
Haematocrit	0.38	0.35–0.53

Questions

- **Can you make a preliminary diagnosis based on the FBC and ESR? What are the abnormalities?**
- **What are the potential causes of the diagnosis?**
- **What are your next steps?**

Chapter summary

- The principal disease of red blood cells is anaemia.
- Anaemia may result from invasion of the bone marrow (e.g. by cancer cells) or suppression (by drugs such as chemotherapy).
- Lack of vital nutrients such as iron and certain vitamins can lead to anaemia.
- Excessive body iron can also lead to disease.
- Disease in other organs (such as the liver, kidney, intestine, and reproductive organs) can also lead to anaemia.

Discussion questions

5.1 Why is the bone marrow important in investigating anaemia?

5.2 Describe the different forms of disease of iron metabolism.

5.3 What are the most important vitamins in red cell production and what are the consequences of their deficiency?

5.4 What are the most common causes of the anaemia of chronic disease?

Further reading

- Barra M, Stahl S, and Hellmann H. ***Vitamin B_6 and its role in cell metabolism and physiology***. *Cells* 2018: **7**: 84; doi:10.3390.
- Casu C, Nemeth E, and Rivella S. ***Hepcidin agonists as therapeutic tools***. *Blood* 2018: **131**: 1790–4.
- Dailey HA and Meissner PN. ***Erythroid heme biosynthesis and its disorders***. *Cold Spring Harbour Perspectives in Medicine* 2013: **3**: a011676.
- Davalia V, Hamilton MS, and Molloy AM. ***Guidelines for the diagnosis and treatment of cobalamin and folate disorders***. *British Journal of Haematology* 2014: **166**: 496–513.
- Green R. ***Vitamin B_{12} deficiency from the perspective of a practicing hematologist***. *Blood* 2017: **129**: 2603–11.
- Horos R and von Lindern M. ***Molecular mechanisms of pathology and treatment in Diamond Blackfen anaemia***. *British Journal of Haematology* 2012: **159**: 514–27.
- Kafina MD and Paw BH. ***Intracellular iron and heme trafficking and metabolism in developing erythroblasts***. *Metallomics* 2017: **9**: 1193–203.
- Knutson MD. ***Iron transport proteins: gateways of cellular and systemic iron homeostasis***. *Journal of Biological Chemistry* 2017: **292**: 12735–43.
- Means RT. ***Pure red cell aplasia***. *Blood* 2016: **128**: 254–9.
- Muckenthaler MU, Rivella S, Hentze MW, and Galy B. ***A red carpet for iron metabolism***. *Cell* 2017: **168**: 344–61.
- Pantopoulos K. ***Inherited disorders of iron overload***. *Frontiers in Nutrition* 2018: **5**: 1–11.
- Surendran S, Adaikalakoteswari A, Saravanan P et al. ***An update on vitamin B12-related gene polymorphisms and B12 status***. *Genes Nutrition* 2018: **13(2)**: doi 10.1168/s12263-018-0591-6.
- Thomas SW, Hinchcliffe RF, Briggs C, Macdougall IC, Littlewood T, and Cavill I. ***Guideline for the laboratory diagnosis of functional iron deficiency***. *British Journal of Haematology* 2013: **161**: 639–48.

- **Weiss G, Ganz T, and Goodnough LT. *Anaemia of inflammation*. *Blood* 2019: 133: 40–50.**
- **http://www.who.int/nutrition/events/2017-meeting-haemoglobin-concentrations-anaemia-29novto1dec/en/ (accessed 19 October 2020).**

Answers to self-check questions and case study questions are provided as part of this book's online resources.

Visit www.oup.com/he/moore-fbms3e.

6

Anaemia 2: Haemolysis

Andrew Blann and Sukhjinder Marwah

Having established the basics of red cell function in Chapter 4, and having described some aspects of anaemia in Chapter 5, we now move to complete our study of the principal diseases of erythrocytes by looking at the remaining causes of anaemia. Whilst some of the different types of anaemia we examined in Chapter 5 were due to a shortage of essential factors (such as iron), in this chapter we focus on the destruction of red blood cells, a process called haemolysis. This destruction can be classified as external factors acting on essentially healthy red cells, or destruction of red cells with defects of haemoglobin, the cell membrane, or in metabolic enzymes.

Learning objectives

After studying this chapter, you should confidently be able to:

- understand the meaning of haemolysis;
- list the different causes of the inappropriate destruction of red blood cells;
- explain the relationship between gene mutation and anaemia;
- describe the major features of the haemoglobinopathies;
- outline how mutations in genes for membrane components and enzymes can lead to anaemia;
- suggest how the laboratory can diagnose these conditions.

6.1 Introduction to haemolytic anaemia

In Chapter 5, we saw how many diverse disease processes can act on red cells to cause anaemia. In this chapter, we will see how anaemia may follow from the attack and destruction (**haemolysis**) of red cells, defined as any condition shortening the red cell lifespan

haemolysis
The inappropriate destruction of red cells.

to less than 120 days, perhaps down to 40 or 50 days. Haemolysis can be broadly classified into two processes. In the first, perfectly good red cells are attacked by an outside agent or agents–so-called *extrinsic* defects. In these situations, there is nothing fundamentally wrong with the red cell itself, but it is destroyed because of some external pathological process, such as drugs, toxins, an autoantibody, or an infection. Alternatively, there may be *intrinsic* defects within the red blood cell itself, due to pathogenic gene mutations, that lead to its premature destruction. These red cells are marked out and destroyed by the body itself because there is something intrinsically wrong with them. This may be due to damage to, or loss of integrity of, the red cell membrane, the absence of certain enzymes, or perhaps to an abnormal type of haemoglobin (Table 6.1). A further complication of the latter type of damage is that those cells with certain intrinsic defects are often unable to carry as much oxygen whilst they are able to evade destruction.

haemolytic anaemia
Anaemia that follows the destruction (lysis) of red blood cells.

Thus, the **haemolytic anaemias** (that is, anaemia due to (premature) destruction of the cell) are generally caused by a shortening of the lifespan of the red cell and failure of the bone marrow to be able to compensate for this reduced lifespan. If bone marrow production matches the rate of red cell destruction, this is called compensated haemolytic disease. The causes of the destruction of the red cell may, in turn, be due to one of two markedly different pathological processes.

Whatever the aetiology, these damaged cells are detected and removed by cells of the reticuloendothelial system, mostly within the spleen, although this can also occur in the bone marrow and liver (all described as extravascular haemolysis). Alternatively, cells may be destroyed whilst in the blood (intravascular haemolysis), possibly by physical forces, attack by phagocytes, and/or antibodies and complement. In the laboratory, haemolysis is investigated by assessing

TABLE 6.1 A classification of haemolytic anaemias.

An extrinsic pathological process acting on a healthy red cell	Antibodies autoantibodies alloantibodies
	Mechanical destruction
	Drug-induced haemolysis
	Infections: malaria
Intrinsic defect of the red cell	The haemoglobinopathies sickle cell disease thalassaemia compound haemoglobinopathy
	The membrane spherocytosis elliptocytosis stomatocytosis ovalocytosis pyropoikilocytosis paroxysmal nocturnal haemoglobinuria
	Enzymes glucose-6-phosphate dehydrogenase deficiency pyruvate kinase deficiency

the red blood cells themselves, and also by looking for evidence of an increased rate of their destruction in the blood, such as free haemoglobin and haem, possibly complexed to haptoglobin and haemopexin respectively, and lactate dehydrogenase (LDH). The latter is present in many cells, and increased levels in the circulation are a general-purpose marker of cell damage.

The clinical consequences of haemolytic anaemia are not simply those related to the inability of the damaged red cells to deliver oxygen. All forms of haemolytic anaemic also bring an increased risk of thrombosis, in contrast to the micronutrient deficiency anaemias of Chapter 5, where thrombosis is encountered far less often. Several of the products of haemolysis have the potential to promote thrombosis, especially if the haemolysis is acute and so overwhelms the natural anticoagulant defences of the body. For example, excess iron and haem can promote cytotoxic reactive oxygen species to damage and activate the endothelium and platelets. Accordingly, patients are at risk of deep vein thrombosis and pulmonary embolism, and if these have occurred, the patient will probably need to be anticoagulated.

Cross reference

Red blood cell function is described in Chapter 4, haemopoiesis in Chapter 3.

6.2 Immune-mediated and other 'extrinsic' causes of haemolytic anaemia

In this section, we will specifically discuss immune-mediated and other extrinsic mechanisms of haemolysis. Immune haemolysis can occur as a consequence of the presence of two main types of antibody:

alloantibodies are produced in response to the immune recognition of foreign red cells that have been introduced either via a blood transfusion or are present due to pregnancy (following the mixing of maternal and foetal blood at delivery, or by exposure to paternal HLA molecules);

autoantibodies are produced when a patient's immune system recognizes their own red cells as foreign and so mediates their destruction.

These two processes cause alloimmune (an 'external' cause) and autoimmune (an 'internal' cause) of haemolysis, respectively. These antibodies are most likely to be of IgG and/or IgM isotypes, the haemolysis itself being effected by a complex group of molecules called **complement**, or phagocytosis by cells of the reticuloendothelial system (including macrophages, often within the spleen). Macrophages of the reticuloendothelial system, in particular, express receptors on their cell surface, which recognize the IgG and the complement components covering the sensitized cells, thus enabling the process of phagocytosis to occur. In the first instance, portions of the sensitized red cell membrane are removed, reducing the surface area of the red cells whilst maintaining their volume. This process results in the formation of spherocytes. These spherocytes become prone to increased sequestration in the spleen, which leads to their shortened lifespan. This type of premature destruction is described as extravascular haemolysis.

complement

A series of defensive proteins that normally assemble on the surface of pathogens such as bacteria, helping to cause their destruction.

6.2.1 Alloimmune haemolysis

The introduction of foreign red cells into an individual (deliberately, as in a blood transfusion or accidentally, as in a mother's absorption of blood from her neonate at childbirth) may initiate an immune response if those foreign cells bear antigens that will be recognized by that individual. If present, this incompatibility may have serious clinical consequences.

Blood transfusion

The objective of this process is to provide the patient with a mass of red cells to alleviate a particular clinical problem (most likely, a profound anaemia). However, donor red cells must be checked for compatibility with a sample of serum or plasma from the recipient. It is very likely that both the AB and Rh blood groups will have been determined in both donor and recipient, but the recipient may have antibodies to any one (or more) of the numerous other minor blood groups (such as Kell, Duffy, and Kidd—see section 4.2.5) that, if present, may cause a severe transfusion reaction.

It is important to recognize that any red cell antibody can cause haemolysis when it combines with its respective antigen *in vivo*. Indeed, the most clinically significant cases of alloimmune haemolysis, which can cause major morbidity and even a fatal outcome, are the result of red cell transfusions where the ABO blood groups of the donor and recipient are incompatible. This is a situation unique to this blood group system because of the unexpected presence of a potent preformed antibody in most individuals, depending on their blood group. This type of alloimmune haemolysis most often involves IgM antibodies, complement activation, and red cell destruction within the vascular system itself, i.e. intravascular haemolysis. However, in several different situations, many damaged or abnormal cells are removed in the spleen, so that removal of this organ (splenectomy) is often a successful treatment and prolongs the life of these cells, resulting in an improved clinical picture (that is, less anaemia). These transfusion reactions may also be caused by preformed antibodies in the recipient that have been stimulated by, and so recognize, minor blood groups.

In order to avoid immune-mediated haemolytic transfusion reactions, which inevitably result in the extravascular or intravascular lysis of donor red cells, compatibility testing procedures are performed in the transfusion laboratory, where donor cells are exposed to recipient serum (possibly containing antibodies) and vice versa in a cross-match. However, in most cases only the donor cells are exposed to recipient serum to determine whether the latter has any antibodies. This process, the **indirect antiglobulin test** (IAT), has a number of steps, which are simplified as follows:

indirect antiglobulin test (IAT)

A commonly used blood transfusion technique to determine the presence of an antibody or antibodies in a sample of serum/plasma that react with antigens on the surface of red blood cells.

1. Donor red cells are washed free of serum/plasma and resuspended in physiological saline.
2. Recipient serum/plasma is mixed with the donor red cells for a fixed period of time to enable any antibodies in the serum/plasma to react with the cells.
3. Excess recipient serum/plasma is washed away in physiological saline.
4. The donor cells (now having possibly bound antibodies) are mixed with an anti-human globulin (AHG) that will cross-link those cells bearing antibodies (if any). The AHG is generally itself an antiserum produced by animals themselves immunized with human globulin.
5. If antibodies in the recipient serum/plasma have bound the donor red cells, the AHG will cause them to agglutinate, indicating that the cells of the donor are incompatible with the recipient and therefore should not be transfused. Conversely, lack of an agglutination implies absence of antibody-coated donor cells that are therefore transfusable.

Figure 6.1 summarizes these steps. The IAT has a number of variants that may be used to detect certain types of antibodies or in certain other circumstances. Since the body had a temperature of 37°C, this is often the temperature of choice for the cell incubations. However, some reactions at room temperature (20–22°C; see Box 6.1 and section 6.2.2) may be important, other

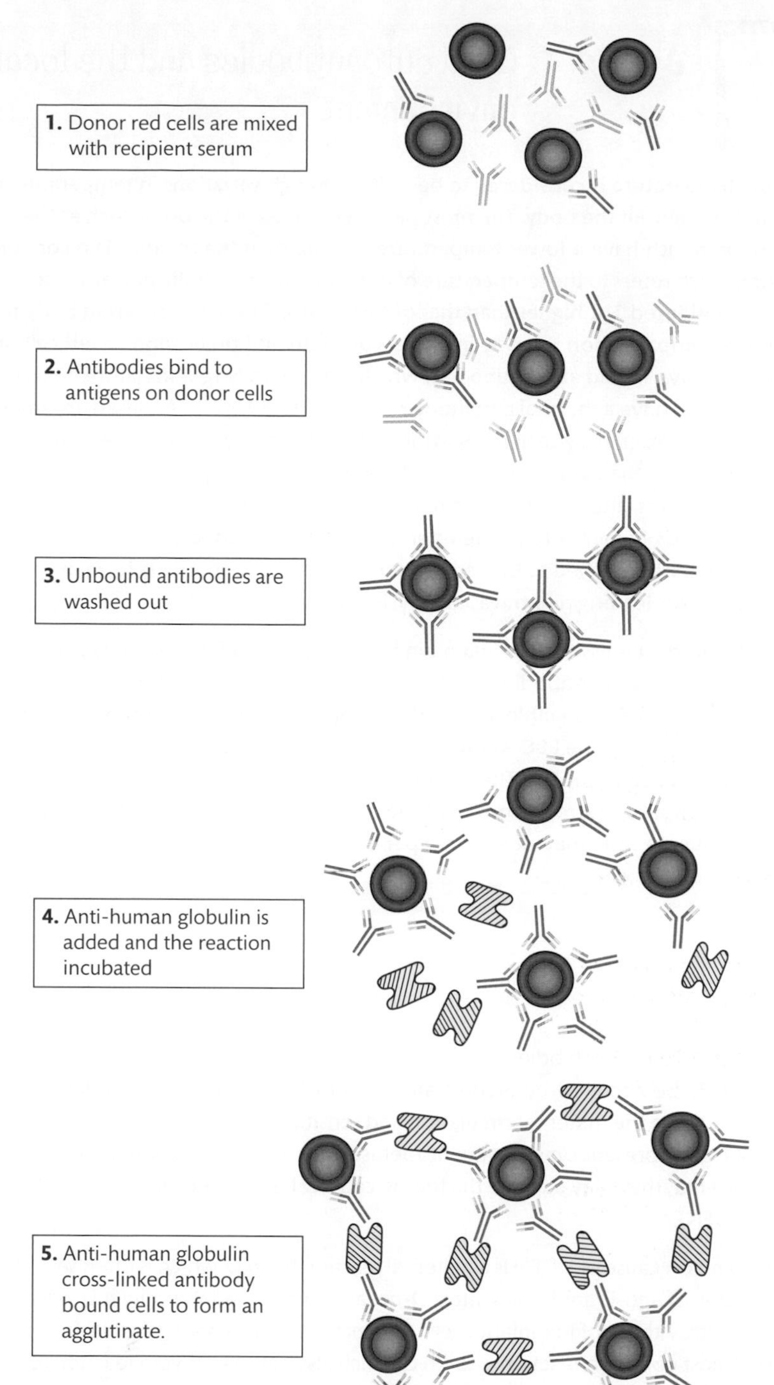

FIGURE 6.1
Simplification of the IDT: see text for details. N.B. the sizes of the antibodies, red cells, and AHG are not to scale. For simplicity, antibodies are represented as IgG—they may be IgM.

incubations being performed in low-ionic-strength saline or with added albumin (as would be present within blood vessels). Different forms of AHG preparations are also available to detect different types of antibodies, but also complement components (mostly anti-C3d). The closely related direct antiglobulin test is discussed in section 6.2.2.

Cross reference

Blood group systems, complement components, and their role in haemolysis are discussed briefly in Chapter 4, and in further detail (with more on antiglobulin testing) in the *Transfusion and Transplantation Science* text in this series.

BOX 6.1 Cold autoantibodies and the local environment

Our core temperature is considered to be 37°C, although variations in temperature are recorded throughout the body. The most peripheral areas of the body, such as the ears, armpits, or mouth have a lower temperature reading than the rectum. The core temperature, which refers to the temperature of the liver—the metabolic powerhouse of the body—is considered 1°C higher than that of the rectum. These variations in body temperature can impact upon the way in which cold autoantibodies bind to red cells and lead to haemolysis. Cold autoantibodies, which are generally IgM in nature, react optimally at 4°C but have a thermal amplitude ranging up to 30–32°C. A low environmental temperature, as frequently seen in the winter months in the UK, will cause cold autoantibodies to bind to the patient's own red cells within the cooler peripheral circulation of the nose, fingertips, and ears and start to fix complement. As the coated red cells move centrally towards higher 'internal' temperatures, the IgM dissociates, leaving complement components attached to the red cell surface. These components form membrane attack complexes leading to intravascular lysis of red cells.

In the laboratory, it is essential to maintain blood specimens from patients with known or suspected cold autoantibodies at 37°C prior to performing a full blood count (FBC) analysis. Failure to do so will lead to red cell agglutination within the specimen and abnormal FBC results. The FBC will show a low red cell count, a normal or raised MCV, and a raised MCH and MCHC. The blood film will contain red cell agglutinates. In the blood transfusion laboratory, strong cold autoantibodies may cause ABO typing anomalies, in which case tests may need to be performed at 37°C to confirm ABO type and antibody identification.

Pregnancy

Although we develop antibodies to various blood groups spontaneously, antibodies to others can only be acquired by blood transfusion or by pregnancy. In the latter, neonatal blood can pass into the maternal circulation and sensitize her, so that she may develop antibodies to antigens present on subsequent foetuses she is carrying. Should these antibodies cross the placenta, they may damage the foetus, causing haemolytic disease of the foetus and newborn (HDFN).

The most common cause of HDFN is a maternal reaction to Rh D, which is often severe, justifying the injection of anti-D antibodies into a D-negative woman whose child is D-positive, in an attempt to destroy those of her infant's cells that may immunize her against its future siblings. The second most common reaction is to Rh c, which also causes a severe reaction (causing 46 deaths per 100,000 pregnancies) whilst an anti-Kell reaction may also be common, causing mild-to-severe disease and depressing erythropoiesis. Consequently, anti-D, anti-c, and anti-Kell antibodies are often sought with laboratory tests. All other minor blood group reactions are either rare (n = 56) or very rare. The IAT can help determine whether the mother is indeed producing antibodies to her child's blood group antigens, some of which may be those of the father.

Whilst not generally thought of as being pathological, women bearing several children to the same man often, due to repeated sensitizations, develop alloantibodies to his human leucocyte antigen (HLA) molecules, present on neonatal white bloods that presumably enter her circulation at childbirth. These antibodies were of great importance in the early days of the dissection of the major histocompatibility system and its component HLA loci.

6.2.2 Autoimmune haemolysis

Patients with plasma autoantibodies directed against their own red cells' antigens, and who have a reduced haemoglobin concentration, are considered to have autoimmune haemolytic anaemia (AIHA). With around 20 per million presenting annually, the disease is uncommon, and red cell destruction can occur with or without complement activation and fixation.

Approximately 80% of patients with AIHA possess warm-reactive IgG autoantibodies that react optimally at 37°C and, as such, are classified as having warm AIHA. The remainder exhibit cold-reactive autoantibodies, and are generally of an IgM type, which react optimally at 4°C. Characteristically, however, these antibodies have a wide thermal amplitude, still being able to react at temperatures of 30–32°C. Those patients that exhibit cold-reactive antibodies are classified as having cold AIHA.

The AIHAs have a wide variety of causes. Those with unknown causes are *primary* or *idiopathic* conditions, whereas those with an identifiable cause are described as *secondary* conditions. Both warm and cold AIHA can be secondary to lymphoproliferative conditions such as chronic lymphocytic leukaemia or lymphoma (both discussed in detail in Chapter 12), autoimmune disease, or drugs. Warm AIHA may also be seen in other systemic autoimmune diseases such as systemic lupus erythematosus (SLE) or ulcerative colitis, in solid non-haemopoietic tumours, or following the use of drugs such as α-methyldopa. Cold AIHA may be secondary to infections such as *Mycoplasma pneumoniae* or infectious mononucleosis (Epstein–Barr virus). (See also Box 6.2.)

Both warm and cold AIHA result in the reduction of the red cell lifespan. Warm AIHA is often accompanied by spherocytes and polychromasia on the peripheral blood film, and hyperbilirubinaemia (the latter reflecting red cell destruction). Cold AIHA is accompanied by red cell agglutination on the peripheral blood film, and a factitiously raised mean cell volume (MCV) and mean cell haemoglobin concentration (MCHC) in the FBC. Clinically, patients will demonstrate signs and symptoms consistent with anaemia and jaundice. Usually patients with cold AIHA do not demonstrate splenomegaly or hepatosplenomegaly, although these are prominent features in warm AIHA.

The principal laboratory test to determine the presence of an autoantibody against a red cell antigen is the **direct antiglobulin test** (DAT), which is essentially the second stage of the IAT. The patient's red cells are washed free of other plasma proteins and a suspension of washed cells are mixed with some AHG. Should the red cells be coated with antibodies, the AHG will cross-link them, forming an agglutinate (Figure 6.2). As with the IAT, different preparations of AHG and different conditions for the test can be used to determine the nature of the antibody (e.g. IgG class, IgM, warm or cold, etc.). We will now consider the mechanisms of red cell destruction in warm and cold AIHA, and mechanisms of treatment.

Warm AIHA

Warm AIHA is associated with the production of autoantibodies of the IgG class, with or without the involvement of complement component 3. These autoantibodies recognize 'self' antigens expressed on the red cell surface, leading to red cell sensitization and removal by the reticuloendothelial system. The reduction in red cells leads to reduced oxygen delivery to the tissues and a corresponding rise in erythropoietin synthesis and secretion by the peritubular cells of the kidney. This increased erythropoietin results in increased red cell production and

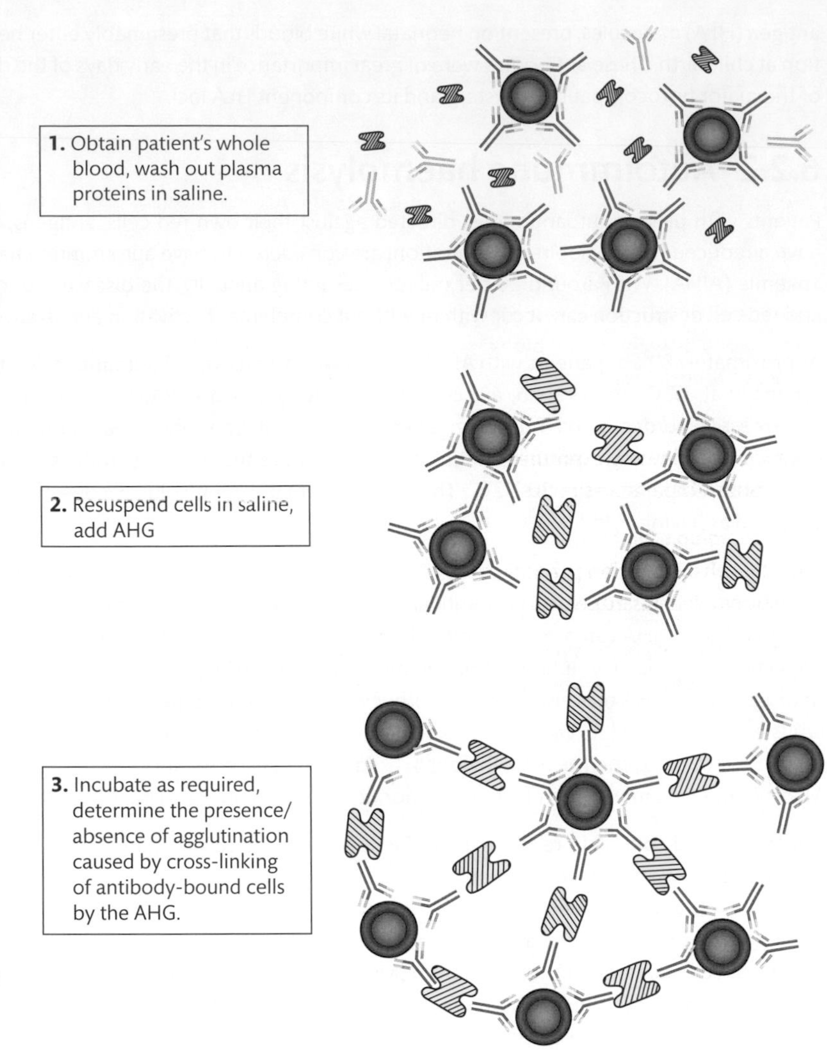

FIGURE 6.2
Simplification of the DAT: see text for details. N.B. the sizes of the antibodies, red cells, and AHG are not to scale. For simplicity, antibodies are represented as IgG—they may be IgM.

release from the bone marrow, resulting in reticulocytosis, with polychromasia apparent on the peripheral blood film.

In symptomatic patients, red cell transfusion may be required to maintain oxygen delivery to the tissues. However, the compatibility testing of these patients is often difficult and requires specialized serological techniques to uncover any alloantibodies that could potentially cause haemolytic transfusion reactions but are being masked by the patient's autoantibody. The antigen involved is often a member of the Rh blood group family, or a glycophorin, and both may involve a drug. Likely causes include autoimmune disease or lymphoma, and a drug, and both may be idiopathic.

Patients may also be administered glucocorticoid corticosteroids to try to improve their red cell lifespan. These immunosuppressive glucocorticoids reduce the expression of IgG-specific receptors on macrophages of the reticuloendothelial system, thereby reducing phagocytosis of these sensitized cells. As a consequence, sensitized cells are retained for a shorter period in the splenic vasculature and can contribute to oxygen delivery to systemic tissues. Glucocorticoids induce remission in approximately 80% of patients, although this remission is maintained in only 20–35% of patients.

BOX 6.2 Nomenclature

Although the naming of particular processes, diseases, and conditions seems to be a mystery to many students, there is in fact logic in the system which can be decoded by breaking down the formal name into its constituent parts. For example:

***Autoimmune haemolytic anaemia (AIHA)*—'auto' means 'one's self', 'immune' implies an antibody, 'haem' = 'of red blood cells', 'lytic' means 'bursting' or 'breaking', and 'anaemia' is essentially low levels of haemoglobin. Hence AIHA = low levels of haemoglobin caused by bursting of red blood cells, itself caused by an antibody to 'self' made by the patient themselves.**

The closely related *alloimmune haemolytic anaemia*—where a patient has developed an antibody following exposure to red cells from somebody else (allo- means a different member of the same species).

***Microangiopathic haemolytic anaemia (MAHA)*—describes anaemia resulting from red blood cell destruction (= haemolytic), which involves disease (= pathic, from pathology) of the small (= micro) blood vessels (= angio).**

***Paroxysmal cold haemoglobinuria (PCH)*—paroxysmal means intermittent, or every now and then; cold is fairly obvious; haemoglobinuria means haemoglobin in urine. Put it all together and PCH becomes a condition where every now and then, when it is cold, the urine goes red because of the presence of haemoglobin. The abbreviation PCH seems to be close to another: *paroxysmal nocturnal haemoglobinuria (PNH)*—which translates to occasional red urine (because of the haemoglobin) during the night. However, both these pictures may also be due to some forms of renal disease.**

As a second-line therapy, splenectomy is considered for the management of these patients and can induce a response in 50% of patients. The removal of the organ primarily responsible for the destruction of these sensitized cells will increase red cell survival. However, splenectomy is not without its risks. Primarily, patients are at increased risk from infection by encapsulated microorganisms, and patients should be vaccinated against pneumococcus, *Haemophilus influenzae* b, and meningococcal serogroup C, and should be prescribed lifelong penicillin or erythromycin.

Other immunosuppressive agents including azathioprine and cyclophosphamide may be indicated, although some, such as cytotoxic drugs, should be used with caution, as their usage is associated with the development of secondary malignancies. In other patients, a monoclonal antibody specific for the B-cell antigen (CD20) called rituximab (anti-CD20) may be indicated. Rituximab is often used for the treatment of B-cell malignancies, but can be used in AIHA by removing the B cells that produce the autoantibody. Removal of these B cells will improve red cell survival, as the production of red cell autoantibodies is inhibited.

Cold AIHA

Cold AIHA (which accounts for 13–15% of all cases of AIHA) is associated with red cell autoantibodies of the IgM type and the patient is often described as having cold haemagglutinin disease or cold agglutinin disease (CAD). Aetiology may be an IgM monoclonal gammopathy, Waldenstrom's macroglobulinaemia, and infection with *Mycoplasma pneumoniae*. Indeed, some consider primary CAD to be a B-cell lymphoproliferative disorder. Autoantibodies (often directed towards blood group I and i antigens) can be of very high levels and may be almost monoclonal-like in nature: the DAT is positive. They are characterized by having a wide thermal range such that they can often still react just above 30°C (but not at 37°C). However, they do not cause a major problem, providing the patients are kept warm. By contrast, patients suffer symptoms in cold conditions, where the temperature of the peripheral areas of the body, such as the finger tips, nose, and ear lobes, may drop to a level that allows the IgM autoantibody to bind to the red cells in these regions and to activate complement. Red cell agglutination may cause some vasculature occlusion in these areas, leading to acrocyanosis (blue coloration of the extremities, mostly the hands, indication hypoxia). When the red cells return to the core body temperature from the extremities, the sensitizing IgM antibodies dissociate from the red cell surface. However, as the IgM autoantibodies have been effective at binding complement components to the red cell surface, and as complement is most active at body temperature, the completion of this chemical cascade can occur, resulting in the formation of membrane attack complexes and intravascular haemolysis of the red cell. As a result, the patient may experience bouts of haemoglobinuria, haemosiderinuria, and possibly jaundice following exposure to the cold. Some extravascular haemolysis will also occur. As with warm AIHA, cold AIHA/CAD can be treated with chloroambucil/cyclophosphamide, rituximab alone, or in combination with the purine analogue fludarabine or prednisone.

The other rarer type of cold AIHA is paroxysmal cold haemoglobinuria (PCH, not to be confused with paroxysmal nocturnal haemoglobinuria, PNH). This transient condition is mostly associated with viral infections and lymphoma and is caused by an unusual 'biphasic haemolysin' complement-binding IgG autoantibody (the Donath–Landsteiner antibody). In this case, the antibody normally has an anti-P specificity. The mechanism of red cell destruction is similar to above: when exposed to cold conditions, the patient will often experience acute episodes of intravascular haemolysis. The haemolysis is self-limiting, but may require transfusions to correct the resulting anaemia.

6.2.3 Other 'extrinsic' causes of haemolytic anaemia

These types of anaemia cover a broad spectrum of pathologies, including mechanical and drug-mediated, and those caused by infections and animal toxins. Red blood cells may also be destroyed as a consequence of other diseases. Examples of the latter include thrombotic thrombocytopenic purpura, malignant hypertension, haemolytic–uraemic syndrome (HUS), and pre-eclampsia/eclampsia.

HUS is a triad of acute renal failure, thrombocytopenia, and microangiopathic haemolytic anaemia (MAHA), and is described in two forms: typical and atypical. The typical variant is inevitably linked to an enteropathogenic strain of the common intestinal bacteria *E. coli* that produces a Shiga-like toxin, and a diarrhoea is common, with around 25% of patients developing renal insufficiency. In the absence of an infectious agent, the disease is described as atypical, hence aHUS, and instead a likely aetiology is of complement dysregulation with excess complement activation. The aetiology may be genetic or acquired, although mutations of genes that encode complement, or antibodies that alter their function, account for around two-thirds of cases. The thrombocytopenia aspect of HUS gives rise to a considerable cross-diagnosis with thrombotic thrombocytopenia purpura (TTP), where deficiency of the serine protease ADAMTS-13 is common. However, MAHA may be present in non-HUS settings, and may be associated with physical damage to red cells either on abnormal surfaces (such as an artificial heart valve) or by damage caused by red cells travelling through fibrin strands deposited in capillary microvessels. These blood vessel problems may also be caused by **disseminated intravascular coagulation** (DIC), probably the most severe and acute coagulopathy. Red cells may also be adversely affected by high levels of urea (i.e. uraemia) and of bilirubin (i.e. hyperbilirubinaemia).

Cross reference

ADAMTS-13 is discussed in Chapters 13 and 16, and DIC is discussed in depth in Chapter 16.

Mechanical damage

Mechanical causes of haemolytic anaemia include prolonged and/or unaccustomed exercise or physical activity, such as may be present in soldiers after a long route march (hence, the term 'march haematuria') or endurance sports such a running a marathon. The same effect can be caused by a malfunctioning artificial heart valve. But whatever the cause, otherwise healthy red blood cells are destroyed by a physical force.

Drug-induced haemolytic anaemia

Haemolytic anaemia can occur as a result of a particular drug being potentially toxic, possibly by accidental or deliberate overdosage, and also by the induction of a hypersensitivity response. Drug effects are moderately common precipitants, accounting for 10–20% of haemolytic anaemias. Mechanisms may include direct chemical destruction of the membrane, or perhaps the generation of an increased oxidative stress by reactive oxygen species, which may lead, for example, to lipid peroxidation and oxidation of sulphydryl groups of the membrane and to the oxidation of ferrous (Fe^{2+}) iron in haemoglobin, forming methaemoglobin, where the iron is in the ferric (Fe^{3+}) state and is markedly less able to transport oxygen, if at all (when all four iron atoms in the ferric state).

Another cause is the development of an antibody that is directed against a drug, such as penicillin, cephalosporin, or tetracycline. The mechanism for cell destruction is such that if the drug binds or otherwise sticks to the cell membrane, this will be recognized and bound by the antibody, which will then render the entire cell more susceptible to haemolysis by the body's immune system. This mechanism may also be called 'neoantigen', or 'immune complex'.

An alternative type of drug-induced immune haemolytic anaemia occurs when the drug itself stimulates the production of an autoantibody to a component of the normal red cell membrane, but the drug itself does not directly take part in the reaction. Perhaps the best example of such a drug that induces this type of haemolytic anaemia is alpha-methyldopa (a drug used in the treatment of hypertension).

Alternatively, the drug itself may bind directly to the red blood cell, and this combination induces the production of an antibody towards the combination of the drug and the red cell. Examples of drugs which induce these antibodies include the blood pressure drug hydrochlorothiazide, the antibiotic rifampicin, levodopa, and procainamide. It could rightly be argued that this is really antibody-mediated haemolysis, but it is clearly primarily a drug reaction and the DAT is generally positive.

Infections

Infection can lead to haemolytic anaemia via several routes. A good example of this is *Clostridium perfringens* septicaemia, which is associated with acute haemolysis.

A high temperature, such as may be present in response to influenza, may also cause red cell destruction. This is likely to be semi-physiological, as a small number of red cells (perhaps those that are aged) are likely to be more sensitive to brief adverse changes in temperature, such as that associated with a fever. This explains the frequent reports of more yellow-coloured urine during and shortly after an episode of influenza.

Malaria is a major health and economic problem in many countries, and worldwide causes perhaps one million deaths annually in the 300–500 million people infected. Accordingly, although a major component of Chapter 7, we are justified in describing key aspects here. Severe infections with the causative parasite—*Plasmodium*—are also likely to precipitate a haemolytic anaemia, in addition to other laboratory and clinical changes, such as a raised erythrocyte sedimentation rate (ESR) and fevers, respectively. These occur partly because the parasite-loaded cell is detected as being abnormal, and so is eliminated. Haemolytic anaemia can also develop as the new red cells lyse releasing merozoites, facilitating the spread in parasites between adjacent red cells.

The parasite is very likely to have driven the development of haemoglobinopathy (as those heterozygous for sickle haemoglobin are relatively protected from infection) and the absence of Duffy blood group structures (hence Fy(a-b-)) in malaria-endemic areas. Notably, blood group O seems to protect against severe malaria. Interaction between CD147 (also known as basigin, coded for at 19p13.3) and RAP2, a parasite-secreted rhoptry protein, is essential for the invasion of the erythrocyte by *P. falciparum*, whilst *P. vivax* expresses a tryptophan-rich molecule that binds to red cell membrane component Band 3.

Cross reference

The cell biology and haematology of malaria and its causative organisms are discussed in detail in Chapter 7.

Key Points

Antibodies constitute a crucial and efficient aspect of our defence against microorganisms such as bacteria. However, if misdirected, they can cause considerable harm, as in the case of antibody-mediated haemolysis. Red blood cells are more susceptible to this type of damage, as they have a highly specialized, yet delicate membrane.

SELF-CHECK 6.1

What are the different types of antibody-mediated haemolytic anaemia?

6.2.4 The role of the laboratory in haemolytic anaemia

An FBC, reticulocyte count, and blood film are essential initial investigations in haemolytic anaemia. In antibody-mediated haemolytic anaemia, IAT and DAT are mandatory, and measurement of serum immunoglobulins, electrophoresis, and immunofixation will probably be needed to determine hypergammaglobulinaemia. Examination of the bone marrow may be required to probe for a very likely cause, non-Hodgkin B-cell lymphoma, which in one study was present in 75% of patients with an apparently primary CAD. There is likely to be a reduced haemoglobin and, if the reticulocyte count is grossly raised, there may also be a raised MCV. Increased numbers of reticulocytes reflects increased bone marrow activity, as it is responding to the anaemia, and on the blood film these larger cells may be more blue-tinged and so contribute to **polychromasia**. If the disease is severe and prolonged, there may be an additional reaction from the bone marrow. This can be demonstrated by the presence of nucleated red blood cells, reflecting (like reticulocytosis) a reactive increase in **erythropoiesis**. Bone marrow examination is infrequently required (except in AIHA, in order to exclude malignancy) but, if so, it is likely to demonstrate erythroid hyperplasia. In malaria, the anaemia is normocytic and normochromic, and a 'thick' and 'thin' blood film are needed to visualize the parasites within red cells by conventional light microscopy. A high MCHC (>370 g/L) is often found as cell dehydration decreases the efficiency of parasite invasion. The parasite consumes haemoglobin, the undigested haem residue being deposited as haemozoin.

polychromasia
A finding on blood film that translates as 'many colours'. In practice, there will be red blood cells of a normal colour, but others (reticulocytes) with a blue tinge.

erythropoiesis
The process of the development of red blood cells.

schistocytes
Fragments of red blood cells that are evidence of haemolysis.

spherocytes
Acquired or inherited disorder of red cells which have lost their discoid shape to become spheroid.

The pathological basis of haemolytic anaemia is the destruction of red blood cells. This is often reflected on the blood film by red cell fragments called **schistocytes**. However, the damage to the red cell may leave the cell intact, but instead change its characteristic discoid shape to that of a sphere—identifiable on the film as **spherocytes**. The blood film may give other clues as to the cause of the haemolytic anaemia. For example, a raised white blood cell count in leukaemia

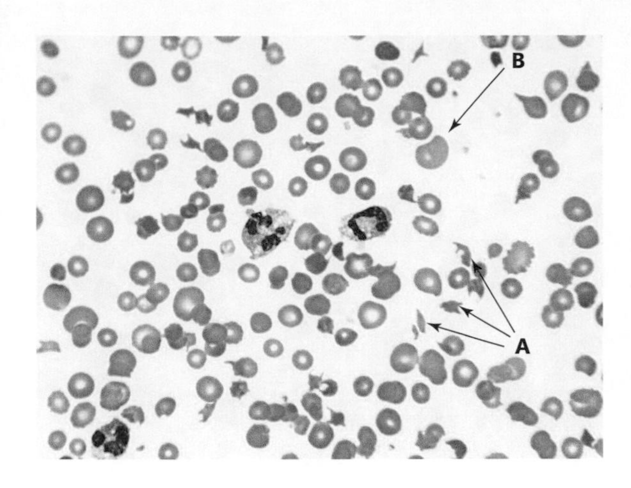

FIGURE 6.3
Schistocytes and marked polychromasia. Destruction of red blood cells in haemolytic anaemia produces fragments–schistocytes (arrows A). If the anaemia is prolonged, the bone marrow may respond by increasing the release of reticulocytes–juvenile red cells markedly larger than mature cells (arrow B). The blue tinge of reticulocytes leads to the descriptive term 'polychromasia'. There are also three neutrophils present. (Magnification ×400.)

or low platelets in idiopathic thrombocytopenic purpura are both diseases where AIHA may occur. Figures 6.3, 6.4, and 6.5 illustrate changes in the blood film that are commonly found in haemolytic anaemia.

Since all these haemolytic diseases are characterized by the destruction of red blood cells, the results of this destruction should be detectable in the circulation. However, in many cases, the red blood cell destruction may occur within organs such as the spleen (that is, it is extravascular). If so, there may not necessarily be a great deal of evidence of red cell destruction in the blood, other than the anaemia itself, although the spleen itself may be enlarged (that is, it exhibits splenomegaly). However, in intravascular haemolysis, there should be abundant evidence of red cell destruction, such as schistocytes.

In Chapter 5, we learned of the fate of the red blood cell and how one sign of this is bilirubin, which (broadly speaking) makes our plasma and urine yellow(ish) and our faeces brown. Thus, in haemolysis, increased unconjugated bilirubin, haemoglobin, and haptoglobin in the plasma, and increased urobilinogen in the urine, are all likely. In severe cases, there may also be free haemoglobin in the urine. In cases of chronic intravascular haemolysis, haemosiderin is stored in the tubular cells of the kidney when these cells are sloughed off; haemosiderin may be detected in the urine when stained with Perls' stain.

As mentioned, HUS is characterized by acute renal failure (marked by raised urea and creatinine and a low estimated glomerular filtration rate), thrombocytopenia, and a MAHA that is DAT-negative. Where HUS is associated with diarrhoea, this may be due to a toxin from *E. coli*

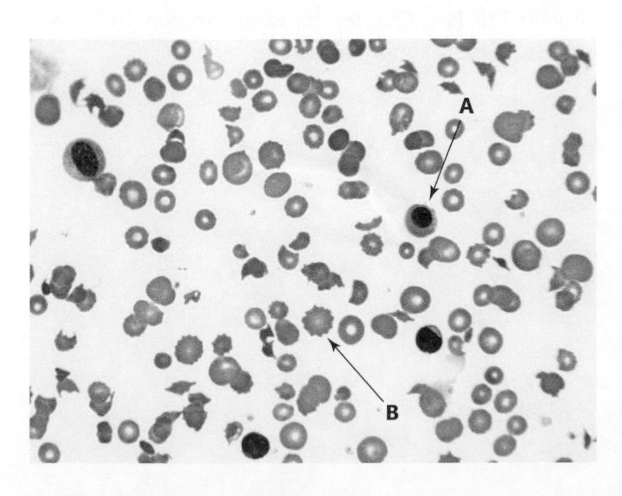

FIGURE 6.4
Schistocytes, polychromasia, and a nucleated red blood cell (arrow A). This film from a patient with DIC illustrates many features of haemolytic anaemia with schistocytes, polychromasia, and a burr cell (echinocyte) (arrow B). There are also two lymphocytes and, on the top left, a metamyelocyte. (Magnification ×400.)

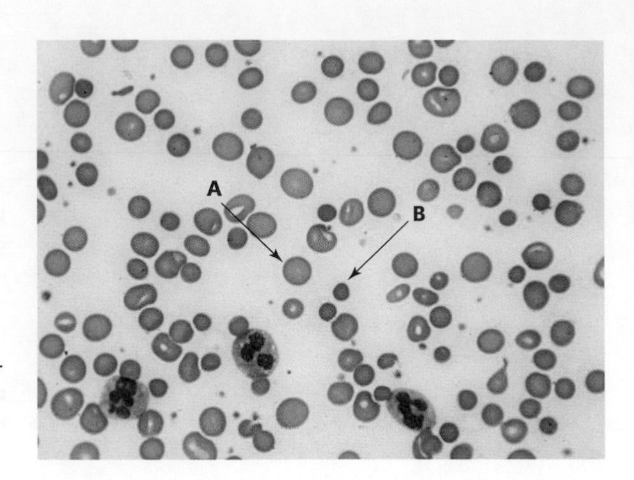

FIGURE 6.5
Spherocytes. This film from a patient with AIHA exhibits spherocytosis (arrow A). Some cells are very small, that is, are microspherocytes (arrow B). There is also a modest degree of polychromasia alongside three neutrophils. (Magnification ×400.)

or other pathogens. Atypical HUS is often characterized by dysregulation of complement components, calling for the measurement of complement components C3, C4, factor H, and factor I: the latter may be measured by serology in an immunology laboratory. Decreased expression of CD46 on mononuclear leucocytes should be sought by fluorescence flow cytometry (CD46 functions to suppress complement activation, so in its absence the complement pathway is hyperactive and attacks the cell membrane, leading to haemolysis). Finally, all cases of suspected aHUS should be tested for low or absent levels of ADAMTS-13 to distinguish aHUS from TTP, which may be due to a constitutive deficiency or to the acquisition of an autoantibody.

Cross-reference
HUS and TTP are discussed in chapter 16.

SELF-CHECK 6.2

What non-antibody factors external to the red cell cause haemolytic anaemia?

CASE STUDY 6.1 Interpretation of a blood film

This blood film is from a patient with TTP (see Chapter 16, where section 16.2.2 is relevant). Study Figures 6.3–6.5 and describe any abnormal features.

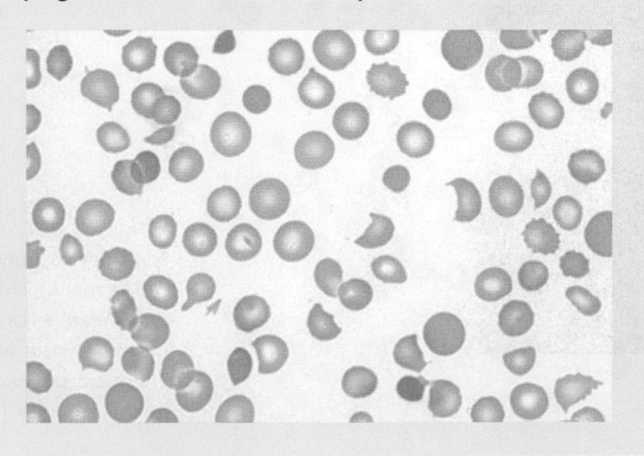

The red cell cytogram

As discussed in Chapter 4, a developing aspect of the flow cytometry of red cells is that of the cytogram. This is an image of nine squares defined by the upper and lower reference levels of the MCH and the MCV. Figure 6.6(a) shows the cytogram profile of a normal population of red cells, where the vast major of cells are clustered tightly in the central quadrant. Contrast this with the profile of a patient with AIHA (Figure 6.6(b)), where the plot of red cells describes an ellipse running from the upper left quadrant (high MCV, but low MCH) to the lower right quadrant (low MCV, but high MCH). Figure 6.6(c) is from a patient with megaloblastic anaemia, whilst Figure 6.6(d) is from a patient with alcoholic liver disease. In both these latter cases, the cloud extends vertically into the upper middle quadrant, and so represents cells with an increased MCV, that is, macrocytes. However, red cells in the former profile are relatively right-shifted and so carry more haemoglobin than those of the liver disease profile, where the cells are clustered more to the left of the cytograms, suggesting a lower overall MCH. The plots offer the practitioner an instant profile of the size and haemoglobin content of the red cell pool, but interpretation can be difficult and demands experience.

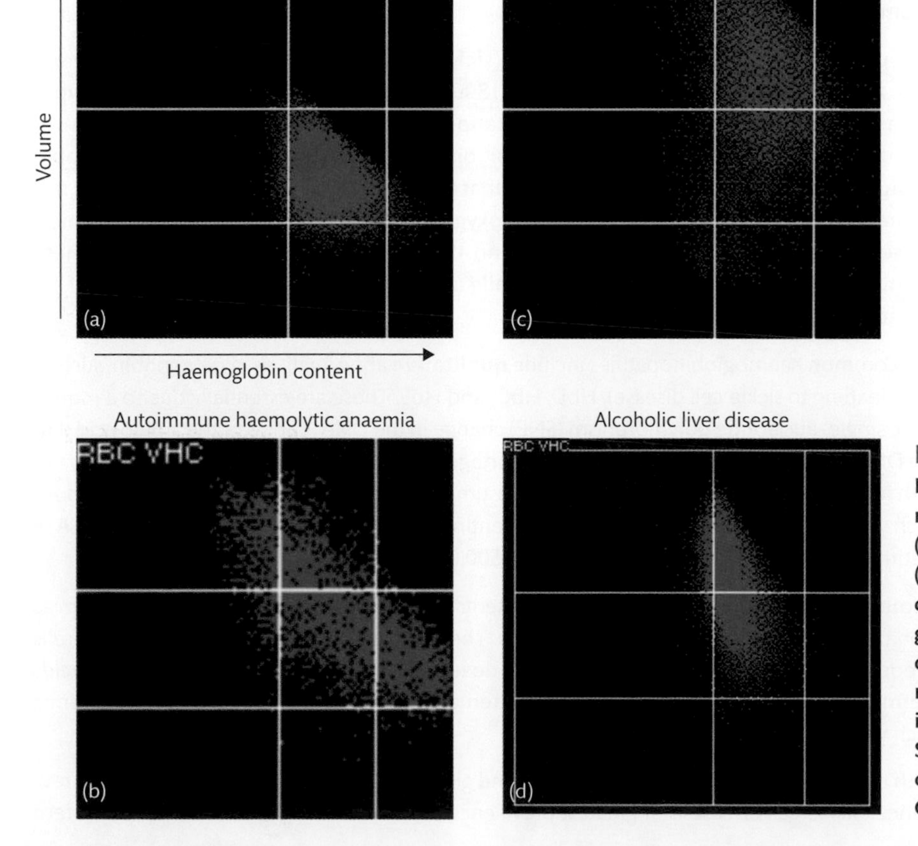

FIGURE 6.6
Red cell cytograms of (a) a normal red cell profile; (b) AIHA; (c) megaloblastic anaemia; and (d) alcoholic liver disease. The cell populations in each disease group demonstrate a greater degree of dispersion than the normal profile, with more cells in the upper three quadrants. See text for discussion. Figures courtesy of, and with thanks to, Graham Gibbs MSc CSci FIBMS.

6.3 Haemoglobinopathy

A second (and very much larger) group of haemolytic anaemias arise from mutations in various genes. These mutations can be classified as those that act on haemoglobin (the haemoglobinopathies, a significant public health problem in several parts of world), on the molecules that make up the cell membrane, and on intracellular enzymes (these two will be considered in sections that follow). Whilst it is certainly the case that the vast majority of these conditions have a hereditary component, they may also arise as *de novo* mutations in the newborn of parents who are themselves subsequently shown to be healthy.

One source suggests that the figure of 5.2% of the global population (i.e. 409 million people) carrying a significant haemoglobin variant is conservative. This figure of 52,000/million dwarfs all other genetic disease. Given that the rate varies considerably by subtype of haemoglobinopathy and by geography, in certain populations these figures will be markedly higher.

6.3.1 An introduction to haemoglobinopathy

Section 4.3.1 of Chapter 4 has highlighted the genetics and structure of haemoglobin. In the adult, globin genes (*HBA* and *HBB*) code for different combinations of globin proteins such as alpha-globin and beta-globin. In the healthy adult, the different variants of globin molecules lead to different types of haemoglobin, such as HbA, HbA_2, and HbF. However, mutations in the genes give rise to different types and amounts of haemoglobin synthesized, which are the cause of a certain type of anaemia. This type of haemoglobin abnormality is called **haemoglobinopathy**.

sickle cell disease
A qualitative haemoglobin disorder characterized in the laboratory by a chronic haemolytic anaemia, and in the clinic by infection and microvascular occlusion.

thalassaemia
A quantitative haemoglobinopathy caused by a mutation in alpha, beta, or both genes that leads to a reduction, or even abolition, in the expression of alpha- or beta-globin.

The principal two haemoglobinopathies (**sickle cell disease** and **thalassaemia**) are almost always inherited, and, to date, over 1800 variants have been identified. Both types of haemoglobinopathy result from the mutation of a healthy globin gene, with a consequent alteration in the amino acid composition of, or the amount of, a particular globin chain. This, in turn, influences the ability of that haemoglobin to transport oxygen. It is this failure to provide the tissues with sufficient oxygen for their needs that leads to the clinical consequences such as venous ulceration and the symptoms of anaemia. Some of these haemoglobin variants are asymptomatic, while others may severely affect the quality of life of patients.

The common haemoglobinopathies include **qualitative** abnormalities in beta-globin, such as HbS (leading to sickle cell disease), HbD, HbC, and HbE. These are essentially due to a monobasic single-nucleotide polymorphism (SNP) change in the DNA, and so of the amino acid that the DNA encodes. Importantly, there is no shortage of globin protein itself; the abnormality lies with its component amino acids. It has been estimated that 230,000 affected children are born each year in Sub-Saharan Africa alone, representing 80% of the global total, with (in the USA) a lifetime discounted inpatient cost of almost $500,000.

Another type of haemoglobinopathy represents a **quantitative** defect—the thalassaemias alpha (α) and beta (β) being good examples. These are called quantitative because there is a reduced mass of globin protein produced: dozens (and perhaps hundreds) of amino acids are missing. It is estimated that 70,000 children are born with various forms of thalassemia each year.

Both types of haemoglobinopathy can be found globally, but there are areas of increased prevalence. Sickle cell disease is of greatest prevalence in West Africa, the Caribbean, the eastern

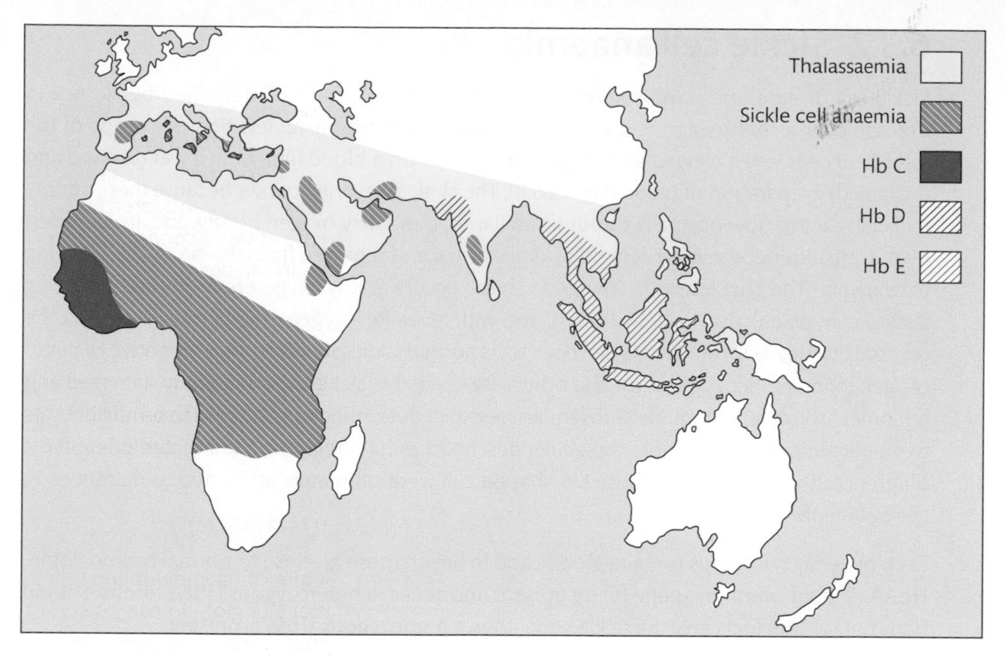

FIGURE 6.7
Distribution of haemoglobinopathy in Europe, Africa, and Asia.

Certain forms of haemoglobinopathy are localized to precise locations (such as HbC to West African and HbD to West India), or have a wide distribution (such as sickle cell anaemia across central Africa, and thalassaemia across Southern Europe, part of North Africa, and most of South and South East Asia). Not shown is sickle disease in the Americas, where it is not indigenous, but can be accounted for historically by the slave trade.

Mediterranean, the Middle East, and South East Asia. Thalassaemia is also found in those individuals who originate from the Mediterranean, the Middle East, and throughout South East Asia, and also in Southern China and Thailand (Figure 6.7).

However, the link between the genetics of the disease and problems with quality of life that the person may have is complex. Nevertheless, we can classify everyone with a haemoglobinopathy by the number and type of genes that are abnormal. Recall that we all have four genes for alpha-globin and two genes for beta-globin (as explained in Chapter 4). If, for example, both beta-globin genes carry a mutation (such as may cause sickle cell disease), then the person is said to be homozygous. By contrast, someone with only one mutated gene but a normal second gene is said to be heterozygous. These heterozygotes are often called 'carriers', and the condition a 'trait'; such subjects typically have few (if any) clinical symptoms and so a reasonably healthy life. It is also possible to have a combination of mutated genes, such as one mutation in a gene for alpha-globin and also a mutation in a gene for beta-globin, or perhaps a different mutation in each of the two beta-globin genes. These points will be revisited in detail in the sections that follow.

SELF-CHECK 6.3

What are the principal differences between the two major types of haemoglobinopathy?

6.3.2 Sickle cell anaemia

Sickle cell disease, the most frequent haemoglobinopathy, was first described by Herrick in Chicago over a century ago. It was so named because of the characteristic sickle shape of the red blood cells when viewed under light microscopy on a blood film, or in a wet (unfixed and unstained) preparation of blood (Figure 6.8). The clinical severity follows because these abnormal cells do not flow normally through capillaries, they carry oxygen poorly, and are detected by the reticuloendothelial system (RES) as abnormal and removed from the circulation, leading to anaemia. The characteristic change in shape occurs at low oxygen tension (that is, during a state of hypoxia), during dehydration, and with fever. Re-oxygenating blood can lead to the reversal of a newly formed sickle cell back to its normal state. However, after repetitive cycles of oxygenation and deoxygenation, the otherwise reversible sickle is permanently damaged and becomes an 'irreversible sickle'. Subsequent research determined the disease to be attributable to characteristics of the globin molecule, described as haemoglobin S (HbS); the phenotypic differences that characterized a sickle-shaped cell were ultimately attributed to mutations in the beta-globin gene.

Each of us has two genes for beta globin, and in health these give rise to normal haemoglobin, HbAA. Should one sickle gene (S) be present and active, a heterozygous HbAS results. Should both beta globin loci carry the sickle gene, then a homozygous HbSS is present.

Epidemiology

Haemoglobin S (sickle haemoglobin) is the most common haemoglobinopathy worldwide. Each year, over 275,000 babies worldwide are born with the disease, with perhaps 1% of all births in Africa being affected. These data extrapolate to 6 million affected people in Africa and 70,000–100,000 in the USA. The frequency of the HbS gene (including both heterozygotes and homozygotes) is found in different racial, geographical, and ethnic groups as follows: West African 1:5, Afro-Caribbean 1:10, Asian 1:50, Mediterranean and the Middle East 1:100. Areas where HbS is common run parallel with areas where falciparum malaria is endemic, leading to the hypothesis that HbS had been selected for in evolutionary terms as a protective mechanism against malaria. In the UK, more than 12 500 people have a clear sickle cell disorder. Over 240,000 are seemingly 'healthy' carriers, and each year 1 in 200 neonates are born with sickle cell disease. In the USA, perhaps 8% of African-Americans are heterozygous for this mutation.

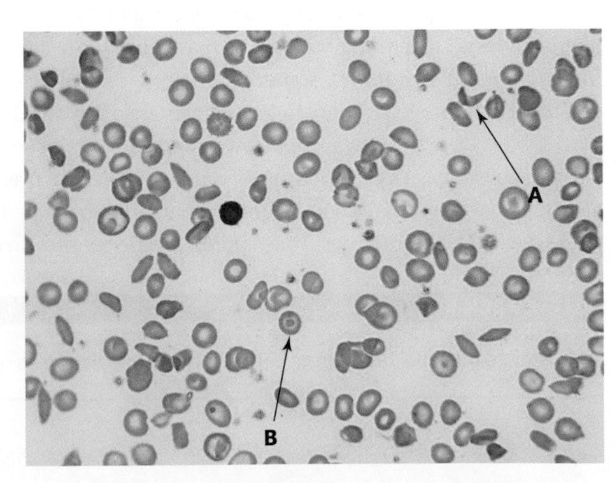

FIGURE 6.8

Sickle cell disease. There are numerous cells that are not round, but instead are oblong, curved, or oval shaped. These are sickle cells: a striking example is in the top right corner (arrow A). There are also several target cells (arrow B) and a lymphocyte. (Magnification ×400.)

Pathophysiology

The genetic basis of the most common form of the disease is a mutation in the beta gene (*HBB*) for one of the major protein components of haemoglobin; that is, beta-globin. Mutations in the alpha gene leading to abnormal alpha-globins are rare. So what is the nature of these mutations at the molecular level? Normally, a triplet of nucleotide bases in the globin gene (guanine, adenine, and guanine: GAG) codes for a globin molecule with the amino acid glutamic acid at position six. However, in sickle cell disease the beta-globin gene is mutated by the substitution of a thymine for the adenine, so the triplet becomes guanine, thymine, guanine: GTG. Consequently, the amino acid valine is substituted at position 6 in the globin chain instead of glutamic acid. This gives the beta-globin molecule, and so the whole haemoglobin molecule, a different shape and, hence, a reduced ability to carry oxygen. This is illustrated in Figure 6.9.

Under normal oxygenation conditions (such as in well-oxygenated arterial blood, especially near the lungs), HbS is fully oxygenated and the haemoglobin remains in the soluble form. However, after deoxygenation, perhaps in the tissues where oxygen is given up, HbS loses its solubility, and becomes polymerized into long, rigid chains called tactoids that distort the red cell into the characteristic sickle shape seen under the microscope. Conditions that predispose to sickling include hypoxia, acidosis, and increased temperature. Polymerization of HbS is reversible and, after re-oxygenation, the sickled red cell may revert to a normal red cell shape. However, after numerous cycles of oxygenation and deoxygenation, the red cell eventually becomes irreversibly sickled. Normal-appearing (i.e. unsickled) HbSS cells are two to three times stiffer than normal-appearing HbAA cells, whilst sickled cells are about twice as stiff as non-sickled cells. However, sickling and stiffness are not the only abnormalities—HbSS red cells are dehydrated (contributing directly to polymerization and stiffness), and have decreased nitric oxide, ATP, and antioxidant capacity, the latter resulting in oxidative damage to many cellular components.

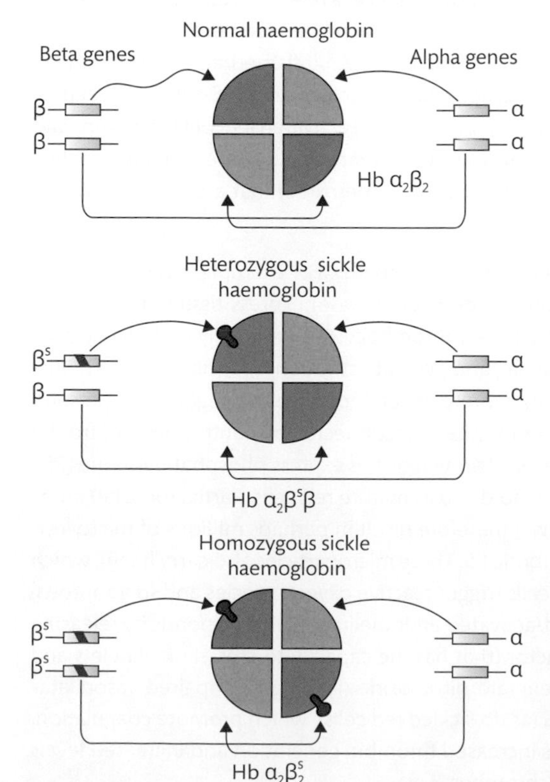

FIGURE 6.9

Relationship between beta-globin gene mutations and haemoglobin S. In health (top panel), alpha and beta genes each produce globin molecules that combine to form normal haemoglobin ($\alpha_2\beta_2$). In heterozygous sickle cell disease (middle panel), one beta gene carries a mutation that generates sickle globin (β^s). The other beta gene generates normal beta globin, resulting in ($\alpha_2\beta^s\beta$). In homozygous sickle cell disease (lower panel), two beta genes each carry a mutation that generates sickle globin, resulting in ($\alpha_2\beta_2^s$).

These permanently damaged elongated rigid red cells contribute to vaso-occlusion and obstruct small blood vessels so that other red cells cannot deliver oxygen to the tissues, leading to infarction. Sickle cells not only lose deformability, but also develop increased 'stickiness', which promotes adhesion to the endothelial cells and increases the risk of thrombosis. These red cell changes lead to a chronic haemolytic anaemia, because the sickled cells are detected as abnormal and so are eliminated by the reticuloendothelial system, most often through the activity of macrophages in the liver and spleen.

In heterozygotes (that is, those with haemoglobin AS; HbAS), the presence of a residual amount of normal haemoglobin (such as HbF and HbA_2) moderates not only the extent of the polymerization, but also its reversal. Thus, sickle cell carrier status leads to a clinical picture that is generally asymptomatic, although there are reports of occasional renal disturbances, and painful crises may occur at times of physiological stress, such as in unpressurized aircraft and during anaesthesia. HbAS red cells do not sickle unless the oxygen saturation is <40%, a level rarely achieved in the circulation.

Inflammation and thrombosis

The pathophysiology of sickle cell disease is not simply about the anaemia brought about by the abnormal haemoglobin and deformed red cells: it extends to inflammation (that may be systemic) and to coagulopathy. As regards the former, there is raised CRP, IL-1β, IL-3, IL-6, TNF-α, endothelin-1, and prostaglandin E2, and a higher neutrophil, monocyte, and eosinophil count, all of which are activated. Activated neutrophils (with increased expression of adhesion molecules such as CD11b/CD18 and CD64) are associated with increased mortality, acute chest syndrome, and stroke. Large, less-deformable white cells are more likely to be adhesive towards the endothelium and exhibit enhanced chemotaxis, and so may also contribute to veno-occlusive disease. If marked, an anaemia will be linked to a reticulocytosis: these cells may contribute to vascular occlusion not only because they are large, but also because they express high levels of adhesion molecules CD36 and integrin α4β1. An inflamed/activated/damaged endothelium also expresses increased levels of adhesion molecules such as E-selectin, that promote leucocyte recruitment. This accumulation of (probably) activated leucocytes may initiate an ischaemia/reperfusion injury, whereby enzymes such as xanthine oxidase generate reactive oxygen species (e.g. the superoxide radical and hydrogen peroxide) that are cytotoxic to the endothelium.

Many of the features described above promote coagulopathy. For example, activated endothelial cells and monocytes (both present in sickle cell disease) express tissue factor, which may initiate coagulation via factor VII. There is a thrombocytosis with activated platelets, as demonstrated by increased soluble P-selectin, and platelet activation is further enhanced during veno-occlusive crises, and significantly more platelet-monocyte aggregates are present. Acute lung injury and veno-occlusive crises may be interconnected by neutrophils, neutrophil extracellular traps (NETs), and haemolysis. Sickle reticulocytes express phosphatidylserine (PS), which promotes thrombosis, and continue to do so as mature red cells. Partial (or total) intravascular haemolysis of a sickled red cell will therefore result in perhaps millions of microvesicles/microparticles expressing pro-thrombotic PS. These microparticles also carry haem, which when transferred to luminal endothelial cells trigger reactive oxygen species and so apoptosis. This may (in part) account for a damaged/activated endothelium, which responds by releasing increased amounts of von Willebrand factor (that has the capacity to cross-link platelets and so generate microthrombi), failing to generate nitric oxide (leading to impaired vasodilatation), and exposing increased levels of PS (as do sickled red cells), which promote coagulation. Further evidence of hypercoagulability is increased thrombin generation and increased levels of D-dimers, indicating a high burden of thrombus.

Clinical features

Despite the singular nature of the genetic lesion, patients with sickle cell anaemia present with a very broad phenotype—ranging from few, if any, clinical complications to severe crises, even as infants—and they may require blood transfusion. This heterogeneity extends to the variability of the inflammatory response, which (in theory) ought not be directed by haemoglobin, so that a substantial non-haemoglobin component must be present and active. A multiplex transcriptome (i.e. total mRNA) analysis of a 31-gene signature taken from peripheral blood mononuclear cells can differentiate two clusters of patients with different mortality rates. This tool has a better predictive value than existing markers or clinical scores alone, but a combination of the gene score and clinical markers of tricuspid regurgitation velocity, white blood cell count, history of acute chest syndrome (features such as chest pain on breathing [not cardiac], cough, and crackles heard using a stethoscope) and haemoglobin level was best at predicting patients with the worse potential outcomes.

The symptoms of anaemia are mild in relation to the low haemoglobin levels, and this is due to a right shift in the oxygen dissociation curve (that is, when HbS releases oxygen to tissues more readily when compared to normal HbA). We discuss the oxygen dissociation curve in section 6.5.2. Disease severity is not simply a matter of the number of mutated genes present (two genes in the case of homozygous disease HbSS, one gene in the case of heterozygous disease HbAS), as the clinical consequences of the disease vary considerably between patients. The major clinical aspects of sickle cell disease are:

Painful vaso-occlusive crises During a crisis, irreversibly sickled cells cause blockage of blood vessels, which in turn initiates tissue hypoxia, and leads to ischaemia with bone and muscle pain. This debilitating process can occur in a variety of organs, including bone (hips, shoulders, and vertebrae), the lungs, and the spleen. These complications may be further enhanced by infection, exposure to cold temperature, hypoxia, strenuous exercise, dehydration, emotional disturbances, and pregnancy. The most serious vaso-occlusive crisis is of the brain (a stroke—which occurs in 7% of all patients) or spinal cord. The 'hand and foot' syndrome is one of the first presentations of the disease in young children and may lead to the formation of digits of varying lengths.

Visceral sequestration crisis Sickling of red cells in the liver and spleen, and sequestration in the lungs is partly responsible for the acute chest syndrome, although infection and infarction may exacerbate the condition.

Aplastic crises This may occur as a result of folic acid deficiency or parvovirus infection and is characterized by a sudden drop in haemoglobin level, resulting in the patient requiring a blood transfusion.

Haemolytic crises These present with an elevated rate of haemolysis with a fall in haemoglobin, and an increase in the reticulocyte count.

Other clinical features commonly include venous thromboembolism (particularly pulmonary embolism) and lower leg ulcers (which may be linked to deep vein thrombosis), which occur as a result of vascular stasis and local ischaemia. The spleen is often enlarged in infancy and early childhood due to trapping of sickle cells, but is later reduced in size as a result of infarcts. However, stroke is the most severe complication in children. Other clinical complications include retinopathy, priapism, chronic liver damage, gall stones, kidney damage, and osteomyelitis. Treatments (described in more detail in section 6.3.8) include red cell transfusions, hydroxycarbamide (which increases HbF by perhaps 10%, although there may be mild neutropenia, but also reduced veno-occlusive disease, retinopathy, acute chest syndromes, transfusions, and admissions to hospital), and haematopoietic stem cell transplantation. These features are summarized in Figure 6.10.

With increased medical care over the decades, it is becoming clear that sickle cell disease brings other serious disease over time. A study of 7512 patients in England found an increased incidence of haematological cancer and some solid tumours, compared to patients hospitalized with minor conditions. In California, a study of 6423 patients found fewer cases of solid tumours, but an increased frequency of leukaemia, principally acute lymphoblastic leukaemia (ALL) and acute myeloid leukaemia (AML). Curiously, beta-thalassaemia trait brings a protection from arterial thrombosis in males, not in females.

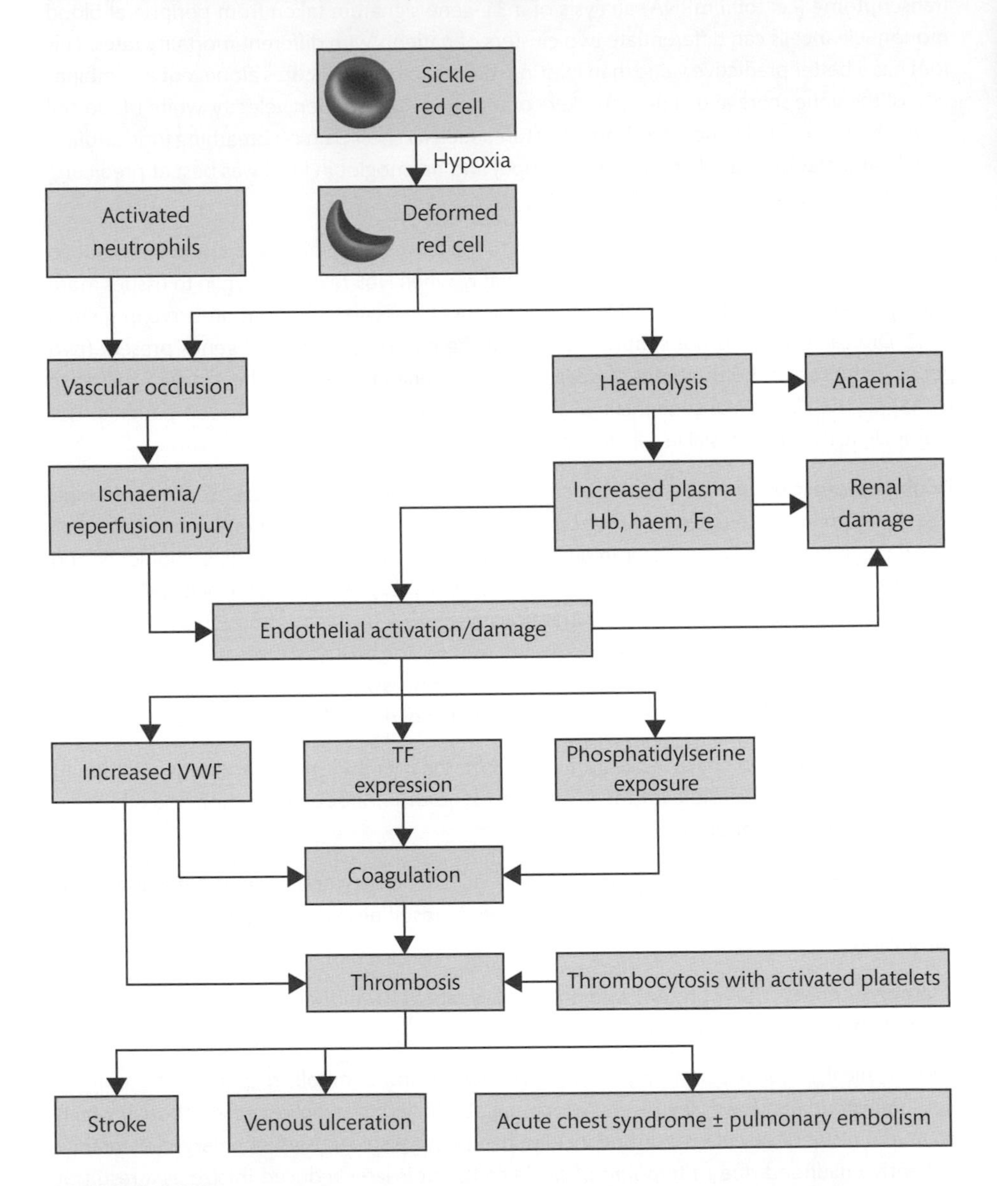

FIGURE 6.10

Relationship between pathophysiology of sickle cell disease and clinical events. In the presence of adequate oxygen, the sickle cell retains its normal shape, but when hypoxic, becomes deformed, leading to vascular occlusion (with trapped activated neutrophils) and haemolysis, the latter leading to anaemia. These changes contribute to endothelial cell activation and/or damage, which, with the release of procoagulants, promote a hypercoagulable state. These lead to thrombosis, which become manifest clinically as stroke, venous ulceration, an acute chest syndrome (possible with pulmonary embolism) and deep vein thrombosis. TF = tissue factor.

6.3.3 Other qualitative beta-globin gene disorders

Haemoglobin E disease

Haemoglobin E (HbE) disease is the second most common haemoglobinopathy worldwide, and is commonly found in South East Asia (including Cambodia, Laos, Malaysia, and Thailand), where the frequency varies from 8% to 50%. This variant is caused by replacement of glutamic acid by lysine at position 26 on the beta-globin chain, leading to β^e. This haemoglobin also has a higher incidence in malarial areas, leading to the suggestion that HbE heterozygotes may also have some protection from this disease. Homozygous HbE disease results in a mild haemolytic anaemia with target cells, and a compensatory erythrocytosis. The heterozygous state is asymptomatic, but the film shows microcytosis.

Haemoglobin C disease

Haemoglobin C (HbC) is the third most common qualitative haemoglobinopathy. Like haemoglobin S, it is also caused by a mutation in the beta gene, which leads to a different amino acid at position 6 of the globin protein chain, β^C. However, in HbC, the amino acid lysine is substituted for glutamic acid, leading to a different type of abnormality. Perhaps the highest frequency of this haemoglobin worldwide is in parts of West Africa, such as Ghana, where the incidence may reach 20%. Approximately 2–3% of African-Americans are heterozygous carriers, such that 1 in 5000 live births in this population will be a homozygote.

Haemoglobin C has a lower solubility, resulting in a crystalline structure that in turn decreases red cell deformability. In the homozygous state (that is, HbCC) there is an enlarged spleen and a mild haemolytic anaemia. Like haemoglobin E, the blood film is dominated by target cells, but these are small, and there may also be some microspherocytes. In the heterozygous (carrier) state (that is, with HbAC) the FBC generally reports normal haemoglobin and the film only a few target cells.

Haemoglobin D disease

Haemaglobin D (HbD) disease may be described as a collection of different mutations in the beta-globin gene. The variant with the highest frequency occurs in the western (Punjab) region of India and Pakistan (hence, it is often described as haemoglobin D_{Punjab}) and is caused by substitution of glutamine for glutamic acid at the 121 position on the beta-globin chain. In the homozygous state, there is a mild haemolytic anaemia with target cells, but haematological abnormalities are generally absent in the heterozygous HbAD form.

Other haemoglobinopathies

Several hundred mutations in alpha and beta genes, all of which give rise to abnormal alpha- and beta-globin, have been described. One such mutation, haemoglobin G Philadelphia, results from a mutation in an alpha gene, but generally does not have clinical implications. The vast majority of this group are family-specific and, as such, have less impact upon public health. Table 6.2 summarizes key aspects of the qualitative beta gene haemoglobinopathies.

SELF-CHECK 6.4

What are the molecular determinants of the different qualitative beta-globin haemoglobinopathies?

6.3.4 Thalassaemia

This group of diseases is characterized by the reduced synthesis of either or both globin chains (hence 'quantitative' defect), often by complete deletions in the relevant DNA, or SNPs leading to premature stop codons or missense sequences that produce a shortened transcript. This results in a serious abnormality in the intact haemoglobin molecule, which leads to poor oxygen transport and, ultimately, the clinical signs and symptoms associated with this disease. In contrast to sickle cell disease, those red cells carrying the abnormal haemoglobin do not go markedly out of shape (as do sickle cells), but there are often morphological changes. However, they are still detected as defective and so are removed from the blood, leading to anaemia.

Epidemiology

Thalassaemia is predominantly found in the eastern Mediterranean region, the Middle East, the Indian subcontinent, and South East Asia—all regions where sickle cell disease is also highly prevalent. Like sickle cell disease, this reinforces the notion that thalassaemia carriers are protected against malaria. Alpha-thalassaemia is frequently present in Chinese people, whilst beta-thalassaemia is often found in Cypriots, Asians, Chinese, and Afro-Caribbeans. In south Italy and Greece, 5–10% of the population is heterozygous for beta-thalassaemia, whereas in Thailand, the gene frequency for the various forms of alpha-thalassaemia reaches 25%. In its most severe form, thalassaemia has been estimated to affect 68,000 births annually.

Pathophysiology

The thalassaemias can be classified as alpha (α)-thalassaemia, in which alpha-globin chain synthesis is reduced (or completely absent), and beta (β)-thalassaemia, in which there is a reduced level of beta-globin chain synthesis, or it is entirely absent. Rarely, there can be defects in both beta and delta genes. Serious clinical disorders result only when two or more of the various genes are affected. Thalassaemia major is the most severe clinical form of the disease and patients inevitably require regular blood transfusions in order to survive. In contrast, thalassaemia carrier status is the mild form of the disease and many patients are asymptomatic, such that they do not generally require a blood transfusion. Between the two are a collection of conditions described as thalassaemia intermedia (Table 6.2).

Cross reference

Globin chains are discussed in Chapter 4, section 4.3.1 and Figure 4.14.

The alpha-thalassaemias

Chromosome 16 carries two copies of the alpha-globin gene (*HBA*). Consequently, as we each carry two copies of chromosome 16, we have four alpha-globin genes overall. SNPs (as can cause sickle cell disease) in particular sections of alpha-globin genes are rare. Instead, the vast majority of mutations are deletions. If both sets of alpha genes are malfunctioning, no alpha-globin chains are produced (α^0-thalassaemia), whereas if only one of the pair of genes is aberrant, the synthesis of globin chain is reduced (α^+-thalassaemia). The clinical severity of any particular thalassaemia depends on the total number of aberrant genes; in many cases, the genes have simply become deleted altogether.

A *one* gene mutation, represented by –/α/α/α, in which alpha-globin output is 75% of normal, is generally clinically silent and without an anaemia. However, this carrier status may give rise to more serious disease if an individual has a child with another such carrier.

Mutations in *two* genes, either –/–/αα or –/α/–/α, may arise from deletions of the two α genes from the same chromosome whilst the two genes of the other chromosomes are intact (that is, –/–/αα). An alternative form is present when one of the two α genes on each chromosome has become deleted, leaving the remaining gene functional (that is, –/α/–/α). This is described as alpha-thalassaemia minor, or as alpha-thalassaemia carrier. In both cases, alpha-globin production is 50% of normal.

Deletion or functional inactivity of *three* of the four α-globin genes, that is, –/–/–α, is haemoglobin H disease. It presents clinically as a type of thalassaemia intermedia and is characterized by levels of alpha-globin 25% of normal.

Complete loss of all *four* alpha-globin genes (–/–/–/–, thus no alpha-globin) results in haemoglobin Bart's, also called hydrops foetalis syndrome, where the foetus is unable to make either normal foetal (α_2/γ_2) or adult A (α_2/β_2) haemoglobin. Hb Bart's is incompatible with life, and death usually occurs *in utero* or soon after birth.

Cases involving both delta- and beta-globin, and complexes of gamma-, delta-, and beta-globin, do occur but are rare. Impaired alpha-globin leads to excess gamma-globin and beta-globin chains, which form unstable and physiologically afunctional tetramers. In haemoglobin Bart's this becomes gamma (γ_4), whilst in haemoglobin H the tetramer is of beta-globin (β_4). Although almost all alpha-thalassaemia is caused by gene defects, it may also be a consequence of a chronic myeloid disorder, usually a myelodysplastic syndrome.

Cross reference

Myelodysplastic syndromes are discussed in depth in Chapter 11.

The beta-thalassaemias

There are more than 200 different mutations of the beta-globin genes (present on chromosome 11) which result in the reduced or absent synthesis of beta-globin chains. The defective production of beta-globin chains leads to the increased synthesis of gamma-globin chains, which pair with alpha-globin chains, and so to increased levels of foetal haemoglobin (HbF, $\alpha_2\gamma_2$). However, the excess free alpha-globin forms insoluble aggregates that damage (via reactive oxygen species) and induce apoptosis in erythroid precursors, leading to ineffective erythropoiesis. The anaemia caused by the latter promotes erythropoietin production, which further stimulates ineffective erythropoiesis in a vicious cycle. Furthermore, excess alpha-globin within the red cells hastens their destruction.

These mutations may be within the gene complex or on nearby promoter or enhancer regions. As with alpha-thalassaemia, the severity of the disease, at both clinical and laboratory levels, depends on the number and character of the abnormal genes. Three broad categories of disease are recognized:

Thalassaemia major This is a consequence of the inheritance of two different mutations, which leads to interference with β-globin chain production and includes both β^0-thalassaemia (where the globin chains are absent) and β^+-thalassaemia (where globin chains are partially present). In homozygous beta-thalassaemia, either no beta-globin chain exists (β^0), or only small amounts (β^+) are synthesized, leading to severe anaemia. Hypertrophy of the ineffective bone marrow is associated with skeletal changes, and there is hepatosplenomegaly.

Thalassaemia intermedia This may be caused by a variety of genetic defects, and so is often classified clinically, as opposed to genetically or haematologically. In one manifestation, there is homozygous beta-thalassaemia, but effects of the anaemia are countered by a greater synthesis of haemoglobin F than usual. In another form, there are mild defects in the synthesis of beta chains. The coexistence of alpha-thalassaemia carrier status improves the haemoglobin levels by decreasing the degree of chain imbalance.

Thalassaemia minor Heterozygotes for beta-thalassaemia are generally asymptomatic. As such, they are described as *thalassaemia minor* (or *trait*), which includes β^0-thalassaemia carrier and β^+-thalassaemia carrier status. These conditions are not generally associated with anaemia (although levels of haemoglobin may be low) because the remaining non-mutated β-globin gene is able to synthesize sufficient beta-globin to permit oxygen carriage. Consequently, clinical presentations are mild. This picture is also seen in alpha-thalassaemia.

Clinical features

Medical care in all cases of thalassaemia focuses on disease severity, regardless of genotype. Deletions in one or two alpha genes are asymptomatic. Deletions in three alpha genes, giving rise to HbH, leads to jaundice, hepatosplenomegaly, leg ulcers, gall stones, and folate deficiency. Most patients will not need blood transfusions, although iron overload can be a problem, probably due to increased intestinal absorption.

In beta-thalassaemia major, anaemia dominates after the age of three months, when the switching from γ- to β-chain synthesis occurs. However, in milder forms, there may be no symptoms up to the age of four years. Children generally fail to thrive, show recurrent infections, pallor, and mild jaundice. Enlargement of the liver and spleen is due to trapping of the excessively damaged red cells. To compensate for the anaemia, extramedullary haemopoiesis becomes dominant. Expansion of the bones caused by intensive marrow hyperplasia leads to thalassaemic facial features such as bossing (swelling of the bone of the jaw) leading to enlarged maxilla, a 'hair on end' appearance on X-ray, and thinning of the cortex of the bones, leading to easy fractures.

In a severe thalassaemia, causing a profound anaemia, blood transfusions are necessary to maintain quality of life, or even life itself, and may be required every six weeks. However, a major complication of these repeated transfusions is the build-up of iron in organs such as the liver, heart, and endocrine organs, which can lead to irreversible damage to these organs, although fortunately there are treatments to remove this excess iron. This is developed in more detail in section 6.3.7. During infancy, the anaemic child is prone to bacterial infection from *Pneumococcus*, *Haemophilus*, and *Meningococcus* species, amongst others. Transfused patients are at risk from virus transmission associated with blood transfusion, such as hepatitis C and B. However, overall, beta-thalassaemia carrier status provides a benign and a generally asymptomatic condition. Regardless of the type of thalassaemia, there is poor growth.

Table 6.2 summarizes major features of the two principal forms of haemoglobinopathy.

6.3.5 Compound and other haemoglobinopathies

It is clear that different haemoglobin gene mutations can occur in various different combinations, and that these different combinations effectively create hybrids of quantitative and qualitative defects, or indeed two different quantitative defects such as in alpha- and beta-thalassaemia. Given the overlapping geographical areas where these mutations are endemic, these associations are to be expected, and arise because of the asymptomatic nature of many types of heterozygous disease.

Combined qualitative defects include those arising from the combination of haemoglobin S and haemoglobin C (that is, HbSC), and of haemoglobin S with haemoglobin D (HbSD). The compound heterozygosity of HbSC may be designated $\alpha_2\beta^s\beta^c$. Combined quantitative and qualitative defects may also be present, such as a combination of beta-thalassaemia and sickle cell disease. Indeed, both HbS/β^0 and HbS/β^+ are known. Because of the dominance of the sickle haemoglobin, all of these are clinical sickling conditions, and the clinical picture resembles homozygous sickle cell disease (e.g. with splenomegaly). The heterozygous disease of combined haemoglobin E and β^0-thalassaemia resembles homozygous β^0-thalassaemia both clinically and haematologically, and so is managed in the clinic as thalassaemia major.

TABLE 6.2 Major aspects of the common haemoglobinopathies.

Sickle cell disease	Thalassaemia
The most common haemoglobinopathy, and also the most common single-gene defect worldwide	Can be quantitative or qualitative defects in α- or β-globins, and heterozygous or homozygous in both globins
A qualitative defect in β-globin that leads to an abnormality in the red cell under hypoxic conditions, and so a change in shape to that of a sickle	Sequential deletions in the four α-globin genes leads to increasingly severe disease; failure of all four leads to hydrops foetalis and is incompatible with life
Can be heterozygous (HbAS, carrier status) or homozygous (HbSS)	Mutations in β-globin genes is also semi-quantitative, the most severe being in both genes (homozygous)
Clinically very variable, with different phenotypes presenting with the same genotype, implying effects of other genes/environment	Clinical disease is classified as thalassaemia major, intermedia, or minor, depending on the extent of the mutations and the phenotype of the patient

In some cases, deletion of β gene(s), δ and β genes, or δ, β or γ genes occurs. In rare conditions, unequal crossing-over of globin genes leads to the production of haemoglobin Lepore. This hybrid has globin chains of δ-globin at one end fused with β-globin at the other end. If present in both sets of genes (that is, as a homozygous state) the clinical presentation is of thalassaemia intermedia, whilst the heterozygous state is less severe and is clinically akin to a carrier status.

Another form of mutation in both delta and beta genes leads to impaired production of both species of globin molecules, hence δβ-thalassaemia. However, foetal haemoglobin (HbF, $\alpha_2\gamma_2$) is increased, giving at least some (limited) capacity to carry oxygen and the production of alpha- and gamma-globin is generally unaffected. Indeed, in homozygous δβ-thalassaemia, only haemoglobin F is synthesized. This leads to the syndrome termed 'hereditary persistence of foetal haemoglobin' (HPFH) and presents clinically as thalassaemia intermedia. However, there are other forms of HPFH caused by deletions, point mutations, or cross-over of beta and gamma genes. Indeed, measurement of the level of HbF can be very informative (Table 6.3). Figure 6.11 illustrates globin molecule interactions in thalassaemia and HPFH.

TABLE 6.3 HbF levels in haemoglobinopathy.

HbF level	Indicative clinical picture
<1%	Normal result
1–5%	Present in approximately 30% of beta-thalassaemia carriers; many other conditions, for example, compound haemoglobinopathy
5–20%	Some cases of beta-thalassaemia trait, compound heterozygotes, some cases of heterozygous HPFH, δβ-thalassaemia
15–45%	Most cases of heterozygous HPFH, some cases of beta-thalassaemia intermedia
>45%	Beta-thalassaemia major, some cases of beta-thalassaemia intermedia, neonates
>95%	Homozygous HPFH, some neonates (especially if premature)

HPFH = hereditary persistence of foetal haemoglobin.

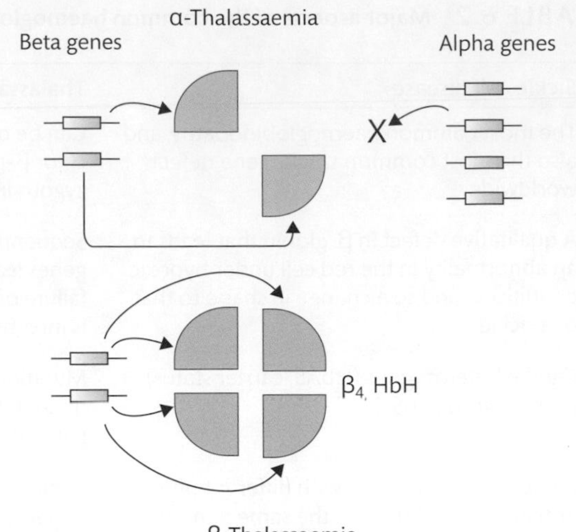

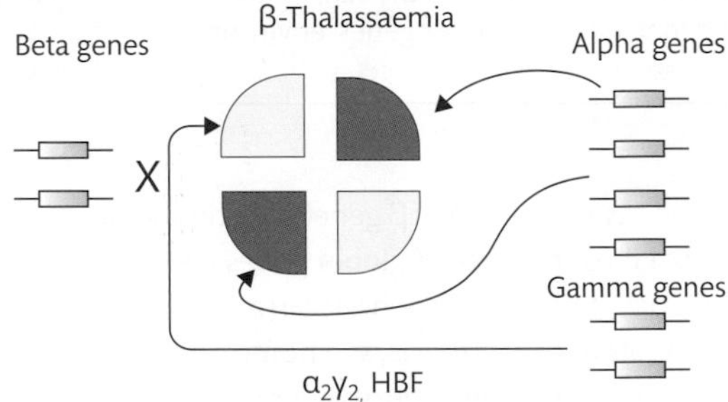

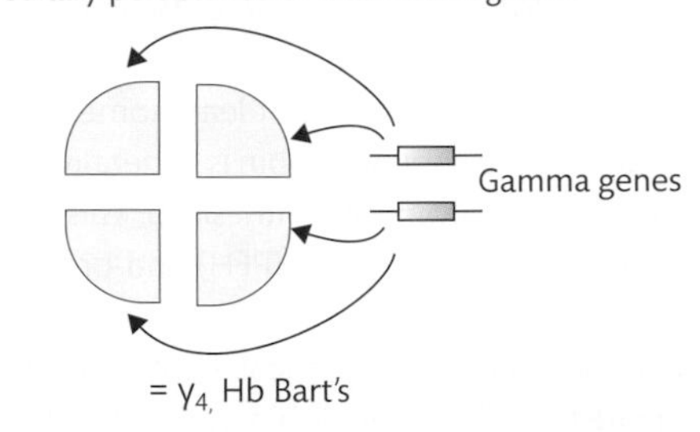

FIGURE 6.11

Molecular variants of haemoglobins in thalassaemia and HPFH. In the most severe form of alpha-thalassaemia (top panel), the four alpha genes fail to produce alpha-globin; the beta genes are unaffected, but the product (02β2) is incompatible with life. However, the two normal beta genes can produce sufficient beta-globin to combine to form a homogeneous haemoglobin (β4) called HbH.

In beta-thalassaemia (third panel), beta genes fail to produce beta-globin. However, functioning alpha genes and gamma genes still produce alpha-globin and gamma-globin, which together form functional haemoglobin F (HbF, α2γ2), which retains oxygen-carrying capacity.

HPFH is present when all alpha and beta genes fail to produce alpha- and beta-globin. The gamma genes are intact and functioning, and generate sufficient gamma-globin to form a haemoglobin molecule where the only globin component is gamma-globin (γ4).

SELF-CHECK 6.5

What are the laboratory differences between the alpha- and beta-thalassaemias?

6.3.6 Laboratory tests in haemoglobinopathy

As in the investigation of any anaemia, an FBC (including reticulocytes) and blood film are essential, and many also find the ESR useful. However, a range of other tests are specific for the haemoglobinopathies, many of which are based on properties of the abnormal globin molecules and of the genes themselves. These are essential in confirming a preliminary diagnosis, in monitoring treatment, and also have a place in screening at-risk populations. Dedicated haemoglobinopathy methods include two 'wet' preparations, electrophoresis (capillary, cellulose, alkaline, and acid), microcolumn analysis for HbA_2, high-performance liquid chromatography (HPLC), estimation of HbF (Betke method), immunoassay, and the analysis of DNA coding for globin chains. Non-specific tests of haemolysis include serum, aspartate aminotransferase and bilirubin, plasma haemoglobin, and red cell microparticles.

Wet preparations

These tests are so named because they are performed directly on whole blood. They are also very quick, cheap (in terms of reagents), and require simple equipment (a light microscope or a water bath). However, they are prone to error.

The simplest test relies on the way that phenotypically normal-looking red cells adopt the sickle phenotype when hypoxic (as they would in the circulation). The test relies on a reducing agent that effectively lowers the oxygen content of a whole-blood sample. A small volume of blood is mixed with an equal volume of the oxygen-consuming reducing agent (such as sodium metabisulphite) in a physiological buffer and is immediately placed in a counting chamber for viewing under a light microscope. As the reducing agent reduces the oxygen levels, cells carrying haemoglobin S will adopt the characteristic sickle shape. The test must be performed with positive and negative controls—respectively, a blood sample known to be from a patient with sickle cell disease and the proband's blood mixed with simple buffer alone (which should not induce hypoxia).

A similar test is of the solubility of sickle cells. Under deoxygenated conditions, sickle haemoglobin becomes insoluble in phosphate buffer supplemented with saponin as a haemolyser and sodium dithionite as a reducing agent. A subject's blood is again mixed with the buffer, then incubated for a short period, and placed in front of a grid with defined lines. As the red cells lyse, the solution becomes clear, implying a negative result for sickle cell disease. However, if the blood solution is opaque or turbid (due to the red cells being intact and in suspension), as defined by an inability to see the lines through the test tube, then the result is deemed positive (Figure 6.12).

This test can be extended by centrifugation of the tubes at 1200 g for five minutes. In the presence of sickle haemoglobin, the buffer solution gives a clear dark-red or purple coloration at the top of the tube, with a thin 'scum' of precipitated protein and red cell stroma, whilst the solution below will be pink or colourless. In the absence of sickle haemoglobin, there is no such coloured ring. False positives may be due to red cell debris, erythrocytosis, increased paraprotein, and unstable haemoglobin(s). This can be corrected by washing the red cells.

A false negative may be due to deterioration of the dithionite, unstable saponin, anaemia, and a post-transfusion sample. High levels (10–20%) of HbF, which counter the polymerization of the sickle haemoglobin, can also give a false negative. Accordingly, the test is unreliable in neonates during the first few months of life.

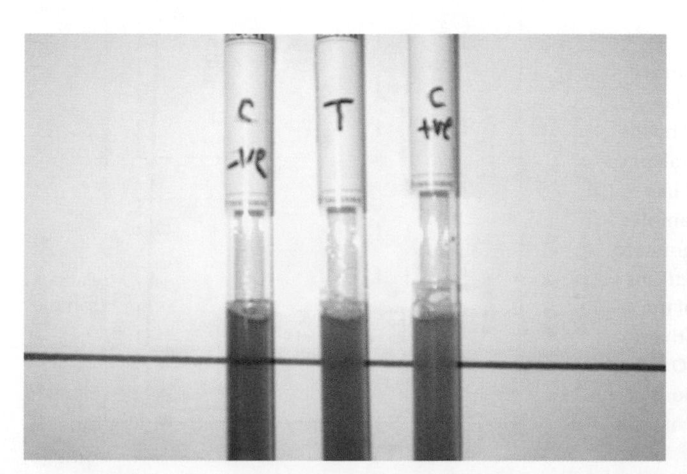

FIGURE 6.12
The sickle solubility test. When sickle positive blood (AS, on the right) is mixed with saponin and a reducing agent, the resultant solution is cloudy, whereas with normal blood (AA on the left) the solution is clear. This property is used to define sickle status. NEG = Negative for sickle status; POS = Positive for sickle status. Figure curtesy of H Bibawi, Tayside NHS Trust.

Electrophoresis

The separation of proteins can be achieved by exploiting the way they carry a net charge. If a sample is placed in an electric field, it will migrate, with its final resting position being dependent on the charge it carries. This process of movement in an electric field is called electrophoresis.

Haemoglobin has a net positive charge, as conferred by the particular combination of amino acid side chains that are present in the molecule. However, different haemoglobin variants are composed of different amino acids, and so carry slightly different electrical charges. As a result, these variants will migrate to different positions on a gel. Different types of electrophoresis (e.g. those employing different pH, or using different physical supports such as cellulose acetate or agarose) can be employed to detect different haemoglobin species.

Cross reference

We learn more about electrophoresis in the *Biomedical Science Practice* volume in this series.

An additional electrophoresis method (using a different buffer and urea, and at pH 6.3 or pH 8) can be used to separate different alpha- and beta-globin chains. Of course, haemoglobin is found within red cells, not in plasma. Therefore the haemoglobin must be released from the red cell by lysis, creating a lysate.

Alkaline (pH 8.4–8.7) electrophoresis When performed on cellulose acetate, this method is simple, reliable, and rapid. Within this pH range, haemoglobin in the red cell lysate is negatively charged and will move towards the anode. However, different types of haemoglobin will separate from haemoglobin A and form their own line of identity. At the practical level, with each set of samples, there must be a control sample containing haemoglobins A, F, S, C, D, and Punjab. However, haemoglobins C and E comigrate to the same position, as do haemoglobins S, D, and G. Haemoglobin Lepore has an electrophoretic motility similar to that of haemoglobin S. Consequently, these haemoglobin variants cannot be differentiated, so that other techniques are demanded. The separation pattern of common haemoglobin variants, and some unknowns, is shown in Figure 6.13.

Acid (pH 6.0–6.3) electrophoresis The broad principle of acid electrophoresis is the same as that for alkaline electrophoresis, but it is best performed on agarose. At an acidic pH, the

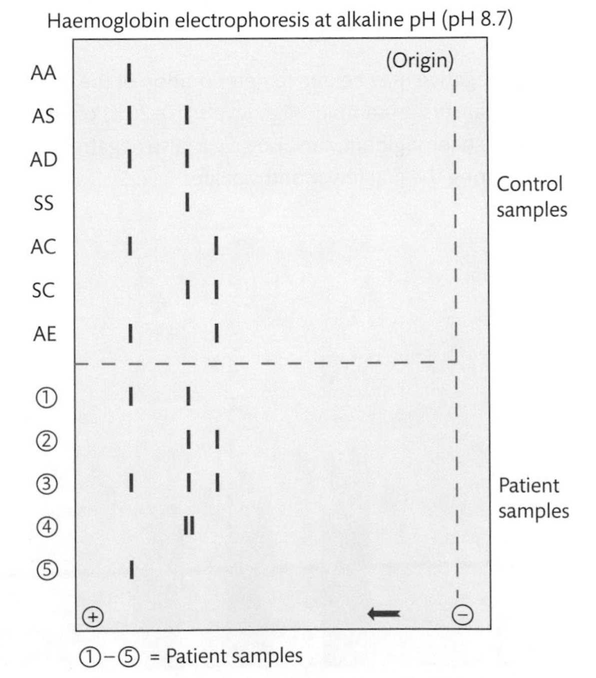

FIGURE 6.13

Haemoglobin electrophoresis patterns. Samples of haemoglobin are placed on the right-hand (anode) edge of the electrophoresis plate and are driven over to the left-hand side (towards the cathode) by the electrical charge. The migration distance depends on their overall physicochemical make up, and so differs between different types of haemoglobin. This figure consists of control samples and patient samples. The topmost trace, AA, is of normal haemoglobin A, which migrates to a fixed position as a single band. Other traces show different patterns according to their composition—HbAS shows two bands, one of the A globin, and the other of the S globin. Thus the trace of HbSS (that is, indicative of homozygous sickle cell disease) is of a single line. Other bands indicate haemoglobins AD, AC, SC, and AE by combinations of different individual lines. The lower traces, 1–5, are the subject of Case study 6.2.

different haemoglobin variants migrate to different positions than they do at an alkaline pH, giving the opportunity to distinguish between alternative phenotypes. For example, haemoglobins S and D comigrate at alkaline pH, as do haemoglobins C and E, but at acid pH they migrate to different positions. Consequently, acid electrophoresis is effectively a back-up method. As before, known positive controls must be run as part of each assay.

Immunoassay

Commercially prepared kits known as HemoCards are available for the detection of common haemoglobins S, C, D, E, and A_2 with sensitivity down to 5–10%. However, the use of the HemoCard kit in the laboratory is limited due to availability, unreliability, and the poor quality of kits. By contrast, HbF can be measured by immunodiffusion and by enzyme-linked immunosorbent assay (ELISA).

HbF quantitation

Foetal haemoglobin (HbF) is a tetramer of two alpha-globin molecules and two gamma-globin molecules (that is, $\alpha_2\gamma_2$). A dominant variant in the foetus, levels fall in the initial few months of life, such that it is present as perhaps 0.5–0.8% of total adult haemoglobin. However, increased levels are present in various haemoglobinopathies and as such are an aid to diagnosis and the monitoring of treatment (Table 6.3).

HbF levels below 12% can be accurately quantified by HPLC or the alkali denaturation method (also called the Betke method, developed some 50 years ago). These have replaced the traditional acid elution method of Kleihauer. Although Betke's alkali method is accurate down to 0.1%, it is labour-intensive, and involves handling a cyanide reagent and an open blood sample. HbF greater than 12% should be confirmed by alternative method.

One of the disadvantages of these tests is that they measure total HbF in a collection of cells, not in individual cells. Consequently, they are unable to distinguish the HbF of a δβ-thalassaemia carrier—in which the distribution of HbF is usually heterocellular (that is, present or absent from particular cells)—from HPFH, in which the distribution is pancellular (that is, present in every single cell). Instead, these two patterns can be determined by staining a blood film with an antibody to HbH, and then staining with a second antibody conjugated to a visualizing signal such as fluorescein isothiocyanate. The qualitative and quantitative differences in the distribution of HbF can be detected by an appropriate microscope set up for fluorescence. HbF-bearing cells can also be detected by flow cytometry.

High-performance liquid chromatography (HPLC)

This method is quite possibly the 'gold standard' technique for determining haemoglobin variants and thalassaemias. Although the capital investment and running costs are considerable when compared to electrophoresis, the method can be automated (giving high throughput), is frugal of sample (requiring as little as 5 μL of blood), and is accurate, rapid (perhaps five minutes), reproducible, and reliable. The method is particularly suited to a high throughput laboratory that may be required to perform antenatal screening for haemoglobinopathy.

The use of HPLC for the determination of haemoglobin variants relies on the different ionic properties of different haemoglobins. Positively charged haemoglobin within the red cell lysate is separated by adsorption on to a negatively charged stationary phase in the column, typically an amino acid coated onto a silica particle resin. The mobile phase (such as a salt solution), featuring an increasing concentration of cations, detaches the bound haemoglobin protein from the anionic-binding amino acid, and the optical density of the eluted solution (which carries the haemoglobins) is measured spectrophotometrically. This result is converted into a chromatogram that shows all the haemoglobin peaks in a particular sample. Each type of haemoglobin molecule carries a distinctive net charge (as we noted above). Consequently, each haemoglobin exhibits a unique retention time on the column, which can be exploited to facilitate haemoglobin identification. This is illustrated in Figure 6.14.

As with gel or cellulose acetate electrophoresis, positive controls of known haemoglobin variants must be present in each batch. An additional aspect of this technique is that it can also provide levels of glycosylated haemoglobin (HbA_{1C}), a valuable tool in the diagnosis and management of diabetes.

Isoelectric focusing (IEF)

This technique, related to electrophoresis, also relies on the different electrical properties of haemoglobin variants. IEF plates are made from either polyacrylamide or agarose and both contain amphoteric molecules with various isolectric points. These molecules are used to establish a pH gradient in the gel, generally running between pH 6 and pH 8.

Isoelectric focusing exploits the way that different proteins have different isoelectric points—that is, the pH at which they carry no net charge. As we noted earlier, when discussing electrophoresis, a charged protein will migrate when placed in an electric field. If this migration occurs through a pH gradient, however, the protein will stop moving once it reaches the point in the gel at which it carries no net charge.

Different haemoglobins have different net charges, with the result that different haemoglobin molecules have unique isoelectric points on the gel (that is, they will migrate to different extents). Once completed, the gel can be stained and the position of the various haemoglobin species can be quantified by densitometry. The densitometry peaks can be superimposed on traces of known variants to assist in the diagnosis.

Like HPLC, IEF demands capital investment, but advantages include the ability to separate many different variants, clear demarcation due to sharp bands, the ability to separate haemoglobins D and G from S, and the requirement for only a small blood sample, so that it is suitable for neonatal haemoglobinopathy investigation and screening.

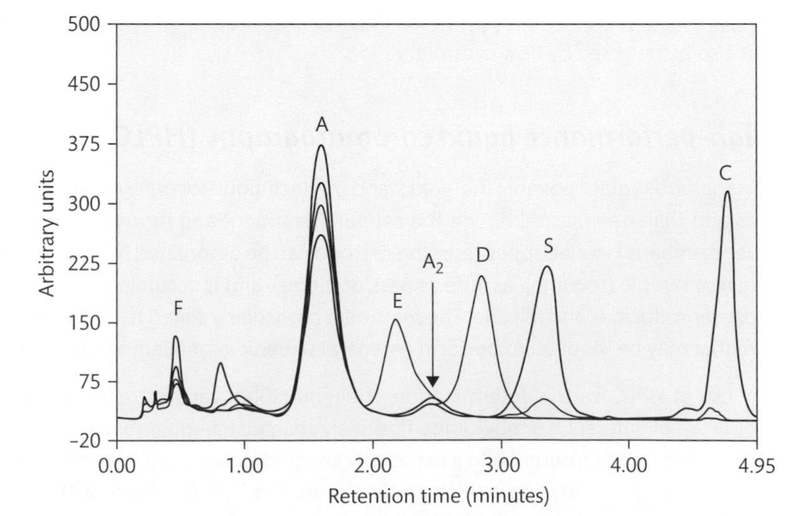

FIGURE 6.14
HPLC chromatograms. HPLC is an additional tool to investigate a presumed haemoglobinopathy. The principle is allied to that of electrophoresis, where different types of globins can be identified by their combined physical and chemical properties. The figure is a composite of a collection of different globins shown together to illustrate how each can be found along the spectrum from left to right. The 'height' of the peak is proportional to the amount of the particular globin present in a sample.

DNA analysis

This specialized method, inevitably found in a reference laboratory, is commonly used for the confirmation of haemoglobin variants, alpha (^{0})-thalassaemia carriers, and prenatal diagnosis of serious disorders of haemoglobin synthesis, and is becoming more popular as technical advances are made. It can also be used for foetal diagnosis of haemoglobinopathy. The general area of molecular genetics is intensely technical and so very demanding of skilled scientists, who are very likely to be based in regional referral laboratories. The basic technique of DNA analysis may be modified according to the mutations expected given the particular ethnicity of the patient, and are used to define the exact nature of a particular gene mutation predominantly in sickle cell disease, beta-thalassaemia, and alpha-thalassaemia.

SELF-CHECK 6.6

What laboratory tests can give the most precise and comprehensive information about the protein nature of a suspected haemoglobinopathy?

6.3.7 Laboratory findings in haemoglobinopathy

The essential tests for studying haemoglobinopathy are the FBC and blood film. The consistent finding from the FBC in haemoglobinopathy is a reduced haemoglobin and a reduced MCV, indicating a microcytosis. If the patient is symptomatic, this makes the diagnosis of microcytic anaemia. However, the sensitivity and specificity of a reduced MCV is poor, as a common alternative diagnosis is iron deficiency, which should be considered and excluded, either by iron studies (such as serum ferritin, iron, transferrin, transferrin saturation) or additional haemoglobinopathy testing.

A reactive erythrocytosis may also be present as the bone marrow attempts to maintain the ability of the blood to deliver oxygen to the tissues. Further evidence of this stress erythropoiesis is a reticulocytosis. These cells are larger than mature erythrocytes, and are certainly larger than microcytes. Consequently, the MCV (a global score of red cell volume) may not be far below the bottom of the reference range. Thus an increased red cell distribution width (RDW) may be present, and this should be detected and reported by the haematology autoanalyser.

Sickle cell disease

Simple screening of this disease is most rapidly achieved by the sickling test, although this will need to be confirmed with other tests such as HPLC. Patients with sickle cell anaemia have a haemolytic anaemia, with a broad range of haemoglobin concentration in the range 60–90 g/L and elevated reticulocyte count of perhaps 10–20%. The haematocrit ranges between 0.18 and 0.30 L/L.

Haemoglobin content of homozygous sickle cell disease (HbSS) will report >80% HbS, with variable levels of HbF (5–15%) but a normal (trace) amount of HbA_2. In HPLC, the chromatogram shows a small peak at the HbF retention time and a large peak in the HbS window. In heterozygous disease (HbAS) there is generally an excess of HbA (50–60%) compared to HbS (perhaps 35–45%). This picture can be contrasted with the relatively common combined defect, sickle/beta-thalassaemia, wherein electrophoresis reveals that 60–90% of the haemoglobin is HbS, whereas 10–30% is HbF. However, if the patient retains some alpha gene activity, then HbA may be present. Haemoglobin A_2 is generally moderately elevated in sickle/beta-thalassaemia disease. In seeking to determine the degree of haemolysis, markers such as LDH, aspartate aminotransferase, bilirubin, and plasma haemoglobin all correlate inversely with total haemoglobin and HbF.

Thalassaemia

General hospital haematology laboratories do not have one single test to confirm an alpha-thalassaemia carrier. Instead, multiple investigation data together with patient ethnic data are used to indicate probable alpha-thalassaemias. The same aspects of screening and prevention apply to other haemoglobinopathies, such as HbS, C, D, E, Lepore, and thalassaemia. In alpha-thalassaemia carrier status (when one or two alpha genes are deleted), the haemoglobin may be normal but produced at slightly reduced levels, providing an asymptomatic phenotype. The MCV (<77 fL) and MCH (<27 pg) may both be reduced, but the red cell count may be raised ($>5.5 \times 10^{12}$/L). Measurement of the relative rates of α- and β-synthesis may support a diagnosis of α-thalassaemia.

When three alpha genes are deleted, HbH disease is present. Haemoglobin is typically 60–110 g/L. In the neonate, electrophoresis or HPLC shows Hb Bart's (γ_4) up to 25%, the remainder being HbA ($\alpha_2\beta_2$) and HbF ($\alpha_2\gamma_2$) with a small amount of HbH (β_4). As the γ-chain production switches to β-chain synthesis, HbH (β_4) gradually replaces Hb Bart's, and HbF also disappears. In adults, the haemoglobin pattern is HbA with a smaller amount of HbH, which may range from 5% to 25%. HbA_2 levels are slightly decreased and the HbF level is normal or slightly increased. The high levels of beta-globin can form deposits or precipitates within the red cell, and this can be detected on a blood film using a supra-vital stain such as Brilliant cresyl blue.

Overall, the beta-thalassaemias demonstrate a broad range of clinical and haematological variability due to the heterogeneous nature of the molecular defects that affect beta-chain production. Heterozygotes for beta-thalassaemia (carriers) are generally asymptomatic, but have microcytic, hypochromic red cells, reduced MCV (typically 65–75 fL), reduced haemoglobin (90–110 g/L), and reduced MCH (<27 pg). There is also likely to be raised HbA_2 (>3.6%).

Patients with thalassaemia intermedia present with a very broad phenotypic picture, ranging from symptomless to the requirement of blood transfusions, and subsequent iron chelation therapy. Patients with haemoglobin ranging from 60–100 g/L generally do not require blood transfusion. The red cells are very microcytic and hypochromic, and there is increased erythropoiesis. These patients may demonstrate some bone deformity, enlarged spleen and liver, and features of iron overload.

Severe beta-thalassaemia major is associated with a profound anaemia (haemoglobin 20–60 g/L), reduced MCV (<65 fL) and MCH, and raised reticulocyte count. Investigation of the bone marrow is generally unnecessary (unless a differential diagnosis needs to be excluded) but, if performed, should show hypercellularity with erythroid hyperplasia. Haemoglobin electrophoresis or HPLC demonstrates the absence (in β^0-thalassaemia) or almost complete absence (β^+-thalassaemia) of normal adult haemoglobin (HbA), with almost all the circulating haemoglobin being foetal haemoglobin (HbF) and varying level amounts of HbA_2, ranging from low to slightly raised. Globin-chain synthesis studies show no synthesis of the β-chain.

The blood film in haemoglobinopathy

The blood film provides two sets of clues that are useful in diagnosis: the specific and the non-specific. The only finding on a blood film specific for haemoglobinopathy is the sickle cell, inevitably present in hypoxic blood from a person with homozygous sickle cell disease (Figure 6.8). However, the intracellular changes brought about by abnormal haemoglobin in severe thalassaemia may also induce the cells to adopt the sickle phenotype.

There are many non-specific aspects of abnormal red cell morphology in the haemoglobinopathies. The presence of schistocytes—see Figures 6.3 and 6.4—is evidence of red cell destruction, the hallmark of many haemolytic anaemias, and so is not specific for haemoglobinopathy. Similarly, microcytes are present in both haemoglobinopathy and in iron-deficient anaemia. Other non-specific morphological changes in haemoglobinopathy include anisocytosis, poikilocytosis, and basophilic stippling, the latter also being found in lead poisoning (Figure 5.16). Target cells are a common finding in haemoglobinopathy (especially in HbC disease), but are also likely to be present in iron deficiency, in liver disease, and following splenectomy. An additional finding following splenectomy, and also in splenic atrophy, are Howell-Jolly bodies. Appearing as a small dark dot (Figure 6.15), Howell-Jolly bodies are an example of an **inclusion body**—a pathological finding within a red blood cell. Other inclusion bodies are described in Table 6.4.

The more severe the disease, then the more marked these changes can be. For example, in severe beta-thalassaemia major, the blood film may well demonstrate marked morphological changes with anisocytosis, hypochromic microcytic cells, basophilic stippling, and nucleated red blood cells (normoblasts) (Figure 6.16). Nucleated red blood cells indicate a hyperactive or stressed erythropoiesis, although these cells are also found in numerous conditions, such as a myeloid malignancy. Conversely, in clinically mild conditions such as sickle cell trait, there

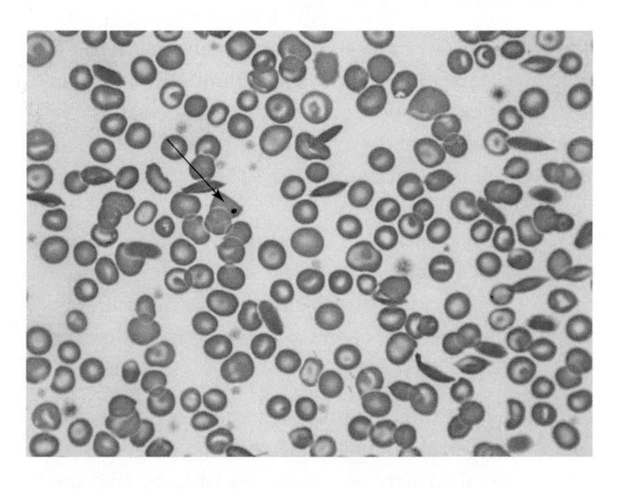

FIGURE 6.15
Sickle cells, target cells, and a Howell-Jolly body (arrowed, left of centre). This film is from a patient with compound haemoglobin S and thalassaemia. There are abundant sickle cells, many target cells, and a single Howell-Jolly body. The latter implies splenic atrophy or post splenectomy. (Magnification ×400.)

TABLE 6.4 **Red blood cell inclusion bodies.**

Basophilic stippling	Denatured RNA
Heinz bodies	Oxidized denatured haemoglobin
Howell-Jolly bodies	Formed from DNA remnants and often found after splenectomy
Nuclei	Nucleated red blood cells
Pappenheimer bodies	Particles of iron-rich protein found in siderocytes
Parasites	As are present in malaria
Cabot rings	Microtubule remnants of the mitotic spindle
Haemozoin	Brown pigment formed from the digestion of haemoglobin by a malarial parasite

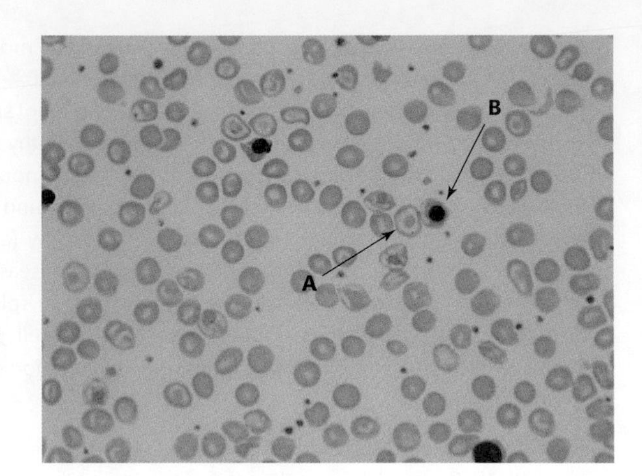

FIGURE 6.16
Target cells (arrow A) and nucleated red blood cells (arrow B). The anaemia in this case of severe beta-thalassaemia is so profound that the overactive bone marrow has released nucleated red blood cells into the circulation. Compare the density of the nuclear staining in these red cells with that of the lymphocyte on the lower margin. (Magnification ×400.)

should be no abnormalities on a blood film, although beta-thalassaemia will always have blood film abnormalities. (See also Case study 6.1.)

Non-specificity and other haemoglobinopathies

A common theme of this section is the clustering of various laboratory findings (such as a low MCV and target cells) in the different haemoglobinopathies, even if compound. For example, both HbS/beta-thalassaemia and HbSC are associated with prominent target cells, basophilic stippling, and a microcytosis. Furthermore, the electrophoresis and HPLC pattern in compound HbS/thalassaemia may resemble HbSS, as both will lack HbA. Target cells and spherocytes are also commonly found in both homozygous and heterozygous HbC and E diseases. In all cases, there will be microcytosis and hypochromia. These two variants comigrate on cellulose-acetate electrophoresis gels but can be separated using acid-citrate agar electrophoresis. In compound HbSF, 70% of the haemoglobin is HbS and 30% is HbF. However, a notable feature of HbC disease is of intracellular crystals of the abnormal haemoglobin, which become more prominent in a hypertonic solution.

Although not generally considered classical haemoglobinopathies, a number of other haemoglobin variants justify a mention. A small number of variants readily take up oxygen, but are reluctant to give it up at the tissues. These are called high-oxygen-affinity variants. The consequences of these mutants are tissue hypoxia, and a reactive raised red cell count in response. Conversely, in low-oxygen-affinity variants, oxygen is taken up poorly, and haemoglobin tends to dissociate into its subunits, again resulting in poor oxygen distribution. Despite the rarity of these conditions, they provide the opportunity to probe the relationships between oxygen uptake, erythropoietin, and erythropoiesis.

Haemoglobin M has a tendency to form methaemoglobin, the molecular basis of which stems from substitutions in either alpha- or beta-globins. The biochemical consequence of this is that the iron remains in the ferric form and so cannot bind oxygen, leading to cyanosis in the patient. However, methaemoglobin may also be found in red cell enzyme deficiency diseases, and in the presence of certain drugs (such as antituberculosis agents and sulphonamide antibiotics) and chemicals (such as silver nitrate and amyl nitrate). Although detectable on electrophoresis, HbM can also be determined by spectrophotometry.

Around 25% of abnormal haemoglobins are described as unstable, defined as species that easily denature and form precipitates as Heinz bodies, leading to variable degree of haemolytic anaemia. Jaundice, discoloured urine due to the excretion of dipyrroles, and occlusive thrombi are common, often after a precipitating factor such as an infection and after a splenectomy. In one case report, Heinz bodies were present in 1–2% of red cells, rising to 30–40% post-splenectomy. The most frequent variant is haemoglobin Koln, characterized by an SNP causing a valine to methionine transition at position 98 in the beta-globin chain. Other unstable haemoglobins are those of Zurich (a histidine to arginine transition at position 63 of beta-globin), Olmsted (leucine to arginine at position 141), and Wein (tyrosine to aspartate at position 130).

Iron

In order to maintain a haemoglobin level of 90–100 g/L in those with beta-thalassaemia major disease, regular transfusion of concentrated red cells is required, possibly at a rate of two to four units of red cells every four to six weeks, but this is very variable and is driven symptomatically and by other treatments. As described above, patients with other forms of haemoglobinopathy may also need regular transfusions. The problem is that each 500 ml of transfused blood contains 200–250 mg of iron and, since the human body has a very limited means to excrete iron, regular blood transfusion eventually leads to iron overload and deposition of iron in various organs (heart, pituitary), leading to their dysfunction. A further issue is that the ineffective erythropoiesis (expansion of immature erythroblasts and apoptosis of mature erythroblasts) in these conditions will be aggravated by iron overload. In the presence of high iron stores, hepcidin levels will be low, and so the importation of dietary iron will proceed without restriction, adding to the overload. In beta-thalassaemia mice, administration of minihepcidins (mimetics of the natural product) are effective in decreasing serum iron and increasing mature red cell. This may become a new therapy in this disease, and in others where iron levels are high, such as polycythaemia vera. A second potential pathway focuses on erythroferrone, and holds distant promise. The combination of peripheral hypoxia (a consequence of the anaemia) and ineffective erythropoiesis increases the production and secretion of erythropoietin, which can prompt erythroblasts to produce erythroferrone. The latter may, in turn, lead to the suppression of hepcidin. Ablation of high levels of erythroferrone mRNA in a mouse thalassaemia model restores hepcidin levels and reduces serum and liver iron, leading to hopes of the value of suppressing erythroferrone in our patients.

However, at present, treatment options for iron overload are limited. After the patient has received about 10–15 units of blood and the serum ferritin level exceeds 1000 μg/L, iron chelation therapy needs to be seriously considered. Although serum ferritin is the marker of choice in assessing iron levels, it is an acute-phase protein and is raised in viral hepatitis and other inflammatory disorders. Therefore, caution in interpretation is required. Removal of iron normally involves overnight infusion of an iron-chelating agent such as desferrioxamine, together with vitamin C, to increase iron excretion, on five to seven nights per week. Deferiprone, an orally active chelator, may be administered in conjunction with desferrioxamine to remove iron from the organs. Combined treatment is more effective than deferiprone alone in reducing serum ferritin. Treatment is also required for iron overload-associated complications such as of the heart, liver, and endocrine systems.

Apart from the iron problem, and that of an increasing likelihood of alloimmunization, blood transfusion seems likely to promote the low level of inflammation seen in patients with haemoglobinopathy, and in transfused beta-thalassaemia patients, transfusions are linked to increased leucocyte apoptosis.

Cross reference

The dangers of iron overload are described in Chapter 5.

SELF-CHECK 6.7

What are the principal findings in the FBC from a patient with a haemoglobinopathy?

SELF-CHECK 6.8

What are the principal findings in the blood film from a patient with a haemoglobinopathy?

6.3.8 Management of haemoglobinopathy

Generally, the patient should avoid known precipitants of sickle cell crisis, such as dehydration, infection, anoxia, strenuous exercise, and the cooling of the skin surface. During sickle cell crises the patient will need to take rest, keep warm, and rehydrate with oral fluids or (if seriously dehydrated) intravenous normal saline. We have already reviewed the role of blood transfusion and of the excess iron it brings. Animal models to reduce iron overload include exogenous apotransferrin (which decreases erythroferrone and the expression of transferrin receptors, suppresses hepcidin, and part reverses ineffective erythropoiesis) and inhibition of haemoxygenase by tin protoporphyrin IX (which decreased liver iron recycling and erythropoietin, increased hepcidin, and alleviated ineffective erythropoiesis). There is no simple effective treatment of haemoglobinopathy: in both forms, bone marrow transplantation, perhaps from an HLA-matching sibling, may offer the prospect of a permanent cure if carried out early in life. The success rate is over 80% in well-chelated young children, provided that liver complications do not arise. Of the remaining treatments, several strands come to the fore.

Several treatments are directed towards increasing the level of HbF. Of these, oral hydroxycarbamide (hydroxyurea) is a common choice to reduce both the frequency and duration of sickle cell crisis. This drug increases HbF synthesis, decreases intracellular HbS levels by increasing the MCV, lowers the white cell count, and reduces the adhesiveness between the sickle cell and the endothelium. New agents, such as butyrates designed to inhibit histone deacetylase, are in development, whilst drugs designed for other purposes (such as beta-blockers for hypertension) may be useful. In this group is pomalidomide, a drug used in myeloma that also induces HbF by decreasing levels of repressors of foetal globin gene expression. Erythropoietin is also used to promote HbF levels. It is also possible that increasing levels of HbA_2 may be of benefit in sickle cell disease.

Increased platelet activation in sickle cell disease and thalassaemia brings a risk of venous thromboembolism and possible oral anticoagulation. For those who have already suffered a deep vein thrombosis or pulmonary embolism, oral anticoagulation is mandatory. Vaccinating patients against pneumococcal and meningococcal organisms may reduce the frequency of infection, whilst oral penicillin is recommended to compensate for splenic atrophy. A vaccination against hepatitis B is also given, since these patients may require blood transfusion at some stage. Splenectomy is considered at the age of five years to reduce the blood requirement, and is followed by oral penicillin therapy for life, as subjects are susceptible to infections. A consequence of splenectomy (often offered to prolong the life of the abnormal red cells) is the appearance on the blood film of Howell–Jolly bodies within red cells, although these bodies (being nuclear remnants) may also be present in any splenic pathology, and so are common in the haemoglobinopathies.

Our understanding of the molecular basis of the haemoglobinopathies has benefited immensely from molecular genetics, and continues to do so, especially when combined with advances in stem cell biology. As outlined above, one therapeutic goal is to increase HbF. Expression of the gene itself, *HBG2*, which is located within the beta-globin cluster, is regulated by two upstream elements, *BCL11A* and *MYB*. Knock-out of *BCL11A* in mice leads to increased HbF, and in our species, deletion mutations leading to loss of function have the same effect, prompting the concept that this may be a possible therapeutic goal. This approach may be viable, as advances in direct gene editing with lentiviral vectors, CRISPR-Cas technology, forced chromatin loops, and zinc finger nucleases can correct abnormalities in human sickle stem cells *in vitro*, and in murine models, both leading to normal red cells. The observation that the co-inheritance of alpha- and beta-thalassaemia leads to a milder phenotype leads to the concept that engineered knock-down or other mechanism for suppressing alpha-globin production may be beneficial. Most cases of HPFH (with high levels of HbF—5–20% in carriers) are caused by a mutation in the promotor region of the γ-globin gene. This mutation can be excised and inserted into an erythroid cell line, which then generates substantial expression of HbF.

Pain is a common and recurring issue, and the patient-driven requirement for analgesia ranges from aspirin to morphine. Curiously, other treatments, such as ticlopidine (an antiplatelet agent), hydroxycarbamide (hydroxyurea), and L-glutamine (a presumed antioxidant) also bring pain relief. Acute interventions for painful crises include urea, methyl-prednisolone (as an anti-inflammatory agent), inhaled nitric oxide (presumed to vasodilate), tinzaparin, arginine, and magnesium. Ongoing clinical trials include drugs to reduce red cell adhesion (such as selectin inhibitors), statins, and other anticoagulants and antiplatelets (rivaroxaban, dalteparin, ticagrelor, and prasugrel). Case study 5.3 describes anaemia in a young woman.

6.3.9 Prenatal diagnosis and prevention of haemoglobinopathy

The severity of many haemoglobinopathies has prompted initiatives aimed at their prevention. Almost all sickle cell disease and thalassaemias are genetically inherited conditions, and the affected patients (carriers) are at risk of transmitting the disease to their descendants.

Under the current NHS guidelines, hospitals within a high-prevalence area perform universal haemoglobinopathy screening for all antenatal attenders (when a first pregnancy), regardless of ethnic or racial origin. Mothers who are tested positive for significant haemoglobinopathy are invited to call the father for screening. If both parents are carriers for a haemoglobinopathy, there is a 25% chance that the newborn may have the homozygous disease during each pregnancy (Figure 6.17). Such couples, with appropriate genetic counselling, would be given the informed choice for investigation and termination.

If both partners are diagnosed with a haemoglobinopathy, then there is a one-in-four chance that the foetus is homozygous, or doubly heterozygous, and a one-in-two chance that the foetus is a carrier. In at-risk couples, prenatal diagnosis is offered using either DNA (from a chorionic villous or amniotic fluid sample) or foetal blood. The foetal DNA is amplified by using polymerase chain reaction (PCR) and the DNA mutation(s) are detected. In homozygous disease of the foetus, the couple should be counselled and, if appropriate, termination may be offered. Authoritative guidelines pertinent to UK practice were published in January 2010.

Key Points

The epidemiology, molecular and cell biology, and pathology of the haemoglobinopathies are well characterized. However, less well understood is the impact that the disease has upon the haematology and clinical course of the disease for each individual. Both these phenotypic aspects can differ widely in two individuals with apparently the same genotype, indicating that there is still much we have to learn.

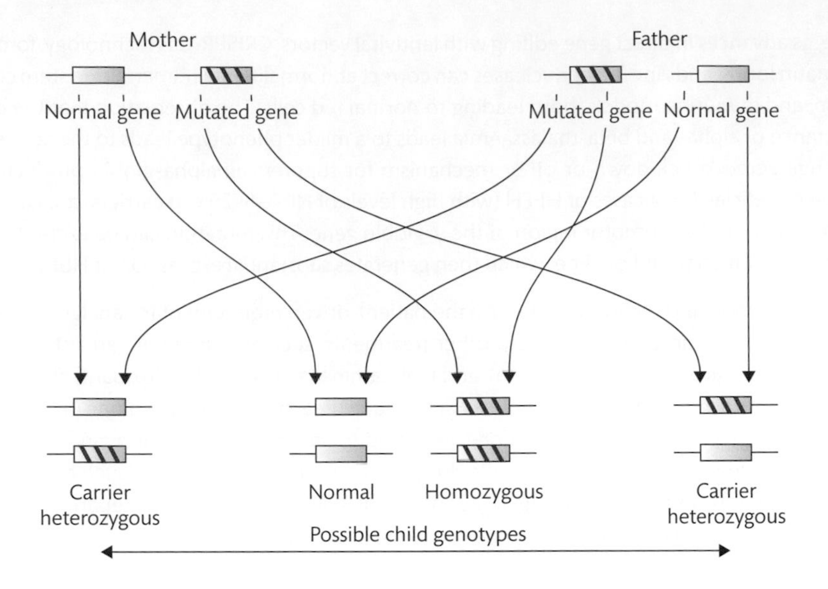

FIGURE 6.17

Simple inheritance of a single gene defect. Inheritance of a haemoglobinopathy (such as sickle cell disease) obeys simple Mendelian genetics. Each parent has two genes, but in the present illustration, each has a normal gene and a mutated gene that gives rise to an abnormal haemoglobin. Thus each parent is a heterozygote and so is a carrier of the disease. Either parent can pass on only one of their genes to their offspring, which then has two genes, one from each parent.

On the bottom far left, the child has inherited a normal gene from the mother and an abnormal gene from the father. Thus, like its parents, it is also a heterozygote and a carrier of the disease. To the right, the offspring has inherited a normal gene from each parent and so is a (normal) homozygote. The offspring third from the left has inherited two mutated genes, one from each parent, and so is a homozygote and will fully express the genotype (such as sickle cell disease). On the bottom right, the offspring has inherited a normal gene from the father and an abnormal gene from the mother, and so is a heterozygote carrier.

CASE STUDY 6.2 Anaemia in a young woman

A 20-year-old female recently moved, with her family, to the UK from the Far East. Following a few weeks' acclimatization and recovery from jet lag, it became clear to her family that she was consistently tired and lethargic, more so than her siblings. On examination she had no symptoms of infection (e.g. fever, sweating) or aches and pains, and did not report heavy menstrual bleeding. The FBC and ESR were as follows. How should this subject be investigated?

Index	Result	Reference range (female) (see front of book)
Haemoglobin	105 g/L	118–148 g/L
RCC	6.0×10^{12}/L	3.9–5.0 $\times 10^{12}$/L
MCH	17.5 pg	26–33 pg
MCV	55 fL	77–98 fL
MCHC	318 pg/L	330–370 pg/L
Hct	0.33	0.33–0.47
Reticulocytes	150×10^{9}/L	25–125 $\times 10^{9}$/L
ESR	15 mm/hour	<10 mm/hour

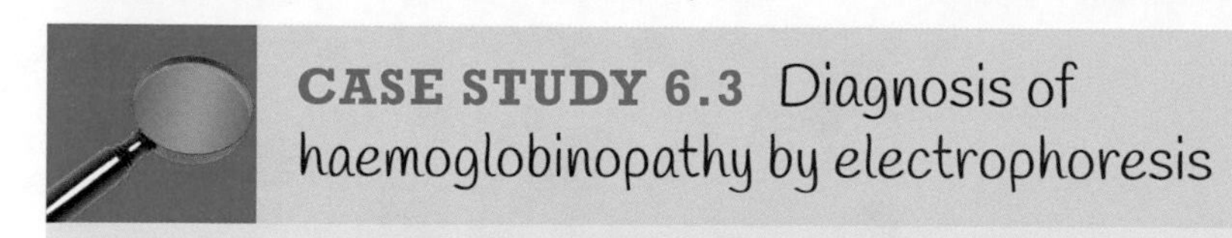

Haemoglobin electrophoresis is a useful tool in determining the phenotype of various haemoglobinopathies. Consider Figure 6.13. The top section consists of patterns from subjects with known haemoglobin patterns—they, therefore, are the positive controls. The lower five samples are from patients with a suspected haemoglobinopathy. Study the patterns of these five samples and compare them with the patterns of the positive controls. In this way, you can determine the haemoglobin phenotype of the five patients.

6.4 Membrane defects

The three principal causes of haemolytic anaemia relating to abnormalities in the red blood cell membrane are hereditary spherocytosis, hereditary elliptocytosis, and PNH. Conditions less frequently encountered include hereditary stomatocytosis, hereditary pyropoikilocytosis, and hereditary ovalocytosis. There can be no complete understanding of these diseases without knowledge of the structure of the membrane, as is presented in Chapter 4, where Tables 4.2–4.4 and Figures 4.6–4.8 apply.

6.4.1 Hereditary spherocytosis (HS)

Hereditary spherocytosis (Figure 6.18) is the most common hereditary haemolytic anaemia in North Europeans and in 75% of cases is inherited in an autosomal-dominant manner, with variable clinical presentations ranging from a severe neonatal haemolytic anaemia to an asymptomatic state. It is present in about one in 2000 individuals of northern European ancestry. Several mutations are known to cause HS, the most common being in *ANK1* that codes for ankryn, followed by *SLC4A1* coding for Band 3, although others are known (in *SPTA1* coding for α-spectrin, *SPTB* for β-spectrin, and *EPB42* for protein 4.2). The link is that the particular gene defect affects the proteins involved in the vertical interaction between the skeleton membrane and lipid layer of the red cell (that is, ankyrin, α- and β-spectrins, Band 3, and protein 4.2). The relationships between these molecules is described in Chapter 4, where Figures 4.7 and 4.8 are relevant. Notably, HS cells present increased Lu glycoprotein to laminin in blood vessel walls, thus promoting adhesion. The molecular basis of this may result from failure of the intracellular portion of Lu interacting with spectrin.

Part of the membrane that is not supported by skeleton is lost, causing the cells to lose sections of membrane and become more and more spherical. These abnormal red cells have a considerably reduced lifespan (between 6 and 20 days, compared to 120 days in health), and are unable to pass through the splenic microcirculation. Consequently, they are eliminated.

Clinical features include anaemia and jaundice. These features may be present at any age throughout the lifespan, but can be compounded if associated with Gilbert's disease, a heritable enzyme defect associated with bilirubin metabolism. Patients with spherocytic anaemia benefit from splenectomy, which may increase their haemoglobin level to normal. Patients are further supported by folic acid supplement to avoid folate deficiency.

The FBC from an HS patient can be expected to show an increased reticulocyte count (5–20%), low haemoglobin (e.g. 70 g/L), reduced MCV, raised MCH, and raised MCHC (such as

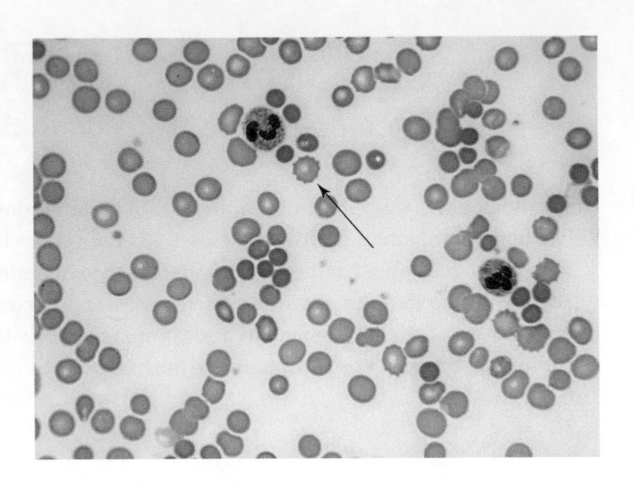

FIGURE 6.18
Hereditary spherocytosis. There is considerable variation in the sizes of the red cells in this film (anisocytosis), with many cells being small. These are not microcytes, but spherocytes, as they are well stained and so have a high component of haemoglobin. There are also two neutrophils, and a burr cell slightly below and to the right of the upper neutrophil (arrowed). Note also the lack of schistocytes. (Magnification ×400.)

360 g/L or more). The red cell distribution width is increased (>14%) in most patients. The blood film shows dense microspherocytes, with no central pallor, as shown in Figure 6.18. However, recall that spherocytes may also be present in other disease, such as AIHA (see Figure 6.5). The laboratory can differentiate between these two manifestations of spherocytosis as the DAT will be positive in AIHA but negative in HS. Other special tests in HS include impaired spectrin phosphorylation and decreased red cell deformability. Polyacrylamide gel electrophoresis of a preparation of red cell membranes will display an absence of particular spectrins.

6.4.2 Hereditary elliptocytosis (HE)

In its classical form, this red cell abnormality is most frequently found in Europeans and is inherited as an autosomal-dominant characteristic. However, in parts of Africa (especially those where malaria is endemic), the incidence may reach one or two in 100. Elliptoid red cells may also be caused by mutations in *SPTA1* and *SPTB*, but also by mutations in *EPB41* at 1p35.3, coding for protein 4.1, or in *SCL4A1* at 17q21.31, coding for Band 3.

On a blood film, the phenotype shows a broad spectrum of abnormal red cell morphology, ranging from slightly oval cells to extreme elliptocytosis. Most patients do not have a haemolytic anaemia and there is no correlation between haemolysis and the degree of elliptocytosis. The underlining membrane abnormality is a deficiency of protein 4.1 leading to the failure of spectrin heterodimers to assemble into heterotetramers. However, the same elliptoid phenotype is also present in some defects of α-spectrin, β-spectrin, and glycophorin C.

Most cases of elliptocytosis are heterozygous with no haemolysis, but a few cases of homozygous hereditary elliptocytosis present with a severe haemolytic anaemia. The FBC may show a normal-to-slightly-low haemoglobin, whilst the blood film may show typically that 80% of cells are oval, as are present in Figure 6.19. However, in cases where numbers are not large, an alternative diagnosis such as acquired elliptocytosis (such as that seen in iron deficiency) may be considered (Figure 6.20).

6.4.3 South East Asian ovalocytosis

This disease, which may be viewed as a variant of elliptocytosis, is commonly found in Indonesia, the Philippines, and Malaysia, where the prevalence may be 30%. The abnormal red cell shape is caused by mutations in *SLC4A1* that lead to errors in membrane component Band 3 (Figure 6.21) This molecule is both a component of the cytoskeleton and also the

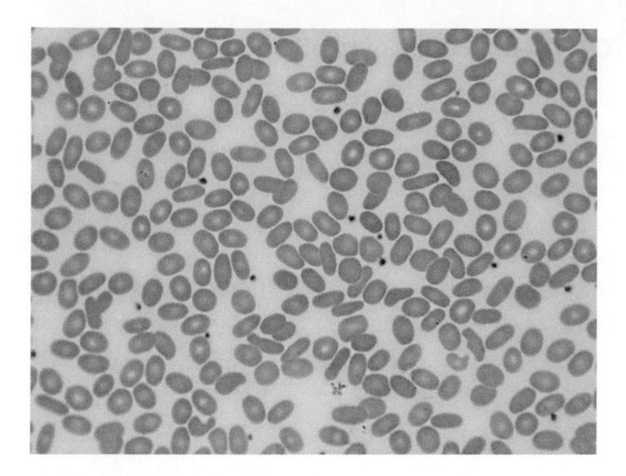

FIGURE 6.19
Hereditary elliptocytes. Well over half of the cells in this film are elliptical–hence elliptocytosis. (Magnification ×400.)

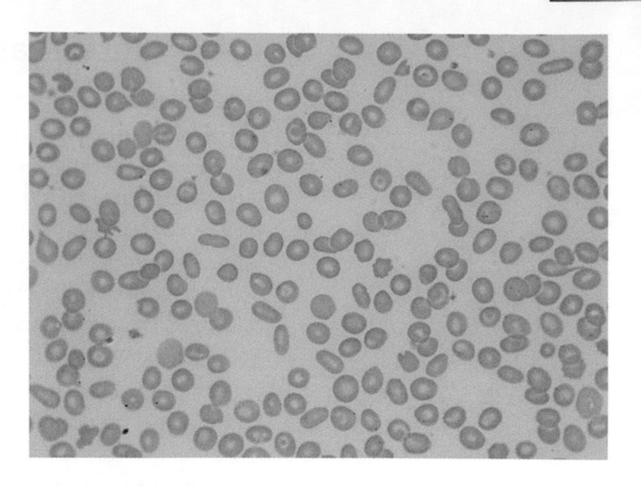

FIGURE 6.20
Iron deficiency with elliptocytes. There are many excellent examples of elliptocytes, but relatively few compared to Figure 6.19; in iron deficiency we tend to refer to these elliptocytes as pencil cells. Note also several target cells and microcytes, as may be expected in this condition. (Magnification ×400.)

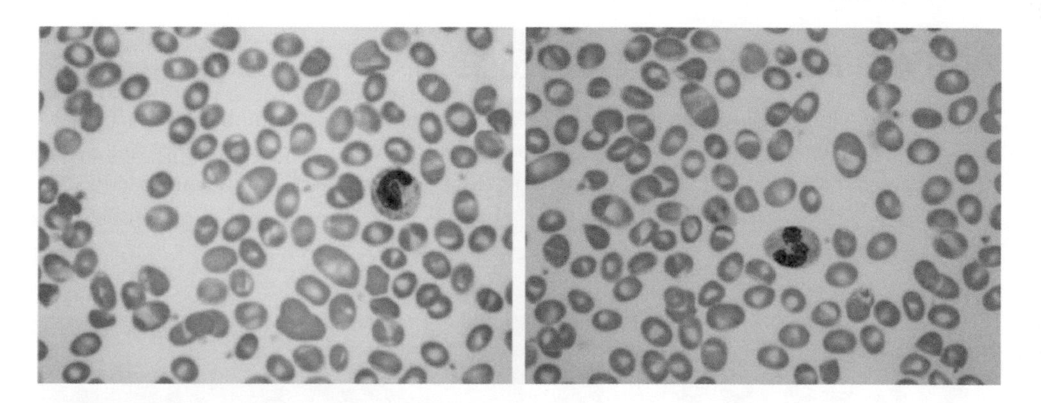

FIGURE 6.21
South East Asian ovalocytosis. The films show maro-ovalocytes, some cells having more than one stoma.
https://onlinelibrary.wiley.com/doi/epdf/10.1002/ajh.23379.

chloride-bicarbonate anion exchanger. Red cells from a homozygote for this mutation showed increased expression of CD44, CD147, and calreticulin, suggesting a defect in reticulocyte maturation.

A consequence of these membrane changes is that the cell is exceptionally rigid, a characteristic that is detected by cells of the reticuloendothelial system (as for HS cells), leading to their elimination, often in the spleen. Like many hereditary membrane, enzyme, and haemoglobin diseases, the high frequency in this part of the world has almost certainly arisen as a protection against malaria, possibly around 10,000 years ago.

6.4.4 Hereditary stomatocytosis (HSt)

HSt describes several autosomal-dominant conditions where the transport of cations (notably sodium and potassium) is impaired. This imbalance leads to osmotic misbalance, and so shape, and (depending on exact pathophysiology) cells can be overhydrated or underhydrated. It is characterized on a blood film by stomatocytes, often with an oblong bar of central pallor (Figure 6.22). The lesion is generally due to abnormalities in membrane components, those of Band 3 being the most common, although it may also be caused by mutations in *PIEZO1* and *KCNN4* (more of which follows). However, some cases of HS and South East Asian ovalocytosis (SEAO) also have defects in this molecule. Red cells are large and osmotically fragile with a low MCHC, which leads to a mild-to-moderate macrocytic anaemia. A variant of HSt is cryohydrocytosis, where there is haemolysis and a cold-induced leak of cations, particularly potassium, and so a mild pseudohyperkalaemia.

6.4.5 Hereditary pyropoikilocytosis

This rare autosomal-recessive disorder is characterized on the blood film by microcytes and bizarre poikilocytes with marked red cell fragmentation (schistocytes) (Figure 6.23), and clinically by a severe haemolytic anaemia. These cells, with a defect in alpha spectrin, show exceptional heat sensitivity, and their spectrin cytoskeleton denatures at a lower temperature (45–46°C) instead of the expected 49°C. The disease is strongly linked with hereditary elliptocytosis, present in up to one-third of family members.

6.4.6 Hereditary xerocytosis

This condition encompasses a number of clinically heterogenous syndromes in which erythrocytes exhibit decreased total cation and potassium content that is not accompanied by the expected increase in sodium and water. Most cases are associated with mutations in *PIEZO1*

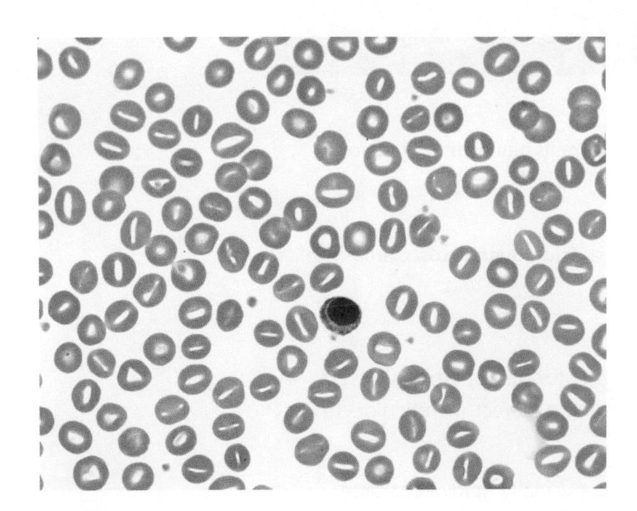

FIGURE 6.22
Hereditary stomatocytosis. Almost all red cells have a central oblong bar. A nucleated red blood cell is also present. https://hereditarystomatocytosiskw.weebly.com/

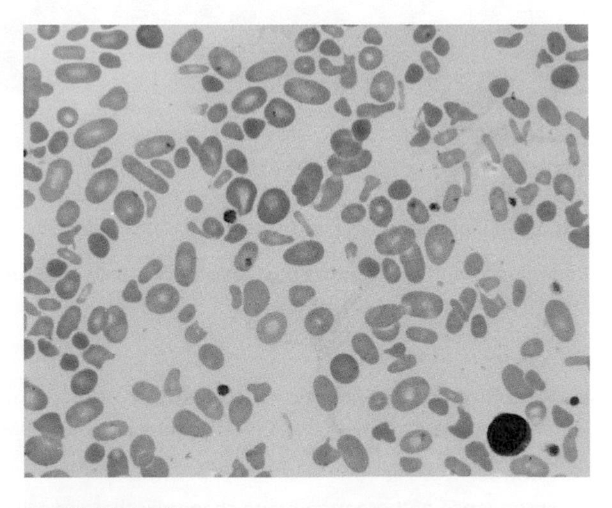

FIGURE 6.23
Hereditary pyropoikilocytosis. There is marked anisocytosis with numerous rod-shaped elliptocytes and bizarre red cell forms; also a lymphocyte. Kindly provided by M Brereton and J Burthem: www.haematologyetc.co.uk.

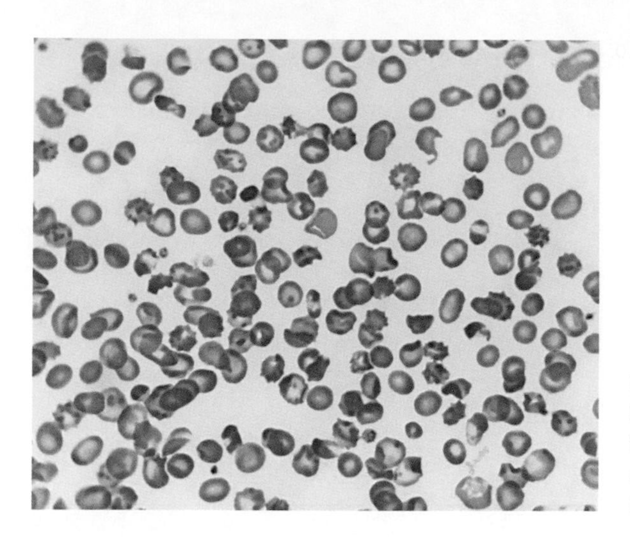

FIGURE 6.24
Hereditary xerocytosis. There is anisopoikilocytosis with frequent dehydrated and crenated cells, schistocytes, and some target cells. Kindly provided by M Brereton and J Burthem: www.haematologyetc.co.uk.

(at 16q24.3), that codes for a pore-forming membrane structure, the Gardos channel. Mutations in *PIEZO1*, mostly at the COOH-terminal of the pore region protein of the channel, ultimately lead to the dehydration of the cell. An alternative cause of defects in the channel is a dominantly inherited mutation in *KCNN4*, found on 19q13.31. Laboratory findings for this dominantly inherited disorder include increased MCHC and decreased osmotic fragility, the film showing a number of abnormalities (Figure 6.24). Clinical consequences include a mild-to-moderate haemolytic anaemia, with hydrops foetalis, thrombosis, gallstones, and iron overload.

6.4.7 Tests of a weakened membrane

Perhaps the simplest and most physiological test of a weak membrane is an increased osmotic fragility. Equal volumes of blood are added to a series of buffered hypotonic sodium chloride solutions, ranging from 0.1% to 0.9% (the latter being physiological). The suspensions are allowed to stand for 30 minutes at room temperature, during which time some cells (the weakest) will lyse. The degree of haemolysis in each test tube is recorded spectrophotometrically and is expressed as a percentage of haemolysis at each sodium chloride concentration when compared to the positive control (that is, distilled water). Interpretation relies on the marked increased fragility of the red cell population; the test can also be useful in other red cell diseases (Figure 6.25). Extension of the incubation test to 24 hours can also be useful in defining other pathologies, such as of enzyme deficiencies.

In an alternative test, the acidified-glycerol lysis test, red cells are suspended in a slightly acidified phosphate-buffered sodium chloride-glycerol reagent and the time taken for 50% haemolysis to occur is measured. Interpretation relies on a shortened haemolysis time in HS compared to normal red cells. The cryohaemolysis test employs a similar principle—namely, to stress the cells and observe the result. The stress in this case is to warm the cells to 37°C, then to transfer them promptly to an ice bath for an additional incubation, at which point weakened cells, such as those from a patient with HS, will lyse.

The fluorescent probe eosin-5-maleimide (EMA) binds to normal red cells, and may be detected by a fluorescence activated cell scanner, but does so less avidly if there are abnormalities in membrane components spectrin, protein 4.2, or Band 3. An authoritative 2012 guideline recommends the use of the EMA and cryohaemolysis tests in screening for HS: the osmotic fragility test is not recommended for routine use, as 25% of patients have a normal fragility profile. These features are summarized in Table 6.5.

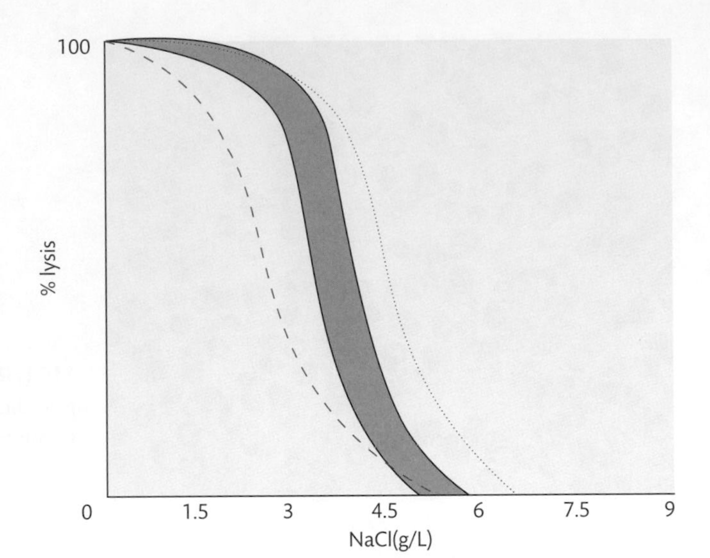

FIGURE 6.25
Osmotic fragility. Red blood cells are suspended in different concentrations of sodium chloride for 30 minutes at 37°C, during which time the weakest will lyse in the most hypotonic saline solutions. The profile of normal red cells, and most enzymopathies, is present as the solid bar. Red cells in sickle cell disease, thalassaemia, and iron deficiency show a left shift (dashed line), those from hereditary spherocytosis show a right shift (dotted line). In HE, the pattern is slightly right shifted.

TABLE 6.5 Laboratory tests in membrane disorders.

Diagnosis	Osmotic fragility test	Acid-glycerol lysis-time test	Cryohaemolysis test	EMA-binding test
HS	Increased fragility	Shortened lysis time	Increased lysis	Decreased fluorescence
HPP	*	*	*	Decreased fluorescence
Overhydrated HSt	Increased fragility	*	*	Increased fluorescence
Dehydrated Hst	Decreased fragility	Normal	*	Increased fluorescence
SEAO	*	*	Increased lysis	Decreased fluorescence

HS = hereditary spherocytosis; HPP = hereditary pyropoikilocytosis; HSt = hereditary stomatocytosis; SEAO = South East Asian ovalocytosis; * = no published data. Modified from Bolton-Maggs et al. (2011).

6.4.8 Paroxysmal nocturnal haemoglobinuria (PNH)

We close this section with a disease whose pathological basis lies not in abnormal intracellular molecules, but in contrast, in the absence of a single molecule that spans the membrane and that is linked to two other molecules. PNH is caused by a rare acquired mutation in the *PIG-A* gene (present at Xp22.2) in haemopoietic stem cells, leading to the defective synthesis of glycosyl phosphatidylinositol (GPI) on the cell surface. Untreated, the disease is fatal in perhaps one-third of subjects within five years of diagnosis, although treatment with a monoclonal antibody (eculizumab, directed towards complement component C5) is a very effective treatment. This agent virtually eliminates intravascular haemolysis, as shown by a rapid decline in LDH after administration of the drug, often eliminating the need for blood transfusions. However, most patients continue to have low levels of haemolysis and so have a mild-to-moderate anaemia.

GPI is an anchor that supports the integrity of several proteins, including two molecules, CD55 and CD59, which are important in resisting the activation of complement. As a consequence, the red cells become sensitive to complement-mediated intravascular haemolysis. The result of this sensitivity is the constant clinical feature of haemoglobinuria. This can give rise to iron deficiency, which may exacerbate the anaemia. Treatments include prednisone, folic acid, oral iron, and blood transfusions. Since CD55 and CD59 are also present on platelets and white cells, other clinical features include recurrent thrombosis of large veins, with intermittent abdominal pain. Thus, laboratory findings include anaemia (ranging from mild to severe), leucopenia, and thrombophilia (for which anticoagulants, such as warfarin, may be required).

The typical FBC from a patient with PNH demonstrates mildly reduced haemoglobin, with an increase in reticulocytes, mild leucopenia ($<2.5 \times 10^9$/L), and thrombocytopenia ($<50 \times 10^9$/L). Consequently, PNH should be excluded in patients with pancytopenia. Serum LDH is inevitably raised, and indeed, normal levels essentially refute the diagnosis.

A number of diagnostic tests are available. The acidified-serum lysis test (Ham test) involves suspending red cells in normal serum which has been acidified to pH 6.5–7.0, followed by incubation at 37°C, and subsequent examination for haemolysis. This red cell destruction will be induced by activation of the alternative pathway of complement activation. Similarly, for the sucrose haemolysis test, red cells are incubated in isotonic solutions of low ionic strength with a small amount of serum present in the mixture. Lysis is induced by activation of the classical pathway of the complement system. Interpretation of <5% haemolysis is inconsequential, 5–10% haemolysis is borderline, but >10% is consistent with PNH. For both these tests, good positive and negative controls are essential. However, fluorescence flow cytometry analysis for the presence of CD55 and CD59 (and perhaps C3) at the cell surface is rapidly becoming the method of choice for the diagnosis and management of the disease.

Cross reference

Fluorescence flow cytometry and immunophenotyping are considered in detail in Chapter 10.

SELF-CHECK 6.9

How would you go about distinguishing between hereditary spherocytosis and hereditary elliptocytosis?

The astute reader will have noted that the pathophysiology of several diseases in this section (aHUS, typical HUS, PNH, CAD, and TTP) may all be linked to complement. Although primarily the remit of the immunologist, the haematologist will be aware of the importance of this pathway, even though it is not generally part of their routine practice. For further details, see the textbook ***Immunology*** in this series. The review by Brodsky is recommended.

6.5 Principal enzyme defects

Although there are many enzyme defects, the principal abnormalities are deficiencies of glucose-6-phosphate dehydrogenase (G6PD) and pyruvate kinase (PK). Deficiency of both these enzymes offers partial protection from malaria, explaining their marked geographic links. Full details of the metabolic roles of these enzymes are presented in Chapter 4, section 4.3.2, where Figure 4.15 and nearby text apply. Quite possibly the third most relevant enzyme, pyrimidine-5-nucleotidase (P5NT), degrades pyrimidine nucleotides from messenger and ribosomal RNA to cytidine and uracil. In the deficiency of this enzyme, partially degraded nucleotides build up within the cell, detectable by conventional staining of a blood film as basophilic stippling. Lead is a reversible inhibitor of this enzyme, and this, as well as the inherited autosomal-recessive enzyme deficiency, results in a chronic, moderately severe haemolytic anaemia. The method for detecting this abnormality is relatively simple, calling for the comparison of the absorbance at 280 nm (where cytidine nucleotides absorb) with that of 260 nm (where adenine, guanine, and uridine absorb). In P5NT deficiency, levels of cytidine will be increased, thus reducing the A260/A280 ratio.

6.5.1 G6PD deficiency

The enzyme glucose-6-phosphate dehydrogenase (G6PD, coded for by *G6PD* at Xq28) is responsible for removing hydrogen from its substrate, the metabolic intermediate glucose-6-phosphate, which becomes 6-phosphogluconate. The hydrogen is received by NADP, which becomes NADPH. In turn, NADPH passes the hydrogen to oxidized glutathione (abbreviated as GS^-, but present as a dimer linked by the sulphur atoms, that is GS-SG), which is then converted to reduced glutathione (GSH). Reduced glutathione is a crucial antioxidant which (alongside other enzymes such as catalase, superoxide dismutase, and peroxiredoxins) counters the damaging and toxic effects of oxygen within the red cell by effectively neutralizing or quenching it with the hydrogen.

In patients with G6PD deficiency, the hydrogen is unavailable to regenerate levels of GSH, which subsequently decrease, leading to a fall in the antioxidant capacity of red cells. A further consequence of the lack of NADPH is the inability to convert levels of methaemoglobin (high levels of which lead to hypoxia and so to headache, dyspnea, and cyanosis) to haemoglobin. Reduced activity of this pathway also leads to damage to the membrane, and ultimately to haemolysis. Figure 6.26 shows a short metabolic pathway illustrating this biochemistry.

As the gene for G6PD is found on the X-chromosome, inheritance is sex-linked. The deficiency condition affects up to 1% of the world's population, but is considerably higher (13%) in those of West African descent. The gene has a number of isoforms, some of which are loss-of-function mutations, leading to G6PD deficiency. This condition shows marked clinical heterogeneity,

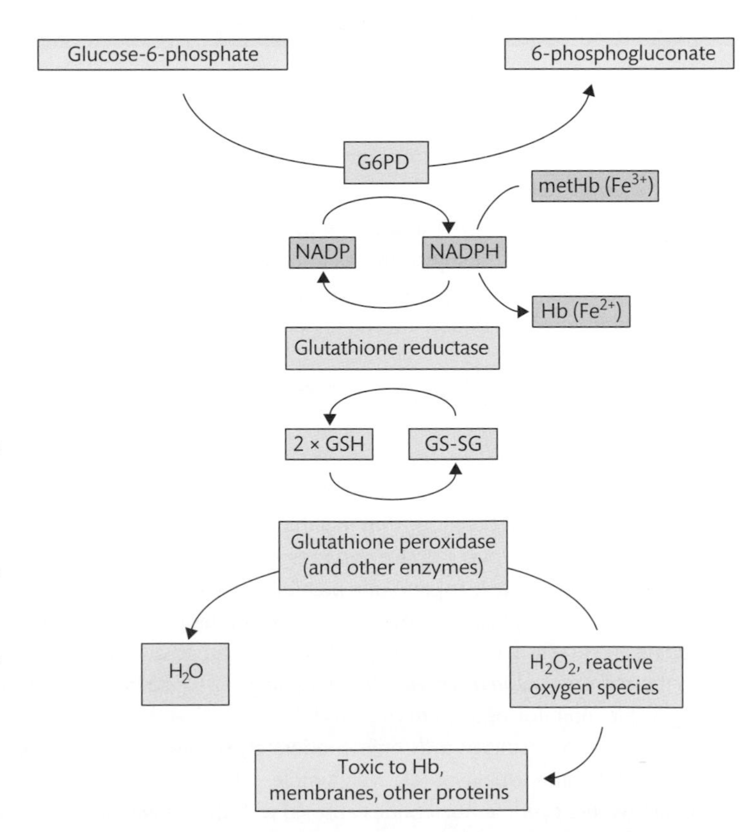

FIGURE 6.26

Biochemical pathway illustrating the effects of glucose-6-phosphate dehydrogenase (G6PD) deficiency. At the top, G6PD catalyses the transfer of a hydrogen atom from glucose-6-phosphate to NADP. Two hydrogen atoms are required for the reduction of GS-SG to two molecules of GSH, and ultimately these atoms convert the dangerous hydrogen peroxide to water. Lack of the enzyme, i.e. G6PD deficiency, leads to failure to generate NADPH, and so failure to reduce GS-SG to 2GSH. Hence no GSH is available to defend against toxic oxygen species, leading to cell and molecular damage. A further aspect of G6PD deficiency (on the right) is the inability to convert metHb to 'standard' Hb. G6PD = glucose 6 phosphate dehydrogenase; NADP = nicotine adenine dinucleotide phosphate; GSH = reduced glutathione; GS-SG = oxidized dimer of glutathione, H_2O_2 = hydrogen peroxide.

with symptoms including sensitivity to different drugs. The most common consequence of G6PD deficiency is drug-induced haemolysis.

The oxidant drugs that can cause haemolysis include antimalarials (such as primaquine), antibiotics (chloramphenicol), analgesics (aspirin), and antihelminths (nitrodazole). The ingestion of fava beans also precipitates acute crises of haemolytic anaemia, with haematuria and pain. In contrast to these acute crises, in 'steady-state' G6PD deficiency there is chronic haemolytic anaemia, and therefore jaundice. Treatment for G6PD deficiency is to treat the cause: avoidance of the precipitating factor(s) (stopping the ingestion of fava beans and agents that promote haemolysis), resolution of an infection, and in severe cases, the consideration of a blood transfusion. There is no replacement therapy.

In the absence of haemolytic crisis, the FBC is normal, although during acute intravascular exacerbations, the blood film demonstrates features of haemolysis, such as red cell ghosts (without haemoglobin—'bite' and 'blister' cells) and polychromasia. Heinz bodies (haemoglobin denatured by the high levels of oxidants) may be seen in reticulocytes.

Commercial screening and assay kits are available for the assessment of G6PD deficiency, and these tests are based on the measurement of the levels of NADPH. The reaction mixture is the substrate glucose-6-phosphate, NADP, oxidized glutathione (GS-SG), and buffers, In the presence of G6PD, NADP is reduced to NADPH, which fluoresces under long-wave ultraviolet light. Specimens from normal healthy individuals fluoresce brightly, indicating a normal G6PD activity. However, specimens from G6PD-deficient samples show reduced or no fluorescence because of the absence of NADPH. Intermediate degrees of fluorescence indicate a specimen from heterozygotes, or patients with a mild G6PD variant. Reticulocyte cells demonstrate higher G6PD activity. An alternative (but rarely used) method is to exploit the reduction of methaemoglobin to haemoglobin by NADPH. This calls for a reaction mixture of sodium nitrite, dextrose, and methylene blue. Although of longer duration than the enzymatic method, it has the advantage of being cheaper, and as the end point is a change in colour in the visible section of the spectrum (red/brown), calls for a standard white light spectrophotometer.

6.5.2 Pyruvate kinase (PK) deficiency

Pyruvate kinase (PK, coded for by *PKLR* on chromosome 1q21) participates in the generation of ATP by transferring a phosphate group from the metabolic intermediate phosphoenolpyruvate to ADP, the remaining product being pyruvate. Consequently, a lack of activity of this enzyme (most often due to point or frameshift mutations in *PKLR*, of which more 160 have been identified) leads to reduced intracellular ATP, this failure most likely being due to a loss-of-function mutation in the relevant gene. Curiously, there is no association between PK enzyme activity and genotype. Deficiency of PK is inherited in an autosomal-recessive manner and is the most common enzyme deficiency in the Embden–Meyerhof pathway, with an estimated prevalence of 1 per 20,000 in the Causasian population. Reduced levels of ATP lead to a mild-to-moderate chronic non-spherocytic haemolytic anaemia (haemoglobin typically 40–100 g/L) and an increased reticulocyte count. The blood film shows poikilocytes, and may show some echinocytes, and a reticulocytosis is common. Reduced red cell survival and chronic haemolysis result in increased iron turnover—increased ferritin and iron overload is found in 60% of untransfused PK-deficient patients and there is low hepcidin, which correlates with haemoglobin. Clinical consequences include splenomegaly, jaundice, gall stones, iron overload (present in 45% and 48%, respectively, of the cohort), thrombosis, and osteopenia. Splenectomy is common as it is all haemolytic anaemia.

A further consequence of PK deficiency is that—as the substrate phosphoenolpyruvate is not consumed—levels rise, and, in turn, cause a rise in levels of another metabolite in the metabolic pathway, 2,3-diphosphoglycerate (2,3-DPG). In Chapter 4, we discussed the importance of this molecule in the movement of oxygen in and out of the red cell in the lungs and in the tissues. Thus, in PK deficiency, high levels of 2,3-DPG shift the oxygen dissociation curve to the right, which effectively reduces the clinical symptoms of anaemia in comparison to the haemoglobin concentration. Downstream of PK, pyruvate is converted to lactate by the enzyme LDH, which requires reduced NADH to donate a proton (Figure 6.23). This requirement for a proton donor (NADH and NADPH) has a parallel with the donation of protons to GS-SG in the G6PD pathway of Figure 6.27. Insufficient availability of NADH will therefore fail to convert pyruvate to lactate, so the levels of the latter build up to eventually saturate the pathway and so inhibit the generate of ATP, as in PK deficiency.

The laboratory confirmation of PK deficiency is relatively simple and precisely follows this biochemistry. Red cells are lysed, and the lysate fed a cocktail of substrates: ADP and phosphoenolpyruvate, with buffers, LDH, and NADH. If PK is present, it will generate products of pyruvate and ATP, and the NADH will be converted to NAD. The disappearance of NADH can be monitored with a standard spectrophotometer by measuring the change in the absorbance of light at 340 nm. Specimens from healthy individuals demonstrate reduced or no fluorescence. In contrast, PK-deficient samples show bright fluorescence.

Table 6.6 summarizes the pathology and laboratory aspects of membrane and enzyme defects.

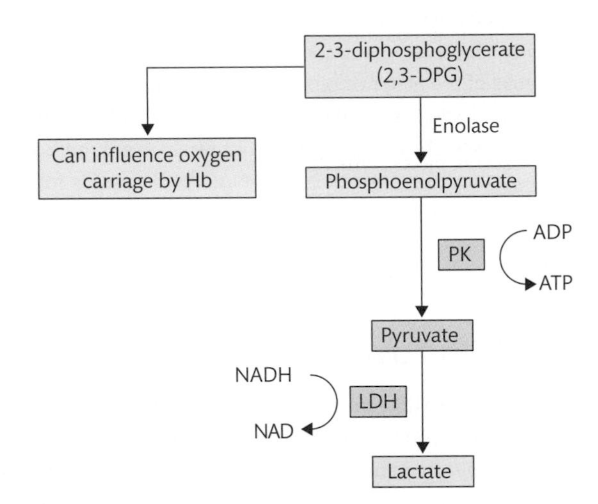

FIGURE 6.27
Place of pyruvate kinase (PK) in red cell metabolism. PK catalyses the transfer of a phosphate group from phosphoenolpyruvate to ADP, giving ATP and pyruvate. In PK deficiency, there is impaired ATP generation and the build-up of 2,3-DPG, the latter possibly interfering with oxygen transport. LDH = lactate dehydrogenase, NAD = nicotinamide adenine dinucleotide.

TABLE 6.6 Common membrane and enzyme defects leading to haemolysis.

Condition	Nature of the pathology
Hereditary spherocytosis	Defects in the internal cytoskeleton leading to abnormal cell shape
Hereditary elliptocytosis	Defects in the internal cytoskeleton leading to abnormal cell shape
Paroxysmal nocturnal haemoglobinuria	Failure to anchor key protective molecules to the cell membrane
Glucose-6-phosphate dehydrogenase deficiency	Poor regeneration of reduced glutathione results in increased oxidant activity and failure to convert metHb to Hb
Pyruvate kinase deficiency	Reduced ability to generate ATP, build-up of 2,3 diphosphoglycerate

Key Points

The small number of well-characterized membrane and enzyme defects provide the haematologist with an excellent opportunity to link a gene defect with a cellular abnormality, and thus a clinical syndrome.

Chapter summary

- Antibody-mediated haemolytic anaemia may be classified as being caused by autoantibodies or alloantibodies.
- External causes of haemolytic anaemia include drugs, infections, and mechanical processes.
- The principal haemoglobinopathy conditions are sickle cell disease and thalassaemia.
- Defects in the red cell membrane, as in hereditary spherocytosis, hereditary elliptocytosis, and PNH, all cause haemolysis.
- The principal enzyme defects leading to haemolysis are G6PD deficiency and PK deficiency.

Discussion questions

6.1 What are the key features of antibody-mediated haemolytic anaemia?

6.2 How can the different haemoglobinopathies be detected in the laboratory?

6.3 What is the cellular link between the membrane diseases and the phenotype of the red cells on a blood film?

6.4 Describe the common and unique features of the laboratory methods for detecting G6PD deficiency and PK deficiency.

Further reading

Bolton-Maggs P, Langer JC, Tittensor P, and King MJ. ***Guidelines for the diagnosis and management of hereditary spherocytosis—2011 update***. *British Journal of Haematology* 2012: **156**; 37–49.

Brodsky RA. ***Complement in haemolytic anaemia***. *Blood* 2015: **126**: 2459–65.

Camaschella C and Nai A. ***Ineffective erythropoiesis and regulation of iron status in iron loading anaemias***. *British Journal of Haematology* 2015: **172**: 512–23.

Conway EM. ***HUS and the case for complement***. *Blood* 2015: **126**: 2085–90.

Hoban MD, Orkin SH, and Bauer DE. ***Genetic treatment of a molecular disorder: Gene therapy approaches to sickle cell disease.*** *Blood* 2016: **127**: 839–48.

Mohandas N. ***Inherited hemolytic anemia: A possessive beginner's guide.*** *Hematology—American Society of Hematology Education Program* 2018: **1**: 377–81.

Mohandas N and An X. ***Malaria and human red blood cells.*** *Medical Microbiology and Immunology* 2012: **201**: 593–8.

L'Acqua C and Hod E. ***New perspectives on the thrombotic complications of haemolysis.*** *British Journal of Haematology* 2015: **168**: 175–85.

Luzzatto L and Seneca E. ***G6PD deficiency: A classic example of pharmacogenetics with ongoing clinical implications.*** *British Journal of Haematology* 2014: **164**: 469–80.

Zhang D, Xu C, Manwani D, and Frenette PS. ***Neutrophils, platelets, and inflammatory pathways at the nexus of sickle cell disease pathophysiology.*** *Blood* 2016: **127**: 801–9.

Answers to self-check questions and case study questions are provided in the book's online resources.

Visit www.oup.com/he/moore-fbms3e.

7

Blood-borne parasites

Gary Moore

In this chapter, you will be introduced to the major blood-borne parasites and their life cycles. You will see how morphological characteristics are important for accurate diagnosis and meet other laboratory tests available to biomedical scientists for the detection of blood parasites.

Learning objectives

After studying this chapter, you should confidently be able to:

- name the main blood-borne parasites that cause disease in humans;
- understand parasite life cycles;
- appreciate the importance of the recognition of parasite morphology;
- describe the appearances of parasites in Romanowsky-stained blood films;
- describe additional laboratory tests for parasite detection.

7.1 Introduction

Parasitism is a form of **symbiosis** where the parasitic organism benefits from the association to the detriment of the host organism. The other forms of symbiosis are mutualism, where both organisms benefit, and commensalism, where one organism benefits but the other is unaffected.

The main types of **endoparasite** organisms that affect humans are protozoa and **helminth** worms. Although they commonly infect the intestines, they also infect other sites, such as the blood, brain, eyes, liver, and kidneys. In this chapter, we will be concerned only with parasites that have life-cycle stages detectable in peripheral blood. Biomedical scientists play a crucial role in the detection and identification of blood-borne parasites, which is integral to their practice of assessing stained peripheral blood films for morphological abnormalities. Arthropods such as head lice and scabies mites are **ectoparasites**; the most likely haematological effect of infection with these parasites is eosinophilia.

symbiosis
Close interaction between different species.

endoparasite
A parasite that lives within the body of the host.

ectoparasite
A parasite that lives on the body of the host.

TABLE 7.1 Blood-borne human endoparasites.

Phylum	Species	Resultant disease
Protozoa	*Plasmodium falciparum* *Plasmodium vivax* *Plasmodium malariae* *Plasmodium ovale* *Plasmodium knowlesi*	Malaria
	Babesia bovis *Babesia microti* *Babesia divergens*	Babesiosis
	Trypanosoma brucei gambiense *Trypanosoma brucei rhodesiense*	African trypanosomiasis (sleeping sickness)
	Trypanosoma cruzi	American trypanosomiasis (Chagas disease)
	Leishmania donovani *Leishmania infantum*	Leishmaniasis (Kala-azar)
Nematoda	*Wuchereria bancrofti* *Brugia malayi* *Brugia timori*	Lymphatic filariasis
	Loa loa	Loa loa filariasis
	Mansonella perstans *Mansonella ozzardi*	Mostly asymptomatic

Cross reference

Eosinophilia in response to certain parasitic infections is discussed in further detail in Chapter 8.

Like many parasites, blood-borne parasites target specific organs and species for different stages of their life cycles. The nature of each life cycle and the target organs involved dictate the clinical symptoms of each infection. The main blood-borne parasitic organisms of humans and the resultant diseases are outlined in Table 7.1.

7.2 Malaria

Malaria is a serious and sometimes fatal vector-borne infectious disease, the vector being mosquitoes. Rarely, malaria infection occurs through transfusion of parasitized blood, transplantation of infected bone marrow, or by placental transmission (congenital malaria). It is widespread in tropical and subtropical areas, including areas of Africa, Asia, and the Americas. Between 300 and 500 million cases of malaria are diagnosed every year, of which between 1 and 3 million will die from the disease. The majority of deaths are of young children in Sub-Saharan Africa.

There are four main species of protozoa from the *Plasmodium* genus which cause malaria in humans: *P. falciparum*, *P. vivax*, *P. malariae*, and *P. ovale*. A fifth species, *P. knowlesi*, predominantly causes fatal malaria in long-tailed macaques (*Macaca fascicularis*) but is known to cause human malaria in most countries in South East Asia. Infection by *P. falciparum* accounts for approximately 80% of all cases. *P. falciparum* malaria is more severe than the other main forms and is the cause of 90% of deaths from this disease, although *P. knowlesi* infection can lead to illness as severe as that caused by *P. falciparum*.

Key Points

Only female mosquitoes of the *Anopheles* genus can act as vectors to transmit malarial parasites to humans. There are more than 450 known species of *Anopheles* mosquitoes, of which approximately 100 can transmit malaria to humans. *Anopheles gambiae* is the best known because of its significant role in the transmission of *P. falciparum*.

SELF-CHECK 7.1

Name the five parasites that cause malaria in humans.

7.2.1 Life cycle of malarial parasites

When taking a blood meal from a human, an infected *Anopheles* mosquito injects thousands of motile, spindle-shaped cells called **sporozoites** into the circulation. The life cycle is illustrated in Figure 7.1. The sporozoites are the stage of the parasite life cycle resulting from sexual reproduction in the midgut of the mosquito. The sporozoites infect the blood for a maximum of a few hours before they migrate to the liver. Those that are not destroyed by phagocytes infect the parenchymal cells of the liver (hepatocytes). The sporozoites divide and mature into **schizonts** by an asexual reproductive process of multiple fission called 'pre-erythrocytic **schizogony**'.

Over the next one to two weeks multiplication occurs inside each schizont such that each will contain thousands of the next phase of the life cycle, the **merozoite**, which consists of a single nucleus and a narrow ring of cytoplasm. Some of the sporozoites of *P. vivax* and *P. ovale* instead differentiate into hypnozoites, which are detailed in Box 7.1. The merozoites exit the hepatocytes contained within structures called **merosomes**, which consist of hundreds of merozoites encased in host membrane. Inside the merosome, the merozoites are protected from phagocytic attack. The merosomes travel to the lungs, where they lodge in pulmonary capillaries and the membranes disintegrate over the next four to six days. This liberates parasites into the bloodstream, where they begin the process of erythrocyte invasion by attaching to specific receptors on red cell membranes:

- *P. falciparum* binds to glycophorins A, B, and C;
- *P. vivax* and *P. knowlesi* bind to the Duffy antigen;
- the receptors for *P. malariae* and *P. ovale* are unknown.

Merozoites initially attach to red cells via any point on the merozoite surface, a process involving their surface-coat filaments. You can see in Figure 7.2 that the merozoite then reorientates to bring its apical pole into direct contact with the red cell membrane. Remarkably, the red cell membrane itself cooperates in this reorientation by partially 'wrapping round' the merozoite to facilitate the repositioning.

Once orientated, a closer membrane-to-membrane adhesion forms between the two cells, which is called a tight junction. At this point, a slight indentation in the red cell membrane appears, into which the apical cone of the merozoite is inserted. Unlike the initial membrane contacts, this adhesion is irreversible and the merozoite is now committed to entering this red cell.

A series of molecular events then occur in order for the merozoite to gain entry into the red cell. The adhesion zone around the tight junction moves to cover the entire merozoite surface,

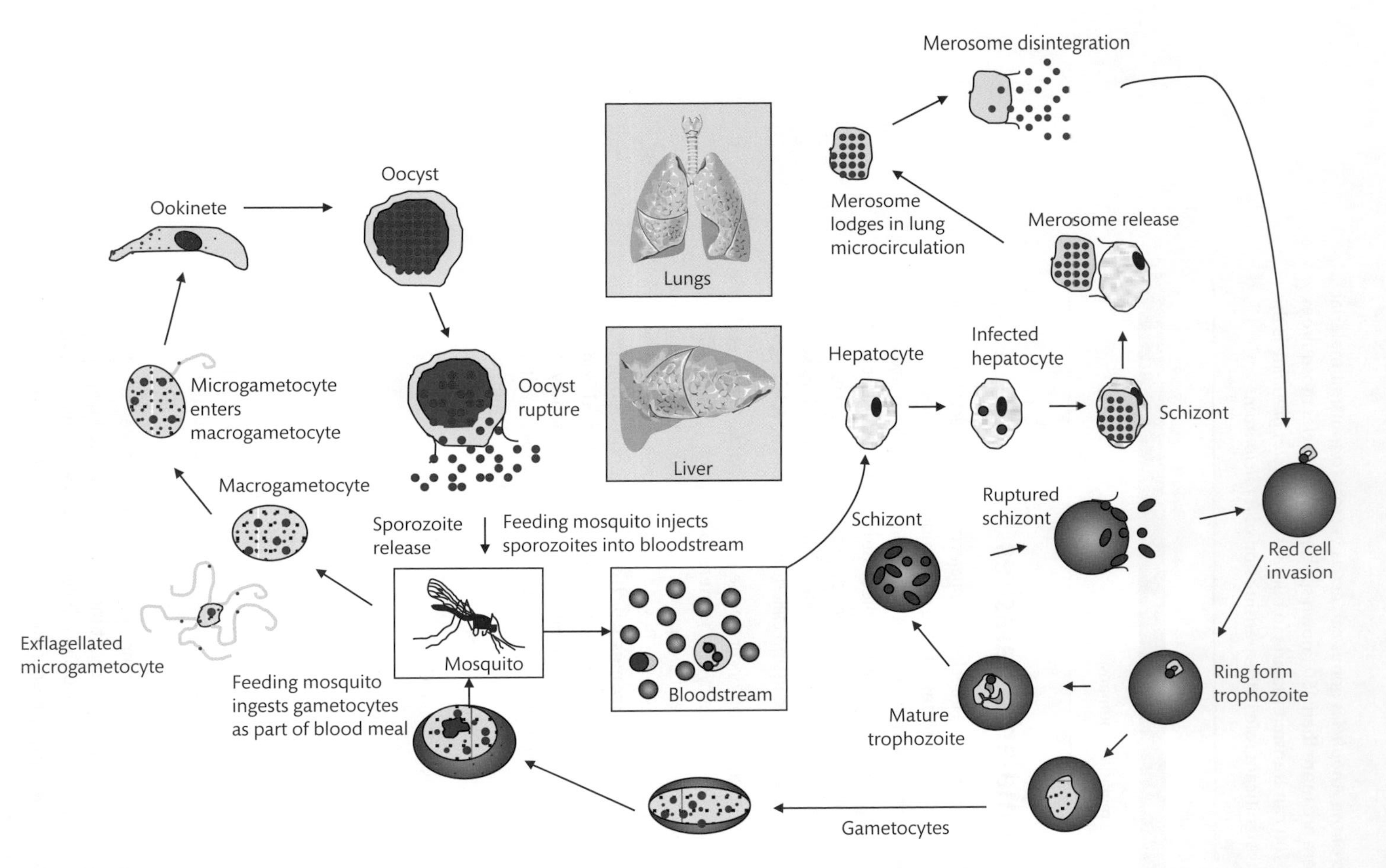

FIGURE 7.1

Life cycle of malarial parasites. An infected mosquito injects sporozoites into the bloodstream of the human host. The sporozoites migrate to the liver and invade hepatocytes, whereupon they enter shizogony and multiply to form merozoites. The merozoites exit the hepatocytes encased in a merosome, which travels to the lungs and lodges in the lung microcirculation. The merosome membrane disintegrates, liberating merozoites to invade host red blood cells and progress through the trophozoite stages and then enter schizogony. Rupture of the schizont releases merozoites, which infect further red cells. Some trophozoites transform into gametocytes, which are ingested by another mosquito when it feeds on the infected human host. The microgametocytes exflagellate and then fertilize the macrogametocytes to form the ookinete which embeds in the gut wall and develops into an oocyst. Multiple cell divisions occur in the oocyst to generate sporozoites, which are released from the oocyst to travel to the mosquito's salivary glands and are injected into another human host to perpetuate the life cycle.

BOX 7.1 Dormant malaria

In *P. vivax* and *P. ovale*, some sporozoites differentiate into **hypnozoites** as well as merozoites. Hypnozoites can remain dormant in hepatocytes for up to 30 years. Once 'reactivated' (by an, as yet, undetermined mechanism), they grow and undergo exo-erythrocytic schizogony to generate a wave of merozoites that invade the blood and produce a clinical relapse. Recent reports have suggested that *P. falciparum* and *P. malariae* may occasionally have a dormant stage, but *P. knowlesi* does not.

so facilitating its entry into the red cell. Merozoites contain an organelle called the **rhoptry**. The contents of the rhoptries are released to interact with inner membrane red cell lipids. Local alteration of the red cell membrane architecture follows, which produces an invagination to create the **parasitophorous vacuole membrane** (PVM). You can see in Figure 7.2 that the PVM then surrounds the merozoite to form the **parasitophorous vacuole** (PV), which remains attached to the red cell membrane via the junction at the posterior end of the PV. The red cell membrane and PVM close and the parasite is now locked inside the red cell. The junction then fuses with the PVM and severs the connection with the red cell membrane so that the parasite is free inside the red cell. (See also Box 7.2.)

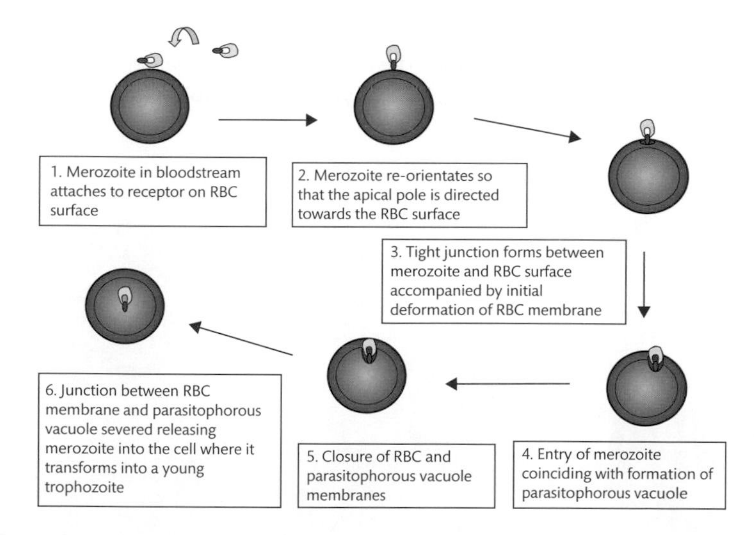

FIGURE 7.2

Red cell invasion by a malarial merozoite. Dense bodies inside the merozoite then fuse with the plasma membrane of the merozoite and release their contents into the PV, which triggers flattening of the merozoite and its transformation into the next stage of the life cycle, the trophozoite. The trophozoites are the feeding stage of the life cycle and ingest haemoglobin and other contents of the red cell cytoplasm. Plasmepsin enzymes degrade up to 80% of the haemoglobin in the red cell, but the parasite only uses about 15% of the amino acids derived from this digestion for protein synthesis. The parasites create new transport pathways to export the excess amino acids from the red cell to prevent rupture of the red cell and thus death of the parasites before they are sufficiently mature to enter the next stage of the life cycle. Digestion of the haemoglobin generates a characteristic brown pigment called haemozoin, which accumulates as the parasites mature.

haemozoin
Product of haemoglobin digestion by malarial parasites comprising polymerized insoluble haem residues.

BOX 7.2 Malaria species and red cell age

P. vivax **and** ***P. ovale*** **are commonly found in young or immature red cells.** ***P. malariae*** **tends to be found in ageing red cells, whilst** ***P. falciparum*** **and** ***P. knowlesi*** **are non-specific.**

On Romanowsky-stained blood films, young trophozoites appear inside red cells as characteristic ring forms with a chromatin dot, which you can see in Figure 7.3. They develop into large trophozoites whose cytoplasm has a more amorphous, amoeboid shape (Figure 7.4).

Most large trophozoites undergo multiplication by erythrocytic schizogony to form schizonts containing merozoites. When mature, the schizonts rupture and break open the red cell to release the merozoites, which then infect further red cells to perpetuate the erythrocytic cycle. The release of merozoites from red cells tends to be synchronized and is responsible for the cyclical nature of fevers associated with malarial infections. The period between fevers varies between species and is usually in the region of two to three days.

After several generations, some of the trophozoites develop into male **microgametocytes** or female **macrogametocytes**, which are typically oval or banana-shaped. You can see gametocytes in stained blood films in Figure 7.5. If the infected human is then bitten again by a mosquito vector, it takes up gametocytes in its blood meal which travel to the mosquito's stomach to undergo the sporogenic or sexual phase of the life cycle. Within 20 minutes of arriving in the stomach, the alteration in pH stimulates the microgametocytes to extend up to eight slender flagella in a process called **exflagellation**. The exflagellated microgametocytes then fertilize the macrogametocytes to form elongated motile **zygotes** called **ookinetes**. (See also Box 7.3.)

zygote
Derived from the Greek for 'joined', it is the unicellular product of joining male and female genetic material—the product of fertilization.

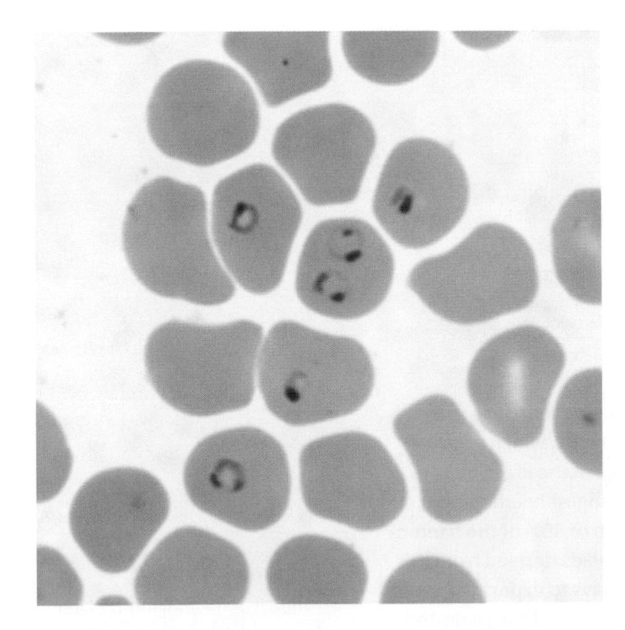

FIGURE 7.3
Ring-form trophozoites of *P. falciparum* on a Giemsa-stained thin film. (Magnification ×500.)

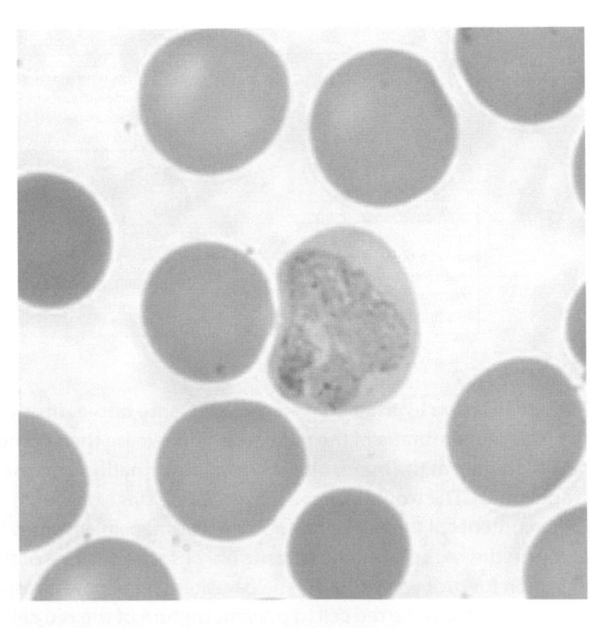

FIGURE 7.4
Large trophozoite of *P. vivax* on a Giemsa-stained thin film. (Magnification ×500.)

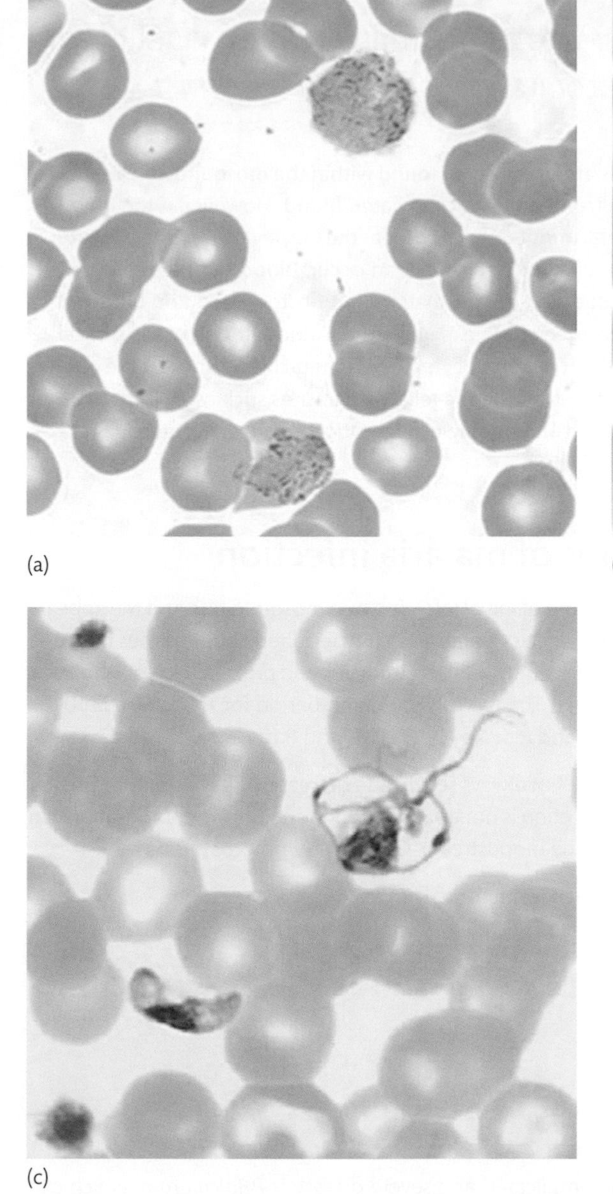
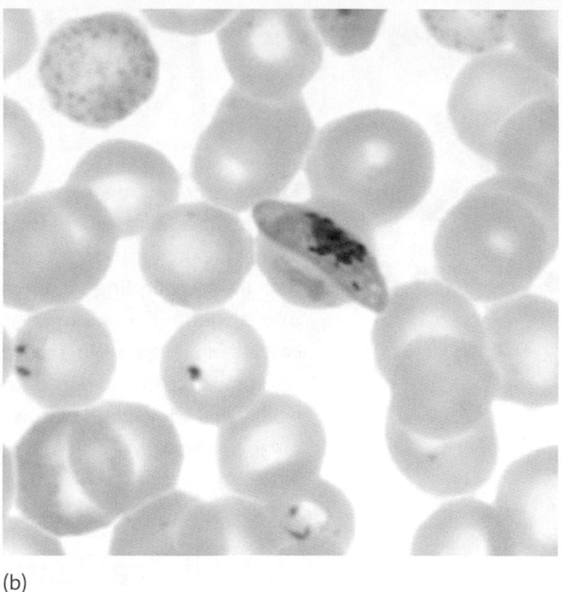

(a) (b) (c)

FIGURE 7.5

Gametocytes of *P. vivax* and *P. falciparum* on Giemsa-stained thin films. (Magnification ×500.) (a) Gametocytes of *P. vivax*; (b) gametocyte of *P. falciparum*; (c) crescent-shaped gametocyte of *P. falciparum* and another undergoing exflagellation.

The ookinetes that survive the immune response penetrate and escape the midgut and then embed themselves in the gut wall close to the exterior. A few even migrate onto the surface. Once embedded, they develop into **oocysts**, which are the spore phase of the life cycle. Multiple cell divisions occur in the oocysts to generate large numbers of small, elongated sporozoites. Oocyst rupture releases the sporozoites, which migrate to the mosquito's salivary glands, where they are injected into the human host at the mosquito's next meal to perpetuate the life cycle.

SELF-CHECK 7.2

Outline the malarial parasite life cycle.

BOX 7.3 Artefactual finding of exflagellated microgametocytes in blood

Exflagellated microgametocytes are usually only found within the mosquito and are not seen in blood films made from fresh, non-anticoagulated blood. However, if the blood sample is exposed to air for several minutes the pH rises as the CO_2 level falls to equilibrate with the surrounding air and *in vitro* exflagellation can occur. Blood samples collected into ethylene diamine tetra-acetic acid (EDTA) that are left unstoppered and unmixed can attain a high enough pH to precipitate exflagellation. Although it is of no clinical significance to find exflagellated microgametocytes in peripheral blood, there is the possibility that they could be mistaken for other flagellated parasites such as *Trypanosoma cruzi*. A gametocyte of *P. falciparum* undergoing exflagellation is shown in Figure 7.5(c).

7.2.2 Clinical features of malaria infection

parasitaemia
Quantitation of the number of parasitized red blood cells (normally expressed as a percentage).

The incubation period between the time of initial infection by a mosquito bite and the appearance of clinical symptoms and appearance of parasites in the peripheral blood varies between species. This is due to differences in the length of the erythrocytic and pre- and exo-erythrocytic cycles and the degree of **parasitaemia**. The typical incubation period for *P. knowlesi* is 10–12 days, *P. falciparum* 7–14 days, *P. vivax* and *P. ovale* 12–17 days, and *P. malariae* 18–40 days.

The symptoms of malaria infection are not always severe and dramatic and are easily dismissed as relatively trivial. However, if the infection is not treated, sudden and drastic deterioration can occur as the parasites rapidly invade the bloodstream. Common symptoms are:

- influenza-like symptoms, including fever and sweating;
- headache;
- weakness;
- dizziness;
- nausea and vomiting;
- anaemia.

The main species associated with complicated and severe disease is *P. falciparum*, which can induce additional symptoms, such as:

- neurological signs—delirium, coma, convulsions, focal signs;
- intense muscle spasms;
- oliguria (low urine output), renal failure;
- pulmonary oedema (fluid on the lungs), laboured breathing, cough;
- jaundice, liver failure;
- hypoglycaemia;
- diarrhoea;
- circulatory collapse.

BOX 7.4 Worldwide malaria monitoring

The World Health Organization reports on the current malaria status of countries throughout the world and on the sensitivity of malarial parasites to particular anti-malarial drugs. Travel clinics and doctors can then use these data to provide the most appropriate prophylaxis for travellers.

BOX 7.5 Malaria as a medicine?!

Before the widespread availability of penicillin in the 1940s, there was no drug available to cure syphilis. Every patient with the disease died once it entered the terminal stage affecting the brain. The Austrian psychiatrist Julius Wagner von Jauregg received a Nobel prize in 1927 for his work on curing syphilis with pyrotherapy. He discovered that patients whose disease had progressed to neurosyphilis could be cured by being infected with a mild strain of malaria. The prolonged high fevers and elevation of body temperature associated with malaria infection were able to kill the causative spirochaete bacterium of syphilis, *Treponema pallidum*. Patients underwent about ten bouts of fever before their malaria was subsequently treated with quinine. The results were remarkable, with many patients being completely cured of the physical and mental effects of their illness and able to return to normal life. Understandably, the advent of widespread penicillin availability made 'treatment' with malaria obsolete.

The first-line prophylaxis and treatment for malaria for many years has been chloroquine, and prior to that, quinine. Parasites are increasingly developing resistance to chloroquine and newer drugs are being used, such as proguanil, atovaquone, mefloquine, and doxycycline. No antimalarial drug is 100% effective and it is not uncommon to prescribe a combination of drugs. Different drugs may be prescribed for prophylaxis, depending on the area someone is travelling to; for instance, if the area is known to have chloroquine-resistant strains of parasite, mefloquine may be prescribed instead. A drug called primaquine can be effective against hypnozoites (see also Box 7.4). Box 7.5 gives a fascinating insight into how clinical aspects of malaria infection were once used to combat another disease.

7.2.3 Haematological and other features of malaria

The anaemia of malaria

The periodic release of merozoites from schizonts within infected red cells together with the accompanying red cell rupture causes a haemolytic anaemia, which can be life-threatening. In fact, malaria is the most common cause of haemolytic anaemia in humans. However, the degree of anaemia cannot be explained by this alone and malarial anaemia has multiple aetiologies.

The **reticuloendothelial system** removes parasitized red cells from the circulation by extravascular phagocytosis. The process of *pitting* in the spleen where red cell parasites (and other red cell inclusions such as Heinz bodies) are removed results in loss of red cell membrane in the area where the inclusion is removed. Although this does not necessarily result in the immediate

reticuloendothelial system
Contains phagocytic cells located in reticular connective tissue, lymph nodes, and spleen that are able to ingest bacteria, immune complexes, and other foreign bodies.

destruction of the red cell, it does reduce its lifespan. Parasitized red cells have reduced deformability that can lead to enhanced clearance by the spleen. Increased reticuloendothelial and splenic function can lead to enhanced clearance of non-parasitized red cells. It is a normal function of the spleen to remove senescent, or ageing, red cells from the circulation and the **splenomegaly** arising from malaria infection exaggerates this normal function.

splenomegaly
Enlarged spleen.

dyserythropoiesis
Abnormal red cell development.

pro-inflammatory cytokines
Regulatory cellular-communication signalling molecules that favour and promote inflammation.

A low reticulocyte response to severe haemolytic anaemia is often a remarkable finding in the severe malarial disease of *P. falciparum* infection. This inadequate response to anaemia arises from **dyserythropoiesis**, which occurs due to the presence of **pro-inflammatory cytokines**, such as tumour necrosis factor-α (TNF-α) and interleukin-12 (IL-12), and suppression of erythropoiesis by haemozoin. Dyserythropoiesis can be seen in infections with the other *Plasmodium* species and can persist for some weeks after the acute phase of the infection.

Malaria occurs in many areas where chronic nutritional anaemias and haemoglobinopathies are endemic, so it often exacerbates a coexisting anaemia. The anaemia of malaria itself is normocytic and normochromic, but the classic morphological appearances of a nutritional anaemia or haemoglobinopathy do not exclude the possibility of coexistent malaria infection. The anaemia can worsen if patients infected with *P. falciparum* develop Blackwater fever, which is detailed in Box 7.6. Box 7.7 describes how you can contract malaria without even visiting a malarial country.

SELF-CHECK 7.3

What are the causes of the anaemia of malarial infection?

BOX 7.6 Blackwater fever

Patients with *P. falciparum* malaria who have previously been infected with that species can present with sudden intravascular haemolysis followed by fever and haemoglobinuria, a syndrome termed Blackwater fever. It is commonly seen in non-immune individuals who have been in malarious countries for less than a year and have had inadequate doses of quinine for prophylaxis and/or treatment. Quinine itself can be a precipitating factor and Blackwater fever almost disappeared after 1950, when quinine was replaced by chloroquine as the treatment of choice. The syndrome reappeared in 1990, when quinine was reintroduced due to chloroquine resistance. More recently, cases have been described associated with newer drugs similar to quinine, such as halofantrine and mefloquine. Other factors that may precipitate an attack of Blackwater fever are cold, sun exposure, trauma, fatigue, pregnancy, and X-ray treatment of the spleen.

It is thought that the haemolysis occurring as a result of schizogony stimulates the formation of haemolysin antibodies that destroy red cells. The patients become hypersensitized as a result of repeated *P. falciparum* infection such that a subsequent heavy infection and administration of quinine, together with other precipitating factors, generates a sudden and massive output of haemolysin antibodies which cause severe, acute intravascular haemolysis. Patients present with pallor due to rapidly developing anaemia, nausea, jaundice, fever, renal failure due to tubule necrosis and haemoglobin deposition in the tubules, and urine that is black or dark red in colour from the excretion of free haemoglobin. Individuals with glucose-6-phosphate dehydrogenase (G6PD) deficiency are more likely to develop Blackwater fever as a complication of malaria infection.

Other haematological changes

The white cell count can be raised in severe malarial infection, but it is often within the reference range. Some patients develop a leucoerythroblastic picture, whilst others may have neutrophilia, atypical lymphocyte morphology, monocytosis, eosinopenia, or a reactive eosinophilia during the recovery phase.

Mild thrombocytopenia is common in malaria infection, resulting from a combination of platelet activation, increased pooling/clearance in the spleen, and reduced platelet lifespan due to immune responses. The platelet count can drop to around 100×10^9/L and sometimes below 50×10^9/L in severe disease. In addition, blood coagulation can be activated in malaria infection, particularly with *P. falciparum*, as a result of the following:

- the cytokine storm produced in response to the infection, which can activate blood-clotting mechanisms;
- alterations to the red cell membrane so that coagulation-promoting phospholipids are expressed on the surface;
- adherence of parasitized red cells to deep tissue capillary endothelium, which damages the endothelial cells and leads to activation of blood coagulation.

Reduced synthesis and increased activation of blood coagulation factors can lead to mild elevations of the prothrombin time and activated partial thromboplastin time. Fibrinogen concentration can be reduced, normal, or elevated. Factor XIII, which is responsible for stabilizing fibrin clots, and antithrombin, a regulator of blood coagulation, can both be reduced in malaria due to their consumption.

Cerebral malaria

Only *P. falciparum* invades the central nervous system to cause **cerebral malaria**. It is an acute, widespread disease of the brain accompanied by fever and can be fatal within 24–72 hours if not treated. It is caused by adherence of both parasitized and non-parasitized red cells to the cerebral microvasculature, leading to blockage of cerebral blood flow. It is also hypothesized that a malarial toxin is released that stimulates macrophages to release cytokines which induce uncontrolled production of nitric oxide in the brain. The nitric oxide diffuses through the blood–brain barrier and affects nerve cell function in a similar way to anaesthetics and alcohol, resulting in a state of reduced consciousness. Children are more vulnerable to cerebral malaria than adults.

Cerebral malaria has three key symptoms:

1. impaired consciousness with non-specific fever;
2. generalized convulsions and neurological abnormalities;

BOX 7.7 Airport malaria

Mosquitoes have been known to enter aeroplanes and be transported to parts of the world where malaria does not occur naturally, and then infect someone at the airport who did not actually travel to a malarial region. In favourable climatic conditions, the mosquito may travel beyond the airport and infect people further afield. Mosquitoes can survive in baggage and infect someone when released from the baggage, not necessarily the person who had travelled.

3. coma that persists for 24–72 hours, in which the patient is initially rousable and then unrousable.

Immediate treatment is necessary and includes chemotherapy with quinine to interfere with the parasite's digestion, antipyretics to reduce fever, and anticonvulsants.

Knob formation

Knobs are conical protrusions of the membranes of red cells infected by *P. falciparum*, *P. ovale*, and *P. malariae* and are not visible by light microscopy. Most of the research has been done on *P. falciparum*-infected cells. The knobs are involved in mediating the cytoadherence of *P. falciparum*-infected red cells to the vascular endothelium and the rosetting of non-infected cells around the adhered infected cells. The knobs form junctions between red cell and endothelial cell membranes. Cytoadherence may be a mechanism for *P. falciparum* to avoid destruction in the spleen.

A number of *P. falciparum* proteins occur on the red cell surface or associated with the red cell cytoskeleton in relation to knob formation. Table 7.2 lists the proteins and their functions.

7.2.4 Genetic protection from malaria

Genetic defects of haemoglobin and red cells are common in all parts of the world where malaria is prevalent, mainly because carriers are afforded a degree of protection from *P. falciparum* infection. These defects include: haemoglobinopathies such as haemoglobins S, C, E, and thalassaemias; red cell membrane defects such as South East Asian ovalocytosis (SEAO) and some forms of elliptocytosis; and metabolic abnormalities such as G6PD deficiency. It is

TABLE 7.2 ***P. falciparum*** **knob-associated proteins.**

Protein	Abbreviation	Function
Histidine-rich protein 1*	HRP-1	Critical for structural formation of the knob
Histidine-rich protein 2	HRP-2	Released into plasma to suppress lymphocyte function
P. falciparum erythrocyte membrane protein 1	PfEMP 1	Expressed on knob surface; cytoplasmic region interacts with HRP-1
P. falciparum erythrocyte membrane protein 2**	PfEMP 2	Anchor for PfEMP 1
P. falciparum erythrocyte membrane protein 3	PfEMP 3	Probably involved in knob formation via interaction with red cell cytoskeleton
Ring-infected erythrocyte membrane surface antigen	RASA	Cytoskeleton binding
Sequestrin	–	Recognition protein for CD36 on endothelial membrane
Rosettin	–	Binds endothelial receptors and ABO antigens

* HRP-1 is also known as knob-associated histidine-rich protein (KAHRP).

** PfEMP 2 is also known as mature erythrocyte surface antigen (MESA).

thought that natural selection has been responsible for elevating and maintaining their gene frequencies where malaria is endemic, and is probably why haemoglobinopathies are the most common single-gene disorders in the human population.

Sickle cell trait is the commonly quoted example of genetic protection from malaria. A variety of mechanisms have been proposed to explain the protective effect against *P. falciparum* infection, such as attenuating the infection by reducing intracellular oxygen tension, or targeting parasite-infected cells for splenic clearance due to reduced deformability of cells containing haemoglobin S and parasites. It is clear though that HbAS-containing cells are just as likely to be infected as cells containing HbAA. It has been proposed that red cells of homozygous haemoglobin C individuals resist lysis at the merozoite release stage.

More recently, atypical display of PfEMP1 has been reported on cells in patients heterozygous and homozygous for haemoglobin C and with sickle cell trait, leading to reduced cytoadhesion and rosetting, and thus protection from cerebral malaria. Cytoadhesion can occur in other organs. The existence of variant malarial surface antigens may also precipitate an enhanced immune response and affect the early development of naturally acquired immunity.

Red cells of patients with thalassaemia and G6PD deficiency are highly sensitive to oxidant stress and provide a poor environment for *P. falciparum* parasites. Modified surface-antigen expression may also operate in some thalassaemias.

SEAO is a hereditary disorder of red cell structure, due to a deletion in the Band 3 gene, that is widespread in parts of South East Asia. The ovalocytic red cells are rigid and resistant to invasion by various malarial parasites.

Absent or mutant Duffy antigens on red cell surfaces provide protection against *P. vivax* because it binds the Duffy antigen at the red cell invasion stage of the life cycle via its Duffy binding-like erythrocyte-binding protein (DBL-EBP). The same is not generally considered to be true for *P. falciparum* because it can use more than one red cell antigen as a receptor and the parasite has four DBL-EBPs. However, about 10% of Melanesians living in the northern provinces of Papua New Guinea are negative for the Gerbich antigen located on glycophorin C, and there is evidence to suggest that it may provide some degree of protection against *P. falciparum* infection.

Cross references

You can find explanations of white cell abnormalities in Chapter 8.

You were introduced to blood coagulation in Chapter 2 and can find more details about its activation in Chapter 13.

7.2.5 Malaria prevention

Anopheles mosquitoes feed at night and only one bite is needed to infect the human host. Some 90% of malaria cases occur in Sub-Saharan Africa, with young children at greatest risk because they have yet to develop natural immunity, which can take up to ten years.

Education is key to engendering the appropriate lifelong behaviours to reduce infection. Sleeping under nets suspended above a bed can be an effective barrier against the mosquitoes reaching humans to feed on, although the nets need to be maintained and used properly to be effective. More recently, nets impregnated with pyrethroid insecticides have become available. The insecticide is either incorporated within or bound around the net fibres. As well as providing a physical barrier, insecticide-treated nets (ITN) deter mosquitoes from feeding and even drive them from their indoor resting places. A treated net with large holes can provide protection as effective as an intact, properly used untreated net. However, the insecticide does not last indefinitely and the nets need to be re-dipped in insecticide every six months. Long-lasting insecticidal nets (LLIN) are now available that remain effective for about three years, but they are inevitably more expensive, albeit more cost-effective in the long term. Unfortunately, neither nets nor insecticides are sufficiently widely available or affordable, and fewer than half the people in Africa who need them have access.

Another strategy to prevent being bitten by mosquitoes in the home is indoor residual spraying (IRS), which involves covering the internal walls with an insecticide that kills and repels mosquitoes. Whilst nets and IRS can be very effective, it is important to remember that some *Anopheles* mosquitoes only bite outdoors. Even those that will bite indoors may enter the house, feed, and leave without resting on any of the indoor walls. The insecticide dichlorodiphenyltrichloroethane (DDT) has been used as an outdoor residual spraying strategy, but it affects other wildlife and DDT resistance is now common in mosquitoes. (See also Box 7.8.)

A different approach to reducing malaria has been to target the aquatic but air-breathing larval stage of the mosquito's life cycle. In 1897, Major Ronald Ross discovered that *Anopheles* mosquitoes were the vectors for malaria. He initiated attempts to reduce mosquito numbers by coating the surfaces of ponds and marshes with oil to prevent the larvae from taking in air through their breathing tubes. Box 7.9 describes how the word 'malaria' was derived from the larvae's living conditions.

chitin
The principal component of arthropod exoskeletons.

Water has been treated with chemicals that can prevent larvae from metamorphosing into adults, impair **chitin** synthesis, or disrupt nerve function. Not surprisingly, there are problems with toxicity to other wildlife and humans. Some habitats lend themselves to the reduction of larvae numbers by introducing predatory fish into the water, such as the common carp (*Cyprinus carpio*) in rice fields and the killifish (*Aphanius dispar*) into man-made containers. Introduction of *Bacillus thuringiensis israelensis* and *Bacillus sphaericus* bacteria into the water kill the larvae because toxins on the bacterial spore coat poison the larvae's stomach. Predatory fish and pathological bacteria can reduce larvae population density by 98–100%, but the fish are only effective for between two weeks and one year, and the bacteria for a maximum of ten weeks.

Expanded polystyrene beads have been used in man-made reservoirs such as water tanks and wells. They form a floating layer on the water surface that blocks oviposition (egg laying). However, the beads can blow away if exposed to wind, and it has been known for local people to collect the beads to manufacture into jewellery.

Environmental modification can contribute to a reduction in malaria infection. Strategies include siting housing away from breeding sites, raising houses on poles because the mosquitoes tend to fly low, covering windows and doors with screening and being vigilant in repairing cracks in walls, restricting outdoor human activity to between sunrise and sunset, and diverting vectors to other mammalian hosts (zooprophylaxis).

An important dimension to environmental manipulation is water management. Strategies include flushing streams and canals because larvae prefer still water, intermittent irrigation in agricultural fields, flooding or temporarily de-watering man-made or natural wetlands, or altering salinity. Planting trees can contribute to draining marshy land.

BOX 7.8 Chrysanthemums and malaria prevention

Pyrethroids are a group of man-made insecticides similar to the natural pesticide pyrethrum, which is derived from the dried flowers of *Chrysanthemum cinerariaefolium*. So-called pyrethrum daisies have been used as an insecticide for over 2000 years; the generic name for the six active compounds derived from them is pyrethrin. Since it is naturally produced, it decomposes rapidly in sunlight and is considered one of the most environmentally safe insecticides, with very low mammalian toxicity.

BOX 7.9 What does 'malaria' mean?

The word 'malaria' is derived from the Italian words *mala* and *aria* meaning bad air. Before the discovery that malaria is caused by a parasitic infection, it was believed that breathing in the foul-smelling air from swamps and latrines was responsible for the disease. It was a mere coincidence that the stagnant water that provided a breeding ground for mosquitoes also frequently contributed to bad air.

Pharmaceutical malaria prevention

You met examples of the main drugs for prophylaxis and treatment earlier in this chapter. The development of drug-resistant parasites and insecticide-resistant mosquitoes is causing an increased burden on health services and even the economic stability of the countries that are worst affected. Alternatives to current treatments and prevention strategies are badly needed. An area that attracts considerable attention is that of a vaccine for malaria, which would be of particular value in individuals who have yet to develop natural immunity, such as young children, travellers, pregnant women, and people who live in endemic areas but are not now regularly exposed to the infections.

Development of vaccines to malaria has proven difficult to date because the parasites are very good at evading the immune response. The parasites enter human cells very quickly and are effectively hidden from the immune response for most of their time inside the human host. As you have seen, they express certain antigens which are recognizable by the immune system, and thus vaccines, but they have the ability to vary their antigen expression. The many different life-cycle stages mean that a vaccine to just one stage will be ineffective. Because it is difficult to make a vaccine that targets the entire organism, most attempts at vaccine production have been directed at subunits.

It has proven difficult to create a vaccine that will precipitate an effective immune response. Some vaccines have been shown to generate antibody formation, but the subsequent processes of the immune response were not initiated. Natural immunity builds up through early life as individuals encounter and develop resistance to different strains of parasite, so people in endemic areas are not effectively immune until adulthood. Therefore it is clearly possible to be immune to malaria, but developing a single vaccine remains elusive. Interestingly, natural immunity doesn't completely destroy the parasites as they can still grow in people who are clinically immune.

7.2.6 Laboratory detection of malaria

Detection of malarial parasites in peripheral blood and species identification is achieved primarily by light microscopy of Romanowsky-stained blood films. This approach has been the mainstay of malaria detection for decades and remains the 'gold standard'. Supplementary assays are available to affirm microscopical findings or for use as initial screening tests.

Light microscopy

Thick and thin blood films should be prepared, stained, and examined for all cases being investigated for malaria. Thin films are conventional wedge films. Thick films are prepared on a separate slide by spreading a drop of blood in a circular motion to cover an area with a diameter of about 1 cm. The films should be made with minimal delay because morphological alteration of parasites occurs in stored EDTA-anticoagulated blood.

Thin films can be stained with a conventional Romanowsky stain such as Leishman's or May Grünwald/Giemsa, although the pH of 6.8 does not permit sufficiently intense staining of some important structures that aid species identification, such as **Schüffner's dots**. Therefore, use of pH 7.2 stains is recommended. Some laboratories stain thin films with their usual preparation as it is usually adequate to demonstrate malarial parasites, and then follow up with Leishman's or Giemsa staining of a separate slide at pH 7.2 if necessary.

Schüffner's dots are multiple, small, brick-red dots inside red cells infected with *P. vivax* or *P. ovale*. The dots are composed of invaginations of the red cell membrane called caveolae complexed to vesicles, forming structures called caveola–vesicle complexes. The dots may not be present in red cells containing the smaller, young ring-form trophozoites. The dots can be darker in *P. ovale* infection and are referred to as **James's dots**.

Thick films are not fixed so that when stained with **Field's stain**, the haemoglobin elutes from the red cells. It is crucial that the film is perfectly dry, otherwise unfixed cellular material will flake off during staining. Giemsa stain can also be used for thick films. Thick and thin films are shown in Figure 7.6 and microscopical appearances of a *P. falciparum* infection on a thick film are shown in Figure 7.7.

Microscopical examination of the thick film is used as the first screening tool because the larger volume of blood used increases the likelihood of finding the parasites, especially in scanty infections. Parasite density can be as low as 1 infected cell per 100,000 red cells and would take in the region of an hour to detect on a thin film, whereas a five-minute examination of a thick film would examine the equivalent amount of material. The limitation of thick films is that the red cell lysis distorts parasite morphology and so thin films are needed to determine the species and assess parasitaemia.

It is crucial to differentiate the species because *P. falciparum* infection can lead to complications and be fatal. Species are identified based mainly on trophozoite appearances, merozoite numbers in schizonts, gametocyte morphology, size and shape of infected red cells, and the presence of Schüffner's dots. Fortunately, a number of morphological features are either specific for *P. falciparum* infection or very rare in other infections. These features are:

- smaller ring forms than in other species;
- double chromatin dots, which can be present in ring forms;

FIGURE 7.6
Stained thick film (top) and thin wedge film (bottom). Royal Perth Hospital, Government of Western Australia. CDC/ Steven Glenn, Laboratory & Consultation Division.

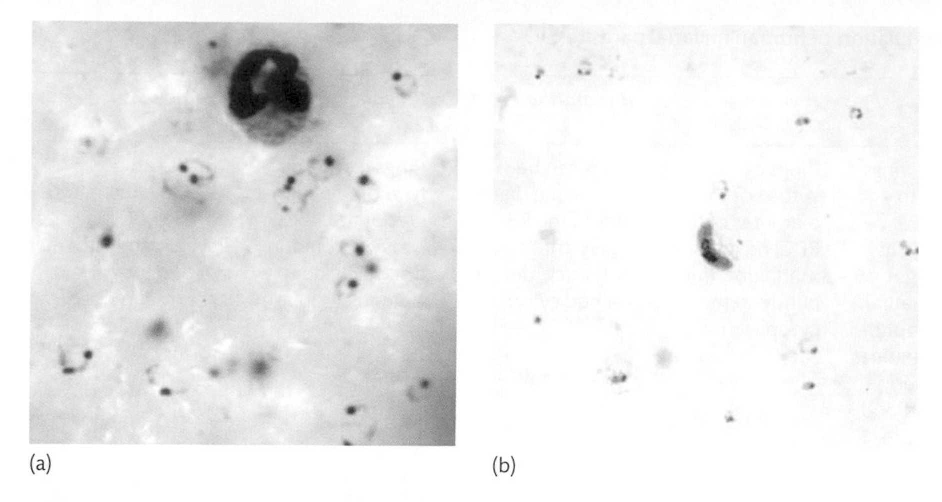

(a) (b)

FIGURE 7.7

Field's-stained thick films. (a) High-power field of *P. falciparum* infection showing numerous ring forms and a neutrophil; (b) low-power field of *P. falciparum* infection showing ring forms and a characteristic crescent-shaped gametocyte. Note that red cells are not visible. Division of Parasitic Diseases and Malaria, Centers for Disease Control and Prevention, USA.

- multiply infected cells, which are more common;
- the absence of Schüffner's dots;
- the presence of Maurer's clefts/dots;
- gametocytes that may be crescent-shaped.

Table 7.3 describes the important morphological features of the malarial parasites of humans. It is important to bear in mind that mixed infections do occur and clear identification of the presence of one species does not preclude the presence of another. The most commonly encountered stages are the ring-form trophozoites, which appear as a ring of blue cytoplasm with red to purple chromatin. The larger trophozoites often contain malaria pigment. Figures 7.8, 7.9, 7.10, 7.11, and 7.12 show examples of microscopic appearance of life-cycle stages on thin films of the malarial parasites of humans.

Maurer's clefts appear as irregular red/mauve dots inside red cells infected with *P. falciparum*. Figure 7.13 demonstrates that Maurer's clefts are larger and fewer in number than Schüffner's dots. They are newly constructed clefts in the red cell cytoplasm that are continuous with the PVM. Maurer's clefts are involved in protein/antigen sorting and trafficking.

SELF-CHECK 7.4

What are the main morphological features that allow differentiation between *Plasmodium* species on Romanowsky-stained blood films?

Estimation of parasitaemia

The percentage parasitaemia must be estimated when *P. falciparum* or *P. knowlesi* infection is detected because the severity of the infection can affect treatment decision-making. Quantification is performed by counting the number of parasitized cells examined from

TABLE 7.3 Morphological differentiation of human malarial parasites.

Life cycle stage/ other morphology	*P. falciparum*	*P. vivax*	*P. malariae*	*P. ovale*	*P. knowlesi*
Ring-form trophozoite	Delicate, small rings with scanty cytoplasm; can have double chromatin dots; older rings may be stippled (Maurer's clefts/dots); trophozoites are sometimes found at the periphery of the red blood cell (RBC) and are referred to as accolé forms	Rings can be ⅓ to ½ of the diameter of the RBC; heavy chromatin dot; thin, faintly stained cytoplasm	Rings can be ⅓ to ½ of the diameter of the RBC; heavy chromatin dot; thick, deeply stained cytoplasm	Rings can be up to ½ of the diameter of the RBC; heavy chromatin dot; thick, deeply stained cytoplasm	Rings can be up to ½ of the diameter of the RBC; delicate cytoplasm with one or two prominent chromatin dots; occasional accolé forms
Large trophozoite	Not normally seen	Amoeboid vacuolated	Compact or band shaped scattered dark granules	Band forms rare; scanty, dark brown granules	Compact cytoplasm and occasional band forms coarse, dark-brown pigment
Mature schizont	Occasionally seen in peripheral blood; 16–30 merozoites; numerous chromatin masses	Parasite can fill entire RBC; 12–24 merozoites in a rosette; fine, central pigment; numerous chromatin masses	Parasite can fill entire RBC; 6–12 merozoites in a rosette; central, coarse clump of pigment; few chromatin masses	8–12 merozoites in a rosette; few chromatin masses	Parasite can fill entire RBC; up to 16 merozoites with large nuclei, clustered around mass of coarse, dark-brown pigment; occasional rosettes; mature merozoites appear segmented
Gametocytes	Crescent or sausage-shaped; central nucleus surrounded by darkly pigmented granules	Round or oval-shaped; fills entire RBC; evenly distributed pigment	Round or oval-shaped; scattered pigment	Round or oval shaped; smaller than *P. vivax*	Round or oval-shaped; may almost fill RBC; diffuse or scattered pigment
Stages present in peripheral blood	Ring-form trophozoites; gametocytes; occasionally schizonts	All stages	All stages	All stages	All stages
Schüffner's dots	Not present	Can be seen in all stages except early ring-form trophozoite	Not present	Can be seen in all stages except early ring-form trophozoite	Light stippling, similar to Schüffner's dots
Multiply infected RBC	Common	Occasional	Rare	Rare	Fairly common
RBC size and shape	Normal	Up to twice as large as normal; normal or oval-shaped	Normal or smaller size; normal shape	RBC usually normal size or slightly enlarged; RBC frequently oval; fimbriated (ragged) edges often seen	Normal or smaller size; normal shape

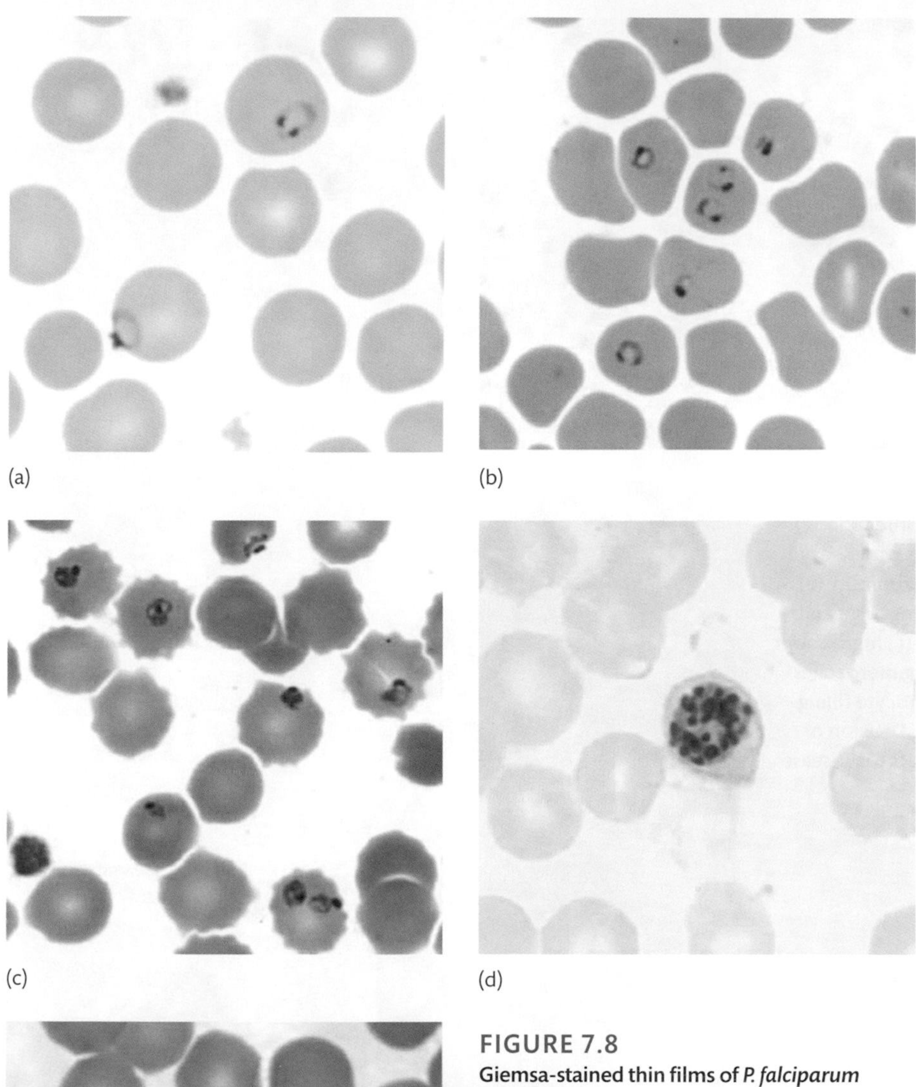

FIGURE 7.8

Giemsa-stained thin films of *P. falciparum* infection (high power). (a) Characteristic thin, delicate rings (note the double chromatin dot–headphone appearance–in the infected red cell at the top and the accolé form in the infected red cell at the bottom); (b) note the multiply infected red cell; (c) *P. falciparum* trophozoites are rarely seen in peripheral blood–older, ring forms are referred to as the trophozoites, which tend to have more dense cytoplasm than in younger rings; (d) mature schizont containing merozoites; (e) crescent-shaped gametocyte. Division of Parasitic Diseases and Malaria, Centers for Disease Control and Prevention, USA.

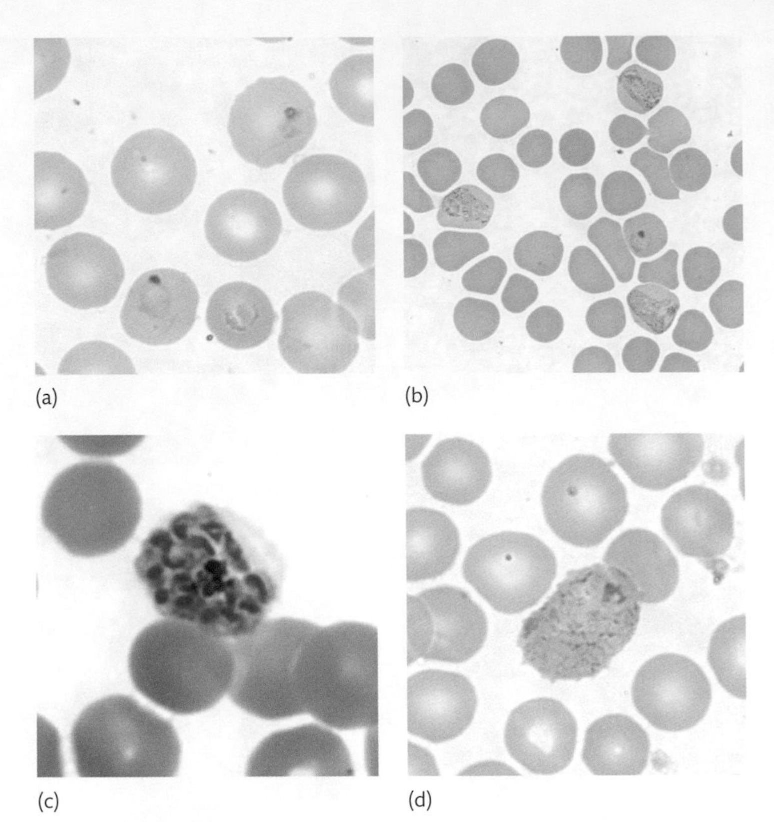

FIGURE 7.9
Giemsa-stained thin films of *P. vivax* infection. (a) High-power field showing large rings with heavy chromatin dots but faintly stained cytoplasm; (b) medium-power field showing one ring form and three large, amoeboid trophozoites; (c) high-power field with mature schizont containing merozoites virtually filling the red cell; (d) gametocyte filling the entire red cell, which is enlarged. Division of Parasitic Diseases and Malaria, Centers for Disease Control and Prevention, USA.

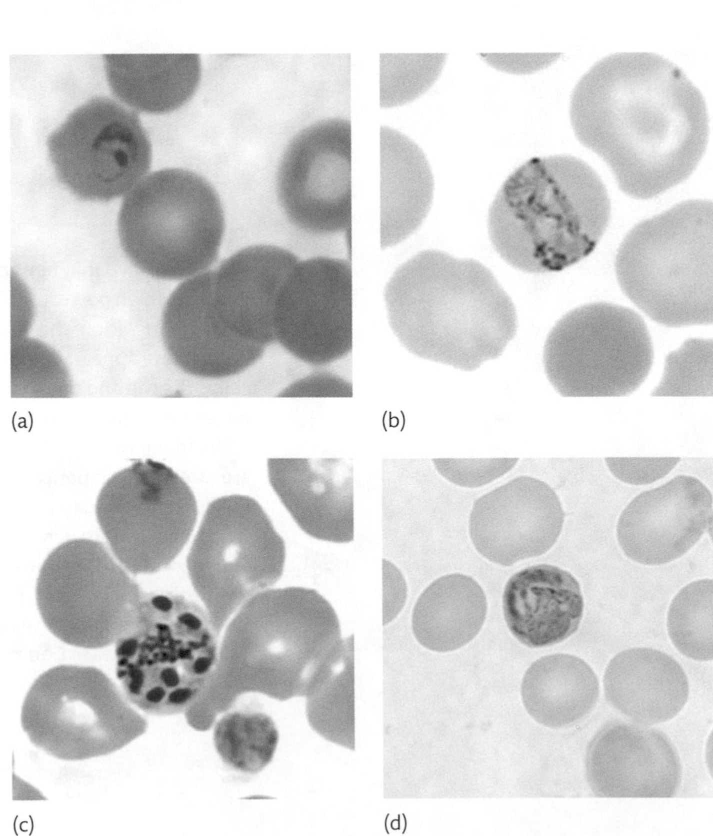

FIGURE 7.10
Giemsa-stained thin films of *P. malariae* infection (high power). (a) Ring form with heavy chromatin dot and thick, deep-staining cytoplasm; (b) band form trophozoite; (c) mature schizont with fewer merozoites than *P. falciparum* or *P. vivax*, central clump of pigment; (d) gametocyte. Division of Parasitic Diseases and Malaria, Centers for Disease Control and Prevention, USA.

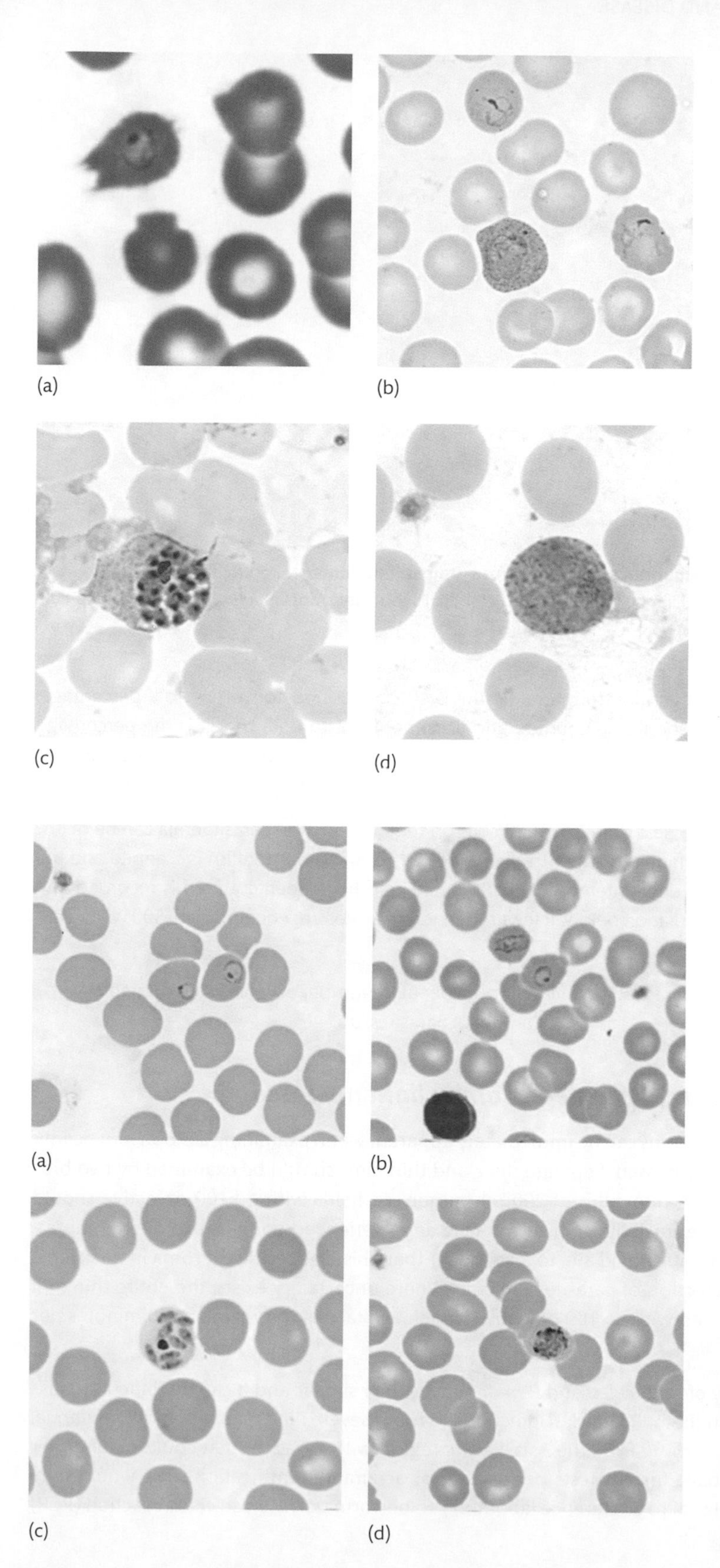

FIGURE 7.11
Giemsa-stained thin films of *P. ovale* infection (high power). (a) Ring form with heavy chromatin dot and thick, deep-staining cytoplasm; (b) ring forms and a developing trophozoite; (c) Schizont containing merozoites with large nuclei; (d) gametocyte. Division of Parasitic Diseases and Malaria, Centers for Disease Control and Prevention, USA.

FIGURE 7.12
Giemsa-stained thin films of *P. knowlesi* infection (high power). (a) Ring forms; (b) ring form and a developing, band form trophozoite; (c) mature schizont (note the pigment has concentrated into a single mass); (d) spherical gametocyte. Division of Parasitic Diseases and Malaria, Centers for Disease Control and Prevention, USA.

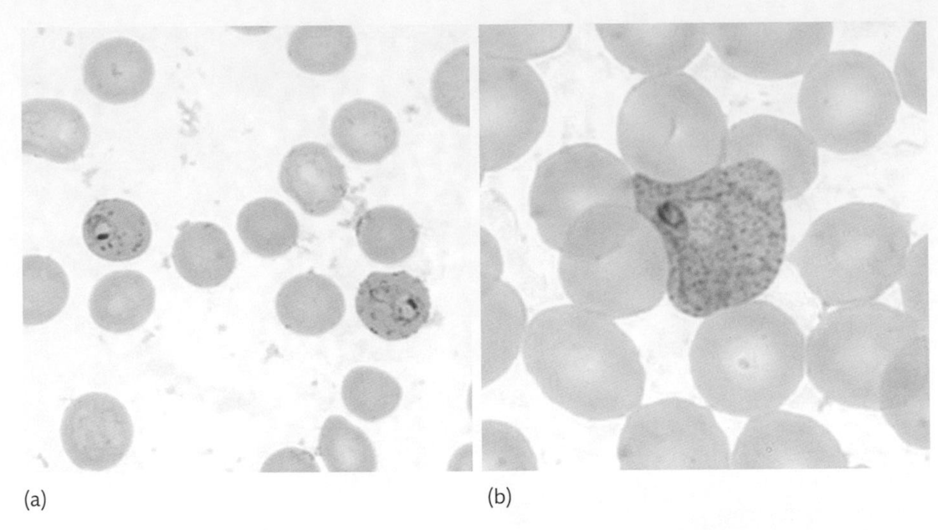

FIGURE 7.13
Maurer's clefts and Schüffner's dots on Giemsa-stained thin films. (Magnification ×500.) (a) Maurer's clefts in *P. falciparum* infected red cells; (b) Schüffner's dots in a *P. vivax* infected red cell.

different areas of a thin film from a minimum of 1000 cells and converting to a percentage. Only asexual stages should be counted and it is important to note that it is the percentage of parasitized cells that is being estimated and not the number of parasites per 100 red cells. A parasitaemia of >5% is considered to be a severe infection and a medical emergency.

If the parasite count is less than 1 in 1000 cells on the thin film, the parasitaemia can be quantified from the thick film in relation to the white cell count. The thick film is scanned and the number of parasites encountered when 200 white cells have been counted is recorded and entered into the calculation below. If fewer than ten parasites are encountered, 500 white cells should be counted.

$$\text{(Number of parasites} \times \text{White cell count per } \mu\text{L)/(Number of white cells counted)} = \text{Number of parasites/}\mu\text{L}$$

Limitations of malaria detection by light microscopy

Whilst marked infections can be immediately apparent when examining a blood film, scanty infections are easily missed. Separate thick and thin films should be examined by two biomedical scientists. A minimum of 200 oil-immersion fields with a ×100 objective should be examined in the thick film, which will take an experienced scientist up to ten minutes. Less experienced staff should aim to take longer than this. The residual stroma in thick films can make the detection of parasites difficult. Where uncertainty exists, the entire thin film should be examined with a ×100 objective, which will take in the region of 30 minutes per operator.

The morphology of *P. knowlesi* and *P. malariae* are very similar and it can be difficult to distinguish between them. *P. knowlesi* infection is a more severe illness than that of *P. malariae*, so any patient returning from the Asia-Pacific region with parasites that could be of either species should have further tests performed for accurate identification. This would normally involve referral to a reference laboratory to perform polymerase chain reaction (PCR) analysis.

Key Points

Cells containing parasites are inevitably heavier than non-parasitized cells, so it is worth looking at the tails and edges of a thin film, where larger and heavier cells can predominate.

Despite these precautions, false-negative reports will inevitably occur and it is good practice for at least three separate films taken during episodes of fever to be examined. Even if all three are negative, a diagnosis of malaria cannot be entirely excluded, particularly if the patient has taken antimalarial drugs. Conversely, a positive finding does not prove that symptoms are due to malaria or malaria alone because asymptomatic malaria is not uncommon in adults from endemic areas.

Malaria pigment can persist in peripheral blood mononuclear cells, which can be a useful diagnostic indicator where parasites cannot be found but the clinical suspicion is high. White cell differential histograms on automated analysers can exhibit abnormal patterns in this scenario and indicate the need for evaluation of white cell morphology.

Key Points

Sequestration of *P. falciparum* in deep capillaries can render the peripheral blood 'temporarily negative', which is another reason for taking a repeat sample of blood for the detection of malarial parasites. If an infection is present, parasites will be released into the blood after schizogony and potentially be detected in a repeat sample.

In scanty infections, where very few parasites can be found, it may not be possible to determine the species with certainty, in which case it is best that the patient is treated as for *P. falciparum* infection. Examining 1000 rather than 200 high-power fields increases the chances of finding more parasites and reduces the chances of false negatives.

Care must be exercised during the apparently straightforward process of staining as parasites can be washed off, particularly *P. falciparum* gametocytes. Staining slides in bulk can even result in the transfer of parasites from one slide to another.

Artefacts on a stained blood film can lead to reporting false-positive results. A common problem is that platelets superimposed onto red cells can appear to be inside the red cell, rather than on it. Moving the microscope stage up and down slightly is a good way of trying to differentiate between objects on red cells or within them. If an object is intracellular, it will be in the same focal plane as the red cell, and therefore both will be in focus or out of focus together. If an object is on top of a cell, the red cell and the object will not be in the same focal plane, and therefore will not be in focus together. Platelets can be confused for trophozoites because they are a similar size and colour. Precipitated stain superimposed onto red cells can also lead to misidentification, as can other chromatoid body inclusions resulting from severe anaemia.

Key Points

Films from all positive cases should be sent to a reference laboratory for confirmation of species identification.

SELF-CHECK 7.5

What are the main limitations of detecting malarial parasites by light microscopy?

Automated malaria detection

The time-consuming nature of light microscopy for malaria detection, coupled with the high degree of skill and experience required, has prompted attempts to automate malaria microscopy with image recognition technology. Image analysis software and machine learning methods, including deep learning, are applied to digital images of blood films so that the technology can subsequently recognize infected red cells, distinguish parasites from artefacts and other cellular material (i.e. platelets), and even achieve supervised classification of *Plasmodium falciparum*, in clinical samples. Applications exist for smartphones to link to a basic magnifying device, or even a regular microscope, to recognize the presence of malarial parasites and calculate percentage parasitaemia. Another cell phone system employs polarized light to detect hemozoin as a marker of malarial infection. Such systems are not yet in widespread use, but further research and refinement of automated digital analysis have the potential to supersede light microscopy, especially in laboratories where throughput of parasite detection is high.

Supplementary assays

Alternatives to microscopy for malaria detection serve a variety of purposes. They can be useful when an individual is relatively inexperienced at microscopy, or an entire department lacks significant experience because their geographical location generates low demand. Work pressures, such as a biomedical scientist working alone on call, can prevent adequate time being spent on microscopy. Outside of pathology laboratories where malaria is diagnosed 'in the field', simple and rapid tests that do not require microscopes, staining equipment, or sophisticated analysers can be extremely valuable.

Rapid and simple to perform, lateral-flow immunochromatographic techniques that detect parasite antigens have been available since the early 1990s. They are available in a variety of formats (such as dipsticks, strips, cards, wells, and cassettes), but all operate to the same basic principle. The blood sample can be anticoagulated whole blood, plasma, or direct from a fingerprick—which is ideal for testing 'in the field'. The blood is added to a buffered solution containing a haemolysing agent and one or more antibodies against malaria antigens that are labelled with a marker which can be visualized by the naked eye, such as colloidal gold. The antibodies complex with their target antigens if present, then migrate by capillary action along the test strip until they encounter separate immobilized capture antibodies directed against each target antigen in specific sections of the strip. There is a further antibody directed against the labelled antibody, which is the final antibody in the sequence, acting as a control to indicate that the procedure itself has worked. The strip is washed with buffer to remove haemoglobin. You can see in Figure 7.14 that any malaria antigens that have been labelled and captured will manifest as coloured lines. The tests are qualitative and cannot assess parasitaemia.

Antibodies are used that target proteins specific for *P. falciparum*, such as HRP-2 (PfHRP-2), or panmalaria proteins present in all species, such as *Plasmodium* aldolase or *Plasmodium* lactate dehydrogenase (pLDH), which are enzymes in the glycolytic pathway of *Plasmodium* species. *P. falciparum* LDH (PfLDH)-specific antibodies are also available.

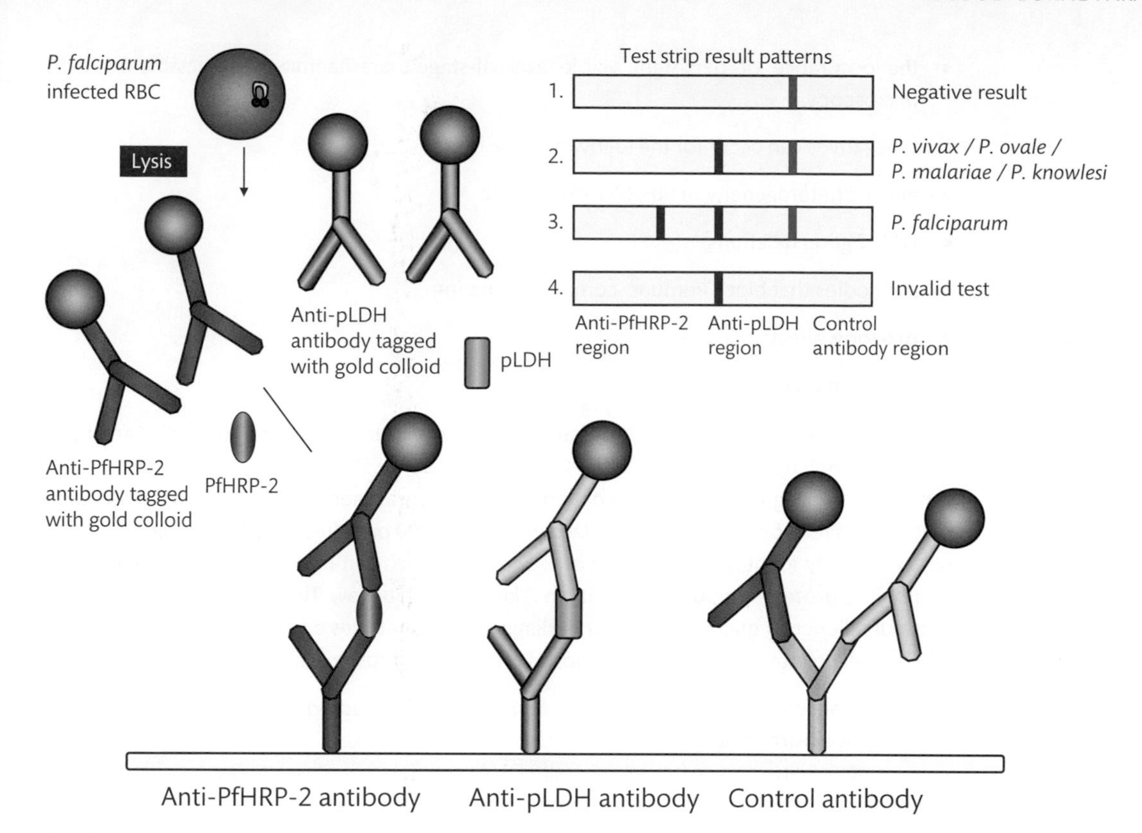

FIGURE 7.14

Immunochromatography for detection of malarial antigens. The patient in this example is infected with *P. falciparum*. Lysis of the red blood cells releases the *P. falciparum* specific antigen PfHRP-2 and the antigen that is present in all human-infecting species, pLDH. The lytic buffer also contains antibodies to PfHRP-2 and pLDH labelled with colloidal gold. The antibodies complex with their target antigens and migrate by capillary action along the test strip until they encounter separate regions of immobilized capture antibodies directed against each target antigen. An antibody directed against the labelled antibodies is present at the end of the strip as a control to ensure the antibodies have fully migrated. You can see that test strip 1 is negative because there is no colour formation in the regions of PfHRP-2 and pLDH antibodies. The band of colour in the control region indicates that the assay had worked. Test strip 2 has a band of colour in the pLDH region but not the PfHRP-2 region, indicating non-falciparum malaria, but it cannot differentiate between the other species. Test strip 3 has bands in all three positions and indicates *P. falciparum* infection, and can also reflect a mixed infection where *P. falciparum* is not the only species present. Although there is a band of colour in the pLDH region of test strip 4, the lack of colour in the control region indicates that the test may not have worked and should be repeated.

Although the rapid immunochromatographic tests appear relatively straightforward in their design and operation, they are not without their limitations and cannot be considered a replacement for microscopy. False positives can occur for the following reasons:

- cross-reacting antibodies such as rheumatoid factor;
- PfHRP-2 can cross-react with non-falciparum malaria;
- PfHRP-2 can persist after parasites have been cleared from the blood;

- the occurance of persistent viable asexual-stage parasitaemia undetectable by light microscopy.

False negatives can occur for the following reasons:

- genetic heterogeneity of PfHRP-2 expression;
- *HRP-2* gene deletions;
- antibodies that block immune-complex formation;
- **prozone** effect;
- unknown causes.

prozone
Concentration of antibody or antigen is so high that the optimal concentration for maximal reaction with antigen is exceeded and binding is reduced or does not occur.

Clearly, the rapid diagnostic tests (RDT) are not totally reliable in detecting falciparum and non-falciparum malaria, although neither is microscopy. Some studies have demonstrated >95% sensitivity, although this has been in patients with high parasitaemia. Sensitivity to the presence of malaria is reduced when the parasitaemia is below 100 parasites per μL of blood. They cannot specifically identify the presence of *P. knowlesi*, which can manifest with the antibody to panmalaria protein, although sensitivity to *P. knowlesi* can be low. The rapid tests are, therefore, a useful adjunct to malaria detection in diagnostic laboratories and have a place in the field where malaria may otherwise be diagnosed on clinical symptoms alone.

Another type of assay that is used as a screening test backed up by microscopy is the **quantitative buffy coat** (QBC) method. Capillary blood is taken into a glass haematocrit tube containing acridine orange (to stain parasite DNA) and potassium oxalate (as an anticoagulant). A cylindrical float is inserted into the tube, which is then centrifuged to separate the cells according to their densities so that they form the discrete bands you can see in Figure 7.15. The bands would normally be small but the presence of the float, which occupies 90% of the bore of the

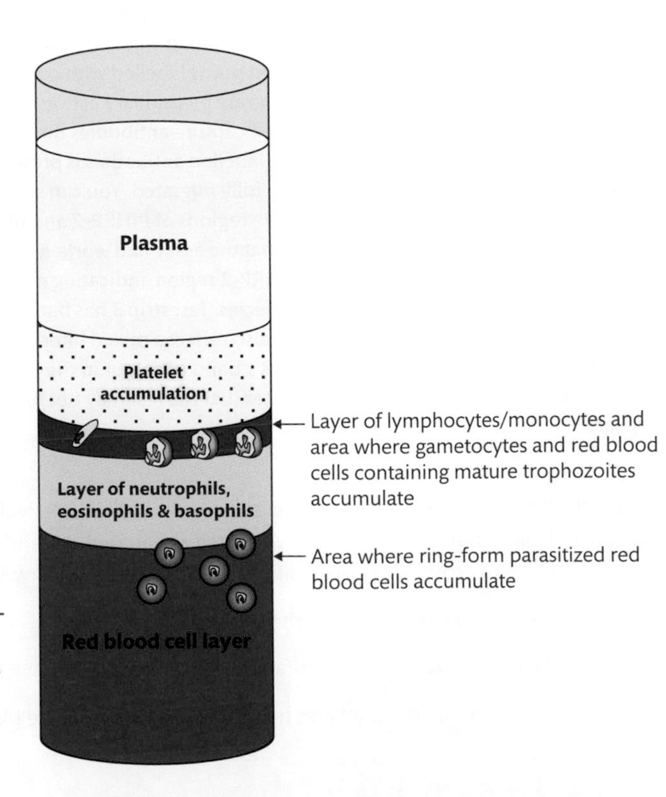

FIGURE 7.15
Distribution of blood components and parasites in the glass haematocrit tube of the quantitative buffy coat technique for detection of malaria.

tube, forces the blood components to the periphery, thus significantly enlarging the areas occupied by each band. The tube is then placed on a holder and examined under a light microscope with an ultraviolet light adapter. Red cells do not contain DNA, but red cells containing parasites will fluoresce because the parasite DNA takes up the acridine orange stain. Red cells containing parasites are less dense than unparasitized cells so they form a discrete band between the non-infected red cells and the white cell layer. The parasite nuclei appear as bright specks of green light and the cytoplasm as yellow-orange against a background of non-fluorescing red cells.

However, there are limitations to the QBC assay: it is almost impossible to differentiate between species, quantification is not possible, filarial worms can be misidentified as malaria, and visual artefacts can lead to false-positive reporting by inexperienced operators.

Parasite nucleic acid detection by PCR techniques have been described to detect and differentiate malarial parasites. They are tenfold more sensitive than microscopy and more reliable in species differentiation, especially in mixed infections, but are not suitable for rapid diagnostics.

Assays are available that detect antibodies against malarial parasites produced by the immune response of the infected patient. The antibodies can persist in the circulation for several months after the infection is over, so the tests do not necessarily demonstrate a current infection.

Abnormalities in cytograms/scattergrams generated from flow cytometry-based haematology full blood count analysers can be an aid in detecting malaria. For instance, unclassified spots extending from the neutrophil area towards eosinophils, dual eosinophil, or neutrophil populations, or overlap between neutrophil and eosinophil populations can occur, probably due to haemozoin ingested by neutrophils causing them to be classified as eosinophils. Pseudoeosinophilia not confirmed by microscopical examination has been reported in up to 39% of cases. Monocytes that have ingested haemozoin can also affect their recognition by flow cytometry. Lymphocyte volume standard deviation and monocyte volume standard deviation can be used to reflect cellular anisocytosis, likely due to the presence of activated monocytes responding to the infection. The information available varies between analysers from different manufacturers and accuracy may vary according to malarial parasite species and parasite load. Consequently, these parameters are not routinely used or recommended for malaria detection but represent a potential future direction.

SELF-CHECK 7.6

What tests are available to supplement peripheral blood microscopy in the detection of malaria and what are their main uses?

7.3 Babesiosis

Babesiosis is an intraerythrocytic non-tropical parasitic infection predominantly described in the USA and Europe. It is caused by protozoa of the genus *Babesia* and transmitted by bites from the deer or black-legged tick (*Ixodes scapularis*) in the USA, which is also the vector for the

BOX 7.10 Another name for babesiosis

In Northern USA, babesiosis occurs mostly in Long Island and the islands off the coast of Massachusetts, where it is sometimes called the malaria of the Northeast.

bacterial infection Lyme disease, and the sheep or castor bean tick (*Ixodes ricinus*) in Europe. *Babesia* organisms have been known to be transmitted from human to human by blood transfusion and via the placenta.

Babesiosis in Europe is mainly due to infection with *B. divergens*, and in the USA, *B. microti* and *B. duncani*. (See also Box 7.10.)

7.3.1 Life cycle of babesiosis parasites

The main host of *Babesia* species is the white-footed mouse (*Peromyscus leucopus*), where the life cycle is similar to malaria in humans apart from the lack of a hepatic phase. An infected tick vector injects sporozoites into the mouse bloodstream when it takes a blood meal, whereupon the sporozoites invade red blood cells and undergo trophozoite formation. Unlike malarial parasites, invading parasites are not encased in a parasitophorous vacuole. Asexual reproduction occurs to form more merozoites, which lyse the red blood cells when released to infect other red cells. When another tick ingests infected cells most of the parasites are destroyed in its gut, but some differentiate into Ray bodies, which divide to form four gametes. Male and female gametes fuse to form a motile zygote called a kinete. In some species, the kinetes invade tick ovaries and are transmitted transovarially to the next generation of ticks. Sporogony occurs in the salivary glands to allow infection of the next host by the tick nymphs. Transovarial transmission does not occur in *B. microti*. Instead, immature ticks are infected whilst feeding on a parasitized host, the parasites invade the ticks' salivary glands, multiply, and are then passed on to the next host at feeding.

Humans are occasional hosts to *Babesia* species. A feeding tick introduces sporozoites into the bloodstream which then invade red blood cells and undergo asexual reproduction by binary fission. Multiplication and release are responsible for the clinical manifestations of infection, but there is little, if any, transmission to ticks that feed on parasitized humans.

7.3.2 Clinical features of babesiosis

Many patients with *Babesia* infections are asymptomatic or have mild influenza-like symptoms, which are ignored and the infection resolves spontaneously. More severe cases present within one to four weeks of exposure with symptoms similar to malaria, such as fever, shaking, chills, fatigue, and (haemolytic) anaemia. The haemolytic anaemia is not as severe as malaria because *Babesia* species do not demonstrate periodicity as their release from red blood cells is not synchronized. Babesiosis commonly presents between May and September as a 'summer flu' because this is the period of nymph feeding. It is important that clinicians ascertain whether the patient has visited tick-infested areas, such as those in the USA where there is a significant deer population. Patients rarely report a tick bite because the nymph stage of *Ixodes scapularis* is very small and mistaken for a small freckle. Severe and more persistent cases, which can lead to respiratory distress syndrome and death, tend to occur in very young children, the elderly, immunocompromised patients, and individuals who have had their spleen removed. Cases of babesiosis in Europe mainly involve splenectomized patients in whom the disease is far more severe, with >50% of patients dying. The clinical presentation can be complicated by coexisting Lyme disease and its own typical symptoms such as erythema migrans skin rash.

7.3.3 Laboratory detection of babesiosis

The parasites are detected on thick and thin Giemsa-stained blood films. The erythrocytic ring forms adopt oval, round, or pear-shaped rings and are easily mistaken for *P. falciparum* early trophozoites. Similarities with *P. falciparum* include normal red cell size, multiply infected cells,

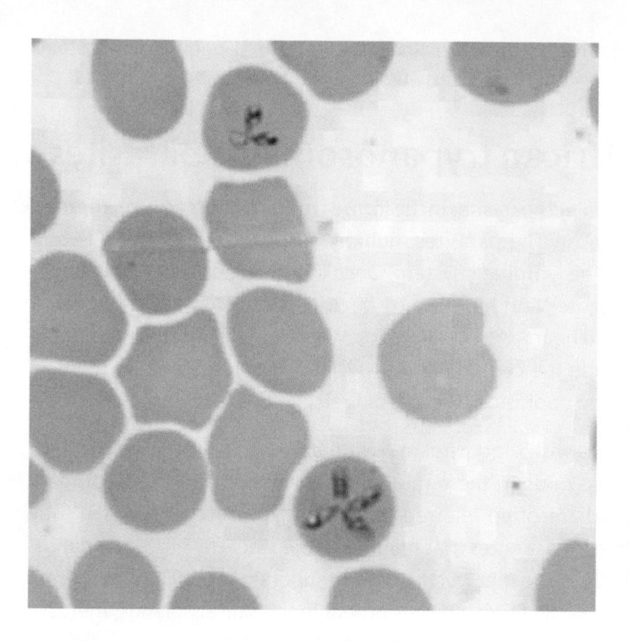

FIGURE 7.16
Red cells infected with *B. microti* on Giemsa-stained thin film. (Magnification ×500.) Note the tetrad of four pear-shaped ring forms forming the characteristic Maltese cross formation.

small rings, and the absence of Schüffner's and James's dots. The main diagnostic criterion to differentiate *Babesia* species from *P. falciparum* is the presence of groups of pear-shaped rings forming tetrads that are referred to as **Maltese cross formations**, which you can see in Figure 7.16. Other clues are a lack of schizonts, gametocytes, Maurer's clefts, and pigment.

Serological tests to identify antibodies against *Babesia* organisms are available and can detect low-level infections that may be missed by microscopy. The most commonly used are immunofluorescent assays that employ *B. microti* parasitized red blood cells as the antigen source. They can also aid differentiation between malaria and babesiosis in patients who may be at risk of contracting both infections. *Babesia* organisms can also be detected using PCR techniques.

SELF-CHECK 7.7

Compare and contrast the human-infective stages of the life cycles of malarial and babesiosis parasites.

7.4 Trypanosomiasis

Human **trypanosomiasis** is a vector-borne parasitic infection caused by haemoflagellate protozoa of the genus *Trypanosoma*. There are two types of the disease in humans, African and American trypanosomiasis, which are fatal if left untreated.

7.4.1 African trypanosomiasis (sleeping sickness)

In Africa, *T. brucei gambiense* and *T. brucei rhodesiense* are transmitted by the tsetse fly (*Glossina* species). The distribution of African trypanosomiasis is determined by the ecological limits of tsetse flies and covers most of Sub-Saharan Africa, although the two trypanosome subspecies have more specific distributions.

T. brucei gambiense is found in the central and western regions of Africa and causes a chronic infection lasting years, referred to as West African trypanosomiasis. *T. brucei rhodesiense* is found in southern and eastern regions and causes an acute illness lasting several weeks, referred to as East African trypanosomiasis.

7.4.2 Life cycle of African trypanosomiasis parasites

metacyclic
A biochemical term for the extension of a cyclic group by another cyclic group.

Whilst feeding, an infected tsetse fly injects parasites in the form of **metacyclic trypomastigotes** into the skin tissue of the human host. The parasites multiply in the vicinity of the bite and then pass into the bloodstream via the lymphatic system. Once in the bloodstream, they transform into long, slender trypomastigotes and replicate by binary fission. They are transported to other sites in the body, such as the heart and other organs, where they replicate in the tissue fluid. Later in the disease, they invade the central nervous system, and it is the tissue damage in the brain that leads to lethargy and confusion (the sleeping sickness) and eventual death.

A feeding tsetse fly becomes infected with bloodstream trypomastigotes which transform into procyclic trypomastigotes in the fly's midgut; these multiply by binary fission. The trypomastigotes are the flagellated form and are often referred to as the **trypanosomes**. When they leave the gut, they transform into epimastigotes that migrate to the tsetse fly's salivary glands, where they multiply and transform into the metacyclic trypomastigotes that will infect the next human host and perpetuate the life cycle. During development in the salivary glands, metacyclic trypomastigotes initiate the expression of metacyclic variant surface glycoproteins (VSGs). Differential activation of VSG genes in the bloodstream occurs to evade waves of antibodies, although it appears that the metacyclic stage aids the generation of population diversity. Humans are the main mammalian host for *T. brucei gambiense*, whereas *T. brucei rhodesiense* mainly infects cattle, and wild game animals and humans are less common, incidental hosts.

7.4.3 Clinical features of African trypanosomiasis

A bite from a tsetse fly is extremely painful and causes a small indurated (hardened) lesion that can persist for several days. The marked inflammatory reaction (a chancre) resulting from parasite multiplication at the injection site can last for up to three weeks. Fever occurs when the parasites enter the bloodstream and can be accompanied by sweating, shivering, and an increased pulse rate. Early stages of the disease are associated with **lymphadenopathy** and anaemia, haemorrhages, and **petechiae** may occur. The anaemia occurs primarily as a result of removal of immune-complex coated red cells by phagocytosis. The bleeding tendency arises from thrombocytopenia, vascular injury, and coagulopathy. In the later stages, when parasites invade the brain, mental dullness, apathy, excessive sleeping, and incontinence can occur; although despite the common name of the disorder, drowsiness is not always present.

lymphadenopathy
Swollen/enlarged lymph nodes.

petechiae
Red, pinpoint-sized haemorrhages of small capillaries in the skin or mucous membranes.

7.4.4 American trypanosomiasis (Chagas disease)

American trypanosomiasis occurs in Mexico, Central America, much of South America, and occasionally in the Southern USA. The infective organism, *T. cruzi*, is transmitted by nocturnal triatomine bugs such as the assassin bug (*Triatoma infestans*). Infection has also been described via blood transfusion, organ transplantation, contaminated food, the placenta, and even breast milk. (See also Box 7.11.)

7.4.5 Life cycle of *T. cruzi*

The bugs tend to feed around the edges of the mouth and eyes of the host and deposit metacyclic trypomastigotes on the skin surface through defaecation. The trypomastigotes enter the skin either by directly penetrating the conjunctiva or membranes of the nose and mouth

BOX 7.11 Affectionate bugs

The group of triatomine bugs that spread Chagas disease are known as kissing bugs because of their predilection for feeding on people's faces.

or from being rubbed into the skin by scratching of the bite site by the human host. Once inside cells, such as those in the subdermal layer or phagocytic cells, the parasites develop into **amastigotes**, multiply by binary fission, and develop into trypomastigotes that then enter the bloodstream. The trypomastigotes do not replicate in the bloodstream and migrate to tissues such as heart and skeletal muscle, nerves, and smooth muscle of the gut. Once inside the cells, the trypomastigotes differentiate into amastigotes which multiply to form pseudocysts containing up to 500 amastigotes. The amastigotes differentiate into more trypomastigotes, which rupture the cells and burst into the bloodstream to travel and infect other cells in a cycle of cell infection, replication, and release, or are ingested by a feeding bug.

Inside the bug, trypomastigotes transform into **epimastigotes** and multiply in the mid-gut and differentiate into metacyclic trypomastigotes in the hind-gut where they can be transferred to the skin surface of a human host upon defaecation.

7.4.6 Clinical features of American trypanosomiasis

The disease begins with an acute phase immediately after infection that can last a few weeks or a few months. It is usually asymptomatic, although there is a swelling at the site of parasite entry, a chagoma, which may be accompanied by fever, anorexia, lymphadenopathy, mild hepatosplenomegaly (enlarged liver and spleen), and myocarditis (inflammation of heart muscle). Inflammation of conjunctiva is termed Romaña's sign and is a recognized marker of Chagas disease.

In most patients, the acute phase resolves into an asymptomatic chronic phase that can last for years, or even decades, before a symptomatic chronic phase evolves. The most serious manifestation is heart disease in the form of arrhythmias and heart enlargement, which occurs in about 30% of patients. A smaller number develop gut abnormalities, such as loss of peristalsis and enlargement of the oesophagus and colon.

7.4.7 Laboratory detection of trypanosomiasis

The trypanosomes in the circulating blood phase can be detected by examining a sample of fresh anticoagulated whole blood, or the buffy coat, under a microscope for motile parasites, a so-called wet preparation. Giemsa-stained thick and thin films can be examined to directly visualize the parasites, and where necessary, identify the species. *T. cruzi* can be confused with *T. rangeli*, which is not known to be pathogenic in humans, and *T. brucei gambiense* and *T. brucei rhodesiense* are morphologically indistinguishable.

You can see in Figure 7.17 that on stained blood films trypanosomes appear with an undulating membrane, central nucleus, anterior flagellum, and posterior **kinetoplast**. They are between 14 and 33 μm in length. *T. cruzi* trypanosomes often adopt a characteristic C-shape.

kinetoplast
Mitochondrial DNA.

The QBC method used for the detection of malarial parasites is also used for trypanosome detection and is the method of choice for African trypanosomiasis. Serological tests are

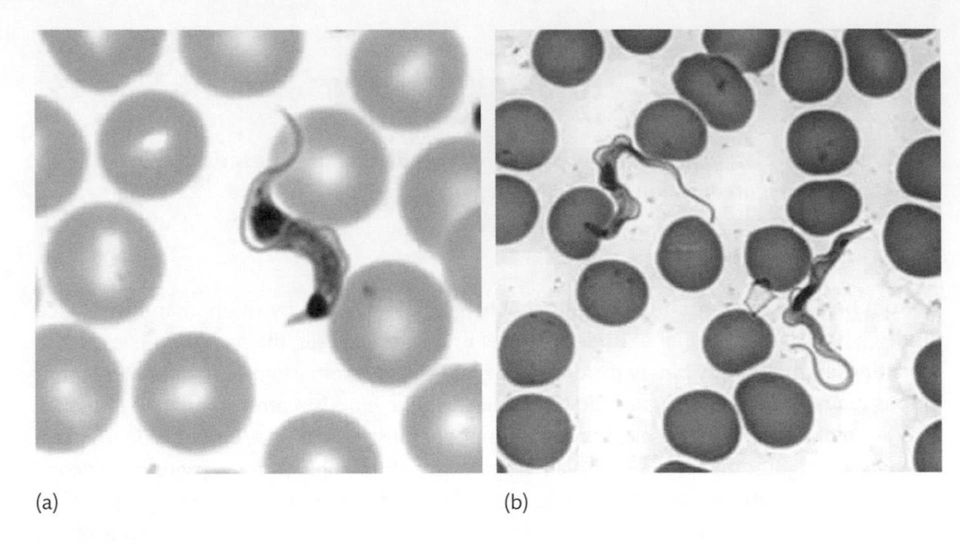

FIGURE 7.17
Trypanosomes on Giemsa-stained thin films. (Magnification ×500.) (a) *T. cruzi*; (b) *T. brucei*.

available that are used in mass population screening for *T. brucei gambiense* in endemic areas of Africa. The card agglutination test for trypanosomiasis (CATT) employs freeze-dried purified, formaldehyde-fixed bloodstream-form of *T. brucei gambiense* variable antigen type LiTat 1.3, stained with Coomasie Blue. Whole blood, plasma, or serum are added to the reagents, which form macroscopic agglutination if antibodies to the antigen are present. An assay based on indirect agglutination is available whereby antibodies in the test sample agglutinate latex beads with *T. brucei gambiense* surface antigens coupled to the surface. Immunochromatograhic RDTs similar to those for malaria are also available for *T. brucei gambiense* detection. In situations where there is a low parasitaemia, *xenodiagnosis* can be performed by allowing an uninfected vector bug to feed on the patient's blood and then examining the bug's gut contents for the presence of parasites. PCR can be used for detection and species identification.

SELF-CHECK 7.8

How are trypanosomes detected in the laboratory?

7.5 Leishmaniasis

Flagellated protozoa of the genus *Leishmania* that are transmitted by night-feeding female sandflies (*Phlebotomus* species) cause visceral and cutaneous leishmaniasis in humans. Only the visceral form, also known as **kala-azar**, has protozoal life-cycle stages associated with haemopoietic tissues—two species of parasite, *L. donovani* and *L. infantum*, are causative. More than 90% of cases of visceral leishmaniasis occur in India, Bangladesh, Nepal, Sudan, and Brazil. It is the most severe form of leishmaniasis and can be fatal if left untreated.

7.5.1 Life cycle of *L. donovani* and *L. infantum*

The infective stage of the life cycle, metacyclic **promastigotes**, are injected into the human host bloodstream by a feeding sandfly. Promastigotes actively invade neutrophils, which are then phagocytosed by macrophages, whereupon the promastigotes invade the macrophages.

Once inside the macrophage, the promastigotes transform into amastigotes and replicate by simple division inside the **phagolysosome** of the macrophage—the very organelle designed to destroy foreign bodies. Promastigotes express surface lipophosphoglycan to resist destruction in the hostile lysosomal environment at the point where they infect the macrophage. Amastigotes do not express lipophosphoglycan and survive in the phagolysosome because their metabolism has evolved to be optimal in an acidic environment. Sheer numbers of amastigotes produced by replication cause the macrophage to rupture and release parasites to invade further macrophages and other organs or to be ingested by feeding sandflies. Once inside the sandfly's stomach, the amastigotes transform into flagellated promastigotes that reproduce asexually before migrating to the sandfly's proboscis to be passed on at the next meal.

phagolysosome
Membrane-enclosed vesicle formed from the fusion of a lysosome (organelle containing digestive enzymes) and a phagosome (vacuole formed around a foreign body inside a phagocytic cell).

7.5.2 Clinical features of visceral leishmaniasis

The clinical picture varies depending on the genotype of the infective organism and the host's immune response. Many infected individuals remain asymptomatic because they are able to preferentially activate T lymphocytes; these produce gamma-interferon that is protective against progression to severe disease. Those whose disease does progress preferentially activate T lymphocytes that produce interleukin-4, which promotes an antibody response but a lesser cellular protective response.

The incubation period typically lasts between two and six months, after which the onset of fever can be sudden. The acute phase can be accompanied by diarrhoea, joint pain, anorexia, and bleeding gums which are followed by progressive muscle wastage, fever, anaemia, and hepatosplenomegaly. Some patients have lymphadenopathy. Death is often caused not by the parasite itself but by **opportunistic infections** such as pneumonia, tuberculosis, and dysentery because of a weakened immune system. In immunocompromised patients, such as those with human immunodeficiency virus (HIV) infection, visceral leishmaniasis behaves like an opportunistic infection.

opportunistic infection
An infection caused by organisms that do not normally cause disease in the presence of a competent immune system.

Development of a normochromic, normocytic anaemia is common, with haemoglobin levels as low as 70 g/L. Splenomegaly can give rise to anaemia, leucopenia, and thrombocytopenia, and liver dysfunction can affect blood coagulation due to the reduced production of coagulation factors.

Some patients who recover can develop a secondary form of the disease called post kala-azar dermal leishmaniasis (PKDL), which can occur up to 20 years later. It is more commonly associated with *L. donovani* infection. PKDL manifests as small, hypopigmented lesions on the face, which gradually enlarge and then spread over the body. The lesions can coalesce to form swollen, disfiguring plaques that resemble leprosy. Blindess may follow if they spread to the eyes.

7.5.3 Laboratory detection of visceral leishmaniasis

The most common method for detecting *L. donovani* and *L. infantum* parasites is by light microscopy of a bone marrow aspirate. The parasites can also be demonstrated in splenic aspirate, lymph nodes, liver biopsy, and peripheral blood buffy coat.

Romanowsky staining reveals the amastigotes as round or oval organisms about 2–3 μm in length with pale blue cytoplasm and a large red nucleus. The cytoplasm exhibits a deep-red or violet, rod-shaped kinetoplast close to the nucleus. The amastigotes are contained within monocytes and macrophages, which you can see in Figure 7.18.

Rapid immunochromatography assays are available to detect parasite antigens, and PCR techniques are available to detect and speciate the parasites. Serological testing for antibodies is

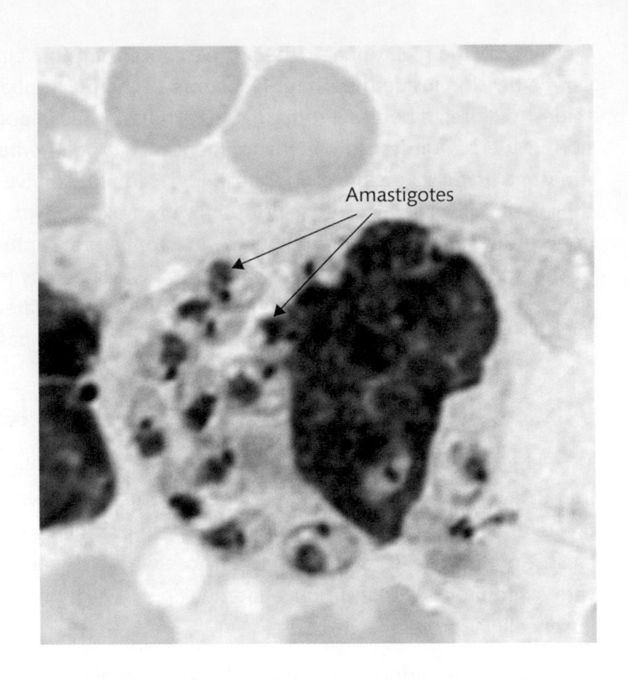

FIGURE 7.18
Monocytes containing *Leishmania* amastigotes on Giemsa-stained skin scraping. (Magnification ×500.)

available and widely used in the field, but not all patients who are infected will develop the clinical disease and require treatment. Antibodies can remain after a patient is cured so these tests cannot confirm cure or indicate re-infection because the tests do not detect the parasites themselves. Immunocompromised patients will not generate antibodies and therefore will be negative in an antibody detection test even if they are infected.

SELF-CHECK 7.9

How is visceral leishmaniasis detected in the laboratory?

7.6 Filariasis

Filarial worms are nematodes (roundworms), and eight species are known to infect humans (causing **filariasis**): *Wuchereria bancrofti*, *Brugia malayi*, *B. timori*, *Loa loa*, *Mansonella perstans*, and *M. ozzardi*, which have blood-inhabiting larvae; and *M. streptocerca* and *Onchocerca volvulus* that do not. Of the peripheral blood-detectable organisms, only *W. bancrofti*, *B. malayi*, *B. timori*, and *L. loa* are associated with significant symptomatic disease in humans. *Dirofilaria* species cause incomplete infection as they are unable to mature into adults in a human host.

W. bancrofti, *B. malayi*, and *B. timori* are transmitted to humans by mosquitoes of the genera *Culex*, *Anopheles*, *Mansonia*, and *Aedes*. Vectors for *L. loa* are mainly the deerfly (or mangofly) species of *Chrysops silacea* and *C. dimidiata*.

7.6.1 Life cycle of *W. bancrofti*, *B. malayi*, and *B. timori*

The life cycles of *W. bancrofti*, *B. malayi*, and *B. timori* are virtually identical except for transmittance by different species of vector mosquitoes. An infected mosquito introduces third-stage filarial larvae (microfilariae) into the host bloodstream when it takes a blood meal. The microfilariae moult twice and migrate to the lymphatic vessels and lymph nodes, particularly in the arms, legs, scrotum, and breast. Here they mature into adult worms, mate, and produce more microfilariae. Mature female *W. bancrofti* are between 80 and 100 mm in length, those of the *Brugia* species being about half that length. Adult male worms are about half the size of females. An adult female can release 50,000 active, immature microfilariae into the peripheral blood each day. The microfilariae of *W. bancrofti* are about 250 μm in length, whilst those of *Brugia* species are about 200 μm in length. The presence of the adult worms in the lymphatic system causes the vessels to dilate, resulting in the lymph fluid travelling more slowly. Microfilariae appear in the peripheral blood within 6–12 months of the adult worms becoming established in the lymphatics, a situation termed microfilaraemia.

Microfilariae circulate in the peripheral blood and can show periodicity, depending on the feeding habits of vectors in the region. Many vectors feed at night and it is advisable to take blood samples for diagnostic purposes when microfilariae will be circulating in larger numbers. A feeding mosquito ingests circulating microfilariae, which shed their outer coat as they penetrate the gut wall of the mosquito. They then migrate to the mosquito's flight muscles in the thorax and enter cells where they transform to first-stage larvae, moult to form second-stage larvae, and then elongate to form the infective third-stage larvae. The third-stage larvae leave the flight muscles, pass through the haemocele (blood space in arthropods), and enter the mouth parts to be transmitted at the next blood meal.

7.6.2 Clinical features of lymphatic filariasis

W. bancrofti, *B. malayi*, and *B. timori* cause lymphatic filariasis in humans. The presence of the worms in the lymphatic system and the resultant tissue damage lead to swelling, scarring, bacterial infections, and a lung condition called **tropical pulmonary eosinophilia**. There are recurrent bouts of fever and patients experience heat, redness, and pain over infected lymphatic vessels. The reduced function of the lymphatic system causes fluid to accumulate and cause swellings (termed lymphoedema) in the arms, legs, breasts, and scrotum. The recurrent infections and inflammation can lead to hardening and thickening of the skin, which is termed elephantiasis because the swollen limbs with thick, hard skin resemble those of elephants. Elephantiasis can lead to severe disfigurement, decreased mobility, and long-term disability. Additional features of *W. bancrofti* infection are fluid accumulation in the scrotum, a hydrocele, and chyluria, which is turbid urine containing emulsified fat or pus.

tropical pulmonary eosinophilia
Nocturnal cough, wheezing, fever, and eosinophilia arising from marked sensitivity to microfilariae in the lungs.

7.6.3 Life cycle of *L. loa*

During a blood meal, an infected fly introduces third-stage larvae onto the skin surface of the human host. The larvae then enter the skin through the bite wound and mature into adults in the subcutaneous tissues, which takes about one year. Adult female worms are 40–70 mm long and males around 30 mm. The adults mate and produce microfilariae of a similar size to those of *W. bancrofti*. The *Chrysops* flies are diurnal feeders so the microfilariae are present in the peripheral blood during the day, and are mainly found in the lungs when not circulating. A feeding fly ingests microfilariae, which moult and migrate from the gut to thoracic muscles via the haemocele. The microfilariae develop into first-, second-, and third-stage larvae, the latter being the infective stage that migrates to the fly's proboscis to infect a host at the next meal.

7.6.4 Clinical features of Loa loa filariasis

Some patients can be asymptomatic, whilst severe infections can result in encephalitis (brain inflammation), cardiomyopathy (inflamed heart muscle), and kidney failure.

The main clinical sign is puffy, diffuse subcutaneous swellings occurring predominantly on the limbs or face termed **Calabar swellings**, so-called because they were first recorded in 1895 in the coastal Nigerian town of Calabar. The swellings are red, itchy, and painful and arise from immune/allergic responses to worms migrating through subcutaneous tissues, to their metabolic products, or to dead worms. *L. loa* are sometimes referred to as the African eyeworm because they can be seen migrating across the conjunctiva. This does not cause loss of vision but can be irritating and painful; it takes about 15 minutes for a worm to cross an eyeball. Adult worms can cause hydrocele formation in the scrotum.

7.6.5 Laboratory detection of filariasis

Microfilariae can be isolated from peripheral blood, urine, hydrocele fluid, or biopsies. Blood specimens should be obtained based on the periodicity of the microfilariae in peripheral blood. The best time for phlebotomy in nocturnal periodicity is between 10 p.m. and 4 a.m. Parasites can be concentrated in blood samples by either passing through a filter or using Knott's technique of formalin centrifugation. Thick and thin blood films can be stained with Giemsa or haematoxylin.

Features of the sheath, cephalic space, and the presence/arrangement of nuclei in the tail (caudal nuclei) enable species identification. Pathogenic microfilariae are sheathed and non-pathogenic species are not. Not all sheaths stain with Giemsa, but they can be visualized by staining with haematoxylin. The main features of Giemsa-stained microfilariae of different species are outlined in Table 7.4 and examples of microfilariae appearances are shown in Figure 7.19. (See also Box 7.12.)

Rapid immunochromatography assays are available to detect *W. bancrofti* antigens, and PCR techniques are available to distinguish between the two main causative organisms of lymphatic filariasis, *W. bancrofti* and *B. malayi*.

SELF-CHECK 7.10

How are microfilariae differentiated in the laboratory?

TABLE 7.4 Morphological features of microfilariae in Giemsa-stained blood films.

Species	Sheath	Cephalic space	Column of caudal nuclei
W. bancrofti	Stains faintly	Short	Does not extend to tip of the tail
B. malayi	Bright pink	Long	Extends to tail tip; two isolated terminal nuclei separated by a constriction
B. timori	Does not stain	Longer than *B. malayi*	Greater number of single-file nuclei towards the tail than *B. malayi*
L. loa	Does not stain	Short	Long column of single nuclei extending to tip of tail
M. perstans	No sheath	Short	Extends to tip of a blunt tail
M. ozzardi	No sheath	Short	Does not extend to tip of the tail

BOX 7.12 A practical tip

Microfilariae can concentrate in the tails of a wedge blood film and eosinophilia is often present. In view of the occurrence of asymptomatic filariasis, a biomedical scientist encountering an unexpected eosinophilia should consider that parasites may be present and check the tails and edges for the presence of microfilariae.

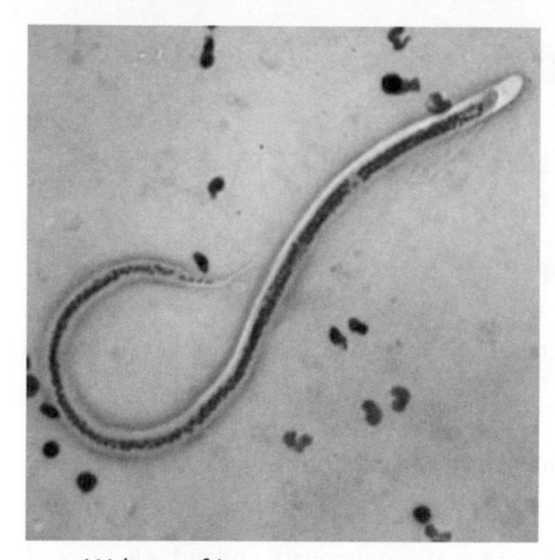

W. bancrofti
(Thick film; x500 magnification)

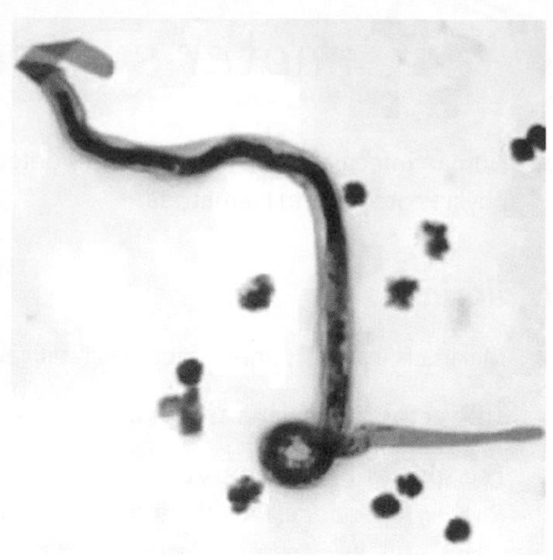

B. malayi
(Thick film; x500 magnification)

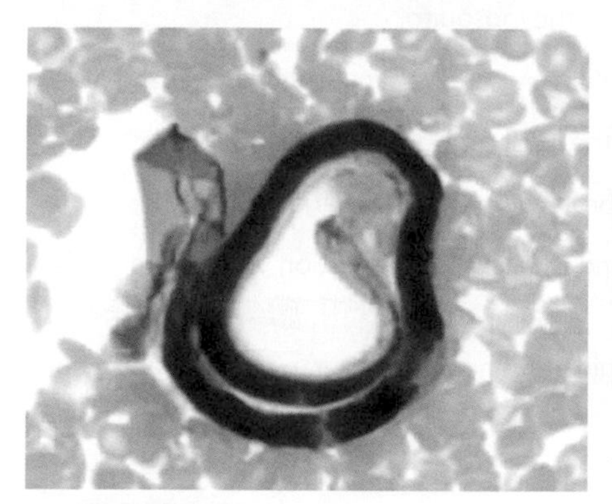

B. malayi
(Thin film; x500 magnification)

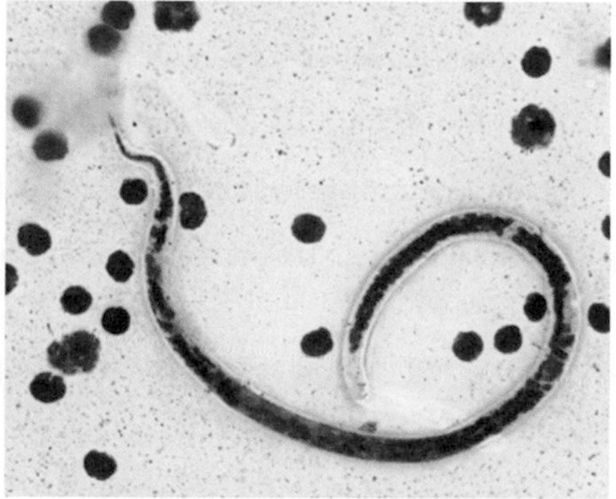

B. timori
(Thick film; x500 magnification)

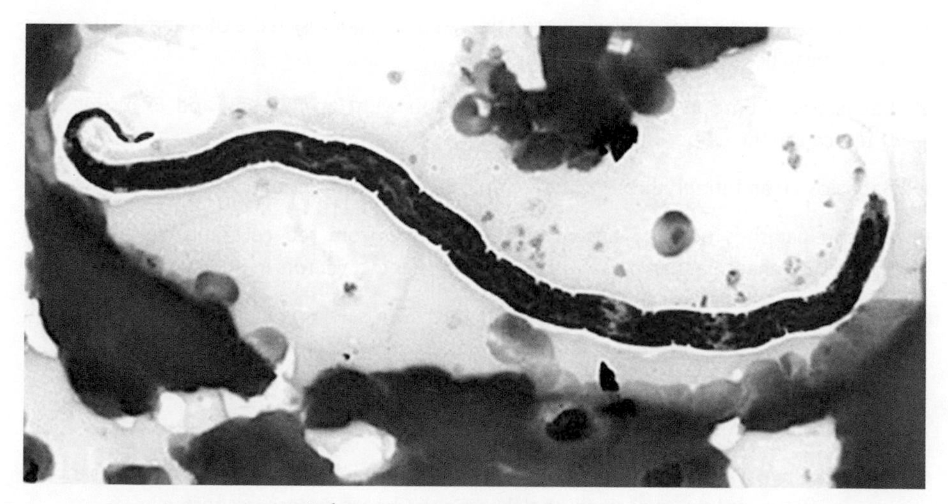

L. loa
(Thin film; x500 magnification)

FIGURE 7.19

Romanowsky-stained microfilariae. Note the absence of red cells in the thick films and the presence of white cell nuclei.

Chapter summary

- Human infective parasites with stages detectable in blood and bone marrow belong to the phyla Protozoa and Nematoda.

Malaria

- Malaria is a serious and sometimes fatal vector-borne infectious disease.
- The vectors for malaria are female *Anopheles* mosquitoes.
- The species of the protozoan genus *Plasmodium* that infect humans are: *P. falciparum*, *P. vivax*, *P. malariae*, *P. ovale*, and *P. knowlesi*.
- Once inside the human host, the parasites enter hepatic, pulmonary, and red cell stages of the life cycle and reproduce asexually.
- The parasites undergo sexual reproduction in the mosquito.
- Synchronized release of parasites from red cells causes fever and haemolytic anaemia.
- *P. falciparum* and *P. knowlesi* infections are the most serious and potentially fatal.
- Some genetic disorders afford a degree of protection against malaria.
- The mainstay of laboratory detection and species differentiation of malaria is light microscopy.
- Supplementary diagnostic tools include rapid immunochromatography, quantitative buffy coat, PCR, and antibody assays.

Babesiosis

- Humans are incidental hosts to babesiosis parasites.
- The vectors are the deer or black-legged tick (*Ixodes scapularis*) in the USA and the sheep or castor bean tick (*Ixodes ricinus*) in Europe.
- Babesiosis in Europe is mainly due to infection with *B. divergens*, and *B. microti* and *B. duncani* in the USA.
- Babesiosis is an intraerythrocytic, non-tropical, parasitic protozoan infection.
- *Babesia* parasites undergo unsynchronized release from red cells and cause a milder haemolytic anaemia than malaria. They do not re-infect vectors in human infection.
- The parasites are detected on thick and thin Giemsa-stained blood films. They have similarities to *P. falciparum* ring forms but Maltese cross formations are diagnostic.

Trypanosomiasis

- There are two types of this parasitic protozoan disease—African and American trypanosomiasis, otherwise known as sleeping sickness and Chagas disease, respectively.
- In Africa, *T. brucei gambiense* and *T. brucei rhodesiense* are transmitted by the tsetse fly (*Glossina* species). In the Americas, *T. cruzi* is transmitted by nocturnal triatomine bugs.

- African trypanosomes travel in the bloodstream to organs such as the heart and replicate in the tissue fluid, and later invade the brain to cause the sleeping sickness.
- Triatomine bugs deposit parasites in their faeces onto the skin surface and the parasites enter the skin where they multiply and enter the bloodstream to travel to other tissues for further replication and ingestion by vectors.
- Trypanosomes are detected in blood samples by wet preparations and light microscopy. Quantitative buffy coat, xenodiagnosis, and PCR can also be used.

Leishmaniasis

- Protozoa of the genus *Leishmania* transmitted by female sandflies (*Phlebotomus* species) cause visceral and cutaneous leishmania in humans.
- Only the visceral form, also known as kala-azar, has life cycle stages associated with haemopoietic tissues and is due to infection by *L. donovani* or *L. infantum*.
- Parasites injected into the human host invade and divide in macrophages.
- The immune response of the host dictates the course of the disease.
- The parasites are detected by light microscopy of mainly bone marrow aspirates, but also splenic aspirate, lymph nodes, liver biopsy, and peripheral blood buffy coat.
- Rapid immunochromatography assays, PCR techniques, and serological testing for antibodies can also be used.

Filariasis

- The nematode parasites *W. bancrofti*, *B. malayi*, and *B. timori* cause lymphatic filariasis and are transmitted to humans by mosquitoes of the genera *Culex*, *Anopheles*, *Mansonia*, and *Aedes*.
- Lymphatic filariasis is characterized by adult worms in the lymphatic system leading to swelling, scarring, bacterial infections, tropical pulmonary eosinophilia, and elephantiasis.
- The nematode parasite *L. loa* causes Loa loa filariasis and is transmitted to humans mainly by the deerfly species of *C. silacea* and *C. dimidiata*.
- Adult *L. loa* live in subcutaneous tissues.
- Some individuals with Loa loa filariasis are asymptomatic, whilst others develop the clinical disease, characterized by Calabar swellings.
- Adult filarial worms release microfilariae into the bloodstream to infect other tissues and to be ingested by feeding vectors.
- Microfilariae are detected microscopically from isolates of peripheral blood, urine, hydrocele fluid, or biopsies. Thick and thin blood films are stained with Giemsa or haematoxylin.
- Microfilariae are speciated based on features of the sheath, cephalic space, and presence/arrangement of nuclei in the tail (caudal nuclei).
- Rapid immunochromatography assays and PCR techniques are available to distinguish between *W. bancrofti* and *B. malayi*.

Discussion questions

7.1 Why are blood-borne parasites an evolutionary success?

7.2 Vector eradication can only ever be an unachievable social and medical ideal. Discuss.

7.3 Why does microscopy in the hands of an experienced scientist remain the mainstay of parasite detection and identification?

Further reading

- Bailey JW, Williams J, Bain BJ, Parker-Williams J, and Chiodini PL. ***Guideline: The Laboratory Diagnosis of Malaria***. British Society for Haematology, 2013. Available at: http://onlinelibrary.wiley.com/doi/10.1111/bjh.12572/pdf.
- Bates I and Ekem I. ***Haematological aspects of tropical diseases***. In: Hoffbrand AV, Higgs DR, Keeling DM, and Mehta AB (eds). ***Postgraduate Haematology***, 7th edn (pp. 854–69). Wiley Blackwell, Oxford, 2016.
- Hoffman SL, Abdalla SH, and Pasvol G (eds). ***Malaria: A Haematological Perspective***. Imperial College Press, London, 2004.
- Poostchi M, Silamut K, Maude RJ, Jaeger S, and Thoma G. ***Image analysis and machine learning for detecting malaria***. *Translational Research* 2018: **194**: 36–55.
- ***http://www.malariasite.com***.
- Website of Centers for Disease Control and Prevention (CDC), Division of Parasitic Diseases and Malaria. Extensive fact sheets on laboratory practice in detection of blood-borne parasites. Available at: http://www.cdc.gov/dpdx/az.html.

Answers to self-check questions and case study questions are provided as part of this book's online resources.

Visit www.oup.com/he/moore-fbms3e.

8

White blood cells in health and disease

Gavin Knight

The following chapter outlines the main types of white blood cell (WBC) encountered in the peripheral blood, their development, structure, and function.

The species of WBC that you will learn about in this chapter are neutrophils, lymphocytes, monocytes, eosinophils, basophils, and their precursors. Each of these species of WBC has a very particular function, and we can often suggest a differential diagnosis for a patient based on the way in which the WBCs are represented in the full blood count and their appearance using microscopy.

At the most basic level, WBCs are responsible for immunity through phagocytic means, through the production of antibodies which help in the destruction of foreign bodies entering our system, or through cytotoxic mechanisms.

phagocytosis
The process of the ingestion and destruction of foreign and unwanted material, such as bacteria and effete red blood cells. Phagocytosis is performed by phagocytes, principally neutrophils and monocyte macrophages.

Broadly speaking, lymphocyte species—called B cells and T cells—are involved in adaptive immunity because they adapt to the specific antigens encountered, whereas other WBC species form part of our innate immune system, which treats all foreign antigens in the same way.

Learning objectives

After studying this chapter, you should confidently be able to:

- explain the structure-function relationship of the different species of WBC;
- describe the processes of maturation and differentiation of WBC species;
- draw the typical morphological characteristics of the different species of WBC and recognize these in a Romanowsky-stained blood film;
- use the correct terminology to describe quantitative abnormalities in the WBC count;
- describe some of the common causes of quantitative abnormalities in the WBC count.

8.1 Introduction

White blood cells (WBCs), also known as leucocytes, are cellular components of the immune system. The concentration of WBCs in the peripheral blood is usually maintained within tight limits, with a reference range of 4.0–10 × 10^9/L. (See also Box 8.1.) **Leucocytosis** occurs when a white cell count exceeds the upper limits of the reference range, whereas a white cell count below the lower limits of the reference range is described as **leucopenia**. WBCs are produced as part of the haemopoietic process, as outlined in Chapter 3. Briefly, the pluripotent stem cell, under the influence of a number of growth factors and cytokines, undergoes the process of differentiation and maturation. Pluripotent stem cells can produce cells of erythroid, megakaryocytic, granulocytic, monocytic, and lymphoid lineages. However, this chapter is only concerned with the white cell components: granulocytes (neutrophils, eosinophils, and basophils), monocytes, and lymphocytes. The initial stages of white cell development occur in the bone marrow (BM), with maturation of all lineages—except a subset of lymphocytes called T lymphocytes—completing within the BM micro-environment. T lymphocytes require a stage of maturation within the thymus and migrate to this area of lymphoid tissue during development. With the exception of T lymphocytes, immature cells tend to remain within the BM—which acts as a reservoir to subsidize the peripheral blood population if necessary. Because the myeloid precursors are held within the BM, the BM contains many more myeloid than erythroid cells, where a ratio of between 2:1 and 12:1 in a normal adult BM can be expected. In cases of acute leukaemia, the process of maturation is disrupted, often resulting in the cessation of maturation at a particular intermediate stage—this process is called maturation arrest. This process is outlined in section 8.2.1.

Cross reference

More detail regarding haemopoiesis can be found in Chapter 3.

8.2 Granulocytes

8.2.1 Granulocyte maturation

Granulocytes are WBCs that contain granules within their cytoplasm, and include neutrophils, eosinophils, and basophils. The process of granulocytic maturation involves a number of different stages, each relating to characteristic changes in cellular size, granularity, staining characteristics, and nuclear structure. The process of normal granulocyte differentiation and maturation occurs within the bone marrow, as shown in Figure 8.1. The earliest stage of concern in this chapter is called the **myeloblast** stage.

Myeloblasts

In a normal bone marrow, up to 4% of cells are myeloblasts. Myeloblasts should not be apparent in the peripheral blood. These cells have a high **nucleocytoplasmic ratio**, i.e. the nucleus of the cell is almost as large as the cell itself, and a fine, open, lacy chromatin pattern is apparent, accompanied by prominent **nucleoli**. The cytoplasm of the cell stains a light shade of blue. The size of a myeloblast is 16 ± 4 μm. **Granulopoiesis** describes the production of granulocytes. In granulopoiesis, myeloblasts have the potential to produce all three of the granulocyte lineages. When maturation arrest occurs at this stage and the percentage of myeloblasts within the BM equals or exceeds 20%, a diagnosis of acute myeloid leukaemia can be made.

nucleolus (plural nucleoli)

An area of the nucleus composed of genes that encode ribosomal RNA.

granulopoiesis

The growth, differentiation, and maturation of granulocytes.

In the next chapters we discuss leukaemias in greater detail and their association with an increased number of myeloblasts in the BM and peripheral blood. Three examples of myeloblasts can be seen in Figure 8.2. In Figure 8.2(a), we can see the typical characteristic of a high nucleocytoplasmic ratio; the nucleus occupies the majority of the volume of the cell. Note also

BOX 8.1 Adult WBC count reference ranges

Total WBC count:	$4.0–10.0 \times 10^9$/L
Neutrophil:	$2.0–7.0 \times 10^9$/L
Lymphocyte:	$1.0–3.0 \times 10^9$/L
Monocyte:	$0.2–1.0 \times 10^9$/L
Eosinophil:	$0.02–0.5 \times 10^9$/L

Some laboratories report their absolute cell counts as × 10^3/μL and it is important to recognize the units used by your laboratory.

It is also important to recognize that reference ranges in your laboratory may be different—when interpreting results from patients in your laboratory, you must use your own local reference ranges.

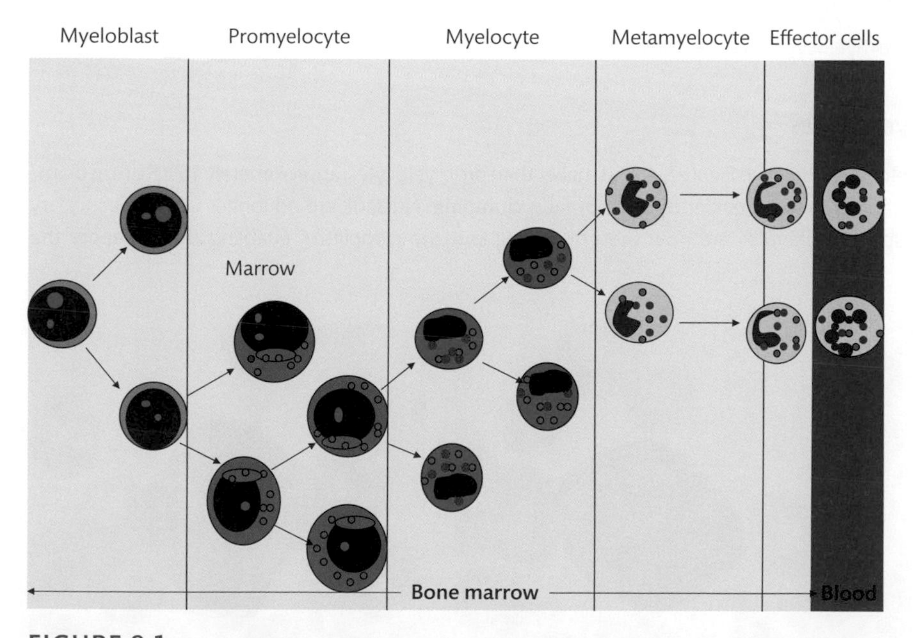

FIGURE 8.1

Granulocyte maturation occurs within the BM. Myeloblasts are shown on the far left. Cell division and maturation results in the production of larger promyelocytes. Promyelocytes demonstrate primary granules, but do not express species-specific characteristics. Promyelocyte division and maturation results in the myelocyte stage. Myelocytes contain species-specific granules and so morphologically can be identified as either neutrophil, eosinophil, or basophil lineages. Myelocytes undergo further mitotic divisions before maturing into metamyelocytes. Metamyelocytes mature into effector cells without any further rounds of mitosis.

the open chromatin configuration and prominent nucleolus, both of which demonstrate the active nature of the DNA within this cell. In Figure 8.2(b), the nucleus occupies less volume than in the previous example, although here we can appreciate the basophilic (blue) coloration imparted on the cytoplasm by the Romanowsky cytochemical stain. This basophilic staining characteristic demonstrates the high mRNA content in the cell's cytoplasm. Figures 8.2(b) and (c) both show cytoplasmic Auer rods, markers of a myeloid malignancy.

Cross reference
Acute myeloid leukaemias and blast percentages are further discussed in Chapters 9, 10, and 11.

The next step in granulocytic development is the promyelocyte stage.

Promyelocytes

Promyelocytes, as shown in Figure 8.3, are significantly larger than myeloblasts, measuring 20 ± 5 μm, although they demonstrate a lower nucleocytoplasmic ratio and occasional nucleoli. The cytoplasm has a deep blue coloration in comparison to the paler myeloblast, and contains primary granules. Primary granules are retained in neutrophils, eosinophils, and basophils, and the function of these granules is identical in each of the cells outlined here. The nucleus of the promyelocyte is also slightly indented. Once myeloblasts have matured into promyelocytes, their specific lineage is set, and they are then destined to follow the neutrophil, eosinophil, or basophil developmental pathways.

Promyelocytes should not be present in the peripheral blood. As we will see in Chapter 9, an increase in the number of promyelocytes in the absence of other intermediate stages of maturation is associated with a subtype of acute myeloid leukaemia called acute promyelocytic leukaemia.

Cross reference
Chapter 9 gives more details on the development of acute leukaemia.

As promyelocytes mature, they become myelocytes.

Myelocytes

Myelocytes, shown in Figure 8.4, are smaller than promyelocytes, approximately 15 ± 5 μm in diameter, and they show evidence of chromatin clumping. Nucleoli are no longer visible. Secondary (or specific) granules are now present throughout the cytoplasm, enabling us to identify the

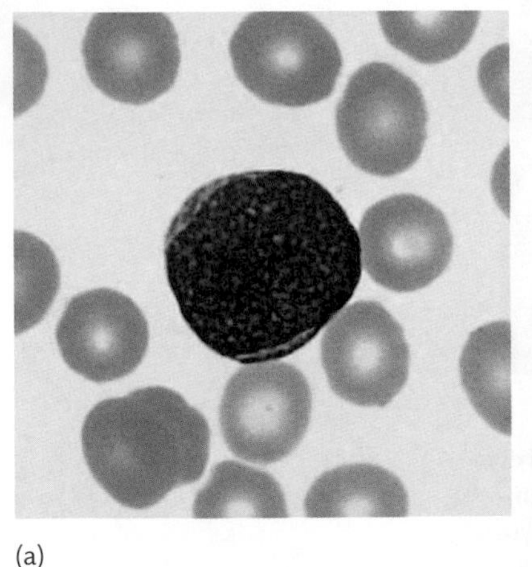
(a)

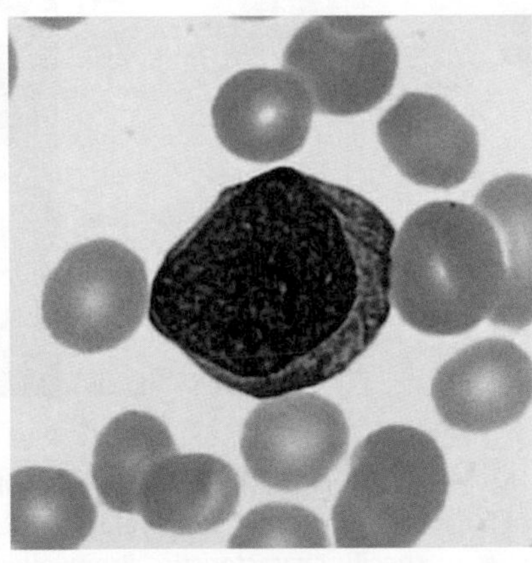
(b)

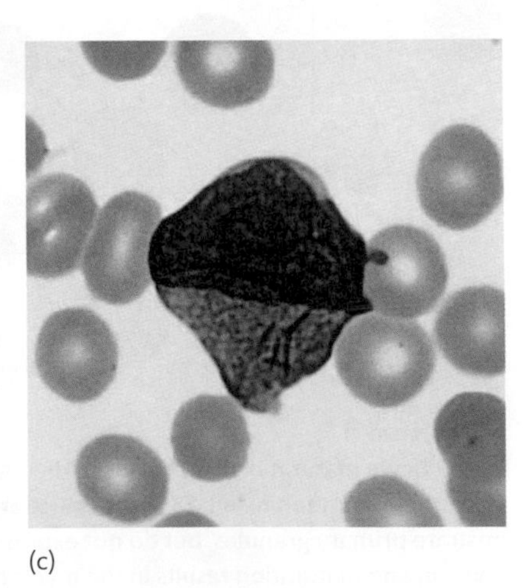
(c)

FIGURE 8.2
Myeloblasts of differing morphology. In all cases shown here, these myeloblasts are in a patient diagnosed with acute myeloid leukaemia. Note the key morphological features of: high nucleocytoplasmic ratio, open chromatin configuration (a), basophilic cytoplasm; Auer rods (b) and (c).

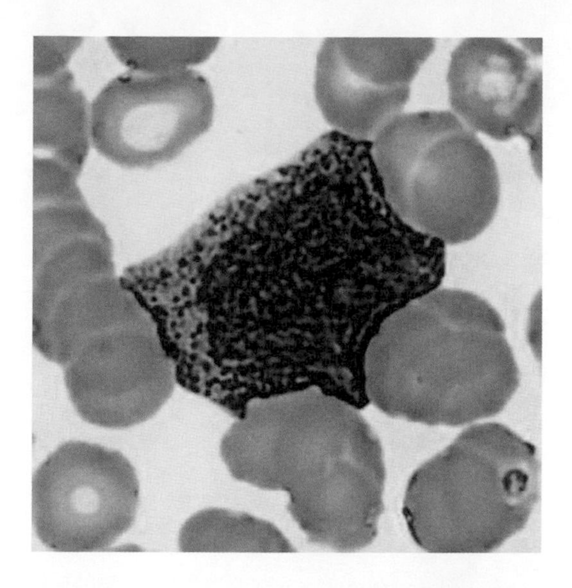

FIGURE 8.3
Promyelocyte. This cell demonstrates a high nucleocytoplasmic ratio with coarse cytoplasmic granules evident. The chromatin does not have the same open, lacy pattern noted in the myeloblast, although a prominent nucleolus is still evident.

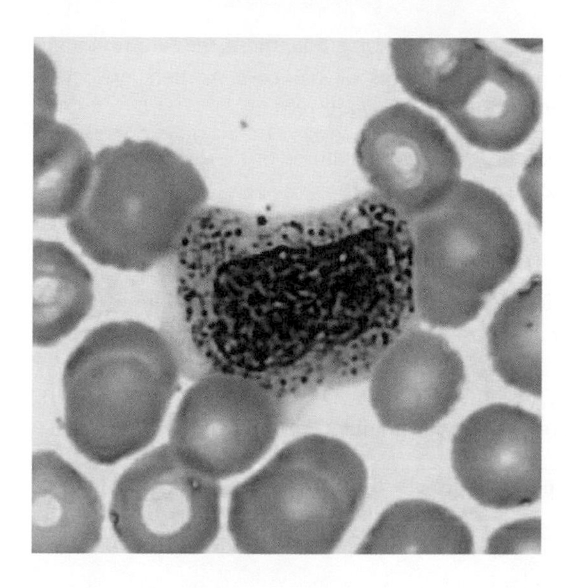

FIGURE 8.4
Myelocyte. The next stage in granulopoiesis. Here the chromatin is becoming increasingly condensed and there is now no evidence of nucleoli. Primary and secondary granules are clearly evident at this stage and the nucleocytoplasmic ratio is lower.

granulocytic lineage to which these myelocytes belong. Secondary granules are so called because they are produced after the primary granules, and they possess a cell-specific function. Gradual changes in the structure of the myelocyte results in the metamyelocyte stage of maturation.

The myeloblast, promyelocyte, and myelocyte stages of maturation are all mitotic stages and are therefore referred to as components of the *mitotic pool*. At each of these stages, identical daughter cells are produced in a symmetrical fashion, so the larger the number of cells entering the cell cycle at the myeloblast stage, the greater the number of cells will become effector (mature) cells.

Metamyelocytes

Metamyelocytes, shown in Figure 8.5, contain distinctive primary and secondary granules. The nucleocytoplasmic ratio is much lower in metamyelocytes and the nucleus is kidney-shaped. As the metamyelocyte matures, the nucleus becomes increasingly curved, so that it almost

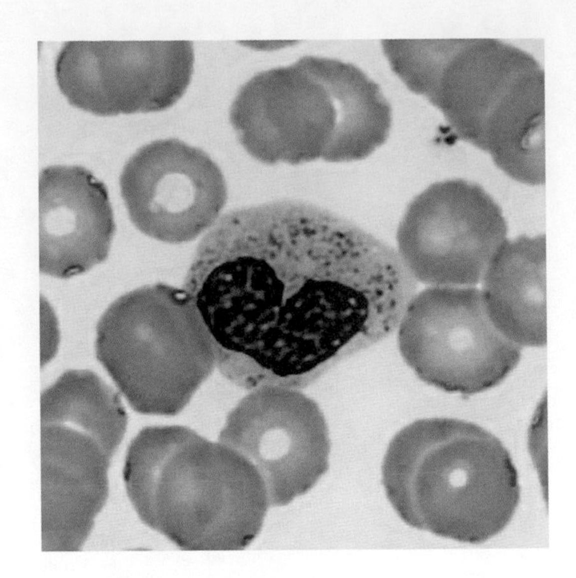

FIGURE 8.5
Metamyelocyte. The features we expect to see in a neutrophil are really becoming evident now. The chromatin is completely clumped and the cytoplasm is much paler. The nucleus has become indented in preparation for segmentation.

band form (also called 'stab cell')
An immature granulocyte with a horseshoe-shaped nucleus.

effector
A fully functional, mature cell.

resembles a horseshoe. Once the nucleus represents a horseshoe, these cells are no longer considered metamyelocytes, but are termed '**band forms**' or 'stab cells'. As metamyelocytes are unable to undergo mitosis, they form part of the *post-mitotic pool*.

Finally, the nucleus becomes segmented and the mature features of the **effector** cell—the neutrophil, eosinophil, or basophil—become apparent.

Each of the stages of granulocyte maturation has an associated expression of cluster of differentiation (CD) markers that allow these stages to be identified using immunophenotyping techniques. This is particularly useful in disease states, such as acute myeloid leukaemia, where maturation arrest occurs at a particular stage or where developmental abnormalities occur,

TABLE 8.1 **Normal CD markers expressed through granulocyte (neutrophil) maturation.**

Haemopoietic stem cell	Myeloblast	Promyelocyte	Myelocyte	Metamyelocyte
CD34 + +	CD34 + +			
	HLA-DR + +			
	CD117 +	CD117 +/-		
	CD13 +	CD13 +	CD13 Dim	CD13 +
	CD33 Dim	CD33 +	CD33 +	CD33 +
	MPO -	MPO +	MPO +	MPO +
		CD65 +	CD65 +	CD65 +
		CD15 +\-	CD15 +	CD15 +
			CD11b +/-	CD11b +
				CD35 Dim
				CD16 +

such as dysplasia. Table 8.1 provides the basic immunophenotype for each of the developmental stages in the granulocytic lineage.

A summary diagram showing the main stages of granulocyte maturation, and the variations in the morphology associated with these stages is shown in Figure 8.6.

Cross reference

Immunophenotyping is covered in more detail in Chapter 10.

SELF-CHECK 8.1

List the different stages of granulocyte maturation.

8.2.2 Neutrophils

In a healthy individual, early or intermediate stages of granulocyte maturation are not to be expected within the peripheral blood, so only mature neutrophils, eosinophils, and basophils should be found. Typical neutrophil morphology is shown in Figure 8.7. It is estimated that 10^{11} neutrophils are produced every day within the BM, providing a peripheral blood range in the order of 2–7 × 10^9/L. Neutrophils are a primary cellular component of the innate immune system, and so, in reactive conditions such as inflammation or infection, immature neutrophils, called band forms, become increasingly apparent in an attempt by the innate immune system to remove the cause. Where increased numbers of granulocytes are produced, we may see changes in granulocyte morphology, as shown in Box 8.2.

A **leukaemoid reaction**, where the leucocyte count exceeds 50 × 10^9/L and all stages of myeloid maturation (including myeloblasts) are present, may be found in the following: in very severe infections; as a response to certain cancers (carcinoma, melanoma, sarcomas, etc.); use of pharmaceuticals such as corticosteroids, minocycline, and myeloid growth factors; haemorrhage; haemolysis; and following alcoholic intoxication. A leukaemoid reaction is very difficult to distinguish from chronic myeloid leukaemia and may be impossible to differentiate from chronic neutrophilic leukaemia. A leukaemoid reaction comprises polyclonal cells, whereas true leukaemia is composed of monoclonal cells. Providing the patient survives the inducing event, their leukaemoid reaction will resolve and blood counts will return to normal.

Cross references

Chronic myeloid leukaemia and chronic neutrophilic leukaemia are discussed in detail in Chapter 11.

Clonality is discussed in Chapter 9.

SELF-CHECK 8.2

Define the term 'leukaemoid reaction'.

As granulocytes mature, they are **sequestered** into the BM storage compartment where they remain acting as a *reserve* to fight acute infection. As granulocytes are needed, they are released from this compartment into the peripheral blood, where they will circulate prior to extravasation (leaving the blood vessels to the tissues). Historically, it was believed that neutrophils only circulate in the blood for six to eight hours, although there is now mounting evidence that this part of their lifecycle could last for almost five-and-a-half days, Following release from the BM into the peripheral blood, neutrophils are either maintained in a free-flowing state, and can be enumerated as part of the full blood count, or **marginate**, primarily within vessels of the liver, spleen, and BM. When required, granulocytes are directed to the site of infection through the process of **chemotaxis**. Granulocytes can then adhere to the vascular endothelium prior to entering the tissues. Adherence, or margination, is predominantly mediated by adhesion molecules such as L-selectin on leucocytes and E-selectin on endothelial cells. Marginating neutrophils can easily enter the tissues from their position on the vascular wall, where they can be involved in inflammation and phagocytosis.

sequester

To remove or separate.

chemotaxis

The process whereby chemical signals (e.g. complement components C3a and C5a) and bacterial products (such as lipopolysaccharide) disseminate, forming a concentration gradient for granulocytes to follow.

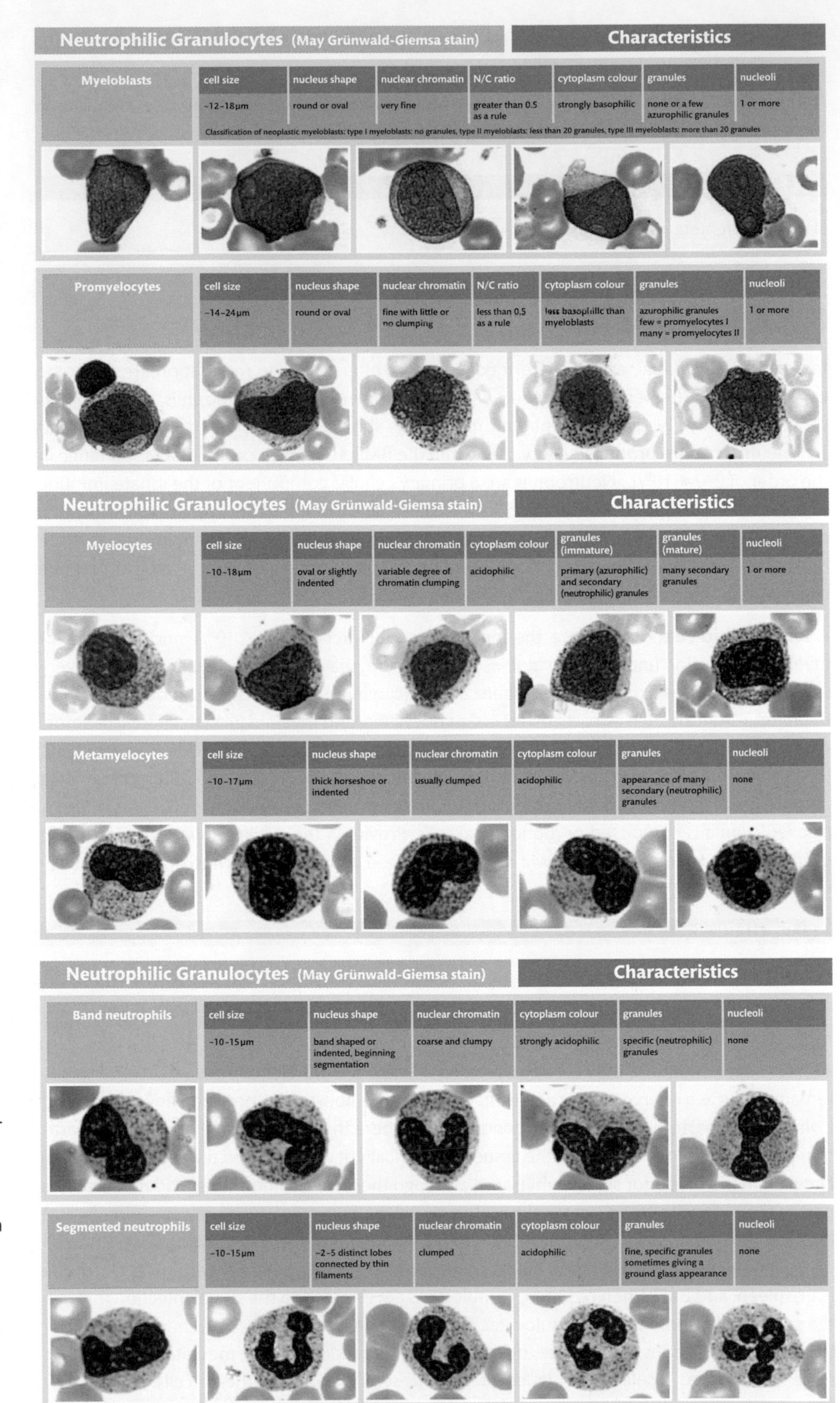

FIGURE 8.6
A summary of the morphological variations we may see during the process of neutrophil granulopoiesis, increasing in maturity from left to right from row to row. © Sysmex Europe GmbH 2015. All texts, images, and published information are under copyright of Sysmex Europe GmbH or published with the consent of their respective owners.

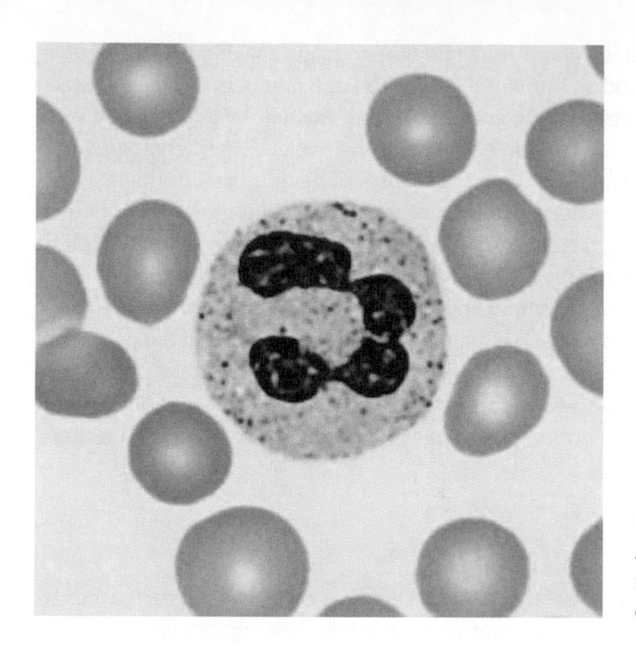

FIGURE 8.7
A mature neutrophil. Three nuclear lobes are joined by thin chromatin bridges, and cytoplasm shows azurophilic granules. Image courtesy of Gordon Sinclair, Royal Marsden Hospital.

Granulocytes, particularly neutrophils, can survive in the tissues for two to three days prior to cell death. The movement of granulocytes between the different compartments—BM and reserve, peripheral blood, and tissues—is a process of dynamic balance. As more granulocytes are required within a particular compartment, so more are released from reserves. The greater the pressure on the reserves, the greater the stimulus for proliferation within the BM compartment.

In an acute setting, where tissue damage has occurred, neutrophils will be stimulated to enter the tissues from their marginating position. Rapidly, peripheral blood neutrophils will marginate to replace those entering the tissues, and will subsequently be replaced by the BM reserve. Margination can happen very quickly, before the BM can begin to synthesize a greater number of neutrophils. In this instance, a temporary reduction in the neutrophil count, called **neutropenia**, becomes apparent as all the neutrophils from the reserve have marginated in preparation for fighting the tissue infection. This neutropenia is short-lived, and as more neutrophils are synthesized, so they are released into the peripheral blood causing a rise in the neutrophil count. A neutrophil count in excess of the reference range is called

BOX 8.2 The dynamic morphology of granulocyte maturation

An important concept to remember when examining the morphology of immature WBCs is that the characteristic stages of development—myeloblast, promyelocyte, myelocyte, metamyelocyte, band form, and effector cell—are formed through a dynamic process. Each stage merges into another, so cells will often share features between different stages of maturation.

a **neutrophilia**. These neutrophils will not have had time to mature within the BM reserve, appearing in the peripheral blood as immature cells with large numbers of granules—they will be **left-shifted** and demonstrate **toxic granulation**. An example of a left-shifted neutrophil is shown in Figure 8.8. Note the curved shape of the nucleus when compared to a mature, segmented neutrophil. There are a range of causes of neutrophilia including: infection, inflammation, malignancy, the administration of specific myeloid growth factors such as granulocyte colony-stimulating factor (G-CSF), haemorrhage, and splenectomy. Each of these possible causes (also known as the differential diagnosis) should be considered where a patient's neutrophil count exceeds the upper limit of the reference range. Where a left-shift occurs and promyelocytes, myelocytes, and metamyelocytes may be found in the peripheral blood, this can be a feature of both reactive and malignant disease. However, the presence of toxic granulation, Döhle bodies (discussed later in this section), cytoplasmic vacuolation, and an increased monocyte count with normal eosinophil and basophil levels suggest a reactive neutrophilia rather than malignancy.

left-shift
This describes an increase in the number of immature neutrophils within the peripheral blood.

toxic granulation
Occurs when neutrophils have been synthesized to fight an infection. The primary granules are more abundant in preparation for fighting infection and have a higher concentration of acid mucosubstances, which have prominent azurophilic-staining characteristics.

lyonization
Involves 'switching off' an X-chromosome in females through chromatin condensation. This chromatin condensation prevents the transcription and translation of genes contained within the additional X chromosome.

Neutrophil lobes

A typical neutrophil has between three and five nuclear lobes connected to each other by thin strands of heterochromatin. A population of neutrophils showing an average of two nuclear lobes or fewer is considered **hyposegmented**. Conversely, a population of neutrophils where in excess of 3% of the total neutrophils show five nuclear lobes or more is considered **right-shifted** or **hypersegmented**, shown in Figure 8.9.

Small regions of nuclear material protrude from the nucleus of neutrophils in females and some males. This excess nuclear material, called **drumsticks**, is consistent with an inactivated or **lyonized** X-chromosome.

It is important to make the distinction between left-shifted neutrophils and hyposegmented neutrophils. Left-shifted neutrophils are often a consequence of infection, inflammation, or pregnancy and denote an increased number of immature neutrophils or progenitors (myelocytes, metamyelocytes, and occasionally promyelocytes and myeloblasts) within the peripheral blood.

Cross reference
Hypersegmented neutrophils are introduced in the context of megaloblastic anaemias in Chapter 5.

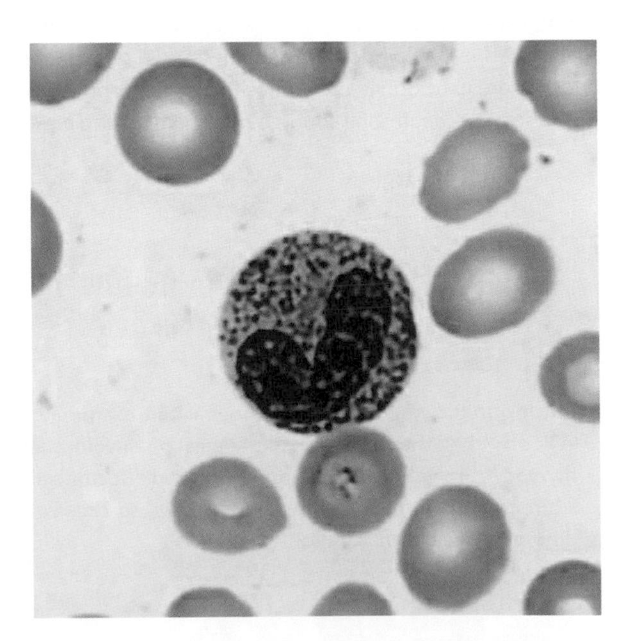

FIGURE 8.8
Left-shifted neutrophil with toxic granulation. The white cell is a band form. Image courtesy of Gordon Sinclair, Royal Marsden Hospital.

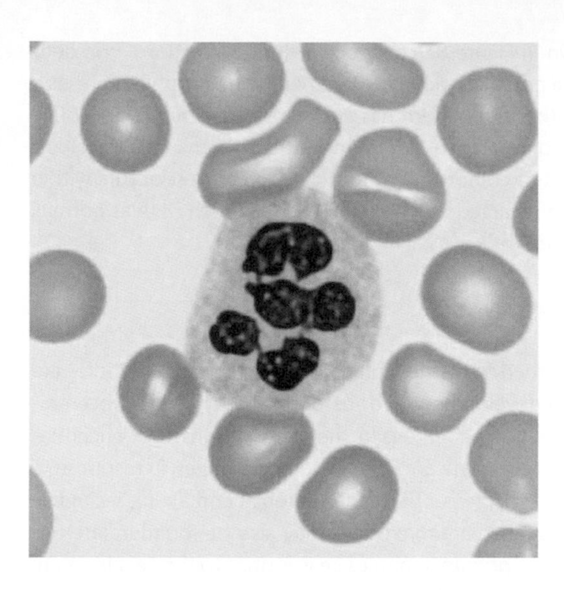

FIGURE 8.9
A hypersegmented neutrophil in a patient with megaloblastic anaemia. Image courtesy of Gordon Sinclair, Royal Marsden Hospital.

The **Pelger–Hüet anomaly** describes hyposegmented neutrophils with a singular or bilobed nucleus. The Pelger–Hüet anomaly may be inherited; most commonly, though, it will be seen as an acquired phenomenon, called the pseudo-Pelger–Hüet anomaly. In pseudo-Pelger–Hüet, the nucleus is usually bilobed, held together by a thin strand of chromatin, in what is referred to as the *pince-nez* configuration (named after circular spectacles with a pinching nose-bridge introduced in the 1840s). Pseudo-Pelger–Hüet neutrophils may also have a nucleus with a single lobe, often found in myelodysplastic syndrome. Examples of Pelger–Huet neutrophils are shown in Figure 8.10.

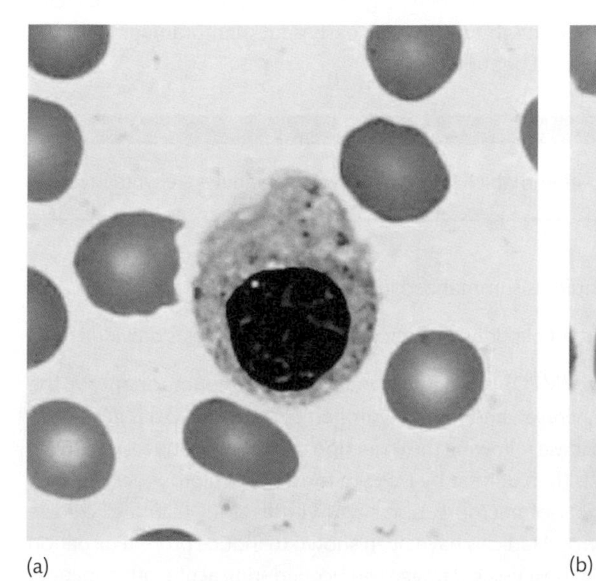

(a)

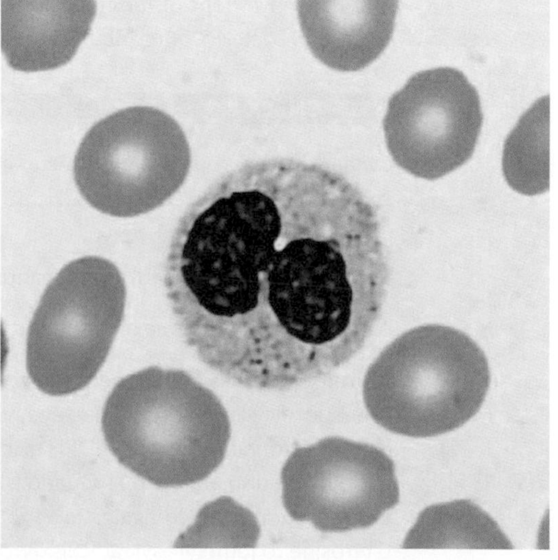

(b)

FIGURE 8.10
Pelger–Hüet phenomenon. (a) A single-lobed (dysplastic) neutrophil; (b) the characteristic pince-nez configuration associated with pseudo-Pelger–Hüet cells. Images courtesy of Gordon Sinclair, Royal Marsden Hospital.

uraemia
The accumulation of urea within the blood.

Hypersegmented neutrophils, as shown in Figure 8.9, may occur during infections, iron deficiency, and in **uraemic** patients. When a population of hypersegmented neutrophils is present, it is always important to consider whether the patient has any signs of vitamin B_{12} or folate deficiency. In patients with vitamin B_{12} or folate deficiency, megaloblastic anaemia develops. In these patients, DNA replication is often compromised and cell-cycle arrest occurs within S phase. Some of the neutrophils will be enlarged and contain twice as much DNA as normal. These enlarged neutrophils with increased DNA are called macropolycytes.

Neutrophil granulation

C/EBPε
CCAAT enhancer-binding protein is a factor important for the transcription of certain genes that control the terminal differentiation of granulocytic precursors. Deficiency of C/EBPε leads to a deficiency of secondary granules.

Primary and secondary granules are formed from the Golgi apparatus. Primary granules are synthesized at the promyelocyte stage of development and are found at the highest concentration within these cells. As the cell matures, and undergoes further cell divisions, the concentration of *primary* granules falls. During the myelocyte and metamyelocyte stages, in response to the myeloid transcription factor **C/EBPε**, synthesis of the antimicrobial containing *secondary* and *tertiary* granules occurs. C/EBPε-inducible antimicrobials may also be found in late primary granules, although formation of these primary granules ceases prior to the development of secondary and tertiary granules. Because the synthetic pathway of secondary granules is longer and occurs following the mitotic stages of development, secondary granules accumulate. More secondary granules are produced, providing the distinctive staining patterns used in the identification of different species of granulocytes and their precursors.

Neutrophil granules contain approximately 300 different types of proteins, which are selectively distributed between the primary, secondary, and tertiary granules. These granules are released at different times, together orchestrating the neutrophil's multifaceted toxic response to foreign objects and inflammation.

exocytosis
Binding of cytoplasmic vesicles to the cell membrane, leading to the release of the vesicle's contents into the extracellular environment.

Tertiary granules are **exocytosed** more readily than secondary granules, with gelatinase-containing granules exocytosed before gelatinase-negative granules. Finally, primary granules are exocytosed. The early release of gelatinase-containing granules is important, as these facilitate the movement of the neutrophil through the vascular basement membrane prior to the release of bacteriostatic and bacteriocidal peptides.

SELF-CHECK 8.3

Outline the stages of granulocyte maturation in which the different types of granules are synthesized.

The identification of neutrophil granules is summarized in Figure 8.11.

A brief summary of some of the important peptides contained within the granules is provided here.

Primary granules ***Myeloperoxidase*** (MPO) is a haem-containing enzyme important for the non-specific elimination of bacteria, viruses, and fungi. Hydrogen peroxide (H_2O_2) is generated by the membrane-bound NADPH oxidase following the reduction of oxygen to superoxide (O_2^-). Superoxide dismutates to H_2O_2 and is then utilized by MPO to form hypochlorous acid (HOCl), an important defence against a variety of microorganisms and viruses. MPO can also oxidize L-tyrosine to form tyrosyl radicals. Tyrosyl radicals have been shown to induce protein oxidation and lipid peroxidation, suggesting that host tissue damage can occur during acute inflammation.

Bacterial permeability-inducing factor (BPI) is a potent peptide with a binding specificity for lipopolysaccharide (LPS), and therefore exhibits specificity for Gram-negative bacteria, increasing bacterial membrane permeability and leading to cell death.

Lysozyme is an enzyme that can degrade the bacterial peptidoglycan cell wall by digesting the β1→4 glycosidic bond between *N*-acetylglucosamine and *N*-acetylmuramic acid residues. Lysozyme is particularly effective at destroying Gram-positive bacteria, although it often requires cofactors such as lactoferrin, H_2O_2, or opsonins when directed against Gram-negative bacteria.

Sulphated mucopolysaccharides provide the azurophilic coloration of the primary granules.

Elastase is a highly destructive enzyme found within the primary granules. It has been shown to be particularly effective at degrading Gram-negative, but not Gram-positive, bacteria. The degradation of fungi and enterotoxins is also considered to be a vital role of elastase. Very high concentrations of elastase are found within primary granules, and are associated with significant tissue injury when released following degranulation.

Acid hydrolases include a wide range of different enzymes, including acid phosphatase, β-galactosidase, β-glucuronidase, arylsulphatase, esterase, and mannosidase.

Cross references

More detail of Gram-positive and Gram-negative bacteria can be found in the *Medical Microbiology* text in this series.

Neutrophil elastase is discussed further in relation to Kostmann syndrome later in this section.

Secondary granules *Lactoferrin* is an important enzyme which plays both a bacteriostatic and a bacteriocidal role. A protein with a similar structure to transferrin, lactoferrin binds iron, thereby preventing uptake by microorganisms. Lactoferrin may work in concert with lysozyme.

Lysozyme, as outlined in the 'Primary granules' section above.

Tertiary granules *Gelatinase* is a matrix metalloproteinase capable of digesting denatured collagen, as well as intact collagen types IV and V. This digestive process enables neutrophils to migrate through the vessel basement membrane and enter the tissue fluid, where they complete their phagocytic function.

Secretory vesicles Secretory vesicles are produced through the process of endocytosis during the final stages of neutrophil maturation in the BM. They contain a variety of plasma proteins. Importantly, the membrane of the secretory vessel expresses a range of important adhesion proteins, alkaline phosphatase, and the complement regulator decay-accelerating factor (DAF). Fusing of these secretory vesicles with the neutrophil membrane enables these proteins to be expressed on the cell surface, without the requirement for neutrophil degranulation.

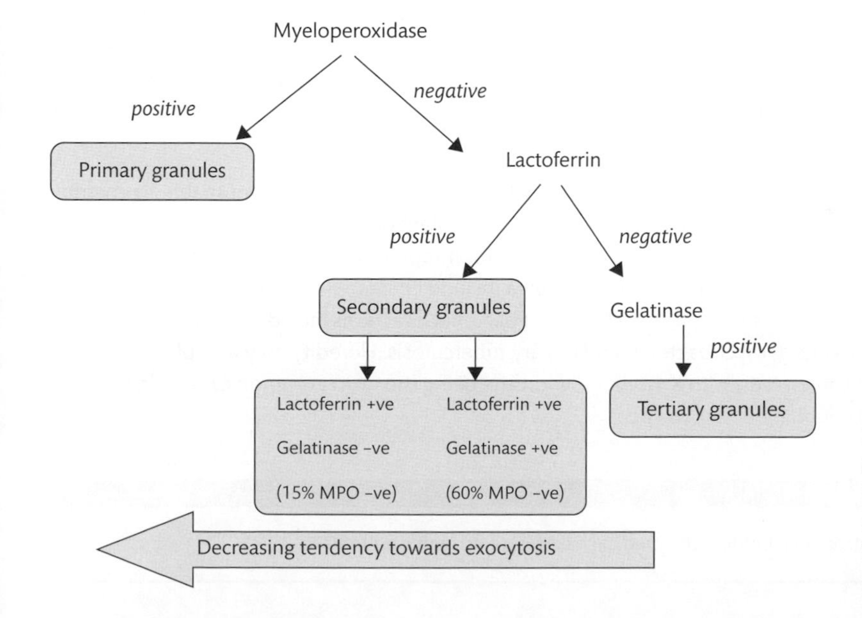

FIGURE 8.11
Differentiation between neutrophil granules based on MPO, lactoferrin, and gelatinase components. Gelatinase may be found in approximately 60% of MPO-negative granules.

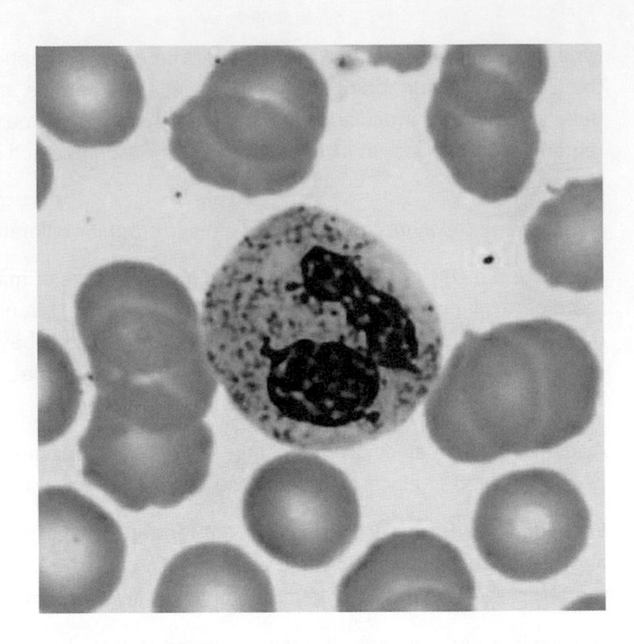

FIGURE 8.12
The neutrophil shows a discrete blue patch in the upper right-hand region of the cytoplasm and a second next to the nucleus. These are Döhle bodies. Image courtesy of Gordon Sinclair, Royal Marsden Hospital.

In cases where toxic granulation is present, light microscopy may reveal the presence of **Döhle bodies**. Döhle bodies (Figure 8.12) are blue-grey in colour, appear as cytoplasmic inclusions, and are composed of remnants of endoplasmic reticular material.

Neutropenia

Neutropenia describes a reduction in the neutrophil count to below the lower end of the reference range. A neutrophil count of less than 0.5×10^9/L is considered serious, as the risk of developing bacterial and fungal infections is high. At this neutrophil level, patients may be considered for isolation.

There are a number of causes of neutropenia, some more common than others, which can be broadly classified as defects in production, accelerated destruction, or sequestration. In the first stages of investigation it is important to establish whether the patient has an isolated neutropenia, i.e. only their neutrophil count is low, or if the patient has **pancytopenia**. Pancytopenia is commonly associated with bone marrow failure and can be an indication or complication of malignancy.

pancytopenia
Describes a reduction in red cells, white cells, and platelets.

Isolated neutropenia can be drug-induced, which is most commonly attributable to chemotherapy, autoimmunity, or can be a consequence of infection. Patients with collagen disorders are particularly prone to neutropenia, where autoantibodies are believed to reduce the lifespan of neutrophils. Viral causes of neutropenia include Epstein–Barr Virus (EBV), cytomegalovirus (CMV), hepatitis A and B, and HIV, whilst bacterial infections include salmonella, brucella, pertussis, rickettsia, mycobacteria, and miliary tuberculosis. Heredity may also play an important role in some cases, with Kostmann syndrome being the most common cause of a reduced neutrophil count. (See also Box 8.3.)

SELF-CHECK 8.4

Define neutropenia. Outline the medical risks for neutropenic patients.

Kostmann syndrome/severe combined neutropenia This is a congenital neutropenia, commonly diagnosed within the first year of life following multiple severe bacterial infections or abscesses. Congenital neutropenia is associated with neutrophil counts typically below 0.2×10^9/L. Monocyte and eosinophil counts are often raised, leading to a white cell count within the reference range.

Kostmann syndrome is largely associated with mutations in the gene encoding neutrophil elastase (*ELA2*), with approximately 60% of all cases demonstrating *ELA2* mutations. These cases are either sporadic or are autosomal-dominant.

In autosomal-recessive neutropenia, as originally described by Kostmann, 30% of all patients demonstrate mutations in *HAX1*, the gene encoding HS-1-associated protein X (HAX). HAX plays an important role in regulating apoptosis. Deficiency of HAX predisposes these cells to apoptosis prior to their release from the BM, resulting in neutropenia.

Cases originally attributable to mutations in the G-CSF receptor have been shown to be somatic secondary mutations associated with Kostmann syndrome, rather than causative.

There is a significant risk of patients with severe combined neutropenia developing myelodysplastic syndrome or acute myeloid leukaemia as part of the disease process. Both of these conditions are associated with significant morbidity and mortality and are discussed in detail in Chapter 11.

Cyclic neutropenia This condition is characterized by oscillations in the neutrophil and monocyte counts. As the absolute neutrophil count (ANC) reaches its trough, often below the lower limit of the reference range, the absolute monocyte count (AMC) peaks, and vice versa, as shown in Figure 8.13. The oscillations are on a 21-day cycle. Bacterial infections are commonly encountered when the ANC reaches its trough. G-CSF can be given to increase the neutrophil count during the trough period, thereby reducing the risk of infection. Mutations have also been identified in *ELA2*.

BOX 8.3 Neutropenia and ethnic variations

A neutrophil count below the lower end of the reference range is considered neutropenia. This corresponds to a value below 2×10^9/L. However, it is important to consider ethnicity when interpreting low neutrophil counts: a condition called benign ethnic neutropenia (BEN) is associated with some ethnic groups. Although there is considerable variation in the literature, up to half of all Africans, 16% of black Ethiopian Jews, 12% of Yemenite Jews, 11% of Arabs, and 5% of African Americans experience BEN. The mechanisms leading to BEN are unclear, although studies have shown that the release of neutrophils from the BM compartment, neutrophil margination, and egress from the peripheral blood may differ between individuals with and without BEN. In these cases, arguably these patients are not truly neutropenic as their neutrophils are present, but they are confined to regions other than the peripheral blood. BEN should be considered a diagnosis in patients from these ethnic groups when there is no evidence of splenomegaly or lymphadenopathy, no other cytopenia, or no presence of any cause of secondary neutropenia. Importantly, BEN is not associated with an increased risk of infection and no monitoring is required.

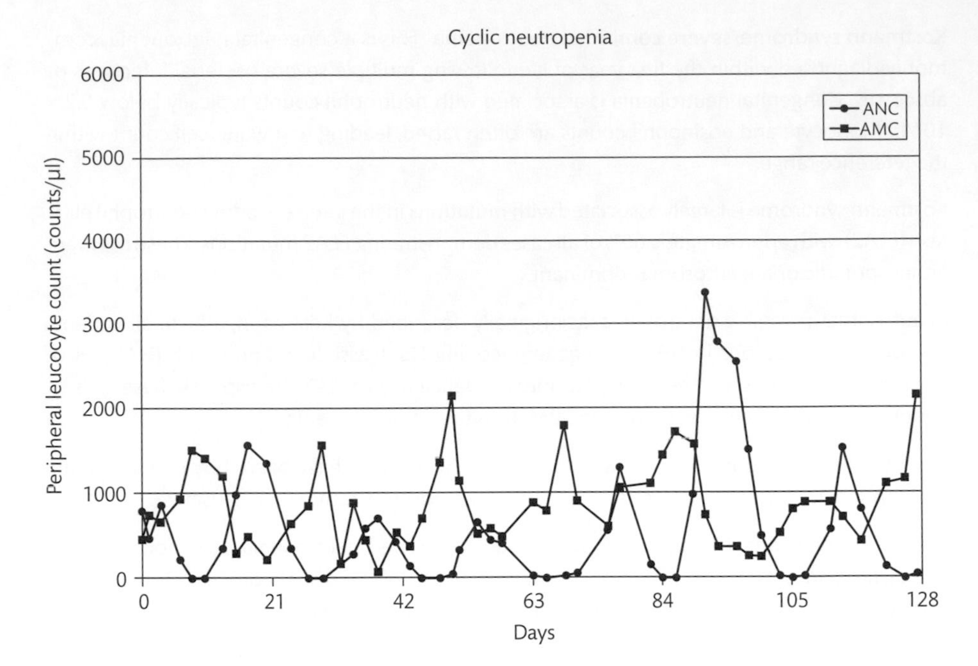

FIGURE 8.13

Oscillations of cyclical neutropenia. At the peak of the neutrophil oscillation, the monocyte concentration is at its lowest, reciprocating as the neutrophil count reaches its nadir.

Based on diagram from Boxer LA. *How to approach neutropenia*. *Hematology* 2012: 174–82.

Autoimmune neutropenia This can be divided into primary and secondary conditions. Primary autoimmune neutropenia is a very rare condition and will not be considered here.

Secondary autoimmune neutropenia is a common complication of drug therapy, malignancy, and can also be secondary to autoimmune diseases such as systemic lupus erythematosus (SLE).

Neutrophil destruction is likely to be due to the coating of neutrophils, called opsonization, with neutrophil-specific antibodies and complement components. These opsonized cells are then detected and removed by phagocytes in the spleen, liver, and BM.

In secondary autoimmune neutropenia, prophylactic antibiotics are prescribed to reduce the risk or severity of infection, and the patient may also be administered G-CSF to improve their neutrophil count. Ciclosporin A has been shown to improve the neutrophil count, although **splenectomy**—a reliable treatment for other autoimmune haematological disorders such as idiopathic thrombocytopenic purpura (ITP)—does not improve the neutropenia.

splenectomy

The procedure of removing the spleen. The spleen is an important component of the reticuloendothelial system, playing a vital role in the removal of senescent red blood cells, bacteria, and virally infected cells. Splenectomy usually increases the lifespan of cells coated in immunoglobulin, and as such is a therapeutic option for individuals with autoimmune diseases of the haemopoietic system.

Neutrophil function

Neutrophils play the primary role in limiting the rate of microbial growth following infection. This antibacterial function is dependent upon both *oxidative mechanisms*, including the generation of hydrogen peroxide (H_2O_2), superoxide (O_2^-), hydroxyl radicals, and nitric oxide, and *non-oxidative mechanisms* tending to be bacteriostatic. During infection, the process of inflammation is initiated. In the first instance, acute inflammation causes an increased permeability of blood vessels, initially mediated by histamine, which is released by basophils, mast cells, and platelets.

Vasodilation The increased vascular permeability initiated by histamine is important in mediating the efflux of neutrophils, monocytes, complement components, and antibodies from the circulation into the peripheral tissues. A second wave of vasodilation occurs in a histamine-independent manner, which is more prolonged and allows for a substantially increased migration of neutrophils from the peripheral blood into the area of infection.

The process of neutrophil migration from the blood to the tissues involves the neutrophil squeezing through the gap junctions between endothelial cells, a process called **diapedesis**, as shown in Figure 8.14. In order to successfully achieve an increased concentration of neutrophils within a particular area of the interstitial compartment, **chemotaxins** are required. Other mediators of vascular permeability include **kinins**, **prostaglandins**, **leukotrienes**, and the basic proteins found in neutrophil- and eosinophil-specific granules. Of note, in the latter stages of inflammation, neutrophils are replaced by other immune cells, including macrophages and lymphocytes. Further detail of this process can be found in the *Clinical Immunology* text in this series.

chemotaxins

Any group of small molecules that can induce chemotaxis.

kinins

A family of proteins that play an important role in inflammation, haemostasis, and pain.

prostaglandin

A lipid-derived potent physiological mediator.

leukotrienes

A type of lipid related to prostacyclins involved in inflammation and allergic reactions.

Chemotaxis This is the process whereby cells move from one place to another when mediated by a chemical stimulus. Chemotaxis (Figure 8.15) is a unidirectional process dependent upon one of five classes of compound:

- products of bacteria;
- products of damaged tissue;
- complement factors C3a and C5a;
- neutrophil-derived products;
- plasma-derived chemicals.

Cross reference

More detail regarding the process of inflammation can be found in the *Clinical Immunology* text in this series.

Each of these compounds is released in a concentration gradient, with the highest concentration closest to the source, and with decreasing concentration the greater the distance from the

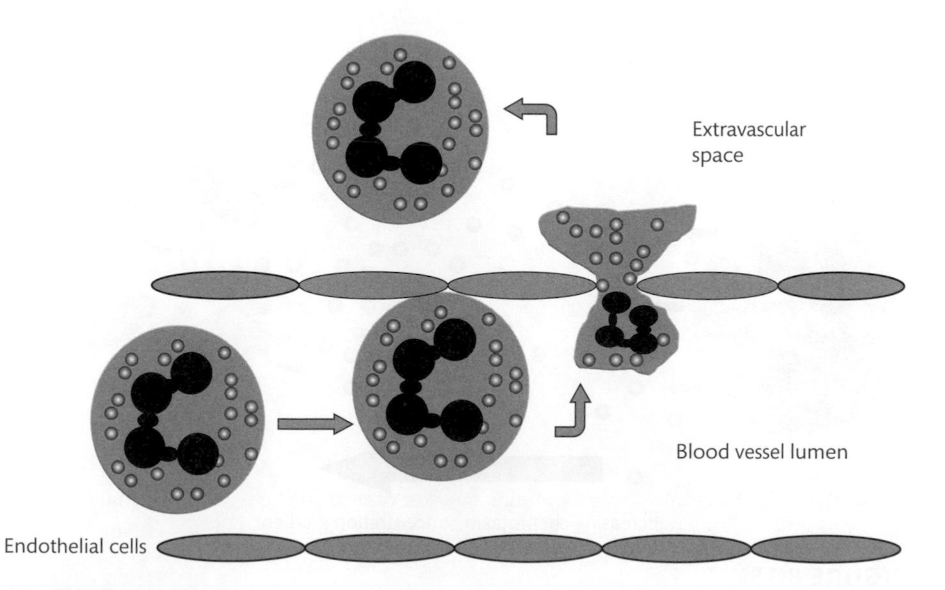

FIGURE 8.14

Diapedesis. Neutrophils adhere to the surface of the vascular endothelium and migrate through the vascular wall via gap junctions between endothelial cells. Diapedesis enables neutrophils to access extravascular tissues to combat infection.

lamellipodia
Lamellipodia are extensions of the cell cytoskeleton, and comprise actin projections which aid cell locomotion.

origin. When neutrophils detect these chemotactic signals, they are able to move in straight lines against the concentration gradient by producing **lamellipodia**, the function of which is to help locomotion-utilizing actin filaments. Repeated contraction and relaxation of these actin filaments is ATP-dependent.

SELF-CHECK 8.5

Describe the process of neutrophil migration from the peripheral blood to a site of infection.

opsonins
Molecules coating the surface of cells which enhance the recognition and destruction of these coated cells by phagocytes.

FcγR
A receptor that recognizes a particular region (called the Fc region) on an IgG antibody.

The interaction between neutrophils and a bacterium This is mediated by receptors on the surface of neutrophils that bind to **opsonins** coating the bacterium. The binding of specific antibodies (e.g. IgG) to the surface of the bacterium enables the neutrophil to interact with the bacterium via the **FcγR** receptor. Expressed on the neutrophil surface, FcγR signalling activates the neutrophil, leading to phagocytosis, degranulation, and **respiratory burst** activity.

Activation of phagocytosis involves the progressive manipulation of the neutrophil membrane, forming a vacuole which surrounds and internalizes the bacterium—called a **phagosome**. Fusion of neutrophil granules with the phagosome produces a phagolysosome, leading to the destruction of the ingested material, as shown in Figure 8.16.

During phagocytosis, the biochemical synthesis of bacteriotoxic compounds within the neutrophil is essential and these products must be directed to the phagosome in order to effectively kill the bacterium. Following phagocytosis, a huge increase in oxygen consumption occurs, called the respiratory burst, which allows the production of hydrogen peroxide and superoxide. Superoxide is generated through the increased availability of NADPH in response to a rising intracellular lactic acid concentration.

The phagocytic activity of neutrophils is clearly demonstrated in Figure 8.17, with this neutrophil-engulfing cryoglobular material.

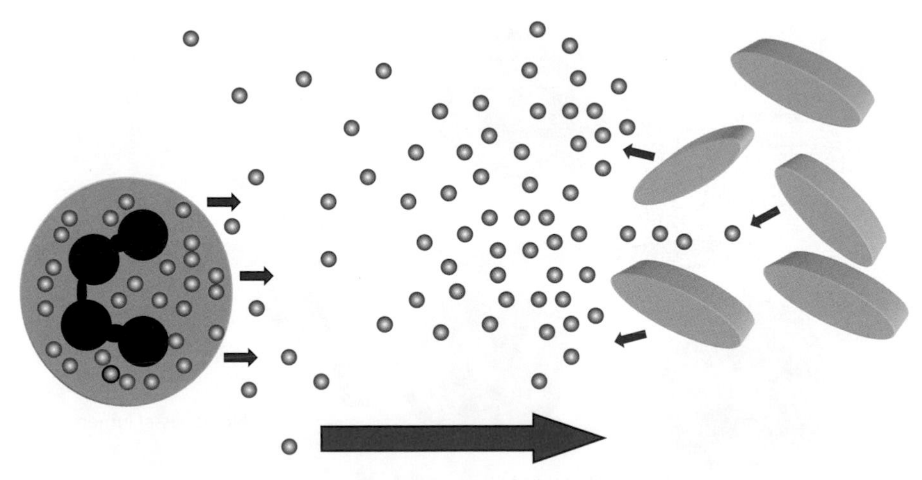

FIGURE 8.15
Bacterial products are released (right) producing a concentration gradient. The highest concentration is at the source, becoming lower the greater the distance from the site of infection. Neutrophils can detect these concentration gradients and move towards the source in order to engage the infection.

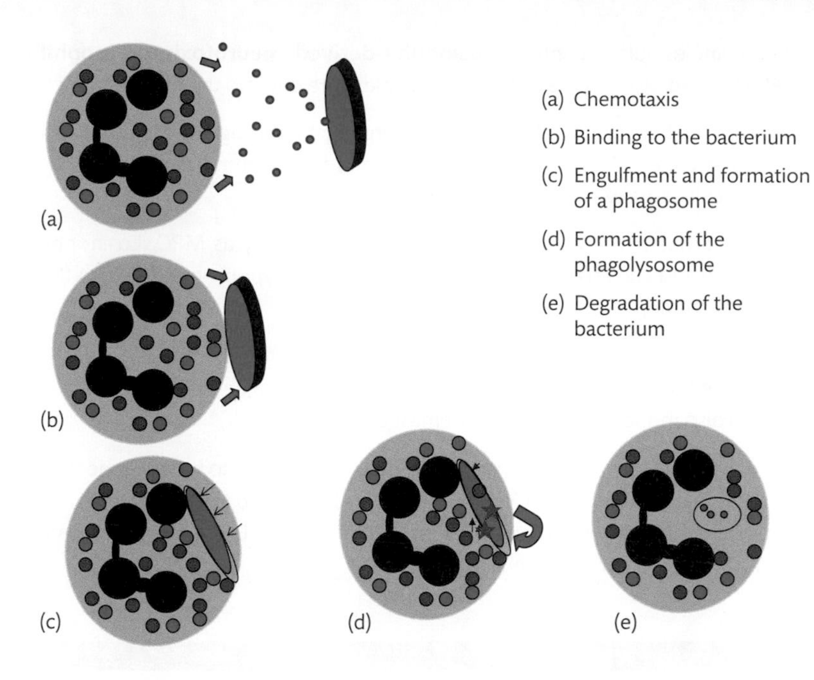

FIGURE 8.16
Formation of a phagolysosome. (a) Initially, neutrophils are attracted to a site of infection via a chemotactic concentration gradient; (b) recognizing the bacterium as foreign, the neutrophil initiates phagocytosis; (c) internalization of the bacterium within a membrane-derived vacuole by the neutrophil produces a phagosome; (d) fusion of lysosomes with the phagosome results in the formation of a phagolysosome; and (e) degradation of the bacterium.

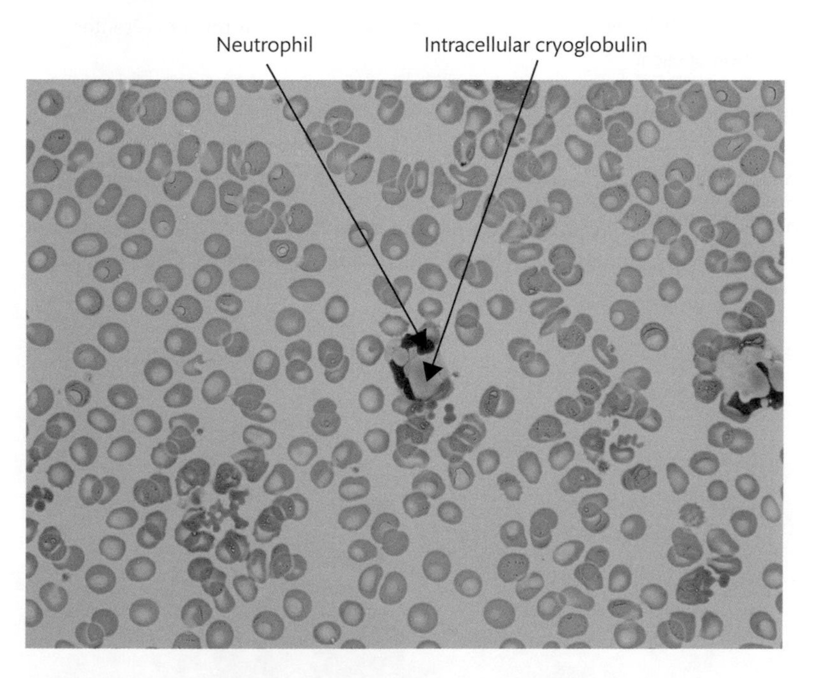

FIGURE 8.17
Neutrophils engulfing cryoglobulins. Cryoglobulin is a type of abnormal protein that precipitates at low temperatures and is usually found in patients with plasma cell-related diseases. In this case, neutrophils have recognized cryoglobulins as abnormal and are utilizing their phagocytic properties in order to remove the cryoglobulin precipitates. In this case, the patient appears thrombocytopenic. Image courtesy of Jackie Warne, Haematology department, Queen Alexandra Hospital, Portsmouth.

8.2.3 Eosinophils

Eosinophils follow the same developmental pathway as neutrophils. However, their concentration in peripheral blood is much lower, with values expected to be within the region 0.02–0.5 × 10^9/L. From the myelocyte stage of maturation, eosinophil-specific granules start to form.

Eosinophils have a bilobed nucleus and large orange–red secondary granules which bind the stain eosin due to the high concentration of bactericidal arginine-rich basic proteins, including **major basic protein (MBP)**, **eosinophil peroxidase**, and **eosinophil cationic protein (ECP)**.

Eosinophil-specific granules also contain **eosinophil-derived neurotoxin/eosinophil protein X (EDN/EPX)**. These proteins are described in a little more detail in the following list:

MBP is known to disrupt the lipid bilayer of parasites and targets cells through interactions with anionic regions on these targets. This disruption results in an increase in membrane permeability and subsequent cell damage.

Eosinophil peroxidase is a haem-containing enzyme of the same family as MPO. Eosinophil peroxidase is bactericidal and can form reactive singlet oxygen and hypobromous acid in the presence of H_2O_2 and bromide.

ECP contains a large number of arginine residues and has ribonuclease (RNase) activity. Independent of its RNase function, ECP is known to be bactericidal and helminthotoxic (toxic to **helminths**) and can also induce degranulation of mast cells.

helminth
A parasitic worm commonly found within the intestines.

EDN/EPX are indistinguishable from one another and are actually believed to be the same protein. EDN/EPX is not restricted to eosinophils, but can also be found in basophils and monocytes. It has the same degree of bactericidal or helminthotoxic activity as ECP or MBP, although as the name suggests, EDN/EPX possesses neurotoxic activity.

SELF-CHECK 8.6

List the contents of eosinophilic secondary granules and provide, in note form, a one-sentence summary of the function of each.

Typical eosinophil morphology is shown in Figure 8.18.

Peripheral blood contains relatively few eosinophils in comparison to the BM and tissues. BM holds significant reserves of eosinophils in the same manner as the neutrophils previously described.

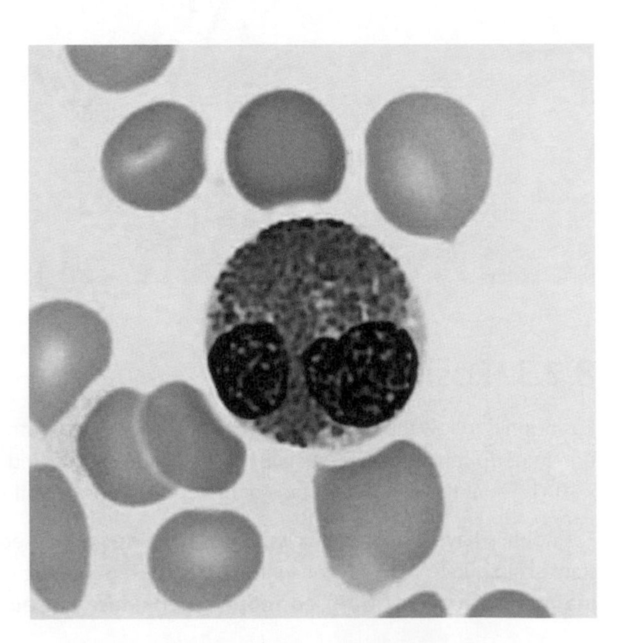

FIGURE 8.18
A typical eosinophil. The nucleus is bilobed and there are distinctive orange (eosinophilic) granules within the cytoplasm.

Image courtesy of Gordon Sinclair, Royal Marsden Hospital.

The development of eosinophils within the BM is regulated by interleukin 3, interleukin 5, and GM-CSF. Interleukin 5, secreted by a subset of lymphocytes called CD4$^+$ T$_H$2 cells, regulates eosinophils specifically and has been demonstrated to prolong their lifespan. This is an important feature, enabling eosinophils to fulfil their function in limiting helminth infections.

Cross reference

T$_H$2 cells are discussed in more detail in the T-cell development section of this chapter (in section 8.4).

Eosinophils are known to be important mediators of allergic (hypersensitivity) reactions. In hypersensitivity reactions, eosinophils counteract the effects of basophils and mast cells. A number of compounds, including prostaglandins, inhibit the release of vasoactive amines from basophils and mast cells. Eosinophils also play a key role in the destruction of helminth parasite infections, including filariasis, hookworm, schistosomiasis, and trichinosis.

Individuals infected with the *Schistosoma* parasite, who have the disease schistosomiasis, demonstrate eosinophilia (meaning an increased number of eosinophils) in their peripheral blood. Following the opsonization of *Schistosoma* with IgG antibodies, eosinophils bind to the Fc fragment of the IgG using FcγR (receptors specific for the Fc fragment of IgG antibodies), inducing eosinophil degranulation and subsequent death of the helminth. MBP, eosinophil peroxidase, ECP, EDN, and **Charcot-Leyden crystal protein** have all been shown to have a regulatory effect on helminth infections.

Charcot-Leyden crystal protein

A type of enzyme found within eosinophils and basophils that has the propensity to form crystals. Evidence of crystals in body fluids indicates allergy.

Eosinophilia describes an eosinophil count greater than 0.5×10^9/L—usually attributable to allergy or inflammation. However, in some situations an eosinophilia could be the consequence of Hodgkin lymphoma, chronic eosinophilic leukaemia, or idiopathic hypereosinophilic syndrome. In cases of prolonged eosinophilia, there is a significant risk of tissue damage following eosinophil degranulation. This tissue damage is characterized by tissue fibrosis and may ultimately lead to organ failure. In patients with prolonged eosinophilia, it is important to establish the cause and treat appropriately in an attempt to limit organ damage secondary to fibrosis. Figure 8.19 shows a blood film taken from a patient demonstrating eosinophilia.

8.2.4 Basophils

Basophils are the least common leucocyte in the peripheral blood, with a concentration of $0.02–0.1 \times 10^9$/L. The nuclear structure of basophils is indented or comprises two nuclear lobes, and their cytoplasm is rich in purple-black granules, as shown in Figure 8.20. The granules are generally so large and numerous that they obscure the outline of the nucleus.

Basophils, accompanied by mast cells, are important in both allergic and anti-helminth responses, and when ligated by IgE, crosslinking of FcεRI (IgE receptor) occurs. This leads to the

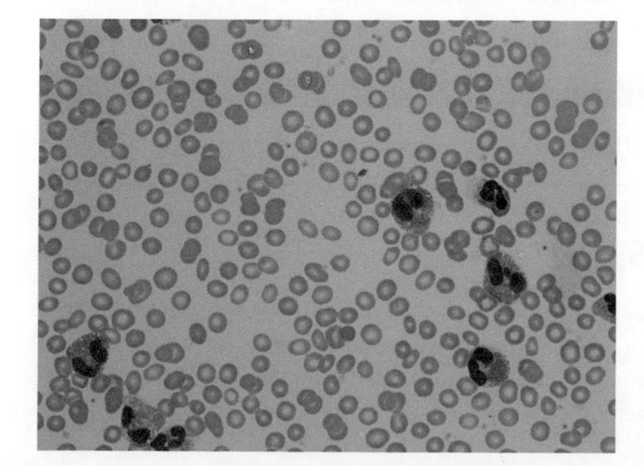

FIGURE 8.19

A patient with a severe allergy showing eosinophilia. Note the prominent orange granules within the cytoplasm of the eosinophils. Image courtesy of Jackie Warne, Haematology department, Queen Alexandra Hospital, Portsmouth.

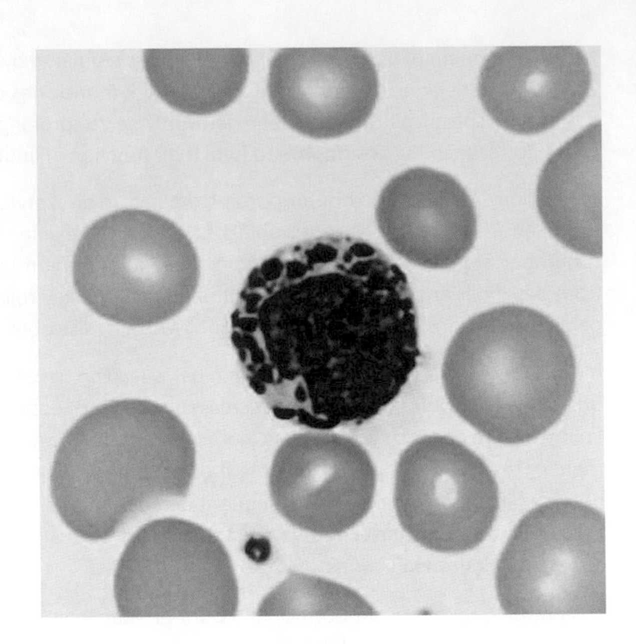

FIGURE 8.20
A normal basophil. Note the purple-black granules within the cytoplasm and the indented shape of the nucleus. Image courtesy of Gordon Sinclair, Royal Marsden Hospital.

activation of basophils, secretion of a range of signalling chemicals including IL-4, and either an immediate hypersensitivity or delayed-type reaction and the initiation of T_H2, $CD4^+$ T-cell differentiation.

basophilia
A term used to describe an increased number of basophils in the blood, beyond that of the upper limit of the reference range. Basophilia is also used to describe the deep blue staining appearance of cell cytoplasm, for example in reactive lymphocytes, plasma cells, or nucleated red blood cells.

A basophil count of greater than 0.1×10^9/L is described as **basophilia**. A basophilia may be found in cases of acute hypersensitivity reactions, sometimes during viral infection, and may also be associated with haematological malignancies—in particular, chronic myeloid leukaemia (CML). Advancing CML is associated with progressive basophilia.

IL-3 is a critical element in the basophilia observed in helminth infections and is believed to primarily originate from the T-cell response to these infections. It is also clear that basophils themselves can synthesize IL-3, further enhancing the concentration of basophils within the peripheral blood, and simultaneously expanding the concentration of mast cells, dendritic cells, eosinophils, and monocytes.

Cross reference
For further details about CML, please refer to section 11.4 on myeloproliferative neoplasms section in Chapter 11.

The basophil-specific granules contain acid mucopolysaccharides such as heparin, and large quantities of histamine. Degranulation of basophils results in the release of histamine and acute inflammation. As previously described, the release of histamine plays an important role in increasing vascular permeability.

SELF-CHECK 8.7

Briefly describe the biological role of basophils.

8.3 Monocytes

Monocytes are found in the peripheral blood at a concentration of $0.2–1.0 \times 10^9$/L. The process of monocyte maturation is much simpler to follow than that of the granulocyte. Morphologically, monoblasts cannot readily be distinguished from the myeloblast, although as monoblasts mature they characteristically appear more like the monocytes we recognize in the peripheral blood.

SELF-CHECK 8.8

Outline the different stages of monocyte development.

A promonocyte is a large BM-derived cell with a large, indented nucleus. Unlike promyelocytes, promonocytes tend not to demonstrate nucleoli and appear agranular using light microscopy. It takes approximately five days for the formation and release of a monocyte into the peripheral blood. Monopoiesis is outlined in Figure 8.21.

Monocytes have a large, highly indented, or C-shaped nucleus accompanied by greyish-blue cytoplasm and small numbers of azurophilic granules, the contents of which play an important role in the degradation of ingested particles. Typical monocyte morphology is demonstrated in Figure 8.22.

The immunophenotype of normal macrophages and their precursors is shown in Table 8.2.

Monocytes are not effector cells, but rather a short (two or three days) blood-borne stage in the development of tissue macrophages. In cases of infection, a monocytosis will often accompany a neutrophilia. The neutrophils act as the first line of defence against foreign pathogens, and monocytes are recruited to support neutrophils in the removal of pathogens and debris within the tissues to enable healing to occur.

Once monocytes enter the tissues, they mature to become macrophages, or **histiocytes**. Examples of macrophages include Kupffer cells of the liver, lung alveolar macrophages, and intraglomerular mesangial cells. Tissue macrophages continue synthesizing DNA and undergo mitosis. After several divisions, they can no longer undergo mitosis, and at this stage are considered to be mature macrophages. Mature macrophages may undergo the process of **endomitosis** until they produce *giant cells*, often found in areas of chronic inflammation. Unlike monocytes in the peripheral blood, macrophages may survive for months in the tissues.

endomitosis
A process involving the replication of a cell's nucleus in the absence of cytokinesis (cell division).

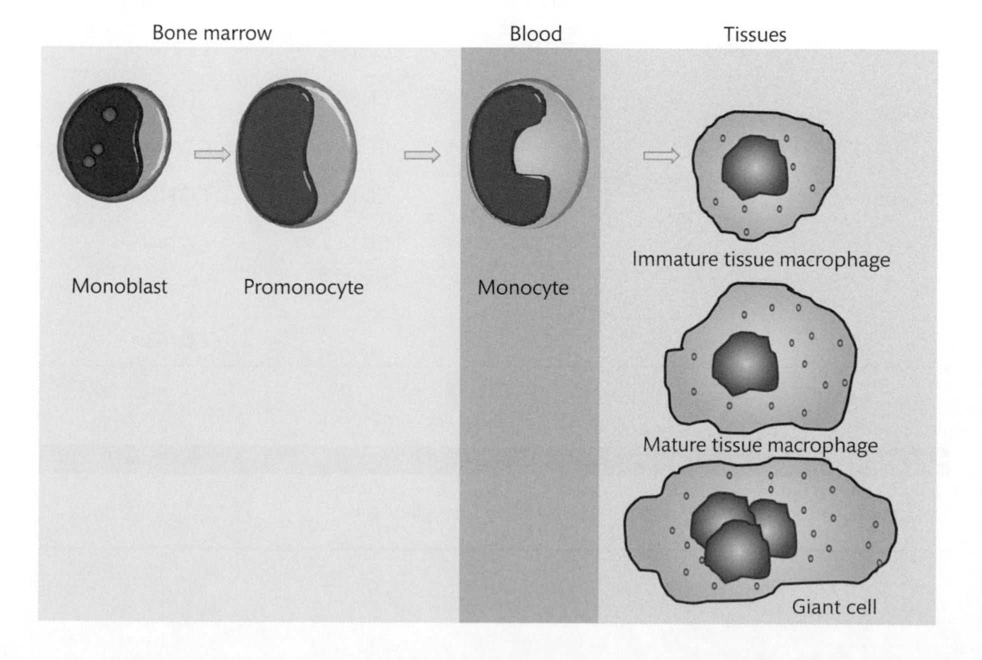

FIGURE 8.21
Monopoiesis. Monocytes begin the developmental process as monoblasts within the BM. Mitosis and maturation result in the production of larger promonocytes. Promonocytes mature further into monocytes, which are typically found within the peripheral blood. Monocytes are an intermediate stage of maturation, and following exit from the blood vessel lumen into extravascular tissue, monocytes mature into tissue macrophages following a number of mitotic cycles.

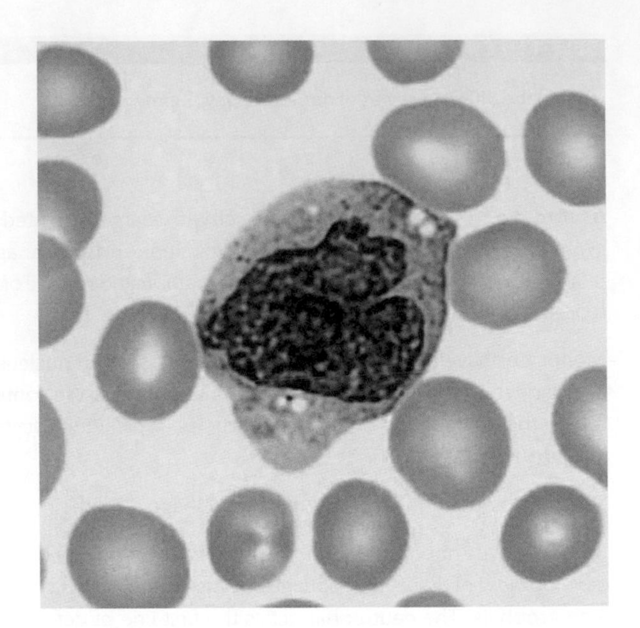

FIGURE 8.22
Peripheral blood monocyte. This monocyte has a characteristic c-shaped nucleus with pale blue cytoplasm. The cytoplasm in this monocyte contains vacuoles. Image courtesy of Gordon Sinclair, Royal Marsden Hospital.

TABLE 8.2 Normal monocyte lineage CD markers expressed during maturation.

Haemopoietic stem cell	Monoblast	Promonocyte	Monocyte	Macrophage
CD34 + +v	CD34 +			
	$CD4^+$	$CD4^+$	$CD4^+$	$CD4^+$
	CD13 +	CD13 +	CD13 +	CD13 +
	CD33 +	CD33 +	CD33 +	CD33 +
	HLA-DR +	HLA-DR +	HLA-DR +	HLA-DR +
	MPO -	CD15 +	CD15 +	CD15 +
		CD36 +	CD36 +	CD36 +
		CD64 +	CD64 +	CD64 +
		CD11b +	CD11b + +	CD11b +
		CD14 +	CD14 + +	CD14 + +
				CD16 +
				CD163 +

SELF-CHECK 8.9

Describe the process through which monocytes become giant cells.

8.4 Lymphocytes

Lymphocytes are mononuclear cells of variable size and are the second most abundant type of leucocyte found in the peripheral blood, accounting for 20–40% (1.0–3.0 × 10^9/L) of nucleated cells. A typical example of a small lymphocyte is shown in Figure 8.23. Approximately 75% of circulating lymphocytes are T cells, although morphologically, differentiation between T cells and B cells cannot be accomplished reliably using Romanowsky-stained blood films, and therefore immunophenotyping should be performed when identification of these subspecies of lymphocytes is required. Between 10 and 15% of lymphocytes are B cells and the remaining 10–15% are **large granular lymphocytes (LGL)**, as shown in Figure 8.24.

Cross reference

More detail regarding immunophenotyping can be found in Chapter 10 of this book.

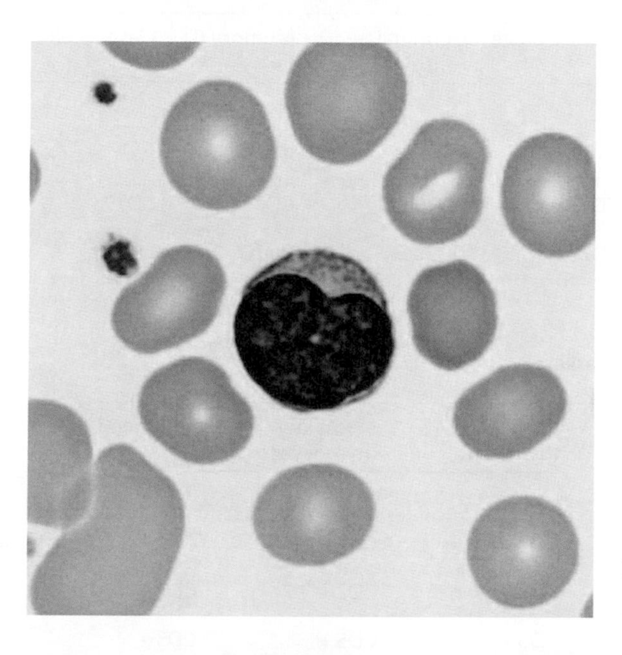

FIGURE 8.23
A single lymphocyte. The nucleus occupies the majority of the cell, with limited cytoplasm surrounding the nucleus. Image courtesy of Gordon Sinclair, Royal Marsden Hospital.

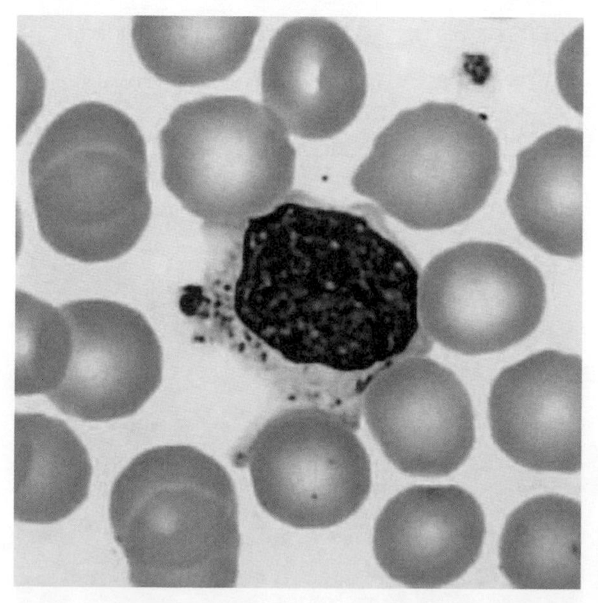

FIGURE 8.24
Large granular lymphocyte (LGL). The cytoplasm appears voluminous, punctuated with discrete vivid violet granules. © Sysmex Europe GmbH 2015. Image is under copyright of Sysmex Europe GmbH or published with the consent of its owner.

The immunophenotype of normal B-lineage and T-lineage cells is shown in Table 8.3 and Table 8.4, respectively.

Small lymphocytes have a round nucleus and a thin, often described as 'scanty', rim of cytoplasm with condensed chromatin. *Large lymphocytes* tend to have abundant cytoplasm with a more open, almost lacy, chromatin configuration. The cytoplasm of lymphocytes has a blue coloration when exposed to Romanowsky stain. Following activation, the cytoplasm becomes basophilic.

TABLE 8.3 Normal B-cell CD marker expression.

Pro-B	Pre-B	Immature B	Mature naive B	Germinal centre B	Memory B	Plasma cell
TDT	TDT	TDT				
PAX5	PAX5	PAX5	PAX5	PAX5	PAX5	
	CD79a	CD79a	CD79a	CD79a	CD79a	CD79a
		CD20	CD20	CD20	CD20	

TABLE 8.4 Normal T-cell CD markers.

Prothymocyte	Subcapsular thymocyte	Cortical thymocyte	Medullary thymocyte	Peripheral T cell
TDT	TDT	TDT		
CD7 +	CD7 +	CD7 +	CD7 +	CD7 +
	CD2 +	CD2 +	CD2 +	CD2 +
	CD5 +	CD5 +	CD5 +	CD5 +
	CD3 cyto	CD3 surface	CD3 surface	CD3 surface
	CD4 *and* CD8	CD4 or CD8	CD4 or CD8	CD4 or CD8
		CD1a		

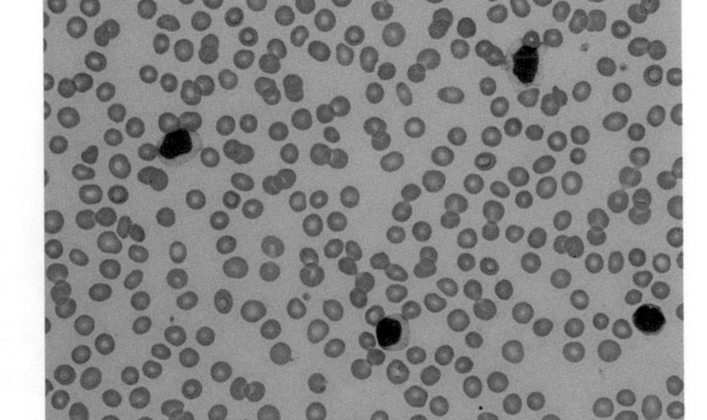

FIGURE 8.25
NK-cell leukaemia. Large numbers of circulating NK cells can be seen. Note the voluminous cytoplasm with the cytoplasmic granules when compared with the normal lymphocyte in Figure 8.16. Image courtesy of Jackie Warne, Haematology department, Queen Alexandra Hospital, Portsmouth.

Between 10 and 15% of lymphocytes are called LGLs because they are larger than the average lymphocyte and include cytoplasmic azurophilic granules. The LGL population includes *natural-killer (NK) cells*, important in antibody-dependent, cell-mediated cytotoxicity, and also in the destruction of infected cells.

Figure 8.25 shows a photomicrograph of a blood film of a patient diagnosed with NK-cell leukaemia. Although the morphology is not quite representative of a normal NK cell, cytoplasmic granules are present within the cytoplasm of these lymphoid cells.

A raised peripheral blood lymphocyte count (called **lymphocytosis**) is often associated with viral infections, but may also be seen in bacterial infections and some haematological malignancies. Following their activation, B and T lymphocytes are described as *'reactive'* or *'atypical'* in morphology. These cells are polyclonal and are a completely normal response to a range of stimuli. Following activation, small, resting lymphocytes develop the capacity to undergo mitosis and acquire the appearance of immature cells—large, with open chromatin configuration and increased DNA synthesis. In the case of B cells, activated lymphocytes begin to synthesize and secrete immunoglobulin, leading to the development of cytoplasmic basophilia. Antibody synthesis is the first step in the process leading to **plasma cell** formation—antibody-secreting cells localized within the lymphoid tissues. T cells often become atypical in morphology following viral infection of B cells, as seen commonly in the EBV-caused condition, infectious mononucleosis (glandular fever). A range of causes of atypical lymphocyte morphology is outlined in Table 8.5 below. (See also Box 8.4.)

plasma cell
Antibody-secreting terminal stage in B-cell maturation following exposure to antigen.

Other morphological changes, illustrated in Figure 8.26, are seen in lymphocytes following infection and include plasmablasts, as discussed in the B-cell maturation section in section 8.4.1 below, and Mott cells. Both of these types of cell are reactive, of B lineage, and involve the synthesis of substantial quantities of immunoglobulin. Mott cells, in particular, are morphologically very impressive and contain numerous spherical cytoplasmic inclusions, called *Russell bodies*, which are composed of immunoglobulin. Similar inclusions, which overlie the nucleus or are invaginated into it, are called *Dutcher bodies*. As well as being notable in infection, Mott cells are also seen in a number of haematological malignancies, particularly those referred to as plasma cell dyscrasias, which are outlined in Chapter 12.

Cross reference
Plasma cell dyscrasias are outlined in Chapter 12.

A reduced lymphocyte count (called lymphopenia) is a relatively common finding, usually associated with acute disease such as trauma or infection, and may also be found in patients with immunosuppression. It is important to establish the cause of the lymphopenia, and to consider whether the patient is immunosuppressed.

8.4.1 Lymphocyte development

Comprehensive coverage of **lymphopoiesis** is provided in the *Clinical Immunology* text in this series, although the development of lymphocytes will be addressed here in sufficient detail to provide a background for haematologists.

lymphopoiesis
The growth, differentiation, and maturation of lymphocytes within primary lymphoid organs.

Lymphocytes provide the body with *adaptive immunity*. Adaptive immunity allows the body to develop a specific, tailored response to foreign antigens.

The process of lymphocyte development begins at the stem cell stage within the foetal liver and adult BM. The haemopoietic stem cell differentiates into *common lymphocyte progenitors* (CLP). Exposure of CLP to a variety of lymphoid-specific cytokines selects for either a population of B cells (so-called because B cells were discovered in birds in the Bursa of Fabricius) or T cells. T cells are so-called because they require a stage of maturation within the thymus, whereas B cells develop within the BM. Because the BM and thymus are the initial sites for the development of B and T cells, they are called **primary lymphoid organs**.

Cross references
More detail regarding the process of lymphopoiesis can be found in the *Clinical Immunology* text in this series.

More information on immunophenotyping can be found in Chapter 10 of this book.

TABLE 8.5 Causes of atypical lymphocytes.

Infection	Drugs and toxic reactions	Autoimmunity
Epstein–Barr virus	Hydantoin drugs	Rheumatoid arthritis
Cytomegalovirus	Para amino salicylic acid	Idiopathic thrombocytopenia purpura
Toxoplasma	Phenothiazine	Systemic lupus erythematosus
Q fever	Organic arsenicals	Autoimmune haemolytic anaemia
Rubella	Lead	Chronic hepatitis
Roseola	Trinitrotoluene	Agammaglobulinaemia
Herpes simplex	Diaminodiphenyl-sulfone	
Haemorrhagic fever		**Malignancy**
Herpes zoster	**Post-perfusion syndrome**	
Rickettsial pox		Hodgkin lymphoma
Rubeola	**Immunizations**	
Mumps		**Idiopathic disorders**
Adenovirus	**Radiation**	
Influenza		Sarcoidosis
Tuberculosis	**Hormonal stress**	Carcinomatous neuropathy
Varicella		Guillain-Barre syndrome
Syphilis	Adrenaline	Myasthenia gravis
HIV (1 and 2)	Addison's disease	Acute disseminated encephalomyelitis
Hepatitis (A and B)	Glucocorticoid deficiency	
Listeria monocytogenes	Hypopituitarism	**Graft rejection**
Mycoplasma pneumonia	Thyrotoxicosis	

Adapted from Simon MW. ***The atypical lymphocyte***. *International Pediatrics* 2003: **18(1)**: 20–2.

BOX 8.4 Describing WBC numerical abnormalities

It is always important to use the correct terminology when describing quantitative abnormalities of WBCs.

The prefix to the noun always denotes the species of WBC being described, whereas the suffix (emboldened here) denotes the quantitative abnormality.

So for an increased cell number, cell names ending in -phil become -philia, whereas cell names ending in -cyte become -cytosis:

Neutro**philia** = increased neutrophils

Eosino**philia** = increased eosinophils

Baso**philia** = increased basophils

Monocy**tosis** = increased monocytes

Lymphocy**tosis** = increased lymphocytes

Leucocy**tosis** = increased number of white cells.

For a decrease in cell numbers, the suffix becomes -penia:

Neutro**penia** = decreased neutrophil count

Eosino**penia** = decreased eosinophil count

Baso**penia** = decreased basophil count

Monocyto**penia** = decreased monocyte count

Lympho**penia** = decreased lymphocyte count

Leuco**penia**—decreased white cell count.

B-cell maturation

The first recognizable stage in the development of a B cell is the **progenitor (Pro)B cell**. The development of pro-B cells to **pre-B cells** requires interaction with BM stromal cells. The pre-B cell shows evidence of immunoglobulin M (IgM) heavy-chain synthesis within the cytoplasm. Immunoglobulins are composed of two identical heavy chains and two identical light chains. As pre-B cells mature, light chains are synthesized and combine with the heavy chains,

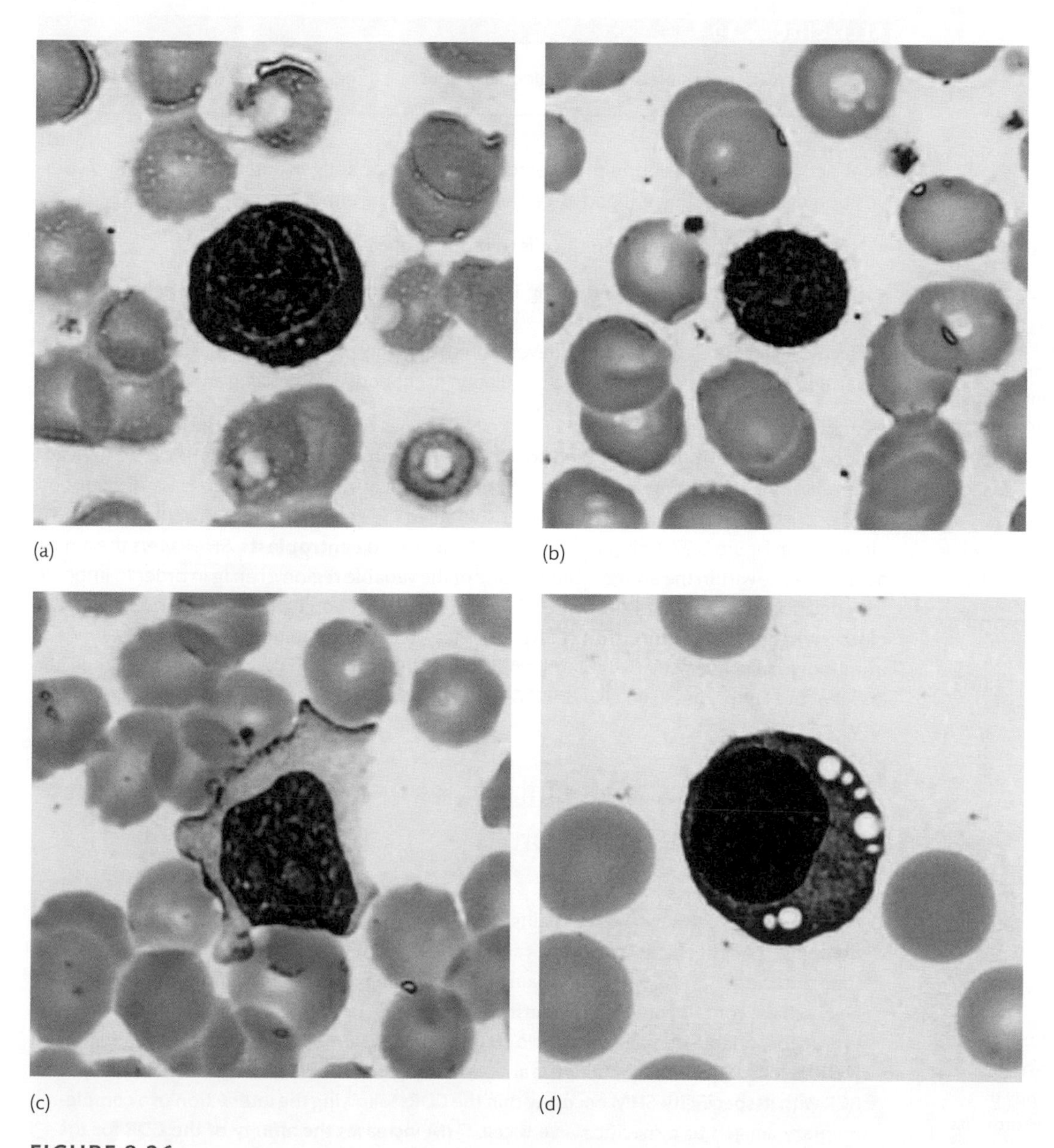

FIGURE 8.26

A range of morphological changes associated with infection. (a) and (b) are both reactive lymphocytes showing a high nucleocytoplasmic ratio and pronounced cytoplasmic basophilia; (c) shows an atypical lymphocyte with scalloped edges, most commonly associated with infectious mononucleosis; and (d) a plasma cell with small Russell bodies present within the basophilic cytoplasm. Images courtesy of Gordon Sinclair, Royal Marsden Hospital.

B-cell receptor (BCR)
The abbreviation BCR should be used with caution as it is context specific. In the context of immunology BCR represents the B-cell receptor, whereas in cytogenetics BCR represents breakpoint cluster regions.

forming a complete surface-bound, **B-cell receptor (BCR)**. At this stage, the cell is called an **immature B cell**. Further BCR changes to the B cell result in the surface expression of IgD. Co-expression of IgM and IgD surface receptors is found on **mature B cells** released from the BM into the peripheral blood. These mature B cells are called **naive B cells** (also unactivated or virgin B cells) because they have yet to encounter antigen. Stages of B-cell maturation can be determined by measuring the expression of different proteins on the cell surface by flow cytometry (immunophenotyping).

SELF-CHECK 8.10

Describe the pattern of surface immunoglobulin expression throughout B-cell development.

variable region (or 'v region')
A region of an antibody that binds to an epitope on an antigen. All antibodies secreted by a particular plasma cell have the same variable region, although antibodies synthesized by different plasma cells have dissimilar variable regions.

B cell activation depends upon the interaction of a specific antigen and the structure and composition of that activating antigen. B-cell activation can be either:

- thymus-dependent—requiring T cells to produce effector B cells; or
- thymus-independent—not requiring T cells for the production of effector B cells.

The activation of B cells results in the development of highly proliferative behaviour, enabling the expansion of the activated B cell into a population of antibody-synthesizing and -secreting B cells.

centroblasts
Highly proliferative cells found within the early germinal centre following antigenic stimulation.

class-switching recombination (CSR)
The process of generating different types of immunoglobulin.

As activated B cells proliferate, certain Ig-specific genes encoding the **variable region** of the BCR undergo point mutation—called **somatic hypermutation (SHM)** (see Box 8.5). The variable region of an Ig generates antigen-binding diversity within a specific class of antibody. SHM occurs within a region of lymphoid tissue called the germinal centre. A germinal centre is illustrated in Figure 8.27. Cells undergoing SHM are called **centroblasts**. SHM alters the amino acid sequence within the antigen-binding site of the variable region of an Ig in order to improve the specificity of the BCR for the stimulating antigen. In addition to SHM, centroblasts undergo **class-switching recombination (CSR)**, a process necessary for generating different types of immunoglobulin (particularly IgG and IgA). Combining CSR and SHM generates highly specific and reactive antibodies. (See also Box 8.6.)

BOX 8.5 Antibodies and somatic hypermutation

An antibody contains two heavy chains and two light chains, each of which is composed of domains. The light chains (known as κ and λ, based on their structure) contain a single constant domain (C_L) and a variable domain (V_L). Depending on the class of antibody produced, heavy chains contain three or four constant domains (C_H) and a single variable domain (V_H). Within the variable domains of the two chains, the **complementarity-determining region** (CDR) is found. This CDR contains the antigen-combining site and provides the antibody or BCR with its specificity. SHM occurs within the CDRs following the interaction of a complementary antigen to a specific naive B cell. SHM increases the affinity of the CDR for the antigen, improving the effectiveness of the immune response.

complementarity-determining region
The region of an antibody that complements the structure of its associated antigen.

Assessing V_H hypermutation plays an important role in establishing prognosis in certain B-cell malignancies. In B-cell chronic lymphocytic leukaemia (B-CLL), malignant lymphocytes showing evidence of SHM have been correlated with a good prognosis when compared with patients failing to show evidence of SHM.

Cross references
For more information regarding CLL, see Chapter 12.
Apoptosis is outlined in Chapter 9.

Following SHM and CSR, **centrocytes** (as they are called at this stage) migrate to regions of secondary lymphoid organs where **follicular dendritic cells (FDCs)** are concentrated. FDCs express antigen, and are one of a number of different types of **antigen-presenting cell (APC)**. B cells expressing receptors with the highest affinity for the antigens presented on the FDC will be selected, expanding the population of B cells specific for that antigen. B cells expressing non-functional receptors or receptors with a weak affinity for the presented peptide will undergo **apoptosis**.

centrocyte
A small, non-dividing cell found within the germinal centre. The nucleus contains a cleft.

follicular dendritic cell (FDC)
A cell with long, branching processes found within lymphoid follicles. FDCs are important for enabling B-cell maturation.

antigen-presenting cell (APC)
Any cell that expresses MHC class II and can present antigenic material to elicit a T-cell response.

apoptosis
The process of programmed cell death.

SELF-CHECK 8.11

Describe the process of SHM and explain how this process alters the affinity of antibodies for their target antigens.

Following positive selection by FDCs, plasma cells are produced—secreting clones of antigen-specific antibodies. The lifespan of plasma cells is approximately 14 days, and in order for long-lived immunity to develop, a proportion of centrocytes must mature into **memory cells**. Memory cells provide a long-term record of previously encountered antigens, allowing a secondary immune response to occur following interaction with a previously recognized antigen. Indeed, recurrent exposure to the same antigen promotes reactivation of memory B cells and differentiation into transient cells called **plasmablasts**. Plasmablasts have undergone CSR, secrete IgG, or IgA, and peak in concentration within the blood six or seven days after infection or immunization. Plasmablast concentration appears to decrease to normal levels within two to three weeks following acute viral infection, although this varies according to the duration of the infection.

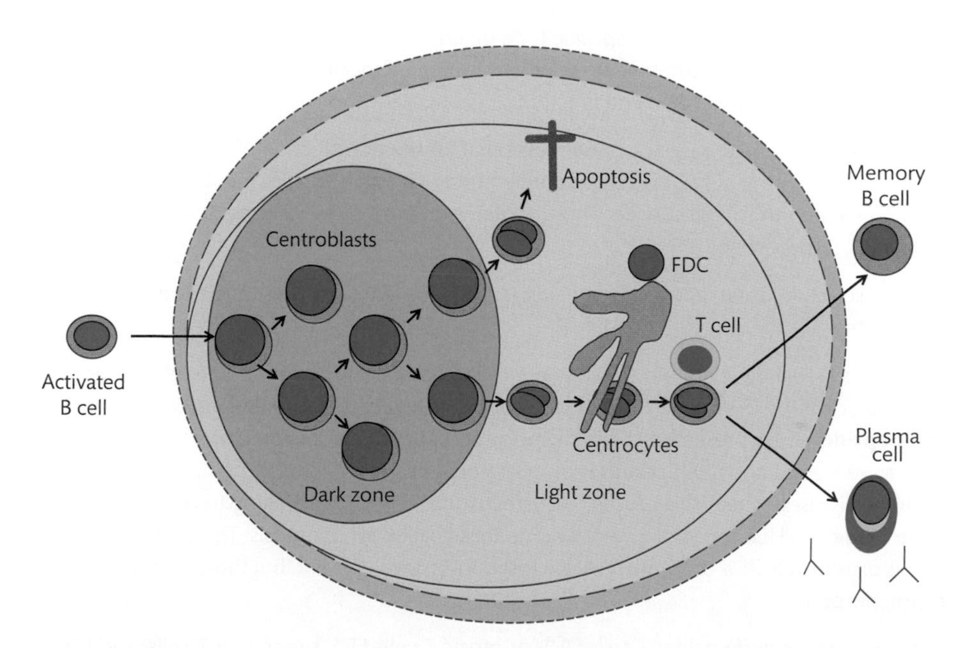

FIGURE 8.27
Structure of a germinal centre. Activated B cells enter the germinal centre from the T zone. Inside the germinal centre, the B cells become larger and, now called centroblasts, undergo repeated rounds of cell division and hypermutation in the variable region of an Ig within the dark zone. Following cell division, centrocytes enter the light zone where—if they encounter FDCs—they survive; otherwise, they are signalled for apoptosis. A final encounter with antigen-specific T cells is required before memory cells and plasma cells are produced.

BOX 8.6 The germinal centre

Germinal centres are found within lymphoid follicles. Following B-cell activation within an intrafollicular area, the B cell will enter a primary lymphoid follicle and begin to proliferate. Within three days, a germinal centre forms and acts as a 'nursery' for the developing centroblasts. The germinal centre is composed of two distinctive anatomical sites: the *dark zone* and the *light zone*. Within the dark zone are actively proliferating centroblasts. As these centroblasts reduce their rate of proliferation, they migrate to the light zone. Within the light zone, non-proliferating centrocytes (as the developing centroblasts are called at this stage) interact with large numbers of FDC and T_H cells (helper T cells, described in the next section). Centrocytes failing to appropriately interact with human leucocyte antigen (HLA)-expressed antigenic peptides undergo apoptosis, ensuring only effective centrocytes remain. Finally, the FDC-selected centrocytes mature into either antibody-secreting plasma cells or memory B cells.

T-cell development

T cells play a central role in the regulation of the immune system. Immature T cells leave the bone marrow or foetal liver and enter the peripheral blood. From here, immature T cells enter the thymus and are called **thymocytes**, where they will develop the surface expression of mature T-cell CD markers. On the thymocyte surface, specific molecules are expressed to enable T-cell function. T-cell receptors (TCRs) are expressed on the T-cell surface following the selection of variable region genes encoding the TCR. T-cell receptors show diversity matched only by the BCR, with each T cell specific for a different antigen. The ability to select different variable region genes and combine these provides antigen-specific diversity, enabling an adaptive immune system to develop and function.

Accompanying the TCR is CD3, a protein essential for the transduction of signals into the cell following the binding of the TCR to its complementary antigen. In addition, CD4 and CD8 are co-expressed on the T-cell surface. These proteins interact with HLA class II and HLA class I molecules, respectively.

Thymic selection then occurs, producing single positive $CD4^+$ or single positive $CD8^+$ thymocytes.

Not all T cells will develop into peripheral T cells (those found within the peripheral blood). The interaction of $CD4^+$ and $CD8^+$ thymocytes with each class of HLA molecules enables **self-peptides** to be presented to the thymocytes. In order to function correctly, thymocytes need to recognize and interact with HLA, but should not recognize self-peptides—as in autoimmune diseases, where this process is dysregulated. Thymocytes failing to bind to HLA, or those binding to HLA too strongly—or avidly—are signalled for apoptosis. This process ensures that approximately 75% of thymocytes are lost during development, but those that survive are 'fit-for-purpose'.

self-peptides
These are short protein sequences presented to developing lymphocytes in order to inactivate self-reactive lymphocytes.

T-cell subtypes include helper T cells (T_H), cytotoxic T cells (T_C), suppressor T cells, and LGLs, each having different functions. These functions are outlined below:

Helper T cells are subdivided into two types: T_H1 and T_H2. T_H1 cells promote the action of macrophages and another subtype of T cell called cytotoxic T cells (T_C). Interleukin 2 (IL-2) and gamma-interferon (IFN-γ) are secreted by T_H1 cells modulating cell-mediated immunity.

B-cell activation and maturation is mediated by the T_H2 subsets through the secretion of IL-4 and IL-10.

Cytotoxic T cells express surface CD8 (abbreviated to $CD8^+T_C$), which interacts with HLA class I. HLA class I is expressed on most cells of the body, and therefore provides an excellent tissue non-specific way of presenting viral peptides to $CD8^+T_C$ cells in order to induce cellular destruction. In contrast, HLA class II is expressed on cells of the immune system only.

The cytoplasm of T_C contains granules which are directed to the site of receptor binding following interaction between HLA class I-presented peptides and CD8. Degranulation of T_C causes the release of perforins and serine proteases called granzymes, causing osmotic lysis and apoptosis of infected cells.

We have previously described the morphological appearance of LGLs. Of this LGL population, natural killer (NK) cells have an important function. NK cells express neither TCR nor BCR, but they are able to recognize antigens without being restricted in their function by antigen-specificity. Being cytotoxic T cells, NK cells have the ability to destroy infected cells through a process called *antibody-directed cellular cytotoxicity* (ADCC). Receptors to IgG antibodies, called FcγR—expressed on the NK cell surface—bind to antigen-bound IgG antibodies. Internal granular rearrangement within the NK cell focuses granules onto the area in which the antibody is bound, initiating degranulation and releasing perforins and granzymes.

Cross reference

Interactions between B cells and T cells are discussed in greater detail in the *Clinical Immunology* text in this series.

SELF-CHECK 8.12

Outline the role of cytotoxic T cells (T_C) in controlling bacterial infections.

Interactions between B and T cells

B cells are particularly efficient APCs. The expression of antigenic peptides by HLA class II activates antigen-specific $CD4^+$ T_H cells. Binding to HLA class II-presented peptides enables $CD4^+$ T_H cells to secrete a range of cytokines which transform and activate B cells. Co-activation of the B cell through the interaction of CD40 and T cell expressed CD40L (L = ligand) is required to prevent B-cell apoptosis.

Now that lymphocytes and their subsets have been considered in some detail, we should examine some of the causes of numerical, morphological, and functional abnormalities of lymphocytes.

8.4.2 Infectious mononucleosis

First described by Nil Filatov in the 1880s and called infectious mononucleosis (IM) by Sprunt and Evans in 1920, IM is a benign **lymphoproliferative** disorder characterized by an EBV-mediated infection of B cells. EBV is a member of the human herpesvirus family that utilizes the CD21 receptor on the lymphocyte membrane to enter B cells. Infected B cells express viral antigens on their surface, initiating a rapid $CD8^+$ cytotoxic T-cell response and latterly—following B-cell activation—the production of EBV-specific antibodies.

lymphoproliferative

Inducing cells of lymphoid origin to proliferate at a rate greater than that usually observed in healthy individuals.

EBV epidemiology

EBV infection occurs commonly in two defined age ranges: children aged 1–6 years and individuals aged 14–20 years. EBV is spread through the saliva of infected individuals, and viral particles have also been identified in other bodily fluids. IM is commonly referred to as 'the kissing disease' because of the association between viral spread and this popular pastime amongst teenagers and young adults. It appears that EBV infection in young children is probably contracted from parents or siblings, and individuals requiring medical interventions including blood transfusion, stem cell transplantation, and solid organ transplantation can also be at risk of infection. Worldwide, approximately 80–90% of adults are **seropositive**, suggesting historical EBV infection. However, IM is not recognized so readily in under-developed parts of the world, probably because the age of primary infection is much lower than in developed countries.

seropositive
Identification of specific antibodies within an individual's serum.

Following an acute phase, EBV infection becomes latent, with the virus persisting in the patient for the remainder of their life. The immune system prevents the re-emergence of the virus, although when immunosuppression occurs, re-emergence of the EBV infection can be detected.

Symptoms associated with infectious mononucleosis

Typically, symptomatic patients present with fever, malaise, and chills, which later develop into sore throat, **pharyngitis**, **exudative tonsillitis**, **palatal petechiae**, stiff neck, abdominal pain, and **lymphadenopathy**, although patients may also present with a slowly developing malaise and fatigue. In approximately 50% of cases, splenomegaly is apparent. The frequency of typical symptoms associated with IM is shown in Table 8.6. Patients exhibiting splenomegaly should avoid strenuous activity due to the risk of splenic rupture—a severe and potentially life-threatening complication of IM. Approximately 50% of patients will be asymptomatic.

pharyngitis
Inflammation of the pharynx caused by an infection.

exudative tonsillitis
Enlarged red tonsils covered in white patches.

palatal petechiae
Small blood spots within the oral cavity.

lymphadenopathy
Enlarged lymph nodes.

Differential diagnoses include CMV, toxoplasmosis, streptococcal pharyngitis, and human immunodeficiency virus (HIV). These infections will be discussed in greater detail later in this section.

Laboratory investigations

The first step in the investigation of a patient with suspected IM is a full blood count, blood film, and heterophile antibody test. The full blood count will reveal a leucocytosis, often ranging from 12 to 25 × 10^9/L, predominantly composed of lymphocytes. Approximately 75% of

TABLE 8.6 Frequency of signs and symptoms associated with infectious mononucleosis.

Signs and symptoms	Frequency (%)
Sore throat	95
Cervical lymphadenopathy	80
Fatigue	70
Upper respiratory symptoms	65
Headache	50
Decreased appetite	50
Fever	47
Myalgia	45
Hepatitis (usually subclinical)	75

lymphocytes are T cells. Lymphocytosis will usually occur after one week following the appearance of symptoms, peaking during the second or third week and lasting for up to eight weeks. Lymphocyte morphology is referred to as 'atypical' or 'reactive', and scalloping of the lymphocyte membrane around neighbouring red cells can be seen. Figure 8.28 shows a typical blood picture for a patient with IM as described above. The patient may also be neutropenic and thrombocytopenic.

Patients with IM develop **heterophile antibodies** called Paul-Bunnell antibodies. Heterophile antibodies do not react with human red cells, although they do cause agglutination of sheep, ox, and horse red blood cells. Other types of heterophile antibody, called **Forssman antibodies**, are not associated with IM, although they do occur in response to a range of infectious agents. Differentially, Forssman antibodies bind to guinea-pig kidney cells but not ox red cells, whereas IM heterophile antibodies react in the opposite fashion, binding to ox red cells but not guineapig red cells, see Table 8.7. The identification of non-Forssman heterophile antibodies provides a diagnosis of IM.

Forssman antigen
A glycosphingolipid found on cell membranes in many different species. Antibodies directed against Forssman antigens are caused by a number of infectious agents, although they are not associated with infectious mononucleosis.

Testing for heterophile antibodies involves comparing the agglutination of horse red cells following mixing predetermined volumes of patient plasma with ground guinea-pig kidney cells and ox red cells. In patients with IM, stronger reactions are seen in the test containing guinea-pig kidney cells as Paul-Bunnell (PB) antibodies *are not* adsorbed by these cells. The ox cells *do* adsorb PB antibodies, leaving fewer antibodies available to agglutinate the horse red cells. The opposite reaction occurs in the presence of Forssman antibodies.

A number of scientific reagent manufacturers have now developed tests for heterophile antibodies using latex agglutination techniques to differentiate between PB antibodies,

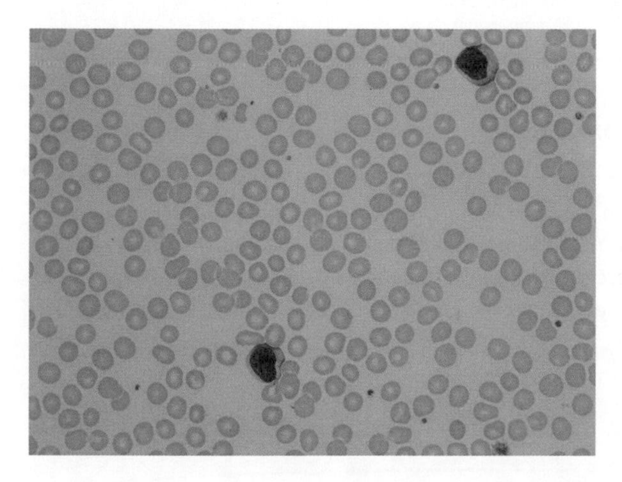

FIGURE 8.28
Blood film for a patient diagnosed with infectious mononucleosis. Two atypical lymphocytes are demonstrated, each with basophilic cytoplasm and one with membrane scalloping around juxtaposing red blood cells. Image courtesy of Sarah Bruty and Amelia Fitzpatrick, Haematology department, Southampton General Hospital.

TABLE 8.7 **The different patterns of antibody absorbance between heterophile antibodies.**

Cell origin	Antibody reaction	
	Forssman	Paul-Bunnell
Ox cells	Negative	Positive
Guinea-pig cells	Positive	Negative

Forssman antibodies, and others shown to react with multiple species. The specificity of latex agglutination tests means that differential absorption is no longer required to remove non-PB heterophile antibodies prior to testing, making latex agglutination tests much quicker and easier to use. Latex agglutination test results are shown in Figure 8.29. (See also Case study 8.1.)

It is important to recognize the limitations of the tests available. In patients infected with EBV under the age of four years, 50% fail to demonstrate PB antibodies. In older patients, 71–91% of patients exhibit PB antibodies. It is important to try to detect these antibodies at the appropriate stage in the infection to reduce the risk of obtaining a false-negative result. During the first week of infection, the false-negative rate could be as high as 25%, dropping to 5–10% in the second week. In the third week, false-negative results could still be as high as 5%.

Antibodies synthesized against the EBV-mediated antigens—*viral capsid antigen* (VCA) and *EBV nuclear antigen* (EBNA)—are far more specific for diagnosing EBV infection, and are particularly useful in patients where there is a strong clinical indication of IM, but a negative PB antibody test. As indicated in Table 8.8, the acute phase of IM is associated with the presence of IgM antibodies called anti-VCA (VCA-IM), up to four months following infection. In the 3–12 months post infection, IgG anti-VCA (VCA-Ig) will be detectable and, once present, these IgG antibodies persist for life. Anti-EBNA will be present after the first six to eight weeks and will also be present for life.

SELF-CHECK 8.13

Describe the findings you would expect from a patient referred to a haematology laboratory with suspected infectious mononucleosis.

Treating infectious mononucleosis

Patients diagnosed with IM are generally directed to rest and take analgesics such as paracetamol, aspirin, or ibuprofen to alleviate the symptoms. In approximately 30% of cases, IM is complicated with a concurrent β-haemolytic streptococcal infection. In these cases, the antibiotic amoxicillin may be prescribed, although there is a risk of skin rashes attributable to the formation of circulating immune complexes. These rashes resolve following withdrawal of the antibiotic.

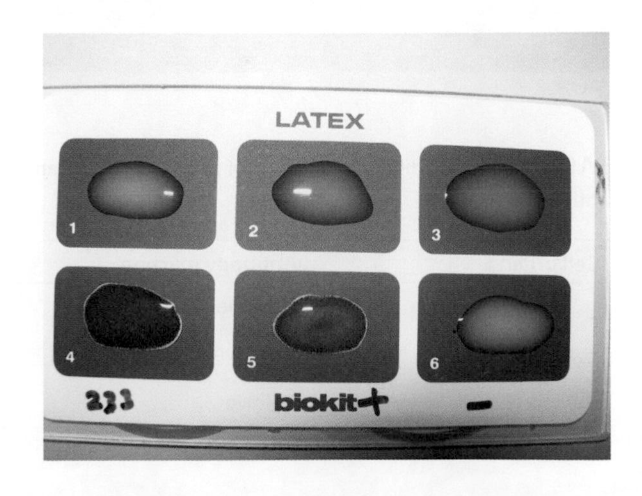

FIGURE 8.29
Results from a Biokit latex agglutination text. The two marked (+) and (−) are positive and negative controls. Note the patient next to the positive control (position 4). Clearly agglutination has occurred following the addition of patient plasma to the latex. This patient is diagnosed with infectious mononucleosis. All other patients on this slide are negative. Image courtesy of the Haematology department, Royal Bournemouth Hospital.

CASE STUDY 8.1 The laboratory investigation of a student with lethargy and sore throat

Patient history

- A 19-year-old male university student presents to his GP complaining of a two-week history of sore throat, stiff neck, and lethargy.
- Clinical examination revealed mild pyrexia with cervical lymphadenopathy and pharyngeal inflammation. Mild splenomegaly was noted through abdominal palpation.
- FBC and blood film were performed.

Results

FBC

Index	Result	Reference range
WBC	11.36	(RR 4–10 × 10^9/L)
Neutrophils	1.50	(RR 2–7 × 10^9/L)
Lymphocytes	8.18	(RR 1–3 × 10^9/L)
Monocytes	1.11	(RR0.2–1.0 × 10^9/L)
Eosinophils	0.03	(RR 0.02–0.5 × 10^9/L)
Basophils	0.54	(RR 0.02–0.1 × 10^9/L)
RBC	4.77	(RR 4.3–5.7 × 10^{12}/L)
Hb	143	(RR 133–167 g/L)
MCV	82	(RR 77–98 fL)
Plt	111	(RR143–400 × 10^9/L)

Blood film

- Atypical lymphocytes noted. Platelet numbers appear reduced. No evidence of platelet clumps or fibrin strands.

Significance of results

The patient has a normocytic normochromic blood picture, although the white cell count is raised and platelet count is reduced. The reversed ratio between neutrophils and lymphocytes, where neutrophil numbers no longer exceed lymphocyte numbers, is suggestive of a viral infection, although other causes of a lymphocytosis, including lymphoma, need to be excluded. The blood film demonstrates lymphocytes of an atypical morphology—where voluminous cytoplasm with membrane scalloping will be observed. The reduced platelet count initially observed in the full blood count is confirmed using the blood film and could be the result of immune-mediated destruction or sequestration in the patient's enlarged spleen. Based upon the age of the patient and his presenting symptoms, a diagnosis of infectious mononucleosis is likely. Laboratory staff decided to use a Biokit latex agglutination test to confirm their suspicions. A positive result was observed, confirming the suspicion of infectious mononucleosis. The patient was instructed to rest and abstain from strenuous activity until resolution of his splenomegaly. Analgesia using paracetamol and ibuprofen was recommended to help alleviate the painful symptoms.

TABLE 8.8 Staging EBV infection by enzyme immunoassay antibody results.

Stage of infection	Time after onset of illness	VCA IgM	VCA IgG	EBNA-1 IgG
EBV naive	n/a	Negative	Negative	Negative
Acute primary infection	0–3 weeks	Positive	Negative or positive	Negative
Subacute infection	4 weeks–3 months	Positive	Positive	Negative
Convalescent infection	4–6 months	Negative or positive	Positive	Negative or positive
Past infection	>6 months	Negative	Positive	Positive

EBNA = EBV nuclear antigen; EBV = Epstein-Barr virus; VCA = viral capsid antigen.

From Balfour HH, Dunmire SK, Hogquist, KA. ***Infectious Mononucleosis***. *Clinical & Translational Immunology* 2015:**4**(2);e33.

8.4.3 Cytomegalovirus

CMV is a human β-herpesvirus transmitted through saliva, breast milk, sexual contact, placental transfer, and organ (including BM) transplantation. Up to 70% of the population in industrialized nations is seropositive for CMV, although this figure is much higher in low socio-economic groups, homosexuals, and in developing countries. As with EBV, following the primary infection, CMV becomes latent, persisting lifelong and being controlled by the host's immune system. Dysregulation of the immune system can result in the re-emergence of CMV infection, with the largest reservoir of the virus comprising the myeloid compartment of the bone marrow.

immunocompetent
Describes a fully functioning immune system.

Immunocompetent, seronegative patients exposed to CMV generally exhibit many of the same symptoms as IM, outlined earlier. The majority of patients tend not to demonstrate pharyngitis, tonsillitis, or splenomegaly to the same extent, and in some cases describe the symptoms as 'flu-like'. In teenagers and older children, there is no benefit in distinguishing these conditions. However, when encountered in pregnancy it is important to differentiate between the diagnosis of EBV, CMV, and toxoplasmosis, since CMV and toxoplasmosis can affect foetal development.

In pregnancy, CMV can cross the placenta, leading to congenital CMV. CMV is the most common congenital infection in the developed world, being identified in up to 6% of live births. In order to classify CMV as a congenital infection, it must be isolated within the first three weeks of life, otherwise CMV is considered the product of a postnatal infection.

Approximately 17% of patients with congenital CMV are asymptomatic, although those demonstrating clinical features may show hepatosplenomegaly, microcephaly (small head), jaundice, intrauterine growth restriction, thrombocytopenia, and petechiae. Some of these features may not be present at birth, but develop later in childhood.

immunodeficiency
Any inherited or acquired deficiency in the immune system.

Acquired CMV in **immunodeficiency** patients results in significant morbidity and mortality. Those at greatest risk are seronegative patients receiving an organ for transplantation from a seropositive donor. If the recipient is also seropositive, then reactivation of the virus can occur. Consideration must be given, especially for patients following stem cell transplantation, that blood is typed as CMV-negative to prevent infection by CMV.

8.4.4 Human immunodeficiency virus 1 (HIV-1)

The retrovirus HIV-1 is the causative agent of acquired immunodeficiency syndrome (AIDS). First reported in 1981 in the United States following a number of cases in which patients developed rare opportunistic infections or rare skin cancers, AIDS quickly spread throughout the developed and undeveloped world. Common complications include pneumocystis pneumonia (PCP) caused by *Pneumocystis jirovecii*, Kaposi sarcoma, and primary central nervous system lymphoma. It was estimated that in 2019, 105 200 people were living in the UK with HIV. With the introduction and improvement of antiretroviral drugs, the mortality rate from AIDS-related complications in the UK has steadily declined from approximately 340 in 2009 to 270 in 2018.

HIV-1 can be transmitted from one individual to another through unprotected heterosexual or homosexual intercourse, from infected blood products, from intravenous drug abuse, or from an HIV-1 infected mother to her infant.

HIV-1 predominantly infects $CD4^+$ T cells, and are called T-tropic strains, although $CD4^+$ monocytes or macrophages can also be infected by M-tropic strains, through the binding of the

CCR5 chemokine receptor. The lipid bilayer of the viral envelope expresses a transmembrane glycoprotein (gp41) associated with gp120. Gp120 binds to the CD4 ligand, and additional binding to the chemokine receptor CXCR4 enables the virus to enter host T cells. Once internalized, the viral RNA is reverse-transcribed to DNA, called a provirus, through the action of viral reverse transcriptase. The provirus is incorporated into the host's genome, with DNA replication enabling the synthesis of new virions. The expression of virions leads to the lysis of the infected T cell and the release of the virions into the patient's blood, enabling further infection of T cells.

Following infection by HIV-1, many patients are commonly asymptomatic. Those showing symptoms may present with fever, rash, and lymphadenopathy, all of which spontaneously resolve in a few weeks. Following a protracted chronic phase, the viral load—the number of copies of viral RNA in the patient's plasma—increases and the number of $CD4^+$ T cells declines. Generally, AIDS is diagnosed when the patient's $CD4^+$ count drops below 200 cells/mm^3. Patients often succumb to opportunistic infection unless treated with one of the many effective antiretroviral agents designed to combat HIV-1.

8.4.5 Toxoplasmosis

Toxoplasma gondii is a protozoan parasite reportedly infecting approximately one-third of the world's population, and accounting for 1–2% of patients with suspected IM. Commonly, infection occurs following the ingestion of oocysts from cat faeces, contaminated and undercooked meat, or water. Approximately 10–20% of infected individuals are symptomatic, and can be diagnosed with toxoplasmosis. Clinical features are similar to those found in IM, the most common of which is cervical lymphadenopathy.

Congenital toxoplasma infection can occur, but is only of risk to foetuses of seronegative mothers as *Toxoplasma gondii* does not cross the placenta in seropositive mothers. Clinical symptoms of congenital toxoplasma are variable, including neurological symptoms, encephalitis, anaemia, jaundice, rash, and petechiae.

Immunocompromised patients may be infected following a transplant from an infected individual; but, more commonly, infections are associated with the reactivation of a latent infection following the suppression of T lymphocytes. In immunocompromised patients, symptoms include pneumonitis, encephalitis, and myocarditis which, if left untreated, are associated with significant morbidity and mortality.

8.5 Plasma cells

Plasma cells are antibody-secreting cells found within the BM and lymphoid organs. In a healthy individual, plasma cells would not be expected to be seen in the peripheral blood. However, occasionally during acute infections, and in some types of haematological neoplasm (including plasma cell myeloma), plasma cells may be evident on a peripheral blood film.

Plasma cells vary in size and shape, but typically contain an eccentric nucleus within deeply basophilic (blue) cytoplasm. The nucleus contains clumped chromatin and is surrounded by a prominent Golgi zone (a pale area) called a perinuclear halo. Figure 8.30 shows plasma cell morphology in a patient diagnosed with plasma cell leukaemia.

Cross reference

Plasma cell neoplasms are discussed in greater detail in Chapter 12.

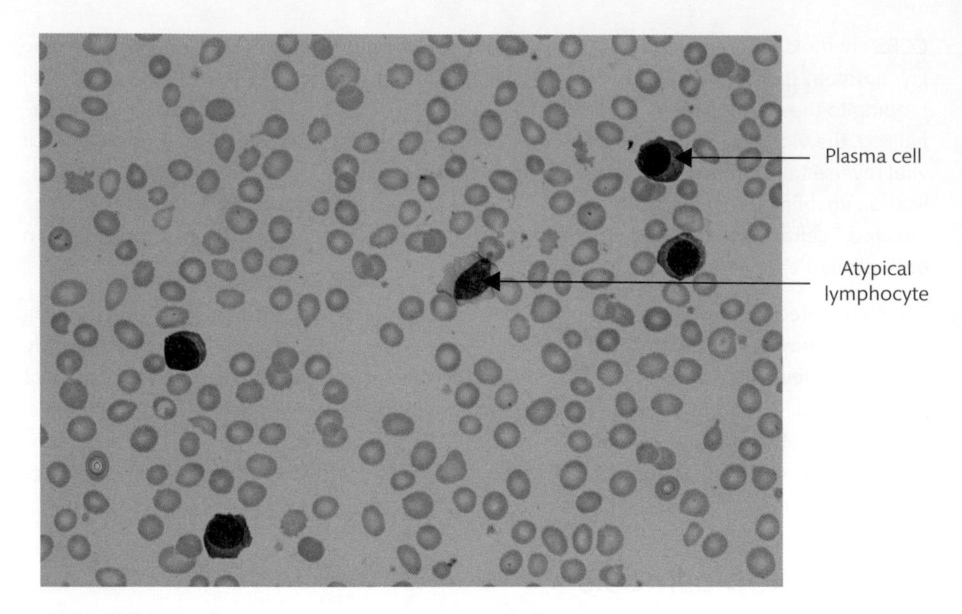

FIGURE 8.30
Plasma cells in the peripheral blood of a patient with plasma cell leukaemia. Four plasma cells are present, accompanied by an atypical lymphocyte in the centre. Note the presence of the deeply basophilic cytoplasm and perinuclear halo, characteristic features of plasma cells. Image courtesy of Jackie Warne, Haematology department, Queen Alexandra Hospital, Portsmouth.

Chapter summary

In this chapter we have:

- determined the structure of granulocytes and the morphological characteristics of each species of WBC;
- considered the identity of blast cells, their structural characteristics, and the significance of identifying them in the blood;
- examined the mitotic and post-mitotic pools of granulocytes and their impact on maintaining the peripheral blood granulocyte concentration;
- outlined the granular contents of granulocytes and their impact on white cell function;
- considered the structural and morphological changes that occur to white cells in a reactive state;
- considered the processes of monopoiesis and lymphopoiesis and the variety of effects mature cells of these lineages determine;
- discussed some of the causes and consequences of abnormal WBC counts;
- examined the methods of investigating suspected cases of infectious mononucleosis.

Discussion questions

8.1 Consider the structure of the germinal centre. In combination with the information available in Chapter 12 and your further reading, critically discuss the relationship between the germinal centre and named lymphoid malignancies.

8.2 Compare and contrast the processes of monopoiesis and granulopoiesis.

8.3 Compare and contrast the effects of named bacterial and viral infections on the WBC count. Explain how the white cell differential and further haematological investigations can aid your diagnosis.

Further reading

- Bain BJ. *Blood Cells: A Practical Guide*, 5th edn. Blackwell Publishing Ltd, Oxford, 2015.
- Balfour HH, Dunmire SK, and Higquist KA. *Infectious mononucleosis*. *Clinical & Translational Immunology* 2015: **4**: e33, doi:10.1038/cti.2015.1.
- Hoffbrand AV, Moss PAH, and Pettit JE (eds). *Essential Haematology*, 6th edn. Blackwell Publishing Ltd, Oxford, 2011.
- Owen JA, Stranfords, and Punt J. *Kuby Immunology*, 7th edn. W.H. Freeman and Co., New York, 2013.
- Klein, C. *Kostmann's disease and HCLS-1-associated protein X-1 (HAX1)*. *Journal of Clinical Immunology* 2017: **37**: 117–22.
- Klein U and Dalla-Favera R. *Germinal centres: Role in B-cell physiology and malignancy*. *Nature Reviews Immunology* 2008: **8**: 22–3.
- Swerdlow SH, Campo E, Harris NL, Jaffe ES, Pileri SA, Stein H, Thiele J, and Vardiman JW (eds). *WHO Classification of Tumours of Haematopoietic and Lymphoid Tissues, vol 2*. IARC, Lyon, 2017.

Answers to self-check questions and case study questions are provided as part of this book's online resources.

Visit www.oup.com/he/moore-fbms3e.

PART 3

Haematological Malignancies

9

An introduction to haematological malignancies

Gavin Knight

The study of cancers of the haemopoietic system is called **haemato-oncology**. The term 'haematological malignancy' is used to describe a range of cancerous conditions of the haemopoietic system and allows us to differentiate cancers of this system with cancers of other tissues. Haemato-oncology is a fascinating subject providing scientists with the opportunity to work at the cellular, molecular, or genetic level, enabling clinicians to piece together a wide variety of data for accurate diagnosis.

Throughout this chapter, we will work through some of the cellular, molecular, and genetic mechanisms responsible for the development of these malignancies. Whilst it is invariably exciting to begin the investigation of a newly presenting haematological malignancy, it is essential to be mindful that every sample represents a patient and our actions should always be measured, respectful, and logical.

Learning objectives

After studying this chapter, you should confidently be able to:

- describe the main differences between the myeloid and lymphoid malignancies;
- describe a number of conditions found in malignant haematology;
- describe the role of the cell cycle in cell proliferation;
- discuss the possible consequences of a dysregulated cell cycle;
- critically discuss the role of the cancer stem cell in leukaemogenesis and lymphomagenesis;
- describe the role of epigenetic processes as mechanisms of inducing malignant change;

- explain the structure of chromosomes, and make appropriate use of chromosomal nomenclature;
- discuss the role of different types of mutation in oncogenesis;
- discuss the concept of clonality in relation to malignant haematology;
- discuss the role of environmental factors in oncogenesis.

9.1 A background to haematological malignancies

The haemopoietic system is responsible for the development of the effector cells we see in the peripheral blood: red blood cells, white blood cells (neutrophils, eosinophils, basophils, monocytes, and lymphocytes), and platelets through the process of **haemopoiesis**. The process of haemopoiesis begins with pluripotent stem cells, which have the capacity to develop into any one of the mature effector cells through their interaction with different growth factors and cytokines. Pluripotent stem cells also have the capacity for **self-renewal**, a process that ensures a sufficient supply of stem cells to enable the future production of mature effector cells.

haemopoiesis
This describes the process of producing the cellular constituents of the blood: red blood cells, white blood cells, and platelets.

self-renewal
A normal property of stem cells enabling the stem cell population to be regenerated without necessarily undergoing differentiation of daughter cells into specific cell lines.

oncogenic
The process of producing cancer.

dysregulation
Describes the abnormal function of one or more regulated processes.

maturation arrest
Occurs in malignant cells when they are unable to develop beyond a particular intermediate stage of maturation. This failure in development prevents the production of a population of effector cells.

apoptosis
The process of programmed cell death. One of the functions of apoptosis is to prevent acquired genetic mutations being passed on to daughter cells.

If particular cancer-causing (**oncogenic**) genetic mutations are acquired by an individual stem cell, the process of stem cell maturation can become dysregulated. **Dysregulation** of haemopoiesis can result in:

- the failure to produce effector cells;
- an accumulation of cells at a particular intermediate stage of maturation (**maturation arrest**);
- an increased cellular proliferation rate;
- a failure of **apoptosis**.

Not every haematological malignancy displays all of these features.

Capability of stem-cell self-renewal can be maintained in maturing populations of daughter cells. By inheriting the ability to self-renew, a population of cells with acquired genetic changes can rapidly accumulate and replace the normal haemopoietic components of the bone marrow. Cancer stem cells will be considered in greater depth later in this chapter.

Now that we have briefly considered some of the features of haematological malignancies, the range of these malignancies should be examined in a little more depth, starting with leukaemias and lymphomas before moving on to myelodysplastic syndromes, multiple myeloma, and then myeloproliferative neoplasms.

SELF-CHECK 9.1

Outline the effects of a dysregulation of haemopoiesis.

9.1.1 Leukaemia and lymphoma

Leukaemia describes the presence of malignant haemopoietic cells within the peripheral blood or bone marrow. These malignant cells may be either *myeloid* (red blood cells, neutrophils, monocytes, eosinophils, basophils, platelets, or their precursors) or *lymphoid* (B-lineage, T-lineage lymphocytes, plasma cells, NK cells, or their precursors).

Lymphomas are *only* lymphoid in origin and are largely restricted to the lymphoid organs (spleen or lymph nodes), although we sometimes encounter patients with extranodal lymphomas (those originating outside the lymphoid organs). Often, lymphomas enter a **leukaemic phase**, where malignant lymphoid cells, which were restricted to a particular lymphoid organ, overspill into the bone marrow and peripheral blood.

leukaemic phase
Describes the presence of malignant cells, derived from a lymphoma, having entered the peripheral blood from their associated lymphoid organ(s).

Leukaemias can be considered as either *acute* or *chronic*. Acute leukaemia is associated with an accumulation of immature cells called **blasts**—comprising at least 20% of the nucleated cells of the bone marrow with evidence of maturation arrest. Acute leukaemias are also clinically aggressive, leading to death relatively rapidly if untreated. Chronic leukaemias are less aggressive and have blasts below 20%. Maturation arrest does not occur in chronic leukaemias, although the cells often fail to undergo apoptosis, allowing the malignant cells to accumulate.

Cross reference
Chapter 3 deals with normal haemopoietic processes.

Cancer Research UK statistics for the period 2014–16 have been compiled to show the total number of male and female cases of leukaemia reported and their relative rates per 100,000 of the population. Figure 9.1 shows that males are more commonly affected throughout life, although rates between males and females start to diverge from the age of 40 years onwards. The greatest differences in rates between males and females is in the 90 years-plus group. This data represents leukaemia in general, irrespective of acute or chronic, or any further subdivisions.

Cross reference
Chapter 11 examines acute and chronic myeloid leukaemias.

Lymphomas can be described as low grade or high grade. Low-grade lymphomas are associated with a lower cellular proliferation rate than high-grade ones, and low-grade lymphomas *tend* to progress slowly. Interestingly, low-grade lymphomas are more difficult to treat than high-grade lymphomas even though, if left untreated, high-grade lymphomas will cause death relatively quickly.

blast
A stage of differentiation of a blood cell as it passes from the stem cell stage to the mature cell stage that should only be found within the bone marrow in low numbers. Blasts may be further characterized by their lineage; e.g. myeloblasts are of the myeloid lineage, lymphoblasts are of the lymphoid lineage.

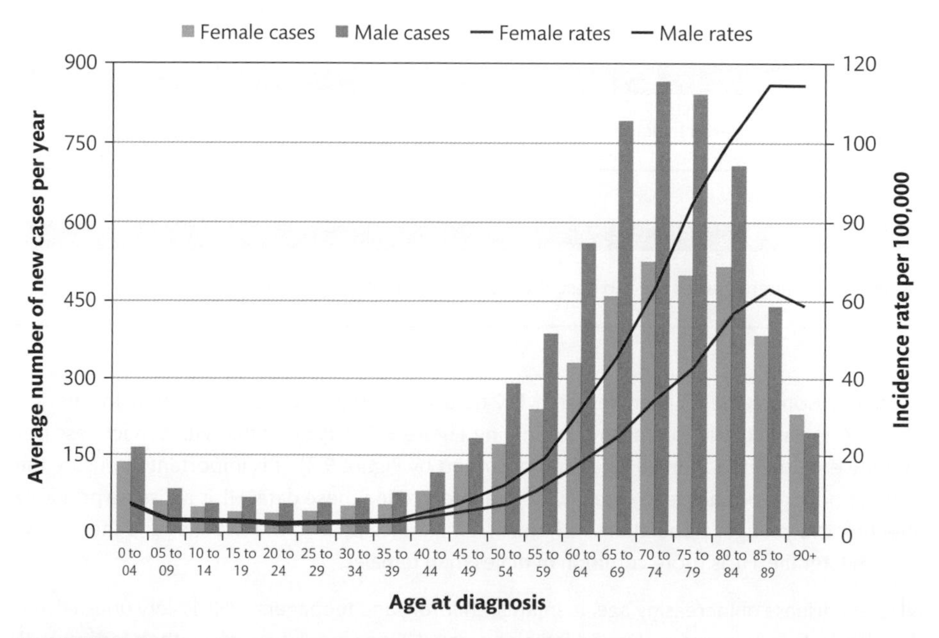

FIGURE 9.1
Cancer Research UK data showing the incidence of leukaemia during 2014–16 stratified according to age and sex. Obtained with permission from Cancer Research UK: https://www.cancerresearchuk.org/health-professional/cancer-statistics/statistics-by-cancer-type/leukaemia/incidence#heading-One accessed January 2020.

Cross references

For more information on lymphoid malignancies please see Chapter 12.

For more information on myeloid malignancies please see Chapter 11.

Cross reference

More detail on HL and the Hodgkin and Reed-Sternberg cell can be found in Chapter 12.

Additionally, we need to consider whether a particular case of lymphoma is a *Hodgkin lymphoma* (HL) or *non-Hodgkin lymphoma* (NHL). The name non-Hodgkin lymphoma describes a large group of solid lymphoid (lymphomatous) tumours which are very diverse in nature. This contrasts with Hodgkin lymphomas, which are relatively restricted in their pathology and are associated with a characteristic mutated B-lineage cell called the **Hodgkin and Reed-Sternberg cell**.

The presence of lymphoblasts in the bone marrow and peripheral blood indicates acute lymphoblastic leukaemia (**ALL**). In cases where there is clear evidence of primary lymphoblastic disease from a lymphoid organ or extranodal tissue, the term 'lymphoblastic lymphoma' should be used.

Key Points

Becoming familiar with the terminology for leukaemias often takes some practice, and you will find some language variations between American and British textbooks. When discussing acute and chronic lymphoid leukaemias, it is important to use the correct suffix for the lineage in question. If you are discussing an acute leukaemia of lymphoid origin, the suffix -blastic must be used, and so the name would be acute lymphoblastic leukaemia. In the chronic form, you should use the suffix -cytic, so the disease would be termed chronic lymphocytic leukaemia. Large numbers of blast cells are evident in acute leukaemias, as denoted by the name. In chronic leukaemias, the cells show maturation, and this is denoted by the suffix -cytic.

There are also variations in the myeloid lineage. Myeloid leukaemias may be described as myeloid, myelogenous, myeloblastic, or in America as non-lymphoblastic leukaemias. The best option here is to use myeloid, as this can be used to describe either acute or chronic forms of the disease.

SELF-CHECK 9.2

List the four main types of leukaemia outlined so far.

SELF-CHECK 9.3

Outline the different grades used to describe the behaviour of lymphoma cells.

A comparison of the Cancer Research UK data for Hodgkin versus NHL from the period 2014–16 is particularly interesting. Examining Figure 9.2, it is clear that HL is much less frequently encountered than NHL (as demonstrated by Figure 9.3). It is important to ensure the axes for both graphs are read carefully before interpreting these data. HL is a disease primarily affecting the younger population, with the first incidence peak in the twenties and the second peak in later life. HL is more common in males than females.

NHL is a disease of increasing age. In infants, children, and teenagers, NHL is very unusual and the distribution between males and females is equal. Once individuals reach their forties, NHL is increasingly found in males more than females, with a gradually increasing incidence until the early fifties. Beyond the early fifties, there is a rapid increase in the incidence of NHL.

9.1.2 Myelodysplastic syndrome (MDS)

Myelodysplastic syndrome describes a range of heterogeneous blood diseases restricted to the myeloid lineage. In MDS, the predominant feature is an increased number of precursor cells within the bone marrow (bone marrow **hypercellularity**) and a concurrent reduction in the number of cells circulating in the peripheral blood (peripheral blood **cytopenia**). The premature death of immature haemopoietic cells within the bone marrow is called **ineffective haemopoiesis**. Clinically, when patients only exhibit signs of anaemia, MDS may be expressed as a low haemoglobin concentration with enlarged red cells (*macrocytic anaemia*). This anaemia will fail to respond to conventional treatments—vitamin B_{12}, folate, or iron—and is therefore considered a **refractory anaemia**. More severe symptoms, including recurrent infections, will be experienced if the patient has a severe reduction in the number of circulating mature white cells with a concurrent increase in the number of blast cells within the bone marrow and peripheral blood.

MDS used to be termed 'pre-leukaemia' because of the perceived increased risk of transformation to an acute leukaemia in these patients. Pre-leukaemia is now a redundant term, since most patients with MDS do not develop acute leukaemia.

hypercellularity
An increase in the number of cells within the bone marrow.

cytopenia
A reduction in one or more cell lines within the peripheral blood. If all cell lines (red cells, white cells, and platelets) are reduced, this is termed pancytopenia.

ineffective haemopoiesis
Describes an accumulation of immature cells within the bone marrow, the vast majority of which undergo apoptosis before maturing into effector cells.

SELF-CHECK 9.4

Outline the common findings associated with MDS.

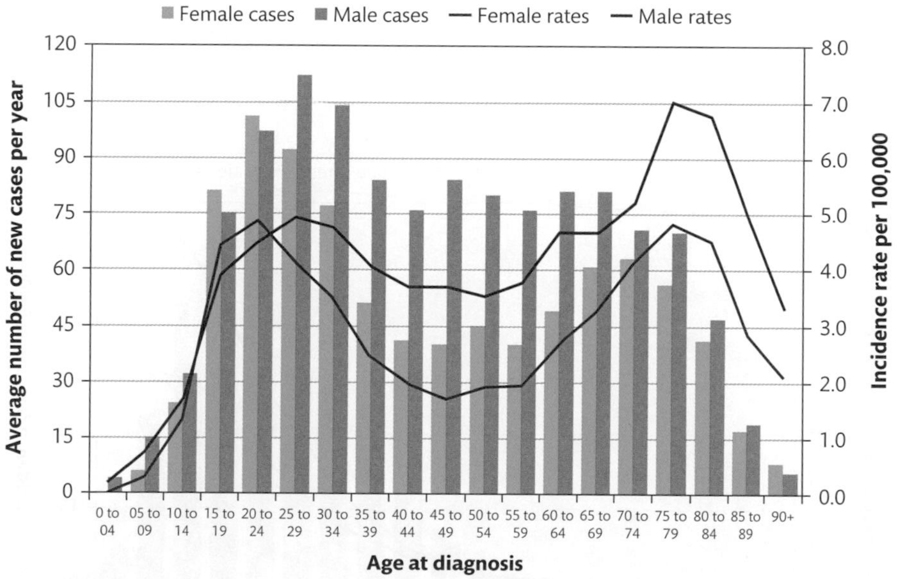

FIGURE 9.2
Cancer research UK data showing the incidence of HL during 2014–16 stratified according to age and sex. Obtained with permission from Cancer Research UK: https://www.cancerresearchuk.org/health-professional/cancer-statistics/statistics-by-cancer-type/hodgkin-lymphoma/incidence#heading-One accessed January 2020.

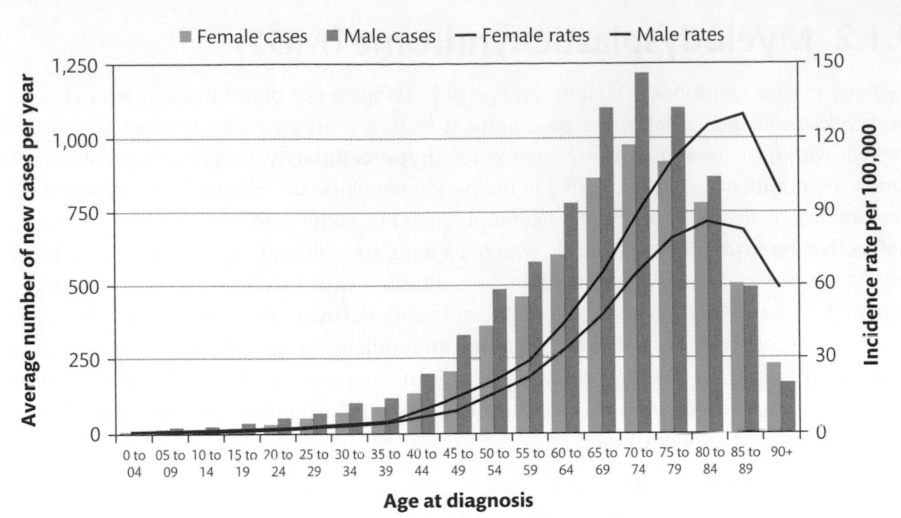

FIGURE 9.3

Cancer Research UK data showing the incidence of NHL during 2014–16 stratified according to age and sex. Obtained with permission from Cancer Research UK: https://www.cancerresearchuk.org/health-professional/cancer-statistics/statistics-by-cancer-type/non-hodgkin-lymphoma/incidence#heading-One accessed January 2020.

9.1.3 Multiple myeloma

Multiple myeloma (MM) is a malignancy of the lymphoid compartment and increases in incidence beyond 35 years of age. Also shown in Figure 9.4, there is a gradual increase from the early forties, at which point the incidence rises markedly. Myeloma is more common in males than females until the late eighties are reached. Only when patients reach their nineties does the number of cases in females exceed that in males.

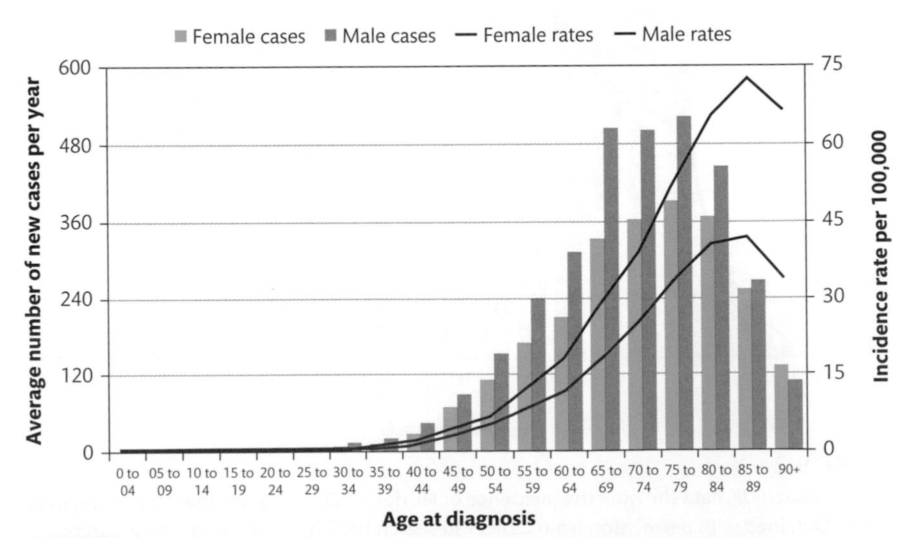

FIGURE 9.4

Cancer Research UK data showing the incidence of myeloma during 2014–16 stratified according to age and sex. Obtained with permission from Cancer Research UK: https://www.cancerresearchuk.org/health-professional/cancer-statistics/statistics-by-cancer-type/myeloma/incidence#heading-One accessed January 2020.

Following stimulation by an antigen, the stimulated B cell will undergo a maturation process whereby either memory B cells (cells which enable a more rapid response to a repeat infection from the same organism), or antibody-synthesizing and -secreting plasma cells, are produced. MM is associated with the proliferation of malignant plasma cells. These malignant plasma cells accumulate in the bone marrow, causing myelosuppression and the activation of osteoclasts (phagocytic cells of the bone involved in bone remodelling). This increase in osteoclast activity causes a breakdown of bone, leading to spontaneous (pathological) fractures and punched-out lesions apparent on X-ray films. MM is characterized by the production of monoclonal antibodies, components of which may damage the kidney if present in the plasma at excessive concentrations.

Cross references

The importance of the micronutrients (iron, vitamin B_{12}, and folate) is explained in Chapter 5.

More detail can be found regarding the classification and pathogenesis of MDS in Chapter 11.

SELF-CHECK 9.5

With which type of cell is multiple myeloma associated?

Cross references

Multiple myeloma and associated bone disease is discussed in further detail in Chapter 12.

Polycythaemia is discussed in detail in Chapter 11.

Myelofibrosis and essential thrombocythaemia are discussed in further detail in Chapter 11.

9.1.4 Myeloproliferative neoplasms

The collective term 'myeloproliferative neoplasms' refers to a range of haematological conditions affecting the myeloid lineage associated with an increased rate of proliferation. Three key disorders are contained under the umbrella of myeloproliferative neoplasms, including:

- **polycythaemia**—a monoclonal proliferation of red cell precursors and an accumulation of mature red cells causing an increased red cell mass. Increases in white cells and platelets may also be apparent;
- **myelofibrosis**—proliferation of fibroblasts within the bone marrow resulting in a modification of the architecture of the bone marrow microenvironment and redistribution of haemopoietic precursors;
- **essential thrombocythaemia**—inappropriate megakaryocyte proliferations causing a sustained increased platelet count.

In order to understand how haematological malignancies develop and exert a clinical effect, we need to examine some of the mechanisms through which a population of malignant cells can develop and expand.

In the next sections we will discuss the processes controlling cell maintenance, growth, and proliferation in normal cells, and then examine how, when these processes malfunction, malignancy can develop.

9.2 Signal transduction

Signal transduction is the cellular process by which signals from the cell's environment regulate DNA transcription and translation, cell proliferation, and apoptosis. Figure 9.5 demonstrates an example of the signal transduction process. Notice how, in this case, an increase in the transcription and translation of target genes is induced as a consequence of intracellular signalling. (See also Box 9.1.)

Signal transduction begins with an initiating factor, for example a growth factor (or ligand) binding to its receptor. This binding causes a conformational change, or dimerization, of receptors—leading to their activation through **phosphorylation**. Once a receptor has been activated, a biochemical cascade is initiated and activation of second messengers (also called signal transduction intermediates) directs the signal to its target. We refer to this process as an amplification cascade: it begins with what is essentially a very small stimulus, but results in an

phosphorylation

The process of adding a phosphate group (PO_4) to an organic molecule.

Cross reference

Each of the lymphoid malignancies is outlined in Chapter 12.

increasing number of additional intermediates becoming involved, causing a *cascade* through the *amplification* of the initial stimulus.

Inhibition of the signal transduction cascade can cause the activation of transcription factors, or, conversely, may cause transcription factor repression. The consequence of this activation or inhibition of transcription factors is the control of gene transcription and translation. The products of these genes then play a specific role in controlling the cell cycle—for example, by inducing proliferation through the activation of the cell cycle.

Now that the concept of signal transduction has been considered, we should examine one of the processes controlled by the signal transduction mechanism—the *cell cycle*.

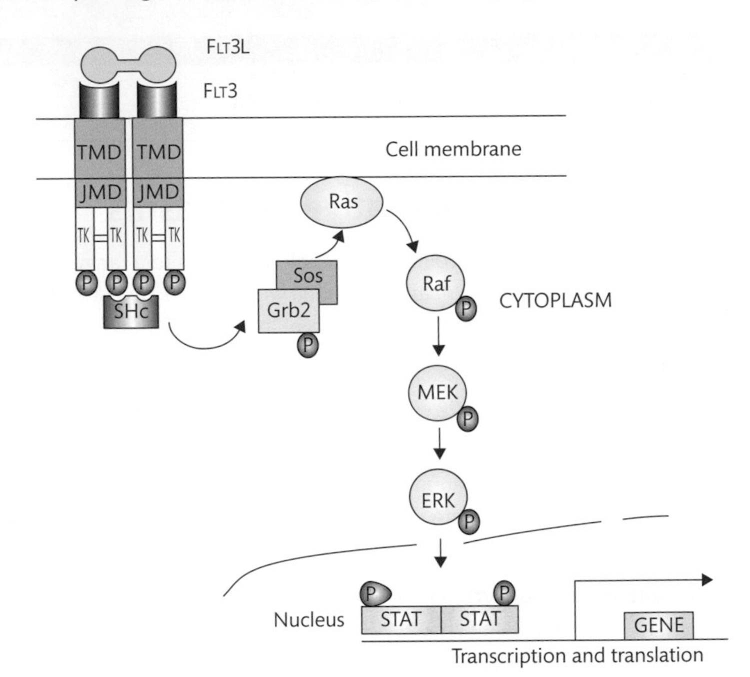

FIGURE 9.5
A signal transduction cascade initiated by the binding of the FLT3 ligand (FLT3L) to its receptor. Subsequent dimerization of the FLT3 receptor occurs leading to the phosphorylation (P) and activation of tyrosine kinase (TK) residues. Phosphorylation of TK induces downstream signalling events and the activation of STAT (Signal Transducer and Activator of Transcription), leading to the transcription and translation of FLT3L inducible genes. TMD = transmembrane domain; JMD = juxtamembrane domain.

BOX 9.1 NFκB signalling

Dysregulation of signal transduction pathways is an important process in the development of haematological malignancies. The Nuclear Factor κB (NFκB) pathways relevant to haematology will be used for illustration below.

As you work through Chapter 12, you will encounter a number of haematological malignancies associated with the over activity of NFκB pathways.

NFκB is a family of transcription factors containing DNA-binding, dimerization, nuclear transactivation, and inhibitor-binding domains. Usually located within the cytoplasm, and associated with inhibitors, NFκB is activated through two pathways—classical and alternate—and is active in many signalling pathways. The *classical pathway* is activated by pro-inflammatory cytokines or the interaction of B-cell or T-cell receptors with their ligands, whereas the *alternate pathway* is activated by the TNF cytokine family. The multimeric complexes that result from the activation of these signal transduction pathways are structurally different, although once activated, these complexes can translocate from the cytoplasm to the cell nucleus, where they induce the transcription of a range of target genes.

NFκB signalling is essential for the development and maturation of B and T cells by suppressing apoptosis. Suppression of apoptosis is particularly important as B cells progress through the germinal centre following activation in an antigen-dependent manner to facilitate the development of highly specific memory B cells and plasma cells.

Constitutive activation of NFκB leads to the accumulation of cells through NFκB-initiated abnormal cell cycling and apoptotic failure; and is associated with a range of haematological malignancies including: follicular lymphoma; MALT lymphoma; activated B-cell type, diffuse, large B-cell lymphoma; anaplastic, large cell lymphoma; and in some EBV-related cases of HL.

SELF-CHECK 9.6

Outline the importance of signal transduction in regulating cell behaviour.

9.3 An introduction to the cell cycle

We have already established that both cellular proliferation and the failure of apoptosis have a role in the development of malignancies. Before moving on to consider specific details of oncogenesis, we should first consider how cells reproduce and try to understand some of the control mechanisms in place. Understanding the mechanisms controlling normal cellular reproduction can aid our understanding of how cells can become malignant when these control mechanisms malfunction.

The process of cellular reproduction comprises a series of phases which, when integrated, form a cycle—called the cell cycle. Figure 9.6 provides an outline of the main phases of the cell cycle. A single cycle usually produces two identical daughter cells from one 'parent' cell. Before the cycle is ready to produce daughter cells, sequential activation and termination of each phase of the cycle must be completed. The phases collectively producing the cell cycle include two gap phases abbreviated to G1 and G2, a synthesis phase (S), and mitosis (M). An additional phase, in which the cell 'rests' may follow mitosis, and is referred to as the quiescent stage (G0). These phases are organized into the following order to ensure the cell is fully prepared to reproduce:

$$G1 \rightarrow S \rightarrow G2 \rightarrow M(G0)$$

During G1 phase, protein synthesis occurs, preparing the cell for the synthesis of DNA in S phase of the cell cycle. G1 is the longest phase of the cell cycle, completing in approximately nine hours. Chromosomes are present as single **chromatids**.

chromatids
Subunits of chromosomes separated by a centromere. Chromatids become chromosomes following division of the centromere during mitosis.

S phase follows G1, and is involved in the synthesis of DNA. This synthesized DNA forms a duplication of all DNA within the cell. This process lasts approximately five hours, with mostly active (or early replicating) genes duplicating first, followed by largely inactive (or late-replicating) genes.

G2 phase then follows, allowing the cell to make the final preparation for cell division.

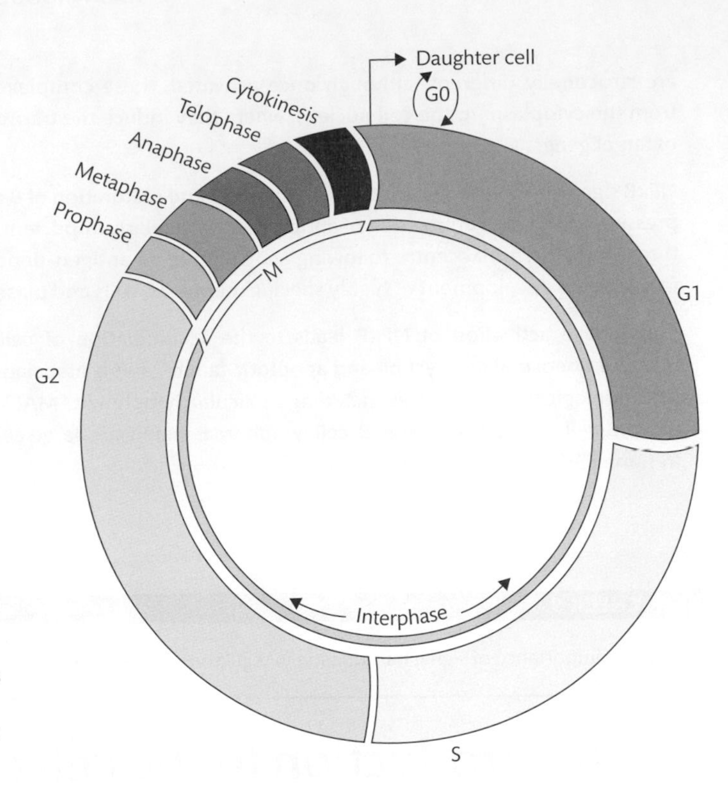

FIGURE 9.6

A brief outline of the main stages of the cell cycle. The cycle contains two gap phases (G1 and G2), a single synthesis phase (S), and mitosis (M). G1, S, and G2 phases constitute interphase, with the different phases of mitosis (prophase, metaphase, anaphase, telophase, and cytokinesis) leading to the production of two daughter cells. G0 represents an optional quiescent or 'resting' phase.

G1, S, and G2 phases are collectively called interphase. Mitosis also includes a number of distinct phases: prophase, metaphase, anaphase, telophase, and finally cytokinesis. During mitosis, separation of duplicated chromosomes and division of the cytoplasm occurs, producing two distinct diploid cells.

SELF-CHECK 9.7

List the main phases of interphase and provide a brief description of each.

Cross reference

Metaphase is discussed in relation to chromosomal structure and cytogenetic analysis in Chapter 10.

The cell cycle includes a number of regulators that induce or inhibit specific phases of the cycle. Between them, these regulators ensure that each phase begins and ends correctly, and that each phase takes an appropriate amount of time to complete.

Whilst the DNA is replicating, the replicating strand of DNA is checked for mutations. If mutations are detected, cell cycle regulators are activated and cell cycle arrest is induced.

Following cell cycle arrest, DNA repair enzymes will be activated; although, if repair is unsuccessful, apoptosis is induced. Inducing DNA repair or apoptosis prevents acquired DNA mutations from passing from one cellular generation to the next, thereby protecting the genome. Unrepaired mutations in the cell cycle machinery are associated with a wide variety of haematological malignancies.

Let's now explore the concept of the cell cycle in a little more detail.

9.3.1 An introduction to cell cycle control

To proliferate and become committed to the cell cycle, cells require particular signals to initiate mitosis. These signals are termed 'mitogenic' and indirectly activate a family of cell cycle control proteins called **cyclin-dependent kinases** (CDKs). The concentration of CDKs remains constant throughout the cell cycle; however, their activity is regulated by another family of proteins called **cyclins**. One group of cyclin family members is called cyclin D (which actually comprises cyclins D1, D2, and D3), the transcription of which is induced through mitogenic signals (e.g. through Ras signalling). Particular cyclins will bind specifically with CDKs, and CDKs will not function unless they are bound to their appropriate cyclin partner. Table 9.1 shows the main CDK-binding partner for each of the cyclins. Just as cyclin–CDK interactions push the cell cycle forward, control mechanisms are in place to inhibit these cyclin–CDK complexes, ensuring that each phase of the cell cycle completes at the appropriate time.

cyclin-dependent kinases
A group of enzymes, requiring cyclins, involved in the transfer of phosphate groups from donors to acceptors to enable the progression of the cell cycle.

Initiation of the cell cycle

Prior to the initiation of a new cell cycle, the cycle is maintained in an inactive state–achieved through the actions of two families of cyclin kinase inhibitor (CKI):

- inhibitors of CDK4 (INK);
- CDK inhibitory proteins (cip/kip).

These inhibitors form the CKI–CDK–cyclin complex. Binding of a **mitogen** (a growth factor, for instance) to its binding site or receptor results in modification to this CKI–CDK–cyclin complex, allowing an interaction between a subgroup of cyclins—the D-type cyclins—and their corresponding CDKs. The remodelling of this complex ensures that initiation of the cell cycle occurs in response to external requirements.

mitogen
Any chemical or substance capable of inducing mitosis.

The cyclin D–CDK4 and the cyclin D–CDK6 complexes then interact with an important regulatory protein called retinoblastoma (pRb). Figure 9.7 outlines the process through which pRb is modified to allow cell cycle progression to occur. When the cell cycle is arrested, for example in the absence of cyclin D–CDK4, pRb binds to a **transcription factor** called E2F-1. By binding with E2F-1, pRb acts as a docking station, preventing E2F-1 from associating with its binding site in the promoter region of target genes. E2F-1 should only be released when appropriate signals are received by pRb. Because E2F-1 cannot bind to its binding site, E2F-1-sensitive genes are repressed, and so transcription of these genes cannot occur. Increasing intracellular concentrations of cyclin D–CDK4 and cyclin D–CDK6, in response to mitogenic activation, add phosphate groups to pRb, in a process called phosphorylation. This phosphorylation reduces the binding capacity of pRb for E2F-1, releasing E2F-1 transcription factors, and activating E2F-1-dependent genes. The release of E2F-1 promotes transcription of a range of genes, including the gene encoding cyclin E. Another protein bound to pRb is an enzyme called **histone deacetylase** (HDAC). The localization of HDAC to pRb and the promoter region of target genes causes modification of the chromatin structure within this region, thus inhibiting genes within the area.

transcription factor
A protein which plays a regulatory role in the transcription of particular genes.

histone deacetylase
A class of enzyme that removes acetyl groups from histones, thereby inhibiting DNA transcription.

SELF-CHECK 9.8

Describe the role of the retinoblastoma protein pRb.

TABLE 9.1 Key cyclins, their binding partners, and typical expression within the cell cycle.

Cyclin	CDK-binding partner(s)	Typical expression in the cell cycle
Cyclin D (D1, D2, D3)	CDK4 or CDK6	Mid G1
Cyclin E	CDK2	Late G1
Cyclin A	CDK2	S phase
Cyclin B	CDK1	Late G2/M

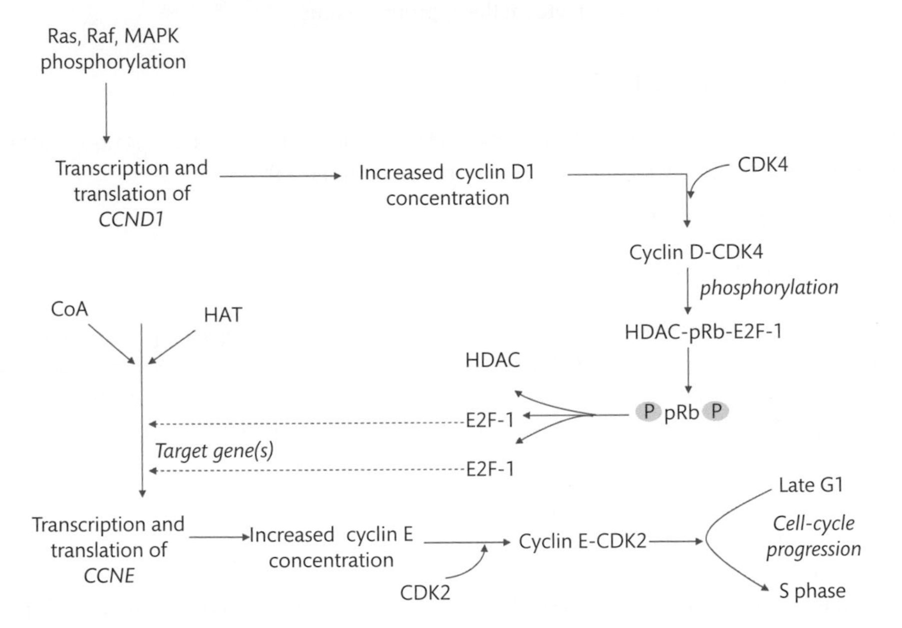

FIGURE 9.7
The induction of cyclin D via the mitogenic activation of Ras, Raf, or MAPK. Transcription of the cyclin D1 gene (*CCND1*) is induced, allowing translation of cyclin D1, thereby increasing its intracellular concentration. Cyclin D1 associates with CDK4 producing the cyclin D-CDK4 complex. This complex phosphorylates pRb, releasing transcription factor E2F-1 and HDAC. Histone acetyltransferase (HAT) and coactivators (CoA) replace HDAC, and in the presence of E2F-1 allow the transcription and translation of *CCNE*. Cyclin E will then form a complex with CDK2, allowing cell cycle progression from late G1 to S phase.

Cross reference
HDAC and other epigenetic modifiers are discussed in greater detail in section 9.8 later in this chapter.

Activation of cyclin E–CDK2 occurs in late G1 phase, just prior to cells entering S phase, and induces factors necessary for S-phase progression.

Cell cycle progression

Cyclins A, B, B2, and B3 are mitotic cyclins and are involved in the progression of the cell cycle through S and G2 phases, although they are actually destroyed during mitosis itself. Cyclin A–CDK2 concentrations peak in early G2 following initial expression in late G1. Cyclin

A-CDK2 is an essential component of the S phase, as DNA replication is dependent upon its activity. In addition, cyclin A-CDK2 has been shown to play an inhibitory role in G1/S-phase transition.

Cell cycle completion

Transition between G2 and M phases is largely controlled by cyclin B and CDK1 (also called cell division cycle 2). Cyclin B-CDK1 is initially expressed in early G2 phase, with a peak in concentration at the G2/M transition. Mitosis can only complete once cyclin B-CDK1 has been degraded. Once mitosis has completed, the cell cycle of the daughter cells may then begin.

Mitosis involves the condensation of duplicated sister chromatids across the equator of the cell. These sister chromatids are separated through the formation of a spindle complex and the associated microtubule network. Each chromatid, now called a chromosome, is localized to the opposite poles of the cell, allowing two genetically balanced daughter cells to be produced. This forms the basis of inheritance and, in the case of haematological malignancies, allows for abnormal genes to be passed on to subsequent generations and the production of a population of malignant cells. Dysregulation of the cell cycle through alterations in cell cycle machinery has been reported in a number of cases of haematological malignancy, and some of these are explored in Chapters 11 and 12. An outline of some of the most common abnormalities in haematological malignancies is provided in Table 9.2.

TABLE 9.2 Common cell cycle control abnormalities in haematological malignancies.

Cell cycle component	Mechanism	Malignancy	Comments
RB1 (pRb)	Various	ALL	No prognostic implication
	Deletion	HL	Occurs in approximately 16% of cases
	Deletion	AML	Occurs in up to 55% of cases
	Mutation	CML	A product, rather than a cause of disease progression
CCND1 (cyclin D1)	Dysregulated	MCL	t(11;14)(q13;q32) *CCND1-IGH*
		Myeloma	As MCL (above)
		MGUS	As MCL (above)
CCND2 (cyclin D2)	Upregulation	CML	A downstream effector of BCR-ABL1
CCND3 (cyclin D3)	Upregulation	Myeloma	t(6;14)(p21;q32) *CCND3-IGH*
CCNE (cyclin E)	Upregulation	AML	Approx. 25% of cases, but depends on subtype
	Upregulation	CLL	No data

Key: ALL = acute lymphoblastic leukaemia; AML = acute myeloid leukaemia; CLL = chronic lymphocytic leukaemia; CML = chronic myeloid leukaemia; HL = Hodgkin lymphoma; MCL = mantle cell lymphoma; MGUS = monoclonal gammopathy of undetermined significance.

9.4 Apoptosis

Apoptosis is a completely normal cellular process involving three main pathways: the *extrinsic*, *intrinsic*, and *perforin/granzyme* pathways. The *extrinsic* pathway occurs in response to external cell signals, in the form of 'death receptor' **ligands** that interact directly with death receptors of the tumour necrosis factor (TNF) receptor superfamily. This receptor family is involved in a number of cellular processes, including regulation of cell survival, cell death, cellular differentiation, and immunological response. Principal ligands and receptors include FasL/FasR, TNFα/TNFR1, and Apo3L/DR3, although others are also important. Extracellular (or extrinsic) signals are passed into the cell via an intracellular 'death domain', coordinating the cellular response to these apoptotic stimuli. Although the final details of the intracellular interactions vary according to receptor type, a multimeric cytoplasmic complex forms a death-inducing signalling complex (DISC), containing an important proenzyme—procaspase 8. Cleavage of procaspase 8 to its active form—caspase 8—leads to the activation of the execution phase of apoptosis and cell death. The execution **caspases** (caspases 3, 6, and 7) activate cytoplasmic endonucleases leading to the degradation of nuclear and cytoplasmic material. The process of apoptosis is independent of inflammation and follows reproducible morphological changes to the structure of the cell, including cell shrinkage, cytoplasmic blebbing, chromatin condensation, and nuclear and DNA fragmentation.

ligand
A molecule that will bind to another. For example, a growth factor that binds to its receptor.

caspase
As cysteine-dependent, aspartate-directed proteases, caspases form an important part of the intracellular cascade leading to apoptosis.

The *intrinsic* pathway acts independently of death receptors and is dependent on mitochondrial signalling. The cellular response to the detection of DNA damage, withdrawal of extracellular survival signals, radiation, toxins, free radicals, and viral infections, induces the intrinsic apoptotic pathway, although, in the context of this text, DNA mutation and failure of DNA repair in response to mutation are of particular importance. Detection of cellular damage results in the activation of key tumour suppressors, such as p53, and the downstream activation of their tumour suppressor substrates. Activation of Bcl-2 family members leads to permeabilization of the outer mitochondrial membrane and to the subsequent release of pro-apoptotic proteins into the cytoplasm. Cytochrome c, Smac/DIABLO, and HtrA2/Omi are released, resulting in a pro-apoptotic intracellular environment. Smac/DIABLO and HtrA2/Omi are negative regulators of endogenous apoptotic inhibitors (i.e. inhibitors of inhibitors) whilst cytochrome c is necessary for the formation of the **apoptosome**. The apoptosome activates caspase 9, leading to the activation of the execution pathway. The intrinsic apoptotic pathway is also controlled by major tumour suppressors, including ATM/ATR and p53 proteins. As we will see later in the malignancy section of this text, mutation or altered regulation of these tumour suppressors can lead to a failure of apoptosis. Failure of apoptosis promotes the development of cancers, evolution of existing cancers, and resistance to a number of commonly used chemotherapeutic agents. Look at Figure 9.8 to note the similarities and differences between intrinsic and extrinsic apoptotic pathways.

apoptosome
A complex produced from apoptotic protease activating factor 1 (Apaf1), following its interaction with cytochrome c, and caspase-9.

The *perforin/granzyme* pathway is mediated by cytotoxic T cells in response to antigen-expressing cells. Cytotoxic T lymphocytes bind to infected cells and degranulate. Release of perforin from these granules leads to the formation of pores within the plasma membrane of targeted cells. Through these pores, granzymes A and B, which originate from cytotoxic T lymphocytes, are projected, leading to activation of caspases, DNA fragmentation, and apoptosis.

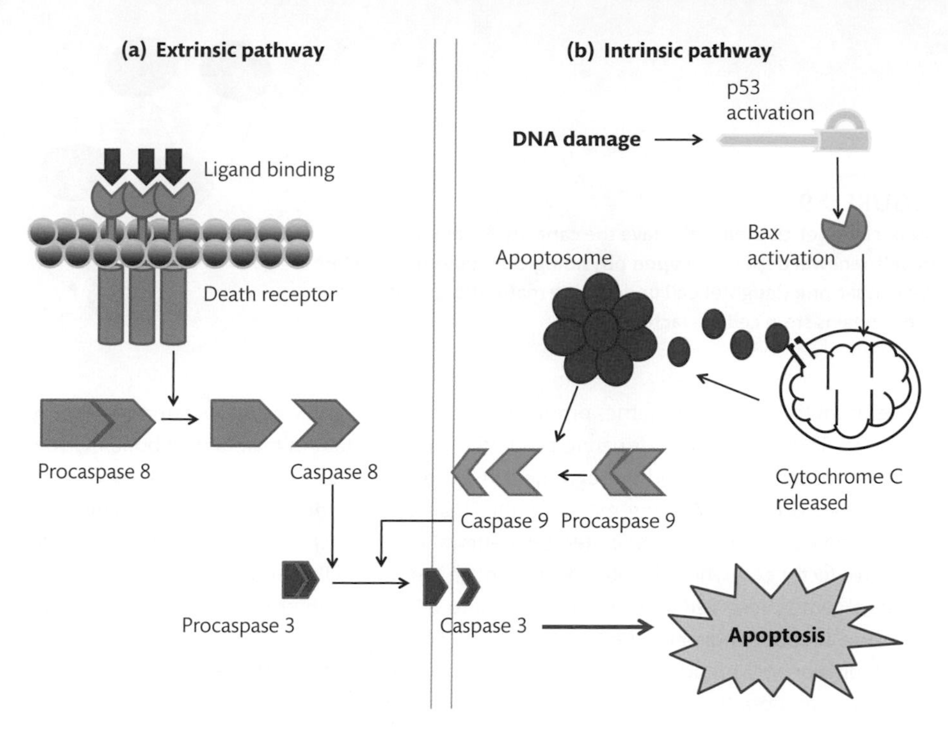

FIGURE 9.8

Basic summary of extrinsic (a) and intrinsic (b) apoptosis. (a) Binding of death ligand, such as Fas ligand, results in clustering of receptors and signalling through the death domain. Cleavage of the pro-enzyme caspase 8 to its active form–caspase 8–leads to the activation of executioner caspase 3 and finally apoptosis. (b) Detection of cellular damage, such as DNA mutation, leads to the activation of cell checkpoint mechanisms and, in this case, p53. Upregulation of Bax (a Bcl-2 family member) and the formation of pores within the mitochondrial membrane follow. The release of important pro-apoptotic molecules, such as cytochrome c, into the cytoplasm results in apoptosis.

9.5 Cancer stem cells

Stem cells are immunophenotypically CD34-positive (CD34+) precursor cells with the potential to differentiate and mature into many different types of effector cell, whilst undergoing the process of self-renewal. Cancer stem cells tend to be interspersed with more mature cancer cells, which lack the ability for self-renewal. Determination of the self-renewal characteristics of these cells is demonstrated through transplantation into immunodeficient mice. Cancer cells with the ability to populate the mouse with tumour cells are considered stem cells, whilst the more mature cancer cells fail to induce cancer.

Haemopoietic stem cells have been reported as occurring in approximately 1 in 10,000 nucleated bone marrow cells. However, there is now evidence to suggest that these stem cells are more numerous than previously thought. These stem cells can produce any of the myeloid or lymphoid lineages constituting the haemopoietic system, whilst maintaining the stem cell pool. Stem cell division is either symmetric, producing two identical daughter cells (the typical

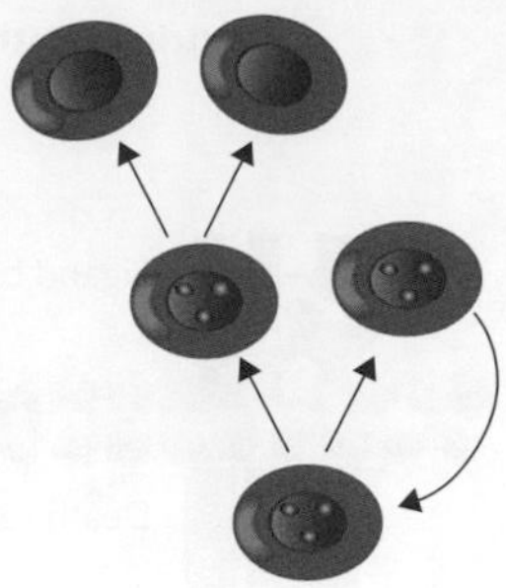

FIGURE 9.9

Stem cell kinetics. Stem cells have the capacity for differentiation or self-renewal depending upon physiological requirements. Here we can see one daughter cell undergoing maturation, whilst the other retains stem cell characteristics.

model of mitosis) or asymmetric, producing dissimilar daughter cells—as demonstrated by Figure 9.9. These apparently disparate properties are important in maintaining bone marrow viability and can be achieved through the process of asymmetric cell division in response to local mitotic drivers. Following mitosis, rather than two identical daughter cells being produced, one daughter cell with limited self-renewal capacity will differentiate and mature to form an effector cell, whilst the other will maintain the stem cell phenotype, allowing the stem cell pool to be preserved. The selection pressure on the daughter cells should be sufficient to ensure that, depending on the body's physiological requirements, symmetric or asymmetric cell division will predominate. For example, following **myeloablative therapy** and stem cell transplantation, the transplanted stem cells may undergo symmetric division in the first instance to repopulate the bone marrow with stem cells. Following repopulation, asymmetric division may occur, producing a range of effector cells, whilst conserving the stem cell pool. (See also Box 9.2.)

myeloablative therapy

The process whereby the bone marrow is destroyed by high doses of chemotherapeutic agents in preparation for stem cell transplantation. Stem cell transplantation provides a potential cure for patients receiving this therapy.

BOX 9.2 Stem cell immunophenotype

Stem cells express a specific surface immunophenotype that enables identification and quantification using flow cytometry. The most important of the markers is CD34, and enumeration of this marker is useful in ensuring that sufficient stem cells have been harvested prior to myeloablative therapy and stem cell transplantation.

The common haemopoietic stem cell phenotype is: CD34+, CD38–, CD117+. Lineage-specific markers are typically absent (Lin–).

The three common markers that should be assessed to characterize stem cells are considered in more detail below:

***CD34*: a heavily glycosylated, monomeric, cell-adhesion molecule rich in O-linked carbohydrates and sialic acid. Approximately 42% of bone marrow blasts express CD34;**

***CD38*: a transmembrane, type-II glycoprotein expressed on early B and T lymphocytes, activated T lymphocytes, NK cells, and plasma cells;**

***CD117*: also known as c-Kit, CD117 is a transmembrane glycoprotein and receptor tyrosine kinase, the ligand of which is stem cell factor. CD117 is expressed on approximately 4% of normal marrow precursors, with the exception of lymphoid-lineage cells. In the lymphoid lineage, CD117 is expressed on a small subset of NK cells and early T-cell precursors.**

Cancer stem cells have been associated with myeloid and acute lymphoid malignancies, although there is less evidence linking them with mature lymphoid malignancies. With myeloid malignancies it is difficult to ascertain whether stem cells acquire a malignant phenotype whilst retaining their self-renewal characteristics, or partially differentiated cells undergo genetic mutation reactivating the stem cell property of self-renewal.

In lymphoid malignancies, mature lymphoid cells have the potential to become self-renewing, highly proliferative cells following stimulation by antigen as they enter the germinal centre. In fact, the potential for self-renewal is maintained until the memory B-cell stage. Activation of this proliferative programme could lead to the expansion of malignant lymphocytes in the absence of 'true' cancer stem cells, and appears to be the explanation for why cancer stem cells are not found in lymphoid malignancies with a mature phenotype. By assessing variable heavy-chain hypermutation in malignant B cells, it is possible to establish the stage of maturation in which oncogenesis has occurred. Cells showing evidence of somatic hypermutation are derived from germinal centre or post-germinal centre cells, whilst those failing to show somatic hypermutation are usually derived from pre-germinal centre cells.

Cross reference

More detail regarding germinal centres can be found in Chapter 8.

Oncogenesis through the accumulation of mutations at the post-germinal centre stage is thought to result in multiple myeloma and Hodgkin lymphoma.

Enumeration of cancer stem cells demonstrates the vast differences in the number of stem cells found in different types of cancer. In acute myeloid leukaemia, the abundance of CD34+ CD38− cancer stem cells has been reported to be as low as 0.02–1%, whereas solid tumours are associated with much larger numbers of cancer stem cells. One explanation for the low numbers of cancer stem cells present in haemopoietic malignancies could be the large numbers of cytogenetic translocations associated with these malignancies inducing stem cell-like behaviour in the absence of CD34 expression.

9.6 Chromosomes and nomenclature

Chromosomes are passed from one generation of a cell to the next. The DNA-histone composite found within chromosomes is called **chromatin**, and is composed of an assembly of nucleosomes. Figure 9.10 demonstrates the composition of the nucleosomes and their interaction with one another. Double-stranded DNA forms a nucleosome when wrapped around a histone octamer. Four histone proteins (H2A, H2B, H3, and H4) are duplicated to form the octamer. Each nucleosome contains 147 base pairs of DNA, with small segments of linker DNA attaching one nucleosome to the next. A single molecule of histone H1 may bind with the linker DNA and the nucleosome, causing stabilization of this structure.

SELF-CHECK 9.9

Describe the components of a nucleosome.

The structure of chromatin varies in different stages of the cell cycle, with chromosomes in metaphase being composed of densely packed chromatin (not involved in the transcription process). It is only during metaphase that chromosomes are visible using light microscopy.

Whilst chromosomes can vary in size, shape, and DNA content, they all have a very similar structure. By understanding the normal structure of chromosomes, we can begin to appreciate the significance of chromosomal changes in haematological malignancies.

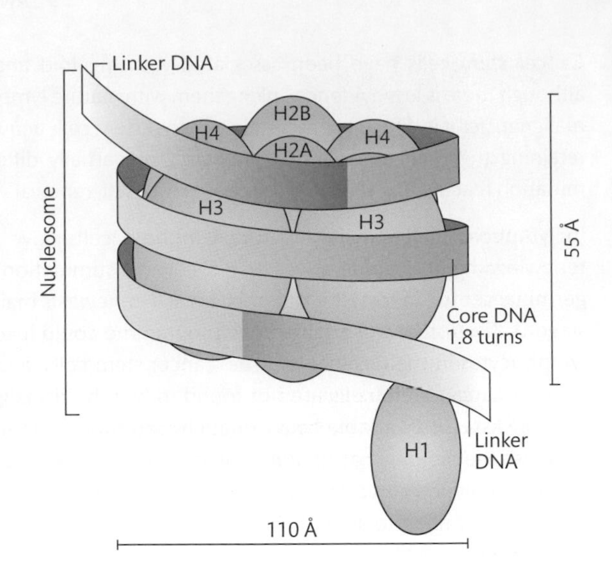

FIGURE 9.10
The structure of a nucleosome. Chromatin is formed by the assembly of nucleosome subunits. Nucleosomes contain eight histones, each wrapped in double-stranded DNA. Each nucleosome attaches to the next by linker DNA, stabilized by histone H1. Reprinted by permission from Macmillan Publishers Ltd: Georgopoulos K. *Haematopoietic cell-fate decisions, chromatin regulation and ikaros*. *Nature Reviews Immunology* 2002: 2: 162–74.

9.6.1 Chromosomal structure

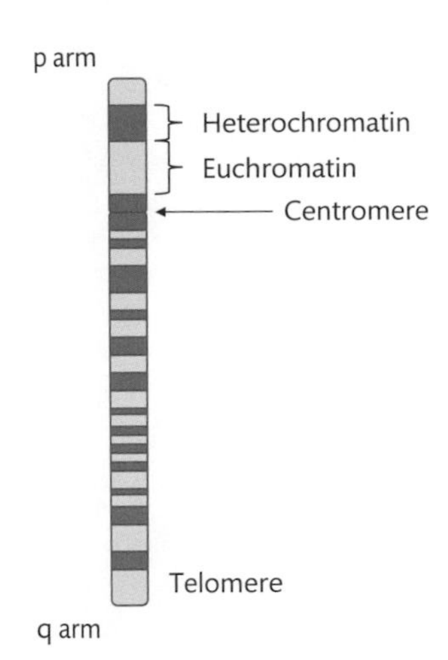

FIGURE 9.11
The basic structure of a chromosome, including an example of G-banding pattern.

Chromosomes have three key, easily identifiable structural elements, as illustrated in Figure 9.11. The most prominent of these is the **centromere**, which joins the two identical (sister) chromatids, produced after DNA replication. It is required for the formation of the kinetochore, where the fibres that pull each of the chromatids to opposite sides (poles) of the cell during anaphase attach, allowing the accurate passage of DNA from one generation to the next.

Either side of the centromere are the two arms. The shorter of the two arms is called the **p-arm**. The p-arm is so called because it is '*p*etite'. The larger of the two arms is called the **q-arm**, so designated because it is next to 'p' in the alphabet. The tip of each chromosome arm has a special DNA structure and is called the telomere.

Now we have established the macro structure of the chromosome, let us consider the micro structure. By applying a variety of different stains to metaphase chromosomes, a **banding pattern** is generated. A brief summary of these banding patterns can be found in the following Method box. The cytogenetic bands can then be used to describe the location of specific genes. Refer to Figure 9.11 for further clarification of chromosomal structure and banding patterns.

banding pattern
A method of identifying chromosomes based upon their appearance following staining.

Each of the landmark bands is numbered from the centromere, in the direction of the chromosome tip, the telomere. To add further specificity, in less-condensed chromosomes bands may divide into sub-bands. In Chapter 11, we will discuss a particular type of leukaemia called acute promyelocytic leukaemia (APL). The locus (plural loci) for one of the genes associated with APL, *RARA*, is 17q12. 17q12 is a specific direction to a particular place within the genome. The locus designation 17q12 tells us that *RARA* can be found on chromosome 17, on the long arm (q), band 12. A normal genome consists of 46 chromosomes and these banding pattern techniques allow the identification of very specific locations on chromosomes and explain complex chromosomal information in a simple, effective manner.

Cross reference
Acute promyelocytic leukaemia is discussed in detail in Chapter 11.

The most important thing to remember about chromosomal nomenclature is that it is universal. All cytogeneticists report and discuss their results according to the International System for Human Cytogenetic Nomenclature (ISCN). This means that communication regarding patients and scientific research can cross geographical borders, and regardless of the country, the terminology is standardized.

METHOD Generating banding patterns in cytogenetic investigations

- A variety of different methods can be used to produce banding patterns on chromosomes of interest. It is important when considering these techniques to remember that metaphase chromosomes are required. Using these techniques, cells are encouraged to progress through the cell cycle, and are then treated with a spindle inhibitor such as colchicine in order to arrest cell division in metaphase. The number of bands produced depends on the level of condensation of the chromatin; shorter chromosomes show fewer bands but retain the main 'landmark' bands.
- The most widely reported of the stains is Giemsa, which provides a *G-banding* pattern. The dark areas are composed of heterochromatin whilst light areas are composed of euchromatin. If you refer to Figure 9.11 you will notice the distribution of heterochromatin and euchromatin. Note that Giemsa stain is also one of the key stains used to assign blood cell morphology because of its robust and reproducible staining characteristics.
- *Q-banding* uses a fluorescent stain, such as quinacrine, to produce a distinct banding pattern. However, this pattern is not visible to the naked eye, and requires fluorescent microscopy to identify the appropriate banding patterns.
- *C-banding* involves the treatment of metaphase chromosomes with an acid and then a base, prior to a final staining procedure using Giemsa. C-banding stains the centromere.
- *R-banding*, or reverse banding, provides a banding pattern opposite to that achieved by G-banding. Giemsa stain is still used with this technique, although in contrast to G-banding, the chromosomes are initially heated, modifying the uptake of this stain by the chromatin. The dark areas are composed of euchromatin, whilst light areas are heterochromatin.

9.6.2 Chromosomes in pathology

Many of the genes involved in leukaemias and lymphomas have been identified following recurring chromosomal rearrangements causing changes in the banding pattern. However, each chromosome band contains many genes and there are now much more sensitive techniques available to establish gene location, such as fluorescence *in situ* hybridization (FISH). FISH is a molecular cytogenetic technique that can be used on interphase or metaphase chromosomes and is discussed in greater detail in Chapter 10.

Cross reference

Cytogenetic analysis is further considered in Chapter 10.

SELF-CHECK 9.10

How does G-banding help with the reporting of a particular gene locus?

The next sections examine different types of mutation at the genetic and chromosomal level in order to demonstrate how these aberrations can result in haematological malignancies.

9.7 An introduction to genetic mutation

Variations in the sequence of a particular gene between individuals are commonplace, provide us with our genetic diversity, and are called **polymorphisms**. The different forms of a specific gene giving rise to polymorphisms are called **alleles**; the alleles for a given gene are found at the same locus, irrespective of their sequence. If an individual has inherited two different copies of an allele at a particular locus, they are called **heterozygous**, and if two identical alleles are inherited at any given locus, they are considered **homozygous** for that particular gene.

Key Points

The name of a gene should always be written in capital letters and italicized in order to differentiate between the gene and its product.

***Example*: The gene encoding retinoblastoma is referred to as *RB1*, whereas its protein product is abbreviated to pRb.**

Non-germline mutations—those that are acquired and found only in somatic cells—occur throughout our lives and are, for the large part, detected by our DNA repair enzymes during the cell cycle. These mutations are either repaired, or the cell harbouring the mutation undergoes apoptosis. As we age, the likelihood of acquiring genetic mutations increases, and acquired mutations are commonly associated with malignancy. There are two particular families of genes that, when mutated, are associated with the development of cancer. These families of genes are:

- **proto-oncogenes**—encode proteins that are often responsible for cell growth, proliferation, and survival. Mutations within proto-oncogenes may result in a protein product with a gain-of-function. Once mutated, proto-oncogenes are called *oncogenes*;
- **tumour suppressor genes**—play an important role in controlling the cell cycle and preventing tumour development. An example of a tumour suppressor gene is *RB1*, which encodes pRb. In Section 9.3.1 we considered the role of pRb in the inactivation of the transcription factor E2F-1, thereby inhibiting the cell cycle—a characteristic role of a tumour suppressor. Tumour suppressor genes (TSGs) are also called *anti-oncogenes*. Mutations within these genes often result in a loss of function. The result of this loss of function is analogous to removing the brakes from the cell cycle.

Cross reference

For specific examples of types of chromosomal abnormalities in leukaemias and lymphomas, refer to Chapters 11 and 12.

Section 9.6 considered the role of chromosomes in the packaging of genetic material. The respective roles of proto-oncogenes and tumour suppressor genes have been outlined above, and we will examine the ways in which the function of these genes can be altered or lost. First, we will consider the possible types of mutation affecting our genes at the nucleotide level, and will then consider major structural alterations of chromosomes. As demonstrated in Chapters 11 and 12, these mutations have an important role in the development (pathogenesis) of haematological malignancies.

In the majority of cases, haematological malignancies are associated with mutations in at least two classes of allele. Mutations in *class I* alleles result in cells gaining a growth advantage, following the activation of signal transduction pathways such as the RAS–MAPK pathway (we will encounter hairy cell leukaemia in Chapter 12, which is associated with a mutation in this pathway). To consolidate these mutations, alteration in *class II* alleles leads to the altered expression of principal transcription factors and results in disordered development. However, technological advances using next-generation sequencing, and more recently from the Cancer Genome Atlas project, have allowed us to further delineate these class I and II mutations into nine classes, based upon the specific biological activity of the gene products affected. Current stratification of functional group mutations is shown in Figure 9.12.

<table>
<tr><th rowspan="2">Analysis</th><th>Before 2008</th><th>2008-12</th><th>From 2013</th><th></th></tr>
<tr><th>Cytogenetic and molecular genetic analysis</th><th>Next-generation sequencing approaches</th><th>The Cancer Genome Atlas project</th><th>Prevelence in AML (%)</th></tr>
<tr><td rowspan="9">Functional groups</td><td rowspan="4">Class I: activated signalling—e.g. FLT3, KIT, RAS mutations</td><td rowspan="3">Class I: activated signalling—e.g. FLT3, KIT, RAS mutations</td><td>Class 1: transcription factor fusions—e.g. (8;21), t(16;16), t(15;17), MLL fusions</td><td>18</td></tr>
<tr><td>Class 2: nucleophosmin 1, NPM1 mutations</td><td>27</td></tr>
<tr><td>Class 3: tumour suppressor genes—e.g. TP53, WT1, PHF6 mutations</td><td>16</td></tr>
<tr><td rowspan="3">Class II: transcription and differentiation—e.g. t(8;21), t(16;16), t(15;17), CEBPA, RUNX1 mutations</td><td>Class 4: DNA-methylation-related genes: DNA hydroxymethylation—e.g. TET2, IDH1, IDH2 DNA methyltransferases eg, DNMT3A</td><td>44</td></tr>
<tr><td rowspan="5">Class II: transcription and differentiation—e.g. t(8;21), t(16;16), t(15;17) CEBPA mutations</td><td>Class 5: activated signalling genes—e.g. FLT3, KIT, RAS mutations</td><td>59</td></tr>
<tr><td>Class 6: chromatin-modifying genes—e.g. SXL1, EZH2 mutations, MLL fusions, MLL partial tandem duplications</td><td>30</td></tr>
<tr><td rowspan="3">Class III: epigenetic modifiers —e.g. TET2, DNMT3A, ASXL1 mutations</td><td>Class 7: myeloid transcription factor genes —e.g. CEBPA, RUNX1 mutations</td><td>22</td></tr>
<tr><td>Class 8: cohesin-complex genes—e.g. STAG2, RAD21, SMC1, SMC2 mutations</td><td>13</td></tr>
<tr><td>Class 9: spliceosome-complex genes—e.g. SRSF2, U2AF35, ZRSR2 mutations</td><td>14</td></tr>
</table>

FIGURE 9.12

The classification of mutational classes as proposed from the Cancer Genome Atlas Project. Reprinted from Meyer SC and Levine RL. *Translational implications of somatic genomics in acute myeloid leukaemia*. *The Lancet Oncology* 2014: 15: e382–94. Copyright © 2014, with permission from Elsevier.

SELF-CHECK 9.11

Briefly, compare and contrast tumour suppressor and proto-oncogenes.

9.7.1 Types of mutation

The sequence of nucleotides in a gene determines the sequence of amino acids in the polypeptide encoded. In turn, the sequence of amino acids in the polypeptide determines the structure and function of the polypeptide itself. Consequently, a mutation in a nucleotide sequence has a direct impact upon the polypeptide it encodes; it is so specific that a change in just one nucleotide may be sufficient to render its polypeptide product dysfunctional.

The process of polypeptide synthesis (translation) requires a gene to first be transcribed into messenger RNA (mRNA). A sequence of three nucleotides in mRNA forms a *codon*, with each codon indirectly encoding a particular amino acid (or a signal denoting the beginning or end of transcription). Sequential codons form a so-called 'reading frame', the sequence of codons which is 'read' to determine the sequence of amino acids in the polypeptide product. Figure 9.13 demonstrates this complementary binding of the anti-codon to the codon, the production of sequence-specific polypeptides, and the consequences of the different types of mutation (as outlined below).

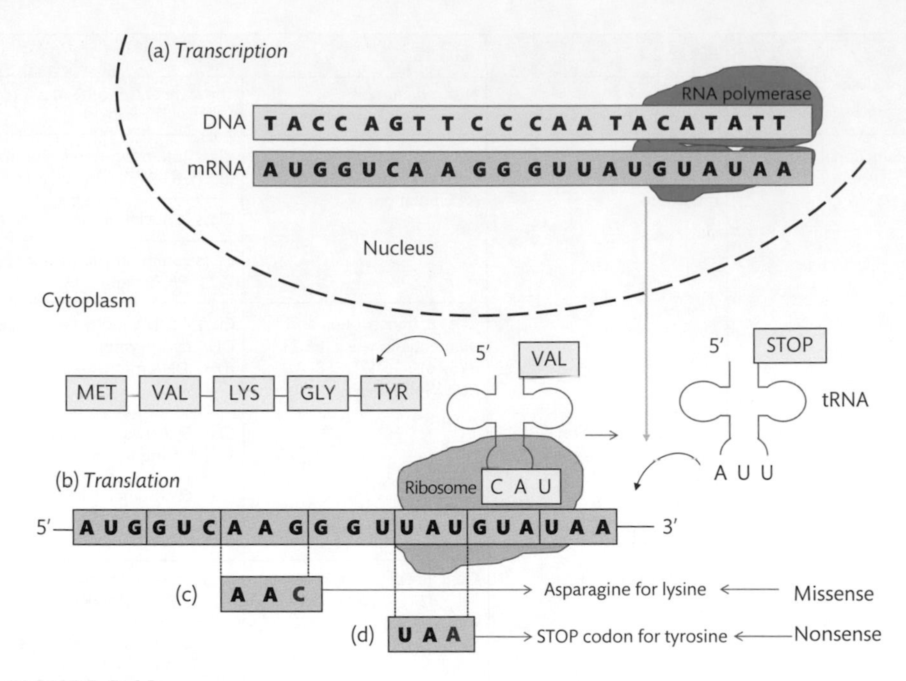

FIGURE 9.13

A simplified outline of transcription and translation. (a) mRNA is transcribed from the DNA template in the 3′ to 5′ direction by RNA polymerase (green). mRNA (purple) leaves the nucleus to interact with ribosomes in the cytoplasm (b). Following binding to the AUG start codon, tRNA anti-codons bind, in a complementary fashion, to the remaining codons in the mRNA sequence, allowing the addition of specific amino acids to the polypeptide chain. Each amino acid binds to the next by a peptide bond. This transcription process is mediated by the ribosome (yellow). Nucleotide substitutions can alter the amino acid composition of the polypeptide. (c) A missense mutation–the nucleotide exchange of cytosine for guanine–changes the codon specificity from lysine to asparagine. (d) Exchanging the nucleotides uracil for adenine, in this instance, represents a nonsense mutation. A stop codon is formed in place of tyrosine, producing a shortened (truncated) polypeptide.

The alteration of a single nucleotide in a gene sequence is called a **point mutation**. A **missense mutation** encodes a different amino acid following a single nucleotide exchange, altering the original codon. By contrast, a **nonsense mutation** generates a codon not encoding an amino acid at all. Nonsense mutations prevent further transcription beyond the mutation and are therefore referred to as stop codons.

Missense mutations may be clinically silent (have no pathological affect) or may alter the structure-function relationship of the protein product. An alteration in the size or charge of the amino acid may change the conformational structure of the protein when folded into its tertiary structure, thereby altering its activity. A nonsense mutation fails to encode an amino acid and invariably results in shortened (or truncated) polypeptides. If the sequence downstream from this nonsense mutation encodes a functional domain, this domain will not be transcribed, and so the protein product loses its function.

The deletion or addition of a nucleotide (or nucleotide sequence) is likely to disrupt the reading frame, causing the synthesis of a structurally and/or functionally abnormal protein.

Let's look at some illustrative examples.

In the following example, the sentence is constructed of a number of three-letter words or syllables, each representing a codon. Each of these codons will encode an amino acid, providing we can understand each word included in the sentence. The first sentence (a) makes sense, is explanatory, and should be considered normal (wild-type) DNA. In (b), we see the substitution of one of the letters (nucleotides) for another. Whilst the essence and structure of the sentence is preserved, the meaning has changed slightly. We can still encode an amino acid from the codon HAT (because we can still make sense of the word), but this will not necessarily be the *intended* amino acid. This is an illustration of a missense mutation. In (c), a nucleotide has been inserted into the sentence. We have tried to maintain the structure of the sentence by keeping to three-letter words. However, we can no longer understand the sentence as we now have a **frameshift mutation**. Translation will occur to the point at which the codon no longer makes sense (in this case ALT). As ALT does not make transcriptional sense, a premature stop codon is formed, and transcription ceases, resulting in a shortened protein product. This example illustrates the addition of nucleotides to a sequence, although it could also be reversed (d) to illustrate the loss of nucleotides.

(a) THE CAT ATE ALL HIS DIN NER

(b) THE HAT ATE ALL HIS DIN NER

(c) THE CAT ATE ALT LHI SDI NNE R

(d) THE CAT ATA LLH ISD INN ER.

In addition, duplications of genes can occur, potentially resulting in an overexpression of the gene product. If a proto-oncogene is overexpressed in this way, the cell harbouring this duplication may have a growth or survival advantage, especially if the duplicated gene encodes a cyclin or an anti-apoptotic protein.

This section has so far considered the translation of polypeptides from mRNA following acquired DNA mutations, and the mechanisms through which the polypeptide product may be dysfunctional. Section 9.7.2 examines acquired alterations within the chromosomal constitution of a cell, and how these alterations can affect polypeptide synthesis.

SELF-CHECK 9.12

List each of the main types of DNA mutation and provide a brief description.

9.7.2 Quantitative and qualitative chromosomal abnormalities

Chromosomal abnormalities can be either quantitative (numerical) or qualitative (structural), many of which can have a profound effect on the host cell. In this section, we will consider the way in which numerical chromosomal abnormalities are reported, before examining some of the structural abnormalities found in malignancies. The most common of qualitative chromosomal abnormalities in haematological malignancies are translocations, although deletions and inversions are also frequently seen. There is also a range of other chromosomal abnormalities; for more information on these, a specialized cytogenetics textbook should be consulted.

Let us now consider the chromosomal complement of normal and malignant cells, by considering how a patient's karyotype is established.

Karyotype

Quantitative and qualitative chromosomal abnormalities can be identified quite easily through karyotypic studies. The **karyotype** is a description of the chromosomal complement of a cell—if abnormal, then the number of cells in which the abnormality was found should be written inside square brackets. In order to assess an individual's karyotype, chromosomes (autosomes and sex chromosomes) are harvested from cells of a particular tissue, stained, and during analysis are placed in decreasing order of size, location of the centromere, and chromosomal staining characteristics (also called the banding pattern). This arrangement of chromosomes is called a karyogram.

From a karyogram, alterations in the number of chromosomes should be readily apparent. **Ploidy** is the term used to describe the number of homologous chromosomes within a cell. Normal somatic cells are **diploid**: they contain *two pairs* of each autosome plus two sex chromosomes. A normal somatic cell contains 46 chromosomes and is described as **euploid**.

If the chromosomal complement is abnormal, having one or more chromosomes missing or gained in an otherwise chromosomally normal (euploid) cell, this is described as **aneuploidy**. Examples of aneuploidy include monosomies and trisomies. The loss of a single chromosome results in a **monosomy** for that particular chromosome. The abbreviation for the loss of a chromosome is the minus sign (−). The karyotype for a female demonstrating a loss of chromosome 7 would read 45,XX,−7. The gain of a particular chromosome is called a **trisomy**, which is abbreviated to a plus sign (+). For a male patient with a trisomy 12, the karyotype would be reported as 47,XY,+12. For more comprehensive information regarding ploidy, please refer to Table 9.3.

Aneuploidy is frequently associated with haematological malignancies, and can be used to indicate the prognosis for a patient. For example, patients with an acute myeloid leukaemia (**AML**) with the loss of a chromosome 5 or 7 (−5 or −7) are considered to have a poor prognosis.

Examples of karyograms for two patients diagnosed with chronic lymphocytic leukaemia are shown in Figures 9.14 and 9.15.

The condition in which cells contain multiple sets of homologous chromosomes, also seen in some haematological malignancies, is called **polyploidy**.

Let us now examine some of the structural (qualitative) chromosomal abnormalities associated with haematological malignancies.

TABLE 9.3 **Nomenclature for describing cellular chromosomal content.**

Ploidy	Number of chromosomes	Comments
Near haploid	23-9	As would normally be found in a germ cell (n = 23)
Low hypodiploid	30-9	
High hypodiploid	40-5	
Diploid	46	Normal chromosomal constitution for a somatic cell
Pseudo-diploid	46	Normal total chromosomal number, but composition is abnormal
Low hyperdiploid	47-50	
High hyperdiploid	51-65	
Hypo/hypertriploid	69 ± (66-80)	Approximately three times the haploid (n = 23) number
Near tetraploid	92 +/−11 (81-103)	Approximately four times the haploid number of chromosomes

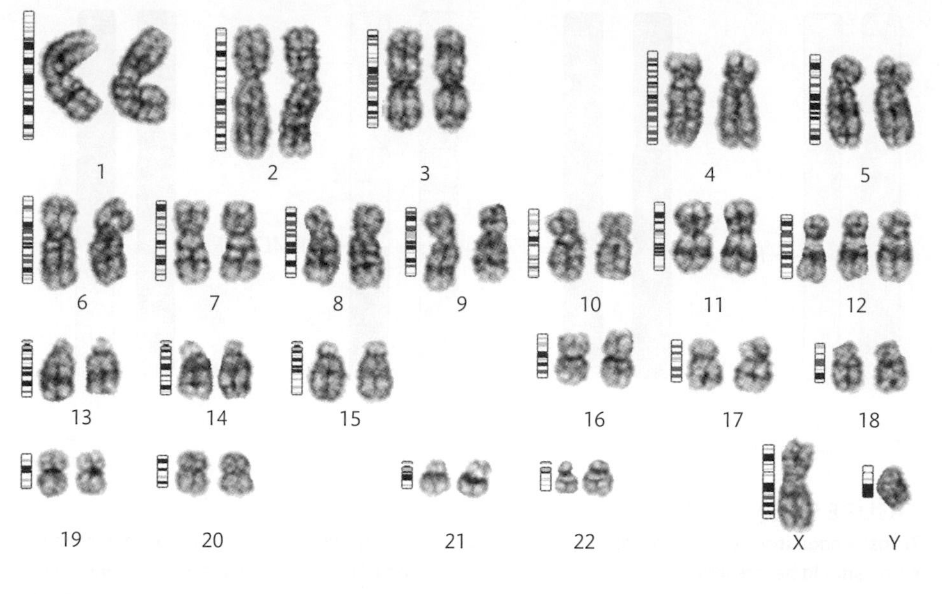

FIGURE 9.14
Karyogram for a male patient with chronic lymphocytic leukaemia. Chromosomes are aligned according to decreasing size and are matched into pairs according to banding pattern. This patient has an additional copy of chromosome 12 (trisomy 12) and deletion of the long arm of chromosome 13 (del13q). Image courtesy of Anne Gardiner, Cytogenetics department, Royal Bournemouth Hospital.

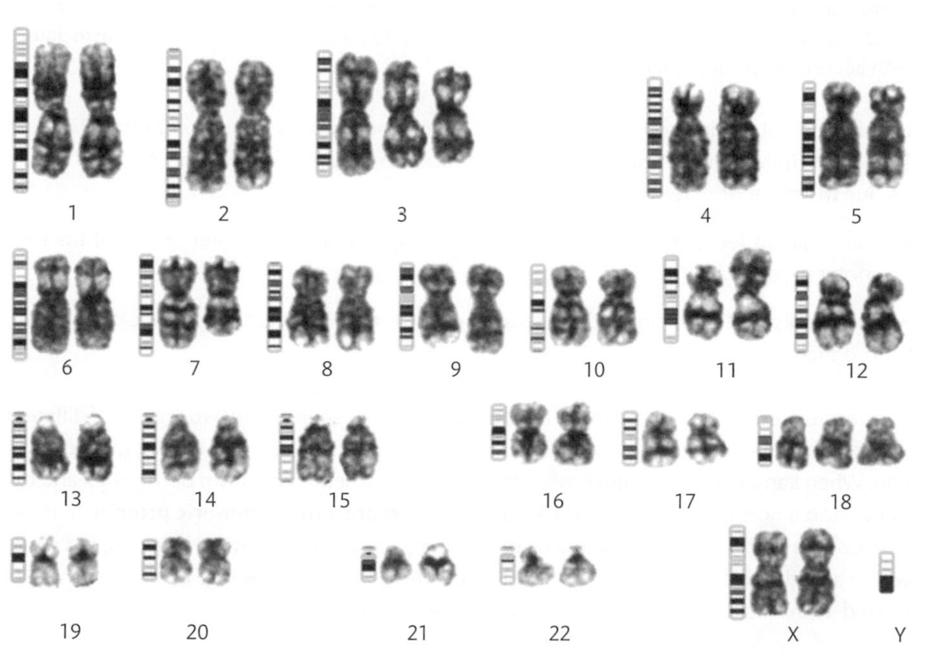

FIGURE 9.15
Karyogram for a female patient diagnosed with chronic lymphocytic leukaemia. This patient has additional copies of chromosome 3 (trisomy 3) and chromosome 18 (trisomy 18) plus t(7;11)(q21;p13). Image courtesy of Anne Gardiner, Cytogenetics department, Royal Bournemouth Hospital.

Translocations

Translocations in their simplest form comprise the exchange of genetic material between chromosomes. Look at Figure 9.16, which shows the mechanics of chromosomal translocation. In most situations, the exchange of chromosomal material results in both chromosomes acquiring new genetic material. We call this exchange a reciprocal translocation. This exchange can have a number of effects on the genes within the translocated region.

First, the translocation may be clinically silent—it has no effect on the host cell. Second, translocations can result in the *overexpression of a gene*. This is usually due to the translocated gene becoming separated from its original promoter (the control region of a gene which regulates transcription), and being controlled by a stronger promoter located upstream of the

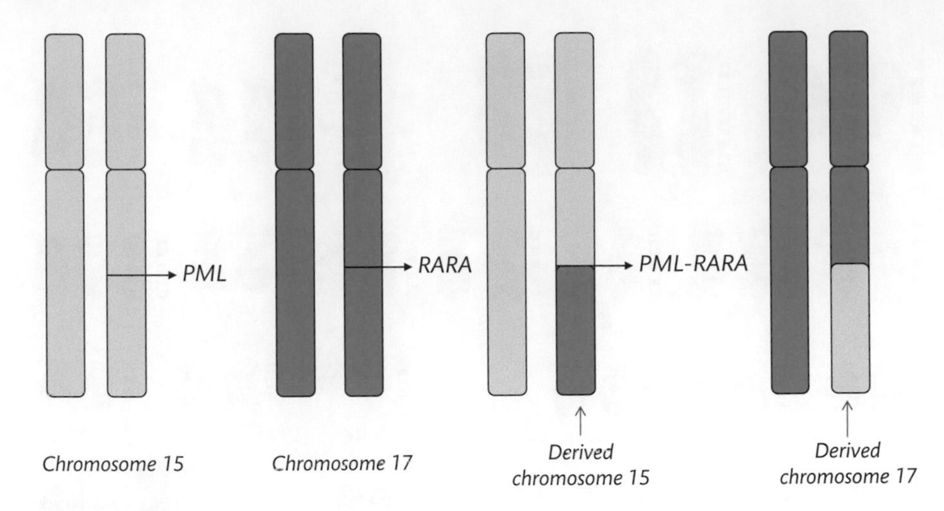

FIGURE 9.16

Translocation involving chromosomes 15 and 17–t(15;17)(q22;q12). Two copies of each chromosome should be present in each diploid cell, but only one of each chromosome is involved in the translocation process. The *PML* gene is located on chromosome 15 (15q22) and *RARA* is located on chromosome 17 (17q12). The derived chromosome 15 harbours *PML-RARA* and is associated with acute promyelocytic leukaemia.

translocation site. The stronger promoter has an enhanced ability to induce transcription and translation. Third, translocations can result in an *underexpression of a gene or its activity*. This can be due to a number of reasons including:

- translocation leading to transcriptional control by a weaker promoter or loss of the promoter region;
- the **breakpoint** occurring within a specific gene, leading to a shortened (truncated) protein product.

breakpoint
An area of a chromosome, usually within a gene, where a region of chromosomal material is lost or exchanged with another chromosome.

Cross reference
Examples of fusion proteins can be found in Chapters 11 and 12.

Furthermore, due to the formation of a fusion (or chimeric) gene where two regions of different genes are brought together by a translocation, the *protein product may acquire a novel function*. When transcribed, the mRNA will comprise retained sequences from both genes, and the translation process will form a product called a **fusion protein** or **chimeric protein**. In many cases, some of the functional domains of both original proteins may be retained, but fused within a single molecule. In the majority of cases, the fusion protein will be unable to complete its wild-type function, or this function may be altered in some way.

Key Points

In order to describe a translocation, a lower case 't' is used followed by two pairs of brackets. Within the first pair of brackets, the numbers of the chromosomes involved in the translocation are presented in numerical order. Each chromosome is separated by a semicolon e.g. t(9;22).

Within the second set of brackets breakpoints are included in an order that allows us to relate the chromosomes to their particular breakpoints. Again, each breakpoint is separated by a semicolon.

Example: A specific translocation between chromosomes 9 and 22 found in chronic myeloid leukaemia would be abbreviated as follows: t(9;22)(q34;q11). From this abbreviation, we can determine that the breakpoints are 9q34 and 22q11. The description of each part of a karyotype is separated by a comma, so the karyotype for a male with this translocation would be 46,XY,t(9:22)(q34;q11).

Whilst translocations can have a variety of effects on the behaviour of a cell, providing we know the function of the polypeptide transcribed from a particular locus, deletions have a more predictable effect and will be considered next.

Deletions

Another common cytogenetic feature found in haematological malignancies is the **deletion**. Deletions, as the name suggests, involve the loss of chromosomal material, including any genes found within that deleted region. A deletion may involve an entire chromosome, or part of a chromosome, either of which may have an overall effect on cell function. For example, if a key gene responsible for the regulation of the cell cycle (such as a tumour suppressor gene) is located within this deleted region, cell cycle control may be lost. Using appropriate chromosomal nomenclature, deletion is abbreviated as 'del'.

There are several types of deletion. Consider the structure of a chromosome in relation to Figure 9.17. A deletion occurring *within* a chromosome is called an **interstitial deletion**. Such deletions have two breakpoints and result from the chromosome being wrongly repaired by the cell's DNA repair machinery. However, if there is only one breakpoint and the rest of the chromosome arm is missing, we have a **terminal deletion**. Small deletions found within a chromosome (often submicroscopic) are called **microdeletions**.

Inversions

Subtle changes in the banding pattern within a chromosome may indicate an inversion. Inversions also need two chromosome breaks and are the result of the portion of chromosome being rotated before being rejoined, causing the 5′ region to face in a 3′ direction and vice versa. Figure 9.18 demonstrates the mechanics of chromosomal inversions. Two types of inversion can occur: a **paracentric inversion** and a **pericentric inversion**. A paracentric inversion is one in which a region restricted to one arm (p or q) is rotated through 180°. A pericentric inversion involves part of both arms of the chromosome, so also includes the centromere and, unlike a paracentric inversion, may change the location of the centromere. An example of a pericentric inversion is inv(16)(p13;q22) and this is found in a particular subtype of acute myeloid leukaemia.

So far, we have considered how gene expression can be altered through the process of mutation, and the gene sequence and chromosomal complement of a cell altered. We have discussed that DNA mutations, if undetected by the cell's DNA repair machinery, can be passed on to daughter cells, causing an accumulation of cells derived from the mutated progenitor. Epigenetic processes also allow constant modification of gene expression. Epigenetic processes are discussed in section 9.8.

Table 9.4 provides a comprehensive summary of the important translocations and inversions included within this book.

Cross reference

Inv(16)(p13;q22) is discussed in further detail in Chapter 11.

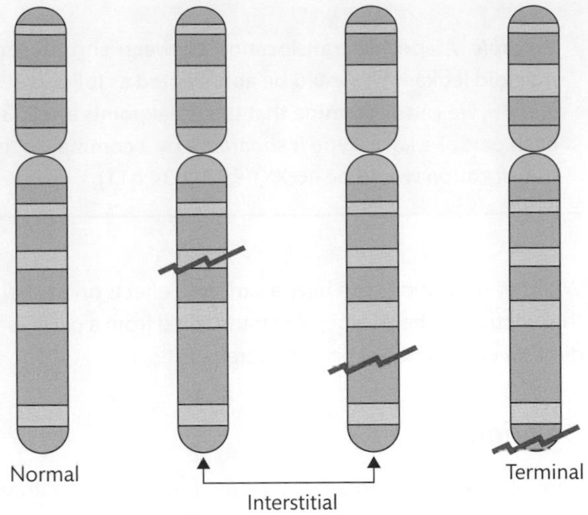

FIGURE 9.17
Chromosomal deletions. Zig-zagged lines represent chromosomal breakpoints. Left: normal chromosomal structure with the p- and q-arms separated by the centromere. Middle: examples of interstitial deletions. Right: a terminal deletion causes the loss of chromosomal material from the tip.

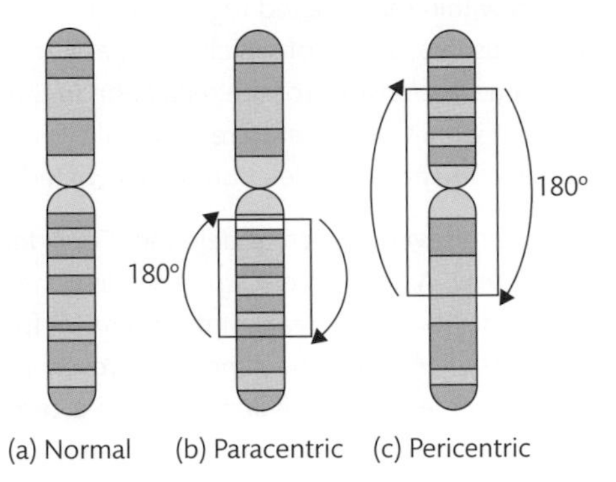

FIGURE 9.18
Chromosomal inversions. (a) A normal chromosome is presented for comparison of the G-banding patterns. A paracentric inversion (b) is restricted to a single arm (in this case the q-arm) and shows a reversal of the banding pattern within the boxed region as a consequence of the 'rotation' of this chromosomal material. The pericentric inversion (c) involves both the p- and the q-arms and describes the rotation of chromosomal material around the centromere. The pericentric inversion also reverses the banding pattern.

TABLE 9.4 **A selection of cytogenetic translocations and inversions considered in this textbook.**

Chromosomal abnormality	Product	Associated pathology
t(8;21)(q22;q22)	*RUNX1–RUNXT1*	AML with maturation
inv(16)(p13.1;q22)	*CBFB–MYH11*	Acute myelomonocytic leukaemia
t(16;16)(p13.1;q22)	*CBFB–MYH11*	Acute myelomonocytic leukaemia
t(15;17)(q22;q12)	*PML–RARA*	Acute promyelocytic leukaemia
t(11;17)(q23;q21)	*PLZF–RARA*	Acute promyelocytic leukaemia
t(11;17)(q13;q21)	*NuMA–RARA*	Acute promyelocytic leukaemia
t(5;17)(q35;q21)	*NPM–RARA*	Acute promyelocytic leukaemia
t(9;11)(p22;q23)	*MLLT3–KMT2A*	AML
t(6;9)(p23;q34)	*DEK–NuP214*	AML with maturation; acute myelomonocytic leukaemia
inv(3)(q21;q26.2)	*GATA2, MECOM*	AML; B-lymphoblastic leukaemia/lymphoma

Chromosomal abnormality	Product	Associated pathology
t(1;22)(p13;q13)	*RBM15-MKL1*	Acute megakaryoblastic leukaemia
t(v;11q23)	*v-KMT2A*	AML; B-lymphoblastic leukaemia/ lymphoma
t(9;22)(q34;q11)	*BCR-ABL1*	Chronic myeloid leukaemia; B-lymphoblastic leukaemia/lymphoma
t(12;21)(p13;q22)	*ETV6-RUNX1*	B-lymphoblastic leukaemia/lymphoma
t(5;14)(p31;q32)	*IL3-IGH*	B-lymphoblastic leukaemia/lymphoma
t(1;19)(q23;p13.3)	*TCF3-PBX1*	B-lymphoblastic leukaemia/lymphoma
t(14;18)(q32;q21)	*IGH-BCL2* *MALT1-IGH*	Follicular lymphoma; MALT lymphoma
t(11;14)(q13;q32)	*CCND1-IGH*	Myeloma; MGUS; mantle cell lymphoma
t(4;14)(p16.3;q32)	*MMSET-IGH*	Myeloma
t(14;16)(q32;q23)	*IGH-CMAF*	Myeloma
t(6;14)(p21;q32)	*CCND3-IGH*	Myeloma
t(14;20)(q32;q11)	*IGH-MAFB*	Myeloma
t(11;18)(q21;q21)	*API2-MALT1*	MALT lymphoma
t(1;14)(p22;q32)	*BCL10-IGH*	MALT lymphoma
t(2;17)(p23;q32)	*CLTC-ALK*	ALK positive large B-cell lymphoma
t(8;22)(q24;q11)	*MYC-IGK*	Burkitt lymphoma
t(2;8)(p12;q24)	*MYC-IGL*	Burkitt lymphoma
t(8;14)(q24;q32)	*MYC-IGH*	Burkitt lymphoma
t(2;5)(p23;q35)	*ALK-NPM*	Anaplastic large cell lymphoma ALK positive
t(1;2)(q21;p23)	*TPM3-ALK*	Anaplastic large cell lymphoma ALK positive

SELF-CHECK 9.13

A male patient's malignant cells are shown, using a variety of cytogenetic techniques, to have the following abnormalities in twenty cells examined:

47 chromosomes.

Deletion of chromosome 8.

Addition of chromosome 7.

Addition of chromosome 12.

Translocation of chromosome 15 long arm—band 22, and chromosome 17 long arm—band 12.

Write this patient's karyotype in abbreviated form.

9.8 Epigenetics

Epigenetics explores how gene expression can be altered whilst maintaining wild-type DNA sequences. Epigenetic processes are entirely normal, although, when they occur inappropriately, can predispose to malignancy. An excellent definition of epigenetic processes was proposed by Adrian Bird of the Institute of Cell Biology in Edinburgh as '... *the structural adaptation of chromosomal regions so as to register, signal or perpetuate altered activity states* ...'. Epigenetic processes can involve region-specific DNA sequences or post-translational modification of histone proteins. **Hypermethylation** (the addition of methyl (CH_3) groups), a process whereby genes may be 'switched off' to repress DNA transcription, is an important epigenetic event. Epigenetic processes also describe the removal of methyl groups, resulting in **hypomethylation**. **Acetylation** will be considered here as an important regulator of transcription.

A range of other epigenetic processes may also be utilized in transcriptional regulation, including phosphorylation—the addition of phosphate groups (PO_4^{3}) to target proteins; and histone ubiquitination—the addition of the protein ubiquitin to histone molecules H2A and H2B. These processes are beyond the scope of this book.

We will now examine hypermethylation, hypomethylation, and acetylation in turn.

9.8.1 Hypermethylation

A family of enzymes called **DNA methyltransferases** (DNMTs) attach methyl groups to **CpG islands** throughout the genome, as illustrated in Figure 9.19. DNMTs are found to be overexpressed in many haematological malignancies.

CpG islands are regions of DNA rich in cytosine and guanine nucleotide repeats, found within the promoter regions of approximately 40% of all mammalian genes. The transcriptionally repressive process of hypermethylation is thought to be mediated by the binding of specific proteins containing a methyl-binding domain (MBD) to methylated CpG islands. The localization of MBD-containing repressors and co-repressors inhibits DNA transcriptional machinery, preventing the production of mRNA. One of the key repressive enzymes to be localized in this fashion is HDAC. As such, hypermethylation enables the repression of specific genes.

In the context of haematological malignancies, hypermethylation of tumour suppressor genes results in their inactivation. Such inactivation of tumour suppressor genes can potentially cause the subsequent accumulation of mutations within the genome of the cell, attributable to a reduction in the intracellular concentration of tumour suppressor proteins. The pattern of methylation is passed to subsequent generations of daughter cells, silencing tumour suppressor genes in a progenitor cell, which in turn will cause silencing of the tumour suppressor in all subsequent

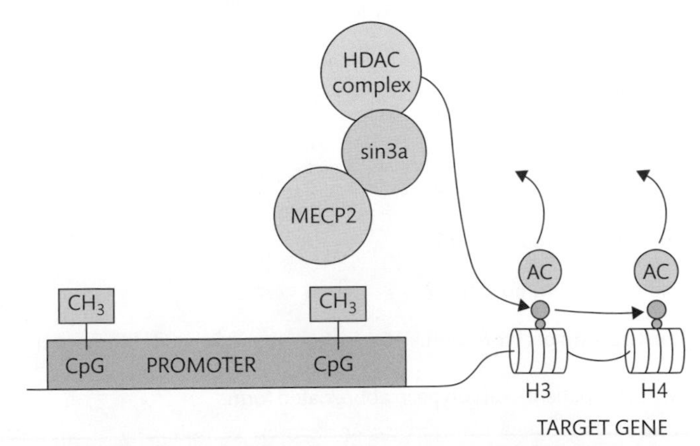

FIGURE 9.19
Epigenetic processes. Methylation of cytosine within CpG islands allows the assembly of an inhibitory complex involving HDAC and sin3A. This inhibitory complex causes chromatin remodelling by deacetylating histone H3 and H4 lysine residues. Transcriptional activators that normally bind to acetylated lysine fail to localize to this region, causing transcriptional repression. CpG = CpG island within the promoter region; MECP2 = methyl-CpG-binding protein 2; HDAC = histone deacetylase; AC = acetylation.

generations of cells. This heritable silencing of tumour suppressors increases the probability of further mutations accruing within this lineage, predisposing these cells to malignant transformation.

The following mechanisms outline the process of silencing tumour suppressor genes through hypermethylation:

- *biallelic methylation* of a particular tumour suppressor gene means that both copies of that gene are silenced, rendering the host cell deficient in a particular tumour suppressor protein;
- *monoallelic hypermethylation with simultaneous mutation of the accompanying tumour suppressor allele* is a scenario where a single copy of the tumour suppressor gene is silenced by methylation, but the alternative allele has been mutated in some way, inhibiting the function of the tumour suppressor protein. This situation also causes the host cell to become deficient in a particular tumour suppressor protein.

Hypermethylation has the same effect as a tumour suppressor deletion or mutation: loss of function. Pro-apoptotic genes, cell cycle regulators, and DNA repair genes are examples of other families of genes that may be silenced through epigenetic processes.

Now that the role of hypermethylation in gene silencing has been considered, we should examine the role of hypomethylation on gene transcription.

9.8.2 Hypomethylation

Whilst the importance of hypermethylation can be seen in the repression of specific genes, another epigenetic mechanism may result in an overexpression of certain genes and their products. This mechanism is called hypomethylation. Hypomethylation, or demethylation, is the opposite of hypermethylation and is a process through which the methyl groups attached to DNA are removed. This demethylation results in gene activation, as methyl-binding proteins can no longer localize within the promoter region of target genes. The failure of this localization subsequently inhibits binding to a range of transcriptional repressors and co-repressors which inhibit DNA transcription and translation. Often in haematological malignancies, widespread hypomethylation occurs (called global hypomethylation), the result of which is the activation of a number of oncogenes, including cyclins and anti-apoptotic genes—allowing transcription and translation to occur. Overexpression of oncogenes can play an important role in oncogenesis.

An indication of some of the effects of hypo- and hypermethylation are shown in Figure 9.20.

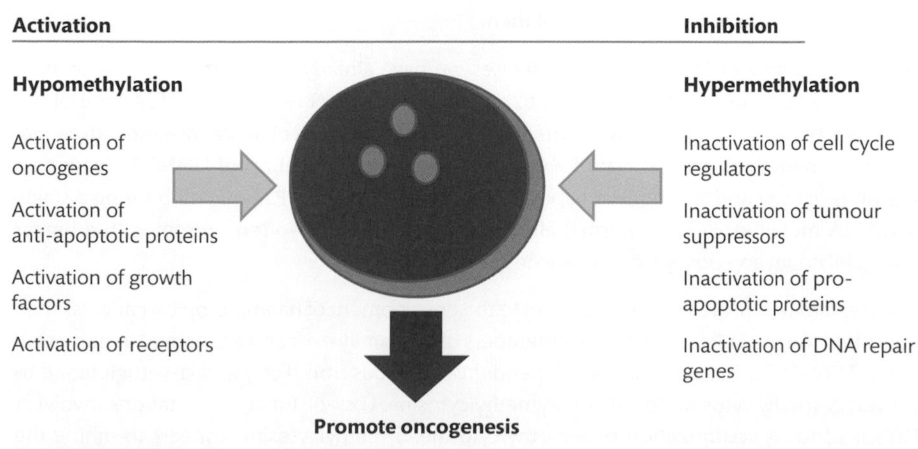

FIGURE 9.20
Potential consequences of hyper- or hypomethylation in oncogenesis.

9.8.3 Acetylation

Acetylation involves the addition of acetyl-Coenzyme A (CoA)-derived acetyl groups (–$COCH_3$) to histone lysine residues in the presence of the enzyme histone acetyltransferase (HAT). Acetylated lysines, usually associated with histones H3 and H4, are unevenly distributed throughout chromatin, although are associated with the 'open' chromatin structure **euchromatin.** The term 'euchromatin' describes regions of chromatin possessing active genes. These regions are therefore subject to DNA transcription. Euchromatin is also associated with hypomethylated CpG islands, ensuring transcriptional repressors fail to localize in this region. In contrast, **heterochromatin**, the 'closed' or 'clumped' chromatin, retains hypoacetylated histones and is associated with hypermethylated CpG islands, both of which repress transcription and translation.

In hypermethylated regions, the localization of large inhibitory protein complexes prevents transcription and translation through a variety of mechanisms (refer to Figure 9.19). Localization of HDAC removes acetyl groups from H3- and H4-associated lysine residues. Co-localization of sin3a (bridging HDAC to specific transcription factors) and mi-2 (involved in the nucleosome remodelling and histone deacetylation complex (NuRD)) enables chromatin remodelling and the inhibition of transcription.

The removal of acetyl groups from lysine is believed to inhibit transcription and translation in two ways:

- the positive charge of the histone-associated lysine residues is regained following neutralization by the acetyl groups. This positive charge induces chromatin clumping, preventing access of the DNA transcriptional machinery to target genes;
- acetylated lysines associated with H3 and H4 act as binding sites for proteins containing a **bromodomain**. These bromodomain-containing proteins are largely considered to be involved in the regulation of transcription and translation. Removal of the bromodomain-binding site causes gene repression.

bromodomains
Originally identified in the *Drosophila* protein brahma. A bromodomain comprises a sequence of 110 amino acids derived from a four-helix bundle. When the helices interact, they produce a hydrophobic binding pocket with a conserved amino acid sequence that recognizes acetyl lysine.

9.8.4 Mutation in epigenetic regulators

Whilst we have seen that class I and class II mutations can lead to the development of haematological cancers, only about 50% of cases of myeloid cancers harbour mutations in class I alleles. There is now building evidence to show that mutations in epigenetic regulators provide a significant contribution to the development of these malignancies, both in concert with class I/II mutation, and also independently of them.

DNMT is a family of DNA methyltransferases we have already encountered in this section. These enzymes transfer methyl groups to cytosine within CpG islands. DNMT3A, one of the members of the DNMT family, is mutated in around 22% of cases of acute myeloid leukaemia, and these mutations are associated with a poor prognosis. Mutation of DNMT3A leads to a loss of its activity and subsequent hypomethylation of cytosines. Patients harbouring a single DNMT3A mutation—i.e. one normal allele, one mutated allele—often present with a higher white cell count and possess other cancer-causing mutations.

Another family of proteins which can lead to the development of haematological cancers when perturbed are the TET proteins. Three members of this family exist and are named in numerical order, TET1–TET3. TET enzymes are dependent on ferrous iron (Fe^{2+}) and α-ketoglutarate to convert 5-methylcytosine to 5-hydroxymethylcytosine. Loss-of-function mutations involving TET2 lead to an accumulation of 5-methylcytosine. 5-methylcytosine appears to inhibit the binding of negative regulatory proteins to DNA, leading to protein overexpression,

IDH1 and IDH2 are two related enzymes that possess overlapping activity with TET2. The conversion of isocitrate to α-ketoglutarate is dependent upon both IDH1 and IDH2 activity, and, as you will have noted from the paragraph above, α-ketoglutarate is an important cofactor for TET proteins. As such, the aberrant methylation patterns noted in IDH mutants is similar to those observed in TET mutants. These TET mutations have the potential to contribute to malignant transformation of haemopoietic precursors. It has also been shown that, when mutated, IDH converts α-ketoglutarate to 2-hydroxyglutarate, which is elevated in cases of AML where mutant IDH exists.

We now recognize that mutations in Janus kinase 2 (JAK2) (V617F), associated with myeloproliferative neoplasms, also play a role in epigenetic modification. The V617F, gain-of-function, mutation allows aberrant re-localization of mutated JAK2 to the nucleus, with phosphorylation of tyrosine 41 on histone H3. The overall effects of this modification include enhanced gene activity and clinical features consistent with myeloproliferative neoplasms. Furthermore, t(15;17), rearrangements involving 11q23, and the core binding factor (CBF) leukaemias: t(8;21), t(16;16) and inv(16) also lead to abnormal epigenetic profiles and transcriptional repression, which are known to contribute to the development of haematological malignancies.

Cross reference

More detail on the CBF leukaemias is given in Chapter 11.

Here we have considered some of the genetic and epigenetic mechanisms responsible for altering cell behaviour and ultimately increasing the probability of oncogenesis. Significant progress is now being made in developing our understanding of the range of epigenetic modifications we see in cancer and the identification of key components within these pathways which can be targeted with new classes of drugs.

To put these concepts into the context of clonal disorders, section 9.9 provides a definition of clonality, and the ways in which we can establish whether a population of cells is a malignant clone. That section will conclude with discussing the acquisition of additional mutations which can cause an alteration in the behaviour of the clonal subpopulation. With this in mind, let's have a look at clonality.

SELF-CHECK 9.14

Briefly discuss the roles of hypomethylation, hypermethylation, and acetylation in the control of transcription and translation.

9.9 Clonality

Unrepaired mutations occurring within a single stem cell or haemopoietic progenitor will be passed on to all subsequent generations of cells derived from that mutated cell, as shown in Figure 9.21. As these daughter cells share the same genetic or molecular features, although they are distinct from the wild-type population of cells, this population of mutated cells is called a clone. Haematological malignancies are clonal disorders and, as such, the malignant cells all share the same basic genetic or molecular abnormalities. The remainder of this section describes some of the ways that are used to determine the clonality of a particular group of cells.

9.9.1 Cytogenetic analysis

Cytogenetically, a population of cells can be called a clone if a minimum of two cells share an additional chromosome, or a minimum of three cells share the same deletion in the absence of any other abnormal cytogenetic features. Two or more cells need to show the same translocation to constitute a clone.

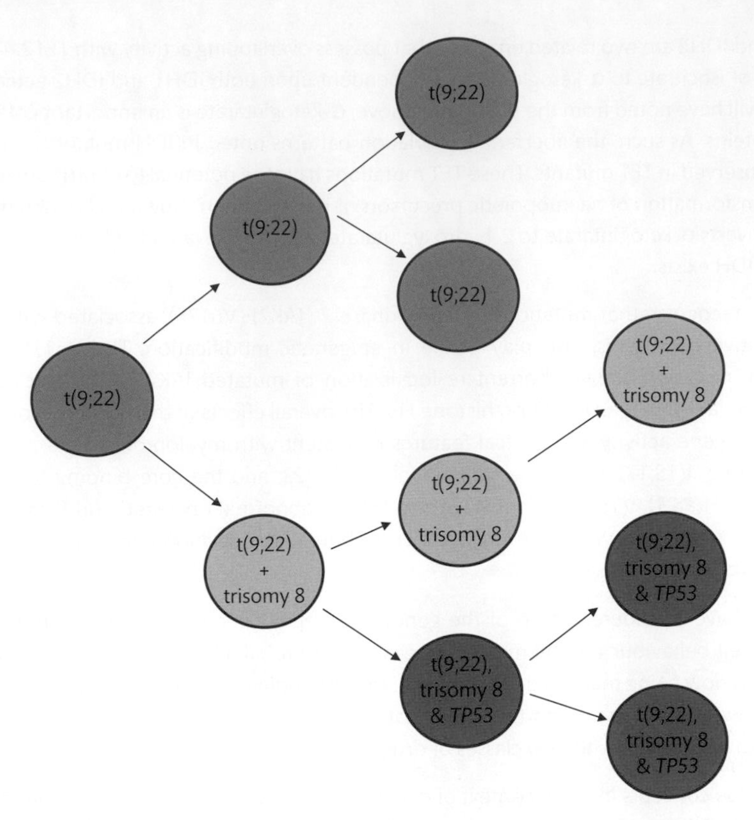

FIGURE 9.21

A clonal population of cells. All cells in this figure are derived from a single mutated stem cell harbouring t(9;22), typically found in CML. Daughter cells also acquire additional mutations, forming subclones (represented by the different colours). These subclones accumulate mutations, altering their behaviour in a process called clonal progression. In this example, trisomy 8 is acquired in a cell already harbouring t(9;22). One of the daughter cells of this subclone acquires a *TP53* mutation, resulting in the loss of a valuable tumour suppressor–forming another subclone with a greater survival advantage and likely resistance to therapy.

A good example of a clonal disorder is chronic myeloid leukaemia (CML). Following the extraction of a bone marrow aspirate for cytogenetic analysis, the cultured cells will be investigated for a specific chromosomal abnormality called the **Philadelphia chromosome**. This chromosome is known as der(22) as it is derived (der) from a reciprocal translocation, although there is a net loss of chromosomal material from the long arm of chromosome 22. Specifically, the translocation t(9;22)(q34;q11) producing der(22) will be investigated and should be found in all the malignant cells within the myeloid lineage. Morphologically, CML appears as an accumulation of mature and maturing granulocytes. There are generally all stages of maturation of neutrophils present, with an increase in basophils and eosinophils also occurring, especially as the disease progresses. Platelet defects will also be apparent. Even though CML appears morphologically heterogeneous, this condition is clonal because all malignant cells possess the Philadelphia chromosome.

Philadelphia chromosome
This denotes the derived chromosome 22 der(22), not to be confused with t(9;22). The translocation *process* is abbreviated to t(9;22), not the chromosome product.

light-chain restriction
An overexpression of either kappa or lambda light chains. This overexpression of a single type of light chain would be the product of a monoclonal population of B cells.

9.9.2 Immunophenotyping

Using immunophenotyping, it can be rather difficult to establish whether a population of cells is clonal. When a B-lineage malignancy is suspected, we can investigate membrane-bound immunoglobulins specifically to assess kappa (κ) or lambda (λ) **light-chain restriction**. A single

B cell will express a single type of light chain, either kappa or lambda, but never both. Figure 9.22(a) shows a monoclonal population of B lymphocytes, with kappa light-chain restriction. We would expect approximately 65% of cells to express kappa light chains, with the remainder expressing lambda. Generally speaking, kappa or lambda light chains found in excess of those ranges are described as light-chain restricted and therefore clonal. Ratios within the reference range suggest a polyclonal population of B cells have produced these light chains, which is a normal response to infection.

If aberrant markers are expressed on a range of malignant cells, for example the T-cell antigen CD7 expressed on myeloid cells, as shown in Figure 9.23, this is also an indication of a clonal population. However, selective loss or gain of antigen expression often makes the immunophenotyping of aberrant markers an unreliable process.

Cross references

Cytogenetic analysis is discussed in greater detail in Chapter 10.

See the *Medical Microbiology* text in this series for more information on immune response to infection.

A more in-depth outline of CML can be found in Chapter 11.

More detail regarding the process of immunophenotyping and an explanation of CD markers can be found in Chapter 10.

9.9.3 Morphology

It is important to recognize that the cells constituting a clonal population do not necessarily look the same (homogeneous morphology)—and in malignancies such as CML they appear heterogeneous. Investigations such as cytogenetic analysis or polymerase chain reaction (PCR) should be completed to determine clonality.

9.9.4 DNA mutations and their products

Genetic mutations play an important role in oncogenesis, often resulting in a structurally and functionally abnormal protein product. If the cell population shares an identical mutation, these cells are clonal. An example of a mutated protein is the JAK2 receptor, derived from the valine to phenylalanine substitution at position 617 (V617F) associated with a number of myeloproliferative diseases, including polycythaemia vera.

Cross references

Some of the methods used to identify DNA mutations and their products are outlined in Chapter 10.

Myeloproliferative diseases are discussed in greater detail in Chapter 11.

Cross reference

JAK2 mutations are discussed in greater detail in Chapter 11.

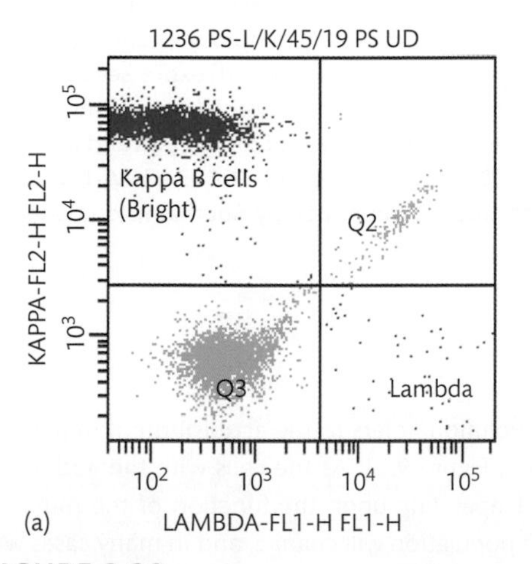

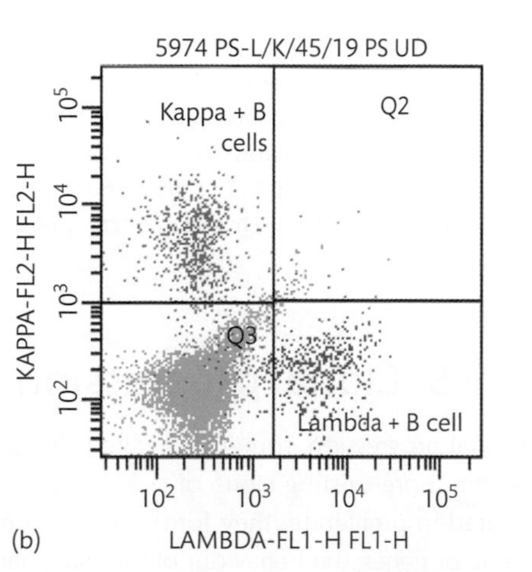

FIGURE 9.22

Flow cytometric evaluation of light chain status. (a) Patient shows only kappa light chains (red), quadrant 1 (Q1). The patient has a clonal population of B cells which are kappa light-chain restricted. (b) A polyclonal B-lymphocyte population in which both kappa and lambda light chains are detected. Cells in Q2 and Q3 are considered negative for kappa and lambda. Images courtesy of Department of Immunology, Southampton General Hospital.

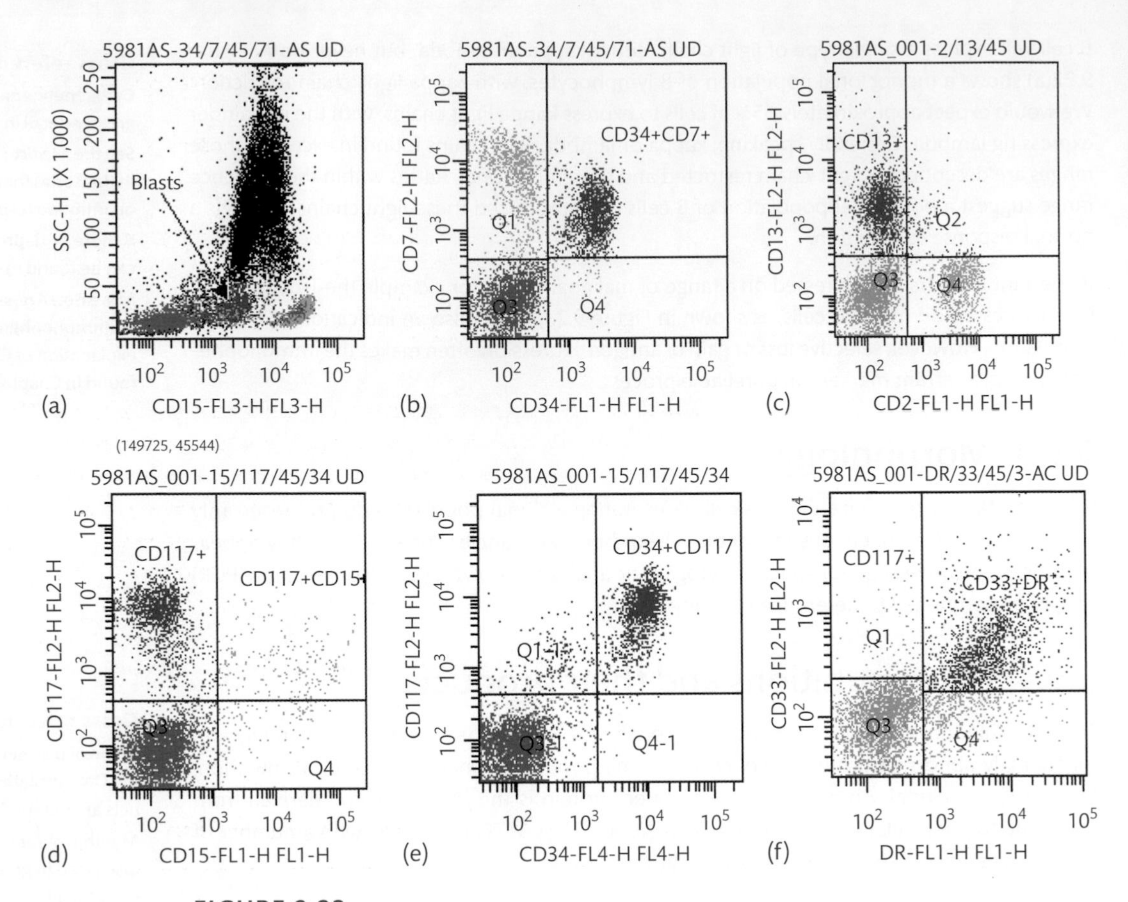

FIGURE 9.23

Aberrant expression of CD7 in acute myeloid leukaemia (AML). CD7 is a T-cell marker, but may be found abnormally expressed on cells of a myeloid lineage in AML. (a) Shows the general distribution of cells within the sample, the target population is shown in red; (b) cells co-express CD34 and CD7; (c) cells show bright expression of CD13 (a marker of granulocytes and monocytes), but do not express CD2 (a marker of T-cells); (d) blast population (red) expresses CD117, the stem cell factor receptor, but not CD15 (another marker of granulocytes); (e) confirms that cells co-express CD34 and CD117; (f) cells co-express CD33 (a marker of monocytes and myeloid progenitor) and HLA-DR. The blast phenotype is: CD34+, CD117+, CD13+, CD33+, HLA-DR+, CD7+ and CD2−, CD15−. Images courtesy of Department of Immunology, Southampton General Hospital.

9.9.5 Clonal progression

Clonal progression, sometimes called clonal evolution, refers to the acquisition of mutations within a pre-existing clone of cells, as shown in Figure 9.21. As the cells with the additional mutation proliferate, they form a sub-clone. Depending upon the function of the mutated gene or genes, the behaviour of the malignant population will change, and in many cases will become more aggressive. Markers of aggressiveness include the spread of malignant cells to other sites (metastasis), or resistance to previously effective therapy. For example, mutation of the tumour suppressor gene *TP53* is associated with a poor prognosis because radiotherapy and many chemotherapeutic agents activate the p53 protein through the DNA damage they induce. This mutation renders these *TP53*-mutated cells resistant to many forms of traditional anti-cancer therapy. With wild-type *TP53*, following chemotherapy or radiotherapy, p53 would be activated, and as a consequence, the apoptotic machinery would be initiated, causing death of the malignant cells.

We have now considered all the important changes that can occur within a cell to cause malignant transformation, but how many of these mechanisms need to occur before a cell will transform? Section 9.10 considers this in relation to the oncogenic mechanisms discussed thus far.

SELF-CHECK 9.15

Provide, using your own words, a definition of a clone.

9.10 Mechanisms of oncogenesis

This section considers the development of haemopoietic tumours. The development of leukaemia is called **leukaemogenesis**, and the same process resulting in lymphoma is called **lymphomagenesis**. The process through which myeloma develops is called **myelomagenesis**.

The development of cancer is a consequence of the loss of a cellular control mechanism, possibly causing an increase in the proliferation rate, or the loss of apoptotic control. Ultimately, the result is an inherent genomic instability within the tumour cells, predisposing the malignant clone to further genetic alteration and clonal progression.

Clonal progression usually heralds an evolution in the behaviour of the tumour cells and can cause resistance to commonly used therapeutic agents. However, the cause of the change from a normal to a malignant cell is unlikely to be the response to a single mutational event. Rather, oncogenesis is due to a multistep process causing a gradual reduction in cellular control. As we have previously discussed, alterations in proto-oncogenes are likely to cause gain-of-function mutations, and mutations in tumour suppressor genes cause a loss of cellular control. Mutations in both tumour suppressor genes and proto-oncogenes are common in haematological malignancies, but will a single mutation in a single gene cause cancer? Undoubtedly, the answer to this question is 'no'.

Take, for example, a mutation in a tumour suppressor gene. Tumour suppressors have a number of functions, but the endpoint of their action is to ensure that the cell cycle completes appropriately, without the incorporation of any mutations within the replicated DNA sequence. Mutating a tumour suppressor gene on its own will not cause cancer. Instead, alterations in tumour suppressor genes (TSGs) will increase the likelihood of *additional* mutations occurring in other areas of the genome, and will then go unchecked because of the initial mutation. For example, the loss of a tumour suppressor may be the first mutation within a cell. If this TSG mutation is followed by a mutation in either the same cell or in that cell's progeny by an activating (or gain-of-function) mutation within a proto-oncogene, uncontrolled proliferation will occur. Once uncontrolled proliferation occurs, additional mutations are likely, leading to clonal progression.

This multistep process of oncogenesis was proposed by Ernest Knudson in 1971 and is called *Knudson's hypothesis*. This hypothesis states that a minimum of two steps (or two hits) are required within a single cell to cause cancer, and this has been repeatedly demonstrated in a wide range of tumours.

The acquisition of somatic mutations is a proven way of developing genetically unstable cells leading to a haematological malignancy. However, this model largely describes oncogenesis in an ageing population, where the accumulation of somatic mutations occurs over time. Clearly, this process does not explain all types of haematological malignancy. Undoubtedly, there is a heritable component to many haematological malignancies, and section 9.11 examines some of those conditions leading to an increased risk of developing malignancy, but these also abide by Knudson's hypothesis.

SELF-CHECK 9.16

Using appropriate examples, explain Knudson's hypothesis.

9.11 Inheritance and leukaemia

A wide range of heritable conditions can predispose individuals to haematological malignancies. A number of these heritable conditions can result in childhood leukaemias. Some of these conditions may be associated with the inheritance of a single inactivating mutation within a TSG, leaving the individual open to further mutations in accordance with Knudson's multi-hit hypothesis. Others, such as Down syndrome, Fanconi syndrome, and Diamond–Blackfan anaemia are associated with alternative mechanisms of oncogenesis, including failure of DNA repair and immunodeficiency. Table 9.5 demonstrates the range of familial conditions associated with a predisposition to haematological malignancies. This section briefly introduces a selection of inherited factors.

9.11.1 Trisomy 21

Trisomy 21, commonly called Down syndrome, is associated with an increased risk of leukaemia. The highest risk is associated with patients below four years of age, with the risk progressively decreasing beyond this age. Children with trisomy 21 commonly undergo a brief period in which the production of their myeloid cells within the bone marrow becomes grossly abnormal (**transient abnormal myelopoiesis**, TAM). When examined in the laboratory, TAM is virtually indistinguishable from acute myeloid leukaemia. The key differentiating feature is that TAM spontaneously remits within two to three months. Approximately 25% of patients with TAM will later go on to develop acute megakaryoblastic leukaemia (cancer of the platelet precursors)—a particularly rare subgroup of AML. Individuals who have a mosaic genotype for trisomy 21, i.e. a mixture of cells with either two or three copies of chromosome 21, are also at risk of TAM. In these mosaic cases, only the trisomic cells are implicated in TAM, indicating that the additional copy of chromosome 21 is essential for this haematological syndrome to manifest itself. Currently, the reason for the association between trisomy 21 and leukaemogenesis is unknown.

Cross reference
Trisomy 21 is discussed in the context of a myeloid proliferation related to Down syndrome in Chapter 11.

9.11.2 Fanconi syndrome

Fanconi syndrome or Fanconi anaemia (FA) is associated with progressive bone marrow failure caused by genetically unstable progenitor cells and consequential reduction in peripheral blood cell counts of all three lineages (pancytopenia): red cells, white cells, and platelets.

The risk of developing AML is 15,000-fold higher in FA patients in comparison to the normal population. Largely inherited in an autosomal-recessive fashion (i.e. the patient must have two copies of the mutated gene in order for the disease to manifest itself), the precise biology of the disease is complex, involving at least 22 genes. The products of these genes are associated with the FANC/BRCA DNA repair pathway, the failure of which leads to the accumulation of DNA mutations and a propensity to develop cancer.

Cross reference
Fanconi anaemia and Diamond-Blackfan anaemia were introduced in Chapter 5.

Cross reference
For more information on myelodysplastic syndromes, refer to Chapter 11.

TABLE 9.5 Hereditary diseases associated with the development of haematological malignancies.

Syndrome	Inheritance pattern	Malignancy
DNA repair		
Ataxia telangiectasia	Recessive	BCL, TCL, T-ALL, T-PLL
Bloom	Recessive	ALL, AML, lymphoma
Fanconi	Recessive	AML
Nijmegen breakage	Recessive	lymphoma, leukaemia
Rothmund-Thomson	Recessive	AML
Seckel	Recessive	AML
Werner	Recessive	AML
Conditions leading to secondary mutations		
Amegakaryocytic thrombocytopenia	X-linked	AMML
Diamond-Blackfan	Dominant and recessive	AML
Familial platelet disorder	Dominant	AML
Kostmann	Dominant and recessive	AML, AMoL
Schwachman-Diamond	Recessive	ALL, AML, AMML, AMoL, EL, JMML
Tumour suppressor		
Li-Fraumeni	Dominant	ALL, Hodgkin, Burkitt, CML B-CLL
Neurofibromatosis	Dominant	AML, JMML
Immunodeficiency		
Common variable immunodeficiency	Dominant and recessive	BCL
Severe combined immunodeficiency disease	Recessive	BCL
Wiskott-Aldrich	X-linked	ALL, Hodgkin lymphoma
X-linked lymphoproliferative syndrome	X-linked	EBV-linked BCL

Key: ALL = acute lymphoblastic leukaemia; AML = acute myeloid leukaemia; AMML = acute myelomonocytic leukaemia; AMoL = acute monocytic leukaemia; BCL = B-cell lymphoma; B-CLL = B-cell chronic lymphocytic leukaemia; CML = chronic myeloid leukaemia; EBV = Epstein-Barr virus; EL = erythroleukaemia; JMML = juvenile myelomonocytic leukaemia; T-ALL = T-acute lymphoblastic leukaemia; TCL = T-cell lymphoma; T-PLL = T-prolymphocytic leukaemia.

Adapted from Segal GB and Lichtman MA. ***Familial (inherited) leukaemia lymphoma, and myeloma: an overview***. *Blood Cell, Molecules, and Diseases* 2004: **32**: 246-61.

9.11.3 Diamond–Blackfan anaemia

Diamond-Blackfan anaemia (DBA) is a particularly interesting congenital disease as it is the first condition to be associated with the mutation of a ribosomal protein, RPS19. The mechanisms responsible for DBA-associated pathologies are incompletely understood. The gene encoding RPS19 is located at locus 19q13, and mutations within this gene are found in 25% of patients. Alternative gene loci have also been implicated in DBA pathogenesis, including 19 other genes that encode ribosomal proteins. Additionally, three mutated genes have been identified in DBA (or DBA-like diseases) independent of those ribosomal proteins, including *TSR2* (involved in ribosome biogenesis), *GATA1* (an erythroid transcription factor) and a single report of a mutation within the erythroid growth factor (erythropoietin). DBA is associated with an increased risk of MDS, AML, colon cancer, osteosarcoma, and cancers of the genitourinary system. Although DBA is clinically and biologically heterogeneous, it usually presents within the first year of life with a failure of red cell production (erythropoiesis), an anaemia associated with enlarged red cells (a macrocytic anaemia), and a significant reduction in the number of developing red cells within the bone marrow (erythroid aplasia).

So far, some of the heritable factors causing a predisposition to oncogenesis have been considered. Now we should examine some of the environmental factors demonstrated to play a role in oncogenesis, and in particular the development of haematological malignancies.

9.12 Environmental causes of haematological malignancies

So far, we have examined some of the molecular and cytogenetic causes of cancer. Later chapters will demonstrate the consequences of the mutations already introduced, and how their very real impact on cell proliferation and survival can ultimately cause malignant transformation. Sometimes mutations will occur through a very traceable series of events, such as the inheritance of a dysfunctional TSG. In other situations, the cause may not be so straightforward, and we need to look into alternative hypotheses to explain the causes of these tumours.

It is important to be aware of environmental factors. Whilst it is very easy to view ourselves as discrete autonomous units, in fact we are the opposite, although we may not be conscious of it. Our bodies are constantly adjusting and adapting to environmental stimuli. The vast majority of these environmental factors are totally harmless, although in some situations, these may be oncogenic. In this section, we will look at some of the environmental factors associated with haematological malignancies.

9.12.1 Ionizing radiation

It is well established that individuals surviving the detonation of atomic bombs in 1945 in Nagasaki and Hiroshima had an increased risk of developing haematological malignancies. The first type of malignancy to manifest itself following irradiation from an atomic bomb detonation is leukaemia. Individuals can also be exposed to ionizing radiation through treatment for solid tumours, or occupational exposure in medical radiology or as a radiation scientist (such as those working in nuclear processing plants). It is believed that Marie Curie and her daughter, Irene, both died from leukaemia following their work with radiation. Health risks are substantially lower as a consequence of recent changes in working practices within these environments, as occupational exposure has been reduced.

The most common types of leukaemia associated with atomic bomb detonations are acute leukaemias and chronic myeloid leukaemia. The Life Span study for atomic bomb survivors has provided the majority of the information used to shape our understanding of these malignancies following exposure to radiation. Recent work completed in Nagasaki suggests that two other haematological conditions may follow atomic bomb exposure: myelodysplastic syndrome and monoclonal gammopathy of undetermined significance (MGUS)—a clonal disorder of plasma cells, identified by an excess production of a single type of immunoglobulin in the absence of any clinical symptoms—two conditions usually affecting the elderly. The group, investigating in excess of 50 000 atomic bomb survivors, suggests an increased incidence of these conditions in survivors. The closer the survivor to detonation, the greater the risk of developing haematological malignancies in later life. It seems likely that irradiated stem cells predispose individuals to haematological malignancies later in life.

9.12.2 Non-ionizing radiation

There is a great deal of public concern and media attention directed towards the issue of non-ionizing radiation in the form of extremely low-frequency electromagnetic fields (ELF-EMF) associated with overhead power cables. The particular concern is that these low-frequency fields can induce DNA damage, especially in children, causing childhood leukaemia. In 2005, Draper et al. published a paper examining 29 081 children with cancer, including 9700 with leukaemia. Draper noted an increased risk of leukaemia (although not other childhood cancers) associated with children living within 200 metres of a high-voltage power line compared with those living at distances greater than 600 metres from the source. However, the group concede that these results may be due to chance, or secondary to a mechanism other than ELF-EMF. To date, there is no convincing evidence of a genuinely increased risk of childhood leukaemia due to ELF-EMF.

Cross references

For more information on myelodysplastic syndromes, refer to Chapter 11.

Late effects of therapy in leukaemogenesis are discussed in detail in Chapter 11.

9.12.3 Late effects of therapy

Many patients treated with radiotherapy (ionizing radiation) and chemotherapy, for example for solid organ tumours, may subsequently develop haematological malignancies, most commonly leukaemias. These leukaemias are called **'secondary' leukaemias** because they arise as a consequence of exposure to a known **mutagen**. By contrast, spontaneously occurring malignancies are called ***de novo* leukaemias** as they have no traceable cause. These are further discussed in Chapters 11 and 12.

mutagen

A chemical substance known to induce DNA mutations.

Chemotherapeutic agents, for example alkylating agents or topoisomerase II inhibitors, work by inducing DNA damage in cells with a high proliferation rate, resulting in cell death through the activation of apoptotic pathways. However, because these therapeutic agents are so toxic, they tend to have innocent bystander effects where non-malignant cells (such as haemopoietic progenitor cells) accrue mutations and become precancerous. Patients developing secondary acute myeloid leukaemia following radiotherapy and chemotherapy commonly have a preceding MDS, which later transforms into AML.

At the forefront of concern for these so-called 'late effects' cancers are patients treated for childhood leukaemia. The most common type of childhood leukaemia is common acute lymphoblastic leukaemia (c-ALL) which accounts for 26% of all childhood cancers in the United Kingdom, and is associated with a five-year survival rate of 80%. These patients tend to have an increased risk of late-effect cancers because of their survival potential due to their young age. That is, their young age increases the likelihood that they will survive ALL—however, if they do survive, they then have a greater chance of developing a late-effect cancer when they get older.

9.12.4 Infectious agents

There are a number of infectious agents, viral and bacterial, associated with haematological malignancies. This section outlines a number of the more common types and their association with malignant transformation.

Epstein–Barr virus (EBV)

EBV is a member of the human herpesvirus family and is the cause of infectious mononucleosis (glandular fever). EBV was the first virus to be associated with human cancer, having been derived from the tissue of patients with Burkitt lymphoma. In some individuals, EBV is associated with a number of haematological malignancies, although this does not necessarily mean that anyone exposed to EBV (or, indeed, who has suffered from glandular fever) is destined to develop cancer.

EBV can infect a wide range of cells, including B and T lymphocytes. When naive B cells are infected, some will migrate to germinal centres—found within lymph nodes and important in the process of B-cell maturation—where they will develop into plasma cells or memory B cells. These memory B cells contain residual viral DNA, and cause the patient to become a carrier of EBV DNA. Over time, carrier status can result in malignant transformation and the onset of B-cell lymphoma, particularly in patients with ineffective immune systems. Although not an exhaustive list, these lymphomas include HL and primary central nervous system lymphoma. EBV-positive tumours are more commonly found in immunocompromised than in immunocompetent patients.

Cross references

Germinal centres are discussed in greater detail in Chapter 8.

More detail regarding EBV can be found in Chapter 8.

Human T-lymphotrophic virus-1 (HTLV-1)

HTLV-1 can be isolated from most cases of adult T-cell leukaemia (ATL) and is commonly found in Africa, the Caribbean, and Japan. The mechanism of transformation is particularly complex, and is thought to involve the inactivation of TSGs, the activation of cell cycle regulators, and the upregulation of anti-apoptotic proteins. The end result is a clone of T cells with an increased proliferation rate and an inability to undergo apoptosis. This cellular behaviour increases the probability of mutations and can result in malignant transformation. Of those patients infected with HTLV-1, up to 3% may develop ATL.

Human herpesvirus-8 (HHV8)

This is a member of a group of viruses that includes herpes simplex 1 and 2, cytomegalovirus (CMV), and EBV. Herpesviruses persist for the life of the host and affect between 60 and 90% of adults.

Human herpesviruses comprise linear, double-stranded DNA within a protein capsid. Initially infecting epithelial cells, following primary infection the virus enters a latent period within the periaxonal sheath of a variety of nerves. Reactivation of the virus involves migration along sensory neurones to the mucocutaneous interface, a process that enables the virus to avoid immune recognition.

HHV8 is recognized as an important cause of Kaposi sarcoma in HIV patients, and can also infect CD19+ B cells, leading to the development of B-cell lymphomas associated with immunosuppression.

Cross references

HHV8 is discussed in relation to large B-cell, lymphoma-associated Castleman disease and primary effusion lymphoma in Chapter 12.

Marginal zone lymphomas are discussed in greater detail in Chapter 12.

Helicobacter pylori

Gastric *Helicobacter pylori* infections are associated with the development of gastritis, peptic ulceration, and the development of gastric carcinoma. Infection by *Helicobacter pylori* is also strongly associated with the development of a marginal zone B-cell lymphoma of mucosa-associated lymphoid tissue (MALT) type. MALT lymphoma is a form of marginal zone lymphoma, which comprises transformed B lymphocytes. MALT lymphoma accounts for approximately 8% of all non-Hodgkin lymphoma. Evidence suggests that the mechanism of lymphomagenesis is due to chronic inflammation caused by infection, as supported by the fact that, in low-grade cases, eradication of *Helicobacter pylori* results in a resolution of the MALT lymphoma. One study revealed that in uncomplicated, low-grade cases, 79% of patients were cured by eradicating the *H. pylori* infection. However, patients with tumours extending beyond the mucosa and submucosa (i.e. those with invasive properties) have a less favourable outcome when treated with antibiotics and require traditional cytotoxic therapy.

9.12.5 Benzene

Many polycyclic and aromatic hydrocarbons are carcinogenic. Benzene is a well-established leukaemogen, having been associated with leukaemia since as early as 1928. In order to become toxic, benzene must first be metabolized in the liver, primarily by cytochrome P4502E1 (CYP2E1). The outcome of this process is to generate phenol, which is then further metabolized to catechol, hydroquinone, and 1,2,4-benzenetriol. 1,4-benzoquinone is subsequently generated by the degradation of catechol (in which bone marrow peroxidases have been implicated). These metabolic derivatives are thought to bind to various intracellular proteins and enzymes such as topoisomerase II to influence DNA indirectly. It is this impact on DNA which is believed to have a leukaemogenic effect.

The association between benzene and leukaemia is strong, although inconsistent between subtypes. AML has the strongest association, as repeatedly demonstrated, whereas chronic lymphocytic leukaemia (CLL) has a far weaker association. The correlation with ALL and CML is ambiguous and requires further, more comprehensive, investigation. Individuals associated with exposure to high levels of benzene include those employed in the petrochemical, shoe, rubber, and paint industries.

Chapter summary

In this chapter we have:

- established the role of the cell cycle in the development of daughter cells;
- examined the definitions and main subgroups of haematological malignancy;
- examined the mechanisms involved in oncogenesis;
- discussed the role of the cancer stem cell in leukaemogenesis and lymphomagenesis;

- established that a population of identical cells, all derived from the same mutated parent, constitutes a clone;
- outlined the process of clonal progression;
- identified the link between loss-of-function mutations in TSGs and gain-of-function mutations in proto-oncogenes established in the context of Knudson's multi-hit hypothesis;
- discussed cytogenetic terminology and introduced chromosome structure in relation to haematological malignancies;
- outlined a range of mutations at the genetic level and linked these to abnormalities in the polypeptide product potentially leading to a malignant phenotype;
- examined the role of epigenetic processes as genetic modifiers, and linked these processes to the alteration of the expression level of a range of different proteins;
- examined some of the heritable and environmental causes of cancer and established links between these causes and genetic mutation.

Discussion questions

9.1 Using appropriate examples, critically discuss the effect of chromosomal abnormalities on cell cycle control and the subsequent development of haematological malignancies.

9.2 Abnormal DNA methylation patterns influence tumour suppressor function in haematological malignancies. Discuss this statement.

9.3 Critically discuss the roles of infectious agents in lymphomagenesis.

Further reading

- Battle E and Clevers H. ***Cancer stem cells revisited***. *Nature Medicine* 2017: **23(10)**: 1124–34.
- Biswas S and Rao CM. ***Epigenetics in cancer: Fundamentals and beyond***. *Pharmacology and Therapeutics* 2017: **173**: 118–34.
- Chaudry SF and Chevassut TJT. ***Epigenetic guardian: A review of the DNA methyltransferase DNMT3A in acute myeloid leukaemia and clonal haematopoiesis***. *BioMedical Research International* 2017. Article ID 5473197, 13 pages, https://doi.org/10.1155/2017/5473197.
- Chen J, Odenike O, and Rowley JD. ***Leukaemogenesis: More than mutant genes***. *Nature Reviews Cancer* 2010: **10**: 23–36.
- Imbert V and Peyron JF. ***NF-KB in hematological malignancies***. *Biomedicines* 2017: **5(27)**: doi:10.3390/biomedicines5020027.
- Inoue S, Lemonnier F, and Mak TW. ***Roles of IDH1/2 and TET2 mutations in myeloid disorders***. *International Journal of Hematology* 2016: **103**: 627–33. https://doi.org/10.1007/s12185-016-1973-7.

- Medinger M and Passweg JR. ***Acute myeloid leukaemia genomics.*** *British Journal of Haematology* 2017: **179**: 530–42.
- Schaffer LG, McGowan-Jordan J, and Schmid M (eds). ***ISCN: An International System for Human Cytogenetic Nomenclature***. S. Karger, Basel, 2013.
- Segal EB and Lichtman MA. ***Familial (inherited) leukaemia, lymphoma and myeloma: An overview***. *Blood Cells, Molecules and Diseases* 2004: **32**: 246–61.
- Watson IA, Takahashi K, Futreal PA, and Chin L. ***Emerging patterns of somatic mutations in cancer***. *Nature Reviews* 2013: **14**: 703–18.

Answers to self-check questions and case study questions are provided as part of this book's online resources.

Visit www.oup.com/he/moore-fbms3e.

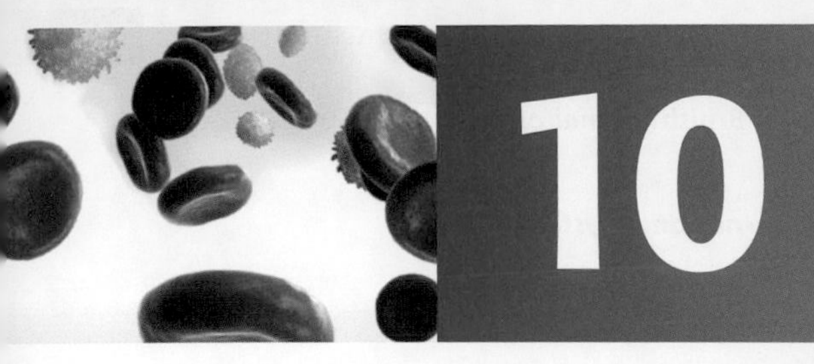

10

The laboratory investigation of haematological malignancies

Gavin Knight

As a biomedical scientist, it is often exciting to be involved in the investigation of a new haematological malignancy. Getting 'hands on' in the examination of blood films and other laboratory-generated data in these cases can be particularly rewarding. However, it is essential that we remain mindful of the fact that every sample represents a patient who is likely to be unwell, emotional, or intimidated by the process of investigation. In every instance, we must remember that these patients are human beings and it is our duty to treat the sample with respect, regardless of the scientific principles to be learnt from the case. Many of these patients will continue to be treated at the hospital in which they present, and over time we can observe changes in their results as a consequence of the treatment regimens they are prescribed. The investigations we complete and the results generated have an important bearing on the patient's management. It is often regarded that working in a laboratory distances you from a patient. However, this is not the case. Instead, you are part of a multidisciplinary healthcare team and your actions as a biomedical scientist can have a profound impact on patient care.

Whilst it is important to recognize that the haematology laboratory plays an important role in the diagnosis of haematological malignancies, it operates as a component of a much larger range of facilities that contribute to patient investigation and diagnosis. Radiological investigations such as computer tomography (CT), magnetic resonance imaging (MRI), positron emission tomography (PET), ultrasound (US), and skeletal scintigraphy (bone scans) can be critical for initial investigations, and a range of other pathology disciplines including biochemistry and histology also play significant roles in investigation, diagnosis, and monitoring. This chapter outlines some of the diagnostic procedures followed when a patient

presents to a medical professional with symptoms that are suggestive of haematological malignancy and therefore require laboratory investigation.

The chapter begins by discussing some of the most common presenting symptoms associated with malignancies, putting them into a biological and physiological context. We then examine the range of laboratory tests often required for a definitive diagnosis. Whilst the focus of this chapter is on haematology, we also discuss a range of molecular biological and cytogenetic investigations essential for the diagnosis and classification of haematological malignancies. Biochemical, microbiological, and histological investigations have been omitted in the interests of focus, but can be explored in the companion titles, *Clinical Biochemistry*, *Medical Microbiology*, and *Histopathology*.

Learning objectives

After studying this chapter, you should confidently be able to:

- discuss the common symptoms experienced by patients presenting with a haematological malignancy and explain the physiological cause of these symptoms;
- describe the investigative process in establishing the diagnosis of a specific haematological malignancy;
- describe the common findings from the initial full blood count and blood film analysis associated with malignant haematological conditions;
- compare and contrast a bone marrow trephine with a bone marrow aspirate and outline the different stains that can be used to provide diagnostic information;
- describe some of the typical cytogenetic abnormalities associated with a range of haematological malignancies, and outline some of the techniques available for their identification;
- discuss the role of named molecular techniques in the investigation of haematological malignancies.

10.1 Patient presentation

Before we begin examining the laboratory procedures for diagnosing and monitoring haematological malignancies, it is important to consider some of the symptoms exhibited by patients when they present to the medical team. The range of these symptoms is particularly broad, and any combination can be present in any one patient. It is also important to realize that these symptoms can be explained using normal physiological principles. These principles will be discussed in the context of the symptoms outlined below.

The symptoms associated with malignant disease include:

- pallor;
- lethargy;
- pharyngitis;
- recurrent infections;
- easy bruising;

- pyrexia;
- night sweats;
- bone pain;
- flu-like symptoms;
- lymphadenopathy;
- splenomegaly;
- hepatomegaly.

Finally, it is important to note that some patients may be **asymptomatic**, with a haematological malignancy indicated from the results generated from a routine full blood count investigation. Let us now consider each of these symptoms in turn.

lethargy
Tiredness or fatigue.

normocytic normochromic anaemia
A particular form of anaemia associated with red cells of a normal size and coloration. Although there can be a number of causes, in the context of a malignancy the most likely causes are anaemia of chronic disease or myelosuppression.

Pallor (a pale coloration to the skin) and **lethargy** are often attributable to anaemia, with an associated reduction in oxygen delivery to the tissues. This reduction in oxygen delivery is a consequence of a reduced haemoglobin concentration or a reduction in the number of circulating red cells. A **normocytic normochromic anaemia**, where the red cells appear of normal size and normal coloration (haemoglobinization), is most frequently encountered with a malignancy.

Pharyngitis and *recurrent infections* are the consequence of a dysfunctional immune system. As we shall see later on in this chapter, although the white blood cell (WBC) count may be raised, this should not be used to imply that the white cells are functional. This raised WBC count (**leucocytosis**) may be due to circulating leukaemic cells. There is a common misconception by students that a raised WBC count as part of a haematological malignancy is a consequence of the body fighting infection. In fact, the raised WBC count is usually due to either circulating malignant cells or non-malignant cells displaced by clonal cells within the bone marrow.

Easy bruising is caused by two main effects:

1. a reduced platelet count (thrombocytopenia);
2. dysfunctional platelets.

A combination of the two may also be apparent. In either case, platelets are unable to form a primary platelet plug, which is essential to prevent leakage of blood from blood vessels. In a patient with a normal platelet count, day-to-day bumps and scrapes go unnoticed, but for patients with thrombocytopenia or dysfunctional platelets, easy bruising occurs. Easy bruising may take the form of **ecchymoses** or **petechiae**.

ecchymoses
Bruises caused by the leakage of blood from blood vessels.

Each of the symptoms outlined above is associated with a condition called **myelosuppression**—the failure of the myeloid component of the bone marrow to produce red cells, white cells, and platelets as a consequence of either a therapeutic agent or a malignant clone. Myelosuppression was initially believed to be the consequence of malignant stem cells competing with normal stem cells for their microenvironment. Although this is partly true, it has been demonstrated that malignant blasts actually secrete cytokines which suppress the proliferation of normal blasts. The consequence of this suppression is that mature effector cells will not be produced, causing a reduction in all myeloid cell lines (pancytopenia).

Cross reference
Normocytic normochromic anaemias were first introduced in Chapters 5 and 6.

Pyrexia (a rise in the body's core temperature) and *night sweats* are often symptoms associated with malignant B lymphocytes and their precursors in response to the synthesis and secretion of cytokines. In particular, interleukin (IL)-1, tumour necrosis factor-α (TNFα), and IL-6 are associated with fever and weight loss. During an infection, the normal physiological response is for white cells to increase in number and secrete a range of cytokines to:

- potentiate the white cell colony;
- stimulate the hypothalamus to increase the body's temperature;
- stimulate the proliferation of haemopoietic stem cells;
- stimulate the proliferation of B cells and promote switching between different types of antibody (isotype-switching).

Increases in white cell numbers will cause an increase in the concentration of circulating cytokines and potentiate their physiological role. However, the white cell count is unlikely to rise as a consequence of elevated cytokine levels because the cytokines produced by the malignant blasts have a myelosuppressive effect on the normal blasts of the bone marrow.

Bone pain can have a number of causes. First, an increase in the bone marrow **cellularity**, as a consequence of large numbers of malignant cells, can compress the nerves running through the bone, causing pain. Second, in myeloma, malignant plasma cells synthesize and secrete osteoclast-activating factors, promoting the degradation of bone. The result of bone degradation is the gradual loss of bone density and, when examined on X-ray, **punched-out lesions** throughout the skeleton become apparent.

cellularity
The proportion of cells within the bone marrow.

punched-out lesions
Caused by the catabolic effects of osteoclast-activating factors on bony structures.

In lymphoma, the spleen or a lymph node may be the primary site of the malignancy. These sites tend to become enlarged and can be felt (palpated) during a clinical examination. *Lymphadenopathy* (one or more enlarged lymph nodes) is often painless. The nodes feel rubbery, and can become quite large in a short space of time. *Splenomegaly* (an increase in the size of the spleen) is often found in splenic lymphoma, although this is not always the case. Splenomegaly is associated with a wide range of illnesses and haematological malignancies. In cases of splenomegaly, the left side of the abdomen can become severely distended and puts pressure on the other visceral organs. Thrombocytopenia (a reduction in a patient's platelet count) may be apparent because platelets accumulate in the enlarged spleen and may result in easy bruising. There may be a concurrent increase in the size of the liver, **hepatomegaly**, due to reduced blood flow through the enlarged spleen and pooling of blood in the liver. A combination of hepatomegaly and splenomegaly is called **hepatosplenomegaly**.

Cross reference
Myeloma is discussed in more detail in Chapters 9 and 12.

When a patient presents to a doctor with any of these symptoms, the doctor will take a thorough medical history and perform a clinical examination. If an obvious cause is not apparent, then more information is needed before a diagnosis can be made, so blood tests are requested. Although a range of blood tests may be requested for the first-line investigation of these symptoms, this chapter will focus on those most pertinent to haematology. The first investigation we will consider is the full blood count.

Cross reference
The full blood count is described in Chapter 2.

SELF-CHECK 10.1

Outline the main presenting symptoms associated with a haematological malignancy.

10.2 The full blood count

quantitative
Types of data that deal with numerical values; for example, the number of platelets in one litre of blood.

qualitative
Non-numerical data; for example, the ability of a platelet to function correctly.

The full blood count (FBC) is a broad measure of the red cells and their contents, white cells, and platelets. The FBC provides us with an enormous amount of **quantitative** data, although it has limited value as it does not provide **qualitative** data, and as such is unable to demonstrate how the measured cells are functioning. The FBC is also non-diagnostic in the large majority of cases. The FBC fails to give a reason for particular indices being outside their reference range. Therefore, in response to an abnormal FBC, the results should be considered in the context of the patient's clinical details and further investigations requested.

Although the FBC can provide important information about haematological malignancies, the presence of a malignancy will not always cause abnormal FBC results. In localized malignancies, such as occurs in some lymphomas, it is possible that the patient may have a normal FBC. Normal FBC results may also be found in the early stages of malignancies, before the bone marrow has been compromised.

10.2.1 Key FBC findings

You may expect to see changes in the following FBC parameters in some patients with haematological malignancies. It is important to note that every patient is different, and some may have **co-morbidities** (from more than one disease process), which may alter the blood profile.

co-morbidities
Two or more co-existing diseases.

Cross reference
More information on lymphomas can be found in Chapters 9 and 12.

Red cell count

The red cell count may be either reduced or increased. A *reduced* red blood cell (RBC) count could be a consequence of myelosuppression or may be attributable to bleeding, especially in patients with thrombocytopenia or abnormal haemostasis. As mentioned in section 10.1, a normocytic normochromic anaemia is commonly associated with malignancy, and whilst the red cells will look normal, they may be reduced in number within the peripheral blood. By contrast, the RBC may be *increased* in polycythaemia vera, where a clonal proliferation of red cell precursors increases the number of circulating red cells. Furthermore, in some B-cell malignancies, autoantibodies directed against red cell antigens may lead to a reduced red cell count, and alterations in the mean corpuscular haemoglobin concentration, as explained below.

Cross reference
More detail on polycythaemia vera can be found in Chapter 11.

Mean cell volume (MCV) and haemoglobin (Hb)

Whilst a normocytic normochromic anaemia is usually associated with haematological malignancies, a macrocytosis (raised MCV) may be apparent in some patients. A macrocytosis is commonly found in patients receiving cytotoxic therapy, or those with myelodysplastic syndrome as a consequence of **cell cycle arrest**. In conjunction with the raised MCV, the haemoglobin concentration is often normal or slightly low.

cell cycle arrest
Occurs when the G1-S-G2-M transitions become disrupted. Arrest usually occurs in late G1 or S phases as a consequence of the activation of tumour suppressor proteins in an attempt to initiate DNA repair. Alternatively, it can be seen during S phase, where synthesis of DNA is disrupted.

In diseases such as polycythaemia, an iron deficiency may develop because of the increased production of red cells and the use of venesection to reduce the haematocrit. This will eventually lead to a microcytosis (low MCV), reduced haemoglobin concentration, and a gradual reduction in the red cell count.

Raised mean corpuscular haemoglobin concentration (MCHC)

The MCHC is calculated from the average haemoglobin concentration of each of the red cells counted. In patients with autoimmune haemolytic anaemia (AIHA), a condition often associated with B-cell malignancies, red cells lose their biconcave disc shape and become spherical (called spherocytes). Part of the red cell membrane is removed by macrophages in response to membrane-bound autoantibodies, whilst the volume of the cell stays the same. This reduction in cell size accompanied by a maintained cell volume causes the MCHC to increase.

White blood cell count

The white cell count represents the number of both mature and immature white cells. In the case of a haematological malignancy, an increased WBC count may be the result of malignant cells circulating in the peripheral blood, or attributable to normal cells being displaced by the malignant clone. When examining malignant cells, the raised WBC count may represent an increased proliferation rate, a failure of apoptosis, or both.

Cross reference

More information about clonality can be found in Chapter 9.

A reduced WBC count may also be found particularly as a consequence of myelosuppression. In myelofibrosis, an increased or decreased WBC count may be seen, depending upon whether the fibrotic marrow has displaced immature cells to the peripheral blood or suppressed leucocyte development.

An important practice point to remember is that even though the WBC count may be within the reference range, this does not necessarily mean these white cells are normal and indeed, the relative proportions of the different species of WBCs may be markedly abnormal.

Abnormal differential

The differential count examines the range of species of WBC: neutrophils, lymphocytes, monocytes, eosinophils, basophils, and their precursors. Each of these should be found within a defined reference range, as determined by testing a representative group of the local population. An abnormal differential is commonly found in haematological malignancies, broadly representing the events occurring within the bone marrow. In lymphoid malignancies, a **reversed ratio** may be found. Normally, the neutrophil count is higher than the lymphocyte count, but in lymphoid malignancies the reverse is often found. A reversed ratio may also occur in viral infections and in children up to the age of seven years, so it is important not to jump to diagnostic conclusions, but to consider all findings logically and perform further investigations as necessary.

reversed ratio

A lymphocyte count that is higher than the neutrophil count.

Platelets

In haematological malignancies, thrombocytopenia is the most commonly identified platelet abnormality, and can be attributable to myelosuppression. However, in splenomegaly, thrombocytopenia can also be evident due to pooling of platelets within the spleen. A raised platelet count (thrombocytosis) may be associated with myeloproliferative conditions such as polycythaemia vera, and is a mandatory finding in **essential thrombocythaemia** (for which there are specific diagnostic criteria). Thrombocytosis may be a reactive phenomenon mediated by inflammatory cytokines.

essential thrombocythaemia

A clonal proliferation of megakaryocytes leading to an increased platelet count.

Whilst this section focuses on findings associated with haematological malignancies, these can also be associated with benign conditions. A raised WBC count may be found in people with bacterial infections, although it may be decreased following a viral infection. Platelets may be decreased following a haemorrhage, and increased during an inflammatory illness. AIHA may not always be associated with malignancy; it can be a complication of another autoimmune disease or infection, and the MCHC may be increased following treatment with certain types of drugs, or following a transfusion reaction. In some types of haematological malignancy, the FBC may be normal.

In cases where abnormal FBC results have been generated or a specific request has been made by a doctor, the next step is to examine a peripheral blood film. The blood film can give us important information regarding the appearance (**morphology**) of the cells, and some of the features apparent on the film may be of diagnostic use. The blood film can also help confirm the legitimacy of the FBC results.

Cross reference

More information regarding essential thrombocythaemia can be found in Chapter 11.

TEST	RESULT	ABN	NORMALS	UNITS
WBC		123.5	(4 - 10)	x10.e9 /L
RBC	4.54		(4.5 - 5.5)	x10.e12 L
HGB	140		(130 - 170)	g/L
HCT	0.41		(0.40 - 0.50)	L/L
MCV	89.7		(83 - 101)	fL
MCH	30.9		(27 - 32)	pg
MCHC	344		(315 - 345)	g/L
CHCM	352		(-)	g/L
PLT		19	(150 - 400)	x10.e9 /L
#NEUT		118.5	(2 - 7)	x10.e9 /L
#LYMPH	2.1		(1 - 3)	x10.e9 /L
#MONO	1.0		(0.0 - 1)	x10.e9 /L
#EOS		1.0	(0 - 0.5)	x10.e9 /L
#BASO		15.7	(0 - 0.1)	x10.e9 /L
#LUC		0.9	(0 - 0.5)	x10.e9 /L
MPXI	8.1		(-10 10)	

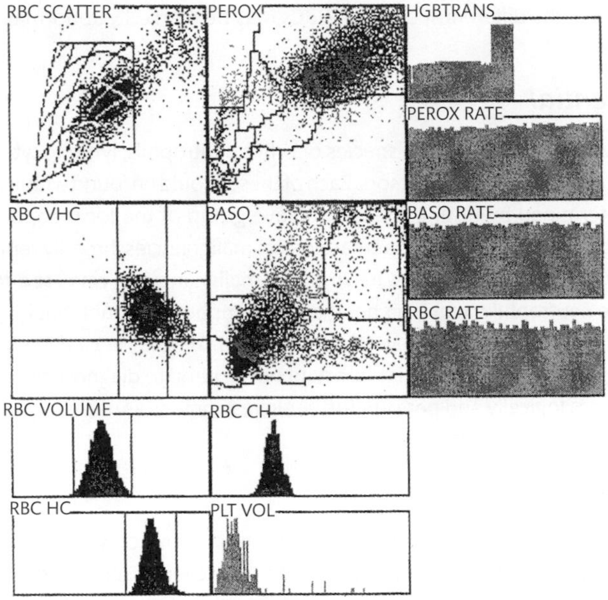

FIGURE 10.1

A full blood count taken from a patient diagnosed with acute promyelocytic leukaemia (APL). Note the patient's high white cell count and low platelet numbers.

SELF-CHECK 10.2

List examples of data provided by a full blood count analysis.

Figure 10.1 shows the presenting FBC from a patient who was later diagnosed with acute promyelocytic leukaemia (APL). You will learn more about this disease later on in this chapter and in more detail in Chapter 11. If you look at the patient's WBC count, you can see the value exceeds the reference range with a value of 123.5 × 10^9/L. When we examine the differential count, i.e. the different species of white cell within the sample, we can see that the absolute neutrophil count (absolute being denoted by the # preceding the name of the white cell) is also very high (118.5 × 10^9/L). The patient's lymphocyte and monocyte counts are within range, although eosinophils and basophils are also raised. LUCs (large unstained cells) are classified as such because their true identity, other than the fact that they can be described as large and unstained, cannot be determined by the analyser, Because of the increased LUCs, and the very abnormal differential, it is likely that the analyser has incorrectly identified the species of white cell, and so a manual differential count should be performed. Also of importance is the very low platelet count—thrombocytopenia. All of the patient's red cell parameters are within the reference range. It is essential in this case to examine the morphology of this patient's cells under the microscope and to perform a manual differential count before further investigations can be requested and a formal diagnosis made.

10.3 Blood film

A blood film should always be examined when a patient has a suspected haematological malignancy. Clinical laboratories have their own guidelines to determine the point at which a blood film examination is indicated.

The indices provided by the FBC should be representative of the cells seen when using a microscope, although in some cases the FBC findings may be inaccurate. For example, **dysplastic** changes in white cells alter their morphology and physical properties to the point that the analyser cannot determine their species. In these instances, the blood film is an important mechanism for identifying the true nature of the FBC results. Some of the limitations of using a FBC alone are outlined below.

- The FBC may suggest that a patient's platelet count is low, but it is considered good practice to examine a blood film to check for **platelet clumps** and fibrin strands, either of which would artificially reduce a patient's platelet count *in vitro*. Where there is an unexpectedly low platelet count, the sample should be checked for a clot before the blood film is made.
- A patient's MCV may be within the reference range. However, when examining a blood film, two distinct populations of red cells, small and large (**dimorphic**), may be distinguished. When each population of red cells has their volume averaged, the mean sits within the reference range. An indication of this may come from a raised red cell distribution width (RDW). *Note*: the RDW is often measured as part of an automated FBC, but is not always reported.
- The technological methods used to determine the identity of WBCs are unreliable when there are signs of dysplasia, or when large numbers of immature cells are present. This could result in misclassification of a particular species of WBC.

dysplasia
The abnormal development or maturation of cells, tissues, or organs.

platelet clumps
Aggregates of platelets within a FBC, demonstrable on a blood film. Often artefactual, facilitated by the presence of ethylene diamine tetra-acetic acid (EDTA) in blood tubes.

dimorphic
Two distinct populations of red cells in the same FBC sample.

The peripheral blood film also allows examination of the structural features of the cells, and identification of any inclusion bodies present. We can obtain many diagnostic hints to aid in the classification process by examining: the cytoplasmic granular contents of granulocytes; the size and haemoglobinization of red cells; the size and number of platelets; the nuclear configuration (nucleoli and chromatin structure) of white cells; and the presence of immature forms.

10.3.1 Blood film examination

Romanowsky stain should be used to obtain the most information from the blood film. According to the International Committee for Standards in Haematology (ICSH), Romanowsky stain should be composed of azure B (a basic stain) and eosin Y (an acidic stain). These stains bind to acidic and basic intracellular or extracellular components of blood, respectively.

By using Romanowsky stain, we would expect to see cell nuclei staining purple, with nucleoli staining a much lighter blue. The cytoplasm of the cell differs according to the function of the cell or its stage of maturation, and therefore its chemical composition. Table 10.1 demonstrates the typical cytoplasmic staining of the different haemopoietic cells following Romanowsky staining.

Ideally, when examining the blood film for the first time, a low-power lens should be used so that the distribution of cells can be ascertained. Any cellular clumps or fibrin strands should be easily recognizable at low power.

Once the low-power examination is complete, a higher-powered lens (×40 magnification) should be used. This more detailed examination should enable the appreciation of cell structure and identification of key cell lines. Any immature cells otherwise retained within the bone marrow should be easily recognizable. In cases of malignancy, a differential count of 200 WBCs should be completed to determine the proportions of the different cells and their precursors. Precursor cells found within the peripheral blood would be counted on an ad hoc basis, but as part of the differential count. The term **leucoerythroblastic blood picture** is used to describe a blood film containing immature white cells and immature red cells.

leucoerythroblastic blood picture
A blood film showing nucleated red cells and immature white cells.

Auer rods
Eosinophilic primary granules are abnormally assembled in rod-like structures and are a marker of myeloid malignancies.

The cytoplasmic contents of granulocytes and the nuclear lobes of polymorphs should be readily appreciable. Examination of the shape, size, structure, and chromatin distribution of the nucleus and identification of nucleoli in lymphocytes is of utmost importance. Should blasts be present, evidence of **Auer rods** should be sought to assist with malignancy classification. Figure 10.2 shows typical morphological characteristics associated with myeloid blast cells.

TABLE 10.1 Typical Romanowsky staining characteristics of erythroid, myeloid, and lymphoid cells.

Lineage	Cell	Cytoplasm	Granulation
Erythroid	Early erythroblast	Dark blue	
	Late erythroblast	Light blue	
	Reticulocyte	Bluish tinge	
	Red blood cell	Dark pink	
Myeloid	Myeloblast	Pale blue	None
	Promyelocyte	Blue	Azurophilic
	Myelocyte	Pink	Lineage-specific granules (see mature cells)
	Metamyelocyte	Pink	Lineage-specific granules
	Neutrophil	Pink/orange	Purple
	Eosinophil	Pink	Red/orange
	Basophil	Blue	Purple/black
	Monocyte	Grey/blue	
Lymphoid	Lymphocyte	Blue	
	Plasma cell	Dark blue	

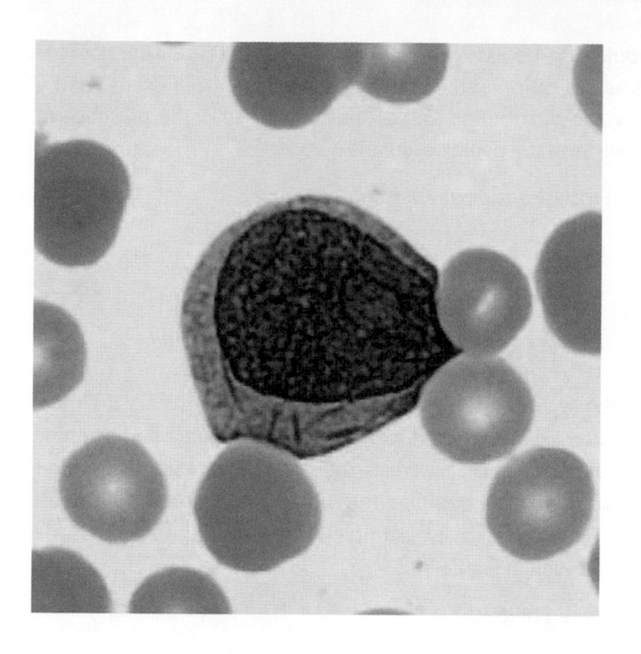

FIGURE 10.2
Typical blast cell morphology. This peripheral blood film shows an example of a myeloblast. The blast contains a prominent nucleolus and Auer rods within the cytoplasm. The presence of Auer rods is conclusive evidence of the myeloid lineage of these cells. The high nucleocytoplasmic ratio, open chromatin configuration, and the presence of nucleoli are characteristics of blast cells. Image courtesy of Gordon Sinclair, Royal Marsden Hospital.

Identification of Auer rods confirms a malignancy to be of myeloid origin, although the absence of Auer rods does not suggest a lymphoblastic malignancy. Only 50% of myeloid malignancies exhibit Auer rods within the cytoplasm of blast cells.

Experience in blood film examination is essential in order to obtain the greatest amount of information from the patient sample. Often, alterations in blood cell morphology can make the identification of individual cells, particularly WBCs, very difficult. Experience of examining blood films is crucial when identifying features of a particular cell's lineage. The difficulty in identifying cells could lead to the incorrect diagnosis being made, or important diagnostic features being overlooked. Whenever the opportunity arises, this should always be taken to gain valuable experience in blood film examination.

Cross reference
See Chapter 8 for the morphology of white blood cells.

SELF-CHECK 10.3

What is the significance of identifying Auer rods in a population of blasts?

Whilst a blood film is important for the examination of peripheral blood cells, a bone marrow sample is a key diagnostic tool. The bone marrow is the site of haemopoiesis and is informative of the pathological process taking place.

10.4 Bone marrow assessment

10.4.1 Bone marrow samples

Whilst blood film examination can give us a great deal of information about a patient's disease and can aid in a diagnosis, we need to assess the patient's bone marrow to fully appreciate the nature of their disease. With the exception of T-lymphocyte production, the process of haemopoiesis occurs within the bone marrow. It is therefore important to examine the bone marrow to obtain as much information as possible about the patient's pathology.

In cases involving a suspected haematological malignancy of a lymph node (or group of lymph nodes), samples from the lymph node(s) should be used for primary diagnostic purposes.

However, the bone marrow should also be assessed to allow the evaluation of the haemopoietic tissue, as this will give us an indication of the presence of malignant cell infiltrates; the architecture of the bone marrow; the distribution and proportions of cells; the presence of abnormal inclusions within the cells; and the overall cellularity.

The following samples can be used to assess the patient's bone marrow:

- aspirate;
- trephine;
- trephine imprint/roll.

The sternum, around the area of the second intercostal space, or the iliac crests are the most commonly chosen sites for bone marrow aspiration. A combination of aspirate and trephine dictates the use of either the anterior or posterior iliac crests, utilizing the same skin incision for both procedures, whereas a lone bone marrow trephine *must* be performed on the iliac crest because of the risk of puncturing the thoracic cavity. Examples of the needles used for bone marrow aspiration and biopsy can be seen in Figure 10.3.

Each of these samples will be considered below.

Routine bone marrow assessment involves the collection of a bone marrow aspirate and, often, a trephine biopsy. It is essential if the aspirate gives a '**dry tap**' or is **haemodiluted** that a trephine is obtained for diagnostic purposes.

dry tap

An unsuccessful bone marrow aspirate which yields either no bone marrow material or insufficient amounts for analysis. A dry tap could be caused by poor technique, but is more likely to be due to bone marrow fibrosis or hypercellularity.

haemodilute

When aspirated contents of the bone marrow are excessively diluted with peripheral blood.

Bone marrow aspirate

Samples collected from a bone marrow aspiration can be used for morphological examination, Perls' staining (to assess iron stores), immunophenotyping, and cytogenetic and molecular genetic testing. An image of a well-produced bone marrow aspirate slide is shown in Figure 10.4. All of these investigations are important in order to provide as much information on the patient's condition as possible. Differential cell counts can be performed easily on a well-stained bone marrow (BM) aspirate as the cellular morphology is well preserved. A minimum of 500 cells should be counted for the BM differential, with the trails of BM behind the fragments being the best position to find cells originating from the BM. A BM fragment at the tip of an aspirate slide is shown in Figure 10.5. Outside this localized area, the differential may start to include cells from peripheral blood contaminating the BM from the aspiration, and may

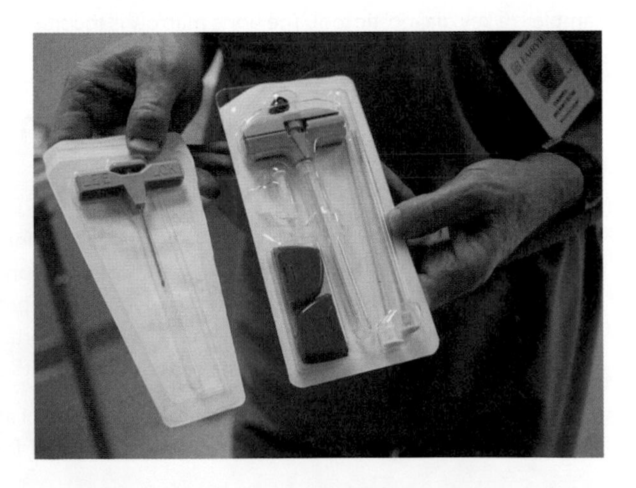

FIGURE 10.3
Bone marrow aspirate (left) and trephine biopsy (right) needles. The aspirate needle is used for the withdrawal of liquid bone marrow to assess morphology and perform differential counts and other downstream tests, whilst the trephine biopsy needle is used to remove a core of bone marrow tissue, thus preserving the architectural arrangement of the marrow. Image courtesy of Thirteen of Clubs, licensed under the Creative Commons Attribution Share Alike 2.0 Generic license.

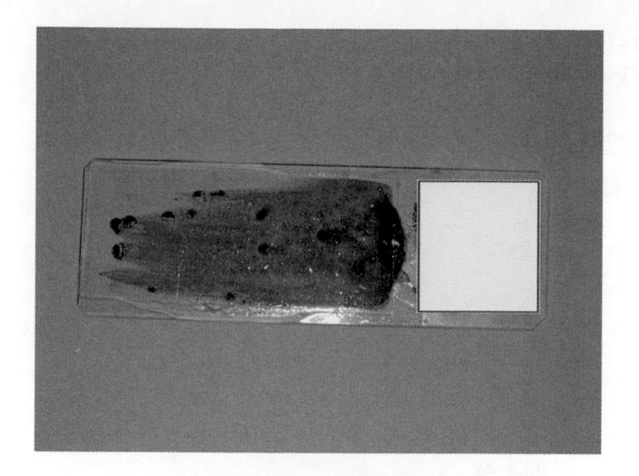

FIGURE 10.4
Bone marrow aspirate. A macroscopic view showing the distribution of the marrow on the slide. The densely stained region on the right is the origin. Tails are to the left. Bony particles can be seen at the tips of the tails. Image courtesy of Dr Robert Corser, Haematology department, Queen Alexandra Hospital, Portsmouth.

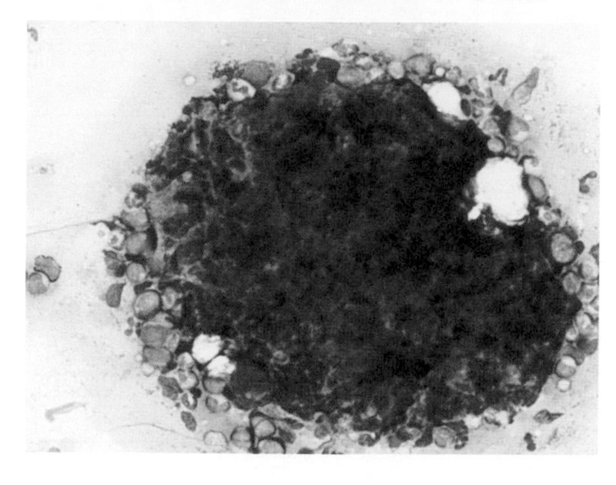

FIGURE 10.5
High magnification of bony particle. Large numbers of cells are contained within the fragment, and these can be used to estimate BM cellularity. Image courtesy of Dr Robert Corser, Haematology department, Queen Alexandra Hospital, Portsmouth.

therefore be inaccurate. Table 10.2 shows the reference ranges for a BM differential count for an individual aged between 21 and 56 years of age.

Bone marrow aspirates should be examined using low power, intermediate, and then high power. At low power, an assessment of the BM cellularity should be made along with the identification of abnormal cells within the marrow. Megakaryocytes and macrophages should also be examined for abnormalities. At high power, the maturation stages of myeloid and erythroid lines should be appreciated and a differential count completed. The myeloid:erythroid ratio should be calculated and areas of necrosis determined. Finally, iron stores should be assessed using Perls' stain.

An initial (baseline) BM aspirate is a useful tool, as it can be used diagnostically, but it also allows clinicians to compare subsequent BM aspirates to the original following treatment. The baseline marrow therefore allows us to assess whether a particular disease is progressing or improving following treatment.

Bone marrow trephine

In comparison with the BM aspirate, the trephine is a core biopsy of marrow in which the architecture is maintained (see Figures 10.6 and 10.7). Bone marrow trephines do not need to be performed in all cases, although, once obtained, the sample should be fixed and prepared prior to staining with a minimum of haematoxylin and eosin (H&E) and a reticulin stain. Romanowsky stain can also be used, and a high-quality Giemsa stain can provide much more

TABLE 10.2 Reference ranges for haemopoietic constituents in bone-marrow-derived samples, for patients between 21 and 56 years of age.

Cell	Mean value (%)	Range (%)	Comments
Myeloblasts	1.4	0–3	WHO classification of acute leukaemia ≥20%
Promyelocytes	7.8	3.2–12.4	'Blast-equivalents' in leukaemia diagnosis
Myelocytes	7.6	3.7–10	Mitosis does not occur beyond this stage
Metamyelocytes	4.1	2.3–5.9	
Band forms	–	–	
Neutrophils	34.2	23.4–45	
Eosinophils	2.2	0.3–4.2	
Basophils	0.1	0–0.4	
Monocytes	1.3	0–2.6	
Erythroblasts	25.9	13.6–38.2	Myeloblast equivalent in erythroleukaemia
Lymphocytes	13.1	6–20	
Plasma cells	0.6	0–1.2	Increased in myeloma and other plasma cell diseases
Myeloid:erythroid ratio	2:4	1.3–4.6	This is a ratio, and as such is not reported in percentage terms

METHOD Bone marrow aspirate

Bone marrow aspiration begins with the rupture of the sinusoidal membrane. The fine needle enters the BM and liquid marrow is withdrawn using a syringe. The liquid nature of the BM is a consequence of peripheral blood leaking into the BM from damaged blood vessels.

Once aspirated, the fluid can be smeared onto a slide, dried, fixed, and stained using Romanowsky stain.

Excess aspirate can be aliquoted into EDTA anticoagulant and used to make additional films, but EDTA can alter cellular morphology and cause the loss of subtle diagnostic clues. The films are prepared for interpretation in the haematology laboratory and the turnaround time can be as little as an hour. EDTA-anticoagulated BM can be forwarded on for immunophenotyping analysis if necessary.

information than an H&E stain, including the identification of cell lineage and the detection of fibrosis (supplemented using a reticulin stain). (See also Box 10.1.)

In contrast to the aspirate, the trephine is prepared by the histopathology laboratory primarily because of the requirement for specialized techniques and equipment. The turnaround time for the trephine biopsy is approximately two days. A trephine biopsy is essential for patients with suspected myeloproliferative disorders, particularly when a BM aspirate has provided a

FIGURE 10.6
Bone marrow trephine. This is a core biopsy in which the BM architecture has been maintained. Note the long, thin appearance of the specimen, which has been removed from the trephine needle, and finally processed by the histology department. Image courtesy of Dr Robert Corser, Haematology department, Queen Alexandra Hospital, Portsmouth.

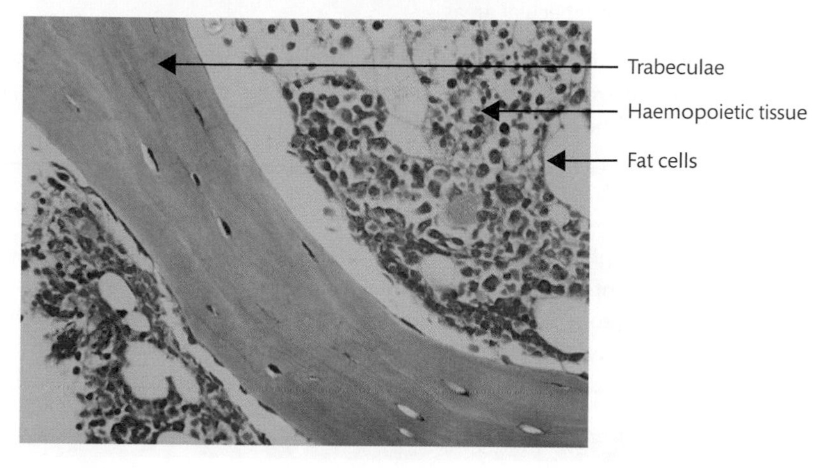

FIGURE 10.7
Microscopic evaluation of the BM structure can be assessed using a BM trephine. The BM architecture is maintained, allowing cell-to-cell interactions and distribution to be assessed. Image courtesy of Dr Robert Corser, Haematology department, Queen Alexandra Hospital, Portsmouth.

BOX 10.1 Bone marrow trephine

A BM trephine is the removal of a core of BM tissue, measuring a minimum of 16 mm, from the iliac crest. The trephine biopsy maintains the architecture of the BM and allows the relationship between different cell types within the BM to be assessed; for example, immature erythrocytes with macrophages or the localization of immature myeloid cells within the marrow.

Once collected, the biopsy is placed in formalin and labelled appropriately with the patient's details. To summarize the process of preparation, the sample is transported to the histopathology laboratory for processing where it undergoes a process of decalcification to remove bony particles contained within the tissue. The tissue is then ready for embedding prior to cutting sections for staining. Once stained, the proportions of different cell species and their interrelationship can be assessed.

dry tap, as is often the case in myelofibrosis. In lymphoid malignancies with suspected BM involvement, a trephine biopsy is important for providing information regarding the distribution of cells infiltrating the marrow. This may help dictate when therapy is initiated, and is also an important baseline for comparison with post-treatment trephines.

Trephine imprint or the trephine roll

The trephine imprint is often indicated in the absence of a BM aspirate. Following biopsy, the trephine is rolled over the surface of a glass slide, depositing BM material, see Figure 10.8. The trephine biopsy is then placed in fixative ready for transportation to the histopathology laboratory. The trephine roll, however, can be processed by the haematology laboratory in the same manner as the BM aspirate. The trephine roll can then be used to complete a differential count prior to the receipt of the processed trephine biopsy. The turnaround time can be as little as an hour.

Haematological malignancies often disrupt the normal distribution of cellular elements within the BM. Granulocytes and their precursors differentiate and mature next to bony trabeculae, and appear to diffuse away from the trabeculae—towards the centre of the bone, where many of the marrow sinusoids (blood vessels) are located—as they mature. Erythroblasts form discrete colonies with other red cell precursors and, along with megakaryocytes, are often found in small islands associated with sinusoids. Lymphocyte aggregates are commonly found, although are not associated with paratrabecular areas.

The dissociation between normal developmental patterns and those seen in cases of BM involvement with a range of malignancies often accounts for the presence of immature haemopoietic cells within the peripheral blood.

Key Points

When examining BM trephines, it is important to consider the normal distribution pattern of haemopoietic elements. In the case of normal lymphoid aggregates, they are absent from paratrabecular regions. However, as is discussed in Chapter 12, when BM involvement occurs in follicular lymphoma, paratrabecular infiltration is commonly found.

SELF-CHECK 10.4

Compare and contrast the bone marrow aspirate and the trephine.

When examining a BM slide, one of the important features to assess is the cellularity. This gives an indication of the integrity of the BM and its capability for haemopoiesis.

FIGURE 10.8
A trephine imprint. When a BM aspirate cannot be taken, a trephine is indicated. It is often useful to obtain an imprint of the trephine surface prior to histological processing of the trephine so the morphology of the cells can be fully appreciated. Image courtesy of Dr Robert Corser, Haematology department, Queen Alexandra Hospital, Portsmouth.

10.4.2 Bone marrow cellularity

Examination of a patient's bone marrow can provide an indication of cellularity—the percentage of cells within the area of the marrow being examined. The trephine is particularly good for being able to assess cellularity, although bone fragments contained within the tails of an aspirate can also provide a good indication of the BM cellularity when a trephine is unavailable.

Bone marrow comprises yellow and red components. Yellow marrow is composed of adipose tissue, whereas red marrow is haemopoietic tissue. When the haemopoietic activity of the marrow reduces, replacement of red marrow with yellow marrow occurs; reciprocally, an increase in haemopoietic activity replaces yellow marrow with haemopoietic tissue.

It is absolutely essential when assessing BM to consider the age of the patient. Bone marrow in normal individuals progressively becomes less cellular with age. Up to three months of age, 100% cellularity can be expected. From three months to ten years, this gradually reduces to approximately 80%. At 30 years, the marrow will comprise approximately 50% cells, and at 70 years, this decreases to approximately 30%. Generally, for an adult, BM is considered **normocellular** within the range of 40–70%.

A percentage of cells within the BM greater than 70% is termed **hypercellular**, whereas a reduction in the percentage of cells within the marrow to less than 40% is termed **hypocellular**.

Let us now consider the process of BM infiltration.

normocellular
The relationship between haemopoietic and adipose tissue within the BM, where the proportion of cells within the BM falls within the age-appropriate range for the patient.

hypercellular
Denotes an increase in the size of the haemopoietic compartment of the BM or an increase in the number of cells within the marrow. These cells do not have to be haemopoietic in origin, but can be metastatic.

hypocellular
A reduction in the size of the haemopoietic component of the marrow and an increase in the size of the yellow marrow (fatty marrow) compartment.

SELF-CHECK 10.5

What is the generally accepted reference range for bone marrow cellularity?

SELF-CHECK 10.6

What is the name given to an increased or reduced cellularity, and what are the respective percentage cut-off points of each?

10.4.3 Patterns of infiltration

The term 'bone marrow infiltration' refers to a process whereby malignant cells originating from outside the BM become deposited within the BM environment. It is important to determine patterns of infiltration, particularly in patients with suspected or confirmed lymphoproliferative conditions, but also in metastatic diseases such as breast or prostate cancers, where there is evidence of BM involvement due to the effects these malignant cells have on haemopoiesis. By determining the pattern of infiltration through the examination of a BM trephine, we can obtain important information to assist diagnosis and the formulation of a prognosis. Four main types of infiltrative pattern occur:

- paratrabecular;
- interstitial;
- nodular;
- diffuse.

Paratrabecular infiltration: malignant cells accumulate along the periphery of the **trabeculae**, displacing myeloid precursors and fatty tissue. Figure 10.9 shows typical paratrabecular patterns

trabeculae
Plural; singular, trabecula. Bony processes which extend from the outer bony tables into the marrow cavity.

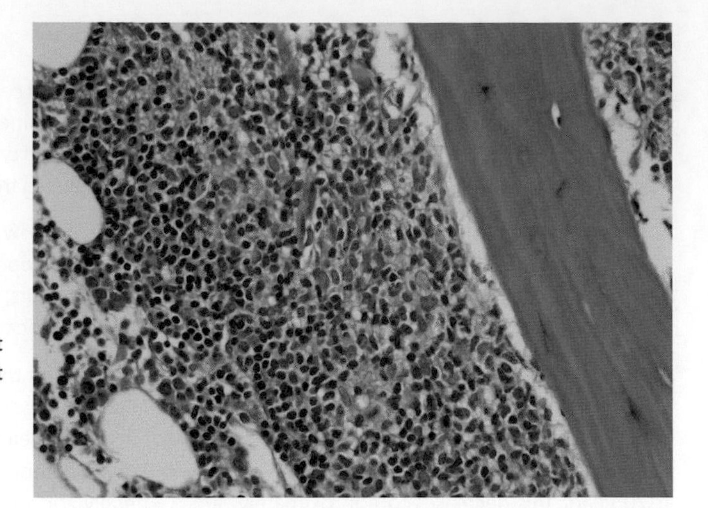

FIGURE 10.9
Paratrabecular infiltration of the BM by lymphoid cells. The trabecula (solid pink structure on the right), runs the length of the microscopic field. The swathe of cells running adjacent to the trabecula are malignant lymphoid cells. From a patient with follicular lymphoma. Reprinted from St J Thomas J. *The diagnosis of lymphoid infiltrates in the bone marrow trephine. Current Diagnostic Pathology* 2004: 10: 236–45, with permission from Elsevier.

of infiltration. Paratrabecular infiltration is commonly found in patients with BM involvement in a particular type of lymphoproliferative disease called follicular lymphoma.

Interstitial infiltration involves the accumulation of malignant cells between the tissues of bone marrow, i.e. between the haemopoietic elements. Fat spaces remain intact, as shown in Figure 10.10.

Nodular infiltrates are small, rounded accumulations of tumour cells within the BM, as shown in Figure 10.11. These can occur in any location within the marrow and have a limited effect on haemopoiesis.

Diffuse bone marrow infiltration occurs throughout the BM. The architecture and cellular interactions of normal haemopoietic elements are disrupted by the sheets of malignant cells accumulating within the marrow environment, as shown in Figure 10.12.

Cross reference
More detail about follicular lymphoma can be found in Chapter 12.

Here, the methods for assessing the morphology of haemopoietic cells, both in the peripheral blood and the BM, have been established. The different types of BM sample and patterns of infiltration within these samples have also been considered. We now need to examine the different methodologies available for identifying cell lineages.

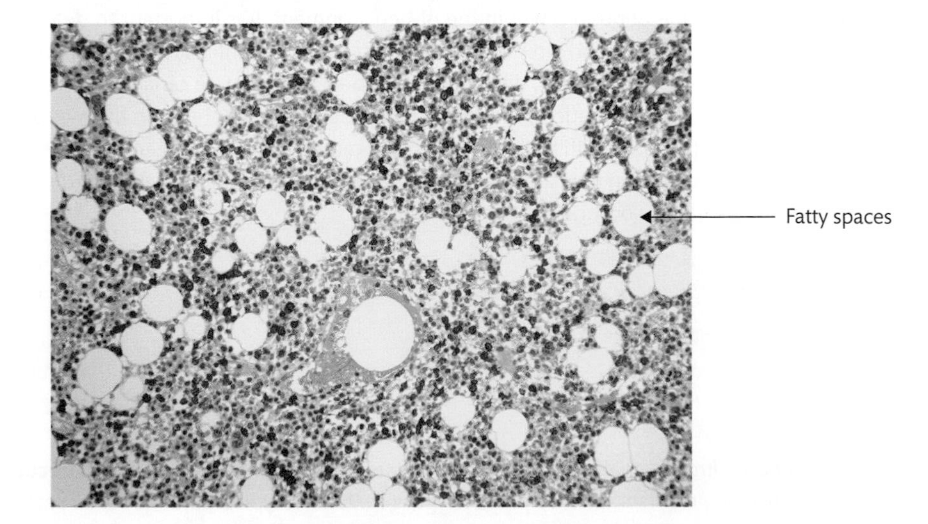

FIGURE 10.10
Interstitial infiltration. Here the BM structure is intact, with few, widely interspersed, clumps of malignant lymphocytes. The fatty spaces and haemopoietic tissue are all intact. Kindly provided by Professor SA Pileri.

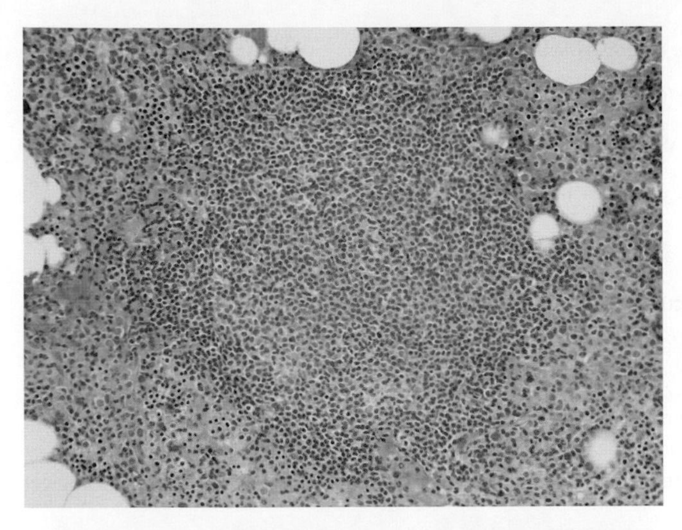

FIGURE 10.11
Nodular infiltration. All haemopoietic elements are maintained, except for discrete areas where malignant lymphoid cells accumulate. Fatty spaces are intact. Reprinted from St J Thomas J. ***The diagnosis of lymphoid infiltrates in the bone marrow trephine***. *Current Diagnostic Pathology* 2004: **10**: 236–45, with permission from Elsevier.

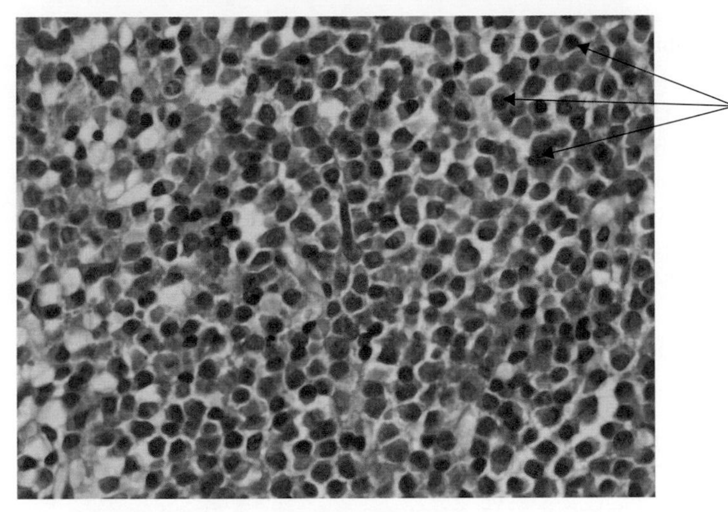

FIGURE 10.12
Diffuse bone marrow infiltration. All normal haemopoietic tissue is replaced by malignant lymphoid cells. Haemopoiesis will be compromised. Reprinted from St J Thomas J. ***The diagnosis of lymphoid infiltrates in the bone marrow trephine***. *Current Diagnostic Pathology* 2004: **10**: 236–45, with permission from Elsevier.

SELF-CHECK 10.7

Define bone marrow infiltration.

10.5 Cytochemistry

Cytochemistry is used to confirm the presence or absence of particular markers within cells. This can assist in the identification of cell lineage. For example, particular enzymes, such as myeloperoxidase, may be present in myeloid cells but absent from lymphoid cells. Being able to confirm the presence or absence of these enzymes helps us to establish cell lineage and to classify haematological malignancies appropriately.

Although a wide range of different cytochemical stains are available, this section will focus on a small selection, including:

- myeloperoxidase (MPO);
- Sudan black B (SBB);

- alpha-naphthyl acetate esterase (ANAE);
- chloroacetate esterase (CAE);
- periodic acid-Schiff (PAS);
- acid phosphatase (AP).

These cytochemical stains are summarized in Table 10.3.

This section does not focus on the chemistry of these reactions, but rather the endpoint required for interpretation. Please refer to Figures 10.13(a–e) in accordance with each of the sections below.

Cytochemistry utilizes microscopic techniques to assess the characteristics of cells following exposure to a range of cytochemical stains. By far the most significant of these is Romanowsky stain, and this should be the platform upon which the diagnosis of a particular haematological

TABLE 10.3 Cytochemical staining characteristics of haemopoietic progenitors and mature cells.

Stain	Lineage	Colour	Cell/maturation stage	Reaction	Comments
MPO	Myeloid	Brown	Early myeloblast	– ve	*Negative in:*
			Late myeloblast	–/+ ve	ALL, AMKL, and
			Promyelocytes	+ + ve	undifferentiated leukaemia
			Myelocytes	+ ve	
			Metamyelocytes	+ ve	
			Neutrophils	+ ve	
			Eosinophils	+ ve	
			Basophils	+ ve	
			Monocytes	+/– ve	
SBB	Myeloid	Black	*as above*		
ANAE	Myeloid	Red/brown	Monoblasts	+ ve	
			Granulocytes	+ ve	
CAE	Myeloid	Bright blue	Myeloblasts	– ve	Increases with maturation
			Granulocytes	+ ve	
PAS	Myeloid/lymphoid	Pink/red	Erythroblasts	– ve	Leukaemic erythroblasts + ve
			Maturing/mature RBCs	– ve	Eosinophil/basophil granules – ve
			Immature granulocytes	– ve	
			Mature granulocytes	+ ve	
			Monocytes	+/– ve	
			Megakaryocytes	+ + ve	
			Platelets	+ + ve	
AP	All cells	Red granules	T cells	+ + ve	Good for distinguishing B/T cells
			B cells	+ ve	Leukaemic monocytes + + ve
			Granulocytes	+ + ve	
			Megakaryocytes	+ + ve	
			Plasma cells	+ + ve	

Key: ALL = acute lymphoblastic leukaemia; AMKL = acute megakaryoblastic leukaemia; – ve = negative; + ve = positive; –/+ ve = occasionally positive; +/– ve = occasionally negative; + + ve = strongly positive.

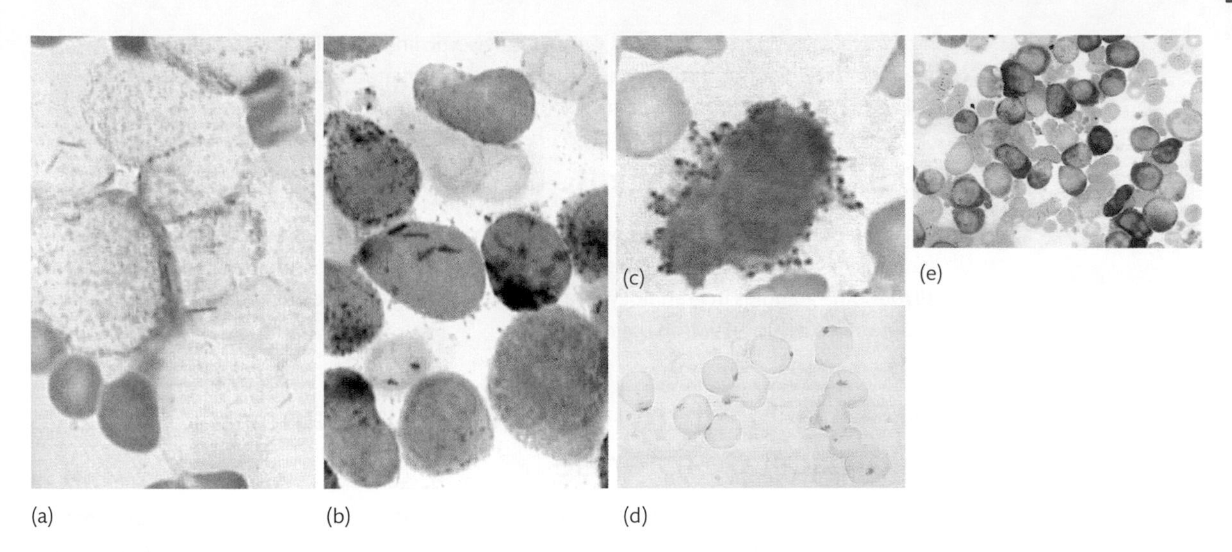

FIGURE 10.13

(a) Myeloperoxidase staining of myeloblasts. Cytoplasm shows diffuse staining of granules and Auer rods are apparent; (b) myeloblasts stained with Sudan black B; (c) periodic acid-Schiff staining of this dysplastic micromegakaryocyte demonstrates pink cytoplasm and granules; (d) acid phosphatase providing distinct red staining in lymphoblasts in a case of T-cell acute lymphoblastic leukaemia; (e) combined esterase α-naphthyl acetate esterase (ANAE) and chloroacetate esterase (CAE). ANAE provides a dark brown coloration to leukaemic monocytes, whereas CAE stains granulocytes bright blue. Reproduced by permission from Lewis et al. *Dacie and Lewis Practical Haematology*, 9th edn. © Elsevier.

malignancy is based. However, other cytochemical stains can prove useful in characterizing cells otherwise difficult to categorize microscopically using Romanowsky stain alone.

Except where Auer rods are present, using Romanowsky stain to differentiate between myeloblasts and lymphoblasts is very difficult and is subject to error. Without utilizing alternative techniques, there is a risk of miscategorization of the cell, and possible misclassification of the malignancy. Differentiation between monoblasts, erythroblasts, and megakaryoblasts also requires the application of alternative techniques, since these cells are morphologically similar, although they differ in their cytoplasmic contents. This cytoplasmic variation can be exploited by using cytochemical stains to produce different colours, which can be interpreted and quantified. However, the use of cytochemistry is also limited, and distinguishing between undifferentiated blasts and megakaryoblasts cannot be achieved using cytochemistry alone. The inability to differentiate between these blasts was one of the important considerations with the implementation of immunophenotyping in the diagnosis of haematological malignancies.

Cross reference

Section 10.3 gives details of blood film examination.

Now we should examine the main cytochemical stains in greater detail.

10.5.1 Myeloperoxidase

Myeloperoxidase is a member of the peroxidase family of enzymes and is found primarily within the azurophilic granules of granulocytes (neutrophils, eosinophils, and basophils), but *may* also be found within monocytes. A positive reaction utilizing the chromogen 3,3′-diaminobenzidine (DAB) provides a brown granular inclusion within the cytoplasm. Early myeloblasts are negative for MPO, but as they mature, they become MPO-positive. Strong reactions with DAB are seen in promyelocytes and myelocytes. Cells of monocytic lineage will also stain with DAB. A negative reaction does not distinguish between undifferentiated leukaemia, acute lymphoblastic leukaemia, or acute megakaryoblastic leukaemia.

10.5.2 Sudan black B

Sudan black B provides comparable information to MPO and can therefore be used to identify granulocytes and *some* monocytes. SBB is a lipophilic stain which binds to the lipid membrane surrounding granules in granulocytes and some monocytes. A positive reaction causes a black and granular inclusion within the cytoplasm.

10.5.3 α-Naphthyl acetate esterase and chloroacetate esterase

Both of these enzymes are part of the esterase group and can be used singly, or in combination with one another. ANAE is useful in the identification of normal and leukaemic monoblasts. ANAE is also useful to identify, through a red/brown coloration, granulocytes in myelodysplastic syndromes and in leukaemia. CAE is useful for identifying granulocytic maturation, as positivity (demonstrated by a bright-blue coloration) increases with maturation. Myeloblasts are generally negative for CAE.

10.5.4 Periodic acid-Schiff

PAS reacts with the 1,2-glycol groups found in carbohydrates, predominantly in glycogen, to produce a pink to bright-red reaction. Normal erythroid cells and their precursors are negative when exposed to PAS, although erythroblasts in erythroleukaemia are generally positive. Cells of all other lineages are likely to show a reaction. Immature granulocytes are negative, although become increasingly positive throughout the maturation process. Cytoplasmic PAS reactions are seen in eosinophils, whilst eosinophil granules are negative. Basophils are often positive, although basophil granules do not stain with PAS. Monocytes may or may not be positive, with variable reactions from fine to coarse granulation. Megakaryocytes are strongly positive, as are platelets, and normal lymphocytes may show granular positivity.

Leukaemic lymphoblasts may be strongly positive, and malignant plasma cells may show variable staining results.

10.5.5 Acid phosphatase

The acid phosphatase reaction produces red granules or a diffuse pattern in all haemopoietic cells. The degree of the reaction is particularly important when discriminating between B and T cells, with the T cells showing the strongest reaction. Acid phosphatase has historically been of value in identifying T-cell leukaemias because of the strong reactive pattern when compared to the B-cell lineage. Acid phosphatase is also of value in identifying acute myeloid leukaemia (AML) with monocytic involvement where strong cytochemical reactions are apparent.

SELF-CHECK 10.8

Provide a brief outline of the reaction patterns and colours associated with the myeloperoxidase, Sudan black B, and Periodic acid-Schiff stains.

As we have determined, cytochemistry is important for identifying cell lineage in haematological malignancies. However, cytochemistry has been largely superseded by the introduction and use of immunophenotyping as the main method for the identification of haemopoietic lineages.

Now that we have considered some of the commonly used cytochemical stains and their application in the differentiation of haemopoietic cell lineages, the methodology of immunophenotyping and its value in the diagnosis of haematological malignancies will be examined.

Cross reference

The World Health Organization classification of tumours of haematopoietic and lymphoid tissues is discussed in detail in Chapters 11 and 12.

10.6 Immunophenotyping

Although immunophenotyping has a wide range of applications, we will briefly examine the role of immunophenotyping and its application to haematological malignancies. Either peripheral blood or a BM aspirate preserved in EDTA anticoagulant can be used for immunophenotyping analysis.

The *Bethesda International Consensus recommendations of the immunophenotypic analysis of hematolymphoid neoplasia by flow cytometry* addresses the use of immunophenotyping in the diagnosis and monitoring of haematological malignancies. These recommendations largely complement the 'WHO classification of tumours of haematopoietic and lymphoid tissues', and also aim to reduce redundancy in the investigative process by using evidence-based practice to reduce the number of unnecessary requests for immunophenotyping.

Figure 10.14 summarizes the Bethesda group's recommendations for the appropriate use of immunophenotyping in the context of haematological malignancies. Sections in blue bubbles

FIGURE 10.14
A summary of the Bethesda group's recommendations for the applications of immunophenotyping, where malignancy is in question. All cases highlighted in blue would form an appropriate basis for immunophenotypic investigation. Isolated scenarios in yellow do not require immunophenotypic investigation.

are considered appropriate requests for immunophenotyping, whilst those highlighted in yellow are not considered to be valid reasons for immunophenotyping requests.

The process of immunophenotyping involves the use of **monoclonal antibodies (mAbs)** directed against intracellular or extracellular **cluster of differentiation (CD) antigens**, of which 350 are currently recognized. Monoclonal antibodies directed against specific CD antigens are called *anti*-CD, followed by the respective number of the CD antigen, for example anti-CD19. CD antigens are expressed on all cells; some CD antigens may be expressed across all lineages, and others are much more specific, being expressed on a single lineage, or at a particular stage of maturation. If you look at Figure 10.15 you will see an illustration of mAbs binding to two different types of CD marker, allowing immunological identification of the cell.

Panels of mAbs directed against different CD antigens provide a very powerful tool for identifying cells accurately, regardless of the cell's morphological features.

In order for the antibodies to bind intracellular antigens, fixation procedures (to prevent cell lysis) and permeablization procedures (to allow the mAb to enter the cell) are necessary.

Monoclonal antibodies are joined to **fluorochromes**, molecules which emit light at a particular wavelength when excited by a laser. The light emitted from each fluorochrome can be

monoclonal antibodies (mAbs)

These are manufactured to recognize a particular epitope on a specific antigen. Although mAbs produced by different manufacturers may have the same name, they may recognize a different epitope on the specified antigen.

cluster of differentiation (CD) antigen

The standardized notation used to describe a range of molecules associated with the differentiation and maturation of cells. CD markers can be measured using epitope-specific mAbs.

fluorochrome

A chemical which, when excited by light of a particular wavelength, will emit light of a different but predictable wavelength, measurable using a photodetector.

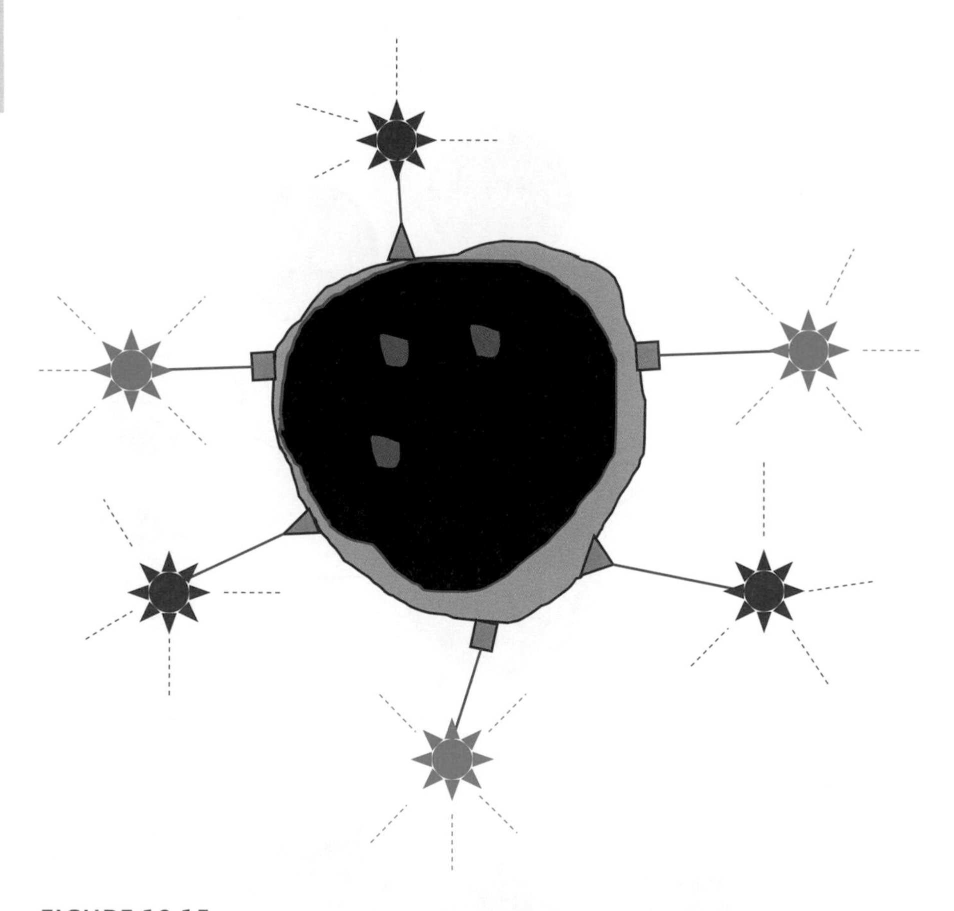

FIGURE 10.15

An illustration of two discrete fluorochrome conjugated mAbs of differing specificity (as suggested by the different-shaped surface antigens) binding to the surface of a malignant blast. An analyser with multiple channels would be able to detect the fluorescence from these different fluorochromes simultaneously, providing evidence of antigenic co-expression.

BOX 10.2 Monoclonal antibodies

The technique to produce mAbs was devised by Milstein and Köhler and reported in 1975. Monoclonal antibody production involves fusing normal-immunoglobulin (Ig) -producing cells, stimulated to produce Ig following sensitization with an appropriate antigen, with myeloma cells. Normal Ig-synthesizing cells do not survive in culture conditions for long, and myeloma cells do not reliably produce Ig, although they survive continuously under culture conditions. This technique produces 'hybridomas' that continually produce antibodies directed to the epitopes expressed by the sensitizing antigen. These hybridomas can be separated by producing single-cell colonies which will expand to form a clonal population, and the specificity of the antibody can be determined. These colonies can be cultured to constantly produce reagent mAbs. For any given epitope, the corresponding mAbs will all have the same structure, specificity, and binding strength (avidity), meaning that these mAbs can be used reproducibly for diagnostic and research purposes.

The seminal findings, in which the synthesis of anti-sheep RBC antibodies were generated, were reported in Köhler and Milstein, 1975.

measured using a detector, and is used to identify whether or not a particular CD marker is expressed by a specific population of cells. We can measure a number of fluorochromes simultaneously by using a flow cytometer with multi-channel capabilities. In order to do this, mAbs joined to fluorochromes emitting light at different wavelengths must be selected. We can use the data generated to assess whether or not a particular set of antigens are co-expressed on particular cells, which can help the identification of the cell. (See also Box 10.2.)

A wide range of fluorochromes have been manufactured; some of the most common are listed in Table 10.4 with their relative excitation and emission wavelengths.

The data generated using immunophenotyping can help to establish whether a particular condition is malignant or reactive by distinguishing between a monoclonal or polyclonal cell population. If the cells are malignant, we can use immunophenotyping to ascertain whether they are myeloid or lymphoid, undifferentiated, or express markers of both myeloid and lymphoid lineages. Immunophenotyping is also important to allow us to define the stage of maturation arrest within the malignant cell population. In a lymphoid population, immunophenotyping can establish whether the cells are of B or T lineage, or express markers of both.

Cross references

See Chapter 9, section 9.9: Clonality.

Chronic lymphocytic leukaemia (CLL) is discussed in detail in Chapter 12.

SELF-CHECK 10.9

What sort of information can be provided through the use of immunophenotyping?

TABLE 10.4 Approximate maximum excitation and emission wavelengths for commonly used fluorochromes in immunophenotyping.

Fluorochrome	Excitation max (nm)	Emission max (nm)
Allophycocyanin (APC)	650	660
Fluorescein isothiocyanate (FITC)	490	520
Peridinin chlorophyll protein (PerCP)	488	578
Phycoerythrin (PE)	564	576
Rhodamine	550	573
Texas Red	596	615

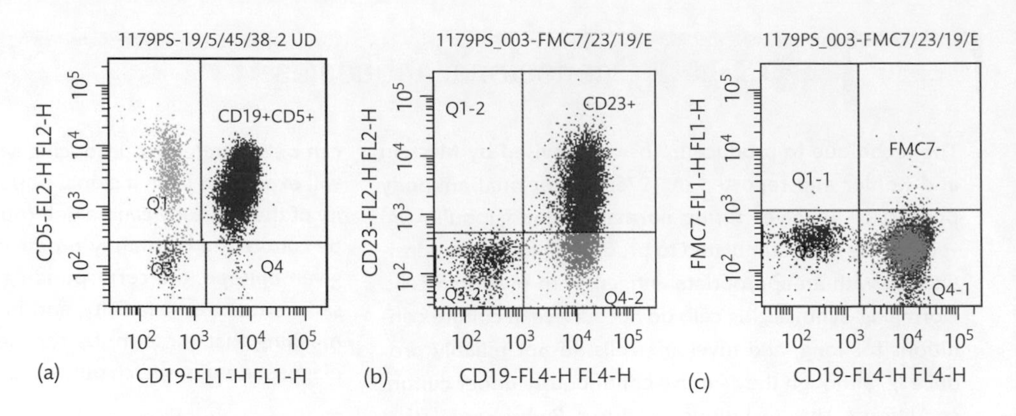

FIGURE 10.16

The typical B- chronic lymphocytic leukaemia (CLL) phenotype. Each scattergram is divided into quadrants—cells within the quadrant on the bottom left (in this case blue) are always antigen negative. (a) Cells are tested against anti-CD5 (*y*-axis) and anti-CD19 (*x*-axis). There is a population of cells that is CD5 positive and CD19 negative (Q1) which correlates with a T-cell population; CD19 positive and CD5 negative cells are shown in Q4. The population of cells highlighted red are the malignant B cells and are CD5 positive and C19 positive (Q2). (b) CD19 positive cells have been selected and, again in quadrant (Q2) show co-expression of CD23. (c) CD19 positive cells rarely co-express FMC7, as shown by the small number of cells within Q2 in this scattergram. The results show CD5+ve, CD19+ve, CD23+ve, FMC7−ve. Immunophenotyping data courtesy of Department of Immunology, Southampton General Hospital.

There is no single CD marker that signals malignancy. CD markers are expressed on normal cells, often in a lineage-restricted manner. Aberrant expression of CD markers, or a co-expression of markers, can suggest an abnormality; the range of markers as a whole needs to be assessed to establish a diagnosis. For example, as shown in Figure 10.16, in B-cell CLL the typical markers are CD5+ve, CD19+ve, CD23+ve, FMC7−ve. Using this combination of CD markers, B-CLL can be confidently diagnosed, even though, individually, each of these is a normal CD marker.

10.6.1 Limitations of immunophenotyping

The immunophenotyping technique has its limitations. In certain malignancies involving mature cells, such as in chronic phase chronic myeloid leukaemia (CML), the expression of CD antigens is consistent with mature myeloid cells and therefore immunophenotyping is of no benefit in the diagnosis of this condition. However, there is value in using immunophenotyping to establish the lineage of blasts in people with CML in blast crisis. Although approximately 70% of cases of CML in blast crisis become AMLs, specific subtypes—for example, erythroleukaemia or lymphoblastic leukaemias—may develop and are subject to different types of therapy and prognoses. According to the British Committee for Standards in Haematology (BCSH) guidelines for the diagnosis and management of adult myelodysplastic syndromes, there are still no specific immunophenotypic markers to support the diagnosis of myelodysplastic syndrome (MDS). Although abnormal expression of certain markers may be apparent, immunophenotyping in MDS is currently only recommended for research purposes or blast cell enumeration. As can be seen in Chapter 11, blast cell enumeration is particularly important for classifying certain types of MDS and for staging the disease.

Whilst immunophenotyping is a very powerful tool, and it is relatively simple to perform the preparatory work and sample processing, analysis of the results may be rather more difficult. The ability to analyse results correctly and formulate a diagnosis requires a great deal of training and experience. New flow cytometers are expensive pieces of equipment to purchase, and mAbs conjugated to fluorochromes (primary mAbs) can also be costly.

Now that some of the general considerations of immunophenotyping have been examined, we should concentrate on how some of the technical considerations affect the way immunophenotyping is used in practice.

Cross references

The pathogenesis of CML and its disease course are discussed in Chapter 11.

See Chapter 11–An introduction to classification systems: myeloid neoplasms.

See Chapter 12–An introduction to classification systems: lymphoid neoplasms.

10.6.2 Technical considerations for immunophenotyping

The BCSH has set out guidelines for the use of immunophenotyping for the laboratory investigation of haematological malignancies of myeloid and lymphoid origin. Revised guidelines were published in 2014 which relate to all aspects of the delivery of a flow cytometry and immunophenotyping service for diagnostic and monitoring purposes. These guidelines do not explicitly state which antibodies should be used or which panels of antibodies are required for diagnostic purposes, although they do direct the reader to guidance on the construction and composition of antibody panels. Broadly, panel selection depends upon clinical suspicion or other laboratory data, for example, as might be indicated by morphology using Romanowsky and, in some cases, cytochemical stains. Based upon suspicion, a first-line panel should be used which broadly tells us whether the patient has an acute leukaemia, a lymphoid malignancy, or plasma cell disorder, Following this initial screen, classification antibody panels can be used to stratify diseases into particular diagnostic entities, as shown in Figure 10.17.

The first stage of analysis is to identify the population of cells to be analysed. Analysis is achieved using two particular measurements: **forward scatter** (FSC) and **side scatter** (SSC). These are plotted on a scatter graph. Forward scatter represents the size of each of the cells in the sample and is measured based on the amount of light (generated by a laser) diffracted by the cell membrane. FSC is plotted on the *x*-axis. SSC, plotted on the *y*-axis, is detected at 90° to the path of the light from the source and represents the internal complexity of each cell. The data representing cell size and internal complexity can then be used to identify the cell species according

METHOD Preparation for immunophenotyping

- Using peripheral blood as an example, once the panel of choice has been selected, the red cells are lysed, and the remaining cells washed to remove red cell stroma.
- The sample is thoroughly mixed and known volumes of blood are aliquoted into separate labelled tubes. These are then incubated with the range of mAbs from the selected panel.
- Following incubation, the contents of each tube are washed to remove excess antibody, which would otherwise interfere with the results (e.g. by providing a false-positive result).
- The sample is now ready for analysis.
- Information regarding the number of tubes to be tested and the specificity of the mAbs contained within each tube should be inputted into the analyser software prior to analysis.
- The analyser will provide a prompt when the analysis of each tube is complete and the next tube can begin.
- The amount of time taken for measuring fluorescence depends upon the number of cells (events) counted.
- Once complete, the data can be analysed.

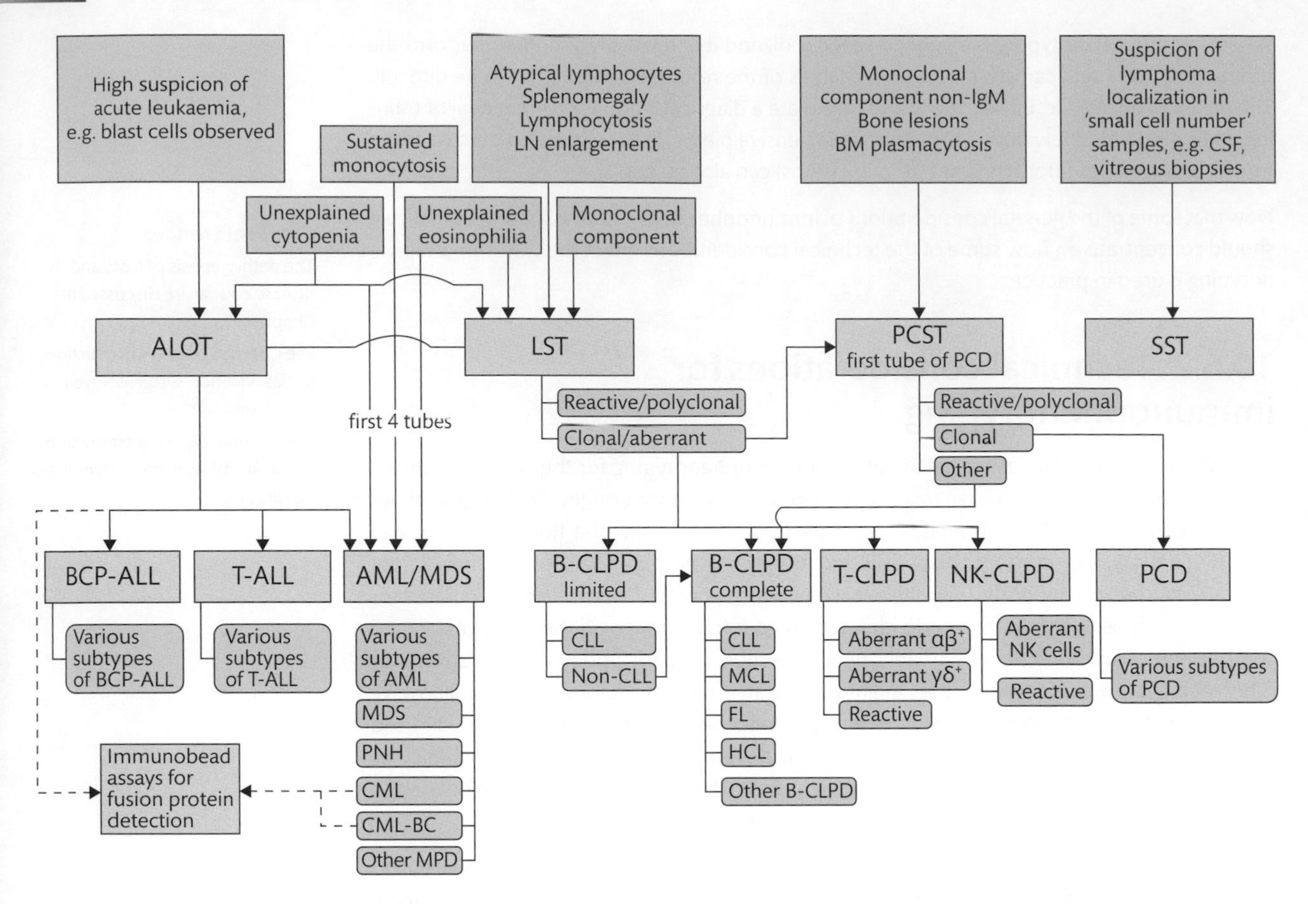

FIGURE 10.17

Flowchart diagram of the EuroFlow strategy for immunophenotypic characterization of haematological malignancies. On the basis of several entries of clinical and laboratory parameters, haematological malignancies are screened using a limited screening panel (i.e. typically one single tube) prior to appropriate and comprehensive characterization using extended antibody combinations. ALOT = acute leukaemia orientation tube; AML = acute myeloid leukaemia; BC = blast crisis; BCP = B-cell precursor; BM = bone marrow; CLL = chronic lymphocytic leukaemia; CLPD = chronic lymphoproliferative disorders; CML = chronic myeloid leukaemia; CSF = cerebrospinal fluid; FL = follicular lymphoma; HCL = hairy cell leukaemia; LN = lymph node; LST = lymphoid screening tube; MCL = mantle cell lymphoma; MDS = myelodysplastic syndrome; MPD = myeloproliferative disorders; PCD = plasma cell disorders; PCST = plasma cell screening tube; PNH = paroxysmal nocturnal haemoglobinuria; SST = small sample tube.

Reprinted by permission from Macmillan Publishers Ltd: van Dongen J, Lhermitte L, Böttcher S et al. ***EuroFlow antibody panels for standardized n-dimensional flow cytometric immunophenotyping of normal, reactive and malignant leukocytes***. *Leukemia* 2012: **26**: 1908–75, https://doi.org/10.1038/leu.2012.120, copyright 2012.

to the structural properties of that cell. For example, neutrophils are relatively large, with a high SSC due to their internal complexity (highly granular with a multi-lobulated nucleus).

Cross references

See Chapter 11—An introduction to classification systems: myeloid neoplasms.

See Chapter 12—An introduction to classification systems: lymphoid neoplasms.

SELF-CHECK 10.10

What is the difference between FSC and SSC? How does this enable us to identify different populations of cells?

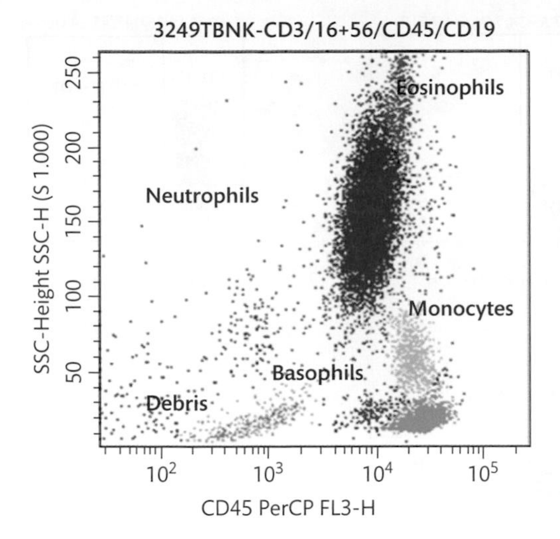

FIGURE 10.18
CD45 can be used in place of FSC against SSC to enable leucocytes to be identified according to CD expression and cellular internal complexity. Neutrophils are coloured red, eosinophils purple, monocytes turquoise, and basophils dark blue. The green population are lymphocytes. Immunophenotyping data courtesy of Department of Immunology, Southampton General Hospital.

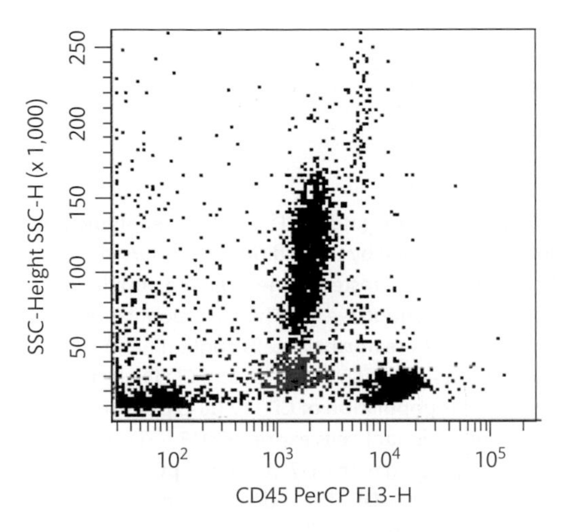

FIGURE 10.19
CD45 expression to identify a population of blasts. Due to the size and internal complexity of blasts, they tend to overlap the distributional patterns of lymphocytes and monocytes when FSC and SSC are used alone. When CD45 is used in place of FSC, the blast population (coloured red) becomes a discrete entity due to the poor expression of CD45 on the blast surface.

Analysis begins with the assessment of anti-CD45, a pan white cell marker, used to exclude any remaining red cell debris or platelets from the analysis; see Figure 10.18. Anti-CD45 is substituted for the FSC measurement on the x-axis. Blast expression of CD45 is significantly lower than that of normal lymphocytes and monocytes, allowing separation of these three mononuclear populations, as shown in Figure 10.19. The use of FSC and SSC alone may allow an overlap between the blast population and the lymphocyte and monocyte populations because of the similar light-scattering characteristics of this heterogeneous population of blasts.

Once the appropriate cell population has been selected, a gate can be applied. **Gating** is a very powerful tool for identifying and subsequently classifying a range of haematological malignancies according to the combined antigenic expression on a particular population of cells. Gating can be performed on the FSC/SSC plot, where a population of cells is selected based on size and internal complexity, although there is a risk that this population will be contaminated by cells of another lineage, as previously highlighted in the context of blasts. Using CD45 provides an excellent way of selecting a pure population of cells because the cell populations are clearly separated.

gating
The process of selecting a particular population of cells on which to base further immunophenotyping analysis.

Once a reaction has been determined between the mAb and the target antigen, to call these cells 'antigen-positive' we need to establish whether or not the cells actually express the

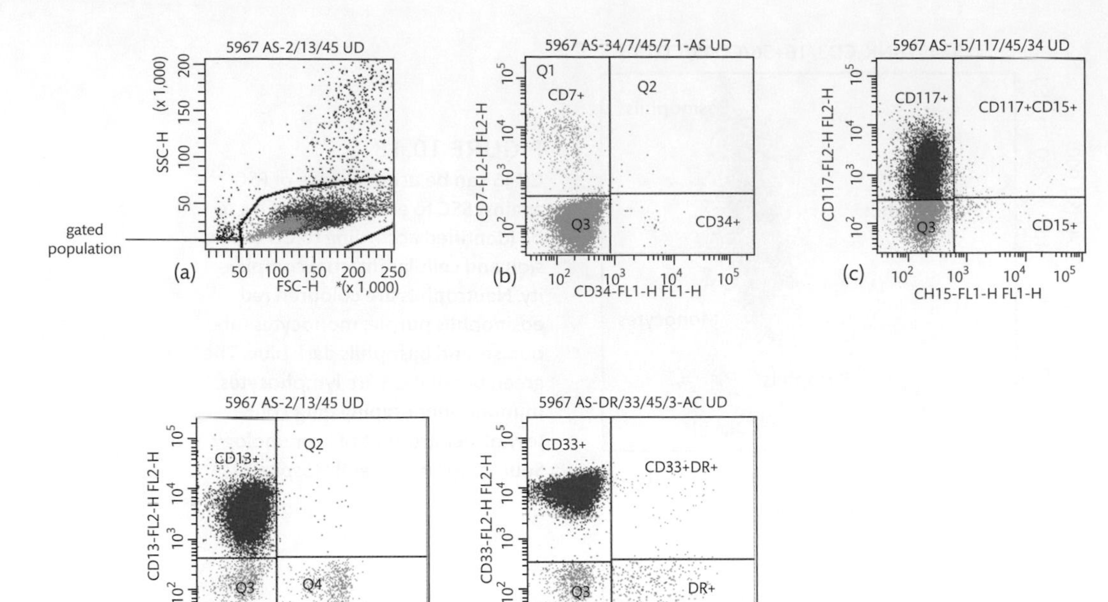

Figure 10.20

Acute promyelocytic leukaemia (APL). (a) Cells of interest are gated. All immunophenotypic data is analysed in relation to this gated population. As we can see by these different colours, this is not a homogeneous population of cells, although we are interested in the large red population of cells in this analysis. Quadrant 3 (Q3) is negative throughout a–e. (b) There is a population of CD34 negative, CD7 positive cells (Q1), although these are normal T cells contaminating the malignant cell population. The majority of the other cells are CD7 negative and CD34 negative (Q3). CD34 is a stem cell marker. (c) There is a clear population of CD117 (also called c-kit) positive cells (Q1), which are CD15 negative. (d) The malignant cells express CD13 (Q1), but not CD2 (Q2). CD2 is a pan T-cell marker and is co-expressed with CD7. (e) The population of interest (red), Q1, expresses CD33, but does not express HLA-DR. The red population has the immunophenotype: CD13+ve, CD33+ve, CD34–ve, HLA-DR–ve, CD7–ve, CD117+ve, typical of APL. Immunophenotyping data courtesy of Department of Immunology, Southampton General Hospital.

antigen at a sufficient density. If no fluorescence is measured, this clearly indicates the cells do not express a particular antigen, although weak fluorescence also needs to be considered.

Having established the pattern of antigen expression on the gated population, we can compare these data with pre-established criteria for different types of malignancy to ascertain the correct diagnosis. For example, we know that if a population of cells are shown to have the following phenotype: CD13+, CD33+, CD117+, CD34– and HLA-DR–, as shown in Figure 10.20, this is characteristic of acute promyelocytic leukaemia.

The identification of lineage is of vital importance in being able to establish the correct diagnosis for a patient with a haematological malignancy. For example, if a population of cells express CD13, CD33, and CD34, we can be confident that these cells are of the myeloid lineage, and because of the expression of CD34, a stem cell marker, these cells are myeloblasts. Lack of CD34 expression implies these cells are beyond the 'blast' stage of maturation.

Certain antigens, such as CD38 and Zap-70 in B-CLL, act as prognostic markers, which are of importance when deciding upon therapeutic intervention, communication between healthcare professionals, and keeping the patient informed of their disease. Immunophenotypic analysis is an important tool in diagnosis and prognostication, although additional diagnostic and prognostic information can be elucidated using cytogenetic analysis.

Cytogenetic testing is a specialized procedure, unavailable in the majority of hospital pathology departments. Specialist referral centres are often used to provide this service, and it is important to have an awareness of the procedures in place and the data obtainable using these techniques, as they are of increasing importance in the field of haemato-oncology.

Cross reference

Zap-70 is discussed in greater detail in Chapter 12.

SELF-CHECK 10.11

What is the purpose of gating?

10.7 Cytogenetic analysis

The term 'cytogenetics' is a fusion of cytology (the study of cells) and genetics, the study of genes. Cytogenetics is the process of studying genetic material at the cellular level and is important in assessing heritable, congenital, and acquired genetic diseases. A considerable amount of scientific and clinical research shows that assessing the cytogenetic profile of patients with haematological malignancies is vital for:

- diagnosing particular haematological malignancies;
- selecting appropriate therapies (in specific cases);
- assessing a patient's prognosis;
- identifying clonal progression.

Indeed, elements of the WHO classification are dependent on cytogenetic analysis to form an accurate diagnosis and enable treatment selection.

The tissue required to complete a cytogenetic analysis depends upon the histological origin of the malignancy and the location of abundant tumour cells. If, for example, we have a patient with a suspected malignancy *localized* in a lymph node, cytogenetic analysis of that lymph node should be performed. Performing a cytogenetic analysis of the BM in this case would not be of value, even if a BM sample has been received.

A BM aspirate usually provides good-quality cells for culture and hence the assessment of patients with acute and chronic leukaemias and for patients with BM infiltration (section 10.4.3). Whether BM or peripheral blood is to be used for the analysis, the sample should be collected into preservative-free, lithium-heparin tubes. (See also Box 10.3.)

Cross references

See Chapter 11—An introduction to classification systems: myeloid neoplasms.

See Chapter 12—An introduction to classification systems: lymphoid neoplasms.

See Chapter 9—An introduction to haematological malignancies, section 9.9: Clonality.

Section 10.4 discusses bone marrow infiltration.

See Chapter 9—An introduction to haematological malignancies for more on karyotype.

See Chapter 9—An introduction to haematological malignancies for more on cytogenetics.

The requirements and procedures for cytogenetic analysis tend to vary between centres, but the patient's **karyotype** will probably be established to identify any numerical chromosomal abnormalities. Comparative genomic hybridization (CGH) can be used to rapidly screen cells for copy number abnormalities whereas G-banding and fluorescence *in situ* hybridization (FISH) are most commonly used to establish any numerical or structural chromosomal abnormalities associated with malignant cells.

karyotype

The chromosomal complement for a cell, or population of cells.

Although an important and commonly used technique, G-banding does have its limitations. It can be impossible to identify microdeletions (submicroscopic deletions of chromosomal material) or some types of translocation, especially when the banding pattern of the chromosomes

BOX 10.3 Cytogenetic techniques

Using peripheral blood as an example, the technique requires approximately 5–10 ml of lithium-heparin anticoagulated whole blood.

- Cultures containing ~1 × 10^6 WBC/ml culture medium (typically RPMI) are set up under sterile conditions, and incubated at 37°C. Depending upon the target cells, a mitogen may be added to stimulate cell division.
- To harvest the cells from the culture, either colchicine or other spindle inhibitors are added to induce cell cycle arrest in metaphase.
- When it is thought that an appropriate number of cells have accumulated in metaphase, a hypotonic solution is added to burst the red cells and swell the cultured cells.
- The cells are then fixed and dropped gently onto microscope slides. The chromosomes spread to produce metaphases, which are then stained and analysed microscopically. At least 30 cells arrested in metaphase should be examined for cytogenetic abnormalities.

has not been disrupted. Translocations unidentifiable using standard microscopic techniques are called **cryptic translocations**. In cases involving cryptic translocations, more powerful techniques may be used. A good example of a cryptic translocation is t(12;21)(p13;q22), found in childhood acute lymphoblastic leukaemia (ALL), where the regions of chromosome exchange have an identical banding pattern to the normal chromosomes 12 and 21, and so cannot be detected using G-banding and light microscopy. Molecular cytogenetic techniques provide more reliable data to establish the cytogenetic profile of patients with haematological malignancies.

cryptic translocation

This occurs when the original morphology and banding pattern of a chromosome is maintained following translocation. This means the translocation may go undetected.

Comparative genomic hybridization (CGH) is a relatively fast screening technique used to identify changes in chromosome copy number, as outlined in Figure 10.21. Once CGH analysis is completed and changes in copy number are established, more specific techniques—such as FISH or DNA sequencing—are used to identify the precise chromosomal defect.

CGH does not require cell culturing prior to analysis. Tumour DNA is extracted, conjugated with a green fluorescent probe, and mixed with normal DNA, which is conjugated to a red fluorescent probe. The green- and red-labelled DNA is then hybridized to preprepared human metaphase chromosomes,

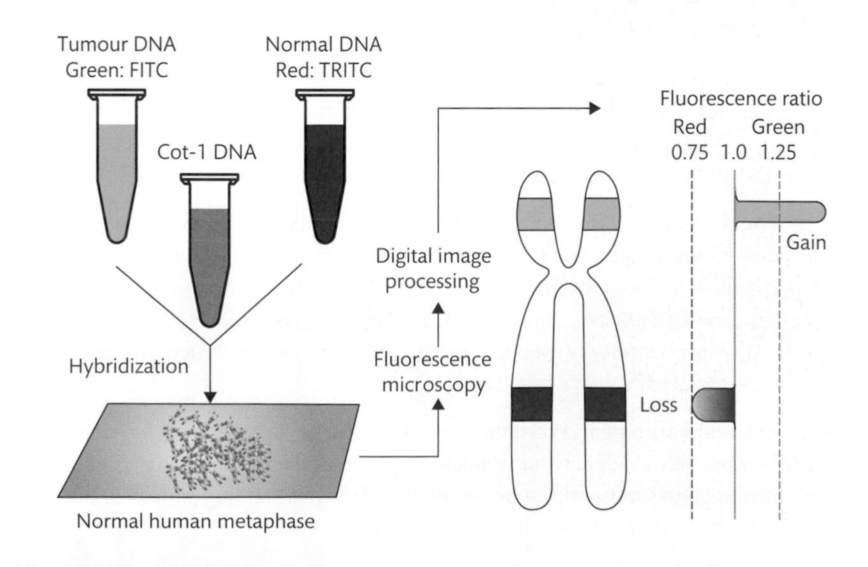

FIGURE 10.21
Comparative genomic hybridization. Tumour DNA is labelled green and normal DNA labelled red. These are mixed in a 1:1 ratio and hybridized to metaphase chromosomes. The ratio of green to red fluorescence is measured. Values >1 are suggestive of chromosomal gains whilst <1 suggest losses. Reproduced with permission from Reis-Filho JS et al. ***The molecular genetics of breast cancer: the contribution of comparative genomic hybridization.*** *Pathology, Research and Practice* 2005: **201(11)**: 713–25. Doi: 10.1016/j.prp.2005.05.013.

and the ratio of green to red fluorescence measured. A ratio of greater than one demonstrates chromosomal gain, whereas a ratio of less than one indicates a loss of chromosomal material.

Fluorescence in situ *hybridization* (FISH) uses fluorescent probes to bind in a complementary fashion to the particular chromosomal segments to be investigated. FISH can enable very accurate, rapid analysis for particular qualitative chromosomal abnormalities, and has the advantage that either metaphase chromosomes or interphase nuclei can be used as the target. FISH is an important technique for assessing individuals with cryptic translocations and those with **complex karyotypes**. By using one or more probes labelled with different fluorochromes, it is possible to examine different chromosomal segments, which allows us to identify deletions, additions, and translocations relatively easily, as shown in Figures 10.22 and 10.23. Identifying identical chromosomal abnormalities in a number of cells can signal clonality, as described in Box 10.4. Unlike G-banding it is not a 'global technique' examining the whole genome; it only provides an answer to the specific question asked by the use of particular probes.

complex karyotype
Describes the situation where more than three acquired cytogenetic abnormalities are found within a population of tumour cells.

Consider the different conditions in the following list of cytogenetic markers associated with different types of haematological malignancy. This is by no means an exhaustive list, although

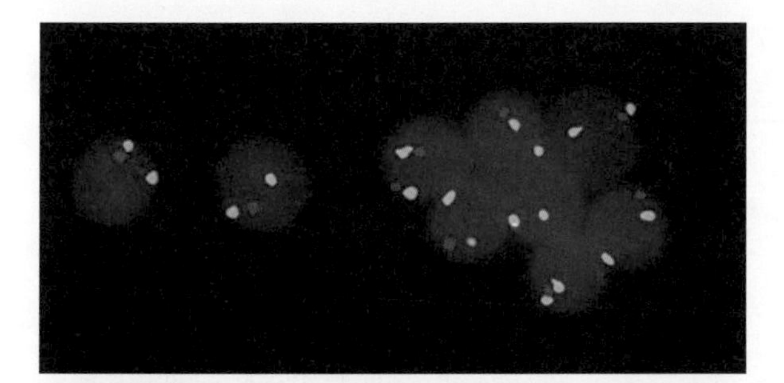

FIGURE 10.22
Fluorescence *in situ* hybridization, using a probe for *ATM* locus 11q23 (red), and one for the centromere of chromosome 11 for this CLL patient. There should be equal numbers of red and green fluorescent spots within each cell. There are fewer red spots, indicating the loss of *ATM* in the majority of these interphase cells. Courtesy of Cytogenetics laboratory, Royal Bournemouth Hospital.

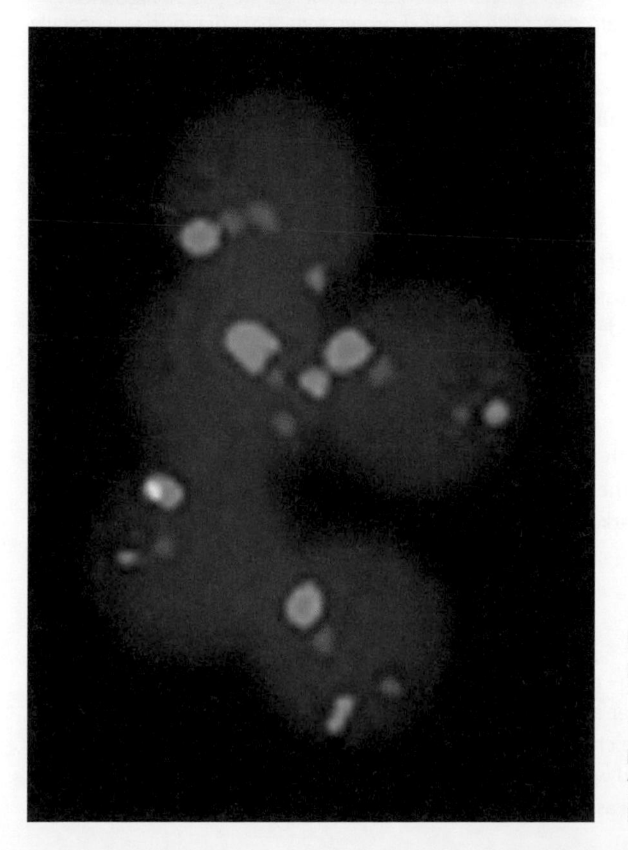

FIGURE 10.23
Fluorescence *in situ* hybridization for the same patient as in Figure 10.22, but with two different probes. A probe for the *TP53* locus, 17p13 (red), and the centromeric probe (green) shows equal numbers of green and red fluorescent spots. There is no evidence of *TP53* loss in this patient. Courtesy of Cytogenetics laboratory, Royal Bournemouth Hospital.

BOX 10.4 Cytogenetics and clonality

Identifying a monoclonal population of cells by immunophenotyping suggests that these cells are malignant, whereas a polyclonal population of cells is likely to be normal or reactive. If cytogenetic analysis shows a chromosomal addition or structural abnormality, excluding a loss, *two cells must share that abnormality* to be considered clonal. If the loss of a chromosome is detected using cytogenetic analysis, *a minimum of three cells must share this abnormality* in order for a clonal disorder to be confirmed.

it demonstrates some of the important cytogenetic findings in a range of different malignant conditions. In some cases, these abnormalities have been linked with prognosis. Pre-treatment cytogenetic data is one of the most important prognostic factors for patients with AML and ALL.

SELF-CHECK 10.12

What is the purpose of performing cytogenetic investigations on patients with a suspected haematological malignancy?

10.7.1 Cytogenetic abnormalities in malignant conditions

AML

Cytogenetic abnormalities can help direct the medical professional to a diagnosis, especially in certain cases where a particular cytogenetic event is closely associated with a specific disease. For example, t(15;17)(q22;q12) is diagnostic for APL, and for this particular translocation, targeted therapy in the form of **all-*trans* retinoic acid (ATRA)** is available. Identification of t(15;17) is associated with a good prognosis and survival prospects because of the success of ATRA in treating these patients.

all-*trans* retinoic acid (ATRA)
A vitamin A derivative which interacts with the PML-RARα fusion protein in acute promyelocytic leukaemia and, at pharmacological concentrations, can induce differentiation of cell cycle arrested promyelocytes, allow their maturation, and the induction of apoptosis.

hyperdiploidy
Human cells containing in excess of 46 chromosomes.

Other cytogenetic abnormalities in AML associated with a good prognosis include t(8;21)(q22;q22), inv(16)(q13;q22), and t(16;16). Intermediate prognostic markers include a trisomy 8, trisomy 21, and the majority of fusion partners associated with the breakpoint 11q23. Markers of a poor prognosis in AML include t(6;9)(p23;q34), inv(3)(q21;q26), del(5q), and del(7q).

ALL

For ALL, markers associated with a good prognosis include **hyperdiploidy** (more than 46 chromosomes), but particularly the presence of 51–65 chromosomes (high hyperdiploidy) and t(1;19)(q23;p13). By contrast, markers of a poor prognosis include t(9;22)(q34;q11) and t(4;11)(q21;q23).

MDS

In MDS, patients with either a 5q− or 7q− are associated with better survival prospects (a good prognosis) than those with either a trisomy 8 (associated with an intermediate prognosis) or those with the loss of the entire chromosome 5 or 7 (markers of a poor prognosis). Because of the good prognosis associated with 5q−, the WHO guidelines contain a subcategory for the classification of MDS—MDS with associated del(5q) or 5q− syndrome.

CLL

When del13q14 is the only chromosome abnormality detected in CLL, it is associated with a good prognosis, as is a normal karyotype. Del13q14 is not associated with the loss of the tumour suppressor gene *RB1*, but rather with the loss of two members of a **microRNA** (miRNA) family of genes found telomeric to *RB1*. In particular, *miR-15a* and *miR-16-1* are deleted, which downregulates the anti-apoptotic protein BCL-2. Loss of miR-15a and miR-16-1 allows overexpression of BCL-2, protecting the cell from apoptosis.

microRNAs
Small fragments of RNA, 20–23 nucleotides in length, which bind in a complementary fashion to mRNA to inhibit translation.

Trisomy 12 may be associated with either a good or poor prognosis, whereas 11q23 abnormalities (loss of *ATM*; Figure 10.22) and 17p deletion (loss of *TP53*) are associated with a poor prognosis. Deletion of either of these genes and mutation of the remaining allele reduces the patient's response to DNA-damaging drugs often used in the treatment of CLL, so alternative therapies should be used.

CML

Identification of the Philadelphia chromosome der(22) or the translocation t(9;22)(q34;q11) is important to enable the appropriate therapeutic selection for patients with CML. This translocation results in the production of a BCR–ABL1 fusion protein with increased tyrosine kinase activity, causing the signal transduction intermediates interacting with this tyrosine kinase to be 'switched on'. This will lead to an increased transcription and translation of the genes activated by this signal transduction cascade, causing CML. Patients with this translocation have a better prognosis than those with atypical CML in which t(9;22) is absent. Individuals with t(9;22) can be treated with the signal transduction inhibitor, **imatinib mesylate**, which prevents activation of the downstream signal transduction intermediates, reversing the molecular effects of BCR–ABL1. The introduction of signal transduction inhibitors is an exciting move towards targeting malignant diseases according to gene expression and cytogenetic data. We now employ three other signal transduction inhibitors for the treatment of CML, **dasatinib**, **nilotinib**, and **bosutinib**, which are more effective than imatinib at killing CML cells in a highly targeted manner.

Cross references
See Chapter 9—An introduction to haematological malignancies—for more information on karyotypes.

For more information on CML, please see Chapter 11.

imatinib mesylate
A signal transduction inhibitor which can be used to target the ATP-binding pocket conserved on the Abl transcript of BCR–ABL1. Inhibition of BCR–ABL1 inhibits the phosphorylation of downstream effector molecules, reinstating apoptosis and overriding the other dominant effects of BCR–ABL1. Imatinib can also inhibit c-kit and the platelet-derived growth factor receptor (PDGFR).

Non-Hodgkin lymphoma (NHL)

NHL is a broad term used to describe a range of different subtypes of lymphoma that are epidemiologically, clinically, and biologically distinct from Hodgkin lymphoma. Particular cytogenetic abnormalities are associated with certain subtypes and aid in their diagnosis. Cytogenetic analysis can be performed on malignant lymphocytes obtained from peripheral blood, BM, or cells aspirated from a lymph node.

In follicular lymphoma, t(14;18)(q32;q21) is found in more than 90% of cases, or less commonly t(2;18)(p11.2;q21) and t(18;22)(q21;q11.2). Overexpression of BCL-2 is due to the translocation of this gene to the immunoglobulin heavy-chain gene locus (*IGH*), or the κ or λ light-chain loci on chromosomes 2 and 22, respectively, although t(2;18) and t(18;22) are rarely seen. Overexpression of BCL-2 prevents apoptosis in cells where this is expressed.

In 80% of cases of Burkitt lymphoma, t(8;14)(q24;q32) causes the translocation of *MYC* to the *IGH* locus. This leads to overexpression of the transcription factor c-Myc and subsequent B-cell proliferation.

In mantle cell lymphoma, t(11;14) (q13;q32) is found in over 70% of cases and induces overexpression of *BCL-1*/cyclin D1 through the translocation of this gene with the *IGH* locus. Overexpression of cyclin D1 pushes the cell cycle from G1 to S phase, driving proliferation in these cells. Rarely, in cases without a t(11;14), other cyclins associated with different translocations have been found to be overexpressed.

Cross references

See Chapter 12—An introduction to classification systems: lymphoid neoplasms.

Thrombophilia is considered in greater detail in Chapter 15.

In certain situations, the tests outlined so far are insufficient for us to be able to obtain the appropriate diagnostic and prognostic information for a particular disease. In these cases, further information can be obtained using molecular techniques.

10.8 Molecular techniques

The use of molecular techniques has improved diagnosis, the selection of therapeutic agents, and provides important prognostic information for a range of haematological malignancies. The polymerase chain reaction (PCR) and variants, DNA sequencing technologies, and gene expression profiling all provide important information to help treat and monitor disease. Of course, not all molecular techniques are available in every laboratory, but there are referral centres and centres of excellence providing these services, allowing us to utilize their expertise in this field. These techniques are also used for research into a wide variety of pathologies, not just haematological malignancies, although a greater understanding of the pathogenesis of haematological malignancies is due largely to the development and availability of these techniques.

Some of the methods used to provide diagnostically and prognostically useful information to haematologists are outlined below. PCR is a basic technique, the product of which can be used for a wide range of important molecular techniques. PCR can be utilized for the investigation of *TP53* polymorphisms in malignancies to the identification of individuals with Factor V Leiden mutations in the investigation of thrombophilia.

10.8.1 Polymerase chain reaction

This technique is extremely important in amplifying particular segments of DNA and can be used to identify specific mutations within the sequence being studied. Once a region of DNA has been selected, providing the sequence of the ends is known, the process of amplification can begin.

The PCR technique involves the use of a single-stranded DNA template, oligonucleotide primers and deoxynucleotides (to excess), and a DNA polymerase to generate multiple copies of a target sequence. A series of cycles of heating (to 95°C) to separate the complementary DNA strands, and cooling (to 50°C) to anneal the oligonucleotide primers, is used. The addition of deoxynucleotides in the presence of the heat-tolerant DNA polymerase, *Thermus aquaticus* (*Taq*), results in the extension of the oligonucleotide primers and the formation of a complementary strand to the template DNA during the extension phase at 72°C. The cycle can be repeated many times, each time generating successively greater copy numbers of the target sequence for analysis.

If a particular region of DNA is of interest within a tumour, this tissue can be compared to non-tumour cells (wild-type cells) from the same patient. For example, if tumour cells from a patient with B-cell CLL were being examined, DNA could be extracted from neutrophils for comparison. Neutrophil DNA would be considered wild-type, and any differences between the two could be considered to be due to the disease process.

restriction enzymes

These enzymes recognize specific nucleotide sequences and can cut DNA wherever the sequence occurs. This selective cutting allows fragments to be formed which can be indicative of mutations within a specified sequence through the loss or gain of these restriction sites.

wild-type DNA

Denotes germline DNA. Mutations found within tumour cells when compared to wild-type DNA are considered to be acquired.

The next step could involve the addition of a **restriction enzyme**, which recognizes specific nucleotide sequences. If a particular sequence is present, the DNA will be cut at that point, producing two fragments. Using electrophoresis, we can visualize the products of the PCR and the effects of the restriction enzyme. By ascertaining whether the DNA has been cut, compared to a **wild-type region of DNA**, we can infer the presence of particular point mutations within this region.

Although PCR is an excellent technique to facilitate further investigation into the molecular biology of particular malignancies, the following techniques have played a pivotal role in our understanding of the biology of malignant disease. DNA sequencing technology allows us to examine the nucleotide sequence of entire genes. Knowing the sequence of a particular gene can assist in the selection of restriction enzymes for screening tests or can aid our understanding of the nature of gene mutations in both health and disease.

10.8.2 DNA sequencing technology

The use of DNA sequencing technology has dramatically improved our understanding of a range of cancers, including B-cell CLL. The importance of DNA sequencing technologies will be illustrated using the example of somatic hypermutation. Elucidating the DNA sequence for the immunoglobulin heavy-chain hypervariable region (IgHV) determines whether the B cell has undergone somatic hypermutation. Somatic hypermutation is usually defined as less than 98% homology with the germline DNA sequence.

Somatic hypermutation is an entirely normal process within B cells, and follows an encounter with an antigen. These somatic mutations cause an increased strength of antigen binding and enhance the specificity for that antigen by the antibody. If somatic hypermutation has occurred, the B cell has passed through a lymph node-derived germinal centre. If the DNA sequence demonstrates an unmutated IgHV region, this suggests that the B cells have not passed through the germinal centre.

Correlating IgHV mutational status with clinical outcome indicates that patients demonstrating hypermutated IgHV have a significantly better prognosis than those with unmutated IgHV genes. Patients with unmutated IgHV are likely to have a more aggressive disease, require treatment earlier, and have a reduced survival rate.

Cross reference

Germinal centres are introduced in Chapter 8.

Sequencing technology has been employed to identify point mutations in cellular receptors which, when mutated, can lead to a growth advantage of those cells expressing these mutant receptors. In Figure 10.24, we can see a point mutation, G1849T, demonstrated using sequencing technology, in the JAK2 receptor leading to the V617F activating amino acid substitution.

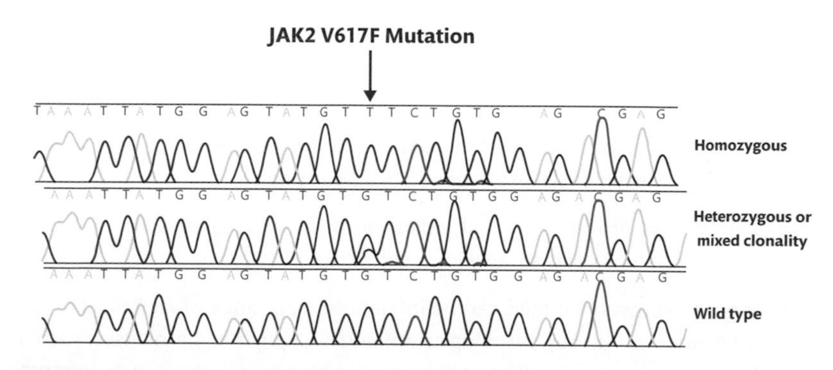

FIGURE 10.24

Gene sequencing of exon 14 in JAK2 granulocyte genomic DNA. The wild-type sequence (bottom row) is 'normal' with a guanine (G) shown below the arrow. The top row shows homozygous mutation where guanine has been replaced by thymine. The middle row shows either heterozygous cells or a mixed clone as shown by the guanine and thymine peaks below the arrow. Republished with permission of American Society of Hematology, from David P Steensma et al. *The JAK2 V617F activating tyrosine kinase mutation is an infrequent event in both 'atypical' myeloproliferative disorders and myelodysplastic syndromes*. *Blood* 2005: 106(4): 1207–09. Doi: https://doi.org/10.1182/blood-2005-03-1183. Copyright 2005.

DNA sequencing is an important, but expensive and technically demanding, technique only available to a relatively small number of laboratories. Surrogate markers (such as Zap-70 in the case of B-CLL, measured using immunophenotyping) can be used as a more accessible marker of IgHV mutational status without the cost implications of sequencing. Gene expression studies were used to identify Zap-70 and its association with unmutated CLL, as shown in Figure 10.25.

The molecular biological methods mentioned so far are available, albeit indirectly, to a large number of laboratories within the UK. The final technique in this chapter—the use of gene expression studies, or the gene microarray—is only available as a research tool at present, although it has provided an abundance of information to help us understand the mechanics of malignancy. In Chapters 11 and 12, we discuss classification systems, and the techniques available to accurately and reproducibly categorize diseases according to a framework of guidelines as outlined by the WHO. These classification guidelines are likely to expand over the next few years to incorporate data from gene expression studies, as this technology becomes more readily available and more pathobiological information is uncovered.

10.8.3 Gene expression studies

The use of gene expression studies, or the DNA microarray, is increasing within the scientific community, although it is still largely a research, rather than a diagnostic tool. Although the principle behind the technique is fairly simple, practically, the technique can be rather difficult to employ. The data generated is indicative of the expression level of a particular gene or set of genes within a disease. By using the term expression level, we are focusing on the amount of mRNA produced when compared to controls. Expression can be normal, low, or high.

The DNA sequence of a number of target genes is obtained and incorporated into an array in the form of probes. These probes are arranged in a carefully recorded order, and inputted onto an analytical database. Separately, patient samples are processed to yield mRNA. This mRNA is then converted to complementary DNA (cDNA) (using reverse transcriptase) to which a fluorochrome is attached. This fluorochrome-labelled cDNA is added to, and incubated with, the DNA probes. Any complementary sequences shared between the DNA probe and the cDNA will result in the binding (annealing) of the complementary sequences. After washing off the excess cDNA, the microarray can be analysed, with the intensity of fluorescence being measured photometrically. The intensity of fluorescence will be proportional to the concentration of labelled cDNA binding to its complementary probe: if the cDNA is present in high quantities (mirroring the high levels of the original mRNA), the fluorochrome will generate an intense signal. By contrast, low levels of cDNA will generate a weak signal.

Cross references

See Chapter 9, section 9.3—Introduction to the cell cycle, and section 9.8—Epigenetics.

Lymphomas are discussed in greater detail in Chapter 12.

Gene expression studies provided the first indication that Zap-70 expression may be linked with the prognosis of patients diagnosed with B-cell CLL.

The results of gene expression studies can be influenced by a range of factors, although this book focuses on some of the key pathological influences. Epigenetic processes, such as the hypermethylation of genes, or the hypoacetylation of histones, will result in gene repression and downregulation, since mRNA will be synthesized either at a reduced rate, or will fail to be synthesized. Reduced or absent mRNA decreases the fluorescence generated from this technique and can be interpreted as underexpression of a gene. The hypermethylation of *CDKN2A* and *CDKN2B* causes the loss of expression of the INK cell cycle regulators p16 and p15, respectively. The loss of p16- and p15-mediated cell cycle control ensures dysregulation of the cyclin D-CDK4/6 complex and the hyperphosphorylation of pRb. *CDKN2A* and *CDKN2B* silencing through methylation or deletion occurs in approximately 60% of cases of non-Hodgkin lymphoma, and results in dysregulation of the cell cycle.

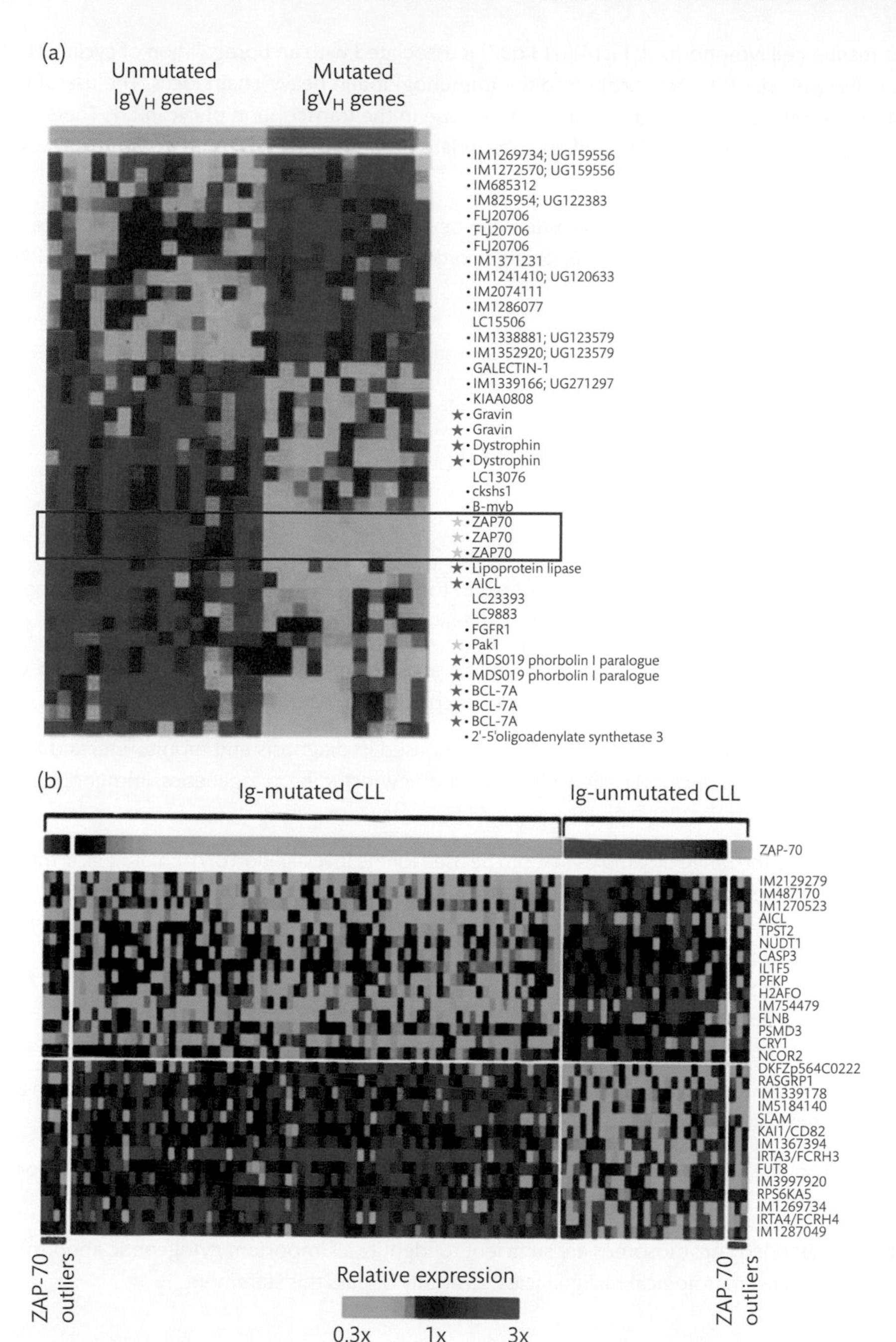

FIGURE 10.25

Classifier lists distinguishing B-cell chronic lymphocytic leukaemia (B-CLL) with mutated and unmutated Ig heavy chain on the basis of gene expression. Note that ZAP-70 (in box) is uniformly underexpressed in CLL harbouring IgV$_H$ mutated genes and is overexpressed in cases with unmutated IgV$_H$ genes. Reproduced with permission from Hubank M. *Gene expression profiling and its application in studies of haematological malignancy*. *British Journal of Haematology* 2004: 124: 577–94.

In mantle cell lymphoma, t(11;14)(q13;q32) is associated with an upregulation of cyclin D1 as a consequence of its translocation to the immunoglobulin heavy-chain locus. The use of the strong promoter in this region causes an increase in the transcription of cyclin D1. Therefore, using gene expression studies and an appropriate probe for cyclin D1 mRNA, overexpression would be demonstrated.

The results from gene expression studies can be consolidated by many other molecular techniques, including PCR and electrophoresis, in order to investigate the cause of abnormal mRNA expression.

Chapter summary

In this chapter, we have:

- examined the laboratory work involved in the investigation, diagnosis, and monitoring of patients with suspected haematological malignancies;
- outlined commonly associated symptoms and discussed their causes in relation to the underlying molecular and biochemical mechanisms of haematological malignancies;
- discussed a variety of laboratory techniques used in diagnosis and monitoring, and have focused on those closely allied with haematology, including cytogenetics, immunophenotyping, and molecular biology;
- discussed the range of samples that can be used for the investigation of haematological malignancies, including peripheral blood, BM aspirates, trephines, and lymph node biopsies.

Discussion questions

10.1 Compare and contrast the value of cytochemical and immunophenotypic methods in the investigation and diagnosis of haematological malignancies.

10.2 Compare and contrast the BM aspirate and BM trephine. Explain which sample is appropriate for the assessment of lymphocyte infiltration of haemopoietic tissue.

10.3 'G-banded chromosomes are sufficient to identify all important cytogenetic abnormalities in haematological malignancies.' Critically discuss this statement.

Further reading

- **Bain BJ. *The bone marrow aspirate in healthy subjects*. *British Journal of Haematology* 1996: 94: 206–9.**
- **Bain BJ, Clark DM, and Wilkins BS. *Bone Marrow Pathology,* 4th edn. Wiley-Blackwell, Oxford, 2009.**
- **Bain BJ, Bates I, Laffan MA, and Lewis SM. *Dacie and Lewis Practical Haematology*, 11th edn. Churchill Livingstone, London, 2011.**

- Goodwin S, McPherson JD, and McCombie WR. ***Coming of age: Ten years of next generation sequencing technologies.*** *Nature Reviews Genetics* 2016: **17**: 333–51.
- Johansson U, Bloxham D, Couzens S, Jesson J, Morilla R, Erber W et al. ***Guidelines on the use of multicolour flow cytometry in the diagnosis of haematological neoplasms.*** *British Journal of Haematology* 2014: **165**: 455–88.
- Köhler G and Milstein C. ***Continuous cultures of fused cells secreting antibody of predefined specificity.*** *Nature* 1975: **256**: 495–7.
- Schaffer LG, McGowan-Jordan J, and Schmid M (eds). ***ISCN: An international system for Human Cytogenetic Nomenclature.*** S Karger, Basel, 2013.
- Shendure J and Ji H. ***Next-generation DNA sequencing.*** *Nature Biotechnology* 2008: **26(10)**: 1135–45.
- Van Dongen JJM, Lhermitte L, Bottcher S, Almeida J, Van der Velden VJH, Flores-Montero J et al. ***Euroflow antibody panels for standardized n-dimensional flow cytometric immunophenotyping of normal, reactive and malignant leukocytes.*** *Leukaemia* 2012: **26**: 1908–75.
- Weiss MM, Hermsen MAJA, Meijer GA, van Grieken NCT, Baak JPA, Kuipers EJ, and van Diest PJ. ***Comparative genomic hybridisation.*** *Molecular Pathology* 1999: **52**: 243–51.

Answers to self-check questions and case study questions are provided as part of this book's online resources.

Visit www.oup.com/he/moore-fbms3e.

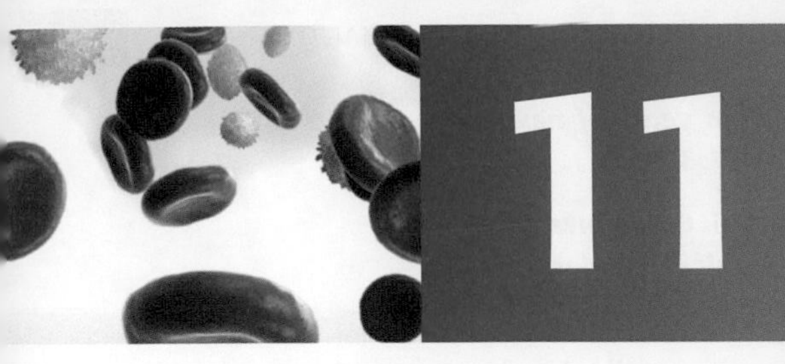

11

An introduction to classification systems: myeloid neoplasms

Gavin Knight

The discovery of leukaemia is generally attributable to Hughes Bennett and Rudolph Virchow independently in 1845. The first description of leukaemia was actually published in 1844 by the French physician Alfred Donné in his text *Cours de microscopie complémentaire des études médicales, Anatomie microscopique et physiologie des fluides de l'Economie*. In more than a century-and-a-half since this initial discovery, we now recognize a wide range of different types of haematological malignancy, not just leukaemia.

Chapter 9 examined the basic theories of haemato-oncology and established that leukaemia can be subdivided into acute and chronic forms. In this chapter, we examine a range of haematological neoplasms, including leukaemias, and explain how we can subdivide these into myeloid or lymphoid lineages and then further subdivide them according to various cellular, molecular, genetic, and clinical features. This chapter focuses on myeloid-related neoplasms, whereas Chapter 12 examines those of lymphoid origin.

Learning objectives

After studying this chapter, you should confidently be able to:

- describe the importance of classification systems and understand the basis of their development;
- critically discuss the types of haematological neoplasm recognized by the World Health Organization (WHO) system;
- describe the common translocations identified in haematological neoplasms and, where appropriate, their role in classification;

- explain the molecular mechanisms underpinning the development and behaviour of selected haematological neoplasms;
- discuss the classification of myelodysplastic syndromes (MDS);
- compare and contrast the disease entities classified as myeloproliferative neoplasms (MPN) and recognize the importance of MDS/MPN overlap conditions;
- using appropriate examples, justify the reasons why the techniques outlined in Chapter 10 are used for the diagnosis of myeloid neoplasms.

11.1 Why is classification important?

It is important to recognize that each neoplastic disease varies with its origin, clinical behaviour, morphology, cytogenetic profile, and molecular make-up. In addition, the way a disease responds to therapeutic agents varies between different types of cancer and, based on our knowledge of the molecular basis of cancer, some therapies are more appropriate for treating a particular disease when compared to others. The best way of ensuring that a patient is prescribed the appropriate medication is to obtain a correct diagnosis and to ensure that, as scientists and clinicians, we have a universal language to describe the disease. The terminology assigned to a particular disease in the UK has to be standardized to have the same meaning throughout the world. This can be achieved via a clear framework allowing us to pigeonhole specific diseases.

From a research perspective, it is important that scientists are able to confidently both investigate and report internationally recognized diseases. Having a standardized diagnostic framework can ensure that research can be reported and interpreted globally. An appropriate classification system is important because it allows the subdivision of patients with different prognoses, ultimately helping to direct future research and therapeutic options.

Ultimately, classification systems need to be dynamic. We need to be able to change different aspects of these systems as technology and scientific understanding develops. For example, prior to the introduction of the World Health Organization (WHO) approach to classification, the French-American-British (FAB) system primarily utilized blood cell morphology for classification purposes. Our classification systems evolved as our understanding improved. At present, the WHO system does not recognize results from gene-expression analysis, although in the future this may be an important consideration for classification to truly delineate different disease states and to provide a wholly personalized approach to treatment and monitoring.

Let's now consider the main classification systems frequently encountered in your studies and in the clinical haematology laboratory.

11.2 The main classification systems

The predecessor to the WHO classification system, the FAB system, provided the cornerstone of classification of leukaemias and myelodysplastic syndromes from 1976 to 2001. FAB used cellular morphology and the arbitrary value of 30% blasts to differentiate between acute leukaemias and other myeloproliferative conditions. Patients were diagnosed with acute leukaemia when their blasts exceeded 30% of nucleated cells—the **blast threshold**. In these patients, more aggressive therapy was indicated than those with <30% blasts. FAB is now considerably outdated and was replaced by the WHO classification system at the turn of the century. Focus is now directed to the revised fourth edition of the WHO classification system, which was most recently updated in 2016.

blast threshold
The minimum percentage of blasts required to diagnose a leukaemia as acute. WHO currently specifies 20% blasts to make this diagnosis.

11.2.1 The World Health Organization classification system

The WHO system for the classification of haematological malignancies revolutionized the way in which these malignancies are reported (Box 11.1). Evidence-based practice is the cornerstone of the WHO system, which initially uses morphological assessment for classification purposes, but then utilizes immunophenotype, cytogenetic, and molecular genetic data to further subgroup patients into their respective disease groups. Using the WHO system, the biology behind the disease (pathobiology) is also considered, with a parallel between the malignant cell and normal cell counterpart being drawn where appropriate. In addition, therapy-related malignancies (i.e. those leukaemias secondary to previous chemo- or radiotherapeutic procedures), and those considered to be a progression of myelodysplasia, are classified as particular disease entities. This clarification is important, as these secondary conditions often have a less favourable prognosis when compared with primary (*de novo*) malignancies.

Nine main *myeloid subgroups* have been established by the WHO system and include:

1. acute myeloid leukaemia and related precursor neoplasms;
2. myeloproliferative neoplasms;
3. myeloid and lymphoid neoplasms with eosinophilia and gene rearrangement;
4. myelodysplastic/myeloproliferative neoplasms;
5. myelodysplastic syndromes;
6. mastocytosis;
7. acute leukaemias of ambiguous lineage;
8. blastic plasmacytoid dendritic cell neoplasm;
9. myeloid neoplasms with germline predisposition (these will not be discussed in this chapter).

BOX 11.1 *History of the WHO classification system*

The WHO classification system of haematological malignancies represents the collaboration between the European Association for Haematopathology and the Society for Haematopathology. Prior to publication, a Clinical Advisory Committee was developed with over 40 international panel members to ensure the system was valid. The first WHO version, published in 2001, based on the revised European and American classification of lymphoid neoplasms (REAL), now applies to all haematological malignancies. In 2014, a Clinical Advisory Committee convened to revise the fourth edition of the WHO classification and, considering the substantial evidence base available since the 2008 revision, made a number of modifications to refine the classification to improve clinical utility. The revised fourth edition was published in 2016.

It has also established six lymphoid subgroups:

1. precursor lymphoid neoplasms;
2. mature B-cell neoplasms;
3. mature T-cell and NK-cell neoplasms;
4. Hodgkin lymphoma;
5. immunodeficiency-associated lymphoproliferative disorders;
6. histiocytic and dendritic cell neoplasms.

In 2016, the update to the fourth WHO version was published, which has refined a number of the disease entities through the recombination, subdivision, or reclassification of previous disease entities to more accurately predict disease behaviour and response to treatment. The classification includes the introduction of a number of provisional disease entities which require appropriate validation before being fully bona fide discrete disease entities.

SELF-CHECK 11.1

What are the main myeloid diseases categorized by the WHO system?

We will now examine each of the myeloid groupings in turn and consider the main diseases that constitute them.

11.2.2 Acute myeloid leukaemia and related neoplasms

The WHO group recognizes six distinctly different subcategories of acute myeloid leukaemia (AML) and related precursor neoplasms:

- AML with recurrent genetic abnormalities;
- AML with myelodysplasia-related changes;
- therapy-related myeloid neoplasms;
- AML not otherwise specified (NOS);
- myeloid sarcoma;
- myeloid proliferations related to Down syndrome.

The WHO classifies AML based on accumulation of 20% myeloblasts or more in the peripheral blood or bone marrow—the blast threshold—*or* the presence of the following cytogenetic or molecular genetic abnormalities irrespective of the blast percentage:

- t(8;21)(q22;q22.1);
- inv(16)(p13.1;q22);
- t(16;16)(p13.1;q22);
- t(15;17)(q22;q12)/*PML-RARA*.

In addition, the following recurrent cytogenetic and molecular genetic features are also considered important in the WHO classification, although each requires blasts of 20% or higher to be classified as acute leukaemia:

- t(9;11)(p21.3;q23.3);
- t(6;9)(p23;q34.1);
- inv(3)(q21.3;q26.2);
- t(3;3)(q21.3;q26.2);
- t(1;22)(p13.3;q13.3);
- AML with mutated *NPM1*;
- AML with biallelic mutation of *CEBPA*;
- two provisional entities, including AML with *BCR–ABL1* and AML with mutated *RUNX1*.

More details regarding NPM1 and CEBPA can be found in Box 11.2.

Individuals harbouring these mutations usually have very specific morphological features which are highly suggestive of the mutations listed above, although a definitive diagnosis should not be made without access to the appropriate cytogenetic and/or molecular genetic data.

In the cases of both *NPM1* and *CEBPA*, there is evidence that these lesions can work independently of any other to cause leukaemia.

A diagnosis of *acute myeloid leukaemia with recurrent genetic abnormalities* is based on the presence of one of eleven recurrent genetic abnormalities, and two provisional entities, identified as providing a predictable clinical outcome for these patients. If features of multilineage dysplasia are present in the blood or bone marrow findings of a patient with AML, the disease will be classified as *AML with myelodysplasia-related changes*. AML with myelodysplasia-related

BOX 11.2 NPM1 *and* CEBPA *in leukaemia*

***NPM1* mutations are associated with approximately 30% of cases of acute myeloid leukaemia and are generally found independently of the other recurrent genetic abnormalities featured in this section. *NPM1* encodes the nucleolar protein nucleophosmin. Mutation of *NPM1* leads to relocalization of nucleophosmin from its usual sites of activity, including the nucleolus, nucleus, and cytoplasm. In these cellular sites, nucleophosmin is involved in the formation of ribosomes and their movement, stabilization of tumour suppressor proteins, and centromere stabilization. Relocation and restriction of nucleophosmin to the cytoplasm leads to its association with novel binding partners, a gain of function, and contribution to oncogenesis.**

***NPM1*-associated leukaemias are called '*AML with mutated NPM1*', or '*NPMc*$^+$ *AML*', where c$^+$ indicates nucleophosmin expression is restricted to the cell's cytoplasm.**

Neutrophil development is controlled by the product of the *CEBPA* gene—the CCAAT/enhancer-binding protein. This protein usually functions as a homodimer, although it can also form heterodimers with alternative family members. Research has shown that patients with AML harbouring *CEBPA* mutation to both alleles—abbreviated to *CEBPA*^{DM} (where DM denotes double mutation)—have superior survival than those with single mutation. For this reason, *CEBPA*^{DM} is now incorporated into the formal classification.

changes predominantly recognizes secondary myeloid leukaemias following a myelodysplastic course or cases showing a myelodysplastic syndrome (MDS)-related karyotype.

Also recognized and classified are acute myeloid leukaemias, MDS, and myeloproliferative disorders secondary to previous exposure to chemotherapeutic agents or ionizing radiation in a clinical setting. *Therapy-related myeloid neoplasms* tend to have characteristic patterns of presentation and prognosis.

Acute myeloid leukaemia not otherwise specified contains acute myeloid leukaemias that fail to show any features consistent with a diagnosis of the other subtypes. Future amendments to this category are likely to occur as the nature of evidence-based practice furthers our level of understanding of these diseases at the genetic or molecular levels.

Myeloid proliferations related to Down syndrome is another important category that relates particularly to the processes of transient abnormal myelopoiesis and Down syndrome-related acute megakaryoblastic leukaemia. Both these conditions are associated with mutations in the *GATA1* transcription factor, and should therefore be treated separately to non-Down syndrome-related myeloid diseases.

Reading below, you will see that the main categories within the AML section of the WHO guidelines are outlined. Where appropriate, the molecular basis of the disease is introduced.

11.2.3 AML with recurrent genetic abnormalities

This particular subgroup includes four main genetic abnormalities identified through clinical research as having either a good or poor prognosis. The classification of AML with recurrent genetic abnormalities should only be applied to *de novo* malignancies, even if the mutations described below are encountered in cases of secondary leukaemia. The use of standard cytogenetic tests, as outlined in Chapter 10, is usually sufficient to identify these abnormalities. However, in the case of AML with *PML-RARA*, this may be caused by cryptic translocation—i.e. a translocation that is present, although it cannot be visualized by karyotyping—or through a complex rearrangement. Regardless, the presence of *PML-RARA* signals a good prognosis and is the defining feature of this particular disease entity.

The identification of recurrent cytogenetic abnormalities should always be considered in the first instance in order to successfully classify AML.

AML with t(8;21)(q22;q22)—RUNX1-RUNX1T1

The translocation t(8;21)(q22;q22) involves the fusion of *RUNX1* (also known as CBFα), found on the long arm of chromosome 21 with the functional co-repressor *RUNX1T1*, on chromosome 8.

Figure 11.1 depicts chromosomes 16 and 21, and the location of the genes *CBFB* and *RUNX1*, respectively. Working through Figure 11.1 you can see that in a normal situation, the protein product RUNX1, acts as a molecular bridge allowing the binding of core binding factor-β (CBFβ) to the CBF binding site. This, in the presence of other transcription factors which are localized to the CBFα/CBFβ complex, allows for the active transcription of target genes.

In Figure 11.2, the N-terminal region of CBFα is fused with the majority of the RUNX1T1 protein forming a chimeric (fusion) protein, preventing the formation of a stable DNA-binding complex involving CBFβ. Instead, functional repressors become involved with RUNX1-RUNX1T1, inhibiting transcription and translation of target genes and causing a failure of cell maturation and differentiation.

Cross reference
Knudson's hypothesis is outlined in Chapter 9.

The RUNX1-RUNX1T1 chimeric protein is insufficient to be independently leukaemogenic, but requires a secondary event to promote leukaemogenesis—as predicted by Knudson's hypothesis.

AML with inv(16) (p13.1;q22) or t(16;16)(p13.1;q22): CBFB-MYH11

The pericentric inversion inv(16) and translocation, t(16;16) are both involved with the disruption of the *CBFB* gene, and the consequent function of CBFβ protein product is disrupted. If you look at Figure 11.3, you will see that, through this inversion or translocation process, *CBFB* is fused with the gene encoding the smooth muscle myosin heavy chain (SMMHC)-*MYH11*. The chimeric protein product, CBFβ-SMMHC, recruits a range of transcriptional repressors and histone deacetylases to the CBF binding site. The consequence of recruiting these repressors is that the transcription and translation of target genes is prevented. A number of studies have demonstrated that inv(16) or t(16;16) alone are insufficient to be leukaemogenic, although they can contribute to a malignant phenotype in the presence of a second mutation. (See also Box 11.3.)

Acute promyelocytic leukaemia with PML-RARA

Approximately 98% of patients with acute promyelocytic leukaemia (APL) with *PML-RARA*, possess t(15;17)(q22;q12). The t(15;17)(q22;q12) is associated with the translocation of the promyelocytic leukaemia gene (*PML*), so-called because it was first identified in patients with this disease, with the retinoic acid receptor α-(*RARA*) gene. *PML* encodes a multimeric protein that is necessary for assembling macromolecular PML nuclear bodies. These nuclear bodies, ranging in number between 1 and 30 per nucleus, have diverse functions involving self-renewal, DNA repair, and apoptosis, to name but a few. Conversely, the retinoic acid receptor α is part of the nuclear receptor superfamily whose function is ligand-dependent transcription factors. If you refer to Figure 11.4 you will see that in a normal situation, retinoic acid, when bound to its receptor (RARα/RXR), induces the dissociation of transcriptional repressors and the binding of transcriptional activators. In Figure 11.5 we can see that t(15;17)(q22;q12) produces the chimeric protein PML-RARα. PML-RARα interacts with a variety of co-repressors, including histone deacetylase (HDAC), inhibiting the retinoic acid-responsive genes necessary for granulocytic maturation beyond the promyelocyte stage, as shown in Figure 11.6. It has also been suggested

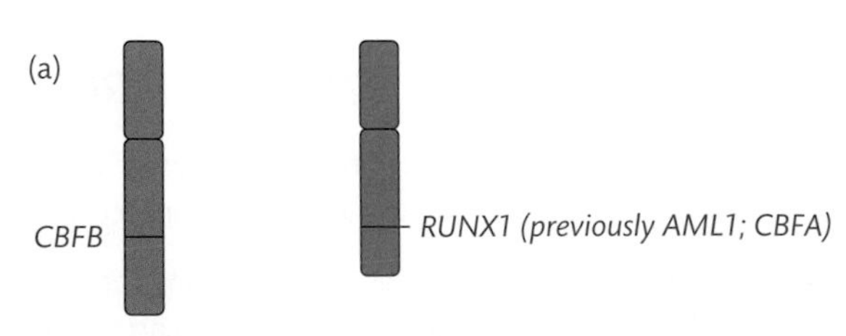

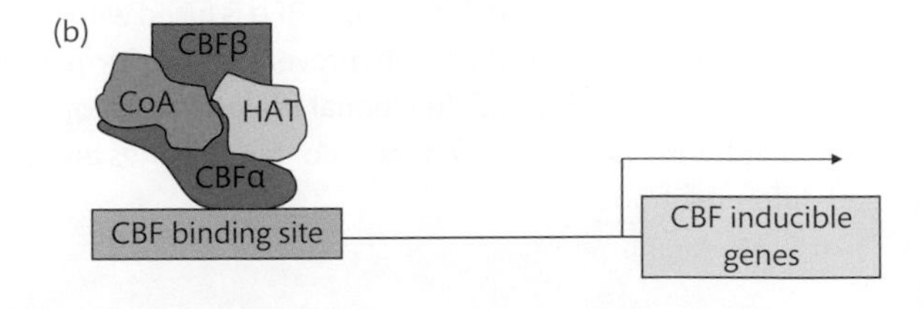

FIGURE 11.1
(a) The locus for the CBFα gene (*RUNX1*) is situated on chromosome 21q22 and CBFβ (*CBFB*) is located on chromosome 16q22. (b) The binding of CBFα to the CBF binding site permits the subsequent binding of CBFβ. This complex enables histone acetyltransferase (HAT) and co-activators (CoA) to bind, allowing histone modification and transcription and translation of CBF inducible genes.

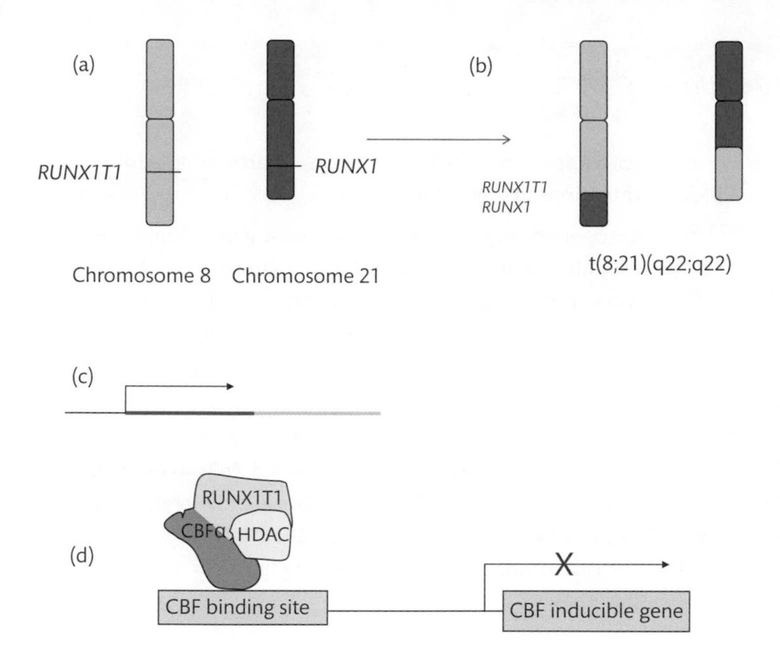

FIGURE 11.2
Repression of the CBF complex through the fusion of *RUNX1* (encoding RUNX1/CBFα) with the transcriptional repressor RUNX1T1. Translocation of *RUNX1* found on chromosome 21 (21q22) and *RUNX1T1* on chromosome 8 (8q22) (a and b) produce a fusion gene (c), which is transcribed and translated into the chimeric protein CBFα-RUNX1T1. CBFα-RUNX1T1 binds to the CBF binding site (d), but instead of enabling transcription and translation of target genes through the binding of CBFβ and transcriptional activators, transcriptional repressors are recruited, inhibiting maturation and differentiation of cells containing this translocation.

that the methylating enzymes Dnmt1 and Dnmt3a are recruited by PML-RARα, causing transcriptional repression through promoter methylation.

The range of retinoic acid responsive genes involved in the pathogenesis of APL has yet to be elucidated, although the transcription factor PU.1, necessary for granulocytic differentiation, can be indirectly inhibited by PML-RARα. Inhibition of PU.1 blocks differentiation, leading to an accumulation of myeloid precursors. Treatment with all-*trans* retinoic acid (ATRA) and

Cross reference

Epigenetic modification and methylation is explored in Chapter 9.

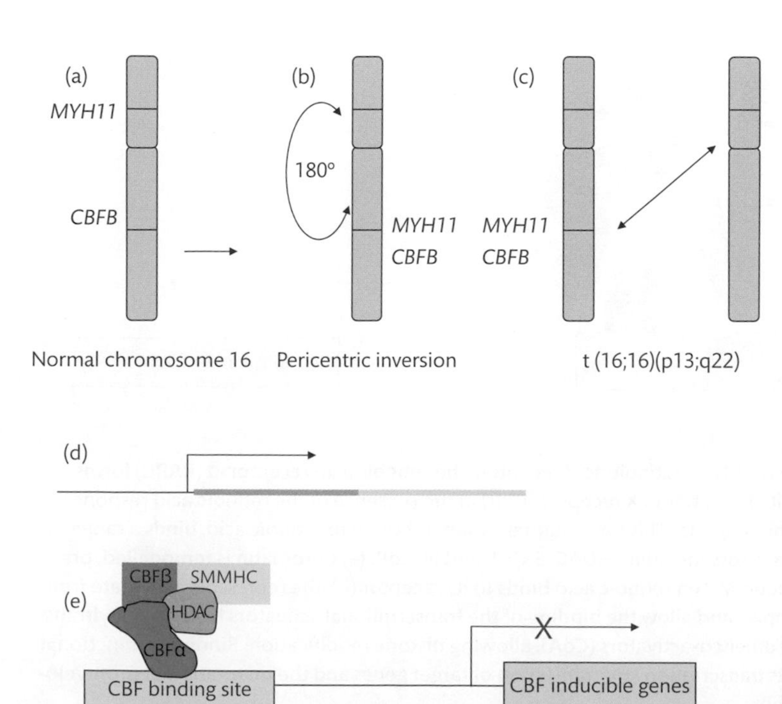

FIGURE 11.3
Disruption of the CBF pathway. The normal chromosome 16 (a) contains the gene encoding CBFβ (16q22) and the gene encoding the smooth muscle myosin heavy-chain gene (*MHY11*) (16p13.1). Pericentric inversion (b), rotation of chromosomal material involving the centromere, or translocation (c), results in *CBFB* and *MYH11* becoming fused. (d) Transcription and translation of the fusion protein follows. (e) *CBF*β-SMMHC is able to recruit potent transcriptional co-repressors (HDAC and other co-repressors) to the CBF complex, inhibiting maturation and differentiation of cells containing this chimeric protein.

BOX 11.3 *CBF leukaemias*

Acute myeloid leukaemias associated with disruption of CBFα (RUNX1) via the *RUNX1-RUNX1T1* fusion, or CBFβ, via the *CBFB-MYH11* fusion are called CBF leukaemias or CBF-AML. In adults with AML over the age of 60 years, approximately 7% of all cases of AML will be associated with CBF disruption. In younger patients, this figure is higher, with approximately 7% of cases of AML in this group associated with *RUNX1* alone, and between 5 and 8% of cases associated with *CBFB* abnormalities.

It is well documented that t(8;21), inv(16), and t(16;16) are insufficient to induce leukaemia alone; a number of co-operating mutations have been identified which support the development of a leukaemic phenotype. Mutations in *FLT3*, *KIT*, *NRAS*, and *KRAS* are associated with CBF-AML, with approximately 90% of all cases harbouring additional mutations in these genes.

Cytogenetically, additional chromosomal abnormalities may also be identified, including deletion of 9q and trisomy 8 accompanying t(8;21) while trisomies of chromosomes 8, 21, and 22 all commonly accompany inv(16).

In terms of treatment, CBF leukaemia is associated with a good prognosis. Patients respond very well to anthracyclins and cytarabine-based induction chemotherapy, with long-term remissions achieved following repetitive cycles of high-dose cytarabine. However, balancing the benefits of high-dose chemotherapy with the serious complications of toxicity is difficult in these patients.

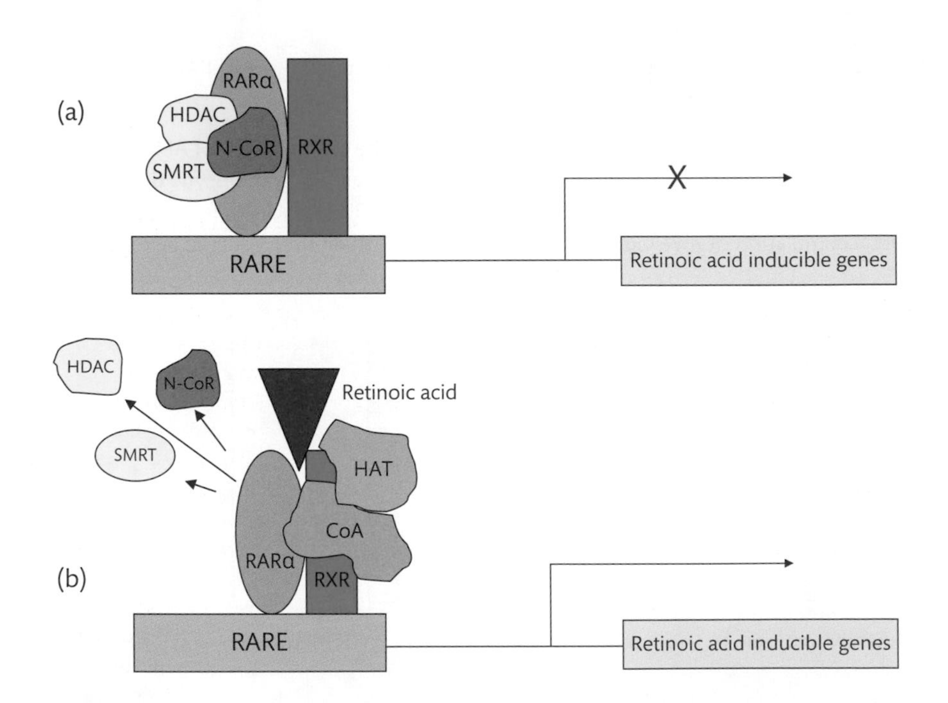

FIGURE 11.4
The action of the wild-type retinoic acid receptor. The retinoic acid receptor-α (RARα) forms a heterodimer with the retinoid X receptor (RXR) in the presence of the retinoic acid response element (RARE) binding site. This heterodimer, when unbound to retinoic acid, binds a range of functional co-repressors including HDAC, SMRT, and N-CoR. (a) Chromatin is remodelled, preventing transcription. When retinoic acid binds to its receptor (b), the repressors dissociate from the receptor complex and allow the binding of the transcriptional activators histone acetyltransferase (HAT) and other co-activators (CoA), allowing histone modification. Binding of functional activators permits transcription and translation of target genes and the maturation of promyelocytes to neutrophils.

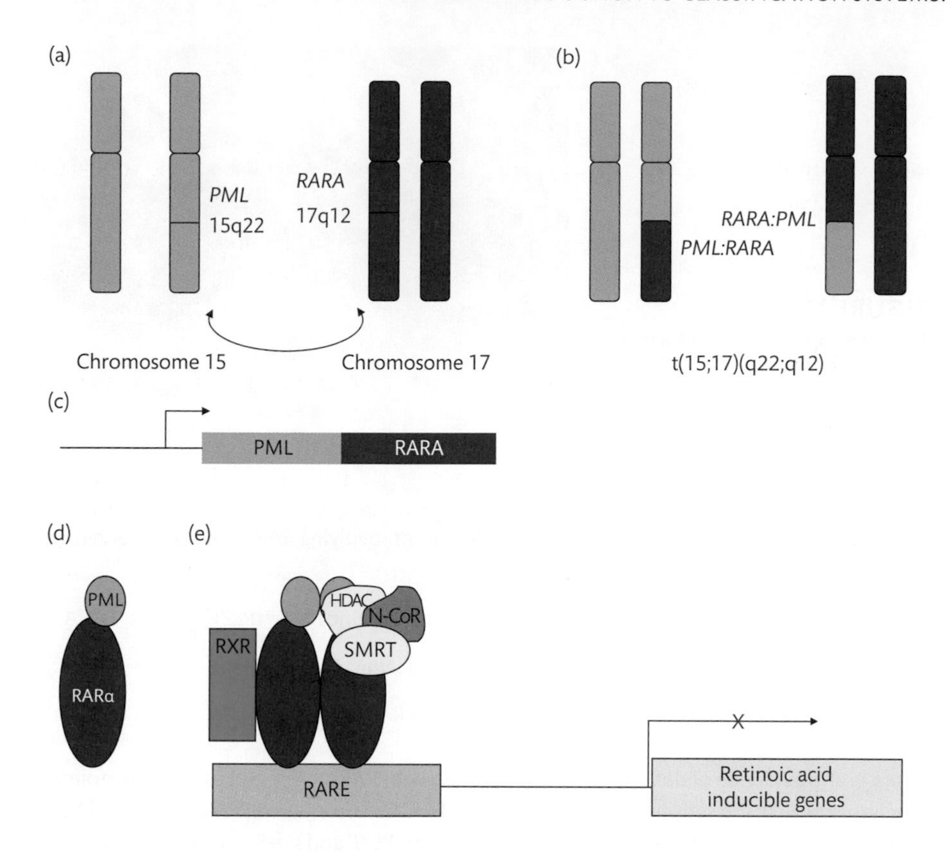

FIGURE 11.5

The inhibitory complex generated by t(15;17) in acute promyelocytic leukaemia (APL). A balanced and reciprocal translocation occurs between the long arms of chromosomes 15 (blue) and 17 (purple). This translocation (a) involves the genes *PML* (on chromosome 15), encoding the PML protein, and *RARA* (on chromosome 17), encoding the RARα protein, being combined. This combination produces the fusion gene *PML:RARA* (b). The translocation process is described as t(15;17)(q22;q12). Wild-type PML nuclear bodies are found in most tissues and belong to the nuclear matrix which is involved in the regulation of DNA replication, transcription, and epigenetic silencing. PML induces post-translational modification of target proteins, modifying their binding partners. Following the fusion of these genes, protein functions are dysregulated and displaced. Transcription of this fusion gene (c) produced a fusion protein (d) incorporating both protein products PML and RARα into a single protein structure. PML-RARα retains important functional domains (e) including the retinoic acid response element (RARE) binding domain, the RXR (retinoid X receptor) binding domain in RARα, and the PML homodimerization domain. PML binds transcriptional repressors, HDAC, N-CoR and SMRT, with high affinity, and these become localized to the RARE via PML-RARα binding. Chromatin modification via the co-repressors prevents transcription and translation of target genes, in this case resulting in a maturation arrest at the promyelocyte stage of development.

combinations of chemotherapy, including ATRA and arsenic trioxide (As_2O_3), restores granulopoiesis by removing the block of differentiation and maturation, and allows the production of mature neutrophils to continue. Indeed, arsenic trioxide is particularly effective at degrading PML and PML-RARα through the process of **sumoylation**. Arsenic binding to the fusion protein promotes polymerization of PML, enhanced binding of SUMO, and subsequent modification and degradation.

sumoylation

This is a post-translational modification that results from the binding of SUMO. SUMO is a *S*mall *U*biquitin like *MO*difier that can attach to lysines on target proteins to alter their function or lead to their degradation.

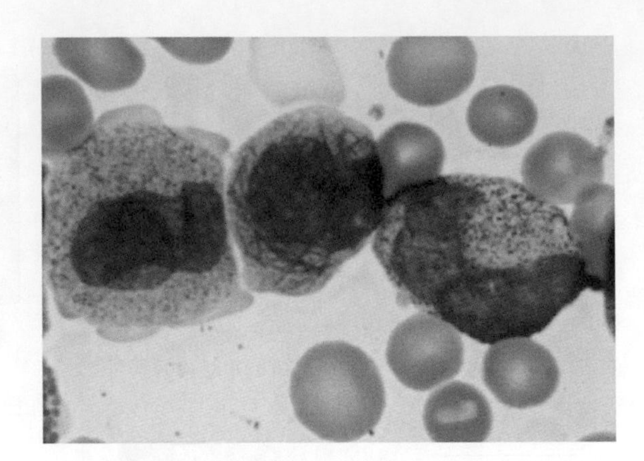

FIGURE 11.6
Typical malignant promyelocytes associated with acute promyelocytic leukaemia. © PR. H. Piguet/ CNRI/Science Photo Library.

In addition to t(15;17), a number of other translocations involving *RARA* have also been identified. The most common of these include:

- t(11;17)(q23;q21), fusing the promyelocytic leukaemia zinc finger (*PLZF*) gene to *RARA*;
- t(11;17)(q13;q21), fusing nuclear matrix associated (*NuMA*) to *RARA*;
- t(5;17), in which nucleophosmin (*NPM)* is fused to *RARA*.

There is also evidence of der(17), which fuses stat5b with RARα. Differences in these molecular mechanisms alter the sensitivity of these cells to ATRA. Cells containing *PML*, *NPM*, and *NuMA* fusions are sensitive to ATRA therapy, whilst those with *PLZF* and *stat5b* fusions are insensitive.

If you consider all cases of AML, approximately 30% will harbour t(8;21)(q22;q22), inv(16) (p13;q22), t(16;16)(p13;q22), or t(15;17)(q22;q12).

AML with t(9;11)(p21.3;q23.3)–MLLT3-KMT2A

The cytogenetic abnormalities t(8;21), inv(16), and t(16;16) are very specific. Here, two genes are involved, and fusion of the two produces a chimeric protein with an altered structure–function relationship.

The original WHO classification for AML with recurrent genetic abnormalities included any primary AML with 11q23 abnormalities. 11q23 abnormalities are a much more heterogeneous group of mutations than the others within this classification, with the only constant being that 11q23 is involved. There are over 50 different partners which have been documented to combine with the lysine (K) methyltransferase gene (*KMT2A)*, although the most common found in AML is MLLT3 in t(9;11), present in approximately one-third of cases. (See also Box 11.4.)

AML with t(6;9)(6p23;9q34)–DEK-NUP214

The translocation resulting in the formation of the *DEK-NUP214* (previously known as *DEK-CAN*) fusion gene is associated with a poor prognosis. A rare condition, accounting for between 0.5 and 4% of all cases of AML, t(6;9) is associated with an aggressive malignancy refractory to common chemotherapeutic agents, with a survival time of less than 12 months.

AML with t(6;9) should only be classified as such following comprehensive cytogenetic analysis—even though the genotype is considered predictable by the discrete morphological

BOX 11.4 *Histone-lysine N-methyltransferase 2A protein activity*

The histone-lysine N-methyltransferase 2A protein produced from *KMT2A* was previously identified in mixed-lineage leukaemias and was called MLL. Histone-lysine N-methyltransferase 2A confers histone methyltransferase activity leading to the methylation of lysine (K) 4 in histone H3 (H3K4me). This enzyme undergoes proteolytic cleavage, and is recruited to a multi-molecular complex to enhance its rate of activity and, in addition to the transfer of methyl groups to H3K4, also acts as an acetyltransferase, particularly to lysine 16 in histone 4 (H4K16ac). A range of other functions associated with histone-lysine N-methyltransferase 2A and chromatin remodelling lead this complex to be an effective epigenetic modifier.

characteristics usually demonstrated. Peripheral cell counts demonstrate pancytopenia, whilst bone marrow investigations reveal a hypercellular marrow with an absolute basophilia and signs of unilineage or multilineage dysplasia.

Internal tandem duplications (ITD) of *fms*-related tyrosine kinase (FLT3-ITD) are found in approximately 70% of patients harbouring t(6;9) and 20–25% of all cases of AML. The combination of t(6;9) and FLT3-ITD is associated with a higher white cell count, blast percentage, and worse prognosis. Although initially believed to be passenger mutations, i.e. those that occur during the course of disease, evidence now suggests FLT3-ITD are important in driving malignancies and are therefore potential drug targets for these conditions. An example of the ITD associated with FLT3 receptor structure is shown in Figure 11.7.

internal tandem duplications (ITD)
These arise from the duplication of sequences from within a gene. In relation to FLT3, ITD of the juxtamembrane region of the gene results in constitutive activation of FLT3.

AML with inv(3)(q21.3;q26.2) or t(3;3)(q21.3;q26.2); GATA2, MECOM

The paracentric inversion inv(3)(q21.3;26.2) or t(3;3)(q21.3;q26.2) results in the relocalization on the *GATA2* enhancer to *MECOM*. The net effect of this is a reduction in the transcription and translation of *GATA2* and an increase in *MECOM* activity. There are six members of the GATA family of transcription factors, which is divided into two discrete groups. GATA 1–3 are involved in the haemopoietic lineages, amongst others, whereas GATA 4–6 are found in cells of the heart, lung, liver, and gut. The expression of GATA 2 is notable in early haemopoietic progenitors. *MECOM* encodes a nuclear DNA-binding protein, which, when dysregulated through enhanced transcription and translation, results in its overexpression and an increased proliferative potential in blast cells harbouring this mutation.

Morphologically, this subgroup of AML demonstrates significant platelet abnormalities including, in some cases, thrombocythaemia and giant hypogranular platelets, with a background of anaemia. Neutrophil dysplasia including hypogranulation and the Pelger–Huet anomaly are also common features.

Patients with this chromosomal abnormality have a poor prognosis.

AML (megakaryoblastic) with t(1;22)(p13.3;q13.1)–RBM15-MKL1

An uncommon cytogenetic abnormality, t(1;22), is associated with infant and childhood acute megakaryoblastic leukaemia, accounting for approximately 10% of all childhood AMLs.

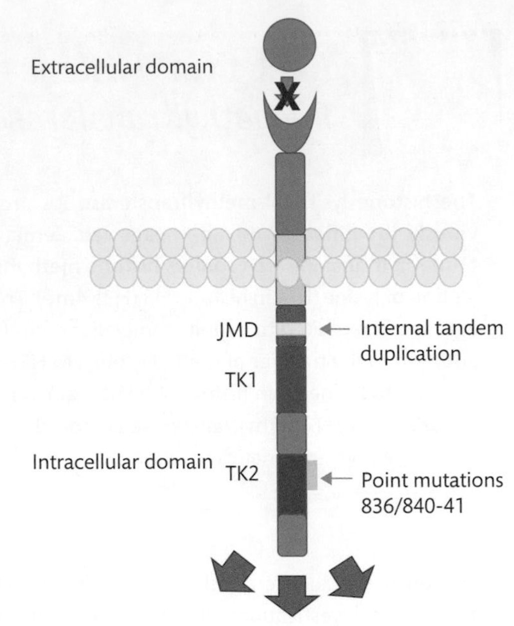

FIGURE 11.7

ITD within the juxtamembrane domain (JMD) domain of the receptor FLT3. ITD and point mutations within the tyrosine kinase 2 (TK2) domain at positions 836 and 840–41 activate FLT3, leading to aberrant signalling from this receptor.

The gene *RBM15*, also called *OTT* following its identification as a fusion partner in the translocation One Twenty Two, encodes the RNA-binding motif protein 15 found in high concentrations in haemopoietic progenitors. Wild-type RBM15 is thought to be involved in the inhibition of differentiation and, in addition, inhibits proliferation. When dysregulated through the fusion with MKL1, also called megakaryoblastic leukaemia-1 (*MAL*, *BSAC*, or *MRTF-A*), RBM15 maintains megakaryoblasts in an immature state.

Experimentally, the knock-down of *RBM15* is associated with an increased number of marrow megakaryocytes; and, accompanied with the inhibitory apoptotic inhibitor MKL, it is a likely candidate that contributes to the pathogenesis of acute megakaryoblastic leukaemia harbouring t(1;22)(p13.3;q13.1).

Morphologically, patients with t(1;22) have increased numbers of megakaryoblasts, some of which are small, and demonstrate cytoplasmic blebbing. Marrow fibrosis is a common feature, as with all pathologies demonstrating megakaryoblast/megakaryocyte accumulation.

SELF-CHECK 11.2

Under what conditions would a blast percentage of less than 20% signal a new diagnosis of acute myeloid leukaemia?

SELF-CHECK 11.3

List the genetic abnormalities considered of diagnostic value in classifying acute myeloid leukaemia.

AML with BCR-ABL1 (provisional entity)

This is a rare condition, comprising <1% of all acute myeloid leukaemias, that must be differentiated from chronic myeloid leukaemia (CML) in blast crisis (CML-BC), mixed phenotype acute leukaemia with t(9;22)(q34.1; q11.2) and B-lymphoblastic leukaemia/lymphoma with t(9;22)(q34.1;q11.2); *BCR-ABL1*. Accompanying *BCR-ABL1*, there appears to be concurrent loss of *IKZF1* and/or *CDKN2A*. *IKZF1* encodes Ikaros family zinc finger protein 1, which is a critical tumour suppressor involved in chromatin remodelling that is deleted in 70% of acute lymphoblastic leukaemias (ALL). *CDKN2A* encodes two proteins, $p16^{INK4A}$ and $p14^{ARF}$, both of which are key tumour suppressors involved in cell cycle control. Additional mutations are found in *IGH* and *TCR* genes. Unlike CML-BC, AML with *BCR-ABL1* has no history of blood disorder and is not associated with splenomegaly or basophilia. Bone marrow cellularity is approximately 80%, compared to the 100% usually identified in CML-BC, and lymphoid markers are often expressed when immunophenotyping is utilized. Signal transduction inhibitors (STI) such as imatinib, dasatinib, and nilotinib can be employed to target *BCR-ABL1* in this condition.

Cross reference

Please see chronic myelogenous leukaemia, *BCR-ABL1* positive in the myeloproliferative neoplasm section of this chapter.

SELF-CHECK 11.4

In what other conditions does *BCR-ABL1* feature?

AML with mutated RUNX1 (provisional entity)

RUNX1 encodes the DNA-binding component—RUNX1, also called CBFα—to the core binding factor transcription factors. These factors are essential for haemopoiesis. *RUNX1* is mutated in up to 15% of *de novo* MDS and up to 30% of cases of therapy-related MDS and AML. AML with mutated *RUNX1* are *de novo* cases of AML, with mutations that directly affect either the DNA-binding component or prevent the interaction between CBFα and its binding partner CBFβ. To be classified as a case of AML with mutated *RUNX1*, MDS-associated cytogenetic abnormalities must be excluded. Blast cells harbouring *RUNX1* demonstrate mutation arrest prior to the promyelocyte stage of maturation. Prognosis for AML patients with mutated *RUNX1* is poor.

SELF-CHECK 11.5

Briefly explain the function of t(15;17)(q22;q12) in leukaemogenesis.

Once recurrent cytogenetic abnormalities have been considered, we should identify the presence of dysplasia. Approximately 30% of cases will have been classified with recurrent cytogenetic abnormalities, but an appropriate designation needs to be found for the remainder of cases with AML.

11.2.4 Acute myeloid leukaemia with myelodysplasia-related changes

This subgroup of the AML classification can, unlike the recurrent genetic abnormalities subgroup, occur in *de novo* or secondary leukaemias.

A patient may be diagnosed with acute myeloid leukaemia with myelodysplasia-related changes (AML–MRC) in the following circumstances:

1. AML arising from a previous MDS or MDS/myeloproliferative disorder;
2. AML with a specific *MDS-related* cytogenetic abnormality;
3. AML with multilineage dysplasia.

However, if multilineage dysplasia is identified and coexists with either *NPM1* mutation or biallelic mutation of *CEBPA*, these mutations are prioritized and this condition is classified accordingly, as a recurrent genetic abnormality. In the case of AML with multilineage dysplasia, two or more of the myeloid lineages must exhibit signs of dysplasia in ≥50% of the cells in the affected cell lineages in order to fulfil the term multilineage, and the blast count must exceed the 20% threshold to be considered an acute leukaemia. If the 20% threshold is not exceeded, the patient is likely to be diagnosed with one of the following:

- MDS with multilineage dysplasia (MDS-MLD);
- MDS with excess blasts (MDS-EB);
- chronic myelomonocytic leukaemia (CMML);
- atypical chronic myeloid leukaemia (aCML).

These conditions are addressed later in this chapter.

Examples of dysplasia might include:

- megakaryocytes with a reduced number of nuclear lobes (hypolobulated megakaryocytes) or those with widely dispersed lobes;
- abnormally small megakaryocytes (micromegakaryocytes);
- hypolobulated neutrophils;
- hypogranular neutrophils;
- multinucleated erythroblasts;
- **dyskaryorrhexis**.

dyskaryorrhexis
Abnormal bursting of the cell nucleus.

AML with multilineage dysplasia is associated with an unfavourable prognosis; patients may be resistant to standard therapies, and survival rates are low.

SELF-CHECK 11.6

Describe some of the dysplastic features you may expect to see in the blood or bone marrow of a patient with an acute leukaemia or myelodysplastic syndrome with multilineage dysplasia.

We have now considered whether the patient has a recurrent genetic abnormality, or if they show signs of myelodysplasia-related changes. Evidence of either of these allows us to apply the appropriate classification, as highlighted above. However, if neither recurrent genetic abnormalities nor myelodysplasia-related changes are present, we should consider *therapy-related myeloid neoplasms*.

11.2.5 Therapy-related myeloid neoplasms

This classification requires the patient to have been previously treated with, or exposed to:

- alkylating agents;
- ionizing radiation;
- Topoisomerase II inhibitors.

By definition, the disorders included in this group are largely attributable to previous medical interventions and are therefore largely considered **iatrogenic**. However, there is growing evidence that the therapy-related myeloid neoplasms are more complex than previously thought. It is now believed that haemopoietic stem cells harbouring pre-existing mutations, which are then exposed to cytotoxic therapy for the treatment of another condition, are positively selected for. These cells then have the opportunity to increase in number, gradually replace normal progenitors, and repopulate the marrow. The risk of developing therapy-related myeloid neoplasms (t-MN) varies according to the nature of the primary cancer, the age of the patient, and the type of the cytotoxic intervention administered. Evidence suggests that up to 4% of patients with myeloma, approximately 8% of patients with CLL, and 10% of patients with non-Hodgkin lymphoma (NHL) will develop t-MN. A summary of the main classes of drugs, with examples, can be seen in Table 11.1.

iatrogenic
A term used to describe a disease or condition caused by medical intervention.

Haematological malignancies that fall within this category include therapy-related (t-) acute myeloid leukaemia (t-AML), myelodysplastic syndrome (t-MDS), and myelodysplastic/myeloproliferative neoplasms (t-MDS/MPN). Most patients present with dysplastic features and, in the case of alkylating agents or ionizing radiation, abnormalities of chromosome 5 or 7. Interestingly, abnormalities of chromosomes 5 and 7 are thought to lead to the development of a **contiguous gene syndrome**, which causes an imbalance in the number and activity of those sites.

contiguous gene syndrome
The manifestation caused by the deletion of a number of genes that are situated next to one another and are deleted simultaneously.

Alkylating agents

Alkylating agents are a group of compounds with electrophilic properties which add alkyl groups to biologically important molecules (a reaction called **alkylation**). Of particular interest is the mechanism by which alkylating agents interact with DNA. The result of this interaction may be a failure of mitosis, preventing cell division. Inhibition of cell division is of particular importance when treating malignancies, as preventing cell proliferation and inducing apoptosis can reduce a patient's **tumour burden**.

alkylation
The transfer of an alkyl group (C_nH_{2n+1}) from one molecule to another.

tumour burden
The size of an individual's tumour or the number of tumour cells involved in a particular malignant case.

The mode of action is dependent upon the properties of the alkylating agent. Some alkylating agents are monofunctional, where a single strand of DNA is alkylated, whereas bifunctional agents can cause cross-linking between two strands of DNA. Examples of alkylating agents include cyclophosphamide, melphalan, chlorambucil, and nitrosoureas. Figure 11.8 demonstrates the action of alkylating agents.

Ionizing radiation

Leukaemias are associated with this particular form of radiation. **Ionizing radiation** is known to cause a range of different cancers, and the dose of radiation has been well established. Exposure to ionizing radiation even at doses as low as <0.2 **Grays** has been associated with leukaemogenesis. Exposure to ionizing radiation can be through medical treatment, occupation, or in some cases, environment.

ionizing radiation
Radiation in the form of either high-energy waves or particles which, when interacting with atoms, can remove electrons, thus producing a charge (ion).

Gray
The unit of measurement for the absorbed dose of radiation (i.e. the amount deposited in the mass of a particular material). One Gray = one joule per kilogram.

Both alkylating agents and ionizing radiation have a latent period of five to ten years prior to the development of t-MN.

Topoisomerase II inhibitors

Topoisomerase II (Topo II) is an enzyme intrinsically involved in the cell cycle through its ability to make cuts in double-stranded (ds) DNA. These cuts not only enable transcription, replication, and recombination to occur, but they also allow for supercoiled DNA to relax by allowing

TABLE 11.1 Principal drug classes associated with therapy-related myeloid neoplasms with mechanisms of action.

Drug class and name	DNA damage
Alkylating agents	
Monofunctional	
Nitrosourea N	Base lesions
Procarbazine	Replication lesions
Dacarbazine	Bulky adducts
Temozolamide	
Bifunctional	
Cyclophosphamide	Base damage
Ifosphamide	DNA cross-links
Micomycin C	Replication lesions
Melphalan	Bulky adducts
Chlorambucil	Double stranded breaks
Cisplatin	
Carboplatin	
Topoisomerase inhibitors	
Etoposide	Double-stranded breaks
Doxorubicin	Single-stranded breaks
Daunorubicin	Replication lesions
Epirubicin	
Mitoxantrone	
Camptothecin	
Antimetabolites	
Azathioprine	Base damage
Fludarabine	Replication lesions
Cladribine	
5-Fluorouracil	

Adapted with permission from Sill H, Olipitz W, Zebisch A, Schulz E, and Wolfer A. ***Therapy-related myeloid neoplasms: pathobiology and clinical characteristics***. *British Journal of Pharmacology* 2011: **162**: 792–805.

the passage of one segment of dsDNA through another. Figure 11.9 demonstrates the process of Topo II-mediated DNA cleavage and transport. The transit of DNA molecules through one another is permitted by the induction of a transient cleavage complex involving a strand of DNA, which is rejoined (re-ligated) following the passage of the second DNA molecule.

Administration of Topo II inhibitors (also known as Topo II poisons) is associated with large numbers of DNA breaks, although the precise mechanism depends upon the particular type of drug used. In the case of *etoposide* and *teniposide*, Topo II is prevented from rejoining (re-ligating) the dsDNA breaks it initiates, causing the formation of multiple permanent DNA

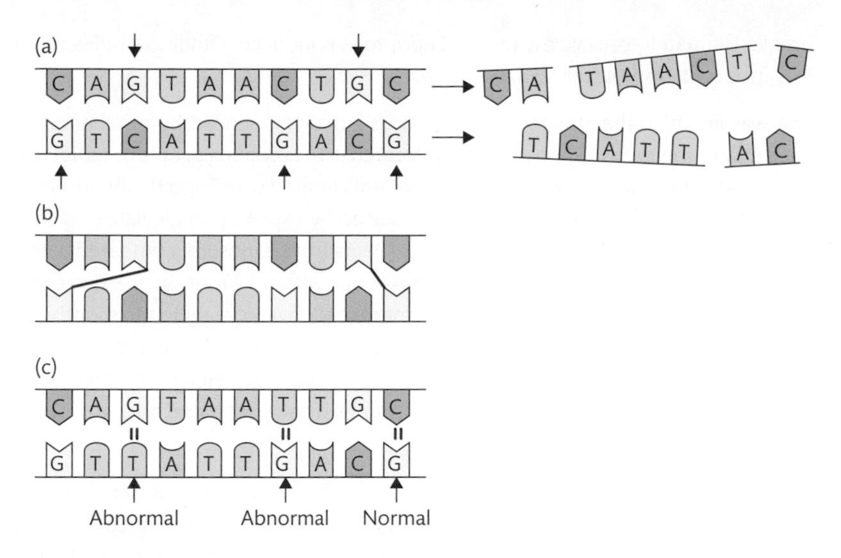

FIGURE 11.8

Mechanisms of DNA disruption through alkylating agent activity. Alkylating agents bind to guanine (G) and cause modification as outlined in (a), (b), and (c). (a) Alkylated guanine (represented by yellow icons) is cleaved by repair enzymes, causing DNA fragmentation; (b) cross-linking guanine bases between DNA strands (as shown) or within strands prevents DNA strand separation, inhibiting DNA transcription; (c) aberrant binding of guanine to thymine (T) instead of cytosine (C) results in the incorporation of point mutations into the DNA template.

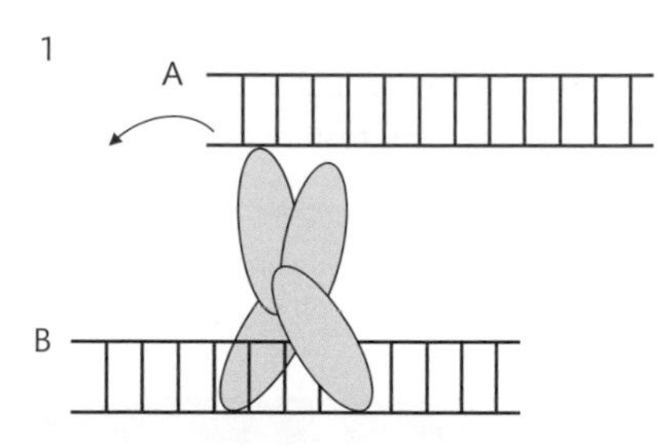

Topoisomerase II binds to DNA segment B (purple).

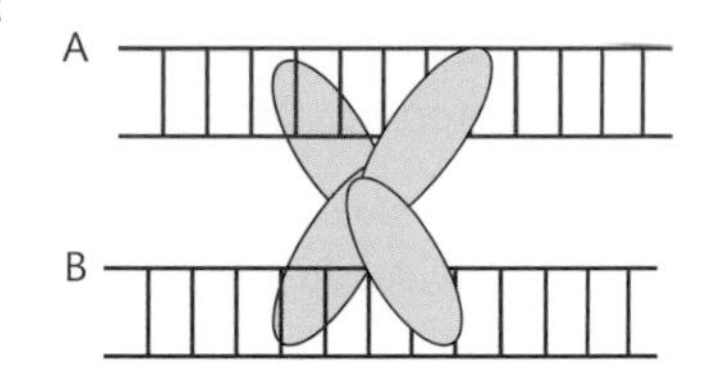

In addition, topoisomerase II now binds to supercoiled DNA segment A (red).

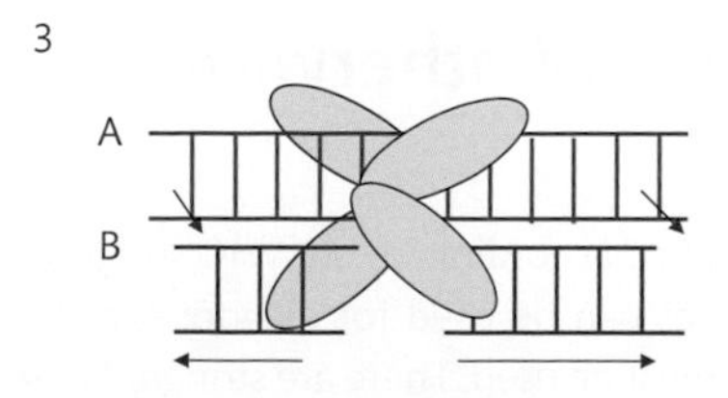

Topoisomerase II cleaves DNA strand B and begins to move supercoiled strand A towards strand B.

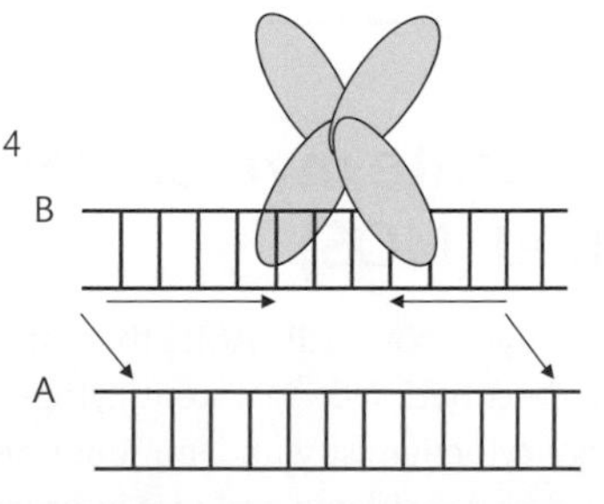

DNA strand A passes through strand B, reducing the mechanical stress caused by supercoiling. Topoisomerase re-ligates strand B.

FIGURE 11.9

(1) Topoisomerase II binds to double-stranded (ds) DNA (B). (2) Whilst bound to B, topoisomerase II binds to a second dsDNA, which is considered supercoiled. In order to reduce the tension in the DNA coil, helix A needs to pass through helix B. (3) In order to mediate this DNA transfer process, strand B needs to be cleaved. Reversible cleavage is achieved by topoisomerase II, enabling strand A to pass through prior to re-ligation (4).

breaks. Ultimately, cell cycle arrest and apoptosis is induced. Other examples of Topo II inhibitors include doxorubicin and daunorubicin.

The way in which the malignancy presents is largely influenced by the therapy to which the patient was previously exposed. Patients treated with alkylating agents or radiotherapy present as an acute leukaemia or MDS, whereas patients treated with Topo II inhibitors present purely with a leukaemia. Leukaemias and MDS, caused by exposure to alkylating agents or ionizing radiation, generally occur within five to ten years from exposure, compared with exposure to Topo II inhibitors, which often leads to leukaemia within 12 months of exposure. Survival following a diagnosis of t-MN is poor, with a median survival of eight to ten months, and five-year survival at 10%. In approximately 20–30% of cases, acute leukaemia may become apparent between one and five years following exposure. Patients will often present with signs of dysplasia in one or more blood cell lines and commonly display signs and symptoms associated with peripheral blood cytopenias.

Leukaemias associated with alkylating agents or ionizing radiation tend to demonstrate monocytic or myelomonocytic lineage and have a poor prognosis compared to those caused by Topoisomerase II inhibitors. In addition, patients who have been exposed to Topo II inhibitors are much more likely to possess 11q23 translocations because of the functional activity of the Topo II enzyme in causing dsDNA breaks. Alternative translocations associated with Topoisomerase II inhibition include t(15;17)(q22;q12), inv(16)(q13;q22), and abnormalities associated with 21q22. In most cases, treating patients with these abnormalities should follow the same protocol employed for the *de novo* equivalent, although evidence suggests that therapy-induced t(8;21) and inv(16) benefit from alternative therapeutic intervention.

It should also be noted that, in addition to alkylating agents, ionizing radiation, and Topoisomerase II inhibitors, *anti-metabolites* are also considered to be leukaemogenic. These drugs are incorporated into DNA, preventing normal cell cycling, thus resulting in apoptosis.

For the majority of other cases not fitting into the categories already discussed, the WHO classification system provides an area in which to deposit these cases: *acute myeloid leukaemia not otherwise specified.*

SELF-CHECK 11.7

Compare and contrast the causes of secondary leukaemia associated with different therapies.

11.2.6 Acute myeloid leukaemia not otherwise specified (NOS)

This section largely covers the AMLs that cannot be classified according to the WHO subgroups previously described. No consistent cytogenetic features can be used for diagnosis, and so morphology, cytochemistry, and immunophenotyping must be used. There are stringent criteria that must be considered and met in order to provide an appropriate classification for leukaemias in this subgroup. A summary of the frequencies associated with each of these subtypes of AML NOS is shown in Table 11.2.

11.2.7 Myeloid sarcoma

Myeloid sarcoma describes the accumulation of myeloblasts within one or more anatomical sites outside the bone marrow, leading to the destruction of the normal tissue within that region. Myeloid sarcoma can occur in conjunction with established AML or other myeloproliferative or

TABLE 11.2 AML NOS. The approximate frequencies of these conditions have been outlined, although in some cases it is difficult to obtain precise figures, as these vary between reports.

Subtype	Approximate frequency (%) cases of AML
AML with minimal differentiation	5
AML without maturation	10
AML with maturation	30–45
Acute myelomonocytic leukaemia	15–25
Acute monoblastic and monocytic leukaemia	5–8
Acute monocytic leukaemia	3–6
Pure eythroid leukaemia	<5
Acute megakaryoblastic leukaemia	3–5
Acute basophilic leukaemia	<1
Acute panmyelosis with myelofibrosis	Very rare

MDS and can frequently involve lymph nodes. Approximately 50% of cases are asymptomatic, whilst the remainder may be associated with organ dysfunction caused or through the patient experiencing localized pain. Due to the extramedullary nature of this condition, radiographic assessment is frequently used to identify the location and extent of the lesion from which biopsy material can be recovered. Diagnosis involves histological examination of biopsy material and the use of immunohistochemistry to determine the immunophenotype of the malignant cells. Treatment is aggressive to maximize the chances of survival, and bone marrow transplantation is often indicated.

11.2.8 Myeloid proliferations related to Down syndrome

The incidence of Down syndrome (DS) is approximately 1 in every 700 births and is associated with a constitutional trisomy 21. Infants and children with Down syndrome have a predisposition to acute lymphoblastic leukaemia, with a 20-fold higher risk of developing the disease in comparison to the general population. These infants are also predisposed to AML, and have a 500-fold increased risk of developing acute megakaryoblastic leukaemia (AMKL).

The development of DS-AMKL is known to be a multi-step process, as indicated in Figure 11.10, although the precise genetic requirement in this disease has yet to be determined. There is a requirement for trisomy 21, although we do not know which genes present on chromosome 21 are important in the pathogenesis of DS-AMKL. A Down syndrome critical region (DSCR) has been identified as being important in determining the clinical phenotypes of patients with DS, and there are candidate genes present within this region that could be involved in leukaemogenesis.

Cross reference

For further information on trisomy 21 and Down syndrome, please refer to Chapter 9.

Importantly, mutations within the haemopoietic transcription factor GATA1 have been identified, and are always present in cases of DS-AMKL. GATA1 mutations are not associated with non-DS-AMKL. GATA1 is an essential transcription factor required for erythropoiesis and

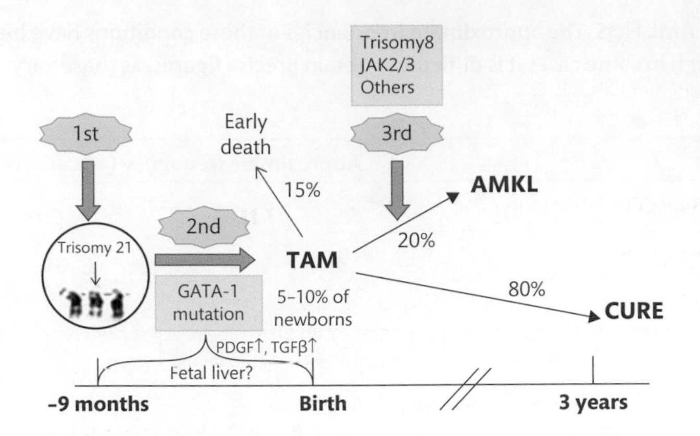

FIGURE 11.10

Multi-step model of myeloid leukaemogenesis in DS. Trisomy 21 enhances the proliferation of foetal liver megakaryo-erythroid progenitors via PDGF and/or TGF beta. The acquisition of GATA1 mutation further enhances the clonal proliferation of immature megakaryoblasts diagnosed at birth as TAM. GATA1 mutations are necessary but insufficient for the development of AMKL. Additional genetic events such as trisomy 8 or JAK2/3 mutations have been proposed in progression from TAM to AMKL. Redrawn from Kazuko Kudo. *Myeloid Leukemia associated with Down Syndrome*, in Subrata Dey (ed.), *Down Syndrome*, InTech. (2013), http://www.intechopen.com/books/down-syndrome/myeloid-leukemia-associated-with-down-syndrome.

megakaryopoiesis. The loss of GATA1 function results in the accumulation of abnormal megakaryocytes within the liver and bone marrow and the failure of these megakaryocytes to produce platelets.

Mutation of *GATA1* is an *in utero* event, with these mutations clustering in exon 2 of *GATA1*. Point mutations, insertions, and deletions have all been documented and produce a premature stop codon. Transcription and translation of mutated *GATA1* produces a truncated transcription factor called GATA1s, in which the transcriptional activation domain is absent.

Recently, *JAK3* mutations have been identified in a small group of patients with DS-AMKL, and are associated with increased signal transducer and activator of transcription (STAT) signalling and cellular proliferation.

The multi-step process of myeloid leukaemogenesis in DS is relatively well described. Between 5 and 10% of infants with trisomy 21 develop a transient leukaemia, also called *transient abnormal myelopoiesis* (TAM) and *transient myeloproliferative disease* (TMD). All these cases studied express GATA1s. In the vast majority of cases, this transient leukaemia resolves spontaneously within three months. In approximately 20% of patients, by the age of four years, AMKL develops following a myelodysplastic stage. This is probably due to residual cells harbouring the GATA1s mutation developing additional mutations and leading to a full leukaemic phenotype.

Clinical and laboratory findings include a raised white cell count with an increased number of megakaryoblasts. The patient may demonstrate thrombocythaemia or thrombocytopenia and show signs of hepatomegaly. Hepatomegaly is due to the retention of neonatal hepatic haemopoiesis and the accumulation of malignant megakaryoblasts within the liver. Abnormal cytokines released by the megakaryoblasts induce fibrotic changes within the liver.

Patients with DS have an increased sensitivity to cytarabine and anthracyclines. Low doses of cytarabine have been shown to cure up to 80% of patients with DS-AMKL.

SELF-CHECK 11.8

Why is the *GATA1* gene important in Down syndrome?

11.2.9 Blastic plasmacytoid dendritic cell neoplasm

Plasmacytoid dendritic cells (pDC) are usually found in clusters within T-cell-rich areas of lymphoid tissue and are important mediators of innate and adaptive immunity. pDC are responsible for the production of large quantities of interferons α and β in response to viruses, or their nucleic acids. Expression of Toll-like receptors on the pDC surface allows recognition of viral RNA and certain types of DNA, leading to the secretion of interferon. Release of interferon leads to a coordinated response by a range of cells within the immune system. Additional features of pDC include their plasma cell-like morphology and their ability to differentiate into classical dendritic cells following appropriate stimulation. pDC only make up approximately 0.5% of peripheral blood cells, and following their release from the bone marrow they enter the lymphoid tissue, where they are retained.

The origin of these cells is unclear and further investigation is required to establish their definitive origin. This malignancy is associated with an accumulation of pDC precursors, occurring usually, but not exclusively, in elderly patients with a median age between 60 and 70 years. This malignancy is highly aggressive, with a median survival of approximately 12 months. Cutaneous lesions with bone marrow, lymphoid, and peripheral blood involvement are common. Prior to diagnosis, comprehensive immunophenotypic analysis should be completed to ensure the accurate classification of this disease. Up to 20% of patients are shown to have a concurrent myelodysplasia which progresses to acute myelomonocytic leukaemia (AMML).

Although these blastic pDC appear lymphoid in origin, immunophenotyping does demonstrate the expression of a number of myeloid markers. In addition, the association with a progressive MDS leading to AMML has resulted in its classification by the WHO as a myeloid malignancy.

11.2.10 Acute leukaemias of ambiguous lineage

Seven main diseases are classified as acute leukaemias of ambiguous lineage:

1. acute undifferentiated leukaemia;
2. mixed-phenotype acute leukaemias (MPAL):
    a. MPAL with t(9;22)(q34;q11.2); *BCR-ABL1*;
    b. MPAL with t(v;11q23); *KMT2A* rearranged;
    c. MPAL, B/myeloid, NOS;
    d. MPAL, T/myeloid, NOS;
    e. MPAL, NOS, rare types;
3. acute leukaemias of ambiguous lineage, NOS.

Acute undifferentiated leukaemia is a separate disease entity to MPAL as phenotypically there are no lineage-specific markers expressed on the former, either cytoplasmically or on the cell surface. Conversely, mixed-phenotype leukaemias show lineage-specific phenotypic antigens, although a combination of lineage-specific antigens are expressed, either within the cytoplasm or on the cell surface. For example, blasts with a B- or T-lineage phenotype accompanied by cytoplasmic myeloperoxidase or monoblastic markers would be sufficient to diagnose mixed-phenotypic acute leukaemia.

The presence of *BCR–ABL1* is an indicator of a poor prognosis and is of relevance to this classification category if an underlying CML can be excluded. There is evidence that including a signal transduction inhibitor, such as imatinib, dasatinib, or nilotinib, in the treatment strategy may afford some benefit to patients harbouring this genetic aberration.

Patients demonstrating 11q23 rearrangement also have a poor prognosis. *MLL/KMTA* can be considered to be a promiscuous gene as it has a number of fusion partners. In this classification, 'v' has been used in the abbreviation t(v;11q23) to demonstrate variable translocation partners, rather than specifying each in turn.

A convenient algorithm to guide the process of classifying the main acute myeloid subgroups is shown in Figure 11.11.

11.3 Myelodysplastic syndrome

Myelodysplastic syndrome (MDS) is diagnosed when one or more of the myeloid cell lines displays signs of dysplasia. Generally, signs of dysplasia will be seen in 10% or more of cells within a particular lineage, when peripheral blood and bone marrow slides are examined. MDS is a disease usually associated with a hypercellular bone marrow, but with the patient showing variably low peripheral blood counts. The patient's blast percentage may be normal or raised, but will not exceed the critical 20% value diagnostic for AML.

It is important when assessing patients' blood films for signs of dysplasia that optimal staining procedures are followed using a good-quality Romanowsky stain. Samples should be less than two hours old to ensure cellular morphology is maintained. Samples older than two hours start to show EDTA changes, making the assessment of dysplastic features difficult. Figure 11.12 provides some morphological examples of dysplasia that may be indicative of MDS.

Evaluation of the patient's bone marrow cytogenetic profile is important for prognostication purposes. Particular emphasis should be made on *TP53* mutation, which signals a poor prognosis in all MDS groups and, in particular, indicates deleterious outcomes in the otherwise favourable diagnosis of MDS with isolated del(5q). *TP53* encodes the critical tumour suppressor p53 and is located at 17p13.1. The mutation of *TP53* is associated with resistance to a number of cytotoxic agents.

Recently, *SF3B1* mutation has been adopted to facilitate the diagnosis and prognostication of MDS with ring sideroblasts. The presence of *SF3B1* mutation is associated with a good prognosis and signals a lower acceptable threshold of 5% ring sideroblasts in this condition. In the absence of *SF3B1* mutation, the threshold of 15% ring sideroblasts is required to make this diagnosis. *SF3B1* encodes an RNA splicing factor in the 3b protein complex, located at chromosome 2q33.1.

Any case of MDS secondary to either the use of alkylating agents or radiotherapy should be considered as a therapy-related neoplasm and called t-MDS. (See also Box 11.5.)

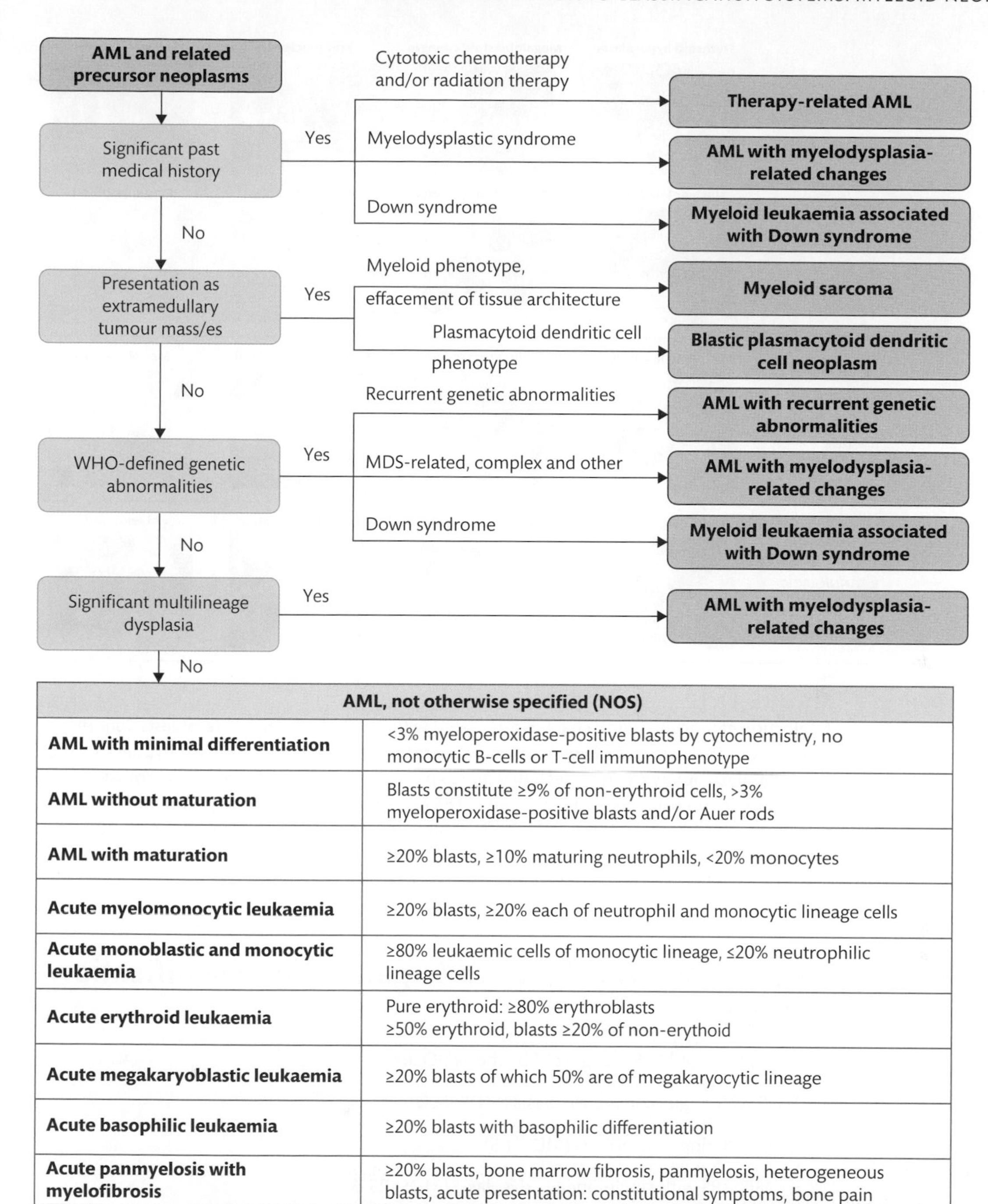

AML, not otherwise specified (NOS)	
AML with minimal differentiation	<3% myeloperoxidase-positive blasts by cytochemistry, no monocytic B-cells or T-cell immunophenotype
AML without maturation	Blasts constitute ≥9% of non-erythroid cells, >3% myeloperoxidase-positive blasts and/or Auer rods
AML with maturation	≥20% blasts, ≥10% maturing neutrophils, <20% monocytes
Acute myelomonocytic leukaemia	≥20% blasts, ≥20% each of neutrophil and monocytic lineage cells
Acute monoblastic and monocytic leukaemia	≥80% leukaemic cells of monocytic lineage, ≤20% neutrophilic lineage cells
Acute erythroid leukaemia	Pure erythroid: ≥80% erythroblasts ≥50% erythroid, blasts ≥20% of non-erythoid
Acute megakaryoblastic leukaemia	≥20% blasts of which 50% are of megakaryocytic lineage
Acute basophilic leukaemia	≥20% blasts with basophilic differentiation
Acute panmyelosis with myelofibrosis	≥20% blasts, bone marrow fibrosis, panmyelosis, heterogeneous blasts, acute presentation: constitutional symptoms, bone pain

FIGURE 11.11

An algorithm can be utilized for the sub-classification of acute myeloid leukaemias. A series of four key questions can be applied to the case under investigation. The answer to each of these questions helps the clinician reduce the differential diagnosis and consider further investigations until a definitive diagnosis is found. Reprinted, with modification, by permission from Macmillan Publishers Ltd: Patel K, Khokhar F, Muzzafar T et al. *TdT expression in acute myeloid leukemia with minimal differentiation is associated with distinctive clinicopathological features and better overall survival following stem cell transplantation*. *Modern Pathology* (2013) 26: 195–203. Doi: https://doi.org/10.1038/modpathol.2012.142, copyright 2013.

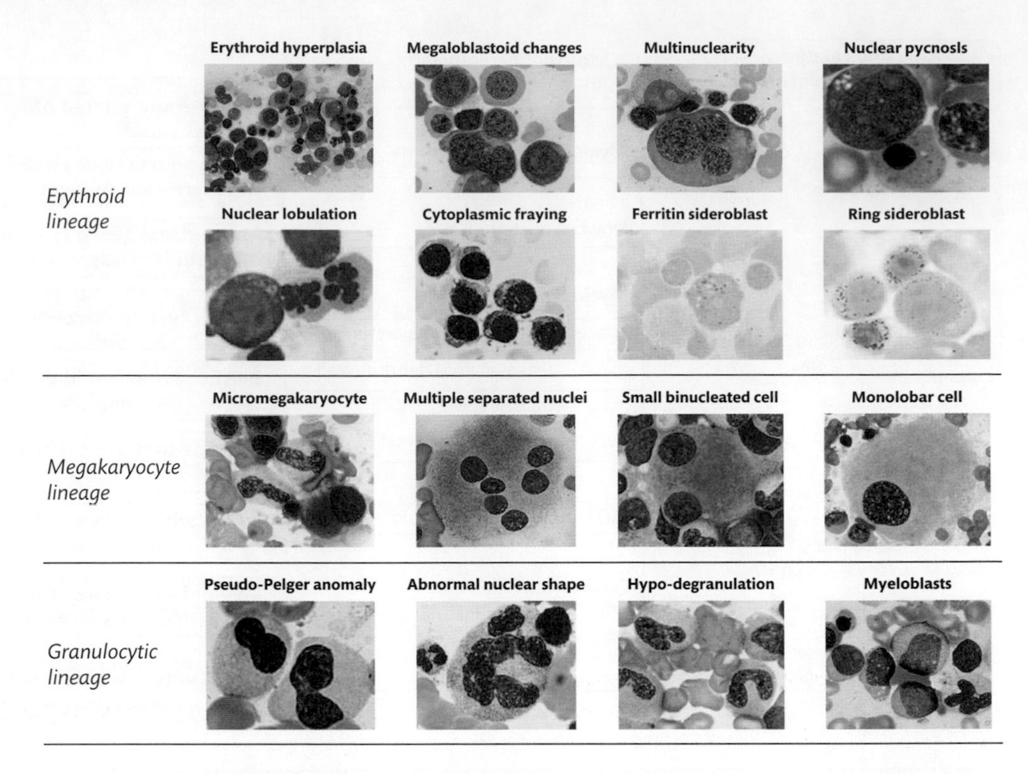

FIGURE 11.12
A selection of morphological changes associated with dysplasia. 1 = ring sideroblasts demonstrated using Perls' stain; 2 = hypogranulation. Care must be taken to ensure that hypogranular or agranular neutrophils are not artefactual due to poor staining. Reproduced with permission from Cazzola M, Della Porta MG, and Malcovati L. ***The genetic basis of myelodysplasia and its clinical relevance***. *Blood* 2013: **122(25)**: 4021–34. Doi: https://doi.org/10.1182/blood-2013-09-381665.

11.3.1 The World Health Organization classification of MDS

The main types of MDS recognized by the WHO are:

- MDS with single-lineage dysplasia (MDS-SLD);
- MDS with ring sideroblasts (MDS-RS):
    - MDS-RS with single-lineage dysplasia (MDS-RS-SLD);
    - MDS-RS with multilineage dysplasia (MDS-RS-MLD);
- MDS with multilineage dysplasia (MDS-MLD);
- MDS with excess blasts (MDS-EB):
    - MDS with excess blasts 1 (MDS-EB-1);
    - MDS with excess blasts 2 (MDS-EB-2);

- Myelodysplastic syndrome unclassified (MDS-U):
  - with 1% blood blasts;
  - with single-lineage dysplasia and pancytopenia;
  - based on defining cytogenetic abnormality;
- MDS associated with isolated del(5q) (5q-syndrome);
- refractory cytopenia of childhood (provisional entity).

Table 11.3 provides a convenient quick reference guide for diagnosing subtypes of MDS.

BOX 11.5 *International Prognostic Scoring System-Revised (IPSS-R)*

The International Prognostic Scoring System (IPSS) was developed to establish associations between patient blast count, cytogenetic features, levels of cytopenia, and patient prognosis. However, now we have more up-to-date data available regarding the parameters which can be used to measure prognosis, this system has been revised (R). The IPSS-R score should be calculated at the point of diagnosis only, with each feature (outlined below) being allocated a score between 0 and 4. Between 0 and 2.0, these scores are staged in increments of 0.5, whereas integers between 2.0 and 4.0 are allocated. The important parameters are shown in the table below:

	Score						
Parameter	0	0.5	1.0	1.5	2.0	3.0	4.0
Cytogenetics (*)	V. good	-	Good	-	Intermediate	Poor	V. poor
BM blasts	≤2.0%	-	<2.0–<5.0%	-	5.0–10.0%	>10.0%	-
Haemoglobin g/dL	≥10.0	-	8.0–<10.0	<8.0	-	-	-
Platelets (×10^9/L)	≥100	50–<100	<50	-	-	-	-
Abs. neutrophil count (×1010^9/L)	≥0.8	<0.8	-	-	-	-	-

To complete the cytogenetics row in the table above and to determine the appropriate score to be allocated, only selected karyotypes or abnormalities are considered relevant. These are outlined in the table below:

Prognostic group	Score	Cytogenetic abnormality (*) used to score IPSS-R
Very good	0	-Y (males), del(11q)
Good	1.0	Normal karyotype, del(5q), del(12p), del(20q), double including del(5q)
Intermediate	2.0	del(7q), +8, +19, i(17q), any other single or double independent clones
Poor	3.0	-7, inv(3)/t(3q)/del(3q), double including -1/del(7q), complex; 3 abnormalities
Very poor	4.0	Complex >3 abnormalities

(Continued)

Once this cytogenetic data has been inputted, it is then possible to determine the patient's projected survival. This is shown in the table below:

IPSS-R prognostic outcome		
Risk category	**Risk score**	**Median survival (years)**
Very low	≤1.5	8.8
Low	>1.5–3.0	5.3
Intermediate	>3.0–4.5	3.0
High	>4.5–6.0	1.6
Very high	>6.0	0.8

As you can see, patients with the very lowest scores have an expected median survival of almost nine years, compared to those in the highest risk category, where median survival is less than one year from diagnosis.

A useful online calculator that can be used to determine IPSS-R score and category can be found at: http://www.mds-foundation.org/ipss-r-calculator/. Age adjustment is also possible using this calculator for a more precise estimate of prognosis.

The WHO has also released a prognostic scoring system (WPSS) based upon the WHO classification system and this is also recommended for use by the British Committee for Standards in Haematology. Unlike the IPSS-R above, this system can be applied at any point in the patient's disease process.

TABLE 11.3 **The WHO method for diagnosing and classifying myelodysplastic syndrome. Characteristic blood and bone marrow findings that help inform the diagnosis have been outlined.**

Type	Dysplastic lineages	Cytopenias*	Ring sideroblasts in erythroid elements of BM	Blasts	Cytogenetics
MDS-SLD	1	1 or 2	RS <15%/(<5% with *SF3B1* mutation)	BM <5%, PB <1%, no Auer rods	Any, unless fulfils all criteria for MDS with isolated del(5q).
MDS-MLD	2 or 3	1–3	RS <15%/(<5% with *SF3B1* mutation)	BM <5%, PB <1%, no Auer rods	Any, unless fulfils all criteria for MDS with isolated del(5q).
MDS-RS					
MDS-RS-SLD	1	1 or 2	≥15%/(≥5% with *SF3B1* mutation)	BM <5%, PB <1%, no Auer rods	Any, unless fulfils all criteria for MDS with isolated del(5q).
MDS-RS-MLD	2 or 3	1–3	≥15%/(≥5% with *SF3B1* mutation)	BM <5%, PB <1%, no Auer rods	Any, unless fulfils all criteria for MDS with isolated del(5q).
MDS with isolated del(5q)	1–3	1–2	None or any	BM <5%, PB <1%, no Auer rods	Del(5q) alone or with one additional abnormality except -7 of del(7q)
MDS-EB					
MDS-EB-1	0–3	1–3	None or any	BM 5–9%, PB 2–4%, no Auer rods	Any

Type	Dysplastic lineages	Cytopenias*	Ring sideroblasts in erythroid elements of BM	Blasts	Cytogenetics
MDS-EB-2	0-3	1-3	None or any	BM 10-19%, PB 5-19%, OR Auer rods	Any
MDS-U					
With 1% PB blasts	1-3	1-3	None or any	BM <5%, PB = 1%**, no Auer rods	Any
With SLD and pancytopenia	1	3	None or any	BM <5%, PB <1%, no Auer rods	Any
Defining cytogenetic abnormality	0	1-3	<15% ***	BM <5%, PB <1%, no Auer rods	MDS defining abnormality
Refractory cytopenia of childhood	1-3	1-3	None or any	BM<5%, PB<2%	Any

* Cytopenia = haemoglobin <100g/L, platelets <100 × 10^9/L, absolute neutrophil count <1.8 × 10^9/L (although caution required with some ethnic groups, who commonly have lower neutrophil counts). PB monocytes must be >1 × 10^9/L.

** 1% PB blasts must have been recorded on at least two separate occasions.

*** Cases with ≥15% ring sideroblasts are classified as MDS-RS-SLD.

MDS with single-lineage dysplasia (MDS-SLD)

MDS-SLD describes a single cell line demonstrating dysplastic features in 10% or more in any one of: red cells, granulocytes (neutrophils), or platelets. For the purposes of identifying dysplasia in megakaryocytes, the WHO recommends evaluating the morphology of a minimum of 30 megakaryocytes.

MDS with ring sideroblasts (MDS-RS)

MDS-RS is subdivided according to the number of myeloid lineages demonstrating signs of dysplasia. In cases where a single lineage is affected, the designation MDS-single-lineage dysplasia (SLD) is necessary. Where more than one lineage expresses dysplastic change, MDS-multilineage dysplasia (MLD) is used. The bone marrow shows ≥15% ring sideroblasts when stained with Perls' stain, unless there is an associated mutation of *SF3B1*. *SF3B1* mutation is considered an early event in MDS and when identified concurrently with ring sideroblasts, the threshold of ring sideroblasts in MDS-RS is reduced to ≥5%.

MDS with multilineage dysplasia (MDS-MLD)

This group demonstrates dysplastic features in >10% of cells in two or more cell lineages. The peripheral blood film shows cytopenias in two or more cell lineages.

The WHO has subdivided the MDS-EB category into two, based on clinical observations and correlating these findings with the blast count and the presence of Auer rods.

MDS with excess blasts-1 (MDS-EB-1)

MDS-EB-1 includes patients with peripheral blood blasts <5%, and 5-9% bone marrow blasts. Dysplastic features may be seen in one or more cell lineages. Patients diagnosed with MDS-EB-1 have a more favourable prognosis than those with MDS-EB-2.

MDS with excess blasts-2 (MDS-EB-2)

This group is an extension to MDS-EB-1, with 5–19% blasts in the peripheral blood, and 10–19% blasts in the bone marrow. If Auer rods are present, providing the blast count is <20%, patients are diagnosed with MDS-EB-2, regardless of any other features. Interestingly, a number of authors have questioned the validity of including the presence of Auer rods in a poor prognostic group, as Auer rods have been shown to be an independent marker of a good prognosis.

The diagnosis of the erythroid/myeloid subtype of acute erythroid leukaemia (Table 11.2) was previously based upon a requirement of ≥50% erythroid with blasts in ≥20% of non-erythroid cells. Adjustment of the thresholds for diagnosis now requires the number of myeloblasts to be reported as a percentage of total nucleated marrow cells, rather than of non-erythroid cells only. This change in calculation will reduce the overall percentage of myeloblasts to <20% in some cases, meaning these patients no longer qualify for a diagnosis of AML. The more appropriate diagnosis in these cases is MDS-EB.

Myelodysplastic syndrome unclassifiable (MDS-U)

The importance of MDS-U should not be underestimated just because it contains unclassifiable cases. Over the coming years, with further clinical research and the emergence of improved availability of new technologies, we should be better equipped to provide a more appropriate classification for these cases. Since the last classification, MDS-U has been refined to include the following entities: MDS-U with 1% blood blasts, MDS-U with single-lineage dysplasia and pancytopenia, and MDS-U based on a defining cytogenetic abnormality.

MDS associated with isolated del(5q)

These patients are usually female and typically present with a normal or raised platelet count, a macrocytic anaemia, and normal white cell count. The bone marrow morphology is specific for 5q- syndrome, with mononuclear or binuclear megakaryocytes and erythroid hypoplasia, allowing a prediction of the patient's cytogenetic profile. Cytogenetic evaluation is essential for these patients to confirm the 5q- status and to provide a definitive diagnosis. The 5q- syndrome is associated with a good prognosis. Recent evidence suggests that *RPS14*, which produces part of the 40S ribosome subunit, is deleted in these patients, causing **haploinsufficiency**. Loss of *RPS14* reduces protein production within the cell harbouring 5q-. (See also Case study 11.1.)

haploinsufficiency
Occurs when one of a pair of genes on homologous chromosomes is silenced through deletion or mutation. The intracellular concentration of the 'normal' gene product is insufficient to complete its intended biological role.

Refractory cytopenia of childhood

These are an extremely rare group of diseases which are largely attributable to heritable syndromes, the most notable of which is Down syndrome. Only primary MDS are considered in this category under the provisional classification of refractory cytopenia of childhood (RCC). It is important when investigating these patients to consider the broad differential diagnoses including infectious agents, vitamin deficiencies, and inherited bone marrow failure syndromes.

RCC covers a range of MDS, the behaviour of which is different to MDS in adults. The majority of cases are associated with a hypocellular marrow, in contrast to the hypercellular marrows seen in adults. As the WHO considers RCC a provisional entity, no further discussion regarding this pathology will be made.

CASE STUDY 11.1 A 58-year-old female patient with macrocytic anaemia

Patient history

- A 58-year-old part time social worker visited her GP for a routine check-up following a few months of feeling tired. There were no other remarkable medical complaints.
- Physical examination showed modest signs of anaemia, including general pallor of the mucous membranes and mild tachycardia on exertion. The patient is a non-smoker and denies recreational use of drugs or alcohol.
- Full blood count (FBC) and blood film were initially performed.

Results 1

Blood film

The blood film revealed red cell morphology consistent with the FBC results. Red cells appeared macrocytic. Leucocytes and platelets appeared morphologically normal.

- Further examinations by the GP demonstrated normal liver function tests, creatinine levels, B_{12} and folate, and thyroid hormones.

Index	Result	Reference range
WBC	7.35	(RR 4–10 × 10^9/L)
Neutrophils	4.1	(RR 2–7 × 10^9/L)
Lymphocytes	2.2	(RR 1–3 × 10^9/L)
Monocytes	0.7	(RR 0.2–1.0 × 10^9/L)
Eosinophils	0.3	(RR 0.02–0.5 × 10^9/L)
Basophils	0.05	(RR 0.02–0.1 × 10^9/L)
RBC	3.6	(RR 3.9–5.0 × 10^{12}/L)
Hb	76	(RR 118–148 g/L)
MCV	102	(RR 77–98 fL)
Plt	398	(RR 143–400 × 10^9/L)

Significance of results 1

The isolated macrocytic anaemia is a cause for concern. The GP has requested tests to investigate all common causes of a macrocytosis (data not shown). Accompanied with the examination of the blood film, the GP has been able to exclude liver disease, kidney failure, B_{12}/folate deficiency (a cause of megaloblastic anaemia), hypothyroidism, and bone marrow compensation to bleeding (polychromasia), and has excluded excess alcohol intake through questioning. As a consequence, this patient was referred for haematology consultation.

- Initial physical examination by the haematologist provided agreement with the GP's findings. The patient did not complain of recent infections or bleeding and admitted her increased lethargy is her only concern. The haematologist requested a bone marrow aspirate, trephine biopsy, and cytogenetic analysis.

Results 2

- The bone marrow revealed normocellular marrow with trilineage haemopoiesis. Increased numbers of megakaryocytes were apparent, with dysmegakaryopoiesis noted. Of particular importance was the presence of monolobed megakaryocytes. Myeloblasts were reported at 2%.
- Cytogenetic analysis revealed an isolated deletion of 5q [del(5)q22q35] in 20/20 metaphase chromosomes.

Significance of results 2

The bone marrow morphology is clearly indicative of 5q– syndrome, which was confirmed by cytogenetic analysis. Results were inputted into the IPSS-R calculator, providing a score of 2.5 (low-risk category). According to the IPSS-R calculator, 38% of MDS patients fall within this category, which is associated with a median survival of 5.3 years. There is a low risk of transformation to acute myeloid leukaemia. (See also Box 11.6.)

BOX 11.6 *Guidelines for the diagnosis and management of MDS*

The 2014 BCSH guidelines outline the diagnostic process that should be employed where MDS is suspected. A comprehensive history should be taken to determine whether there is a family history of MDS or AML. It should also be determined whether patients have received prior chemotherapy or radiotherapy, as these treatments can induce dysplastic change and lead to *therapy-related myeloid neoplasms*. It is also necessary to ascertain whether the patient has had a history of recurrent infections or bleeding, which can point to the duration of the disease, particularly when white cells (neutropenia) or platelets (thrombocytopenia) are involved. Physical examination should pay particular attention to any abnormal physical features, which could be associated with congenital disorders of the bone marrow, and any obvious signs of infection, bleeding, or bruising. Finally, the patient should be assessed for signs of splenomegaly.

From a laboratory perspective, patients will be required to undergo full blood count, particularly to assess red cell count, haemoglobin, mean cell volume, platelet count, and white cell differential. Neutrophil and monocyte counts are particularly important components of the white cell count when MDS is suspected. Reticulocytes should be enumerated, and will usually be inappropriately low for the degree of anaemia. A blood film should be assessed to confirm the full blood count results and to assess signs of dysplasia within the peripheral blood. Bone marrow aspirates and trephines should be performed to consolidate the data obtained from the blood film, also to allow the enumeration of blasts and to assess for signs of fibrosis. Immunocytometric enumeration of CD34 may be useful to determine blast percentages, although care should be taken, as not all blasts express CD34. Perls' stain should be used to evaluate iron stores and to allow the enumeration of any ring sideroblasts. Biochemical tests—including lactate dehydrogenase, to measure cell turnover and ineffective erythropoiesis; ferritin to assess iron levels and to provide valuable prognostic information, as hyperferraemia is associated with poorer outcome; and β2 microglobulin—should be performed.

Although this process is important to follow, the majority of cases will be determined, according to the WHO classification, by morphology and cytogenetic results.

Now that the myelodysplastic syndromes have been considered, we will examine the myeloproliferative neoplasms. We will then explore the myeloproliferative neoplasms overlapping with MDS. Myelodysplastic/myeloproliferative diseases exhibit dysplastic features and a higher cell count, since these cells have a larger proliferative component than purely MDS.

11.4 Myeloproliferative neoplasms

Myeloproliferative neoplasms have a blast level below 20% and are generally associated with an increased bone marrow cellularity. Typically, peripheral cell counts are raised—particularly in the early phases—but in later stages bone marrow fibrosis can occur. Fibrosis compromises haemopoiesis, leading to peripheral blood cytopenias.

Of the conditions outlined below, polycythaemia vera (PV), essential thrombocythaemia (ET), and primary idiopathic myelofibrosis (PMF) can morph into one another throughout their natural course. Where myelofibrosis follows PV or ET, the diagnosis of post-PV myelofibrosis or post-ET myelofibrosis can be made, respectively. Criteria for the diagnosis of myelofibrosis secondary to PV and essential thrombocythaemia have been clearly outlined by the WHO.

One of the most important mechanisms of oncogenesis in PV, PMF, and ET is the acquisition of a point mutation in the JAK2 tyrosine kinase. Found in approximately 90% of cases of PV, 50% of PMF patients, and 30% of ET, the substitution of amino acids valine for phenylalanine at position 617 results in the constitutional activation of this tyrosine kinase and enhanced sensitivity to specific growth factors, including Epo, GM-CSF, Tpo, IL-3, and stem cell factor.

The *JAK2* allele burden is higher in PV than ET, and this has been shown to influence the expression of a polycythaemic phenotype. In 70–85% of patients diagnosed with ET, or where JAK2 and MPL (thrombopoietin receptor) mutations are not present, gain-of-function mutations in exon 9 of the **calreticulin** (*CALR*) gene can be identified. Indeed, it is evident that *CALR* are important driver mutations associated with MPN stem cells, which do not change during the course of the disease and enhance signalling through the MPL–JAK–STAT pathway.

calreticulin
A protein, associated with multiple functions, including calcium storage within the endoplasmic and sarcoplasmic reticulum, which functions as an endoplasmic reticulum-based chaperone protein that is involved in the clearance of misfolded proteins prior to their export to the Golgi.

Key Points

JAK2 is a receptor-associated tyrosine kinase involved in the signalling from a number of cytokine receptors, including EpoR. JAK2 also acts as a chaperone facilitating the trafficking of EpoR to the Golgi for post-translational modification prior to receptor expression on the cell surface.

Binding of cytokines to their JAK2-associated receptor induces the formation of either homo- or heteromeric receptors and the clustering of JAK2. Receptor-associated JAK2 kinases then phosphorylate one another (in a process called transphosphorylation), and tyrosine residues within the cytokine receptor undergo phosphorylation. STAT molecules associate with the receptor and are phosphorylated, leading to the activation of downstream signalling pathways. Of particular importance is the association between JAK2 and the STAT pathway. Subsequent dimerization of STAT and localization within the nucleus results in increased transcription and translation of STAT-inducible genes.

JAK2 is composed of seven JAK homology domains (JH1–7). JH1 comprises the tyrosine kinase domain and JH2 the pseudokinase or tyrosine kinase-like domain. The pseudokinase domain negatively regulates JH1, controlling phosphorylation and activation of JAK2. Wild-type V617, located within JH2, preserves the inactive conformation of a critical structure called the activation loop. V617F modifies the structure of JH2, preventing inhibition of JH1, allowing continuous signalling utilizing the newly positioned activation loop.

11.4.1 Polycythaemia vera

Polycythaemia vera (PV) is characterized by an increase in an individual's red cell mass >25% above the patient's mean predicted value. In half of patients, thrombocytosis is a feature and two-thirds of patients demonstrate neutrophilia. Surrogate markers, including patient haematocrit (discussed later in this section) and haemoglobin concentration, have been adopted, although these should be interpreted in the context of the patient's clinical examination and medical history.

Cross reference
Polycythaemia vera is contrasted with erythrocytosis in Chapter 4.

PV is a clonal erythroid disorder, and so all defective cells will be the product of a mutated progenitor and nucleated cells will harbour any cytogenetic abnormalities present in that progenitor cell. The vast majority of patients harbour *JAK2* V617F mutations. Other patients demonstrate variant mutations within JAK2, leading to constitutional activation of this receptor and increased tyrosine kinase signalling.

In the absence of clear molecular evidence of clonality, the difficulty with diagnosing PV is determining whether the erythrocytosis is the consequence of a clonal population of red cells, or another event raising the red cell count and haematocrit, causing secondary erythrocytosis.

METHOD Calculating a patient's red cell mass

In nuclear medicine, red cell mass is measured using sodium radiochromate (^{51}Cr) or sodium pertechnetate (^{99m}Tc)-labelled red cells as per the International Committee for Standardization in Haematology (ICSH) guidelines. The value obtained should be compared to the patient's predicted RCM, and is calculated as follows:

For males: [(1486 × surface area) − 825] mL

For females: [(1.06 × age) + (822 × surface area)] mL.

oximetry
A non-invasive method for determining the amount of oxygen within arterial blood.

To exclude secondary erythrocytosis, serum erythropoietin should be measured and is expected, in PV, to be low for the degree of erythrocytosis; pulse **oximetry** should be used to investigate for insufficient oxygen in the blood (hypoxaemia), and would be expected to fall within the reference range in PV; investigations such as high-performance liquid chromatography (HPLC) or isoelectric focusing (IEF), to exclude high-affinity haemoglobins, should be completed; and renal pathology should be excluded, as reduced blood flow through the glomerulus may induce erythropoietin (Epo) production, and consequently raise the red cell count. Additionally, in some cases of liver disease, Epo-like peptides may be synthesized, leading to an increased red cell count, so these should be investigated.

A major diagnostic criterion is the haematocrit. This threshold result is stratified according to *JAK2* mutation and is indicative of PV (as in Table 11.4).

The combination of a high haematocrit or raised red cell mass, plus evidence of *V617F*, provides the diagnosis of PV. In cases where *V617F* is absent, a raised haematocrit or raised red cell mass (RCM) is necessary, plus the exclusion of any cause of secondary erythrocytosis or another acquired genetic abnormality within the haemopoietic cells (excluding *BCR–ABL1*). A combination of two or more of the following can also be used in the absence of an additional acquired genetic abnormality: prolonged thrombocytosis, neutrophilia, or low serum Epo concentration. (See also Box 11.7.)

11.4.2 Primary myelofibrosis

Primary idiopathic myelofibrosis (PMF) is a clonal disorder characterized by the proliferation of immature myeloid precursors. Mutation of the pluripotent stem cell precursor is believed to induce the proliferation of erythroblasts, megakaryoblasts, and their progeny. Megakaryoblasts are thought to play a key role in the development and potentiation of the myelofibrotic phenotype. Megakaryoblasts can stimulate fibroblast proliferation via basic fibroblast growth factor b (b-FGF) and transforming growth factor β (TGF-β), leading to the development of fibrosis.

TABLE 11.4 Stratification of haematocrit according to sex and *JAK2 (V617F)* mutational status.

	V617F positive	*V617F* negative
Males	>0.52	>0.48
Females	>0.60	>0.56

BOX 11.7 PV guidelines

Comprehensive guidelines outlining the strategy employed in investigating and diagnosing PV are provided by the British Committee for Standards in Haematology (BCSH); reference details can be found in the Reference section at the end of this book.

In the early stages, PMF is characterized by a leucoerythroblastic blood picture featuring teardrop poikilocytes. An increased white cell count composed primarily of raised neutrophils, often with raised eosinophils and basophils, is a feature. Extramedullary haemopoiesis is commonly associated with PMF, and an associated hepatosplenomegaly is also frequently encountered. In keeping with the megakaryocytic component of this disease, a raised platelet count is often observed in the early stages. As the disease progresses, thrombocytopenia develops, and the white cell count drops as haemopoiesis becomes compromised by the increasing marrow fibrosis. In up to 10% of patients, PMF transforms to AML.

Primary myelofibrosis can be subdivided into two disease processes:

- PMF, prefibrotic/early stage;
- PMF, overt fibrotic stage.

Both forms of PMF are associated with megakaryocytic atypia, including changes to the nucleocytoplasmic ratio, nuclear structure, and staining intensity and their distribution within the marrow. Investigation of key driver mutations including *JAK2*, *CALR*, and *MPL* is necessary and, when absent, evidence of mutations to show a clonal disorder is required. Although common, minor diagnostic criteria include anaemia that cannot be attributed to another cause, a WBC $\geq 11 \times 10^9$/L, splenomegaly, and raised lactate dehydrogenase (LDH).

Prefibrotic and overt fibrotic stages can be differentiated from one another by the presence of marrow fibrosis. In prefibrotic PMF, ≤grade 1 reticulin/collagen fibrosis (scattered or loose network of fibres) is required compared with ≥grade 2 fibrosis (diffuse and dense fibrosis) in the overt fibrotic stage. Examples of fibrotic change and associated grading can be seen in Figure 11.13 where comparisons between haematoxylin and eosin and Gomori trichrome stain can be seen. Where a diagnosis of prefibrotic PMF is considered, it is necessary to exclude essential thrombocythaemia (ET) from the differential diagnosis as ET has a more favourable prognosis than prefibrotic PMF.

11.4.3 Essential thrombocythaemia

Essential thrombocythaemia (ET) is a clonal disorder of megakaryocytes, characterized by a persistently raised platelet count of $>450 \times 10^9$/L in the absence of any reactive condition. In addition to the presence of JAK2 V617F, up to 10% of patients will harbour a clonal mutation of *MPL*, the gene encoding the thrombopoietin receptor. In a number of cases, this results in the amino acid tryptophan being replaced by another, usually leucine, at position 515 (W515). MPL mutations increase thrombopoiesis independently of environmental stimulation, thus leading to a growth advantage in those cells harbouring these mutant receptors. Reactive conditions should be identified via thorough clinical examination or a raised erythrocyte sedimentation rate (ESR) or C-reactive protein (CRP). Previously, ET was suggested with a platelet count $>600 \times 10^9$/L for a duration of two months or more, but the WHO has amended to the lower value of $>450 \times 10^9$/L based on some patients experiencing the complications of ET at platelet counts

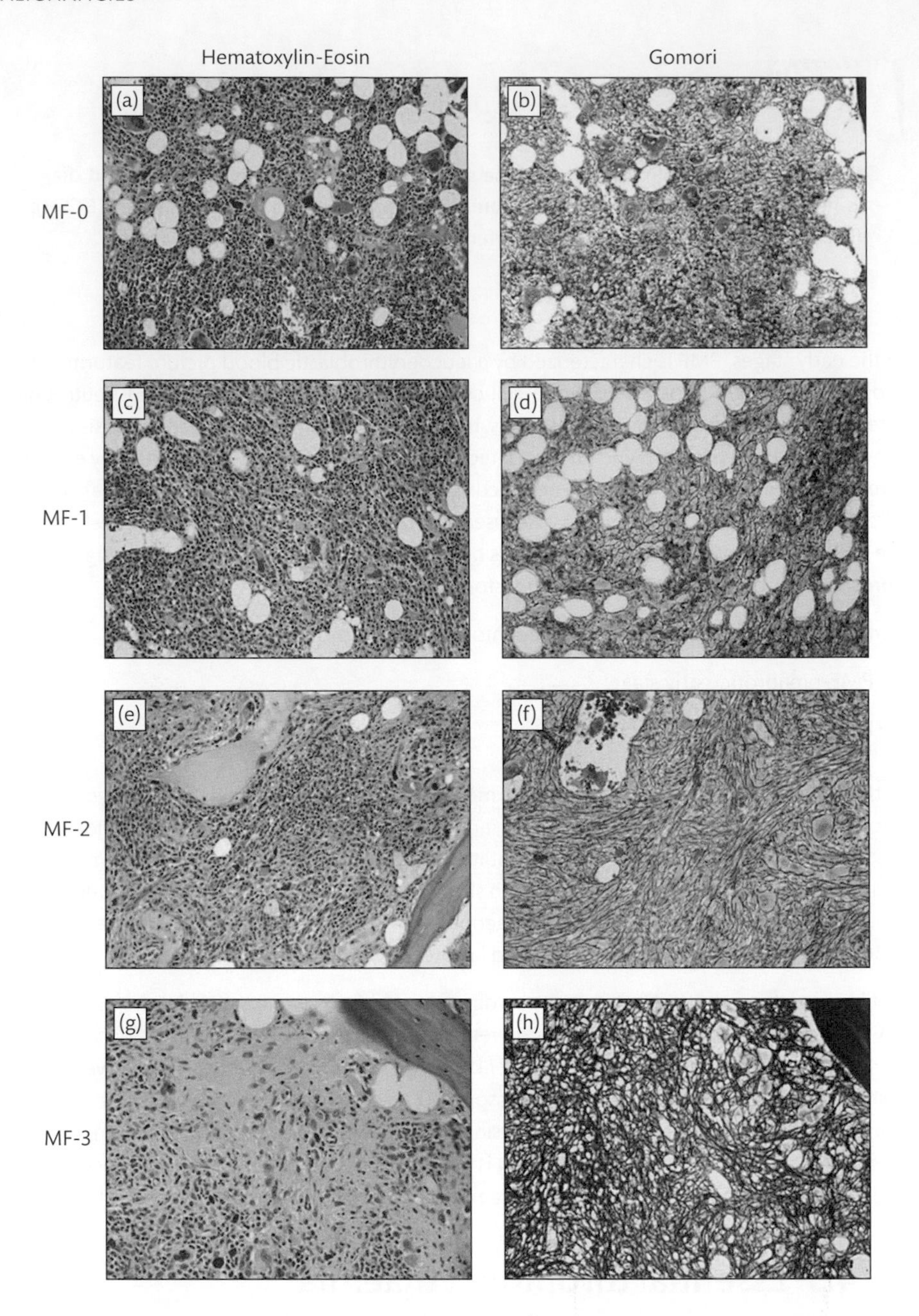

FIGURE 11.13

Examples of bone marrow fibrosis (haematoxylin-eosin and Gomori, × 20) and its grading. (a, b) Pre-fibrotic primary myelofibrosis (MF-0): scattered linear reticulin fibres with no intersections; (c, d) early-stage primary myelofibrosis (MF-1): loose network of reticulin fibres with many intersections; (e, f) fibrotic stage primary myelofibrosis (MF-2): diffuse and dense increase in reticulin fibres, with extensive intersections and occasionally with focal bundles of collagen; (g, h) fibrotic stage primary myelofibrosis (MF-3): diffuse and dense increase in reticulin fibres, with extensive intersections and coarse bundles of collagen. Reproduced with permission from Gianelli U, Vener C, Bossi A et al. *The European consensus on grading of bone marrow fibrosis allows a better prognostication of patients with primary myelofibrosis*. *Modern Pathology* 2012: 25: 1193–202. Doi: https://doi.org/10.1038/modpathol.2012.87.

below 600 × 10^9/L. Other myeloproliferative disorders should also be excluded where practical, although in some patients ET may coexist with another disease entity.

Diagnosis relies on the exclusion of diseases which may cause a secondary thrombocytosis. Iron deficiency should be excluded, as this is associated with a reactive increase in platelet count.

It is also important to exclude the presence of the Philadelphia chromosome and BCR-ABL1 to ensure CML is not missed.

Detailed investigations should be completed to ensure the patient has none of the morphological or cytogenetic features of any other myeloid neoplasm. The best diagnostic tool for this is bone marrow examination, allowing for the morphology of megakaryocytes and other haemopoietic progenitors to be fully appreciated. Megakaryocytes may vary in shape and size, ranging from small to giant, accompanied by either scant or voluminous cytoplasm. Hyperlobulated nuclei may be apparent.

SELF-CHECK 11.9

Outline the interrelationship between the myeloproliferative disorders.

Cross reference

Bone marrow assessment is outlined in Chapter 10.

11.4.4 Chronic myelogenous leukaemia, BCR-ABL1 positive

Chronic myelogenous leukaemia (CML), also known as chronic myeloid leukaemia or chronic granulocytic leukaemia (CGL), is the most clinically researched haematological malignancy in history. CML was the first malignant disease to be associated with a particular chromosomal abnormality, the Philadelphia chromosome, and the first to have significant reported success with a molecular targeted therapy—imatinib mesylate. Following the successes of imatinib as a first-line therapy for CML, significant advances have been made to improve the range of drugs available to treat this condition. Second-generation tyrosine kinase inhibitors including nilotinib, dasatinib, and bosutinib are now available and can be used as second- and third-line treatments. Treatment failure often occurs following the acquisition of point mutations within BCR-ABL1 prohibiting imatinib binding. Nilotinib is approximately 30-fold more potent than its predecessor, imatinib, which affects both the dose required and its effectiveness.

In 1960, Nowell and Hungerford made the association between the typical morphological features we see in patients diagnosed with CML and a minute chromosome, the shortened chromosome 22. In 1973, the origin of this shortened chromosome 22 became clear, when Rowley identified t(9;22), the Philadelphia translocation. The precise breakpoints to this translocation have been identified as: t(9;22)(q34;q11). (See also Box 11.8.)

BOX 11.8 *First report linking malignant disease with an acquired chromosomal abnormality*

Nowell and Hungerford's original findings were the first linking a malignant disease to an acquired chromosomal abnormality. The minute chromosome reported, using microscopic examination of cells derived from seven patients diagnosed with chronic granulocytic leukaemia, was later identified as chromosome 22, and is now abbreviated to der(22): the Philadelphia chromosome.

CML is a clonal disorder of pluripotent stem cells causing proliferation of myeloid precursors, and their differentiation into mature effector cells. These effector cells are likely to be qualitatively abnormal, so they are functionally impeded. CML is not associated with maturation arrest, but there is a failure of apoptosis. In the case of CML, an increased proliferation rate accompanied by apoptotic failure results in an increased white cell count. In many cases, the red cell count and platelet count are also increased.

The defining feature of CML is the Philadelphia (Ph) chromosome or the presence of the fusion product of t(9;22)(q34;q11), BCR-ABL1. Using cytogenetic analysis, 95% of patients demonstrate the Ph chromosome. A further 2.5% of patients have a masked (cryptic) translocation undetectable using conventional cytogenetic techniques, despite evidence of BCR-ABL1 using molecular techniques. The remaining 2.5% are considered to be Ph chromosome negative, have no molecular evidence of BCR-ABL1 and therefore do not have CML, despite a morphological blood picture indicative of this disease.

Atypical morphology in the absence of the Ph chromosome can be considered a myelodysplastic/myeloproliferative disease that is discussed in section 11.5.2.

The key fusion gene found within the Philadelphia chromosome der(22), *BCR-ABL1*, is formed from the proto-oncogene *ABL1*, originally located on chromosome 9—encoding a non-receptor tyrosine kinase—and the *BCR* (breakpoint cluster region) gene. The actual physiological role of the BCR protein has yet to be effectively demonstrated. The Philadelphia translocation can produce protein products of three sizes: p190 (190 kDa), p210 (210 kDa), and p230 (230 kDa), depending upon the site of the breakpoint within the *BCR* gene—p210 is by far the most common of the three fusion products found in CML. Figure 11.14 shows a simplified version of the formation of the different-sized fusion proteins.

The clinical course of CML can be subdivided into three distinct phases: chronic phase, accelerated phase, and blast crisis. Each of these phases has distinctive associated clinicopathological features, although with the effectiveness of the tyrosine kinase inhibitors used in clinical practice, accelerated phase is becoming a less frequent manifestation of this condition.

The majority of patients who present are in *chronic phase*. Chronic phase is usually associated with a hypercellular bone marrow with peripheral leucocytosis, and is responsive to therapy. The majority of patients at presentation have splenomegaly or hepatosplenomegaly, and this can be reduced in chronic phase using standard chemotherapeutic agents. Figure 11.15 shows

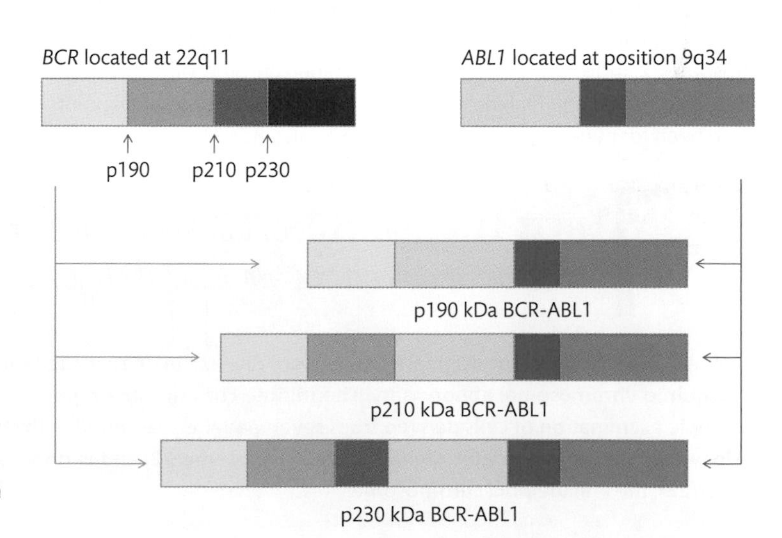

FIGURE 11.14
The formation of BCR-ABL1 fusion proteins of differing size. *ABL1* remains intact, but different breakpoints within *BCR* can result in different-sized fusion genes and proteins.

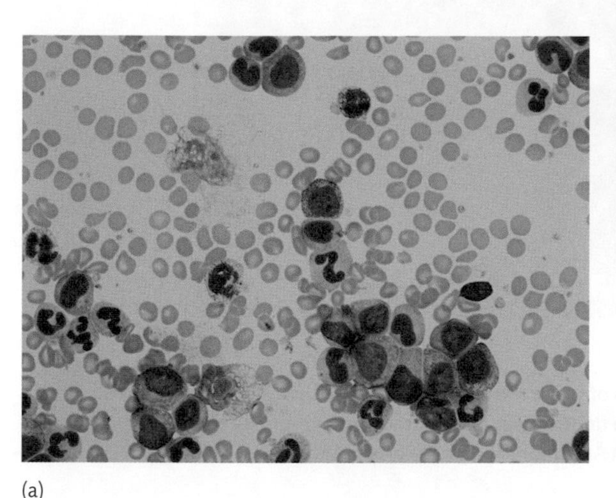

(a)

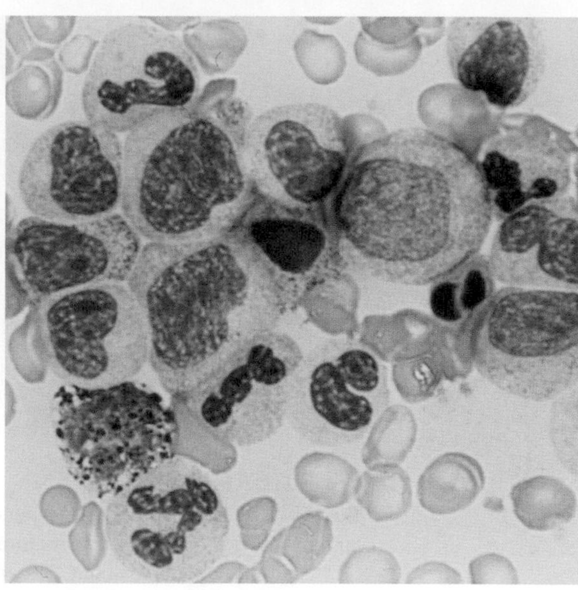

(b)

FIGURE 11.15

(a) CML in chronic phase. Large numbers of myeloid cells are present within the peripheral blood. Varying degrees of maturation are present, although many are in the process of maturing. (b) High power image of CML cells. Image (a) courtesy of Jackie Warne, Haematology department, Queen Alexandra Hospital, Portsmouth. Image (b) from Provan D, Baglin T, Dokal I, and de Vos J. *Oxford Handbook of Clinical Haematology*, 4th edn, Oxford University Press, Oxford, 2015. By permission of Oxford University Press.

a peripheral blood film from a patient in chronic phase. Note the large numbers of differentiated, maturing cells. There is no evidence of maturation arrest.

As the disease progresses to *accelerated phase*, it becomes more difficult to treat. Indeed, for patients receiving tyrosine kinase inhibitors, the manifestations of accelerated phase require refinement. Accelerated phase is currently associated with an increasing blast count, although this remains less than 20%, and the basophil count equals or exceeds 20% in the peripheral blood. Leucocytosis becomes difficult to control, and is accompanied by either a persistent thrombocytopenia ($<100 \times 10^9$/L) or a thrombocytosis (in some cases $>1000 \times 10^9$/L). The presence of thrombocytopenia is part of the disease process, and is not secondary to **cytoreductive therapy**. Additional cytogenetic abnormalities, including additional Philadelphia chromosomes, trisomy 8, isochromosome 17q, trisomy 19, or complex karyotype may be found, suggesting disease progression.

cytoreductive therapy
A form of therapy effective at reducing high blood cell counts. An example of cytoreductive therapy is hydroxycarbamide.

Blast crisis describes the transformation to acute leukaemia. The blast count is 20% or higher, and immunophenotyping required to determine the lineage of the acute leukaemia. It is important to note that the extramedullary (outside the bone marrow) infiltration of blasts may occur, which will not be obvious from bone marrow examination, and can still lead to a diagnosis of blast crisis. Furthermore, evidence of lymphoblasts in the peripheral blood or bone marrow before the 20% blast threshold is reached should trigger close monitoring of the patient for lymphoblastic transformation. Figure 11.16 shows a case in which CML has transformed to acute leukaemia. (See also Case study 11.2.)

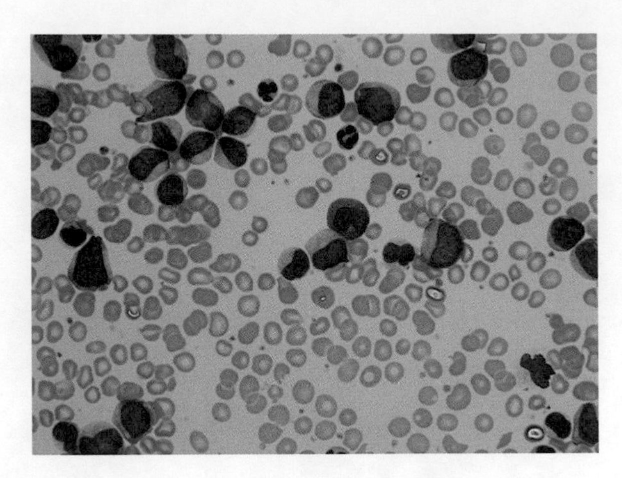

FIGURE 11.16
CML in blast crisis. Large numbers of blasts are apparent in the peripheral blood. Note the prominent nucleoli. Some mature neutrophils are apparent. Image courtesy of Jackie Warne, Haematology department, Queen Alexandra Hospital, Portsmouth.

Cross reference
CML is briefly introduced in Chapter 9.

SELF-CHECK 11.10

Give a brief account of the different phases of chronic myeloid leukaemia.

11.4.5 Chronic neutrophilic leukaemia

Chronic neutrophilic leukaemia (CNL) is an exceptionally rare myeloproliferative disorder associated with a neutrophil leucocytosis in the absence of infection, inflammation, or confounding myeloid malignancy.

CNL is a disorder that, until recently, was classified through the process of exclusion. However, mutation of the gene encoding the granulocyte colony-stimulating factor (G-CSF) 3 receptor CSF3R- has now been identified in the majority of cases. G-CSF provides proliferative and survival signals for granulocytes, with mutation of CSF3R replicating this in a ligand-independent manner. The most common mutation, *CSF3R*T618I – enables an accurate diagnosis, although other CSF3R-activating mutations may otherwise be present. Patients with CNL demonstrate a peripheral blood leucocyte count of >25 × 10^9/L with >80% neutrophils or band forms. Less than 10% of the peripheral blood leucocytes are immature granulocytes. The bone marrow appears hypercellular, with increased granulopoiesis evident. In the absence of CSF3R-activating mutations, there should be no evidence of any cytogenetic abnormalities characteristic of other myeloid diseases, including: the Philadelphia chromosome and its BCR–ABL1 fusion product (diagnostic for CML), nor rearrangement of *PDGFRA*, *PDGFRB*, *FGFR1*, and *PCM1-JAK2*, which signal the presence of other myeloid malignancies. Other myeloproliferative disorders and MDS should also be excluded. The majority of patients demonstrate hepatosplenomegaly.

Cross reference
Multiple myeloma is discussed in Chapter 12.

Historically, a number of cases of CNL have been associated with multiple myeloma (discussed in chapter 12), and recent evidence suggests that the neutrophilia may be driven by malignant plasma cells in an inflammatory process, rather than it being an independent clonal neutrophilic manifestation.

CASE STUDY 11.2 An evolving case of chronic myeloid leukaemia

Patient history

- A 64-year-old male, diagnosed with CML 14 months previously, returned to his haematologist following a marked decline in his physical well-being. He noted an uncomfortable mass in his abdomen and increasing shortness of breath on exertion.
- Physical examination by the haematologist noted splenomegaly.
- FBC and blood film were initially performed.

Results 1

(Please note that not all WBC species are included below.)

Blood film

Increased numbers of neutrophils and their precursors were apparent with concurrent absolute basophilia and occasional blasts noted. Examination of the platelets and red cells revealed thrombocytopenia and anisopoikilocytosis, respectively.

Index	Result	Reference range
WBC	82.3	(RR 4–10 × 10^9/L)
Neutrophils	38.2	(RR 2–7 × 10^9/L)
Basophils	2.4	(RR 0.02–0.1 × 10^9/L)
Blasts	0.85	(RR <0.02 × 10^9/L)
Promyelocytes		0.9 × 10^9/L
Myelocytes		4.5 × 10^9/L
Metamyelocytes		8.2 × 10^9/L
Band forms	20.6	(RR 0.1–0.8 × 10^9/L)
Hb	101	(RR 133–167 g/L)
MCV	86	(RR 77–98 fL)
Plt	98	(RR 143–400 × 10^9/L)

Significance of results 1

Checking these results against the patient's records (not shown) it appears to be a significant change in the patient's haematological parameters. The white blood cell (WBC) count has more than doubled (from 14.3 × 10^9/L), thrombocytopenia is now apparent, and the patient has a notable splenomegaly. The reduced platelet count could be a consequence of the enlarged spleen sequestering the platelets. The increased blast percentage is a cause for concern and so a bone marrow biopsy and immunophenotyping analysis was requested.

Results 2

Bone marrow

Bone marrow analysis revealed a hypercellular marrow with a 17% blasts, 7% basophil, and left-shifted myelopoiesis. Micromegakaryocytes were frequent.

Flow cytometry

Key immunophenotyping demonstrated 17% of cells with a B-lineage phenotype expressing CD34, CD19, CD10, and TdT.

Significance of results 2

These further investigations indicate that this patient has progressed from chronic phase to accelerated phase CML. The patient's blast percentage is now 17% and immunophenotyping confirms the presence of CD34 positive blast cells. Importantly, the critical 20% blast threshold has not yet been exceeded. The presence of B-lineage markers shows that the patient is now in the process of transforming to an acute lymphoblastic leukaemia. Cytogenetic analysis should be performed to assess for additional chromosomal lesions, which can be used to predict disease outcome and therapeutic options.

11.4.6 Chronic eosinophilic leukaemia, not otherwise specified

A raised eosinophil count is a common feature of allergic reactions, parasitic infections, and malignancy to name but a few, and it is important to consider these when attempting to make a diagnosis of chronic eosinophilic leukaemia (CEL) or idiopathic hypereosinophilic syndrome (HES).

A significant overlap between CEL and HES exists and the WHO has made an attempt to allow differentiation between them. The defining feature common to both conditions is a raised eosinophil count >1.5 × 10^9/L for six months or more.

All reactive causes of an eosinophilia should be excluded, as should any malignant condition known to cause a raised eosinophil count, either directly (as part of the malignant clone), or indirectly (through chemical mediation). An abnormal T-cell population should also be excluded; that could promote a sustained eosinophilia through the activity of interleukin-5 (IL-5).

CEL is diagnosed when a clonal population of eosinophils is identified, or there is an increase in myeloblasts (below 20%) accompanied by a persistently raised eosinophil count. If the clonal nature of the eosinophils cannot be established, and the blast percentage is not raised, then a diagnosis of idiopathic HES is appropriate.

11.4.7 Myeloproliferative neoplasms, unclassifiable

As the name suggests, chronic myeloproliferative disease, unclassifiable (CMPD-U) should be used when we are unable to classify a thoroughly investigated CMPD as a specific disease entity.

The presence of the Philadelphia chromosome and the fusion product BCR-ABL1 should be excluded, otherwise the condition should be considered CML. It is important that CMPD-U is not used for inadequately investigated cases.

Now that we have considered myeloproliferative neoplasms and MDS, we should consider overlap conditions that show both myelodysplastic changes and are myeloproliferative.

11.5 Myelodysplastic/ myeloproliferative diseases

This category includes diseases that demonstrate dysplastic features, a hypercellular bone marrow, and increased peripheral cell counts. Included within this category are:

- chronic myelomonocytic leukaemia (CMML);
- atypical chronic myeloid leukaemia (aCML), *BCR-ABL1*$^-$;
- juvenile myelomonocytic leukaemia (JMML);
- MDS/MPN with ring sideroblasts and thrombocytosis (MDS/MPN-RS-T);
- MDS/MPN, unclassifiable.

Each of the MDS/MPN neoplasms is associated, to some degree, with mutation of *SETBP1*. Located on chromosome 18, the protein product, SETBP1, is involved in chromatin modification. Somatic mutation of SETBP1 enhances its stability compared to the wild type protein and enhances its binding to target DNA promoter regions. Mutant SETBP1 enhances HOXA9 and HOXA10 activity, which increases cell self-renewal. Furthermore, mutant SETBP1 promotes greater proliferative potential than the wild type protein.

Somatic SETBP1 mutation is a poor prognostic factor and is found in:

- CMML—15% of cases;
- atypical CML—25–33% of cases;
- JMML—8% of cases;
- MDS/MPN-RS-T—13% of cases.

We will now examine each of these conditions in more depth.

11.5.1 Chronic myelomonocytic leukaemia

CMML is characterized by a hypercellular bone marrow and a normal or increasing white cell count. A sustained peripheral monocytosis (>1 × 10^9/L) is apparent, with monocytes accounting for ≥10% of the leucocyte count with no obvious reactive cause. Monocytes are precursors to tissue macrophages and usually increase in number during bacterial infections or following tissue damage to remove cell debris. Bacterial infection and tissue damage should be excluded prior to a diagnosis of CMML being made. Variable signs of dysplasia should be apparent in the myeloid series. Cytochemical stains, including ANAE, can be used to identify cells of a monocytic lineage, where cells will stain red/brown, whilst CAE can assist in differentiating dysplastic granulocytes staining bright blue. There should be no evidence of the **Philadelphia chromosome**, or the BCR-ABL1 transcript (otherwise suggestive of a diagnosis of CML).

Philadelphia chromosome
This denotes the derived chromosome 22 der(22) not to be confused with t(9;22). The translocation process is abbreviated to t(9;22).

CMML should be separated into proliferative or dysplastic types, defined by the WBC. Proliferative type CMML is associated with a WBC ≥13 × 10^9/L, whereas dysplastic type has WBC counts <13 × 10^9/L. CMML is further subdivided into three groups based on the number of blood and bone marrow blasts (the term 'blast' in this instance includes the promonocyte stage of maturation):

- CMML-0 includes peripheral blood blasts <2% with <5% marrow blasts;
- CMML-1 comprises 2–4% peripheral blood blasts and <5–9% bone marrow blasts;
- CMML-2 demonstrates 5–19% peripheral blood blasts and 10–19% bone marrow blasts.

Patients with a higher blast percentage have a poorer prognosis and are at greater risk of transforming to AML.

11.5.2 Atypical chronic myeloid leukaemia, *BCR-ABL1* negative

Although the name of this disease suggests that it is a variant of CML, atypical CML (aCML) is caused by a separate pathological process. Atypical CML is not associated with the Philadelphia chromosome, and does not show evidence of BCR-ABL1 at the molecular level. Furthermore, common mutations associated with myeloproliferative neoplasms including *JAK2*, *CALR*, and *MPL*, are not associated with aCML. However, although non-diagnostic for the condition, mutations of *ASXL1* (20–70%), *SETBP1* (10–48%), *NRAS* (10–30%), *KRAS* (8–10%), *JAK2* (4–8%),

TET2 (16–30%), and *SRSF2* (12–40%) are the most common. Cooperative mutations in *SETBP1* and *ASXL* occur in 48% of patients.

Within both the peripheral blood and bone marrow, aCML often shows leukocytosis, predominantly attributable to left-shifted neutrophilia accounting for ≥10% of the leucocyte population with evidence of dysgranulopoiesis. Basophils and monocytes should account for <2% and <10% leucocytes, respectively.

Cross reference

Please compare to chronic myelogenous leukaemia, *BCR-ABL1* positive in section 11.4.4.

The prognosis for patients diagnosed with aCML is particularly poor, with a median survival of 15 months. The majority of patients succumb to complications of bone marrow failure, including overwhelming infections, complications of anaemia, or bleeding. Up to 20% of patients will transform to acute myeloid leukaemia.

11.5.3 Juvenile myelomonocytic leukaemia

This is a rare myelodysplastic/myeloproliferative condition associated with infants and young children, accounting for approximately 2% of all childhood leukaemias. Characterized by a proliferative neutrophil and monocytic compartment, juvenile myelomonocytic leukaemia (JMML) can be differentiated from CML through the absence of the Philadelphia chromosome and the lack of BCR-ABL1.

In the vast majority of cases, dysregulation of Ras leading to its hyperactivity can be identified. Some 35% of patients have gain-of-function mutations of *PTPN11*, which encodes Shp2, a protein tyrosine phosphatase; 35% of patients have gain-of-function mutations in either *NRAS* or *KRAS*; and 15% of patients have inactivating mutations of *NF1*. Wild-type NF1 activates a GTPase which negatively regulates Ras.

Current guidelines for the classification of JMML include a monocyte count ≥1 × 10^9/L. Blasts must not exceed 20%, whilst *BCR-ABL1* must be absent. The total WBC may be more than 10 × 10^9/L and in some cases the patient's HbF percentage may be higher than expected for their age. The patient will demonstrate splenomegaly. Immature granulocytes may be present in the peripheral blood, and these may be hypersensitive to GM-CSF. Genetic studies must be conducted with somatic mutation of *PTPN11*, *KRAS*, or *NRAS* and monosomy 7 investigated. Furthermore, mutation of *NF1* should be explored, with evidence of germline mutation of *CBL* sought.

11.5.4 MDS/MPN-RS-T

The median age of diagnosis of this complex overlap syndrome is 70–75 years. MPS/MPN-RS-T is a rare neoplasm associated with dyserythropoiesis, anaemia, and bone marrow-derived ring sideroblasts exceeding 15% of erythroid precursors. A sustained platelet count ≥450 × 10^9/L is required, with bone marrow examination revealing large, atypical megakaryocytes. It is important to exclude essential thrombocythaemia and reactive thrombocytosis prior to a diagnosis of MDS/MPN-RS-T. Approximately 85% of patients possess *SF3B1* mutation—although this has no impact on the requirement for the presence of 15% ring sideroblasts, as applies in MDS-RS-SLD and MDS-RS-MLD. *SF3B1* mutation is often coincidental with the following: *JAK2*V617F (approx. 50%), *TET2* (approx. 25%), *ASXL1* (approx. 20%), *DNMT3A* (approx. 15%), and *SETBP1* (approx. 10%). Mutation of *ASXL1* and *SETBP1* have both been associated with a poorer prognosis.

11.5.5 Myelodysplastic/myeloproliferative disease, unclassifiable

Myelodysplastic/myeloproliferative disease, unclassifiable (MDS/MPD-U) should not be considered a subgroup where inadequately investigated cases are placed, but rather those conditions with a proliferative element showing dysplasia in one or more cell lines.

MDS/MPD-U may be misclassified in cases where patients have been administered chemotherapy, which would induce dysplastic change, or growth factors, which cause myeloproliferation. Care must be taken to exclude both scenarios prior to a diagnosis being made.

Cytogenetic features characteristic of specific disease entities, for example the Philadelphia chromosome and BCR-ABL1 fusion product, should be excluded, with patients classified according to the WHO criteria dictated by those cytogenetic abnormalities.

11.6 Mastocytosis

Mast cells are important effectors in both innate immunity and in IgE-associated immune responses. Increased numbers of mast cells showing clonal gene rearrangements and a propensity to accumulate in one or more organ systems is called mastocytosis. The main categories of mastocytosis, named according to the organ system within which the clonal mast cells are identified and their behaviour, are listed below:

- cutaneous mastocytosis:
  - diffuse cutaneous mastocytosis;
  - maculopapular mastocytosis;
  - cutaneous mastocytoma;
- systemic mastocytosis:
  - indolent systemic mastocytosis;
  - smouldering systemic mastocytosis;
  - systemic mastocytosis with an associated haematological neoplasm;
  - aggressive systemic mastocytosis;
  - mast cell leukaemia;
- mast cell sarcoma.

In addition there are a number of extremely rare subvariants to these mast cell diseases.

Commonly, a point mutation within the *c-KIT* gene is identified, leading to a substitution of valine for aspartic acid at position 816 (ASP816VAL or D816V) within the KIT receptor. The ligand for KIT is stem cell factor (SCF). D816V is an activating mutation causing enhanced and inappropriate signalling through KIT in a ligand-independent manner. Other KIT mutations are documented, and these tend to be associated with the mastocytosis variants.

There is a range of mast cell disorders demonstrating mutations within platelet-derived growth factor receptor *(PDGFRA)* and *PDGFRB*, but these should be considered as distinct entities and classified as myeloid or lymphoid neoplasms with eosinophilia and abnormalities of *PDGFRA*, *PDGFRB*, or fibroblast growth factor receptor 1 (*FGFR1*) (outlined in section 11.7).

Depending upon the subtype of mastocytosis present, there are variable clinical and laboratory features, so bone marrow analysis is important to assess the number and distribution of mast cells. This can also assist in establishing prognosis. The spleen, liver, and lymph nodes are also commonly affected. If mast cells exceed 20% of nucleated cells within the bone marrow, mast cell leukaemia should be diagnosed. Mast cells are likely to feature in the peripheral blood where a diagnosis of mast cell leukaemia is suspected, with levels of 10% of the total WBC count not uncommon.

11.7 Myeloid and lymphoid neoplasms with eosinophilia and gene rearrangements

The diseases contained within this category are diverse and uncommon. The main diseases are listed below:

- myeloid and lymphoid neoplasms with platelet-derived growth factor receptor A gene (*PDGFRA*) rearrangement;
- myeloid neoplasms with platelet-derived growth factor receptor B gene (*PDGFRB*) rearrangement;
- myeloid and lymphoid neoplasms with fibroblast growth factor receptor-1 gene (*FGFR1*) rearrangement;
- myeloid and lymphoid neoplasms with *PCM1-JAK2*.

The provisional entity *PCM1-JAK2*, formed through t(8;9)(p22;p24.1), has been included in the 2016 WHO classification, although further research is required before this is made an established discrete disease entity.

The majority of cases demonstrate a degree of eosinophil involvement, with the eosinophil count generally over 1.5×10^9/L. The malignant progenitor cell is thought to be a stem cell capable of forming myeloid or lymphoid lineages, and this is reflected in the classification title for these diseases. This would explain why in some cases there is an overt lymphoblastic leukaemia accompanied by eosinophilia. Care must be taken with this diagnosis to ensure the eosinophilia is a consequence of genetic mutation, rather than aberrant growth factor release from the malignant lymphoid cells.

Although four main disease entities are classified under this heading, many of those possessing *PDGFRA* or *PDGFRB* mutations comprise chronic eosinophilic leukaemia (CEL). Patients demonstrating the cryptic *FIP1L1-PDGFRA* fusion through deletion of the intervening chromosomal segment 4q12, or *PDGFRA* rearrangements, are included within the category 'Myeloid and lymphoid neoplasms with *PDGFRA* rearrangement'.

Patients generally possess mutations at the locus 5q31-33 involving the *PDGFRB* gene. Translocations are the common finding, unlike mutations involving 4q12, where cryptic deletion is commonly present. A number of these cases also produce CEL. Other cases may represent a variety of myeloproliferative diseases but these commonly demonstrate eosinophilia.

Abnormalities with the gene *FGFR1* are associated with translocations involving the 8p11-12 locus. These translocations give rise to a range of both myeloid and lymphoid diseases. (See also Box 11.9.)

BOX 11.9 *Growth factors*

Platelet-derived growth factor receptor

The receptor for platelet-derived growth factor (PDGF) contains α- and β-subunits which dimerize following the binding of PDGF. PDGF has mitogenic activity, i.e. mitosis can be induced when it is bound to its receptor. The binding of PDGF to PDGFR initiates the dimerization of PDGFRα- and PDGFRβ-subunits and autophosphorylation of intracellular tyrosine kinase domains. This phosphorylation induces the activation of intracellular signalling. Mutations within PDGFRα or PDGFRβ can switch this tyrosine kinase on, leading to inappropriate signalling.

Fibroblast growth factor receptor-1

This receptor is involved in intracellular signalling, inducing proliferation, cell growth, and differentiation in response to the binding of fibroblast growth factor (FGF). The extracellular region is involved in ligand binding, whilst the intracellular region contains a tyrosine kinase domain, which, when activated, is important in initiating signal transduction pathways. The translocations involving *FGFR1* induce overactivity of the receptor, causing dysregulation of cell growth and proliferation.

Chapter summary

In this chapter, we have:

- outlined the importance of appropriate classification systems;
- examined the morphological, genetic, and molecular events commonly associated with different types of myeloid malignancy;
- explored specific chromosomal abnormalities in detail to understand their causal relationship, or association, with specific disease entities;
- considered the importance of the IPSS-R system for prognostication of MDS;
- dismantled many of the subgroups adopted by the WHO system, and examined the diseases that comprise these groups in order to gain an understanding for the reasoning behind the classification process;
- examined the importance of using evidence-based practice to provide a framework for diagnostic and treatment strategies;
- explored, where relevant, current UK guidelines in relation to the diagnosis of myeloid malignancies.

Discussion questions

11.1 In the light of the recent developments in molecularly targeted chemotherapeutics, critically discuss the value of classifying specific diseases according to the WHO category *AML with recurrent genetic abnormalities*.

11.2 Critically discuss the investigative steps required for the investigation of a patient in CML blast crisis.

11.3 Critically discuss the distinction between MDS and MDS/MPN.

Further reading

- Arber DA, Orazi A, Hasserjian R, Thiele J, Borowitz MJ, Le Beau MM, Bloomfield CD, Cazzola M, and Vardiman JW. ***The 2016 revision to the World Health Organization classification of myeloid neoplasms and acute leukaemia***. *Blood* 2016: **127(20)**: 2391–405.
- Cazzola M, Della Porta, MG, and Malcovati L. ***The genetic basis of myelodysplasia and its clinical relevance***. *Blood* 2013: **122(25)**: 4021–36.
- Elliott MA and Tefferi A. ***Chronic neutrophilic leukemia: 2018 update on diagnosis, molecular genetics and management***. *American Journal of Hematology* 2018: **93**: 578–87.
- Hong M and Guangsheng H. ***The 2016 revision to the World Health Organization classification of myelodysplastic syndromes***. *Journal of Translational Internal Medicine* 2017: **5(3)**: 139–43.
- Killick SB, Carter C, Culligan D, Dalley C, Das-Gumpta E, Drummond M et al. ***Guidelines for the diagnosis and management of adult myelodysplastic syndromes***. *British Journal of Haematology* 2014: **164**: 503–25.
- McMullin MF, Harrison CN, Ali S, Cargo C, Chen F, Ewing G et al. ***A guideline for the diagnosis and management of polycythaemia vera. A British Society for Haematology Guideline***. *British Journal of Haematology* 2019: **184(2)**: 176–91.
- Montalban-Bravo G and Garcia-Manero G. ***Myelodysplastic syndromes: 2018 update on diagnosis, risk-stratification and management***. *American Journal of Hematology* 2018: **93(1)**: 129–47.
- Pardanani A, Lasho TL, Laborde RR, Elliott M, Hanson CA, Knudson RA, Ketterling RP, Maxson JE, Tyner JW, and Tefferi A. ***CSF3R T618I is a highly prevalent and specific mutation in chronic neutrophilic leukemia***. *Leukaemia* 2013: **7**: 1870–3.
- Patnaik MM and Tefferi A. ***Refractory anemia with ring sideroblasts (RARS) and RARS with thrombocytosis (RARS-T)–'2017 update on diagnosis, risk-stratification, and management'***. *Hematology* 2017: **92(3)**: 297–310.
- Reilly JT, McMullin MF, Beer PA, Butt N, Conneally E, Duncombe AS et al. ***Use of JAK inhibitors in the management of myelofibrosis: A revision of the British Committee for Standards in Haematology Guidelines for Investigation and Management of Myelofibrosis 2012***. *British Journal of Haematology* 2014: **167**: 418–38.

- Shwartz LC and Mascarenhas J. ***Current and evolving understanding of atypical chronic myeloid leukaemia***. *Blood Reviews* 2019: **33**: 74–81.
- Swerdlow SH, Campo E, Harris NL, Jaffe ES, Pileri SA, Stein H, Thiele J, and Vardiman JW (eds). ***WHO Classification of Tumours of Haematopoietic and Lymphoid Tissues***. IARC, Lyon, 2008.
- Swerdlow SH, Campo E, Harris NL, Jaffe ES, Pileri SA, Stein H, Thiele J, and Vardiman JW (eds). ***WHO Classification of Tumours of Haematopoietic and Lymphoid Tissues, volume 2***. IARC, Lyon, 2017.
- Tefferi A. ***Primary myelofibrosis: 2019 update on diagnosis, risk-stratification and management***. *American Journal of Hematology* 2018: **93**: 1551–60.
- Valent P, Akin C, and Metcalfe D. ***Mastocytosis: 2016 updated WHO classification and novel emerging treatment concepts***. *Blood* 2017: **129(11)**: 1420–7.

Answers to self-check questions and case study questions are provided as part of this book's online resources.

Visit www.oup.com/he/moore-fbms3e.

12

An introduction to classification systems: lymphoid neoplasms

Gavin Knight

In Chapter 11, we examined the principles underpinning the classification of haematological neoplasms, emphasizing the 2016 revision of the World Health Organization (WHO) classification system for myeloid diseases. In this chapter, we apply the same principles to the classification of lymphoid neoplasms. This chapter cannot consider all these malignancies in detail, but it will provide the pathobiological background or a brief description of the most common and interesting of cases.

Learning outcomes

After studying this chapter, you should confidently be able to:

- outline the main differences between precursor and mature cell neoplasms;
- describe the key recurrent cytogenetic abnormalities associated with precursor lymphoid neoplasms;
- identify key genetic mutations associated with selected lymphoid disease states and indicate how these mutations contribute to leukaemo- and lymphomagenesis;
- critically discuss some of the common mature B-cell neoplasms and outline the key behaviours of each;
- differentiate between the main plasma cell neoplasms and explain the key clinical and laboratory findings in each;
- discuss the pathogenesis of plasma cell myeloma-related bone disease;
- describe the common T-lineage malignancies and provide an outline of the role of ALK dysregulation;
- compare and contrast Hodgkin and non-Hodgkin lymphoma.

12.1 The World Health Organization (WHO) classification of tumours of haematopoietic and lymphoid tissues

The World Health Organization (WHO) classification, first published in 2001, enhanced the already popular Revised European American classification of Lymphoid neoplasms (REAL). Following the detailed 2008 update, the WHO system was again revised in 2016 and is the most comprehensive and clinically representative classification to date. The WHO classification utilizes the following data in order to enable the appropriate classification of malignancies.

- *Morphology*—this is still the cornerstone of lymphoma diagnosis, although it is important to recognize that lymphomas of different origin may share morphological features, whilst demonstrating different clinical behaviour.
- *Immunophenotype*—this is an important adjunct to morphology and provides quantitative information useful in determining B-, T-, or NK-cell lineage. Immunophenotype also helps identify the maturation stage of the lymphoma cells; for example pre-germinal centre, germinal centre, or post-germinal centre. As we saw in Chapter 10, immunophenotype can identify surrogate markers that help us predict disease outcome, and is critical for identifying cells that express combinations of lymphoid and myeloid markers.
- *Genotype*—a large number of non-random translocations are identifiable in lymphoma, some of which are diagnostic for particular subtypes; for example, follicular lymphoma is associated with t(14;18). A section entitled *B-lymphoblastic leukaemia/lymphoma with recurrent genetic abnormalities* describes aggressive conditions with typical cytogenetic characteristics that define particular disease entities.
- *Normal cell counterpart*—this is important to recognize because we can predict the behaviour of the disease, to a certain extent, by knowing the properties of the normal cellular equivalent.
- *Site of origin*—this should be considered when reporting the lymphoma findings. Is the lymphoma of nodal or extranodal origin (gastrointestinal tract, central nervous system, skin, etc.)? This information can help predict clinical outcome through disease behaviour.
- *Clinical aggressiveness and prognosis*—these are also considered and play an important role in choosing the appropriate therapy and keeping the patient informed of their condition.

The WHO classification of lymphoid neoplasms is outlined in Tables 12.1–12.6.

This chapter aims to outline some of the most commonly encountered or interesting lymphoid disease entities, but does not attempt to provide a comprehensive review of all the conditions outlined in the WHO classification.

The WHO has also included a number of 'provisional disease entities' within the lymphoid group, although these are not considered in detail in this book. These diseases have been allocated the 'provisional status' due to a lack of scientific and clinical literature that can provide an appropriate evidence base for their distinctive classification. Further research should address this deficiency of data, and attempt to provide much-needed clarity in these areas.

First, we will consider acute lymphoblastic leukaemias and lymphoblastic lymphomas, before moving on to the specific criteria used by the WHO for the classification of mature lymphoid neoplasms.

TABLES 12.1–12.6 The range of lymphoproliferative neoplasms recognized in the WHO classification system, 4th edition, volume 2 (2017 publication).

SELF-CHECK 12.1

Outline the features considered important by the WHO for the classification of lymphoproliferative diseases.

TABLE 12.1 Acute lymphoblastic lymphomas and leukaemias are incorporated into precursor lymphoid neoplasms.

B lymphoblastic leukaemia/lymphoma
B lymphoblastic leukaemia/lymphoma, not otherwise specified
B lymphoblastic leukaemia/lymphoma with recurrent genetic abnormalities
  B lymphoblastic leukaemia/lymphoma with t(9;22)(q34.1;q11.2); *BCR-ABL1*
  B lymphoblastic leukaemia/lymphoma with t(v;11q23.3); *KMT2A* rearranged
  B lymphoblastic leukaemia/lymphoma with t(12;21)(p13.2;q22.1); *ETV6-RUNX1*
  B lymphoblastic leukaemia/lymphoma with hyperdiploidy
  B lymphoblastic leukaemia/lymphoma with hypodiploidy (hypodiploid ALL)
  B lymphoblastic leukaemia/lymphoma with t(5;14)(q31.1;q32.3); *IL3-IGH*
  B lymphoblastic leukaemia/lymphoma with t(1;19)(q23;p13.3); *TCF3-PBX1*
  B lymphoblastic leukaemia/lymphoma BCR-ABL1 like
  B lymphoblastic leukaemia with IAMP21
T lymphoblastic leukaemia/lymphoma
Early T-cell precursor lymphoblastic leukaemia
NK lymphoblastic leukaemia/lymphoma

TABLE 12.2 Malignancies associated with mature cell morphology and phenotype are classified as mature B cell malignancies.

Chronic lymphocytic leukaemia/small lymphocytic lymphoma
Monoclonal B-cell lymphocytosis
B-cell prolymphocytic leukaemia
Splenic marginal zone lymphoma
Hairy cell leukaemia
Splenic B cell lymphoma/leukaemia, unclassifiable
  Splenic diffuse red pulp small B cell lymphoma
  Hairy cell leukaemia variant
Lymphoplasmacytic lymphoma
IgM monoclonal gammopathy of undetermined significance (MGUS),
Heavy-chain diseases
  Alpha heavy-chain disease
  Gamma heavy-chain disease
  Mu heavy-chain disease
Non-IgM monoclonal gammopathy of undetermined significance (MGUS)
Plasma cell myeloma
Plasma cell myeloma variants
  Smouldering myeloma
  Non-secretory myeloma
  Plasma cell leukaemia

(Continued)

TABLE 12.2 ***(Continued)***

Plasmacytoma
- Solitary plasmacytoma of the bone
- Extraosseous plasmacytoma

Monoclonal immunoglobulin deposition diseases
- Primary amyloidosis
- Light-chain and heavy-chain deposition diseases

Plasma cell neoplasms with associated paraneoplastic syndrome
- POEMS (polyneuropathy, organomegaly, endocrinopathy, monoclonal protein, skin changes)
- TEMP syndrome

Extranodal marginal zone lymphoma of mucosa-associated lymphoid tissue (MALT lymphoma)

Nodal marginal zone lymphoma
- Paediatric nodal marginal zone lymphoma

Follicular lymphoma
- Testicular follicular lymphoma
- In situ follicular neoplasia
- Duodenal-type follicular lymphoma

Paediatric type follicular lymphoma

Large B-cell lymphoma with *IRF4* rearrangement

Primary cutaneous follicle centre lymphoma

Mantle cell lymphoma
- Leukaemic non-nodal mantle cell lymphoma
- In situ mantle cell neoplasia

Diffuse large B-cell lymphoma (DLBCL), not otherwise specified

T-cell histiocyte-rich large B-cell lymphoma

Primary DLBCL of the central nervous system

Primary cutaneous DLBCL, leg type

EBV^+ DLBCL, not otherwise specified

EBV^+ mucocutaneous ulcer

DLBCL associated with chronic inflammation

Fibrin-associated DLBCL

Lymphomatoid granulomatosis

Primary mediastinal (thymic) large B-cell lymphoma

Intravascular large B-cell lymphoma

ALK^+ large B-cell lymphoma

Plasmablastic lymphoma

Primary effusion lymphoma

HHV8-associated lymphoproliferative disorders
- Multicentric Castleman disease
- $HHV8^+$ DLBCL, not otherwise specified
- $HHV8^+$ germinotropic lymphoproliferative disorder

Burkitt lymphoma

Burkitt-like lymphoma with 11q aberration

High-grade B-cell lymphoma
- High-grade B-cell lymphoma with *MYC* and *BCL2* and/or *BCL6* rearrangements
- High-grade B-cell lymphoma, not otherwise specified

B-cell lymphoma, unclassifiable, with features intermediate between DLBCL and classical Hodgkin lymphoma

TABLE 12.3 Categories that form the spectrum of WHO-defined Hodgkin lymphomas.

Nodular lymphocyte predominant Hodgkin lymphoma
Classical Hodgkin lymphoma
Nodular sclerosis classical Hodgkin lymphoma
Lymphocyte-rich classical Hodgkin lymphoma
Mixed-cellularity classical Hodgkin lymphoma
Lymphocyte-depleted classical Hodgkin lymphoma

TABLE 12.4 Range of mature T-cell and NK-cell neoplasms.

T-cell prolymphocytic leukaemia
T-cell large granular lymphocytic leukaemia
Chronic lymphoproliferative disorder of NK cells
Aggressive NK-cell leukaemia
EBV^+ T-cell and NK-cell lymphoproliferative diseases of childhood
Systemic EBV^+ T-cell lymphoma of childhood
Chronic active EBV infection of T- and NK-cell type, systemic form
Hydroa vacciniforme-like lymphoproliferative disorder
Severe mosquito bite allergy
Adult T-cell leukaemia/lymphoma
Extranodal NK/T-cell lymphoma, nasal type
Intestinal T-cell lymphoma
Enteropathy-associated T-cell lymphoma
Monomorphic epitheliotropic intestinal T-cell lymphoma
Intestinal T-cell lymphoma, not otherwise specified
Indolent T-cell lymphoproliferative disorder of the gastrointestinal tract
Hepatosplenic T-cell lymphoma
Subcutaneous panniculitis-like T-cell lymphoma
Mycosis fungoides
Sézary syndrome
Primary cutaneous $CD30^+$ T-cell lymphoproliferative disorders
Lymphoproliferative disorders
Lymphomatoid papulosis
Primary cutaneous anaplastic large cell lymphoma
Primary cutaneous peripheral T-cell lymphomas, rare subtypes
Primary cutaneous gamma delta T-cell lymphoma
Primary cutaneous CD8 positive aggressive epidermotropic cytotoxic T-cell lymphoma
Primary cutaneous acral CD8+ T-cell lymphoma
Primary cutaneous $CD4^+$ small/medium T-cell lymphoproliferative disorder
Peripheral T-cell lymphoma, not otherwise specified
Angioimmunoblastic T-cell lymphoma and other nodal lymphomas of T-follicular helper (TFH) cell origin
Angioimmunoblastic T-cell lymphoma
Follicular T-cell lymphoma
Nodal peripheral T-cell lymphoma with TFH phenotype
Anaplastic large cell lymphoma, ALK^+
Anaplastic large cell lymphoma, ALK^-
Breast implant associated with anaplastic large cell lymphoma

TABLE 12.5 Immunodeficiency-associated lymphproliferative disorders.

Lymphoproliferative diseases associated with primary immune disorders
Lymphomas associated with HIV infection
Post-transplant lymphoproliferative disorders (PTLD)
Non destructive
Polymorphic PTLD
Monomorphic PTLD (B- and T/NK-cell types)
Monomorphic B-cell PTLD
Monomorphic T/NK-cell PTLD
Classical Hodgkin lymphoma PTLD
Other iatrogenic immunodeficiency-associated lymphoproliferative disorders

TABLE 12.6 Rare WHO-classified histiocytic and dendritic cell neoplasms.

Histiocytic sarcoma
Tumours derived from Langerhans cells
Langerhans cell histiocytosis
Langerhans cell sarcoma
Indeterminate dendritic cell tumour
Interdigitating dendritic cell sarcoma
Follicular dendritic cell sarcoma
Inflammatory pseudotumour-like follicular/fibroblastic dendritic cell sarcoma
Fibroblastic reticular cell tumour
Disseminated juvenile xanthogranuloma
Erdheim-Chester disease

12.2 WHO classification of precursor lymphoid neoplasms

Key Points

In American texts, acute lymphoblastic leukaemia (ALL) is referred to as acute lympho-*cytic* leukaemia. This terminology is not employed in the United Kingdom and should be avoided. ALL is associated with an accumulation of lymphoblasts rather than mature lymphocytes, and its name should reflect this.

Lymphoblastic malignancies are thought to be derived from haemopoietic stem cells which harbour mutations ultimately leading to maturation arrest, self-renewal, and failure of apoptosis. In the context of ALL, the concept of the cancer stem cell seems to be controversial, with recent studies suggesting that some lymphoblastic cells possess stem cell characteristics, although fail to express the stem cell marker CD34.

Cross references

Cancer stem cells are outlined in Chapter 9.

CD34 is outlined in Chapter 9.

Cytogenetic techniques and immunophenotyping are both outlined in Chapter 10.

Common features of patient presentation are outlined in Chapter 10.

Using molecular techniques, we can identify clonal rearrangements in immunoglobulin or T-cell receptor genes demonstrating differentiation into a particular lymphoid lineage. Additional identification of cellular subtype or normal cell counterpart is of limited value in the selection of therapies, and so immunophenotypic analysis should often focus on identifying the following key features:

- T-cell phenotype;
- mature B-cell phenotype;
- B-cell precursor phenotype.

Approximately 88% of lymphoblastic malignancies are of B-cell lineage, with the remainder being of T-cell lineage.

It is important to identify the aberrant expression of myeloid antigens on malignant lymphoid cells, as these can be useful in differentiating this clone from the normal lymphoid population following treatment, thus enabling the appropriate assessment of minimal residual disease.

The accumulation of lymphoblasts showing minimal differentiation in the bone marrow and peripheral blood is the key finding to the diagnosis of ALL. Examples of lymphoblast morphology are shown in Figure 12.1. In the majority of patients, in excess of 20% lymphoblasts will be found in the peripheral blood. Immunophenotyping and cytogenetic analysis are crucial in providing an accurate diagnosis, especially in the light of the WHO classification, and are important in the process of prognostication.

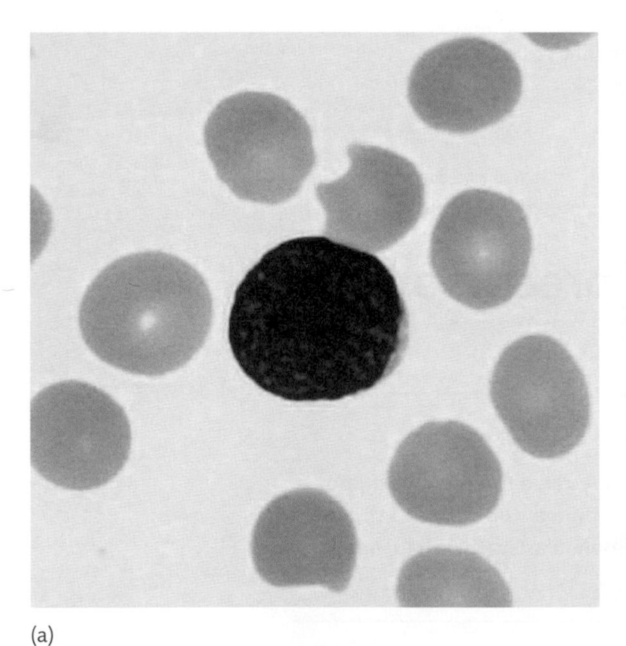

(a)

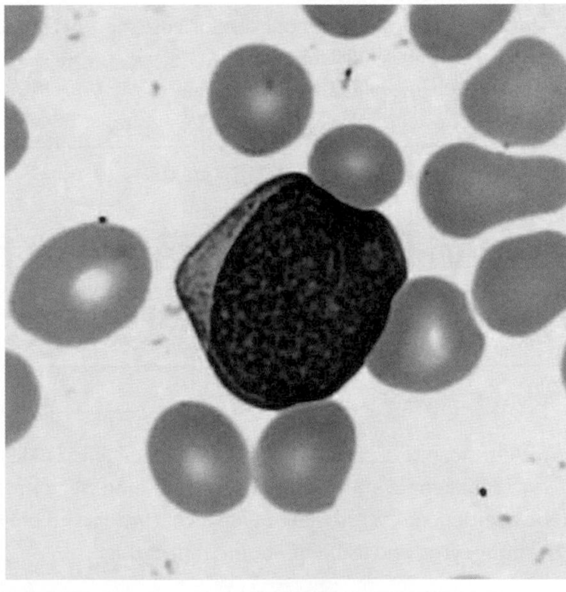

(b)

FIGURE 12.1

Examples of lymphoblast morphology associated with a diagnosis of acute lymphoblastic leukaemia. (a) The lymphoblast shows a high nucleocytoplasmic ratio—a tiny amount of cytoplasm is shown to the left of the cell structure—and open chromatin configuration; (b) lymphoblast with a slightly lower nucleocytoplasmic ratio, prominent nucleolus, and open chromatin configuration. Note the absence of cytoplasmic granules. Auer rods will not be seen in these blasts, as Auer rods signal cells of myeloid origin.

ALL is much more common in children than in adults, and age is an important factor in prognostication. Fewer than 5% of children have a heritable predisposition to lymphoblastic malignancies, with the remainder demonstrating a range of acquired genetic abnormalities implicated in the pathogenesis and progression of ALL and lymphoblastic lymphoma (LBL).

The highest incidence of ALL occurs in children between two and five years of age. There is a subsequent gradual decline followed by an increasing incidence after the age of 40 years and another peak between 80 and 84 years. Children between the ages of one and nine years have the best prognosis, with approximately 80% of patients within this age group being cured.

Overall, leukaemia-free survival in adults is approximately 34%, although this can be further stratified according to particular age brackets, and by cytogenetic composition. Between 35 and 40% of patients aged 20–50 years can expect a leukaemia-free survival, but this drops significantly to 10–20% in patients aged over 60 years. Complete remission rates are as high as 95% in children and 40–60% in adults over 50 years of age.

Clinical findings are diverse, with the majority associated with bone marrow failure. There is a risk of testicular involvement in males and the development of central nervous system (CNS) disease in both sexes. A lumbar puncture followed by cytological investigation of cerebrospinal fluid for the presence of lymphoblasts is an important additional investigation to identify CNS involvement.

The white blood cell (WBC) count is variable in childhood cases. Approximately 45% of patients have a WBC count less than 10.0×10^9/L and 50% of cases have a WBC in excess of 50.0×10^9/L. Thrombocytopenia is present in approximately 70% of patients, and this can be associated with bleeding complications. Anaemia is atypical at presentation in younger patients. In adults, the WBC count is raised in over 60% of cases and approximately 80% of patients are anaemic.

The classification of ALLs, and their division into appropriate subgroups, has become possible through the use of immunophenotyping and molecular biological techniques. The specificity and reproducibility of immunophenotyping to distinguish between myeloid and lymphoid lineages, and the ability to further differentiate lymphoid lineages into B and T lineages, relies upon the identification of particular **lineage-restricted** cluster of differentiation (CD) markers.

lineage-restricted
Refers to CD markers expressed by only one cell lineage. Lineage restriction allows us to confidently identify malignant cells. An example is CD20, a B-lineage-restricted marker.

The WHO classification system includes a 'recurrent genetic abnormalities' section, mirroring that included for the myeloid malignancies. This section recognizes non-random, reproducible cytogenetic translocations which can be used to predict disease behaviour. Not all recognized translocations are included in this section, as the vast majority are not associated with disease behaviour, morphology, or prognosis.

Based on a cell's immunophenotype, the WHO separates lymphoblastic leukaemias according to B-cell, T-cell, and NK-cell neoplasms, as shown in Table 12.7.

The WHO system does not attempt to stratify these conditions any further than these broad groups, since further subclassification has no therapeutic value.

The WHO has divided precursor lymphoid neoplasms into two broad groups based on whether the cells are characterized as B lineage or T lineage. Included within this section are lymphoblastic leukaemias and lymphoblastic lymphomas. These two entities are combined because these diseases are clinically very similar. Lymphoblastic leukaemias are associated with an accumulation of lymphoblasts within the bone marrow and peripheral blood, whereas lymphoblastic lymphomas tend to be site-restricted as a lymphomatous lesion.

In Chapter 11, we examined the classification of myeloid malignancies and established the role of classification, where possible, according to the presence of defined recurrent genetic abnormalities. The WHO has adopted this system for lymphoblastic leukaemias/lymphomas

TABLE 12.7 The key immunophenotypic features used to distinguish between the acute lymphoblastic leukaemias as recognized by the WHO.

WHO classification	Immunophenotype
B-lymphoblastic leukaemia/lymphoma	TdT+ HLA-DR+ CD79a+/- CD10+
T-lymphoblastic leukaemia/lymphoma	TdT+ cytoplasmic CD3+ CD2+/- CD4+/- CD5+/- CD7+/- CD8+/- CD10+/-
Mature B-cell (Burkitt lymphoma/leukaemia)	CD19+ CD20+ CD22+ CD10+ CD5- CD23- TdT- BCL6+ BCL2-

Key: BCL = B-cell lymphoma; HLA-DR = human leucocyte antigen—DR; TdT = terminal deoxynucleotidyl transferase.

Cross reference
Immunophenotyping is discussed in detail in Chapter 10.

Cross reference
Acute myeloid leukaemia with recurrent genetic abnormalities is discussed in detail in Chapter 11.

of B-cell lineage where these genetic abnormalities can predict clinical outcome. Named cytogenetic aberrations are considered under the overarching section *B-lymphoblastic leukaemia/lymphoma with recurrent genetic abnormalities.*

B-lymphoblastic leukaemias or lymphomas failing to demonstrate any of the named recurrent genetic abnormalities are combined within another group—*B-lymphoblastic leukaemia/lymphoma not otherwise specified (NOS).* Whilst this group should not be used to classify inadequately investigated patients, it is likely that the composition of cases within this group will change in subsequent editions of the WHO classification as more is discovered about this particular group of lymphoid malignancies.

According to the revised fourth edition of the WHO classification, T-lymphoblastic leukaemias and lymphomas have no defining cytogenetic abnormalities that can be used for classification purposes, and therefore these T-lineage lymphoblastic conditions are combined under the heading of *T-lymphoblastic leukaemia/lymphoma* regardless of their cytogenetic profile. However, it is highly likely that over the course of the next few years, evidence will accumulate enabling the identification of clinically useful recurrent genetic abnormalities for T-lymphoblastic malignancies.

We will now consider the WHO classification of the range of lymphoblastic leukaemia/lymphoma entities in some detail.

12.2.1 B-lymphoblastic leukaemia/lymphoma with recurrent genetic abnormalities

In this section, seven cytogenetic abnormalities are recognized as recurrent, disease-defining entities. Where appropriate, the molecular basis of these translocations in leukaemogenesis is discussed.

B-lymphoblastic leukaemia/lymphoma with t(9;22)(q34.1;q11.2); BCR-ABL1

The fusion protein BCR-ABL1 is associated with approximately 30% of cases of adult ALL and 3% of childhood cases. This translocation is uniformly associated with a poor prognosis in cases of ALL, even though in CML t(9;22) is associated with a good prognosis. The size of the BCR-ABL1 peptide is 190 kDa in childhood ALL, due to breakpoints occurring in the minor **breakpoint cluster region (BCR)**, with a larger product of 210 kDa associated with adult ALL and chronic myeloid leukaemia (CML). Abnormal signalling initiated by the mutated Abelson (ABL) tyrosine kinase is responsible for dysregulated cell growth, differentiation, and failure of apoptosis.

breakpoint cluster region (BCR)
An area of a gene where chromosomal breakages are particularly common and well defined.

The cell surface expression includes CD10+, CD19+ and TdT+. CD25 has been linked with cases of ALL harbouring t(9;22), although the presence of CD25 should not be used as a surrogate marker for this translocation. The frequent aberrant expression of the myeloid antigens CD13 or CD33 has also been reported in cases harbouring t(9;22)(q34.1;q11.2). As in CML, the use of imatinib mesylate has shown some success in targeting the tyrosine kinase activity of the BCR-ABL1 peptide.

A provisional entity is included in the 2016 version of the WHO classification, called *B-ALL, BCR-ABL1-like*, that represents lymphoblastic leukaemias with a similar gene-expression profile to B-lymphoblastic leukaemia/lymphoma with t(9;22)(q34.1; q11.2) *BCR-ABL1*, although alternative translocations are present. Often, these alternative translocations are not identifiable through routine karyotyping, and have a deleterious impact on patient outcomes. Translocation of *CRLF2* (cytokine receptor-like factor 2) occurs in approximately 50% of cases, half of which also exhibit mutant *JAK1* or *JAK2*.

Cross reference
The translocation t(9;22) is discussed in greater detail in Chapter 11.

B-lymphoblastic leukaemia/lymphoma with t(v;11q23.3); KMT2A rearranged

This group only contains 11q23 rearrangements. Cases involving deletion are not included in this category, as they have not consistently been associated with predictable disease behaviour.

KMT2A rearrangements are associated with a poor prognosis, with age also seeming to be an important factor. In paediatric cases of ALL, patients under the age of one year harbouring *KMT2A* rearrangements have a worse prognosis than patients over the age of one year.

Interestingly, in the research arena, cases of ALL harbouring *KMT2A* abnormalities have a distinctive gene expression profile that seems to enable these *KMT2A*-related diseases to be identified without the need for cytogenetic analysis. As this finding has only been reported in a handful of cases, replacing cytogenetic analysis *should not* be considered appropriate in the classification of cases suspected of harbouring *KMT2A* rearrangement at present. The most common of the 11q23 rearrangements, t(4;11)(q21;q23.3), results in the fusion of *KMT2A* to

AFF1. Cases harbouring 11q23 abnormalities tend to be associated with a higher WBC count and CNS involvement.

B-lymphoblastic leukaemia/lymphoma with t(12;21)(p13.2;q22.1); ETV6-RUNX1

This is the most common cytogenetic abnormality associated with paediatric ALL, occurring in up to 25% of B-ALL cases between the ages of two and ten years. As a consequence of t(12;21)(p13.2;q22.1) ETV6, a member of the E26 transformation specific (ETS) family of transcription factors, is fused to RUNX1—a component of the core binding factor complex. This translocation has similar molecular consequences to t(8;21)(q22;q22), one of the recognized translocations associated with acute myeloid leukaemia (AML) outlined in *AML with recurrent genetic abnormalities*.

Cross reference

AML with recurrent genetic abnormalities is discussed in Chapter 11.

ETV6 contains a **helix-loop-helix domain**, essential for homodimerizing with other ETV6 proteins and heterodimerizing with other members of the ETS family. An ETS domain enables ETV6 to bind to DNA.

ETV6 can bind to a number of transcriptional repressors, including histone deacetylase (HDAC), thereby suppressing target gene transcription and translation and functioning as a likely tumour suppressor. The fusion of ETV6 with RUNX1 results in the production of a fusion protein containing the ETV6-derived helix-loop-helix domain and repressor binding regions combined with the entire RUNX1 structure. The net result of this translocation is the inappropriate localization of transcriptional repressors and co-repressors to RUNX1-inducible genes. This in itself is insufficient to be oncogenic, but additional genetic lesions could lead to a malignant phenotype.

Additional evidence suggests that the helix-loop-helix region can bind the wild-type ETV6, inhibiting ETV6's tumour suppressor function. In the majority of cases, however, the wild-type ETV6 located on the other chromosome 12 is deleted.

Considerable evidence shows that acquisition of t(12;21)p13.2;q22.1 is an initiating event resulting in the formation of a pre-leukaemic clone and that subsequent, secondary, mutations are required to develop paediatric B-ALL.

t(12;21)(p13.2;q22.1) is associated with a good prognosis, with over 90% of children cured.

B-lymphoblastic leukaemia/lymphoma with hyperdiploidy

These cases typically possess between 50 and 66 chromosomes per cell, and are associated with a good prognosis. Care is required to ensure that any hyperdiploid cells detected are derived from true hyperdiploidy and are not the consequence of abnormal cell division in an otherwise hypodiploid case. On occasion, hypodiploid cases appear as a mixed and unbalanced population of hypodiploid and hyperdiploid cells—caused by endoreplication. However, when quantified, the hyperdiploid cells predominate which can lead to misdiagnosis. **Endoreplication** is the process of chromosomal duplication in the absence of cytokinesis and is also relevant in the following section. B-lymphoblastic leukaemia/lymphoma with hyperdiploidy is common within the one-to-ten-year age range, and is associated with a low WBC count, with cells predisposed to spontaneous apoptosis.

Endoreplication

The process of chromosomes replicating in the absence of cytokinesis.

In excess of 90% of children with hyperdiploidy may be cured.

B-lymphoblastic leukaemia/lymphoma with hypodiploidy (hypodiploid ALL)

In cases of hypodiploid ALL, chromosome number is less than 45 per cell and is associated with a poor prognosis. Hypodiploidy is subdivided as follows:

- high hypodiploid—40-45 chromosomes;
- low hypodiploid—31-39 chromosomes;
- near haploid—24-30 chromosomes.

Occasionally, *masked hypodiploidy* may occur in otherwise low-hypodiploid and near-haploid populations, where the hypodiploid population may be dwarfed by a deviant hyperdiploid or near-triploid population formed through aberrant endoreplication.

In low-hypodiploid ALL, approximately 90% of cases possess *TP53* mutation that results in the loss of function of p53. These mutations are not seen in near-haploid ALL, although approximately 70% of near-haploid cases demonstrate Ras dysfunction or abnormal receptor tyrosine kinase signalling and the worst prognosis in this hypodiploidy group.

B-lymphoblastic leukaemia/lymphoma with t(5;14)(q31.1;q32.3) (IL3-IGH)

This is a particularly rare entity, accounting for less than 1% of cases of ALL. The translocation t(5;14)(q31;q32) is considered a discrete entity, as it is associated with an increased number of circulating eosinophils. These eosinophils are not part of the malignant clone, but rather the consequence of an induced inflammatory response secondary to the translocation. The *IL3* gene, localized on 5q31, is translocated to the IgH locus on 14q32, resulting in the upregulation of interleukin-3 (IL-3) expression as a consequence of the stronger IgH promoter. IL-3 then stimulates eosinophil differentiation from common myeloid precursors, resulting in a peripheral blood eosinophilia. At the time of writing, there are no conclusive survival data for this particular subtype.

B-lymphoblastic leukaemia/lymphoma with t(1;19)(q23;p13.3);TCF3-PBX1

TCF3 encodes two transcription factors, E12 and E47, which bind to the regulatory elements of a range of genes, including *IGK*—the gene encoding Igκ. E12 and E47 are essential for normal lymphopoiesis and control the rate of transcription of a number of important genes actively involved in the process of lymphocyte development. *PBX1* is a **homeobox** gene, encoding a transcription factor essential for lymphopoiesis from the haemopoietic stem cell stage to the pro-B-cell stage of maturation.

homeobox
Genes which encode transcriptional regulators which are expressed at particular times and in particular places within an organism—for example, during embryonic development or in cell differentiation.

The E12 and E47 transcriptional activation domains are kept intact in this fusion product, and are localized to the PBX1 DNA-binding domain. It is likely that the fusion of these proteins results in the expression of atypical proteins within cells harbouring this translocation, leading to maturation arrest and leukaemogenesis.

Generally, t(1;19)(q23;p13.3) is associated with 6% of B-ALL cases and confers a good prognosis.

Cross reference
TCF3 is discussed in further detail in the context of Burkitt lymphoma later in this chapter at section 12.3.2.

SELF-CHECK 12.2

Outline the reasons why eosinophilia occurs in cases of leukaemia harbouring t(5;14)(q31.1;q32.3).

12.2.2 T-lymphoblastic leukaemia/lymphoma

T-ALL is an aggressive form of leukaemia, accounting for approximately 15% of childhood and 25% of adult ALLs. T-LBL accounts for 85–90% of lymphoblastic lymphomas. T-ALL/T-LBL is often aggressive and is associated with a poor prognosis. Patients often have a high white cell count and are slightly older than their patient counterparts with B-ALL.

There is often significant bone marrow and peripheral blood involvement, commonly accompanied by an apparent mass within the thoracic cavity (**mediastinum**), hepatosplenomegaly, and lymphadenopathy. In excess of 25% blasts are usually found within the bone marrow in T-ALL, although this is not the case in T-LBL, where the blasts are usually confined to a bulk lesion outside of the bone marrow microenvironment.

mediastinum
This defines the area within the thorax between the lungs containing the heart, trachea, oesophagus, and thymus.

Immunophenotyping can often be used to identify the normal cell counterpart based on the expression of different combinations of T-cell antigens on the cell surface.

A variety of genetic and chromosomal abnormalities are associated with T-lymphoblastic leukaemia/lymphomas, although these are not related to particular disease subgroups at the time of writing.

Now that we have considered the precursor neoplasms, we should go on to look at the mature B-cell neoplasms.

12.3 Mature B-cell neoplasms

Mature B-cell neoplasms demonstrate the expression of lineage-specific markers and follow a less aggressive course than the precursor lymphoid neoplasms. It is important to recognize that, in these cases, the diseases can still be aggressive and are associated with significant **morbidity** and **mortality**.

morbidity
The state of being diseased, or unhealthy, or in poor health.

mortality
Untimely death or an increased risk of death.

An understanding of the concept of the germinal centre is important in understanding the pathogenesis of these diseases. Recognizing the terms *centroblast*, *centrocyte*, *mantle zone*, *marginal zone*, *memory B cell*, and *plasma cell* is important in understanding the origin of the normal cell counterpart. In this section, the normal cell counterpart has not been explicitly outlined, as is the case in the original fourth edition WHO publication, although where critical in understanding the basic concepts of the disease, the normal cell counterpart is included.

The role of cytogenetic analysis is critical in understanding B-cell lymphoid neoplasms, the majority of which are associated with discrete non-random chromosomal translocations. The mechanisms in place for controlling somatic hypermutation and immunoglobulin class-switching often become dysregulated in B-cell neoplasms, and it is thought that dysregulation of this normal mutational machinery is responsible for the large numbers of cytogenetic abnormalities associated with lymphoid malignancies. Where appropriate, the genetic or cytogenetic basis of mature B-cell neoplasms will be discussed in relation to discrete disease entities.

Cross reference
Germinal centres are outlined in Chapter 8.

12.3.1 B-cell prolymphocytic leukaemia

B-cell prolymphocytic leukaemia (B-PLL) is a rare, aggressive, B-cell lymphoproliferative disorder accounting for 1–2% of all chronic lymphoid leukaemias. Although historically B-PLL was believed to share developmental pathways with chronic lymphocytic leukaemia/small lymphocytic lymphoma (CLL/SLL), this is no longer the case. However, occasionally the differentiation between CLL/SLL with increased prolymphocytes, mantle cell lymphoma in leukaemic phase, and genuine B-PLL can prove challenging. Judicious morphological examination, Matutes score (see CLL/SLL section for further details) and genetic analysis will clarify the diagnosis in the majority of cases.

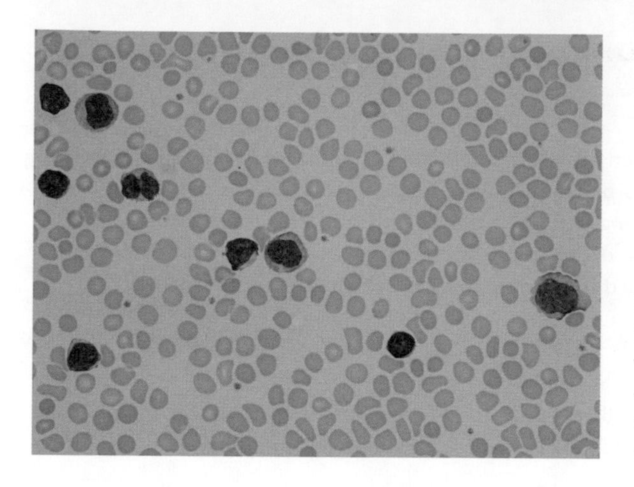

FIGURE 12.2
Prolymphocytic leukaemia. The lymphocytes here are pleomorphic, with typical prolymphocytes present. The chromatin is not as clumped in these cases as in CLL, with many demonstrating prominent nucleoli. Image courtesy of Jackie Warne, Haematology department, Queen Alexandra Hospital, Portsmouth.

Patients present at a mean age of 70 years, with splenomegaly, absence of lymphadenopathy, and a very high WBC count (>100 × 10^9/L). Splenic biopsy shows infiltration of the spleen in both white and red pulp. The threshold for the diagnosis of B-PLL is 55% prolymphocytes in the peripheral blood, although in most cases this exceeds 90%. Figure 12.2 shows typical prolymphocytes in a case of B-PLL.

pleomorphic
Variability in size and shape of cells.

The bone marrow from many of these patients shows a marked infiltration by prolymphocytes, with either an interstitial/diffuse or diffuse distribution pattern which disrupts normal haemopoiesis. This infiltrative pattern is present in excess of 50% of patients at the time of diagnosis and manifests itself clinically as anaemia and thrombocytopenia. Other patients may show a nodular distribution of prolymphocytes within the bone marrow, which does not disrupt normal haemopoiesis.

The B-PLL immunophenotype is strongly positive for: CD19, CD20, and CD22, surface membrane immunoglobulin (smIg), FMC7 (an epitope of the CD20 antigen), and CD79b. CD5 is variably expressed, whilst CD23 is often negative.

Cytogenetically, a number of abnormalities are associated with B-PLL. Mutations involving chromosome 14 occur in excess of 50% of cases, and t(11;14)(q13;q32) has been described. It is now recognized that cases resembling B-PLL but harbouring t(11;14) are actually a variant of mantle cell lymphoma rather than B-PLL. Key mutations include the deletion or mutation of 17p13, leading to the loss of *TP53* or p53 dysfunction, deletion of 13q14, and the associated loss of *RB1*, and deletion of 11q23and *ATM* loss are notable.

Cross reference
Mantle cell lymphoma is discussed in section 12.3.9.

Cross reference
Chronic lymphocytic leukaemia is discussed later in this chapter at section 12.3.3.

There is no correlation between prognosis in B-PLL and the prognostic indicators in B-CLL/SLL. Overall survival is in the range of 30–50 months from diagnosis.

12.3.2 Burkitt lymphoma

Burkitt lymphoma (BL) was first identified by Denis Burkitt, working in Uganda (see Box 12.1). He noted the presence of a significant deformity in the jaws and abdomen of a number of children. Following collaboration with Anthony Epstein in 1961, the herpesvirus, later named the Epstein–Barr virus (EBV), was identified within the tumour cells. BL became the first human malignancy to be associated with an oncogenic virus.

The BL identified in these cases was certainly endemic BL, although we can now recognize three subtypes as outlined below:

Endemic BL—this occurs mainly in children aged between four and seven years in equatorial Africa and Papua New Guinea, and presents as large tumours of the jaw or abdomen, which are

BOX 12.1 Burkitt lymphoma

In 1972, Denis Burkitt published the paper describing his findings of an unusual sarcoma in 38 children during 7 years based at Mulago hospital, Kampala, Uganda. Although Christiansen published a paper describing a patient with similar clinical features in 1938, Burkitt and colleagues did an exceptionally thorough investigation of these children, leading to the recognition of a discrete disease entity, which later bore his name.

sarcoma
A tumour that arises from connective tissue.

invariably EBV-positive. Endemic BL accounts for up to 75% of all malignancies in children in equatorial Africa or 40% of all childhood cases of non-Hodgkin lymphoma worldwide.

Sporadic BL—again, this occurs mainly in children but has a weaker association with EBV, with up to 20% of cases in the West demonstrating evidence of EBV within tumour cells. Geographically, poorer, tropical countries have a higher association with EBV in sporadic BL and it should, therefore, be considered a heterogeneous disease. The incidence of sporadic BL is approximately two per million.

Immunodeficiency BL—this, the most common BL, occurs most in approximately 30–40% of cases of non-Hodgkin lymphoma in patients infected by HIV. In approximately 30% of cases, EBV will be identified in tumours. Development of immunodeficiency BL is indicative of progression to acquired immune deficiency syndrome (AIDS).

BL tumour cells typically have a very rapid proliferation rate, with the tumour cells often having a doubling time of 24–48 hours. Immunostaining using the proliferation marker Ki-67 provides a Ki-67 index of 95% (see Ki-67 Key Points box below for an explanation), demonstrating that the majority of cells are actively proliferating. Accompanying this proliferative behaviour is an increased rate of cell death, leading to an increase in the number of macrophages within the tumour. These macrophages scavenge the cellular debris produced by the high rate of cell death. Histological investigation of the tumour reveals large numbers of macrophages interspersed with tumour cells—demonstrating what has often been described as a 'starry sky appearance'.

Key Points

Ki-67 is a protein marker used for determining the number of cells within a tissue progressing through the cell cycle. Visualization of Ki-67 can be accomplished by immunostaining utilizing monoclonal antibodies. The most commonly used monoclonal antibody for the purposes of Ki-67 identification is called MIB-1 (Molecular Immunology Borstel-1). G1, S, G2, and M phases all express Ki-67, and only quiescent cells (those in G0) fail to express the Ki-67 antigen. The percentage of cells within a tissue expressing Ki-67 provides the 'Ki-67 index' or the proliferation fraction. Ki-67 is useful to measure because it tells us how quickly cells divide and enables us to predict tumour growth. The Ki-67 index can also be used as a prognostic marker.

Figure 12.3 shows an example of typical BL cells within the peripheral blood. Note the presence of cytoplasmic vacuoles and deeply basophilic cytoplasm.

The tumour cell immunophenotype includes surface expression of IgM and is accompanied by the B-cell markers CD19, CD20, CD22, and CD79a. In addition, the germinal centre markers

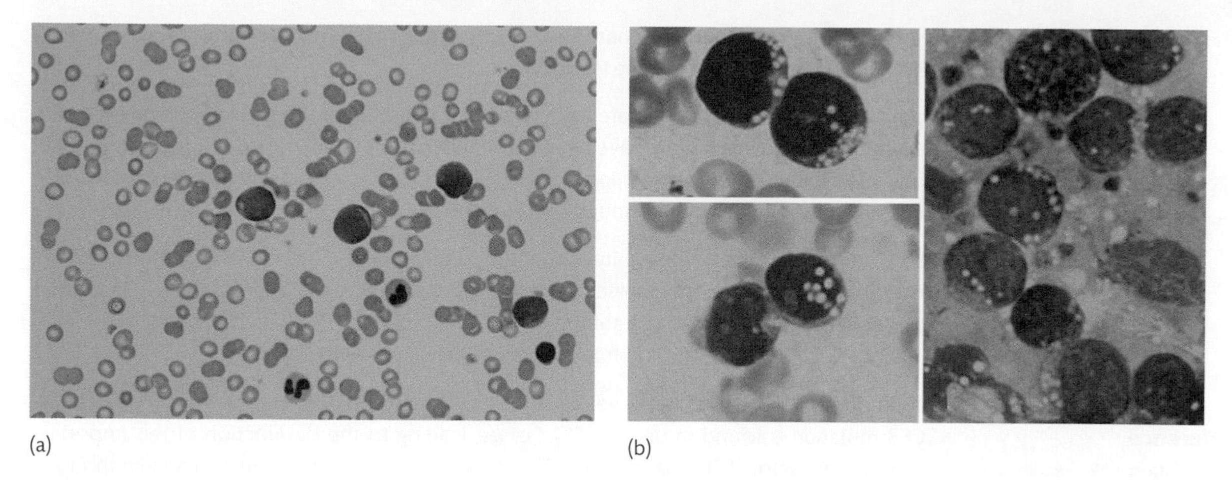

FIGURE 12.3
Burkitt lymphoma (a) and (b). In BL, the lymphoblasts have a high nucleocytoplasmic ratio with deep cytoplasmic basophilia and vacuolation. The nucleus contains fine open chromatin, with numerous nucleoli. Image (a) courtesy of Jackie Warne, Haematology department, Queen Alexandra Hospital, Portsmouth. Image (b) from Longmore M, Wilkinson I, Baldwin A, and Wallin E. *Oxford Handbook of Clinical Medicine*, 9th edn. Oxford University Press, Oxford, 2014. By permission of Oxford University Press.

BCL6, CD10, and CD38 are expressed and are accompanied by molecular evidence of somatic hypermutation. The immunophenotypic data coupled with evidence of somatic hypermutation indicate that BL is caused by mutated germinal centre-derived B cells.

Typical cytogenetic findings in Burkitt lymphoma

Cytogenetically, three translocations are associated with BL, all involving the upregulation of the proto-oncogene *MYC*. These translocations are t(8;14)(q24;q32), t(8;22)(q24;q11), and t(2;8)(p12;q24). Accounting for up to 80% of cases of BL, t(8;14) utilizes the *IGH* promoter, whereas the kappa and lambda genes are located on chromosomes 22 (5%) and 8 (15%), respectively. Variable breakpoints occur within these translocations, although the end result—the upregulation of c-Myc—is a constant finding.

Although the vast majority of cases of BL harbour translocations involving *MYC*, these translocations are not exclusively found in BL, and are therefore *not* diagnostic for BL.

The role and dysregulation of c-Myc

The role of c-Myc in cellular control is both complex and diverse. c-Myc has been implicated in the control of cell growth, proliferation, induction of apoptosis, and metabolism. Dysregulation of c-Myc following mutation or translocation can lead to modification of all of these functional properties.

Research has shown that in non-malignant cells, the tumour suppressors $p14^{ARF}$ and p53 are upregulated following the overexpression of c-Myc. Upregulation of these tumour suppressors leads to cell cycle arrest and the induction of apoptosis in cells where c-Myc is dysregulated. Overexpression of c-Myc and inactivating mutations of p53 are common findings in patients with BL, with up to 35% of cases harbouring *TP53* mutation, thereby facilitating oncogenesis by removing this important negative feedback mechanism.

Approximately 38% of BL cases are associated with mutation in the gene encoding cyclin D3, *CCND3*. A range of mutations have been identified in this oncogene, from single-point mutations to deletions of up to 41 amino acids, resulting in its increased stability, activity, and propensity to drive the cell cycle through the G1-S transition.

In approximately 17% of cases, inactivating mutations of the cell cycle regulator *CDKN2A*, encoding p16 and p14arf, are present. These cell cycle inhibitors are part of the INK (inhibitors of kinase) family and are important for suppressing cyclin D-cdk4 activity. Loss of p16/p14arf activity results in increased G1-S transition and a cellular growth advantage.

Cross reference

TCF3 dysregulation is also recognized as a B-lymphoblastic leukaemia/lymphoma recurrent genetic abnormality as part of the t(1;19)(q23;p13.3) mechanism covered in section 12.2.1.

Mutations in TCF3 and its negative regulator ID3 are also found commonly in all types of BL. TCF3 mutation is found in up to 25% of cases, leading to the dysfunction of this important transcription factor. TCF3, as shown in Figure 12.4, plays an instrumental role in lymphopoiesis and is necessary for both B- and T-cell development. Normally, ID3 forms heterodimers with TCF3, preventing its binding to DNA recognition sequences. However, mutation of ID3 prevents the formation of these negative regulatory heterodimers, allowing enhanced TCF3 activity. Enhanced TCF3 activity appears to increase the expression of BL B-cell receptors and promotes signalling via the PI3K pathway.

Cross reference

The cell cycle and its regulation are outlined in Chapter 9.

Interestingly, the overexpression of c-Myc downregulates the activity of the CDK inhibitor p27. Cyclin E-CDK2 activity is controlled by p27, resulting in the progression of the cell cycle through G1 into S-phase. c-Myc also activates CDK4. When accompanied by cyclin D, cyclin D-CDK4 phosphorylates pRb, releasing the transcription factor E2F-1, leading to the increased synthesis of cyclin E.

The overexpression of c-Myc is thought to be a prerequisite for BL lymphomagenesis, although the mutation of *MYC* alone is not thought to be sufficient to induce malignant transformation. Therefore, a number of events, including *MYC* dysregulation, seem important.

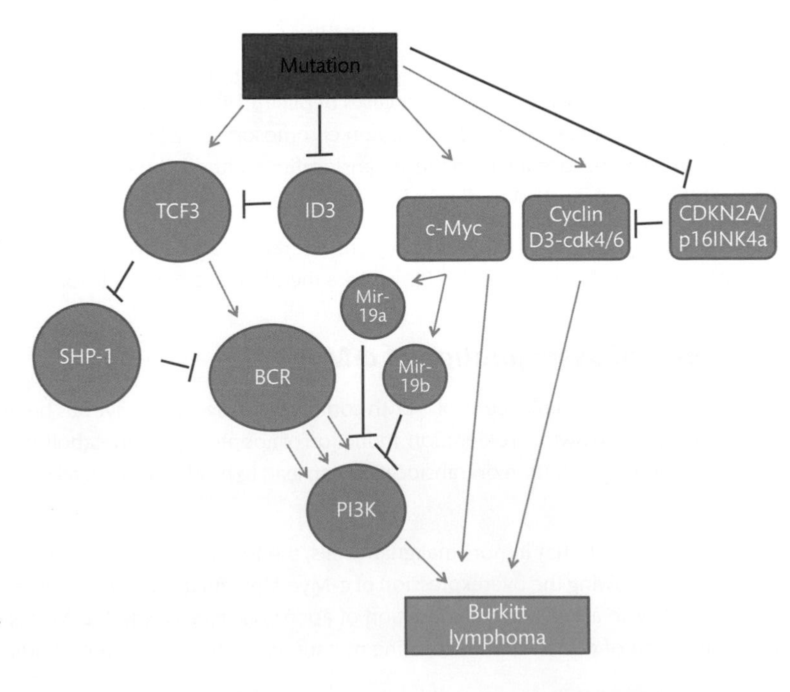

FIGURE 12.4

Mutation occurs at a number of levels leading to BL. Mutation of TCF3 leads to enhanced B-cell receptor (BCR) signalling, thus promoting PI3K signal transduction. Furthermore, enhanced TCF3 activity downregulates the negative regulator of BCR, SHP-1. This also leads to enhanced BCR activity. In BL, we also see loss of function mutations in the TCF3-negative regulator ID3. Loss of ID3 also leads to enhanced TCF3 signalling. Loss and mutation of *CDKN2A/p16INK4a* leads to removal of the negative regulators of cyclin D3-CDK4/6, and when accompanied by mutation of cyclin D3, results in enhanced cell cycle transition from G1-S. Finally, mutation of c-Myc promotes a BL phenotype directly, and also by upregulating a family of microRNAs which indirectly upregulate PI3K through their direct inhibition of a PI3K inhibitor.

Burkitt lymphoma prognosis

The prognosis for patients with BL is very good. Largely, endemic BL is considered to have the highest mortality, but this is due to endemic BL occurring in Africa and equatorial regions, where a poor standard of health care is expected. The use of high-dose combination chemotherapeutic agents leads to rapid cell lysis, and the risk of **tumour lysis syndrome**. Up to 90% of individuals with BL can be cured using appropriate combination chemotherapy.

Key Points

Tumour lysis syndrome is considered to be a medical emergency and is a consequence of the release of a number of intracellular components following the lysis of tumour cells. Patients will have raised plasma potassium (hyperkalaemia), phosphate (hyperphosphataemia), and uric acid (hyperuricaemia) concentrations, accompanied by reduced calcium (hypocalcaemia) levels and often acute renal failure.

SELF-CHECK 12.3

Describe the role of Ki-67 in evaluating BL-cell behaviour. Describe the molecular mechanisms that might be responsible for the typical Ki-67 index.

12.3.3 Chronic lymphocytic leukaemia/small lymphocytic lymphoma

Chronic lymphocytic leukaemia (CLL) and small lymphocytic lymphoma (SLL) are considered variations of the same disease. CLL, as its name suggests, is composed of malignant lymphocytes in leukaemic phase (i.e. within the peripheral blood), whereas SLL is composed of malignant lymphocytes in one or more bulk lesions, a peripheral blood lymphocyte count $<5 \times 10^9$/L and an absence of cytopenias. The morphology of the lymphocytes and their immunophenotype are identical in each case. CLL/SLL is a disease of the elderly, occurring predominantly in Westernized societies. In the UK, the incidence is approximately 3 per 100,000, occurring mainly in those over 50 years of age.

The diagnosis of CLL, the process of which is signposted in Box 12.2, is largely dependent on a lymphocyte count in excess of 5.0×10^9/L for three months or more. Malignant lymphocytes appear homogeneous or monomorphic, with a thin rim of cytoplasm and a nucleus containing clumped chromatin. **Smear cells** (also known as smudge cells or Gumprecht shadows)—cells that have ruptured and smeared across the slide in response to mechanical damage from producing the blood film—are frequent features. Figure 12.5 shows a photomicrograph of a typical case of B-CLL. In the vast majority of cases, with lymphocyte counts in excess of 30.0×10^9/L, immature lymphocytes

BOX 12.2 *Guidelines*

The British Committee for Standards in Haematology (BCSH) has guidelines for the diagnosis and management of CLL published on its website https://b-s-h.org.uk/guidelines/

called **prolymphocytes** will account for up to 5% of the white cells. Approximately 50% of cases of CLL are diagnosed following referral for a full blood count for an unrelated clinical situation.

Key Points

A common complication of CLL is the production of autoantibodies directed against red cells. During the diagnostic process, it is important that patients should have a direct antiglobulin test (DAT) performed to determine whether or not red cell autoantibodies are present. It is important that the DAT is monitored throughout the duration of the disease. DATs are performed within blood transfusion laboratories and use antihuman globulin (AHG) to cause agglutination of red cells sensitized with antibodies or complement components.

The presence of red cell agglutination following incubation of the patient's red cells with the patient's plasma in the presence of AHG suggests that autoantibodies have been produced against the patient's red cells.

Autoantibodies directed against neutrophils and platelets can also occur in patients with CLL.

Cross reference

The direct antiglobulin test is illustrated and described in Chapter 6.

As CLL progresses, a normocytic normochromic anaemia may be found, but patients may also develop an autoimmune haemolytic anaemia following the production of autoantibodies directed against red cells or their precursors. Red cell aplasia can occur as a consequence of human parvovirus infection, leading to a profound anaemia requiring blood transfusion support.

CLL/SLL is typified by the immunophenotype: CD5+, CD19+, CD20+, CD23+, CD22–, smlg, FMC7, and CD79b may or may not be expressed. Other causes of malignant lymphocytosis, including mantle cell lymphoma and splenic marginal zone lymphoma, may appear morphologically identical, but can be distinguished on the basis of their immunophenotype. A scoring system, using immunophenotyping outlined in Table 12.8, was developed by Matutes et al in 1994 in order to help differentiate CLL from other lymphoproliferative disorders, and is still

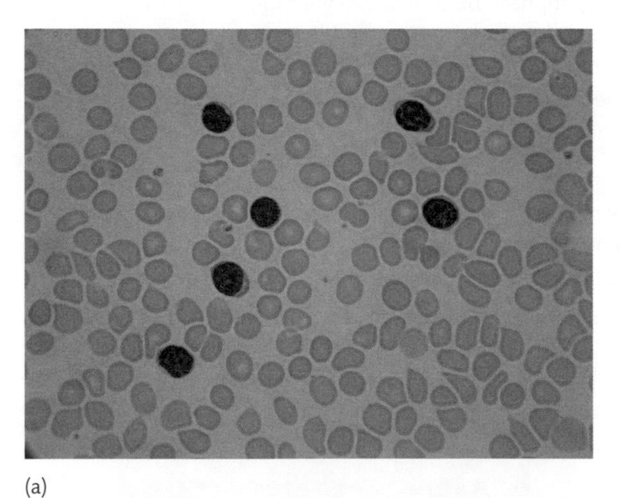
(a)

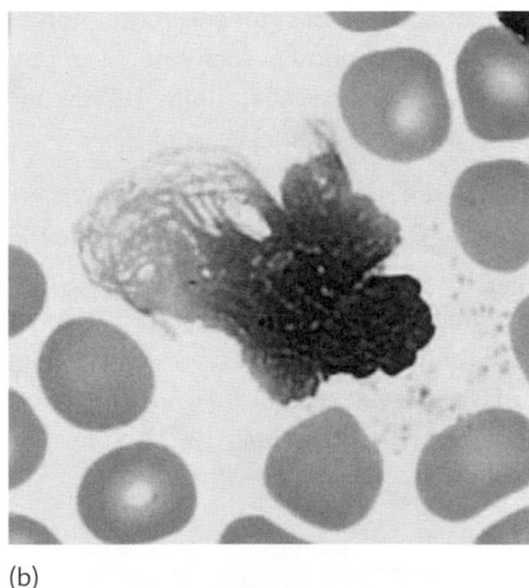
(b)

FIGURE 12.5
(a) A case of chronic lymphocytic leukaemia. Note the increased number of mature lymphocytes. The cytoplasm is pale blue and the nucleus is composed of clumped chromatin. (b) A CLL smear cell produced by mechanical damage of abnormally fragile cancer cells induced by the slide preparation process.

TABLE 12.8 **CLL immunophenotyping (Matutes) score.**

CD marker	Score	
	1	0
CD5	Positive	Negative
CD23	Positive	Negative
FMC7	Negative	Positive
SmIg	Weak	Moderate/strong
CD22/CD79b	Weak/negative	Moderate/strong

Key: score >3 = CLL; <3 = other B-cell malignancy.

used in many centres today. CLL will achieve a score of three or above, whereas other lymphoproliferative diseases should score less than three.

Examination of a bone marrow trephine is an essential component of the diagnostic process and indicates the level of bone marrow infiltration by malignant lymphocytes. A diffuse pattern of infiltration is associated with a poor prognosis, as it indicates advanced disease. The sheets of malignant lymphocytes occupying the bone marrow compromise haemopoiesis leading to anaemia, neutropenia, and thrombocytopenia.

Cross reference

Characteristic immunophenotyping results for CLL can be found in the immunophenotyping section of Chapter 10.

Significant progress has been made in correlating disease behaviour and patient survival with the molecular and cytogenetic events occurring within CLL/SLL. The identification of two distinct CLL pathologies based around *IGHV* mutational status has aided our understanding of disease behaviour. One CLL subgroup shows evidence of *IGHV* somatic hypermutation, whilst the other possesses unmutated *IGHV*. *IGHV* hypermutated cases are associated with a better prognosis than those cases failing to demonstrate hypermutated *IGHV*.

Zeta-associated protein 70 (Zap-70) is a tyrosine kinase signalling molecule not usually found in normal circulating B lymphocytes. The association of Zap-70 with unmutated *IGHV* has enabled routine laboratories to quantify Zap-70 using immunophenotyping technologies to provide data in the absence of performing complex and expensive *IGHV* sequencing. However, this is still a relatively demanding technique to set up, and many laboratories have opted not to test Zap-70 using their own equipment.

Cytogenetic studies have demonstrated that up to 80% of patients exhibit chromosomal abnormalities, the majority of which can be used for prognostication.

Del13q14.3, occurring in approximately 55% of cases, is associated with a good prognosis. Patients with this deletion, as an isolated genetic abnormality, have been shown to have a better prognosis than patients demonstrating a normal karyotype. Deletion of 13q14.3 results in the loss of *DLEU2*, which encodes micro-RNAs MiR-15a and MiR-16-1. When intact, these micro-RNAs negatively regulate the expression of *CCND1*, *CCND3* (genes encoding cyclins D1 and D3, respectively), *CDK6*, and *BCL2*. The tumour suppressor *RB1*, also found at locus 13q14, remains intact in these cases. Patients with del13q14.3 have a reported survival of 132 months, compared with 120 months for patients with a normal karyotype.

Deletion of 17p13 and the associated loss of its product, p53, occurs in approximately 5–8% of untreated cases and is universally associated with a poor prognosis, with survival as low as 36 months in these patients. In patients with the loss of a single *TP53* allele through deletion, approximately 80% show mutation of the remaining allele. For those without *TP53* deletion, between 4 and 7% of patients harbour mutations within at least one copy of this gene.

In addition, the loss of 11q23, the locus for the tumour suppressor gene ataxia telangiectasia mutated (*ATM*), is also associated with a poor prognosis and is reported in approximately 25% of untreated cases and in 10% of early-stage disease. Isolated 11q23 abnormalities are associated with a bulky lymphadenopathy, rapid progression, and survival of approximately 84 months.

Additional, recently identified, recurrent mutations associated with a poor prognosis affect *NOTCH1*, *SF3B1*, and *BIRC3*. These are introduced in a little detail below.

Approximately 10% of cases of newly diagnosed CLL harbour mutations in *NOTCH1*, which is usually involved in the differentiation of B cells into plasma cells, and the development of T-lineage lymphocytes. Mutation of *NOTCH1* in CLL is associated with the loss of function of its **PEST domain**. The inhibitory effects usually provided by the PEST domain are lost, thus modifying cellular behaviour. Between 15 and 20% of progressive cases of CLL harbour *NOTCH1* mutations, indicating that this is a passenger mutation that occurs later on in the life cycle of the malignancy, rather than an initial, founding mutation. NOTCH1 mutations appear to be associated with a worse prognosis than those without this mutation.

PEST domains

These are composed of proline, glutamic acid, serine and threonine and provide proteolytic signals to enhance protein degradation.

spliceosomes

These are complex macromolecular structures composed of small nucleoproteins, involved in the removal of introns from pre-mRNA.

SF3B1 is mutated in approximately 10% of CLL cases at diagnosis and 17% of patients with progressive cases of CLL. Playing an important role in **spliceosomes**, mutation of *SF3B1* is thought to contribute to cell cycle progression, enhanced cell survival, and failure of apoptosis, and is associated with a poor clinical course and outcome.

BIRC3 mutations, occurring in approximately 4% of newly diagnosed CLL, are another reproducible finding in CLL and result in enhanced MAP3K14 and constitutive NFκB signalling. *BIRC3* mutations are found only in CLL and splenic marginal zone lymphoma (SMZL) B-lymphoid malignancies and are associated with a poor prognosis. Other important prognostic markers that will be assessed in the routine diagnostic laboratory include CD38. CD38 expression is associated with a poor prognosis, but expression of CD38 is thought to vary over the course of the disease. Lymphocyte doubling time (LDT) is an important indicator of disease progression and prognosis. CLL/SLL lymphocytes have a low proliferative rate, with high lymphocyte numbers being a consequence of accumulation following the failure of apoptosis, rather than rapid proliferation. However, patients with an LDT of less than 12 months have a worse prognosis than those with an LDT in excess of 12 months. Beta 2-microglobulin (β_2m) and lactate dehydrogenase (LDH) can easily be measured by the clinical biochemistry department, and patients showing a raised β_2m and LDH have a worse prognosis than those with a normal β_2m and LDH. β_2m forms part of the major histocompatibility complex (MHC) class I structure expressed on B cells, whilst LDH is a useful marker of cell turnover and tumour size.

Cross reference

SF3B1 has significant involvement in myelodysplastic syndrome with ring sideroblasts with single-lineage dysplasia (MDS-RS-SLD) and MDS-RS with multilineage dysplasia (MDS-RS-MLD), with further information included in section 11.3.

SELF-CHECK 12.4

Describe how CLL has been subdivided into two different pathologies.

12.3.4 Monoclonal B-cell lymphocytosis

Almost all cases of CLL/SLL are preceded by monoclonal B-cell lymphocytosis (MBL). Progression from MBL to CLL/SLL or other low-grade B-cell lymphoma occurs in approximately 1% of cases per annum. Since 2010, it has become apparent that MBL can be divided into two types based upon cell count. These are:

- low-count MBL (MBL^{Lo});
- high count MBL (MBL^{Hi}).

The critical threshold differentiating these MBL subgroups is 0.5×10^9/L: MBL^{Lo} is associated with clonal lymphocytes $<0.5 \times 10^9$/L, whereas MBL^{Hi} demonstrates clonal lymphocytes $\geq 0.5 \times 10^9$/L. Some variation in immunophenotype between the two is noted. Immunophenotypic analysis demonstrates one of three phenotypes:

- CLL-like phenotype (CD5+, CD19+, CD20 (dim) CD23+);
- atypical CLL phenotype (CD5+, CD19+, CD20 (bright) or CD23-);
- non-CLL type phenotype (CD5-, CD20+).

MBL^{Lo} provides limited evidence of progression to CLL/SLL, with follow-up currently not indicated, whereas MBL^{Hi} is more likely to progress to CLL/SLL, albeit slowly, and does require annual follow-up. However, Orfao et al (2017) have provided evidence to show that progression from MBL^{Lo} to MBL ^{Hi} is possible so practice may change as our understanding improves.

12.3.5 Diffuse large B-cell lymphoma

Diffuse large B-cell lymphoma (DLBCL) is a heterogeneous group of B-cell malignancies, accounting for up to 40% of all cases of NHL. This group of disorders is so broad, containing both common and rare DLBCL variants, that they cannot all be discussed in depth within this textbook. Included is a broad summary of each type, but you are encouraged to read around the subject using current literature reviews and research articles to further your knowledge.

These malignancies are designated as large B-cell lymphomas because they demonstrate large B-cell morphology. The B cells are approximately twice the size of normal B lymphocytes and grow in a diffuse pattern when examined using histological techniques. Approximately 50% of cases are nodal, with the remainder occurring extranodally.

DLBCL appears to be derived from antigen-experienced cells that have had some association with the germinal centre. The majority of cases show evidence of *IGHV* somatic hypermutation, and some cases (especially the germinal centre type outlined below) show intraclonal variation of *IGHV*, suggesting ongoing somatic hypermutation. Intraclonal variation of *IGHV* sequences is consistent with the intra-germinal-centre phenotype. Molecular profiling of DLBCL lymphocytes has allowed scientists and clinicians to allocate malignant cells to two key DLBCL groups:

- germinal centre B-cell like (GCB)
- activated B-cell like (ABC).

Particular chromosomal and molecular abnormalities, survival profiles, and cells of origin have been established for each of these conditions, and these are summarized in Table 12.9.

The WHO classification has successfully stratified DLBCL into a number of well-recognized variants and subgroups to help the clinician with diagnosis, treatment, and prognostication, and it also recognizes the molecular variants highlighted above. Unfortunately, the vast majority of cases, although subdivided to an extent, are still classified as *DLBCL not otherwise specified*.

Routine clinical use of molecular profiling for DLBCL into GC and non-GCB groups is not widely available. It is therefore necessary to employ immunohistochemical (IHC) techniques to differentiate these conditions. Although concordance between approaches is not absolute, IHC is still the method of choice. The Hans classifier, utilizing CD10, BCL6, and MUM1 expression, can provide clinically useful differentiation between the types, as shown in Table 12.10, although it should be noted that other diagnostic algorithms are in use.

TABLE 12.9 Chromosomal translocations and five-year survival figures associated with the GCB and ABC molecular subtypes of DLBCL.

Molecular subtype of DLBCL	Origin	Typical chromosomal abnormalities	Five-year survival
Germinal centre B cell	GC	t(14;18)(q32;q21)Genomic gains of 2p and 2q 6q21-22 abnormalities	59%
Activated B cell	Post-GC (non-GCB)	Gains of 3q and 18q21-22 Loss of 6q21-q22 Constitutive activation of NFκB	30%

TABLE 12.10 Differentiation of GC and NGC DLBCL according to the Hans algorithm.

	CD10*	BCL6*	MUM1/IRF4*
GC	Positive	Negative	Negative
GC	Negative	Positive	Negative
NGCB	Negative	Negative	Negative
NGCB	Negative	Positive	Positive

GC = germinal centre; NGCB = non-germinal-centre B cell; * = >30% cells should express the marker.

Other DLBCL disease entities include:

T-cell/histiocyte-rich large B-cell lymphoma—This accounts for approximately 10% of all DLBCL, and is associated with a poor prognosis.

Primary diffuse large B-cell lymphoma of the CNS (CNS DLBCL)—this is a particularly interesting subtype as it occurs within the brain, spinal cord, or eye, but systemic disease is not evident. As lymphoid tissue is absent in the brain, and the malignant lymphocytes show evidence of *IGHV* somatic hypermutation, it remains unclear how these tumours develop and from where the malignant cells originate. Associated with a poor prognosis, CNS DLBCL can be stratified according to the immune status of the patient. Immunocompromised patients—most commonly those with human immunodeficiency virus (HIV)—have an EBV-positive tumour. In contrast, immunocompetent patients possess tumours driven by a different, currently unclarified mechanism.

Primary cutaneous DLBCL, leg-type—this is a rare disease occurring predominantly in the elderly, and is a typically aggressive disease with only 50% of patients surviving beyond five years.

EBV-positive diffuse large B-cell lymphoma, NOS—this is always EBV-driven and often occurs in patients over the age of 50 years. Formerly, this condition was called *EBV-positive DLBCL of the elderly*, although recent evidence has shown that this condition is also found in younger patients, albeit infrequently. Consequently, the name of this entity was refined in the 2016 WHO revision. Although the prevalence of EBV infection in DLBCL is unknown, there does appear to be geographical variation, with approximately 15% cases involving EBV in Asia and South America compared to 5% in the West. Survival of up to two years is reported.

DLBCL associated with chronic inflammation—this form of DLBCL only seems to occur following a period of 10–20 years' chronic inflammation. The proliferation of transformed B cells seems to be driven by EBV-infected cells in these cases. This is an aggressive lymphoma with mortality approaching 50% at five years.

Lymphomatoid granulomatosis—this is an EBV-driven B-cell malignancy primarily affecting the lung, although other sites may also be involved. The median survival is approximately two years.

Primary mediastinal (thymic) large B-cell lymphoma (PMBL)—this is considered a disease in its own right, although it is still consistent with the DLBCL classification. PMBL *has* been outlined as a distinctive molecular entity along with GC- and ABC-type DLBCL, as listed above.

Cross reference

Anaplastic lymphoma kinase (ALK) is discussed in greater detail in relation to anaplastic large cell lymphoma, ALK-positive in section 12.4 on mature T-cell and NK-cell neoplasms.

Intravascular large B-cell lymphoma—this is an aggressive disease associated with a poor prognosis. Malignant B cells accumulate within the lumen of blood vessels in a variety of organs, although infrequently malignant cells may be identified in a full blood count (FBC) and peripheral blood film.

ALK-positive large B-cell lymphoma—the presence of anaplastic lymphoma kinase (ALK) protein defines this rare DLBCL subtype first described in 1997 by Delsol et al in the journal *Blood*. This disease is associated with mutations of 2p23—the ALK locus commonly expressed as t(2;17)(p23;q23). Although typically aggressive, variable survival patterns have been reported, especially when paediatric cases are considered, as these appear to have a more favourable prognosis than adult cases.

Plasmablastic lymphoma—this is a very rare and aggressive form of DLBCL occurring predominantly in immunocompromised patients, especially those harbouring HIV. EBV tends to be co-expressed with HIV in most cases, and survival is reported to be less than 12 months from presentation.

HHV8$^+$DLBCL, NOS—this is an aggressive DLBCL, which—as its name suggests—is associated with human herpesvirus-8 (HHV8). Frequently this condition also demonstrates EBV co-infection. Rarely seen in the FBC and peripheral blood film, this condition is typically found in patients with HIV, or elderly immunocompetent patients with HHV8. Survival is less than six months from diagnosis.

Primary effusion lymphoma—this is a rare malignancy commonly arising in body cavities such as the pleural, peritoneal, and pericardial spaces, and is predominantly associated with HIV-positive individuals, the immunocompromised, and the elderly. All cases are associated with HHV8, although in immunocompromised patients EBV may also be co-expressed. Symptoms are dependent upon the site of the lymphoma. Patients with peritoneal localization present with abdominal distension, whilst those with pericardial or pleural disease exhibit dyspnoea. Typically, there is no evidence of lymphadenopathy. Prognosis is poor, and the majority of patients die within six months from initial presentation.

To capture a broader spectrum of aggressive B-cell lymphomas, *High grade B-cell lymphoma, with MYC and BCL2 and/or BCL6 translocations* now captures the so-called double- (MYC+ *and* BCL2+ *or* BCL6+) or triple hit (MYC+ *and* BCL2+ *and* BCL6+) lymphomas, of which the former (2008, 4th edn) classification of *B-cell lymphoma, unclassifiable* formed a part. Double-and triple-hit lymphomas account for up to 10% cases of DLBCL and are associated with a poor prognosis. It is noteworthy that MYC+, BCL2+, BCL6– and MYC+, BCL2–, BCL6+ phenotype neoplasms, although both double-hit lymphomas that are classified as the same condition, appear to behave differently and therefore are likely candidates for differentiation into separate disease entities in the future.

High grade B-cell lymphomas, NOS—this captures aggressive B-cell lymphomas that fail to express MYC, do not demonstrate BCL2 and/or BCL6 aberration, and cannot be classified as another type of aggressive B-cell malignancy.

SELF-CHECK 12.5

List the subtypes of DLBCL associated with viral infection.

12.3.6 Follicular lymphoma

Follicular lymphomas (FL) comprise approximately 30% of all non-Hodgkin lymphomas and, as its name suggests, in the vast majority of cases cell growth follows a follicular pattern. In some cases, a combination of follicular and diffuse cell growth is present. Within the diffuse regions of growth, sclerosis will often be present.

As with normal lymphoid follicles, in FL a combination of large, non-cleaved highly proliferative centroblasts are present accompanied by smaller, cleaved, non-proliferative centrocytes. Additionally, FL cells are admixed with normal T cells, follicular dendritic cells, and macrophages, which support FL cells through their microenvironmental interaction with the malignant cells. Microenvironment is important for promoting the growth and survival of FL cells, whilst protecting these cells from the host immune system.

Sometimes there will be evidence of FL cells within the peripheral blood. These cells are often small with pale cytoplasm, with a proportion of cells showing a nuclear cleft. Figure 12.6 shows an example of FL cells in the peripheral blood of a patient in leukaemic phase. Note the small cell size and the distinctive nuclear clefts.

FL is graded according to the number of centroblasts and centrocytes found in the biopsy, as measured when examining the tissue using high-power fields (hpf). Grading ranges from 1 to 3, based on the number of centroblasts present per hpf. Grade 3 is further subdivided into 3A and 3B according to the presence or absence of centrocytes in the assessed follicles. The greater the number of centroblasts, the higher the assigned grade. The immunophenotype of FL cells is CD19+, CD5−, CD10+.

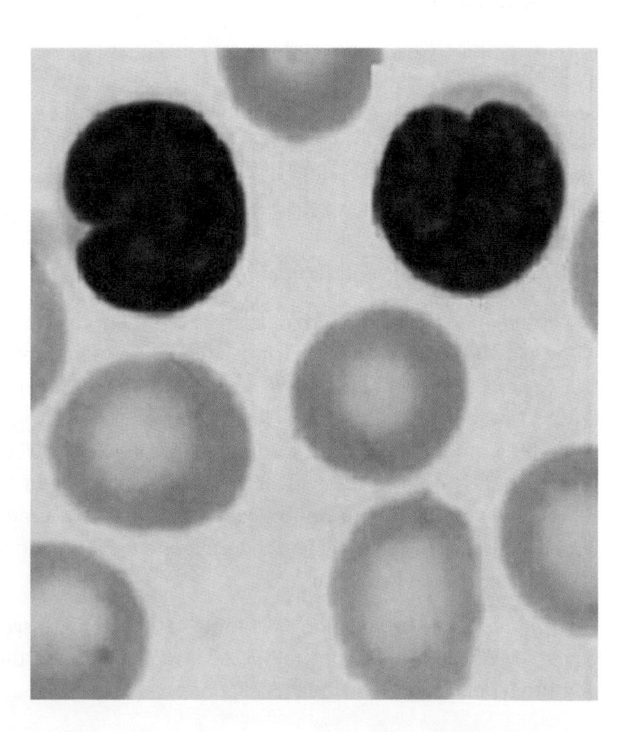

FIGURE 12.6
Follicular lymphoma. Predominantly composed of small lymphocytes; a thin rim of cytoplasm is a typical feature. Image from Longmore M, Wilkinson I, Baldwin A, and Wallin E. *Oxford Handbook of Clinical Medicine*, 9th edn. Oxford University Press, Oxford, 2014. By permission of Oxford University Press.

TABLE 12.11 **Common genetic abnormalities associated with follicular lymphoma.**

Locus/chromosome	Aberration	Frequency (%)
3q26	Gain	31-50
7p21	Gain	16-34
8q24	Gain	16-36
1p13-p31	Loss	29-52
6q23-27	Loss	23-38
9p21	Loss	18-31
11q22-q23	Loss	21-59
13q11-q13	Loss	22-55
13q14-q34	Loss	43-51
17p13pter	Loss	21-45
12	Trisomy	25

Key Points

t(14;18)(q32;q21) has been identified in B cells of normal individuals, implying that this translocation alone is insufficient to cause lymphomagenesis, conforming to Knudson's hypothesis. The frequency of t(14;18) appears to increase with age, and higher levels of this translocation are seen in smokers compared to non-smokers.

The translocation t(14;18)(q32;q21) is believed to be the first of the multiple hits necessary to induce malignancy, is present in up to 90% of cases of grade 1 and 2 FL, although is found in up to 70% of grade 3A and between 15 and 30% of grade 3B FL. This translocation leads to the over-expression of the anti-apoptotic protein BCL2. Following translocation, *BCL2* is moved from its original locus 18q21 to the *IGH* locus at 14q32. BCL2 is now under the transcriptional control of the *IGH* promoter, leading to BCL2 upregulation resulting in continued cell survival. However, we recognize that t(14;18) alone is insufficient to cause FL, as 50-70% of individuals without FL also harbour this translocation. Additional somatic mutations within t(14;18) harbouring cells, including losses at 1p36, 6q, and chromosomal gains of 7, 18, and X are common, as are the additional chromosomal abnormalities shown in Table 12.11. Furthermore, there are a number of additional genes that are shown to play a role as 'secondary hits' in FL, including those with tumour suppressor function: *EPHA7* (70%), *TNFAIP3* (2-26%), *TP53* (<5%); histone modification: *MLL2* (89%), *CREBBP* (33%), *MEF2B* (15%), *EP300* (9%); and apoptosis regulation: *FAS* (6%).

In normal B cells, BCL2 is often overexpressed during Ig rearrangement, meaning that apoptosis is not initiated, ensuring the cells survive. Following Ig rearrangement, BCL2 should be repressed enabling apoptosis of poorly functional/non-functional B cells to occur.

Disease progression is common in FL and although delaying the beginning of treatment in some patients in favour of a watch-and-wait policy is common practice, at autopsy up to 70% of patients show evidence of high-grade transformation. These high-grade malignancies are very aggressive and often fail to respond to therapy. The majority of patients survive for less than 12 months following transformation.

SELF-CHECK 12.6

Outline the role of t(14;18)(q32;q21) in the pathogenesis of follicular lymphoma.

12.3.7 Hairy cell leukaemia

Hairy cell leukaemia (HCL) accounts for approximately 2% of all lymphoid leukaemias, with patients presenting at a median age of 50 years. HCL is associated with the accumulation of malignant lymphocytes with a mature B-cell phenotype.

Patients present invariably with splenomegaly and the FBC shows pancytopenia. Even though the patient is pancytopenic, an absolute monocytopenia is a typical finding in HCL. HCL is often a difficult disease to diagnose because low numbers of leukaemic cells are found in the peripheral blood. Morphologically, these B cells have an oval or indented nucleus and fine hair-like projections—attributable to abnormalities in the organization of the cell cytoskeleton—protruding from the cell surface. These are the defining morphological feature of the disease. Figure 12.7 shows typical peripheral blood film features of HCL.

The spleen, liver, and bone marrow become infiltrated by hairy cells, but the lymph nodes tend to be spared. Synthesis and secretion of basic fibroblast growth factor (bFGF) and tumour growth factor-β1 (TGFβ1) stimulate fibroblast proliferation within the bone marrow, resulting in marrow fibrosis. This bone marrow fibrosis can lead to the aspirate producing a dry tap, so a trephine biopsy is a useful diagnostic tool in cases of HCL. The combination of bone marrow infiltration and fibrosis is thought to be predominantly responsible for pancytopenia at presentation, although cell sequestration by the enlarged spleen accompanied by haemodilution is also likely to reduce peripheral blood counts, contributing to anaemia, bruising, and opportunistic infection.

In over 85% of cases, HCL cells show evidence of somatic hypermutation, indicating these cells have progressed through the germinal centre; and HCL cells express IgM, IgG, and IgA on the cell surface in 40% of patients. This unusual expression of surface immunoglobulin is found in no other lymphoproliferative disorder, and suggests HCL is derived from cells undergoing

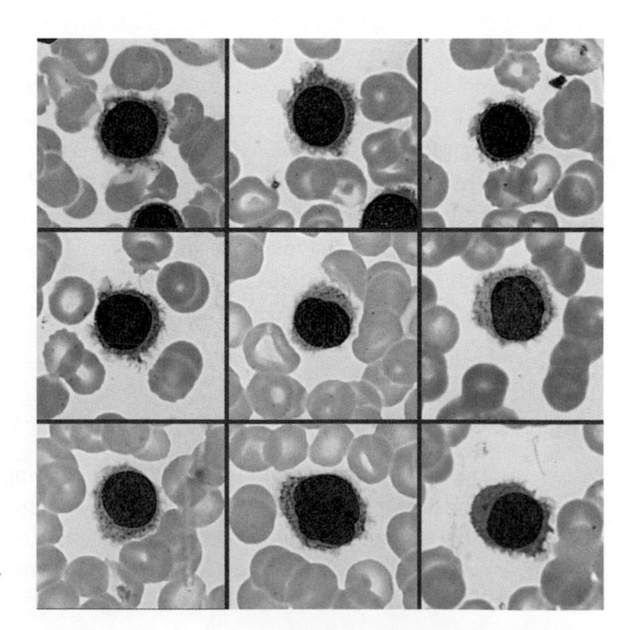

FIGURE 12.7
Hairy cell leukaemia. Various examples of the cytoplasmic projections that are characteristic of hairy cell leukaemia. Panel courtesy of © Sysmex Europe GmbH 2015. Images are under copyright of Sysmex Europe GmbH or published with the consent of their respective owners.

isotype-switching. Gene expression profiling suggests that HCL cells are actually derived from mutated memory B cells and as such have completed their transit through the germinal centre.

Mutation of the signal transducer B-Raf is one of the key molecular features of HCL. B-Raf is a serine-threonine kinase which forms part of the mitogen-activated protein kinase (MAPK) pathway. Missense mutation at position 1799 of the *BRAF* gene, 7q34, (exon 15) leads to the replacement of valine with glutamic acid at position 600 (V600E) of the protein. This amino acid substitution results in constitutive B-Raf activation and enhanced cell survival, proliferation, and differentiation, as shown in Figure 12.8. Although *BRAF* mutations have been found in many cancers, V600E appears to be uniquely associated with classic HCL compared to other B-cell malignancies, including HCL variants. Alternatively, mutation of *MAP2K1*—the gene encoding the signal transducer MEK—is found mutually exclusively to V600E in classic and variant HCL. MEK, a component of the MAPK signalling pathway, is phosphorylated by B-Raf, resulting in enhanced transcription and translation of target genes. Mutation of *MAP2K1* results in the loss of MEK's negative regulatory regions and catalytic core, leading to enhanced signalling and, in some cases, resistance to MEK inhibitors. Together, this indicates that classic HCL and variant HCL are discrete disease entities. The revised fourth edition of the WHO classification includes a number of provisional disease entities, including HCL-variant to account for these differences.

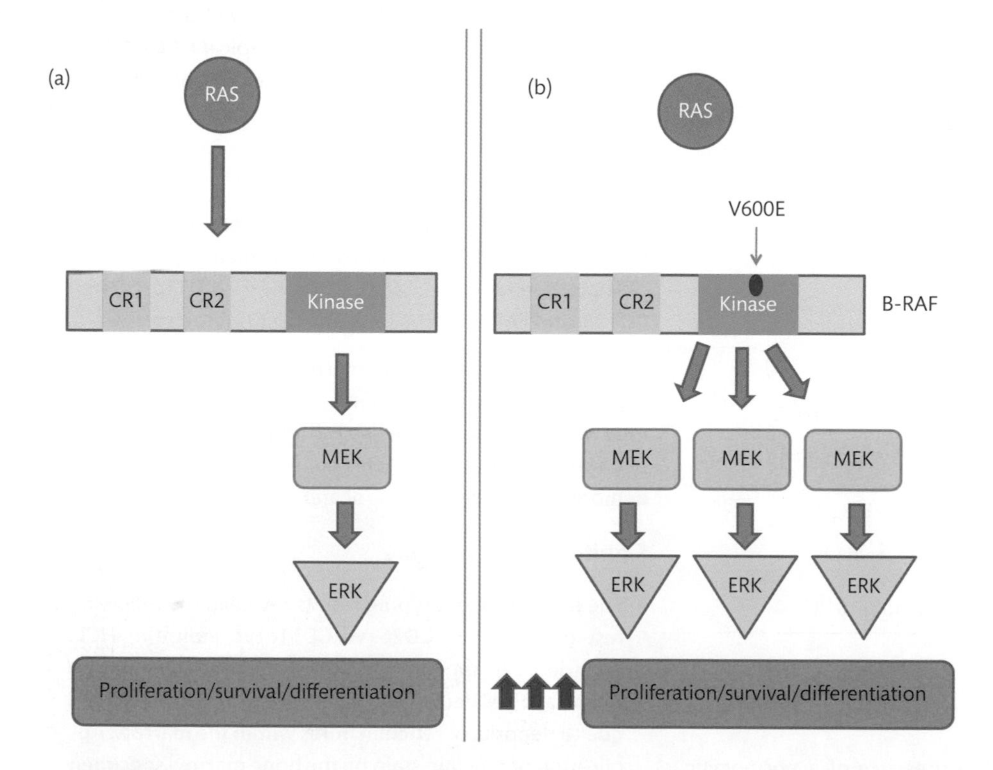

FIGURE 12.8

Normal Ras signalling leads to B-Raf activation, modification of the kinase domain, and downstream signalling events, resulting in the transcription and translation of target genes promoting proliferation, survival, and differentiation. (a) The presence of target genes promoting proliferation, survival, and differentiation; (b) the presence of V600E within the kinase domain leads to constitutive B-Raf activation and enhanced downstream signalling completely independent of exogenous signals. Mutant B-Raf signalling leads to enhanced proliferation, survival, and differentiation.

SELF-CHECK 12.7

What is the normal HCL cell counterpart? Explain how you have made this determination.

CASE STUDY 12.1 A 62-year-old male patient presents with sore throat and left-sided abdominal pain

Patient history

- A 62-year-old male presented to his GP with left-sided abdominal pain of three weeks, fatigue, and recurrent pharyngitis.
- Physical examination showed the spleen was palpable 10 cm below the costal margin. The liver and lymph nodes appeared normal.
- FBC and blood film were initially performed.

Results 1

Index	Result	Reference range
WBC	2.65	(RR 4–10 × 10^9/L)
Neutrophils	0.8	(RR 2–7 × 10^9/L)
Lymphocytes	1.6	(RR 1–3 × 10^9/L)
Monocytes	0.0	(0.2–1.0 × 10^9/L)
Eosinophils	0.2	(RR 0.02–0.5 × 10^9/L)
Basophils	0.05	(RR 0.02–0.1 × 10^9/L)
RBC	3.6	(RR 3.9–5.0 × 10^{12}/L)
Hb	97	(RR 118–148 g/L)
MCV	94	(RR 77–98 fL)
Plt	110	(143–400×10^9/L)

Blood film

The blood film confirmed the presence of a normocytic normochromic blood picture with a degree of polychromasia. Neutrophils appeared normal, although reduced in number. Occasional atypical mononuclear cells noted with hair-like projections protruding from the cell membrane. No monocytes seen. Platelets were reduced, although morphologically normal.

Significance of results 1

The upper left quadrant pain is associated with splenomegaly, which was demonstrated as part of the clinical examination. The recurrent pharyngitis is likely linked with neutropenia. The presence of monocytopenia given the signs of infection is a cause for concern, and given the splenomegaly and atypical mononuclear cells, hairy cell leukaemia (associated with monocytopenia at diagnosis) should be considered in the differential diagnosis. Other possible causes of the symptoms include splenic lymphoma with villous lymphocytes (SLVL), B-PLL, and atypical HCL. The GP referred to a haematologist who requested bone marrow examination and immunophenotyping.

Results 2

Immunophenotyping was performed on a peripheral blood sample, which demonstrated a small population of cells positive for SmIg, CD20, CD22, CD25, CD11c, CD103, CD123, and FMC7. CD5, CD10, and CD23 were negative.

Attempts to acquire a bone marrow aspirate resulted in a dry tap. Bone marrow trephine produced a 20 mm specimen with patchy mononuclear cell infiltration with reduced but normal-appearing haemopoiesis. Mononuclear cells demonstrated a fried-egg appearance.

Significance of results 2

- The immunophenotyping results revealed the following results: CD103+ve, CD25+ve, CD11c+ve, indicating HCL. The fried-egg appearance of mononuclear cell infiltrate is typical of HCL cells, whilst the dry tap would have been due to deposits of reticulin fibres within the marrow. Application of reticulin stain on the bone marrow specimen would have shown reticulin deposits, which are not uncommon in HCL.
- Selection of a purine nucleoside analogue such as pentostatin plus anti-CD20 therapy provides excellent results for patients with HCL.

12.3.8 Lymphoplasmacytic lymphoma

Lymphoplasmacytic lymphoma (LPL) is a heterogeneous lymphoid malignancy associated with the accumulation of small B lymphocytes, plasmacytoid cells, and plasma cells within the bone marrow, lymph nodes, or spleen. It is derived from a mutated post-germinal-centre B cell prior to differentiation into a plasma cell. Figure 12.9 shows a blood film containing a number of small lymphoid cells in a case of LPL. A subtype of LPL called Waldenström **macroglobulinaemia** (WM) occurs in approximately 95% cases of LBL and is diagnosed following cytological findings associated with LPL infiltration of the bone marrow, often accompanied by mast cell infiltration, and evidence of an IgM paraprotein.

macroglobulinaemia
Very high levels of large proteins circulating in the blood, for example, IgM antibodies.

Studies examining the *IGHV* genes associated with LPL/WM have shown that somatic hypermutation has occurred and therefore these cells have passed through the germinal centre. Furthermore, there is a lack of intra-clonal *IGHV* sequence variation in individual tumours—suggesting the cells, rather than being of a late germinal centre origin, are of a post-germinal centre origin.

Complicating features occur in patients expressing high levels of IgM paraprotein, where this can result in an increased blood viscosity (called **hyperviscosity**), associated with cardiovascular complications in some patients. Occasionally, the paraprotein precipitates and aggregates at low temperatures, forming **cryoglobulins**. In some cases, the paraproteins have autoreactivity with host antigens, leading to autoimmune complications.

Immunophenotypically, these cells express CD19, CD20, and surface immunoglobulin (SmIg). In WM, this is restricted to IgM, although IgG or, rarely, IgA may be expressed in other LPL. CD138 expression represents plasmacytic differentiation and should be positive in a subset of cells in LPL/WM.

In approximately 90% of cases of WM, a point mutation in *MYD88*, leading to an exchange between leucine and proline at amino acid 265 (L265P), is detected. *MYD88* encodes an adaptor molecule involved in the inflammatory signalling pathway by linking the IL-1 receptor (IL-1R)—or Toll-like receptor family members—to IL-1R-associated kinases (IRAK). Activation of this pathway leads to enhanced signalling of nuclear factor kappa B (NFκB), MAPK, or AP-1, thus promoting survival. In WM cells, this activated adaptor pathway links Bruton's tyrosine kinase (BTK)—a B-cell receptor-associated enzyme—to downstream effectors, which enhances cell survival. L265P MYD88 mutations are also found in approximately 50% of IgM, producing monoclonal gammopathy of undetermined significance (MGUS). IgM MGUS progresses to WM, or other B-cell lymphoproliferative neoplasms (less frequently), at a rate of approximately 1.5% per year.

Cross reference
IgM monoclonal gammopathy of undetermined significance is discussed in section 12.3.11.

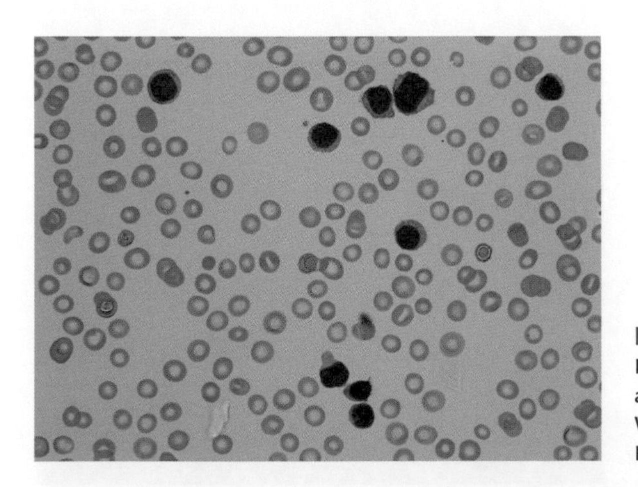

FIGURE 12.9
Lymphoplasmacytic lymphoma. Lymphoid cells are numerous and demonstrate cytoplasmic basophilia. Image courtesy of Jackie Warne, Haematology department, Queen Alexandra Hospital, Portsmouth.

Deletions of chromosome 6q21 are found in approximately 42% of patients, whilst trisomy 4 occurs in approximately 20%. On a number of occasions, t(9;14)(p13;q32) has been reported in LPL/WM, although we now know that t(9;14) (p13;q32) is associated with a range of lymphoid malignancies and is not disease specific. t(9;14)(p13;q32) results in the fusion of PAX5 to the IgH promoter, leading to dysregulation of *PAX5*. *PAX5* encodes the B-cell-specific activator protein (BSAP), a transcription factor, involved in early B-cell development.

LPL/WM is associated with a median survival of approximately 60 months.

12.3.9 Mantle cell lymphoma

Mantle cell lymphoma (MCL), originally described in 1992, comprises approximately 5% of all non-Hodgkin lymphomas. MCL is associated with a poor prognosis, although in younger patients with new treatment regimes, survival exceeding ten years is documented.

Most patients present with bone marrow involvement, hepatosplenomegaly, and lymphadenopathy. Pancytopenia or leucocytosis may feature at presentation and there may be evidence of malignant B cells within the peripheral blood. Figure 12.10 shows the typical appearance of a mantle cell in the peripheral blood of a patient in leukaemic phase. Immunophenotypically, MCL cells express a phenotype very similar to the CLL phenotype: CD5+, CD19+, CD20+; but FMC7, CD10, CD20, CD23, and BCL6 are typically not expressed. IgM and IgD are present on the MCL cell surface, helping to identify the pre-germinal-centre origin of these cells.

SELF-CHECK 12.8

Compare and contrast the immunophenotype of CLL and MCL.

Approximately 95% of cases harbour t(11;14)(q13;q32), fusing *CCND1*, encoding cyclin D1, with the IgH gene. Cyclin D1 is important in controlling the G1 phase of the cell cycle. Patients failing to express t(11;14) demonstrate translocations involving the other G1 cyclins, cyclin D2 or cyclin D3, leading to their overexpression. Overexpression of the D-type cyclins pushes these malignant cells through G1 to the S phase of the cell cycle, resulting in increased cell proliferation.

Sox11 aberration has been identified as a reliable marker that can consistently help identify MCL independently of cyclin D1 levels. Sox11, a neural transcription factor, is aberrantly expressed in MCL and promotes enhanced growth and survival. Sox11 is considered a promiscuous transcription factor, binding to a number of genes, leading to the activation of a variety of pathways. In the

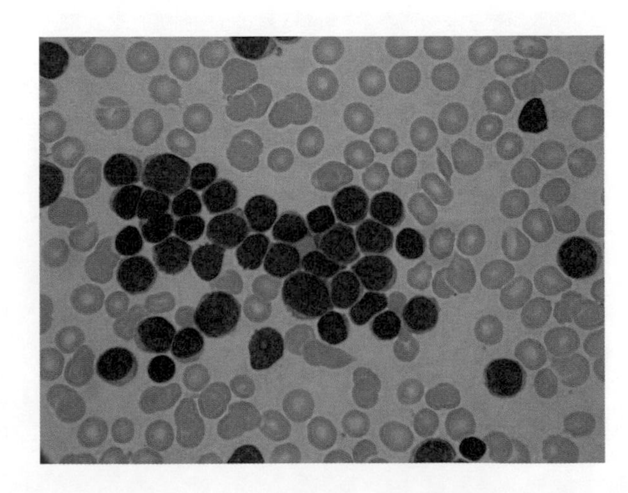

FIGURE 12.10

Mantle cell lymphoma. Large B cells within the peripheral blood during leukaemic phase. The mantle cells each have a large nucleus with a prominent nucleolus and pale cytoplasm.

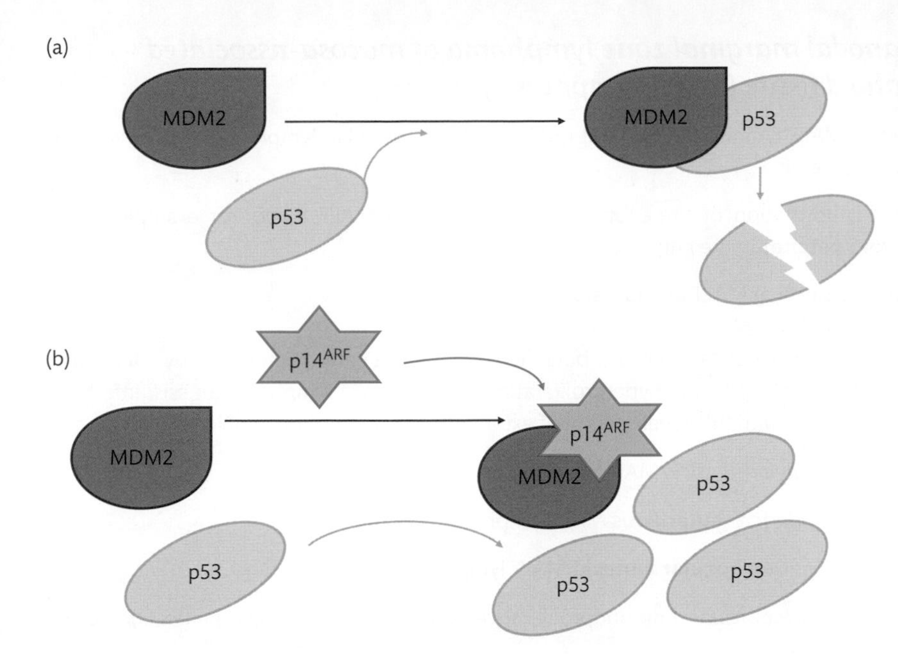

FIGURE 12.11
Regulation of p53 by MDM2 and p14^{ARF}. (a) MDM2 is a negative regulator of p53. Interaction between MDM2 and p53 leads to the inhibition and degradation of p53. (b) p14^{ARF} inhibits MDM2, allowing p53 to accumulate within the cell. Inhibition of p14^{ARF} by deletion or epigenetic silencing allows MDM2 to accumulate–inhibiting p53-dependent tumour suppressor functions.

majority of cases, *in situ* mantle cell neoplasm harbours t(11;14) and overexpresses cyclin D1. In addition to cyclin D1 amplification, cells *with* aberrant Sox11 (Sox11^{+}) fail to enter the germinal centre, retain unmutated *IGHV* genes, and progress to classic MCL. Additional mutations, including those of *TP53*, result in an aggressive blastoid subtype. Conversely, *in situ* mantle cell neoplasm, harbouring t(11;14) *without* aberrant expression of Sox11 (Sox11^{-}) enter the germinal centre, undergo *IGHV* hypermutation, and become leukaemic, non-nodal, MCL. These cells, although initially indolent, may progress rapidly should further somatic mutation of key genes, such as *TP53*, occur.

Additional genetic mutations involving the tumour suppressor ATM occur in 40–75% of cases, although—unusually—ATM aberration is not associated with an inferior outcome in these patients. However, mutations and dysregulation of the ARF/MDM2/p53 and p16^{INK4A}/CDK4 pathways in MCL are associated with aggressive disease and poor outcomes. Figure 12.11 outlines the interactions between p14^{ARF}, p53, and MDM2.

In a normal situation, MDM2 is responsible for the sequestration and degradation of p53. p14^{ARF} is a negative regulator of MDM2, and the homozygous silencing of p14^{ARF} removes the negative regulator of MDM2, increasing MDM2/p53 interactions. The net result of this interaction between MDM2 and p53 is a loss of p53 activity within p14^{ARF} mutated cells. In addition, the loss of p16^{INK4A} results in the uncontrolled interaction between cyclin D1 and its associated kinase CDK4. The association of cyclin D with CDK4 leads to progression of the cell cycle through G1 into S phase. These tumour suppressor and proto-oncogene mutations are associated with an increased proliferation rate, more aggressive disease, poor response to therapy, and reduced survival.

12.3.10 Marginal zone lymphomas

Marginal zone lymphomas (MZL) are a diverse group of lymphoid malignancies, and according to the WHO, can be subdivided principally into:

- extranodal MZL of mucosa-associated lymphoid tissue (MALT lymphoma);
- nodal MZL.

Each of these conditions will be considered in turn.

Extranodal marginal zone lymphoma of mucosa-associated lymphoid tissue (MALT lymphoma)

Extranodal MALT lymphomas account for 5–8% of non-Hodgkin lymphomas and develop as a consequence of:

- the transformation of pre-existing previously normal MALT tissue, for example in the **Peyer's patches** in the gut;
- the acquisition of MALT in unusual sites.

Peyer's patches
A group of lymph nodes in the wall of the ileum.

A significant number of cases are attributable to chronic antigenic stimulation, by infection or inflammation, leading to the hyperproliferation of B cells and lymphomagenesis. Infectious agents associated with the development of extranodal MALT lymphoma include:

- *Helicobacter pylori*—gastric MALT lymphoma;
- *Borrelia burgdorferi*—**cutaneous** MALT lymphoma;
- *Chlamydia psittaci*—**ocular adnexal** MALT lymphomas;
- *Camphylobacter jejuni*—immunoproliferative small intestine disease (IPSID)/α-chain disease,

cutaneous
Pertaining to the skin.

ocular adnexa
The accessory structures of the eye; they include the eyelids, lacrimal glands, orbit, and paraorbital areas.

and may also occur in autoimmune conditions such as:

- Hashimoto's thyroiditis;
- Sjögren's syndrome.

Following the isolation of the causative pathogen, treating patients with specific antibiotics can, in a significant number of cases, induce regression of the MALT lymphoma and lead to cure.

Typically, MZL cells are morphologically heterogeneous—representing a wide variety of marginal zone-derived B-lineage cells responding to antigen. Immunophenotypically, these cells express surface and cytoplasmic immunoglobulin. IgM is more frequently expressed than IgG. B-cell antigens CD19, CD20, CD22, CD79a, and CD79b are expressed, whilst CD3, CD5, CD10, CD11c, and CD23 are not expressed in these malignancies.

Approximately 60% of cases of extranodal MALT lymphoma demonstrate cytogenetic abnormalities. The most frequent translocation, occurring in approximately 30% of cases, is t(11;18)(q21;q21), although t(1;14)(p22;q32) and t(14;18)(q32;q21) are also found. We will now look at each of these translocations in turn.

t(11;18)(q21;q21) This translocation fuses the *API2* (apoptosis inhibitor-2) gene, located at 11q21 with the *MALT1* gene, locus 18q21—leading to the production of the fusion gene *API2-MALT1*. API2 is a member of the inhibitor of apoptosis (IAP) gene family, whilst MALT1 has been identified as a **paracaspase**, activated following the formation of a complex with BCL10. As a consequence of the activation of MALT1 via BCL10-binding, **ubiquitination** of the inhibitor of nuclear factor κB (IκB) occurs, leading to increased NFκB signalling. This in turn results in the overexpression of a number of NFκB regulated genes, many of which are involved in apoptotic inhibition. Interestingly, one of the downstream targets of NFκB is API2, the expression of which is increased following signalling via API2-MALT1. In addition, increased NFκB signalling via API2-MALT1 mimics B-cell receptor signalling, which usually follows an encounter with antigen. In this case, increased NFκB signalling leads to an upregulated proliferative drive, clonal growth, and survival of B cells.

paracaspase
A caspase-related protein.

ubiquitination
Involves the addition of ubiquitin monomers to a peptide, which allows for degradation of the peptide via the proteasome.

SELF-CHECK 12.9

Outline the relationship between NFκB and API2-MALT1.

t(1;14)(p22;q32) The translocation t(1;14)(p22;q32) is encountered less frequently than t(11;18) but has a similar biological role in tumour progression. The gene encoding BCL10 is situated at locus 1p22 and, as explained above, is *normally* associated with the activation of NFκB following the formation of a BCL10-MALT1 complex. In t(1;14), *BCL10* fuses with the immunoglobulin heavy-chain gene (*IGH*)—*IGH-BCL10*, resulting in the upregulation of BCL10 through the control of the strong IGH promoter. BCL10 has an important role in promoting B-cell survival following signalling via the BCR.

t(14;18)(q32;q21) Translocation t(14;18)(q32;q21) involves the fusion of *MALT1* with *IGH*, leading to the overexpression of MALT1 and the localization of BCL10 within the perinuclear region of malignant B cells. Occurring in 15-20% of cases, the consequence of IGH-MALT1 is the constitutional activation of NFκB signalling as described for t(11;18).

Key Points

MALT lymphoma-related t(14;18)(q32;q21) should not be confused with the translocation involving the same chromosomal breakpoints in follicular lymphoma (FL). In FL, *BCL2* is involved, whereas in MALT, *MALT1* is rearranged. *BCL2* is located approximately 5 Mb telomeric to *MALT1*, but is still considered the same region according to banding pattern.

Nodal marginal zone lymphoma

Nodal MZL is a rare malignancy, accounting for less than 2% of all lymphoid malignancies, and is most frequently found in patients aged 50-62 years. Largely restricted to lymph nodes, nodal MZL cells are rarely found in the bone marrow or peripheral blood. The majority of patients present with disseminated disease involving cervical and abdominal lymph nodes. The five-year survival is between 50 and 70%.

A paediatric form of this condition is also notable, although is not considered in any detail in this chapter.

12.3.11 Plasma cell neoplasms

Plasma cells are post-germinal-centre antibody-secreting cells. A wide range of plasma cell neoplasms are recognized, and in this section we will look at some of the most common. Figure 12.12 provides a brief summary of the range of plasma cell neoplasms recognized by the WHO. Only plasma cell myeloma and related disorders, MGUS, extraosseous plasmacytoma, and solitary plasmacytoma of bone are discussed in this section.

Plasma cell myeloma

Plasma cell myeloma, also called multiple myeloma, is a malignancy of plasma cells with a median age at presentation of 70 years, and a 50% survival rate of approximately five years.

Patients usually present with bone pain or pathological fractures and demonstrate signs and symptoms of anaemia, recurrent infections, and renal failure. The guidelines for the diagnosis of plasma cell disorders, including myeloma are outlined in Box 12.4. Serum electrophoresis demonstrates a monoclonal band, called an **M-protein** (monoclonal protein) or paraprotein,

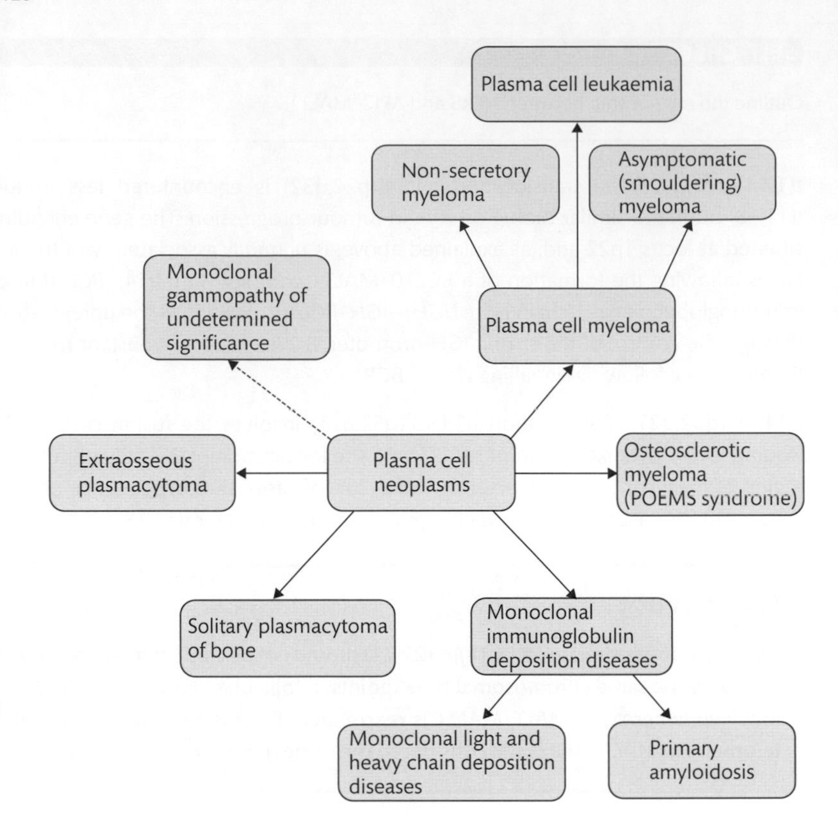

FIGURE 12.12
A chart to show the different types of plasma cell neoplasm and the way in which some of these conditions inter-relate. These conditions are all represented in the WHO classification system, although they are not all discussed in this text.

within the electrophoretogram. M-proteins are either whole immunoglobulins or just the free light-chain component. Urinary analysis shows paraproteins, called **Bence-Jones proteins**, in approximately 80% of cases. (See also Box 12.3.)

Figure 12.13 shows an electrophoretogram for serum electrophoresis for a wide range of patients. If you look along the bottom row, you will see patients 5, 7, and 8 all show distinct bands. Patient 5 has an additional band within the β-region, whilst patients 7 and 8 have dark bands within the γ-region. These bands are paraproteins. Figure 12.14 shows the densitometric analysis of the electrophoretogram for patient 5. Densitometry is used to quantify the size of the protein bands. As you can see, peak size matches the banding pattern. In this patient, the paraprotein is hidden within the β-region.

BOX 12.3 *Bence-Jones proteins*

Although Bence-Jones proteins (BJP) in multiple myeloma were first described by Dr Henry Bence-Jones in 1847, the nature of BJP was first investigated and reported by Edelman and Gally in 1962. BJP were identified as the light-chain components of immunoglobulin, comprising the M-protein. Heavy chains are too large to pass through the glomerular filter.

Identification of BJP can be performed from urinary samples, and the monoclonal nature of these proteins can be established using immunofixation.

Paraproteins are excreted in the urine before accumulating in the plasma. Only when renal excretion is compromised do paraproteins accumulate in the plasma.

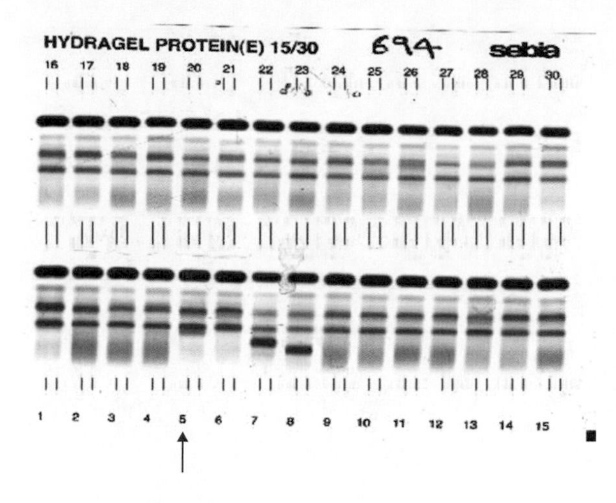

FIGURE 12.13

Serum electrophoresis. Of particular interest are patients 5, 7, and 8, all of whom show additional bands. Patient 5 shows a band within the β-region and patients 7 and 8 show distinctive bands within the γ region. Densitometry and immunofixation are performed to attempt to quantify and identify the species of abnormal proteins. Image courtesy of Rajee Goswami, Southampton General Hospital.

Figure 12.15 demonstrates the result of immunofixation for patient 5. The electrophoretic strip on the left demonstrates a match to the original in Figure 12.13. We can identify from this immunofixation that this patient has an IgA κ paraprotein.

Investigation of suspected plasma cell myeloma should include the following:

- FBC;
- urea and electrolytes (U&Es);
- serum calcium;
- serum creatinine;
- serum electrophoresis;
- erythrocyte sedimentation rate (ESR);
- blood film;
- bone marrow biopsy.

In addition, radiologic investigations should be included to identify the presence of **osteolytic lesions**. Although X-ray-based analysis of bone health has long been a staple investigation of plasma cell-based diseases, it is recognized that X-ray-based radiologic evaluation of patients with suspected or confirmed myeloma or other plasma cell neoplasm is of limited clinical utility. This is due to the relatively poor sensitivity of X-rays compared to other available radiologic approaches. Bone lesions of <5 mm in size are likely to be missed due to the poor sensitivity of X-rays, whilst active disease and response to therapy are poorly determined using these radiologic approaches. As such, the European Myeloma Network and the British Society for Haematology guidelines recommend three approaches to assess bone involvement, active disease, and response to therapy in these patients.

Whole-body low-dose computerized tomography (WBLDCT)—this provides an initial evaluation of bone disease. Being more sensitive than X-ray and able to detect lesions of <5 mm, this allows bone disease and osteolytic lesions to be detected earlier, providing greater value for patients with asymptomatic disease.

Whole-body MRI—this is excellent for establishing bone marrow infiltration and involvement and is considered the gold standard imaging approach for this type of assessment.

18 Fluorine-fluoro-deoxy-glucose positron emission tomography (FDG-PET)—this provides evidence of active disease by differentiating cells and tissue, such as normal and cancer cells, based upon their relative glucose usage. Cancer cells have higher glucose requirements than

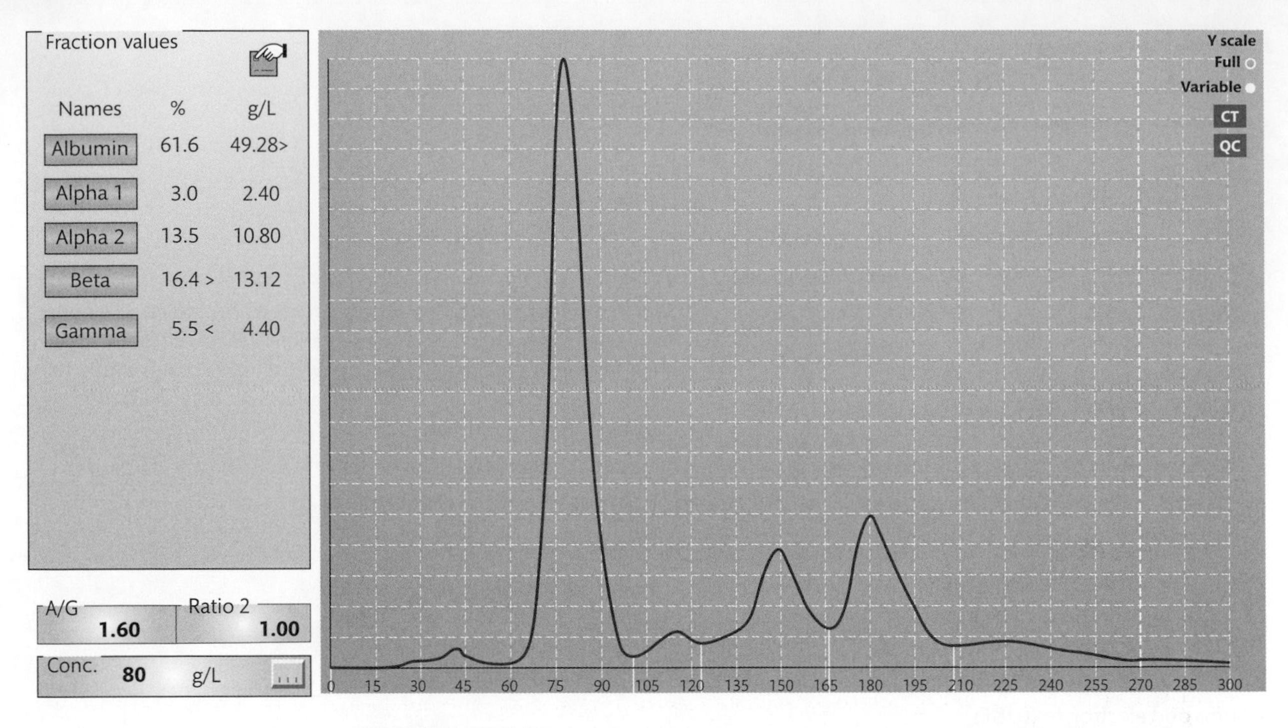

FIGURE 12.14

Results of densitometric analysis. Patient 5 showed a total protein concentration of 80g/L. The major species, albumin, gives a large peak at position 80. The beta region (position 150) is increased in size, whereas the gamma region (position 180) is reduced, as indicated in the fraction values on the left. The increased protein band visualized on the electrophoretogram (see Figure 12.13) is confirmed within the beta region using this technique. Higher levels of paraprotein are expected in myeloma, and it is likely this patient is receiving treatment. Image courtesy of Rajee Goswami, Southampton General Hospital.

normal cells and can be differentiated when FDG-PET is employed. Consequently, reduction in the size of tumour burden secondary to successful treatment is detectable and can be monitored using this approach.

The FBC may show a low haemoglobin concentration and a normal mean cell volume, the typical findings of a normocytic normochromic anaemia. Neutropenia may also be present, which may explain the occurrence of recurrent infections in a proportion of these patients.

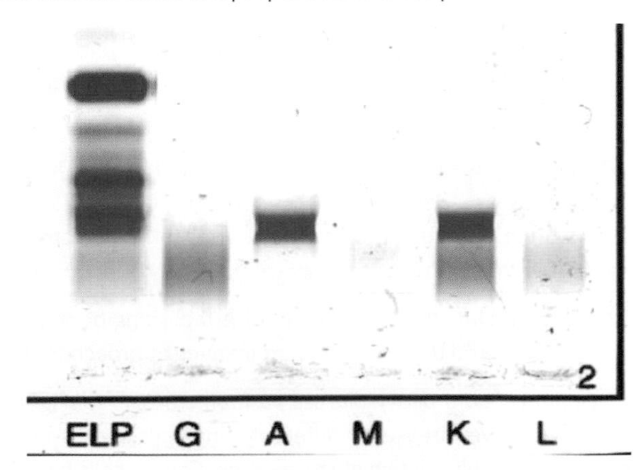

FIGURE 12.15

Using immunofixation, a distinct monoclonal band is identified as IgA, also demonstrating κ light chain restriction. The electrophoretic strip (labelled ELP) on the left shows the distribution of proteins in the patient's plasma. Image courtesy of Rajee Goswami, Southampton General Hospital.

TABLE 12.12 M-proteins identified in plasma cell myeloma.

M-protein	Frequency (%)
IgG	53
IgA	22
Light chain	20
Non secretory	3
IgD/IgE	1.5
IgM	0.5

Prior to the microscopic examination of a blood film, the macroscopic examination of the blood film following Romanowsky staining is usually the first indication of increased protein content in the plasma. The blood smear will have a deep-blue coloration—a consequence of the Romanowsky stain interacting with the fixed paraprotein on the blood film. Microscopically, paraprotein will be apparent as **background staining**. In cases with a paraprotein, rouleaux (red cells stacking upon one another) formation will be apparent on the blood film. Absence of these staining characteristics does not exclude a diagnosis of myeloma, as these are all a consequence of raised serum paraprotein levels.

background staining
The uptake of Romanowsky stain by plasma-derived paraprotein fixed on the microscope slide.

In patients with a raised plasma paraprotein concentration, the ESR will be raised.

Biochemical analysis will show a raised serum calcium concentration, called **hypercalcaemia**—caused by increased bone degradation in a number of these patients (discussed later in this section). Serum creatinine will be raised in patients with renal impairment, secondary to increased paraprotein concentrations and hypercalcaemia. In addition, serum electrophoresis demonstrates an M-protein which can be present within any region. The frequencies of the different M-proteins are outlined in Table 12.12. The use of immunofixation—a technique used to identify proteins within a sample using monoclonal antibodies—will enable the monoclonal or polyclonal nature of the proteins to be established. Total protein levels may be increased in patients with plasma cell myeloma. Conversely, there may also be a decrease in the other immunoglobulins—**immune paresis**. Immune paresis can be seen in up to 90% of patients with plasma cell myeloma.

immune paresis
This describes the reduction of all immunoglobulins in a patient's serum, except for the paraprotein.

SELF-CHECK 12.10

Outline the key laboratory findings associated with plasma cell myeloma.

Cross reference
FISH is introduced in Chapter 10.

Cytogenetic changes in plasma cell disorders

Cytogenetic abnormalities demonstrated by fluorescence *in situ* hybridization (FISH) are found in more than 90% of patients. Approximately 40–50% of cases possessing chromosomal translocations involve 14q32, combining the IgH locus with:

- 11q13—*CCND1*—cyclin D1;
- 4p16.3—FGFR3/MMSET (fibroblast growth factor receptor 3/multiple myeloma SET domain);
- 16q23—*CMAF*—cellular **musculoaponeurotic fibrosarcoma** proto-oncogene;

musculoaponeurotic
Pertaining to a fibrous or membranous sheath that connects muscle to bones.

fibrosarcoma
Tumour developing in the fibrous connective tissue.

- 6p21–*CCND3*–cyclin D3;
- 20q11–*MAFB*–c-MAF-related gene.

Monosomy 13 or del(13q14) occurs in approximately 15–40% of patients and trisomy 3, 5, 7, 9, 11, 15, 19, and 21 are also common features. Up to 10% of patients with an abnormal karyotype demonstrate gains of chromosome 1q. In approximately 30–40% of cases, patients demonstrate *RAS* mutations, p16 gene methylation in 20–30%, *MYC* rearrangements in 10–15%, and *TP53* mutations in 5–10%. These mutations either induce an increase in proliferative capacity within the malignant cells or a dysregulated range of tumour suppressor proteins.

Myeloma-related bone disease A frequent finding in patients with plasma cell myeloma and related diseases is bone disease, with areas containing osteolytic or 'punched-out lesions' apparent on X-ray analysis, as shown by the skull X-ray in Figure 12.16. This bone disease is responsible for the bone pain often experienced by patients, and also accounts for the high frequency of pathological fractures associated with plasma cell neoplasms.

Malignant plasma cells, as shown in the bone marrow aspirate in Figure 12.17, are often associated with areas of active bone resorption. Bone remodelling is an essential process for maintaining bone architecture, involving both bone resorption and bone deposition through the balanced action of osteoclasts and osteoblasts, respectively. Increased activity of osteoclasts,

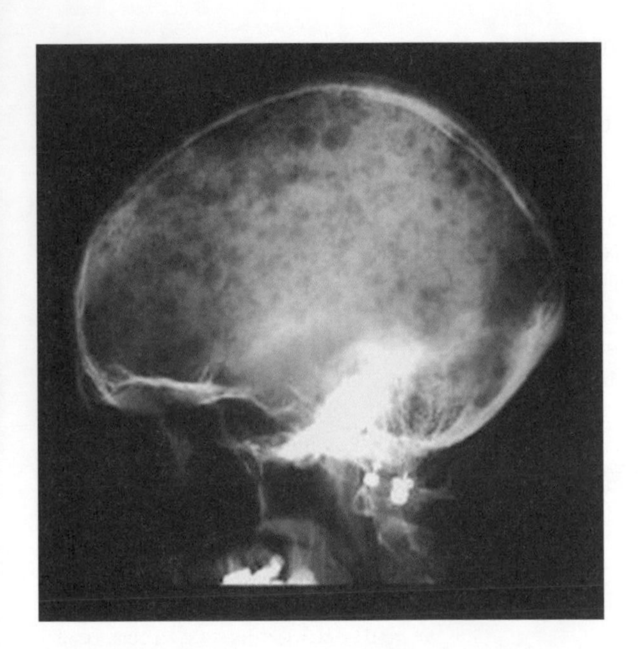

FIGURE 12.16
X-ray of the skull from a patient diagnosed with myeloma. The light patches in the skull, called punched-out lesions, represent enhanced osteolytic activity of osteoclasts, which have been activated by the myeloma cells resident within the bone marrow in these areas. Image from Provan D, Baglin T, Dokal I, and de Vos J. ***Oxford Handbook of Clinical Haematology***, 4th edn. Oxford University Press, Oxford, 2015. By permission of Oxford University Press.

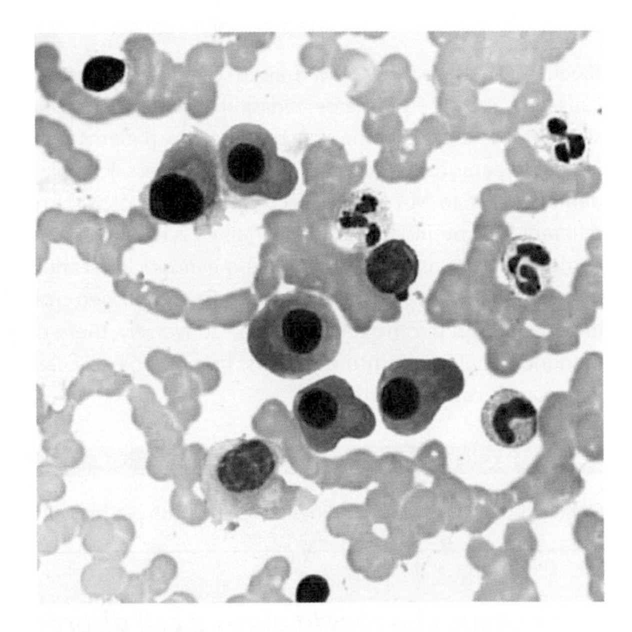

FIGURE 12.17
Malignant plasma cells within a bone marrow aspirate. Note the tissue architecture has been disrupted by the bone marrow aspiration process. Malignant plasma cells have a small, condensed nucleus and basophilic cytoplasm, accompanied by a white, perinuclear halo. Image from Provan D, Baglin T, Dokal I, and de Vos J. ***Oxford Handbook of Clinical Haematology***, 4th edn. Oxford University Press, Oxford, 2015. By permission of Oxford University Press.

and a relative reduction in osteoblast numbers and activity during the later stages of myeloma, leads to a loss of localized bone density and the accumulation of osteolytic lesions.

Secretion of osteoclast-activating factors (OAFs) is essential in the development of myeloma-associated bone disease. Chemical signalling between plasma cells and normal osteoclasts results from localization of these plasma cells to areas in close proximity to osteoclasts. There are two important processes involved in myeloma-related bone disease, including the upregulation of macrophage inflammatory protein-1α (MIP-1α) and dysregulation of the *osteoprotegerin/Receptor activator of nuclear factor κB ligand/Receptor activator of nuclear factor κB* (OPG/RANKL/RANK) pathway. These processes are illustrated in Figure 12.18.

MIP-1α is a chemokine thought to be responsible for the proliferation of osteoclasts in regions occupied by malignant plasma cells. MIP-1α levels increase in two ways:

- MIP-1α is synthesized by malignant plasma cells;
- MIP-1α synthesis and secretion can be induced from osteoblast-like cells in response to IL-1 and tumour necrosis factor-α(TNFα)—both synthesized and secreted by malignant plasma cells.

MIP-1α binds to the chemokine receptors CCR1 and CCR5 on the surface of osteoclasts, and induces proliferation and activation of these cells and uncoupling of the bone remodelling balance.

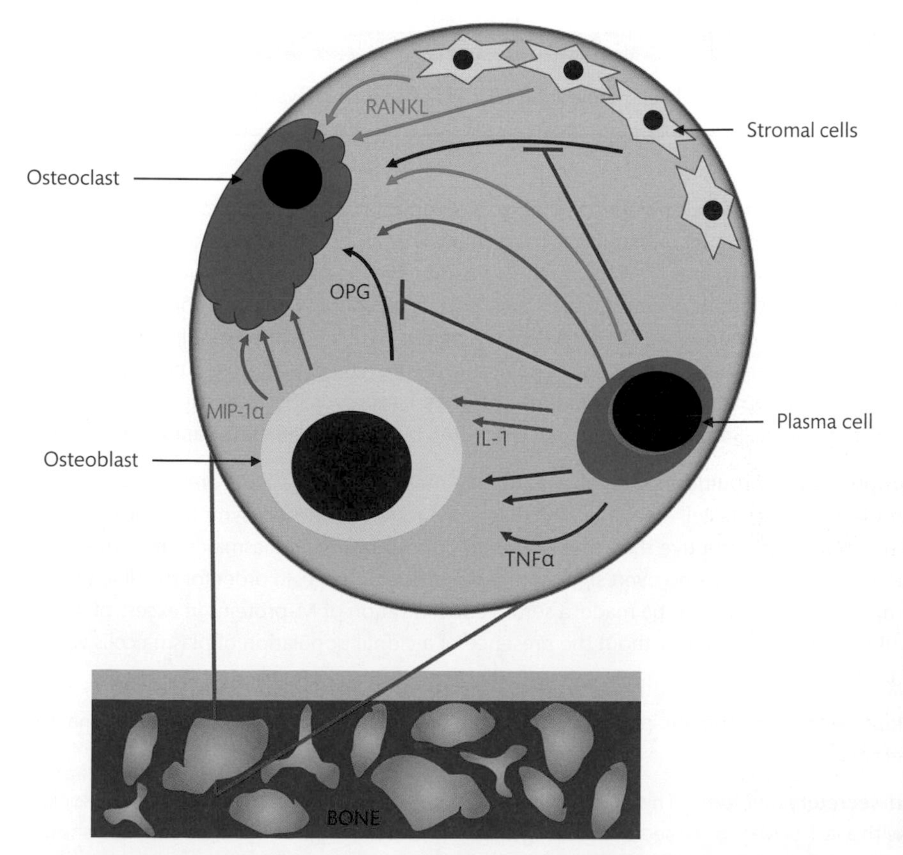

FIGURE 12.18
Pathogenesis of myeloma-related bone disease. Malignant plasma cells accumulate within the bone marrow and interact with a variety of cells. The secretion of IL-1 (brown arrows) and TNFα (purple arrows) stimulates osteoblasts to synthesize and secrete macrophage inflammatory protein -1α (MIP-1α) (blue arrows). MIP-1α is also synthesized by the plasma cells and increases osteoclast proliferation. Receptor activator of nuclear factor κB ligand (RANKL) (green arrows) is secreted by malignant plasma cells and bone marrow stromal cells, resulting in an increase in the number of osteoclasts and their activity in the local area. Osteoprotegerin (OPG) (black arrows), a negative regulator of RANKL secreted by osteoblasts and bone marrow stromal cells, is inhibited in myeloma. The net result is an increase in irreversible bone resorption.

RANK is a member of the TNF superfamily of receptors and is expressed on osteoclasts. The binding partner for RANK–RANKL (RANK ligand)–is expressed on the surface of osteoblasts, and interaction between RANKL and RANK results in:

- increased osteoclast differentiation;
- increased osteoclast activity;
- inhibition of osteoclast apoptosis;
- increased bone resorption.

The expression of soluble osteoprotogerin (OPG) by bone marrow stromal cells and osteoblasts inhibits the interaction between RANK and RANKL, thereby limiting the differentiation and activation of osteoclasts. In plasma cell myeloma, there is a decrease in the synthesis of OPG and, importantly, an increase in the expression of RANKL, not only by bone marrow stromal cells, but also by the malignant plasma cells. This upregulation of RANKL and downregulation of OPG favours the osteolytic activity of osteoclasts, leading to a net loss of bone density in those areas.

The diagnosis of plasma cell myeloma must be considered in all patients demonstrating an M-protein. There are two distinct types of myeloma: symptomatic plasma cell myeloma and asymptomatic (or smouldering) plasma cell myeloma. The most important feature in discriminating between the two is the presence of organ or tissue damage, including hyper*c*alcaemia, *r*enal insufficiency, *a*naemia, and *b*one lesions–described by the acronym **CRAB**. It is important to recognize that these symptoms are not specific to plasma cell myeloma and patients must be fully investigated prior to a diagnosis and subsequent treatment.

SELF-CHECK 12.11

Describe the process through which myeloma-related bone disease develops.

Symptomatic plasma cell myeloma Patients show evidence of CRAB and harbour a clonal population of plasma cells, either in the bone marrow or as a plasmacytoma. In most patients, at least 10% of the nucleated cells within the bone marrow are plasma cells, but this figure is not considered an important diagnostic threshold. Specific threshold concentrations of M-protein within the plasma or urine are not considered as part of the classification requirement because these are so variable between patients. Depending on the type of paraprotein produced by the plasma cells, many patients will have plasma IgG levels in excess of 30 g/L or an IgA concentration greater than 25 g/L. Some patients present with CRAB but do not fulfil these paraprotein criteria.

Asymptomatic (smouldering) myeloma Approximately 8% of patients are diagnosed with asymptomatic plasma cell myeloma. The diagnostic requirements of asymptomatic myeloma are much more prescriptive than those required for symptomatic plasma cell myeloma, primarily because there are no overt signs of organ or tissue damage. In order for the diagnosis of asymptomatic myeloma to be made, a serum concentration of M-proteins in excess of 30 g/L should be present with or without the presence of a clonal population of plasma cells within the bone marrow.

Patients with asymptomatic plasma cell myeloma may progress to symptomatic plasma cell myeloma.

Non-secretory myeloma This occurs in approximately 3% of cases of myeloma and is associated with a lack of M-protein secretion. Using immunohistochemical analysis, a distinction should

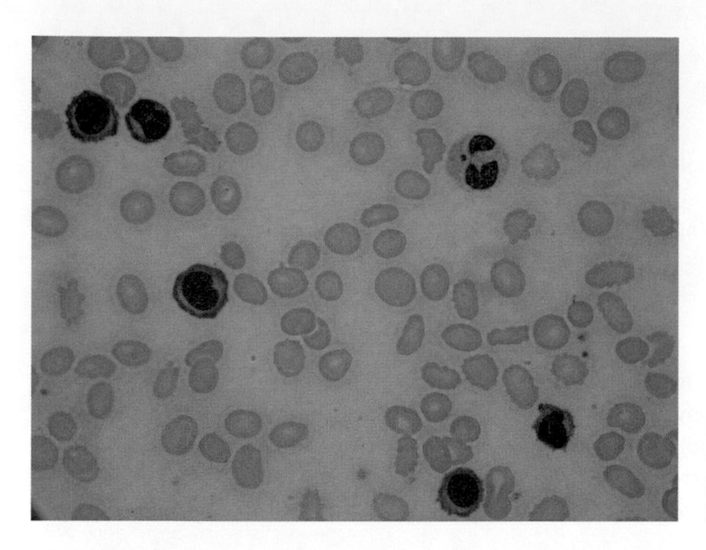

FIGURE 12.19
Plasma cell leukaemia. The presence of circulating plasma cells shows distinctive eccentric nuclei with a pale Golgi zone called the perinuclear halo. Cytoplasm is deeply basophilic.

be made between cases showing cytoplasmic M-protein and cases in which no M-protein is synthesized. The behaviour and treatment of these variants is identical. Careful consideration of a diagnosis of non-secretory myeloma should be given for patients demonstrating clinical signs of myeloma with absence of an M-protein. Bone marrow biopsy and histochemical investigations are indicated to ensure this diagnosis is not missed.

Plasma cell leukaemia Plasma cell leukaemia (PCL) is defined by the presence of:

- 2.0×10^9/L plasma cells or more in the peripheral blood;
- plasma cells comprising 20% of the total WBC count—which is particularly important to consider in patients with a low WBC count.

PCL can be either a primary or secondary disease. Primary PCL occurs in about 5% of patients from presentation with no previous medical history or evidence of plasma cell myeloma. Secondary PCL is the leukaemic phase of an established plasma cell myeloma. Primary PCL is considered to be a separate disease entity due to its different disease behaviour, cytogenetic, and immunophenotypic profile. PCL is associated with considerable extramedullary infiltration and aggressive behaviour.

The peripheral blood cell morphology for a patient with PCL can be seen in Figure 12.19.

Monoclonal gammopathy of undetermined significance (MGUS)

MGUS is a frequently encountered plasma cell disorder within the haematology laboratory that does not always progress to an overt malignancy. IgM-MGUS is differentiated from non-IgM MGUS—principally of the IgG/IgA type—in the revised fourth edition of the WHO guidelines. MGUS is found in approximately 3% of all individuals over the age of 50 years, rising to 5% of individuals aged 70 years and over.

The majority of patients are diagnosed following investigation for a completely unrelated disorder; these patients have an M-protein of less than 30 g/L and no evidence of Bence-Jones proteinuria. There should be an absence of CRAB and FBC indices should be normal.

Approximately 46% of patients demonstrate 14q32 translocations—the most common of which is t(11;14)(q13;q32)—which result in the overexpression of cyclin D1.

The risk of transformation to a malignant disease is approximately 1% per year for IgG MGUS, rising to approximately 1.5% a year for MGUS secreting IgA, and requires the accumulation

BOX 12.4 *UK Myeloma Forum guidelines*

An archive of key guideline documents for the diagnosis, monitoring, and treatment of patients with plasma cell diseases are hosted on the UK Myeloma Forum website. These can be accessed by using the following link. https://www.ukmf.org.uk/guidelines-page/bshukmf-guidelines/.

of additional oncogenic mutations to progress to myeloma. Approximately 50% of MGUS are hyperdiploid, often involving the odd-numbered chromosomes. Secondary cytogenetic events include, but are not restricted to, the loss of chromosome 13 in approximately 30% of cases, amplification of chromosome 1q (25% of cases), activation of *NRAS* (36%), or *BRAF* (27%). Deranged serum-free light-chain ratios (κ:λ) may help predict the risk of progression.

To date, two forms of MGUS have been identified based on the following paraproteins:

- IgM paraprotein;
- IgG or IgA paraprotein.

IgM-related MGUS is thought to be derived from a B-cell population failing to undergo Ig class-switching, although there is evidence of somatic hypermutation. Conversely, IgG- and IgA-related cases are derived from a post-germinal-centre plasma cell.

Should transformation occur, IgM-secreting cells are likely to transform to lymphoplasmacytic lymphoma or Waldenström macroglobulinaemia, in approximately 50% of cases, whilst IgG or IgA MGUS may progress to plasma cell myeloma or **amyloidosis**.

In cases where MGUS is diagnosed, a watch-and-wait strategy is employed, ensuring that patients are monitored every six months to determine whether the disease remains stable or the paraprotein concentration is increasing.

Solitary plasmacytoma of bone and extraosseous plasmacytoma

A plasmacytoma is an accumulation of malignant plasma cells within a particular tissue type. Solitary plasmacytoma of the bone can be defined as '*the presence of a single lytic lesion due to monoclonal plasma cell infiltration with or without soft tissue extension*' whereas extraosseus plasmacytoma is a '*soft tissue mass not in contact with bone*'. The majority of solitary plasmacytoma cases occur in vertebrae, femurs, pelvis, and ribs, whereas in the extraosseous form, the head and neck, gastrointestinal tract and lungs predominate. Solitary plasmacytoma of the bone carries a high risk of evolution to myeloma, with 50% progression at ten years. Evolution of extraosseous plasmacytoma to myeloma occurs in approximately 30% of cases at ten years. Whilst there is a risk of pathological fractures at the site of the plasmacytoma in bony disease, neither disease is associated with other myeloma-related symptoms.

12.3.12 Splenic marginal zone lymphoma

Splenic marginal zone lymphoma (SMZL) is invariably associated with splenomegaly in the absence of lymphadenopathy and occurs in approximately 2% of patients diagnosed with a lymphoid malignancy. The median age of patients at presentation is 65 years. The bone marrow always shows signs of infiltration, with any of the patterns of infiltration occurring—nodular,

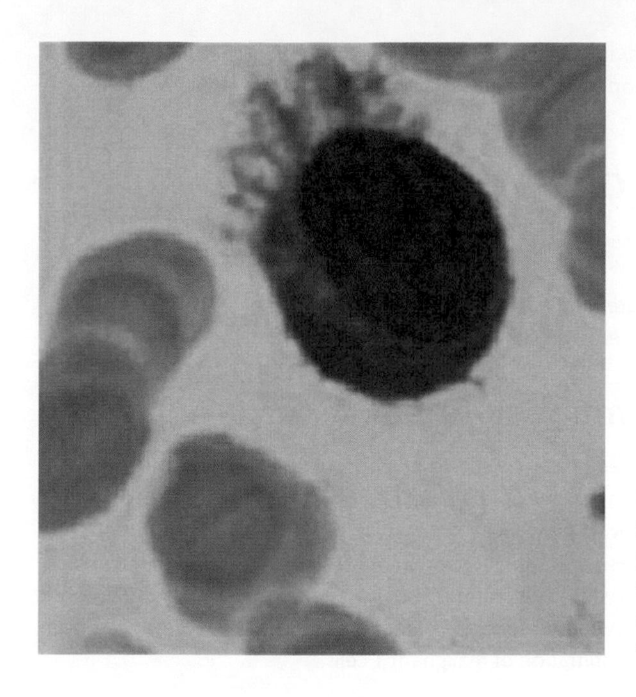

FIGURE 12.20

Splenic marginal zone lymphoma. Heterogeneous cells with villous protrusions of the cell membrane are typical features. Image from Longmore M, Wilkinson I, Baldwin A, and Wallin E. *Oxford Handbook of Clinical Medicine*, 9th edn. Oxford University Press, Oxford, 2014. By permission of Oxford University Press.

interstitial, paratrabecular, or diffuse. Characteristically, an intrasinusoidal pattern of infiltration is noted, helping to differentiate SMZL from all other types of low-grade lymphoma. A bone marrow trephine is usually more useful for the diagnosis of SMZL than a bone marrow aspirate, and often, because the malignant B cells express CD20, anti-CD20 should be used to visualize the extent of bone marrow infiltration by SMZL cells.

Cross reference

Bone marrow assessment is discussed in Chapter 10.

The peripheral blood contains variable numbers of malignant lymphocytes, with an increase in the absolute lymphocyte count occurring in approximately 75% of cases. Villous lymphocytes are apparent in 15% of cases and are demonstrated in Figure 12.20.

Signs and symptoms of anaemia occur in approximately 64%, and severe thrombocytopenia features in 15% of patients. Neutropenia is also often apparent, leading to an increased risk of opportunistic infections. Reduction in the platelet, neutrophil, and red cell counts may be a consequence of splenic sequestration, bone marrow infiltration, or autoimmunity.

Approximately 50% of patients demonstrate a monoclonal band—usually IgM, IgG, or, very rarely, IgA—using serum electrophoresis, but with serum concentrations.

Cytogenetically, there are no characteristic chromosomal abnormalities diagnostic for SMZL, although the most common features involve the rearrangement or loss of chromosome 7 and trisomy 3. Mutations of *TP53* have been variably reported, but only occur in advanced disease.

Clinically, SMZL is a stable disease, with median survival between 8 and 13 years. Approximately 10% of patients transform to an aggressive lymphoma.

T-cell malignancies are far less frequently encountered than B-cell malignancies and do not have the same cytogenetic or molecular predictability as associated with the development of malignant B cells. However, on occasion, working in a haematology laboratory you will encounter T-cell malignancies and you should be aware of some of the important features of the most frequently occurring.

The WHO subdivides the T-cell malignancies into *precursor lymphoid neoplasms* and *mature T-cell and NK-cell neoplasms.* So far we have considered the important precursor neoplasms and the mature B-cell neoplasms, and now we should move on to look at the most common of the mature T-cell and NK-cell neoplasms.

12.4 Mature T-cell and NK-cell neoplasms

The diseases considered below are of a mature T-cell origin and are clearly defined using the WHO guidelines. Only the most frequently encountered T-cell neoplasms are outlined here.

12.4.1 T-cell prolymphocytic leukaemia (T-PLL)

This is an aggressive T-cell malignancy which used to be considered the T-cell counterpart to B-CLL, although this is no longer the case. Approximately 75% of patients present with splenomegaly, and 50% of cases will show signs of hepatomegaly or lymphadenopathy. Importantly for laboratory haematologists, these cells will be found in the FBC and on the peripheral blood film, as seen in Figure 12.21. The white cell count is usually greater than 100.0×10^9/L, with the FBC demonstrating a normocytic normochromic anaemia and thrombocytopenia, both caused by a diffuse bone marrow infiltration of malignant T cells.

The most frequently encountered cytogenetic feature is an inversion or translocation of chromosome 14, particularly involving 14q11 and 14q32.1. Overexpression of *TCL1A* results in an

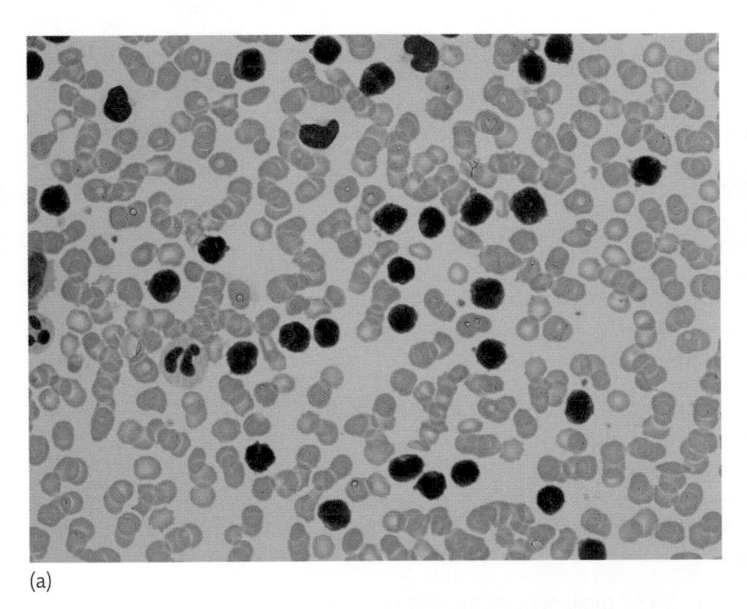

(a)

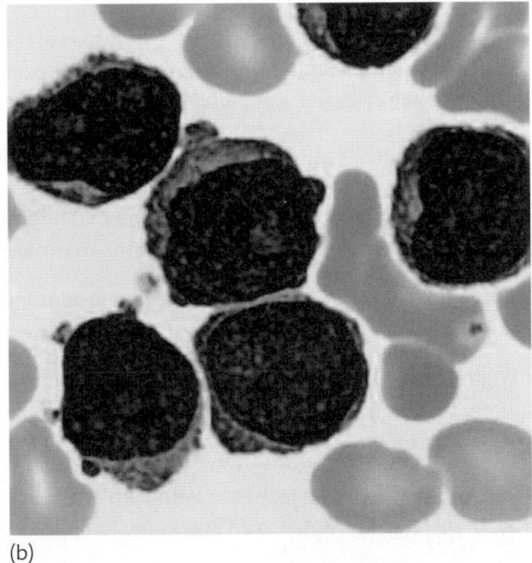

(b)

FIGURE 12.21

T-cell prolymphocytic leukaemia. (a) A high WBC count showing small-sized prolymphocytes. The cells have scant basophilic cytoplasm, with many unusually shaped nuclei. This image was taken from a relatively thick region of the blood film, resulting in the considerable red cell stacking apparent in this image. (b) High power image of T-PLL cells. Image (a) courtesy of Jackie Warne, Haematology department, Queen Alexandra Hospital, Portsmouth. Image (b) courtesy of Gordon Sinclair.

increase in TCL1 expression and activity. TCL1 enhances signalling via the serine-threonine kinase AKT, resulting in increased growth and survival. Immunophenotypically, T-PLL demonstrates surface expression of CD2, CD3, and CD7, although CD1a and TdT are not expressed.

Response to chemotherapy is generally poor and median survival is less than 12 months.

12.4.2 T-cell large granular lymphocytic leukaemia (T-LGL)

T-LGL is associated with an increased number of circulating large granular lymphocytes expressing CD3, CD8, and CD57. Although a number of patients will be diagnosed following an FBC for an unrelated medical complaint, approximately 50% of patients present with splenomegaly, whilst hepatomegaly occurs in 20%. A common feature is an isolated neutropenia (52–84%), which increases the risk of opportunistic infections for patients with this malignancy. Occasionally, anaemia and thrombocytopenia are evident, although these are unusual manifestations. The blood film frequently shows large granular lymphocytes, as shown in Figure 12.22. Autoimmunity is a common complication, with rheumatoid arthritis frequently reported.

Median survival has been reported to be 13 years following diagnosis.

12.4.3 Mycosis fungoides and Sézary syndrome

Although these diseases are considered as *separate entities* by the WHO, they are related and will be considered as variations of a single pathology in this text. Mycosis fungoides (MF) is the most frequently occurring cutaneous T-cell lymphoma, accounting for over 50% of all lymphomas of the skin, where it usually remains restricted. In contrast, Sézary syndrome accounts for only 5% of cutaneous T-cell lymphomas, does have a leukaemic phase, and, as a consequence, can become easily disseminated. Morphologically Sézary syndrome is readily identified within the peripheral blood film. Typically, the cells for both conditions are cerebriform—meaning the nucleus of the cell looks 'brain-like', as demonstrated in Figure 12.23. Sézary syndrome is also associated with **erythroderma** and lymphadenopathy, demonstrating the broad effects this malignancy can have on a range of tissues.

erythroderma
Red and scaling skin caused by inflammation.

Immunophenotypically, both MF and Sézary cells express T-cell markers CD2, CD3, and CD5, but generally fail to express CD8. An increased population of CD4-positive, CD7-negative cells is characteristic of the Sézary phenotype with peripheral blood immunophenotyping.

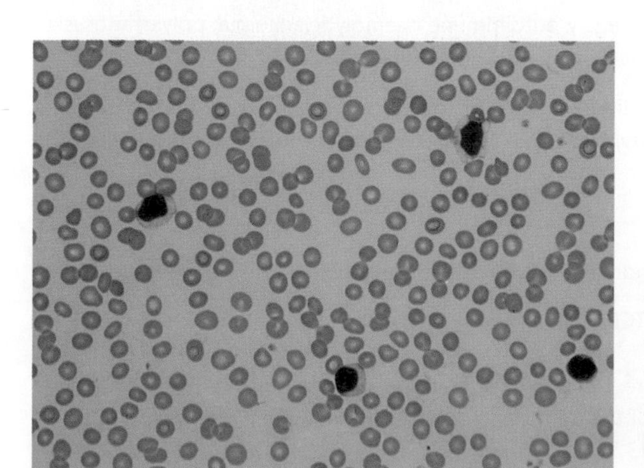

FIGURE 12.22
T-cell large granular lymphocytic leukaemia. Circulating large granular lymphocytes (NK cells) can be seen dominating this blood film. The cytoplasm is light blue, almost clear in places except for some large granules scattered throughout the cytoplasm. Image courtesy of Jackie Warne, Haematology department, Queen Alexandra Hospital, Portsmouth.

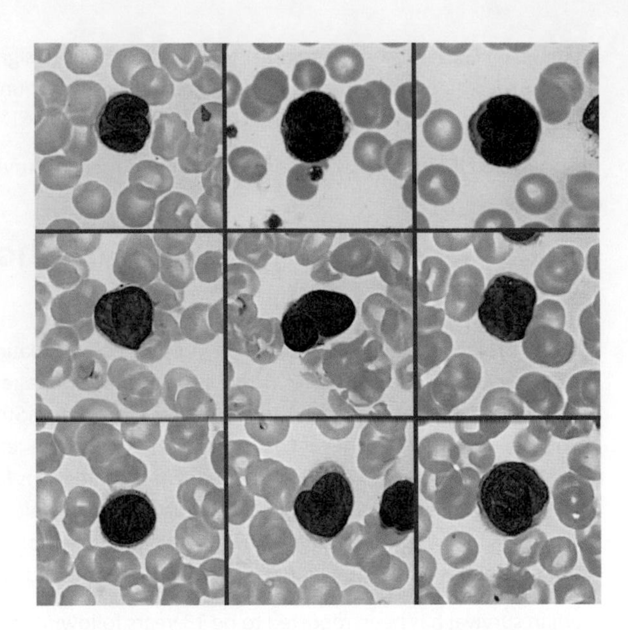

FIGURE 12.23
Sézary syndrome. The large lymphocytes demonstrated contain cerebriform nuclei with condensed chromatin. The cytoplasm is moderately basophilic. © Sysmex Europe GmbH 2015. Images are under copyright of Sysmex Europe GmbH or published with the consent of their respective owners.

Depending on the stage of the disease, patients with MF have been reported to survive for over 33 years following diagnosis, compared with a 5-year survival of 20% for patients diagnosed with Sézary syndrome.

12.4.4 Peripheral T-cell lymphoma, not otherwise specified

These diseases account for the majority of cases (30%) of T-cell lymphoma and are incorporated into the *not otherwise specified* category due to a lack of predictable features of morphology, immunophenotype, and cytogenetics. The diseases included in this section are highly aggressive, with the majority of patients dying within five years following diagnosis.

12.4.5 Angioimmunoblastic T-cell lymphoma (AITL)

Originally considered a reactive process, AITL is the most common of the *T-cell lymphomas*. Occurring most frequently in the elderly, with a median age at diagnosis of 65 years, clinical indicators include hepatosplenomegaly, autoimmune haemolytic anaemia, polyarthritis, skin rash, eosinophilia, and a polyclonal hypergammaglobulinaemia.

According to the revised fourth edition of the WHO classification, a new category called *nodal T-cell lymphoma with TFH phenotype* has been introduced. This includes three disease entities including:

- AITL;
- nodal peripheral T-cell lymphomas (PTCL) with T follicular helper (TFH) phenotype;
- follicular T-cell lymphoma (FTCL).

Only AITL will be considered here. Defined by immunophenotype, these cells should include the expression of two or three of the following normal follicular helper T antigens, including CD279/PD1, CD10, BCL6, CXCL13, ICOS, SAP, and CCR5.

The mutational landscape of AITL is complex. A point mutation within Ras homology family member A (RHOA) occurs in up to 70% of cases, leading to the substitution of glycine for valine at amino acid 17—G17V. This acquired loss of function mutation prevents GTP interacting with RHOA, leading to the dysregulation of downstream signalling pathways. Additionally, loss of TET2 function occurs in approximately 47-83% of cases, DNMTA in 20-30%, and IDH2 in 20-45% of cases. Other than in myeloid malignancies, gliomas, and glioblastomas, the only lymphoid neoplasm to harbour IDH2 abnormalities is AITL.

Lymph node examination reveals tumour cells on a background of neutrophils, eosinophils, plasma cells, and epithelioid histiocytes, typically accompanied by **atrophied** lymphoid follicles.

atrophy
Wasting of a structure, with subsequent change in function.

Coexisting EBV-infected B cells are apparent in the majority of cases. In addition, following the immunosuppression and dysregulation encountered in AITL, the unchecked re-emergence of EBV can occur, leading to an EBV-driven population of immortalized B cells.

Overall five-year survival is approximately 30%, with a median survival reported in most publications of approximately three years.

12.4.6 Anaplastic large cell lymphomas (ALCLs)

These are a group of mature T-cell malignancies sharing common morphological features, CD30 expression, with varied expression of other T-cell markers including CD3, although they differ in their disease behaviour, genetic profiles, and prognosis. ALCLs can be subdivided into the following types:

- systemic ALK$^+$ ALCL;
- systemic ALK$^-$ ALCL;
- primary cutaneous ALCL;
- breast implant-associated ALCL (provisional entity).

This chapter will provide some clarity regarding the systemic ALK$^+$ and ALK$^-$ subtypes.

Systemic ALK$^+$ ALCL

Accounting for 10-20% of childhood lymphomas and 12% of mature T/NK-cell malignancies, this is a relatively uncommon T-cell malignancy associated with heterogeneous morphology. The defining features of ALCL are shared between ALK$^+$ and ALK$^-$ subtypes and include the presence of horseshoe-shaped cells (called hallmark cells) independent of the other morphological features and CD30+ immunophenotype,

The most common cytogenetic feature, occurring in 40-60% of patients, is a translocation involving chromosomes 2 and 5—t(2;5)(p23;q35)—resulting in the fusion of the *ALK* and *NPM* genes, and is associated with a relatively good prognosis. Currently, a number of *ALK* fusion partners have been described, with the second most common being tropomyosin 3 (*TPM3*) as a consequence of t(1;2)(q21;p23). The chimeric product of t(2;5)(p23;q35), NPM-ALK, has **constitutionally** active tyrosine kinase activity leading to enhanced signalling through the JAK3-STAT3 transduction pathway.

constitutionally
Pertaining to an entity's composition.

Nucleophosmin (NPM) is important for the shuttling of ribonucleoproteins from the cell nucleus to the cytoplasm, and is found in both the nucleus and the cytoplasm in cases harbouring t(2;5)(p23;q35). NPM contains an oligodimerization domain which is conserved in the NPM-ALK fusion protein, enabling NPM-ALK dimers to form and constitutional activation of ALK.

Systemic ALK⁻ ALCL

Genetically ALK⁻ ALCL is divided into three forms:

1. DUSP22-*rearranged*—this group has the best prognosis of the three subtypes of ALK⁻ ALCL with similar outcomes to ALK⁺ ALCL. The *DUSP22* locus, 6p25.3, also includes the IRF4 gene and therefore is called the *DUSP22-IRF4* gene region. Approximately 30% of cases involve rearrangement with *FRA7H* via t(6;7)(p25.5;q32.2), resulting in the downregulation of the *DUSP22* product, the negative regulator of the MAP kinase signalling pathway, dual-specificity phosphatase protein 22.
2. TP63-*rearranged*—8% of ALK⁻ ALCL cases involve the rearrangement of the *TP53* homologue, *TP63*, resulting in the production of a fusion protein that exerts a dominant-negative (silencing) effect on p53. *TP63* rearrangements are the defining feature of the ALK⁻ ALCL poor prognostic subgroup.
3. *Triple negative ALCL*—here, rearrangement of DUSP22, TP63, and ALK is not present.

Triple negative ALCL is associated with a prognosis intermediate between the good prognosis of *DUSP22* rearrangement and the poor prognosis associated with *TP63* rearrangement.

Now that the most common of the non-Hodgkin lymphomas have been considered, we will go on to examine Hodgkin lymphomas.

SELF-CHECK 12.12

Describe the role of ALK in lymphomagenesis.

12.5 Hodgkin lymphoma

Hodgkin's disease, later renamed Hodgkin lymphoma (HL), was first identified as a specific disease entity in 1832 by Thomas Hodgkin. In 1898 and 1902, respectively, Reed and Sternberg independently described the mononucleated 'Hodgkin cell' and the multinucleated cells characteristic of HL; these were later described as 'Reed–Sternberg cells' (HRS cells) and are shown in Figure 12.24. Hodgkin lymphoma is a lymphoid malignancy commonly associated with young adults.

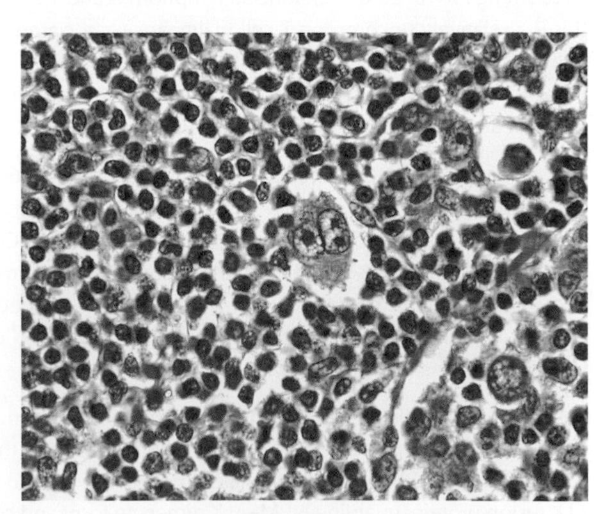

FIGURE 12.24
Typical Hodgkin/Reed–Sternberg cells. Note the owl's eye appearance characteristic of this type of cell. Image courtesy of Ed Uthman MD, licensed under the Creative Commons Attribution Share Alike 2.0 Generic license.

The incidence of Hodgkin lymphoma has remained steady in comparison with other types of lymphoma, and is considered to be relatively treatable. Approximately 40% of patients experience constitutional (or B-) symptoms, including weight loss, night sweats, low-grade fevers, fatigue, and **cachexia**. Up to 70% of patients diagnosed with HL can be considered cured following intensive chemotherapy and radiotherapy.

cachexia
Describes progressive weight loss, anorexia, and a gradual reduction in the patient's body mass.

Hodgkin lymphoma is not considered a single disease entity, but rather a composite of two disorders—classical HL and nodular lymphocyte predominant HL—subdivided according to the morphology and immunophenotype of the malignant cells and the nature of the cellular infiltrate around the malignant cells within lymphoid tissue. Figure 12.25 demonstrates how Hodgkin lymphoma can be subdivided.

Classical HL is further subdivided into four distinct entities; importantly, all classical HLs can be treated in the same way. In contrast, nodular lymphocyte-predominant type requires correct identification, as its higher rate of relapse is an indication for specific therapy.

Compared with other haematological malignancies, the cancer cells in HL only comprise between 0.1 and 1% of the total number of nucleated cells in the affected tissue. The majority of cells within the tumour mass are inflammatory cells, as described in mixed cellularity HL.

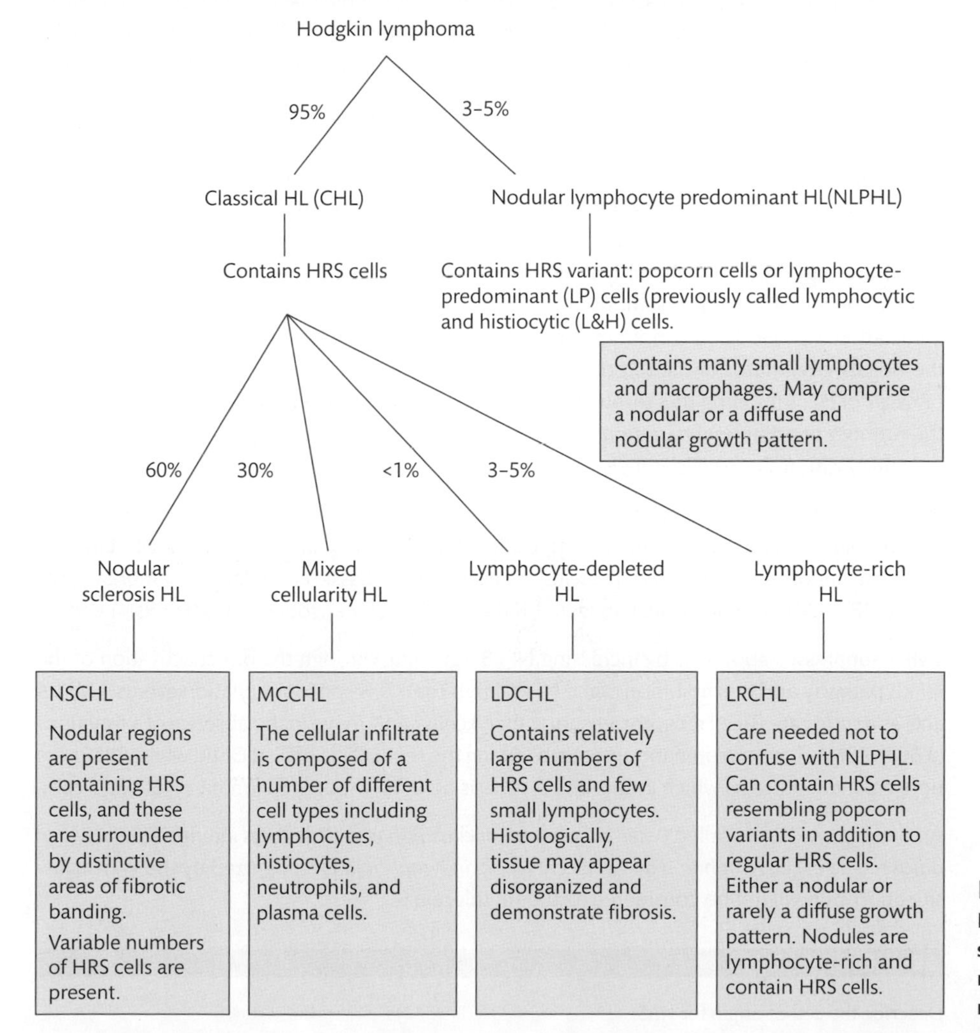

FIGURE 12.25
Hodgkin lymphoma and the subdivision into classical and nodular lymphocyte-predominant (NLPHL) diseases.

HL is associated with dissemination to nearby (contiguous) lymph nodes. Adjacent lymph node regions are rarely uninvolved, as the lymphoma spreads throughout the lymphatic vessels.

12.5.1 What are HRS cells?

HRS cells are derived, in about 98% of cases, from germinal-centre B cells that have undergone somatic hypermutation of *IgV* genes. In approximately 2% of cases, HRS cells are of T-cell origin. Somatic hypermutation is dysregulated in HRS cells, leading to the incorporation of nonsense mutations in immunoglobulin genes in about 25% of cases. As a consequence of these nonsense mutations, B-cell receptors are not expressed on the cell surface. Failure of B-cell receptor expression should induce apoptosis, but in these cells the apoptotic mechanism fails, leading to the survival of dysregulated B cells.

The expression of most B-cell immunophenotypic markers is lost in HRS cells, making their origins difficult to establish in initial scientific investigations. Expression of CD45, CD19, CD20, and CD22 is completely absent from the majority of HRS cells. Furthermore, analysis using DNA microarrays has shown that gene expression is decreased in HRS cells, resulting in difficulty in assigning a normal cell counterpart to HRS cells. In the vast majority of cases, HRS cells continue to express MHC class II, CD40, and other molecules, which facilitate interaction with T_H cells—the most numerous of the infiltrative cells. The expression of CD30 is important in facilitating the interaction between eosinophils and mast cells commonly seen in the HL infiltrate.

A number of cytogenetic and molecular events have been reported in HRS cells, leading to an overexpression of transcription factors and overactivity of signalling pathways. Probably the most important event is increased NFκB signalling, leading to apoptotic failure. In addition, increased signalling through the JAK–STAT pathway increases the number of transcription factors within the nucleus of these cells. Translocations involving the germinal centre regulator BCL-6 are often found in B-cell lymphomas, but are not associated with HRS cells. Studies have identified BCL-6 translocations in some cases of HL expressing LP cells. LP cells and their association with Hodgkin lymphoma are outlined in Figure 12.25. Overexpression of MDM2, the negative regulator of the tumour suppressor protein p53, is found in approximately 60% of cases of HL. Inhibiting p53 through altered MDM2 expression has the potential to increase the number of additional mutations occurring within HRS cells, leading to increased genomic instability within these cells and diverse progression.

In the Western world, EBV can be identified in approximately 40% of classic HL cases, of which the mixed cellularity type has the greatest association. NLPHL is very rarely associated with EBV. The following important EBV-derived proteins are expressed in classic cases: latent membrane protein (LMP)1, LMP2A, EBV nuclear antigen-1 (EBNA-1), and EBV encoded RNAs (EBERs)-1 and -2.

LMP1 suppresses apoptosis by increasing NFκB signalling through the direct activation of the NFκB pathway and the inhibition of the NFκB inhibitor IκB via phosphorylation events. LMP2A acts as a surrogate B-cell receptor, ensuring that B cells can survive in the absence of a functional BCR. EBNA-1 is a maintenance protein allowing the replication of viral DNA, whilst EBERs are non-essential proteins which induce the synthesis of IL-10, thought to inhibit cytotoxic T cells.

An overlap syndrome called *B-cell lymphoma, unclassifiable, with features intermediate between diffuse large B-cell lymphoma and classical Hodgkin lymphoma* is recognized by the WHO classification, but will not be considered further in this text.

SELF-CHECK 12.13

Describe the cell of origin for HRS.

- The WHO classification takes into account a wide range of investigative techniques in order to classify distinct clinicopathological entities.
- Clear cytogenetic abnormalities have been identified and used, where appropriate, to outline distinctive pathologies.
- As scientists and clinicians, we have a vast knowledge of the molecular basis of a number of these diseases which allows us to predict disease behaviour.
- Viruses—including HIV, EBV, and HHV8—are important in the pathogenesis of a number of lymphoid malignancies.

Discussion questions

12.1 How can knowledge of the germinal centre improve our understanding of malignant B-cell behaviour?

12.2 Critically discuss the role of IgH translocations in lymphoid malignancies.

Further reading

- Ansell SM. ***Hodgkin lymphoma: 2016 update on diagnosis, risk-stratification, and management.*** *Journal of Hematology* 2016: **91**: 435–42.
- Chantry A, Kazmi M, Barrington S, Goh V, Mulholland N, Streetly M et al. ***Guidelines for the use of imaging in the management of patients with myeloma***. *British Journal of Haematology* 2017: **178**: 380–93.
- Delsol G, Lamant L, Mariamé B, Pulford K, Dastugue N, Brousset P et al. ***A new subtype of large B-cell lymphoma expressing the ALK kinase and lacking the 2; 5 translocation***. *Blood* 1997: **89(5)**: 1483–90. doi:10.1182/blood.V89.5.1483.
- Fukumoto K, Nguyen TB, Chiba S, and Sakata-Yanagimoto M. ***Review of the biologic and clinical significance of genetic mutations in angioimmunoblastic T-cell lymphoma***. *Cancer Science* 2018: **109**: 490–96.
- Hallek M. ***Chronic lymphocytic leukaemia: 2017 update on diagnosis, risk stratification, and treatment***. *American Journal of Hematology* 2017: **92**: 946–65.
- Inaba H, Greaves M, and Mullighan CG. ***Acute lymphoblastic leukaemia.*** *Lancet* 2013: **381**: 9881. doi:10.10.1016/S0140-6736(12)62187-4.
- Jaffe ES, Harris NC, Stein H, and Isaacson PG. ***Classification of lymphoid neoplasms: The microscope as a tool for disease discovery***. *Blood* 2008: **112**: 4384–99.
- Kalisz K, Alessandrino F, Beck R, Smith D, Kikano E, and Ramaiya N. ***An update on Burkitt lymphoma: A review of pathogenesis and multimodality imaging and assessment of disease presentation, treatment response, and recurrence.*** *Insights into Imaging* 2019. https://doi.org/10.1186/s13244-019-0733-7.

- Klener P. ***Advances in molecular biology and targeted therapy of mantle cell lymphoma***. *International Journal of Molecular Science* 2019: **20**: 4417. doi:10.3390/ijms20184417.
- Matutes E, Owusu-Ankomah K, Morilla R, Garcia Marco J, Houlihan A, Que TH, and Catovsky D. ***The immunological profile of B-cell disorders and proposal of a scoring system for the diagnosis of CLL***. *Leukemia* 1994: **8(10)**: 1640–5.
- Montes-Majarro IA, Steinhilber J, Bonzheim I, Quintanilla-Martinez L, and Fend F. ***The pathological spectrum of systemic anaplastic large cell lymphoma***. *Cancers* 2018: **10**: 107. doi:10.3390/cancers10040107.
- Nieuwenhuijzen N van, Spaan I, Raymakers R, and Peperzak V. ***From MGUS to Multiple Myeloma, a paradigm for clonal evolution of premalignant cells***. *Cancer Research* 2018: **78(10)**: 2449–56.
- Orfao A, Criado I, Rodriguez-Cabarello A, Gutierrez ML, Pedreira CE, Alcoceba M et al. ***Low count monoclonal B-cell lymphocytosis persists after 7 years of follow-up and confers a higher risk of death***. *Blood* 2017: **130(1)**: 4299.
- Rezk SA and Weiss LM. ***Epstein–Barr virus-associated lymphoproliferative disorders***. *Human Pathology* 2007: **38**: 1293–304.
- Safavi S and Paulsson K. ***Near-haploid and low-hypodiploid acute lymphoblastic leukaemia: two distinct subtypes with consistently poor prognosis***. *Blood* 2017: **129(4)**: 420–3.
- Shaffer AL, Young RM, and Staudt LM. ***Pathogenesis of human B cell lymphomas***. *Annual Review of Immunology* 2012: **30**: 565.
- Swerdlow SH, Campo E, Pileri SA, Harris NL, Stein H, Siebert R, Advani R, Ghielmini M, Salles GA, Zelenetz AD et al. ***The 2016 revision of the World Health Organization classification of lymphoid neoplasms***. *Blood* 2016: **127**: 2375–90.
- Swerdlow SH, Campo E, Harris NL, Jaffe ES, Pileri SA, Stein H, Thiele J, and Vardiman JW (eds). ***WHO Classification of Tumours of Haematopoietic and Lymphoid Tissues, volume 2***. IARC, Lyon, 2017.
- Vose JM. ***Mantle cell lymphoma: 2017 update on diagnosis, risk-stratification, and clinical management***. *American Journal of Hematology* 2017: **92**: 806–13.
- Wenzinger C, Williams E, and Gru AA. ***Updates in the pathology of precursor lymphoid neoplasms in the revised fourth edition of the WHO classification of tumours of hematopoietic and lymphoid tissues***. *Current Hematologic Malignancy Reports* 2018: **13**: 275–88.
- Zucca E and Bertoni F. ***The spectrum of MALT lymphoma at different sites: Biological and therapeutic relevance***. *Blood* 2016: **127(17)**: 2082–92.

Answers to self-check questions and case study questions are provided as part of this book's online resources.

Visit www.oup.com/he/moore-fbms3e.

PART 4

Haemostasis in Health and Disease

13

Normal haemostasis

Gary Moore

In this chapter, we will outline the processes that cause blood to clot in a controlled manner and present an overview of the interplay between the various components.

Learning objectives

After studying this chapter, you should confidently be able to:

- explain the importance of effective haemostasis;
- list the major components of haemostatic mechanisms;
- describe the mechanisms of primary haemostasis;
- describe the mechanisms of secondary haemostasis;
- describe the mechanisms of fibrinolysis;
- outline the interplay between the elements of haemostasis.

13.1 Introduction to haemostasis

The mechanisms of haemostasis have a number of crucial functions:

- under normal physiological conditions, to maintain blood contained within the vasculature in a fluid state;
- upon vessel trauma/injury, to limit and arrest bleeding by the formation of a blood clot, whilst at the same time maintaining blood flow through the damaged vessel;
- removal of the blood clot upon completion of wound healing.

13.1.1 Mechanisms of haemostasis

In health, the haemostatic mechanisms maintain the fluidity of circulating blood and are essentially anticoagulant in nature—that is, they retard or prevent clotting. Upon damage to blood vessels, a variety of **procoagulant** mechanisms are initiated, which involve a complex interplay

procoagulant
Physiological or pharmacological mechanism that promotes clotting processes.

between vessel endothelium, circulating platelets, and coagulation factors. A fine balance exists between maintaining blood fluidity and preventing excessive activation of procoagulant pathways that may lead to thrombosis and vascular occlusion. This balance is controlled by positive and negative feedback pathways, localization mechanisms, and inhibitory pathways. In this chapter, we will describe the haemostatic mechanisms as three distinct phases:

- **primary haemostasis**, which involves interactions between blood vessels, platelets, and von Willebrand factor (VWF), leading to formation of the initial barrier to blood loss, the primary haemostatic plug;
- **secondary haemostasis**, whereby the pathways of coagulation biochemistry act in concert to generate fibrin strands which strengthen the clot;
- **fibrinolysis** is a separate biochemical system that degrades the fibrin clot to prevent vascular occlusion and ultimately remove the clot once the wound has healed.

Cross reference

Primary haemostasis, secondary haemostasis, and fibrinolysis are all covered in more detail in dedicated sections of this chapter.

Although we will describe haemostasis under these discrete headings, it is important to recognize that the different phases are in fact integrated, many of the processes occurring simultaneously.

SELF-CHECK 13.1

What are the main roles of haemostasis and what are the mechanisms involved?

13.2 Primary haemostasis

The initial responses after vessel injury are mediated by interactions between the vessel wall and circulating platelets to begin the formation of a clot. In order to understand the events of primary haemostasis, we will first consider some relevant details of the main players: blood vessels, VWF, and platelets.

13.2.1 General blood vessel structure

The walls of the larger blood vessels (i.e. arteries and veins) consist of three layers: the intima, media, and adventitia. The intima is the inner layer, which is lined by a single, continuous layer of endothelial cells that rests on a basement membrane of subendothelial microfibrils. These microfibrils predominantly consist of different types of **collagen** and also some elastin. The middle layer, or media, contains mainly circularly arranged smooth muscle cells, which allow contraction and relaxation of the vessel. The media is separated from the outer layer, the adventitia, by the external elastic lamina, which allows the vessel to stretch and recoil. The adventitia comprises collagen fibres and fibroblasts that protect the vessel and anchor it to its surroundings. It also contains smaller blood vessels and nerves. The structure of large blood vessels is shown in Figure 13.1. Small blood vessels (i.e. capillaries, post-capillary venules, and arterioles) are also lined by a single layer of endothelial cells surrounded by a continuous basement membrane. Pericytes (or Rouget cells) provide support and stability by surrounding the endothelial layer, which is encircled by the adventitia.

collagen

A long protein fibre that connects and strengthens numerous tissues.

13.2.2 Haemostatic roles of blood vessels

The vessel wall has many important haemostatic functions. Blood flow contributes to effective haemostasis, the flow conditions in a given vessel being a direct function of its diameter. Higher flow rates are encountered in the centre of the vessel lumen and can be significantly lower at the endothelium surface. As a result, the relatively larger erythrocytes are more concentrated in the centre of a vessel and the smaller platelets circulate in areas of lower flow near to the

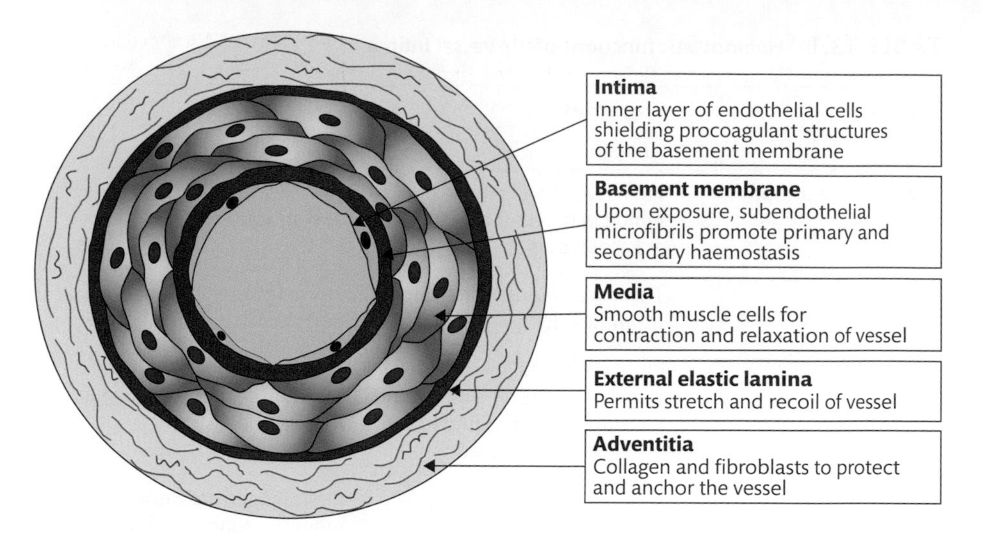

FIGURE 13.1
Large blood vessel structure.

vessel wall. **Shear forces** induced by blood flow affect the reactivity of cells and the crucial haemostatic protein VWF.

shear forces
Forces produced when surfaces are pressed together or move over each other.

Under normal physiological conditions, where there is no vessel trauma or disease, the endothelial layer of the intima is **antithrombotic**. This layer enlists a variety of mechanisms that actively maintain blood fluidity by the synthesis and secretion of agents that inhibit clotting processes. Upon vessel injury, or as a result of certain disease states, endothelial cells switch to functions that actively promote blood clotting: they become **prothrombotic**. The main haemostatic contributions of endothelial cells and the basement membrane are summarized in Table 13.1.

Once exposed to circulating blood, the subendothelium of the intima is strongly **thrombogenic** as it contains structures that promote both primary and secondary haemostasis. Vascular injury causes endothelin-1, which acts as a potent vasoconstrictor, to be released from endothelial cells. **Vasoconstriction** is brought about by the smooth muscle cells of the media and is an important mechanism for reducing or preventing blood loss, especially in the microvasculature. Cells of the adventitia express tissue factor, which is crucial to the initial reactions of the coagulation biochemistry of secondary haemostasis.

thrombogenic
Causing blood to clot; causing thrombosis.

vasoconstriction
Narrowing of the diameter of a blood vessel.

SELF-CHECK 13.2

How do blood vessels contribute to haemostasis?

Cross reference
The significance of the prothrombotic and antithrombotic functions of blood vessels will be discussed later in this section, and also in Chapter 14 when we consider disorders of primary haemostasis.

13.2.3 von Willebrand factor (VWF)

VWF is a large, adhesive glycoprotein produced constitutively in endothelial cells and megakaryocytes. Unlike many of the activated coagulation factors we will encounter in the section on secondary haemostasis, it is not an enzyme. The primary functions of VWF involve binding to cells and molecules and are outlined below:

- binding to subendothelial collagen exposed due to vessel injury;
- binding to a **receptor** on the surface of non-activated platelets, the glycoprotein Ib (GPIb) complex;

receptor
A molecule on the surface of (or in) a cell that acts as a recognition site for other molecules that bind in order to promote specific functions.

TABLE 13.1 Haemostatic functions of the vessel intima.

Function	Properties	
	Anticoagulant	**Procoagulant**
Synthesis/secretion	Prostacyclin and nitric oxide inhibit platelet function and act as vasodilators	von Willebrand factor is integral to platelet function
	Tissue plasminogen activator promotes clot lysis	Tissue factor initiates coagulation reactions
	Thrombomodulin, endothelial protein C receptor, and heparan sulphate contribute to the inhibition of coagulation	Factor V and factor VIII are essential cofactors in coagulation biochemistry
		Plasminogen activator inhibitor-1 inhibits clot lysis
		Thrombospondin and endothelin-1 promote platelet aggregation
Protein and cell binding	Binding of coagulation inhibitors	Adhesion molecules for neutrophils
		Exposure of subendothelial microfibrils to promote platelet binding via VWF
		Binding of some coagulation factors
Coagulation biochemistry	Endothelial protein C receptor binds protein C to allow its activation by the thrombin-thrombomodulin complex, leading to inactivation of activated factors V and VIII	Thrombin-thrombomodulin complex activates a fibrinolysis inhibitor
	Heparan sulphate enhances the inhibition of activated clotting factors by antithrombin	Tissue factor binds factor VII to begin coagulation

monomer

A molecule that can combine with others to form a polymer; the smallest repeating unit of a polymer.

domain

A discrete section that is part of the structure of a protein.

dimer

A molecule composed of two linked subunits.

- binding to a receptor on the surface of activated platelets, the GPIIbIIIa complex;
- binding to factor VIII (FVIII) to protect it from proteolytic degradation in the plasma.

Structure and assembly of VWF

The basic VWF **monomer** consists of 2050 amino acids containing distinct **domains** that have specific functions. The A1 domain binds GPIb and heparin and the A3 domain specifically binds collagen. Binding to GPIIbIIIa is via the C1 domain and the domain for FVIII binding is D′/D3, which can also bind with heparin. This pre-pro-VWF molecule is formed in the endoplasmic reticulum, where the monomers are then linked via C-terminal disulphide bonds to form **dimers**, which are termed the pro-VWF molecules.

Post-translational glycosylation then occurs in the Golgi, and it is interesting to note that this results in VWF being one of few proteins that carry ABO antigens. This has clinical relevance, as populations with blood group O have lower mean levels of VWF than those with blood groups A, B, and AB, which is a result of altered VWF survival *in vivo*.

Processing of the dimers continues in the Golgi, where the formation of N-terminal disulphide bonds facilitates **multimerization** of the dimers—that is, formation of an aggregate of multiple molecules joined by non-covalent bonds. **Multimers** vary in size, and the larger forms are functionally more effective. Processing of the pro-VWF into mature VWF is carried out by the enzyme furin, which also cleaves the propolypeptide known as von Willebrand antigen II (VW AgII) or von Willebrand propeptide (VWFpp) from the pro-VWF dimers. VWFpp is required for multimer formation and also chaperones VWF into storage granules, which in endothelial cells are the **Weibel-Palade bodies**, and in platelets are the **α-granules.**

Cross references

Details of blood groups and their antigens are covered in Chapter 4, section 4.2, and in the *Transfusion and Transplantation Science* volume in this series.

You will meet α-granules later in this chapter when we look at platelet structure and function.

VWF is constitutively released into the plasma from endothelial cells and can also be released from the Weibel-Palade bodies in response to stimuli such as exercise, adrenergic stimulation, and certain drugs such as **desmopressin**. Constitutive secretion only occurs from endothelial cells, as platelets do not release α-granule contents until activated.

multimer
A protein composed of more than one peptide chain.

desmopressin
Synthetic antidiuretic drug that stimulates the release of FVIII and VWF.

Vascular endothelial cells and platelets contain ultra-large VWF multimers that are highly adhesive. However, they are only transiently detectable in plasma as they are cleaved by a circulating protease called **ADAMTS-13**, the acronym for *A* zinc- and calcium-dependent *D*isintegrin and *M*etalloprotease with *T*hrombo*S*pondin type 1 motifs, member 13 (also known as VWF-cleaving protease). ADAMTS-13 degrades the ultra-large multimers into smaller forms ranging in size from 500 kDa to ~20,000 kDa.

SELF-CHECK 13.3

How does von Willebrand factor contribute to primary haemostasis?

13.2.4 Platelets

As presented in Chapter 3 on haemopoeisis, platelets are anucleate fragments of megakaryocyte cytoplasm, each megakaryocyte producing between 2000 and 3000 platelets. Platelets circulate in a dormant state, and, as we will see, are capable of a rapid and dramatic response to vessel injury. This is crucial to effective haemostasis as they form the platform for a number of essential haemostatic mechanisms.

The main functions of platelets are to:

- interact with VWF to form the initial barrier to blood loss;
- allow platelet-platelet interactions to propagate the thrombus;
- provide a negatively charged lipid surface to support key reactions of coagulation biochemistry;
- deliver a variety of haemostatically active molecules to increase their local concentration;
- localize thrombus formation;
- promote vasoconstriction;
- promote vessel repair;
- maintain the molecular integrity of endothelial cell junctions.

Relative to leucocytes and erythrocytes, platelets are small and discoid in shape, are approximately 3.0 μm × 0.5 μm, and have a volume of 7.5–11.5 fL. The shape and small size of platelets, together with conditions of blood flow, result in them circulating towards the edges of blood vessels, where they are ideally placed to rapidly respond to vessel damage. To achieve the functions listed above, platelets have a highly specialized structure which we will now consider in more detail. Figure 13.2 shows you the general structure of a platelet and organization of the features covered in the following sections.

Key Points

Localization mechanisms are important to ensure that a clot forms only where it is needed. We will see the pathological consequences of breakdown of localization processes in Chapter 14.

Platelet membrane

glycoproteins
Proteins with carbohydrate chains.

The platelet membrane is a complex structure containing a variety of components crucial to effective haemostasis. The external coat of the membrane, or glycocalyx, is rich in anchored **glycoproteins**. Some of these glycoproteins act as receptors for molecular stimuli which trigger platelet activation, whilst others facilitate platelet adhesion to subendothelium via VWF, or platelet aggregation. The middle layer is rich in asymmetrically distributed phospholipids that act as a surface for the interaction with specific components of coagulation biochemistry, such as factor II and factor X. The inner layer takes part in translating signals received on the outer coat into chemical messages and physical alterations in platelet structure that accompany activation. Figure 13.3 depicts a schematic of the platelet membrane showing the main receptors you are about to meet in the following sections. Note that some are directly connected to components of the cytoplasm.

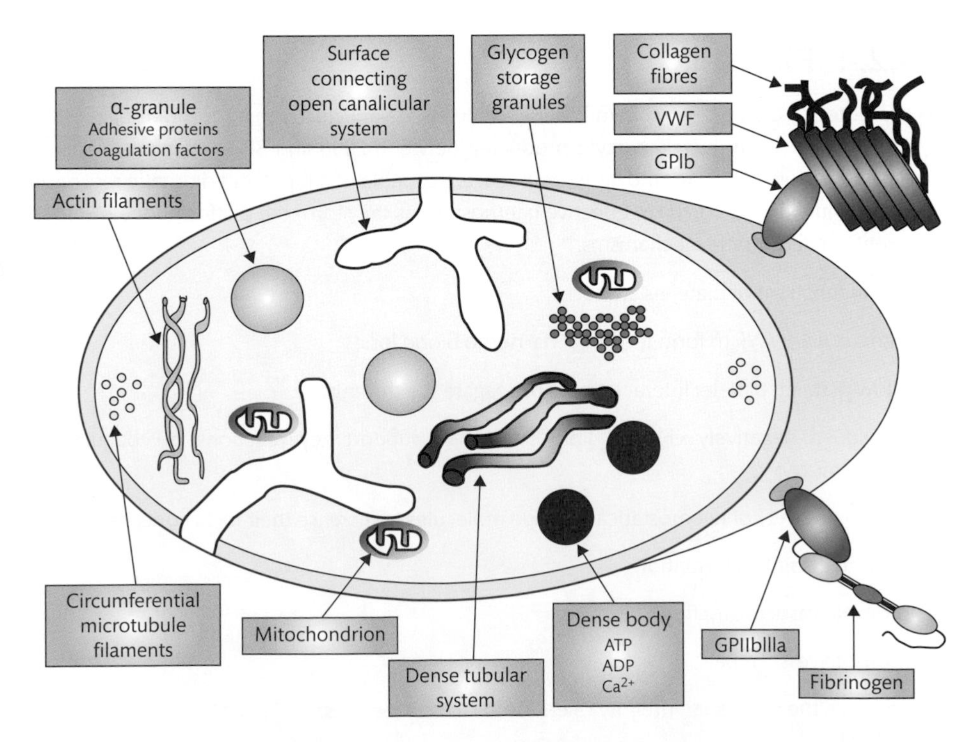

FIGURE 13.2
Platelet ultrastructure. The two main glycoprotein (GP) receptors are shown binding to their main ligands, GPIb binding to VWF and GPIIbIIIa binding to fibrinogen. Dense bodies containing adenine nucleotides and calcium, and α-granules containing adhesive proteins and coagulation factors are present in the cytoplasm. The surface-connected open canalicular system forms channels to allow molecules in and out of the platelet, and the dense tubular system regulates platelet activation. Actin filaments and the microtubules maintain platelet shape, and change shape upon activation. Mitochondria and glycogen-storage granules provide energy.

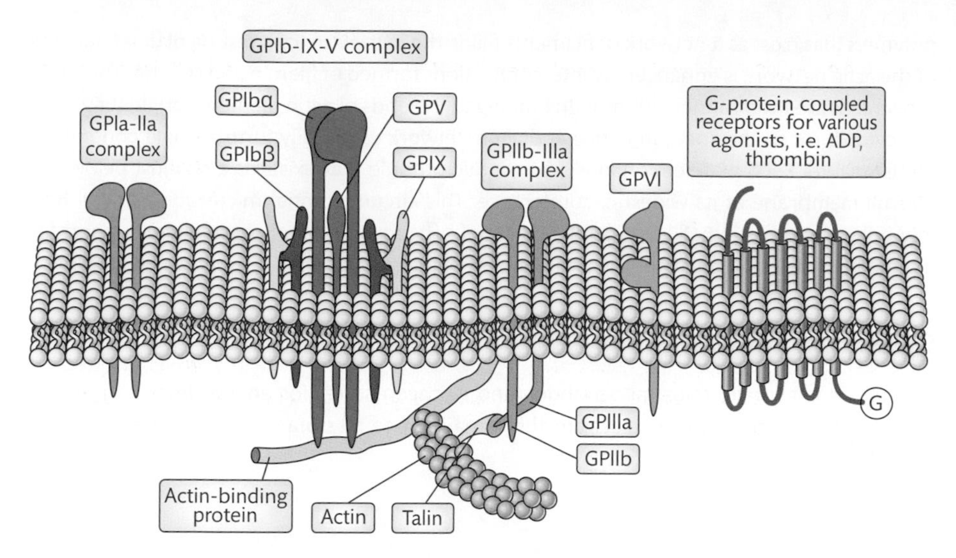

FIGURE 13.3

Platelet membrane receptors. Platelets have receptors for each of the agonists capable of activating resting platelets, such as ADP, thrombin, and thromboxane A_2 (TXA_2). The GPIb-IX-V complex that is responsible for binding platelets to VWF consists of two molecules each of GPIbα, GPIbβ, and GPIX flanking a single molecule of GPV. The GPIIbIIIa complex, responsible for binding to fibrinogen to facilitate platelet-to-platelet aggregation, consists of one molecule of each GP (GP, glycoprotein) with GPIIIa anchored to actin via talin. The GPIaIIa complex and GPVI are involved in binding platelets to collagen. Note that some are directly connected to components of the cytoplasm, such as GPIbα molecules anchored to an actin-binding protein and GPIIIa anchored to actin via talin.

actin-binding protein
Protein that binds actin.

talin
Ubiquitous cytoskeletal protein linking integrins to the actin cytoskeleton.

The **surface-connected open canalicular system** (SCOCS) is a network of channels throughout the interior of the platelet formed from invaginations of the membrane. These channels, or canaliculi, are continuous with the external membrane and significantly increase the surface area of platelet exposed to plasma. This provides a route for molecules to reach deep within the platelet. The canaliculi also serve as conduits for the extrusion of substances from within platelets once activated.

A separate internal membrane system is the **dense tubular system** (DTS), which is derived from the endoplasmic reticulum of the parent megakaryocyte. The channels of this system also pervade the cytoplasm. The DTS regulates platelet activation by sequestering or releasing calcium, and is also the site of prostaglandin synthesis.

Cytoskeleton

Three cytoskeletal systems maintain the discoid shape of resting platelets as they encounter the shear forces generated in flowing blood:

- the membrane skeleton;
- the cytoplasmic actin network;
- the microtubule coil.

The membrane skeleton is composed of elongated spectrin strands that are connected to the membrane. Actin is the most abundant platelet protein, forming 2000-5000 linear actin polymers that exist as a network of filaments filling the cytoplasm of a resting platelet. Rigidity of the actin network is enhanced by interconnections formed of filamin and α-actin. The spectrin strands are interconnected through binding to the ends of actin filaments originating from the cytoplasm, so the spectrin lattice and actin network essentially form a single continuous ultrastructure. Each platelet contains a single microtubule that exists as a coil just below the plasma membrane at its widest circumference. This circumferential microtubule coil, a hollow polymer consisting of 12–15 subfilaments of αβ-tubulin dimers, is the major support for maintaining the discoid shape. The arrangement of the subfilaments is shown in Figure 13.4.

Cytoplasm contents

Platelet cytoplasm contains mitochondria and glycogen stores for energy. Important to the haemostatic functions of platelets are three main types of storage granules which rapidly release their contents upon activation:

- α-granules;
- dense bodies;
- lysosomes.

The α-granules are the most numerous platelet organelle, about 80 per platelet, and the main secretory granule, containing a vast array of predominantly large molecules necessary for haemostasis and wound healing. Upon platelet activation, the molecules are carried to the cell surface for release, some of which adhere to the platelet surface, whilst others are incorporated in the platelet membrane or diffuse into the extracellular fluid.

The **dense bodies**, also termed dense granules or δ-granules, are the smallest platelet granule with a frequency of five to seven per platelet. They are so-called because when viewed with an electron microscope they are electron-dense, which means that they are impermeable to the electron beam in the microscope. They are the storage and secretory organelles for small non-protein molecules necessary for effective haemostasis. The main contents of these organelles, and their functions, are detailed in Table 13.2. Note that some constituents have multiple functions.

Lysosomes are organelles that contain digestive enzymes and are found in most cells. In platelets, they are also termed λ-granules and are few in number. They contain acid hydrolase enzymes, which are assumed to facilitate the digestion of blood clots when they are no longer needed.

13.2.5 The events of primary haemostasis

Primary haemostasis comprises a series of events involving the activation of platelets and their interplay with VWF and subendothelium. These events lead to the formation of a plug of platelets and set the stage for secondary haemostasis.

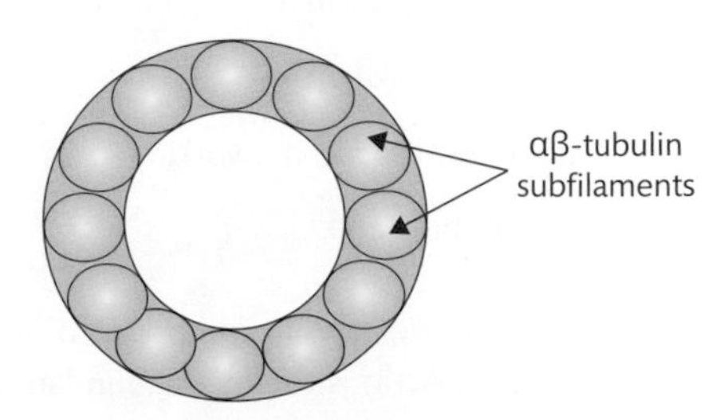

FIGURE 13.4
Substructure of platelet microtubule. The coil consists of 12–15 subfilaments of αβ-tubulin dimers circularly arranged to form a hollow tube.

TABLE 13.2 **Platelet α-granule and dense body constituents and their functions.**

Function	Constituents
α-granules	
Adhesive proteins	VWF (platelet adhesion), fibrinogen (platelet aggregation), fibronectin, vitronectin, fibrinogen (wound healing)
Membrane proteins	GPIIbIIIa, GPIb, P-selectin
Coagulation biochemistry	FV, FVIII, fibrinogen
Fibrinolysis	Plasminogen
Inhibition	TFPI, protein S, α_2-macroglobulin, α_1-antitrypsin, protease nexin II (inhibit coagulation) α_2-antiplasmin, PAI-1 (inhibit fibrinolysis), thrombospondin-1 (inhibits angiogenesis)
Chemokines for leucocyte attraction	Platelet factor 4, β-thromboglobulin
Cellular mitogens	PDGF (smooth muscle proliferation) VEGF-A, VEGF-C (angiogenesis)
Dense bodies	
Vasoconstriction	Serotonin
Platelet activation	ADP, serotonin (weak activator)
Coagulation biochemistry	Calcium ions
Energy for platelet biochemistry	ATP
Membrane proteins	GPIIbIIIa, GPIb, P-selectin
Integrin regulation	Calcium ions, magnesium ions

This table outlines the main constituents of α-granules and dense bodies, but is not exhaustive. VWF = von Willebrand factor; GP = glycoprotein; TFPI = tissue factor pathway inhibitor; PAI-1 = plasminogen activator inhibitor-1; PDGF = platelet-derived growth factor; VEGF = vascular endothelial growth factor; ADP = adenosine diphosphate; ATP = adenosine triphosphate.

Adhesion

The primary haemostatic response begins with vessel injury, which results in disruption of the vessel wall and the consequent exposure of procoagulant stimuli. The initial event following vessel damage is that of **platelet adhesion** to the collagen components of the subendothelium. Platelet adhesive events are mediated by two groups of glycoprotein (GP) surface membrane receptors, the integrin family and the leucine-rich motif (LRM) family. The integrin family consists of GPs comprising non-covalently associated αβ-heterodimers that mediate attachment to the subendothelium and/or platelet-to-platelet cohesion. LRM proteins are defined by the presence of a 24-residue structural motif. Four members of this family–GPIbα, GPIbβ, GPIX, and GPV–form the GPIb-IX-V complex, which serves as the main VWF receptor. This complex is depicted in Figure 13.3.

At the low-shear stress conditions encountered in the venous circulation, platelets adhere to collagen directly via interactions with specific receptors, predominantly GPIaIIa (integrin $\alpha_2\beta_1$) and GPVI.

Under the intermediate-to-high-shear conditions that exist in arteries, arterioles, capillaries, and stenosed (narrowed) vessels, platelet adhesion to collagen is indirect and dependent on the interaction between VWF and the GPIb–IX–V complex, the VWF providing a molecular bridge between the platelet surface and exposed collagen. This interaction can support adhesion in the venous circulation.

Circulating VWF is immobilized via the collagen within the exposed subendothelium and promotes *tethering* of platelets, which you can see in Figure 13.5. Resting platelets (i.e. non-activated) express GPIb–IX–V on their surface, but binding to VWF either does not occur or occurs with low affinity in the normal circulation. This is because the GPIb–IX–V-binding section of circulating VWF is masked. Immobilization of VWF on the collagen surface allows the shear forces of flowing blood to unravel VWF and expose the binding section, thus facilitating binding to GPIb–IX–V and consequent capture of platelets circulating at the edges of the vessel.

Look at Figure 13.6 and you will see that once tethered, platelets *roll* over immobilized VWF in the direction of blood flow as a result of the progressive formation of new VWF–GPIb–IX–V interactions as subsequent sections of the platelet membrane come into contact with the VWF, beginning at the site of tethering.

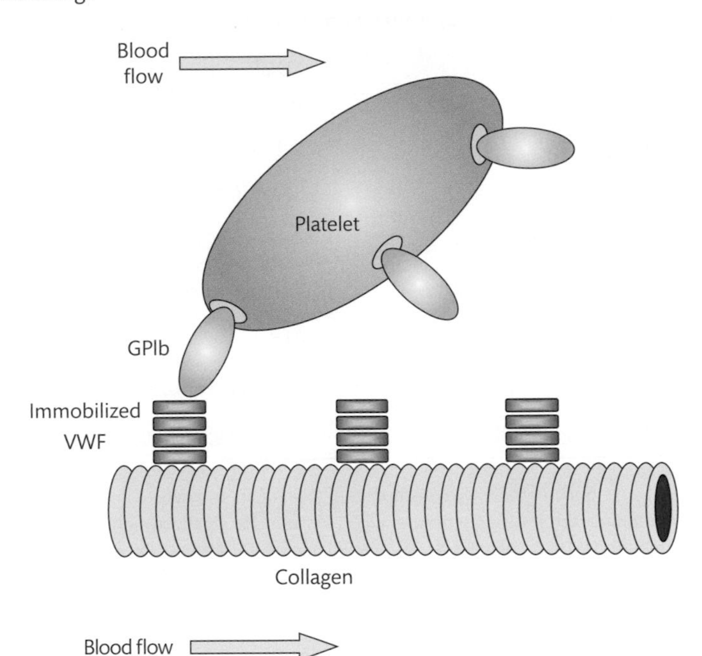

FIGURE 13.5

Platelet tethering via collagen, VWF, and GPIb. VWF anchors to exposed collagen and blood flow facilitates unravelling of VWF to expose the GPIb-binding site. Platelets circulating at the edge of the vessel are then captured and tethered by VWF via binding to the constitutively expressed GPIb.

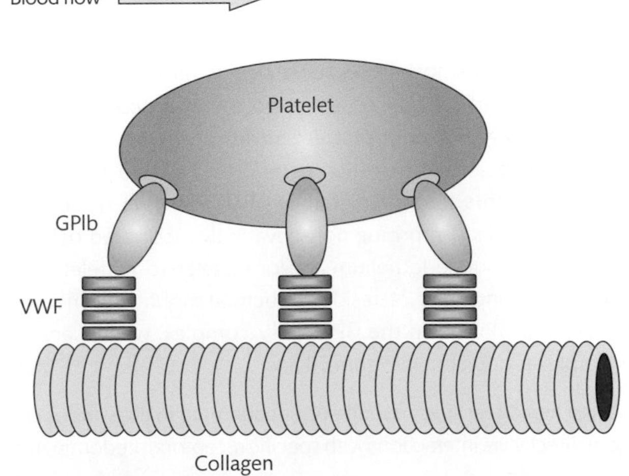

FIGURE 13.6

Platelet rolling. Once tethered via initial VWF–GPIb interactions, blood flow forces the platelet to roll over from the tethering point and promote further VWF–GPIb interactions.

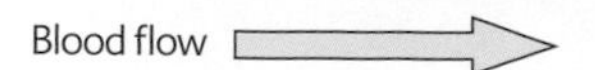

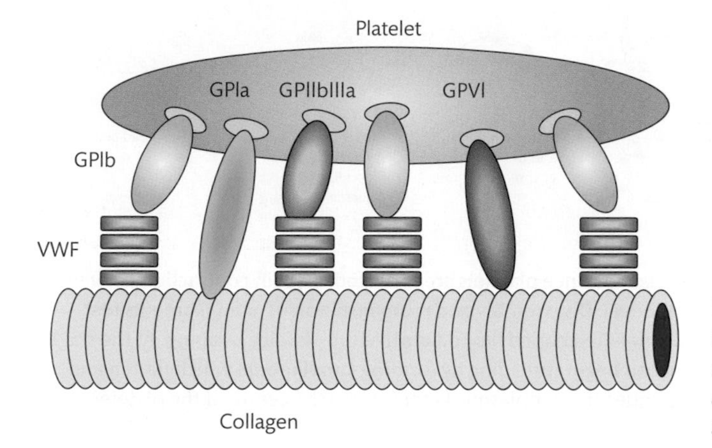

FIGURE 13.7
Stable adhesion of platelets. Stabilization of platelet adhesion to exposed subendothelium occurs via binding to GPs additional to GPIb. GPIa and GPVI bind directly to collagen and GPIIbIIIa binds via VWF.

Interaction with VWF is not stable due to a fast dissociation rate, so *stable adhesion* of platelets to the subendothelium also involves the collagen receptors GPIaIIa and GPVI, the **fibronectin** receptor GPIc′–IIa (integrin $\alpha_5\beta_1$), and the fibrinogen receptor GPIIbIIIa (integrin $\alpha_{IIb}\beta_3$), the latter being able to bind VWF at high shear rates, and also fibronectin and **vitronectin**. The main interactions are shown in Figure 13.7.

fibronectin
A multifunctional glycoprotein involved in cell adhesion, differentiation, growth, and wound healing.

vitronectin
A multifunctional protein that stabilizes plasminogen activator inhibitor 1, contributes to binding of platelets to vascular cell walls, and promotes adherence, migration, proliferation, and differentiation of many different cell types.

ligands
Binding molecules.

Interaction of GPVI with collagen triggers signalling responses that lead to platelet activation, including elevation of cytosolic calcium ions, cytoskeletal actin-filament rearrangements, activation of enzymes (i.e. tyrosine kinases), and activation of GPIIbIIIa. Circulating platelets express GPIIbIIIa on the surface, but there is no binding to plasma fibrinogen or VWF as the integrin maintains very low affinity for these **ligands** in resting platelets. Signalling responses resulting from collagen adhesion that activate GPIIbIIIa via tyrosine kinases are termed *inside-out signalling*. They modulate the adhesive activity of GPIIbIIIa by two distinct processes:

- activation of GPIIbIIIa via its cytoplasmic tails causes a conformational change in the extracellular domain that increases its affinity for ligands;
- clustering of GPIIbIIIa receptors increases their local concentration, thereby promoting binding to multivalent ligands such as fibrinogen.

Binding of fibrinogen and VWF to the extracellular domain of GPIIbIIIa causes further clustering and also conformational changes in the cytoplasmic domains that promote *outside-in signalling*. Signalling cascades are triggered in the cytoplasm that influence cytoskeletal reorganization, full platelet aggregation, granule secretion, and availability of procoagulant phospholipid activity. We will meet other GPIIbIIIa activators shortly, when we explore platelet aggregation.

SELF-CHECK 13.4

What are the main events of platelet adhesion?

Shape change and platelet spreading

Once activated, platelets undergo a change in shape, transforming from discoid to a spherical centre with cytoplasmic extensions that enable closer physical interactions with each other and allow attachment to the vessel wall.

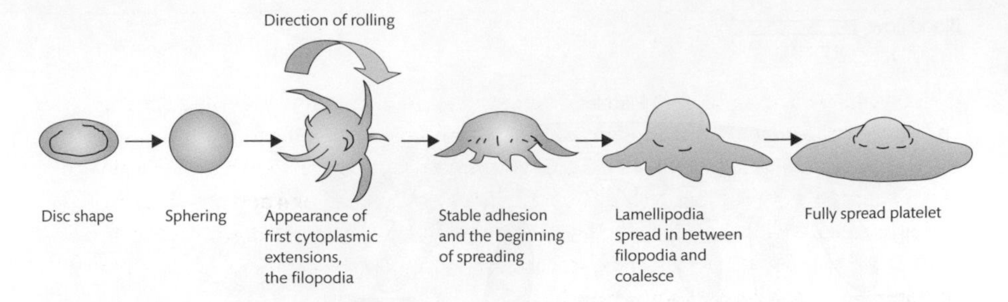

FIGURE 13.8

Platelet shape change and spreading. Once platelets are activated, myosin contraction converts their shape from discoid to spherical. Disassembly and reorganization of actin filaments first results in the formation of thin, needle-shaped filopodia while the platelet is rolling. At the stage of stable adhesion the platelet can begin spreading by forming lamellipodia which flatten and spread the platelet. Spreading squeezes cytoplasmic granules into the centre of the platelet.

Transition from disc to sphere is mediated by the contraction of platelet myosin. Upon activation, the microtubule depolymerizes and the actin filaments of the cytoskeleton undergo a rapid disassembly and reorganization into new structures. This is under the control of a number of actin regulatory proteins, primarily gelsolin, which is activated by the significant increase in intracellular calcium ions mediated by phospholipase C. The reassembly of actin filaments forms two types of cytoplasmic extensions. The first to appear are the needle-shaped *filopodia*, which grow from the periphery of the platelet, and the larger and flatter *lamellipodia*, which fill in the area between the filopodia. Extending these protuberances has the effect of flattening and spreading the platelet which squeezes granules and organelles into the centre of the platelet, giving the characteristic 'fried-egg' appearance when viewed microscopically. The stages of transformation from disc to 'fried egg' are shown diagrammatically in Figure 13.8.

The process of adhesion generates a platelet monolayer that is sufficient to initiate platelet plug formation, but the single layer of cells itself is insufficient to prevent bleeding. We will now turn our attention to events that work towards extension of the forming thrombus.

Release reaction

Centralization of the α-granules and dense bodies, in the presence of elevated levels of calcium ions, leads to fusion of the granules with the membranes that line the SCOCS. The SCOCS acts as a conduit system of internal membrane leading to external secretion of the granule contents into the surrounding medium or onto the surface of platelets. Several of the haemostatic proteins within α-granules are present in relatively higher concentrations in platelets than in plasma. VWF and fibrinogen are 3–4 times more concentrated, and factor V is nearly 30 times more concentrated. The release of granule contents serves to increase their concentration in the area of the forming thrombus, one of the important localization mechanisms of the haemostatic response.

In addition to secretion, fusion of α-granules to the SCOCS and platelet membrane provides a number of crucial surface receptors, in particular, GPIIbIIIa and P-selectin. Although activated platelets already express functional GPIIbIIIa, this fusion provides a significant increase in GPIIbIIIa that magnifies its functions in relation to platelet spreading and aggregation. P-selectin binds to a receptor on leucocytes that initiates the leucocyte adhesion cascade, an essential step in inflammation.

Key Points

Inflammation (the process by which white blood cells and an array of biochemicals protect from infection and foreign substances) and tissue repair are directly linked to haemostasis. Activated platelets release inflammatory mediators such as histamine to increase capillary permeability. Substances such as platelet factor 4 and adenine nucleotides modulate white cell function, whilst others (such as platelet-derived growth factor) contribute to repair mechanisms. You will meet other instances of cross-over with inflammation and repair later in this chapter.

The **serotonin** released from dense bodies functions mainly as a vasoconstrictor, but it is also a weak platelet activator. The prime function of adenosine triphosphate (ATP) is for energy, although it also has a role in platelet activation by triggering calcium influx. You will see in the next section that adenosine diphosphate (ADP) plays a major role in platelet activation.

serotonin
Also known as 5-hydroxytryptamine (5-HT), a hormone that acts as a neurotransmitter in the brain and contributes to haemostasis by initiating vasoconstriction.

Platelet agonists and aggregation

Platelets express surface receptors for a number of ligands that will activate them as they arrive at the scene of vessel damage. This allows activated platelets to be recruited to accumulate on the platelet monolayer. These ligands, often referred to as **agonists** in this context, become available as part of the haemostatic response and contribute to one or more of three main platelet signalling processes:

agonists
Molecules that activate platelet functions via surface receptors.

- increase in cytoplasmic calcium ions;
- reorganization of the actin cytoskeleton;
- suppression of cyclic adenosine monophosphate synthesis.

We have already covered the roles of the first two responses when we considered activation via GPVI, and these ligands provide alternative routes to platelet activation. Cyclic adenosine monophosphate (cAMP) is formed in platelets from ATP and inhibits platelet signalling, and so needs to be suppressed for platelets to become activated.

The main agonists are ADP, thrombin, and **thromboxane A_2** (TXA_2) . Thrombin plays a pivotal role in haemostasis and we will meet it in more detail in the section on coagulation biochemistry. TXA_2 is produced from membrane arachidonic acid as part of the cyclooxygenase pathway, which is initiated upon platelet activation.

thromboxane A_2
A platelet agonist produced from membrane arachidonic acid as part of the cyclooxygenase pathway within platelets.

ADP, thrombin, and TXA_2 activate platelets via binding to surface G-protein coupled receptors (GPCR), which are structurally characterized by an extracellular N terminus attached to seven transmembrane α-helices and an intracellular C terminus. The agonists bind to surface accessible domains, which causes conformational changes in the receptors, leading to C-terminus interaction with, and activation of, cytoplasmic G proteins. The G proteins subsequently activate metabolic pathways that promote platelet activation.

Platelet aggregation is the formation of platelet-to-platelet linkages through the binding of mainly fibrinogen to activated GPIIbIIIa on adjacent cells, allowing the thrombus to grow beyond the initial platelet monolayer.

You can see in Figure 13.9 that the VWF- and collagen-mediated events of platelet adhesion generate the platform for the recruitment of resting platelets, which once activated, provide the platform for secondary haemostasis.

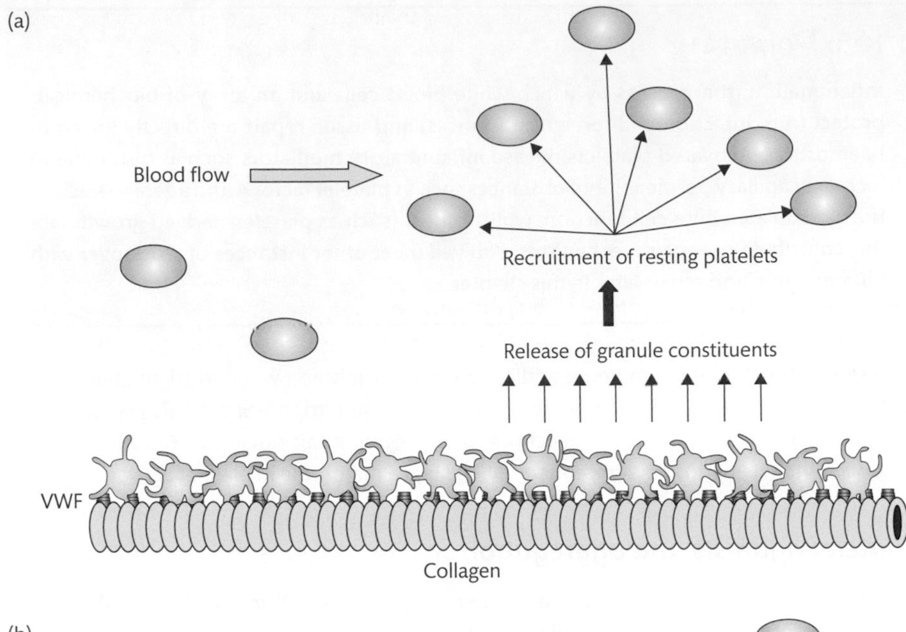

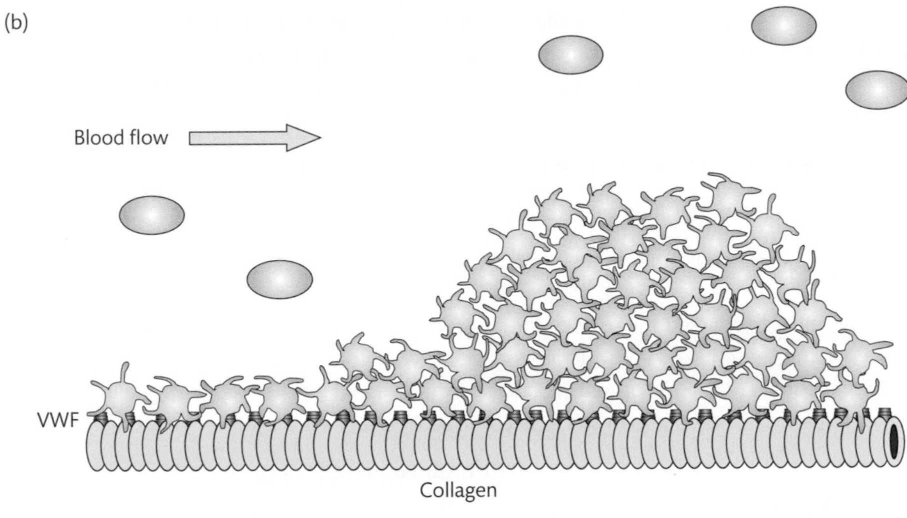

FIGURE 13.9
Activation and recruitment of platelets. (a) Platelets forming a monolayer on the surface of exposed collagen are activated and release agonists into the local circulation to activate platelets arriving at the scene; (b) once activated, these platelets change shape to promote greater physical interactions and express surface GPIIbIIIa to facilitate aggregation via the formation of fibrinogen bridges.

Cross reference

The cyclooxygenase pathway is shown in Figure 17.12 of Chapter 17.

SELF-CHECK 13.5

What are the main platelet events following adhesion?

Procoagulant activity

Critical to an effective haemostatic response is the presentation of negatively charged phospholipids on the surface of activated platelets which act as a platform for the assembly of two multiprotein complexes that are vital to coagulation biochemistry. Strong platelet activation or potent shear stress can induce the release of fragments of platelet membrane called *platelet microparticles*. These fragments express GPIb, GPIIbIIIa, and also phosphatidylserine, which may play a role in supporting coagulation by providing an additional surface for the formation of the multiprotein complexes.

Additionally, exposed vessel adventitia cells and epithelial cells of surrounding tissues constitutively express **tissue factor** (TF) on their surfaces, which can also be expressed on platelet microparticles. TF is crucial to the activation of coagulation biochemistry, and we will now turn our attention to that compartment of haemostasis.

13.3 Secondary haemostasis

In the microcirculation, a platelet plug alone is normally sufficient to stem blood loss, but it needs to be reinforced when forming in larger vessels if it is to be haemostatically effective. This is where the biochemical reactions of coagulation come into play. They converge on soluble fibrinogen to convert it into a meshwork of insoluble fibrin that intertwines with the cellular components of the forming thrombus to form a supporting scaffold.

The terms 'haemostasis' and 'coagulation' are often used interchangeably, yet the latter is but a part of the former. The processes of coagulation comprise a tightly regulated orchestration of coagulation factors, cofactors, and inhibitors that result in the controlled formation of the pivotal enzyme thrombin, which, in addition to activation and regulatory functions, initiates fibrin formation. The main players and their principal roles are outlined in Table 13.3.

13.3.1 Coagulation factors

The coagulation factors are a group of **zymogens** which cooperate in an integrated system of enzyme activation (and inactivation) steps. The substrate(s) of a given activated coagulation factor will be the zymogen of a different coagulation factor, which becomes activated after part of the molecule, the activation peptide, has been cleaved to expose the active site.

zymogen
Inactive precursor of an enzyme, also called a proenzyme.

Coagulation factors are numbered using roman numerals in the order they were discovered and not in the order they were first thought to work. As indicated in Table 13.3, convention has it that zymogens are abbreviated to the letter F plus the roman numeral, and the activated forms are then subscripted with lower case a, for example FX and FXa. The majority of the activation reactions are confined to membrane surfaces because the coagulation factors or their precursors have specific binding sites for phospholipids, calcium ions, or specific receptors or cofactors expressed on cell surfaces, which contribute to the concentration and localization of the haemostatic response.

All the coagulation factors that function as activation enzymes belong to a class of **peptidases** called serine proteases, which are characterized by the presence of a serine residue in the active site of the enzyme.

peptidases
Enzymes that cleave peptide bonds in proteins.

Many of these enzymes require the presence of cofactors to maximize their rate of substrate activation. Examples of cofactors are TF and the activated forms of the coagulation factors V and VIII. Strictly speaking, they are coenzymes, as they are organic (cofactors are inorganic), but it is common for them to be referred to as cofactors and we adopt that convention in these chapters.

Vitamin K-dependence

A group of the coagulation factors, FII, FVII, FIX, and FX, are termed vitamin K-dependent. During their biosynthesis in the liver, a series of post-translational enzymatic reactions include a step that requires vitamin K to function as a cofactor to a carboxylase enzyme. The enzyme + cofactor add carboxyl groups to the γ-carbon of a number of specific glutamic acid (Glu) side chains converting them to γ-carboxyglutamic acid (Gla). The section of the molecule containing them is termed the **Gla domain**.

TABLE 13.3 Key proteins of coagulation biochemistry.

Common name	Abbreviation	Vitamin K-dependence	Mean plasma level (μg/mL)	Half-life (h) in plasma	Main roles of active form
Factor II	FII	Yes	90	60–70	Converts fibrinogen to fibrin, activates PC, FXI, and TAFI
Factor V	FV	No	10	12–15	Cofactor for FXa
Factor VII	FVII	Yes	0.5	3–6	Activates FIX and FX (and FVII)
Factor VIII	FVIII	No	0.1	8–12	Cofactor for FIXa
Factor IX	FIX	Yes	5	18–24	Activates FX
Factor X	FX	Yes	8	30–40	Activates FII
Factor XI	FXI	No	5	45–52	Activates FIX
Fibrinogen	Fib/FGN	No	3000	72–120	Converts to insoluble fibrin clot
Factor XIII	FXIII	No	10	200	Stabilizes fibrin clot
Tissue factor	TF	No	NA	Cellular	Cofactor for FVII/FVIIa
Tissue factor pathway inhibitor	TFPI	No	NA	Cellular	Inhibits FVIIa activation of FX
Antithrombin	AT	No	140	72	Inhibits FIIa, FIXa, FXa, and FXIa
Protein C	PC	Yes	4	6	Inactivates FVa and FVIIIa
Protein S	PS	Yes	10	42	Cofactor for activated PC and TFPI
Protein Z	PZ	Yes	2.2	60	Cofactor for ZPI
Protein Z-dependent protease inhibitor	ZPI	No	4.0	60	Inhibits FXa and FXIa
Heparin cofactor II	HCII	No	90	60	Secondary FIIa inhibitor
α_1-antitrypsin	α_1AT	No	250	120	Inhibits FXIa
α_2-macroglobulin	α_2M	No	2000	36	Back-up to AT in certain conditions
Thrombomodulin	TM	No	NA	Cellular	Activation of PC
Endothelial protein C receptor	EPCR	No	NA	Cellular	Activation of PC

TF, TFPI, TM, and EPCR are membrane-bound and do not have plasma levels or half-lives. TAFI = thrombin activatable fibrinolysis inhibitor. Suffix a to a clotting factor denotes an active form.

The Gla residues bind calcium ions and are necessary for the activity of these coagulation factors. Figure 13.10 shows that binding of calcium alters the conformation of the Gla domain, which exposes hydrophobic phospholipid-binding sites, thus enabling them to interact with membrane surfaces by direct binding to phospholipids.

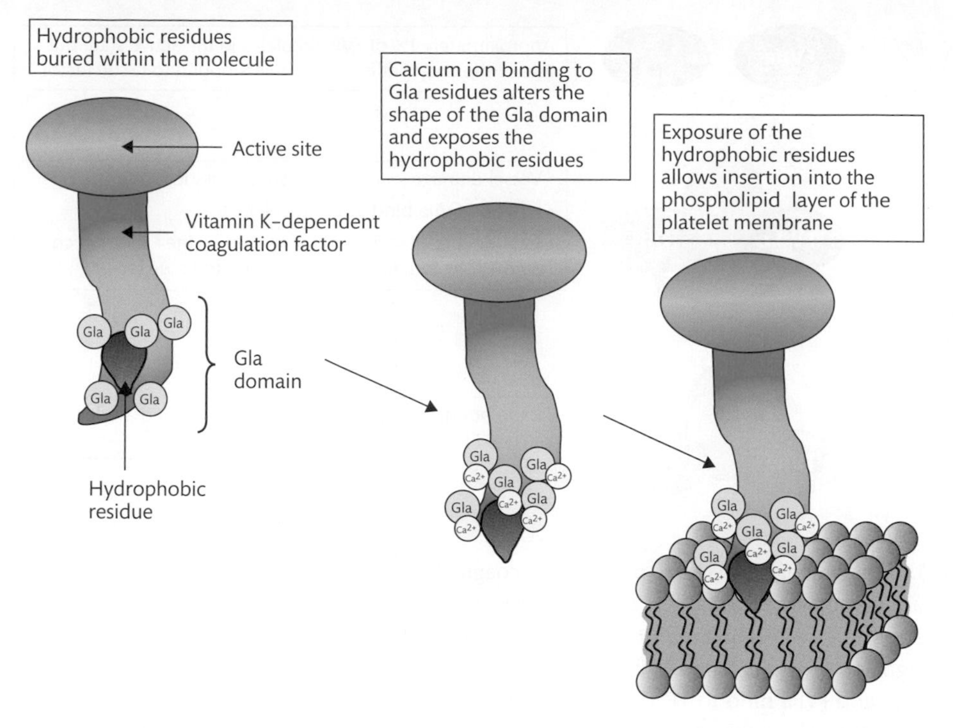

FIGURE 13.10
Gla-domain function in vitamin K-dependent coagulation factors. Gla domains mask hydrophobic residues that are exposed upon binding of calcium ions to the Gla domains. The hydrophobic residues facilitate insertion of the coagulation factor into the phospholipid of the platelet membrane, which ensures localization of the coagulation reactions. Gla = γ-carboxyglutamic acid.

Key Points

Naming the vitamin K-dependent factors is a common examination question, so worth committing to memory. Note in Table 13.3 that protein C, protein S, and protein Z are also vitamin K-dependent. You will meet them in more detail when we look at coagulation inhibitors.

13.3.2 The events of coagulation

The generation of fibrin is essential to stabilize clots in larger vessels. Coagulation begins with an initiation phase, which is then amplified to set the stage for a burst of thrombin generation during a propagation phase.

Initiation

The events that start coagulation biochemistry via TF and activated factor VII are referred to as the **initiation** phase. The sole initiator of coagulation is the membrane-bound protein TF. It is constitutively expressed at biological boundaries such as skin and organ surfaces, where it forms a protective envelope immediately available to initiate blood coagulation in the event of injury or blood loss. It is found in almost all tissues, although, interestingly, not in joints, where severe haemophiliacs have their worst bleeding episodes. TF is not expressed on vascular endothelial surfaces but is abundant in the vascular adventitia.

Initiation of coagulation begins with exposure of subendothelium after vessel damage, which brings circulating blood into contact with the TF of the adventitia and surrounding tissues. TF binds FVII and functions as a cofactor in the activation of FVII, the formation of the TF–FVII complex also serving as a localization mechanism.

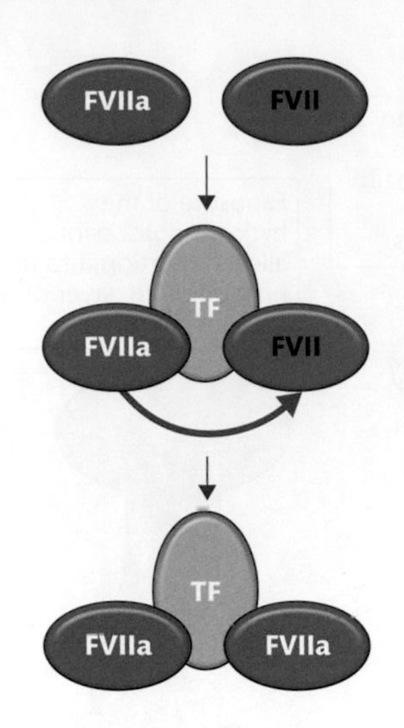

FIGURE 13.11
Stages of FVII activation. Activated and non-activated FVII bind to TF that is exposed upon vessel damage. Binding of FVIIa to TF, its cofactor, bestows full enzymatic activity whereby it can activate zymogen FVII to FVIIa.

Unique amongst the serine proteases of coagulation, not all of FVII circulates as a zymogen: approximately 1% circulates in the active state (FVIIa). Plasma FVIIa has an inefficient active site and thus little **proteolytic** activity unless bound to TF, so at normal blood levels has no significant activity towards its substrates and is unreactive with circulating inhibitors. Both FVII and FVIIa bind to TF, the FVII remaining enzymatically inactive until autoactivated by TF–FVIIa, which induces a conformational change resulting in activation to the functional serine protease. The stages of FVII activation are shown in Figure 13.11. This TF–FVIIa complex is anchored on the cell surface in close proximity to negatively charged phospholipids, facilitating optimal positioning for its substrates. TF expressed on the surface of macrophages, monocytes, fibroblasts, and platelet microparticles can also bind FVII/FVIIa. Figure 13.12 shows that the TF–FVIIa complex then binds FX forming the so-called **extrinsic tenase complex**, which activates the FX to FXa on the subendothelial surface. It also converts some factor IX to factor IXa. The complex is more efficient at catalysing the conversion of FX, so FXa is the initial product.

proteolytic
Enzymes that digest or lyse proteins into smaller sections or amino acids.

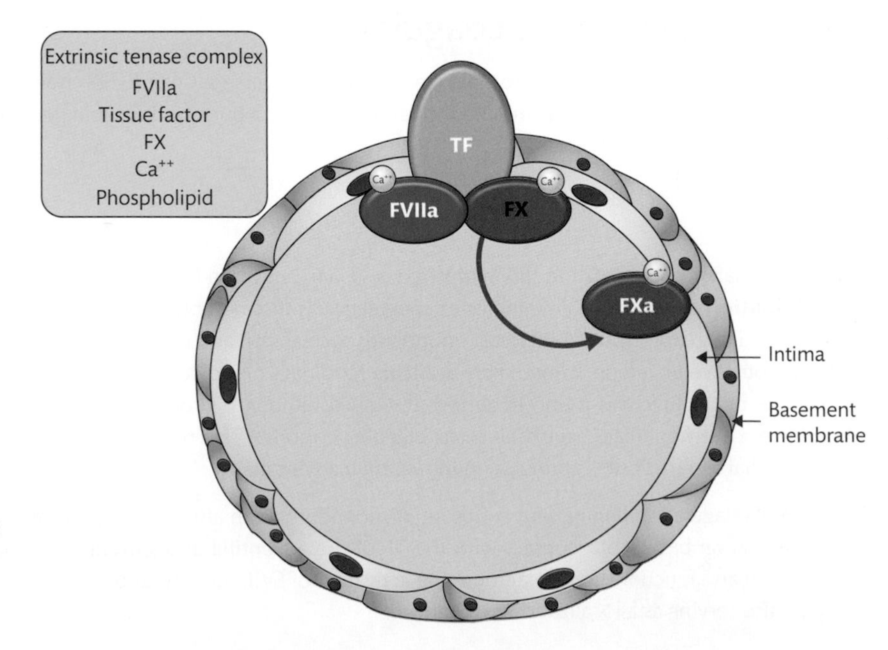

FIGURE 13.12
Formation of the extrinsic tenase complex. The TF–FVIIa complex is bound to the exposed vessel subendothelium and combines with FX to form the extrinsic tenase complex. The product of this reaction is FXa.

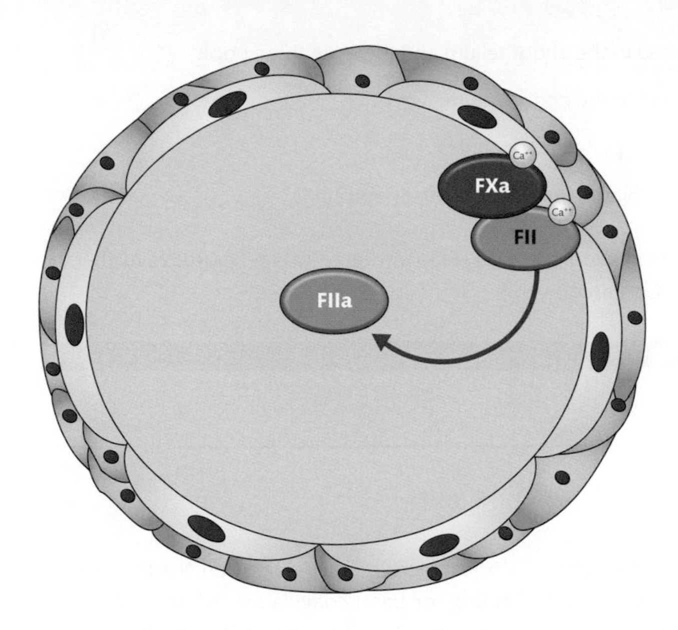

FIGURE 13.13
Generation of trace thrombin. FXa bound to phospholipid via calcium-ion interactions combines with FII (prothrombin) and converts it to FIIa (thrombin). At this stage, the cofactor for FXa (FVa) is unavailable so the reaction rate is slow and only trace amounts of thrombin are formed.

At this stage, the cofactor for FXa is unavailable, so its reaction rate is relatively slow. Nonetheless, the FXa is able to cleave its main substrate, FII (prothrombin), to generate trace amounts of FIIa, otherwise known as thrombin, which is shown in Figure 13.13. Whilst insufficient to initiate significant fibrin formation, this thrombin generation is pivotal to amplification of the coagulation response.

Only small amounts of FXa are formed, as its generation by this mechanism is rapidly downregulated by **tissue factor pathway inhibitor** (TFPI) , which first binds to the active site of FXa via its Kunitz 2 (K2) domain and inhibits it. The TFPI–FXa complex then binds with high affinity to the FVIIa within the TF–FVIIa complex via the K1 domain of TFPI. The K3 domain binds to a cofactor, protein S, which enhances inhibition of FXa by the K2 domain. You will meet protein S again in Chapter 15. This results in a fully inhibited quaternary complex of TF–FVIIa–TFPI–FXa preventing further activation of factor X via TF–FVIIa, as depicted in Figure 13.14.

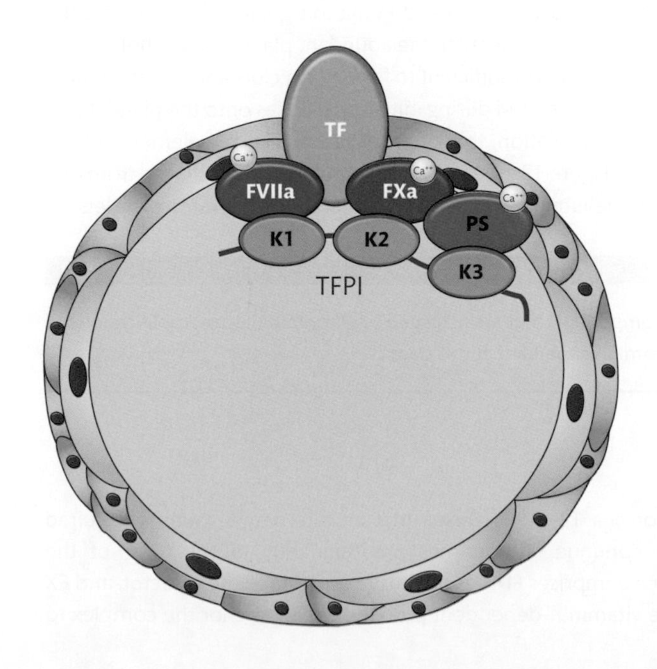

FIGURE 13.14
Downregulation of initiation by TFPI. Only small amounts of FXa are formed at the initiation stage because the TF pathway is downregulated by the inhibitor TFPI, which first binds FXa via its K2 domain and then the TF–FVIIa complex via the K1 domain. The K3 domain binds to the cofactor protein S (PS).

TFPI is predominantly produced in the endothelium and there are three pools:

- approximately 90% is bound to the endothelium;
- between 5 and 10% resides in platelet α-granules;
- a small amount circulates in plasma associated with lipoproteins.

Despite rapid inhibition of the initiation phase via TFPI, sufficient FXa is generated to further the coagulation response before it is shut down.

SELF-CHECK 13.6

How is secondary haemostasis initiated?

Idling

idling
Extravascular priming of coagulation.

Factors II, VII, IX, and X are small enough to leak across spaces between tissues and the vessel wall. This permits the process called **idling**, or basal coagulation, where continuous pre-formation of the TF-FVIIa complex primes coagulation by producing low levels of thrombin in the extravascular space despite the absence of vessel injury. The low level of thrombin promotes faster activation of coagulation upon vessel damage. Progression to clot formation does not occur during idling because larger elements of clotting, such as platelets, fibrinogen, and VWF/FVIII, remain sequestered in the vascular space.

Amplification

The small amount of thrombin (FIIa) generated during initiation then enters an **amplification** phase to facilitate the availability of important cofactors. The trace amount of thrombin is able to activate the protein cofactors FV and FVIII, and also platelets and some FXI. FVa is the cofactor for FXa (the cofactor that was unavailable during the initiation phase), and FVIIIa is the cofactor to FIXa. Activation of more platelets results in exposure of the procoagulant phospholipid surface on a large scale. The availability of FVa and FVIIIa to significantly accelerate the reaction rates of their partner enzymes, together with the abundant platelet phospholipid, set the stage for large-scale thrombin generation sufficient to form fibrin clots and initiate important feedback mechanisms. The FIXa generated during initiation diffuses onto the platelet surface to facilitate the final phase of coagulation, that of propagation. The key elements of the amplification phase are depicted in Figure 13.15. Note that coagulation is transferred from one cell surface, the exposed subendothelium, to another, the membrane of activated platelets.

SELF-CHECK 13.7

What are the three haemostatic components that are activated by thrombin during amplification? Why is only a trace amount of thrombin generated at this stage?

Propagation

The shut-down of FX activation via TF–FVIIa means that an alternative route is required if thrombin generation is to continue. This comes predominantly in the form of the **intrinsic tenase complex** which comprises FIXa as the enzyme, FVIIIa as the cofactor, and FX as the substrate. FIX and FX are vitamin K-dependent proteins and so anchor the complex to

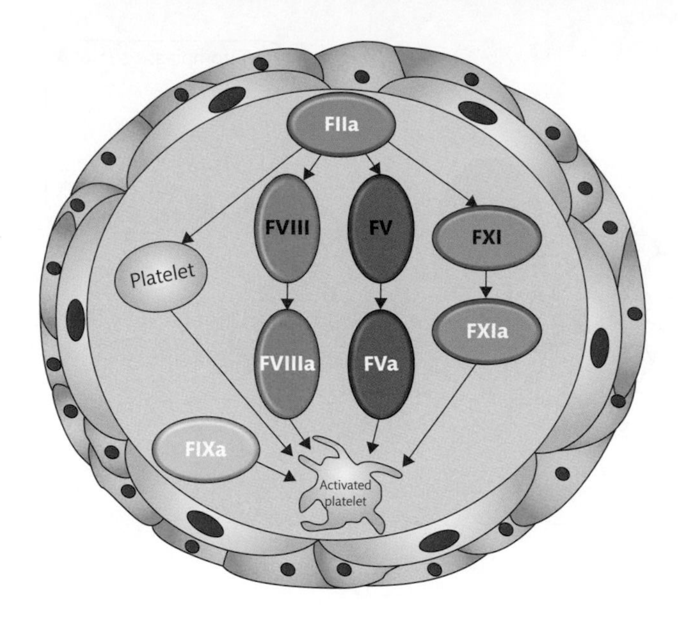

FIGURE 13.15
Amplification phase of secondary haemostasis. The small amount of thrombin generated at initiation is sufficient to activate circulating FV and FVIII and some platelets and FXI. The FIXa generated at initiation migrates to the surface of activated platelets to begin the propagation phase of coagulation.

the platelet phospholipid surface as a result of their interaction with calcium ions via their Gla domains. Activated platelets undergo a membrane reorientation which affects the distribution of phospholipids, the activation process leading to a greater availability of phosphatidylserine to promote the binding of activated coagulation factors. Assembly of the intrinsic tenase complex is shown in Figure 13.16.

The initial source of FIXa comes from the activation of FIX by TF–FVIIa during the initiation phase. Additionally, some of the membrane-bound FXa formed via this route will have activated FIX. Once FVIIIa is formed during the amplification phase, the intrinsic tenase complex becomes the major activator of FX. Binding of FIXa to FVIIIa induces a 10^6-fold increase in its ability to activate FX, the complex being 50 times more efficient than TF–FVIIa in FXa generation. Consequently, more than 90% of FXa is ultimately produced by the intrinsic tenase complex.

Cross reference
You met the role of Gla domains in Figure 13.10.

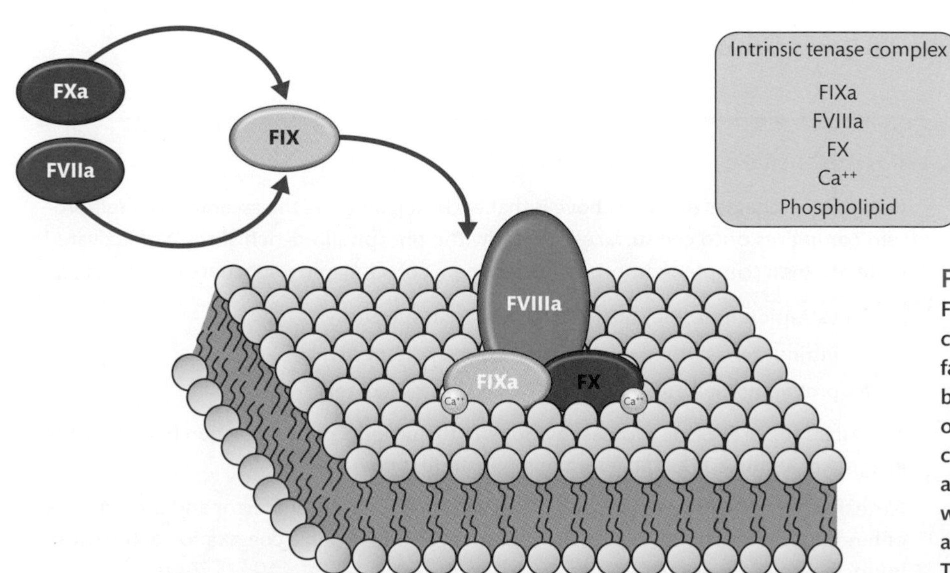

FIGURE 13.16
Formation of the intrinsic tenase complex on the phospholipid surface of an activated platelet. FIXa binds to the phospholipid surface of the platelet via Gla domain–calcium ion interactions, and associates with its cofactor, FVIIIa, which was generated during amplification, and its substrate, FX. The product of the reaction is FXa.

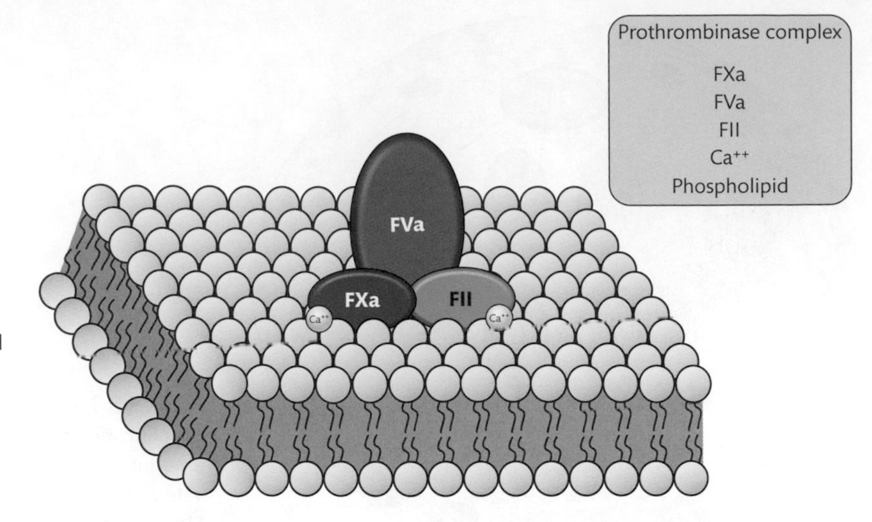

FIGURE 13.17
Formation of the prothrombinase complex on the phospholipid surface of an activated platelet. FXa binds to the phospholipid surface of the platelet via Gla domain-calcium ion interactions, and associates with its cofactor, FVa, which was generated during amplification and released from platelet α-granules and its substrate, FII. The product of the reaction is FIIa (thrombin).

FXa combines with its cofactor and substrate, FVa and FII respectively, anchored to the platelet surface via calcium ions and phospholipid, to form the **prothrombinase complex**, which is shown in Figure 13.17. Along with the FVa generated in plasma, activated platelets provide another source of FV, which is secreted from α-granules and is partially activated, becoming fully activated once in contact with thrombin.

You can see in Figure 13.18 that the product of the prothrombinase complex is thrombin. This enzyme has multiple functions, mainly integral to haemostasis. Its main roles are listed below:

- proteolysis of fibrinogen;
- activation of FV, FVII, FVIII, FXI, and FXIII;
- activation of platelets via protease-activated receptor-1 (PAR-1);
- activation of protein C;
- activation of thrombin-activatable fibrinolysis inhibitor (TAFI);
- tissue repair and development.

Key Points

Key to the processes detailed above is that each step involves the assembly of multiprotein complexes onto cell surfaces, primarily the phospholipid-rich surface of activated platelets. Each complex comprises an enzyme, a cofactor, and a substrate:

- **for extrinsic tenase, they are FVIIa, TF, and FX, respectively;**
- **for intrinsic tenase, they are FIXa, FVIIIa, and FX, respectively;**
- **for prothrombinase, they are FXa, FVa, and FII, respectively.**

The substrates are serine protease zymogens, the product of one reaction becoming the enzyme of another. FIX is an alternative substrate for FVIIa + TF.

Note that the enzyme in each complex is a vitamin K-dependent factor and thus capable of binding to phospholipid surfaces via calcium ions to localize coagulation at the site of injury where appropriate surfaces are exposed.

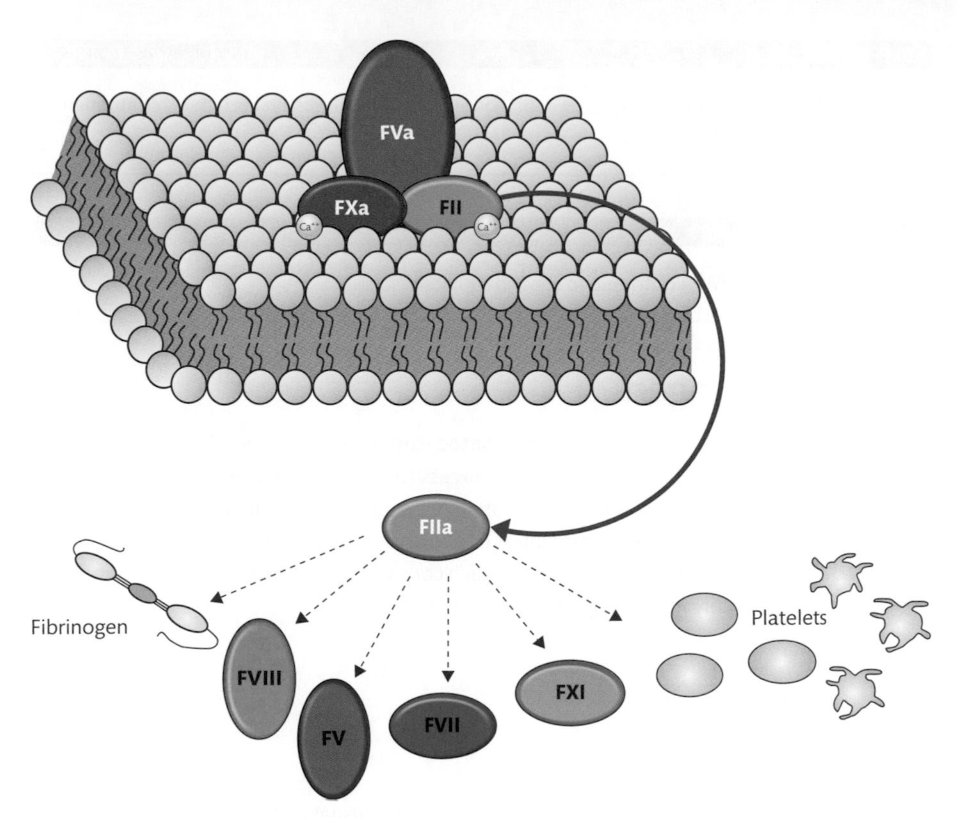

FIGURE 13.18
Burst of thrombin generation. The thrombin generated via the prothrombinase complex activates further components of haemostasis, allowing a positive feedback loop that leads to further thrombin generation. More platelets are activated by thrombin, and fibrinogen can now be converted to fibrin to stabilize the forming clot.

With thrombin now being produced via the intrinsic tenase and prothrombinase complexes, it is available to enter into positive feedback loops to propagate coagulation by activating more FV, FVIII, and FXI. FXI circulates in complex with high-molecular-weight kininogen (HMWK), which is required for binding FXI to phospholipid surfaces. In the presence of calcium ions and HMWK, FXIa activates FIX and can also autoactivate FXI to form more FXIa to propagate this route of FIX activation.

Despite the shutting down of FVII activation by TFPI during the initiation phase, FIXa, FXa, and thrombin are all capable of activating FVII and provide a route for introducing FVIIa into the propagation phase to generate more FIXa, FXa, and of course, FVIIa. Ultimately though, TF availability is reduced as the initial vessel breach is obscured by the forming clot, and thrombin generation continues without TF–FVIIa.

Key Points

You saw in Figure 13.16 how assembly of the intrinsic tenase complex on the surface of an activated platelet localizes the formation of its product, FXa, which then participates in assembly of the prothrombinase complex on the same surface to generate the pivotal enzyme thrombin.

Formation of thrombin via the intrinsic tenase complex and subsequent positive feedback mechanisms produce a burst of thrombin generation sufficient to cleave fibrinogen to form fibrin, and also to continue the feedback processes.

Key Points

Thrombin triggers inflammatory responses such as the production of chemokines to attract white cells to the site of injury.

SELF-CHECK 13.8

Name the three multiprotein complexes of coagulation biochemistry and their component enzymes, substrates, and cofactors.

SELF-CHECK 13.9

What do platelets and VWF contribute to secondary haemostasis?

13.3.3 Fibrinogen

hepatocytes
Liver cells involved in protein synthesis, storage of protein, iron, and vitamins, and detoxification.

Fibrinogen is a glycoprotein synthesized in **hepatocytes** and is present in a high concentration in plasma, (approximately 3.0 g/L), in marked excess to the minimal requirements necessary for normal haemostasis. It is also found in platelet α-granules, where it is not synthesized but taken up from plasma by receptor-mediated endocytosis. Fibrinogen is a symmetrical dimer composed of two identical subunits, each of which consists of three non-identical polypeptide chains that are intricately folded together, these being:

- Aα chains;
- Bβ chains;
- γ chains.

Fibrinogen was the first biological macromolecule to be visualized by electron microscopy. The molecular architecture was revealed to be trinodular, comprising two roughly spherical modules termed the D domains, connected to a central E domain. The E domain contains the N termini of all six chains tethered to each other via disulphide bonds. Two thin, coiled regions stretch out on either side of the E domain, each consisting of one Aα, one Bβ, and one γ polypeptide. Each coil ends in a globular D domain, which contains the C termini of the Bβ and γ chains and part of the Aα chain. The remainder of the Aα chain protrudes from the D domain

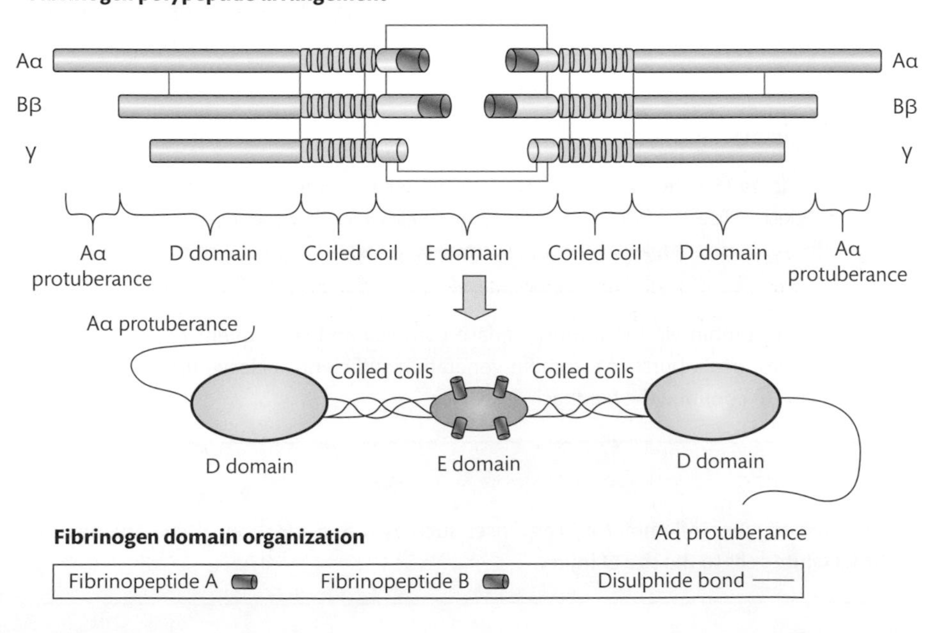

FIGURE 13.19
Fibrinogen structure. Symmetrical dimer structure composed of two identical subunits, each containing three non-identical polypeptide chains: the Aα, Bβ, and γ chains. The N termini of all six chains are tethered together via disulphide bonds in the E domain. Two coiled coil regions stretch out on either side of the E domain, each consisting of one each of the Aα, Bβ, and γ chains. Each coil ends in a globular D domain, which contains the C termini of the Bβ and γ chains and part of the Aα chain. The remainder of the Aα chain protrudes from the D domain as the Aα protuberance.

TABLE 13.4 Functions of fibrinogen.

Process	Functions
Haemostasis	Supports platelet aggregation Parent molecule for fibrin in the formation of an insoluble fibrin clot Clot dissolution
Inflammation	Bridging molecule in cell-cell interactions during inflammatory cell trafficking
Wound healing	Modulates cellular responses Structural support of adhesive cell-cell/cell-extracellular provisional matrix interactions Fibrinogen (and fibrin) found in tumour matrices to support angiogenesis Matrices provide structural scaffold to bind growth factors and support: cell adhesion, spreading, migration, and proliferation

as a long strand. Look at Figure 13.19 to see how the polypeptide chains are arranged and that they are linked at various positions by disulphide bonds. Figure 13.19 translates polypeptide arrangement into domainal structure for ease of interpreting the subsequent related diagrams.

In approximately 15% of circulating fibrinogen molecules one of the γ-chains is a minor variant termed γ′, and molecules where both are γ′ chains comprise about 1%. Molecules containing γ′ chains are called fibrinogen 2.

Fibrinogen is a multifunctional molecule with roles in haemostasis, inflammation, and wound healing that are summarized in Table 13.4. We have covered its role in primary haemostasis and will now look at its function in secondary haemostasis.

SELF-CHECK 13.10

Name the peptide chains of fibrinogen.

Fibrin formation

Fibrin is formed from fibrinogen via the action of thrombin, which cleaves small peptide fragments from the N-terminal ends of the Aα and Bβ chains, which are termed **fibrinopeptide A** (FPA) and **fibrinopeptide B** (FPB), respectively. The fragments are small and constitute just 3% of the mass of fibrinogen. The fibrinopeptides shield polymerization sites on the parent molecule and prevent interaction with complementary structures in other fibrinogen molecules, sometimes referred to as 'knobs' and 'holes', respectively.

Figure 13.20 shows FPA release from a single fibrinogen subunit exposing the E_A polymerization site in the E domain, which joins non-covalently with the unshielded D_a-binding site on the γ chain of the outer D domain of another fibrin monomer/fibrinogen molecule. Note that one portion of the E_A site is at the N terminus of the Aα chain and the other on the Bβ chain. FPB release, which is slower than FPA release, exposes the separate E_B polymerization site of the Bβ chain in the central E domain, which joins non-covalently with an inherently expressed D_b-binding site of a Bβ chain in the D domain of another molecule. Note that the differing chain lengths and positions of the 'holes' result in the molecules overlapping—the so-called half-staggered overlap.

Fibrin molecules can link together through the interaction of the E domain on one fibrin molecule to the D domains on up to four other fibrin molecules. Fibrinopeptide release converts fibrinogen molecules to fibrin monomers and initially leads to the formation of fibrin dimers. The fibrin dimers

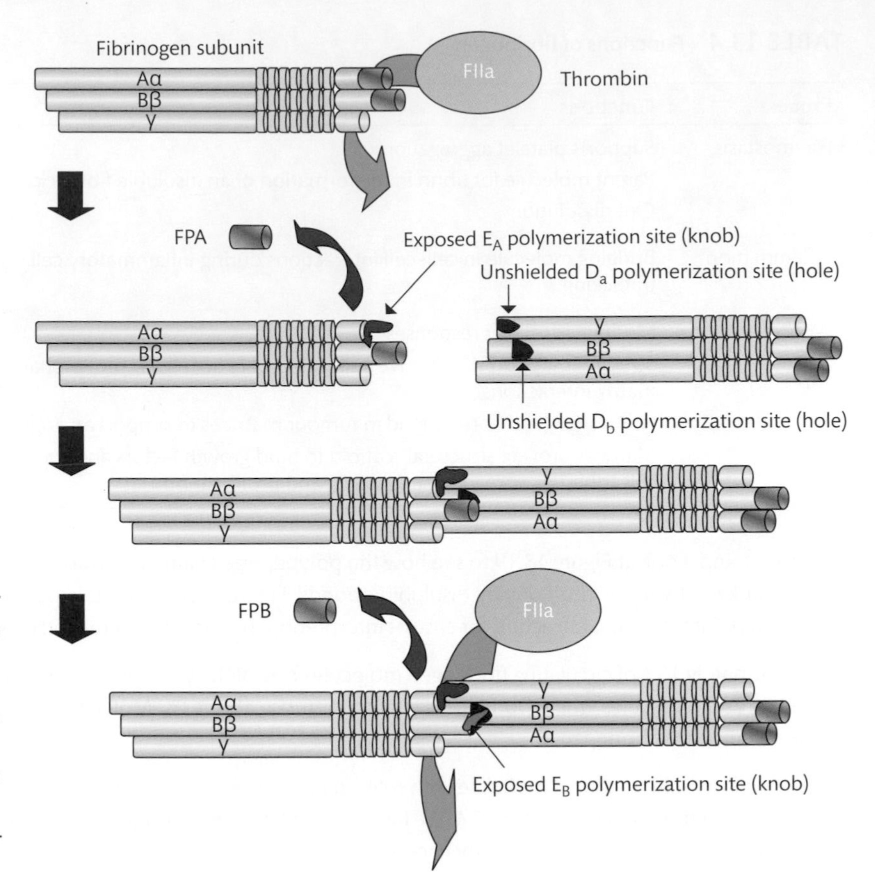

FIGURE 13.20
Fibrinopeptide release. Thrombin cleaves FPA from a fibrinogen subunit to expose the E_A polymerization site in the E domain, which joins with an unshielded D_a-binding site on the γ chain of the outer D domain of another molecule. Thrombin also cleaves FPB to expose a separate E_B polymerization site on the Bβ chain in the central E domain, which joins with the unshielded D_b-binding site of a Bβ chain in the D domain of another molecule. The differing chain lengths and positions of the 'holes' result in the molecules overlapping to form the half-staggered overlap.

are characterized by the overlap structure and are stabilized by the non-covalent interactions between the knobs and holes on the central and outer domains, which you can see represented in domain form in Figure 13.21. At this stage, the fibrin remains soluble. As the thrombin-catalysed fibrinopeptide cleavage continues, rapid polymerization ensues to generate two-stranded fibrin polymers called protofibrils. The fibrin molecules align in the staggered overlap end-to-middle domain arrangement, which facilitates the formation of twisting fibrils. Multi-stranded fibrils are formed as a result of lateral association. Fibril branching to form a clot network is facilitated by formation of bilateral and equilateral branch junctions, which you can see in Figure 13.21.

The initial interactions forming the staggered overlaps are via E_A–D_a associations, release of FPB contributing to lateral associations. After a certain molecular size is reached, fibrin solubility is significantly reduced, leading to formation of an insoluble polymer.

SELF-CHECK 13.11

Outline the molecular events of fibrin formation.

Fibrin stabilization

Despite the intricacies of fibrin formation, the clot requires stabilization if it is to withstand higher blood pressure and shear forces and become fully haemostatically effective. The fibrin clot is stabilized and rendered mechanically stronger by the action of the activated form of FXIII.

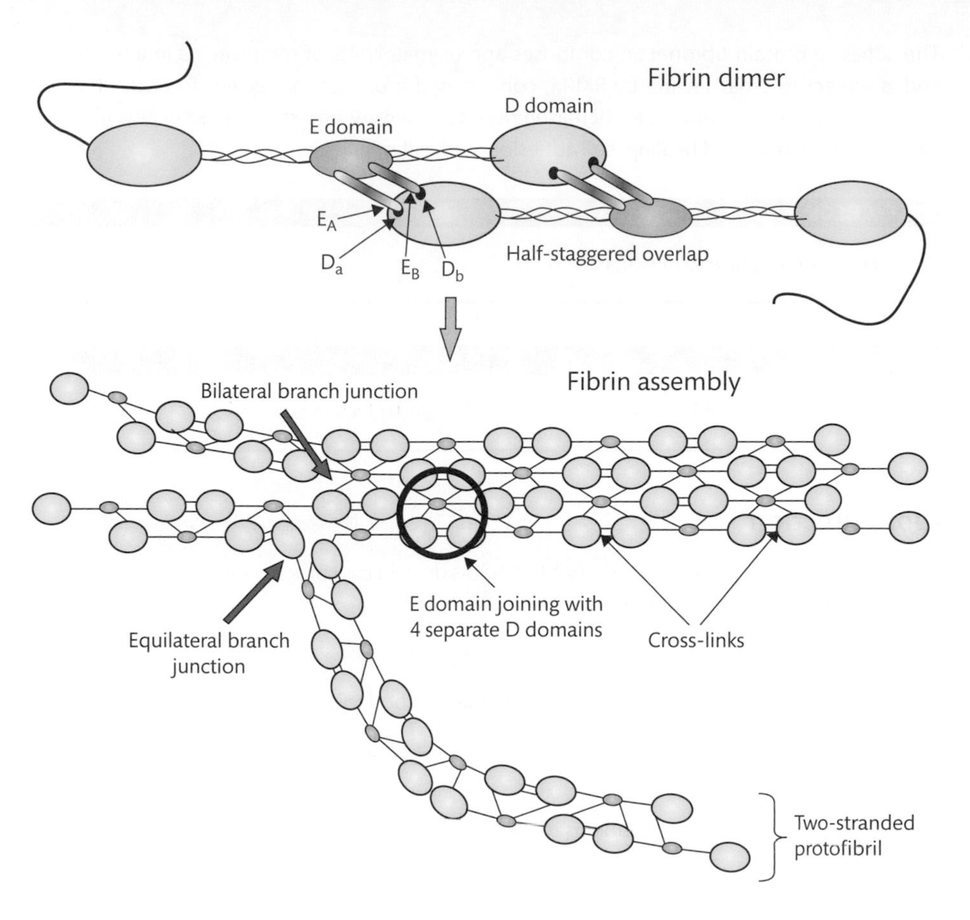

FIGURE 13.21
Domainal representation of fibrin formation. Fibrin dimer formation represents the start of fibrin polymerization; the diagram depicts the associations you met in Figure 13.20 in terms of E-domain and D-domain interactions. Formation of lateral fibrin polymers occurs when E domains join with four separate D domains, indicated by the thick circle. Branching occurs upon formation of bilateral or equilateral junctions.

Plasma FXIII circulates in the form of a tetramer composed of two A subunits, which are the catalytic component, bound to two stabilizing carrier B subunits. The plasma FXIII circulates bound by its B subunits to fibrinogen 2. FXIII also occurs intracellularly in platelets in the form of a homodimer of the two A subunits.

Activation of plasma FXIII is by calcium-dependent thrombin cleavage of the A chains. Thrombin first cleaves off the activation peptide from the A chains, and then in the presence of calcium ions, the B chains dissociate and the A chains transform into the active enzyme (FXIIIa). It is a calcium-dependent transglutaminase that catalyses the formation of intermolecular covalent glutamine–lysine bonds between fibrin molecules (and between fibrin and other proteins). These links are commonly referred to as **cross-links**. Intracellular FXIII is activated without proteolytic cleavage by the elevated calcium levels in thrombin-activated platelets, which causes a conformational activation.

In the early stages of fibrin assembly, FXIIIa catalyses the formation of cross-links between the D domains of the assembling fibrin molecules via γ chains. After the majority of γ chains have been paired, a slower process of cross-link formation between α chains proceeds. Cross-linking can also occur between α chains and γ chains. The stabilized fibrin mesh strengthens the clot by binding platelets together and contributing to their attachment to the vessel wall, mediated by binding to platelet receptors and interactions with other adhesive proteins. FXIIIa cross-links fibrin γ chains to the platelet membrane.

FXIIIa also binds to α_2**-antiplasmin**, the main inhibitor of the enzyme plasmin which is responsible for dissolution of the clot, thereby preventing premature clot destruction.

The adhesive protein fibronectin comprises approximately 4% of the proteins in a fibrin clot and is subject to cross-linking by FXIIIa, connecting fibronectin molecules to each other or fibrin. Fibronectin contributes to increased fibre size, density, and strength. FXIII also plays an important role in wound healing, cell adhesion, and cell migration.

SELF-CHECK 13.12

What is the haemostatic role of FXIIIa?

SELF-CHECK 13.13

What are the main roles of fibrinogen in primary and secondary haemostasis?

Clot retraction

Over a time course of minutes to hours, blood clots draw in by a process termed clot retraction, which serves four crucial functions:

- helping platelet-rich thrombi withstand high shear forces;
- compacting the clot to form a tighter seal to stem blood loss;
- compacting the clot to make it less obstructive to facilitate continued blood flow whilst the wound repairs;
- contraction of the clot, which helps to close the wound.

The process of clot retraction is driven by forces within platelets. GPIIbIIIa is integral to this function, as it acts as a link between cytoplasmic actin filaments and surface-bound fibrin polymers. Additionally, GPIIbIIIa generates the intracellular signals that trigger retraction through activation of tyrosine kinases, which lead to the assembly of an intracellular complex of actin-binding proteins, resulting in tethering of the actin to GPIIbIIIa. FXIIIa also cross-links actin to fibrin and to myosin. Once tethered, actin interacts with cytoplasmic myosin, which is the molecular motor that applies the contractile force on actin filaments. Contracted clots develop a specific structure where a meshwork of fibrin and platelets aggregates on the exterior of the clot surrounding a close packed array of dramatically compressed red cells, called polyhedrocytes due to their altered shape, within the clot.

13.4 Regulation of secondary haemostasis

Although localization mechanisms exist to restrict clot formation to the site of injury, the generation of potentially lethal enzymes such as thrombin in the circulation requires additional regulatory mechanisms. Left unchecked, the thrombin that could be generated from 20 mL of blood could (theoretically) clot all circulating fibrinogen in less than a minute. This does not happen due to the existence of localization mechanisms and a group of physiological anticoagulant substances that cooperate to inhibit coagulation.

Once the coagulation reactions have been activated, a separate series of reactions occur that bring inhibitory mechanisms into play to limit and control the clotting processes. The **naturally occurring inhibitors** of coagulation fall into three main subtypes:

- Kunitz-type inhibitors;
- serine protease inhibitors (serpin);
- components of the protein C system.

We will now look at each of these inhibitors in turn.

13.4.1 Kunitz-type inhibitors

These inhibitors are so-termed because they contain biochemical modules called Kunitz domains. These domains are found throughout nature in many proteins that are usually inhibitors of specific serine proteases. The general mechanism of inhibition is the insertion of a looped structure from the inhibitor into the serine protease module of the enzyme, where it occupies and disrupts the active site.

The two Kunitz-type inhibitors of importance to coagulation are TFPI, which we met earlier, and **protease nexin-2** (PN-2), which has inhibitory activity against FXIa and also FIXa and FXa in the prothrombinase complex. PN-2 circulates at low concentrations in plasma, so has little inhibitory activity against any free FXIa, and the FIXa and FXa in the prothrombinase complex are membrane-bound anyway. However, PN-2 is present in platelet α-granules and this is the likely route for its contribution to the inhibition of coagulation.

13.4.2 Serpins

As most TFPI is membrane-bound and therefore present in low concentrations in plasma, it can only delay the coagulation reactions, and so other inhibitors must come into play to regulate clot formation during subsequent stages. As well as FVIIa, the other main enzymes of coagulation are serine proteases, and human plasma contains a number of serpins that perform anticoagulant functions.

Serpins inhibit their target enzymes using a suicide substrate-inhibition mechanism. The reactive centre sequence of the serpin is recognized by the target enzyme as a substrate-like sequence and cleaves a peptide bond in the serpin. This leads to the formation of a covalent bond between the enzyme and the serpin and results in a massive conformational change in the serpin, and, to a lesser extent, of the enzyme too, which is consequently inactivated. The resultant enzyme–inhibitor complex is irreversible.

The plasma serpins vary in their specificity, some having almost incidental activity against activated coagulation factors, with their main physiological roles being elsewhere in human biochemistry. The serpins that act predominantly as coagulation inhibitors are antithrombin (AT), heparin cofactor II (HC II), and protein Z-dependent protease inhibitor (ZPI).

Antithrombin

There were originally thought to be six antithrombins, but subsequent research has proven all but one to be other entities; thus, what has been termed antithrombin III in many publications is now referred to as just antithrombin.

AT is an important endogenous coagulation inhibitor, its name being something of a misnomer as it has inhibitory activity towards FIXa, FXa, and FXIa as well as thrombin. It also possesses activity against FVIIa in the FVIIa–TF complex, but not free FVIIa.

AT forms stable 1:1 complexes with its target enzymes; that is, one molecule of inhibitor binds to one molecule of target enzyme and neither takes any further part in haemostasis.

It is an inefficient inhibitor in the absence of activating cofactors, and so complex formation without cofactors is progressive. Cofactor activity is provided *in vivo* by heparan sulphate, a **glycosaminoglycan** anchored to the vessel wall by a proteoglycan core and present on the surface of most eukaryotic cells and in the extracellular matrix. It serves to localize and concentrate activated AT on or in the vessel wall. Heparan sulphate binds AT via a unique pentasaccharide sequence, inducing a conformational change in the reactive centre of AT that potentiates its ability to bind and inhibit serine proteases. Once the inactive complexes are formed, heparan sulphate's affinity for AT decreases, allowing dissociation and subsequent rapid clearance of the complexes by the liver.

AT inhibits free FXa and thrombin more efficiently than FXa and thrombin, which are bound to activation assemblies or clots, as they are resistant to AT inactivation in those forms. In this case, AT acts as a scavenger of FXa, and thrombin molecules which diffuse away from complexes and clots, thus playing a crucial role in localizing and limiting coagulation and preventing unnecessary fibrin formation and deposition.

AT plays an important role in regulating the low-level activation of coagulation that occurs normally in the uncompromised circulation. In direct contrast to the protection afforded FXa when part of platelet surface complexes, FVIIa when in complex with TF on a cell surface is more susceptible to AT inactivation and provides a route for AT to regulate TF-induced coagulation.

AT circulates in two isoforms, α-AT and β-AT, the latter being less glycosylated. β-AT has enhanced affinity for heparan sulphate and is more effective than α-AT at inhibiting the thrombin appearing on the wall of injured vessels. The fact that 90% of circulating AT is the α-isoform is probably a reflection of the increased binding capacity and increased vessel wall consumption of β-AT.

AT also adopts non-anticoagulant functions. Its action on thrombin results in regulation of the enzyme's own non-coagulant functions, and thus has anti-proliferative and anti-inflammatory properties. Cleaved and latent forms of AT have potent anti-angiogenic properties.

SELF-CHECK 13.14

How are the anticoagulant roles of AT split between the two isoforms?

Heparin cofactor II

Heparin cofactor II (HC II) inhibits thrombin by forming 1:1 complexes, but it has no activity against other coagulation serine proteases. Binding to dermatan sulphate enhances the inhibitory action of HC II towards thrombin. AT is present in plasma in a two-fold molar excess over HC II, and it appears that the latter has different physiological roles related to thrombin inhibition in extravascular lesions and thrombin regulation during pregnancy.

Protein Z-dependent protease inhibitor

Protein Z-dependent protease inhibitor (ZPI) inhibits FXa via assembly of a calcium ion-dependent tertiary complex containing FXa, ZPI, and the vitamin K-dependent cofactor protein Z (PZ) at the phospholipid surface. PZ enhances the rate of FXa inhibition by ZPI more than 1000-fold. A portion of plasma ZPI circulates in complex with PZ. ZPI also inhibits FXIa, independently of calcium ions, phospholipids, and PZ, by competing with other FXIa inhibitors and FIX for the active site of FXIa. The interaction between ZPI and FXIa generates cleaved, inactive ZPI, which reduces the amount of ZPI available for FXa inhibition.

ZPI can also inhibit FXa contained within the prothrombinase complex although its effect is negated at high prothrombin concentrations and so occurs either prior to FV activation or after local prothrombin consumption. Thus, it seems that ZPI has a role in dampening coagulation by delaying initiation and reducing thrombin generation.

α_2-*macroglobulin*

α_2-macroglobulin (α_2M) is not strictly speaking a member of the serpin superfamily as its effects are not restricted to serine proteases, its binding to activated coagulation factors being away from the serine-active centre. In a normal adult, it contributes to no more than 20% of the inhibition of thrombin and 10% of FXa inhibition. In coagulation, α_2M functions as a 'back-up' or 'fail-safe' serine protease inhibitor, particularly in times of pathological stress, when some of the main inhibitors are overwhelmed. α_2M levels in children can be double those of adults, suggesting a compensatory role for the lower levels of AT in children. Other physiological roles for α_2M include the inhibition of non-coagulation protease enzymes and the regulation of growth factors and immune function.

Other serpins

α_1-antitrypsin has moderate affinity for thrombin and plasmin, although it does not have a major impact in the regulation of thrombin generation. Its prime physiological role is the inhibition of neutrophil elastase. C1-esterase inhibitor plays a minor role in FXIa inhibition, its primary role being inhibition of the complement pathway. The final serpin of relevance to haemostasis is one of the main fibrinolysis inhibitors, α_2-antiplasmin (α_2AP), which we will meet in more detail later in this chapter.

13.4.3 The protein C anticoagulant system

We have seen that the generation of the cofactors FVa and FVIIIa during the amplification phase of coagulation is a pivotal point in clot formation, as their presence facilitates assembly of the tenase and prothrombinase complexes on phospholipid surfaces. As AT has a limited inhibitory effect on FXa and thrombin bound to activation assemblies, the protein C pathway plays a crucial role in the regulation of coagulation on phospholipid surfaces by directly limiting the procoagulant activity of FVa and FVIIIa, and indirectly, thrombin. There are six key players in the pathway.

Protein C

Protein C (PC) is the circulating vitamin K-dependent zymogen of the serine protease **activated protein C** (APC), the enzyme responsible for inactivating FVa and FVIIIa.

SELF-CHECK 13.15

Which clotting factors are inhibited by activated protein C?

Thrombomodulin and thrombin

Thrombomodulin (TM) is a transmembrane receptor present on the endothelial cells of most arteries, veins, capillaries, and lymphatic vessels, as well as most other tissues. The ratio between endothelial cell surface and blood volume is at its highest in the capillaries, so the TM concentration in the microcirculation is >1000-fold higher than in the major vessels. A fascinating paradox in haemostasis is that the pivotal procoagulant molecule, thrombin, when in complex with TM, loses its procoagulant activity and becomes anticoagulant in nature by activating protein C. Thrombin binds with high affinity for TM, which functions as a cofactor for thrombin, producing a >1000-fold increase in the rate of PC activation. Thrombin and TM bind in a 1:1 complex.

Endothelial protein C receptor

Endothelial protein C receptor (EPCR) is a transmembrane receptor on endothelial cells that is abundantly expressed in the larger arterial vessels and at lower levels in veins and capillaries. Although TM enhances activation of PC by thrombin, the conversion is a relatively low-affinity reaction. EPCR augments the activation of PC, providing a further 20-fold increase in the rate of PC activation by the thrombin–TM complex.

Protein S and factor V

APC requires the presence of two cofactors to exert its full anticoagulant effect: **protein S** (PS) and the intact form of FV.

PS is a vitamin K-dependent factor, but is not a serine protease. Approximately 60% of circulating PS is associated in a 1:1 complex with C4b-binding protein (C4bBP) and is unavailable for APC cofactor functioning. The remaining 40% is termed free PS and it is this pool that is available to act as an APC cofactor. Free PS has little or no intrinsic inhibitory activity against FVa and FVIIIa; it acts by binding strongly (via the Gla domain) to negatively charged phospholipids on the surface of activated platelets, and then forming a calcium ion-dependent complex with APC. Whilst the presence of PS alone is sufficient for APC to inactivate FVa, regulation of FVIIIa within the tenase complex requires the synergistic APC cofactor activities of PS and FV.

PS also operates as a non-enzymatic cofactor for TFPI in the inhibition of FXa at the initiation stage of coagulation. Binding of TFPI to PS enables optimal interaction between TFPI and FXa on a phospholipid surface. PS has a role in programmed cell death by binding to phospholipid exposed on apoptotic cells and stimulating phagocytosis.

Activation and function of protein C

The protein C pathway is brought into play as a direct consequence of the generation of thrombin. As the thrombin concentration rises, its high affinity for TM results in much of it binding to TM, primarily on the endothelial surface. This binding promotes three important anticoagulant functions:

- occupation of a specific area of the thrombin molecule by TM prevents the binding of thrombin to its procoagulant substrates and activated platelets;
- binding to TM causes a conformational change in thrombin, altering its substrate specificity, and allowing it to activate PC;
- when bound to TM, thrombin is more susceptible to inhibition by AT than free thrombin, so once thrombin generation ceases, the activation complex stops activating PC. A chondroitin sulphate side chain of TM stimulates this inhibition.

The Gla domains of PC/APC allow binding to negatively charged phospholipids; they are also the site of binding to EPCR. Figure 13.22 shows that PC is localized to the endothelial

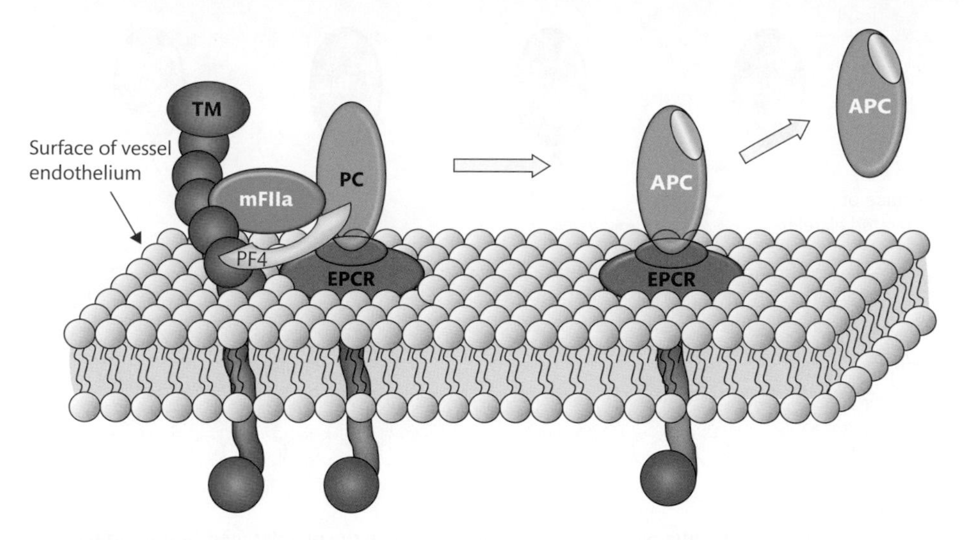

FIGURE 13.22

Activation of protein C on the surface of the vessel endothelium. Protein C (PC) is first localized on the endothelial surface by attachment to endothelial protein C receptor (EPCR) via Gla domains and then locates a thrombin-thrombomodulin complex. Binding of platelet factor 4 (PF4) enhances the affinity of PC for the thrombin-thrombomodulin complex, which activates PC to APC. The APC-EPCR complex then dissociates from the thrombin-thrombomodulin complex but APC does not exert any anticoagulant effect until it subsequently dissociates from EPCR. Note that thrombomodulin (TM) and EPCR are transmembrane proteins. mFIIa = thrombomodulin-modified thrombin.

surface by EPCR, which then moves laterally on the cell surface to locate a thrombin-TM complex. Here, PC is aligned by EPCR for optimal cleavage of an activation peptide, thereby converting the serine protease domain to its active conformation. The platelet α-granule protein platelet factor 4 (PF4) that is released from activated platelets enhances PC activation by binding to the Gla domain and causing a **conformational change** that enhances PC affinity for the thrombin-TM complex. APC is unreactive with its substrates whilst part of the activation complex, and so dissociates. You can see in Figure 13.22 that APC initially remains bound to EPCR, where it is able to contribute to cell signalling processes related to inflammation. APC must also dissociate from EPCR and onto the platelet surface in order to exert its anticoagulant effects.

PS binds with high affinity to negatively charged phospholipids and forms a membrane-bound complex with the APC on the surface of activated platelets. Notice in Figure 13.23 that assembly of this complex orients the active site of APC above the platelet surface, thereby enhancing its activity against FVa and FVIIIa within the forming clot. Without this mechanism, FVa and FVIIIa are protected from APC inactivation by FXa and FIXa, respectively. APC is a highly specialized enzyme that inactivates membrane-bound FVa and FVIIIa by cleaving just a small number of peptide bonds in these homologous molecules, so preventing further participation in effective thrombin generation.

The concentration of FV in plasma is about 100 times greater than that of FVIII, so APC is more likely to inactivate FVa than FVIIIa and mechanisms are required to ensure the inactivation of FVIIIa. The synergistic cofactor effect of FV and PS contributes to this, the effect also being required as the tenase complex is highly efficient and APC + PS alone is insufficient to regulate it. Also, FVIIIa is the preferred substrate for APC.

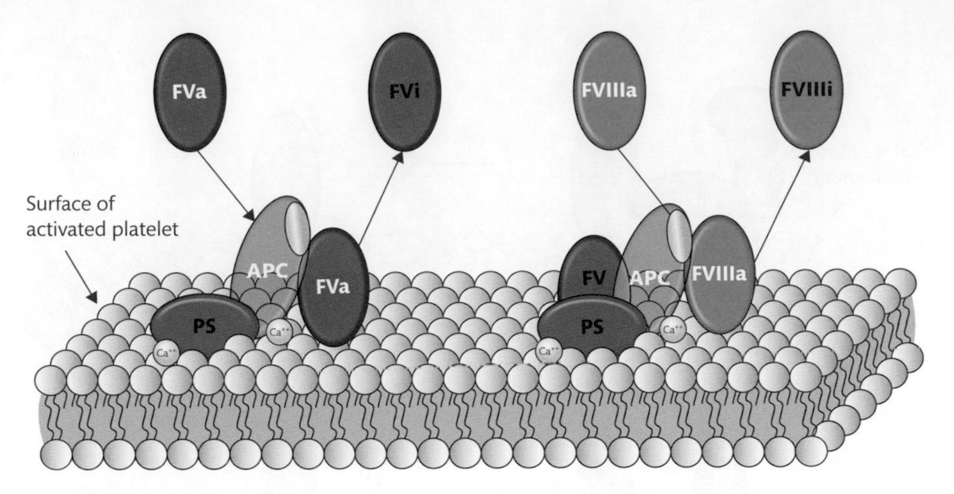

FIGURE 13.23

Inactivation of FVa and FVIIIa by activated protein C (APC) on the platelet membrane surface. APC released from the vessel surface forms a membrane-bound complex with protein S (PS), which orientates the active site of APC above the platelet surface to increase activity against FVa and FVIIIa. The APC–PS complex inactivates FVa and FVIIIa by cleaving a small number of peptide bonds, but it needs the zymogens FV and PS to work in synergy as cofactors for the cleavage of the bonds in FVIIIa.

SELF-CHECK 13.16

Outline the interplay of the components of the protein C system.

Inhibition of the protein C pathway

APC is slowly and progressively inhibited by the protein C inhibitor (PCI; a serpin) through the formation of 1:1 complexes. Free thrombin can cleave free PS and abolish its ability to bind phospholipid and APC.

SELF-CHECK 13.17

Name the vitamin K-dependent factors in haemostasis.

13.5 Fibrinolysis

Fibrin clots cannot grow unchecked or remain in place indefinitely. They are degraded by enzymatic proteolysis through the multi-component mechanisms of fibrinolysis. Like coagulation, it includes zymogen precursors, activators, cofactors, and inhibitors and is a localized, surface-bound process. The main roles of fibrinolysis are:

- to act as the principal defence mechanism against vascular occlusion by preventing fibrin formation in excess of that required to prevent blood loss or rapidly removing excess fibrin;
- removal of a fibrin clot as part of the process of tissue remodelling.

Fibrinolytic activity is always present in plasma, playing a role in the control of the normally occurring low-level activation of coagulation. Fibrin will only accumulate when its production exceeds its destruction.

Key Points

Platelets must also be removed from a clot once it has served its purpose. Interestingly, exocytosed lysosome granules may contribute to clot dissolution, and thrombin can activate platelet apoptosis.

13.5.1 Components of the fibrinolytic system

The pivotal enzyme that degrades fibrin (and fibrinogen) is **plasmin**, which is formed from its zymogen **plasminogen**. Several activators of plasminogen are known, and the system is regulated by inhibitors of plasmin and plasminogen activators.

The interplay between the main players of fibrinolysis is shown in Figure 13.24, which you can refer to as you meet each player in more detail in the following sections.

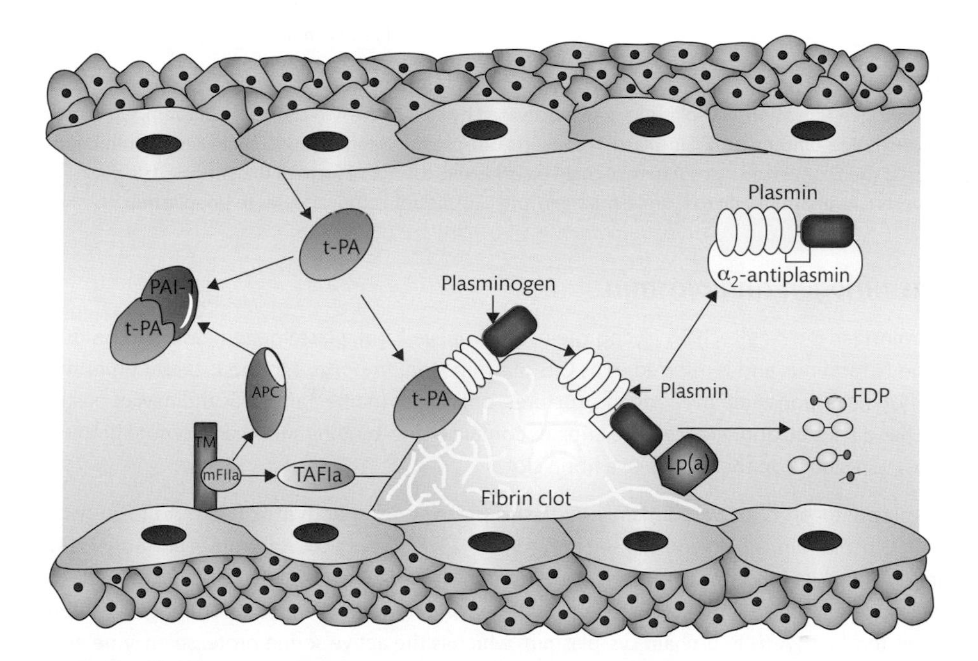

FIGURE 13.24
Interplay of components of fibrinolysis. t-PA released from vessel endothelium rapidly complexes with its inhibitor PAI-1. In the presence of a fibrin clot, t-PA and plasminogen assemble on the fibrin surface via attachment to lysine residues whereby t-PA undergoes a conformational change to facilitate the activation of plasminogen to plasmin, the enzyme that degrades fibrin to its breakdown products. Plasmin activity is regulated by its inhibitor α_2-antiplasmin. Lp(a) acts as a competitive inhibitor by competing for lysine-binding sites on the fibrin surface. The thrombin-thrombomodulin complex activates TAFI and protein C. TAFIa acts as an inhibitor by removing lysine-binding sites but activated protein C impairs the inhibition of t-PA by PAI-1. t-PA = tissue plasminogen activator; PAI-1 = plasminogen activator inhibitor type 1; TM = thrombomodulin; mFIIa = modified thrombin; APC = activated protein C; TAFIa = activated thrombin activatable fibrinolysis inhibitor; Lp(a) = lipoprotein (a); FDP = fibrin degradation products.

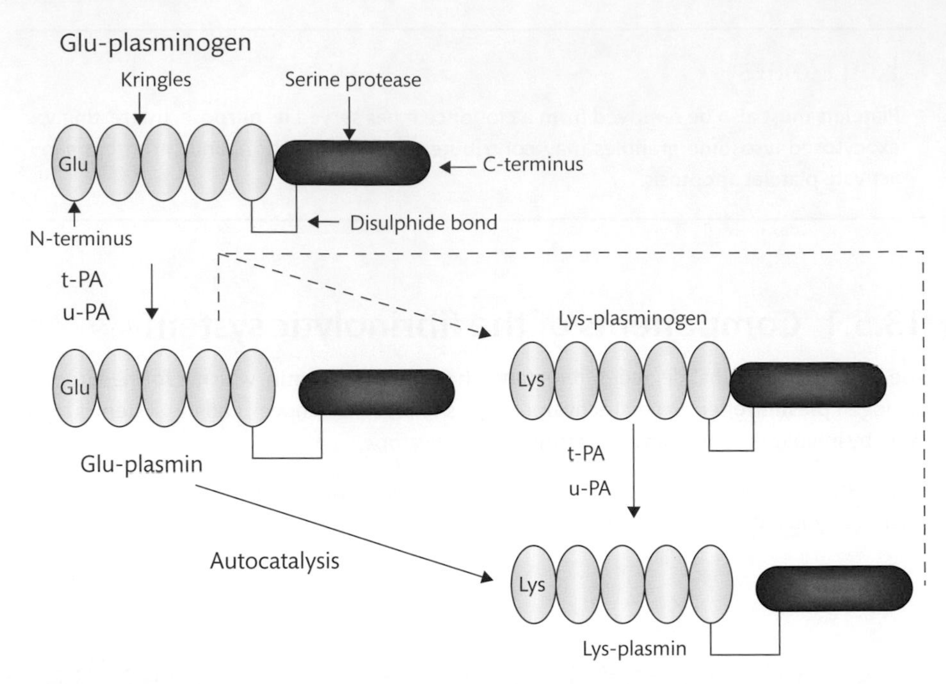

FIGURE 13.25

Plasminogen activation. Glu-plasminogen is activated by t-PA to the two chain Glu-plasmin, which has masked lysine-binding sites. Glu-plasmin converts Glu-plasminogen to single-chain Lys-plasminogen, which is then converted by t-PA to two-chain Lys-plasmin, the active form of the enzyme. Lys-plasmin converts Glu-plasminogen to Lys-plasminogen and Glu-plasmin autocatalyses to Lys-plasmin.

Plasminogen and plasmin

Plasminogen is a single-chain glycoprotein. In its native form, plasminogen has a Glu residue at the N terminus and is referred to as Glu-plasminogen. Important to the localization of the fibrinolytic response are five looped structures, or kringles (named after a Scandinavian pastry that has a knotted appearance). The kringles contain lysine-binding sites that promote binding to lysine residues on the surface of a fibrin clot.

Glu-plasminogen is activated mainly by **tissue plasminogen activator** (t-PA) to form two-chain Glu-plasmin, which you can see in Figure 13.25. However, Glu-plasmin is fibrinolytically ineffective as its lysine-binding sites are masked. Glu-plasmin converts Glu-plasminogen to single-chain lysplasminogen, which has a higher affinity for fibrin. Lys-plasminogen is then converted by t-PA to two-chain Lys-plasmin, which is the active serine protease enzyme as it can bind to fibrin. Lys-plasmin can also convert Glu-plasminogen to Lys-plasminogen, and Glu-plasmin can also autocatalyse to Lys-plasmin.

Plasmin can hydrolyse a variety of substrates by cleaving lysine–arginine bonds. Its main substrates are fibrinogen and fibrin, but it can also hydrolyse FV and FVIII. Plasmin also has a role in inflammation by activating the complement protein, C3.

Tissue plasminogen activator

t-PA is a serine protease and its only known substrate is plasminogen. It is secreted into the circulation by endothelial cells as a single-chain form, and quickly cleared by the liver or inactivated by formation of a 1:1 complex with the fast-acting inhibitor, **plasminogen activator inhibitor type 1** (PAI-1) . Most circulating t-PA is bound to PAI-1.

The rate of t-PA release is markedly increased by a variety of physical and biochemical stimuli, including venous occlusion, strenuous exercise, diet, hypercoagulability, thrombin, vasopressin or its analogues, and some antidiabetic drugs. Free t-PA, however, is an inefficient activator of plasminogen in the absence of fibrin. Assembly of t-PA and plasminogen onto the fibrin surface to form a ternary complex, shown in Figure 13.24, causes a conformational change in the t-PA, so facilitating plasminogen activation. The resultant plasmin not only digests the fibrin, but also cleaves t-PA into a two-chain form which enhances the formation of t-PA–plasminogen–fibrin complexes. Conversion of fibrinogen to fibrin exposes cryptic binding sites for t-PA–plasminogen, so free t-PA has minimal affinity for plasma fibrinogen. The requirement of fibrin as a cofactor for the t-PA–plasminogen complex is a localization mechanism, and demonstrates that fibrin necessarily colludes in its own destruction.

SELF-CHECK 13.18

How is the action of plasmin localized?

Urinary plasminogen activator

Sometimes called urokinase, the **urinary plasminogen activator** (u-PA) is synthesized predominantly in the kidneys and also by fibroblast-like cells in the gastrointestinal tract. It is secreted as a single-chain zymogen termed pro-urokinase and becomes a two-chain serine protease upon activation by kallikrein and plasmin. The one-kringle module in u-PA has no fibrin affinity, and u-PA binds instead to its cell-associated receptor, u-PAR. Thus, its main functions are within tissues where it plays a role in the cellular events of differentiation, mitogenesis, wound healing, and inflammation, and also has a minor role in clot lysis.

Other plasminogen activators

A number of exogenous activators of human plasminogen from animals and microorganisms have been described. Various strains of the bacterium *Streptococcus* produce an extracellular protein called **streptokinase** (SK) , purified versions of which have been used to treat life-threatening thrombotic events such as myocardial infarction by activating plasminogen to destroy the clot. Staphylokinase (SAK) from the bacterium *Staphylococcus aureus* also has thrombolytic properties. SK and SAK form 1:1 complexes with plasminogen, but there is no activation in the absence of fibrin.

A number of snake venoms contain plasminogen activators to exacerbate bleeding in their prey by digesting clots. Examples include TSV-PA from the Chinese green tree viper (*Trimeresurus stejnegeri*), LV-PA from the bushmaster pit viper (*Lachesis muta muta*), and Haly-PA from the Siberian pit viper (*Gloydius halys*).

The saliva of the vampire bat (*Desmodus rotundus*) has a highly potent plasminogen activator (DS-PA) to keep blood in a fluid state while it feeds. DS-PA has a higher affinity for fibrin than t-PA, so there is virtually no plasminogen activation without it and it was trialled as a potential therapeutic.

13.5.2 Inhibitors of plasmin

Similar to thrombin generation, the plasmin-generating potential of plasma is considerable and requires regulatory mechanisms to prevent build-up of a potentially lethal enzyme. The inhibitors affect either plasmin itself or its activators.

α_2-antiplasmin and α_2-macroglobulin

The primary inhibitor of plasmin is the serpin α_2-antiplasmin, which exerts its effect in two ways:

- formation of a stable 1:1 complex with plasmin, which completely inactivates the enzyme;
- retardation of fibrinolysis by masking lysine-binding sites on Glu-plasminogen and thus interfering in the binding to fibrin. Lys-plasmin(ogen), however, has a higher affinity for the fibrin surface and is less susceptible to this effect of α_2-antiplasmin.

The cross-linking of α_2-antiplasmin to the fibrin clot by FXIIIa is a mechanism to resist premature clot destruction. In plasma, the concentration of plasminogen exceeds that of α_2-antiplasmin, which is not the case in a fibrin clot, where the latter is concentrated as a result of the cross-linking. The small amount of plasmin generated during normal physiology is rapidly neutralized by α_2-antiplasmin, but under pathological conditions of extreme fibrinolytic activation (i.e. snake bite, streptokinase therapy) it can become swamped. This is where **α_2-macroglobulin** comes into play, acting as a scavenger inhibitor by forming 1:1 complexes with excess plasmin that retain weak enzymatic activity but are cleared rapidly by the liver.

Thrombin activatable fibrinolysis inhibitor

Also known as carboxypeptidase B, **thrombin activatable fibrinolysis inhibitor** (TAFI) reduces fibrin's ability to act as a cofactor by removing lysine-binding sites for plasminogen and t-PA. Interestingly, TAFI is activated by thrombin only when the thrombin is in complex with thrombomodulin.

Red blood cells

Red cells contain a fibrinogen receptor which can modify fibrin fibre structure and confer some resistance to fibrinolysis resulting from impaired plasminogen activation.

SELF-CHECK 13.19

What are the main mechanisms of plasmin inhibition?

Other plasmin inhibitors

Histidine-rich glycoprotein (HRG) regulates fibrinolysis due to its capacity to bind and block the high-affinity lysine-binding sites of plasminogen. Approximately 50% of circulating plasminogen is reversibly bound to HRG.

Lipoprotein (a) has striking structural homology with plasminogen and acts as a competitive inhibitor for binding sites on fibrin, fibrinogen, and t-PA.

Thrombospondin is a platelet α-granule constituent that has activity against plasmin.

13.5.3 Inhibitors of plasminogen activators

Activators of plasminogen are regulated by their own set of inhibitors. Plasminogen activator inhibitor type 1 (PAI-1) is principally synthesized and secreted by endothelial cells and is present in platelet α-granules and adipocytes (fat cells). Platelets are the major pool of PAI-1. Endothelial synthesis is upregulated by many compounds, including thrombin, lipoproteins, insulin, and pro-insulin. PAI-1 functions as a fast-acting inhibitor of t-PA, u-PA, and, to a lesser

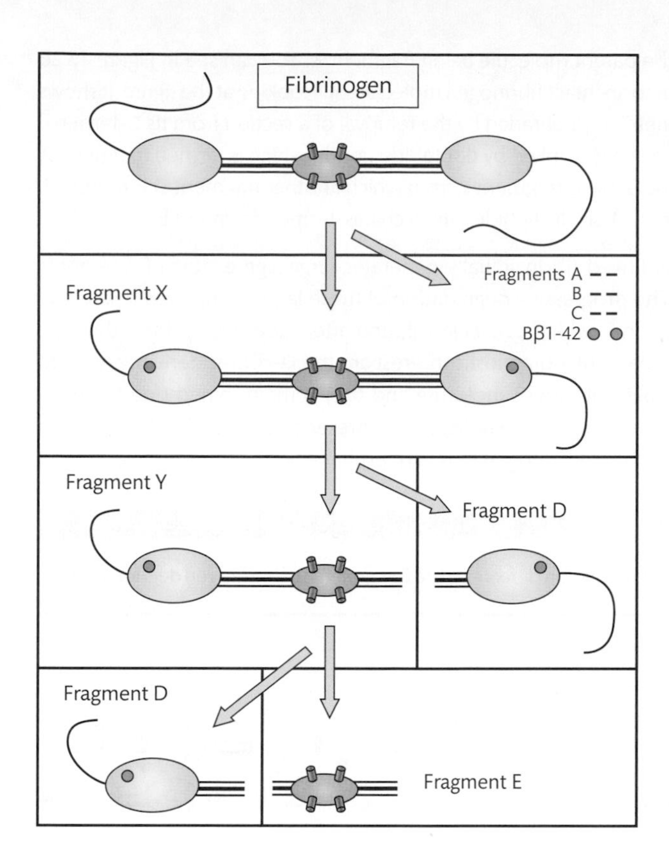

FIGURE 13.26
Fibrinogen degradation by plasmin. The first stage of the plasmin-mediated proteolysis of fibrinogen is the cleavage of fragments A, B, and C from the C-terminal of the Aα chains, followed by cleavage of the Bβ1–42 fragments from the Bβ chains. The remainder of the parent molecule is termed 'fragment X'. The next plasmin attack point leads to the removal of one D fragment; the remainder of the parent molecule is referred to as fragment Y. Removal of the second D fragment from fragment Y leaves the tethered N-termini of all six chains, fragment E.

extent, plasmin, all of which cleave the same bond in PAI-1 and are subsequently inhibited by the formation of a 1:1 complex.

Approximately 80% of plasma PAI-1 is in complex with t-PA; the remaining free PAI-1 is functionally active and stabilized by association with vitronectin. Plasma PAI-1 levels exhibit diurnal fluctuations, with an early morning peak that can halve by the afternoon.

Plasminogen activator inhibitor type 2 (PAI-2) exists in both an intracellular form and a secreted form that vary with respect to glycosylation. Very little PAI-2 is detectable in plasma, except during pregnancy, where it is produced by the placenta. This is because villus cells are the main source of PAI-2, so the presence of a villus-rich organ like the placenta results in increased plasma levels. It is more effective against u-PA than t-PA, but is less effective than PAI-1 against both activators. Its main roles are intracellular, where it can alter gene expression and influence the rate of cell proliferation and differentiation.

APC forms a tight complex with PAI-1 whereby it can no longer inhibit t-PA. It also has an indirect inhibitory effect on TAFI as it downregulates thrombin generation which, in turn, leads to reduced TAFI activation. Interestingly, protein C inhibitor (PCI), the APC inhibitor, is also referred to as **plasminogen activator inhibitor type 3** (PAI-3).

13.5.4 Fibrinogen and fibrin degradation by plasmin

Fibrinogen and fibrin are degraded in a sequential fashion by plasmin to soluble degradation products. Figure 13.26 shows that the first stage of fibrinogen proteolysis is the symmetrical removal of three small peptides from the C-terminal of the Aα chains, termed fragments A, B, and C. This is followed by cleavage of the first 42 amino acids of the Bβ chains generating the Bβ1–42 fragments, the remainder of the parent molecule being fragment X. You can see in Figure 13.26 that fragment X is little different to an intact fibrinogen molecule. Still looking at the figure, it shows that fragment X is then asymmetrically degraded by the removal of a section from its C-terminal end comprising parts of all three chains linked by disulphide bonds, which is termed fragment D. The remainder of the parent molecule is fragment Y, from which another fragment D is removed, leaving the N-terminal portion of all six disulphide-linked chains, termed fragment E.

Plasmin degradation of cross-linked fibrin initially generates large aggregates of fragments termed X and Y **oligomers**. The progressive degradation of these large complexes to the terminal derivatives, DD–E, DD, and E, which occurs in solution after their release from the clot, is depicted in Figure 13.27. Of particular diagnostic interest are the **D-dimer** derivatives. Note that each D fragment has a different parent molecule and they remain joined due to cross-linking, so the detection of D-dimers in plasma indicates the presence of an *in situ* clot, whereas detection of single D fragments from fibrinogen lysis can result from other conditions.

Cross reference

You will meet the use of D-dimers in the diagnosis of thrombotic disease in Chapter 17.

SELF-CHECK 13.20

How do the D fragments of fibrin degradation differ from the D fragments of fibrinogen degradation?

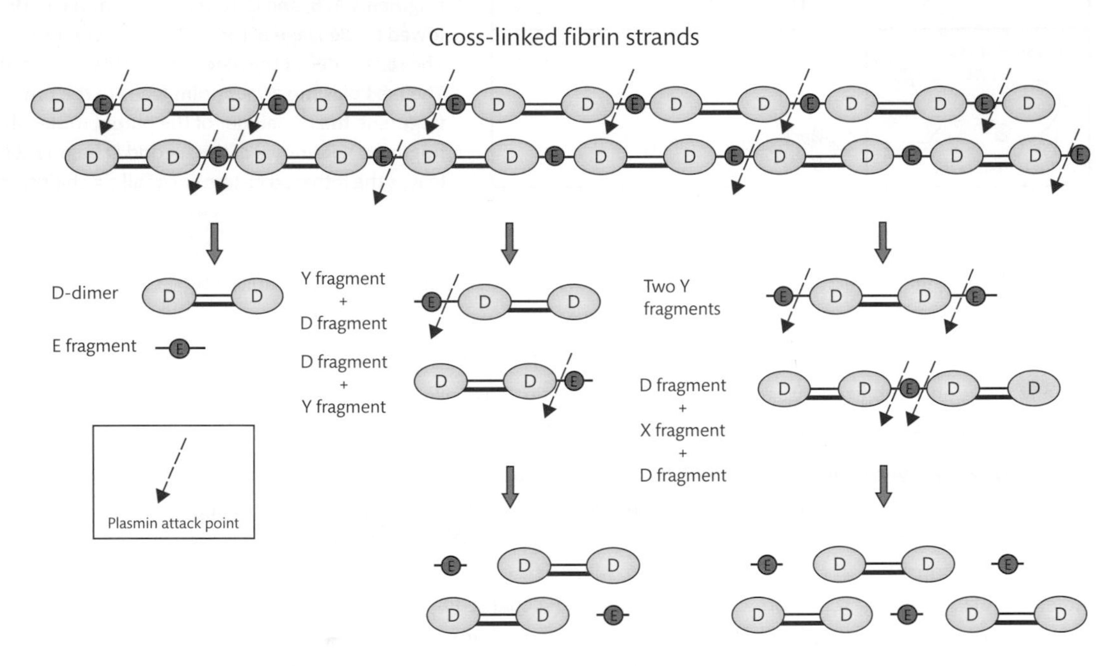

FIGURE 13.27

Fibrin degradation by plasmin. Plasmin degrades fibrin in a similar way to fibrinogen, although the presence of cross-links results in differing products. Degradation of fibrin initially generates large aggregates of fragments termed X and Y oligomers, which are further digested by plasmin to the terminal derivatives DD–E, DD, and E.

Key Points

Inflammation promotes the production of increased levels of a group of proteins called acute-phase reactants, such as the complement proteins. Many of the haemostatic proteins you have met in this chapter are acute-phase reactants, such as FVIII, VWF, fibrinogen, PAI-1, t-PA, and plasminogen.

Chapter summary

Roles

- Haemostasis exists to maintain blood fluidity and to limit and arrest blood loss resulting from injury.
- Mechanisms exist to remove clots once they have served their purpose.

Components

The components of haemostasis are:

- vascular integrity;
- platelets;
- blood coagulation;
- fibrinolysis;
- inhibitory mechanisms.

Primary haemostasis

- Damaged blood vessels expose structures that promote clotting mechanisms.
- VWF is immobilized on collagen fibres and facilitates the adhesion of platelets.
- Adhered platelets become activated and recruit more platelets via the release reactions and aggregation.

Secondary haemostasis

- Damaged blood vessels expose tissue factor, which initiates coagulation biochemistry via FVII/FVIIa to activate FIX and FX.
- Trace levels of thrombin are generated that activate FV, FVIII, and platelets before the pathway is shut down by TFPI.
- FVIIIa acts as cofactor to FIXa, which moves to the phospholipid surface of activated platelets to activate more FX, which, once activated and linked to cofactors, generates more thrombin.
- The thrombin back-activates the coagulation system and converts soluble fibrinogen to insoluble fibrin strands, which are stabilized by cross-linking and then intertwine between platelets to consolidate the clot.

Fibrinolysis

- t-PA activates plasminogen to plasmin once they have both bound to the fibrin surface.
- Plasmin degrades fibrin, generating soluble fragments.

Inhibitory mechanisms

- AT forms 1:1 complexes with thrombin and FXa.
- APC inhibits phospholipid-dependent reactions in the presence of PS and FV by cleaving FVa and FVIIIa.
- PAI-1 inhibits t-PA to regulate plasmin generation.
- α_2-antiplasmin inhibits plasmin.
- TAFI removes binding sites for plasmin on the fibrin surface.

Discussion questions

13.1 How do localization mechanisms contribute to effective haemostasis?

13.2 What might be the physiological advantages of having multiple-component enzyme systems for coagulation and fibrinolysis in preference to the direct generation of fibrin and plasmin?

13.3 Which came first, the coagulation factor or the inhibitor?

13.4 Platelets are the platform for haemostasis; discuss.

Further reading

- **Gomez K and McVey HJ. *Normal haemostasis*. In: Hoffbrand AV, Higgs DR, Keeling DM, and Mehta AB (eds). *Postgraduate Haematology*, 7th edn (pp. 676–98). Wiley Blackwell, Oxford, 2016.**
- **Kattula S, Byrnes JR, and Wolberg AS. *Fibrinogen and fibrin in hemostasis and thrombosis*. *Arteriosclerosis, Thrombosis and Vascular Biology* 2017: 37: e13–e21.**
- **Marder VJ, Aird WC, Bennett JS, Schulman S, and White II, GC (eds). *Hemostasis and Thrombosis: Basic Principles & Clinical Practice*, 6th edn. Lippincott Williams & Wilkins, Philadelphia, PA, 2012.**

  Detailed chapters on different components of haemostasis.
- **Michelson AD, Cattaneo M, Frelinger A, and Newman P (eds). *Platelets*, 4th edn. Elsevier Science, California/London, 2019.**

  Extensive coverage of platelets in health and disease, with coverage of laboratory techniques.
- **Patri B, Sahu AA, Pal S, and Chakravarty S. *The cell based model of in-vivo coagulation: A work in progress*. *International Blood Research and Reviews* 2016: 6: 1–10.**
- **Watson SP, Morgan NV, and Harrison P. *The vascular function of platelets*. In: Hoffbrand AV, Higgs DR, Keeling DM, and Mehta AB (eds). *Postgraduate Haematology*, 7th edn (pp. 699–714). Wiley Blackwell, Oxford, 2016.**

  Excellent detail on fibrinogen.

Answers to self-check questions and case study questions are provided as part of this book's online resources.

Visit www.oup.com/he/moore-fbms3e.

14

Bleeding disorders and their laboratory investigation

Gary Moore and David Gurney

In this chapter, you will be introduced to bleeding disorders that arise from hereditary or acquired abnormalities of coagulation factors, von Willebrand factor (VWF), platelets, and blood vessels. We will then look at the main laboratory tests available to biomedical scientists to identify and characterize bleeding disorders.

Learning objectives

After studying this chapter, you should confidently be able to:

- name a number of bleeding disorders and describe their causes;
- describe the principles and interpretation of coagulation screening tests;
- describe the principles of factor assays and raw data assessment;
- describe the principles of inhibitor screening and measurement;
- describe the laboratory investigation for von Willebrand disease and outline the subclassification;
- describe screening tests for primary haemostatic disorders;
- describe the principles and interpretation of platelet function analysis.

14.1 Bleeding disorders

Bleeding disorders occur when components of haemostasis are deficient to an extent that confers a tendency to bleed excessively. The phrase *deficiency* is used as an all-encompassing term relating to the biological activity of a given substance. Some patients make molecules with normal function, but the deficiency exists because the molecules are manufactured in lower amounts than normal. Other patients manufacture the molecules in normal amounts but with impaired function, so the biological activity is deficient even if the concentration of the molecule is not.

Patients with more severe disorders will experience bleeding episodes that begin in early childhood, often with no obvious precipitating event—these are called spontaneous bleeds. Those with mild disorders are more likely to exhibit excessive bleeding following trauma, surgery, or dental extraction. Common symptoms of bleeding disorders include the following:

- epistaxis (nosebleeds);
- gingival (gum) bleeds;
- bruising;
- purpura (skin haemorrhages appearing as purple spots or patches);
- petechiae (small skin haemorrhages appearing as minute red or purple spots);
- menorrhagia (heavy menstrual bleeding);
- joint bleeds leading to arthritic complications;
- muscle bleeds;
- chronic anaemia.

The type, site, and severity of haemorrhage can give important clues to whether a haemostatic disorder is present, and if so, the areas of haemostasis that are affected. Easy bruising, epistaxis, and menorrhagia are common, but unless persistent and severe, do not necessarily indicate a haemostatic abnormality. Broadly speaking, there are differences between the clinical manifestations of defects in primary and secondary haemostasis. In the former, bruises are small, epistaxis is common and often severe, bleeding from cuts and abrasions can be prolonged and profuse, and bleeding after dental extraction or surgery is immediate. Conversely, patients with factor deficiencies can present with large bruises at unusual sites, epistaxis is uncommon, bleeding from cuts and abrasions is not severe, and bleeding after dental extraction or surgery tends to be delayed. **Haematuria** is common in factor deficiencies, as is **haemarthrosis** in severe haemophilia, but they are rare in primary haemostatic disorders. Haemostatic abnormalities leading to bleeding disorders can involve any area of haemostatic function and can be hereditary or acquired.

haematuria
Presence of blood in urine.

haemarthrosis
Bleeding into joint spaces that can lead to joint damage and disability.

14.1.1 Hereditary bleeding disorders

Probably the most well-known are the haemophilias, yet the most common hereditary bleeding disorder is **von Willebrand disease (VWD)**, which you will meet later in this chapter.

Haemophilia

Deficiency of FVIII is referred to as **haemophilia A** and deficiency of FIX is **haemophilia B**, the latter sometimes referred to as Christmas disease, as it is named after the first boy described with it, Stephen Christmas. Both are X-linked disorders and so occur almost exclusively in males. You will remember from Chapter 13 that the function of FVIII and FIX are directly linked, FVIIIa being a cofactor for FIXa, so deficiencies of one or the other have virtually identical clinical signs and symptoms. Many different genetic mutations have been described that give rise to deficiencies of varying clinical severity and bleeding frequency, which are inversely correlated

TABLE 14.1 **Sub-classification of haemophilia A.**

Classified as	Severe	Moderate	Mild
FVIII level (IU/dL)	<1.0	1.0–5.0	>5.0
Age at presentation	Infancy	<2 years	>2 years
Bleeding symptoms and frequency	Frequent spontaneous bleeds into joints, muscles, and internal organs Severe bleeding after trauma	Much fewer spontaneous bleeds than patients with severe disease Minor trauma can precipitate a bleed	No spontaneous bleeds Bleeding after significant trauma/surgery
Approximate percentage of cases	50	30	20

BOX 14.1 *2nd century AD teachings on haemophilia*

The earliest known description of haemophilia and the recognition that it affects males can be found in the Babylonian *Talmud*, a book of Jewish laws, ethics, customs, and history. The *tana'im*, who were Rabbinic sages, taught that 'If she circumcised her first son and he died, and a second son and he died, she must not circumcise a third one.'

with the FVIII or FIX level of activity. Haemophilia A is subclassified according to FVIII activity, as shown in Table 14.1, which can also be applied to FIX levels in haemophilia B.

Haemophilia A occurs at a frequency of approximately 1:10,000 males, of which about 30% are spontaneous mutations where no family history of a bleeding disorder will be apparent. Haemophilia B has an approximate frequency of 1:50,000 males. (See also Box 14.1.)

A third haemophilia exists, **haemophilia C** (or Rosenthal's disease), which is due to a deficiency of FXI. It is an autosomal disorder of variable clinical severity where the bleeding manifestations do not correlate with FXI levels. It is predominantly, but not exclusively, found in patients of Ashkenazi Jewish heritage. About 8% of Ashkenazi Jews have FXI deficiency.

Haemophilia in females

Some carrier females have a sufficient reduction in FVIII/FIX to require treatment prior to invasive procedures or after major trauma. The main causes of markedly reduced FVIII/IX levels in females are listed below:

- extreme **lyonization** of the FVIII or FIX gene in a carrier (see also Box 14.2);
- hemizygosity (unpaired genes) of X chromosome with mutant FVIII/FIX gene (i.e. XO in Turner syndrome);
- female with Normandy variant VWD;
- true haemophiliac female (parents are a male haemophiliac and a female carrier who are often cousins);
- female with acquired haemophilia.

lyonization
The process by which one copy of the X chromosomes in a female is inactivated.

BOX 14.2 Lyonization

Only one X chromosome is necessary to generate sufficient levels of the gene products. Lyonization occurs to prevent females having twice the levels of males, who only possess one copy of the X chromosome. The inactivation is random in all mammals except marsupials, where it occurs only in the X chromosome derived from the male parent. A clear manifestation of lyonization is seen in tortoiseshell cats, which have black and orange alleles for fur colour on their X chromosomes. Every patch of orange fur occurs because the X chromosomes in those areas that carry the black allele have been inactivated, and vice versa. If a haemophilia carrier undergoes extreme lyonization of her X chromosomes carrying normal genes for FVIII/FIX, she can have sufficiently low levels to present with a symptomatic bleeding disorder.

SELF-CHECK 14.1

What are the main differences between haemophilias A, B, and C?

Von Willebrand disease

This disease was first described in 1926 by Dr Eric Adolf von Willebrand in a population on the Åland Islands, an archipelago province of Finland, and termed 'pseudo-haemophilia'. VWD is now known to be an autosomal bleeding disorder of variable clinical severity resulting from a variety of subtypes of VWF deficiency. It is present in about 1% of the population and classically presents as a mild-to-moderate bleeding disorder with symptoms such as epistaxes, bruising, excessive minor wound bleeding, heavy menstruation, and excessive, yet rarely life-threatening, bleeding after trauma/surgery. Prolonged bleeding after a first dental extraction is often the first manifestation. Because primary haemostasis (i.e. VWF, platelet, and vessel interactions) predominates in small vessels, patients with all but the most severe VWD do not suffer the crippling and painful joint and muscle bleeds seen in people with haemophilias A or B, which are more dependent on effective secondary haemostasis.

In addition to the effects of VWD on primary haemostasis, the reduction in VWF leads to a concomitant reduction in FVIII levels because there is less VWF to protect FVIII from proteolytic degradation in the plasma. In the most severe form of VWD, where VWF is absent, this leads to FVIII levels similar to those of moderate or severe haemophiliacs and exacerbates the clinical condition. However, the FVIII level does not fall to zero, as the patients have functional FVIII genes and continually produce enough FVIII to maintain a baseline of about 1.0–2.0 IU/dL. The full subclassification of VWD is detailed later in this chapter (section 14.2.6), when we consider its laboratory investigation.

SELF-CHECK 14.2

What are the two mechanisms by which a deficiency of VWF can contribute to bleeding symptoms?

Other clotting factor deficiencies

Deficiencies of other clotting factors are much rarer than VWD or haemophilias A and B. Deficiency of FVII has an estimated prevalence of 1:500,000, whilst deficiencies of fibrinogen, FV, FX, and FXI are each about 1:1,000,000. Prothrombin and FXIII deficiencies are rarer still, each occurring with a frequency of about 1:2,000,000. They all demonstrate an autosomal-recessive pattern of inheritance, with the exception of dysfunctional fibrinogens, which tend to be autosomal-dominant. Population frequencies show geographical differences, the recessive disorders being more common in countries where consanguineous marriages are frequent.

An interesting autosomal-recessive disorder presenting as a combined deficiency of FV and FVIII is known, with a prevalence of about 1:1,000,000. It is not, as one might expect, the result of dual inheritance of FV and FVIII mutations, but the deficiency of a chaperone protein for these factors that is involved in their intracellular transport known as **lectin mannose-binding protein 1 (LMAN1)**. The disorder can also be caused by deficiency of the cofactor molecule for LMAN1, **multiple coagulation factor deficiency 2 (MCFD2)**.

The clinical severity of these rarer disorders is variable. Most **dysfibrinogenaemias** (60%) are asymptomatic, whilst about 20% will have bleeding symptoms and 20% thrombotic episodes. **Hypofibrinogenaemia** is usually mild and often not detected until surgery or trauma. **Afibrinogenaemia** gives rise to a severe bleeding disorder. Prothrombin deficiency can be severe when levels are low and undetectable prothrombin levels are probably incompatible with life. Combined FV and FVIII deficiency is usually clinically mild (the levels are not as low as seen in haemophilia itself), as is isolated FV deficiency. The clinical severity of FVII deficiency is variable and does not correlate well with plasma levels, probably because only a small amount of FVIIa is necessary to trigger coagulation. Deficiencies of FX and FXIII tend to be more severe, with patients often suffering the bleeding into joints and muscles seen in haemophilias A and B.

dysfibrinogenaemia
Functional deficiency of fibrinogen.

hypofibrinogenaemia
Low concentration of functionally normal fibrinogen.

afibrinogenaemia
Absence of plasma fibrinogen.

Treatment

The main treatment is replacement therapy with concentrates of the deficient factor. Concentrates can be derived from multiple donations of human plasma, or manufactured utilizing recombinant technology. An advantage of recombinant FVIII and FIX products is that there is no risk of transmitting infectious agents such as hepatitis C or human immunodeficiency virus. Patients who are more severely affected are treated prophylactically, whilst others are treated when a bleed is recognized or suspected. Because FVIII has a half-life of 8–12 hours, most patients with severe haemophilia A need to self-administer their treatment intravenously about three to four times a week for their prophylaxis. The half-life of FIX is longer, at 18–24 hours, so patients with severe haemophilia B will inject themselves with FIX concentrate on average about twice a week as protection against spontaneous bleeding. Even with regular prophylactic factor concentrate infusions, which is demanding on the patients and their families, spontaneous bleeds can still occur that require additional treatment.

The most recent advance in haemophilia treatment has been development of modified FVIII and FIX concentrates known as extended half-life (EHL) products. A variety of methods are employed to extend time of clearance from the circulation to reduce the number of infusions required whilst maintaining equal clinical efficacy to existing concentrates:

- fusion of FVIII or FIX to a plasma protein with a longer half-life, such as albumin or the Fc fragment of immunoglobulin G;
- covalent attachment of one or more chains of polyethylene glycol (PEG), a non-toxic and non-immunogenic polymer, to a therapeutic molecule, which is termed PEGylation. Conjugation of PEG to FVIII or FIX prolongs their half-lives mainly by increasing solubility, protecting against enzymatic digestion and blocking interaction with clearance receptors;
- recombinant single-chain FVIII with a truncated B-domain (native FVIII is a two-chain molecule). This EHL product has improved stability and an increased affinity for VWF, the natural chaperone for FVIII.

You saw in Chapter 13 that FVIII circulates bound to VWF to protect it from proteolytic degradation, so clearance of FVIII occurs mostly in complex with VWF. The EHL FVIII products are also largely regulated by the interaction with VWF, so they only achieve moderate increases of half-lives, in the region of 1.5–2-fold as compared to unmodified FVIII. In contrast, the EHL FIX products achieve four–six-fold increases in half-lives.

A novel approach to treatment of haemophilia A is use of a recombinant, humanized bi-specific antibody called emicizumab that mimics cofactor activity normally provided by FVIIIa. Emicizumab is described as bi-specific because it is directed against both FIXa and FX, and works by bridging FIXa to FX to facilitate activation of FX by FIXa in the absence of FVIIIa. Unlike FVIII, emicizumab does not bind to VWF or phospholipid and does not require activation by thrombin. It is administered subcutaneously, has a half-life of 28 days, and can remain in circulation for months, so some patients do not even need treatment every week.

Treatment of VWD can involve concentrates, but some subtypes respond to a drug called **desmopressin**, which increases the release of VWF from endothelial cells.

14.1.2 Other primary haemostatic disorders

Other than VWD, primary haemostatic disorders occur as a result of a reduction in the numbers or function of platelets, which are covered later in this chapter, or the haemorrhagic vascular disorders.

Hereditary haemorrhagic telangiectasia (HHT) is an autosomal-dominant disorder characterized by fragile blood vessels, which are prone to bleeding as they cannot support the mechanisms of primary haemostasis. Type 1 HHT is caused by a mutation in **endoglin**, which is involved in cytoskeletal organization, and type 2 HHT by a mutation in **activin receptor-like kinase 1** (ALK-1), which controls the maturation phase of **angiogenesis**.

angiogenesis
Blood vessel formation.

Connective tissue disorders can result in fragile blood vessels, such as the collagen abnormalities seen in Ehlers–Danlos syndrome and osteogenesis imperfecta, or the elastic-fibre abnormality of Marfan syndrome.

Acquired disorders of haemostasis

A number of clinical situations exist where bleeding symptoms occur that are not due to hereditary haemostatic abnormalities but are secondary to other disorders, and are thus referred to as acquired bleeding disorders. The main causes of acquired bleeding disorders are summarized in Table 14.2 and are covered in detail in Chapter 16.

SELF-CHECK 14.3

What are the main distinctions between hereditary and acquired bleeding disorders?

14.2 Laboratory investigation of a suspected bleeding disorder

Diagnosis of a bleeding disorder is dependent on the presence of bleeding symptoms in the patient, yet compiling the bleeding history is subjective. Questionnaires are available where the examining clinician gives scores depending on the presence and severity of particular

TABLE 14.2 Acquired bleeding disorders.

Primary disorder	Acquired cause of bleeding
Sepsis, shock, obstetric calamities, trauma, surgery, some malignancies, transplant rejection, recreational drugs, snake bite	Acute disseminated intravascular coagulation (DIC) arising from depletion of platelets, coagulation factors, and fibrinogen
Some malignancies, chronic infections, chronic kidney disease, Kasabach-Merritt syndrome	Chronic DIC (if present, bleeding is far less severe than acute DIC)
Liver disease	Reduced synthesis of coagulation factors and thrombocytopenia
Renal disease	Uraemia and increased prostacyclin release impair platelet function
Vitamin K deficiency	Dietary deficiency or malabsorption reduces synthesis of vitamin K-dependent factors Newborns are vitamin K-deficient and can present with haemorrhagic disease of the newborn
Autoantibodies to coagulation factors or platelets	Reduction in affected coagulation factor or platelet numbers
Old age, prolonged steroid use, vitamin C deficiency	Compromised blood vessel integrity
Amyloidosis (extracellular protein deposition)	Impaired platelet function; amyloid binds FX causing plasma FX deficiency and can interfere with fibrin polymerization
Dilutional coagulopathy	Massive transfusion of stored blood products and/or volume replacement with blood substitutes (i.e. they do not contain coagulation factors and live platelets) can reduce platelet numbers and concentrations of circulating coagulation factors
Drug therapy	Anticoagulant therapy (see Chapter 17); various drugs impair platelet function (see later in this chapter)

symptoms, the so-called bleeding score, which can aid diagnosis and choice of laboratory investigations. When there is clinical suspicion of a bleeding disorder, characterization of the presence and nature of any defect(s) begins with the performance by biomedical scientists of screening tests that will indicate the area(s) of haemostasis affected. Table 14.3 lists the commonly used tests.

14.2.1 Coagulation screening

The investigation of platelets and fibrinolysis is discussed elsewhere, so we will now concentrate on **coagulation screening**. Before we consider the hows and whys of each test, we need to introduce the concept of the **cascade theory of coagulation**. This theory preceded the cell-based model that you saw in Chapter 13, and forms the basis of some screening test designs. Briefly, two separate enzyme pathways were considered to each generate FXa, which then began a third pathway culminating in fibrin generation. The coagulation factors in each pathway were considered to follow strict sequential order, one enzyme activating many more molecules of the next in sequence before being inactivated.

coagulation screening
A set of global tests to assess general coagulation factor status.

The cascade theory is presented in Figure 14.1, where you will notice some factors that were not part of the coagulation model you met in Chapter 13: FXII, prekallikrein (PK), and

TABLE 14.3 Screening tests for defects of haemostasis associated with bleeding.

Area of haemostasis	Tests	Result determinants
Platelets/vessels	Platelet count	Platelet numbers
	Blood film examination	Platelet morphology
	Platelet function screen	Platelet function; VWF
	Bleeding time	Platelet function; VWF; vessel abnormalities
Coagulation	Prothrombin time	Levels of factors II, V, VII, X, and fibrinogen; inhibitors
	Activated partial thromboplastin time	Levels of factors II, V, VIII, IX, X, XI, XII, PK, high-molecular-weight kininogen (HMWK), and fibrinogen: inhibitors
	Mixing tests	Distinguish between factor deficiencies and inhibitors
	Thrombin time	Level of fibrinogen
	Reptilase time	Level of fibrinogen
	Fibrinogen activity	Direct measurement of fibrinogen
Fibrinolysis	D-dimers	Direct measurement of fibrin degradation products (FDPs)
	Dilute clot lysis time, euglobulin clot lysis time, fibrin plate	Mainly t-PA, PAI-1, plasminogen, and α_2 antiplasmin

high-molecular-weight kininogen (HMWK). These are involved in the contact activation of the **intrinsic pathway**, where exposure to the collagen surface partially activates FXII, which becomes fully activated via PK and HMWK before it activates FXI. Deficiencies of the three 'contact factors' do not cause bleeding disorders. As you saw in Chapter 13, they play little part in coagulation *in vivo* and in fact have other roles related to fibrinolysis and inflammation.

In contrast to the cell-based model, the cascade theory does not explain why subjects with contact factor deficiencies are free of bleeding problems, or why the haemophilias are so severe when there is apparently another pathway that can generate FXa. It does not account for why FXI deficiency is less severe than deficiencies of FVIII and FIX when FIX apparently relies on FXIa for its activation. Similarly, the cofactors FV and FVIII are activated by thrombin, yet generation of the latter occurs after FV and FVIII activation in the cascade model.

SELF-CHECK 14.4

Name and outline the three discrete pathways of the cascade theory of coagulation.

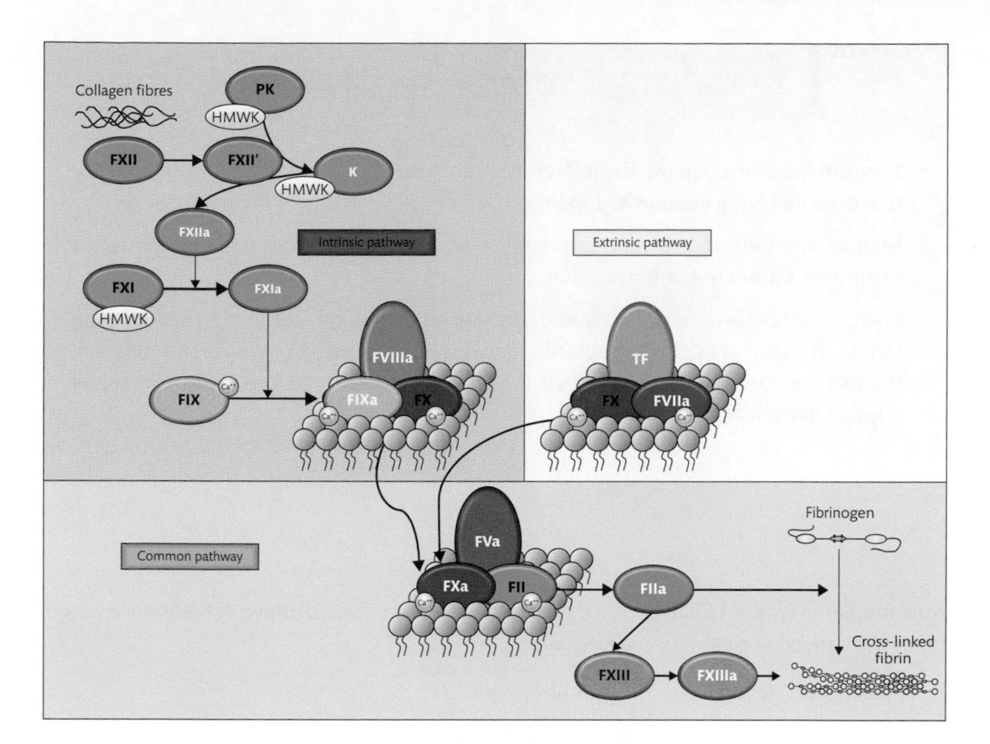

FIGURE 14.1

Coagulation cascade. The intrinsic pathway is activated by the contact of FXII with the negatively charged surface of collagen. This partially activates FXII to FXII', which becomes fully activated by kallikrein, which is generated via its precursor prekallikrein in the presence of high molecular weight kininogen. FXIIa activates FXI in the presence of high molecular weight kininogen. FXIa activates FIX, which combines with FVIIIa, calcium, phospholipids, and FX to form the intrinsic tenase complex, whose product is FXa. The extrinsic pathway involves FVIIa, tissue factor, calcium, phospholipids, and FX, which form the extrinsic tenase complex, which also generates FXa. The FXa from these pathways begins the common pathway by forming the prothrombinase complex in tandem with FVa, calcium, phospholipids, and FII. The product is FIIa (thrombin), which converts fibrinogen to fibrin, which is stabilized by thrombin-activated FXIII.

14.2.2 Screening tests

Coagulation screening tests are designed to isolate specific compartments of the cascade theory to give clues to where an abnormality exists. An hereditary or acquired abnormality of coagulation is termed a **coagulopathy**. You saw in Table 14.3 the coagulation factors that affect each test and we will now look at the design and use of each test.

Prothrombin time

The reagents detailed in Box 14.3 allow the patient's plasma to form a fibrin clot via the **extrinsic pathway** and **common pathway** and the time taken to clot is recorded as the **prothrombin time (PT)**.

The patient's PT, measured in seconds, is compared to a reference range which will vary depending on the analytical technique and type of thromboplastin used. (There are more details on

BOX 14.3 *Prothrombin time design*

- **Thromboplastin** contains tissue factor to 'activate' FVII, and also phospholipid for reactions involving vitamin K-dependent factors. (Which ones will operate here?)
- Manual methods add calcium ions subsequently, whereas automated methods use thromboplastins containing calcium.
- The intrinsic pathway is not activated, as there is no contact factor activator. Although FVIIa can activate FIX, thromboplastin reagents provide such a powerful stimulus that the FXa generated will form a clot via the common pathway before FIX can exert any appreciable effect.

thromboplastin types in Chapter 17.) If the clotting time is elevated above the reference range, it may indicate one or more of the following:

- deficiencies of factors II, V, VII, X, or fibrinogen;
- autoantibodies against the above clotting factors;
- anticoagulant drugs affecting the production of vitamin K-dependent factors (i.e. warfarin);
- anticoagulant drugs directly inhibiting FXa (i.e. rivaroxaban, apixaban, edoxaban);
- anticoagulant drugs directly inhibiting thrombin (i.e. dabigatran, argatroban);
- vitamin K deficiency;
- liver disease;
- disseminated intravascular coagulation (DIC);
- lupus anticoagulants, albeit rarely (see Chapter 15).

SELF-CHECK 14.5

Describe the principle of the PT, and the causes of elevated clotting times with this test.

Activated partial thromboplastin time

The reagents detailed in Box 14.4 allow a patient's plasma to form a fibrin clot via the intrinsic and common pathways—the time taken to clot, measured in seconds, is recorded as the **activated partial thromboplastin time (APTT)**.

The patient's APTT is compared to a reference range, and if the clotting time is elevated above the reference range, it may indicate one or more of the following:

- deficiencies of factors II, V, VIII, IX, X, XI, XII, PK, HMWK, or fibrinogen;
- some subtypes of VWD (due to associated FVIII deficiency);
- autoantibodies against the above clotting factors;

BOX 14.4 *APTT design*

- **Patient plasma is incubated for a set time period, typically two to five minutes, with a contact activator to activate FXII and begin the intrinsic pathway. Commonly used contact activators are kaolin, silica, and ellagic acid.**
- **Although not a significant reaction *in vivo*, FXII is used in this way to activate FXI independently of thrombin generated via the extrinsic pathway. Similarly, the activation of FIX and FX is independent of FVII.**
- **The reagent also contains a 'partial' thromboplastin comprising phospholipids for reactions involving vitamin K-dependent factors (which ones will operate here?) but no tissue factor, thereby preventing participation of the extrinsic pathway.**
- **Incubation with activator only allows the intrinsic pathway to proceed as far as FXIa generation. After the incubation period, calcium ions are added to allow coagulation to proceed to completion.**

- anticoagulant drugs affecting the production of vitamin K-dependent factors, although the APTT is less affected than the PT because there are more non-vitamin K-dependent factors contributing to the clotting time;
- anticoagulant drugs directly inhibiting FXa (i.e. rivaroxaban, apixaban, edoxaban);
- anticoagulant therapy with heparin;
- anticoagulant drugs directly inhibiting thrombin (i.e. dabigatran, argatroban);
- vitamin K deficiency;
- liver disease;
- DIC;
- lupus anticoagulants.

SELF-CHECK 14.6

Describe the principle of the APTT and causes of elevated clotting times with this test.

Cross reference

The end point of clotting-based assays like PT and APTT is a fibrin clot. You saw in Chapter 2 the different techniques available to detect the fibrin clot so that time to clot formation can be determined.

Result interpretation We can narrow down the *in vitro* compartment of coagulation from where a deficiency involving a single factor exists based on whether one or both of the PT and APTT results are abnormal:

PT and APTT elevated:	deficiency exists in the common pathway
PT only elevated:	FVII deficiency
APTT only elevated:	deficiency exists in the intrinsic pathway.

Box 14.5 describes how some rare subtypes of FVII and FX deficiencies can generate unusual screening test results.

BOX 14.5 *Misbehaving molecules*

Some dysfunctional defects of FVII will generate markedly discrepant results, depending on the type of thromboplastin used in the assays. Patients homozygous for FVII Padua, FVII Nagoya, and FVII Tondabayashi will generate markedly elevated PT and markedly reduced FVII activity if tested with rabbit brain-derived thromboplastin, mildly elevated PT and mildly to moderately reduced FVII activity if tested with human recombinant thromboplastin, and normal results with bovine thromboplastin. Some patients are asymptomatic, whilst others have a mild bleeding tendency.

Similarly, some dysfunctional defects of FX will vary in their laboratory presentation, depending on the assay type used. So-called Friuli-like molecules will be abnormal with all assays except those employing Russell's viper venom to activate FX *in vitro*. Padua-like defects tend to react only, or predominantly, in assays based on extrinsic-Xase (i.e. PT), whilst Melbourne-like defects will only or predominantly manifest in assays based on intrinsic-Xase (i.e. APTT). Other defects will be abnormal in all assays except chromogenic FX activity measurement.

These misbehaving molecules are rare subtypes of already rare disorders and should be considered only when standard testing has not identified any abnormalities.

Mixing tests

Factor deficiency can result from a reduced production/increased clearance of the factor, or the presence of antibodies that interfere with function (i.e. inhibitors). If you mix an equal volume of normal plasma with plasma from a patient who has a factor deficiency and repeat the screening test that was initially abnormal, the missing factor will be replaced by that present in the normal plasma and the result will correct into the reference range. However, if the patient has an **inhibitor**, it will interfere with the function of the normal plasma as well as the patient's plasma, and the result will not correct. These simple follow-up tests assist biomedical scientists and medical staff in deciding on the next stages of the diagnostic process. We will meet assays for specific factors and inhibitors later in this chapter, and the *in vitro* inhibitors known as lupus anticoagulants are covered in Chapter 15. A word of caution: some inhibitors are time-dependent and may not manifest without prolonged incubation at 37°C. There are more details about them later in this chapter.

Key Points

The diagnostic process involves marrying clinical and laboratory data.

Thrombin time and reptilase time

The **thrombin time (TT)** is a simple test, whereby thrombin is added to patient plasma to directly convert fibrinogen to fibrin and the time to clot measured in seconds. The patient's TT is compared to a reference range, and if the clotting time is elevated above the reference range, it may indicate one or more of the following:

- hypofibrinogenaemia—acquired causes such as DIC are more commonly encountered than hereditary forms;
- dysfibrinogenaemia—can be hereditary or acquired (i.e. liver disease);

CASE STUDY 14.1 Screening tests in a patient with a factor deficiency

Patient history

- A 14-month-old boy presented with an intramuscular haematoma after falling over in a playground.
- He had no history of bleeding from minor cuts or abrasions.
- There was no family history of bleeding disorders.
- Coagulation screening tests were performed.

Results 1

Index	Result	Reference range
PT (s)	11.9	(RR: 10.0–14.0)
APTT (s)	102.3	(RR: 32.0–42.0)
APTT 1:1 mixing test (s)	38.0	
TT (s)	9.3	(RR: 9.0–11.0)
Fibrinogen (g/L)	3.8	(RR: 1.5–4.0)
Platelet count (× 10^9/L)	250	(RR: 150–400).

Significance of results 1

The PT was normal and excluded deficiencies or inhibitors of factors in the extrinsic and common pathways, and thus, vitamin K deficiency too. It also excluded the presence of anticoagulant drugs that affect PT. The fibrinogen level was normal and the normal thrombin time (TT) additionally excluded the presence of anticoagulant drugs that affect TT. Therefore, the markedly elevated APTT was not due to anticoagulant drugs. The platelet count was normal, although that does not exclude a functional defect. In view of the presentation of a severe bleed in a young male child after relatively minor trauma, and normalization of the APTT upon mixing with normal plasma, a deficiency of a coagulation factor in the intrinsic pathway was suspected. As the child was male, one of the X-linked haemophilias was most likely, but the main factors of the intrinsic pathway were all measured.

Results 2

Index	Result	Reference range
FVIII (IU/dL)	135	(RR: 50–150)
FIX (IU/dL)	2.0	(RR: 50–150)
FXI (IU/dL)	85	(RR: 60–140)
FXII (IU/dL)	121	(RR: 50–150)

Significance of results 2

The FIX level was extremely low, which combined with the sex of the child, and age and severity of bleeding at presentation, was sufficient to diagnose moderate haemophilia B. The normal FVIII excluded haemophilia A and the more severe forms of VWD, and the normal FXI excluded haemophilia C. Although we know that FXII deficiency does not cause a bleeding disorder, it could have been the cause of the child's elevated APTT and it would have been necessary to look elsewhere for the cause of his bleeding. See also Box 14.6

- afibrinogenaemia;
- anticoagulant therapy with heparin;
- anticoagulant drugs directly affecting thrombin (e.g. dabigatran);
- hypoalbuminaemia, amyloidosis, paraproteins, and elevated levels of fibrin/fibrinogen degradation products (FDP) can interfere with fibrin polymerization.

In clinical practice, there are occasions when the laboratory is inadvertently not informed that a patient is receiving therapeutic heparin and the biomedical scientist is presented with an

BOX 14.6 *Haemophilia B variants*

Haemophilia B Leyden is an interesting variant characterized by severe FIX deficiency at birth that is gradually ameliorated from the onset of puberty—probably by the action of testosterone on an androgen-responsive element in the FIX promoter.

The results in Case study 14.1 are typical for a patient with haemophilia. Mutations causing haemophilia Bm result in a FIX molecule with an inhibitory effect on the activation of FX by FVIIa/tissue factor, which is detectable *in vitro* when bovine thromboplastin is used for PT analysis, leading to an elevated PT as well as APTT. The 'm' stands for Martin, the surname of the family in which the variant was first described.

unexpected elevated TT accompanied by an elevated APTT but normal PT. One way to confirm that the results are due to heparin therapy is to perform a **reptilase time (RT)**. The RT is similar to the TT except that the reagent is not thrombin but a thrombin-like enzyme purified from the venom of the common lancehead pit viper snake (*Bothrops atrox*). Heparin anticoagulates by markedly potentiating antithrombin, but because reptilase is a reptile enzyme it is unaffected by human antithrombin, so the RT will be normal. RT can be elevated for the reasons stated above for TT, except for anticoagulant drugs directly affecting thrombin, and of course, heparins. (See also Case study 14.2.)

Since heparin is administered by continuous intravenous infusion, blood samples are often taken from the line in the vein so that the patient does not have to receive a separate venepuncture. Unless the line is properly flushed to remove heparin, blood samples can be contaminated such that the APTT, TT, and even PT (which is rarely affected by therapeutic levels of heparin), are incoagulable. A normal RT in this circumstance would indicate heparin contamination and a repeat sample should be requested. Alternatively, reagents are available that neutralize heparin prior to performing any assays, such as heparinases, protamine sulphate, and hexadimethrine bromide.

Interestingly, the snake venom enzyme only removes fibrinopeptide A from fibrinogen, in contrast to thrombin, which additionally removes fibrinopeptide B. Discrepancies between TT and RT can sometimes indicate a dysfibrinogenaemia if the mutation is in the Bβ chain, as the TT will be abnormal but the RT normal.

SELF-CHECK 14.7

What are thrombin time and reptilase time used for in clinical laboratories?

Direct assays of fibrinogen

In the **Clauss fibrinogen assay**, plasma is diluted (usually one part plasma to nine parts buffer) to dilute out interfering substances like heparin and fibrinogen degradation products. The diluted plasma is clotted with a strong concentration of thrombin so that the clotting times are independent of thrombin concentration over a wide range of clotting times. In view of the thrombin concentration, the dilution of the plasma also prevents clotting times being too short. The clotting time is read off a standard curve and converted to fibrinogen concentration. If the fibrinogen concentration is sufficiently high or low that the clotting time is beyond the linearity limits of the standard curve, the plasma is rediluted to bring the clotting time into the measurable limits and the dilution taken into account in the calculation of the final result. This means that a plasma with a low fibrinogen level may need a 1:5 or 1:2 dilution, resulting in less dilution of interfering factors, which can then generate an erroneous result that underestimates the fibrinogen level.

Automated analysers using photo-optical clot detection can derive a fibrinogen level from the patient's PT data. The optical density changes for a range of dilutions of a plasma with a known fibrinogen level are used to generate a standard curve. The optical density change from the patient's PT curve is read off the standard curve to produce the fibrinogen value. This derived fibrinogen assay is simple and ostensibly free because the result is derived from the raw data of another test. However, there is wide variation in results, depending on the thromboplastin and analyser pairing used. The optical clarity and fibrinogen level of the standard can affect accuracy, many methods generating higher results than Clauss assays.

Immunological methods such as enzyme-linked immunoabsorbent assay (ELISA) can be used to measure the total amount of fibrinogen antigen present, irrespective of function. Dysfibrinogenaemias will have lower functional results than antigenic results, whereas they will be concordant in patients with hypofibrinogenaemia.

Case study 14.2 shows how fibrinogen, TT, and RT assays can contribute to interpretation of PT and APTT.

SELF-CHECK 14.8

What tests can be used to assess fibrinogen levels?

BOX 14.7 *A false sense of security?*

It is important to remember that a coagulation screen (+ platelet count) will not identify all haemostatic abnormalities. A patient's positive clinical history of bleeding but normal coagulation screen and platelet count may be due to the following:

- **bleeding not due to a haemostatic disorder (e.g. mechanical);**
- **disorders of platelet function;**
- **vascular disorders;**
- **FXIII deficiency—stabilization by cross-linking is not necessary for clots to form and persist *in vitro*;**
- **mild factor deficiencies, which may not manifest because reagents tend to be sensitive to deficiencies in the region of 30–40% of normal and below, although this varies from one reagent to another. In some instances of FII, FV, and FX deficiencies, this can result in *either* PT *or* APTT being elevated, which is not the expected laboratory presentation of a common pathway deficiency. For most patients, mild factor deficiencies are of little clinical significance;**
- **milder forms of VWD, which may not reduce FVIII levels sufficient to prolong an APTT;**
- **fibrinolytic disorders.**

CASE STUDY 14.2 Grossly abnormal screening tests

The following results were obtained on a sample of acceptable quality for analysis. No clinical details were available.

Index	Result	Reference range
PT (s)	> 120	(RR: 10.0–14.0)
APTT (s)	> 120	(RR: 32.0–42.0)
TT (s)	> 120	(RR: 9.0–11.0)

Further investigations to explain the deranged results showed the following:

Index	Result	Reference range
Fibrinogen (g/L)	5.9	(RR: 1.5–4.0)
RT (s)	13.3	(RR: 12.0–15.0)
Platelet count (× 10^9/L)	321	(RR: 150–400)

What do the above follow-up results suggest? If, however, the follow-up tests had given the following results what would they indicate?

Index	Result	Reference range
Fibrinogen (g/L)	0.2	(RR: 1.5–4.0)
RT (s)	> 120	(RR: 12.0–15.0)
Platelet count (× 10^9/L)	25	(RR: 150–400)

Key Points

All coagulation screening tests are performed at 37°C.

14.2.3 Pre-analytical variables

Unless care is taken to ensure that samples received in the laboratory are of suitable quality for analysis, performing the tests can generate results that are useless or even dangerous to the patient. A vital role of the biomedical scientist is ensuring that samples are fit for purpose, and Table 14.4 details the main areas that impact on sample quality.

14.2.4 Isolating a factor deficiency

If coagulation screening suggests a **factor deficiency**, assays can be performed to determine the plasma concentration of individual factors. The choice of factors to assay will be informed by the patient's clinical history and which of the screening tests was abnormal.

TABLE 14.4 Pre-analytical variables in haemostasis analysis.

Variable	Effect on results
Phlebotomy	Venous samples should be used whenever possible. Capillary samples can be contaminated/diluted with tissue fluids and may require modified analytical techniques. Stress and exercise prior to venepuncture can increase FVIII, VWF, PAI-1, and t-PA. Venous occlusion with a tourniquet can activate platelets, some clotting factors, and fibrinolysis. Difficult/traumatic venepuncture can activate some clotting factors and lead to shortened PT and/or APTT results that may mask an abnormality. Platelets may also be activated. Some samples may also contain small clots, or be completely clotted, and are unsuitable for analysis. Not checking patient identity can mean taking a good-quality sample, but from the wrong patient.
Sample collection tube with correct anticoagulant	Trisodium citrate (usually 105 mmol/L) is the anticoagulant of choice because its removal of calcium is reversible. Heparin and EDTA directly inhibit coagulation and are unsuitable for coagulation testing.
Sample volume	Trisodium citrate is a liquid anticoagulant, so there is a dilution factor. Nine parts blood is taken into one part citrate. If too much blood is put in the tube, it will be under-anticoagulated and clot quicker when tested. If the sample is underfilled, it will be over-anticoagulated, which may generate erroneously elevated clotting times and falsely suggest an abnormality.
Sample age	Components of haemostasis deteriorate quite rapidly and coagulation screens should be performed within four hours of collection. FV and FVIII are particularly labile. Plasma for most other tests can be stored frozen until analysis. Platelet function analysis can only be undertaken on fresh blood while the platelets are still 'alive'.
Centrifugation	Insufficient centrifugation can result in platelets remaining in the plasma and compromise standardization of phospholipid composition in PT and APTT. Lupus anticoagulant testing is particularly reliant on platelet removal.
Properties of plasma	Haemolysed samples can activate coagulation factors and the colour can interfere with photo-optical clot detection. Similarly for the strong colour of icteric samples. Lipaemic samples can interfere with turbidity recognition in photo-optical clot detection. It can also increase viscosity and interfere with mechanical clot detection.
Interfering substances/contamination	PT and APTT are used to monitor therapeutic anticoagulation but it can interfere with other tests that attempt to isolate specific components based on the assumption that the patient's coagulation is otherwise normal, e.g. lupus anticoagulant testing, activated protein C resistance screening. Other drugs can interfere with specific tests, e.g. aspirin affects platelet function.
Storage	Frozen plasma can be stored between −40°C and −70°C for some weeks without significant loss of activities. Samples cannot be repeatedly frozen and thawed as this leads to progressive loss of activities.

Irrespective of whether a deficiency is due to a dysfunctional molecule or reduced synthesis/increased clearance of a normally functioning molecule, assays assessing biological function in coagulation-based systems will indicate a deficiency. Immunological methods do not detect dysfunction.

As well as investigating patients with a bleeding history, coagulation screens are commonly performed prior to surgery to check that coagulation status is sufficient to cope with the trauma of surgery. The unexpected and incidental preoperative finding of abnormal results may result in cancellation of the surgery until the cause is isolated.

SELF-CHECK 14.9

If a young male patient presented with an intramuscular bleed and the only abnormal screening test was his APTT, which two clotting factors would you assay first and why?

One-stage coagulation factor assays

Bioassays for coagulation factors are usually performed with **one-stage coagulation assays**, so called because the fibrin clot endpoint is generated directly with one set of reagents. The assays use PT or APTT reagents to facilitate *in vitro* coagulation of manipulated patient plasma, as detailed in the Method box below and in Figure 14.2.

Assays for factors II, V, VII, and X are usually performed using PT reagents to facilitate the coagulation reactions. Factors VIII, IX, XI, XII, PK, and HMWK are assayed with APTT reagents.

Assessment of raw data

The plots for the patient and standard, and any controls, should form parallel straight lines. A patient's line that is to the right of that of the standard (i.e. the clotting times are longer at each dilution) indicates that the level of factor being measured is lower than the standard value, and if to the left, it is higher.

To calculate the actual result, the 1:10 dilution of the standard plasma is assigned a potency of 100% because that is the lowest dilution used and represents 100% of calibrator in this

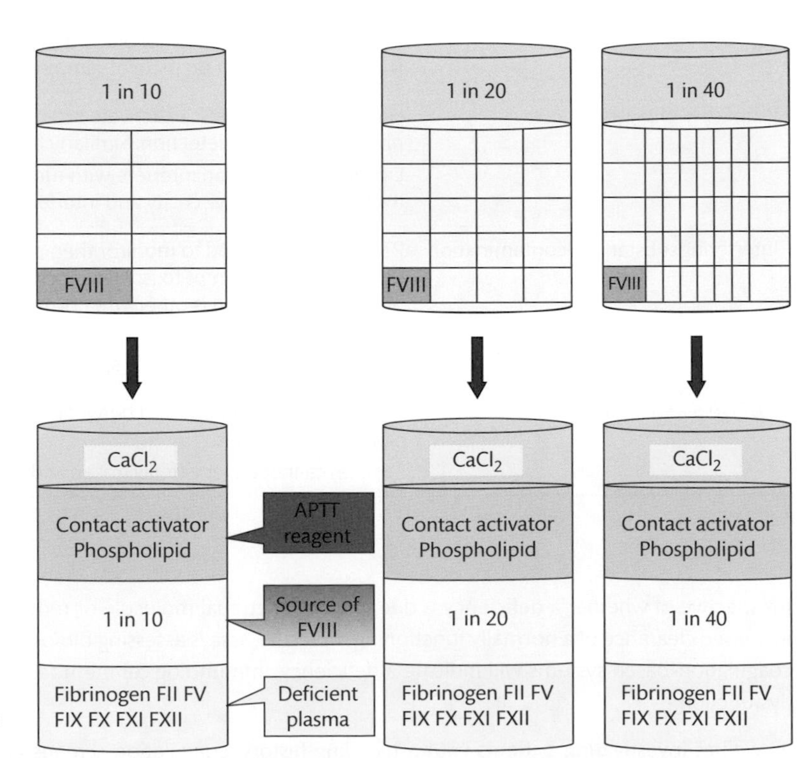

FIGURE 14.2
Principle of a one-stage FVIII assay. Test plasma is doubly diluted to gradually reduce the amount of FVIII present. One volume of each dilution is separately added to an equal volume of FVIII-deficient plasma and an APTT performed on each mixture. Full details are given in the Method box below.

METHOD One-stage factor assays

- At least three doubling dilutions are made of patient plasma in buffer, typically 1:10, 1:20, and 1:40.
- In separate tubes, one volume of each dilution is added to an equal volume of a plasma that is totally deficient in the factor being tested, but otherwise normal.

Manipulating the patient plasma like this means that the only source of the factor being assayed is the patient's plasma, so it becomes rate-limiting.

The levels of all the other coagulation factors will be constant in each tube so the clotting time of the 1:40 will be longer than the 1:20, which in turn will be longer than the 1:10 as a direct result of diluting the patient's plasma. Although all other coagulation factors will be present in the patient plasma, the dilution renders their contribution to coagulation capacity negligible in relation to that supplied by the deficient plasma.

- Clotting times are plotted against dilution factor on semi-log or double-log graph paper and compared to those of a standard plasma, as shown in Figure 14.3.

diluted system. Thus, the 1:20 and 1:40 become 50% and 25%, respectively, and so on if further dilutions are used. This means that if a 1:5 dilution is added, it adopts a value of 200%. Similar to the Clauss fibrinogen assay you met earlier, the standard curves are only linear within a certain range of clotting times and can vary from one batch of deficient plasma to another. Therefore, if the clotting times from a patient's dilutions are beyond those of the standard curve, the patient must be re-assayed using dilutions that generate clotting times within that range.

You can see in Figure 14.3 that the 1:10 dilution of the patient's plasma is read directly from the curve, the 1:20 is multiplied by 2 as it is the 50% dilution, and the 1:40 multiplied by 4 as it is the 25% dilution. Providing that the results for each of the patient's dilutions are within 10% of each other, the mean is calculated to generate the potency of the patient's factor.

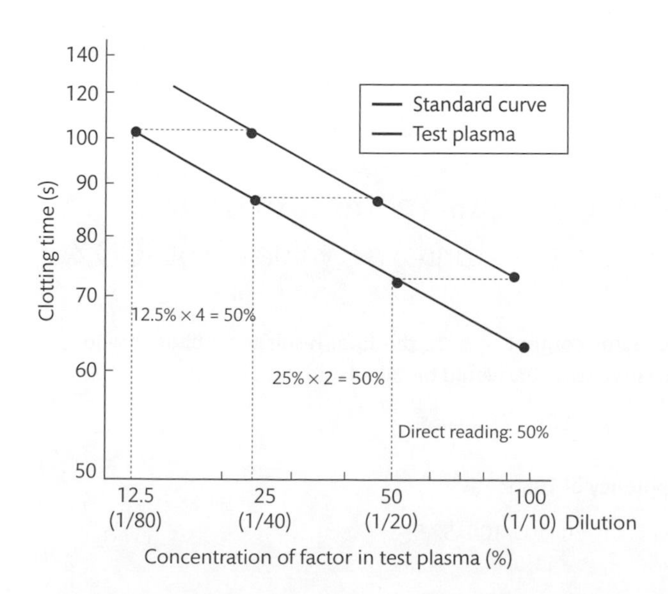

FIGURE 14.3
Graphical plot of standard and test plasma for a one-stage factor assay, a so-called 'parallel-line bioassay plot'.

Although we arbitrarily assign a value of 100% to the standard when plotting the data, its actual value may be more or less than that, which means that all we have done so far is assess the potency of the patient's factor as a percentage of the standard. If the actual value of the standard is indeed 100%, no further calculations are necessary. If the actual value is different, a further calculation is necessary, as shown in the Method box below.

Units

Plasma concentration of coagulation factors varies widely, so rather than measure factor levels in concentration units such as g/L, we report them in units that indicate enzyme potency relative to normality.

So, what exactly do we mean when we report a factor assay result as a percentage, that is to say, a percentage of what? Before the availability of standards with potency assigned as enzyme units, factor levels were reported merely as a percentage of the potency of a pooled normal plasma. Because the pools are prepared from a large number of healthy donors who would have a range of potencies of the factor being measured, the pools were assumed to have a potency of 100%, i.e. the theoretical mean.

Standards are now available with potency assigned as (enzyme) u/mL or u/dL, which have a theoretical equivalence to percentage values, thus: 100% of a given factor is considered equivalent to 1.00 u/mL or 100.0 u/dL.

International standards, or secondary standards calibrated against them, are available for most factors, the units designated as IU/mL or IU/dL.

Assay validity

The reason we assay the patient's plasma at more than one dilution is because there are situations where the line is not parallel to the standard. The main reasons why this occurs are:

- presence of inhibitors;
- low but detectable levels;
- total absence of the factor being measured.

Look at Figure 14.4, which illustrates typical results in the presence of an inhibitor. At the lowest dilution of 1:10, the inhibitor interferes with coagulation and significantly prolongs the clotting

METHOD Calculating potency of a test sample where the standard's value is not 100%

If the standard had a true factor content of 92%, the final result for a patient whose potency from the standard curve was 50% would be calculated thus:

$$50 \times 92/100 = 46\%$$

If the standard had a true potency of 105%:

$$50 \times 105/100 = 52.5\%$$

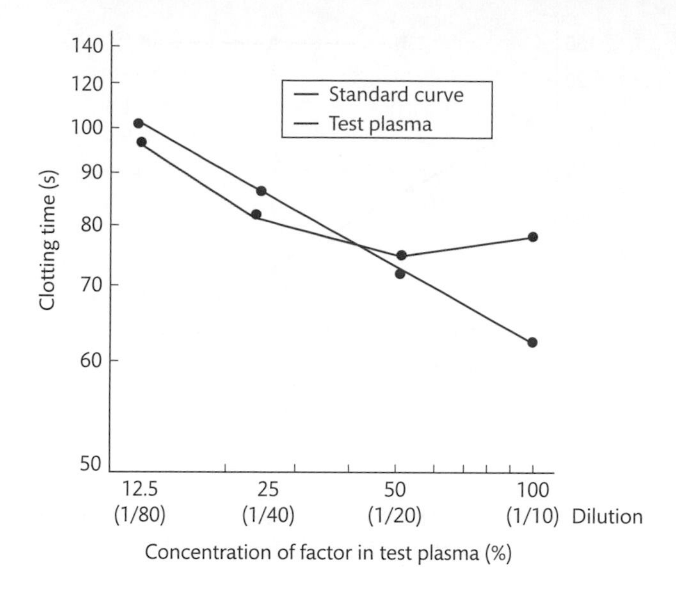

FIGURE 14.4

Graphical plot of standard and test plasma for a one-stage factor assay where the patient has an inhibitor that interferes with linearity and parallelism at low dilutions.

time. The inhibition remains in the 1:20 dilution, although you can see that the effect is less marked because the point on the graph is much closer to the standard line and the point for the 1:40 is below the standard. The higher the dilution, the greater the loss of inhibitory effect, so the calculated factor potency gets higher with each dilution. Notice that this assay included a 1:80 dilution, which, when married with the 1:40 dilution, generated a linear line parallel to the standard. This indicates that the inhibitor had been diluted to such a degree that it no longer interfered and the values from those dilutions can be used to report an accurate result.

The graph in Figure 14.5 has achieved linearity and parallelism at lower dilutions but not in the higher dilution. This is because the factor being measured is present but at a low concentration, so by the time it is diluted 1:40 it is virtually undetectable and linearity is lost. Performing the assay again using lower dilutions (e.g. 1:5) should provide a linear and parallel assay to allow calculation of the coagulation factor level. An apparent decrease in relative potency as samples are further diluted may also indicate an activated sample.

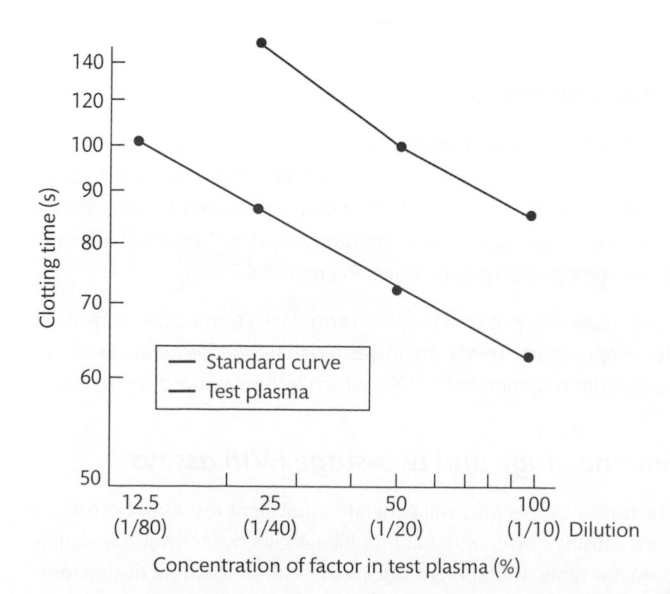

FIGURE 14.5

Graphical plot of standard and test plasma for a one-stage factor assay where a patient has a low level of the clotting factor, such that parallelism and linearity are lost at the higher dilution because the factor has been diluted to an undetectable level.

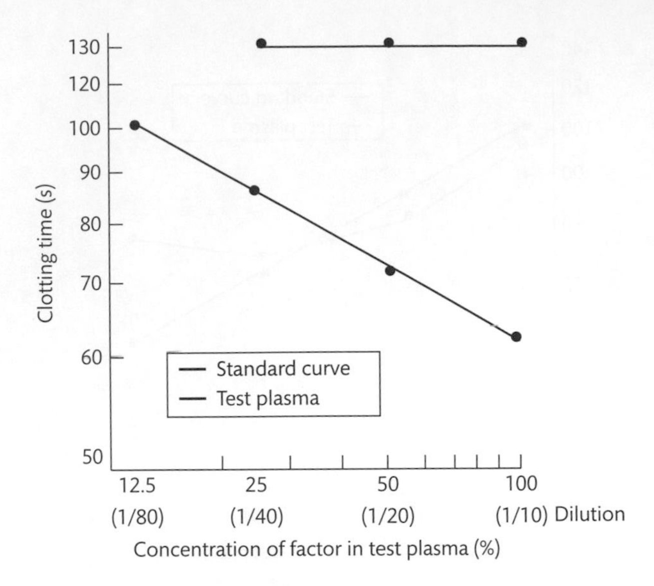

FIGURE 14.6
Graphical plot of standard and test plasma for a one-stage factor assay where a patient has undetectable levels of the factor being assayed.

The plot in Figure 14.6 is interesting because it is linear but not parallel. This is because the factor being measured is absent, so whatever dilution is used, the clotting time is the same because each one is ostensibly a blank. This could also be due to an extremely potent inhibitor.

Factor assays can be performed manually or on automated analysers. Some of the latter perform all the dilutions and calculations for you and display the parallel line graph, allowing the operator to assess validity. Others will only assay a single dilution of patient plasma and/or not display the graph. You can see from Figures 14.4 and 14.5 that if the mean result from each dilution was calculated without reference to the graphical display for assay validity, erroneous results would be reported with the potential for inaccurate diagnosis and inappropriate treatment. *Scrutiny of factor assay graphs is a crucial part of the analytical process.*

SELF-CHECK 14.10

Outline the principle of a one-stage clotting assay to measure FIX.

Two-stage coagulation factor assays

Also available for the measurement of FVIII are **two-stage factor assays**. The first stage involves the generation of FXa in a FVIII-dependent manner. In the second stage, the FXa is either added to normal plasma, phospholipids, and calcium ions to form a clot, or allowed to react with a chromogenic substrate. Clotting-based two-stage assays are rarely used as they are technically demanding. The design for the chromogenic assay is shown in Figure 14.7.

An occasional problem with one-stage assays is that they are sensitive to sample pre-activation that can result from a difficult venepuncture. This is circumvented in two-stage assays as all the patient's FVIII is pre-activated in order to generate the FXa, which is then assayed separately.

Discrepancies between one-stage and two-stage FVIII assays

In the absence of interfering factors, most patients will generate equivalent results in both assay types. However, some mutations causing non-severe haemophilia A give rise to clinically significant discrepancies between the assay types. Mutations reducing the stability of the activated molecule give rise to two-fold, or higher, results from one-stage compared with two-stage assays, some

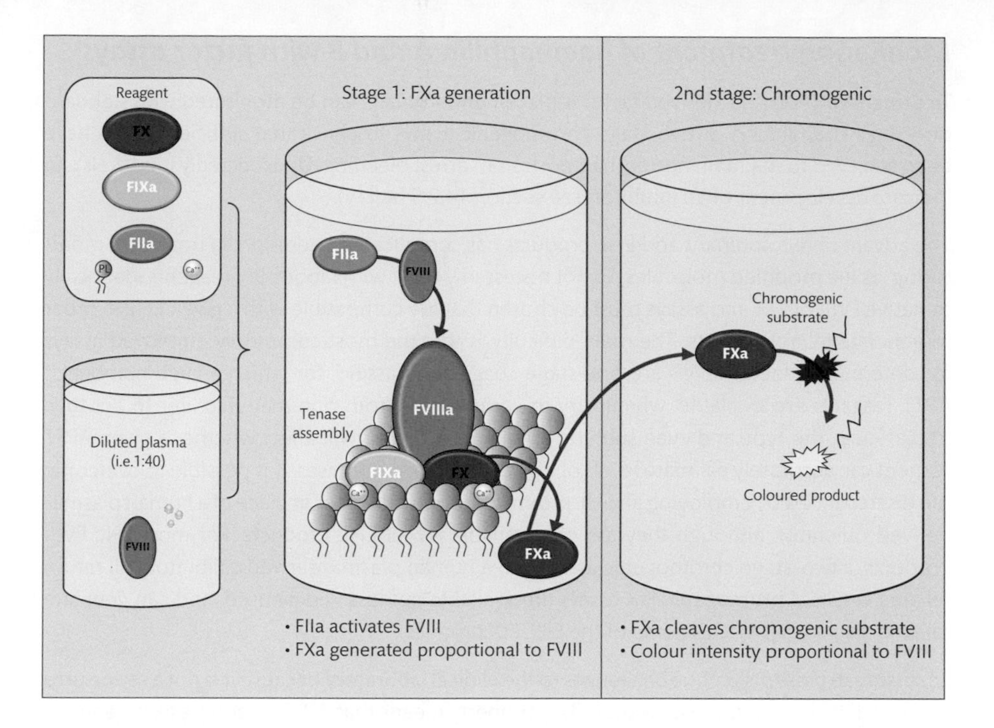

FIGURE 14.7

Principle of the chromogenic two-stage FVIII assay. FXa is generated in a FVIII-dependent manner in the first stage by activating the patient's FVIII with a low level of thrombin (FIIa) and then reacting it with exogenous FIXa, FX, phospholipids, and calcium ions. The low level of thrombin is insufficient to form a fibrin clot in the diluted plasma, and the higher concentrations of other reagent components facilitate preferential formation of FXa within the reaction time. The amount of FXa generated is directly proportional to the amount of FVIII in the patient's plasma. In the second stage of the assay, the FXa is reacted with a chromogenic substrate, the resultant colour intensity being directly proportional to the amount of FXa generated and thus the FVIII level.

even giving FVIII activity results within the one-stage assay reference range. This is because thrombin is only available for a short time in APTT assays before the clot is formed so there is just a small window where (thrombin-activated) FVIIIa is present and thus little time for it to destabilize. In the two-stage assay, thrombin is present in the reagents at the outset, so more FVIIIa will destabilize prior to endpoint generation. The bleeding phenotype correlates with the two-stage result. Less common is the reverse discrepancy where two-stage results can be double the one-stage, arising from mutations that generate FVIII molecules relatively resistant to thrombin activation. Thus, the short period of thrombin availability in one-stage assays results in less activation of FVIII than in the two-stage assay, where thrombin is present for longer to activate the resistant molecules.

Alternative factor assays

Alternatives to standard one-stage and two-stage techniques exist when exogenous enzymes are available that directly activate specific factors. We have met one of these already in the form of the reptilase enzyme that directly converts fibrinogen into fibrin. Venom from the Russell's viper (*Daboia russelii*) contains a fraction that directly activates FX, and venom from the world's most dangerous land snake, the coastal taipan (*Oxyuranus scutellatus*), contains an FII activator. Directly activating a factor at different plasma dilutions in the presence of appropriate cofactors generates clotting times that can be plotted as parallel-line bioassays. A variation of the Russell's viper venom FX assay allows the generated FXa to cleave a chromogenic substrate instead of generating a fibrin-clot endpoint.

Monitoring treatment of haemophilia A and B with factor assays

Treatment with plasma-derived factor replacement products can be monitored with standard one-stage coagulation or two-stage chromogenic assays to check that sufficient levels have been achieved to maintain normal haemostasis or arrest bleeding. Unexpectedly low levels can indicate development of an inhibitor (see section 14.2.5 below).

The advent of recombinant and EHL products has complicated haemophilia treatment monitoring, as the modified molecules do not necessarily react with laboratory reagents identically to native FVIII or FIX and assays must be chosen that are compatible with a given EHL in order to generate reliable results. The main difficulty is that the most commonly employed assays for determining factor levels are one-stage coagulation assays, for which a large number of APTT reagents are available, which vary in composition from one manufacturer to another. In particular, the type and even subtype of contact activator will affect whether a given APTT reagent can accurately estimate levels of a given EHL. In some cases, it is possible to overcome the discrepancies by employing an EHL product-specific calibrator in place of a human plasma-derived calibrator, although they are not available for all EHL products. For most EHL FVIII products, a two-stage chromogenic assay with a human plasma-derived calibrator will return reliable results. Chromogenic FIX assays are available but less widely used, and can generate reliable results with at least some of the EHL FIX products.

Emicizumab presents unique challenges to the clinical laboratory because it is not a serine protease and does not require activation. This property means that APTT screening tests generate short clotting times, and the consequent short clotting times in one-stage coagulation assays lead to overestimation of FVIII levels because the emicizumab masks the activity of any endogeneous FVIII present. Because emicizumab is an antibody that recognizes human coagulation factors, it is necessary to use a chromogenic assay with bovine-derived reagents to measure the patient's own FVIII level. One-stage coagulation assays and two-stage chromogenic assays can be calibrated with emicizumab itself to measure circulating drug levels.

Assays for factor XIII

The standard coagulation screening tests PT, APTT, and thrombin time, and other assays based on them, are unaffected by FXIII deficiency. If screening tests give normal results in a patient whose clinical symptoms indicate a bleeding disorder, separate assays are required to detect FXIII deficiency.

A qualitative clot solubility screening test for FXIII activity does exist, but it is not standardized and will only detect the more severe deficiencies. The principle is based on cross-linked fibrin clots being insoluble in urea and weak acid solutions. The method involves clotting plasma with Ca^{++} (± thrombin), adding a solubilizing agent (i.e. 5 mol/L urea, 2% acetic acid, 1% monochloroacetic acid), incubating at 37°C, and visually examining for persistence or dissolution of the clot over time. A clot formed from normal control plasma with sufficient FXIII will remain for over 24 hours, but the clot will dissolve within an hour where severe FXIII deficiency is present because cross-linking will have been impaired. Mixing tests can distinguish between deficiency and inhibitor.

Where resources allow, a quantitative FXIII activity assay should be the first-line test, the most commonly used being the ammonia release assay. It is based on activating FXIII with thrombin and Ca^{++} so that the FXIIIa cross-links a small peptide substrate to a glutamine-containing oligopeptide, whereby ammonia product is released during the transglutaminase reaction. The ammonia then enters a parallel NAD(P)H-dependent glutamate dehydrogenase catalysed reaction, the decrease in NAD(P)H being measured spectrophotometrically at 340 nm, being proportional to the FXIII activity. Plasma blanks are required to correct for FXIIIa-independent ammonia production, otherwise levels can be overestimated, especially where the FXIII level is low. Fibrinogen is not removed from the plasma because it would result in reduction of FXIII activity, and fibrin formed by thrombin accelerates FXIII activation. To prevent the thrombin from generating fibrin polymers and clotting the reaction mixture, a peptide that inhibits fibrin polymerization is used to hold fibrin monomers in solution. The assay principle is shown in Figure 14.8. Other FXIII activity assays are based on amine incorporation or isopeptidase activity of FXIII.

Key Points

Inherited FXIII deficiency is a rare bleeding disorder that can be a quantitative defect due to decreased synthesis, or to FXIII molecules with functional defects. Umbilical cord bleeding at birth occurs in >80% of patients with severe FXIII deficiency.

Measuring FXIII antigen by ELISA helps distinguish type I deficiency, where activity and antigen will be concordantly reduced, from type II deficiency, where activity will be reduced but the antigen will be normal, or at least significantly higher than activity. Distinguishing between deficiencies of the A and B subunits can be undertaken with ELISA assays specific for either subunit. Performing the quantitative assay on mixing studies can detect neutralizing antibodies in acquired FXIII deficiency, whilst clearing antibodies can be detected in ELISAs using purified FXIII as the capture antigen.

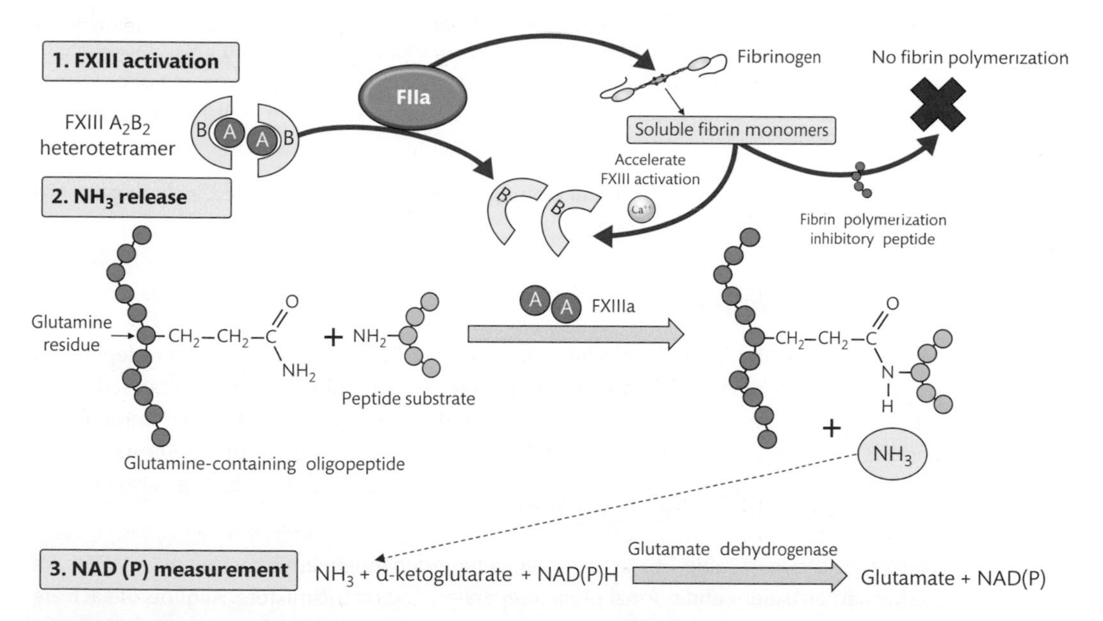

FIGURE 14.8

Ammonia release assay for FXIII activity. 1. Thrombin cleaves the activation peptide from the A-subunits, and in the presence of fibrin and Ca^{++}, the B-subunits separate from the A-subunits to form the active enzyme, FXIIIa, comprising the two activated A-subunits (FXIII-A_2). 2. The FXIIIa cross-links the glutamine-containing oligopeptide to the small peptide by catalysing formation of an isopeptide bond, the product of which is ammonia. 3. Measuring the ammonia level in an NAD(P)-producing reaction catalysed by glutamate dehydrogenase serves as an indirect measurement of FXIII activity.

14.2.5 Detection and characterization of inhibitors

Whilst most deficiencies of coagulation factors will be due to hereditary mutations or acquired causes of reduced production/increased consumption, some patients generate antibodies (inhibitors) that can interfere with coagulation function or facilitate immune-mediated clearance from the circulation of a specific factor.

The most commonly encountered inhibitors in clinical practice are lupus anticoagulants, which can be found incidentally during pre-operative screening in asymptomatic patients. These antibodies interfere with the phospholipid-dependent stages of some coagulation tests, but do not inhibit specific coagulation factors. In symptomatic patients, they are clinically associated with thrombosis and are considered in depth in Chapter 15.

alloantibodies
Antibodies directed against substances that are recognized as foreign to self.

autoantibodies
Antibodies directed against an organism's own tissues or native proteins.

The most common antibodies to specific factors associated with bleeding are those directed against FVIII. Approximately 30% of patients with severe haemophilia A develop **alloantibodies**, which result from treatment with FVIII replacement products because their immune system considers FVIII to be a 'foreign' protein. One molecule of inhibitor combines with one FVIII molecule, the resulting complex having no FVIII activity. Such inhibitors are considered to possess simple kinetics. Approximately 8% of patients with mild or moderate haemophilia A develop inhibitors that are **autoantibodies** with complex kinetics. These inhibitors rapidly inactivate FVIII, but the complex then dissociates, leaving some FVIII activity, and adding further FVIII results in the same residual activity. These are considered complex kinetics and such antibodies can also occur in non-haemophiliacs, who are referred to as having 'acquired' inhibitors. About 12% of patients with severe haemophilia B develop inhibitors, but the overall incidence of inhibitor formation in haemophilia B is only about 3%. (See also Box 14.8.)

The type of gene mutation causing haemophilia can make patients more likely to develop an inhibitor, such as major gene deletions or nonsense mutations. There is a lower incidence of such mutations in haemophilia B compared to haemophilia A, which is why inhibitor formation is less common in haemophilia B. The site of the mutation on the FVIII/FIX molecules can also influence the likelihood of inhibitor development.

Inhibitors directed against all the other clotting factors have been described, including fibrinogen and FXIII, but they are very rare.

Inhibitor screening

Most inhibitors are immediate acting *in vitro*, but FVIII inhibitors are time-dependent because the FVIII has to dissociate from VWF before the inhibitor can act. We can screen for inhibitors by mixing equal volumes of patient and normal plasma and testing immediately and then serially after incubation at 37°C over two hours. Although this is normally done using APTT because most clotting factors contribute to the clotting time, it can also be done using PT, and even Clauss fibrinogen.

The results shown in Table 14.5 are typical of a time-dependent inhibitor. At time 0, APTTs are performed on patient and normal plasma separately and on the mixture. Aliquots of each are incubated and tested again at 30-minute intervals, as well as a fresh mixture prepared from the separately incubated patient and normal aliquots. Results are interpreted as follows:

Tube 1—the normal plasma clotting times are equivalent throughout, indicating no appreciable deterioration over the test period for this plasma.

Tube 2—the patient at time 0 has an elevated clotting time because the inhibitor has already acted on the FVIII *in vivo* and remains elevated throughout.

BOX 14.8 *Acquired haemophilia due to autoantibodies*

Acquired haemophilia due to anti-FVIII antibodies is largely a disease of the elderly, but with a smaller peak associated with pregnancy or post-partum. It is a rare condition, with an incidence of 2:1 million annually. There is a greater likelihood of developing acquired haemophilia in the presence of malignancy or an autoimmune disorder.

Patients present with prominent subcutaneous haematomas (and bleeding elsewhere). The bleeding is often more severe than suggested by the inhibitor titre, although haemarthroses are rare. Acute bleeds are treated with prothrombin concentrate or recombinant FVIIa, and the inhibitors are eliminated with steroids and intravenous immunoglobulin plus low-dose cytotoxic therapy. Results from a patient with acquired haemophilia are shown in Case study 14.3.

Most elderly patients die within two years of diagnosis from comorbid conditions rather than from bleeding.

Tube 3—the clotting time of the incubated mixture increases over time, revealing the progressive nature of the inhibitor, which had no effect on the FVIII in the normal plasma at time 0 but a marked effect by completion of incubation.

Tube 4—performing APTTs on the fresh mixture proves that the abnormality is only expressed when patient plasma is incubated with normal plasma over time.

Key Points

Non-FVIII inhibitors would be demonstrated at time 0 in the mixture and no further incubation would be necessary.

Quantifying an inhibitor

If an inhibitor is suspected, factor assays will reveal which clotting factor is affected, whereupon biomedical scientists can perform a **Bethesda assay** that will measure inhibition of that factor. In haemophiliacs, it will be obvious which factor is affected and it is important to monitor the inhibitor levels as they have a marked effect on treatment. The assay is detailed in the following Method box and in Figure 14.9.

In practice, you are rarely fortunate enough to get a result of exactly 50% FVIII. Therefore, the results for the dilutions whose results are closest to 50% are plotted, as shown in Figure 14.10, to estimate the dilution that would have given exactly 50% residual FVIII and hence the inhibitor level itself. The assay design assumes there is no FVIII in the patient plasma, which is not necessarily the case for inhibitors with complex kinetics. In such instances, patient plasma can be pre-incubated at 56°C for at least 30 minutes, which destroys FVIII, but not the antibodies. The Bethesda assay is also used to quantify FIX inhibitors where extended incubation is unnecessary.

TABLE 14.5 Inhibitor screening results for a progressive inhibitor in APTT.

Time point (min.)	TUBE 1 Normal plasma clotting time (s)	TUBE 2 Patient plasma clotting time (s)	TUBE 3 Incubated 50:50 mixture clotting	TUBE 4 Fresh 50:50 mixture clotting time(s) time(s)
0	30	75	31	-
30	30	76	37	32
60	31	77	45	33
90	32	79	54	34
120	32	80	78	34

METHOD Bethesda assay for FVIII inhibitor

- Doubling dilutions are made of patient plasma in buffer to increasingly dilute out the inhibitor.
- One volume of each dilution is incubated with an equal volume of a normal plasma known to contain 100% FVIII (1.0 IU/mL), for two hours at 37°C.
- The tubes are then put on ice to prevent any further antibody-antigen reactions, and each one assayed for FVIII.

If there is an inhibitor present, more residual FVIII will be present as the dilution factor increases.

- One Bethesda unit (BU) is the amount of inhibitor that will neutralize 50% of a normal plasma containing 100% FVIII after a two-hour incubation at 37°C. Thus, the dilution factor of the patient plasma dilution that has a residual FVIII of 50% can be used to calculate the inhibitor level in BU/mL; i.e. if the neat plasma had 50% residual FVIII, by definition it would contain 1.0 BU/mL. If, however, the inhibitor is more potent than that and the plasma needed diluting, say, 1:10 to obtain 50% residual FVIII, the inhibitor level is 10 × 1.0 BU/mL.

You can see in Figures 14.9 and 14.10 that there was no residual FVIII in the neat plasma and the 1:2 dilution because the inhibitor was sufficiently potent to neutralize all the FVIII in the time frame. Between the 1:4 and 1:32 dilutions there is an increasing amount of FVIII, as the effect of the inhibitor diminishes with increasing dilution. From the 1:64 dilution and above, the dilution of the inhibitor was sufficient to prevent it from acting within the time frame.

SELF-CHECK 14.11

Outline the methods used to detect clotting factor inhibitors.

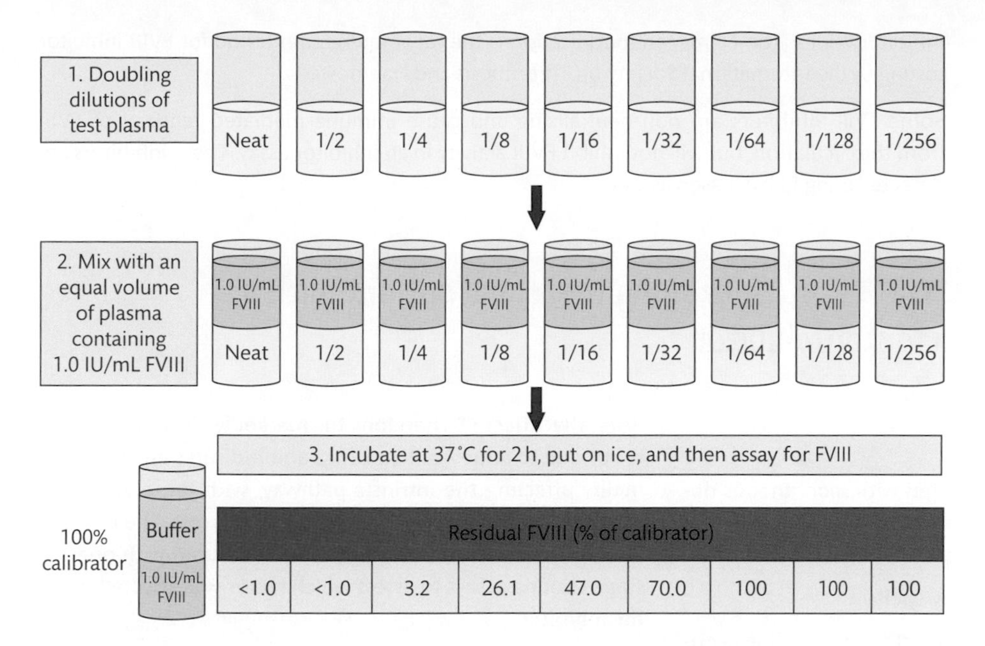

FIGURE 14.9

Principle of the Bethesda assay for FVIII inhibitor quantification. The inhibitor (if present) is increasingly diluted before each dilution is then reacted with an identical amount of FVIII during a two-hour incubation at 37°C. Antibody-antigen reactions are stopped by putting the tubes containing each dilution on ice. Each dilution is then assayed for residual FVIII.

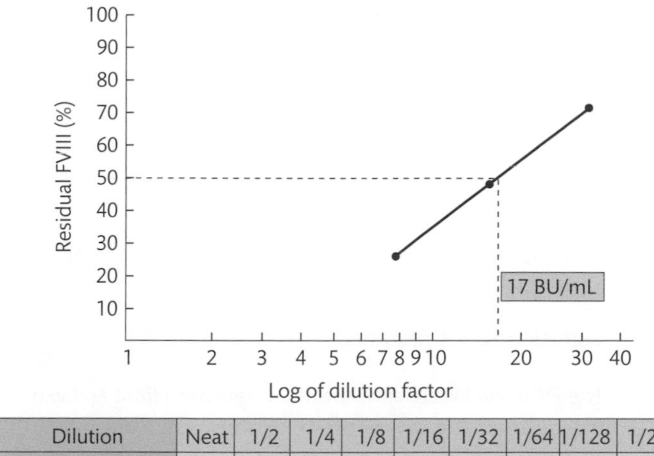

Dilution	Neat	1/2	1/4	1/8	1/16	1/32	1/64	1/128	1/256
Residual FVIII (%)	<1.0	<1.0	3.2	26.1	47.0	70.0	100	100	100

FIGURE 14.10

Graphical plot for Bethesda assay. The log of the three dilutions with residual FVIII closest to 50% are plotted against residual FVIII. The dilution that would have given exactly 50% residual FVIII is read off the curve and is reported as the Bethesda value.

Other inhibitor assays

Changes in pH and protein concentration during the two-hour incubation of a Bethesda assay can affect FVIII stability, and its inactivation increases as the pH increases and at low protein concentration. The **Nijmegen modification** of the Bethesda assay involves direct buffering of FVIII source plasma to counter pH drift, and dilution of test plasma in FVIII deficient plasma to maintain constant protein concentration. A cheaper alternative to FVIII-deficient plasma is 4% bovine serum albumin. These modifications reduce the incidence of false-positive weak inhibitor values. The Nijmegen modification is the recommended method for FVIII inhibitor testing by the International Society on Thrombosis and Haemostasis.

Some FVIII inhibitors are non-neutralizing and cause immune-mediated removal of FVIII from the circulation, but will not inhibit FVIII activity in an inhibitor assay. These inhibitors are detected using ELISA-based assays.

CASE STUDY 14.3 Screening tests in a patient with a coagulation factor inhibitor

Patient history

- **A 76-year-old woman presented with spontaneous development of haematomas on her left leg and both feet during the previous four weeks.**
- **She reported fatigue and dizzy spells.**
- **She had no personal or family history of bleeding problems.**
- **Full blood count (FBC) and coagulation screening tests were performed.**

Results 1

Index	Result	Reference range
WBC (× 10^9/L)	7.7	(RR: 4.0–11.0)
Hb (g/L)	62	(RR: 120–160)
Platelet count (× 10^9/L)	201	(RR: 150–400)
PT (s)	12.1	(RR: 10.0–14.0)
APTT (s)	92.3	(RR: 32.0–42.0)
APTT 1:1 mixing test (s)	58.0	
TT (s)	9.3	(RR: 9.0–11.0)
Fibrinogen (g/L)	3.8	(RR: 1.5–4.0)

Significance of results 1

The reduced Hb was the cause of her tiredness and dizzy spells and reflects the severity of bleeding. The platelet count was normal, but this does not exclude an acquired functional defect. PT was normal and excludes deficiencies or inhibitors of factors in the extrinsic and common pathways, vitamin K deficiency, and anticoagulant drugs that affect PT. The fibrinogen level was normal and the normal TT additionally excludes the presence of anticoagulant drugs that affect TT. Therefore, the markedly elevated APTT is not due to anticoagulant drugs and indicates an abnormality affecting the intrinsic pathway, with the elevated APTT mixing test revealing an inhibitory process. The main factors of the intrinsic pathway were measured with one-stage clotting assays to assess which factor was affected by the inhibitor.

Results 2

Index	Result	Reference range
FVIII (IU/dL)	<1	(RR: 50–150)
FIX (IU/dL)	142*	(RR: 50–150)
FXI (IU/dL)	112*	(RR: 60–140)
FXII (IU/dL)	99*	(RR: 50–150)

*** The parallel line bioassay plots exhibited inhibition at lower dilutions that was abolished at higher dilutions (see Figure 14.4).**

- **Nijmegen Bethesda assay for FVIII inhibitor: 143 NBU/mL (RR: Not detected).**

Significance of results 2

The FVIII level was below the assay detection limit and was the cause of the severe bleeding. Although FVIII inhibitors are progressive and do not always manifest in non-incubated APTT mixing tests, the FVIII inhibitor titre was extremely high and thus able to achieve inhibition in the mixing test. Because the one-stage assays for FIX, FXI, and FXII rely on normal amounts of FVIII, the high titre inhibitor was able to interfere in the factor assays at low sample dilutions. Abolishing the inhibition at higher dilutions permitted accurate quantification and excluded deficiency or inhibition of FIX or FXI as the cause of bleeding. The patient had developed acquired haemophilia due to autoantibodies against endogenous FVIII.

14.2.6 Detection and classification of von Willebrand disease

Approximately 70% of patients with VWD have a quantitative deficiency of normally functioning VWF (type 1) and approximately 25% have a structural/functional defect (type 2). The remainder have severe VWD, where VWF is absent (type 3).

The variable clinical and phenotypic expression of VWD means that standard screening tests generate limited information for the detection of VWF deficiencies:

- PT, TT, and fibrinogen are unaffected;
- APTT will be elevated only if the FVIII is sufficiently reduced.

Platelet count is normal in most subtypes; platelet morphology may be useful when distinguishing from other primary haemostatic disorders.

Other screening tests, the bleeding time, and analysers assessing high shear-dependent platelet function, are often abnormal in VWD, but do not distinguish between subtypes. These are also abnormal in platelet disorders and can be normal in mild VWD. They are covered in more detail later in this chapter.

If VWD is suspected from the clinical history (with or without screening tests), biomedical scientists will then perform assays that assess VWF concentration, function, and structure:

- Patients with type 1 VWD will have concordant functional and concentration assays because they produce low amounts of normally functioning VWF. Patients with type 3 VWD will have concordant results insomuch as the VWF will be absent, irrespective of how it is measured.
- Most patients with type 2 VWD will have functional results <70% of the concentration result.

Measurement of FVIII activity is also included as it aids diagnosis and can have implications for treatment. The lower the FVIII, the more severe the VWD and the risk of bleeding. A normal FVIII level does not exclude VWD, whilst a reduced FVIII is not indicative in itself of VWD.

Assessing VWF concentration

Immunological assays that measure the total amount of VWF protein present, irrespective of function, are used to assess the overall VWF concentration as VWF antigen (VWF:Ag). The most commonly used methods are ELISA or latex immunoassay (LIA), the principles of which you met in Chapter 2.

Assessing VWF function

Three assays are used to assess different aspects of VWF function based on binding to platelets, collagen, or FVIII.

Platelet binding

Binding of VWF to the glycoprotein 1b (GPIb) platelet receptor is measured in the **ristocetin cofactor assay** (VWF:RCo). Ristocetin A is an antibiotic isolated from the soil bacterium *Nocardia lurida* that promotes VWF–GPIb binding, as shown and described in Figure 14.11.

Cross reference

You can find further detail about aggregometers in section 14.3.13 on the investigation of platelet disorders.

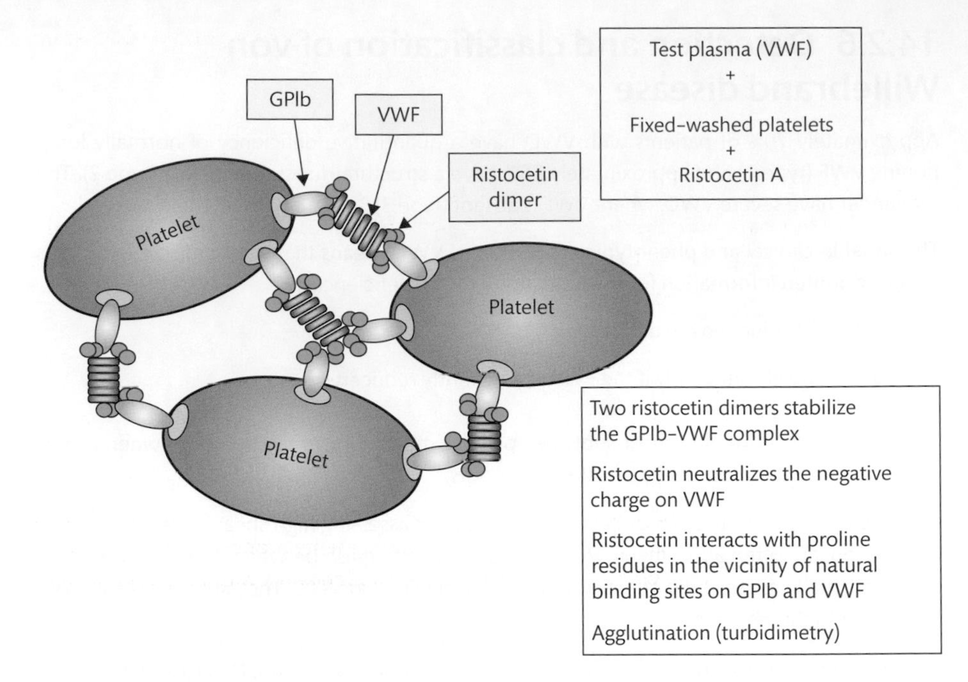

FIGURE 14.11
Principle of ristocetin cofactor assay. Patient plasma is mixed with fixed-washed platelets and ristocetin. Ristocetin dimers stabilize the GPIb–VWF complex by neutralizing the negative charge on VWF molecules and interacting with proline residues in the vicinity of natural binding sites, leading to agglutination of platelets. The amount of agglutination and slope of the resultant graph are proportional to the amount of VWF in the patient plasma.

Dilutions of patient plasma are incubated at 37°C in an aggregometer together with ristocetin and a suspension of platelets that have been washed to remove plasma and then fixed in formalin. The VWF in the patient plasma promotes platelet agglutination, and generates graphical traces with slope values that are directly proportional to the amount of VWF present in relation to its GPIb-binding capacity.

Key Points

Binding of ristocetin on the platelet surface reduces the platelet's negative charge, and similarly on VWF. This reduces electrostatic repulsion between platelets and/or between platelets and VWF, thus allowing VWF to cause agglutination by bridging between platelets. This is in contrast to the platelet agonists you will meet in the following section, which activate platelets to induce the aggregation processes you met in Chapter 13.

New-generation GPIb-binding assays for assessing VWF activity

The VWF:RCo platelet aggregometry assay has been the mainstay VWF activity technique for decades, yet it is a technically demanding manual method with low sensitivity and high inter-assay and inter-laboratory variability. Fully automated assays are now available, and in routine use, that employ latex beads coated with a recombinant fragment of GPIbα through a highly specific monoclonal antibody. In one version of the assay, VWF in the sample binds to the GPIbα fragment via ristocetin and the degree of agglutination is directly proportional to VWF activity, determined by decreased light transmittance due to the agglutination. An alternative assay similarly employs latex-bound GPIbα fragments, but they contain two gain-of-function mutations and agglutination is ristocetin-independent. These assays are technically far more straightforward than the VWF:RCo platelet aggregometry assay, with marked improvements in variability and reliability. Alternative nomenclature is in use to distinguish the new-generation assays from the VWF:RCo platelet aggregometry assay. Ristocetin-triggered GPIb-binding assays are abbreviated to VWF:GPIbR, and gain-of-function mutant GPIb-binding assays are abbreviated to VWF:GPIbM.

Collagen binding

Integral to effective primary haemostasis is the ability of VWF to bind to subendothelial collagen that is exposed upon vessel trauma. This property is assayed in an indirect ELISA, where the patient VWF is captured on an ELISA plate using collagen itself to generate a **collagen-binding activity assay** (VWF:CB) result.

Chemiluminescence assays

The new-generation VWF:GPIbR assay is also available as an automated chemiluminescent immunoassay. It is a two-step assay employing magnetic particles coated with the GPIbα fragments via a monoclonal antibody as the solid phase, and a chemiluminescent detection phase. The first step involves mixing test plasma with the magnetic particles and a buffer containing ristocetin. The VWF in the sample binds to the GPIbα fragments proportionally to its GPIb-binding activity/ability, after which there is magnetic separation and a washing step. The magnetic particles that have captured VWF via the GPIbα fragments are incubated in the second step with an isoluminol-labelled monoclonal VWF antibody. A second magnetic separation and washing step follows, after which two triggers for the chemiluminescent reaction are added, which is measured as relative light units that are directly proportional to the GPIb-binding activity and read from a calibration curve to generate a quantitative result.

Chemiluminescence assays are also available for VWF:CB and VWF:Ag. In the VWF:CB assay, the VWF in the sample is captured with magnetic particles coated with a type III collagen-triple-helical peptide. In the VWF:Ag assay, the magnetic particles are coated with a polyclonal antibody to VWF that will capture VWF molecules irrespective of any functional defects.

FVIII binding

One subtype of VWD, type 2N, has a normal VWF concentration and function with respect to both collagen and platelet binding. However, the VWF binds poorly to FVIII and results in circulating FVIII levels of around 15–30% in heterozygotes, thus mimicking mild haemophilia A.

A family history revealing affected females may be the only clue that it is not true haemophilia, prompting measurement of the **factor VIII-binding capacity** (VWF:FVIIIB). The principle of this ELISA-based assay is illustrated in Figure 14.12.

Identifying structural defects

You were introduced to the multimeric structure of VWF in Chapter 13. For diagnostic purposes, VWF can be split into its component multimers by heating the patient plasma to 56°C and then subjecting it to sodium dodecyl sulphate-polyacrylamide gel electrophoresis (SDS-PAGE), which separates the multimers according to size. The presence or absence of some or all multimers aids diagnosis and classification when assessing results in conjunction with those from the other assays. You can see how VWD is classified in Table 14.6.

Distinguishing between type 2B and pseudo-VWD

High molecular weight (HMW) multimers of VWF have the greatest effect on collagen and platelet binding, so VWF:RCo and VWF:CB assays are sensitive to their absence. In type 2B VWD, the HMW multimers are hyperresponsive and attach themselves to platelets *in vivo*,

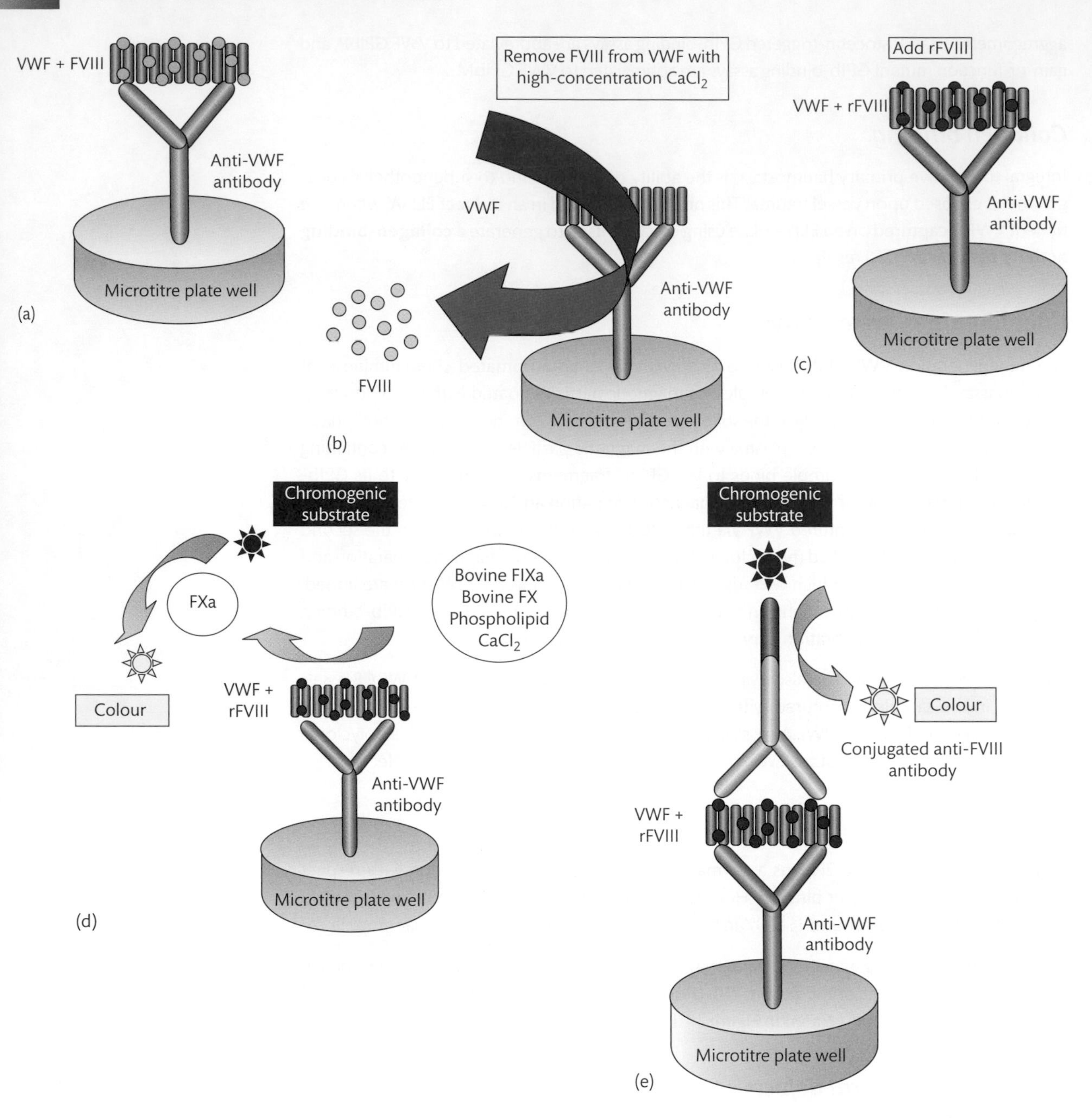

FIGURE 14.12

FVIII-binding assay. (a) The patient's VWF is captured by an anti-VWF antibody. (b) Depending on whether the patient's VWF can bind FVIII, there may be some FVIII bound; however, this will vary between patients so it is removed with a high-concentration calcium chloride solution. (c) Exogenous recombinant FVIII (rFVIII) is then added to standardize the amount of FVIII that is added to each patient's VWF. There are two methods available for determining the amount of FVIII that has bound to the patient's VWF. (d) shows the captured FVIII being reacted with FIXa, FX, phospholipids, and calcium ions to generate FXa, which is subsequently reacted with a chromogenic substrate. The intensity of the coloured product is proportional to the amount of FVIII that bound to the VWF. (e) Alternatively, an enzyme-linked antibody to FVIII (not VWF) is added and reacted with a substrate that produces a coloured product.

TABLE 14.6 Classification of VWD with typical laboratory findings.

Subtype	PFA-100	FVIII	RIPA	VWF:RCo	VWF:CB	VWF:Ag	FVIII binding	VWF:RCo: VWF:Ag ratio	VWF:CB: VWF:Ag ratio	Multimer distribution	Further details
Type 1	N/↑	N/↓	N/↓	↓	↓	↓	N	>0.6	>0.6	All multimers present	Concordant activity and antigen; partial quantitative deficiency of normally functioning VWF due to reduced production, defective release, or increased clearance
Type 2A	↑	N/↓	↓	↓	↓	↓/N	N	<0.6	<0.6	HMW and intermediate forms absent	Discordant activity and antigen; reduced platelet dependent function due to absence of HMW multimers
Type 2B	↑	N/↓	↑	↓	↓	↓/N	N	<0.6	<0.6	HMW forms absent	Discordant activity and antigen; distinguish from pseudo-VWD with RIPA studies, variable thrombocytopenia
Type 2M	↑	N/↓	↓	↓	↓/N	↓/N	N	<0.6	>0.6	All multimers present	Discordant VWF:RCo activity and antigen; decreased VWF-dependent platelet adhesion despite presence of HMW multimers, i.e. dysfunctional HMW multimers
Type 2N	N	↓	N	N/↓	N/↓	N/↓	↓	>0.6	>0.6	All multimers present	Concordant activity and antigen; VWF-dependent platelet function is normal; VWF has reduced affinity for FVIII
Type 3	↓↓	↓↓	↓	↓↓	↓↓	↓↓	N/A	N/A	N/A	Multimers absent	(Concordant activity and antigen); severe bleeding disorder due to markedly reduced or absent VWF; markedly reduced FVIII (~2%)

N = normal; HMW = high molecular weight.

causing the formation of microaggregates and moderate thrombocytopenia. In pseudo-VWD, sometimes called platelet-type VWD, the GPIb is hyperresponsive and removes HMW multimers from the circulation. It is important to distinguish between these subtypes because desmopressin treatment for type 2B is contraindicated, as it releases more hyperresponsive HMW multimers into the circulation and can lead to severe thrombocytopenia. This is done using the **ristocetin-induced platelet aggregation (RIPA) test.**

Subjecting a preparation of patient platelets, suspended in their own plasma, to a fixed concentration of ristocetin (1.2 mg/mL) will generate reduced agglutination compared to normal in most VWD subtypes. Both type 2B and pseudo-VWD are over-responsive and cause agglutination at lower concentrations, where even normal VWF levels are unreactive. To distinguish between type 2B and pseudo-VWD, patient platelets are washed to remove plasma constituents, resuspended in a normal plasma, and then subjected to RIPA.

- If the RIPA remains over-responsive, then the defect must lie in the patient's platelets, as their VWF has been removed and replaced with normal VWF.
- If the RIPA normalizes, the defect was in the patient's plasma. This can be double-checked by washing normal platelets, suspending them in patient plasma, and performing RIPA, which will be over-responsive.

SELF-CHECK 14.12

Outline the principles of assays used to detect and characterize VWD.

Reference ranges

You saw in Chapter 13 that VWF carries ABO blood group antigens which affect *in vivo* survival. The effect is sufficiently significant that we cannot apply a single reference range to the entire population. Individuals with blood group O and no VWF mutation can have VWF:Ag levels as low as 35 IU/dL. For blood groups A and B, the lower thresholds are in the regions of 50 IU/dL and 55 IU/dL, respectively, and for individuals who are AB it is about 65 IU/dL. However, it is more important that the laboratory identifies whether VWF levels are above or below a threshold likely to be a cause or contributor to a bleeding phenotype, irrespective of the distribution of VWF levels in relation to blood group. The British Committee for Standards in Haematology recommends the following in relation to laboratory values and diagnosis of VWD:

- Do not use blood group specific reference ranges.
- When investigating a patient with mucocutaneous bleeding, a diagnosis of VWD can be made when the VWF activity is <0.30 IU/mL.
- Patients with an appropriate bleeding history and VWF activity of 0.30–0.50 IU/mL should be regarded as having primary haemostatic bleeding with reduced VWF as a risk factor, rather than VWD itself.

Further notes on type 1 VWD

Type 1 VWD classically presents in the laboratory with reduced but concordant activity and antigen results and normal multimer distribution, the VWF:Ag levels being in the region of 20–50%. Interestingly, only about 50% of these patients have a VWF mutation, the others having reduced levels for other reasons, such as blood group O. The advent of more sensitive multimer

CASE STUDY 14.4 von Willebrand disease

Patient history

- A 14-year-old female presented with an 18-month history of menorrhagia, easy bruising, and occasional epistaxis.
- She had not been exposed to any surgery in her lifetime.
- Her father had epistaxis as a child and his mother had menorrhagia.
- Coagulation screening tests were performed.

Results 1

Index	Result	Reference range
PT (s)	12.3	(RR: 10.0–14.0)
APTT (s)	44.0	(RR: 32.0–42.0)
APTT 1:1 mixing test (s)	36.0	
TT (s)	10.7	(RR: 9.0–11.0)
Fibrinogen (g/L)	4.0	(RR: 1.5–4.0)
Platelet count (× 10^9/L)	185	(RR: 150–400)

Significance of results 1

The normal PT, TT, and fibrinogen exclude deficiencies or inhibitors of factors in the extrinsic and common pathways, vitamin K deficiency, and many anticoagulant drugs. The bleeding is not due to thrombocytopenia. The mildly prolonged APTT that corrects upon mixing with normal plasma suggests a factor deficiency, although not necessarily a severe one. In view of the sex of the patient and type of bleeding, factor assays and VWF assays were performed.

Results 2

Index	Result	Reference range
FVIII (IU/dL)	35	(RR: 50–150)
FIX (IU/dL)	121	(RR: 50–150)
FXI (IU/dL)	100	(RR: 60–140)
FXII (IU/dL)	99	(RR: 50–150)
VWF:RCo (IU/dL)	27	(RR: 55–200)
VWF:CB (IU/dL)	29	(RR: 55–200)
VWF:Ag (IU/dL)	33	(RR: 55–200).

Significance of results 2

The patient is female so the reduced FVIII, sufficient to mildly but not markedly elevate the APTT, was not due to haemophilia A. VWF activity measured as GPIb and collagen binding was reduced below the 0.30 IU/mL threshold, indicating VWD. Although the VWF:Ag was above 0.30 IU/mL, it was nonetheless reduced, and the threshold is specifically for activity values. The activity and antigen levels were ostensibly concordant, the VWF:RCo:VWF:Ag ratio being 0.82, indicating type 1 VWD. The FVIII was reduced to a similar level to that of its carrier.

analysis techniques has revealed that many patients considered to have more severe type 1 VWD (VWF:Ag <20%) have a slight decrease in HMW multimers or altered subunit structure and discordant activity and antigen results. The majority of these patients do have a VWF mutation and may be better classified as type 2M. Results from a patient with type 1 VWD are shown in Case study 14.4.

An interesting variant of type 1 VWD is the Vicenza subtype. The VWF has reduced survival in plasma, so the VWF:Ag and VWF:RCo levels are very low (<15 IU/dL) and concordant. However, the bleeding symptoms are milder than would be expected from these levels because the VWF in platelets is unaffected. The small amount of VWF that is detected in plasma contains ultra-high molecular weight multimers—because it has only just been released from endothelial cells and has yet to be subjected to cleavage by ADAMTS-13.

You saw in Chapter 13 that von Willebrand antigen II, also known as the VWF propeptide (VWFpp), is cleaved from pro-VWF to release the mature VWF. Because the defect in the Vicenza subtype is reduced survival of mature VWF in plasma, synthesis of VWF is normal, so measurement of VWFpp levels is diagnostically valuable. In a normal individual, the ratio of VWFpp to VWF:Ag is around 1.0, but in a patient with increased clearance, where VWFpp levels are normal but VWF:Ag levels markedly reduced, the ratio is increased. VWFpp assays are commonly ELISAs.

Acquired VWD

Acquired von Willebrand disease (AVWD) is a rare bleeding disorder with similar clinical and laboratory findings to the inherited disorder. It can be caused by specific or non-specific antibodies, adsorption of VWF onto malignant cell clones, hypothyroidism, or loss of high-molecular-weight multimers under conditions of high shear stress such as in **aortic stenosis**. AVWD is usually diagnosed using standard laboratory tests for inherited VWD in the absence of a family history of bleeding. Functional inhibitors can be demonstrated in a modification of the VWF:RCo assay, in which the patient's plasma is mixed with normal plasma to facilitate inhibition of the VWF in the normal plasma by the antibody present in the patient's plasma.

aortic stenosis
Narrowing of the heart valve between left ventricle and the aorta.

Similar to the Vicenza subtype, autoantibodies to VWF can result in increased clearance, so assessing the VWFpp to VWF:Ag ratio can help identify immune-mediated AVWD.

14.3 Diagnosis of platelet disorders

As you saw in Chapter 13, platelets are pivotal to an effective haemostatic response to injury, and comprise the platform for many crucial mechanisms. Thus, defects in both numbers and function, which can be hereditary or acquired, must be detected in the clinical laboratory. Inherited platelet disorders are rare, acquired abnormalities being more commonly encountered in clinical practice.

14.3.1 Thrombocytopenia

Even if platelet function is normal, reduced platelet numbers (thrombocytopenia) compromises both platelet plug formation and the availability of a phospholipid surface for coagulation biochemistry. Additionally, circulating platelets secrete **platelet-activating factor (PAF)**, which contributes to the maintenance of endothelial cell junctions, so reduced PAF secretion in severe thrombocytopenia leads to 'unzipping' of endothelial junctions and bleeding due to compromised vascular integrity.

Thrombocytopenia is defined as subnormal numbers of circulating platelets, the reference range being 150–400 × 10^9/L of blood. In practice, platelet counts >100 × 10^9/L rarely cause bleeding problems, whilst the risk of haemorrhage increases as the platelet count falls. Spontaneous bleeding commonly occurs when the platelet count falls below 20 × 10^9/L. The main causes of thrombocytopenia are outlined in Table 14.7 and are the result of one of three fundamental mechanisms:

- impaired platelet production;
- increased platelet destruction/consumption;
- increased pooling of platelets in the spleen.

Note in Table 14.7 that some disorders have reduced numbers and function and we will now look at disorders of platelet function in more detail.

TABLE 14.7 Thrombocytopenias.

Disorder	Features	Mechanism of thrombocytopenia	Associated with platelet function defect?
Inherited			
Fanconi's anaemia	Progressive bone marrow failure + other congenital anomalies	Impaired production	No
Thrombocytopenia with absent radius (TAR)	Skeletal anomalies + reduced or absent megakaryocytes	Impaired production	No
Bernard–Soulier syndrome	Abnormal GPIb + giant platelets	Impaired production	Yes
MYH9 disorders: May-Hegglin anomaly Sebastian syndrome Fechtner syndrome Epstein syndrome	MYH9 is the gene encoding for the heavy chain of non-muscle myosin White cell inclusions (Döhle-like bodies) + moderate thrombocytopenia and giant platelets	Impaired production	Yes
Grey platelet syndrome	Absent α-granules with mild-to-moderate thrombocytopenia and large platelets	Impaired production	Yes
Congenital amegakaryocytic thrombocytopenia	Almost complete absence of megakaryocytes leading to severe thrombocytopenia	Impaired production	No
Wiskott–Aldrich syndrome	X-linked immune deficiency + eczema + thrombocytopenia with small platelets	Impaired production	Yes
X-linked thrombocytopenia with dyserythropoiesis	Red cell abnormalities + marked thrombocytopenia with giant platelets	Impaired production	No
Montreal platelet syndrome*	Severe thrombocytopenia with giant platelets and spontaneous *in vitro* aggregation	Impaired production	Yes
Type 2B VWD	Abnormal VWF function causes platelet microaggregates leading to moderate thrombocytopenia**	Increased consumption	No
Platelet-type VWD	Abnormal GPIb function causes platelet microaggregates, leading to moderate thrombocytopenia	Increased consumption	Yes
Acquired			
Disseminated intravascular coagulation	Progressive activation and consumption of haemostatic components	Increased consumption	No
Immune thrombocytopenic purpura (ITP)	Accelerated platelet destruction due to autoantibodies to platelets; acute onset in children with spontaneous recovery; insidious onset in adults, where spontaneous recovery is rare	Increased destruction	No

(Continued)

TABLE 14.7 (Continued)

Disorder	Features	Mechanism of thrombocytopenia	Associated with platelet function defect?
Neonatal alloimmune thrombocytopenia (NAIT)	Transplacental passage of maternal antibodies to platelet antigens not shared with the foetus	Increased destruction	No
Pregnancy-associated thrombocytopenia	Incidental; secondary to hypertension or fatty liver, ITP, HELLP syndrome (*h*aemolysis, *e*levated *l*iver enzymes and *l*ow *p*latelets)	Impaired or production or increased destruction/ consumption	No
Megakaryocytic aplasia	Autoimmune suppression of megakaryocyte development	Impaired production	No
Heparin-induced thrombocytopenia	Antibodies to the heparin–platelet factor 4 complex	Increased destruction	No
(Non-heparin) drug-induced thrombocytopenia	Chemotherapy; many other drugs have been described to cause thrombocytopenia, but not in every patient	Impaired production or increased destruction	No
Viral infections	Usually mild thrombocytopenia; can be seen in HIV, rubella, cytomegalovirus infections	Impaired production	No
Nutritional disorders	Deficiencies of vitamin B_{12} and folic acid; alcoholism	Impaired production	No
Thrombotic thrombocytopenic purpura Haemolytic uraemic syndrome	Autoantibodies to ADAMTS-13 lead to increased VWF function and formation of platelet aggregates; rare hereditary form of TTP known as Upshaw–Schulman syndrome (ADAMTS-13 deficiency)	Increased consumption	No
Haemolytic uraemic syndrome	Childhood disorder with haemolytic anaemia, thrombocytopenia, and acute renal failure	Increased consumption	Yes
Bone marrow infiltration	Thrombocytopenia secondary to haemato-oncological disorders	Impaired production	No
Splenomegaly	Pooling of up to 90% of total body platelets in the spleen	Splenic pooling	No

* Recent reports suggest that Montreal platelet syndrome may be a manifestation of type 2B VWD.

** Note that the thrombocytopenia itself is acquired as a result of the inherited type 2B VWD.

14.3.2 Hereditary platelet function disorders

Disorders of platelet function are characterized by the area of function affected:

- surface receptors;
- cytoplasmic granules;
- platelet biochemistry;
- phospholipid exposure.

Key Points

Biomedical scientists need to be aware of the phenomenon of pseudothrombocytopenia. The EDTA used to anticoagulate FBC samples can cause platelets to clump *in vitro* due to the presence of calcium-dependent antibodies that are present in approximately 0.1% of the population. The automated platelet count is low and platelet clumps can be seen on a stained blood film. The antibodies are not clinically significant and an accurate platelet count can be obtained by taking a new sample into a different anticoagulant such as trisodium citrate. This phenomenon can also be seen when patients receive therapeutic antibody preparations such as abciximab. Other antibodies can cause platelets to attach to white cells, typically neutrophils, and is termed 'platelet satellitism' as they are seen to form rosettes around the periphery of the white cell.

14.3.3 Surface receptor disorders

The two most common receptor disorders are **Glanzmann's thrombasthenia** and **Bernard-Soulier syndrome**, which are amongst the most clinically severe of hereditary platelet function defects.

Glanzmann's thrombasthenia—this is characterized by a deficiency or functional defect of the GPIIbIIIa complex. It is classified into three subtypes:

- type I has 0–5% of normal levels of GPIIbIIIa;
- type II has 6–20% GPIIbIIIa;
- 'variant' has 50–100% GPIIbIIIa, but with reduced fibrinogen binding.

The most common clinical features are typical of a primary haemostatic disorder, such as epistaxis, purpura, petechiae, excessive bruising, gum bleeding, and menorrhagia. Platelet counts are normal. Bleeding after trauma/surgery can be severe and pregnancy/delivery markedly increase the risk of bleeding.

Bernard-Soulier syndrome—this results from the absence or decreased expression of the GPIb-V-IX complex on the platelet surface. Patients present with mild to moderate thrombocytopenia and enlarged or giant platelets. Bleeding symptoms are also typical of a primary haemostatic disorder, and are more severe than expected for the degree of thrombocytopenia alone because there is a concomitant functional defect.

Defects in the coupled ADP receptors $P2Y_1$ and $P2Y_{12}$, the collagen receptors GPVI and GPIaIIa, the thromboxane A_2 receptor, and the epinephrine receptor have been described in rare cases.

Key Points

Glanzmann's thrombasthenia and Bernard-Soulier syndrome are both inherited as autosomal-recessive conditions. They have a higher incidence in populations where consanguineous partnerships are common. Heterozygotes tend to be asymptomatic.

SELF-CHECK 14.13

What are the two most common platelet receptor disorders and which receptors are affected in each?

14.3.4 Platelet granule disorders

Platelet function can be defective if granule numbers are reduced, granule contents are deficient, or if release mechanisms fail. Most of these disorders affect dense bodies or α-granules, but rarely affect both. Since the dense bodies and α-granules are storage sites, their defects are also referred to as storage pool disease.

14.3.5 Dense body disorders

Disorders of dense bodies are often part of a more complex congenital disorder, although isolated dense body defects are known. The bleeding phenotype associated with dense body disorders is usually of mild-moderate severity with significant bleeding associated with trauma/surgery. Platelet counts and size are normal. There are three main disorders.

Hermansky-Pudlak syndrome—this is a diverse autosomal-recessive disorder affecting a number of organelles, in particular, **melanosomes** and platelet dense bodies. Consequently, albinism and dense body deficiency are characteristic.

melanosome
Pigment-containing organelle.

Chediak-Higashi syndrome—this is an autosomal-recessive disorder also associated with albinism and dense body deficiency. Additionally, the immune system is affected and large inclusion bodies are seen in white cell precursors in the bone marrow.

Primary dense body deficiency—this is a clinically heterogeneous disorder with an uncertain genetic basis that is not associated with other abnormalities.

14.3.6 Alpha-granule disorders

Defects of α-granules are extremely rare and so it is difficult to generalize about clinical features. There are three main disorders:

Grey platelet syndrome—this is the only disorder associated with a complete absence of α-granules, and thus the levels of their contents are reduced or absent. This results in the characteristic appearance of platelets on Romanowksy-stained blood films as agranular, misshapen, grey blobs. The platelet count is often reduced and platelets can be slightly larger than normal.

Paris-Trousseau syndrome—this is characterized by thrombocytopenia, together with other congenital abnormalities. There are giant α-granules in a percentage of the circulating platelet population that cannot release their contents, and a subpopulation of abnormally matured micromegakaryocytes that lyse upon maturation.

Quebec disorder—this is characterized by increased levels of platelet urinary plasminogen activator (uPA), which activates platelet plasminogen that subsequently degrades α-granule contents.

SELF-CHECK 14.14

Outline the main platelet granule disorders.

14.3.7 Disorders of platelet biochemistry

Abnormalities in platelet biochemistry inevitably impair platelet function. Wiskott-Aldrich syndrome (WAS) is a rare X-linked disorder resulting from defects in the *WAS* gene that encodes for the WAS protein (WASp). WASp takes part in biochemical signalling and cytoskeleton maintenance. Children born with WAS present with bruising, small platelets, and purpura resulting from thrombocytopenia and abnormal platelet function. WAS is a complex disorder, with eczema and immune deficiencies that can be severe.

Deficiencies of platelet enzymes such as cyclooxygenase have also been described.

14.3.8 Disorders of phospholipid exposure

Scott syndrome is an extremely rare bleeding disorder characterized by the reduced exposure of negatively charged phospholipids to facilitate tenase and prothrombinase formation. *Stormorken syndrome* is also rare and is almost the reverse of Scott syndrome. Non-activated platelets express full procoagulant activity, but a reduced response to collagen.

14.3.9 Acquired platelet function disorders

Some systemic disorders affect platelet function, such as renal failure (due to the retention of platelet inhibitory substances), liver disease, and DIC. Abnormal platelet function, as well as thrombocytopenia, is sometimes seen associated with the chronic myeloproliferative disorders, acute myeloid leukaemia, and myelodysplastic syndromes you met in Chapter 11. Significant numbers of patients with myeloma and Waldenström macroglobulinaemia have impaired platelet function due to the interference of function by paraproteins.

Drugs are the most common cause of platelet dysfunction. The humble aspirin many of us use to alleviate headaches irreversibly inactivates the platelet cyclooxygenase pathway, thereby blocking aggregation. Aspirin can therefore be used as an effective antithrombotic drug. Many other drugs are known to affect platelet function to varying degrees, such as:

- non-steroidal anti-inflammatory drugs (NSAIDs)—aspirin is also an NSAID;
- antibiotics;
- cardiovascular drugs;
- psychotropic drugs, such as antidepressants;
- anaesthetics;
- chemotherapeutic drugs;
- anticoagulants.

Not every drug within each category will affect platelet function and their effects tend to be concentration-dependent.

Some foods and food additives can also affect platelet function if ingested in sufficient amounts. Examples include fish oils, green tea, garlic, cumin, turmeric, chocolate, excessive use of vitamin supplements, and the black tree fungus used in Chinese takeaways. In sufficient amounts, these foodstuffs can generate abnormal results in platelet function testing that can complicate diagnostic interpretation. However, abnormal results due to foodstuffs alone are unlikely to translate into a genuine acquired bleeding disorder.

SELF-CHECK 14.15

Name the main causes of acquired platelet dysfunction.

14.3.10 Laboratory investigation for platelet defects

Assessing a patient for a platelet function defect begins with taking a detailed history from the patient and should include the following:

- personal and family history of bleeding episodes;
- severity, frequency, and type of bleeding episodes;
- drugs the patient is taking, detailing both prescribed and over-the-counter drugs;
- dietary intake;
- lifestyle choices, such as smoking and exercise, which can affect platelet function.

Pre-analytical questionnaires are used to rationalize this process and assist the clinician. If they reveal that the patient's diet, lifestyle, or medication could significantly interfere with platelet function, laboratory testing may need to be postponed to allow for a period of abstinence prior to analysis. For an example of a pre-analytical questionnaire, see Figure 14.13, which is used in the authors' laboratories.

14.3.11 Laboratory analysis

The first assays a biomedical scientist will undertake will be to check the size and number of the platelets. This is easily done on the current automated analysers that you met in Chapter 2, although their limitations must be recognized. Analysers that distinguish between cellular components based on size may erroneously place large platelets in the same sector as red cells, whilst small red cells would be placed in the platelet segment. This will lead to falsely reduced or elevated platelet counts.

Compare the platelet count histograms in Figure 14.14. In the normal histogram, you can see there is no cross-over between the population of particles counted as platelets and the threshold for inclusion in the erythrocyte count. The abnormal histogram is from a patient with giant platelets, some of which are so large that they cross the threshold. When the histograms produced by the analyser indicate this is the case, an alternative method is required to obtain an accurate count. Flow cytometric counts can be performed and you will meet them later in this chapter. As you will see later in this section, platelet morphology on Romanowsky-stained peripheral blood films can aid diagnosis.

Questionnaire for all platelet aggregation patients & controls

Question	Response
Are you on:	
• aspirin (within the last 14 days MINIMUM)	
• over-the-counter cold relief medication e.g. Lemsip	
• ibuprofen	
on any of the prescription medications below? COX1 inhibitors (indomethacin, sulfinpyrazone, naproxen), ADP receptor agonists (ticlopidine, clopidogrel), GPIIb/IIIa agonists (Reopro, tirofiban, eptifibatide, integrilin), prostaglandins (cliostazol), tri-cyclic antidepressants (imipramine, amitripyline).	
If the PATIENT has answered YES to the above, please reschedule. If the CONTROL has answered YES, please find another volunteer.	
on any other current prescription medication? Please circle relevant drugs.	
Penicillin, ampicillin, propranolol, atenolol, captopril, perindopril, anaesthesia.	
Are you currently taking vitamin supplements? **If yes, tick or circle if listed below:** Vitamins B_6, C, E all affect platelets.	
Are you taking dietary supplements or herbal remedies? **If yes, please tick or circle if listed below:** Starflower oil, fish oil, Ginko biloba and green tea all have platelet antagonistic ingredients.	
Have you had a take-away within the last 48 hours? Elements in turmeric, cumin, onion, garlic, ginger, clove, black tree fungus & monosodium glutamate all affect platelets.	
Do you drink or smoke? Alcohol, caffeine and smoking have been known to affect platelets.	
Was this a clean draw?	

FIGURE 14.13
Pre-analytical questionnaire for completion prior to platelet function analysis.

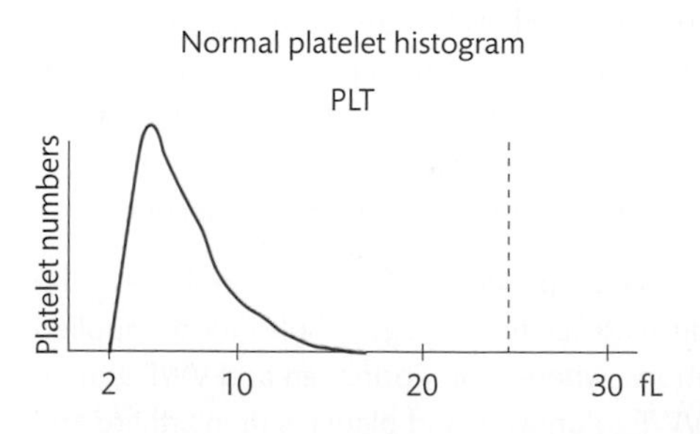

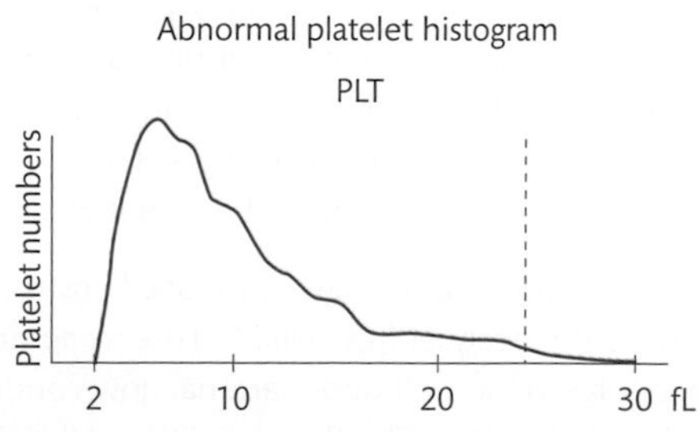

FIGURE 14.14
Platelet count histograms. The vertical dashed line indicates the cell-volume threshold for differentiating between platelets and red blood cells. Any cell above 25 fL is counted as a red blood cell and below 25 fL as a platelet.

14.3.12 Assessment of platelet function—screening tests

Platelet function can be empirically assessed with a mildly invasive technique known as the *bleeding time*. A small uniform incision 1 mm deep is made in the skin of the underside of the forearm. Blood emerging from the wound is carefully removed with blotting paper at 30-second

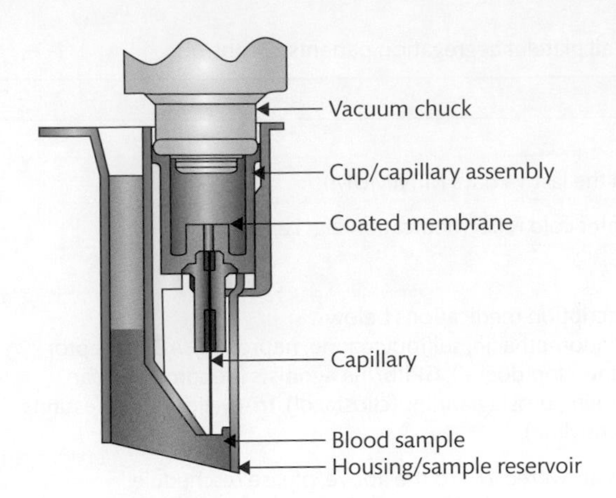

FIGURE 14.15

PFA-100® cartridge. Anticoagulated whole blood is added to the reservoir in the cartridge. Pressure equivalent to that in the arteries passes the blood into a thin-bore capillary towards the agonist-impregnated collagen membrane. Blood passes through the pore in the membrane and the collagen-VWF-platelet interactions of primary haemostasis are initiated. The time taken for the clot to occlude the aperture is reported as the closure time.

intervals and the time is noted when the bleeding stops. The procedure is partly standardized by using blades of a standard width and depth and using a sphygmomanometer to maintain a blood pressure of 40 mmHg in the arm being tested. However, the test is prone to operator variability and relies on the patient keeping still throughout. The small incision only induces trauma in the microcirculation and thus relies almost entirely on fully functioning primary haemostasis. Therefore, blood vessel disorders, platelet number, and/or function abnormalities and some types of VWD will prolong the bleeding time—but haemophilia will not.

In response to the poor reproducibility of the bleeding time test, a number of instrument manufacturers have produced analysers that screen for abnormalities of primary haemostasis. The most widely used in the UK is the platelet function analyser (PFA-100®) from Siemens Healthcare Diagnostics, or its updated version, the Innovance® PFA-200. The PFA-100 cartridge, shown in Figure 14.15, functions by using an equivalent pressure to that in the arteries to pass whole blood from a reservoir through a capillary. At the end of the capillary is a collagen membrane coated with either ADP or epinephrine (adrenaline) that contains a pore with a diameter of 150 μm. The collagen reproduces the exposure of platelets to the subendothelial matrix and the agonists mimic localized activation. The analyser measures the time taken for enough platelets to adhere and aggregate that they occlude the pore in the membrane; the result is reported as closure time and is measured in seconds. Insufficient platelet numbers, reduced platelet function, and some forms of VWD will give elevated closure times, but, clearly, vessel abnormalities cannot be detected.

Other manufacturers have produced instruments employing differing technologies. The Impact-R™ analyser from DiaMed uses *cone-and-plate(let)* technology. Whole blood is applied to a polystyrene well under arterial flow conditions, whereupon fibrinogen and VWF adhere to the well surface. The flow conditions allow VWF to unravel, and platelets then adhere and aggregate. Excess blood is washed off, and the adhered and aggregated platelets are stained. An image analyser quantifies the percentage of the well covered by the aggregates, which represents adhesion, and the average size of the aggregates themselves, representing aggregation. You can see platelet aggregates in a well in Figure 14.16.

Thromboelastography (TEG) is a measure of clot elasticity that assesses the whole dynamic process of haemostasis rather than isolating specific areas, and as such measures the strength of platelet interactions. Whole blood is activated via the contact pathway using Celite or kaolin

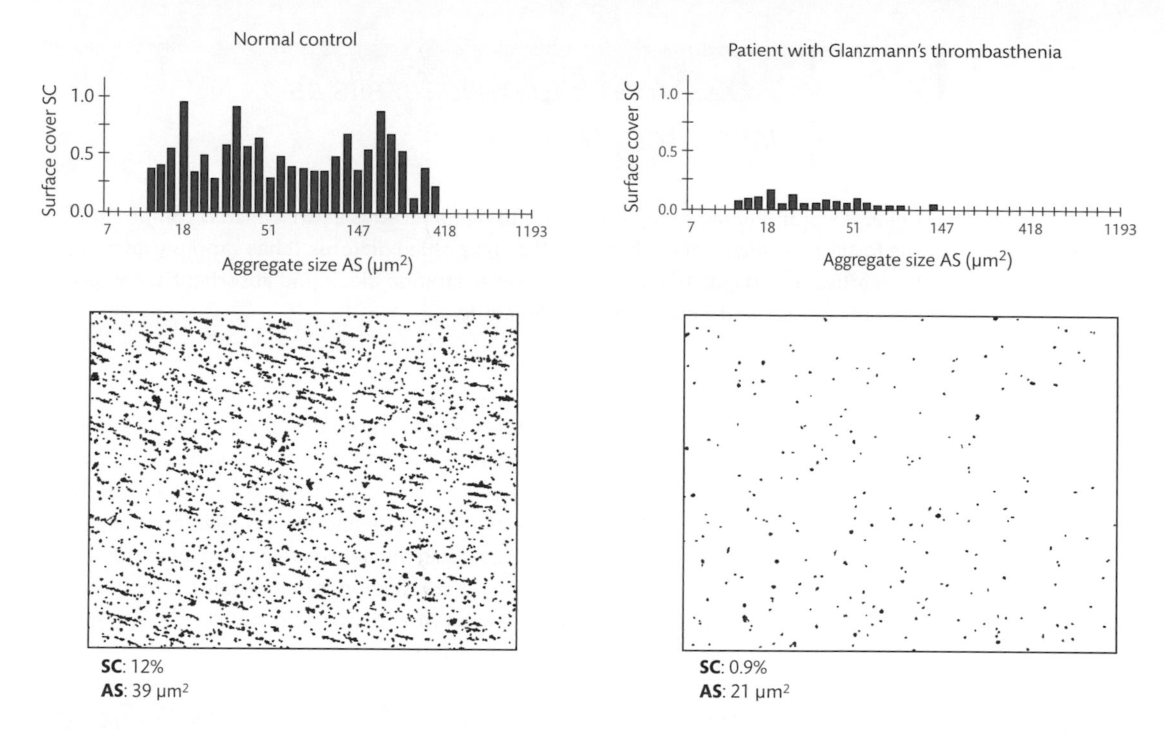

FIGURE 14.16

Result output from a DiaMed Impact-R analyser showing the surface coverage vs aggregate size histograms and the images produced by the test well. In the example on the right, the patient's platelets have largely failed to adhere to the test surface or to aggregate with each other. This gives poor surface coverage and very small aggregates indicative of a platelet function disorder: in this case, Glanzmann's thrombasthenia.

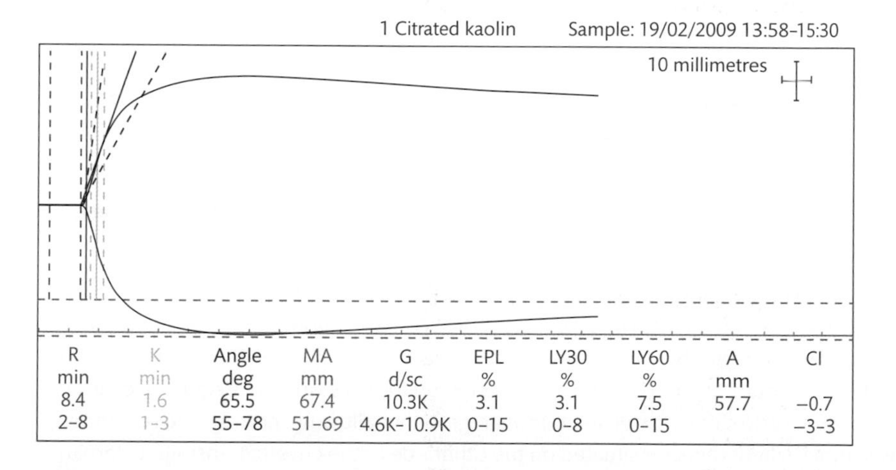

FIGURE 14.17

Output from a kaolin-activated normal whole-blood sample processed on a Haemoscope TEG® analyser, showing: R, the value from addition of blood sample to clot formation; K, time taken to reach a given clot strength; G, measure of clot strength; angle, a measure of the speed/strength of clot formation; MA, the maximum amplitude, which is when the clot formed is at its strongest. LY30 and LY60 measure clot lysis after 30 and 60 minutes, respectively. EPL is the estimated clot lysis measured at a specific time point, and is equivalent to LY30 at 30 minutes and LY60 at 60 minutes. A is the amplitude of the tracing at the latest time point. CI (coagulation index) is a manufacturer's composite value taking into account R, K, MA, and angle.

BOX 14.9 *Explosive fossils as a laboratory tool*

Celite is a brand name for a preparation of diatomaceous earth, which is composed of the fossilized remains of the hard-shelled algae called diatoms. It has various uses other than activating coagulation *in vitro*, such as a filtration aid, liquid absorbent, a component of cat litter, and a component of dynamite.

and placed into a pre-warmed cuvette (see Box 14.9 for more details on Celite). A suspended rotating torsion wire is lowered into the cuvette and the blood is allowed to clot under a low-shear environment resembling sluggish venous blood flow. Fibrin strands interact with activated platelets and attach to the surface of the cuvette and torsion wire, resulting in the clot transmitting its movement to the wire. Changes in resistance are plotted and analysed on computer software to give distinctive 'fingerprints' for different disease states. A modification of this method exists where it is the cuvette and not the torsion wire that rotates. A graphical plot from a TEG analysis is shown in Figure 14.17.

SELF-CHECK 14.16

What first-line screening tests are available when assessing platelet abnormalities?

14.3.13 Assessment of platelet function–diagnostic tests

Because the screening tests cannot distinguish between VWD and platelet function disorders, more specific analyses are required to isolate causes of platelet dysfunction.

Platelet function analysis

Platelet aggregation was devised by Gustav Börn and John O'Brien in the early 1960s. There are two main techniques.

- The first involves producing *platelet-rich plasma* (PRP) by centrifuging the sample at a low speed for an extended time. The supernatant then contains plasma and platelets, the leucocytes and erythrocytes being spun down. PRP is placed in a cuvette in front of a light source. The PRP is constantly stirred to keep the platelets uniformly suspended. Addition of an agonist to the cuvette causes the platelets to aggregate, thus allowing more light to pass through the sample. A light detector situated on the other side of the cuvette transmits absorbance changes to a chart recorder. The amount and rate of fall of absorbance (i.e. increase in light transmission) are a function of agonist concentration and innate platelet function. You can see in Figure 14.18 that a normal aggregation response has different stages.
- The second involves placing a probe with two fine wires in the sample. A current is passed through the wires and as the platelets adhere to the wires in response to agonists the resistance changes. In accordance with Ohm's law, the current changes and this change in electrical signal is converted to an aggregation value. As this method doesn't rely on light passage, whole blood can be analysed for aggregation.

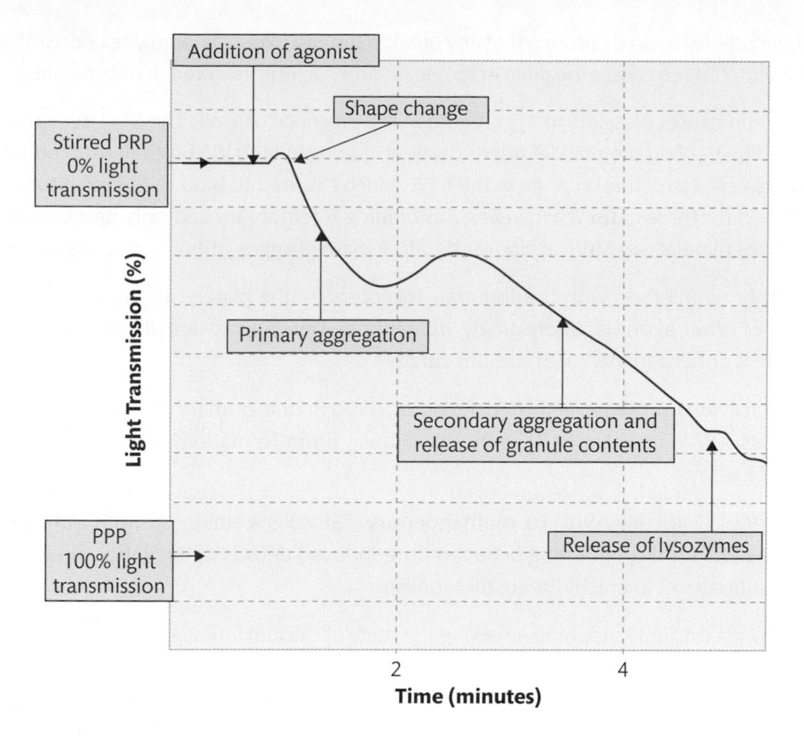

FIGURE 14.18

Platelet aggregation trace showing a normal response to the agonist ADP at a concentration of 2 µmol/L. Assay performed on a BioData PAP-8E aggregometer. The analyser is set to recognize platelet-rich plasma (PRP) as zero light transmission. The patient's platelet-poor plasma (PPP) is used to set the analyser to recognize 100% light transmission. Addition of the agonist causes the platelets to initially change shape from a disc to a spiny sphere. They then enter a phase of primary aggregation where their altered shape promotes interaction by overlapping. This phase can reverse if the stimulus is not strong enough or the platelets are dysfunctional and cannot enter the secondary phase. The release reaction occurs in the secondary phase, causing irreversible aggregation.

A number of agonists are available that cause platelet activation, aggregation, or agglutination via different mechanisms. Different platelet function abnormalities can be identified, depending on the pattern of responses to the panel of agonists. The most common agonists used are ADP, arachidonate, collagen, ristocetin, and epinephrine. Thrombin-receptor agonist peptide (TRAP), U44619, and A23187 are used in more specialist laboratories. The modes of action of each agonist are summarized below.

- ADP activates the $P2Y_1$ and $P2Y_{12}$ receptors on the platelet surface. For ADP to function as an activator, both receptors must be triggered. The $P2Y_1$ receptor triggers an enzyme cascade within the platelet which results in calcium mobilization. Calcium is a potent local platelet activator. $P2Y_{12}$ amplifies the signal via adenylyl cyclase inhibition to increase aggregation and activate secretion.
- Arachidonate is converted into thromboxane A_2 (TXA_2) by intra-platelet cyclooxygenase (COX). TXA_2 then activates platelets and also activates an internal enzyme cascade, which mobilizes calcium.
- Collagen activates GPIa and GPVI. When GPIa is activated, the platelet adheres to the exposed surface. This interaction ensures that the platelet will adhere to exposed collagen from the

subendothelial matrix, ensuring that any break in the vasculature is rapidly sealed. GPVI stimulation by collagen causes the platelet to release other activators stored in the granules.

- Ristocetin causes platelets to agglutinate in the presence of VWF. The VWF receptor complex GPIb–V–IX on the platelet needs divalent cations to bind VWF. When ristocetin is present, there is a structural change in the VWF which causes it to bind to the receptor without the need for these cations. However, this binding is temporary and only produces agglutination of platelets via VWF as opposed to true platelet aggregation.
- Epinephrine interacts with α-adrenergic receptors on the platelet surface, enhancing the effect of other agonists. Interestingly, epinephrine only aggregates platelets *in vitro* if the sample is anticoagulated with sodium citrate.
- TRAP activates protease-activated receptor (PAR) 1, the primary thrombin receptor on platelets. TRAP activates PAR1 without initiating fibrin formation, which would interfere with the assay.
- U46619 (9,11-dideoxy-9α,11α-methanoepoxy $PGF_2α$) is a stable thromboxane **mimetic** and is useful for distinguishing between drug-induced defects of thromboxane generation and disturbances in arachidonate metabolism.
- A23187 is a calcium ionophore, enabling the study of calcium flux across platelet membranes.

Mimetic
A compound that mimics properties of another.

These agonists activate enzyme cascades within the platelet that mobilize the surface receptor GPIIbIIIa (inside-out signalling), which, in turn, binds fibrinogen in a calcium-dependent fashion. Other agonists, such as those derived from snake venoms, are available, but have limited use outside the research arena.

SELF-CHECK 14.17

Outline the principles of platelet function analysis.

Identifying platelet disorders

Platelet aggregation studies are most commonly performed using the light-transmittance method with PRP. Table 14.8 details expected aggregation patterns encountered in the platelet function disorders. Further analyses can be performed to confirm and characterize diagnoses of glycoprotein defects and storage pool disorders.

Glycoprotein defects

Platelet glycoprotein receptors can be identified, or demonstrated to be reduced/absent, using the immunophenotyping method on a flow cytometer you met in Chapter 10 for identifying the cell types in haematological malignancies. The antibodies used are raised against the different platelet surface glycoproteins themselves.

Antibodies raised to the specific platelet receptors GPIb and GPIIbIIIa can be used to place a fluorescent 'tag' on normal platelets, and the number of 'tags' counted using a flow cytometer. This will then give an indication of the number of platelets within the sample that is more accurate than FBC analysers, which utilize the size gating system to count cells. Size gating suffers if the cells you are attempting to count are outside the set gates, such as small red cells or large platelets. Tagging platelets for flow cytometric counting means that distinguishing them is not dependent on size and a more accurate count is produced.

TABLE 14.8 Typical platelet aggregation response patterns in platelet disorders.

Condition	Platelets		Aggregation responses								Comment/further testing
	Count	Size	ADP	Coll	Ri	AA	Epi	TRAP	U46619	A23187	
Glanzmann's thrombasthenia	N	N	0	0	1	0	0	N	N	0	Flow cytometric analysis for GPIIbIIIa
Bernard–Soulier syndrome	Low	Large	N	N	0	N	N	N	N	N	Flow cytometric analysis for GPIb
COX deficiency	N	N	1/N	R	N	R	N	N	N	R	COX analysis
Drug-induced platelet defects; results shown are for aspirin	N	N	1	R	N	R/0	N	N	N	N/R	Stop aspirin (use of questionnaire reduces frequency of this scenario)
Thromboxane deficiencies	N	N	1/N	R	N	R/0	N	N	N/R	N	Platelet nucleotides, platelet biochemistry analysis
Thromboxane receptor defects	N	N	1/N	R	N	R/0	N	N/R	R/1	N	Platelet nucleotides, platelet biochemistry analysis
$P2Y_{12}$ defects	N	N	R/0	N	N	N	N	N	N	N	Receptor analysis, platelet biochemistry analysis
MYH9-RD	Low	Large	N/R	N/R	N	N	N	N	N	N	Genetic analysis, blood film for neutrophil (Döhle) inclusions
PAR1 and PAR4 defects	N	N	N	N	N	N	N	R/0	N	N	Genetic analysis
Ehlers-Danlos syndrome	N	N	N	N	N	N	N	N	N	N	Genetic analysis; other associated clinical manifestations, e.g. albinism, hearing loss, nephritis (note that the bleeding is due to collagen defects, not platelet function defects)
Storage pool defects											
Grey platelet syndrome	Low	N	R/1	N/R	N	N	N	R	N	N	Electron microscopy, alpha-granule content analysis
Quebec platelet syndrome	N	N	N	N	N	N	0	N	N	N	Multimerin analysis, other alpha-granule content analysis
Chediak–Higashi syndrome	N	N	R/1	R	N	N	R	N	N	N	Dense granule analysis, platelet nucleotide analysis
Hermansky–Pudlak syndrome	N	N	R/1	R	N	N	R	N	N	N	Dense granule analysis, platelet nucleotide analysis; other syndrome-associated clinical manifestations, e.g. albinism, prone to infections
Wiskott–Aldrich	Low	Small	R/1	R	N	N	R	N	N	N	Genetic analysis, WASp analysis
Montreal platelet syndrome	Low	Large	R	R	R	N	N	N/R	N	N	Spontaneous aggregation
Membrane anomalies											
Scott syndrome	N	N	N	N	N	N	N	N	N	N	Flow cytometric annexin A5 analysis, genetic testing
Paris-Trousseau syndrome	Low	Large	N/R	N	N	N	N	R/1	N	N	Electron microscopy for large alpha granules; other physical/clinical manifestations
Stormorken syndrome	Low	Large	N	R	N	N	N	N	N	N	Reduction in ATP secretion, spontaneous aggregation in whole blood, flow cytometric annexin A5 analysis

N = normal; 0 = absent; 1 = primary wave only; R = reduced; Coll = collagen; Ri = ristocetin; AA = arachidonate; Epi = epinephrine; COX = cyclooxygenase; TRAP = thrombin receptor agonist peptide; MYH9-RD = myosin heavy-chain-9-related disorders.

Storage granule defects

If defects in the dense bodies are suspected (δ-storage pool disorders) a luminosity technique can ascertain the quantity of the nucleotides ATP and ADP in the dense bodies. The nucleotides must be extracted from platelets by preparing PRP and then adding the following:

- EDTA to prevent any Ca^{2+}- dependent reactions;
- the detergent Triton-X to disrupt platelet membranes and release the cytosol;
- ethanol to precipitate out membrane proteins and stabilize nucleotide concentrations.

This is then centrifuged and the supernatant analysed. The supernatant is incubated with the enzyme *luciferase*, which is extracted from fireflies (*Photinus pyralis*) and its substrate, *luciferin*. The following reaction occurs:

$$\text{ATP + Luciferin} \xrightarrow{\text{Luciferase + Mg}^{++}} \text{Adenyl Luciferin + Light}$$

The amount of light produced is measured and is directly proportional to the amount of ATP in the supernatant. This reaction only measures ATP. The nature of ADP is such that it is unstable and cannot be measured directly. In order to measure the ADP, it is converted to ATP in a separate aliquot of supernatant in the reaction below and the ATP is quantified again using the reaction above.

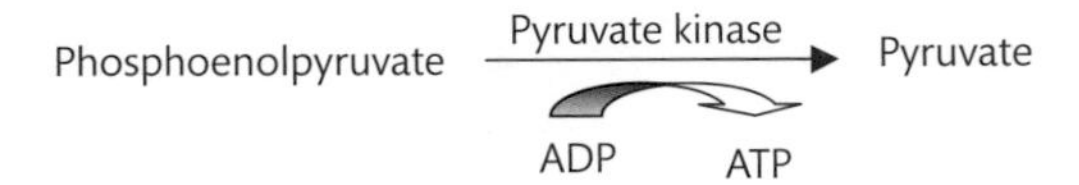

The ATP will be higher in the second result because the reaction tube will contain the ATP already present from the cytoplasm and dense bodies plus the converted ADP. The concentration of ADP is calculated by subtracting the result of the first reaction from that of the second.

About 30% of platelet ATP and most of the ADP is in the dense bodies, the ratio of total ATP:ADP in normal platelets being <2.5. The storage nucleotides are reduced in dense body defects, resulting in an increase in the ATP:ADP ratio. Up to 25% of patients with storage pool disorders may not manifest in standard platelet aggregation testing, so nucleotide quantification is an important part of the diagnostic armoury.

When a defect in α-granules is suspected (α-storage pool disorders), ELISA techniques are available to measure α-granule contents such as VWF, fibrinogen, beta-thromboglobulin, platelet factor 4, and P-selectin.

Lumi-aggregometry

Simultaneous analysis of both aggregation and the release of storage pool contents can be undertaken to detect enzyme defects. The analyser used is termed a lumi-aggregometer and contains two channels. One channel is for standard light impedance-based aggregometry and the other is for a luminometer that measures ATP release over time via the luciferase reaction. The analysis can be performed on whole blood or PRP. An enzyme defect that does not initiate dense body release will demonstrate normal measured levels of nucleotides and a normal ATP:ADP ratio, but reduced release patterns by lumi-aggregometry. Dense body defects will be abnormal in both analyses. Additionally, release defects will give a reduced response to arachidonate in aggregometry, whilst storage pool defects will be normal. Typical lumi-aggregometer tracings are shown in Figure 14.19.

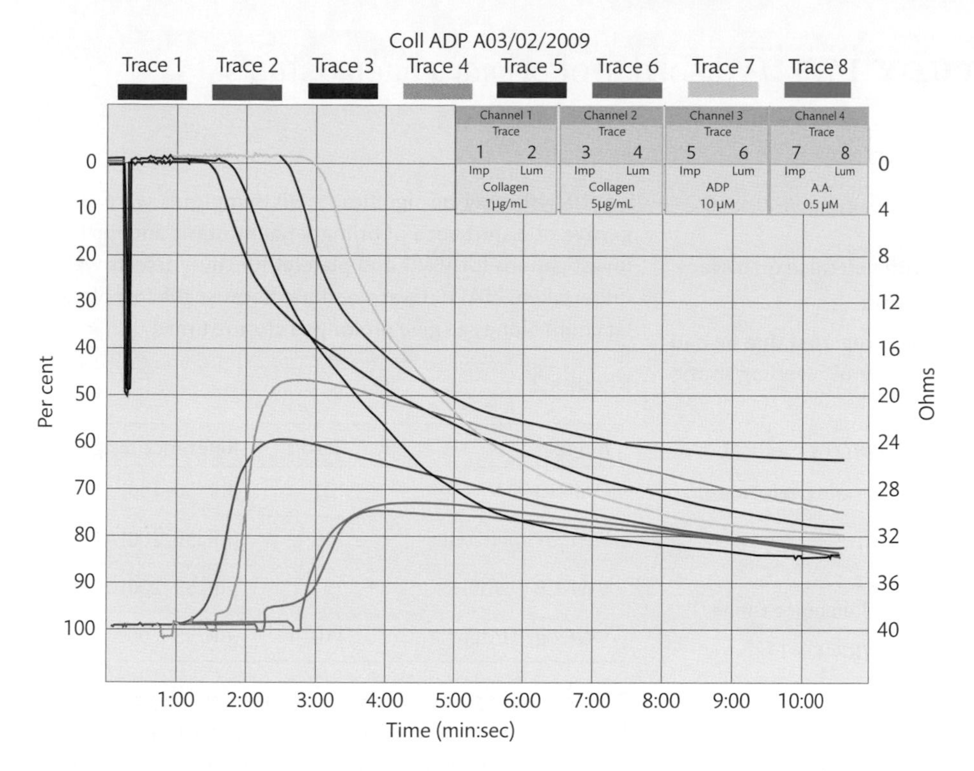

FIGURE 14.19

Normal traces from a Chronolog 700V whole-blood aggregation analyser showing both the whole-blood impedance aggregation trace (Imp) and the release of ADP detected by chemiluminescence (Lum). These traces show that the patient's platelets respond to the addition of varying agonists both by aggregation (top set of curves) and by simultaneously releasing nucleotides (bottom set of curves), to initiate the second phase of aggregation.

Pitfalls of platelet function testing

As you saw earlier (and see Figure 14.13), many external factors interfere with platelet function. The patient should be made aware and requested to minimize their exposure to these elements before analysis takes place.

Most platelet function testing has to be done on fresh blood samples because platelets lose activity over time and cannot be stored. This means the patient must attend the clinic and the analysis be performed within three hours of venepuncture. Similar to coagulation screening, the quality of phlebotomy is important for the generation of clinically useful results. The needle must be of sufficiently wide bore so that it does not activate the platelets in the process of venepuncture. The venepuncture itself should not be traumatic for the patient, otherwise the platelets can become activated and mask a functional defect.

DNA-based diagnosis of inherited platelet disorders

Whilst inherited platelet disorders (IPD) such as Glanzmann's thrombasthenia and Bernard-Soulier syndrome can usually be confidently diagnosed with clear medical histories and standard phenotypic assays, other IPDs are do not always generate clear-cut, standard phenotypic data. Examples include Scott syndrome and mild Stormorken syndrome. Greater use of high-throughput genetic sequencing has the potential to improve diagnostic accuracy for IPDs. Case study 14.5 details the medical history and laboratory results of a patient with a disorder of primary haemostasis.

CASE STUDY 14.5 A disorder of primary haemostasis

Patient history

- A six-year old male presented with petechiae on his face, chest, abdomen, and all four limbs.
- His parents, who were first cousins, reported that he had a history of epistaxis but no history of bleeding in the immediate family.
- He was otherwise well and was not on any medications.
- FBC and coagulation screening tests were performed.

Results 1

Index	Result	Reference range
WBC (× 10^9/L)	10.7	(RR: 4.0–11.0)
Hb (g/L)	126	(RR: 120–160)
Platelet count (× 10^9/L)	88	(RR: 150–400)
PT (s)	11.9	(RR: 10.0–14.0)
APTT (s)	35.7	(RR: 32.0–42.0)
Fibrinogen (g/L)	3.2	(RR: 1.5–4.0)

Blood film examination revealed numerous large and giant platelets.

Significance of results 1

The normal PT and APTT excluded deficiencies or inhibitors of factors in the extrinsic, intrinsic, and common pathways, vitamin K deficiency, and many anticoagulant drugs. Fibrinogen was normal. Although there was a clear thrombocytopenia, the platelet count was not necessarily low enough to cause spontaneous bleeding, yet the large and giant forms may be significant. His symptoms were suggestive of disturbance of primary haemostasis and further investigations for VWD and platelet function defects were undertaken. PFA-100 was not done because the low platelet count alone can give prolonged closure times.

Results 2

Index	Result	Reference range
FVIII (IU/dL)	101	(RR: 50–150)
VWF:GPIbR (IU/dL)	105	(RR: 55–200)
VWF:CB (IU/dL)	99	(RR: 55–200)
VWF:Ag (IU/dL)	110	(RR: 55–200)

Platelet aggregometry revealed normal responses to ADP, arachidonic acid, collagen, and epinephrine, but no response to ristocetin at any concentration.

Significance of results 2

The normal FVIII and VWF parameters exclude VWD, meaning that the absent response to ristocetin in platelet aggregometry was due to a platelet function defect. Since the patient's platelets responded to all other agonists, and given the presence of large and giant platelets, this is a classic presentation of Bernard-Soulier syndrome (BSS), which involves deficiencies in components of the GPIb-V-IX surface receptor complex. BSS is an autosomal recessive disorder, which explains why the parents were asymptomatic as they were both heterozygous carriers. Offspring of consanguineous parents can be at greater risk of autosomal recessive disorders inherited from a common ancestor, with first cousins predicted to share 12.5% of their genes.

14.4 Diagnostic algorithm for bleeding disorders

Now that you have met the screening tests and main diagnostic assays for identifying platelet abnormalities, VWD, and coagulation factor deficiencies, look at Figure 14.20, which shows a generalized algorithmic approach to detecting and characterizing bleeding disorders. Such an algorithm cannot encompass every subtype of each disorder, and instead gives a broader approach that will detect most disorders in the first instance. The algorithm shows a decision

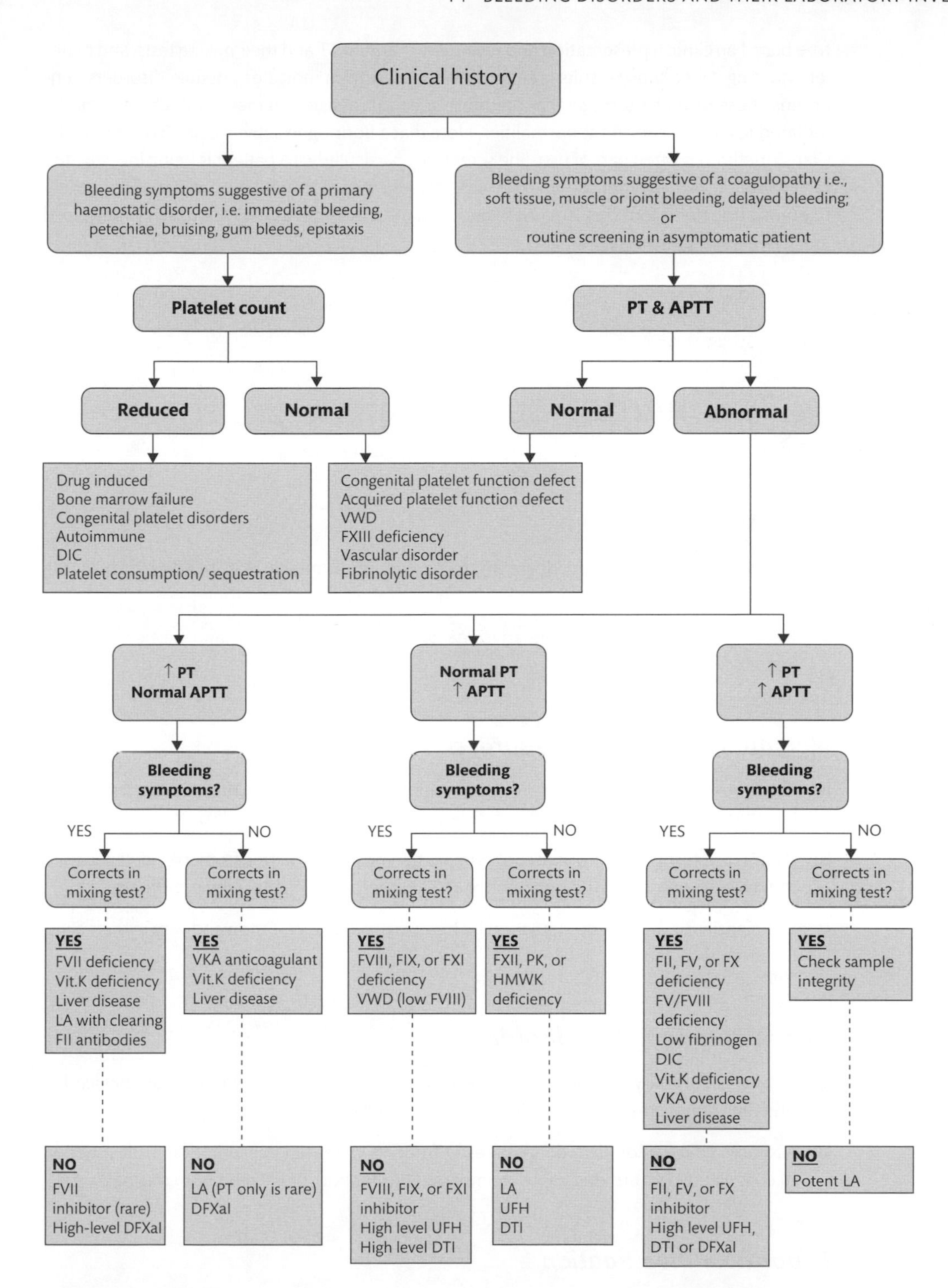

FIGURE 14.20

General diagnostic algorithm for detecting bleeding disorders. APTT = activated partial thromboplastin time; DFXaI = direct activated factor X inhibitor; DIC = disseminated intravascular coagulation; DTI = direct thrombin inhibitor; HMWK = high molecular weight kininogen; LA = lupus anticoagulant; PK = prekallikrein; PT = prothrombin time; UFH = unfractionated heparin; vit. K = vitamin K; VKA = vitamin K antagonist; VWD = von Willebrand disease.

tree based on clinical presentation and results of PT and APTT and their mixing tests, and platelet counting. Once those results are mapped, the algorithm indicates possible disorders generating those result patterns so that the more specialist assays you met in this chapter can be initiated to identify specific abnormalities. Note that a fibrinogen activity assay (commonly the Clauss method) is often part of first-line screening, particularly if a patient is being investigated for bleeding. You also met the PFA-100/200 analysers as a screening procedure for primary haemostatic defects, but they are not included in the algorithm as they are not as widely used as PT, APTT, and platelet counting.

Chapter summary

Clinical considerations

- The nature and sites of bleeding symptoms give important clues to the type of bleeding disorder that may be present.
- Bleeding disorders can involve any area of haemostatic function and be hereditary or acquired.

Hereditary coagulation disorders

- Haemophilia A and B are rare X-linked deficiencies of FVIII and FIX, respectively, and can be life-threatening. Haemophilia C, a deficiency of FXI, is a less severe disorder.
- VWD is a heterogeneous disorder of primary haemostasis resulting from VWF deficiency. In the more severe forms, the reduction in FVIII exacerbates the bleeding symptoms.
- Deficiencies of the other coagulation factors are much less common and vary in their clinical severity.

Acquired coagulation disorders

- A variety of primary disorders can cause secondary disturbances in haemostasis that lead to bleeding symptoms.
- Acquired coagulation disorders can result from an increased consumption or destruction of coagulation factors, decreased or impaired production, or via inhibitory mechanisms.

Laboratory investigation

- Investigation for a bleeding disorder begins with a coagulation screen and platelet assessment.
- Coagulation screening comprises PT and APTT and assessment of fibrinogen activity via TT and/or direct quantification.

- Interpretation of PT and APTT is based on the coagulation cascade theory. Although outmoded in terms of *in vivo* haemostasis, the cascade is used in the diagnostic setting in view of the test principles. PT assesses extrinsic and common pathway function, whilst APTT assesses intrinsic and common pathway function.
- Mixing studies can help direct diagnostic pathways by indicating the possible presence of either factor deficiencies or inhibitors.
- Coagulation screening does not detect all haemostatic abnormalities.
- High-quality results can only be generated from high-quality blood samples.
- Individual clotting factors are quantified by rendering the patient's clotting factor under investigation the rate-limiting factor in specialist assays based on PT or APTT. Scrutiny of the raw data is essential to ensure analytical validity and exclude interfering factors.
- Inhibitors of specific clotting factors are quantified by mixing a range of dilutions of the patient's plasma with a plasma of known factor content and incubating at 37°C for an appropriate length of time. The inhibitor will neutralize the factor to varying degrees because of the increasing dilutions and the value derived from this data.
- Most inhibitors act immediately, but FVIII inhibitors act progressively because the FVIII has to dissociate from VWF before the inhibitor can exert its effect.
- VWF is assayed using a variety of techniques in order to identify the subtypes. Activity is measured by platelet-binding and collagen-binding techniques and total VWF protein by immunological techniques. Some patients require ristocetin dose–response studies. VWF substructure is demonstrated by electrophoresis.
- Thrombocytopenia alone can cause bleeding symptoms, so the platelet count is crucial as an initial investigation.
- Disorders of platelet function can be detected in purpose-designed analysers that screen for abnormalities of primary haemostasis, which includes VWD, but not vessel disorders.
- The mainstay for assessing hereditary platelet function disorders is aggregometry on PRP. Agonists that initiate platelet aggregation via different physiological mechanisms are added to separate PRP aliquots and the responses plotted graphically. The response patterns indicate the nature of any functional abnormalities.
- Follow-up investigations for platelet diagnostics include flow cytometry to indicate the presence/absence of surface glycoproteins, nucleotide quantification, and the release and measurement of α-granule contents.

Discussion questions

14.1 Why are the clinical symptoms of haemophilia A different to those of VWD, taking into account that the latter can be accompanied by reduced FVIII levels?

14.2 What do we gain by using coagulation screening tests that are based on an outmoded theory that does not fully mirror *in vivo* events?

14.3 What are the possible causes of finding an elevated PT and APTT in the presence of a normal TT and fibrinogen level? Justify your answer.

14.4 Assays to quantify FX tend to be based on PT analysis. Why might this be the case when an APTT-based assay would also work?

14.5 Surely a VWF:RCo assay tells us all we need to know about the clinical severity of any VWD subtype, so why bother with all the other tests?

14.6 If platelet function *in vivo* is inextricably linked to interactions with vessel endothelium, do the existing methods for assessing platelet function give clinically relevant information?

Further reading

- **Bain BJ, Bates I, Laffan MA, and Lewis SM (eds). *Dacie and Lewis Practical Haematology*, 12th edn. Churchill Livingstone, Elsevier, London, 2017.**

  Includes sections on coagulation screening, factor and inhibitor assays, and platelet function analysis.

- Chee YL, Crawford JC, Watson HG, and Greaves M. ***Guidelines on the assessment of bleeding risk prior to surgery or invasive procedures***. *British Journal of Haematology* 2008: **140**: 496–504.

- Harrison P, Mackie I, Mumford A, Briggs C, Liesner R, Winter M, Machin S; British Committee for Standards in Haematology. ***Guidelines for the laboratory investigation of heritable disorders of platelet function***. *British Journal of Haematology* 2011: **155**: 30–44.

  Further detail on testing for platelet function disorders.

- Keesler DA and Flood VH. ***Current issues in diagnosis and treatment of von Willebrand disease***. *Research and Practice in Thrombosis and Haemostasis* 2018: **2**: 34–41.

- Laffan MA and Pasi KJ. ***Haemophilia and von Willebrand disease***. In: Hoffbrand AV, Higgs DR, Keeling DM, and Mehta AB (eds). *Postgraduate Haematology*, 7th edn (pp. 715–32), Wiley Blackwell, Oxford, 2016.

- Lentaigne C, Freson K, Laffan MA, Turro E, Ouwehand WH; BRIDGE-BPD Consortium and the ThromboGenomics Consortium. ***Inherited platelet disorders: toward DNA-based diagnosis***. *Blood* 2016: **127**: 2814–23.

- Marder VJ, Aird WC, Bennett JS, Schulman S, and White II, GC (eds). ***Hemostasis and Thrombosis: Basic Principles & Clinical Practice***, 6th edn. Lippincott Williams & Wilkins, Philadelphia, PA, 2012.

  Detailed chapters on different components of haemostasis.

- Mumford AD, Ackroyd S, Alikhan R, Bowles L, Chowdary P, Grainger J, Mainwaring J, Mathias M, O'Connell N; BCSH Committee. ***Guideline for the diagnosis and management of the rare coagulation disorders: A United Kingdom Haemophilia Centre Doctors' Organization guideline on behalf of the British Committee for Standards in Haematology***. *British Journal of Haematology* 2014: **167**: 304–26.

- **Srivastava A, Santagostino E, Dougall A, Kitchen S, Sutherland M, Pipe SW, Carcao M, Mahlangu J, Ragni MV, Windyga J, Llinás A, Goddard NJ, Mohan R, Poonnoose PM, Feldman BM, Lewis SZ, van den Berg HM, Pierce GF; WFH Guidelines for the Management of Hemophilia panelists and co-authors. *WFH guidelines for the management of hemophilia*, 3rd edn. *Haemophilia* 2020, Aug 3. doi: 10.1111/hae.14046 (in press).**

- **Young GA and Perry DJ; for the International Prophylaxis Study Group (IPSG). *Laboratory assay measurement of modified clotting factor concentrates: A review of the literature and recommendations for practice. Journal of Thrombosis and Haemostasis* 2019: 17: 567–73.**

Answers to self-check questions and case study questions are provided as part of this book's online resources.

Visit www.oup.com/he/moore-fbms3e.

15

Thrombophilia

Gary Moore and Ian Jennings

In this chapter, we will define thrombophilia, look at the causes, and discuss the methods used to identify the defects which may be involved.

Learning objectives

After studying this chapter, you should confidently be able to:

- understand what is meant by the term 'thrombophilia';
- know the main causes of thrombophilia;
- distinguish between hereditary and acquired causes of thrombophilia;
- outline the tests used to investigate thrombophilia;
- understand how the presence of one thrombophilia can interfere with the testing for some others;
- understand the importance of accurate result interpretation in thrombophilia diagnosis.

15.1 Introduction

In Chapter 13, you met the normal function of the haemostatic processes which make up the balanced system between pro- and anticoagulant mechanisms to keep the blood flowing through intact vessels, cause a clot to form at the site of vessel damage to prevent blood loss, and to prevent extension of the clot beyond the region of damage. You learnt in Chapter 14 that a reduction in the quantity or quality of the procoagulant clotting factors could lead to a tendency to bleed, such as the haemophilias. In a similar way, defects or deficiencies of the anticoagulant system can lead to a predisposition to thrombosis, which is termed **thrombophilia**, which translates literally to 'clot-loving'. The consequence of thrombophilia for the individual is an *increased risk* of venous thrombosis. This is in contrast to the more severe bleeding disorders where bleeding symptoms are inevitable.

Thrombophilia may be either hereditary or acquired. Some defects which are thought to cause thrombophilia are of uncertain **aetiology**; that is, we do not know whether or not they are caused by some genetic changes, or acquired, or possibly both. These hereditary, acquired, and uncertain causes of thrombophilia are listed in Table 15.1, and will be discussed in detail later.

TABLE 15.1 Causes of thrombophilia.

	Loss of function	Gain of function
Congenital/ inherited causes	Protein S deficiency Antithrombin deficiency Protein C deficiency	FV Leiden Dysfibrinogenaemia Prothrombin G20210A
Mixed/ uncertain	Elevated levels of FIX, XI, Thrombin-activatable fibrinolysis inhibitor (TAFI) Hyperhomocysteinaemia Elevated levels of FVIII APC resistance in the absence of FVL	
Acquired	*Persistent*: antiphospholipid syndrome, age, malignancy, history of venous thromboembolism *Transient*: surgery, immobilization, pregnancy, hormone therapy/oral contraceptive use	

The mechanisms involved in keeping blood in an anticoagulated (fluid) state were introduced in Chapter 13. Figure 15.1 is a reminder of some of the key proteins involved in this process. Central to both procoagulant and anticoagulant mechanisms is the key enzyme thrombin.

15.1.1 Natural anticoagulants

When produced in large amounts from the cleavage of prothrombin, thrombin will in turn cleave fibrinogen to form fibrin, will activate FXIII to stabilize the fibrin clot, and will feed back into the coagulation mechanism to produce further amounts of thrombin. However, when produced in smaller quantities, two mechanisms act to have a negative feedback on the

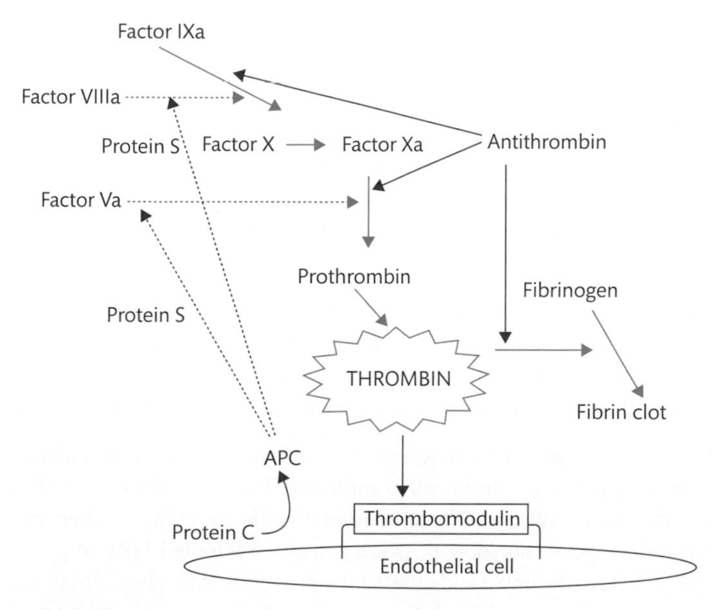

FIGURE 15.1

The central role of thrombin in procoagulant and anticoagulant mechanisms.

Key ⟶ Procoagulant activity

⋯► Anticoagulant/inhibitory activity

stoichiometric
The quantitative relationship between reactants and products in a balanced chemical reaction.

deep vein thrombosis (DVT)
A condition where a blood clot forms in one or more of the deep veins in the body, commonly in a leg.

pulmonary embolism (PE)
A fragment of a clot (the embolus) that travels through the circulation and lodges in the pulmonary artery or one of its branches, causing partial or total blockage. It is potentially fatal.

procoagulant system. First, thrombin will bind to the endothelial membrane-bound protein thrombomodulin to activate protein C. Activated protein C then proteolytically cleaves active factor V and factor VIII, converting these two procoagulant proteins into inactive forms. In addition, antithrombin binds in a 1:1 **stoichiometric** complex with thrombin, rendering it inactive.

Key Points

Thrombophilia is generally taken to relate to venous thrombosis—thrombi forming in the venous system—which can cause deep vein thrombosis (DVT) and pulmonary embolism (PE). Thrombosis may also occur in the arterial system, where a thrombus may cause a myocardial infarction (MI) or stroke. The mechanisms and causes are different in the two systems. Arterial thrombosis will be discussed later in section 15.7.3.

15.2 Deficiency of the natural anticoagulants

Deficiency of the natural anticoagulants antithrombin, protein C, and protein S may be as a result of a mutation causing a reduced concentration of the normally functioning protein in the circulation (**type I defect**), or a mutation affecting the function of the protein (**type II defect**). Deficiencies of these anticoagulants are sometimes termed *loss-of-function* defects.

15.2.1 Antithrombin deficiency

Antithrombin is a naturally occurring inhibitor of several components of the procoagulant system, including thrombin, FIXa, FXa, FXIa, and FVIIa–tissue factor (TF) complex.

In *type I antithrombin deficiency*, the concentration of antithrombin in the circulation is reduced. Antithrombin deficiency is an autosomal-dominant disorder, with levels in heterozygous individuals typically being about 50% of normal. Homozygous type I antithrombin deficiency is not encountered, and is thought to be incompatible with life. Heterozygous deficiency is associated with an up to ten-fold increased risk of **venous thromboembolism** (VTE) compared to normal. Mutations may cause a reduction in the amount of antithrombin produced, or increased destruction; point mutations, frameshift deletions or insertions, and major gene deletions have all been described.

venous thromboembolism
A disorder that encompasses both deep vein thrombosis (DVT) and pulmonary embolism (PE).

Type II antithrombin deficiency occurs when a mutation in the antithrombin gene affects the function of the molecule. Here, assays which measure the activity of antithrombin in plasma will be affected, but normal concentrations of protein may be measured by immunological assays. Type II defects are further subdivided, depending on the site of the mutation and the effect on function.

To understand the different subgroups of type II defects, it is necessary to review the function of antithrombin. In Chapter 13, you were introduced to antithrombin as an inefficient inhibitor of target enzymes in the absence of cofactors. The antithrombin molecule contains a heparin-binding region (D-helix) and a reactive site loop; this loop contains a **scissile** P1–P1′ (Arg393–Ser394) reactive site, which binds to thrombin and other serine proteases. When heparin (added *in vitro* or administered as an anticoagulant) binds to the antithrombin molecule, the reactive site loop undergoes conformational change and exposes the P1–P1′ reactive centre, which then inhibits its target proteases approximately 1000 times more efficiently than in the absence of heparin. *In vivo*, the heparin is provided by heparan sulphate on the surface of endothelial cells which line the blood vessel.

scissile
Easily split or cleaved.

Mutations causing type II antithrombin defects may affect the heparin binding site (type II HBS defects), or the reactive site (type IIRS), or different mutations in other regions of the antithrombin gene may be having varied effects on antithrombin function (type II pleiotropic defects).

15.2.2 Laboratory assays for antithrombin

In order to distinguish between defects of concentration and function, two types of assays are required: immunological and functional.

Immunological assays for antithrombin

Since type II defects will not be detected with assays measuring the amount of protein present in plasma (immunological assays), these assays are generally only useful to distinguish type I and type II antithrombin deficiency.

Radial immunodiffusion assays, Laurell rocket electrophoresis, enzyme-linked immunosorbent assays (ELISA), and latex bead-based immunoassays can all be used to measure antithrombin antigen. Figure 15.2 shows the principle of the Laurell rocket electrophoresis method. Examples of latex-based and ELISA methods are shown later in this chapter (Figures 15.5 and 15.6, respectively).

1. A gel plate is prepared containing antibodies against the target antigen.
2. Plasma is added to wells prepared in the gel and a current is applied to the plate.

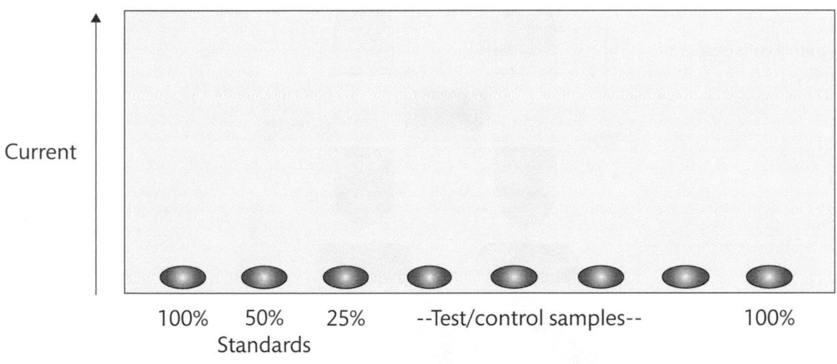

3. Proteins move across the plate and the antigen is captured and precipitated by the antibody. The distance travelled by the antigen is proportional to the quantity of antigen present in the plasma.
4. This can be visualized by staining the plate after drying, and the peak height can be plotted against the percentage of protein present.

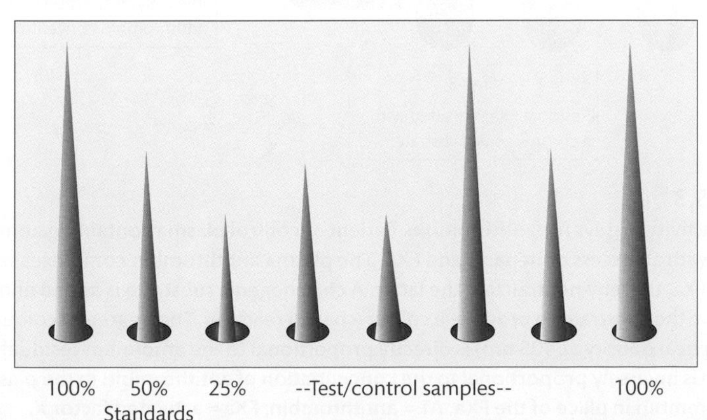

FIGURE 15.2
Principle of immunoelectrophoresis (Laurell rocket electrophoresis).

chromogenic substrate
A substrate that mimics the natural substrate of an enzyme, to which a dye, paranitroaniline (pna), is added. The enzyme will cleave the substrate to release the dye, causing a measurable colour change.

antithrombin Cambridge II
An antithrombin variant arising from an alanine 384–serine (A384S) substitution, with a prevalence of up to 1.14 per 1000 individuals, and associated with a mild increase in thrombotic risk. The mutation modifies, rather than abolishes, serine protease inhibitory activity, and inhibits FXa more efficiently than thrombin.

Antithrombin circulates at a concentration of approximately 125 mg/L (2.3 μmol/L) in normal plasma, although measurement is more frequently determined as a percentage of antithrombin compared to normal.

Functional assays for antithrombin

The principle of antithrombin assays is based on the inhibition of thrombin or factor Xa; you can see how the assays work in Figure 15.3. When an excess of thrombin or factor Xa is added to plasma in the presence of heparin, antithrombin in the plasma will bind to heparin and inactivate some of the thrombin or factor Xa. The thrombin or factor Xa remaining free in the test system can be measured using a **chromogenic substrate**.

The amount of colour change that results from cleavage of the substrate is inversely proportional to the amount of antithrombin in the plasma sample.

Functional antihrombin assays can be performed using human thrombin, bovine thrombin, or factor Xa. The choice of assay is important. Human thrombin also reacts with heparin cofactor II (see Chapter 13); as this can interfere with the measurement of antithrombin, assays employing human thrombin are not usually used. Over-incubation with heparin can mask some type II HBS defects. A relatively frequent, but clinically mild, antithrombin defect, **antithrombin Cambridge II**, may not be detected with anti-Xa-based assays.

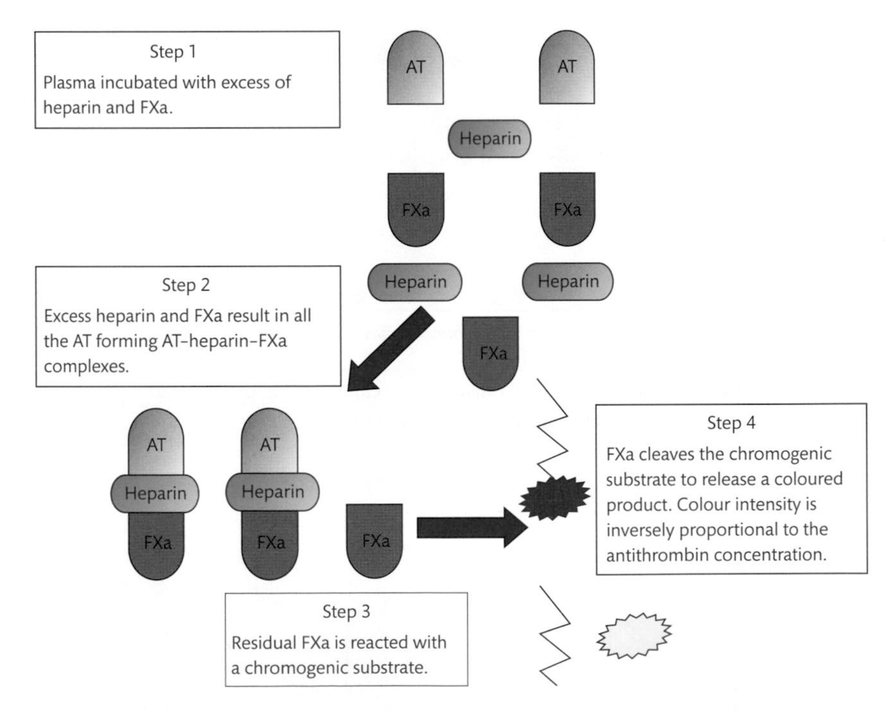

FIGURE 15.3
Functional (activity) assays for antithrombin. Patient or control plasma containing antithrombin is mixed with an excess of heparin and FXa. The plasma antithrombin complexes with the heparin and FXa, thereby neutralizing the latter. A chromogenic substrate is added and residual FXa will cleave the substrate to produce a colour-change reaction. The degree of colour change (change in optical density at 405 nm) is directly proportional to the amount of residual FXa, which in turn is inversely proportional to the concentration of antithrombin in the plasma. Some assays use thrombin in place of the FXa. AT = antithrombin; FXa = activated factor X.

SELF-CHECK 15.1

What is the difference between type I and type II antithrombin deficiency?

SELF-CHECK 15.2

What are the three subtypes of type II antithrombin deficiency?

15.2.3 Protein C deficiency

In Chapter 13, you were introduced to the function of protein C, which is activated by thrombin and then cleaves activated factors V and VIII to prevent these two factors acting to promote clotting. Deficiency of protein C means some of the negative feedback mechanisms in haemostasis are disrupted, and therefore deficiency is associated with an increased risk of thrombosis.

Mutations in the protein C gene can result in type I or type II defects. There is some debate about whether or not protein C deficiency is an autosomal-dominant or autosomal-recessive disorder. Some subjects with genetically confirmed protein C deficiency have protein C levels that fall within the reference range, which can complicate the diagnosis of this defect. Furthermore, age, sex, and plasma lipids can affect plasma protein C levels. Homozygous deficiency of protein C is associated with massive thromboembolic complications shortly after birth, and characteristic symptoms such as **purpura fulminans**. Heterozygosity for protein C deficiency is associated with an up to ten-fold increase in the risk of VTE. In type I protein C deficiency, a concordant reduction in immunological and functional levels of protein C is seen, usually to about 50% of normal. The majority of protein C defects are type I defects, and approximately 60% are caused by missense mutations.

purpura fulminans
A life-threatening disorder characterized by cutaneous haemorrhage and necrosis (tissue death) and disseminated intravascular coagulation (DIC).

15.2.4 Laboratory assays for protein C

Similarly to identifying and subtyping AT deficiencies, assays measuring concentration and function are also required for protein C deficiency.

Immunological assays

Laurell rocket electrophoresis, ELISA, and radioimmunoassays have all been developed for measuring protein C, although for health-and-safety reasons radioimmunoassays are less commonly employed. Concentrations of protein C in plasma are low, approximately 3–5 μg/mL, which makes some assays such as Laurell rocket electrophoresis quite difficult to interpret. For protein C deficiency, there seems to be little difference clinically between type I and II deficiency, so most laboratories only use a functional assay to diagnose a deficiency.

Functional assays: chromogenic and clotting

As we discovered in Chapter 13, protein C is activated by thrombin, and this process is accelerated when thrombin is complexed with membrane-bound thrombomodulin. Assays which measure the function of protein C also require protein C activation, but *in vitro* this is achieved by using a snake venom from the southern copperhead (*Agkistrodon contortrix contortrix*) which directly activates protein C in a similar way to thrombin. The function of activated protein C can then be measured using a chromogenic assay or an assay based on the clotting of plasma. You can see in Figure 15.4 how these two assays differ.

A small proportion (<5%) of type II defects will only be detected using a clot-based assay, such as mutations affecting the binding of protein C to calcium or protein S; however, the clot-based assay is also affected by a number of other factors, such as the presence of lupus anticoagulants or factor V Leiden in plasma, or an increased concentration of factor VIII.

SELF-CHECK 15.3

What activates protein C ***in vivo*** and ***in vitro***?

15.2.5 Protein S deficiency

complement system
An enzyme cascade triggered by antigen-antibody complex formation and resulting in the activation of a series of defence mechanisms, including inflammation and cell lysis.

Protein S acts as a cofactor for protein C by helping to localize activated protein C on cell surfaces so that factors Va and VIIIa are more effectively inactivated, which you saw in Figure 13.23 in Chapter 13. However, protein S is a complex protein, in that approximately 60% of protein S in plasma is bound to **C4b-binding protein**, a component of the **complement system**, and this bound protein S has no significant anticoagulant role. Therefore, measurement of protein S (PS) in plasma must take into account the amount of total PS (both free and bound) and free PS. (See also Box 15.1.)

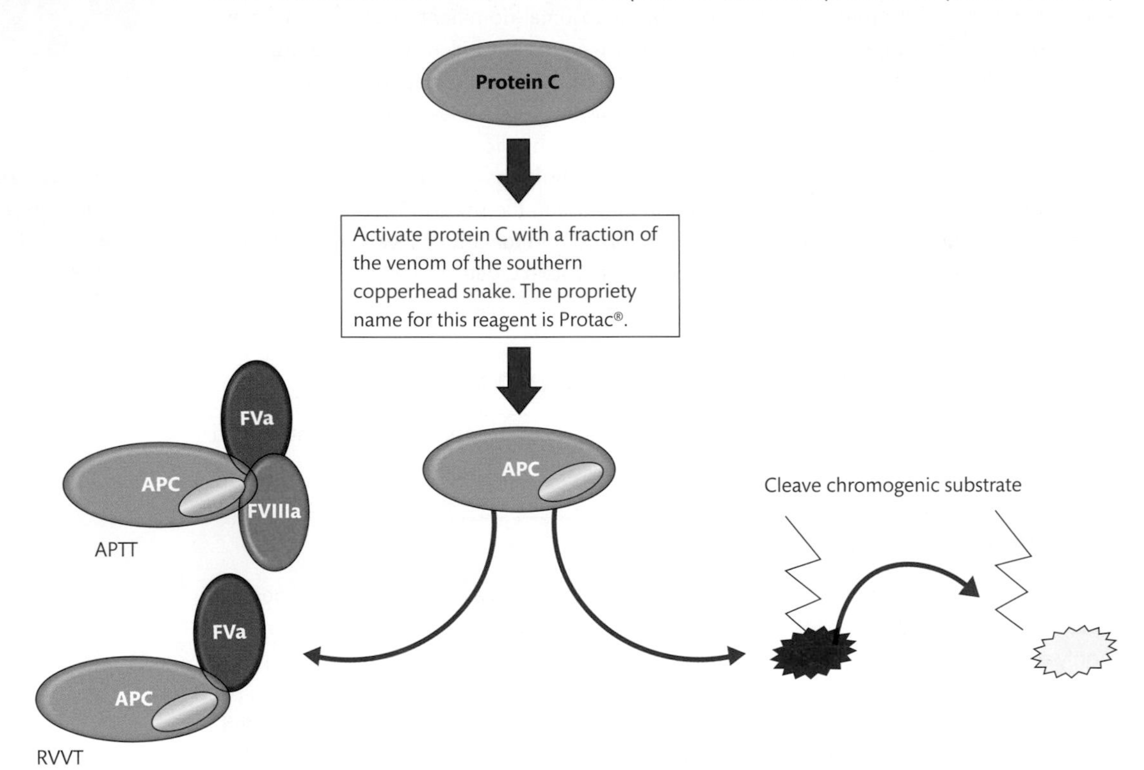

FIGURE 15.4
Functional (activity) assays for protein C. Addition of Protac to plasma causes activation of protein C; addition of a chromogenic substrate specific for protein C will result in cleavage of the substrate and a colour change which is directly proportional to the concentration of protein C in the plasma sample. Alternatively, adding Protac to an APTT will result in prolongation of the clotting time through inactivation of FVa and FVIIIa, the prolongation being directly proportional to the concentration of protein C in the plasma sample. Another clotting test, the Russell's viper venom time (RVVT), can also be used, but as the venom directly activates FX, the activated protein C will only inactivate FVa. APC = activated protein C; APTT = activated partial thromboplastin time; FVa = activated factor V; FVIIIa = activated factor VIII.

Deficiency of protein S is caused by mutations in the gene for protein S, and can be subclassified on the basis of the plasma levels of total and free PS antigen and of PS function (also termed activity). This is slightly more complex than the type I and II deficiencies seen for protein C and antithrombin deficiency. Table 15.2 shows you the levels of the different PS measurements for each subgroup. Heterozygosity for protein S deficiency is associated with an up to ten-fold increase in the risk of VTE, and homozygosity presents similarly to homozygous protein C deficiency.

15.2.6 Laboratory assays for protein S

Protein S can be also measured by Laurell rocket electrophoresis, ELISA, and radioimmunoassays. Latex-based assays have also been developed and are widely used to measure PS in plasma. Total PS can be measured with no special treatment of the plasma. However, some assays used to measure free PS require bound PS to be removed first. To do this, polyethylene glycol (PEG) is added to plasma, which precipitates the bound PS; after centrifugation, the supernatant will contain just free PS. Other immunological methods for PS utilize specific antibodies that allow direct measurement of the free PS without removing bound PS from the plasma. These assays may give greater precision, as a layer of complexity has been removed from the assay. In Figures 15.5 and 15.6 you can see two approaches to the measurement of free PS in plasma.

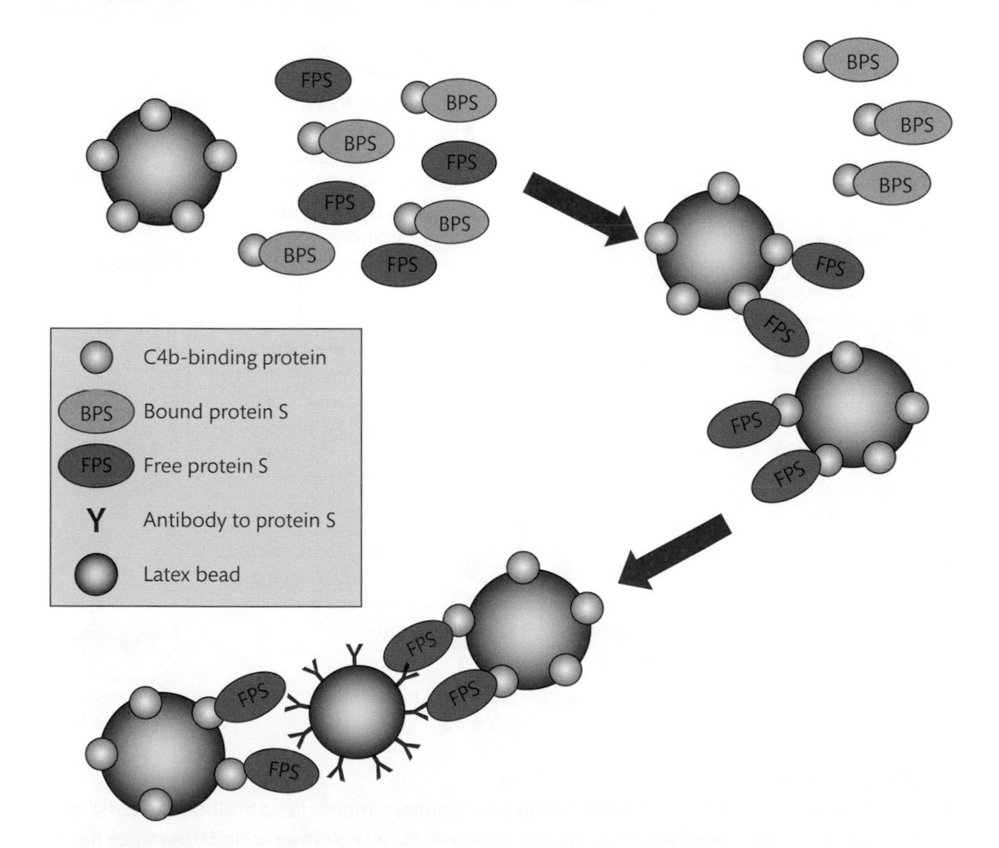

FIGURE 15.5

Free protein S latex immunoassay. Latex beads coated with C4b-binding protein (BP) are added to plasma. Only free protein S in the plasma which is not already bound to C4b-BP binds to the beads. Addition of a second set of beads coated with a monoclonal antibody to protein S results in further binding and agglutination of the beads, which can be measured optically. The turbidimetric changes are directly proportional to the amount of free PS in plasma.

TABLE 15.2 Protein S levels in type I, II, and III deficiency.

	Total PS	Free PS	PS activity
Type I	Reduced	Reduced	Reduced
Type II	Normal	Normal	Reduced
Type III	Normal	Reduced	Reduced

BOX 15.1 *Naming of protein S and protein C*

Protein S is so-called because it was discovered in Seattle, USA. Protein C is so-called because it was the third vitamin K-dependent protein to elute from DEAE-Sepharose, a protein separation technique.

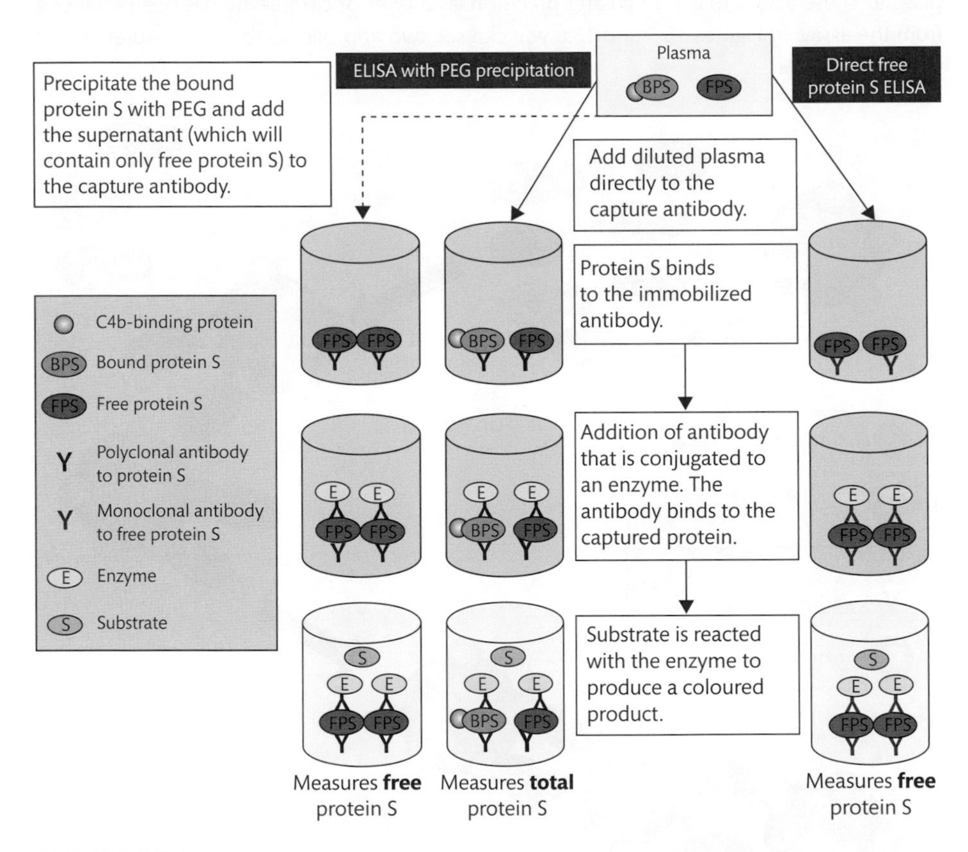

FIGURE 15.6
Protein S ELISA. A microtitre-well plate is coated with either a monoclonal antibody specific for free protein S or a polyclonal antibody against protein S. Plasma is either added directly or has bound protein S removed first by precipitation with PEG. The amount of protein S bound to the antibodies in the wells of the plate is directly proportional to the concentration of protein S in the plasma. An enzyme-linked, anti-protein S antibody is then added and binds to the protein S on the plate. A substrate is added and this is converted by the enzyme to a coloured product; the degree of colour change is directly proportional to the amount of enzyme-linked antibody bound to the protein S, and therefore to the concentration of protein S in the plasma.

Functional PS, or PS activity, cannot utilize a chromogenic assay method, since protein S does not cleave a target substrate. Methods are therefore based on the ability of protein S to prolong the clotting time of plasma to which activated protein C has been added. Figure 15.7 shows you that this can be carried out using different activators of the clotting process. Functional protein S assays may be affected by **activated protein C resistance**, lupus anticoagulants, and increased levels of FVIII in the plasma.

SELF-CHECK 15.4

Why do we measure free protein S rather than total protein S?

15.3 Gain-of-function mutations

In contrast to deficiencies of the natural anticoagulants, some mutations in the genes coding for haemostatic proteins have been identified which either alter the interactions involving these proteins or increase their concentration, causing an increased risk of thrombosis. These are sometimes termed **gain-of-function mutations**.

15.3.1 Factor V Leiden mutation

You saw in Chapter 13 that activated factor V (FVa) is a component of the procoagulant system, acting as a cofactor to FXa to enhance the activation of prothrombin. You also learnt that activated protein C (APC) is a naturally occurring anticoagulant which inactivates FVa. This inactivation occurs through cleavage of FVa at three specific sites on the heavy (aminoterminal) chain: Arg306, Arg506, and Arg679. Cleavage at Arg506 is required for efficient exposure of the other two cleavage sites.

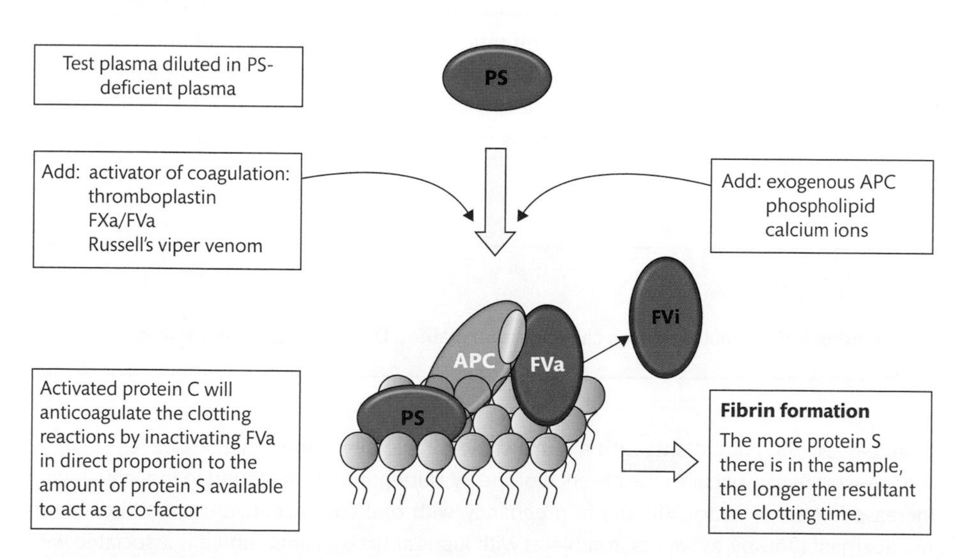

FIGURE 15.7

Functional protein S assays. Test or control plasma is mixed with protein S-deficient plasma (to ensure an excess of all other plasma proteins), and activated protein C is added. The plasma is then clotted by the addition of activators (thromboplastin, FXa, FVa, or Russell's viper venom). The ability of the mixture to prolong the clotting time is directly proportional to the amount of protein S in the plasma sample. PS = protein S; APC = activated protein C; FXa = activated factor X; FVa = activated factor V; FVi = inactivated FV.

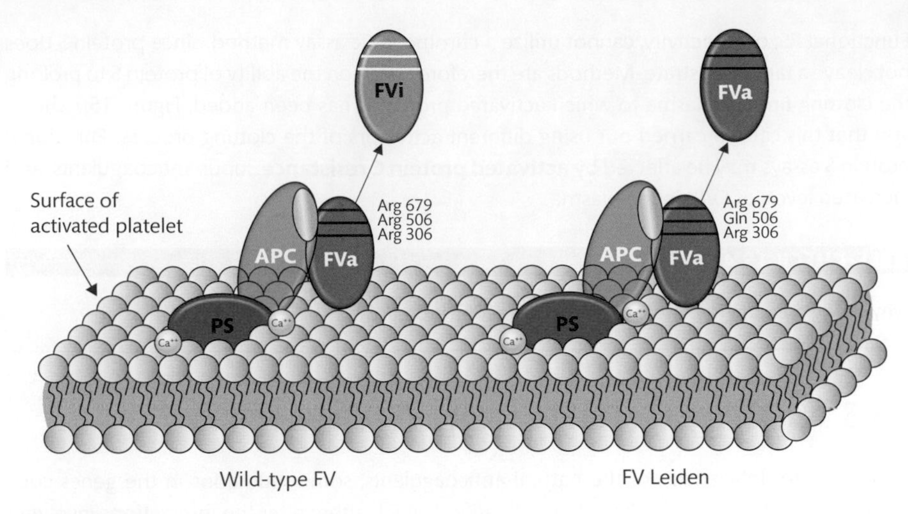

FIGURE 15.8

FV Leiden mutation. Wild-type (normal) factor Va is cleaved by activated protein C and cofactor protein S at positions Arg306, Arg506, and Arg679 to inactivate the active enzyme. Activated protein C is unable to cleave FV Leiden due to a substitution (Gln506) and cannot inactivate factor Va. PS = protein S; APC = activated protein C; FVa = activated factor V; FVi = inactivated FV.

point mutation

A type of mutation that causes a single-base nucleotide in DNA to be replaced with another.

Cross references

You will meet lupus anticoagulants later in this chapter in section 15.6.

Principles of PCR analysis are covered in Chapter 10.

factor V Cambridge

Arg306Thr mutation in the factor V gene, rendering the FV molecule resistant to inactivation by activated protein C.

factor V Hong Kong

Arg306Gly mutation in the factor V gene. It does not cause activated protein C resistance.

factor V Liverpool

Ile359Thr mutation in the factor V gene, rendering the FV molecule resistant to inactivation by activated protein C.

factor V Nara

Trp1920Arg mutation in the factor V gene, rendering the FV molecule highly resistant to inactivation by activated protein C.

A **point mutation** in the gene that codes for the intact factor V molecule, guanine to adenine at nucleotide position 1691, causes an amino acid change at position 506 from arginine to glutamine. You can see in Figure 15.8 that because this is the site that is recognized by activated protein C, the FV molecule formed when this mutation is present cannot be cleaved by APC. We term plasma that contains this mutant factor V 'APC resistant'. As this mutation was first identified in the town of Leiden in the Netherlands, the mutant factor V is called **factor V Leiden**. The mutation is very common, with about 3–5% of the Caucasian population heterozygous for FV Leiden, though it is virtually unknown in African and Oriental populations. There is a 2.5-fold increased risk of thrombosis for carriers of the mutation; individuals who are homozygous for FV Leiden have an 80-fold increased risk of developing thrombosis.

Key Points

FV Leiden is the most common cause of unexplained DVT and PE in Caucasians.

In at least 90–95% of patients with APC resistance, FV Leiden is the cause. However, acquired APC resistance can occur in the absence of the FV Leiden mutation—found in individuals with increased levels of factor VIII and in pregnancy, with oral contraceptive use, and hormone-replacement therapy, as well as in subjects with lupus anticoagulants—and it is associated with increased risk of venous thrombosis.

Two rare mutations in the factor V gene, **factor V Cambridge** and **factor V Hong Kong**, have been described, which cause a change in the arginine at position 306. Both are associated with an increased thrombotic risk. Another rare mutation, **FV Liverpool**, causes a change in the isoleucine at position 359 and leads to mild APC resistance. Conversely, **factor V Nara** has greater APC resistance than FV Leiden.

15.3.2 Testing for factor V mutations that confer APC resistance

A screening test for such FV mutations can be performed by measuring APC resistance in plasma. To do this, a clotting test, such as the activated partial thromboplastin time (APTT), can be performed in plasma with and without APC added. In normal individuals, activated protein C will cleave and inactivate FVa, and a prolonged clotting time will be obtained. In subjects with FV Leiden, activated protein C will be unable to cleave the FVa, and the clotting time with APC will not be as prolonged. Because a number of variables can affect this basic test, a modification using FV-deficient plasma can be used, to make the test almost 100% specific for FV mutations conferring APC resistance.

Screening the FV gene for the mutations should give a definitive result to show whether the patient is heterozygous or homozygous for a FV mutation, or does not have a mutation ('wild-type FV'). Polymerase chain reaction (PCR) and restriction enzyme (RE) digests have been used to detect the FV mutations, although many other techniques applicable to the detection of point mutations can also be used, such as melting curve analysis and fluorescence resonance energy transfer (FRET) techniques.

15.3.3 Prothrombin gene mutation

Soon after the discovery of FV Leiden, a mutation was described in the prothrombin gene that seemed to increase the risk of venous thrombosis. Here, the mutation is located not in the part of the gene that codes for the protein but in a non-translated part of the gene. The mutation, a guanine-to-adenine transition at position 20210 on the gene, causes increased levels of prothrombin in the circulation. As prothrombin is the precursor of thrombin, it is probably not surprising that increases in the amount of prothrombin can cause a two-to-three-fold increased risk of VTE in heterozygotes compared to normal. The increased risk for homozygous subjects is unknown as few data are available, although asymptomatic homozygotes have been described, indicating that this is a mild defect.

It is possible to measure the factor II (prothrombin) activity in patients with this mutation, but the increased level in affected individuals overlaps with the factor II level in normal subjects, so it can be difficult to diagnose on the basis of laboratory clotting methods. Instead, we can search for this mutation using the same genetic analysis methods as for FV Leiden; in some cases, laboratories will use methods that can detect both mutations in the same system (multiplex methods).

15.3.4 Antithrombin resistance

Further prothrombin gene mutations have recently been described that result in reduced reactivity of thrombin with antithrombin, leading to antithrombin resistance. Prothrombin Yukuhashi is due to a substitution of arginine for leucine at position 596 in the gene encoding prothrombin, and prothrombin Belgrade is due to substitution of the same arginine for glutamine. Formation of thrombin–antithrombin complexes are significantly impaired in patients with either mutation, resulting in susceptibility to thrombosis. The prevalence of such mutations is not yet known and they are not tested for routinely. They can be detected using the same genetic analysis methods used for FV Leiden and the prothrombin 20210 mutation. A chromogenic assay has also been described that activates prothrombin with taipan snake venom, the resulting thrombin then being incubated with excess antithrombin and the residual thrombin reacted with a chromogenic substrate to generate a coloured

Cross reference

You will meet taipan snake venom, and other snake venoms, in more detail in section 15.6.7 on testing for lupus anticoagulants.

product. Individuals with normal prothrombin will have their thrombin inhibited by antithrombin and have little residual thrombin, but patients with antithrombin resistance will have considerably more.

SELF-CHECK 15.5

Is APC resistance always caused by the factor V Leiden mutation?

15.4 Other thrombophilias

There are a number of other factors that have been implicated in thrombophilia; some are definitely heritable, others are definitely acquired, and for some it may be difficult to know whether genetics or environmental factors have caused the abnormality.

Fibrinogen—mutations in the fibrinogen gene are rare, but some abnormalities cause dysfibrinogenaemia with an associated increased risk of thrombosis, due to altered polymerization and/or resistance to plasmin digestion. In most cases, an abnormal thrombin time can suggest dysfibrinogenaemia, and an abnormal fibrinogen clotting activity:antigen ratio is confirmatory.

Homocysteine—homocysteine is a non-protein sulphydryl amino acid derived from the metabolism of methionine. Homocysteine is subsequently remethylated to methionine, or undergoes trans-sulphuration to cysteine, depending upon methionine levels. Increased levels can occur in people with folate deficiency, and also in those with hereditary defects. The mechanism by which homocysteine increases thrombotic risk is not known, but it may relate to interference with endothelial cell function. A variety of laboratory methods are available to measure homocysteine, and it is important that plasma is removed from red cells within an hour of collection to avoid falsely raised plasma levels.

High levels of clotting factors—some reports have shown that high levels of certain clotting factors, such as FVIII, FIX, and FXI, may increase the risk of thrombosis. In some cases, high levels of FVIII and thrombosis have been seen in the same family, suggesting that there may be a genetic cause of the defect.

Antiphospholipid syndrome—this is an autoimmune disorder associated with both arterial and venous thrombosis, which is described in section 15.6.

Fibrinolysis—in Chapter 13 you learnt about fibrinolysis, the mechanism by which a fibrin clot is broken down. In theory, failure of this system could lead to an extension of clots formed to repair small areas of vessel damage, and could lead to thrombosis. However, there are so many pre-analytical and analytical variables which affect tests for fibrinolysis, it has not been possible to demonstrate any definite association between a defect in the fibrinolytic system and thrombophilia, although there is evidence that a few PAI-1 polymorphisms, disturbances of t-PA release, and secondary disturbances (such as insulin resistance syndrome) may contribute to thrombophilia.

15.5 Pitfalls of thrombophilia testing

There are many considerations to be taken into account with thrombophilia screening. Sample collection and sample quality are important; as you will see later for lupus anticoagulant testing, some tests require plasma that is centrifuged twice to remove all residual platelets, some tests are very sensitive to the concentration of anticoagulant used, and others require the separation of red cells from plasma within an hour of collection.

Interpretation of data is sometimes straightforward, as in the example in Case study 15.1. However, some defects will cause abnormal results in more than one test. For example, lupus anticoagulants and APC resistance can cause abnormal results in PC and PS clotting-based functional assays in individuals who actually have normal levels of protein C and protein S, which you can see in Case study 15.2. Testing patients soon after a venous thrombosis, or whilst they are pregnant or receiving anticoagulant therapy, can complicate interpretation of the results. Case study 15.3 details the effects of anticoagulant therapy on thrombophilia screening. Therefore, knowledge of the problems and pitfalls of the tests used for thrombophilia screening is an important role of the biomedical scientist.

The mechanisms involved and the interpretation of results in antiphospholipid syndrome are particularly complex, and these are described in section 15.6.

CASE STUDY 15.1 Thrombosis and pregnancy

Patient history

- A 32-year-old woman suffered a deep vein thrombosis during her first pregnancy.
- She reported a family history of thrombosis.
- Coagulation screening and hereditary thrombophilia screening were performed one month after she gave birth.

Results 1

Index	Result	Reference range (see front of book)
PT (s)	11.1	(RR: 10.0–14.0)
APTT (s)	33.0	(RR: 32.0–42.0)
Fibrinogen (g/L)	3.7	(RR: 1.5–4.0)
Antithrombin activity (u/dL)	88.5	(RR: 80–120)
PC activity (u/dL)	43.0	(RR: 77–126)
Total PS antigen (u/dL)	98.0	(RR: 64–135)
Free PS antigen (u/dL)	87.0	(RR: 67–125)
APC resistance test (ratio)	2.25	(RR: >2.0)
FV Leiden mutation	Absent	
Prothrombin G20210A mutation	Absent	

Significance of results 1

It was important that the patient was tested after she had given birth because protein S can fall dramatically in pregnancy, and mild elevations can be seen in protein C. The only abnormality in the battery of tests was a markedly reduced PC activity. It is common practice to repeat any abnormal results, so the PC activity was repeated six weeks later, along with a PC antigen.

Results 2

Index	Result	Reference range (see front of book)
PC activity (u/dL)	45.0	(RR: 77–126)
PC antigen (u/dL)	88.0	(RR: 76–121)

Significance of results 2

The PC activity was proven to be reduced, and to a similar level to that at initial testing. Interestingly, the PC antigen level was normal and almost twice that of the activity result. Remember that antigen assays measure the concentration of the protein irrespective of function, so the normal antigen result accompanied by a much reduced activity result indicates a type II (dysfunctional) deficiency.

CASE STUDY 15.2 Detection of a hereditary thrombophilia following a DVT

Patient history

- A 39-year-old man presented with an unprovoked deep vein thrombosis.
- He had been adopted as an infant and was unable to indicate any family history of thrombotic disease.
- Coagulation screening and hereditary thrombophilia screening were performed.

Results 1

Index	Result	Reference range (see front of book)
PT (s)	12.1	(RR: 10.0–14.0)
APTT (s)	37.0	(RR: 32.0–42.0)
Fibrinogen (g/L)	3.1	(RR: 1.5–4.0)
Antithrombin activity (u/dL)	99.2	(RR: 80–120)
PC activity (u/dL) (clotting-based assay)	49.0	(RR: 74–129)
PS activity (u/dL)	51.0	(RR: 60–135)
APC resistance test (ratio)	1.05	(RR: >2.0)

Significance of results 1

This laboratory only performed functional assays. These data suggest a combined protein C and protein S deficiency, together with an abnormal APC resistance test, which could be due to factor V Leiden. Combined deficiencies, though rare, markedly increase the risk of repeated thrombotic episodes.

However, APC resistance caused by factor V Leiden is known to interfere with clotting-based PC and PS assays. Further investigation at another laboratory gave the following results:

Results 2

Index	Result	Reference range (see front of book)
Antithrombin antigen (u/dL)	104.0	(RR: 75–125)
PC activity (u/dL) (chromogenic assay)	89.0	(RR: 77–126)
PC antigen (u/dL)	99.0	(RR: 76–121)
Total PS antigen (u/dL)	128.0	(RR: 64–135)
Free PS antigen (u/dL)	114.0	(RR: 67–125)
FV Leiden mutation	Homozygous	
Prothrombin G20210A mutation	Absent	

Significance of results 2

These data confirm that protein C levels by chromogenic assay and protein S antigen levels are normal, and the only likely abnormality is the homozygous FV Leiden mutation. It is possible that a type II protein S defect and/or a rare subtype protein C defect are also present, but coexistence of all three defects is very unlikely.

CASE STUDY 15.3 Thrombophilia screening in an anticoagulated patient

Patient history

- A 42-year-old man presented to the Accident and Emergency department with a pulmonary embolism.
- He had no history of VTE, but reported that his father and grandfather both experienced deep vein thromboses in middle age.
- Samples were sent to the laboratory for hereditary thrombophilia screening two weeks later, once he was stabilized on warfarin treatment.

Results 1

Index	Result	Reference range
INR	2.5	(TR: 2.0–4.0)
APTT (s)	44.0	(RR: 32.0–42.0)
Fibrinogen (g/L)	4.9	(RR: 1.5–4.0)
Antithrombin activity (u/dL)	110.0	(RR: 80–120)
PC activity (u/dL) (chromogenic)	21.0	(RR: 74–129)
Free PS antigen (u/dL)	35.0	(RR: 60–135)
Modified APC resistance test (ratio)	3.21	(RR: >2.0)

Significance of results 1

At first sight it appears that the patient is protein C and protein S deficient. However, as you will see in Chapter 17, warfarin works by interfering with production of functionally effective vitamin K-dependent factors, so PC and free PS levels will be reduced anyway whilst the patient is warfarinized. This was indicated on the laboratory report and a repeat sample was requested to be sent when the period of anticoagulation was complete. The patient completed his course of warfarin therapy and samples were sent for his repeat thrombophilia screening. (TR = therapeutic range).

Results 2

Index	Result	Reference range
PT (s)	13.1	(RR: 10.0–14.0)
APTT (s)	39.4	(RR: 32.0–42.0)
Fibrinogen (g/L)	3.2	(RR: 1.5–4.0)
Antithrombin activity (u/dL)	112.0	(RR: 80–120)
Protein C activity (u/dL) (chromogenic)	44.0	(RR: 74–129)
Free protein S antigen (u/dL)	112.8	(RR: 60–135)
Modified APC resistance test (ratio)	3.01	(RR: >2.0)
Prothrombin G20210A mutation	Absent	

Significance of results 2

His free protein S level was now within the reference range. However, although the protein C was higher than before, it was still markedly reduced and the patient was diagnosed with protein C deficiency.

(See also Box 15.2.)

BOX 15.2 *Interference by therapeutic anticoagulants in thrombophilia assays*

You saw in Case study 15.3 that protein C and protein S levels are genuinely reduced in warfarin therapy. Since warfarin also reduces levels of factors II, VII, IX, and X, clotting-based assays can be rendered unreliable by the consequent increase in clotting times. Unfractionated heparin, direct FXa inhibitors, and direct thrombin inhibitors can have a similar effect on clotting times. Antithrombin assays work by inhibition of FXa or thrombin, so patients on direct FXa inhibitors and direct thrombin inhibitors can lead to a reduction in residual enzyme and give rise to a falsely elevated antithrombin activity result.

15.6 Antiphospholipid antibodies

The **antiphospholipid syndrome** (APS) is an autoimmune disorder characterized by one or more episodes of arterial, venous, or small vessel thrombosis, or pregnancy morbidity, accompanied by the persistent presence of one or more types of **antiphospholipid antibody** (APA). In the context of APS, pregnancy morbidity can take the form of unexplained intrauterine death, recurrent unexplained spontaneous abortions, or one or more premature births due to either **pre-eclampsia**, **eclampsia**, or **placental insufficiency**.

Crucial to the diagnosis of APS is the laboratory detection of the different types of APA. The assays detect a heterogeneous group of antibodies that in fact target phospholipid-binding proteins such as **β_2-glycoprotein-I** (β_2GPI) and prothrombin, or protein–phospholipid complexes. Subclassification of APA is based on the assays that are used to detect them:

- antibodies detected in solid-phase assays, usually ELISAs, are **anticardiolipin antibodies** (aCL) and **anti-β_2GPI antibodies** (aβ_2GPI);
- antibodies that interfere with phospholipid-dependent coagulation assays are termed **lupus anticoagulants** (LAs).

The persistent presence of APAs is significantly associated with thrombotic disease, yet for many years it was unclear whether they are coincident, consequent, or causative. Recent evidence suggests that they are indeed causative and that antibodies with specificity for a small **epitope** on domain I of β_2GPI are key players in the thrombotic process. When the antibodies bind to β_2GPI it adopts a new, stretched conformation that allows it to bind to cell-surface receptors for which its normal, circular conformation has poor affinity. Binding of the antibody-β_2GPI complex to vascular endothelial cells causes increased tissue factor expression, and binding to monocytes causes cytokine release, leading to a prothrombotic state.

Perhaps unsurprisingly, there is evidence that APA may directly interfere with haemostasis/coagulation, such as inhibition of the protein C pathway, platelet activation, impairment of fibrinolysis, and disruption of the **annexin A5** anticoagulant shield. The main mechanism of APA-mediated foetal loss is thought to be interference of **trophoblast** invasion/implantation rather than via a thrombotic mechanism, although placental thrombosis leading to **infarction** and placental insufficiency has also been described in APS.

Some patients, for whom the symptoms of APS and the presence of APA are the only manifestations of disease, are said to have primary APS, whereas for others, APS coexists alongside other conditions. Many of these are themselves autoimmune disorders, most patients with 'secondary' APS having a condition called **systemic lupus erythematosus** (SLE). It is possible the APS seen in SLE is not actually secondary, but that both conditions represent different elements of the same process, or at least that SLE provides a pathophysiological setting for development of APS. Alternatively, APS and SLE may be two diseases coinciding in an individual patient.

An important criterion for the diagnosis of APS is the persistence of APA. This is because transient APA can be encountered as a consequence of other disease states and only persist for the duration of the primary disorder—these transient antibodies are very rarely associated with thrombosis. The only way to confirm that an antibody is persistent, and thus more likely to be involved in APS, is to repeat any positive test at least 12 weeks later. The transient antibodies can be associated with the following:

- viral infections, for example human immunodeficiency virus; hepatitis C;
- bacterial infections, for example syphilis, leprosy;

pre-eclampsia
Pregnancy-induced high blood pressure associated with protein in the urine. The only cure is delivery or abortion of the foetus.

eclampsia
One or more convulsions occurring during or immediately after pregnancy.

placental insufficiency
Placental failure to supply nutrients to the foetus and remove toxic waste.

β_2-glycoprotein-I (β_2GPI)
Plasma protein involved in removal of toxic substances and cellular debris.

epitope
A portion or section of a molecule to which an antibody binds.

annexin A5
Annexin A5 forms a shield around negatively charged phospholipid molecules to regulate phospholipid availability to coagulation factors.

trophoblasts
Cells of the outer layer of the blastocyst ('pre-embryo') that attach a fertilized ovum to the uterine wall and develop into the placenta.

infarction
The process that results in the generation of an area of necrotic (dead) tissue caused by the loss of an adequate blood supply.

systemic lupus erythematosus
A chronic autoimmune disease with multiple organ/system involvement that is associated with antibodies directed against cell nuclei.

- parasitic infections, for example malaria;
- lymphoproliferative disorders, for example lymphoma, chronic lymphocytic leukaemia;
- drug exposure, for example some anticonvulsants and some antimalarials.

Antibody profiles are important because patients positive for more than one antibody, and particularly those with triple positivity (LA, and aCL and aβ_2GPI of the same isotype) have a stronger association with thrombotic and obstetric APS and a higher risk of recurrence of thrombosis.

A separate group of patients exists who do indeed possess persistent APA, yet they are clinically well and very few progress to the classic signs and symptoms of APS unless they are triple positive. In some cases, these patients may only be identified through abnormal coagulation screening tests performed for other reasons, such as preoperative screening.

Key Points

The vast majority of lupus anticoagulants detected in children are secondary to infections and are not associated with VTE.

The heterogeneous nature of APA and lack of standardization of the assays makes their detection an interesting diagnostic challenge. We will now turn our attention to the procedures available to the biomedical scientist in the identification of APA.

SELF-CHECK 15.6

What are the clinical criteria for a diagnosis of antiphospholipid syndrome?

15.6.1 Antibodies detected in solid-phase assays

aCL and aβ_2GPI can be detected and quantified using **enzyme-linked immunosorbent assays (ELISA)**. The availability of standard preparations of polyclonal antibodies allows the calculation of aCL results to be reported in immunoglobulin G (IgG) or IgM (anti)phospholipid units—GPLU/mL and MPLU/mL, respectively—in relation to the concentration of the standards. Standards made from monoclonal antibodies are also available and can be used to calibrate ACL or aβ_2GPI assays, but are reported in arbitrary units such as U/mL.

First isolated from heart tissue, **cardiolipin**, or diphosphatidylglycerol, is a phospholipid that makes up about 20% of the inner mitochondrial membrane. It is found in high concentrations in metabolically active cells, such as heart and skeletal muscle, and has been found in some bacterial membranes. aCL produced as part of an autoimmune process, as in APS, are dependent on β_2GPI to bind to cardiolipin in the assay system. aCL are detected using an indirect ELISA using the design described in the Method box that follows.

Despite the availability of purified standards, variations in reagent composition and details of methodology lead to only moderate interlaboratory agreement. This is improved if results are stratified as weak, moderate, or strong positive in relation to the individual method and cut-off, rather than directly comparing quantitative results.

METHOD Indirect ELISA for detecting anticardiolipin antibodies

Unlike a sandwich ELISA, where the antigen is captured by an antibody and then tagged with another, indirect ELISA captures the antigen with an immobilized substance that is known to bind with it.

- **A 96-well microtitre is coated with purified cardiolipin and β_2GPI.**
- **Calibrators containing increasing amounts of purified standard are added to separate designated wells.**
- **Dilutions (in the region of 1 in 100) of the control and test samples are added to separate designated wells. The plate is left to incubate to allow binding of aCL, if present, to the immobilized cardiolipin/β_2GPI.**
- **The plate is washed three times with buffer to remove all unbound material and then coated with a solution of antibody to human immunoglobulins that is attached to an enzyme (i.e. horseradish peroxidase), termed the conjugate. This is left to incubate to allow the conjugate to react with any antigen–antibody complexes that formed during the first stage.**
- **Unbound conjugate is then washed off.**
- **The enzyme is then provided with a substrate, the product of the reaction being coloured. The reaction is given time to generate sufficient colour that can be read spectrophotometrically and then stopped with dilute acid, causing a second colour change.**
- **The colour intensity (referred to as optical density) is in direct proportion to the amount of conjugate bound to the antigen-antibody complex, which itself is proportional to the initial concentration of the aCL in the control and test samples.**

Non- β_2GPI-dependent aCL can be generated as part of the immune response to infection, such as seen in syphilis, but are still detected in standard aCL assays. To circumvent this potential for 'false-positive' aCL results, that is to say non-APS-related antibodies, assays using β_2GPI itself as the capture antigen have been developed and are widely available.

Consequently, assays for aβ_2GPI show greater specificity than aCL for the diagnosis of APS. However, similar problems exist with both methodology and standardization, particularly in relation to the quality of β_2GPI used as the capture antigen. Additionally, antibodies directed to parts of the molecule other than the limited epitope on domain I may be detected that are not due to APS, such as the antibodies to domain V found in leprosy.

The variation between assays means that some patients with β_2GPI-dependent antibodies may be detected in an aCL assay, but not an aβ_2GPI assay, and vice versa, so performing both types of assay increases the chances of detecting clinically significant antibodies. Persistent positivity in just one assay is sufficient to diagnose APS if accompanied by appropriate clinical signs and symptoms, although many patients will inevitably be positive in both.

Typically, the cut-off for normality is in the region of 10–15 U/mL, yet assay variation makes it impossible to define a generic cut-off. Medium-to-high aCL titres have a better correlation with thrombotic APS, but recent evidence suggests that persistent slight elevations are significant in obstetric APS.

SELF-CHECK 15.7

How are aCL and aβ_2GPI detected in the laboratory? What are the main problems with the assays?

Antibodies to prothrombin, or prothrombin in complex with phosphatidylserine, are commonly present in patients with APS and can be detected by indirect ELISA. However, it is not yet clear whether the presence of these antibodies alone is sufficient for a diagnosis of APS and routine measurement is not recommended. Similarly, antibodies to specific phospholipids that are not complexed with protein have been demonstrated in the serum of APS patients, but their presence in isolation does not seem to be associated with thrombotic disease.

15.6.2 Lupus anticoagulants

A subgroup of APA, the lupus anticoagulants (LA), have the ability to interfere with phospholipid-dependent coagulation assays in an inhibitory manner. Their very name is a misnomer as they are present in patients other than those with SLE, the assays are not tests for the diagnosis of SLE, and their anticoagulant properties are purely an *in vitro* phenomenon.

Although aCL are five times more common than LA, it is the latter that have a greater association with thrombosis, pregnancy morbidity, and thrombosis in SLE. The LA can be dependent on β_2GPI or prothrombin for their *in vitro* anticoagulant activity. Many patients have both types, although the β_2GPI-dependent antibodies have an even greater association with the above clinical manifestations and are often those with domain I specificity. (See also Box 15.3.)

Patients with LA do occasionally present with bleeding symptoms, for the following reasons:

- coexistence of antibodies to platelet glycoproteins, leading to thrombocytopenia;
- coexistence of a separate prothrombin antibody that does not cause *in vitro* inhibition but forms **immune complexes** with prothrombin, resulting in its removal from the circulation, causing an acquired deficiency;
- antibodies to other coagulation factors, such as FX, FVII, or FVIIa.

immune complex
The combination of an antibody bound to its target antigen.

LA show significant heterogeneity and can even vary in their *in vitro* behaviour over time in the same patient. Purified standards such as those used in solid-phase assays do not exist, so they are detected by inference due to their ability to inhibit phospholipid-dependent coagulation assays. Antibody heterogeneity and assay variability is considerable, and to such a degree that *no single test is capable of identifying all LA.*

Key Points

Some LA react better in certain types of clotting tests than others, or may be only detectable in one type of clotting test.

National (UK) and international expert guidelines suggest diagnostic strategies to maximize detection rates and include the recommendation to use at least two tests that are of different types. The four main criteria for the laboratory diagnosis of the presence of an LA are outlined in the Method box that follows.

SELF-CHECK 15.8

What are the usual presenting symptoms of patients with lupus anticoagulants and what other presentations are known?

BOX 15.3 *Discovery of lupus anticoagulants*

LA are so called because they were first described by Conley and Hartman in 1952 in two patients with SLE who presented with bleeding and whose plasma exhibited *in vitro* anticoagulant activity. It was suggested that the anticoagulant was associated with the bleeding, although it was in fact due to the thrombocytopenia frequently associated with SLE. A subsequent study by Bowie et al. in 1963 showed that the presence of this aspecific coagulation inhibitor was more frequently associated with thrombosis. The term 'lupus anticoagulant' was proposed nine years later by Feinstein and Rapaport due to the frequency of the anticoagulants in patients with SLE, although we now recognize they are by no means confined to patients with SLE.

METHOD Criteria for the laboratory detection of lupus anticoagulants

- **Prolongation of at least one phospholipid-dependent screening test.**
- **Evidence that the abnormality is phospholipid-dependent.**
- **Evidence that the abnormality is inhibitory in nature.**
- **Ability to distinguish results from other coagulopathies that may mask, mimic, or coexist with LA.**

15.6.3 LA screening tests

A variety of screening tests of different analytical designs are available for detecting LA (detailed in Table 15.3; see also Box 15.4). Virtually all of them incorporate significantly diluted phospholipid to accentuate the inhibitory effect of the LA. They are all based on activating specific parts of the coagulation mechanisms and are thus susceptible to interference by other abnormalities that can lead to false-positive results, such as:

- coagulation factor deficiencies;
- therapeutic anticoagulation;
- non-phospholipid-dependent inhibitors.

Key Points

LA interfere with phospholipid-dependent coagulation tests because they act as competitive inhibitors by competing with coagulation factors for binding to phospholipid. Diluting phospholipid in the assay increases the competition, as there are limited phospholipid molecules available, which therefore increases the likelihood of an LA prolonging the clotting time above the reference range.

BOX 15.4 *Snake venoms*

You will see in Table 15.3 that a number of the assays use snake venoms to activate *in vitro* coagulation at specific points in the pathways. Some of the snakes that provide the venom, which you can see in Figure 15.9, are amongst the most dangerous to man and possess some of the most potent venoms known.

Because we detect LA by inference, and because of potential interfering factors, the case studies in this section are provided to give an insight into the importance of the interpretive role of biomedical scientists in generating accurate and meaningful results and reports. You would not be expected to practise at this level until you reach more senior grades.

Which screening tests should you use?

You will see from Table 15.3 that the tests available to screen for LA fall into three categories, depending on whether the test design is based on directly activating the 'intrinsic', 'extrinsic', or the 'common' pathway. The recommendation to use at least two test types means that using, for instance, two different APTT reagents is insufficient, as some LA are not reactive in assays of that design.

The most common pairing of tests is APTT and dilute Russell's viper venom time (DRVVT). This combination has been repeatedly shown to have high detection rates and one of the three available expert guidelines suggests that no other tests should be used. However, this pairing will not detect all LA, the main reasons for which are listed below.

- Some laboratories use their routine APTT in the reagent pairing for LA detection. Many of these have poor sensitivity to LA compared to APTT reagents specifically formulated to detect LA.
- LA that are potentially detectable in a given assay type (i.e. DRVVT) may not be reactive to the reagents/methodology in local use for that type.
- A small subpopulation of LA are detectable only in extrinsic pathway-based assays.

In cases where the clinical suspicion of APS is high despite negative results, the other two available expert guidelines suggest it can be beneficial to have alternative assays available. Where APTT and DRVVT are the first-line assays, appropriate second-line tests would be an 'extrinsic pathway'-based assay and using APTT/DRVVT reagents from different manufacturers to those used initially. There are resource implications for such a strategy, so a common approach is to send samples to another laboratory which uses different tests/reagents. DRVVT is known to be particularly sensitive to domain I β_2GPI-dependent antibodies and highly correlate with thrombosis and recurrence, and is thus ostensibly indispensable from routine testing. (See also Box 15.5.)

Results are often expressed as ratios that are derived from dividing the patient's clotting time by that of the normal control. This is preferable to using raw clotting times, as it accounts for changes in reagent stability, variation in the clotting time of normal plasma, and operator/analyser variability. Alternatively, the mean clotting time from the reference range can be used.

TABLE 15.3 Screening tests for the detection of lupus anticoagulants (LA).

Screening test	Assay principle/design	Interfering factors	Further information
Activated partial thromboplastin time (APTT)	• 'Intrinsic pathway'-based assay • Coagulation activated by the action of a contact activator on FXII • Phospholipid dependent • Ca^{2+}-dependent	False positives can be due to: • anticoagulant therapy with vitamin K antagonists, unfractionated heparin, direct FXa inhibitors, and direct thrombin inhibitors • non-phospholipid-dependent inhibitor • deficiencies of FII, FV, FVIII, FIX, FX, FXI, PK, HMWK, FXII, or fibrinogen False negatives can occur with weak LA in the presence of elevated FVIII	• Routine APTT reagents do not contain dilute phospholipid and can miss weak/moderate LA • Variation in phospholipid composition leads to inconsistent sensitivity between reagents • Normal result with a routine APTT reagent is insufficient to exclude the presence of an LA • Dilute APTT with LA-sensitive phospholipid composition should be used to screen for LA; phospholipid is dilute but not the contact activator • Rarely affected by fractionated heparin
Kaolin or silica clotting time (KCT/SCT)	• 'Intrinsic pathway'-based assay • Coagulation activated by the action of a contact activator (kaolin or silica) on FXII • No phospholipid added at all • Ca^{2+}-dependent	As for APTT	• Residual plasma lipid and platelets are the sources of phospholipid • Performed on neat plasma and a 1:4 mixture of test:normal plasma (which can dilute LA and lead to false-negative results) • Rarely performed with confirmatory test
Dilute prothrombin time (DPT)	• 'Extrinsic pathway'-based assay • Coagulation activated by the action of thromboplastin on FVII/FVIIa • Dilution of phospholipid achieved by diluting the entire thromboplastin reagent • Ca^{2+}-dependent	False positives can be due to: • anticoagulant therapy with vitamin K antagonists, unfractionated heparin, direct FXa inhibitors, and direct thrombin inhibitors • non-phospholipid-dependent inhibitor • Deficiencies of FII, FV, FX, or fibrinogen	• Dilution of thromboplastin results in dilution of both activator and phospholipid • Considerable reagent variation • Recombinant thromboplastins tend to be more sensitive to LA and exhibit less variation than those from brain/placenta
Activated seven lupus anticoagulant (ASLA) assay	• 'Extrinsic pathway'-based assay • Coagulation activated by the action of recombinant human FVIIa on FX • Recombinant FVIIa used in supraphysiological concentration and works independently of thromboplastin (as it is already activated) • Phospholipid is diluted • Ca^{2+}-dependent	False positives can be due to: • anticoagulant therapy with vitamin K antagonists, unfractionated heparin, direct FXa inhibitors, and direct thrombin inhibitors • non-phospholipid-dependent inhibitor • deficiencies of FII, FV, FVII, FX, or fibrinogen	• Recombinant FVIIa is produced in baby hamster kidney cells and is structurally very similar to human FVIIa • More specific for LA than the other 'extrinsic pathway'-based assay, the DPT • Not widely used because the recombinant FVIIa is expensive

(Continued)

TABLE 15.3 (Continued)

Screening test	Assay principle/design	Interfering factors	Further information
Dilute Russell's viper venom time (DRVVT)	• 'Common pathway'-based assay • Coagulation activated by the action of a purified FX activator from the venom of Russell's viper (*Daboia russelii*) • Venom is diluted • Phospholipid is diluted • Ca^{2+} dependent	As for ASLA	• Most widely used assay for LA • Reagent variation does exist, although less marked than for APTT • Reagent variation due to phospholipid composition and venom heterogeneity (there are a number of Russell's viper subspecies)
Taipan snake venom time (TSVT)	• 'Common pathway'-based assay • Coagulation activated by the action of a purified FII activator from the venom of the coastal taipan (*Oxyuranus scutellatus*) • Venom is diluted • Phospholipid is diluted • Ca^{2+} dependent	False positives can be due to: • anticoagulant therapy with unfractionated heparin and direct thrombin inhibitors • non-phospholipid-dependent inhibitor • deficiencies of FII or fibrinogen	• Taipan venom can activate the PIVKA form of FII and is thus insensitive to the effects of vitamin K antagonist anticoagulation • Insensitive to direct FXa inhibitors because direct FII activation bypasses FXa
Textarin time	• 'Common pathway'-based assay • Coagulation activated by the action of a purified FII activator from the venom of the Australian eastern brown snake (*Pseudonaja textilis*) • Venom is diluted • Phospholipid is diluted • FV and Ca^{2+} dependent See also Box 15.4.	False positives can be due to: • anticoagulant therapy with unfractionated heparin and direct thrombin inhibitors • non-phospholipid-dependent inhibitor • deficiencies of FII, FV, or fibrinogen	• Textarin venom can activate the PIVKA form of FII and is thus insensitive to the effects of vitamin K antagonist anticoagulation • Insensitive to direct FXa inhibitors because direct FII activation bypasses FXa

15.6.4 Confirmatory tests

As you saw in Table 15.3, a screening test may be prolonged for reasons other than the presence of an LA. Thus, we cannot stop the diagnostic process at the stage of an elevated screening test and must progress to other procedures to confirm the nature of the abnormality.

There are two aspects of LA *in vitro* behaviour that we assess in order to confirm that an abnormal screening test is due to an LA, those of *inhibition* and *phospholipid dependence*.

Inhibition

Inhibition is demonstrated by mixing the patient's plasma with an equal volume of a normal pooled plasma (NPP) and then repeating the assay. If the patient's plasma contains an inhibitor, it should continue to interfere with the phospholipid-dependent stages of *in vitro* coagulation, whereas if the abnormality is due to a factor deficiency, the normal plasma will supply the missing factor and generate a normal result. In practice, there is an unavoidable dilution of the LA antibodies in this process, which leads to false-negative mixing studies in a considerable number of cases.

Presence of inhibition can be assessed by calculating the Index of Circulating Anticoagulant (ICA) or assessing the result against a mixing test-specific cut-off. The ICA is calculated as a percentage from clotting times as follows: $[(\text{screen 1:1 mix (s)} - \text{screen NPP(s)})/\text{screen test (s)}] \times 100$. Mixing test cut-offs are always lower than for undiluted plasma due to donor plasmas with clotting times at the extremes of normality being 'normalized' towards the clotting time of the entire pool. This results in increased sensitivity to the presence of inhibition and is more sensitive than ICA.

Phospholipid dependence

Crucial to the demonstration of an LA is confirmation of phospholipid dependence. This is usually achieved by performing the screening test in an identical fashion, apart from markedly increasing the concentration of the phospholipid. This has the effect of swamping the antibody so that competition for phospholipid binding between the LA antibody and clotting factors is reduced, thereby leading to a reduction in the clotting time compared to that of the screening test. Such reagents are referred to as **confirm reagents**.

Other abnormalities—such as factor deficiencies, therapeutic anticoagulation, and non-phospholipid-dependent inhibitors—will prolong the confirmatory test to a degree similar to the screening test because their effects on clotting times are not related to phospholipid concentration. A typical example of positive testing for LA is given as Case study 15.4.

BOX 15.5 *Expert guidelines*

It is common for groups of experts in a given area of medicine to prepare guideline publications to inform regular practitioners of current thoughts on best practice. This is particularly useful in areas such as LA detection, where variations in antibodies, reagents, and clinical practice are marked. The currently available guidelines for LA detection have been produced by the International Society on Haemostasis and Thrombosis, the British Committee for Standards and Haematology, and the Clinical and Laboratory Standards Institute.

(a) (b) (c) (d)

FIGURE 15.9

Snakes whose venom is used in lupus anticoagulant detection. (a) Russell's viper (*Daboia russelli*); (b) coastal taipan (*Oxyuranus scutellatus*); (c) saw-scaled viper (*Echis carinatus*); (d) Australian eastern brown snake (*Pseudonaja textilis*). Russell's viper is the second largest venomous snake in the world and responsible for thousands of human deaths every year across Southern Asia. The coastal taipan is an Australian snake related to cobras and considered by some to be the most dangerous of all land snakes to humans. The saw-scaled viper is found in Africa and Asia and gets its name from the sawing sound it makes by rubbing scales from either side of its body together as a warning sign—it is highly dangerous to man. The Australian eastern brown snake is also related to cobras and is one of the few snakes that both poisons and constricts its prey.(a) © Muhammad Sharif Khan; (b) © Stewart Macdonald/Ug Media; (c) © Muhammad Sharif Khan; (d) © Roger Michael Lowe.

SELF-CHECK 15.9

What are the laboratory criteria for the detection and confirmation of the presence of a lupus anticoagulant?

The two most common methods employed to increase the phospholipid concentration are:

- platelet neutralization procedure (PNP);
- high-concentration phospholipid reagents.

As its name suggests, reagents for the PNP are prepared from platelets, as they contain high concentrations of procoagulant phospholipids. The phospholipids are made available by washing normal platelets and either activating them with a calcium ionophore or repeatedly freezing and thawing them to release the phospholipid by disrupting the platelet membranes. The PNP has been used to confirm phospholipid dependence in APTT, DRVVT, activated seven lupus anticoagulant (ASLA) assay, and Taipan snake venom time (TSVT).

One drawback of using such a reagent prepared from platelets is that it will contain other platelet constituents, in particular **platelet factor 4 (PF4)**. One property of PF4 is that it neutralizes heparin, which can lead to false-positive confirmatory tests in patients receiving therapeutic heparin, as you can see in Case study 15.5.

platelet factor 4 (PF4)
A small cytokine released from activated platelets that modulates the effects of heparin-like molecules, stimulates protein C activation, and is chemotactic for neutrophils and monocytes.

In practice, investigation for LA in patients whose clinical condition requires anticoagulation by unfractionated heparin is rarely indicated, so PNP confirmatory tests remain in use.

An alternative is to use a confirmatory reagent with a high concentration of the same phospholipid preparation that was diluted in the screening test. Such reagents give comparable results to PNP and are not prone to false-positive interpretations in unfractionated heparin therapy and they are now more commonly used than PNP. Such confirmatory reagents are used in APTT, DRVVT, dilute prothrombin time (DPT), and ASLA. The screen and confirm results are interpreted identically to those shown in Case study 15.4.

A variation on this theme is the use of a confirmatory reagent that does not contain concentrated phospholipid, but whose constituent phospholipids are nonetheless insensitive to the presence of LA. This is a function of the relative amounts of the individual phospholipids—phosphatidylcholine, phosphatidylethanolamine, phosphatidylserine, and sphingomyelin. This approach is often adopted in APTT testing for LA where an APTT reagent whose phospholipid composition is sensitive to LA is used as the screening test, accompanied by an LA-insensitive APTT reagent for the confirmatory test. Reagents rich in phosphatidylserine are insensitive to LA, so screening reagents are more sensitive if they have a low phosphatidylserine content.

How do we assess whether the correction of the screen ratio by the confirm ratio is significant?

In theory, the screen and confirmatory tests should give identical results unless an abnormality is phospholipid-dependent. So in a patient with a factor deficiency, both screen and confirm results should be prolonged by a similar degree. Similarly, if we performed a confirmatory test

CASE STUDY 15.4 Typical results for a positive LA screen/confirm by DRVVT

Index	Result	Reference range
DRVVT screen ratio	1.45	(RR: 0.85–1.15)
DRVVT confirm ratio	1.00	(RR: 0.86–1.14)

The screen ratio is markedly above the upper value of the reference range and the confirm ratio is significantly lower and within the reference range. This confirms that the abnormality was corrected by the addition of concentrated phospholipid, i.e. a lupus anticoagulant.

CASE STUDY 15.5 False-positive LA screen in a patient on unfractionated heparin when using PNP in the DRVVT confirmatory test

Index	Result	Reference range
PT (s)	11.1	(RR: 10.0–14.0)
APTT (s)	70.0	(RR: 32.0–42.0)
Thrombin time (s)	101.2	(RR: 9.0–11.0)
Reptilase time (s)	15.1	(RR: 14.2–16.0)
DRVVT screen ratio	1.90	(RR: 0.85–1.15)
DRVVT confirm ratio	1.30	(RR: 0.86–1.14)

The coagulation screening results are indicative of the presence of heparin. The heparin is causing the elevation of the DRVVT screen, but its effect is partly neutralized in the confirmatory test due to the presence of PF4 in the PNP reagent. This can be mistaken for the presence of an LA. It is, of course, possible that a patient on unfractionated heparin does indeed have an LA, but it is impossible to tell in this scenario, so many reagent manufacturers add a heparin neutralizer to both the screen and confirm reagents. However, the neutralizers are only effective up to a certain level of unfractionated heparin, in the region of 1.0 U/mL. If the confirmatory test ratio returns to the reference range, this tells you that the neutralizer has worked, so if the screening test ratio remains elevated, you have found a LA.

on a patient with a normal screening test result, they should give identical results. In practice, however, we have to allow for a degree of analytical error inherent in any method, and partial corrections may be seen. We approach this by applying calculations to assess the degree of correction against a predefined cut-off point. Although much of the research has been done on DRVVT, the calculations can be applied to any assay. There are two calculations recommended by the guideline panels.

Per cent correction of ratio—the degree of correction of the screening test ratio by the confirmatory test ratio is calculated as follows:

$$[(\text{Screening test ratio} - \text{confirmatory test ratio})/\text{Screening test ratio}] \times 100$$

Provided other causes of elevated clotting times have been excluded, in the region of ≥10% correction is considered indicative of the presence of an LA.

The normalized screen/confirm ratio—the screening test ratio is divided by the confirmatory test ratio to generate a third ratio, which is the normalized screen/confirm ratio.

$$\text{Screening test ratio} \ / \ \text{Confirmatory test ratio}$$

A normalized screen/confirm ratio in the region of >1.2 is considered positive for LA. The exact values of cut-offs for either calculation should be derived locally from a set of normal donor plasmas analysed with the reagents and analytical equipment in local use.

Look at Table 15.4, which demonstrates how to incorporate raw data into each of the calculations above in a patient with an LA.

There is no consensus on which calculation is the optimal method. Some laboratories use the calculation recommended by the manufacturers of particular reagents, whilst others adopt the locally preferred calculation for specific assay types. Apart from cases where there is minimal elevation of the screening test above the reference range, correction of the confirm result back into the reference range is strong evidence for the presence of an LA. Some LA are sufficiently potent to achieve a degree of resistance to the 'swamping' effect of high phospholipid reagents and the confirm ratio does not return into the reference range, yet the discordance between the screen and confirm results realizes a positive interpretation from the phospholipid-dependence calculation.

Before you read and work through the case studies that follow, look through the flow chart in Figure 15.10 to understand the diagnostic strategies used to detect LA in the clinical haemostasis laboratory.

15.6.5 Exclusion of other causes of elevated clotting times

Interpretation of the screening and confirmatory procedures we have met so far assumes that no other abnormalities are present, and so the results can be accepted at face value. When another coagulation abnormality coexists, interpreting the result is difficult because you cannot necessarily tell how much each abnormality is contributing to screening test elevation, and, hence, whether the confirmatory test has corrected the effect of an LA.

The starting point for the exclusion of other abnormalities is the coagulation screen, as it will give an immediate indication of the presence of abnormalities. The phospholipid in routine thromboplastin reagents is concentrated, so it is rare for LA to cause an elevation of the prothrombin time. Routine APTT reagents vary widely in their sensitivity for LA, so results will not always be elevated when an LA is present. As you will see in Chapter 17, therapeutic anticoagulants can generate specific result patterns in coagulation screening and they will alter the way in which results are interpreted. Case study 15.6 below demonstrates interpretation.

TABLE 15.4 Worked through examples of calculations for phospholipid dependence on DRVVT testing in a patient with a lupus anticoagulant.

Raw data		
DRVVT screen (test)	60 s	Screen ratio = $\frac{60}{32}$ = 1.88 (Reference range: 0.85–1.15)
DRVVT screen (normal control)	32 s	
DRVVT confirm (test)	31 s	Confirm ratio = $\frac{31}{30}$ = 1.03 (Reference range: 0.86–1.14)
DRVVT confirm (normal control)	30 s	
Calculation for % correction of ratio $\frac{(1.88-1.03) \times 100}{1.88} = 45.2\%$		
Calculation for normalized screen/confirm ratio $\frac{1.88}{1.03} = 1.83$		

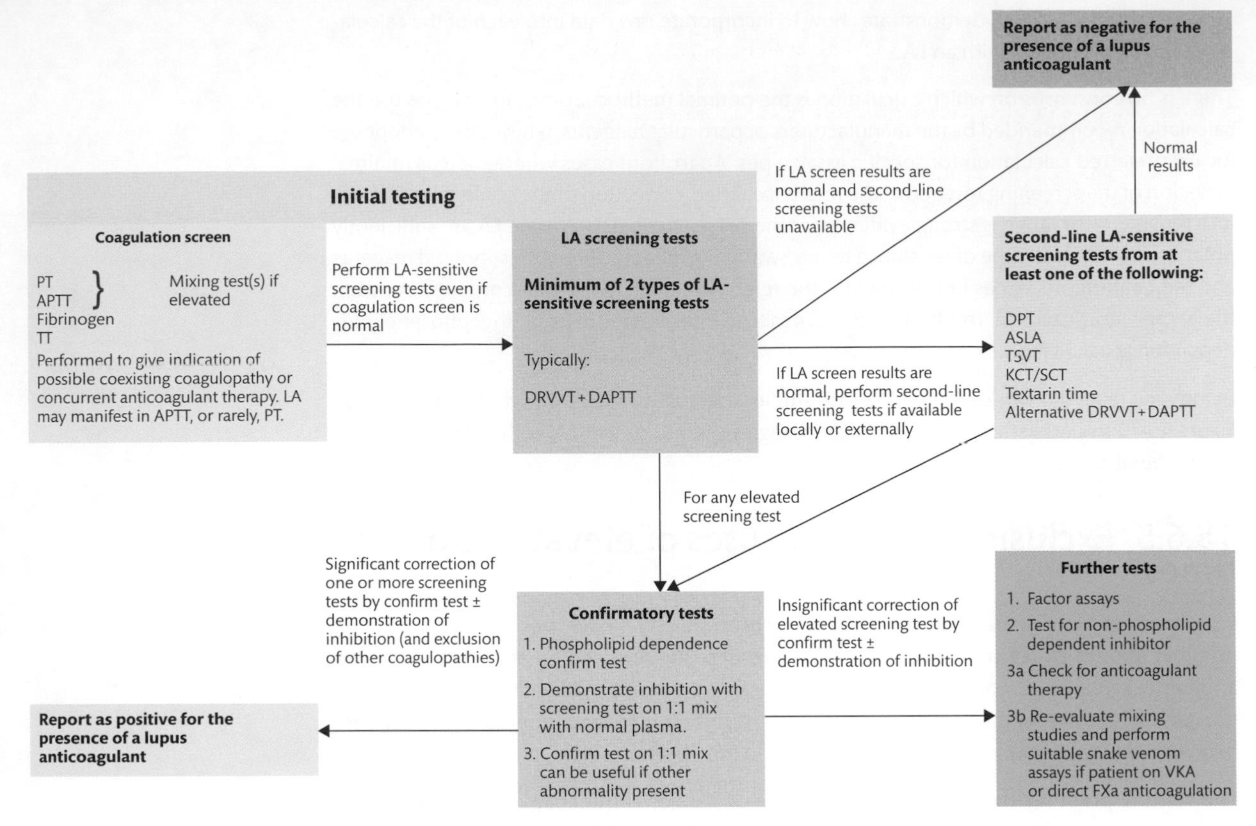

FIGURE 15.10
Flow chart for the laboratory detection of lupus anticoagulants.

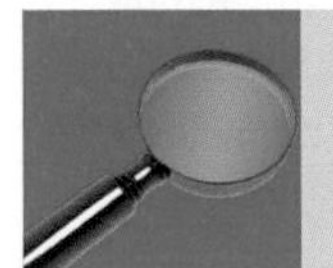

CASE STUDY 15.6 Positive LA screening with a normal coagulation screen

Index	Result	Reference range (see front of book)
PT (s)	12.6	(RR: 10.0–14.0)
APTT (s)	38.2	(RR: 32.0–42.0)
TT (s)	9.5	(RR: 9.0–11.0)
Fibrinogen (g/L)	3.2	(RR: 1.5–4.0)
DRVVT screen ratio	1.70	(RR: 0.85–1.15)
DRVVT confirm ratio	1.11	(RR: 0.86–1.14)
% correction of ratio	35.3	
DRVVT screen ratio on 1:1 mix	1.45	(RR: 0.90–1.11)
DAPTT screen ratio	1.30	(RR: 0.81–1.19)
DAPTT confirm ratio	0.98	(RR: 0.82–1.17)
% correction of ratio	24.6	
DAPTT screen ratio on 1:1 mix	1.24	(RR: 0.84–1.14)

The coagulation screen results show no abnormalities, yet the DRVVT and dilute activated partial thromboplastin time (DAPTT) tests are clearly positive from the elevated screen ratios and significant correction by confirm ratios. Routine coagulation screen reagents are not necessarily LA-sensitive. The DRVVT and DAPTT screening tests on 1:1 mixtures with normal plasma are elevated above the mixing test-specific reference ranges, indicating inhibition. The LA appears more potent with DRVVT than DAPTT. Some antibodies react better in some systems than others.

Case study 15.7 presents with an abnormal coagulation screen, so it is important to find an explanation before taking the LA assays at face value.

CASE STUDY 15.7 A 32-year-old male presented with a deep vein thrombosis

Index	Result	Reference range
PT (s)	13.0	(RR: 10.0–14.0)
APTT (s)	63.1	(RR: 32.0–42.0)
TT (s)	10.1	(RR: 9.0–11.0)
Fibrinogen (g/L)	2.8	(RR: 1.5–4.0)

At first sight, these results suggest a factor deficiency in the intrinsic pathway. Before embarking on time-consuming and expensive factor assays, an APTT on a 1:1 mixture with normal plasma was performed and gave the following result:

APTT1:1 mix(s) 59.1 (RR:32.0-42.0).

Although slightly shorter than the APTT in neat plasma, this result is nowhere near the reference range for undiluted plasma, and suggests the presence of an inhibitor and not a factor deficiency. In view of the thrombotic presentation, LA screening was performed.

Index	Result	Reference range
DRVVT screen ratio	1.01	(RR: 0.85–1.15)
DAPTT screen ratio	2.32	(RR: 0.81–1.19)
DAPTT confirm ratio	1.40	(RR: 0.82–1.17)
% correction of ratio	39.7	
DAPTT screen ratio on 1:1 mix	1.80	(RR: 0.84–1.14)
DAPTT confirm ratio on 1:1 mix	1.11	(RR: 0.85–1.12)
% correction of ratio in mixing tests	38.3	

The DRVVT is normal, yet there is a clear LA in the DAPTT in both neat plasma and the mixing tests. The confirmatory test in neat plasma does not return to the reference range, but significantly corrects the screening test result. The screening test in the mixture gives a lower result than in neat plasma due to a dilution effect. Note that a confirmatory test was also performed on a 1:1 mix. This was because the confirmatory test on undiluted plasma did not return into the reference range and could represent a coexisting factor deficiency, so performing the test on a 1:1 mix can correct it. In view of the patient's thrombotic symptoms, the most likely explanation is that his LA was sufficiently potent to partially resist the swamping effect of the confirm reagent when tested undiluted, but not when diluted. The results demonstrate a phospholipid-dependent inhibitor (i.e. LA) that is reacting only in APTT-based assays.

Key Points

As long as one assay (and its confirmatory tests) demonstrate the presence of an LA, that is sufficient to confirm positivity. Some LA are detectable by only one type of test.

Case study 15.8 demonstrates the importance of performing confirmatory tests for phospholipid dependence and inhibition in order to distinguish LA from other inhibitors.

CASE STUDY 15.8 An inhibitor is demonstrated in the coagulation screen but is it an LA?

Index	Result	Reference range
PT (s)	13.1	(RR: 10.0–14.0)
APTT (s)	110.0	(RR: 32.0–42.0)
APTT 1:1 mix (s)	90.3	(RR: 32.0–42.0)
Fibrinogen (g/L)	3.2	(RR: 1.5–4.0)
DRVVT screen ratio	2.40	(RR: 0.85–1.15)
DRVVT confirm ratio	2.30	(RR: 0.86–1.14)
% correction of ratio	4.2	
DRVVT screen ratio on 1 : 1 mix	1.89	(RR: 0.90–1.11)
DRVVT confirm ratio on 1 : 1 mix	1.92	(RR: 0.90–1.10)
% correction of ratio in mixing tests	−1.6	
DAPTT screen ratio	3.23	(RR: 0.81–1.19)
DAPTT confirm ratio	3.30	(RR: 0.82–1.17)
% correction of ratio	-2.2	
DAPTT screen ratio on 1 : 1 mix	2.87	(RR: 0.84–1.14)
DAPTT confirm ratio on 1 : 1 mix	2.86	(RR: 0.85–1.12)
% correction of ratio in mixing tests	0.3	

The LA tests reveal a non-phospholipid-dependent inhibitor, as there are no significant differences between screen and confirm assays in neat plasma or mixing tests. Further tests gave the following results:

Index	Result	Reference range
TT (s)	>120	(RR: 9.0–11.0)
RT (s)	14.5	(RR: 14.2–16.0)

The patient was receiving unfractionated heparin, the DRVVT and DAPTT confirm reagents were not PNP-based, and neither reagent contained a heparin neutralizer.

CASE STUDY 15.9 A patient was being investigated for thrombophilia, but was not on medication

The following results were obtained:

Index	Result	Reference range
PT (s)	15.0	(RR: 10.0–14.0)
APTT (s)	35.8	(RR: 32.0–42.0)
TT (s)	9.9	(RR: 9.0–11.0)
Fibrinogen (g/L)	3.9	(RR: 1.5–4.0)
DRVVT screen ratio	0.94	(RR: 0.85–1.15)
DAPTT screen ratio	0.99	(RR: 0.81–1.19)

The only abnormality is a slight elevation of the PT, which could indicate a mild FVII deficiency. Measurement of FVII activity gave a normal result and the patient was not on any drugs. Further tests showed the following:

Index	Result	Reference range
DPT screen ratio	1.51	(RR: 0.85–1.14)
DPT confirm ratio	1.10	(RR: 0.88–1.12)
% correction of ratio	27.1	
DPT screen ratio on 1:1 mix	1.30	(RR: 0.90–1.12)

Some LA will not be detected in the most commonly used LA assays, such as DRVVT and DAPTT, and this LA was detected in an extrinsic pathway-based assay.

Case study 15.9 is an example of antibody heterogeneity and reagent variability.

15.6.6 Oral anticoagulation

As you will read in Chapter 17, oral anticoagulant treatment with vitamin K antagonists (such as warfarin) leads to acquired functional deficiencies of factors II, VII, IX, and X. Since this inevitably results in elevated clotting times in screening tests for LA, we have to rely on mixing studies to correct the deficiencies and thus allow LA to manifest, which you can see in Case study 15.10.

Often the mixing studies are negative, but because dilution of the LA can prevent it from prolonging the screening test, you cannot be certain that the testing is truly negative. Thus, a positive result is diagnostic but a negative result does not exclude a weaker antibody.

The situation is different for patients receiving direct oral anticoagulants, since there is no factor deficiency to correct and the inhibitors will affect the normal plasma as well as the patient's plasma. We can use tests that bypass FX for patients receiving direct FXa inhibitors, which you will meet in section 15.6.7, but there are no clotting tests for LA that bypass thrombin. A pre-analytical procedure is available involving brief incubation of plasma with activated charcoal to adsorb DOACs and then centrifugation to pellet particulate matter and permit LA testing on the supernatant. The adsorbent is effective at removing most concentrations of DOACs and there are occasional effects on the residual plasma. LA testing is best postponed in patients receiving direct thrombin inhibitors until their period of anticoagulation is completed.

SELF-CHECK 15.10

Why do we have to use more than one screening test and a battery of confirmatory tests to detect lupus anticoagulants?

15.6.7 Testing for LA in anticoagulated patients using snake venoms

The venoms from the coastal taipan (*Oxyuranus scutellatus*) and Australian eastern brown snake (*Pseudonaja textilis*) contain FII activators, which are capable of activating the abnormal FII produced during warfarin therapy to an altered version of thrombin, termed meizothrombin, that will clot *in vitro*. This means that the TSVT and Textarin times (see Table 15.3) will give normal results in patients anticoagulated with vitamin K antagonists and patients who do not have LA or deficiencies of FII and fibrinogen. Because FII activation bypasses FX, the assays can also be used in patients anticoagulated with direct FXa inhibitors.

The Textarin time utilizes a novel confirmatory test, the Ecarin time, which uses venom from the saw-scaled viper (*Echis carinatus*). Like Taipan and Textarin venoms, the Ecarin venom directly activates FII but in a phospholipid- and calcium-*independent* manner, so it is impossible for LA to interfere with the clotting times. Ecarin venom can also activate the FII formed in vitamin K antagonists therapy. If Textarin and Ecarin venoms are titrated to give similar clotting times with a normal control, dividing the patient's Textarin time by their Ecarin time generates a Textarin/Ecarin ratio, which, if above 1.3, is indicative of the presence of an LA.

The Ecarin time can also be used as a confirmatory test for the TSVT too and is more sensitive to the presence of LA in vitamin K absence/antagonism (VKA) anticoagulated patients than using a PNP confirmatory test. The results are interpreted using the % correction of ratio. In view of the dilution effect in DRVVT and DAPTT mixing studies, some laboratories routinely

CASE STUDY 15.10 Results from a patient receiving warfarin therapy

Index	Result	Reference range (see front of book)
PT (s)	36.7	(RR: 10.0–14.0)
APTT (s)	51.8	(RR: 32.0–42.0)
TT (s)	9.2	(RR: 9.0–11.0)
Fibrinogen (g/L)	4.5	(RR: 1.5–4.0)
DRVVT screen ratio	2.90	(RR: 0.85–1.15)
DRVVT confirm ratio	2.30	(RR: 0.86–1.14)
% correction of ratio	20.7	
DRVVT screen ratio on 1:1 mix	1.54	(RR: 0.90–1.11)
DRVVT confirm ratio on 1:1 mix	1.03	(RR: 0.90–1.10)
% correction of ratio in mixing tests	33.1	
DAPTT screen ratio	1.84	(RR: 0.81–1.19)
DAPTT confirm ratio	1.44	(RR: 0.82–1.17)
% correction of ratio	21.7	
DAPTT screen ratio on 1:1 mix	1.30	(RR: 0.84–1.14)
DAPTT confirm ratio on 1:1 mix	0.99	(RR: 0.85–1.12)
% correction of ratio in mixing tests	23.8	

As described above, even though the neat plasma results show > 10% correction of the screen ratio, they are not necessarily reliable. The screening tests in the mixing studies remain elevated, and the confirmatory tests correct by >10% and back into the reference ranges, indicating that the normal plasma has corrected the factor deficiencies of warfarin therapy and thus allowed the LA to manifest. Similar results would be seen with an isolated hereditary factor deficiency plus LA.

include TSVT with Ecarin time confirmation on all patients receiving VKA or direct FXa inhibitor anticoagulation to improve detection rates.

Pre-analytical variables

Along with the usual considerations for obtaining a high-quality sample, a crucial consideration for LA testing is ensuring that the contamination of the plasma by platelets is minimized, and in any case, the residual platelet count is $<10 \times 10^9$/L. This is because residual platelet material can 'neutralize' LA, particularly the weaker or low-titre antibodies. Most laboratories freeze plasma aliquots in order to batch their LA assays so that the freeze-and-thaw process lyses platelet membranes, thereby releasing procoagulant phospholipid in a similar fashion to the process used to produce PNP reagents. Platelets can be removed from plasma to an appropriate level by double-centrifugation. This involves centrifuging the primary sample at 1500–2000 g for 15 minutes and then carefully separating the plasma to avoid the buffy layer into a secondary tube. This tube is capped and centrifuged at 1500–2000 g for 15 minutes. The now double-centrifuged plasma is separated into a storage tube and frozen, ideally at −70°C or below, until analysis. The normal control plasma must also be similarly platelet-free.

There is variation in analytical performance between different reagents, even those used for the same test but prepared by different manufacturers. This results in the necessity for locally derived reference ranges specific to the reagents and analytical equipment in use. It is insufficient to use manufacturers' stated reference ranges or those quoted by other laboratories or in research papers.

15.7 Thrombosis

15.7.1 Risk of venous thrombosis

Not all subjects with a hereditary or acquired thrombophilia go on to have a thrombosis. However, the risk of thrombosis for an individual with one of these factors is higher than for someone without the defect. As you saw earlier in section 15.2, the risk of thrombosis in someone with heterozygous protein C, protein S, or antithrombin deficiency is up to ten times higher than in an unaffected individual. Since only about 1 in 1000 people per year suffer a thrombosis, and that figure includes all other causes of thrombosis, this risk is still relatively small. However, subjects with thrombophilia are more likely to suffer a thrombosis at a relatively young age (<40 years old). In addition, certain high-risk situations (such as surgery, using the oral contraceptive pill, or hormone replacement therapy) can increase the risk of thrombosis, and subjects who know they have thrombophilia may need to consider alternative options or receive prophylactic anticoagulation at certain times. Identifying heritable thrombophilia in a subject can also have implications for close relatives, who may also wish to be screened.

15.7.2 Treatment of thrombophilia

Because anticoagulants such as heparin and warfarin are associated with a relatively high risk of bleeding, subjects with thrombophilia are not usually treated until they have had a thrombotic episode, or are placed in a high-risk situation, such as needing surgery. At present, it is also difficult to predict which patients will suffer recurrent thrombosis (up to 30% of individuals who have had a VTE will suffer a further event within five years, but individuals with no obvious cause for the first event are at least as likely to suffer a recurrence as those with thrombophilia). At present, therefore, long-term oral anticoagulant therapy is only recommended for subjects with recurrent spontaneous thrombosis, a single thrombosis at an unusual or life-threatening site, a single thrombosis with antithrombin deficiency or LA, or co-inheritance of two or more genetic defects. These recommendations may change as the findings of ongoing studies are reported.

15.7.3 Other types of thrombosis-related disorders

Although investigation for hereditary and acquired thrombophilias represents a considerable part of the workload in more specialized haemostasis laboratories, other disturbances in haemostasis can result in thrombotic disease.

Arterial thrombosis

You learnt in the introduction to this chapter that thrombophilia, and the causes of thrombophilia, generally refer to venous thromboembolic disease. Thrombosis can also occur in the arterial system, but the causes are different. Arterial thrombosis is often preceded by **atherosclerosis**, where the blood vessel is narrowed through deposits of **atheroma**. The composition of the vessel wall and flow characteristics also result in clots forming that are rich in platelets, and treatment of arterial thrombosis will often include antiplatelet therapy. The risk of arterial thrombosis is increased by a number of factors including smoking, **hypertension**, diabetes, and obesity. APS is also associated with an increased risk of arterial thrombosis.

atheroma

An accumulation of lipids, macrophages, and connective tissue causing thickening of the inner wall of the arteries in patches, or plaques.

hypertension

Increased blood pressure.

15.7.4 Other acquired thrombotic disorders

Acquired thrombotic thrombocytopenic purpura (TTP) occurs due to the presence of autoantibodies to ADAMTS-13. This results in reduced regulation of von Willebrand factor (VWF) multimer size, leading to an excess of circulating, highly adhesive ultra-high molecular weight multimers and a thrombotic presentation. Haemolytic uraemic syndrome (HUS) is a clinically similar disorder, but tends to have normal ADAMTS-13 levels. There are congenital forms of both disorders.

A relatively rare complication of anticoagulation with heparin is heparin-induced thrombocytopenia (HIT). These patients make antibodies to heparin-platelet factor 4 complexes, which activate platelets, causing them to aggregate and form blood clots and further reducing the platelet count.

Cross references

You met ADAMTS-13 in Chapter 13.

TTP and HUS are covered in more detail in Chapter 16.

HIT antibodies and their laboratory detection are covered in more detail in Chapter 17.

Chapter summary

- Thrombophilia is an increased risk of venous thrombosis caused by the inheritance of certain genetic defects, and it can also be acquired.
- Inherited causes include deficiency of protein C, protein S, or antithrombin, and mutations in the factor V gene (FV Leiden) or the prothrombin gene.
- Acquired defects, such as lupus anticoagulants, high homocysteine levels, or increased levels of certain clotting factors, are also associated with thrombophilia.
- Different assays are available to measure levels of protein C, protein S, and antithrombin. Some assays measure the level of protein in plasma (immunological assays), whilst others measure the activity of the protein (functional assays).
- Some individuals have defects which only affect the activity of these proteins, termed type II defects. It is important, therefore, to use functional assays to investigate thrombophilia.
- Mutations such as FV Leiden and the G20210A prothrombin mutation may be detected through genetic investigations.
- Not all patients with a heritable thrombophilia will go on to develop thrombosis.

- However, detection of thrombophilia may provide an explanation for a thrombotic event, and may indicate the need for prophylaxis (in the form of anticoagulation) in high-risk situations.
- The tests recommended for inclusion in a *thrombophilia screen* are shown in Table 15.5.
- Antiphospholipid syndrome is an autoimmune disorder characterized by thrombotic disease or pregnancy morbidity accompanied by the presence of persistent antiphospholipid antibodies.
- Antiphospholipid antibodies are subclassified depending on whether they are detectable in solid-phase assays such as ELISA, or in coagulation-based tests.
- Anticardiolipin antibodies and anti-β_2GPI antibodies are detected in solid-phase assays and lupus anticoagulants are detected in coagulation tests.
- Antibodies detected in solid-phase assays can be quantified by comparison with affinity-purified standards.
- LAs are detected by their interference with phospholipid-dependent coagulation assays, and their presence confirmed with tests for inhibition and phospholipid dependence.
- Antibody heterogeneity and reagent variability mean that no single reagent, or reagent type, will detect all LAs. Some antibodies react more strongly in one test than another, and others may only be identifiable by one type of test.

Discussion questions

15.1 Why could it be that having an inherited thrombophilia doesn't inevitably lead to the development of thrombotic disease?

15.2 What is the difference between type I and type II defects?

15.3 In the contexts of hereditary and acquired thrombophilia, what might be gained by measuring a particular analyte by more than one type of analytical method?

15.4 Why do we need multiple screening and confirmatory tests to detect lupus anticoagulants?

Further reading

- **Baglin T and Keeling D. *Heritable thrombophilia*. In: Hoffbrand AV, Higgs DR, Keeling DM, and Mehta AB (eds). *Postgraduate Haematology*, 7th edn (pp. 795–808). Wiley Blackwell, Oxford, 2016.**
- **Cooper PC, Pavlova A, Moore GW, Hickey KP, and Marlar RA. *Recommendations for clinical laboratory testing for protein C deficiency, for the subcommittee on plasma coagulation inhibitors of the ISTH*. *Journal of Thrombosis & Haemostasis* 2020: 18: 271–7.**

  International guideline on protein C testing.

TABLE 15.5 Tests commonly employed in thrombophilia screening.

Thrombophilia screening test	Method	Comments	Further tests if abnormal
Prothrombin time APTT Thrombin time	Routine screening reagents (but be aware of the LA sensitivity of your reagents)	May detect LA (prolonged APTT), dysfibrinogenaemia (prolonged TT), high FVIII (shortened APTT) More importantly, will detect effects of heparin or warfarin, or liver disease affecting vitamin K-dependent clotting factors—protein C and S	Mixing tests with normal plasma, LA screening, fibrinogen assay, factor assays
Antithrombin assay	Amidolytic/chromogenic assay using bovine thrombin or factor Xa as substrate	Not all molecular defects are detected equally by all functional assays	Immunological assay Genetic screening
Protein C assay	Amidolytic/chromogenic assay or clotting assay	Chromogenic assay insensitive to some dysfunctional sub-types Clotting assay affected by FV mutations conferring APC resistance and other variables	If a clotting assay is used, it is important to follow up abnormal results with chromogenic assays and/or immunological assays to exclude a falsely reduced assay Genetic screening may be performed
Protein S assay	Clotting assay or free PS antigen	Clotting assay affected by FV mutations conferring APC resistance and other variables	Free PS antigen (if clotting assay used) Total PS antigen Genetic screening
APC resistance/FV Leiden screening	APC resistance test without FV-deficient plasma, or APC resistance test with FV-deficient plasma, or DNA screening	Test without FV-deficient plasma may detect APC resistance in the absence of FV mutations conferring APC resistance—not specific for FVL	DNA screening if either APC resistance test is abnormal
Prothrombin G20210A mutation screening	DNA screening	FII levels may be elevated, but overlap with reference range precludes screening with a factor assay	
Antiphospholipid syndrome screening	Solid-phase assays Prolongation of phospholipid-dependent coagulation tests (e.g. DRVVT)	Guidelines with expert recommendations exist for LA testing	Evidence of an inhibitor (mixing studies) Confirmation of the phospholipid-dependence (e.g. DRVVT with high-concentration phospholipid) Further tests should be considered if screen is negative
Homocysteine assay	FPIA, ELISA, high-performance liquid chromatography (HPLC), latex immunoassay (LIA)		Some centres also screen for methylene tetrahydrofolate reductase (MTHFR) C677T mutation
Factor VIII assay, other assays	Routine factor assays	May be employed to try and help explain a thrombotic history; however, utility of this test in individual cases is debatable	

- **Devreese KMJ, de Groot PG, de Laat B, Erkan D, Favaloro EJ, Mackie I, Martinuzzo M, Ortel TL, Pengo V, Rand JH, Wahl D, Cohen H. *Guidance from the Scientific and Standardization Committee for lupus anticoagulant/antiphospholipid antibodies of the International Society on Thrombosis and Haemostasis Update of the guidelines for lupus anticoagulant detection and interpretation. Journal of Thrombosis & Haemostasis* 2020: 18: 2828–2839.**

  International guideline on LA detection.

- **Keeling D, Mackie I, Moore GW, Greer IA, Greaves M, and British Committee for Standards in Haematology. *Guidelines on the investigation and management of antiphospholipid syndrome*. *British Journal of Haematology* 2012: 157: 47–58.**

  Guidelines written by the Haemostasis Task Force of the British Committee for Standards in Haematology, with recommendations for LA screening.

- **Ledford-Kraemer MR, Moore GW, Bottenus R, Daniele C, de Groot PG, Exner T, Favaloro EJ, Moffat KA, and Nichols WL. *Laboratory testing for the lupus anticoagulant*. Approved guideline—first edition. CLSI document H60-A. Clinical and Laboratory Standards Institute Wayne, PA, April 2014.**

  Extensive guideline on laboratory detection of LAs.

- **Linnemann B and Hart C. *Laboratory diagnostics in thrombophilia*. *Hamostaseologie* 2019: 39: 49–61.**

  Details on principles and limitations of testing for hereditary thrombophilia.

- **Moore GW, Van Cott EM, Cutler JA, Mitchell MJ, and Adcock DM. *Recommendations for clinical laboratory testing of activated protein C resistance; Communication from the SSC of the ISTH*. *Journal of Thrombosis & Haemostasis* 2019: 17: 1555–61.**

  International guideline on activated protein C resistance testing.

- **Stevens SM, Woller SC, Bauer KA, Kasthuri R, Cushman M, Streiff M, Lim W, and Douketis JD. *Guidance for the evaluation and treatment of hereditary and acquired thrombophilia*. *Journal of Thrombosis and Thrombolysis* 2016: 41: 154–64.**

  Guidance on patient selection for thrombophilia testing.

- **Van Cott EM, Orlando C, Moore GW, Cooper PC, Meijer P, and Marlar R. *Recommendations for clinical laboratory testing for antithrombin deficiency; Communication from the SSC of the ISTH*. *Journal of Thrombosis & Haemostasis* 2020: 18: 17–22.**

  International guideline on antithrombin testing.

Answers to self-check questions and case study questions are provided as part of this book's online resources.

Visit www.oup.com/he/moore-fbms3e.

16

Acquired disorders of haemostasis

Gary Moore

In this chapter, you will be introduced to non-hereditary causes of bleeding and thrombosis. They are therefore referred to as acquired disorders, often arising as a consequence of another, primary disorder, which itself may be hereditary or acquired. We will discuss how the tests you have met so far can be employed to identify and monitor these conditions.

Learning objectives

After studying this chapter, you should confidently be able to:

- name a number of acquired haemostatic disorders and describe their causes;
- know how acquired disorders of haemostasis can affect routine coagulation screening tests;
- know how routine and specialist haemostasis assays can be employed to diagnose acquired disorders of haemostasis;
- understand principles of ADAMTS-13 activity and antibody assays.

16.1 Acquired bleeding disorders

Acquired bleeding disorders occur more frequently than inherited forms and are more common in hospitalized patients. They often consist of multiple coagulation factor deficiencies and can be severe to the point of being life-threatening. Excessive bleeding in a hospitalized patient is often due to recent surgery and can be controlled with a few, expertly placed stitches. However, bleeding can often be accentuated by, or entirely due to, secondary defects in haemostasis that require prompt diagnosis and treatment.

The possibilities of multiple abnormalities means these patients will often generate abnormal laboratory results and variably present with any combination of elevations of prothrombin time (PT), activated partial thromboplastin time (APTT), thrombin time (TT), and reduction in platelet count and/or abnormal platelet function.

16.1.1 Disseminated intravascular coagulation

Disseminated intravascular coagulation (DIC) results from excessive activation of coagulation exacerbated by loss of regulatory and localization mechanisms. This leads to inappropriate, systemic activation of coagulation within the vascular space, causing widespread formation of fibrin, which can lead to microthrombosis and bleeding due to consumption/exhaustion of coagulation factors and platelets.

Acute DIC is predominantly a haemorrhagic disorder, whilst chronic DIC rarely involves bleeding but is more associated with thrombotic manifestations.

Pathogenesis of DIC

The major pathological triggering event for DIC is exposure of the blood to a source of tissue factor that initiates coagulation, leading to thrombin generation. Increased tissue factor expression or availability can occur for the following reasons:

- upregulation of tissue factor expression on vascular endothelial cells and monocytes stimulated by **endotoxin** or **cytokines** (e.g. sepsis);
- release of tissue factor into the circulation following mechanical tissue damage (e.g. brain trauma);
- exposure to malignant cells or **microparticles** that express surface tissue factor (e.g. acute promyelocytic leukaemia, pancreatic cancer).

endotoxin
Lipopolysaccharide bacterial cell wall component that can initiate a potentially catastrophic inflammatory response.

cytokines
Small, hormone-like intercellular mediators with a diverse range of functions, including the stimulation of the immune system in response to an encounter with a pathogen.

microparticles
Small fragments released from endothelial cells, leucocytes, and platelets consisting of a plasma membrane and a small amount of cytoplasm.

Proinflammatory cytokines that upregulate tissue factor expression include tumour necrosis factor and interleukin-1. Other mechanisms that initiate DIC are snake venoms containing fractions that are capable of directly activating specific coagulation factors, and direct activation of factor X by pancreatic, tumour cell, or amniotic fluid enzymes.

SELF-CHECK 16.1

What is the main pathological trigger for DIC?

Acute DIC develops quickly, over hours or days, occurring when sudden exposure of blood to triggers such as tissue factor generates intravascular coagulation, leading to increased fibrin formation within the vessels. The upregulation of tissue factor expression, especially where it is not confined to the endothelium itself, leads to systemic activation of coagulation because localization mechanisms are bypassed. Thus, fibrin formation is not restricted to the surface of activated platelets aggregating at a site of vessel injury, as occurs in normal haemostasis. Even when there is vessel trauma, severe vessel and organ damage can result in release of tissue factor into the circulation and cause systemic activation of coagulation.

In overt DIC, thrombin generation is so excessive and intense that the regulatory systems involving antithrombin, activated protein C, and tissue factor pathway inhibitor (TFPI) become overwhelmed. In any case, excessive thrombin generation leads to formation of thrombin-antithrombin complexes (TAT), resulting in reduced levels of antithrombin due to consumption. Antithrombin levels can be further reduced due to degradation by elastase enzymes released from activated neutrophils, which will also degrade protein C and TFPI. Endothelial damage in DIC, such as that induced by endotoxin and cytokines, can downregulate thrombomodulin expression, and thus impair protein C activation. Additionally, reduced availability of the main cofactor for protein C, protein S, can occur due to elevated levels of C4b-binding protein binding to circulating free protein S, thereby reducing its availability.

Key Points

FVIII, fibrinogen, von Willebrand factor, PAI-1, and C4b-binding protein are all acute phase proteins.

Levels of plasminogen activator inhibitor type 1 (PAI-1) are elevated in DIC, since it is released from damaged or activated vessel endothelium, and it is an **acute phase protein**. This leads to suppression of fibrinolysis and failure to effectively clear intravascular fibrin thrombi. Interestingly, higher mortality in meningococcal sepsis is associated with higher PAI-1 levels.

acute phase protein
A protein whose circulating levels increase in response to acute conditions such as infection, injury, tissue destruction, some malignancies, surgery, burns, or trauma.

ischaemia
Inadequate blood supply to a tissue due to blood vessel blockage.

The excess thrombin generation ineffectively regulated by inhibitors, leading to multiple fibrin clots in the circulation ineffectively cleared by plasmin, results in thrombosis. The fibrin clots can entrap platelets to become larger clots which lodge in the microcirculation, large vessels, and organs, which leads to **ischaemia**, reduced organ perfusion, and end-organ damage and failure. Red blood cells passing through the fibrin meshwork in the microvasculature can be sheared by the fibrin strands, resulting in haemolysis arising from fragmentation of the cells.

Although fibrinolysis is reduced by the elevation in PAI-1 levels, it is not absent and tissue plasminogen activator (t-PA) still binds to plasminogen on the abundant fibrin surfaces to form plasmin. This generates high levels of fibrin degradation products (FDP), which can themselves exacerbate any bleeding by inhibiting thrombin, fibrin polymerization, and platelet function.

If the excessive, unregulated activation of coagulation continues unabated, production of coagulation factors in the liver is ultimately unable to compensate for their consumption and levels begin to decline. This has a particularly marked effect on factors V, VIII, XIII, and fibrinogen. Similarly, megakaryocytes in the bone marrow cannot keep up with platelet consumption and thrombocytopenia ensues. Thrombin-induced platelet aggregation significantly contributes to platelet consumption. Consequently, the most common sign of severe DIC

BOX 16.1 Neutrophil extracellular traps

You saw in Chapter 8 that neutrophils kill pathogens at the site of infection by engulfing and then destroying them. Neutrophils can also prevent circulating pathogens from spreading by expelling neutrophil extracellular traps (NETs), which are predominantly formed of unravelled DNA, that bind pathogens. NETs are lined with granular proteins, such as neutrophil elastase (NE) and myeloperoxidase, to kill the captured extracellular pathogens. Serine proteases in NETs, such as NE, promote fibrin formation to generate microthrombi that further limit pathogen survival and spreading, and tissue invasion. This intravascular fibrin formation as part of the innate immune defence is referred to as immunothrombosis. Platelets also adhere to NETs, leading to activation and aggregation.

Under certain pathological conditions, such as DIC due to sepsis, dysregulation of NET-induced microvascular thrombosis may lead to propagation of clot formation and contribute to venous and arterial thrombosis. NETs drive thrombus formation through tissue factor-mediated initiation of coagulation, contact-pathway initiation of coagulation, recruitment of VWF and subsequent platelet adhesion and aggregation, provision of a fibrin-independent scaffold for red blood cell recruitment, and inhibition of fibrinolysis.

is bleeding due to coagulation factor deficiencies, including reduced fibrinogen, reduced platelet numbers and function, and inhibition by elevated FDP levels. Mechanisms of DIC are summarized in Figure 16.1. Box 16.1 describes how neutrophils can contribute to the thrombosis in DIC.

SELF-CHECK 16.2

Summarize the pathophysiology of acute DIC.

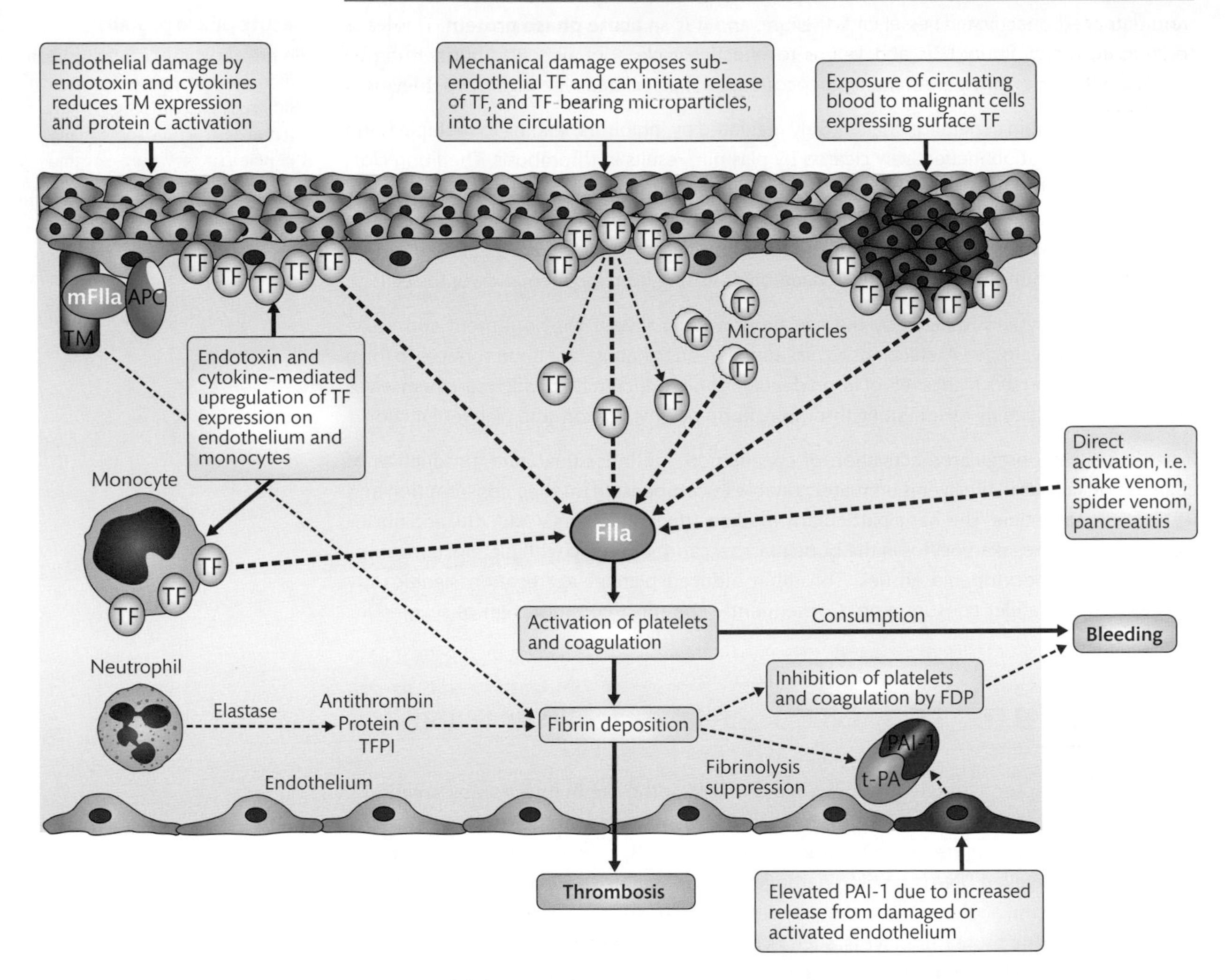

FIGURE 16.1

Mechanisms of DIC. Exposure of blood to a source of tissue factor (TF) initiates coagulation and subsequent thrombin generation. Direct activation can occur via endogenous activators. Excessive thrombin generation overwhelms regulatory molecules, which are further compromised due to degradation by neutrophil enzymes. This loss of regulation can be exacerbated by suppression of fibrinolysis, and clot formation ensues. Consumption of coagulation factors and platelets beyond compensatory capacity leads to bleeding. TM = thrombomodulin; mFIIa = modified thrombin; APC = activated protein C; TFPI = tissue factor pathway inhibitor; PAI-1 = plasminogen activator inhibitor type 1; t-PA = tissue plasminogen activator; FIIa = thrombin; FDP = fibrin degradation products.

Causes of DIC

One of the most commonly encountered causes of acute DIC is sepsis, with trauma and complications of pregnancy also being relatively common. DIC presents with a spectrum of severity that ranges from a chronic disorder with minimal coagulation and haemolysis to a life-threatening disorder encompassing uncontrollable bleeding and major organ failure resulting from small vessel thromboses.

In contrast, chronic DIC develops when blood is continuously or periodically exposed to small amounts of tissue factor, yet the liver and bone marrow are able to compensate. It is less clinically obvious than acute DIC and develops slowly, over weeks or months, and causes excessive clotting but does not progress to bleeding. Chronic DIC is more frequently seen in patients with solid tumours or large aortic **aneurysms**.

aneurysm
Localized, blood-filled dilation (bulge) in a blood vessel resulting from a weakened or injured vessel wall. Risk of rupture increases as an aneurysm grows in size.

The major causes of DIC are listed in Table 16.1.

SELF-CHECK 16.3

What are the main differences between acute and chronic DIC?

Diagnosis of DIC

The variety of causes of DIC, and the fact that severity can worsen over time, make both clinical and laboratory diagnoses challenging. Patients with DIC can experience simultaneous bleeding and thrombotic manifestations, although predominant clinical features can depend on the underlying cause.

The DIC typically seen in sepsis involves severe coagulation activation but fibrinolytic activation is mild due to markedly elevated PAI-1. Therefore, dissolution of multiple microthrombi is inefficient and severe organ dysfunction arises from impaired microcirculation. This can lead to renal, cardiac, and/or pulmonary failure, or skin necrosis and digital gangrene. Bleeding complications are relatively mild. DIC with enhanced fibrinolysis is typically seen in obstetric disorders, acute promyelocytic leukaemia (APML), and prostate cancer. Bleeding is the predominant symptom and can be severe, due to rapid consumption in obstetric disorders and production of plasminogen activators in APML. DIC with a balance between activation of coagulation and fibrinolysis is common in solid tumours. Bleeding and organ dysfunction are relatively uncommon except in advanced cases, although some cancers can progress to DIC with enhanced fibrinolysis, such as prostate cancer.

There is no gold standard or single laboratory test that can establish or exclude a diagnosis of DIC. Diagnosis is based on clinical and laboratory data and scoring systems can be employed for overt and non-overt DIC, examples of which are presented in Tables 16.2 and 16.3.

Laboratory tests in DIC

In acute DIC, laboratory tests that are quick to perform and able to provide evidence of the degree of activation and consumption are essential to effective clinical management. The routine screening tests that you met in Chapter 14, PT, APTT, and fibrinogen activity, fit the bill, as well as platelet counts and markers of fibrin degradation such as D-dimers. DIC is a dynamic situation, so any batch of tests is merely a window in time and regular testing, over hours and days, can be necessary. (See Case study 16.1.)

TABLE 16.1 **Conditions associated with DIC.**

Type of disorder and examples	Main pathophysiological mechanisms
Infection Sepsis or severe bacterial infection Viral infections (e.g. dengue, Lassa, Ebola) Protozoan (e.g. malaria) Fungal (e.g. *Candida*, *Aspergillus*)	Endotoxin release, endothelial damage
Trauma/tissue damage Crush and penetrating injuries (e.g. brain injury) Surgery Thermal injuries (heat and cold) Fat embolism Ischaemia-infarction	Release and exposure of tissue factor
Malignancy Solid tumours Myeloproliferative malignancies Lymphoproliferative malignancies	Release and exposure of tissue factor, release of cancer procoagulants, tumour necrosis factor
Obstetric complications Abruptio placentae Amniotic fluid embolism Eclampsia and pre-eclampsia Retained dead foetus Uterine rupture	Tissue factor release, ischaemia, amniotic fluid enzymes
Vascular and circulatory disorders Kasabach–Merritt syndrome (Giant haemangiomas) Large vascular aneurysims Acute myocardial infarction	Abnormal endothelium
Severe immunological reactions Blood transfusion reaction Heparin induced thrombocytopenia Transplant rejection	Complement activation, tissue factor release
Severe toxic reactions Snake venom Spider venom Pancreatitis	Direct enzyme activation

PT and APTT

Both PT and APTT are used as broad indicators of the ongoing activation and subsequent exhaustion of coagulation factors in DIC. One or both of these screening tests will be elevated at some point in the course of the disorder in about 50–60% of cases. However, nearly half of patients with DIC will generate normal, or even reduced PT and APTT results. This can arise merely because compensatory mechanisms are not yet defeated, or from the presence of circulating activated coagulation factors, such as thrombin or factor Xa, or elevated levels of the acute phase proteins factor VIII and fibrinogen. When APTT is measured by photo-optical endpoint detection,

TABLE 16.2 **Diagnostic scoring system for overt DIC; only for use in patients with an underlying disorder associated with DIC.**

Test	Result	Score
Platelet count	> 100 × 10^9/L	0
	< 100 × 10^9/L	1
	< 50 × 10^9/L	2
Fibrin marker (e.g. D-dimer, fibrin degradation products, soluble fibrin monomer complexes)	No increase	0
	Moderate increase	2
	Strong increase	3
	< 3 s	0
Prothrombin time (in seconds above upper limit of normal)	3–6 s	1
	> 6 s	2
Fibrinogen level	> 1.0 g/L	0
	< 1.0 g/L	1

Score of ≥5 compatible with overt DIC; scoring should be repeated daily.

Score of <5 suggestive for non-overt DIC; scoring should be repeated in one–two days.

TABLE 16.3 **Diagnostic scoring system for non-overt DIC.**

Test/criterion	Result			Score	
1. Risk assessment; is there an underlying disorder known to be associated with DIC?	Yes No	2 0			
2. Major criteria					
Platelet count	>100 × 10^9/L <100 × 10^9/L	0 1	+	Rising Stable Falling	−1 0 1
Fibrin marker (e.g. D-dimer, fibrin degradation products, soluble fibrin monomer complexes)	No increase Elevated	0 1	+	Rising Stable Falling	−1 0 1
Prothrombin time (in seconds above upper limit of normal)	<3 s >3 s	0 1	+	Rising Stable Falling	−1 0 1
3. Specific criteria					
Antithrombin	Normal Reduced	−1 1			
Protein C	Normal Reduced	−1 1			
Thrombin–antithrombin complexes	Normal Elevated	−1 1			

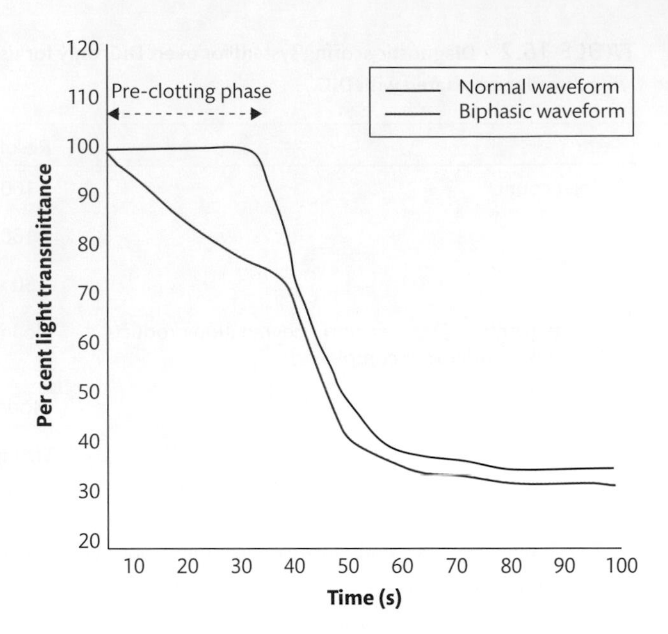

FIGURE 16.2
Normal and biphasic APTT waveforms. The graphs plot decrease in light transmittance against clotting time in seconds. Note that in the biphasic pattern there is an immediate, progressive decrease in light transmittance, even during the pre-clotting phase.

> **Cross reference**
> You met the coagulation screening tests PT and APTT in Chapter 14.

which you met in Chapter 2, an atypical light transmittance profile referred to as the biphasic waveform has been reported in DIC and shown to have positive predictive value. It occurs due to a calcium-dependent formation of a complex between **C-reactive protein** and very-low-density lipoprotein, which increases thrombin generation. A biphasic waveform can occur independently of an elevated APTT. Figure 16.2 compares a normal and a biphasic APTT waveform.

> **C-reactive protein**
> A protein released into the circulation by the liver in response to inflammation. It binds to dead or dying cells and some bacteria to promote clearance.
>
> **acute phase response**
> The physiological response to the onset of infection, inflammation, trauma, and some cancers involving increase in plasma concentration of acute phase proteins, fever, increased vascular permeability, and metabolic changes.

Fibrinogen

Although our knowledge of the aetiology of DIC tells us that fibrinogen can ultimately fall to very low levels, this tends to happen only in the very severe cases of acute DIC. The **acute phase response** will often elevate fibrinogen initially, and despite ongoing consumption, plasma levels can remain normal for some time. Nonetheless, fibrinogen measurement is integral to the investigation and monitoring of DIC and sequential fibrinogen testing is particularly useful.

D-dimers and soluble fibrin

You met D-dimers in Chapter 13 as part of the family of fragments generated by plasmin degradation of fibrin and fibrinogen. Each D domain of a D-dimer fragment is derived from a different parent fibrinogen molecule, bonded together by factor XIIIa-mediated cross links. This means that unlike some other fibrin/fibrinogen degradation products, they can only have originated from a cross-linked fibrin clot, and thus reflect thrombin formation as well as clot dissolution.

> **venous thromboembolism**
> A disorder that encompasses both deep vein thrombosis (DVT) and pulmonary embolism (PE).

Some degree of hyperfibrinolysis is common in DIC and elevated D-dimers will be found. However, many conditions, such as trauma, recent surgery, or **venous thromboembolism**, will generate elevated D-dimers/FDPs. Thus, elevated D-dimers in isolation are not diagnostic of DIC, but can be used in scoring systems and as a useful indicator of the state and course of the DIC process in conjunction with falling platelet counts and changes in PT and APTT.

You saw in Chapter 13 that thrombin cleavage of fibrinopeptides from fibrinogen molecules produces fibrin monomers which polymerize to form insoluble fibrin clots. In the presence of sufficient fibrinogen concentrations, fibrin monomers can form complexes with intact fibrinogen molecules, so-called soluble fibrin monomer complexes (SFMC). Levels of SFMC reflect thrombin generation and can reveal the early-activated state of blood coagulation. SFMC measurement is potentially advantageous over D-dimer because it specifically reflects the action of thrombin on fibrinogen, although SFMCs have a short half-life. SFMC levels have a high sensitivity for diagnosing DIC, but like D-dimers, poor specificity. They can be included in scoring systems yet results vary between assay types, which include enzyme-linked immunosorbent assay (ELISA) and latex immunoassay (LIA).

Cross reference

You met the principles of ELISA and LIA in Chapter 2.

Platelet count

A low platelet count or rapidly developing reduction in numbers is a central, albeit not specific, sign of DIC. The majority of patients with DIC will develop moderate-to-severe thrombocytopenia (platelet count <100 × 10^9/L), whilst up to 50% will develop more severe reductions (platelet count <50 × 10^9/L). Sequential platelet counts are more informative, as an initial result can be within the reference range, and a continuous reduction within the reference range can be an indication of active thrombin generation. Low platelet counts in DIC are associated with increased risk of haemorrhage, particularly when the count falls below 50 × 10^9/L.

SELF-CHECK 16.4

Which rapid, routine haematology tests are used to assess for DIC?

Other tests

Antithrombin and protein C levels are often reduced in DIC and have been shown to have prognostic significance. Relatively rapid and straightforward chromogenic assays are available, but are not in the repertoire of every laboratory. Again, sequential determinations are more informative than an isolated result.

The degree of coagulation derangement can be evaluated by assaying molecular markers of activation. Activation of prothrombin (factor II) yields the prothrombin activation peptide fragment 1 + 2, which can be measured by ELISA, as can fibrinopeptide A, plasminogen-α_2-antiplasmin complexes and TAT. The levels of these markers are elevated in DIC, although the assays take hours to perform and are not available in all diagnostic laboratories.

Blood film examination can reveal red cell fragmentation.

CASE STUDY 16.1 Acute disseminated intravascular coagulation

Patient history

- **A 65-year-old woman attended casualty department complaining of headaches and impairment of recent memory.**
- **She had been successfully treated for bowel cancer three years ago.**
- **CT scan revealed a mass in her brain and she was booked for surgery to remove it.**
- **Standard pre-operative blood tests were undertaken.**

Results 1

- Haemoglobin and platelet count were normal.
- White cell count was 15.9 × 10^9/L (RR: 4.0–11.0).
- PT, APTT, and fibrinogen were all normal.

Post-operative period

- The tumour was excised uneventfully and intra-operative blood loss was minimal.
- Increased bleeding was subsequently noticed from the suture line on her scalp and the catheter drain began aspirating a lot of unclotted blood.
- A second CT scan revealed brain haemorrhage.
- Blood count and clotting screen were repeated.

Results 2

- Haemoglobin was 105 g/L (RR: 120–160).
- Platelet count was 62 × 10^9/L (RR: 150–400).
- Coagulation testing was as follows:

Index	Result	Reference range (see front of book)
PT (s)	45.2	(RR: 10.0–14.0)
APTT (s)	72.9	(RR: 32.0–42.0)
Fibrinogen (g/L)	0.82	(RR: 2.0–4.0)
D-dimers (mg/L)	1.42	(RR: <0.25)

Significance of results

The patient had no evidence of abnormal clotting prior to surgery, so the subsequent deranged coagulation results and low platelet count were acquired as a result of her surgery. The combination of surgery, brain injury, and possibly the tumour itself, led to development of DIC. Consumption of platelets, coagulation factors, and fibrinogen account for the laboratory results and the bleeding. The elevated D-dimers reflect the increase in fibrin generation.

Treatment of DIC

Specific and vigorous treatment of the underlying disorder is key to treating DIC, such as antibiotic administration or surgical drainage, and will spontaneously resolve the DIC in many cases. However, supportive treatment aimed at haemostatic abnormalities can be necessary. This can include transfusion of donor platelets and/or fresh-frozen plasma to combat active bleeding, and even fibrinogen concentrate if fibrinogen deficiency develops. Where thrombosis is the predominant manifestation, anticoagulation with UFH can be used, although monitoring with APTT may be complicated if coagulation factor reduction ensues.

16.1.2 Hepatic disease

The liver is the site of synthesis of all the coagulation factors of secondary haemostasis, plasminogen, and the regulatory inhibitors antithrombin, protein C, protein S, and α_2-antiplasmin. Clearance of activated coagulation factors is also undertaken by the liver. Thus it is unsurprising that liver disease can impact on secondary haemostasis and fibrinolysis, but also primary haemostasis, via different mechanisms that affect platelet number or function. Symptoms can be haemorrhagic or thrombotic.

Liver disease and primary haemostasis

Thrombocytopenia is a common finding in liver disease and can arise from a variety of mechanisms, which are detailed in Table 16.4. Von Willebrand factor (VWF), which is synthesized in megakaryocytes and blood vessel endothelial cells, can be elevated in liver disease. This provides a degree of compensation by enhancing platelet adhesion. Possible mechanisms for the

TABLE 16.4 **Mechanisms of thrombocytopenia in liver disease.**

Cause of thrombocytopenia	Mechanisms
Reduced production	Reduced production of thrombopoietin by the liver Hepatitis C virus-related myelosuppression affecting megakaryocytes Toxic effect of alcohol on megakaryocyte function in alcoholic liver disease Folate deficiency in alcoholic liver disease
Hypersplenism	Splenic sequestration of platelets
Accelerated destruction	Immune-complex-mediated clearance and destruction
Consumption	DIC

elevation in VWF are endotoxin-induced endothelial damage, induction of VWF synthesis in the liver, or reduced ADAMTS-13 synthesis, leading to delayed VWF clearance.

Patients with acute **hepatitis** will often have only a mild thrombocytopenia, in the region of 100–150 × 10^9/L, whilst it can be mild-to-moderate in chronic disease, in the region of 50–150 × 10^9/L. Up to 30% of patients with chronic liver disease will exhibit thrombocytopenia, rising to around 90% in end-stage disease.

hepatitis
Inflammation of the liver, which can be due to infection, autoimmunity, or exposure to toxic substances.

Platelet dysfunction is present in some but not all patients with liver disease, and can be in addition to any reduction in number. Hypoaggregability has been demonstrated, with abnormalities in primary and secondary aggregation to agonists such as adenosine diphosphate, epinephrine, thrombin, and ristocetin. Causes of platelet dysfunction can be extrinsic, such as increased levels of FDPs, ethanol, and oxidized high-density lipoproteins, which are inhibitory. Intrinsic platelet function defects have also been described, such as reduced arachidonic acid availability for thromboxane A_2 generation, acquired nucleotide deficiency, and abnormalities of platelet membrane composition and signalling.

Cross reference
You met platelet function testing for hypoaggregability in Chapter 14 on bleeding disorders.

SELF-CHECK 16.5

What are the causes of thrombocytopenia in liver disease?

Liver disease and secondary haemostasis

Suppression of coagulation factor production due to reduction in **hepatocyte** function leads to a fall in concentrations of most coagulation factors, which can be sufficient to promote bleeding. Extrahepatic production can compensate for levels of some, such as FVIII, FV, and PAI-1. Measured levels of coagulation factors broadly mirror the extent of liver damage, but are poor predictors of bleeding risk in an individual patient, as there may be other causes or contributors.

hepatocyte
Liver cell involved in protein synthesis and storage.

Levels of FVIII and fibrinogen can be increased as part of the acute phase response associated with liver disease. Cytokine release from **necrotic** liver tissue may contribute to increased hepatic and extra-hepatic FVIII synthesis, as well as reduced liver clearance. Unlike FVIII, fibrinogen is exclusively synthesized in the liver, yet the organ has a remarkable capacity to maintain fibrinogen production in both acute and chronic disease. It will often be low in acute disease, and a marked reduction in fibrinogen may herald end-stage liver failure or DIC. Dysfibrinogenaemia can ensue in some patients, due not to inherent structural abnormalities

necrotic
Localized tissue death resulting from insufficient blood flow to the tissue due to injury, radiation, or toxic substances.

such as those encountered in most hereditary forms, but to an increase in sialic acid residues on fibrinogen molecules that impair polymerization and clot stability.

Chronic liver disease is associated with a global reduction in levels of circulating coagulation factors and inhibitors, whilst acute liver failure will manifest an early reduction in levels of FV and FVII, due to their relatively short half-lives, followed by reductions in FII and FX, but not FIX or FXI. Reductions in antithrombin, protein C, and protein S can be encountered in acute and chronic liver disease. To some extent, a rebalancing of the system occurs as the reduction in procoagulant clotting factors is tempered by a concomitant reduction in regulatory factors. Cholestatic liver disease, where bile production by hepatocytes is reduced, leads to impaired absorption of fat-soluble vitamins, including vitamin K, and can result in reduced levels of vitamin K-dependent factors.

SELF-CHECK 16.6

Why are coagulation factor levels often low in liver disease?

Liver disease and fibrinolysis

Tissue plasminogen activator (t-PA) is synthesized in endothelial cells, so levels are not reduced in liver disease. However, t-PA levels are elevated in chronic liver disease due to reduced hepatic clearance, whilst its inhibitor, PAI-1, may be normal or decreased. This can lead to increased fibrinolysis and promote bleeding, whilst the elevated PAI-1 levels in acute liver disease can alter the balance towards reduced fibrinolysis and promote thrombosis.

DIC in liver disease

It is common for patients with advanced liver disease to develop chronic, low-grade DIC. Possible causes include:

- release of procoagulants from damaged hepatocytes;
- release of intestinal endotoxins into the **hepatic portal circulation**;
- reduced hepatic clearance of activated coagulation factors;
- inordinate reduction in levels of antithrombin, protein C, and protein S relative to the degree of hepatic disease.

hepatic portal circulation
Circulation of blood from the small intestine, right-side of the colon, and spleen to the liver via the hepatic portal vein.

Diagnosis of DIC is complicated by the fact that the reduced coagulation factors, fibrinogen and inhibitors, thrombocytopenia and elevated FDPs, can also arise from the multiple defects of chronic liver disease itself.

Coagulation tests in liver disease

Standard coagulation screening tests and platelet counts are normally sufficient to assess possible haemostatic defects in patients with liver disease who are actively bleeding. Chronic liver disease will commonly give rise to elevations in PT and APTT, reflecting a general reduction in production of coagulation factors by the failing liver. Abnormal screening tests alone do not indicate necessity for coagulation factor replacement therapy, which is only required when there is evidence of active bleeding or a patient requires surgery.

An isolated, elevated PT may be the only abnormality in early or mild disease due to the short half-life of FVII causing levels to fall earlier than other factors. The PT has long been used as a test to indicate the degree of hepatic biosynthetic function, although it is, of course, affected by levels of three vitamin K-dependent factors, FII, FVII, and FX. Where vitamin K-related coagulation factor reduction is present or suspected, measuring FV levels is also a sensitive indicator of disease severity because it is not vitamin K-dependent.

Fibrinogen is commonly assayed by the Clauss technique, which is sensitive to reductions in concentration and function, so any dysfibrinogenaemia of liver disease will manifest as well as any quantitative deficiency. Thrombin time and reptilase time may be elevated due to either deficiency type, whilst a fibrinogen antigen assay, typically ELISA, will be reduced in quantitative but not qualitative fibrinogen deficiency. The dysfibrinogenaemia of liver disease is rarely associated with bleeding. D-dimer levels are commonly employed to evidence hyperfibrinolysis and DIC.

Cross references

You met the principles of Clauss fibrinogen, thrombin time, and reptilase time in Chapter 14.

D-dimer measurement is covered in Chapter 17.

Liver transplantation

A patient with chronic or acute liver disease requiring a transplant is likely to have some degree of existing coagulopathy and bleeding can be severe during the transplantation procedure. A mild baseline deterioration of the coagulopathy coincides with removal of the diseased liver. Prior to implantation of the new liver, the anhepatic stage, the coagulopathy worsens as there is no synthesis or clearance of coagulation factors and hyperfibrinolysis ensues due to marked t-PA release and concomitant fall in PAI-1 and fibrinogen. Consumptive thrombocytopenia with DIC is a common occurrence at this stage and massive blood product support is required. Once the new organ is in situ, reperfusion is initially accompanied by release of t-PA and a heparin-like substance from the graft, which can markedly elevate the APTT. Normal haemostasis is restored within a few days if the graft is successful.

16.1.3 Vitamin K

You saw in Chapter 13 that factors II, VII, IX, and X, and proteins C, S, and Z, require dietary vitamin K to facilitate post-translational conversion of glutamic acid (Glu) side chains to γ-carboxyglutamic acid (Gla). These Gla residues promote calcium-dependent attachment to phospholipid surfaces. Vitamin K deficiency results in manufacture of undercarboxylated forms of these proteins and can lead to a bleeding tendency when functional levels of vitamin K-dependent factors fall to around 30% of normal or below. It is only at this point that screening tests like PT and APTT may become abnormal, and earlier detection requires other assays such as direct measurement of vitamin K levels or assays to detect elevated levels of undercarboxylated vitamin K-dependent factors. PT will often be prolonged early, and more severely than APTT, as is seen in warfarin therapy. TT, fibrinogen, and platelet count will be normal unless DIC is present. One-stage factor assays to measure functional levels of factors II, VII, IX, and X will often generate levels below reference ranges before PT or APTT screening tests are elevated.

Cross references

The vitamin K cycle is described in Chapter 17.

One-stage coagulation factor assays are described in Chapter 14.

Deficiency of vitamin K can be due to dietary insufficiency, malabsorption syndromes, fasting, hyperemesis gravidarum, alcoholism, or ingestion of certain drugs (i.e. warfarin, some antibiotics, salicylates).

Vitamin K-deficiency bleeding

The combination of poor passage of vitamin K across the placenta and consequent low vitamin K stores at birth, absent gut flora at birth, and low vitamin K content of human milk conspire to render newborns relatively vitamin K deficient. The deficiency may intensify in the first few days of life without supplement and can lead to bleeding manifestations, which may be severe or even life-threatening. This is referred to as vitamin K-deficiency bleeding (VKDB) or haemorrhagic disease of the newborn (HDN), which occurs in three forms:

Early VKDB manifests within the first 24 hours of life and arises from maternal ingestion of substances that cross the placenta and interfere with vitamin K metabolism, usually pharmaceuticals, such as certain anticonvulsants and antituberculosis drugs, warfarin, and barbiturates. Bleeding typically occurs at sites related to birth trauma, such as cranium, thorax, or abdomen. Early VKDB is very rare and can be easily prevented by maternal vitamin K administration for about two weeks prior to delivery.

Classic VKDB is also rare, occurring in about 1 in 200–400 births without vitamin K prophylaxis for the infant. It manifests within one to seven days of birth, bleeding sites can be gastrointestinal, skin and mucous membranes (i.e. nose and gums), umbilical stump, or after circumcision. Classic VKDB commonly arises in breastfed infants who have not received prophylactic vitamin K, and may also be related to maternal drug usage.

Late VKDB manifests within two weeks to six months of birth, with a peak incidence at three to six weeks, and can also occur in breastfed infants who did not receive vitamin K at birth. It can also be associated with other conditions, such as cholestatic liver disease, cystic fibrosis, coeliac disease, and deficiency of α_1-antitrypsin. Frequency of late VKDB in infants without vitamin K prophylaxis is about 1 in 15,000 births, typically manifesting as intracranial, gastrointestinal, or skin bleeding. Late VKDB is the greatest risk to these children due to sudden bleeding into the central nervous system.

Key Points

Commensal intestinal flora (bacteria) synthesize vitamin K_2 (menaquinone), which supplements vitamin K_1 (phylloquinone) obtained in the diet, primarily from leafy green vegetables. There are about 100 trillion microorganisms in an average human intestine.

SELF-CHECK 16.7

Summarize the three main types of vitamin K-deficiency bleeding and their causes.

16.1.4 Acquired coagulation factor deficiencies due to inhibitors

You met inhibitors of FVIII and FIX in patients with haemophilia A and B in Chapter 14, mainly due to alloantibodies, and also the concept of acquired haemophilia due to autoantibodies. You also met acquired von Willebrand disease in Chapter 14. Far more rarely, inhibitors specific to other coagulation factors arise and can be associated with significant bleeding.

Deficiency of FXI is, of course, the third congenital haemophilia, but inhibitor development is unusual. This is mainly because haemophilia C is commonly less severe than the other two and many patients require little or no blood product therapy unless challenged by surgery or trauma. Inhibitors do occur in patients homozygous for certain mutations. Spontaneous FXI inhibitors have been reported, almost exclusively in patients with underlying autoimmune disorders such as rheumatoid arthritis and systemic lupus erythematosus. Significant bleeding is rare and specific therapy rarely necessary. A spontaneous inhibitor may be suspected upon finding an isolated, elevated APTT that does not correct upon mixing with normal plasma, and can be identified with a Bethesda-type inhibitor assay not requiring the prolonged incubation necessary to detect FVIII inhibitors.

Inhibitors in congenital FV deficiency are also rare. Spontaneous autoantibodies have been reported, mainly after surgery and consequent exposure to the topical haemostatic agents bovine thrombin and fibrin glue, as they both contain bovine FV. Bleeding manifestations are variable, but can be severe. Interestingly, where patients are resistant to treatment with plasma infusion, platelet concentrates can be successful because the FV contained in alpha granules is protected from the circulating inhibitor. Other associations for the development of FV inhibitors include β-lactam antibiotics, blood transfusions, and malignancies. An FV inhibitor will elevate PT and APTT but not TT, and can be quantified in a Bethesda-type inhibitor assay not normally requiring prolonged incubation (see Case study 16.2). Inhibitors to FII and FX will generate similar laboratory results. Antibodies to FII are most commonly encountered in patients with antiphospholipid syndrome, although they tend to be non-neutralizing and ELISA, not Bethesda assays, are required to detect them. Bleeding can be severe. Acquired FX inhibitors have been described, presenting with elevated PT and APTT but not TT, and can be quantified in a modified Bethesda assay.

Development of acquired inhibitors of FVII is associated with septicaemia, pancreatitis, multiple myeloma, and use of drugs such as certain antibiotics. The antibodies can be inhibitory or cause accelerated clearance. The laboratory will find an isolated, elevated PT that does not correct on mixing with normal plasma, and neutralizing antibodies can be detected with a modified Bethesda assay.

Patients treated with bovine thrombin or fibrin glue will very occasionally generate antibodies to bovine thrombin. The patients are asymptomatic, as the antibody does not interfere with human thrombin. Coagulation screening will generate normal PT and APTT, again because the thrombin in the reaction is the patient's own human thrombin, but TT will be elevated, or even unclottable, and reptilase time normal. Clauss fibrinogen measurement is normally unaffected because of the plasma dilution and higher thrombin concentration than is employed in TT reagents. Performing the TT with a human thrombin-derived reagent will generate a normal result and allay any fears of a genuine coagulopathy. More recently, fibrin glue products containing human thrombin have become available.

Acquired autoimmune antibodies that interfere with fibrin polymerization, or are directed against fibrinogen itself, have been described and will present with elevated PT, APTT, TT, and reptilase time, and a reduced Clauss fibrinogen. TT mixing studies can reveal such an inhibitor. Acquired inhibitors to FXIII have been described but PT, APTT, TT, reptilase time, and Clauss fibrinogen are all normal and specific assays for FXIII activity are required.

SELF-CHECK 16.8

Inhibitors to which coagulation factors might be expected to elevate both PT and APTT?

16.1.5 Other acquired bleeding disorders

The following section will introduce you to a variety of clinical situations where haemostasis can be compromised, although regular laboratory testing is not necessarily a major feature in all of them.

Renal disease

Approaching 50% of patients with renal disease will experience haemorrhagic manifestations, such as **purpura**, **menorrhagia**, **epistaxis**, and gut bleeding, although it is rarely severe. Coagulation screening tests such as PT and APTT tend to be normal, as coagulation factors are not reduced and may even be elevated. Severe thrombocytopenia secondary to renal disease is rare and the main haemostatic disturbance is impaired primary haemostasis arising largely from the effects of **uraemia**.

purpura
Red or purple skin discoloration produced by bleeding small blood vessels near the surface.

menorrhagia
Heavy menstrual bleeding.

epistaxis
Nosebleeds.

uraemia
An excess of end products of amino acid and protein metabolism in blood that would normally be excreted in urine, such as urea and creatinine.

CASE STUDY 16.2 Acquired coagulation factor deficiency

Patient history

- A 55-year-old diabetic woman presented with acute gastrointestinal bleeding.
- She had no previous history of bleeding symptoms.
- She was taking antibiotics.
- She was not on any anticoagulant drugs.
- Full blood count and coagulation screen were performed.

Results 1

- Haemoglobin was 60 g/L (RR: 120–160).
- White cell count was 12.1 × 10^9/L (RR: 4.0–11.0).
- Platelet count was 392 × 10^9/L (RR: 150–400).
- Coagulation testing was as follows:

Index	Result	Reference range (see front of book)
PT (s)	84.2	(RR: 10.0–14.0)
PT mixing test (s)	83.1	
APTT (s)	131.7	(RR: 32.0–42.0)
APTT mixing test (s)	129.8	
Thrombin time (s)	11.0	(RR: 10.0–13.0)
Fibrinogen (g/L)	3.0	(RR: 2.0–4.0)

Significance of results 1

- The low haemoglobin was due to blood loss.
- Platelet numbers were normal, so bleeding was not due to thrombocytopenia.
- Markedly elevated PT and APTT with normal TT and fibrinogen suggest a common pathway defect or multiple defects.
- Non-correcting mixing tests suggest a common pathway inhibitor.

Results 2

- Factor assays gave the following results:
- Bethesda assay for FV inhibitor: 209 BU/mL

Index		Result	Reference range
FII	(IU/dL)	78	(RR: 50–150)
FV	(IU/dL)	<1	(RR: 50–150)
FVII	(IU/dL)	120	(RR: 50–150)
FVIII	(IU/dL)	148	(RR: 50–150)
FIX	(IU/dL)	99	(RR: 50–150)
FX	(IU/dL)	92	(RR: 50–150)
FXI	(IU/dL)	65	(RR: 50–150)
FXII	(IU/dL)	72	(RR: 50–150)

Significance of results 2

- Only FV was reduced and was the cause of the bleeding.

The undetectable FV was due to the presence of an acquired antibody to FV that completely inhibited function.

The pathophysiology of impaired primary haemostasis in renal disease is multifactorial and affects endothelium function, VWF, and platelet function. Accumulation of guanidinosuccinic acid in uraemic plasma upregulates endothelial production of nitric oxide, which inhibits platelet function. Plasma VWF levels are normal or elevated, but the high-molecular-weight multimers may be affected by uraemia. Impaired VWF-platelet binding has been reported due to a decrease in glycoprotein Ib (GpIb) complexes on the platelet surface. Other intrinsic platelet abnormalities can be present in uraemic patients, such as abnormal granule secretion, abnormal calcium mobilization, and reduced thromboxane A_2 generation.

The severe anaemia of renal disease disrupts the radial transport of platelets, as there is reduced displacement of platelets to the periphery by red blood cells. This decreases their immediate availability for primary haemostatic interactions on the endothelium. Blood tests of primary haemostatic function are not routinely undertaken.

Cross reference

You met the chronic anaemia of renal disease due to reduced erythropoietin production in Chapter 5.

Coagulopathy of massive transfusion

Massive blood transfusion for major trauma and/or significant bleeding can trigger a complex, multifactorial coagulopathy. The bleeding can occur due to one or more of hypothermia, acidosis, dilutional coagulopathy, increased fibrinolysis, reduced fibrinogen, or impaired platelet function.

Hypothermia and acidosis

Massive, rapid transfusion of blood products can induce hypothermia, commonly considered to be a core body temperature of <34°C. Activated coagulation factors are temperature-dependent enzymes and are slowed down by reduction in body temperature. A patient whose body temperature is reduced to 32°C will have an approximate 50% reduction in their coagulation function. Any attempt to treat bleeding with plasma products will fail unless the hypothermia is reversed, as it will also affect any infused coagulation factors. Hypothermia also causes reversible platelet dysfunction, increases fibrinolytic activity, and can ultimately reduce coagulation factor production.

Metabolic acidosis can occur after massive transfusion resulting from the initial blood loss and **hypoxia**, use of volume replacement solutions that cause dilutional acidosis, and changes that occur in stored blood prior to transfusion. Coagulation enzyme activity is pH-dependent and thus adversely affected. Synergism of hypothermia and acidosis promotes coagulopathy, the combination of all three often being referred to as the 'vicious bloody triad' or 'trauma triad of death'.

metabolic acidosis

Accumulation of acid in blood, leading to pH reduction.

hypoxia

Inadequate oxygen supply.

Key Points

Interestingly, the *in vivo* hypothermia-induced reduction in enzymatic activity of activated coagulation factors is not mirrored in coagulation screening with PT and APTT because the *in vitro* laboratory tests are performed at normal body temperature of 37°C.

Similarly, buffered reagents can compensate for reductions in pH.

Dilutional coagulopathy

Replacement of large amounts of lost blood by volume expanders and red blood cells, both of which lack coagulation factors and functional platelets, inevitably results in reductions of both. The initial cause of blood loss, such as DIC or trauma, can contribute to ongoing coagulation factor and platelet consumption.

Fibrinogen and fibrinolysis

Ongoing systemic fibrinolysis, such as in DIC or advanced liver disease, can exacerbate bleeding by premature clot destruction. Fibrinogen itself may be degraded if regulatory mechanisms fail to prevent plasmin entering the circulation. Dilution, and bleeding-induced loss and consumption, will reduce circulating fibrinogen levels, and thus fibrin formation. Fibrin polymerization can be impaired by interaction with some volume expanders.

Platelet dysfunction

Thrombocytopenia is a major contributor to bleeding due to dilution, yet other factors can additionally impair platelet function. High concentrations of FDPs can bind to GpIIbIIIa and block platelet-fibrinogen bridging, and free plasmin can cleave platelet receptors. Platelets can become 'exhausted' from premature release of granule contents arising from intravascular platelet trauma.

Use of laboratory tests

Notwithstanding the problems assessing contribution of hypothermia and acidosis to a bleeding patient, tests such as PT, APTT, fibrinogen, platelet count, and D-dimer can be useful indicators of dilution, consumption, DIC, and fibrinolysis. They can also be applied to assess plasma and platelet replacement therapy.

SELF-CHECK 16.9

List the possible causes of the coagulopathy of massive transfusion.

Pregnancy

It is normal in pregnancy to see physiological changes in haemostasis, such as elevations of FVIII, VWF, and fibrinogen, and a reduction in protein S. You saw in Table 16.1 that obstetric complications can trigger DIC, which can lead to bleeding, and in some instances may be fatal.

Thrombocytopenia is a common finding in pregnancy, occurring in about 10% of women. The most common cause is gestational or incidental thrombocytopenia arising from increased blood volume (and thus dilution), and increased platelet turnover. The thrombocytopenia tends to be only mild, commonly >70 × 10^9/L, and usually occurs later in the pregnancy. It is rarely associated with bleeding, even at delivery, and will normally resolve within days of delivery. (See also Box 16.2.)

Pregnant patients who develop a platelet count <70–80 × 10^9/L should be suspected of having other causes of the thrombocytopenia, such as the pregnancy-specific pre-eclampsia, HELLP syndrome (*h*aemolysis, *e*levated *l*iver enzymes, and *l*ow *p*latelets) and acute fatty liver of pregnancy. Other conditions that are not specific to pregnancy that lead to thrombocytopenia include immune thrombocytopenia (ITP), viral infections, autoimmune disorders, and of course, DIC. Unlike gestational thrombocytopenia, these situations can lead to bleeding.

BOX 16.2 Twins

Platelet counts tend to be slightly lower if the mother is expecting twins, rather than a single child. It is thought this may be due to increased activation of coagulation in the placenta.

Amyloidosis

Amyloidoses are a group of rare disorders that involve deposition of abnormal protein, called amyloid, in tissues and organs throughout the body. The most common haemostatic abnormality associated with **amyloidosis** is acquired FX deficiency due to adsorption of FX onto the amyloid, leading to a reduction in circulating levels that can cause mild-to-life-threatening bleeding. Both PT and APTT will be elevated in acquired FX deficiency, although bleeding is rare until the FX falls to 10% or less of normal. Reductions in levels of all other coagulation factors, and VWF, have also been described and presumed to also be due to adsorption.

amyloidosis
Deposition of abnormally folded protein in tissues.

Abnormalities in fibrin polymerization occur in amyloidosis due to inhibition by serum amyloid P component. Consequently, elevated TT and reptilase time despite normal or even elevated fibrinogen, by the Clauss method, is a common finding. Increased fibrinolysis has been reported in amyloidosis, possibly due to elevated levels of urokinase. Impaired platelet function is seen in some patients due to amyloid light chains binding to platelet membranes.

Paraproteinaemias

You met B-cell lymphoid neoplasms with circulating **paraproteins**, such as myeloma and Waldenström macroglobulinaemia, in Chapter 12. The paraproteins in some patients can interact with platelets, VWF, or coagulation factors and produce haemostatic abnormalities. Some merely manifest as abnormal *in vitro* testing not associated with bleeding, whilst others will be symptomatic.

paraprotein
Circulating monoclonal immunoglobulin or light chain.

The main causes of bleeding are as follows:

- hyperviscosity causing abnormal vessel wall stress, which can lead to arterial or retinal bleeds;
- inhibitory antibodies to VWF or FVIII;
- non-neutralizing antibodies to VWF or FVIII, leading to increased clearance;
- impaired platelet function, such as antibodies to glycoprotein IIbIIIa;
- inhibition of fibrin polymerization;
- circulating heparin-like anticoagulant.

Vascular abnormalities

You saw in Chapter 13 how blood vessels play important roles in haemostasis, so it is unsurprising that blood vessel abnormalities can compromise haemostasis. The two main conditions in which vascular abnormalities can lead to bleeding are hereditary haemorrhagic **telangiectasia** (HHT) and Kasabach–Merritt syndrome (KMS).

telangiectasia
Visibly dilated (widened) blood vessels near the surface of the skin.

HHT, also known as Osler–Weber–Rendu disease, is an autosomal-dominant disorder giving rise to telangiectasia on the skin and mucous membranes. Large malformations of arteries and veins can arise in solid organs such as lungs, liver, and brain. Recurrent epistaxis is a common symptom, sometimes accompanied by mouth bleeding. Gastrointestinal bleeding can occur, and chronic bleeding can lead to iron deficiency. Some patients develop intracranial bleeding. The patients have normally functioning platelets, VWF, and coagulation factors and the bleeding arises because the abnormally formed blood vessels are fragile. Consequently, testing in the haemostasis laboratory is rarely necessary.

KMS is often present at birth and caused by a vascular tumour. Rapidly growing tumours can trap platelets, probably by adhesion to the abnormally proliferating vascular endothelium,

BOX 16.3 *Psychogenic purpura*

Gardner–Diamond syndrome is an extremely rare disorder characterized by spontaneous, painful subcutaneous purpura (ecchymoses) and is predominantly encountered in psychologically disturbed adult women. It is an autoimmune vasculopathy arising from sensitization to phosphatidylserine, a component of erythrocyte stroma. It manifests clinically as spontaneous development of painful swellings, predominantly in the extremities, which progress to ecchymoses within 24 hours. Standard tests for haemostasis are normal and treatment is psychiatric.

which leads to thrombocytopenia. KMS is defined by the combination of vascular tumour and consumptive thrombocytopenia. The activated platelets cause secondary consumption of VWF and coagulation factors, which can progress to DIC.

An intriguing disorder associated with purpura and an acquired vasculopathy is detailed in Box 16.3.

Collagen disorders

Effective interactions between sub-endothelial collagen, VWF, and platelets are crucial to normal primary haemostasis, so abnormalities in collagen can be associated with a mild bleeding tendency due to generalized soft tissue fragility. The blood vessels themselves can be fragile and compromise haemostasis.

Ehlers–Danlos syndrome (EDS) is a group of inherited connective tissue disorders where bruising is a common feature. The subtype vascular EDS is also associated with spontaneous rupture of major blood vessels. Mild platelet dysfunction has been reported in some patients. Other connective tissue disorders, such as osteogenesis imperfecta and Marfan syndrome, are also associated with bleeding in some cases.

Vitamin C deficiency, or scurvy, impairs normal collagen synthesis and leads to formation of fragile blood vessels which impair primary haemostasis. Bleeding gums and bruising are common symptoms. The loss of collagen and elastin fibres in subcutaneous tissues as people age causes vessel fragility. Blood vessels in the skin are easily damaged by minor trauma and bruising is often seen on hands and forearms.

SELF-CHECK 16.10

What is the pathophysiological commonality between hereditary haemorrhagic telangiectasia and collagen disorders?

16.2 Acquired thrombotic disorders

You saw earlier in this chapter that whilst acute DIC has predominantly haemorrhagic manifestations, thrombosis can also occur, especially in chronic DIC. The antiphospholipid syndrome, an autoimmune cause of acquired thrombosis, was covered in detail in Chapter 15, and the remainder of this chapter will introduce you to three other acquired thrombotic disorders. Note that two of them also have congenital forms of the disorder.

16.2.1 Heparin-induced thrombocytopenia (HIT)

In approximately 5% of patients receiving unfractionated heparin (UFH) therapy, antibodies against heparin-platelet factor 4 complexes develop, resulting in platelet activation. This typically takes place around five days after treatment begins, and may also occur with low-molecular-weight heparin therapy. A marked drop in platelet count is seen, and additionally, both arterial and venous thrombus formation may occur. It is important that heparin therapy is stopped and replaced with an alternative anticoagulant regime.

Cross reference

HIT antibodies and their laboratory detection are covered in detail in Chapter 17.

16.2.2 Thrombotic thrombocytopenic purpura

Von Willebrand factor (VWF) is involved in the normal primary haemostatic mechanisms involving platelet adhesion. VWF subunits combine covalently to form multimers of differing sizes, and larger multimers have a greater facility to promote platelet aggregation. You saw in Chapter 13 how the size of the VWF multimers in plasma is regulated by the metalloprotease ADAMTS-13, which breaks down large multimers of VWF into smaller, less reactive multimers. VWF circulates in a globular confirmation where the ADAMTS-13 cleavage site is masked, so proteolysis of VWF by ADAMTS-13 occurs in situations where VWF unravels in response to shear forces, such as at the point of secretion, when tethered at the site of vessel injury, or during passage though the microvasculature. Deficiency of ADAMTS-13 leads to a condition called **thrombotic thrombocytopenic purpura (TTP)**, which can be either congenital or acquired.

In congenital TTP, also known as **Upshaw-Schulman syndrome**, mutations cause deficiency of ADAMTS-13, which generally affects synthesis or secretion of the enzyme rather than causing production of dysfunctional molecules. Acquired TTP arises from formation of autoantibodies that inhibit normal enzymatic function of ADAMTS-13 and/or bind to ADAMTS-13 and accelerate its clearance from the circulation. The antibodies are usually of the IgG class, but rarely are IgA or IgM class. Each mechanism results in an excess of highly adhesive ultra-large VWF multimers in the circulation, leading to accumulation of VWF-platelet aggregates that occlude the microvasculature and can lead to organ failure. Newly secreted VWF is transiently attached to the endothelial surface, and in ADAMTS-13 deficiency states, ultra-large VWF multimers decorated with platelets can persist on the endothelium. Coagulation screening tests are usually normal or only slightly disturbed, which helps to distinguish TTP from DIC, though thrombocytopenia and a **microangiopathic haemolytic anaemia (MAHA)** will be evident from a full blood count and blood film. Haemoglobin is usually below 105 g/L and mean cell volume (MCV) normal unless there is marked red cell fragmentation, which will be evident from the blood film. The red cell fragments arise from being sheared as they travel past and through the microthrombi, which is the cause of the intravascular haemolysis. Reticulocytes are increased, which can increase the MCV, and nucleated red cells can be seen in the peripheral blood, both indicators of a bone marrow response to the haemolysis. Serum lactate dehydrogenase (LDH) is also elevated due to release from ischaemic tissues and damaged red blood cells. Neurological signs, such as coma, stroke, seizures, and even personality change, are a presenting feature due to formation of thrombi in the cerebral circulation. You can see in Figure 16.3 that deficiency of ADAMTS-13 leads to the presence of circulating uncleaved VWF that promotes clots, or it can bind to the endothelium and promote platelet adhesion and aggregation.

Acquired TTP can also occur secondary to massive endothelial activation resulting from release of large amounts of ultra-large VWF multimers sufficient to overwhelm degradation capacity of their normal or slightly reduced ADAMTS-13 levels. This can occur in conditions such as metastatic tumours, organ transplantation, and in association with drugs such as ciclosporin and interferon alpha.

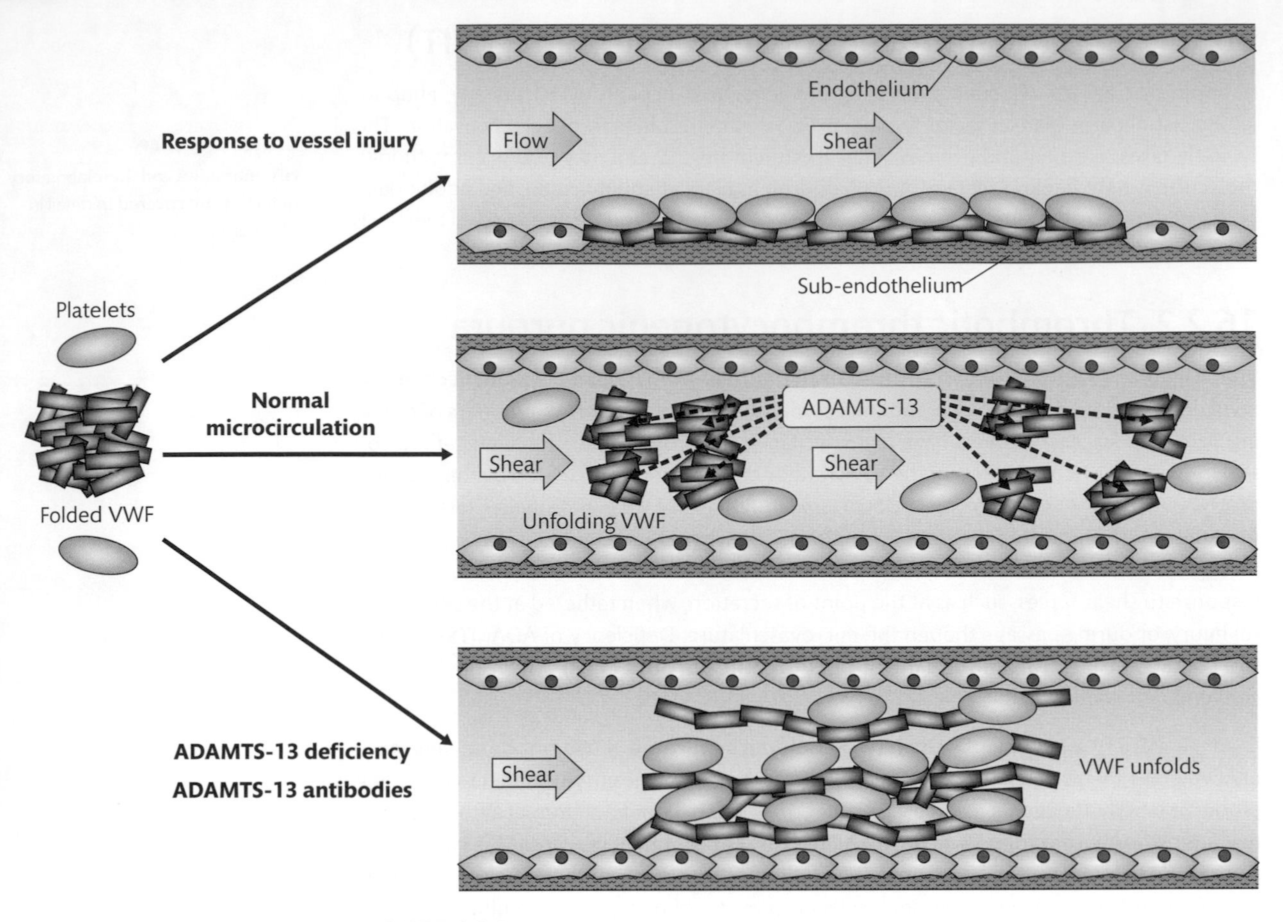

FIGURE 16.3

TTP and microvascaular thrombosis. Exposure of sub-endothelium at sites of vessel injury promotes binding of VWF, which is then rapidly unfolded by flow and shear to promote platelet adhesion. In the normal circulation, transit of VWF through the microcirculation provides brief exposure to high shear stress, where it partially unfolds and permits cleavage by ADAMTS-13, where scissile bonds are exposed. The smaller forms of VWF that subsequently circulate do not promote platelet adhesion and aggregation during normal blood flow. Deficiency or inhibition of ADAMTS-13 results in non-cleavage of unfolded VWF in the microcirculation still containing the highly adhesive ultra-large multimers, thereby inducing platelet adhesion and aggregation.

SELF-CHECK 16.11

What distinguishes the causes of hereditary and acquired TTP?

ADAMTS-13 activity assays

A variety of assays have been described for measuring ADAMTS-13 activity based on degradation of purified, plasma-derived, or recombinant VWF, or of synthetic VWF peptides, by ADAMTS-13 in patient plasma. The resultant VWF cleavage products can then be measured by ristocetin-induced platelet aggregation, electrophoresis, collagen binding, fluorescence resonance energy transfer (FRET) or immunoassay. The platelet aggregation and electrophoretic methods can be complicated and time consuming and the other three assay types are more commonly employed in diagnostic laboratories.

The collagen-binding assay first involves mixing test plasma with a normal pooled plasma as a source of VWF, or purified VWF. The purified VWF does not contain ADAMTS-13, but the normal plasma does, so it is first inactivated with EDTA. Any ADAMTS-13 present in the test plasma will cleave VWF multimers into monomers and the cleavage products are assayed in a collagen-binding assay, which you met in Chapter 14. Collagen preferentially binds multimers, so high levels of binding indicate low levels of ADAMTS-13.

The FRET assay involves addition of VWF73, a recombinant VWF peptide containing the ADAMTS-13 cleavage site, to diluted test plasma. Figure 16.4 shows that the VWF73 is modified so that an amino acid on one side of the cleavage site is labelled with a donor fluorophore, and an adjacent amino acid on the other side of the cleavage site with an acceptor fluorophore. Excitement of the donor fluorophore via exposure to 340 nm light results in transfer of fluorescence resonance energy to the nearby acceptor fluorophore, which quenches emission of fluorescence. Any ADAMTS-13 present in test plasma cleaves the VWF73, thereby separating the donor and acceptor such that quenching does not occur and fluorescence is detected and quantified in a fluorimeter.

There are two immunoassays, also shown in Figure 16.4, a chromogenic ELISA and a chemiluminescent immunoassay (CLIA), both of which also employ the VWF73 peptide. The process of manufacturing the recombinant VWF73 region involves tagging the N terminal with glutathione S-transferase (GST), a property that is used in the design of the assays. The ELISA plate is coated with anti-GST antibody and the GST-VWF73 added to act as a substrate. The resultant antibody–antigen complex is then incubated with test samples whereby the ADAMTS-13 cleaves the immobilized recombinant VWF73, exposing the cleavage site. The plate is washed to remove unbound protein and then incubated with a horse radish peroxidase (HRP)-conjugated monoclonal antibody to N-10, the C-terminal edge residue of the VWF-A2 domain generated after ADAMTS-13 cleavage. The HRP is provided with a substrate, the product of which is coloured, and the degree of colour formation is directly proportional to the ADAMTS-13 activity that cleaved VWF73 and exposed N-10. In the CLIA, magnetic particles are coated with anti-GST antibody and the GST-VWF73 added to act as a substrate. Incubation with test plasma permits any ADAMTS-13 present to cleave the particle-bound VWF73 proportional to its activity. A magnetic separation step and a wash step follow, after which an isoluminol-conjugated monoclonal antibody to the N-10 residue exposed after cleavage is added in a second step, which is followed by magnetic separation and washing. Two triggers of the chemiluminescent reaction are then added, which is measured as relative light units (RLU), which are directly proportional to the ADAMTS-13 activity in the test plasma.

ADAMTS-13 inhibitor assays

Both neutralizing (inhibitory) and non-neutralizing (immune) anti-ADAMTS-13 antibodies have been reported in patients with acquired TTP. Neutralizing antibodies are found in about two-thirds of cases. They are detected by mixing test plasma with normal plasma and then measuring residual ADAMTS-13 activity in a similar design to the Bethesda assay you met in Chapter 14 for detecting antibodies to FVIII.

The remaining one-third of patients with acquired TTP present with non-neutralizing antibodies that bind to and accelerate the clearance of ADAMTS-13 from the plasma. These antibodies are most often detected in an indirect ELISA assay. (See Case study 16.3.)

Cross reference

The principle of indirect ELISA is covered in Chapter 15, with a description of the anticardiolipin antibody assay.

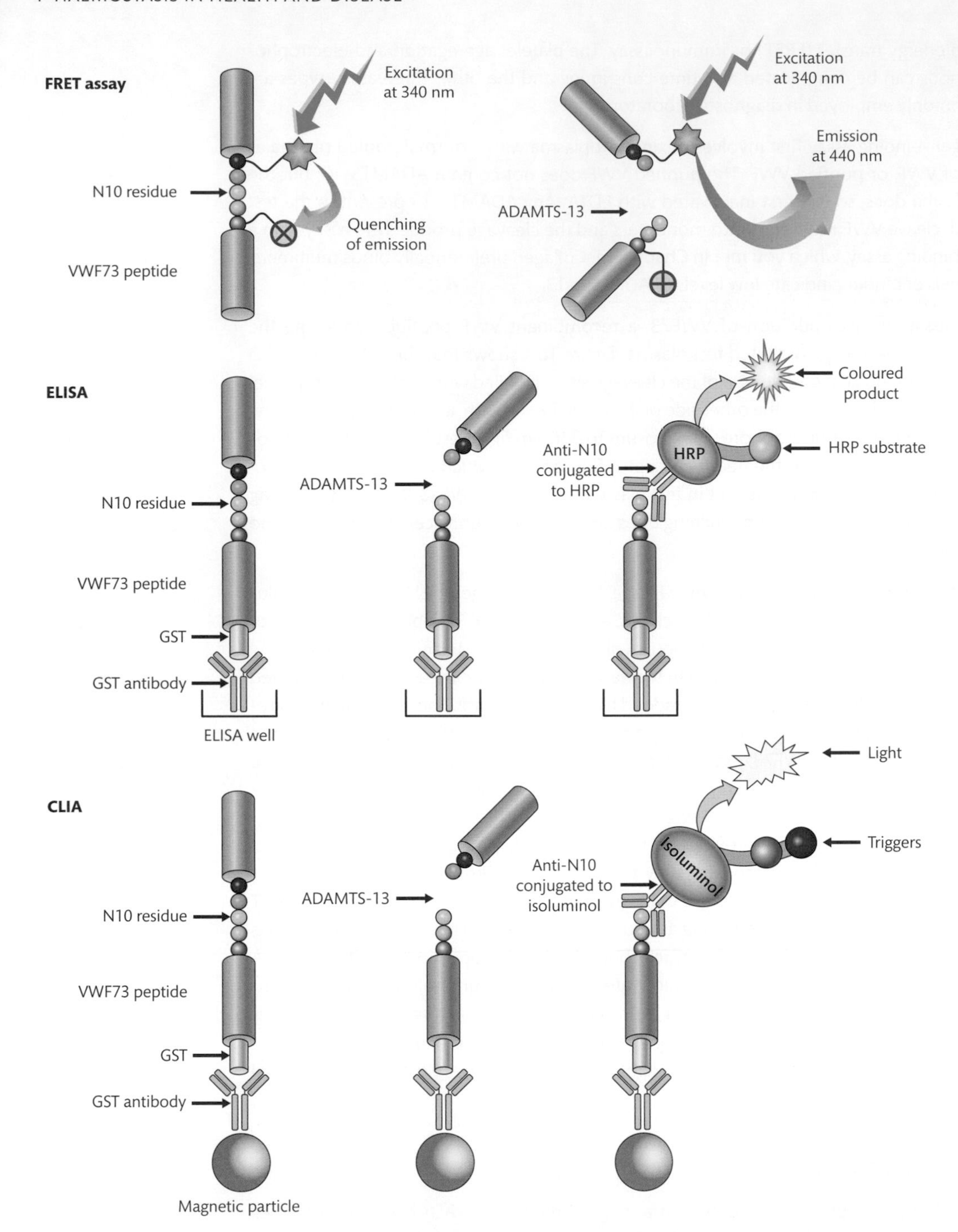

FIGURE 16.4

ADAMTS-13 activity assays. The fluorescence resonance energy transfer (FRET) assay employs the recombinant VWF peptide VWF73 labelled with a donor fluorophore on one side of the ADAMTS-13 cleavage site and an acceptor, quenching fluorophore on the other side. Excitation of the donor produces fluorescence that is quenched if the acceptor is in close proximity. Peptide cleavage by ADAMTS-13 uncouples the fluorophores, resulting in increased fluorescence proportional to ADAMTS-13 activity. In the ELISA assay, VWF73 is anchored to the plate by antibody to GST. Peptide cleavage by ADAMTS-13 exposes the C-terminal edge residue of the VWF-A2 domain, N-10, and an HRP conjugated antibody to N-10 is added and reacted with a chromogenic substrate. Colour intensity of the product is proportional to the ADAMTS-13 activity. The CLIA employs VWF73 anchored to magnetic particles. The VWF73 is cleaved by ADAMTS-13 to expose the N-10 residue, which is recognized by an isoluminol-labelled antibody. Two triggers of the chemiluminescent reaction are then added, which is measured in relative light units that are proportional to the ADAMTS-13 activity.

CASE STUDY 16.3 Acquired TTP

Patient history

- A 40-year-old male presented with a history of recent fever, abdominal pain, vomiting, and diarrhoea.
- A petechial rash was present over much of his body and nose bleeds had begun to occur in the previous two days.
- Deteriorating consciousness was a feature at presentation.
- The haematology laboratory was asked to perform full blood count and coagulation screen.

Results 1

- Haemoglobin was 100 g/L (RR: 120–160).
- Reticulocytes were 135 × 10^9/L (RR: 25.0–80.0).
- White cell count was 10.6 × 10^9/L (RR: 4.0–11.0).
- Platelet count was 36 × 10^9/L (RR: 150–400).
- PT, APTT, and fibrinogen were all normal.
- Blood film examination revealed red cell fragments and confirmed the reduced platelet count.

Significance of results 1

- The low haemoglobin and red cell fragments were due to mechanical haemolysis.
- The elevated reticulocytes reflect the bone marrow responding to the anaemia.
- Platelet numbers were markedly reduced and further tests were needed to confirm or exclude TTP.

Results 2

- ADAMTS-13 activity was <5 % (RR: 70–120).
- ADAMTS-13 Bethesda titre was 9.8 BU/mL (RR: <0.4).
- ADAMTS-13 antibody ELISA was 30 U/mL (RR: <15).
- LDH was 956 U/L (RR: 250–450).

Significance of results 2

- The patient had inhibitory antibodies to ADAMTS-13.
- Marked inhibition of ADAMTS-13 function led to reduced cleavage of high-molecular-weight VWF multimers when VWF was subjected to high shear in the microcirculation, leading to platelet adhesion and aggregation and thrombocytopenia.
- The microvascular thrombosis caused haemolytic anaemia due to red blood cells being damaged and fragmented as they travel through the partially occluded small vessels.
- The elevated LDH was due to release from ischaemic tissues and damaged red blood cells.

16.2.3 Haemolytic uraemic syndrome

A disorder clinically similar to TTP, **haemolytic uraemic syndrome (HUS)**, occurs primarily in infants and young children and occasionally in the elderly. Typical HUS occurs acutely following gastrointestinal infections with *Escherichia coli* 0157:H7, or less commonly, *Shigella dysenteriae*, both of which produce Shiga-toxins. Most patients will have experienced episodes of bloody diarrhoea during the infection. The Shiga-toxin enters the bloodstream and attaches to glomerular endothelial cells because in infants, young children, and the elderly, these cells are rich in the membrane receptor globotriaosyl ceramide (Gb_3). The toxin-receptor complex is endocytosed, which leads to activation of endothelial cells and consequent thrombosis in the renal microvasculature. ADAMTS-13 levels are usually normal. The acute renal failure can be so severe that haemodialysis is required. Neurological signs are far less common than in TTP. The MAHA can also be severe, and the platelet count can fall below 60 × 10^9/L. Typical HUS can be referred to as Shiga toxin-producing *Escherichia coli* HUS, or STEC-HUS.

Familial or atypical HUS (aHUS) accounts for about 5–10% of all cases and results from deficiency or dysfunction of **complement factor H**, a regulator of the complement system that ensures the system is directed against pathogens and not self. The deficiency results in excessive activation of complement component 3 (C3), which is deposited on membranes and leads to renal cell injury. Atypical HUS, which is non-diarrhoea related, is distinguished from TTP by the predominance of renal symptoms and absence of neurological signs. It is commonly associated with the post-partum period or exposure to certain drugs, such as quinine, ciclosporin, and ticlopidine.

SELF-CHECK 16.12

What are the main similarities and differences between TTP and HUS?

16.3 Laboratory parameters to distinguish between DIC, TTP, and HUS

Acute DIC, TTP, and HUS are serious medical conditions associated with activation of haemostasis and MAHA that require prompt diagnosis and treatment. Table 16.5 details typical results of key laboratory parameters for each disorder.

TABLE 16.5 Typical laboratory results in acute DIC, TTP, and HUS.

Parameter	Acute DIC	TTP	HUS
Platelets	Reduced	Reduced	Reduced
Haemoglobin	Often reduced	Reduced	Reduced
Blood film	RBC fragments ±	RBC fragments +++	RBC fragments ++
Prothrombin time	Elevated	Normal	Normal
APTT	Elevated	Normal	Normal
Fibrinogen	Reduced	Normal	Normal
D-dimers	Elevated	Normal	Normal
ADAMTS-13 activity	Can be reduced	Markedly reduced	Normal
Lactate dehydrogenase	Elevated	Elevated	Elevated

16.4 Haemostatic complications of mechanical circulatory support

Mechanical circulatory support (MCS) is employed for patients with severe organ failure to supplement or replace diminished function. Several devices are used for MCS that involve extracorporeal circulation, the main ones being listed below:

- Extracorporeal membrane oxygenation (ECMO) provides non-pulmonary gas exchange in severe respiratory failure, and sometimes in patients with concurrent cardiac failure whilst waiting for organ recovery.

- Left ventricular assist devices (LVAD) are battery-operated mechanical pumps that help a weakened or failed heart pump blood. They are used in patients while their heart recovers from disease or who are waiting for a transplant, or may be long term if a patient is unsuitable for a transplant.
- Cardiopulmonary bypass (CPB) provides a bloodless field for open heart surgery by temporarily taking over heart and lung function.
- Haemodialysis is used in patients with renal failure by undertaking extracorporeal removal of excess fluid, salt, and wastes from the blood.

All these devices remove blood from the patient's circulation, whereupon it comes into constant contact with synthetic, non-biological surfaces in device components such as tubing, membranes, and connectors, leading to activation of primary and secondary haemostasis, and fibrinolysis. This altered-state blood is then returned to the patient's circulatory system and can lead to bleeding, thrombosis, and device failure. Activation of haemostatic mechanisms can be attenuated, but not eliminated, by covering synthetic surfaces with anticoagulants.

Key Points

Patients who require ECMO are already critically ill, so the haemostatic activation arising from contact with MCS device components are superimposed on other haemostatic changes in the patient due to their underlying disease state. Examples include the haemostatic effects of hepatic and renal failure, DIC, and the procoagulant acute phase response in patients with infectious or inflammatory disease.

In particular, ECMO is associated with significant haemostatic challenges compared to CPB, mainly because ECMO can last for days or weeks, whilst CPB is implemented for just the hours of cardiac bypass surgery. Additionally, for patients undergoing CPB, the level of anticoagulation is higher and they are subjected to therapeutic hypothermia and haemodilution. LVAD have a significantly reduced surface area compared to ECMO and CPB to achieve haemostatic activation. The following section discusses haemostatic complications mainly pertaining to ECMO, where bleeding and thrombotic events can coexist in the same patient.

Thrombosis during ECMO

Thrombosis is one of the most common complications of ECMO support and patients are systemically anticoagulated, commonly with unfractionated heparin, to combat it. Despite this, haemostatic activation can lead to deep vein thrombosis, pulmonary embolism, small vessel thrombosis and thrombus deposition in the ECMO circuit, oxygenator, or pump. Heparin-induced thrombocytopenia is also a potential thrombotic complication in patients on ECMO.

A layer of blood proteins, mainly fibrinogen and albumin, adsorb onto the non-biological surfaces of the ECMO circuit very soon after blood comes into contact with them. This protein coating induces platelet adhesion and activation and leads to an increased thrombotic propensity. Activated platelets can initiate coagulation through exposure of tissue factor, giving rise to thrombin generation and formation of fibrin clots. Interestingly, the contact system of FXII, prekallikrein, and high molecular weight kininogen leading to FXI activation, that has a negligible contribution to normal haemostasis, can be activated by contact with the foreign surfaces of the circuit. Not only will the contact system contribute to thrombin generation, but it also activates the fibrinolytic system. The contact of blood with artificial surfaces can also trigger a systemic inflammatory response that leads to elevation in some clotting factors (i.e. FVIII and fibrinogen) and expression of tissue factor on the surface of activated monocytes, facilitating haemostatic activation that can lead to thrombosis and circuit dysfunction or loss. The inflammatory response will downregulate antithrombin, protein C, and fibrinolysis and alter the haemostatic balance in favour of a prothrombotic state. DIC as a result of the underlying illness can contribute to both bleeding and thrombosis, and the extracorporeal support itself can lead to a DIC-like consumptive coagulopathy resulting from tissue injury and contact activation in the artificial membrane. Vessel injury at sites of **cannula** insertion can promote blood clotting.

cannula

A thin tube inserted into a vein or body cavity to administer drugs, drain fluid, or insert a surgical instrument.

Device-induced haemolysis may occur due to mechanical shear stress and can induce a prothrombotic state. Free haemoglobin in the plasma binds nitric oxide, which causes vasoconstriction and increased platelet reactivity. Red blood cell microparticles can induce thrombin generation arising from exposure of phosphatidylserine. Thrombotic events can generate platelet and endothelium-derived microparticles, which also express procoagulant phosphatidylserine, as well as tissue factor.

Cross reference

You will encounter more detail on therapeutic anticoagulation and heparin-induced thrombocytopenia in Chapter 17.

Bleeding during ECMO

Systemic anticoagulation is itself a bleeding risk, and a variety of other mechanisms cause bleeding in patients on ECMO such that it is a significant clinical problem. As discussed earlier in this chapter, DIC can cause life-threatening bleeding. Thrombocytopenia is common in patients receiving ECMO and can be severe, with counts $<50 \times 10^9$/L. Furthermore, the high shear flow in ECMO circuits can cause shedding of the key platelet adhesion receptors, GPIbα and GPVI. Excessive fibrinolysis can also contribute to bleeding. Over time, VWF activity decreases because shear stresses in the ECMO circuit expose the ADAMTS-13 cleavage site and lead to a loss of high-molecular-weight VWF multimers, causing acquired von Willebrand syndrome.

Chapter summary

- Disseminated intravascular coagulation (DIC) results from excessive activation of coagulation exacerbated by loss of regulatory and localization mechanisms.
- DIC can be acute or chronic.
- Liver disease can affect primary haemostasis, secondary haemostasis, and fibrinolysis.
- There are three forms of vitamin K-deficiency bleeding: early, classic, and late.
- Acquired coagulation factor deficiency due to spontaneous antibodies can occur for any of the factors, including fibrinogen.
- Bleeding in renal disease occurs mainly due to disturbances of primary haemostasis.
- Massive blood transfusion can trigger a multifactorial coagulopathy.
- Mild thrombocytopenia without bleeding is common in pregnancy.

- Hereditary and acquired vascular abnormalities can give rise to bleeding due to compromised primary haemostasis.
- Standard coagulation screening tests are the mainstay of laboratory detection and monitoring of most acquired bleeding disorders.
- Hereditary and acquired TTP arises from deficiency of ADAMTS-13 or antibodies to ADAMTS-13, respectively.
- Acquired haemolytic uraemic syndrome is toxin-induced, whilst the familial form arises from deficiency of complement factor H.
- Contact of blood with synthetic surfaces during mechanical circulatory support can activate haemostatic mechanisms and lead to bleeding or thrombosis.

Discussion questions

16.1 Compare and contrast the pathophysiologies of acute and chronic DIC.

16.2 What are the mechanisms of bleeding in hepatic and renal disease?

16.3 Discuss the contribution of antibodies to acquired haemostatic disorders.

16.4 Distinguish between hereditary and acquired thrombotic thrombocytopenic purpura.

16.5 Discuss the disturbed haemostatic balance in patients on ECMO.

Further reading

Doyle AJ and Hunt BJ. *Current understanding of how extracorporeal membrane oxygenators activate haemostasis and other blood components*. *Frontiers in Medicine* 2018: 5: 352.

Iba T, Levy JH, Warkentin TE, Thachil J, van der Poll T, Levi M; Scientific and Standardization Committee on DIC, and the Scientific and Standardization Committee on Perioperative and Critical Care of the International Society on Thrombosis and Haemostasis. *Diagnosis and management of sepsis-induced coagulopathy and disseminated intravascular coagulation*. *Journal of Thrombosis and Haemostasis* 2019: 17: 1989–94.

Key N, Makris M, O'Shaughnessy D, and Lillicrap D. *Practical Haemostasis and Thrombosis*, 3rd edn. Blackwell Publishing Ltd, Massachusetts–Oxford–Carlton, 2017.

Levi M, Toh CH, Thachil J, and Watson HG. *Guidelines for the diagnosis and management of disseminated intravascular coagulation. British Committee for Standards in Haematology*. *British Journal of Haematology* 2009: 145: 24–33.

Mackie I, Mancini I, Muia J, Kremer Hovinga J, Nair S, Machin S, and Baker R. *International Council for Standardization in Haematology (ICSH) recommendations for laboratory measurement of ADAMTS13*. *International Journal of Laboratory Hematology*. 2020 16 July. doi: 10.1111/ijlh.13295 (in print).

Mannucci PM, Peyvandi F, and Palla R. ***Thrombotic thrombocytopenic purpura and haemolytic-uraemic syndrome (congenital and acquired)***. In: Hoffbrand AV, Higgs DR, Keeling DM, and Mehta AB (eds). *Postgraduate Haematology*, 7th edin (pp. 783–94). Wiley Blackwell, Oxford, 2016.

Peyvandi F, Palla R, Lotta LA, Mackie I, Scully MA, and Machin SJ. ***ADAMTS-13 assays in thrombocytopenic purpura***. *Journal of Thrombosis and Haemostasis* 2010: **8**: 631–40.

Roberts LN, Patel RK, and Arya R. ***Haemostasis and thrombosis in liver disease***. *British Journal of Haematology* 2009: **148**: 507–21.

Scully M, Hunt BJ, Benjamin S, Liesner R, Rose P, Peyvandi F, Cheung B, Machin SJ; British Committee for Standards in Haematology. ***Guidelines on the diagnosis and management of thrombotic thrombocytopenic purpura and other thrombotic microangiopathies***. *British Journal of Haematology* 2012: **158**: 323–35.

Wada H, Thachil J, Di Nisio M, Mathew P, Kurosawa S, Gando S, Kim HK, Nielsen JD, Dempfle CE, Levi M, and Toh CH. ***The Scientific Standardization Committee on DIC of the International Society on Thrombosis Haemostasis. Guidance for diagnosis and treatment of DIC from harmonization of the recommendations from three guidelines***. *Journal of Thrombosis and Haemostasis* 2013: **11**: 761–7.

Answers to self-check questions and case study questions are provided as part of this book's online resources.

Visit www.oup.com/he/moore-fbms3e.

17

Haemostasis and anticoagulation

Gary Moore and Jane Needham

In this chapter, you will encounter an outline for the reasons for anticoagulant therapy, mode of action and use of therapeutic agents, laboratory assays for monitoring treatment, and associated clinical services for patient care.

Learning objectives

After studying this chapter, you should confidently be able to:

- explain the reasons for anticoagulant therapy;
- understand the mode of action of anticoagulants;
- understand the importance of anticoagulant management and associated clinical risks;
- know how to monitor anticoagulant therapy in the laboratory;
- be aware of hereditary or acquired conditions that can influence treatment outcomes;
- demonstrate awareness of interacting drugs;
- demonstrate awareness of models of clinical services.

17.1 Reasons for anticoagulation therapy

In Chapter 15, you learnt about the defects and deficiencies that are associated with thrombophilia and can lead to an imbalance in the haemostatic mechanism regulating the activation or inhibition of the haemostatic process. This has the potential of establishing a **hypercoagulable state**, which increases the risk of thrombosis. The formation of a thrombus

hypercoagulable state
An increased tendency towards blood clotting.

results in a firm fibrin clot within veins. Venous thrombosis is a serious condition occurring in approximately 1 in 1000 individuals per year in the United Kingdom and can be fatal.

The exact trigger for a venous thrombotic event is still unclear, but it has been established that certain individuals have a higher disposition to thrombosis. This risk is associated with the following:

- having an inherited or acquired thrombophilia defect;
- medical conditions that can lead to thrombosis secondary to the primary condition;
- events or environment that lead to blood stasis, such as immobility.

The aims of prescribing anticoagulant therapy are to:

- prevent the growth of an existing **thrombus** or **embolus** through the **vasculature**;
- prevent the recurrence of venous thrombotic events;
- prevent the occurrence of a thrombosis in 'high-risk' medical conditions or during surgical procedures and periods of immobility.

Thrombosis can also occur in the arterial circulation and lead to heart attack or stroke. There are important differences between causes and treatment for arterial and venous thrombosis, which you will meet later in the chapter.

thrombus
A blood clot that usually develops in a deep vein of your body, normally in the calf or higher up the leg in the thigh.

embolus
A blood clot or part of a blood clot that migrates through the bloodstream and then lodges in another vessel, causing a blockage.

vasculature
The network of blood vessels through your body or organs.

morbidity
High rate of sickness and medical complications compared to the normal population.

mortality
Increased risk of death compared to the normal population.

stasis
A condition in which the normal flow of blood through a vein is slowed or halted.

17.1.1 Venous thromboembolism

Venous thromboembolism (VTE) is considered to be one disorder, which includes both deep vein thrombosis (DVT) and pulmonary embolism (PE). It is a common medical condition that can occur in individuals with a hereditary thrombotic defect, an associated high disposition to thrombosis, 'spontaneously' with no identified cause or risk factors, or secondary to other clinical conditions, particularly malignancy. VTE has a high rate of **morbidity** and **mortality** if not correctly diagnosed and treated.

The clinical event normally occurs with the formation of a blood clot in a deep vein within the leg, or very rarely in the arm. The exact sequence of events is still not fully understood. Although local venous **stasis** is necessary, this is not thought to be sufficient on its own to cause thrombosis. Increased turbulence of blood around the valves in veins may initiate DVT. It has been suggested that endothelial damage occurs with the migration of white cells through the vessel wall during stasis, causing platelet activation and subsequent thrombus formation; an alternative hypothesis is clot formation due to localized trapped activated clotting factors which are not exposed to the body's natural inhibitors. As there is no rupture to the vessel wall in the formation of a DVT to involve platelets, the composition of the thrombus is predominantly coagulation factors, particularly fibrin. The lower shear rates in the venous circulation compared to the arterial circulation can also contribute to the formation of venous thrombosis.

DVT below the knee is unlikely to cause acute complications. Presenting symptoms include pain and swelling of the calf due to blocked veins. However, a clot formed above the knee has the increased risk of breaking away and forming a life-threatening PE within the lungs, usually resulting in shortness of breath and chest pain, and the need to seek urgent medical attention.

A long-term complication following a DVT, called '**post-thrombotic syndrome**', can occur due to tissue damage following diversion of blood into other veins to avoid the blockage. This presents with a range of symptoms, including calf pain and swelling, with ulceration of the skin in severe cases.

SELF-CHECK 17.1

What are the key aims of anticoagulant therapy?

SELF-CHECK 17.2

Where does a blood clot usually initially form in VTE and where can it move to?

17.1.2 Diagnosis of VTE

The diagnosis of DVT or PE is not without difficulties; in the fairly recent past many patients were diagnosed on clinical symptoms alone. Advances in both laboratory assays and radiological investigations have seen the introduction of clinical assessment scores, which you can see in Tables 17.1(A) and (B), and use of D-dimer assays within a flow-chart algorithm as shown in Figure 17.1. The clinical assessment score, or **Wells' score**, assigns a numerical value to the presence of specific signs and symptoms. The higher the total score, the more likely that a DVT or PE is present, and hence informed decisions can be made about the initiation of diagnostic procedures. If used correctly, this initial assessment reduces the frequency of having to perform invasive procedures, which can themselves have clinical complications, and streamlines and improves diagnosis.

Cross reference

You saw in Chapter 13 that D-dimers are cross-linked D fragments that are breakdown products of fibrin. Elevated levels can indicate the presence of a thrombus.

The assays are very sensitive to the presence of D-dimers, but an elevated result is not specific for VTE because D-dimer can be raised in a number of other clinical situations, i.e. post surgery, malignancy, renal failure. Therefore, it is applied as a **negative predictive index** of DVT and clinicians will rarely initiate imaging tests if there is a normal D-dimer level in conjunction

negative predictive index

The probability that a patient with a normal D-dimer level does not have a DVT or PE.

TABLE 17.1(A) Wells' DVT pre-test probability (PTP) score.

Clinical assessment	Score
Active malignancy (treatment is ongoing, within previous 6 months, or palliative)	1
Paralysis, recent immobility	1
Recently confined to bed >3 days	1
Major surgery within previous 12 weeks	1
Entire leg swollen	1
Swollen calf	1
Pitting oedema (fluid retention) in symptomatic leg	1
Collateral superficial veins (small veins that have grown around a blockage)	1
Previous DVT	1
Alternative diagnosis to DVT as likely or more likely	−2

PTP score <2 DVT unlikely; if D-dimers normal, DVT is excluded, if they are elevated, medical imaging is performed.

PTP score ≥2 DVT likely; perform D-dimers and medical imaging.

TABLE 17.1(B) Wells' PE pre-test probability score.

Clinical assessment	Score
Clinical signs and symptoms of DVT	3.0
Alternative diagnosis to PE is less likely	3.0
Heart rate >100 beats/min.	1.5
Immobilization in previous 4 weeks	1.5
Surgery in previous 4 weeks	1.5
Previous DVT/PE	1.5
Active malignancy (treatment is ongoing, within previous 6 months, or palliative)	1.0
Haemoptysis (coughing up blood/blood-stained sputum from the lungs)	1.0

PTP score <2.0 low probability of PE;

PTP score 2.0–6.0 moderate probability of PE;

PTP score >6.0 high probability of PE.

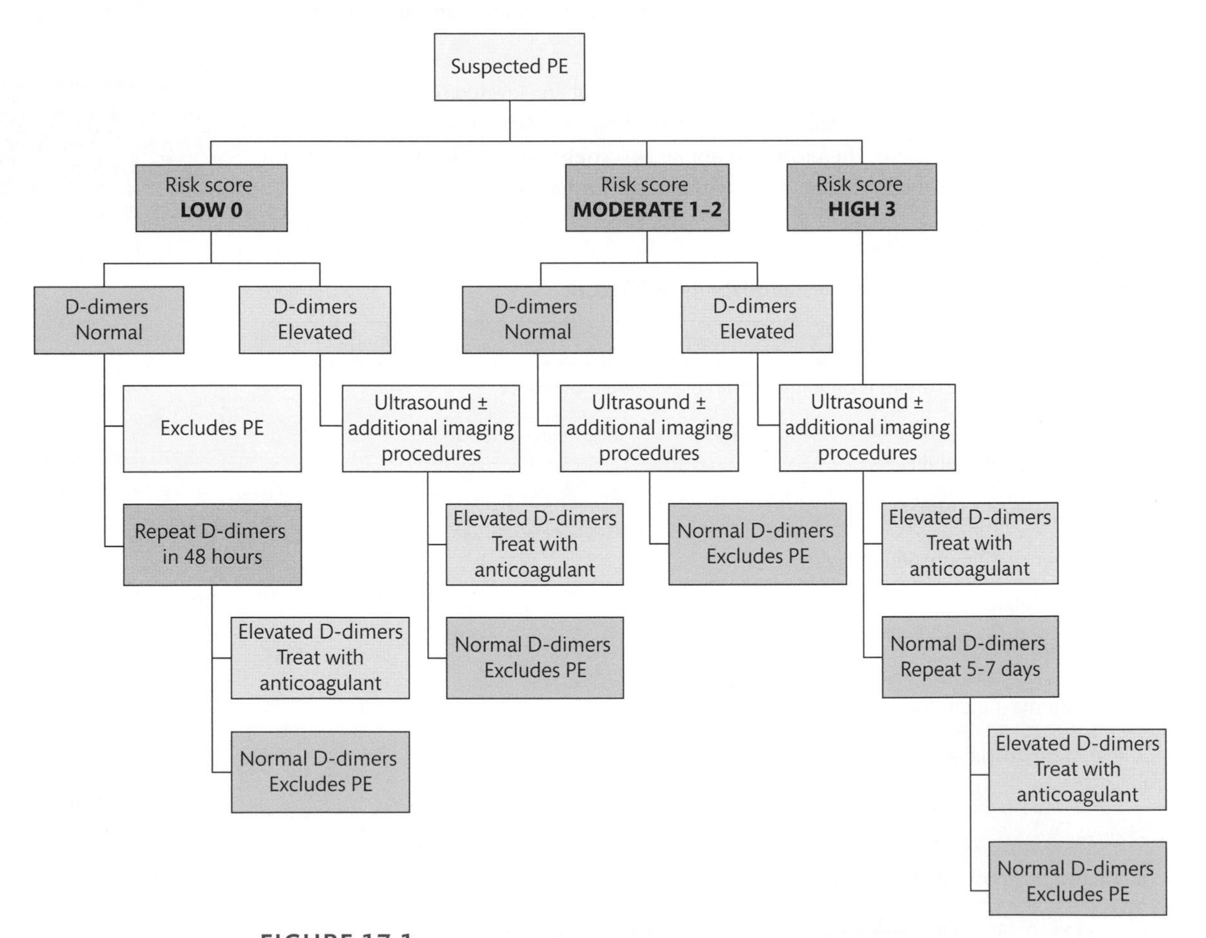

FIGURE 17.1

Example of use of D-dimers within a diagnostic algorithm for suspected PE. The flowchart maps the clinical steps to diagnosis, and treatment with anticoagulant therapy, from initial symptoms, D-dimer result, and medical imaging procedures.

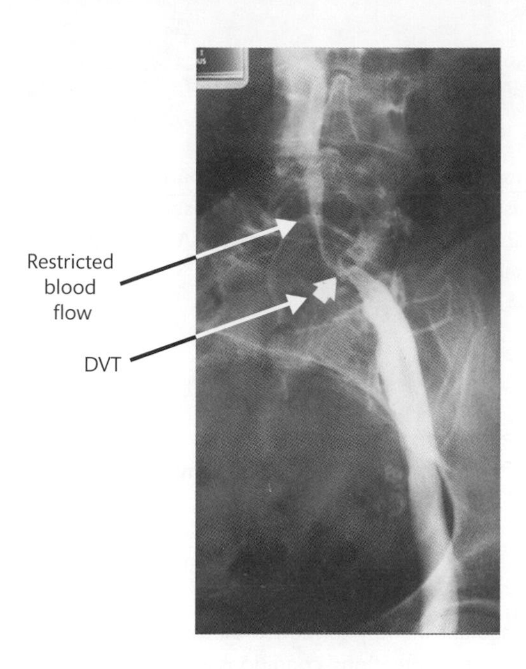

FIGURE 17.2
Example of venograph identifying the presence of a DVT and restricted venous blood flow in the leg. If ultrasound using sound waves to evaluate blood flow in the veins is inconclusive, then venography is performed whereby a dye is injected into a vein in the foot; it then travels the length of the suspect vein to make it visible by X-ray, thus allowing any decrease in blood flow or blockage to be visualized. Reproduced by permission from Zollikofer C and Laerum F. *The peripheral vessels* in *The NICER Centennial Book 1995–A Global Textbook of Radiology*, 2005.

with a low clinical probability score. Radiological investigation, including **venography** and **compression ultrasonography** is normally performed if D-dimers are elevated and/or there is a raised probability score for suspected DVT. A venograph of a blood clot occluding a leg vein is shown in Figure 17.2. (See also Box 17.1.)

Patients with a suspected PE normally have a chest X-ray in conjunction with D-dimer and probability scoring, followed by **pulmonary angiography** and **ventilated-perfusion (V/Q)** lung scanning as appropriate.

Normal levels of D-dimers mean it is highly unlikely that a fibrin clot is present. However, fibrinolysis of a small thrombus may not generate sufficient breakdown products to elevate D-dimers above the reference range. A clot that has been *in situ* for more than ten days may reach a physiological equilibrium, whereby D-dimers that were initially elevated fall back into the reference range. Although a normal D-dimer level can exclude VTE in patients with a non-high clinical probability, D-dimer levels increase with age and specificity decreases in older patients. Diagnostic accuracy is improved if cut-off levels are adjusted for age in patients over 50 years old.

SELF-CHECK 17.3

How is D-dimer testing used to establish a diagnosis of DVT or PE?

17.1.3 Prevention and thromboprophylaxis

Anticoagulant therapy can also be used in certain clinical settings or medical conditions to try and reduce the chance of a thrombotic event occurring and the subsequent risks and complications.

Atrial fibrillation

Atrial fibrillation (AF) is a heart condition where the electrical activity to the atria is disorganized. You can see in Figure 17.3 that AF is characterized by rapid, irregular atrial impulses and ineffective atrial contractions. Stasis of blood in the left atrium contributes to the risk of **stroke** and VTE.

venography
A radiology procedure where a contrasting agent is injected into the vein to enable a blockage to be visualized on the X-ray.

compression ultrasonography
Used for examining soft tissue; provides real-time images in conjunction with venous compression to diagnose DVT.

pulmonary angiography
Performed to assess blood circulation to the lungs by adding a contrast dye through a catheter into a vein and taking X-rays of the lung.

ventilated-perfusion (V/Q) scan
A procedure using small amounts of inhaled or injected radioisotopes to measure the flow of blood and air in the lungs. Images taken capture the flow patterns.

stroke
Occurs when the brain's supply of oxygen is cut off by a blockage or interruption to the blood flow, which causes severe neurological damage.

electrocardiography
Procedure for measuring electrical changes in the heart which take place with each beat.

sinus rhythm
A natural heart (cardiac) rhythm generated through the sinus node (normal rate 60–100 beats/min).

cardioversion
Procedure where the heart is stopped, then an electrical current is applied to shock it into restarting with a normal sinus rhythm.

prophylaxis
Procedure to prevent a condition occurring, rather than to treat or cure it after it has occurred.

atherosclerotic plaque
Deposit of cholesterol and other fatty material that builds up within the vessel wall, normally an artery.

hypertension
A sustained increase in systolic and/or diastolic blood pressure.

atherosclerosis
Narrowing and hardening of the arteries over time. Associated with age, but also other risk factors, e.g. hypertension, high cholesterol.

BOX 17.1 *D-dimer assays*

Measuring D-dimer levels for VTE diagnosis is a common request for a routine haemostasis laboratory. The most commonly used assay type is latex immunoassay (LIA), the principle of which you met in Chapter 2. Other assay types such as ELISA and whole-blood agglutination methods are also used.

AF is diagnosed by **electrocardiography**.

A number of clinical studies have confirmed the benefits of long-term anticoagulant therapy (i.e. warfarin) in these patients, unless they are returned to normal **sinus rhythm** with **cardioversion** intervention.

Thromboprophylaxis

Thromboprophylaxis in the form of anticoagulant therapy to reduce the risk of VTE is prescribed for short-term use in medical or surgical interventions, conditions where research has demonstrated there is an increased risk of thrombosis, or in patients with a previous history of VTE. A risk assessment is performed to ensure the benefits of **prophylaxis** to select the appropriate therapy in individual patients. There is now a national programme aimed at reducing the incidence of hospital-acquired VTE.

Arterial thrombosis

Whereas in VTE the predominant component of the thrombus is coagulation factors, in arterial thrombosis the predominant component is platelets. The main event leading to arterial thrombosis is the rupture of an **atherosclerotic plaque**, which occurs in conditions such as **hypertension** and **atherosclerosis**. (See also Box 17.2.) This results in the rapid adherence of circulating platelets to the exposed collagen via von Willebrand factor and specific platelet surface receptors, which you met in Chapter 13. Following activation, subsequent aggregation

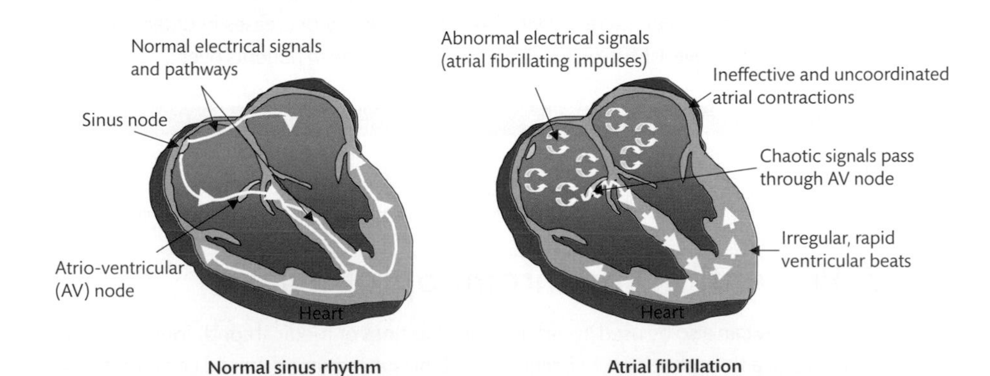

FIGURE 17.3
Atrial fibrillation (AF): identifying area of erratic impulses affecting the normal sinus rhythm within the heart. AF is a heart condition where the electrical activity to the atria is disorganized and is characterized by rapid irregular atrial impulses and ineffective atrial contractions. Stasis of blood (lack of movement) in the left atrium contributes to the risk of stroke and VTE.

is triggered by the generation of **thromboxane A_2** by platelets. This process is normally controlled by **prostacyclin** found in the vascular endothelium, which inhibits platelet aggregation. Thus, pharmaceutical intervention is with antiplatelet therapy, rather than anticoagulant therapy. The differences in the composition of arterial and venous clots are illustrated in Figure 17.4.

thromboxane A_2
A potent inducer of platelet aggregation generated through the arachidonic acid pathway within the platelet.

prostacyclin
A potent inhibitor of platelet aggregation generated through the arachidonic acid pathway within the vessel wall.

SELF-CHECK 17.4

How do the composition and formation of a venous and an arterial thrombus differ?

17.2 Current therapeutic anticoagulant pharmaceuticals for VTE

Pharmaceutical intervention, in the form of anticoagulant therapy, is the mainstay in the treatment of established VTE or as prophylactic treatment in conditions with a high risk of thrombosis, such as AF. These agents inhibit or reduce the normal function of coagulation proteins. Heparin and heparin derivatives are delivered by injection, either **intravenously** or **subcutaneously**, and available oral preparations include **coumarins**, such as warfarin and phenindione, and direct oral anticoagulants (DOACs), such as rivaroxaban and dabigatran.

intravenous
Administration of a drug in liquid form into the body by injection through a vein.

subcutaneous
Administration of a drug in liquid form into the body by injection to just under the skin.

Anticoagulant therapy regimes for VTE have little benefit for the prevention of arterial thrombosis because of its significant platelet component. Until recently, therapy has been limited to aspirin, which inhibits the platelet generation of thromboxane A_2, but recently a new generation of antiplatelet drugs has become available.

Cross references
You will meet the INR system in section 17.4.1 covering laboratory assays.
The clinical management of patients on VKA anticoagulant therapy is covered in the final section (17.5) of this chapter.

17.2.1 Oral anticoagulants

For decades, the mainstay oral anticoagulants have been coumarin derivatives that exert their therapeutic effect by antagonizing the action of vitamin K. By doing so, they inhibit the production of functional vitamin K-dependent clotting factors II, VII, IX, and X, the naturally occurring inhibitor protein C and its cofactor protein S, and protein Z. Although patients on vitamin K antagonist (VKA) **anticoagulation** continue to manufacture the vitamin K-dependent factors, they have impaired function, and are referred to as 'proteins induced by vitamin K absence/antagonism' (PIVKA). VKA anticoagulants need to be carefully monitored using the International Normalized Ratio system to ensure their clinical effectiveness and safety. The most commonly used VKA anticoagulant is **warfarin**. (See also Box 17.3.)

For the majority of conditions, patients receive a warfarin dose that aims to maintain their INR within a therapeutic range of 2.0–3.0, which significantly reduces their thrombotic risk. Those with a higher risk of thrombosis are maintained within a therapeutic range of 3.0–4.0.

BOX 17.2 *Atherosclerotic plaque rupture*

This leads to thrombus formation, which rapidly slows or arrests blood flow, leading to death of the tissues fed by the artery, an event referred to as an **infarction**. When this happens in a coronary artery it is termed a myocardial infarction, otherwise known as a heart attack, and is the most common cause of death in the developed world.

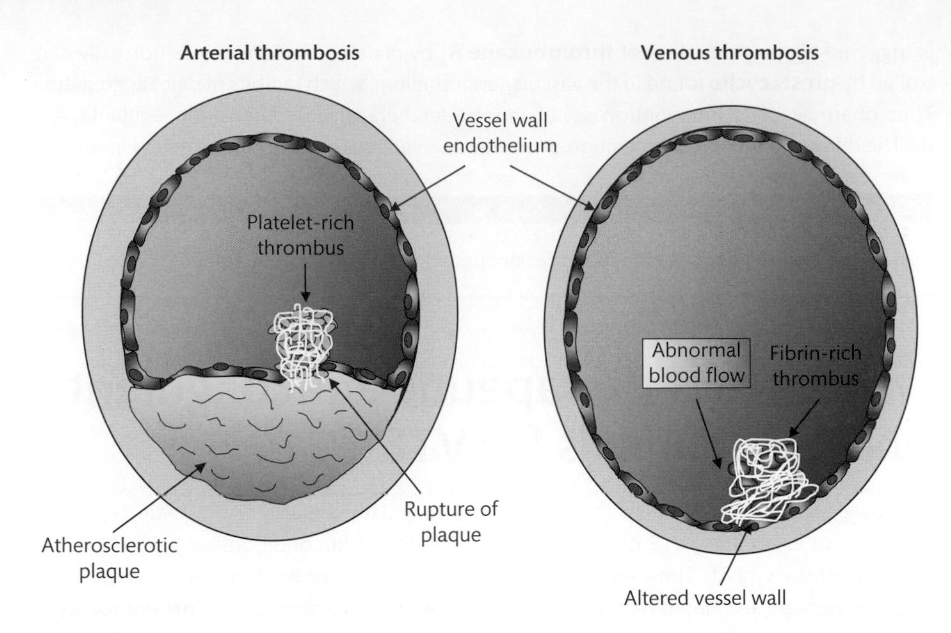

FIGURE 17.4

Differences in clot composition between arterial and venous clots. Arterial thrombosis results from the rupture of an atherosclerotic plaque after a build-up of cholesterol under the arterial wall and thrombus formation with a high platelet composition. Venous thrombosis can result from stasis of the blood or imbalance in levels of pro-coagulant clotting factors and their natural inhibitors, which causes damage to the surface of the vessel wall and subsequent thrombus formation with a high fibrin concentration.

There are three widely used VKA anticoagulant drugs (warfarin, acenocoumarin, and phenoprocoumarin), which are all derivatives of 4-hydroxycoumarin. Warfarin contains one asymmetrical carbon atom and exists in two **enantiomeric** forms (R and S), with chemical synthesis resulting in a **racemic mixture**. Warfarin is readily absorbed from the gastrointestinal tract, with over 90% bound to serum albumin. R-warfarin has a half-life of about 37–89 hours and S-warfarin 21–43 hours, although S-warfarin has two to five times more anticoagulant activity than R-warfarin. Warfarin is removed from the system by oxidation in the liver and the metabolites excreted in the urine.

enantiomeric
Two stereoisomers that are complete mirror images of each other.

racemic mixture
Equal amounts of two stereoisomers of an optically active substance which does not rotate plane-polarized light.

Vitamin K is essential for the formation of biologically active vitamin K-dependent clotting factors. These clotting factors are initially non-functional when first synthesized by hepatocytes in the liver. Vitamin K is an essential coenzyme for the post-translational carboxylation of these factors. The glutamyl residues of these proteins are converted by the liver enzyme carboxylase to gamma-carboxyglutamyl residues and comprise the Gla domain which, in the presence of calcium ions, are essential for anchoring with the negatively charged phospholipids on cellular membranes.

Cross references
You met Gla domains and their function in Chapter 13.
You can find the half-life values of vitamin K-dependent clotting factors in Table 13.3 of Chapter 13.

The concentration of vitamin K-dependent clotting factors is determined by a dynamic equilibrium between synthesis and degradation. Although the rate of synthesis is independent of plasma concentration, the rate of degradation is proportional to the plasma concentration and is influenced by the half-life of the respective clotting factors. The effect of warfarin is to reduce the rate of synthesis of functioning vitamin K-dependent clotting factors, which results in the establishment of a new equilibrium within the plasma. The concentration of coagulation factors will be reduced to the same extent as the rate of synthesis.

BOX 17.3 *A brief history of warfarin*

The basis of warfarin anticoagulant therapy goes back to the winter of 1921 when an epidemic of haemorrhagic disease devastated herds of cattle in Canada. It was found that this problem only occurred when cattle were fed on sweet clover, a new fodder crop. In 1924, the vet Frank Schofield made the observation that the toxic effect was only associated with spoiled sweet clover. Following a further outbreak in Wisconsin in the 1930s, unclotted cattle blood and spoiled sweet clover were sent to the local university for analysis. After many years of research, Karl Paul Link extracted the compound dicoumarol (4-hydroxycoumarin) from clover, which was identified as being responsible for the clotting defect. Damp storage of sweet clover had led to its spoiling by bacterial action which caused the oxidation of coumarin (the compound that gives freshly cut hay its sweet smell), to dicoumarol. Mark Stahmann, a graduate student, went on to develop a large-scale isolation of dicoumarol. Dicoumarol was first used as an anticoagulant in humans by Hugh Roland Butt in 1941, although not with widespread use because of toxicity fears. Link continued to investigate coumarin compounds in a search for an anticoagulant with improved pharmacological properties. Amongst over 150 compounds synthesized was 3-(α-acetonyl-benzyl)-4-hydroxycoumarin, which proved to be well suited for clinical use. As the research was funded by the Wisconsin Alumni Research Foundation (WARF) the compound was named warfarin by combining the organization's acronym with the last four letters of coumarin. It was originally marketed as a rodenticide, for which it is still used. Warfarin was approved for human use in 1954 and, since the 1960s, has been the main therapeutic agent for the prevention of thromboembolic disease. (See also Box 17.4.)

The human body has a limited ability to store vitamin K, so it is recycled by the liver thousandsfold before it is catabolized into inactive degradation products. This recycling process occurs by the reduction of vitamin K to its hydroquinone form KH_2, followed by oxidation to vitamin K epoxide, KO. This reaction is coupled to the carboxylation of glutamate residues to give gamma-carboxyglutamate and the enzymatic reduction of epoxide with NADH as cofactor, regenerating vitamin K to the KH_2 form. Figure 17.5 shows you that warfarin blocks this reaction by competitively inhibiting the enzyme epoxide reductase and hence preventing the reduction of vitamin K epoxide.

Drugs

Acenocoumarin has the same anticoagulant properties as warfarin. The difference between the two medications relates to the activity of the R- and S-enantiomer forms of acenocoumarin. Unlike warfarin, it is the R-enantiomer which has the greater anticoagulant activity.

BOX 17.4 *The first warfarinized VIP*

An early recipient of warfarin was US President Dwight D. Eisenhower, who was prescribed the drug after experiencing a heart attack in 1955.

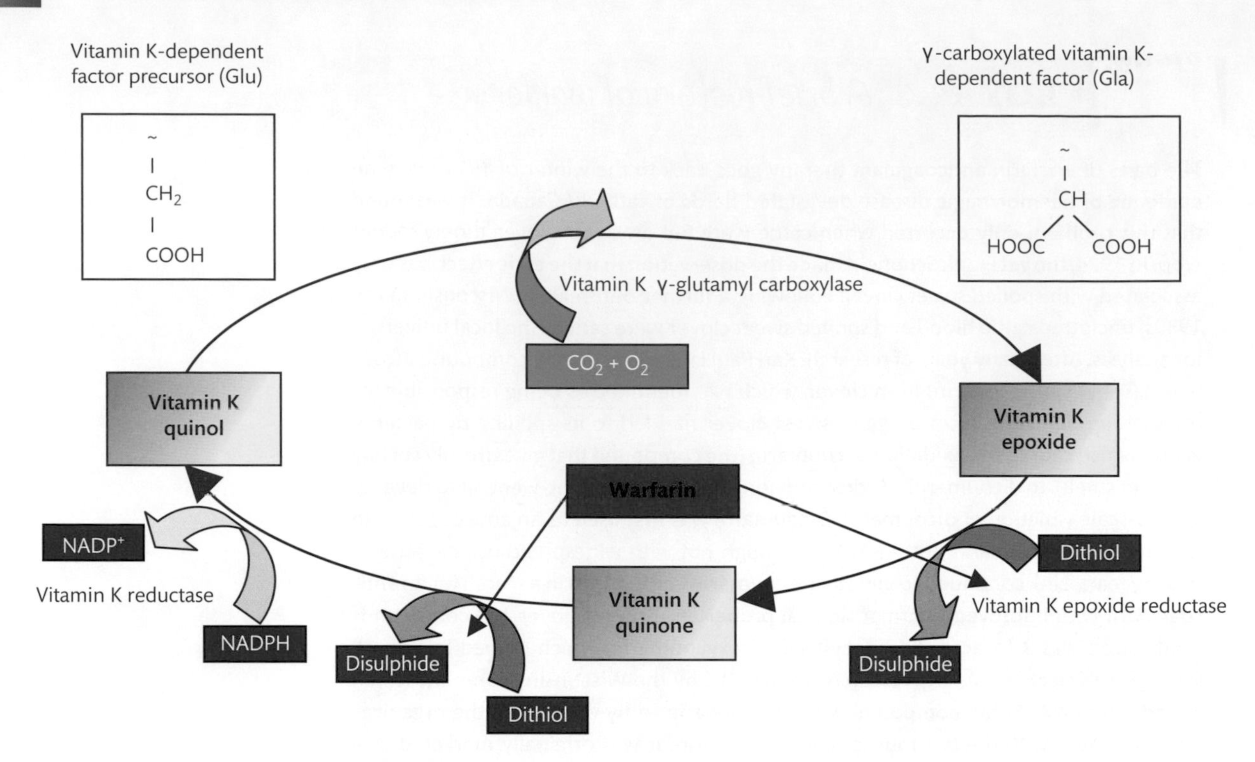

FIGURE 17.5

Coumarin-based drugs and the vitamin K cycle. Coumarins inhibit the action of the vitamin K epoxide reductase enzyme, which results in decreased concentrations of vitamin K and vitamin K hydroquinone. This leads to inefficient carboxylation of vitamin K-dependent factors by the glutamyl carboxylase enzyme.

Phenprocoumarin **Phenprocoumarin** also has the same anticoagulant properties as warfarin; it is in more widespread use in mainland Europe. Phenprocoumarin has a long half-life, 144 hours compared to warfarin's 40 hours. Its clearance from the body is not so dependent on the cytochrome P450 system and there is significant excretion of unchanged drug in urine.

Phenindione **Phenindione**, in the form of dindevan, was once the most commonly prescribed oral anticoagulant; its anticoagulant mode of action is identical to that of coumarin. It has largely been replaced by warfarin due to its severe side effects (including pink urine), and now tends to be used as an alternative therapy in the rare cases of patients experiencing adverse reactions to warfarin, such as skin irritation.

fresh-frozen plasma
Whole blood contains cells and fluid. The fluid component, plasma, contains all the plasma proteins, including coagulation factors. It is prepared at the time of donation when the unit of blood is centrifuged after collection to separate the plasma from the cells. The plasma is removed under sterile conditions and stored frozen to preserve the activity of the coagulation factors. When required, it is thawed and transfused through a vein.

Risks of VKA anticoagulant therapy

There are two main and potentially life-threatening risks of prescribing VKA anticoagulant therapy.

The first is bleeding, which can occur even if treatment is within the therapeutic range. The anticoagulant effect of warfarin or phenindione can be effectively reversed with oral or intravenous vitamin K. The dose needs to reflect the level of the INR and the severity of bleeding. In cases of life-threatening bleeding, immediate reversal of the anticoagulant effect may be achieved with a specific clotting factor concentrate containing factors II, VII, and X, or, if unavailable, **fresh-frozen plasma**.

Key Points

It is crucial to remember at all times that anticoagulant therapy is a balance between hyper- and hypocoagulability. An overdose of anticoagulant can precipitate potentially dangerous bleeding symptoms. Underdosing can allow the thrombotic risk factor(s) to precipitate the thrombotic event that the anticoagulation was prescribed to prevent.

The second potential complication is warfarin-induced skin necrosis, which although rare is a very serious complication of therapy. The incidence is higher in patients with a thrombophilia defect, particularly protein C deficiency. It usually presents during the first three to five days of initiation of therapy, where levels of functional vitamin K clotting factors are falling, but this also includes the vitamin K-dependent naturally occurring inhibitors, namely proteins C, S, and Z. Because protein C has a short half-life compared to most other vitamin K-dependent factors, its level falls sooner, generating a relative protein C deficiency in the early stages of treatment. This can result in the formation of microthrombi in the skin, leading to reduced oxygen supply to surrounding tissue, and ultimately cell death.

You will see later in this chapter (in sections 17.4 and 17.5) that VKA anticoagulants require regular monitoring with laboratory assays because a variety of factors affect how individuals respond to the drug.

SELF-CHECK 17.5

Explain the anticoagulant effect of warfarin.

17.2.2 Direct oral anticoagulants

You will see later in this chapter that although heparins and coumarins are clinically effective, they entail some practical, management, and clinical disadvantages. These issues prompted development of a new generation of orally delivered anticoagulants that exert their effects by direct inhibition of either thrombin or FXa, the DOACs. Regular laboratory monitoring is unnecessary in most clinical circumstances because the anticoagulant effect from a fixed DOAC dose is predictable. Their characteristics compared to VKA anticoagulants are outlined in Table 17.2. When first described, they were often referred to as new/novel oral anticoagulants (NOACs) and you will likely encounter that acronym in any further reading. Their sites of action in coagulation biochemistry relative to heparin and warfarin are shown in Figure 17.6. VKA anticoagulants, predominantly warfarin, remain the most commonly used anticoagulants in the UK, although use of DOACs is increasing.

Direct thrombin inhibitors

Dabigatran is orally administered as the pro-drug dabigatran etexilate, rapidly absorbed and converted to the active form of dabigatran by non-specific esterases in the gut, plasma, and liver. Approaching 80% of the pro-drug is excreted by the kidneys so **bioavailability** is low, about 6.5%, and high doses are required to achieve therapeutically adequate concentrations in plasma. The plasma concentrations and anticoagulant effect are dose-dependent and predictable, peaking within 0.5 to 2 hours of oral administration. The half-life is 8–10 hours

bioavailability
The degree to which a drug or other substance is absorbed or becomes available where it is required physiologically and can act on its target.

TABLE 17.2 Features of direct oral anticoagulants compared to VKA anticoagulants.

Key features	VKA anticoagulants	Dabigatran	Rivaroxaban, Apixaban, Edoxaban
Site of effect	VKOR	Factor IIa	Factor Xa
Oral	Yes	Yes	Yes
Immediate anti-coagulant effect	No	Yes	Yes
Monitoring	Yes	No	No
Dosing	Variable	Fixed dose	Fixed dose
Use in pregnancy	No	No	No
Antidote	Yes	Yes	Yes

if administered as a single dose and 14–17 hours for multiple doses. Dabigatran is a potent, competitive, and reversible direct inhibitor of the active site of thrombin. Dabigatran was first licensed for use in the UK in 2011 and is used in prevention and treatment of VTE, and stroke due to AF. An antidote, idarucizumab, is available as a reversal agent in situations of overdose or drug accumulation. It is a humanized monoclonal antibody that specifically binds to dabigatran and its metabolites to reverse the anticoagulant effect.

Direct inhibitors of FXa

Rivaroxaban is a highly selective, direct inhibitor of FXa that is rapidly absorbed and has bio-availability of 80–100%. The plasma concentrations and anticoagulant effect are dose-dependent and predictable, peaking within two to three hours of oral administration. It is metabolized in the liver and up to two-thirds of the drug is eliminated by the kidneys. The half-life is nine hours in patients with normal renal function. Rivaroxaban was first licensed for use in the UK in 2011 and is used in prevention and treatment of VTE, and stroke due

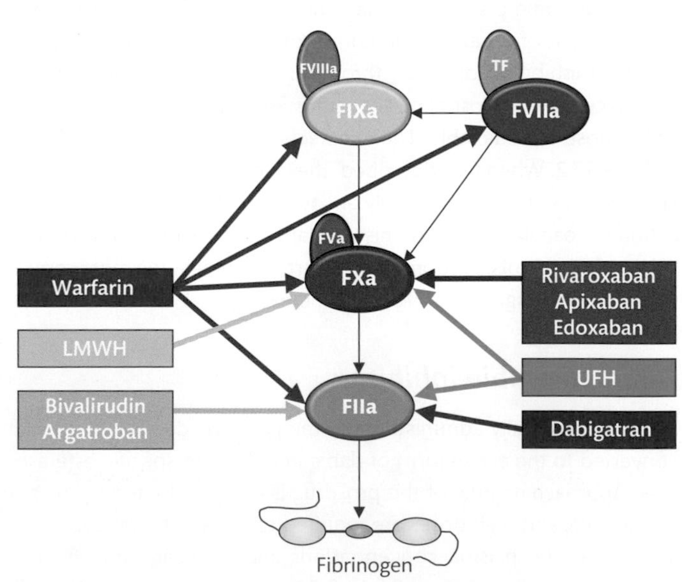

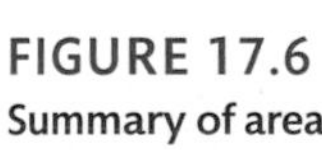
FIGURE 17.6

Summary of areas of coagulation affected by currently used anticoagulant drugs. Warfarin affects the production of vitamin K-dependent factors and thus has an effect at various stages of coagulation biochemistry. Unfractionated heparin (UFH) predominantly affects factors IIa and Xa, whereas low-molecular-weight heparin (LMWH) predominantly affects factor Xa. Rivaroxaban, apixaban, and edoxaban directly inhibit FXa, whilst bivalirudin, argatroban, and dabigatran are direct thrombin inhibitors.

to AF. **Apixaban** is also a direct FXa inhibitor, with oral bio-availability of 50% or greater and is metabolized in the liver. It is mainly excreted by the intestine, with 25% via the kidneys. Peak plasma concentration is reached within three to four hours. The half-life is 9–14 hours. It has been available in Europe since 2012 for prevention and treatment of VTE and stroke prevention in AF. In the UK in 2015, the direct FXa inhibitor **edoxaban** was recommended for prevention and treatment of VTE and stroke prevention in AF. It has oral bioavailability of about 62%. About 50% is excreted by the kidneys, whilst biliary secretion and a small degree of hepatic metabolism account for the rest of clearance. Plasma levels peak 1-2 hours after drug administration and the half-life is 10–14 hours. Another direct FXa inhibitor, betrixaban, is approved as an oral anticoagulant in the USA but not in Europe, mainly due to concerns about an increased bleeding risk.

The antidote for rivaroxaban, apixaban, and edoxaban is andexanet alpha. It is a recombinant modified FXa without the Gla domain, thereby eliminating its ability to bind to phospholipid membranes to form the prothrombinase complex. Consequently, it operates as a decoy molecule that binds the direct FXa inhibitor drugs so that native FXa can enter the prothombinase assembly and participate in coagulation unimpeded.

17.2.3 Parenteral anticoagulants

Heparin—mode of action

The anticoagulant properties of heparin were identified in 1916 by a US medical student, Jay McLean, who was investigating coagulation activation by phosphatides. He discovered that a preparation from liver prolonged the clotting time of plasma; it was subsequently found to be heparin. The chemical structure of this naturally occurring **glycosaminoglycan** was determined by a Swedish chemist, J. Erik Jorpes, in 1936. Following purification, the first use in humans was in 1937, when a heparin-saline solution was injected into the brachial artery, resulting in prolongation of clotting time and no detrimental side effects. Now heparin, one of the oldest drugs, and its derivatives, are some of the most widely used anticoagulants in clinical practice. However, it is only suitable for short-term use, as it is associated with **osteoporosis**—it decreases the bone-forming cells, osteoblasts, and increases the activity of osteoclasts which reabsorb bone. Heparin exerts an immediate anticoagulant effect and is essential in suspected cases of VTE to prevent thrombus extension through the vasculature, as demonstrated in the first clinical trial during the 1960s.

parenteral
Not given orally through the alimentary canal, but administered by injection.

glycosaminoglycan
Group of high molecular weight polysaccharides that includes chondroitin sulphates, dermatan sulphates, heparan sulphate, and heparin.

osteoporosis
A skeletal disease that leads to loss of bone mass and structure, resulting in an increased risk of fractures.

Unfractionated heparin

Unfractionated heparin (UFH) is a heterogeneous mixture of negatively charged sulphated glycosaminoglycans. It is composed of alternating uronic and glucuronic acid resides with polysaccharide chains, ranging in molecular weight from 3000 to 30,000. Heparin is isolated from porcine intestinal mucosa or bovine lung for manufacture into a therapeutic agent. The structure of a heparin subunit is shown in Figure 17.7.

The anticoagulant activity of heparin is contained in just one-third of the molecule, and consists of a pentasaccharide sequence with high-affinity binding properties to antithrombin, the major serpin found in plasma. Heparin brings about a conformational change in antithrombin, which potentiates its action in the region of 1000-fold, significantly enhancing its inhibition of factors IIa (thrombin), IXa, and Xa. Heparin inhibition is most effective against thrombin. You can see in Figure 17.8 that heparin forms a template to which antithrombin and thrombin bind, thus generating a **ternary complex**.

ternary complex
A complex containing three different molecules.

Therapeutic use

The clinical advantage of heparin is that the anticoagulant effect is seen immediately because it is administered intravenously and binds straight away to antithrombin. Intravenous administration means that UFH is only suitable for use within a hospital setting. Dosage depends on the level of anticoagulant effect required and is calculated based on the individual's weight. It can be prescribed either as an eight-hourly injection to provide intermittent cover, or as a continuous infusion. The plasma half-life is about 100 minutes, with little anticoagulant effect seen after six hours. Heparin is partly excreted in the urine or destroyed in the liver.

Monitoring of UFH is essential to ensure clinical effectiveness whilst minimizing the risk of over- or under-anticoagulation, potentially resulting in bleeding or further thrombosis, respectively. In the case of UFH, this is determined by achieving a defined anticoagulant effect *in vitro*, as assessed by prolonging the normal clotting time of the activated partial thromboplastin time (APTT), although there is considerable *in vivo* variation between individuals to the same dosage. This is mainly due to differences in bioavailability (about 50%) as heparin also binds to

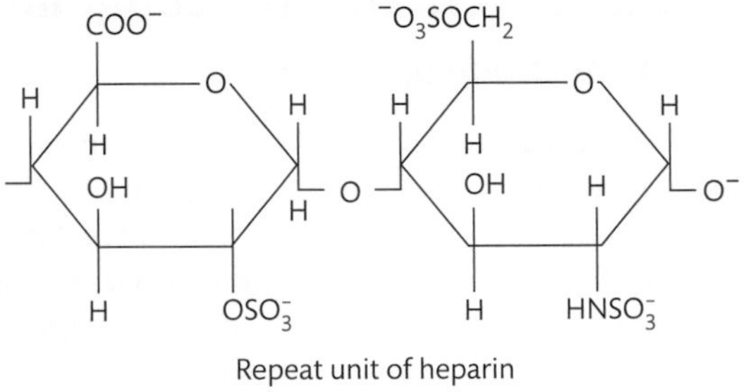

Repeat unit of heparin

FIGURE 17.7
Single unit of unfractionated heparin. Heparin is a mucopolysaccharide structured with alternate units of sulphated glucosamine and glucuronic acid. It is a strongly negatively charged molecule, which facilitates its binding to and inactivation of the positively charged coagulation factors, predominantly IIa and Xa. Reproduced by permission from Quader, Stump, and Sumpio. *Low molecular weight heparins: current use and indications*. *Journal of the American College of Surgeons* 1998.

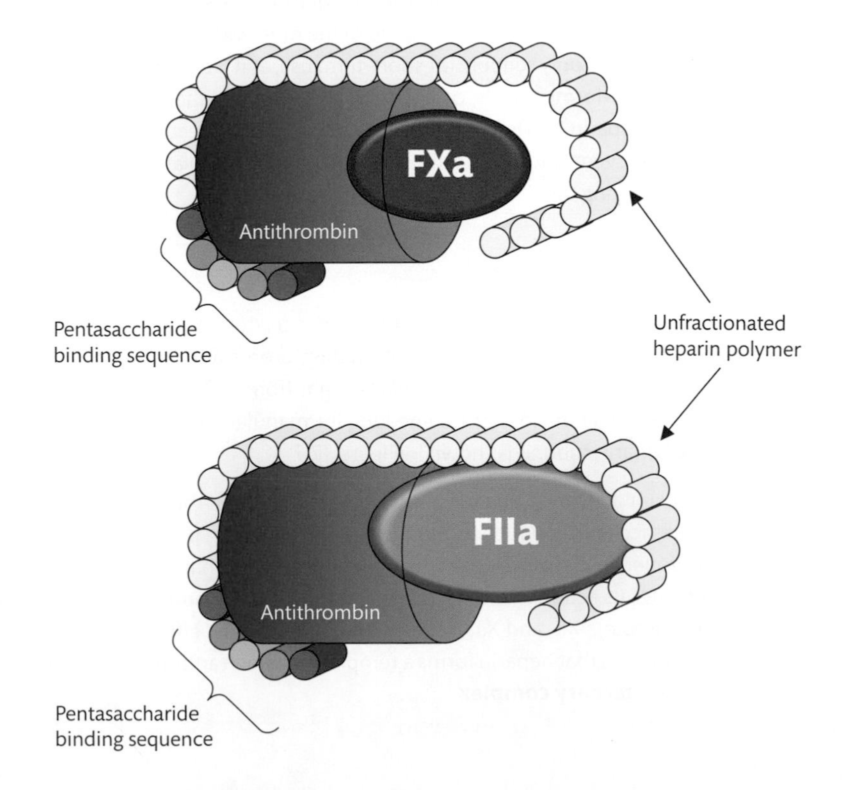

FIGURE 17.8
Illustration of unfractionated heparin (UFH) binding to antithrombin and thrombin or FXa in a template-complex formation. UFH promotes anticoagulant activity by binding to antithrombin and enhancing its major inhibitory effect on factors IIa and Xa. The binding of UFH to antithrombin is through a unique pentasaccharide molecule randomly distributed in the heparin molecule. Binding of the AT-UFH complex to FXa causes a conformational change resulting in the inactivation of FXa. With thrombin binding, however, UFH has to simultaneously bind to AT and thrombin, which requires a longer chain of heparin (saccharide units).

endothelium, macrophages, and plasma proteins, including platelet factor 4 (PF4), fibrinogen, factor VIII, and histidine-rich glycoprotein. It is therefore not possible to administer a fixed dose because individual responses to UFH are unpredictable. There is also considerable *in vitro* variation in APTT reagent sensitivity to heparin; therefore the lower and upper limit of clotting times that identifies effective and safe anticoagulant therapy, termed a **therapeutic range**, needs to be locally established. You can see in Table 17.3 that therapeutic ranges are reagent and analyser combination-specific.

therapeutic range
A concentration range within which a drug is clinically effective, whilst minimizing risk factors and side effects.

Risks of heparin therapy

There are two main and potentially life-threatening risks of prescribing heparin anticoagulant therapy.

The first is bleeding, which can occur even if treatment is within the therapeutic range. It is, of course, more likely if the patient becomes over-anticoagulated. The anticoagulant effect of UFH can be effectively reversed with protamine sulphate, a protein which is strongly basic and neutralizes acidic heparin.

The second is **heparin-induced thrombocytopenia** (HIT). When platelets are activated, they release PF4, some of which binds to the platelet surface. Due to opposite charges, heparin will bind to PF4 on the platelet surface, exposing **neoepitopes**, which are immunogenic and lead to antibody production, as depicted in Figure 17.9. Thrombocytopenia is due to the removal of antibody-coated platelets via the **reticuloendothelial system**.

neoepitope
A localized area of a molecule that is recognized by an antibody; a neoepitope is a structure that is recognizable by an antibody which is only exposed upon formation of a complex.

reticuloendothelial system
The phagocytic system of the body involved in the removal of foreign matter.

HIT can occur in two forms:

- Type I is benign and is associated with an early (within four days), usually mild decrease in the platelet count, which rarely falls below 100×10^9/L. The platelet count quickly returns to normal after stopping heparin, and will often do so even if heparin treatment continues. It occurs as a result of the mild direct activation of platelets by heparin and is not immune-mediated. Type I HIT is associated with no major clinical complications and occurs primarily in the setting of high-dose intravenous UFH.
- Type II is caused by an immune response to the heparin and PF4 complex, and is associated with thrombosis. It is reported to occur in 1–5% of patients on UFH, usually between days five and ten of therapy, and treatment needs to be stopped and anticoagulant therapy changed to a **direct thrombin inhibitor** (DTI), e.g. argatroban (see the later subsection 'Parenteral direct thrombin inhibitors').

TABLE 17.3 Examples of APTT reagent-dependent therapeutic ranges for UFH monitoring based on *in vivo* anti-Xa chromogenic heparin assays.

Reagent + analyser	Therapeutic range APTT ratio
General guidelines	1.5–2.5
Heparin level by anti-Xa assay	0.3–0.7 u/mL
Reagent A + analyser 1	2.5–4.0
Reagent B + analyser 2	2.0–3.0
Reagent C + analyser 1	1.5–3.5
Reagent D + analyser 3	1.8–4.1

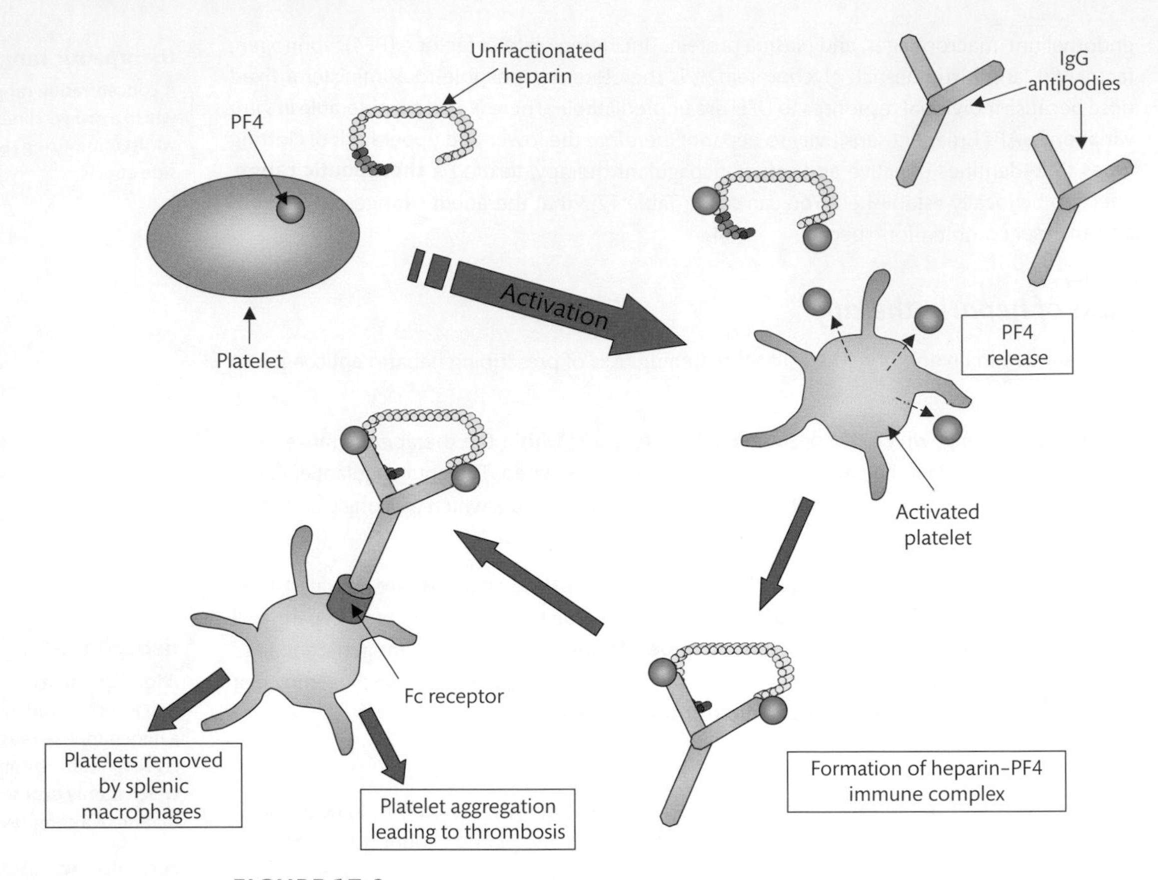

FIGURE 17.9

Pathophysiology of type II HIT. Platelets release PF4 and heparin binds to it, resulting in a PF4–heparin complex. IgG antibodies bind to this complex forming a PF4–heparin–IgG immune complex, which then binds to Fc receptors on the platelet surface inducing platelet activation. This releases more PF4, thus feeding the cycle and resulting in microthrombi and thrombocytopenia.

If the patient has previously been treated with heparin, the thrombocytopenia can occur immediately the drug is administered.

Cross references

You will meet direct thrombin inhibitors later in this section.

Assays for detecting HIT antibodies are covered in section 17.4 Laboratory monitoring of anticoagulant therapy.

Interestingly, bleeding is rarely a problem, but thrombosis can be life-threatening. The immune complexes of PF4-heparin bound to antibody bind to Fc receptors on platelets and endothelial cells. This activates the platelets, which release more PF4 and form platelet aggregates, further reducing the platelet count. Pro-coagulant platelet microparticles are also formed. Excess PF4 that has not bound to heparin binds to endothelial heparan sulphate, which can also generate antibody formation and subsequent endothelial damage by immune complexes. Damaged endothelium can lead to thrombus formation or disseminated intravascular coagulation.

Type II HIT is more common in surgical than medical patients, varies with the dose of heparin, and is more common with bovine- than porcine-derived heparin.

Low-molecular-weight heparin

Low-molecular-weight heparin (LMWH) is commercially produced from UFH by a process of chemical or enzymatic depolymerization, resulting in a heparin with a lower molecular weight than UFH. This process produces LMWH that retains the unique pentasaccharide sequence for specific binding to antithrombin and effective anti-Xa activity, but with reduced ability to inhibit thrombin. You can see in Figure 17.10 that FXa inhibition by LMWH does not require a template mechanism, unlike thrombin, IXa, and XIa, which need both enzyme and antithrombin bound to the same heparin chain.

Due to the reduction in molecular weight, LMWH has:

- reduced affinity for other proteins, thereby improving bioavailability;
- a longer half-life;
- a more predictable anticoagulant response to standard treatment doses;
- reduced side effects relating to bleeding, osteoporosis, or HIT.

LMWH is not readily neutralized by PF4 and has reduced binding to endothelial cells, which increases the half-life to about four hours, and has a bioavailability of 90–100%. Therefore, LMWH can be administered as a single, subcutaneous daily injection. LMWH also interacts less with platelets, which decreases the risk of HIT. Although a number of manufacturers produce a range of LMWH preparations (albeit with variations in properties), with a few exceptions, they are considered clinically identical.

Due to the predictable bioavailability of LMWH, only a few high-risk clinical situations require laboratory monitoring. In most patients, dosage is based on bodyweight and the intensity of anticoagulant effect required. However, due to the targeting of FXa by LMWHs, traditional APTT and thrombin times are relatively insensitive and, therefore, specific anti-Xa assays need to be performed to estimate anticoagulant levels. With the significant benefits outlined, LMWH is now used in the majority of clinical situations requiring immediate, short-term anticoagulation. There are, however, some situations where UFH is preferable to LMWH:

- renal failure where the drug can accumulate;
- patients with replacement heart valves (LMWH is not licensed for use in this group);
- situations where rapid reversal may be required, such as emergency obstetric or invasive procedures.

Cross reference

You will meet the anti-Xa assay for LMWH monitoring in section 17.4 Laboratory monitoring of anticoagulant therapy.

Risks of LMWH therapy

A number of clinical trials have demonstrated a reduced risk of bleeding with LMWH compared to UFH, particularly when used in conjunction with VKA anticoagulants. However, as LMWH is cleared predominantly by the kidneys, it has to be carefully assessed for use in

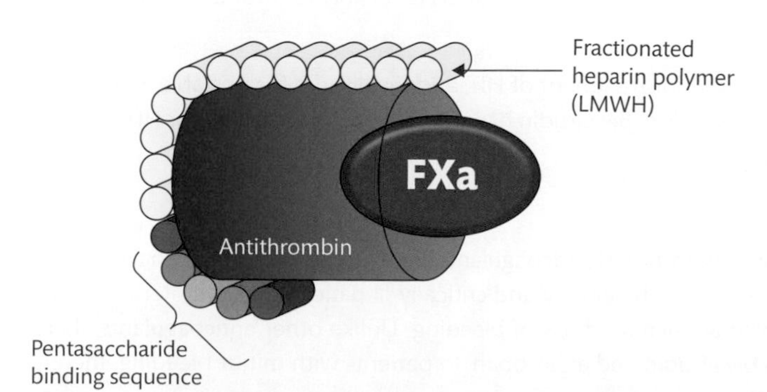

FIGURE 17.10

Illustration of low molecular weight heparin (LMWH) binding to antithrombin and factor Xa in a template-complex formation. LMWHs are fragments of UFH derived from depolymerization processes and are about one-third the size of the original molecule. The LMWH binds to AT through a unique pentasaccharide binding sequence, enhancing the inhibitory activity of AT. However, as LMWH is a much shorter molecule, it exerts its anticoagulant effect predominantly on FXa.

patients with renal failure, in whom it can accumulate in the circulation. If required, LMWH can be reversed with **protamine sulphate**, although not as effectively as UFH, and only about 60% is inactivated.

Danaparoid and fondaparinux

Danaparoid comprises a mixture of heparan sulphate, dermatan sulphate, and chondroitin sulphate, which are all glycosaminoglycan heparin analogues, or heparinoids. It mainly inhibits FXa and is used clinically as an LMWH, although it is clinically distinct from heparin and has different protein-binding properties. It is administered intravenously or subcutaneously. **Fondaparinux** also inhibits FXa and is used clinically as an LMWH, although it is actually a synthetic pentasaccharide virtually identical to the pentasaccharide sequence in heparins, with high affinity for antithrombin. Unlike heparin, fondaparinux has no inhibitory effect on thrombin. It is administered subcutaneously.

SELF-CHECK 17.6

What are the therapeutic differences between unfractionated heparin and LMWH?

Parenteral direct thrombin inhibitors

The most potent natural thrombin inhibitor is **hirudin**, which is extracted from the medicinal leech (*Hirudo medicinalis*) and was the first anticoagulant used in humans, in 1905. Hirudins for therapeutic use are now produced by **recombinant technology** using yeast and administered parenterally, normally by intravenous injection. Although differing slightly in structure from natural hirudin, recombinant (r)-hirudin is still highly specific for thrombin.

recombinant technology
Artificial production of a molecule produced by inserting foreign DNA into a host DNA which will manufacture the product.

You can see in Figure 17.11 that, unlike UFH and LMWH, the DTIs inhibit thrombin directly in a 1:1 irreversible complex independent of any cofactor. Also, unique to DTIs, they can inhibit thrombin that is free, clot-bound, or bound to fibrin degradation products, but have minimal interaction with plasma proteins. This provides a more predictable anticoagulant response to a given dosage, with a bioavailability of nearly 100% for bivalirudin and argatroban via subcutaneous administration. Therefore it is not considered clinically necessary to regularly monitor therapy, although laboratory assays exist that will detect the effects of DTIs if required.

Currently, two hirudin-like DTIs are licensed for use in the UK:

- **bivalirudin**, a synthetic 20-residue peptide analogue of hirudin, which binds reversibly to thrombin and is eliminated predominantly by proteolysis, whilst 20% is renally excreted.
- **argatroban**, a synthetic small molecule, which binds reversibly to thrombin and is cleared by the liver.

Cross reference
You will meet assays for monitoring DTIs in section 17.4.4 covering laboratory assays.

Argatroban is currently licensed for treatment of HIT, and bivalirudin for unstable angina and percutaneous coronary intervention. Bivalirudin has also been recommended for HIT.

Risk factors

Pharmacokinetics of most hirudin-based anticoagulants are very dependent on renal function, which makes predicting dosage in the elderly and critically ill patients difficult. Renal impairment is also associated with an increased risk of bleeding. Unlike other anticoagulants, there is no specific antidote to bivalirudin and argatroban. In patients with minor bleeding, merely

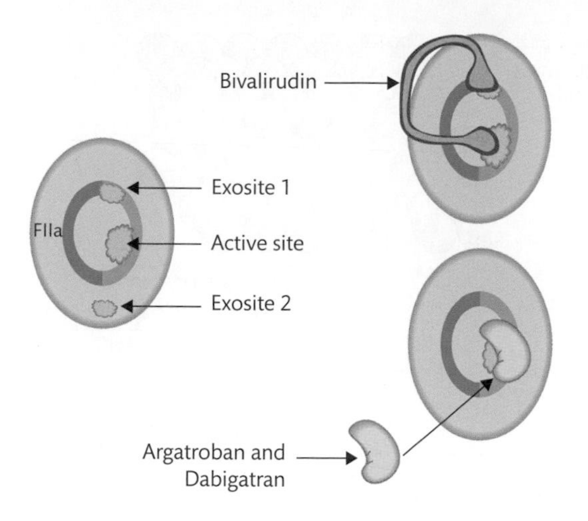

FIGURE 17.11
Binding of direct thrombin inhibitors to thrombin. The active site of thrombin, which binds substrate peptides, exists within a narrow gap. There are two secondary binding sites, exosites 1 and 2, which bind fibrin and heparin, respectively. Bivalirudin is a bivalent inhibitor as it simultaneously blocks the active site and exosite 1. Argatroban and dabigatran are univalent and block only the active site.

stopping therapy is usually effective because these DTIs have a short half-life (approximately 25 minutes for bivalirudin and 50 minutes for argatroban). In cases of major bleeding, haemodialysis or haemofiltration are the only viable methods of reducing the concentration of the circulating anticoagulant.

17.3 Current therapeutic anticoagulant pharmaceuticals for arterial thrombosis

Pharmaceutical intervention for the prevention of arterial thrombosis is focused on inhibiting platelet aggregation. These drugs are now increasingly being prescribed in clinical practice for a wide range of **cardiac** and **vascular** conditions.

- Aspirin binds to cyclooxygenase and inhibits the arachidonic acid pathway, thereby preventing the generation of the platelet agonist thromboxane A_2, which you can see illustrated in Figure 17.12. Risks associated with aspirin include intestinal bleeding and the phenomenon of aspirin resistance (see section 17.4.5).
- Dipyridamole and Cilostazol inhibit phosphodiesterase enzymes, which suppresses degradation of cyclic adenosine monophosphate (cAMP), leading to an increase of cAMP in platelets and blood vessels and consequent inhibition of platelet aggregation and vasodilation.
- Abciximab is a monoclonal antibody that selectively binds with glycoprotein IIbIIIa on the platelet surface, reducing platelet-to-platelet aggregation through fibrinogen.
- Tirofiban is a synthetic version of a small molecule from the venom of the saw-scaled viper (*Echis carinatus*) that inhibits interaction between fibrinogen and glycoprotein IIbIIIa.
- Eptifibatide is a cyclic heptapeptide derived from a component of the venom from the southeastern pygmy rattlesnake (*Sistrurus miliarius barbouri*) that also acts as a glycoprotein IIbIIIa inhibitor.
- Clopidogrel, Prasugrel, Ticagrelor, and Cangrelor are inhibitors of adenosine diphosphate (ADP)-induced platelet aggregation, which act by blocking the binding of ADP to its platelet membrane receptor, $P2Y_{12}$.

cardiac
Relating to the function of the heart and associated medical conditions.

vascular
Relating to the network of blood vessels in the body, namely arteries, veins, and capillaries, and associated medical conditions.

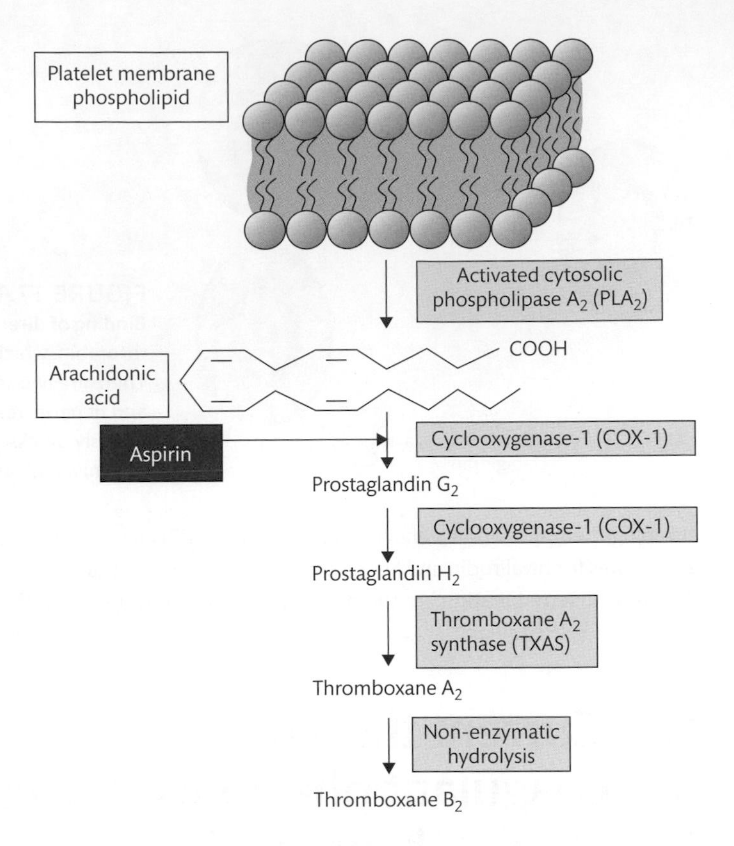

FIGURE 17.12
Anticoagulant effect of aspirin on the arachidonic acid (AA) pathway. Upon platelet activation, cytosolic phospholipase A_2 liberates AA from membrane phospholipids. Once released, AA is oxygenized by cyclooxygenase-1 (COX-1) to prostaglandin G_2 (PGG_2) and then reduced to prostaglandin H_2 (PGH_2), also by COX-1. PGH_2 is then isomerized to thromboxane A_2 (TXA_2) by thromboxane A_2 synthase (TXAS) and TXA_2 is released to function as a platelet activator. TXA_2 is unstable and is rapidly hydrolysed non-enzymatically to its metabolite thromboxane B_2. Aspirin, and other NSAIDs, exert their anticoagulant effect by inhibiting COX-1 activity, thereby preventing the generation of TXA_2 to facilitate platelet activation. The action of aspirin is immediate and irreversible and platelets are affected for the remainder of their lifespan.

Cross reference
Some of the platelet function assays that you met in Chapter 14 for diagnosing bleeding disorders are described here in their role as tools for monitoring antiplatelet therapy.

Monitoring the clinical effectiveness of these medications is not routinely required. However, the impairment of platelet function means that assessment is occasionally necessary, for example prior to surgery to clarify bleeding risk, or to establish the presence of ineffectual therapy (aspirin or clopidogrel resistance).

SELF-CHECK 17.7

What is the anticoagulant effect of aspirin?

17.4 Laboratory monitoring of anticoagulant therapy

The nature of different anticoagulants and test design means that no single test can be used to monitor each drug.

17.4.1 Laboratory monitoring of VKA anticoagulant therapy

The prothrombin time (PT) is the most commonly used laboratory test for monitoring VKA anticoagulant therapy. It was first described by Quick in 1935, before the discovery of VKA anticoagulants, and was initially introduced as a screening test for coagulation defects. As you saw in Chapter 14, it remains in everyday use for that purpose.

The PT measures the time a patient's plasma takes to clot in the presence of optimal concentrations of tissue factor, phospholipids, and calcium ions. Since there is no contact activator in the assay system, it means that factor XII remains dormant and subsequent components of the intrinsic pathway take no part in the generation of factor Xa to begin the common pathway. Therefore, the PT reflects the combined activities of three vitamin K-dependent factors (factors II, VII, and X) and two non-vitamin K-dependent factors (factor V and fibrinogen). Reduced fibrinogen levels will only prolong a PT if they are marked <1.0g/L, and factor V deficiency is extremely rare. Therefore, the major determinants of the clotting times are vitamin K-dependent factors, making PT an ideal tool for VKA anticoagulant monitoring. A modification proposed by Owren in 1947 involves adding components of bovine plasma to the thromboplastin reagent to compensate for variability in factor V and fibrinogen. Consequently, so-called combined reagents can be used specifically for VKA anticoagulant monitoring, versions of which are based on bovine or rabbit thromboplastins. Reagents for the Quick PT without bovine plasma are referred to as plain thromboplastins. Whilst the APTT is also affected by reductions in vitamin K-dependent factors (II, IX, and X), there are more stages involving non-vitamin K-dependent factors that contribute to the clotting time, so it is less sensitive as a VKA anticoagulant monitoring tool.

Standardization

Most laboratory reagents for common analyses are available from a variety of manufacturers. Whilst each type will be broadly similar, the differences in reagent composition can lead to variation in analytical performance. For instance, in the diagnostic setting when investigating for a factor deficiency, the PT of a patient's plasma with thromboplastin A may give a clotting time of 10 s, but give one of 16 s with thromboplastin B. This could result in the patient being considered normal if analysed with the first reagent, but abnormal with the second, leading to the initiation of further investigations to characterize the apparent deficiency. An alternative explanation is that, although the clotting times are different, the reagents are still measuring the same factors, but have different performance characteristics and sensitivities. Thus, the reference range for thromboplastin A is 10–13 s and for thromboplastin B is 15–17 s. As long as reagent- (and analyser-) specific reference ranges are used then these differences can be accounted for and the patient above would be considered normal by both reagents.

The situation is different when using the PT to monitor VKA anticoagulant therapy because the results are not compared to a reference range, but to a therapeutic range, which needs to be the same wherever and however the PT is done. Since it would be impossible to supply every laboratory worldwide with the same thromboplastin reagent to monitor warfarinized patients, alternative measures are used to achieve an acceptable degree of interlaboratory standardization based on the known properties of thromboplastin reagents.

Thromboplastin reagents

Quick's original thromboplastin reagent was prepared from rabbit brain, the brain being an organ rich in both tissue factor and lipid. When PT monitoring of VKA anticoagulant therapy was first introduced, most thromboplastin reagents were prepared locally from human cadaver brains, as they were easily obtained from the hospital mortuary. Other tissues have since been used as a source of thromboplastin such as bovine brain, rabbit lung, and human placenta. (See also Box 17.5.)

As there was no standardization for the local preparation of thromboplastin reagents, variations in analytical performance existed between batches within a given laboratory and from one laboratory to another. The problems were exacerbated by reporting PTs in seconds because a stably anticoagulated patient could have a very different clotting time when tested with different batches of thromboplastin, whether at the same hospital over time or a different hospital using its own reagent. This resulted in unwarranted dose alterations, and thereby increased the risks of bleeding and thrombosis. Commercial supplies of thromboplastin were introduced in the 1950s that relieved hospital staff of the burden of local reagent preparation. Standardization was assumed—but not necessarily delivered—between preparations from alternative manufacturers, and poor interlaboratory agreement remained evident.

The late 1960s saw the availability in the UK of a standardized thromboplastin, the Manchester Comparative Reagent (MCR), a phenolized extract of human brain. PT results were expressed as a PT ratio, where the patient's PT with the MCR was divided by that of the PT of a normal pooled control plasma analysed with the same MCR. Whilst this greatly improved interlaboratory agreement, it was impossible to manufacture enough reagent to supply other countries. An alternative model was needed to standardize PT analysis for VKA anticoagulant monitoring on a worldwide scale and this came in the form of the **International Normalized Ratio** (INR).

The International Normalized Ratio

Cross reference

More details are given on GMNPT later in this section.

The INR was adopted as the World Health Organization (WHO) international PT standardization scheme in 1983. Theoretically, the INR obtained with a given thromboplastin gives the PT ratio that would have been generated if the patient's plasma was analysed with a WHO standard thromboplastin.

BOX 17.5 *Recombinant thromboplastins*

In 1986, native human brain preparations were withdrawn from general use due to potential infection risks during manufacture, particularly in relation to Creutzfeldt-Jakob disease and the human immunodeficiency virus. Rabbit brain thromboplastins were the initial replacement. Thromboplastin reagents containing highly purified recombinant human tissue factor produced in *E. coli* and reconstituted into synthetic phospholipid vesicles became available in the early 1990s. The vesicles must contain phospholipids with a net negative charge to function effectively during *in vitro* coagulation, the most effective being phosphatidylserine. Consequently, recombinant thromboplastins are composed entirely of defined ingredients, making them potentially superior to the relatively crude tissue extract reagents of human or animal tissues.

The INR is derived by first calculating the PT ratio from a previously determined geometric mean normal prothrombin time (GMNPT) for that thromboplastin. The sensitivity of the thromboplastin to the VKA anticoagulant effect relative to the WHO standard is defined by a numerical value, the International Sensitivity Index (ISI). The INR is calculated from the PT ratio and ISI as follows:

$$\mathrm{INR}=\{\mathrm{Patient\,PT\,(s)/GMNPT\,(s)}\}^{\mathrm{ISI}}.$$

Therefore, accurate ISI assignment is crucial to the generation of accurate INRs. ISIs are assigned to thromboplastins after calibration against the WHO standard or a secondary standard that has been calibrated against it.

SELF-CHECK 17.8

How is an INR calculated?

Thromboplastin calibration and ISI assignment

The original WHO International Reference Preparation (IRP), designated 67/40, was a human brain thromboplastin, but it is no longer available. Replacement IRPs were calibrated against 67/40 before supplies were exhausted, and subsequent IRPs have been/will be calibrated against them. Thus, all IRPs will have had their ISIs defined in terms of 67/40. The inevitable degree of uncertainty in the accuracy of ISI assignment to new IRPs is limited by assigning ISIs using large, multicentre calibration exercises. These involve collaboration between 20 laboratories or more from at least 10 different countries. Currently, there are WHO standard preparations for thromboplastins of recombinant human and native rabbit origin. WHO no longer supplies a native bovine preparation and combined thromboplastins are calibrated against the plain rabbit standard.

Any new IRP is calibrated against all existing IRPs. The ISI assigned to the new IRP is taken as the mean ISI obtained from the calibration with each current IRP. Each collaborating laboratory identifies a minimum of 20 healthy individuals and 60 stably VKA anticoagulated patients with INRs spread throughout the range of 1.5–4.5. Each day, fresh plasma samples from two of the healthy individuals and six of the VKA-anticoagulated patients are tested by PT, with each of the current IRPs and the new IRP being calibrated. PTs on each sample are typically performed in quadruplicate. An **orthogonal regression line** is prepared from the natural logarithms of the PTs, which you can see in Figure 17.13.

The orthogonal regression model assumes that a single line can be drawn through both populations, yet you can see in Figure 17.13 that the line may deviate between them. This lack of coincidence can be corrected using Tomenson's correction calculation. The slope of the orthogonal regression line through both populations is used to calculate the ISI of the new IRP.

orthogonal regression line
Generating a best-fit line through a series of points commonly involves ordinary regression, which minimizes the sum of the squares of the vertical distances from each point to the line. Ordinary regression assumes the measurements were made without error. Orthogonal regression allows for random error by minimizing the sum of squares of the perpendicular distances between each point and the line.

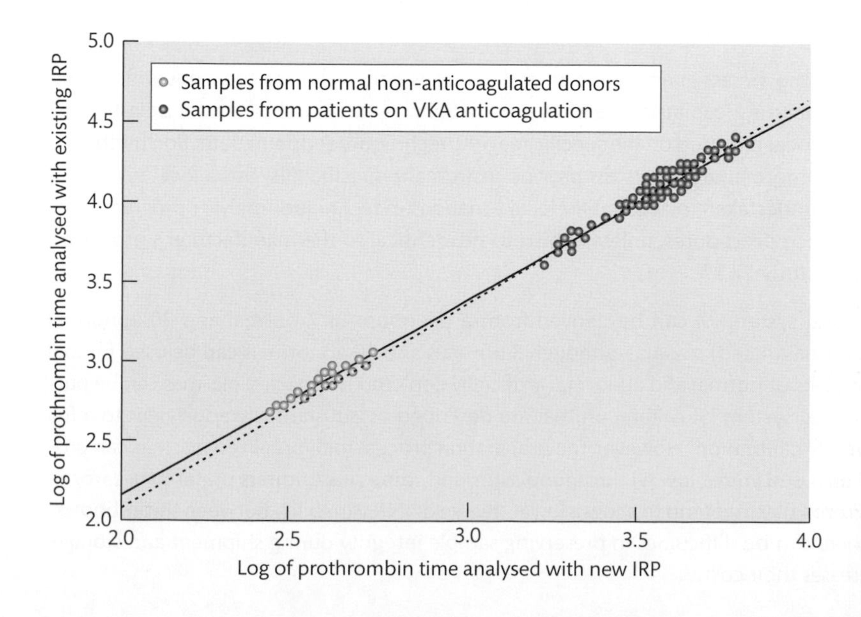

FIGURE 17.13
Orthogonal regression line for ISI calibration prepared by plotting the log of prothrombin times (PT) with the new thromboplastin against the log of PTs with the International Reference Preparation (IRP).

SELF-CHECK 17.9

How is the ISI of a new IRP generated?

National reference preparations and manufacturers' working standards are considered to be secondary standards and should be calibrated against the IRP for the corresponding species. Large, multicentre calibration exercises are impractical for generating secondary standards, but the exercises should be performed in at least two laboratories to minimize the effects of inter-laboratory variation. Secondary standards are used to calibrate manufacturers' commercial reagents. Each reagent batch should be calibrated against a secondary standard from the same tissue of the same species that has been prepared using a similar manufacturing process. These calibrations can be performed using pooled normal plasmas and pooled plasmas from VKA anticoagulated patients. Plasmas artificially depleted of vitamin K-dependent factors are also available. The advantage of using pooled plasmas means that far fewer PT estimations need to be performed to generate the regression line than if fresh individual samples were used. Similarities between the secondary standard and the reagent being calibrated, coupled with the reduction of biological variation between individual samples due to pooling, result in minimal data point scatter about the regression line.

SELF-CHECK 17.10

How is the ISI of a new secondary standard thromboplastin generated?

Local calibration

Cross reference

You met the analytical platforms of tilt-tube techniques, mechanical clot detection, and photo-optical clot detection in Chapter 2.

Commercial thromboplastins are supplied with assigned ISI values in order to contribute to a reduction in the inter-laboratory variation of INR results. However, this is not the full story because there are other variables that impact on raw clotting times apart from thromboplastin variability, principal amongst them being the different analytical platforms used to deliver reagents and detect fibrin clot endpoints. Each platform has innate characteristics and limitations that impact on the clotting time, even if the same plasma sample and thromboplastin are analysed by each method.

Consequently, the ISI assigned by the manufacturer is only valid for their equipment and method(s) of analysis. Hospital laboratories should calibrate each batch of thromboplastin they use to assign a local ISI based on the specific reagent/technique/endpoint detection method(s) in local use. Some manufacturers do provide instrument-specific ISIs, but a local system ISI should still be undertaken to account for local variations in technique, analyser programming, and maintenance procedures, unless known to be identical to the manufacturer's processes. (See also Case study 17.1.)

A reliable local system ISI can be derived from a minimum of 7 normal and 20 abnormal (i.e. VKA-anticoagulated) plasmas, although 3 normals and 10 abnormals can be used for low ISI reagents. Sets of normal and abnormal artificially depleted lyophilized plasmas can be purchased for local system ISI calibration that are described as 'substantially equivalent to a full fresh plasma ISI calibration'. However, the preparation process for lyophilized plasmas can prolong PTs if analysed with a low ISI thromboplastin, and some practitioners prefer to use frozen plasmas. Frozen plasmas tend to show a lower degree of INR variability between thromboplastins, but there can be difficulties in preserving sample integrity during shipment and storage, which increases their cost.

CASE STUDY 17.1 Impact of analytical platform variation on INR generation

As part of a local quality check, plasma from a patient on warfarin had a PT performed by the manual tilt-tube technique and automated photo-optical clot detection using the same thromboplastin. The patient's target INR range was 3.0–4.0. The manufacturer's assigned ISI was 1.05. The results were as follows:

	Patient PT (s)	INR
Tilt-tube (GMNPT: 15 s)	59.0	4.2
Automated analyser (GMNPT: 12 s)	43.2	3.8

Note that the GMNPT and patient PT by each method are quite different. Analysis by the tilt-tube technique would lead to dose alteration, but the patient would have remained on the same dose if clinical decision-making was based on the automated INR.

The thromboplastin was then calibrated for each method and gave an ISI of 1.00 for the tilt-tube technique and 1.04 for the automated method. Recalculation of the INRs gave the following results:

	Patient PT (s)	INR
Tilt-tube (GMNPT: 15 s)	59.0	3.9
Automated analyser (GMNPT: 12 s)	43.2	3.8

There was virtually no difference between the manufacturer's ISI and that generated locally for the automated analysis, probably because the manufacturer calibrated using a similar instrument. The ISI for the manual technique was sufficiently different to make a clinically significant difference to the INR when adopting the manufacturer's ISI, which reverted to a clinically equivalent result when the local ISI was applied.

The plasmas are each assigned a PT value by the manufacturer using an IRP thromboplastin and may be analysed by a manual technique. IRPs are calibrated using manual techniques. The local laboratory performs PTs on each plasma, ideally in quadruplicate and in the same working session, using the local analytical platform(s), and determines the ISI as described above.

An alternative method is to use plasmas that are assigned INR values rather than PT values by the manufacturer. The PTs are analysed using the local analytical platform(s) and the logs of the PTs are plotted against the logs of the reference INR, which you can see in Figure 17.14. Patient INRs are generated from local PT analysis by direct interpolation from the orthogonal regression line, thereby negating the need to determinate the ISI or GMNPT. Look again at Figure 17.14 and you will note that the regression line is generated from only four PT results because each plasma sample is a pool.

Manufacturers of certified plasmas should specify the reagent/analyser combinations for which their products have been shown to operate reliably. Local ISI calibration should ideally be derived using certified PT or INR values derived from the same species of thromboplastin.

SELF-CHECK 17.11

How is a local system ISI generated?

Geometric mean normal prothrombin time (GMNPT)

For the reasons mentioned earlier, the local system for analysing PT will affect the GMNPT as well as the ISI, and so the GMNPT must also be derived locally in order to generate accurate INR results. GMNPT is used in preference to an arithmetic mean because there is a log-normal distribution of PTs in a healthy adult population. Calculating the GMNPT effectively performs a log transformation of the data to normalize the distribution before calculating the mean itself. Although mathematically correct, there is rarely a significant enough difference between the arithmetic and GMNPT to adversely affect INR values, yet most laboratories apply the GMNPT to account for the rare occasions where its benefit might be realized.

A GMNPT is generated from the PTs of a minimum of 20 fresh plasmas from normal adult donors from both sexes and analysed over a period of several days for each analyser and technique in local use. Sophisticated automated analysers will deteriorate over time, even with regular and effective maintenance, so GMNPT may also alter over time. In reality, any drift will be minimal but re-assessment of GMNPT warrants consideration if the internal or external quality control schemes suggest a deterioration in PT/INR performance. A major equipment service may necessitate re-assigning the GMNPT and/or ISI.

Commercially prepared plasmas specifically for GMNPT evaluation are not available, even though in theory GMNPT should be locally assessed for every new thromboplastin batch. PTs for GMNPT assessment should be performed on fresh plasma, not frozen or lyophilized, because routine VKA anticoagulant monitoring is performed on fresh plasma. Few laboratories have the donor availability to do this and tend to rely on the fact that GMNPT drift, if it occurs at all, is

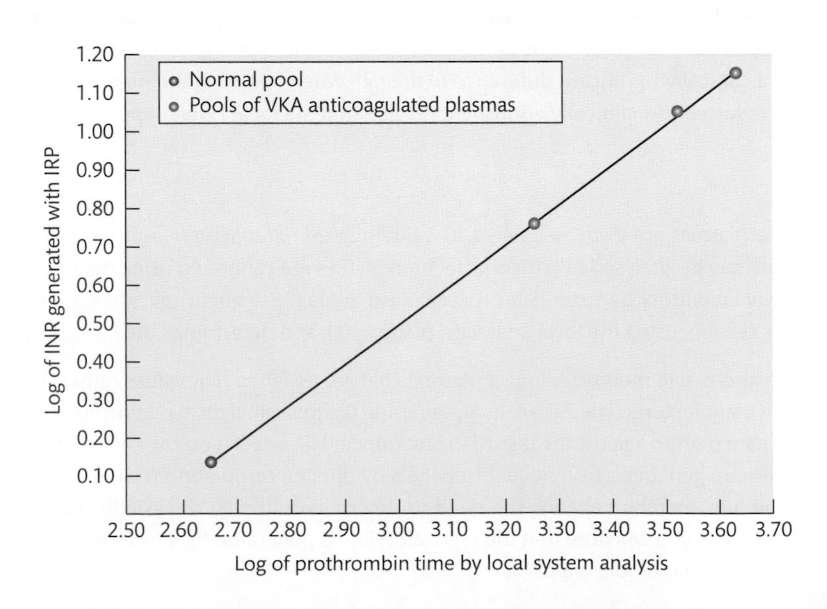

FIGURE 17.14
Local ISI calibration prepared by plotting the log of PTs with the new thromboplastin against the logs of INRs with the reference thromboplastin.

usually minimal. One way round this is to calculate the GMNPT from the PT and INR of a lyophilized or frozen pooled normal plasma obtained for the ISI calibration exercise, as shown below:

$$\text{Theoretical GMNPT} = \frac{\text{PT of normal pooled plasma (s)}}{\text{INR of normal pooled plasma}}$$

Commercial normal plasma pools are usually prepared from a large number of donors, at least 20 and often in excess of 100. Calculating the GMNPT in this way assumes that the PT of such a large pool does indeed equate to the GMNPT, had it been derived from a group of separate donors. Fresh plasmas can only be substituted by lyophilized or frozen plasmas if the laboratory has demonstrated that the mean result of replicate PTs of lyophilized or frozen plasmas analysed by local procedures is identical to that of fresh plasma samples.

SELF-CHECK 17.12

How is a GMNPT generated?

Point-of-care testing

Efficient delivery of an outpatient anticoagulant clinic often requires fast turnaround times of INR results, which are not necessarily achievable via a remote centralized laboratory. A common response to this problem is for biomedical scientists to use **point-of-care test** (POCT) instruments in the environs of the clinic itself. POCT devices are portable, semi-automated analysers that perform only a single type of test, or a very limited repertoire. The reagents and blood samples are loaded manually and endpoint detection is automatic.

Most POCT devices for PT/INR testing are designed to analyse whole-blood samples from capillary punctures, although many can also analyse plasma. The biomedical scientist performs the capillary puncture, often referred to as a finger-prick, and the non-anticoagulated whole blood is immediately applied to a disposable test strip or cartridge containing thromboplastin in a dry state. The blood reconstitutes the thromboplastin and the strip/cartridge is immediately inserted into the analyser for clot detection. Interestingly, fresh capillary blood contains sufficient calcium ions to facilitate *in vitro* coagulation and some strips/cartridges do not contain calcium within the reagent.

A number of POCT instruments for INR generation are available that use a variety of clot detection principles, as summarized below.

- As the whole blood starts to clot, the reduction and eventual cessation of red cell motion is detected using optical pattern recognition by changes in laser interference.
- Integral to the reagent once it has been reconstituted are paramagnetic iron oxide particles. Alternating magnetic fields cause pulsation of the particles, which slows and stops as the blood clots. The changes in particle motion are detected optically.
- A reaction tube containing a magnet rotates slowly. Clot formation changes the position of the magnet and is detected because it removes the magnet from a magnetic field.
- Whole blood enters small-bore channels in a cartridge and is pumped back and forth. Clot formation reduces the oscillation and is detected photo-optically.
- Clot formation generates thrombin, which is reacted with a fluorescent substrate.
- Clot formation is detected by a change in the electrical impedance of blood that occurs when fibrinogen is converted to fibrin.

Performance issues with POCT instruments

Many research studies have been published comparing the analytical performance of POCT analysers between each other and standard laboratory equipment in the context of warfarin monitoring. Some studies have concluded that there is strong agreement between POCT devices and standard analytical techniques, whilst others have reported systematic bias or agreement only up to a certain point within the therapeutic range, typically when the INR is between 3.0 and 4.0. Although the two methods may give statistically different results, the degree of difference may be small enough that decisions on dosing are not markedly affected, in which case the methods can be considered clinically equivalent. If a haematology department uses standard analyses in the centralized laboratory and POCT devices in the anticoagulant clinic, it is important to be aware of the degree of correlation. Analytical parity within the therapeutic range is important because patients may be tested by both methods during a given period of anticoagulation. It is common for POCT results to be checked by standard analysis when INRs are beyond the point where the methods have been shown to agree.

One additional factor that contributes to differences between POCT and standard analysis is the fact that most thromboplastins for POCT are dry reagents. Upon reconstitution of dry reagents, FVII is exposed to thromboplastin in intermediate states, which can affect the clotting time. Liquid reagents used in standard testing contain a stable and fully hydrated thromboplastin immediately available for interaction with FVII. Consequently, recombinant reagents are better suited to POCT analysis because their uniform composition results in a sharp phase transition from the dry state to the liquid state.

Effective ISI calibration of a POCT device requires a calibration exercise comparable to that of a secondary standard. This would need relatively large numbers of normal donors and VKA-anticoagulated patients to provide fresh capillary blood, and is thus a prohibitive exercise for routine hospital laboratories. Consequently, manufacturers assign the ISIs to batches of test strips/cartridges using an IRP, and the ISI is adopted by all users of the instrument and its specific reagent strip/cartridge. To a large extent, this practice in effect generates a local system ISI, although operator and individual device variability are not accounted for.

POCT instruments are necessarily portable and relatively easy to operate. This has led to their operation by non-laboratory personnel such as nurses and pharmacists in clinics and primary-care settings. As long as appropriate training is provided and quality control is monitored by biomedical scientists, this appears to be a safe and effective addition to the armoury of anticoagulant monitoring. Some patients are suited to self-testing and even self-management but they must be carefully selected, and not all patients are willing to do so, or can be successfully trained.

SELF-CHECK 17.13

What are the main differences between performance of INR by routine automated analysis and POCT?

Interlaboratory INR variability

Although we know that a local system ISI corrects for major differences in PT/INR results between systems, a surprising number of laboratories adopt the manufacturer's assigned ISI, irrespective of the analytical platforms in local use. This is often a resource issue, but it may reflect a lack of awareness of the clinical gains to local calibration. Reports from the UK National External Quality Assurance Scheme (UK NEQAS) reveal that interlaboratory agreement remains a problem. Even for those laboratories who do undertake local calibration, there are important variables to take into account, and they are detailed in Table 17.4.

Key Points

Patients whose INR is beyond the therapeutic range or who are unstable (variable INR) are less likely to give good agreement between methods, and these are the patients most likely to be investigated for analytical anomalies.

Even if an effective calibration is undertaken, performance of PTs in the diagnostic/monitoring setting is subject to the effects of pre-analytical variables, which themselves will impact on ISI generation if the samples used for calibration are thus affected.

Use of the arithmetic mean normal PT in place of GMNPT can potentially affect the accuracy of INR results, even if the calibration was well performed. Errors in the generation of GMNPT will affect INR accuracy, such as an unrepresentative donor population, analyser/technical faults, or complications in pre-analytical variables.

Cross references

The types and effects of pre-analytical variables on coagulation testing are detailed in Chapters 2 and 14.

Causes of VKA anticoagulant resistance are detailed in section 17.5.2 later in the chapter.

Other factors interfering with the INR for VKA anticoagulant monitoring

Some patients require immediate anticoagulation and may be given UFH or LMWH therapy until the VKA anticoagulant effect has reached a therapeutic INR value. It is possible for UFH to further prolong a PT and give a false impression of the degree of anticoagulation achieved by the VKA anticoagulant. In practice, it is rare for therapeutic levels of UFH to interfere with

TABLE 17.4 Variables affecting ISI calibration.

Variable	Potential effect on calibration
Calibration against an inappropriate IRP	Using a thromboplastin derived from a different species will impact on raw clotting times and alter the slope of the calibration curve.
Insufficient sample numbers	Calibration has to be based on a sufficiently representative set of donors, otherwise the calibration curve will not equate to the patient population.
Distribution of INRs in the VKA anticoagulated donors	These samples must have INRs distributed within the range of 2.0–4.5 and have representative numbers of low, medium, and high values in order to equate to the patient population. If calibrating with pooled plasmas, it is the responsibility of the manufacturer to ensure an appropriate spread of INRs amongst the donors.
Calculation errors	Mathematical errors will generate an incorrect ISI.
Analyser malfunction	If the calibration is performed on an analyser that is malfunctioning, which is not always immediately obvious to the operator, the calibration will be irrelevant to INRs generated once the fault is repaired.
Operator variability	This is particularly relevant to manual and semi-manual techniques where relative experience and quality of training will impact on competence.

PT analysis because of the powerful stimulus of a significant excess of tissue factor in thromboplastin reagents. Some manufacturers add heparin neutralizers to thromboplastin reagents to further reduce the potential for UFH interference, although they are only effective up to certain UFH concentrations. The main problems are encountered when blood samples are taken from central lines that have not been flushed sufficiently to prevent UFH entering the sample. Use of LMWH, which does not interfere with PT analysis, is becoming more common.

It is rare for lupus anticoagulants (LAs) to cause an elevated PT, because the high concentration of phospholipid in thromboplastin reagents swamps most LAs and prevents them from manifesting in this test. Warfarin is used in the treatment of antiphospholipid syndrome, so patients with clinically significant PT-reacting LA can be difficult to monitor because there is an ever-present risk of overestimation of anticoagulation. Such patients can usually be monitored with an alternative thromboplastin if it is shown to be unresponsive to that patient's LA. However, it should be borne in mind that LA-reactivity characteristics can alter over time in the same patient. Where this is unachievable, one-stage clotting assays for FII or FX can be used, because the additional dilutions integral to the assay design abolish the inhibitory effect of all but the most potent LA. Chromogenic factor X assays are available for patients whose LA overcomes even the dilution effect. A factor X level around 40% of normal equates to an INR of 2 and a level around 15% equates to an INR of 3.

Despite its limitations, the INR system has significantly decreased interlaboratory variability and facilitated the adoption of standard therapeutic ranges worldwide.

Investigating VKA anticoagulant resistance

Individual patients vary in their response to a given VKA anticoagulant dose, hence the need for monitoring therapy by INR. Some may require 0.5 mg daily to achieve a therapeutic INR, whilst others may need 15 mg daily to reach a similar INR. Some patients respond poorly to VKA anticoagulants, even at increasing doses, which can be due to a phenomenon commonly called **warfarin resistance**, since warfarin is the most used and studied VKA anticoagulant.

The patient is given an oral dose of VKA anticoagulant under supervision and serial blood samples are taken over the course of three days for INR determination and direct assay of the VKA anticoagulant. Measurement of VKA anticoagulant concentration over time allows determination of the bioavailability of the VKA anticoagulant and its **half-life** in that patient. Measuring the INR over time in response to a single dose of VKA anticoagulant reveals information about the **pharmacodynamic** response.

half-life
The time taken for the amount of drug in the body to reduce its initial concentration by one-half.

pharmacodynamics
The action, effect, and breakdown of drugs in the body.

Key Points

Apparent warfarin resistance may be due to non-compliance; the dose of warfarin has to be supervised when testing for resistance to ensure that the patient has correctly taken the drug.

Sampling is more frequent on the first day to ascertain the peak warfarin concentration and assess the degree of absorption. Daily measurements are sufficient after that at 48 and 72 hours post-dose in order to calculate the half-life. In the rare cases where the results suggest warfarin malabsorption, a further study using intravenously administered warfarin is necessary to assess warfarin resistance due to causes post-absorption. These regimes tend to be undertaken only in specialist laboratories.

17.4.2 Laboratory monitoring of heparins

The APTT is the most widely used test for monitoring UFH therapy. The APTT is considered to be a sensitive test to the presence of UFH, with some reagents demonstrating prolongation of the APTT when the plasma heparin concentration is as low as 0.1 IU/mL. However, numerous APTT reagents are commercially available, but the variation in their phospholipid composition and concentration results in marked variability in heparin sensitivity.

Similar to warfarin monitoring, patients are assessed for adequate levels of anticoagulation by reference to a therapeutic range. This is often taken as 1.5–2.5 times the APTT of the normal plasma pool if reported in seconds, or an APTT ratio between 1.5 and 2.5. In view of the reagent variability, which is compounded by the type of analytical platform in local use, arbitrary use of these ranges is inappropriate. Instead, individual laboratories should calibrate their APTT therapeutic range to equate to plasma heparin concentrations between 0.3 and 0.7 IU/mL, as measured by a chromogenic factor Xa inhibition assay, or 0.2 and 0.4 IU/mL if measured by protamine sulphate titration. The plasmas used for the calibration should be obtained from patients receiving UFH rather than spiking normal plasma, in order to reflect the effects of patient physiology on circulating UFH.

Another limitation to the use of APTT in UFH monitoring is that the APTT response is affected by other aspects of the patient's haemostasis. In particular, elevated FVIII levels—which can occur postoperatively or associated with acute illness, malignancy, or pregnancy—can reduce the APTT and give the impression that the patient is less anticoagulated than is actually the case. Markedly elevated fibrinogen can have a similar, although less obvious, effect. Conversely, patients on concomitant VKA anticoagulation, with factor deficiencies (e.g. disseminated intravascular coagulation), or LAs may further prolong an APTT and give the impression of a greater degree of anticoagulation than has actually been achieved.

The short plasma half-life of UFH and the possibility of a degree of neutralization by PF4 means that blood samples for monitoring purposes should be centrifuged and analysed within two hours of collection. This is not always logistically possible and can lead to misleading results because the sample may contain significantly less available UFH by the time it is analysed. Taking the sample into a cooled collection tube and keeping it on ice until it reaches the laboratory can reduce these effects. (See also Box 17.6.)

In recognition of the variables that affect UFH monitoring by APTT, use of factor Xa inhibition assays, so-called **anti-Xa assays**, is becoming more common. They are predominantly

BOX 17.6 *APTT and PT ratios*

The APTT ratio is derived from dividing the patient's APTT by that of the normal plasma pool or a mean APTT value from a panel of normal donors. Some laboratories report their diagnostic screening APTT results in ratio format in place of the raw clotting time in seconds. This makes no difference to diagnostic efficacy as long as reference ranges are determined locally.

Similarly, some laboratories report diagnostic PTs as a prothrombin time (PT) ratio, or even an INR. Use of INR in diagnostic screening is contentious, as it was designed to improve analytical parity in the context of the VKA anticoagulant effect, rather than disease states.

chromogenic assays, and thus unaffected by many of the variables that affect clotting-based tests. The requirement to analyse within a short time frame remains, although as long as the sample is cooled and the plasma is separated from the cells in time, it can be stored frozen until analysis.

Some patients who require large doses of UFH to prolong their APTT are said to exhibit **heparin resistance**. This can be due to AT deficiency, short heparin survival *in vivo* associated with large thrombi, elevated FVIII or fibrinogen, circulating activated coagulation factors, or HIT (as a result of PF4 release from aggregating platelets, which neutralizes heparin).

SELF-CHECK 17.14

What are the problems associated with monitoring UFH by APTT?

Anti-Xa assays

Anti-Xa assays use the same principle as the assay you met in Chapter 15 for measuring antithrombin (AT) levels, except that the patient's plasma 'supplies' a different variable. In the antithrombin assay, the variable supplied by the patient's plasma is AT itself, which is reacted with fixed amounts of excess exogenous heparin and factor Xa to form AT–heparin–factor Xa complexes. The residual factor Xa is measured via its reaction with a chromogenic substrate, the intensity of the resultant coloured product being inversely proportional to the patient's AT concentration; i.e. the more AT that was present in the plasma, the more factor Xa will complex with AT–heparin, thereby leaving less residual factor Xa to react with the chromogenic substrate.

In the anti-Xa assay for heparin monitoring, the variable 'supplied' by the patient's plasma is the circulating UFH, which is reacted with fixed amounts of excess exogenous antithrombin and factor Xa, as shown below. In this case, the residual factor Xa that reacts with the chromogenic substrate is inversely proportional to the UFH concentration. So if the UFH concentration is high, there will be less residual FXa and the colour intensity of the final reaction will be low. Results are read from a standard curve.

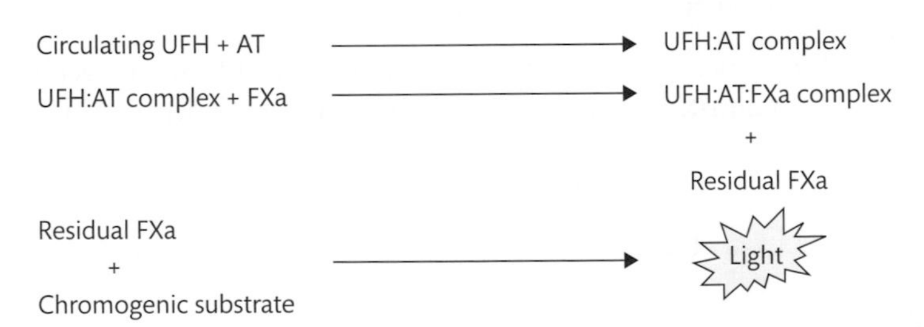

Most patients will have normal levels of circulating AT, but some may be deficient. The anti-Xa assay principle is based on there being an excess of AT, which is provided in the reagent, so the results are unaffected by the patient's own antithrombin levels. Additionally, this ensures that all heparin becomes complexed with antithrombin, which would not necessarily happen if the assay relied on the patient's AT alone, and therefore measures 'free' heparin and provides a truer estimate of the concentration. Some commercially available assays do not employ the AT excess principle, relying instead on the patient's AT alone to complex with the exogenous factor Xa. This is not necessarily inferior because, although methods using excess AT may be more accurate in terms of estimating the actual concentration of circulating heparin, they may not provide a true reflection of the actual degree of anticoagulation. For instance, a patient with reduced antithrombin will generate a heparin concentration value that appears to be within the therapeutic range when measured by an AT-excess assay. However, the heparin will not be providing sufficient anticoagulant effect because the patient has insufficient circulating AT for the heparin to exert its full effect *in vivo*. The APTT, of course, gives a direct indication of the anticoagulant effect. Anti-IIa assays are also available for monitoring UFH, although most laboratories employ anti-Xa because that assay can be used for LMWHs too.

Clot-based assays exist for measuring anti-Xa activity in heparinized patients which rely on the patient's AT to complex with an excess of factor Xa. The residual factor Xa is then reacted with bovine plasma containing prothrombin, factor V, and fibrinogen, which also contains optimal concentrations of calcium chloride and phospholipid to allow clot formation. The clotting time is directly proportional to the heparin concentration; i.e. a high heparin concentration results in less residual factor Xa and hence longer clotting times. Results are read from a standard curve.

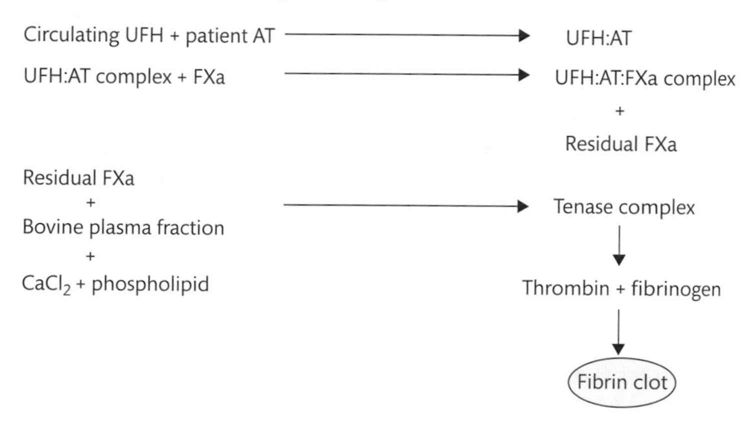

The superior bioavailability and longer half-life of LMWHs means they are administered in lower doses than UFH, and subcutaneously rather than intravenously. Therefore, LMWHs rarely interfere with routine coagulation screening tests and have to be monitored with anti-Xa assays. Routine laboratory monitoring of LMWHs is, however, unnecessary for most patients because there is a predictable dose–response. Situations requiring monitoring include morbid obesity, pregnancy, renal failure, low body weight, young children, and patients on long-term therapy to detect drug accumulation.

SELF-CHECK 17.15

What are the principles of anti-Xa assays for monitoring heparin therapies?

Anti-Xa performance considerations

Standard curves for anti-Xa assays are prepared by spiking normal plasma with heparin at different concentrations and assaying each concentration for anti-Xa activity. A crucial consideration when undertaking anti-Xa assays for heparin monitoring is that UFH preparations from different manufacturers can vary in their biochemical composition. Consequently, different preparations will yield different values for standard curves. Ideal practice dictates that the standard curve should be prepared from the same preparation that the patient is receiving. The variation is greater between LMWHs because the manufacturers use alternative methods of chemical or enzymatic depolymerization of UFH. Generic heparin calibrators are available that can give comparable results to more specific preparations. Anti-Xa assays can be used to measure danaparoid and fondaparinux levels and preparation-specific standard curves must be used for them.

BOX 17.7 Do APTT and anti-Xa assays mirror the full clinical picture in heparin monitoring?

Although LMWHs are considered to have predominantly anti-Xa activity, the variation in manufacture results in some preparations having appreciable anti-IIa activity. These are usually the preparations that may interfere with coagulation screening tests. Measuring only anti-Xa activity may not provide a complete picture of the degree of anticoagulation of UFH or LMWH when anti-IIa activity is also present. Additionally, heparins induce the release of tissue factor pathway inhibitor, which will also contribute to anticoagulation by downregulating the initiation stage of coagulation.

Anti-Xa assays are more complex than APTT to perform by manual techniques. Most hospital laboratories have automated equipment capable of performing anti-Xa assays, yet they are considerably more expensive than APTT and may not be a financially viable proposition for laboratories with a low demand. (See also Box 17.7.)

Thrombin time

The concentration of thrombin used when the thrombin time (TT) is employed diagnostically makes the TT oversensitive to UFH. The TT will often be unmeasurable when the APTT indicates a therapeutic anticoagulant response. The TT can be modified to employ a higher concentration of thrombin for UFH monitoring. However, this requires having two types of TT in the repertoire, one of which has limited applications and results in resource wastage. Use of TT for UFH monitoring is largely historical and APTT is now used by the vast majority of laboratories that opt to monitor with a coagulation screening test.

Some LMWHs have sufficient anti-IIa activity to elevate a TT, and occasionally APTT, but not to such an extent that therapeutic ranges can be applied.

Protamine sulphate neutralization test

Protamine sulphate, which is strongly basic, binds to heparin, which is strongly acidic, and forms a stable salt lacking anticoagulant effect. This test is based on the TT and uses a range of protamine sulphate concentrations that are added to the patient's plasma before the addition of thrombin. The TT normalizes when all the UFH is neutralized, and the concentration of UFH is calculated from the amount of protamine sulphate required to produce this effect. The test can also be used to calculate the amount of protamine sulphate required to reverse the anticoagulant effect of UFH *in vivo*, for instance after cardiopulmonary surgery or haemodialysis. (See also Box 17.8.)

17.4.3 Laboratory detection of heparin-induced thrombocytopenia antibodies

Type II HIT occurs in about 3% of patients treated with UFH, and the haemostasis laboratory plays an important role in confirming the clinical diagnosis. The clinical probability of the presence of HIT II is assessed using the 4Ts scoring system, which you can see in Table 17.5. HIT antibodies are detected in the laboratory by immunological assays and platelet activation assays.

Immunological HIT assays

The most commonly used immunological assays for HIT antibodies are the indirect enzyme-linked immunosorbent assays (ELISA). Antibodies in patient plasma are captured by surface-bound PF4-heparin or PF4-polyvinylsulphate complexes. Binding of PF4 to polyvinylsulphate generates a cryptic autoepitope in the PF4 that is recognizable to HIT antibodies.

Cross references

You met the principle of indirect ELISA in the Method box in Chapter 15, which described its use in the detection of anticardiolipin antibodies.

The ELISAs do not quantify HIT antibodies against a standard curve. Rather, reagents are provided with batch-specific cut-off points for optical density values. Patient samples whose optical density exceeds the cut-off value are reported as positive for the presence of HIT antibodies. If a patient is positive, then the ELISA can be repeated in the presence of a high concentration of heparin. If this inhibits the reaction and causes a >50% reduction in the optical density value, it is considered characteristic of HIT antibodies.

In Chapter 15, you met a similar principle of swamping an antibody with an antigen excess to reduce its *in vitro* effect in the high phospholipid confirmatory tests used in lupus anticoagulation detection.

The ELISA assays have a high sensitivity for the presence of HIT antibodies (~90%), but a lower specificity for HIT II. This is because patients on heparin without clinical HIT II, and even healthy subjects, can generate positive results in the ELISA assays. Clearly, heparin-treated patients can make the antibodies without developing clinical symptoms of HIT II. In healthy patients, the apparent positivity may be due to a conformational change in PF4 upon binding to the microtitre wells that allows antibodies non-specific for PF4-heparin complexes to bind. Alternative immunological

BOX 17.8 *The link between heparin neutralization and fish*

Protamine sulphate was first isolated from salmon sperm, and subsequently from other fish. Patients with fish allergies can react violently to the drug, so it is administered slowly. Protamine sulphate is now primarily produced using recombinant biotechnology.

TABLE 17.5 **The 4Ts scoring for probability of the presence of HIT II.**

Category	2 points	1 point	0 points
Thrombocytopenia	>50% reduction or nadir ≥ 20 × 10^9/L	30–50% reduction or nadir 10–19 × 10^9/L	<30% reduction or nadir <10 × 10^9/L
Timing of reduction in platelet count	Days 5–10 or ≤day 1 with heparin exposure in past 30 days	>Day 10 or ≤day 1 if heparin exposure in past 30–100 days	<Day 4 (no recent heparin exposure)
Thrombosis or other sequelae	Proven thrombosis, skin necrosis, or acute systemic reaction after bolus of heparin	Progressive, recurrent, or silent thrombosis; erythematous skin lesions	None
Other potential causes of thrombocytopenia	Not apparent	Possible reasons	Definite reasons

Score 6–8—high probability of HIT (leads to replacement of heparin with DTI);
score 4–5—moderate risk for HIT (clinical judgement whether to alter treatment);
score 0–3—low risk for HIT.

assays are available that circumvent the latter scenario because their design does not involve binding of PF4–heparin complexes to a solid surface, and these have been shown to have improved specificity. A fluorescence-linked immunofiltration assay (FLIFA) immobilizes antibodies in the patient's serum on a nitrocellulose membrane, which are then detected fluorimetrically with PF4–heparin complexes conjugated with fluorescein isothiocyanate (FITC). An enzyme-linked immunofiltration assay (ELIFA) immobilizes the antibodies in the same way, but detects them using an enzymatic chromogenic reaction using peroxidase-linked, PF4–heparin complexes.

A strongly positive result indicates a greater likelihood of HIT II than a weak positive that minimally or moderately exceeds the cut-off optical density (OD) value. Cut-off values tend to be in the region of 0.4 OD units and results of >1.0 have been associated with an increased risk of thrombosis. Only IgG antibodies require detection.

The above assays are quite time-consuming and can take up to four hours to complete. A rapid gel particle-agglutination technique is available that employs polystyrene beads coated with PF4–heparin complexes that act as the solid phase. The beads are mixed with patient serum in a well containing a gel. Any HIT antibodies present in the patient serum bind to the immobilized PF4–heparin complexes. A secondary anti-human immunoglobulin antibody is used, which acts as a bridge between beads that are coated with HIT antibodies, leading to agglutination. The well is centrifuged for ten minutes and the results are interpreted visually. You can see in Figure 17.15 that if HIT antibodies are present, the red polystyrene particles agglutinate and remain on top of the gel after centrifugation. If HIT antibodies are absent, the polystyrene particles are centrifuged to the bottom of the gel.

Fully automated latex immunoassays employing polystyrene latex particles coated with purified monoclonal antibodies to PF4–heparin are available for performance on automated analysers. They give rapid results and have been shown to be useful as negative predictors of HIT II. A rapid (approximately 30 minutes) chemiluminescence immunoassay is also available that captures IgG HIT antibodies onto magnetic particles coated with PF4–polyvinylsulphate complexes and then goes through the washing and isoluminol labelling steps you met for the von Willebrand factor assays in section 14.2.6 of Chapter 14 and an ADAMTS13 activity assay in section 16.2.2 of Chapter 16.

Platelet activation assays

This group of assays is based on the platelet-activating capacity of PF4–heparin–antibody complexes. HIT antibodies bound to PF4–heparin complexes activate platelets via interaction of the antibody's tail with the FcγIIa receptor on the platelet surface. Patient serum is incubated in the presence of fresh donor platelets and heparin to activate the platelets and cause them to aggregate. The most commonly used method to detect that the platelets have been activated is a standard platelet aggregometer. If a patient demonstrates aggregation with heparin at a concentration of 0.1–0.5 IU/mL that is abolished in a separate test using heparin at a concentration of 100 IU/mL, the presence of HIT antibodies is confirmed. Figure 17.16 shows the aggregation plots from a patient with HIT antibodies.

Other endpoints can be used to detect the platelet activation, such as the measurement of the release of radioactive serotonin from the platelets or ADP release in lumiaggregometry. The serotonin-release assay is considered to be the 'gold standard', but it is technically demanding and use of radioactivity requires additional health-and-safety measures, so it is rarely routinely available. A flow cytometric method has been described where patient serum and donor platelets are incubated together with high and low concentrations of heparin. The low concentration of heparin should activate the platelets in the presence of HIT antibodies, but the activation is abolished with the high concentration. The platelets are then incubated with fluoresceinated

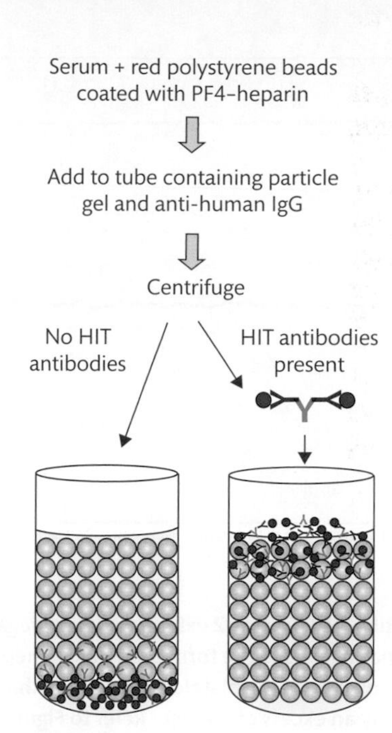

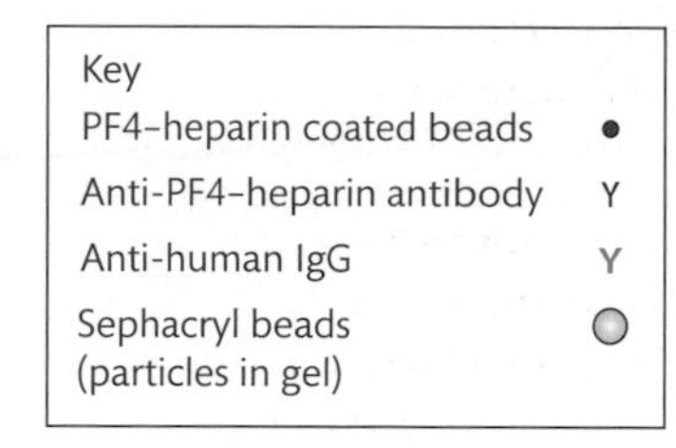

FIGURE 17.15

Gel particle-agglutination technique for the detection of HIT antibodies. Patient serum is incubated with PF4-heparin-coated polystyrene beads at 37°C for five minutes. The serum + bead mixture is then added to a well containing sephacryl beads and anti-human immunoglobulin antibody. The well is then centrifuged for ten minutes. HIT antibodies in the patient serum bind to the PF4-heparin complexes on the polystyrene beads, whereupon the anti-human immunoglobulin antibody binds to the HIT antibodies on different beads, leading to agglutination. When centrifuged, the agglutinated beads are too large to pass through the particles in the gel and remain at the top. If no HIT antibodies are present, no agglutination occurs and the beads are centrifuged to the bottom of the well. The results are read by eye, a red band at the top of the gel being positive and one at the bottom being negative. Weak antibodies can generate a diffuse red colour throughout the gel.

annexin V, which interacts with the anionic phospholipids exposed on the platelet surface after activation. The degree of binding of annexin V is assessed by flow cytometry. Platelets activated by PF4-heparin-antibody complexes can exhibit a 300-fold increase in binding compared to resting platelets. This method has been shown to have comparable diagnostic performance to that of the serotonin-release assay.

The functional assays have a lower sensitivity for HIT antibodies, but a higher probability of identifying cases that have the clinical hallmarks of HIT II. The sensitivity can be improved if the donor platelets are washed to remove potential interfering or masking factors in the donor's plasma, such as IgG, acutephase proteins (i.e. fibrinogen), or variation in calcium concentration. A major influence on test sensitivity is the significant variation between donors of the activation response of their platelets to HIT antibodies.

Some patients may be positive by immunoassay but not platelet activation assay, and vice versa; so both types of assay are necessary for HIT diagnosis.

SELF-CHECK 17.16

What are the two main types of assay used to detect HIT antibodies and what are their basic principles?

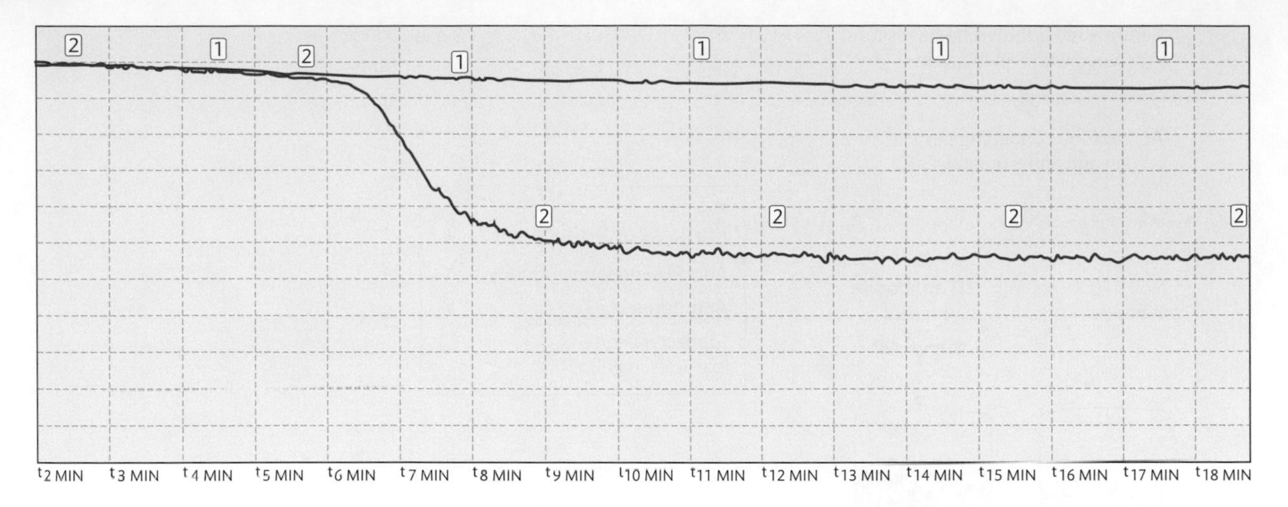

FIGURE 17.16
Platelet aggregation plots for the detection of HIT antibodies. Tracing 2 exhibits a full aggregation pattern because the therapeutic concentration of heparin leads to the formation of PF4-heparin complexes to which the HIT antibodies attach and then activate the platelets. Tracing 1 exhibits no aggregation because the antibodies are swamped by an excess of antigen. Refer to Figure 14.18 in Chapter 14 for details of platelet aggregation plots.

17.4.4 Laboratory assessment of DOACs

Although DOACs do not require regular laboratory monitoring, there are situations where assessment of the degree of anticoagulation may be required:

- when a patient is bleeding;
- following a suspected overdose;
- in patients requiring surgery in whom the drug may still be present;
- when a patient develops renal impairment;
- to assess compliance;
- at extremes of body weight.

pharmacokinetics
The absorption, distribution, metabolism, and excretion of a drug.

Trough levels may be useful in assessing for accumulation in very elderly patients.

Because the DOACs have relatively short half-lives, the results of any laboratory test will depend on when the drug was last taken and interpreted in light of dose, half-life, and factors affecting **pharmacokinetics**. Broadly speaking, peak plasma concentrations for rivaroxaban, apixaban, edoxaban, and dabigatran are in the range 100–400 ng/mL, and for trough concentrations, 10-150 ng/mL.

Cross reference
In Chapter 15, you met the Ecarin time in its role as a confirmatory test for lupus anticoagulants.

Laboratory assessment of direct thrombin inhibitors

Theoretically, the thrombin time should lend itself to assessing oral or parenteral DTI therapy. However, as you saw with UFH, the concentration of thrombin used in standard tests makes the TT oversensitive to DTIs in most reagents.

Although PT will be prolonged at high DTI concentrations, it has low sensitivity to clinical levels and should not be used to assess DTI anticoagulation. Bivalirudin and argatroban can be monitored similarly to UFH by using a therapeutic range of APTT ratios from 1.5 to 2.5 or 3.0. There are also similar reagent considerations because APTT reagents also have variable responses to all DTIs, and the type of analyser used can influence result variability. As you saw for UFH, the sensitivity of APTT reagents to other aspects of the patient's haemostasis compromises their effectiveness in monitoring/assessing DTIs. APTT reagents exhibit a curvilinear response to all DTIs and give a poor indication of anticoagulant response at high and low doses. Thus, a high APTT (i.e. ratio >3.0) can indicate a supratherapeutic level, but underestimate the degree of over-anticoagulation. A normal APTT does not exclude a low drug level, although would normally indicate that a therapeutic level has not been reached. These issues can be resolved by using the Ecarin fraction from the venom of the saw-scaled viper to activate *in vitro* coagulation via prothrombin.

Ecarin has been shown to give a linear response to bivalirudin, argatroban, and dabigatran over a wide range of concentrations and can be used in three ways for monitoring purposes:

- comparing the raw Ecarin clotting time to a therapeutic range;
- using Ecarin clotting times of spiked normal plasma to generate a standard curve which can be used to quantify DTI levels;
- using the thrombin generated from the Ecarin activation of prothrombin to cleave a chromogenic substrate from which a standard curve can be prepared to quantify DTI levels. Thrombin is neutralized by the DTI and the amount of residual thrombin is inversely proportional to the DTI concentration.

The clotting-based Ecarin assays are reliant on sufficient levels of prothrombin in the patient's plasma. An alternative snake venom-based assay for monitoring DTIs is the **prothrombinase-induced clotting time (PiCT)**. The factor V activator from Russell's viper venom activates the patient's factor V, which is then reacted with reagents containing factor Xa, phospholipid, and calcium ions to form the prothrombinase complex and promote clot formation. The time taken to clot is directly proportional to the DTI concentration. A standard curve is generated by spiking normal plasma with different concentrations of the DTI that the patient has received. The PiCT is a stable and reliable assay for levels of DTIs throughout and above therapeutic concentrations, but it can be affected by patient levels of factors II, V, and X below 25% of normal and by lupus anticoagulants. Patients on concomitant oral anticoagulant therapy will have reduced levels of factors II and X and cannot be monitored by PiCT. The assay can also be used for direct FXa inhibitors, although its main use is as a more reliable clotting-based assay than APTT for UFH monitoring.

The most widely used assay in the UK for quantifying DTI levels, particularly dabigatran, is the Hemoclot DTI assay based on a dilute TT. The test plasma is first diluted either 1:2 or 1:8 in buffer and then one volume of the dilution is mixed with two volumes of normal pooled plasma, which supplies prothrombin and fibrinogen. The diluted plasma is incubated to reach 37°C and then a human thrombin calcium reagent is added to clot the plasma. The clotting time is compared to a standard curve to generate the quantitative result.

Laboratory assessment of direct factor Xa inhibitors

There is a linear concentration response of PT to rivaroxaban and apixaban, but marked between-reagent sensitivity, with some reagents giving normal results in patients with therapeutic drug levels. Converting PT to INR should not be employed, as it increases the variability, since ISI is specifically a calibration for a different anticoagulant mechanism. Rivaroxaban, apixaban, and edoxaban will also prolong APTT, which will vary between reagents, but PT is generally more sensitive to the presence of direct FXa inhibitors. Rivaroxaban and edoxaban tend to have a more pronounced effect on PT and APTT than apixaban. Each laboratory should be aware of the relative sensitivity of their PT and APTT reagents to direct FXa inhibitors, since not all of them are suitable for even a crude estimation of the drug levels.

Anti-Xa assays are the method of choice for determining rivaroxaban, apixaban, and edoxaban levels. Similar to UFH, LMWH, danaparoid, and fondaparinux, the standard curves must be preparation-specific; that is, you cannot measure a direct FXa inhibitor from an assay calibrated for any of those anticoagulants. Anti-Xa reagent sets are available specifically for quantifying direct FXa inhibitors that do not contain endogenous antithrombin. Direct FXa inhibitors are antithrombin-independent so it is not an essential component, and it can lead to overestimation, since it possesses innate anti-Xa activity.

17.4.5 Monitoring antiplatelet therapy

The effects of antiplatelet therapy can be monitored by the platelet aggregometry techniques used for detecting platelet function disorders. Although effective at demonstrating an overt physiological response, they may not detect more subtle changes of altered platelet function. Biochemical assays can be used as surrogate markers by measuring receptor changes/occupancy upon drug binding, enzyme function, or changes in surface proteins after activation. The biochemical assays may detect more subtle changes, but may not reflect the degree of anticoagulation. Flow cytometry can be used to detect some of these changes. These techniques are time-consuming and technically demanding, and often require fresh platelets from the patient which cannot be routinely available. Consequently, monitoring of antiplatelet therapy is not routinely undertaken.

The PFA-100 and 200, which you met in Chapter 14 as a screening tool for primary haemostatic disorders, has been used to monitor antiplatelet therapy, particularly aspirin in terms of identifying non-compliance and **aspirin resistance**. Some patients appear under-responsive to standard doses of aspirin according to their PFA-100/200 results with the collagen-epinephrine cartridge. However, it is unclear whether this *in vitro* phenomenon of apparent resistance to the effects of aspirin equates to a genuine clinical failure of aspirin to protect the patient from thrombotic events. Many patients who appear aspirin-resistant from their PFA-100 results demonstrate a reduced response to arachidonic acid by light-transmission platelet aggregometry (i.e. aspirin-sensitive). The standard PFA-100 collagen-ADP cartridge lacks sensitivity to the platelet aggregation inhibitory effect of $P2Y_{12}$ inhibitors and an alternative, the $P2Y^*$ cartridge, can be used for this purpose. The activation of platelets by collagen in the standard collagen–ADP cartridge, together with residual activation potential of the ADP via $P2Y_1$, appears to be sufficient to overcome the $P2Y_{12}$ inhibition in many patients. The $P2Y^*$ cartridge comprises collagen coated with ADP, prostaglandin E1, and calcium chloride. The inconsistent sensitivity of PFA-100 to the effects of antiplatelet drugs is due to the many variables that impact on results, such as platelet count, anaemia, and inherent platelet reactivity to collagen. High levels of von Willebrand factor have been shown to affect PFA-100/200 results and may be the reason why some patients appear to be aspirin-resistant by this method.

True aspirin resistance occurs when the drug fails to completely inhibit thromboxane formation. Thromboxane A_2 is unstable under physiological conditions, so measurement of its metabolite, **thromboxane B_2**, as an indicator of cyclooxygenase activity is the best way to detect clinical aspirin resistance. Thromboxane B_2 can be assayed by ELISA techniques.

A POCT analyser called VerifyNow™ has been designed specifically for the rapid monitoring of antiplatelet agents. Similar to the PFA-100/200, it is a cartridge-based, semi-automated device that tests whole-blood samples. The VerifyNow™ was originally developed to assess the effects of GpIIbIIIa inhibitors by measuring the agglutination of fibrinogen-coated beads by platelets that have been stimulated by the thrombin receptor agonist peptide (TRAP). The TRAP-activated platelets bind to the beads and fall out of solution, and the rate of change in light transmittance is reported as platelet-aggregation units (PAU). Cartridges using arachidonic acid as the agonist are available for aspirin monitoring, the reactivity reported as aspirin reaction units (ARU). The effect of $P2Y_{12}$ inhibitors on the $P2Y_{12}$ ADP receptor is monitored using a cartridge containing ADP as the agonist and reported in $P2Y_{12}$ Reaction Units (PRU). The principle of the VerifyNow™ system is shown in Figure 17.17. An analyser based on whole-blood impedance aggregometry called Multiplate® employs ADP aggregometry to assess effects of $P2Y_{12}$ inhibitors, TRAP aggregometry for GpIIbIIIa inhibitors, and arachidonic acid aggregometry for aspirin.

Direct measurement of drug levels is rarely helpful because of their pharmacological behaviour. For instance, the effects of aspirin are immediate and irreversible and remain long after the drug has been cleared from the plasma. Abciximab binds to platelets very quickly and is released very slowly, so most remains platelet-bound and measurement of plasma levels does not reflect the antiplatelet effect.

Although there are tests that demonstrate inhibition of platelet function by antiplatelet drugs, an important question is whether they are effective, or even necessary. More data from clinical trials are needed to ascertain whether specific limits of laboratory values can be used to alter dosing decisions such has been achieved with the INR in VKA anticoagulant monitoring. Less than 5% of patients are poor aspirin responders, some of which are due to non-compliance, so laboratory monitoring is difficult to justify. Since platelet function testing does not predict clinical events, regular testing does not guarantee effective monitoring. Existing tests can, however, indicate in most patients that the drugs are reaching their targets and exerting the desired pharmacological effect.

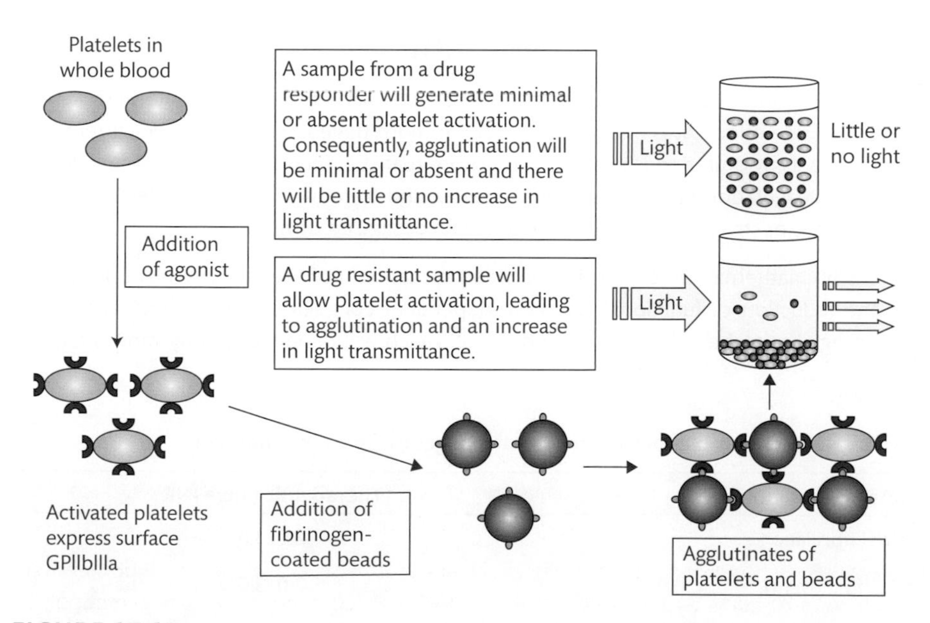

FIGURE 17.17

Principle of the VerifyNow™ system for monitoring antiplatelet agents. The appropriate agonist is added to a whole blood sample to maximally activate the platelets and promote surface expression of the fibrinogen receptor GpIIbIIIa. The blood is mixed with fibrinogen-coated beads and leads to the formation of platelet-bead agglutinates which fall out of solution. Light transmittance through the sample increases in proportion to the degree of agglutination.

17.5 Management of VKA anticoagulant therapy

VKA anticoagulants, normally in the form of warfarin in the UK, have been in use since the 1940s. Their clinical use declined in the 1960s due to concerns over their effectiveness in the management of myocardial infarction. However, improvements in safety due to analytical standardization and the demonstration of improved patient outcomes from better-designed clinical trials, resulted in increased prescribing for medical conditions where anticoagulant therapy is known to be clinically beneficial.

Anticoagulant therapy is considered effective if it prevents stroke and thromboembolism in patients with AF or the recurrence of thrombosis in patients with a confirmed venous thromboembolic event. To achieve this, VKA anticoagulant therapy is routinely monitored by the INR, which you met earlier in section 17.4.1. Adoption of this method by the WHO has resulted in a global improvement in the safety and effectiveness of VKA anticoagulant therapy. Depending on the clinical condition, a patient is assigned an INR target range and the VKA anticoagulant dosage is adjusted to maintain the INR within that range, as detailed in the British Committee for Standards in Haematology (BCSH) guidelines. Therapeutic ranges for the main conditions for which VKA anticoagulant therapy can be used are given in Table 17.6.

Crucial to safe and effective therapy is patient counselling to ensure they understand the reason for therapy, associated risks, and the 'do's' and 'don'ts' whilst on anticoagulation therapy. Good communication links between patients and those managing anticoagulant therapy are essential.

Within each therapeutic range a patient will have a target INR; for instance, a patient with a therapeutic range of 2.0–3.0 will have a target INR of 2.5. If the dosing practitioner aims to keep the patient at the target INR, then they are less likely to fall outside the therapeutic range.

17.5.1 Need for management

Unlike the majority of prescription medication that does not need regular monitoring for effectiveness or side effects, it is essential for safety in those patients prescribed the current VKA anticoagulants. Monitoring is necessary to ensure that therapeutic reduction of functioning vitamin K coagulation factors has been achieved, and to reduce the risk of life-threatening haemorrhage or stroke by timely intervention with dosage adjustment.

There is no mathematical relationship between the amount of VKA anticoagulant prescribed and the reduction of functional vitamin K-dependent coagulation factors. VKA anticoagulant dosage has to be tailored to the individual, which will be influenced by various factors,

TABLE 17.6 Examples of INR therapeutic ranges for specific conditions.

Clinical condition	Therapeutic range INR
Atrial fibrillation	2.0–3.0
VTE (first event)	2.0–3.0
VTE (recurrent on warfarin)	3.0–4.0
Mechanical mitral valve replacement	3.0–4.0
Antiphospholipid syndrome (with VTE)	3.0–4.0

including age, lifestyle, diet, alcohol intake, genetic factors, ethnicity, other medical conditions, and, importantly, interaction of other medication in association with liver function.

VKA anticoagulant clearance from the body occurs in the liver via **cytochrome P450** catalysed oxidation. Many other drugs influence the liver and activity of the various cytochromes, which can increase VKA anticoagulant clearance and reduce the anticoagulant effect, or inhibit clearance and enhance the anticoagulant effect. Other influences of therapy involve drugs that inhibit the intestinal flora that produce vitamin K, notably antibiotics.

cytochrome P450
A large superfamily of haem-containing proteins that take part in electron transfer reactions.

All these factors contribute to the requirement to regularly monitor a patient's INR and adjust the VKA anticoagulant dosage accordingly. National targets exist to maintain patients in the therapeutic range and reduce adverse events.

17.5.2 Treatment risks

The specific risks of haemorrhage and thrombosis associated with anticoagulant medication have been detailed. However, the experience of most anticoagulant clinics is that these risks are exacerbated by other underlying causes. Most common is the cessation or introduction of medication for other clinical conditions without reference to the interaction with VKA anticoagulants. Patients with additional medical conditions that affect diet and the gastrointestinal tract can have adverse effects on vitamin K levels and VKA anticoagulant absorbance. Patient non-compliance, either in taking medication or attending for monitoring, can result in very high INRs and an increased risk of life-threatening **cerebral haemorrhage**.

cerebral haemorrhage
A rupture of a blood vessel in the brain or leakage of blood from a vessel, resulting in impaired oxygen supply to the area and subsequent brain damage of variable severity.

Although rare, there are also some genetic factors that can influence the effectiveness and management of VKA anticoagulant therapy. The term 'warfarin resistance' has been attributed to patients requiring significantly higher-than-average dosage to achieve a therapeutic INR. This has now been identified as a genetic mutation in the vitamin K epoxide reductase enzyme at the warfarin-binding site (vitamin K epoxide reductase complex subunit 1—VKORC1), which reduces its inhibitory effect on vitamin K metabolism. In contrast, the term '**warfarin sensitivity**' has been attributed to patients requiring significantly less than the average dosage to achieve a therapeutic INR. This has now been identified as being due to polymorphism variants in cytochrome P450 CYP2C9, which results in slower breakdown and clearance of the more potent enantiomer S-warfarin.

17.5.3 Drug interactions

Drug interactions have been discussed in other sections of this chapter in relation to the safety and effectiveness of anticoagulant therapy. If clinically indicated, most medication can be prescribed alongside VKA anticoagulants, but close INR monitoring is required to ensure VKA anticoagulant dosage adjustment occurs at the beginning of any interacting effect.

Interactions fall into three groups:

- enhanced VKA anticoagulant effect with a risk of bleeding;
- reduced VKA anticoagulant effect with a risk of thrombosis;
- variable effect, which can enhance or reduce response.

Also requiring consideration is other anticoagulant medication, predominantly antiplatelet drugs. Unless prescribed by a clinician, these should not be taken concurrently with VKA anticoagulants as they increase the risk of bleeding due to the combined effects of impaired platelet function and reduced functional coagulation factors. (See also Case study 17.2.)

CASE STUDY 17.2 Effect of antibiotics on VKA anticoagulant therapy

Patient history

- A 65-year-old man had been on warfarin for four months for atrial fibrillation.
- His target INR therapeutic range was 2.0–3.0.
- He had been compliant and had a stable INR (mean INR: 2.7).
- He then presented to his Accident and Emergency department feeling 'unwell' with 'flu-like' symptoms.
- His blood test results are below in Results 1.

Results 1

- FBC results were as follows:

Index	Result	Reference range (see front of book)
WBC (× 10^9/L)	25.5	(RR: 4.0–11.0)
Haemoglobin (g/L)	125	(RR: 120–160)
MCV (fL)	85	(RR: 80–100)
Platelets (× 10^9/L)	510	(RR: 150–400)
C-reactive protein (mg/L)	129	(RR: 0–5)
INR	2.3	(TR: 2.0–3.0)

Significance of results 1

His white cell count was markedly elevated, which together with the elevated C-reactive protein (a marker of infection/inflammation) suggests an infection. The elevated platelet count was likely a reactive response to the infection. His INR remained within the therapeutic range. He was diagnosed as having bacterial endocarditis, an infection of the heart's inner lining or the heart valves that can damage or destroy the heart valves. He was started on antibiotics and admitted. His INR was checked two days later. There was no evidence of any bleeding.

Results 2

- INR 12.2 (TR: 2.0–3.0).

Significance of results 2

His INR was dangerously high, sufficient to potentially precipitate a severe bleeding event. Antibiotics are known to interact with VKAs, the main mechanism being disruption of intestinal flora that synthesize vitamin K. Loss or significant reduction in these bacteria removes a constant source of vitamin K that contributes to the balance achieved by the ongoing drug dose such that it is tipped in favour of the drug, which consequently exerts a greater effect. Some antibiotics also inhibit cytochrome P450. The patient's warfarin was stopped and he was given 2.0 mg of vitamin K. His INR was checked 24 hours later.

Results 3

- INR 1.8 (TR: 2.0–3.0).

Significance of results 3

His INR was now slightly below the therapeutic range, indicating that the vitamin K dose in conjunction with warfarin cessation had permitted synthesis of vitamin K-dependent factors sufficient to reduce the INR. Warfarin has a half-life of 20–60 hours, so the dose of vitamin K has to be sufficient to account for residual warfarin. Patients in situations such as this require a reduced dose of warfarin, and regular monitoring, until the interacting drug is withdrawn.

17.5.4 Contraindications (liver disease)

Other medical conditions should be carefully considered prior to commencing anticoagulant therapy to assess clinical benefit versus clinical risk. It is particularly important to perform a coagulation screen prior to starting therapy in case there is an underlying coagulation defect or abnormal liver function, which would make achieving a safe stable dosage difficult because coagulation factors are mainly produced in the liver.

17.5.5 Dosing regimes

There are national guidelines both from professional bodies and the National Patient Safety Agency on initial dosage (loading doses) regimes, as well as algorithms to assist with subsequent prescribing. A typical dosing algorithm for a patient with AF is shown in Table 17.7.

Cross reference

The commonly used guideline documents are cited in the Further reading list at the end of this chapter.

The main VKA anticoagulant, warfarin, is manufactured as 0.5, 1.0, 3.0, and 5.0 mg colour-coded tablets. It is recommended that patients are provided with a range of tablets to enable 'fine tuning' of dosage, prevent the need to break tablets in half, and to provide a consistent dosage throughout the week.

Patients fall into two categories:

- patients with VTE or prophylaxis for VTE, where it is critical to obtain a therapeutic INR as soon as possible. LMWH will need to be continued until two INRs are within the therapeutic range;
- patients with AF who can be started on a low dose and then increased gradually until their INR is in the therapeutic range.

SELF-CHECK 17.17

Why must VKA anticoagulation be regularly monitored?

TABLE 17.7 Example of warfarin dosing initiation algorithm for a patient with AF.

INR	Day	Dose (mg)
1.1	1	6
	2	6
	3	3
If, e.g., 2.0–2.5	4	3 mg daily, test INR in 3 days

Duration of VKA anticoagulation

The reason a patient is receiving VKA anticoagulation dictates not just the therapeutic range and target INR, but also the duration of anticoagulant therapy. The British Society for Haematology recommends the following:

- at least six weeks' anticoagulation is recommended after calf vein thrombosis;
- at least three months' anticoagulation is recommended after proximal DVT or PE;
- anticoagulation for three months should be sufficient for patients with temporary risk factors (i.e. combined oral contraceptive, plaster cast) and a low risk of recurrence (i.e. surgery);
- long term anticoagulation should be considered for patients with a first unprovoked, or recurrent unprovoked, proximal DVT or PE.

17.5.6 Service models

Service models by healthcare professionals

Over the past 15 years, service models of care for the provision of anticoagulant therapy have undergone major changes, and continue to do so due to the increasing numbers of patients requiring therapy and today's prolonged life expectancy. There are a number of approaches to service delivery that are very dependent on local needs and geographical area.

Traditionally, patients were seen in a hospital-based, medical consultant-led, face-to-face clinic service. This model of service is still provided in some hospitals, but in most this service has undergone significant change. Although still under consultant clinical direction, the service is delivered by a range of healthcare professionals—i.e. nurses, biomedical scientists, or pharmacists—often working as multidisciplinary teams.

Many hospital-based clinics now offer face-to-face clinics for a patient's initial visit to explain the therapy and provide advice for any problems or queries. Subsequently, patients are either bled in the community (usually provided by their GP's surgery), or attend hospital-based phlebotomy; the samples are then sent to the local haematology laboratory that provides the INR and anticoagulant management service. Information about changes in dose, or maintenance of current dose, is returned to the patient by post, telephone, or text. Many patients are given a yellow booklet where the INR and any dosage changes are recorded at every visit. Patients are advised to keep it with them at all times. Most hospital-based anticoagulant services use a computerized anticoagulant management system as a tool to manage all aspects of the service, including dosage decisions, appointments, length of treatment, and information about interacting drugs. This provides a comprehensive record and audit tool on all aspects of service.

In some areas, GPs provide an anticoagulant management service either in conjunction with the local laboratory to provide INR testing or using one of the POCT devices available for INR testing.

It is important that the same level of quality assurance is applied to all aspects of the service, whether delivered as a hospital- or GP-based service. A flow chart for dosing by healthcare staff is given in Figure 17.18, showing the different models that can be adopted.

Service models by patients

With the increased availability and reliability of POCT devices, there is evidence that self-testing is a viable alternative for patients suitably trained in testing procedures, quality assurance, and procedures for clinical supervision. Relatively small numbers of patients currently elect to self-test and contact their local anticoagulant management service with their INR result, which then provides warfarin dosage instructions. Figure 17.19 presents a flow chart for patient self-testing using POCT devices.

There have been some trials on patients self-testing and self-dosing as more patients seek autonomy and control over their medical condition and treatment. Safety and quality assurance aspects are paramount, and there are now national guidelines on patient self-testing and monitoring to address these issues.

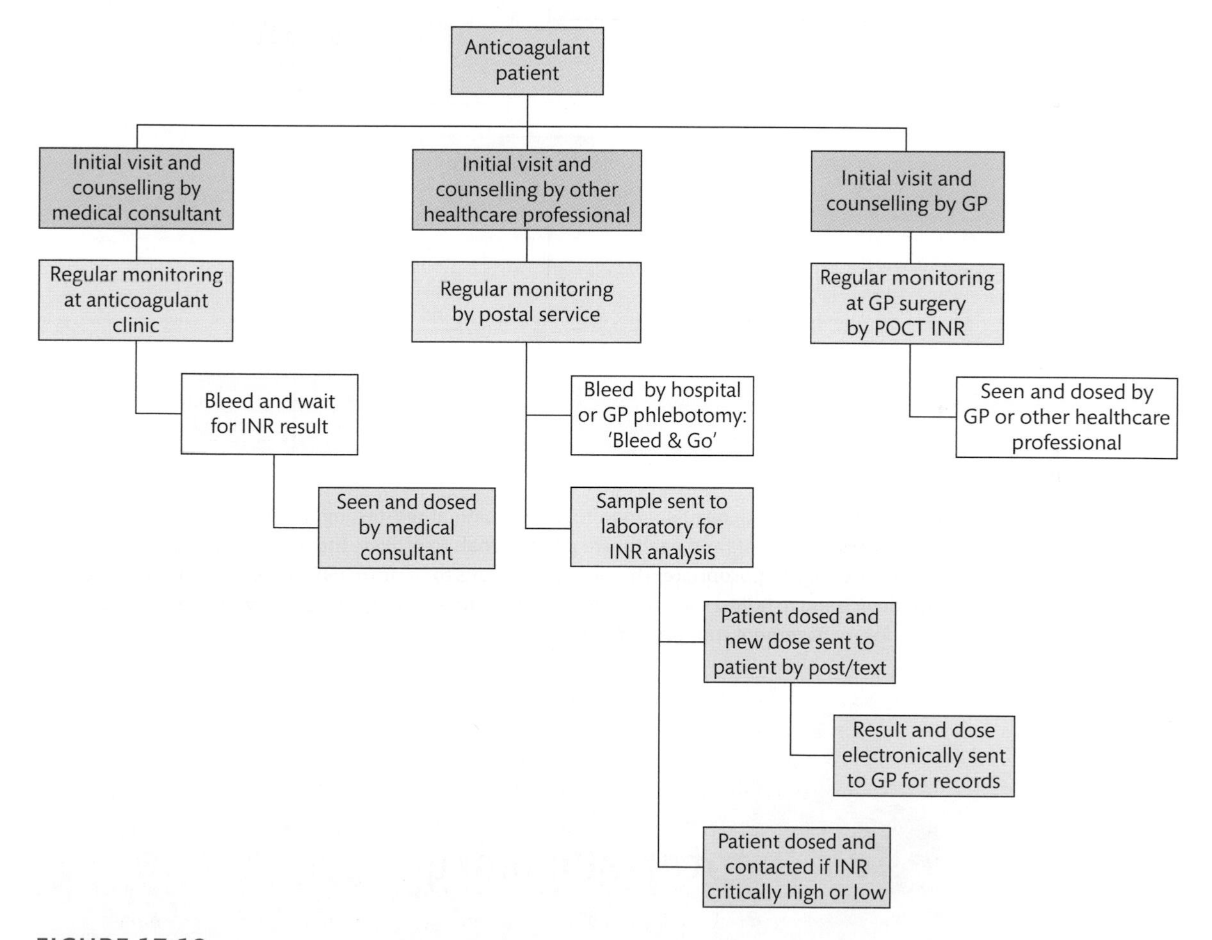

FIGURE 17.18

Service models of anticoagulant management by healthcare professionals. The flowchart illustrates options for the delivery of anticoagulant services provided either by a medical consultant, other healthcare professionals, or by a GP. They range from traditional 'face-to-face' clinics to community-based sample collection, with postal and phone dosing service. It also identifies the integral role of the diagnostic laboratory service within each model.

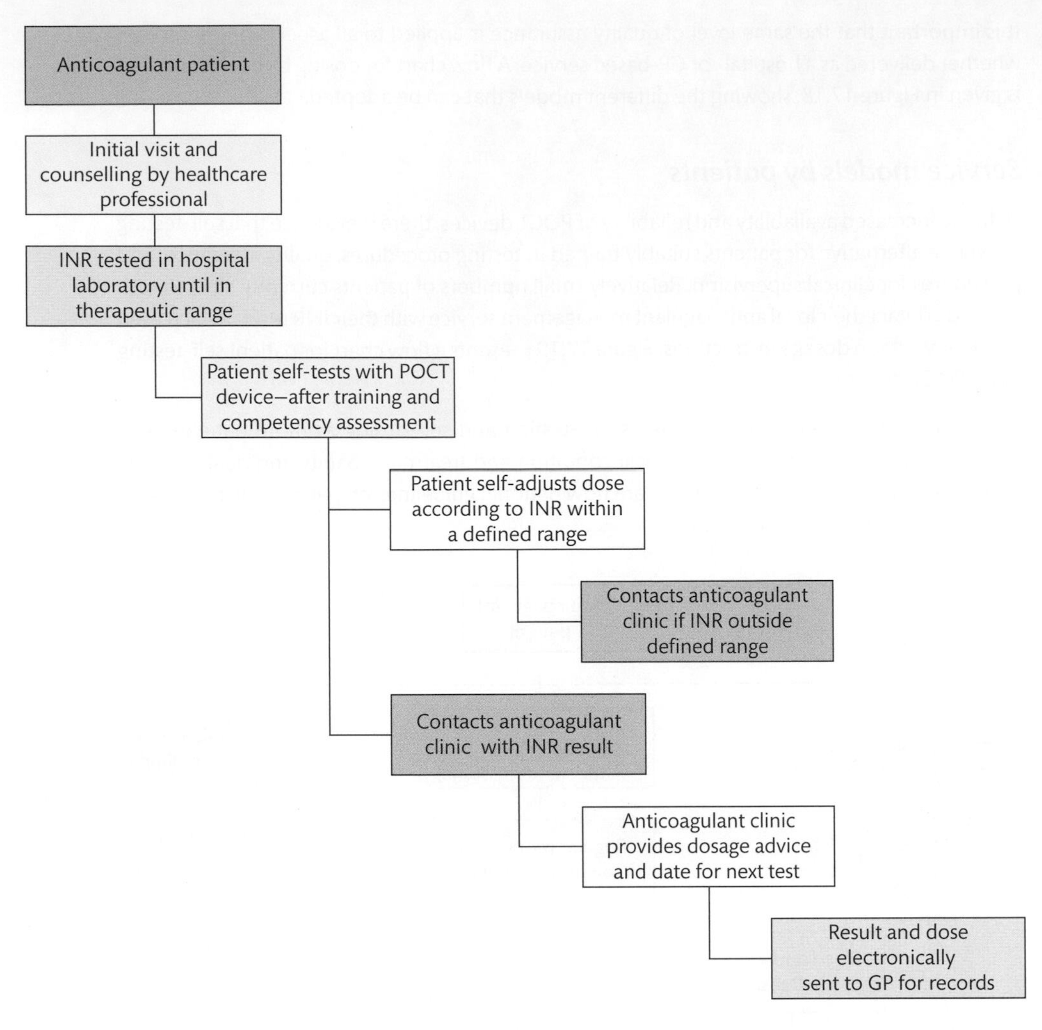

FIGURE 17.19

Service models of anticoagulant monitoring via patient self-testing. The flowchart illustrates procedures for patients and healthcare professionals to provide the option of patient self-testing and self-dosing, if appropriate. This enables patients to perform INR at home using a point-of-care device and to adjust their own dose within defined limits, interacting with the anticoagulant clinic for management support and advice.

Chapter summary

General

- Anticoagulant therapy is provided to prevent the genesis or growth of a thrombus and to reduce the risk of re-thrombosis.
- Aetiology of thrombosis formation and composition is dependent on its occurrence in a vein or an artery.

- Anticoagulants work indirectly or directly against coagulation proteins to reduce the risk of thrombosis.

Diagnosis of VTE

- D-dimer levels can be used as a negative predictive indicator of VTE, and so rapidly exclude a diagnosis of DVT.
- Confirmation of VTE requires radiological investigation.

Anticoagulants

- UFH or LMWH provide an immediate anticoagulant effect, and are primarily used to provide cover until VKA anticoagulant therapy is established.
- Monitoring of VKA anticoagulation is essential to reduce the risk of catastrophic haemorrhage or further thrombosis. Overdosing can lead to the former and underdosing to the latter.
- DOACs are alternatives to VKA anticoagulants that have more predictable effects and do not require regular monitoring

Laboratory monitoring

- VKA anticoagulant therapy is monitored using the INR system.
- Effective ISI calibration at international, manufacturer, and local levels is crucial to generating accurate INR results.
- UFH is usually monitored by APTT, but this test can be markedly affected by concomitant haemostatic abnormalities or anticoagulant therapies. APTT reagents from different manufacturers vary in their sensitivity to UFH, and therapeutic ranges should be locally assigned with respect to UFH levels measured by anti-Xa or protamine sulphate titration assays.
- LMWH rarely needs monitoring and assessment is performed using anti-Xa assays utilizing the same preparation administered to the patient to generate the standard curve. Anti-Xa assays can be used to monitor UFH.
- Type II HIT is a clinically significant complication of heparin therapy. It is detected in the laboratory using immunological assays and platelet activation assays.
- PT and APTT have markedly variable sensitivity to DOACs.
- DTIs can be measured with Ecarin venom-based assays, PiCT, or dilute thrombin time.
- Direct FXa inhibitors can be measured with modified anti-Xa assays.
- Antiplatelet therapy is not routinely monitored. Platelet aggregometry, functional markers, PFA-100/200, thromboxane B_2 levels, VerifyNow™, or Multiplate® can be used when necessary.

Management of anticoagulation

- Genetic mutations in the form of vitamin K epoxide reductase enzyme at the warfarin-binding site (VKORC1) and polymorphism variants in cytochrome P450 CYP2C9 have been identified as causes of anticoagulant management difficulties in some patients.

- A number of service models are available for the delivery of anticoagulant management; currently, most common is a postal service. Use of confidential electronic media is under development.
- New therapeutics may change monitoring and service requirements for the management of VTE in the future.

Discussion questions

17.1 Why do some anticoagulants need to be monitored and not others?

17.2 Why does the site of thrombus formation influence selection of anticoagulant therapy?

17.3 How do laboratory assay principles and limitations affect their effectiveness in monitoring anticoagulant therapy and what measures are taken to overcome or limit these effects?

17.4 What developments could lead to changes in how anticoagulant management services are provided?

Further reading

- Baglin T and Keeling D. ***Antithrombotic agents***. In: Hoffbrand AV, Higgs DR, Keeling DM, and Mehta AB (eds). ***Postgraduate Haematology***, 7th edn (pp. 820–9). Wiley Blackwell, Oxford, 2016.
- Baglin T, Keeling D, Kitchen S; British Committee for Standards in Haematology. ***Effects on routine coagulation screens and assessment of anticoagulant intensity in patients taking oral dabigatran or rivaroxaban: Guidance from the British Committee for Standards in Haematology***. *British Journal of Haematology* 2012: **159**: 427–9.
- Gosselin RC, Adcock DM, Bates SM, Douxfils J, Favaloro EJ, Gouin-Thibault I, Guillermo C, Kawai Y, Lindhoff-Last E, and Kitchen S. ***International Council for Standardization in Haematology (ICSH) recommendations for laboratory measurement of direct oral anticoagulants***. *Thrombosis and Haemostasis* 2018: **118**: 437–50.
- Jennings I, Kitchen D, Keeling D, Fitzmaurice D, Heneghan C, and the BCSH Committee. ***Patient self-testing and self-management of oral anticoagulation with vitamin K antagonists: Guidance from the British Committee for Standards in Haematology***. *British Journal of Haematology* 2014: **167**: 600–7.
- Keeling DM, Mackie IJ, Moody A, and Watson HG (for the BCSH). ***Diagnosis of deep vein thrombosis in symptomatic outpatients and the potential for clinical assessment and D-dimer assays to reduce the need for diagnostic imaging***. *British Journal of Haematology* 2004: **124(1)**: 15–25.
- Keeling D, Baglin T, Tait C, Watson H, Perry D, Baglin C, Kitchen S, Makris M, and British Committee for Standards in Haematology. ***Guidelines on oral anticoagulation with warfarin: 4th edition***. *British Journal of Haematology* 2011: **154**: 311–24.

- Kitchen S, Gray E, Mackie I, Baglin T, Makris M; BCSH committee. ***Measurement of non-coumarin anticoagulants and their effects on tests of haemostasis: Guidance from the British Committee for Standards in Haematology***. *British Journal of Haematology* 2014: **166**: 830–41.

- Moore GW, Thomson GA, and Harrington DJ. ***Warfarin monitoring; standard practice and beyond***. Nova Science Publishers Inc., New York, 2012.

- National Patient Safety Agency. ***Actions that can make anticoagulant therapy safer: Alert and other information***. Available at: http://nrls.npsq.nhs.uk/resources/? entry-id45 = 59814.

  Crucial reading regarding required competencies of healthcare professionals involved in anticoagulant therapy.

- Watson H, Davidson S, and Keeling D. ***Guidelines on the diagnosis and management of heparin-induced thrombocytopenia: 2nd edition***. *British Journal of Haematology* 2012: **159**: 528–40.

Answers to self-check questions and case study questions are provided as part of this book's online resources.

 Visit www.oup.com/he/moore-fbms3e.

PART 5

Case Studies

18

Case studies in haematology

Andrew Blann, Gavin Knight, and Gary Moore

Although the case studies in each of the chapters focus on the particular message of that chapter, haematology and its diseases clearly encompasse many different themes, not merely red cells, white cells, or haemostasis alone. Furthermore, haematology is just one of several biomedical sciences, all of which have their place in patient care. Here we present three cases that will bring together many of these diverse characteristics of our discipline, and demonstrate how these different aspects interact in defining a diagnosis and supporting clinical management.

Learning objectives

After studying these case studies, you should confidently be able to:

- appreciate the multi-faceted nature of haematology;
- outline the part played by haematology in diagnosis;
- give examples of the value of haematology in the management of disease;
- describe how scientists from different disciplines together provide essential information to the clinician.

CASE STUDY 18.1

Day 1

A General Practitioner (GP) visits a 67-year-old woman at home as she is unable to come to the practice. He finds her in bed looking unwell with a persistent fever. Her past medical history includes morbid obesity (body mass index > 40 kg/m^2), an artificial hip joint on the left, and an artificial knee joint on the right. The operations were done at different times in the past three years, but these give occasional pain. The remaining native joints are osteoarthritic.

The combined mobility problems and pain have caused her to be 90% bed-bound for the past four months. Her current medications include amoxicillin for a presumed chest infection and paracetamol for arthralgia. He calls for an ambulance to take her to Accident and Emergency. Examination on arrival finds her lucid but distressed, pyrexial with a temperature of 40°C, tachypnoeic (rapid breathing—25 breaths/minute), tachycardic (pulse rate 105/minute), and with bilateral sacral sores with purulent (pus-like) discharge. Systolic and diastolic blood pressures are 138 mm Hg and 86 mm Hg, respectively, and an electrocardiogram (ECG) demonstrates sinus tachycardia. Both ankles are swollen. Routine haematology and biochemistry bloods are taken and she is assessed for risk of venous thromboembolism (VTE). As this risk is found to be high, she is started on 40 mg of a common low-molecular-weight heparin (LMWH) daily. The routine haematology full blood count (FBC) test results are returned from the path lab, as shown in Table 18.1. Routine biochemistry is unremarkable, all indices being within their reference ranges.

Question CS 1.1 Comment on the results of the FBC.

Based on this, blood is drawn for blood cultures and D-dimers, and she is admitted to a medical ward. Her oral antibiotics, paracetamol, and LMWH are continued. A chest X-ray is ordered and swabs of her sacral sores are sent to microbiology for routine analysis.

Question CS 1.2 Why were these blood tests and chest X-ray ordered, and the sores swabbed?

Day 2

She has a restless night but in the morning feels a little better. The D-dimer result from Day 1 is 575 units/ml (reference range <500 units/mL). She has little appetite but is able to drink, which she does profusely, claiming to be hot and thirsty. An emergency physiotherapist attempts an assessment, but the patient is uncooperative. Nursing staff attempt to get her into a chair but she resists, although she is able to sit up in bed. A catheter is inserted into her urethra, and fluid intake and urine production are monitored. Her breathing is still rapid (24/minute), her heart rate has come down to 80/minute, but she still has a high temperature (39°C). She is taken to the radiology department for a repeat chest X-ray, which is clear. That evening the microbiology results come back, which reports a borderline positive blood culture and positive swabs with a heavy infection of common commensal organisms. The dose of the oral antibiotic is increased and a second-line antibiotic is added. Routine bloods are repeated. A nurse notes that her urine output is high.

Question CS 1.3 Discuss the results of Day 2's investigations.

Day 3

The patient has a poor night and has developed a wheezy cough. During the morning, the previous night's blood and other results are reviewed by the Registrar. Finding persistent tachypnoea, tachycardia, and high temperature, the Registrar orders a second blood culture, another FBC, and a test for haemoglobin A_1c. The patient's sores are dressed and topical antibiotics applied.

Question CS 1.4 Justify the Registrar's actions.

The FBC result from the third day is shown in Table 18.1. The haematology laboratory is trialling some advanced software that flags up cases where there is a linear change in an index over three samples where those results are outside the reference range, and flags the white blood cell count and the neutrophil count. The duty scientist bleeps the Registrar and reports a rising neutrophil leucocytosis, and looks at the film, but is unable to find other abnormalities. The Registrar is concerned. He stops the current oral antibiotics and initiates an intravenous antibiotic in saline every six hours. The haemoglobin A_1c result is 55 mmol/mol (reference range 20–42) which, with the clinical picture (polyuria, polydipsia, morbid obesity) makes a diagnosis of diabetes mellitus, very likely to be type 2 (insulin-resistant). That evening the blood culture result grows positive culture for *Escherichia coli* (*E. coli*). This completely vindicates the Registrar's actions and the diagnosis becomes *E. coli* septicaemia. On the basis of this, the patient is moved to the intensive care unit (ICU). An endocrinology opinion is sought regarding the diabetes.

Day 4

The patient has a bad night with nausea (probably due to the antibiotics), and after four doses of intravenous antibiotics, the patient's temperature is still raised but reduced slightly (38.5°) and she is less troubled by her chest, although there are persistent crackles and some wheezing. Blood pressures, an ECG, fluid balance, and urine output are all normal. She has another two doses of intravenous antibiotics followed by a repeat blood test, including routine biochemistry, C-reactive protein (CRP), and FBC. A sample is also sent to the blood bank for group and save. All the routine biochemistry is once more within the reference range, but the CRP result is 7.5 mg/L (reference range (<5)).

TABLE 18.1 Blood test results.

Test	Reference range	Day										
		1	2	3	4	5	6	7 (1)	7 (2)	8 (1)	8(2)	8(3)
Haemoglobin	118–148	125	124	126	124	125	123	119	100	91	84	70
Red cell count	3.9-5.0	4.21	4.10	4.15	4.08	4.11	4.01	3.92	3.52	3.19	2.70	2.38
MCV	77.98	90.0	88.8	89.0	90.5	89.2	90.5	90.5	89.4	92.0	90.3	88.7
Hct	0.33–0.47	0.379	0.364	0.369	0.369	0.367	0.363	0.355	0.315	0.290	0.244	0.211
MCH	26–33	29.0	30.2	30.4	30.4	30.4	30.7	30.3	28.4	28.5	31.1	29.4
MCHC	330–370	330	341	341	336	341	339	335	318	310	344	331
White cell count	4–10	11.5	13.6	14.9	14.5	15.8	16.8	18.5	21.5	20.8	18.7	20.9
Neutrophils	2–7	7.51	9.45	10.95	10.51	11.77	12.57	14.05	15.58	14.90	14.85	17.8
Lymphocytes	1–3	2.52	2.43	2.15	2.29	2.40	2.45	2.31	2.45	2.01	1.08	1.12
Monocytes	0.2–1.0	0.93	1.21	1.34	1.28	1.14	1.23	1.55	2.50	1.95	1.12	0.69
Eosinophils	0.02–0.5	0.45	0.40	0.35	0.36	0.25	0.21	0.23	0.35	0.31	0.22	0.14
Basophils	0.02–1.0	0.05	0.07	0.05	0.02	0.05	0.09	0.07	0.15	0.07	0.25	0.11
Blasts/atyps	<0.01	0	0.01	0.02	0.04	0.15	0.25	0.30	0.45	1.55	1.18	1.05
Platelets	143–400	190	195	191	180	160	100	75	65	50	55	40
ESR	<10	12	15	18	18	23	25	25	30	28	35	30
PT	11–14	12.5	*	*	*	*	*	13.0	14.5	16.5	17.5	19.2
APTT	24–34	30.9	*	*	*	*	*	31.5	34.3	37.8	42.2	55.4
D-dimers	<500	575	*	*	*	*	*	*	*	1250	1509	*
CRP	<5.0	*	*	*	7.5	*	12.5	15.5	*	37.5	47.5	*
Fibrinogen	1.5–4.0	*	*	*	*	*	*	*	*	1.7	1.2	1.0

Units: haemoglobin g/L; red cell count × 10^{12}/L; MCV = mean cell volume, fL; Hct = haematocrit, proportion; MCH = mean cell haemoglobin, pg; MCHC = mean cell haemoglobin concentration, pg/L; white cell count, neutrophils, lymphocytes, monocytes, eosinophils, basophils, and blasts/atyps all × 10^9/L; ESR = erythrocyte sedimentation rate, mm/hour; PT = prothrombin time and APTT = activated partial thromboplastin time, seconds; D-dimers units/mL; CRP = C-reactive protein, mg/L.

The FBC results are shown in Table 18.1. The intravenous antibiotics are continued and the patient is started on the anti-hyperglycaemia drug metformin, 500 mg twice a day.

Question CS 1.5 What are the salient points of the results of Day 4's FBC, and how would you summarize the overall condition of the patient since her admission?

Day 5

The day passes unremarkably—the antibiotics are continued and an FBC is requested in the late afternoon, just after another dose of intravenous antibiotics. The results (Table 18.1) are bleeped to the Registrar, who increases the dose of the intravenous antibiotic and adds a second. The

nurses changing the dressing on the sacral sores report more redness and exudate (which is blood streaked), an unpleasant odour, and the patient says they are more painful. Paracetamol is increased to its maximum permitted dose.

Question CS 1.6 Why did the Registrar take this course of action?

Day 6

More bloods, including blood cultures and CRP, are taken mid-morning. Results are shown in Table 18.1. The major white cell indices, CRP, and erythrocyte sedimentation rate (ESR) have all increased, and an additional problem is the profound fall in platelets, and so the development of thrombocytopenia. Later, the microbiology biomedical scientist calls to report that the blood culture has grown a moderately intense infection with *E. coli*. This prompts the Registrar to increase the doses of antibiotics to their maximum permitted doses. The nurse changing the dressing reports increased exudation with blood, and the concentrations of the topical antibiotics are increased. Urine output is reduced, despite the same intake of fluids.

Question CS 1.7 What are the implications of the events of Day 6?

Day 7

The patient sleeps well, but wakes in a heavy sweat and thirst, complaining of abdominal discomfort and increasing pain from her sores. A nurse checks the dressings and finds them to be markedly blood stained and increasingly malodorous, and the patient's temperature has risen to 39.5°C. The patient has a prolonged bout of coughing, which is relieved by a salbutamol inhaler. Nevertheless, the patient is still breathing rapidly (25 breaths/minute) and has a marked tachycardia (120 beats/minute). A full panel of blood tests are taken at around 9 a.m. All biochemistry results are within the reference range, although those of renal function are all near to the top of their particular range with a fall in the estimated glomerular filtration rate. Haematology and CRP results are shown in Table 18.1. The blood film reveals some metamyelocytes. The patient complains loudly of being hot and to address this high temperature she is washed down with ice water by nursing staff.

Question CS 1.8 What are the most important results in this profile and what are the implications?

All afternoon of the seventh day the patient complains of increasing abdominal discomfort, but at 6 p.m. she cries out with cramp-like abdominal spasms. Whilst writhing to gain pain relief, the dressings over the sacral sores become loose, revealing a very bloody exudate. The patient has a coughing fit, which causes a nose bleed, and 30 minutes later she collapses into semi-consciousness. Blood pressures are 119/69 mm Hg. The biomedical scientist in the blood bank is asked to prepare some platelet concentrates for possible transfusion. Blood tests are repeated, haematology results of which are shown in Table 18.1.

Question CS 1.9 What has happened in the past ten hours?

Four units of platelets are formally requested and at 10 p.m. they arrive and are transfused. There is a trace of blood in her urine, so the catheter is removed and replaced, and subsequent urine is clear. As a result of the fall in haemoglobin, the blood bank is asked to prepare four units of packed red cells. However, the scientists in the blood transfusion laboratory find that she has a moderately strong titre of an antibody that cross-reacts with all the potential donor red cell packs in the blood bank. An emergency request goes out to the regional blood transfusion service.

Day 8: Early morning

At 4 a.m., bloods are taken for FBC, principally to check the platelet count following the platelet transfusion (Table 18.1). The haemoglobin and red cell count has fallen further, and prompts calls to the blood bank for the requested units of packed red cells. A minor piece of good news is that the total white cell, neutrophil, lymphocyte, and monocyte counts have all fallen but the blasts/atypics have increased (as has the CRP, markedly so), and the blood film shows more metamyelocytes. However, despite the platelet transfusion, the platelet count has continued to fall. This triggers a coagulation panel check. The patient has now lost consciousness and has developed a systemic purpuric rash. Blood pressure has fallen to 114/62 mm Hg.

Question CS 1.10 What are the most important laboratory and clinical findings, and what does this mean for the pathophysiology of the septicaemia?

Day 8: Afternoon

By mid-afternoon, the patient's breathing has deteriorated and she is put on artificial ventilation, and her urine output has fallen dramatically. There is more blood in her urine, the sacral sores continue to discharge blood-stained exudate, and there is some blood draining from her vagina. Blood pressures are now 106/59 mm Hg and the pulse rate is 130 beats per minute. There is a degree of peripheral oedema. Accordingly, a full set of bloods are drawn mid-afternoon (Table 18.1). The serum urea (7.5 mmol/L) is now above the top of the reference range (3.3–6.7 mmol/L), and serum bicarbonate is low (22 mmol/L, reference range 24–29 mmol/L), but all other biochemistry is normal. Eight packs of red cells have been received by the blood bank, and scientists perform urgent crossmatch tests to find any that are compatible with the patient.

Question CS 1.11 What has happened in the past ten hours?

Day 8: Evening

Although the total white cell count has fallen, the neutrophil count continues to rise, and the blood film has a nucleated red blood cell. The developing coagulopathy brings the call for prothrombin complex concentrate from the blood bank. Three units are in stock and are transfused (more are requested from the regional blood transfusion centre), and an hour later a third blood sample that day is taken (Table 18.1). The patient's purpuric rash has noticeably worsened in the past few hours, her temperature has increased to 40.5°C, and she is still markedly tachycardic (125 beats per minute). A biomedical scientist arrives with the requested four units of red cells, but the patient suffers a cardiac arrest and cannot be resuscitated.

Question CS 1.12 Describe the events of the patient's final hours.

Synopsis

The admitting diagnosis was a neutrophil leucocytosis that may have arisen from a bacterial chest infection, but may also be partly caused by the infected sacral sores. However, the positive blood cultures make a clinical diagnosis of septicaemia, calling for intravenous antibiotics. Unfortunately, these are unable to stem the infection, and the clinical picture is worsened by the presumed heparin-induced thrombocytopenia. The diagnosis of diabetes does not add much to the management of the developing septicaemia but may have contributed to its initiation. The thrombocytopenia seems to have been responsible for the presumed gastrointestinal and other minor haemorrhages. To make things worse, disseminated intravascular coagulation (DIC) has developed and consumes not only clotting factors but also platelets, and these cannot be corrected by replacement therapy. In hindsight, the falling platelet count may have been an early sign of the DIC.

Cross reference

Chapter 5 has details of normocytic anaemia due to blood loss, Chapter 8 of the white cell response to severe infection, and Chapter 16 has further details of DIC as an acquired bleeding disorder.

CASE STUDY 18.2

A 43-year-old male, employed as a steel worker, presents to his GP with a sore throat, fever, muscular aches, pains, and a general complaint of feeling tired all the time. Upon examination, the GP notes inflammation of the pharynx with tonsillar exudates, palatial petechiae, slight pallor, and a temperature of 38.9°C. Anterior cervical lymph nodes are tender and enlarged and the patient denies a recent cough. Following application of the Centor criteria, used to identify the likelihood of bacterial throat infection, four points are awarded, and a *Streptococcus pyogenes* infection is suspected. The patient is prescribed a course of phenoxymethyl penicillin 500 mg six-hourly for seven days, with instructions to return if the throat infection fails to resolve.

Following seven days of penicillin, the majority of the patient's symptoms resolve. Although still feeling tired and feverish at times, the sore throat no longer persists. The patient considers this a residual part of his throat infection and gives it no more attention. However, ten days later, his sore throat returns and he notes a pink, non-blanching 'pin prick' rash over the front of his arms and across his legs. He also experiences bleeding from gums on brushing teeth. Concerned, he returns to his GP, having booked an emergency morning appointment.

The GP re-evaluates this patient and notes the return of his tonsillar exudates and palatial petechiae. Physical examination confirms the presence of petechial rash and a fever of 39.6°C. The patient appears pale and has a blood pressure as 105/72 mmHg. Concerned, the GP asked the patient to attend the Accident and Emergency department immediately.

Following a thorough examination of the patient, the Accident and Emergency Registrar requested a throat swab for culture and sensitivity, accompanied by an urgent FBC and coagulation screen. The microbiology results will take 48 hours to return.

Question CS 2.1 Using the correct nomenclature, comment on the initial FBC results (Table 18.2) and outline the logic underpinning the next series of investigations.

TABLE 18.2 Initial blood test results.

Test parameter	Result	Reference range	Units
WBC	12.8	4.0–10.0	$\times 10^9$/L
RBC	3.93	4.5–5.5	$\times 10^{12}$/L
Hb	124	140–170	g/L
Hct	0.38	0.4–0.5	L/L
MCV	95.4	83–101	fL
MCH	31.6	26–33	pg
MCHC	331	330–370	g/L
PLT	18	150–400	$\times 10^9$/L
Neutrophils	3.68	2.0–7.0	$\times 10^9$/L
Lymphocytes	2.60	1.0–3.0	$\times 10^9$/L
Monocytes	0.32	0.2–1.0	$\times 10^9$/L
Eosinophils	0.21	0.02–0.5	$\times 10^9$/L
Basophils	0.03	0.02–0.1	$\times 10^9$/L
Blasts/atyps	5.96	<0.01	$\times 10^9$/L
PT	18	11–14	secs
APTT	38	24–34	secs
Fibrinogen	0.9	1.5–4.0	g/L
D-dimer	1,375	<500	units/mL

WBC = white blood cells; RBC = red blood cells; Hb = haemoglobin; Hct = haematocrit; MCV = mean cell volume; MCH = mean cell haemoglobin; MCHC = mean cell haemoglobin concentration; PLT = platelets; PT = prothrombin time; APTT = activated partial thromboplastin time.

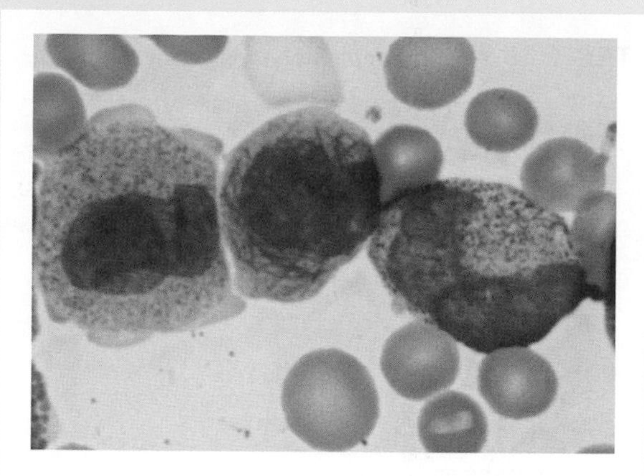

FIGURE 18.1
Typical malignant promyelocytes containing cytoplasmic azurophilic granulation. © PR. H. Piguet/CNRI/ Science Photo Library.

TABLE 18.3 Manual differential.

	Absolute count ($\times 10^9$/L)	Percentage
Neutrophils		11
Lymphocytes		15
Monocytes		1
Eosinophils		0
Basophils		0
Myeloblasts		10
Promyelocytes		59
Myelocytes		2
Metamyelocytes		2
Band forms		0
Total	12.8	100

The patient's initial prothrombin time (PT) and activated partial thromboplastin time (APTT) results are both increased and, in light of the abnormal FBC results, are discussed with the consultant haematologist. Fibrinogen and D-dimer analyses are requested, along with an urgent review of the patient's blood film.

The patient's blood film demonstrates the presence of normocytic normochromic red cells, although schistocytes are apparent throughout the film, indicating a microangiopathy. The film reveals a population of hypergranular promyelocytes with prominent azurophilic granules, as shown in Figure 18.1. Several promyelocytes containing bundles of Auer rods (faggot cells) are noted. Based upon these findings, a manual differential count is performed, with the findings presented in Table 18.3.

Question CS 2.2 Using the percentages obtained from the 100-cell manual peripheral leucocyte count, calculate the absolute count for each of the white cell species present. Based upon the information available, what is your suggested diagnosis?

Now that a presumptive diagnosis had been made from the patient's blood film, the medical team decides to apply the International Society for Thrombosis and Haemostasis (ISTH) diagnostic scoring system to the patient's haemostasis results to determine whether he is suffering from DIC.

Question CS 2.3 Using the ISTH scoring system for DIC, available at https://www.mdcalc.com/isth-criteria-disseminated-intravascular-coagulation-dic, determine whether it is appropriate to apply the ISTH scoring system, and if so whether the patient has an overt or non-overt DIC.

Based upon all the information available, the results are considered a medical emergency. Within two hours of phlebotomy, Accident and Emergency is contacted by the haematology consultant as a matter of urgency. A direct referral to the haemato-oncology ward is made for thorough investigation and commencement of treatment under a consultant haematologist. In the meantime, the peripheral blood specimen undergoes fluorescent flow cytometric analysis to determine the immunophenotype of the patient's peripheral blood leucocytes.

These important data are summarized in Table 18.4.

The data obtained from the immunophenotyping analysis (Figure 18.2) consolidates the findings from the FBC and peripheral blood film, confirming the diagnosis of acute promyelocytic leukaemia (APL).

Following admission, the patient is immediately treated for his potentially life-threatening coagulopathy. A sufficient number of adult therapeutic doses of platelets are administered to maintain his platelet count above 30–50 × 10^9/L, with fresh-frozen plasma and cryoprecipitate infusions to maintain his fibrinogen >1.5–2.0 g/L. Following stabilization, a bone marrow aspirate is obtained. A sample of liquid marrow is forwarded to the cytogenetics department for analysis, whilst the remainder produces a number of bone marrow slides to assess morphology. Following specimen retrieval, the patient immediately commences combination therapy with all-*trans* retinoic acid (ATRA) 45 mg/m^2 in two divided doses to reduce the number of malignant promyelocytes within peripheral blood and bone marrow and to treat the underlying coagulopathy. Routinely, ATRA is given for one to three days prior to the introduction of anthracyclin-based chemotherapy or arsenic trioxide (ATO); however, because this patient had a white cell count >10 × 10^9/L, he is considered high risk, thus ATRA, ATO, and anthracyclin-based chemotherapy are initiated both immediately and simultaneously.

Question CS 2.4 Why is it necessary to ensure the patient is haemodynamically stable prior to the extraction of bone marrow through aspiration?

TABLE 18.4 Immunophenotyping data.

Marker	Result
CD7	Negative
CD13	Positive
CD33	Positive
CD34	Negative
CD117	Positive
HLA-DR	Negative
MPO—not shown	Positive

HLA-DR = human leucocyte antigen-DR; MPO = myeloperoxidase.

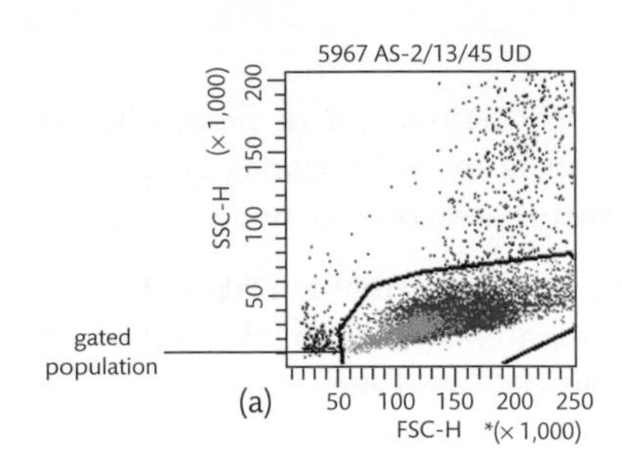

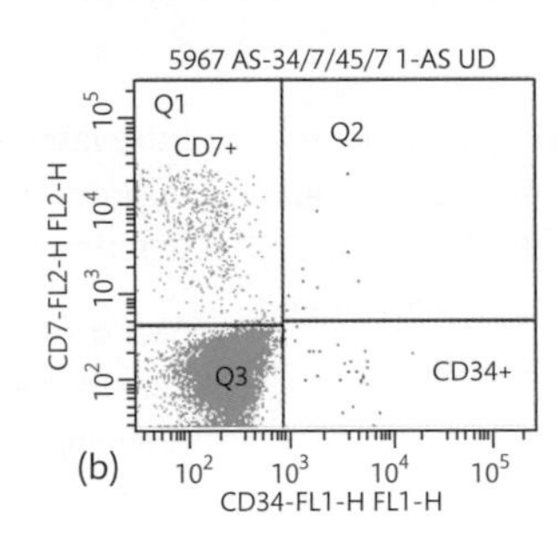

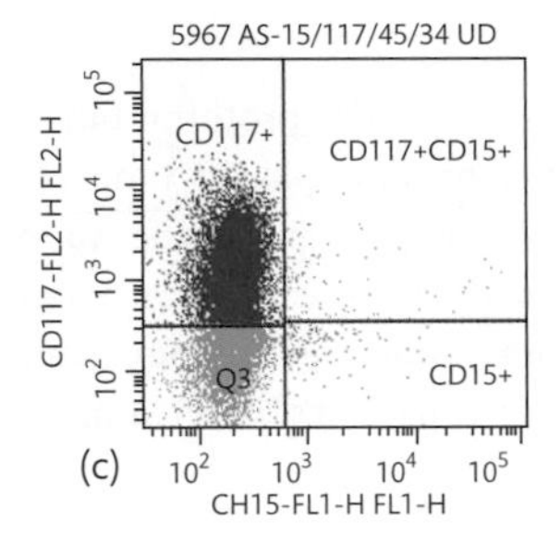

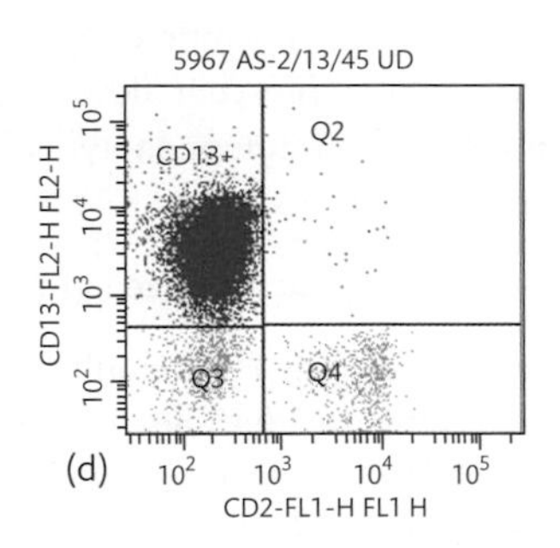

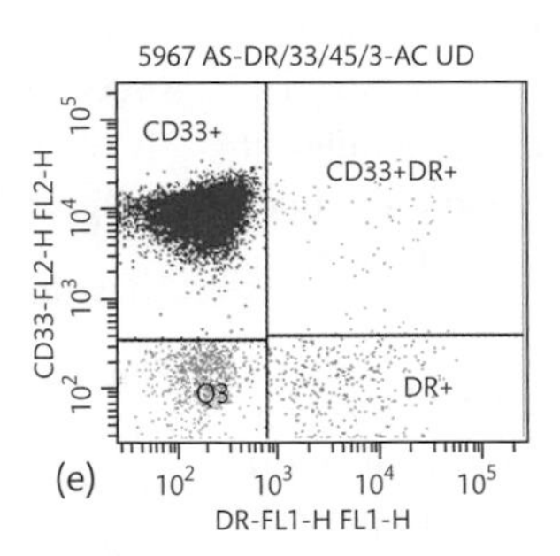

FIGURE 18.2

Fluorescence flow cytometry. (a) Cells of interest are gated. All immunophenotypic data are analysed in relation to this gated population. Quadrant 3 (Q3) is negative throughout frames a–e. (b) There is a population of CD34 negative, CD7 positive cells (Q1), although these are normal T cells contaminating the malignant cell population. The majority of the other cells are CD7 negative and CD34 negative (Q3). (c) There is a clear population of CD117 (also called c-kit) positive cells (Q1), which are CD15 negative. (d) The malignant cells express CD13 (Q1), but not CD2 (Q2). CD2 is a pan T-cell marker and is co-expressed with CD7. (e) The population of interest (red), Q1, expresses CD33, but does not express HLA-DR. Immunophenotyping data courtesy of Department of Immunology, Southampton General Hospital.

TABLE 18.5 **Initial microbiology results.**

Pathogenic microorganism	Antibiotic sensitivity	Zone diameter (mm)	BSAC reference range guideline zone diameter (mm)			BNF antibiotic recommendation
			Resistant (≤)	Intermediate	Sensitive (≥)	
β-haemolytic streptococcus identified.	Penicillin	22	19	–	20	First choice
	Clarithromycin	23	19	20–21	22	Second choice
	Erythromycin	23	19	20–21	22	Alternative second choice
Streptococcus pyogenes						

BSAC = British Society for Antimicrobial Chemotherapy; BNF = British National Formulary.

Using a Jenner Giemsa Romanowsky stain, the bone marrow is found to be hypercellular with 78% promyelocytes. These promyelocytes are morphologically heterogeneous, with a high nucleocytoplasmic ratio, bilobed nuclei, and prominent azurophilic granules. A small proportion of the promyelocytes contain bundles of Auer rods, consistent with faggot cells. Erythroid and megakaryocytic lineages appear normal.

Further results

Cytogenetic analysis employing dual fusion PML/RARA probes confirms the presence of the PML-RARA fusion associated with t(15;17)(q22;q12), providing results concordant with reptilase time-polymerase chain reaction (RT-PCR) techniques. Both provide a definitive diagnosis for acute promyelocytic leukaemia (APL) and predict a positive response to the selected induction chemotherapy.

Forty-eight hours following the initial request for culture and sensitivity, the microbiology results are returned, as shown in Table 18.5.

Question CS 2.5 What do these microbiology results tell us about the patient's initial presenting sore throat?

Synopsis

This is a case of APL. Most cases have a typical presentation, with a white cell count of <10 × 10^9/L, although in some cases, considered high risk, the white cell count may be elevated. There is also a microgranular variant APL, of which this is not an example. This patient demonstrates a normocytic normochromic anaemia which often accompanies malignant states, although in this case, this is overlaid with a microangiopathy. Microangiopathy follows an acquired coagulopathy leading to intravascular red cell destruction and is demonstrated through the presence of the schistocytes within the peripheral blood and, in this case, correlates with the deranged haemostasis results. These results are considered the sequelae following the release of profibrinolytic agents from malignant promyelocyte granules, thus leading to a hyperfibrinolytic state. Whilst this patient scores positively using the ISTH algorithm for DIC, patients with APL do not demonstrate typical DIC. The APL coagulopathy is more biochemically aligned with hyperfibrinolysis than typical DIC, which can be considered the result of consumption of all clotting factors, natural coagulants, platelets, and hyperfibrinolysis. This patient was fortunate to only experience a petechial rash and minor gingival bleeding, whereas

Cross reference

Chapter 5 has details of the anaemia of cancer, Chapters 9–11 have details of the pathophysiology and diagnosis of leukaemia.

others often present with more significant bleeding, where haemorrhage in brain, lungs, and gastrointestinal tract would not be considered uncommon. Nevertheless, the bleeding diathesis associated with APL is considered a medical emergency and, if left untreated or is mismanaged, will lead to a very quick death. The introduction of ATRA and ATO have provided considerable survival advantage to patients with APL, with the identification of t(15;17) (q22;q12) now signalling a good prognosis, compared with the pre-ATRA era, when survival rates were poor.

This patient underwent induction, consolidation, and maintenance therapy as per standard protocols and required the administration of supportive measures to meet clinical needs. The patient responded well to therapy, is now in molecular remission, and is monitored every three months for residual disease.

CASE STUDY 18.3

A 49-year-old man who works as a warehouse supervisor presents to his GP with sharp, low back pain which has been present since a fall from the steps leading to his front door six weeks ago. The pain has been getting worse, and for the last four weeks has been virtually constant. He admits that he has been feeling noticeably more tired than usual since just before the fall, although it was not particularly marked and he put it down to age and needing a holiday. He also comments on having had a series of headaches recently, which is unusual for him. On examination, he is in considerable pain, with slight pallor but otherwise apparently fairly fit for his age. There is no fever, lymphadenopathy, hepatosplenomegaly, abdominal masses, or signs of bruising. He is prescribed analgesia and the GP arranges for the patient to have blood taken for an FBC, ESR, and a routine biochemistry profile.

Question CS 3.1 Why did the GP request those tests?

The patient has his blood taken that day and the GP receives the results the next day. The results are shown in Table 18.6. A blood film is examined and confirms the normal white cell differential. White cells and platelets are morphologically normal. The red cells are normochromic and normocytic, without significant anisocytosis, and no polychromatic cells are seen, although this is difficult to gauge, since there is marked rouleaux present. The blood film report also comments that blue background staining is present.

Question CS 3.2 What are the pertinent results?

Upon receiving the results, the GP immediately refers the patient to the specialist haemato-oncology team at the local hospital for further investigation.

Question CS 3.3 What prompted the GP to make that decision?

The patient is seen in the haemato-oncology clinic and the Consultant Haematologist initiates further blood tests, urine tests, bone marrow aspirate and biopsy, and radiographic investigations. X-rays of his back show a small, punched-out lesion in the third lumbar vertebra and destruction of the symphisis pubis. In the biochemistry laboratory, serum protein electrophoresis reveals a monoclonal band in the gamma region and IgG-kappa paraprotein is identified by immunofixation of both serum and urine. The bone marrow aspirate reveals infiltration with 42% plasma cells. Bone marrow trephine confirms plasma cell infiltration. In the histology laboratory, fluorescence *in situ* hybridization shows t(11;14)(q13;q32). Results from other blood tests are shown in Table 18.7. Based on the clinical findings and these results, a diagnosis of multiple myeloma is made.

Question CS 3.4 What laboratory and clinical features make that diagnosis, and what are the potential differential diagnoses?

The patient is treated with the VAD chemotherapy regimen (vincristine, doxorubicin, and dexamethasone). He achieves complete remission after three cycles of VAD therapy and then undergoes autologous transplantation

TABLE 18.6 FBC, ESR, and routine biochemistry results at presentation.

Parameter	Result	Reference range	Units
Hb	98	133–167	g/L
RBC	3.1	4.3–5.7	$\times 10^{12}$/L
Hct	0.27	0.35–0.53	L/L
MCV	88	80–98	fL
MCH	31.6	26–33	pg
Reticulocytes	28	25–125	$\times 10^{9}$/L
Platelets	158	150–400	$\times 10^{9}$/L
WBC	7.84	4.0–11.0	$\times 10^{9}$/L
Neutrophils	4.4	2.0–7.0	$\times 10^{9}$/L
Lymphocytes	2.9	1.0–3.0	$\times 10^{9}$/L
Monocytes	0.4	0.2–1.0	$\times 10^{9}$/L
Eosinophils	0.1	0.02–0.5	$\times 10^{9}$/L
Basophils	0.04	0.02–0.1	$\times 10^{9}$/L
ESR	95	>10	mm/first hour
Urea	8.9	1.7–8.3	mmol/L
Sodium	138	135–145	mmol/L
Potassium	4.4	3.5–5.0	mmol/L
Creatinine	90	45–85	μmol/L
Calcium	3.00	2.15–2.55	mmol/L
Total bilirubin	12	0–20	μmol/L
Alanine aminotransferase	25	4–45	IU/L
Aspartate aminotransferase	15	10–40	IU/L
Alkaline phosphatase	89	35–129	IU/L
Albumin	48	40–52	g/L

Hb = haemoglobin; RBC = red blood cells; Hct = haematocrit; MCV = mean cell volume; MCH = mean cell haemoglobin; MCHC = mean cell haemoglobin concentration; WBC = white blood cells; ESR = erythrocyte sedimentation rate.

with an enriched population of CD34+ stem cells, the quality of which will be checked by fluorescence flow cytometry.

Three years later, he begins experiencing severe back pain again and an FBC, bone marrow aspirate, and biopsy are performed, which confirm relapse of the multiple

TABLE 18.7 Additional testing at referral.

Parameter	Result	Reference range	Units
Total serum protein	88	64–78	g/L
Albumin	41	40–52	g/L
IgG	47	8–20	g/L
IgM	0.2	0.5–2.0	g/L
IgA	0.4	0.8–5.0	g/L
β_2 microglobulin	2.91	0.8–2.0	mg/L
Lactate dehydrogenase	342	130–310	IU/L
Plasma viscosity	2.1	1.1–1.5	cP
Prothrombin time	12	10–13	s
APTT	34	30–40	s
Fibrinogen (Clauss)	2.3	2.0–4.0	g/L

IgG = immunoglobin G; IgM = immunoglobin M; IgA = immunoglobin A; APTT = activated partial thromboplastin time.

TABLE 18.8 FBC results at follow-up.

Parameter	Result	Reference range	Units
Hb	125	133–167	g/L
RBC	3.9	4.3–5.7	$\times 10^{12}$/L
Hct	0.35	0.35–0.53	L/L
MCV	90	80–98	fL
MCH	32.1	26–33	pg
Reticulocytes	35	25–125	$\times 10^9$/L
Platelets	135	150–400	$\times 10^9$/L
WBC	8.3	4.0–11.0	$\times 10^9$/L
Neutrophils	5.1	2.0–7.0	$\times 10^9$/L
Lymphocytes	2.2	1.0–3.0	$\times 10^9$/L
Monocytes	0.7	0.2–1.0	$\times 10^9$/L
Eosinophils	0.2	0.02–0.5	$\times 10^9$/L
Basophils	0.1	0.02–0.1	$\times 10^9$/L

Hb = haemoglobin; RBC = red blood cells; Hct = haematocrit; MCV = mean cell volume; MCH = mean cell haemoglobin; WBC = white blood cells.

myeloma, and he is started on pomalidomide therapy. The FBC results are shown in Table 18.8.

Three days later he begins to bleed from the biopsy site, and reports blood upon brushing his teeth. Blood samples are taken for coagulation screening. The results are shown in Table 18.9.

Question CS 3.5 Discuss the results from Tables 18.8 and 18.9, the latter with respect to the patient's bleeding symptoms. What tests would you do next, and why?

To discern the nature of the inhibitor, the laboratory performs factor assays and lupus anticoagulant screening. The results are shown in Table 18.9.

Based on these data, further coagulation investigations are performed, and the results are shown in Table 18.10.

Question CS 3.6 How would you interpret the results from Table 18.10? How do you know the inhibitor is not a lupus anticoagulant? What testing would you do next, and why?

TABLE 18.9 Coagulation screening after bleeding from biopsy site.

Parameter	Result	Reference range	Units
Prothrombin time	45	10–13	s
Prothrombin time 1:1 mixing test	42	10–13	s
APTT	72	30–40	s
APTT 1:1 mixing test	70	30–40	s
Fibrinogen (Clauss)	2.1	2.0–4.0	g/L
Thrombin time	10	9–11	s

APTT = activated partial thromboplastin time.

A Bethesda assay based on residual FX is performed on a range of dilutions of the patient's plasma. The raw data are given in Table 18.11.

TABLE 18.10 Follow-up testing from deranged coagulation screen.

Parameter	Result	Reference range	Units
One-stage clotting assays			
Factor II	121	50–150	IU/dL
Factor V	111	50–150	IU/dL
Factor VII	95	50–150	IU/dL
Factor VIII	150	50–150	IU/dL
Factor IX	131	50–150	IU/dL
Factor X	<1	50–150	IU/dL
Factor XI	110	60–140	IU/dL
Lupus anticoagulant assays			
DRVVT screening test	2.61	0.83–1.17	Ratio
DRVVT confirmatory test	2.63	0.85–1.15	Ratio
DRVVT mixing test	2.10	0.90–1.10	Ratio
DAPTT screening test	1.99	0.80–1.20	Ratio
DAPTT confirmatory test	1.97	0.88–1.18	Ratio
DAPTT mixing test	1.78	0.85–1.15	Ratio

DRVTT = dilute Russell's viper venom time; DAPTT = dilute activated partial thromboplastin time.

TABLE 18.11 FX Bethesda assay raw data.

Dilution of patient plasma	Neat	1:2	1:4	1:8	1:16	1:32	1:64	1:128	1:256	1:512
Residual FX (%)	<1.0	<1.0	<1.0	5.0	28.0	52.0	75.0	100	100	100

Question CS 3.7 Use the graphical method illustrated in Figure 14.9 to calculate the FX Bethesda result.

Cyclophosphamide plus prednisolone are initiated in an attempt to eradicate the inhibitor and pomalidomide is continued for the multiple myeloma. Acute bleeding episodes are managed with activated prothrombin complex concentrates.

Synopsis

The main presenting clinical feature was significant back/bone pain. Although this is typical of myeloma, back pain can have entirely mechanical causes, or may be due to an arthritis (which, incidentally, may also be linked to a normocytic anaemia and a modestly raised ESR), so that jumping to a conclusion of haematological malignancy would be premature. In view of the mild malaise, the GP was right to initially prescribe analgesia and request preliminary blood tests. Those blood tests raised concerns of more than just mild anaemia, and the patient required assessment by a specialist. A large battery of diagnostic testing was initiated, including radiography and cytogenetics, to generate sufficient information to differentiate between a number of potential pathologies. One such alternative diagnosis is temporal arteritis, perhaps the only other possible explanation for a grossly elevated ESR in the absence of an overt inflammatory state. However, this alternative diagnosis would have been rapidly dismissed once the other blood results were to hand. The patient was diagnosed with multiple myeloma, a B-cell neoplasm. Rarely, multiple myeloma can be complicated by bleeding due to hyperviscosity, perivascular amyloidosis, acquired coagulopathy, or thrombocytopenia. At presentation, this patient had a normal platelet count and coagulation screen and no bleeding symptoms. Despite successful chemotherapy and autologous stem cell transplantation, the patient relapsed, and exhibited bleeding symptoms upon challenge. His coagulation screening indicated the presence of an inhibitor, which subsequent analysis revealed to be an inhibitor of FX function. His coagulation was normal at presentation so this represented generation of an acquired autoantibody, possibly from a new, antibody-producing B-cell clone.

Cross reference

Chapter 5 has details of the anaemia of cancer, Chapter 12 details of malignant myeloma, and Chapter 14 of the investigation of bleeding disorders.

Chapter summary

It is all too easy to focus on particular finer points of haematology and, indeed, of any of the pathology disciplines. However, patents almost never present with a single pathological problem, and a holistic approach is required.

- The first case illustrates the concerted efforts of scientists in haematology, blood transfusion, microbiology, and biochemistry who all (possibly unknown to each other) are providing vital support to our colleagues in the intensive care unit fighting to save the life of the patient.
- The second case also illustrates the importance of scientists with different skills, not only in haematology, but also in microbiology, immunology, and cytogenetics. The laboratory will continue to be crucial in checking for any return of the leukaemia.
- In the third case, haematologists and biochemists together made the diagnosis of myeloma. They would also have been heavily involved in the transplantation and subsequent follow-up, and provided key information in subsequent long-term follow-up. The skills of the coagulation laboratory were essential in determining the precise cause of the haemorrhage.

Answers to case study questions are provided as part of this book's online resources.

Visit www.oup.com/he/moore-fbms3e.

Glossary

α-granules Platelet storage granules containing numerous large molecules necessary for haemostasis and wound repair.

α_2-antiplasmin Serine protease inhibitor that inactivates the enzyme plasmin, which degrades fibrin clots.

α_2-macroglobulin A protease inhibitor able to inactivate numerous protease enzymes, including plasmin and thrombin.

Acenocoumarin An oral anticoagulant.

Acetylation The addition of acetyl groups to histone lysine residues to induce an open chromatin configuration.

Acquired von Willebrand disease (AVWD) VWF deficiency that is secondary to other disease states (i.e. non-hereditary).

Actin-binding protein Proteins that bind actin.

Activated partial thromboplastin time (APTT) A coagulation screening test for global assessment of intrinsic and common pathways.

Activated protein C (APC) The activated form of the naturally occurring inhibitor of the activated forms of factor V and factor VIII.

Activated protein C resistance A hereditary or acquired abnormality where the anticoagulant action of normal activated protein C is reduced by other factors, e.g. factor V Leiden.

Activin receptor-like kinase 1 (ALK-1) An endothelium bound growth factor receptor involved in angiogenesis.

Acute phase protein A protein whose circulating levels increase in response to acute conditions such as infection, injury, tissue destruction, some malignancies, surgery, burns, or trauma.

Acute phase response The physiological response to the onset of infection, inflammation, trauma, and some cancers involving increase in plasma concentration of acute phase proteins, fever, increased vascular permeability, and metabolic changes.

ADAMTS-13 An enzyme that degrades ultra-large VWF multimers.

Aetiology The cause(s) of a disorder/disease.

Afibrinogenaemia The absence of plasma fibrinogen.

Agonists Molecules that activate cell functions via surface receptors.

AIHA (Autoimmune haemolytic anaemia) Anaemia that follows the destruction of red cells by an inappropriate antibody produced by the patient themselves.

Alkylation The transfer of an alkyl group (C_nH_{2n+1}) from one molecule to another.

ALL Acute lymphoblastic leukaemia or acute lymphocytic leukaemia.

All-*trans* retinoic acid (ATRA) A vitamin A derivative which interacts with the PML–RARα fusion protein in acute promyelocytic leukaemia and, at pharmacological concentrations, can induce differentiation of cell cycle-arrested promyelocytes and allow their maturation and induction of apoptosis.

Alleles Different forms of a specific gene.

Alloantibodies Antibodies directed against substances from a member of one's own species that are recognized as foreign to self.

Alpha antiplasmin Prime inhibitor of plasmin.

Alpha macroglobulin 'Back-up' inhibitor for antithrombin and α_2-antiplasmin.

Alpha- and beta-spectrin The major structural components of the matrix that provide 'skeletal' support for the structure of the red blood cell.

Amastigote Asexual replication phase of *T. cruzi*, *L. donovani*, and *L. infantum*.

AML Acute myeloid leukaemia.

Amplification The activation of various haemostatic components by the trace thrombin formed at initiation of coagulation.

Amyloidosis Deposition of abnormally folded protein in tissues.

Anaemia The principal disease of red blood cells, characterized by reduced haemoglobin and oxygen-carrying capacity, and symptoms such as tiredness and lethargy.

Aneuploidy The numerically abnormal chromosomal complement of a cell.

Aneurysm Localized, blood-filled dilation (bulge) in a blood vessel resulting from a weakened or injured vessel wall. Risk of rupture increases as an aneurysm grows in size.

Angiogenesis Blood vessel formation.

Angiogenic The promotion of blood vessel formation.

Anisocytosis The variation in the size of the red blood cells when viewed by light microscopy.

Annexin A5 A phospholipid-binding protein with inhibitory activity towards blood coagulation.

Anti-β2 GPI antibody (aβ_2GPI) A type of antiphospholipid antibody detected in solid-phase assays.

Anticardiolipin antibody (aCL) A type of antiphospholipid antibody detected in solid-phase assays.

Anticoagulant A physiological or pharmacological mechanism that retards clotting processes.

Anticoagulation The clinical practice of minimizing the risk of thrombosis with anticoagulants.

Antigen-presenting cell (APC) A cell that expresses MHC class II molecules in conjunction with antigenic material to elicit a T-cell response.

Antiphospholipid antibodies A heterogeneous group of autoantibodies that bind to phospholipid, complexes of phospholipid and protein, or phospholipid-binding proteins. Persistent presence of such antibodies is associated with thrombotic disease.

Antiphospholipid syndrome (APS) A disorder of haemostasis that predisposes patients to arterial or venous thrombosis and pregnancy morbidity, which occur in the presence of autoimmune antiphospholipid antibodies.

Antithrombin A naturally occurring inhibitor which predominantly acts on thrombin and activated factor X.

Antithrombin Cambridge II An antithrombin variant associated with mild increase in thrombotic risk.

Antithrombotic A molecule, process, or therapy that prevents or interferes with the processes that form a blood clot.

Anti-Xa assay Assays based on FXa inhibition that are used to measure levels of UFH, LMWHs, and direct FXa inhibitors.

Aortic stenosis Narrowing of the heart valve between left ventricle and the aorta.

Apixaban Orally administered anticoagulant that directly inhibits activated factor X.

Aplastic anaemia An anaemia associated with pancytopenia and generally caused by suppression and/or damage to the bone marrow that results in low numbers of red blood cells and haemoglobin.

Apoptosis The process of programmed cell death involving the ordered removal of organelles and cells. One of the functions of apoptosis is to prevent acquired genetic mutations being passed on to daughter cells.

Apoptosome Is composed of cytochrome c and apoptotic protease activating factor 1 (Apaf1) and is used to initiate the execution pathway by activating caspase 9.

Apotransferrin The carrier protein transferrin when free of iron is called apotransferrin.

Argatroban A direct thrombin inhibitor derived from hirudin.

Aspirin resistance A condition where the platelets of some patients appear under-responsive to standard doses of aspirin, although evaluation of the phenomenon is method dependent.

Asymptomatic A disease is considered asymptomatic if a patient is a carrier for a disease or infection but experiences no symptoms. A condition might be asymptomatic if it fails to show the noticeable symptoms with which it is usually associated. Asymptomatic infections are also called subclinical infections. The term 'clinically silent' is also used.

Atherogenesis The process of the development of atherosclerosis, often involving blood vessel damage, thrombosis, and hypercholesterolaemia.

Atheroma The accumulation of lipids, macrophages, and connective tissue causing thickening of the inner wall of the arteries in patches, or plaques.

Atherosclerosis The narrowing and hardening of the arteries over time leading to impaired blood flow. Associated with age but also other risk factors, e.g. hypertension, high cholesterol.

Atherosclerotic plaque A deposit of cholesterol and other fatty material that builds up within the arterial wall.

Atrophy Wasting of a structure, with subsequent loss of function.

Auer rods Eosinophilic primary granules abnormally assembled into rod-like structures and used as a marker of myeloid malignancies.

Autoantibodies Antibodies that are directed against an organism's own tissues or native proteins.

β2 glycoprotein-I (β_2GPI) Plasma protein with procoagulant and anticoagulation properties.

B-cell receptor (BCR) The term BCR should be used with caution, as it is context specific. In the context of immunology, BCR represents the B-cell receptor, whereas in cytogenetics, BCR represents breakpoint cluster regions.

Babesiosis Parasitic protozoan infection by *Babesia* organisms.

Background staining Uptake of Romanowsky stain by plasma-derived paraprotein fixed on the microscope slide.

Band 3 Component of the membrane of the red cell that provides recognition, transport, and anchorage sites.

Band form An immature granulocyte with a horseshoe-shaped nucleus.

Banding pattern A method of identifying chromosomes based upon their appearance following staining.

Basophil A granulocyte characterized morphologically by large violet or black granules that may partially or completely obscure the bilobed nucleus.

Basophilia An increase in the number of basophils, beyond the upper limit of the reference range.

BD Vacutainer® A glass tube with an inbuilt vacuum that draws blood directly from the vein without the use of a syringe. Vacutainer is a brand name; there are other types of evacuated tubes.

Bence-Jones protein Paraproteins found in the urine.

Bernard-Soulier syndrome A bleeding disorder characterized by defects of the GPIb-IX-V platelet surface receptor complex.

Bethesda assay The method used to quantify an inhibitor that requires prolonged incubation to detect progressively acting FVIII inhibitors.

Bilirubin The major breakdown product of haem: high plasma levels cause the clinical sign of jaundice.

Biliverdin A primary breakdown product of haem that is converted into bilirubin.

Bioavailability The degree to which a drug or other substance is absorbed or becomes available where it is required physiologically and can act on its target.

Bivalirudin A direct thrombin inhibitor derived from hirudin.

Blackwater fever A sudden intravascular haemolysis followed by fever and haemoglobinuria in patients infected with *P. falciparum*.

Blast An early stage of differentiation of a blood cell as it passes from the stem cell stage to the mature cell stage that should only be found within the bone marrow in low numbers. Blasts may be further characterized by their lineage; e.g. myeloblasts are of the myeloid lineage, lymphoblasts are of the lymphoid lineage.

Blast threshold The minimum number of blasts required to diagnose a leukaemia as acute. The World Health Organization (WHO) currently specifies 20% blasts to make this diagnosis.

Blood coagulation The process where specialized proteins interact to form a clot (in conjunction with platelets).

Blood film A stained smear of a drop of blood that, when viewed through a microscope, provides additional information on red blood cells, white blood cells, and platelets.

Blood transfusion The science of ensuring the safe transfer of blood and other substances from one person to another.

Bone marrow Soft tissue located inside hollow bones responsible for production and maturation of blood cells.

Bone marrow aspiration The process of obtaining a sample of liquid bone marrow.

Bosutinib A next-generation signal transduction inhibitor employed for the treatment of chronic myeloid leukaemia.

Breakpoint Describes an area of a chromosome, usually within a gene, where a region of chromosomal material is lost or exchanged with another chromosome.

Breakpoint cluster region (BCR) An area of a gene where DNA breakages tend to occur.

Bromodomains Originally identified in the *Drosophila* protein brahma. A bromodomain comprises a sequence of 110 amino acids derived from a four-helix bundle. When the helices interact, they produce a hydrophobic binding pocket with a conserved amino acid sequence that recognizes acetyl lysine.

Cannula A thin tube inserted into a vein or body cavity to administer drugs, drain fluid, or insert a surgical instrument.

C/EBPε CCAAT enhancer binding protein, a factor important for the transcription of certain genes.

C-reactive protein A protein released into the circulation by the liver in response to inflammation. It binds to dead or dying cells and some bacteria to promote clearance.

C4b-binding protein A component of the complement system that forms a high-affinity complex with protein S.

Cachexia Describes progressive weight loss, anorexia, and a gradual reduction in the patient's body mass.

Calabar swellings Characteristic subcutaneous swellings of *Loa loa* filariasis.

Calreticulin A protein, associated with multiple functions, including calcium storage within the endoplasmic and sarcoplasmic reticulum and the regulation of gene transcription by nuclear hormone receptors through nuclear localization.

Carbaminohaemoglobin A subtype of haemoglobin characterized by binding to carbon dioxide.

Carboxyhaemoglobin A species of haemoglobin formed by binding with carbon monoxide.

Cardiac Relating to the function of the heart and associated medical conditions.

Cardiolipin A phospholipid found in high concentrations in metabolically active cells used as a capture antigen for the detection of anticardiolipin antibodies in the diagnosis of antiphospholipid syndrome.

Cardioversion Procedure of atrial fibrillation where heart is stopped, then an electrical current is applied to shock it into restarting with a normal sinus rhythm.

Cascade theory of coagulation Compartmentalized *in vitro* model of coagulation comprising three converging pathways that involve a series of enzyme and cofactor reactions to ultimately generate fibrin.

Caspases A family of cysteine proteases that play essential roles in apoptosis (programmed cell death).

CD34 A molecule of the surface of stem cells that is a marker of these cells and so allows their identification and collection (see also Cluster of differentiation (CD) antigen).

Cell cycle arrest Occurs when the G1–S–G2–M transitions become disrupted. Arrest usually occurs in late G1 or S phases as a consequence of the activation of tumour suppressor proteins in an attempt to initiate DNA repair. Alternatively, it can be seen during S phase, where synthesis of DNA is disrupted.

Cellularity The proportion of cells within the bone marrow.

Centrifuge A commonplace but essential item of laboratory equipment for separating solid material (e.g. blood cells) from liquids (e.g. serum or plasma).

Centroblasts Highly proliferative cells found within the early germinal centre following antigenic stimulation.

Centrocyte A small, non-dividing cell found within the germinal centre. The nucleus contains a cleft.

Centromere A prominent structure that joins two identical sister chromatids.

Cerebral haemorrhage A rupture of a blood vessel in the brain or leakage of blood from a vessel resulting in impaired oxygen supply to the area and subsequent brain damage of variable severity.

Cerebral malaria Invasion of the central nervous system by cytoadherence of *P. falciparum* infected red cells to the vascular endothelium and rosetting of non-infected cells around the adhered infected cells.

Charcot-Leyden (crystal protein) A type of enzyme found within eosinophils and basophils that has the propensity to form crystals. Evidence of crystals in body fluids indicates allergy.

Chemotaxins Any group of small molecules that can induce chemotaxis.

Chemotaxis The process whereby chemical signals, for example, complement components C3a and C5a, and bacterial products, such as lipopolysaccharide, disseminate, forming a concentration gradient for granulocytes to follow.

Chemotherapy (treatment using drugs) Generally held to be treatment of a cancer with cytotoxic drugs, although the use of any drug, such as an antibiotic, is also chemotherapy.

Chimeric protein Following translocation, the protein product obtained containing elements of both fused genes.

Chitin The principal component of arthropod exoskeletons.

Chromatids Subunits of chromosomes separated by a centromere. Chromatids become chromosomes following division of the centromere during mitosis.

Chromatin The DNA–histone composite found within chromosomes.

Chromogenic substrate A substrate that mimics the natural substrate of an enzyme, to which a dye, such as paranitroaniline is added. The enzyme will cleave the substrate to release the dye, causing a measurable colour change.

Class-switching recombination (CSR) The process of generating different types of immunoglobulin.

Clauss fibrinogen assay Assay based on the thrombin time that quantifies fibrinogen activity.

Clone A cell, group of cells, or organism descended from and genetically identical to a single common ancestor.

Cluster of differentiation (CD) antigen The standardized notation used to describe a range of molecules associated with the differentiation and maturation of cells. CD markers can be measured using epitope-specific monoclonal antibodies.

Coagulation The process where specialized proteins interact to form a clot.

Coagulation factors Circulating inert enzymes and cofactors that become activated to operate in concert to ultimately generate fibrin in a blood clot.

Coagulation screening A group of tests that assess global functionality of plasma clotting factors.

Coagulopathy Hereditary or acquired abnormality of blood coagulation.

Cobalamin The chemical basis of vitamin B_{12} and related molecules.

Collagen A long protein fibre that connects and strengthens numerous tissues.

Collagen-binding activity assay (VWF:CB) An enzyme-linked immunosorbent assay (ELISA assay) to quantify the collagen-binding capacity of VWF.

Colony-forming unit (CFU) A lineage-specific stem cell, responsible for the production of only one type of blood cell.

Colony-stimulating factor (CSF) A cytokine that participates in the regulation of haemopoiesis by acting on precursor cells to induce their differentiation into colonies of themselves or into the next stage of differentiation.

Common pathway Compartment within the cascade theory of coagulation involving a sequence of enzyme activation reactions starting with activation of FII (prothrombin) by FXa and cofactors of FVa, phospholipid, and calcium ions. The activated FII (thrombin) converts fibrinogen to fibrin.

Co-morbidities Two or more co-existing diseases.

Complement A series of inflammatory and defensive proteins that normally assemble on the surface of pathogens such as bacteria, helping to cause their destruction. However, inappropriate complement activation can occur on the cells of the body (such as red blood cells), also contributing to their destruction.

Complement factor H A regulator of the complement system that ensures the system is directed against pathogens and not against self.

Complement system An enzyme cascade triggered by antigen–antibody complex formation and resulting in the activation of a series of defence mechanisms, including inflammation and cell lysis.

Complementarity-determining region (CDR) The region of an antibody that complements the structure of its associated antigen.

Complex karyotype Describes the situation where more than three acquired cytogenetic abnormalities are found within a population of tumour cells.

Compression ultrasonography Used for examining soft tissue; provides real-time images in conjunction with venous compression to diagnose deep vein thrombosis (DVT).

Confirm reagent A reagent used to demonstrate phospholipid dependence when the presence of a lupus anticoagulant is suspected.

Conformational change Alteration in the structure and shape of a protein that can alter biochemical properties.

Constitutionally Pertaining to an entity's composition.

Contiguous gene syndrome The manifestation caused by the deletion of a number of genes that are situated next to one another and are deleted simultaneously.

Continuing professional development (CPD) Practical and educational activities that maintain up-to-date and informed professional practice.

Coumarin A plant toxin that is the precursor for a number of anticoagulants, notably warfarin.

CpG islands Regions of the genome that are composed of dinucleotides of cytosine and guanine.

CRAB Hyper*c*alcaemia, *r*enal insufficiency, *a*naemia, and *b*one lesions.

Cross-links Covalent bonds between adjacent fibrin molecules introduced by activated FXIII.

Cryoglobulins Proteins that precipitate and aggregate at low temperatures.

Cryptic translocation This occurs when the original morphology and banding pattern of a chromosome is maintained following translocation. This means the translocation may go undetected.

Cutaneous Pertaining to the skin.

Cyclins A number of regulatory proteins whose concentration oscillates throughout the cell cycle.

Cyclin-dependent kinases (CDKs) A group of enzymes, requiring cyclins, that are involved in the transfer of phosphate groups to enable the progression of the cell cycle.

Cytochemistry A technique that uses chemical stains that react with cytoplasmic components and so define different cells in the blood and bone marrow.

Cytochrome P450 A large superfamily of haem-containing proteins that take part in electron transfer reactions.

Cytokines Small, hormone-like intercellular mediators with a diverse range of functions, including the stimulation of the immune system in response to an encounter with a pathogen.

Cytopenia A reduction in one or more cell lineages within the peripheral blood. If all lineages (red cells, white cells, and platelets) are reduced, this is termed pancytopenia.

Cytoreductive therapy A form of therapy effective at reducing high blood cell counts. An example of cytoreductive therapy is the use of hydroxycarbamide.

2,3-diphosphoglycerate (2,3-DPG) A red blood cell metabolite product of anaerobic respiration but which may also partially regulate in oxygen carriage by haemoglobin.

D-dimer Terminal degradation product of cross-linked fibrin.

Dabigatran Orally administered anticoagulant that directly inhibits thrombin.

Danaparoid Parenterally administered anticoagulant containing a mixture of heparin analogues.

Dasatinib A next-generation signal transduction inhibitor employed for the treatment of chronic myeloid leukaemia.

***De novo* leukaemia** A primary leukaemia; of unknown cause.

Deep vein thrombosis (DVT) A condition where a blood clot forms in one or more of the deep veins, commonly in the inner thigh or lower leg.

Deletion Loss of chromosomal material.

Dense bodies Platelet storage granules containing numerous small molecules necessary for haemostasis.

Dense tubular system Platelet internal membrane system facilitating internal platelet biochemistry.

Deoxyhaemoglobin Haemoglobin that is not carrying oxygen.

Desmopressin Synthetic antidiuretic drug that stimulates release of FVIII and VWF.

Diamond–Blackfan anaemia The principal congenital cause of pure red cell aplasia.

Diapedesis The movement of neutrophils, between the vascular endothelial cells, from the blood into the tissues.

Dimer A molecule composed of two linked subunits.

Dimorphic A population of two distinct populations of red cells in the same full blood count (FBC) sample.

Diploid Chromosomal complement of a normal somatic cell, containing 22 pairs of autosomes and 2 sex chromosomes.

Direct antiglobulin test (DAT) A laboratory test to search for an antibody on the surface of red blood cells. Commonly used in the diagnosis of AIHA.

Direct oral anticoagulants (DOACs) Drugs that act directly on certain coagulation molecules (thrombin or Factor Xa).

Direct thrombin inhibitor (DTI) Anticoagulant drug that interacts directly with thrombin without the need for cofactors.

Disseminated intravascular coagulation (DIC) A potentially life-threatening condition resulting from excessive activation of coagulation plus loss of control and localization mechanisms.

Divalent metal transporter-1 The molecule at the surface of the enterocyte through which iron passes into the body.

DNA methyltransferases (DNMTs) A family of enzymes responsible for the transfer of methyl groups to DNA, leading to gene silencing.

Döhle bodies Cytoplasmic remnants of endoplasmic reticular material; blue-grey in colour.

Domain A discrete section that is part of the structure of a protein.

Drumstick Small nuclear protrusion often expressed by neutrophils on a blood film and found in females (and some males) comprising a condensed X chromosome.

Dry tap An unsuccessful bone marrow aspirate which yields either no bone marrow material or insufficient amounts for analysis. A dry tap could be caused by poor technique, but is more likely to be due to bone marrow fibrosis or hypercellularity.

Dyserythropoiesis Abnormal red cell development.

Dysfibrinogenaemia Functional deficiency of fibrinogen.

Dyskaryorrhexis Abnormal bursting of a cell's nucleus.

Dysplasia The abnormal development or maturation of cells (dysplastic), tissues, or organs.

Dysplastic This refers to cells that have abnormal maturation characteristics.

Dysregulation Describes the abnormal function of one or more regulated processes within a cell.

Ecchymoses Bruises caused by the leakage of blood from blood vessels.

Eclampsia One or more convulsions occurring during or immediately after pregnancy.

Ectoparasite A parasite that lives on the external surface of the host.

Edoxaban Orally administered anticoagulant that directly inhibits activated factor X.

EDTA An anticoagulant chemical that, when added to whole blood, allows a full blood count to be undertaken.

Effector Describes a fully functional, mature cell.

Electrocardiography A procedure for measuring electrical changes in the heart which take place with each beat.

Electrophoresis The migration of dispersed particles (perhaps molecules) relative to a fluid under the influence of an electric field.

Embden–Myerhof glycolytic pathway A complex series of metabolic reactions whereby energy (in the form of ATP, NADH, and NADPH) is generated from glucose.

Embolus A blood clot or part of a blood clot that migrates through the bloodstream and then lodges in another vessel, causing a blockage.

Enantiomeric Two stereoisomers that are complete mirror images of each other.

Endoglin A membrane glycoprotein involved in angiogenesis.

Endomitosis Replication of the nuclear genome in the absence of cell division, which leads to elevated nuclear gene content and polyploidy.

Endoparasite A parasite that lives within the body of the host.

Endoreplication The process of chromosomes replicating in the absence of cytokinesis.

Endothelial cell A specialized cell that forms the internal lining of the blood vessels and sinuses: damage/dysfunction promotes thrombosis and hypertension.

Endothelial protein C receptor (EPCR) Transmembrane receptor that binds protein C to enhance its activation.

Endotoxin Lipopolysaccharide bacterial cell wall component that can initiate a potentially catastrophic inflammatory response.

Enzyme-linked immunosorbent assay (ELISA) An immunoassay that employs an enzyme linked to an antibody or antigen as a marker for the detection and quantification of proteins/antibodies.

Eosinophil A granulocyte characterized morphologically by large red-brown granules that may partially or completely obscure the bilobed nucleus.

Eosinophil cationic protein (ECP) A bactericidal and helminthotoxic protein that can also induce degranulation of mast cells.

Eosinophil-derived neurotoxin/eosinophil protein X (EDN/EPX) Is not restricted to eosinophils, but can also be found in basophils and monocytes.

Eosinophil peroxidase An enzyme forming reactive singlet oxygen and hypobromous acid in the presence of H_2O_2 and bromide ion.

Epidermotropic A preference to existing within the epidermis.

Epimastigote Replication phase of *T. cruzi* in the vector gut.

Epistaxis Nosebleeds.

Epitope A portion or section of a molecule to which an antibody binds.

Eryptosis The process of red blood cell death and haemolysis.

Erythroblast or normoblast Blast of the erythrocyte (red blood cell) pathway.

Erythroblastic island A highly specialized microenvironment in the bone marrow where erythropoiesis takes place.

Erythrocyte A red blood cell, carrier of haemoglobin.

Erythrocyte sedimentation rate A physical property of blood defined by the rate of settling of the blood by gravity.

Erythrocytosis Increased numbers of circulating red blood cells, generally the consequence of the bone marrow's response to loss of mature red cells.

Erythroderma Red and scaling skin caused by inflammation.

Erythroferrone A hormone produced by erythroblasts that inhibits the production of hepcidin, so promoting iron uptake.

Erythropoiesis The process of the development of red blood cells.

Erythropoietin A growth factor, generally derived from the kidney, that promotes the development of red blood cells.

Essential thrombocythaemia A clonal proliferation of megakaryocytes leading to an increased platelet count.

Euchromatin Describes regions of DNA with an open chromatin structure possessing active genes.

Euploid Containing 46 chromosomes.

Exflagellation Extension of flagella from a malarial microgametocyte.

Exocytosis Binding of cytoplasmic vesicles to the cell membrane leading to the release of the vesicle's contents into the extracellular environment.

External Quality Assessment (EQA) Where a central agency supplies all registered laboratories with blood samples for analysis with locally employed techniques to enable performance comparisons.

Extramedullary haemopoiesis Haemopoiesis occurring outside the bone marrow, such as in the liver, spleen, and lymph nodes.

Extravascular haemolysis Haemolysis outside the blood vessels, generally in the liver and spleen.

Extrinsic pathway A compartment within the cascade theory of coagulation involving activation of FX by FVIIa and cofactors of tissue factor, phospholipid, and calcium ions.

Extrinsic tenase complex Enzyme/cofactor/substrate complex (FVIIa/TF/FX + phospholipid and calcium ions) that generates the initial activated factor X.

Exudative tonsillitis Enlarged red tonsils covered in white patches.

Factor deficiency Reduced concentration and/or function of a coagulation factor that may predispose to a bleeding tendency.

Factor V Cambridge Arg306Thr mutation in the factor V gene, rendering the FV molecule resistant to inactivation by activated protein C.

Factor V Hong Kong Arg306Gly mutation in the factor V gene. It does not cause activated protein C resistance.

Factor V Leiden A mutation in the factor V gene, rendering the FV molecule resistant to inactivation by activated protein C.

Factor V Liverpool Ile359Thr mutation in the factor V gene, rendering the FV molecule resistant to inactivation by activated protein C.

Factor V Nara Trp1920Arg mutation in the factor V gene, rendering the FV molecule highly resistant to inactivation by activated protein C.

Factor VIII-binding capacity (VWF:FVIIIB) A chromogenic/ELISA assay to quantify the FVIII binding capacity of VWF.

Faggot cells Immature granulocytes that contain bundles of Auer rods. These are commonly seen in acute promyelocytic leukaemia.

Fanconi's anaemia (FA) The principal congenital cause of pancytopenia, also linked with skeletal and other abnormalities.

FcγR A receptor that recognizes a particular region (called the Fc region) on an IgG antibody.

Ferritin The main storage protein for iron: can be present in cells or in the serum.

Ferroportin A component of the cell membrane of enterocytes, macrophages, and hepatocytes controlling the export of iron into the circulation.

Fibrin Insoluble polymer strands derived from fibrinogen that form part of a blood clot and strengthen it.

Fibrinogen Fibrin precursor and ligand that facilitates platelet–platelet aggregation and forms a mesh that is the basis of the thrombus.

Fibrinolysis Regulated mechanism of fibrin clot destruction.

Fibrinopeptide A Peptide that shields polymerization site on fibrinogen Aα chain.

Fibrinopeptide B (FPB) Peptide that shields polymerization site on fibrinogen Bβ chain.

Fibroblast A cell found in the bone marrow and numerous other sites in the body that synthesizes many products such as collagen, but also maintains structural integrity and contributes to wound healing and tissue repair.

Fibronectin A multifunctional glycoprotein involved in cell adhesion, differentiation, growth, and wound healing.

Fibrosarcoma A tumour developing in the fibrous connective tissue.

Field's stain A rapid Romanowsky stain used on unfixed thick films of peripheral blood for malarial parasite detection.

Filariasis Parasitic nematode infections by *W. bancrofti*, *B. malayi*, *B. timori* (lymphatic filariasis), and *L. loa* (Loa loa filariasis).

Flow cytometer Machine that analyses cells treated with fluorescent dyes and moving in a liquid stream past a laser beam. Analysis is based on the size, granularity, and fluorescence of the individual cell.

Flow cytometry A technique that identifies blood and bone marrow cells according to size and granularity. Also used in conjunction with fluoresceinated antibodies to determine the presence of different molecules on the cell surface or in the cytoplasm.

Fluorochrome A chemical which, when excited by light of a particular wavelength, will emit light of a different but predictable wavelength, measurable using a photodetector.

Fluorescence flow cytometry (FFC) An extension of flow cytometry that identifies the presence of different molecules on the cell surface or in the cytoplasm.

Folate Essential micronutrient requirement in the development of the red blood cell.

Follicular dendritic cell (FDC) Cells with long, branching processes found within lymphoid follicles. FDCs are important for enabling B-cell maturation.

Fondaparinux Parenterally administered synthetic anticoagulant that directly inhibits activated factor X.

Forssman antibody An antibody directed against Forssman antigen.

Forssman antigen A glycosphingolipid found on cell membranes in many different species. Antibodies directed against the Forssman antigens are caused by a number of infectious agents, although are not associated with infectious mononucleosis.

Forward scatter A flow cytometric measurement to determine the size of a cell.

Frameshift mutation An alteration in the codon structure, changing the reading frame.

Fresh-frozen plasma A therapeutic that contains plasma proteins including coagulation factors and is used clinically to restore haemostasis.

Full blood count (FBC) The most common and important laboratory test in haematology; it provides information on red blood cells, white blood cells, and platelets.

Fusion protein Following translocation, the protein product containing elements of both fused genes. Also called chimeric protein.

G0 The part of the cell cycle when it is not actively proliferating but is resting.

Gain-of-function mutations Mutations that result in increased concentrations or interactions with target molecules.

Gastric parietal cells Cells of the luminal wall of the stomach which synthesize and secrete the vitamin B_{12} carrier molecule intrinsic factor, essential for the absorption of the vitamin.

Gating The process of selecting a particular population of cells being analysed by flow cytometry on which to base further immunophenotyping analysis.

Geometric mean normal prothrombin time (GMNPT) Locally derived mean value from a minimum of 20 healthy donors used in calculation of PT ratios and INRs.

Gla domain Calcium binding portion of vitamin K-dependent factors comprising γ-carboxyglutamic acid residues.

Glanzmann's thrombasthenia Bleeding disorder characterized by defects of the GPIIbIIIa platelet surface receptor.

Globin That part of the haemoglobin molecule that is protein.

Glucose-6-phosphate dehydrogenase (G6PD) A metabolic enzyme involved in the generation of NADPH.

Glutathione An amino acid that provides protection against toxic reactive oxygen species.

Glycophorin A and glycophorin C Components of the membrane of the red cell that provide recognition, transport, and anchorage sites.

Glycoproteins Proteins with carbohydrate chains.

Glycosaminoglycan Group of high molecular-weight polysaccharides, commonly adopting structural functions, that includes chondroitin sulfate, dermatan sulfate, heparan sulfate, and heparin.

Granules Small bodies within the cytoplasm of granulocytes that contain bioactive molecules such as enzymes.

Granulocyte A white blood cell with granules in the cytoplasm. Neutrophils, eosinophils, and basophils all have prominent granules (and an irregular nucleus); monocytes may have granules but this is rare.

Granulopoiesis The growth, differentiation, and maturation of granulocytes.

Gray The unit of measurement for the absorbed dose of radiation (i.e the amount deposited in the mass of a particular material). One Gray = one joule per kilogram.

Growth factors Cytokines produced by one type of cell that initiate or promote the growth or differentiation of another cell.

Haem The non-protein part of the haemoglobin molecule that contains iron.

Haemarthrosis Bleeding into joint spaces that can lead to joint damage and disability.

Haematinic Nutrients required for effective red cell and haemoglobin production.

Haematocrit (Hct) The proportion of blood that is made up of red blood cells.

Haemato-oncology The study of cancers of the haemopoietic system.

Haematuria Presence of blood in urine.

Haemochromatosis The pathological condition of iron overload.

Haemodilute When aspirated contents of the bone marrow are excessively diluted with peripheral blood.

Haemoglobin A specialized iron-containing protein within the red blood cell that transports oxygen.

Haemoglobinopathy Disease (such as thalassaemia and sickle cell disease) resulting from mutations in the globin genes and so abnormal haemoglobin synthesis.

Haemolysis The inappropriate destruction of red blood cells that commonly leads to anaemia.

Haemolytic anaemia The consequence of inappropriate destruction of red blood cells in the circulation and/or liver and spleen, perhaps due to a defect in the red cells themselves, or due to a pathological condition acting on otherwise healthy cells.

Haemolytic uraemic syndrome (HUS) Thrombotic microangiopathy occurring mainly in young children following gastrointestinal illness accompanied with diarrhoea.

Haemopexin A plasma protein that complexes with, and thus removes, haemoglobin and free haem from the circulation.

Haemophilia Hereditary bleeding disorder caused by a deficiency in a clotting factor.

Haemophilia A FVIII deficiency.

Haemophilia B FIX deficiency.

Haemophilia C FXI deficiency.

Haemopoiesis The process of producing the cellular constituents of the blood: red blood cells, white blood cells, and platelets.

Haemoproliferative disorder A condition characterized by inappropriately increased numbers of circulating blood cells and their precursors.

Haemorrhage Excessive bleeding caused by a breakdown in haemostasis.

Haemorrhagic disease Disorders that lead to excessive bleeding.

Haemosiderin A granular and storage form of iron normally present in cells such as those of the liver, spleen, and bone marrow.

Haemostasis The interplay of cellular and molecular processes that maintain blood fluidity and also generate blood clots at sites of injury, regulate clot formation, and degrade clots.

Haemoxygenase A key enzyme in the recycling pathway of the haem molecule that generates iron, carbon monoxide, and biliverdin.

Haemozoin Product of haemoglobin digestion by malarial parasites comprised of polymerized insoluble haem residues.

Half-life The period of time for the amount of drug in the body to reduce its initial concentration by one-half.

Haploinsufficiency Occurs when one of a pair of genes on homologous chromosomes is silenced through deletion or mutation. The intracellular concentration of the 'normal' gene product is insufficient to complete its intended biological role.

Haptocorrin An additional plasma carrier of vitamin B_{12}.

Haptoglobin A plasma protein that complexes with, and thus removes, haemoglobin and free haem from the circulation.

Health and Care Professions Council (HCPC) The government body that regulates professionals in a variety of healthcare disciplines to ensure they are fit to practice.

Helicobacter pylori A bacterium commonly found in the stomach and a likely cause of disease such as gastritis and stomach cancer. These in turn may contribute to micronutrient malabsorption and thus anaemia.

Helix-loop-helix domain Contained by a family of transcription factors, called helix-loop-helix proteins, these domains facilitate the activation of dimers from inactive monomers.

Helminth A parasitic worm commonly found within the intestines.

Heparin A natural anticoagulant often added to blood to allow additional analyses of white blood cells.

Heparin resistance Some patients require large doses of UFH to prolong the APTT due to AT deficiency, short heparin survival *in vivo* associated with large thrombi, elevated FVIII or fibrinogen, circulating activated coagulation factors, or HIT.

Heparin-induced thrombocytopenia (HIT) Reduction in platelet count associated with heparin and LMWH therapy. Type I disease is benign and results from the activation of platelets by high-dose UFH. Type II disease is caused by an immune response and associated with thrombosis.

Hepatic portal circulation Circulation of blood from the small intestine, right side of the colon, and spleen to the liver via the hepatic portal vein.

Hepatitis Inflammation of the liver, which can be due to infection, autoimmunity, or exposure to toxic substances.

Hepatocyte Liver cell involved in protein synthesis and storage.

Hepatomegaly Increase in the size of the liver.

Hepatosplenomegaly Increase in the size of the spleen and liver.

Hepcidin A liver-derived molecule that regulates the movement of iron into the blood.

Hereditary haemochromatosis (HH) The principal congenital cause of iron overload, most often due to a mutation (termed C282Y) in the gene for the iron regulator HFE.

Hereditary haemorrhagic telangiectasia (HHT) Fragile blood vessel disorder.

Heterochromatin Contains hypoacetylated histones and a closed chromatin configuration. Genes within these areas are switched off.

Heterophile antibody An antibody that reacts with antigens expressed on the surface of cells derived from an unrelated species.

Heterozygous Inheritance of two different copies of an allele at a particular locus (compare Homozygous).

High molecular weight (HMW) multimers The largest VWF subunits; the main contributors to GPIb and collagen-binding capacity.

High-pressure [/performance] liquid chromatography Column chromatography technique used to separate, identify, and quantify compounds.

Hirudin Direct thrombin inhibitor extracted from leech saliva.

Histidine-rich glycoprotein (HRG) A regulator of fibrinolysis.

Histiocyte A type of macrophage found within connective tissue.

Histone deacetylase (HDAC) A class of enzyme that removes acetyl groups from histones, thereby inhibiting DNA transcription.

Hodgkin and Reed-Sternberg cell Characteristic mutated B cell associated with Hodgkin lymphoma.

Homeobox Genes that encode transcriptional regulators which are expressed at particular times and in particular places within an organism—for example, during embryonic development or in cell differentiation.

Homeostasis The maintenance of stable physiological systems.

Homozygous Inheritance of two identical copies of an allele at a particular locus.

Human leucocyte antigen (HLA) Specialized cell membrane proteins that help defend us from infection with viruses but severely restrict the success of transplants.

Hypercalcaemia Raised plasma calcium concentration.

Hypercellular Denotes an increase in the size of the haemopoietic compartment of the bone marrow or an increase in the number of cells within the marrow.

Hypercellularity Increase in the number of cells within the bone marrow.

Hypercoagulable state An increased tendency towards blood clotting.

Hyperdiploidy Human cells containing in excess of 46 chromosomes.

Hypermethylation The process of switching off genes through the addition of methyl groups.

Hypersegmented Greater than 3% of the neutrophil population showing in excess of five nuclear lobes.

Hypersegmented neutrophil A principal blood film sign of vitamin B_{12} deficiency. The nucleus of the normal neutrophil may have perhaps three or four lobes; in vitamin B_{12} deficiency the number of lobes may rise to seven or eight.

Hypertension A sustained increase in systolic and/or diastolic blood pressure.

Hyperviscosity Increased blood viscosity.

Hypnozoite A dormant stage in the life cycles of *P. vivax* and *P. ovale*.

Hypocellular A reduction in the size of the haemopoietic component of the marrow and an increase in the size of the yellow marrow (fatty marrow) compartment.

Hypochromic On a blood film, a red blood cell lacking in colour and therefore very likely to be lacking in haemoglobin.

Hypofibrinogenaemia Low concentration of functionally normal fibrinogen.

Hypomethylation The process of removing methyl groups from DNA, leading to gene activation.

Hyposegmented A neutrophil with two nuclear lobes or fewer.

Hypoxia Inadequate oxygen supply.

Iatrogenic Describes a disease or condition caused by medical intervention.

Idling Extravascular priming of coagulation.

Imatinib mesylate A signal transduction inhibitor which can be used to target the ATP binding pocket conserved on the Abl transcript of BCR-ABL but which can also inhibit c-Kit and the platelet-derived growth factor receptor (PDGFR).

Immature B cell An early B cell, more mature than the pre-B cell, that expresses intact IgM.

Immune complex The combination of an antibody bound to its target antigen.

Immune paresis Reduction of all immunoglobulins in a patient's serum, except for the paraprotein.

Immunocompetent Fully functioning immune system.

Immunodeficiency Any inherited or acquired inability to mount a fully effective immunological response to a microbial pathogen.

Immunophenotyping The process of determining the characteristics of a group of cells by probing for the presence of cell-based molecules using monoclonal antibodies conjugated to fluorochromes: synonymous with FFC analysis.

Inclusion body An abnormal body within the red blood cell generally associated with a particular pathology.

Indirect antiglobulin test (IAT) A laboratory test to search for an antibody on the surface of red blood cells. Commonly used in the diagnosis of AIHA.

Ineffective haemopoiesis A state in which the bone marrow produces a large number of precursor cells, which perish (usually via apoptosis) before entering the peripheral blood.

Infarction The process that results in the generation of an area of necrotic (dead) tissue caused by the loss of an adequate blood supply.

Infection The presence of sufficiently high numbers of a microorganism that invoke clinical symptoms and provoke a defensive response.

Inhibitor In haemostasis, inhibitors are pathological antibodies that interfere with function of haemostatic molecules or initiate their immune-mediated removal from the circulation.

Initiation Events that start coagulation biochemistry via tissue factor and activated factor VII.

Interleukin (IL) A cytokine that passes instruction messages from one leucocyte to another.

Internal quality controls (IQC) Samples of known values that are analysed simultaneously with patient samples to monitor performance of an assay.

Internal tandem duplications (ITD) Arise from the duplication of sequences from within a gene. In relation to FLT3, ITD of the juxtamembrane region of the gene result in constitutive activation of FLT3.

International Normalized Ratio (INR) A system established by the WHO to standardize prothrombin time (PT) reporting for patients receiving oral anticoagulants.

International Sensitivity Index (ISI) Calibrated numerical value defining the sensitivity of a thromboplastin to the oral anticoagulant effect relative to the WHO standard.

Interstitial deletion Loss of chromosomal material from within a chromosome.

Intravascular haemolysis Haemolysis occurring in the circulation.

Intravenous Administration of a drug in liquid form into the body by injection into a vein.

Intrinsic factor (IF) A small glycoprotein exported from gastric parietal cells into the lumen of the stomach to enable the transport of vitamin B_{12} to the large intestines and facilitate its absorption.

Intrinsic pathway Compartment within the cascade theory of coagulation involving a sequence of enzyme activation reactions starting with contact activation of FXII leading to activation of FX via PK, HMWK, FXI, FIX, FVIII, and cofactors of phospholipid and calcium ions.

Intrinsic tenase complex Enzyme/cofactor/substrate complex (FIXa/FVIIIa/FX + phospholipid and calcium ions) that generates activated factor X on the platelet surface.

Ionizing radiation Radiation in the form of either high-energy waves or particles which, when interacting with atoms, can remove electrons, thus producing a charge (ion).

Iron-deficiency anaemia An anaemia that results from insufficient iron being delivered to the bone marrow.

Ischaemia Inadequate blood supply to a tissue due to blood vessel blockage.

JAK2 An enzyme activated by the binding of erythropoietin to its receptor that ultimately results in cell proliferation and differentiation.

James's dots Multiple small, brick-red dots inside red cells infected with *P. ovale* that are darker than the Schüffner's dots seen inside red cells infected with *P. vivax*.

Jaundice A yellow coloration of the skin caused by high levels of bilirubin that may in turn be a sign of excessive haemolysis and/or liver damage.

Kala-azar Alternative name for visceral leishmaniasis.

Karyotype Description of the chromosomal complement of a cell.

Kinetoplast Mitochondrial DNA.

Kinins A family of proteins that play an important role in inflammation, haemostasis, and pain.

Köhler illumination The mechanism to set up a microscope ready for use.

Lamellipodia Extensions of the cell cytoskeleton; they comprise actin projections which aid cell locomotion.

Large granular lymphocytes (LGL) A population of large lymphocytes containing cytoplasmic granules.

Lectin Mannose Binding Protein 1 Intracellular transport protein for FV and FVIII.

Left-shift Describes an increase in the number of immature neutrophils within the peripheral blood.

Lethargy Tiredness or fatigue.

Leucocyte An alternative name for the white blood cell.

Leucocytosis An increase in the white blood cell count.

Leucoerythroblastic blood picture A blood film showing nucleated red cells and immature white cells.

Leucopenia Describes a white cell count below the lower limit of the reference range.

Leukaemia A haemoproliferative disorder usually characterized by inappropriately increased numbers of white blood cells and their precursors in the circulation. There may also be abnormalities in red blood cells and platelets.

Leukaemic phase Describes the presence of malignant leukaemic cells that have entered the peripheral blood from their associated lymphoid organ(s).

Leukaemogenesis The development of leukaemia.

Leukaemoid reaction A white cell count greater than 50×10^9/L with all stages of maturation present within the peripheral blood.

Leukotrienes A type of lipid related to prostacyclins involved in inflammation and allergic reactions.

Ligands Molecules that bind other structures, such as cell surface receptors.

Light-chain restriction An overexpression of either kappa or lambda light chains. This overexpression of a single type of light chain would be the product of a monoclonal population of B cells.

Lineage-specific growth factors Growth factors that act on only one set of cells and not on any other, e.g. on cells of the red cell series and not on cells of the granulocyte series.

Lineage-specific stem cell A stem cell that will give rise to only one type of blood cell, e.g. red blood cells alone, not red blood cells and/or granulocytes.

Lineage-restricted A particular immunophenotypic marker that is only expressed on one particular cell lineage.

Lipoprotein (a) Competitive inhibitor of plasmin.

Low molecular weight heparin (LMWH) A depolymerized form of unfractionated heparin that retains anticoagulant activity, mainly towards FXa, but with more predictable and favourable pharmacokinetics.

Lupus anticoagulant (LA) A type of antiphospholipid antibody specifically detected in coagulation assays.

Lymphadenopathy Swollen or enlarged lymph nodes, possibly the consequence of infection or a malignancy such as lymphoma.

Lymphoblast A blast of the lymphocyte pathway that ultimately develops into the mature lymphocyte.

Lymphocyte A small white blood cell with a round and regular nucleus that occupies approximately 95% of the cell. It has immunological properties, such as antibody production or destruction of cells infected with viruses.

Lymphocytosis A raised lymphocyte count above the upper end of the reference range.

Lymphomagenesis The development of lymphoma.

Lymphomas Malignancies of lymphoid cells largely restricted to the lymphoid organs (spleen or lymph nodes), although sometimes encountered in patients with extranodal lymphomas (those originating outside the lymphoid organs).

Lymphopenia A reduction in the number of lymphocytes, below the lower end of the reference range.

Lymphopoiesis The growth, differentiation, and maturation of lymphocytes within primary lymphoid organs.

Lymphoproliferative Inducing cells of lymphoid origin to proliferate at a rate greater than that usually observed in healthy individuals.

Lyonization Switching off an X chromosome in females through chromatin condensation. This chromatin condensation prevents the transcription and translation of genes contained within the additional X chromosome.

Lysosomes Organelles that contain digestive enzymes.

Macrocyte A large red blood cell.

Macrocytic Describes an anaemia where the MCV is above the top of the reference range.

Macrogametocyte Female malarial gametocyte.

Macroglobulinaemia Large amounts of big proteins circulating in the blood, for example IgM antibodies.

Macrophage A specialized white blood cell that is found in the tissues and derived from circulating monocytes.

Major basic protein (MBP) Disrupts the lipid bilayer of parasites and target cells through interactions with anionic regions on targets.

Malaria An infectious disease found in tropical and subtropical regions. It is caused by parasites of *Plasmodium* species that are carried by mosquitoes.

Maltese Cross formations Tetrad of *Babesia* ring form trophozoites.

Marginate Adhere to the vascular endothelium.

Maturation arrest Occurs in malignant cells when they are unable to develop beyond a particular stage of maturation. This failure in development prevents the production of a population of mature and functional effector cells.

Mature B cell Surface expression of both IgM and IgG, these cells are yet to encounter antigen. Also called naive B cells.

Maurer's clefts Irregular red/mauve dots inside red cells infected with *P. falciparum* that are larger and fewer in number than Schüffner's dots.

Mean cell haemoglobin (MCH) The amount of haemoglobin inside the 'average' red blood cell.

Mean cell haemoglobin concentration (MCHC) The concentration of haemoglobin inside the 'average' red blood cell.

Mean cell volume (MCV) The size of the 'average' red blood cell.

Mediastinum This defines the area within the thorax between the lungs containing the heart, trachea, oesophagus, and thymus.

Megakaryoblast The precursor cell to the megakaryocyte.

Megakaryocyte The bone marrow cell that gives rise to platelets.

Megaloblastic anaemia A type of anaemia defined by the presence of large erythroblasts (megaloblasts) in the bone marrow and generally caused by lack of vitamin B_{12}.

Megaloblasts Literally 'large blast cells' present in the bone marrow, often as a result of lack of vitamin B_{12} or folate.

Melanosome Pigment-containing organelle.

Menorrhagia Heavy menstrual bleeding.

Memory cell Provides a long-term memory of previously encountered antigens thus facilitating the development of a secondary immune response.

Merosome Structure within which malarial merozoites exit hepatocytes.

Merozoite The stage of the malarial parasite life cycle resulting from schizont multiplication.

Metabolic acidosis Accumulation of acid in blood leading to pH reduction.

Metacyclic A biochemical term for the extension of a cyclic group by another cyclic group.

Metacyclic trypomastigotes Infectious stage of trypanosome life cycle that expresses variant surface antigens.

Metamyelocyte The final stage of the development of a granulocyte before it becomes a mature polymorphonuclear leucocyte.

Methaemoglobin A subtype of haemoglobin characterized by the presence of iron in its oxidized state.

Microangiopathic haemolytic anaemia (MAHA) Loss of, or damage to, red blood cells through destruction caused by factors in the small blood vessels, leading to haemolysis and anaemia.

Microcyte A small red blood cell.

Microcytic Describes an anaemia where the MCV is below the bottom of the reference range.

Microdeletion A submicroscopic deletion within a chromosome.

Microgametocyte Male malarial gametocyte.

Micronutrients Essential minerals and vitamins absorbed from the diet and required in trace amounts for the correct function of a cell or biochemical process.

Microparticles Small fragments released from endothelial cells, leucocytes, and platelets consisting of a plasma membrane and a small amount of cytoplasm.

MicroRNAs (miRNA) Small fragments of RNA, 20–23 nucleotides in length, which bind in a complementary fashion to mRNA to inhibit translation.

Mimetic A compound that mimics properties of another.

Missense mutation Encodes a different amino acid following a single nucleotide exchange.

Mitogen Any chemical or substance capable of inducing mitosis.

Mixing tests Mixing patient plasma with normal plasma and performing coagulation tests can indicate the presence of a factor deficiency or inhibitor.

Monoblasts and promonocytes Immature cells that are bone marrow precursors of mature monocytes.

Monoclonal antibodies (mAbs) These are manufactured to recognize a particular epitope on a specific antigen. Although mAbs produced by different manufacturers may have the same name, they may recognize a different epitope on the specified antigen.

Monocyte A large white blood cell whose nucleus is regular and generally round, and which occupies 60–80% of the cell. It has immunological and secretory properties and, in the tissues, develops into a macrophage.

Monomer A molecule that can combine with others to form a polymer; the smallest repeating unit of a polymer.

Mononuclear leucocyte A white blood cell with a round and regular nucleus. Generally, these cells do not have granules in their cytoplasm.

Monosomy Loss of a particular chromosome.

Morbidity High rate of sickness and medical complications compared to the normal population.

Morphological Pertaining to the external appearance of cells.

Morphology The shape or appearance of a cell.

Mortality Increased risk of death compared to the normal population.

M-protein A monoclonal protein or paraprotein.

Multimer A protein composed of more than one peptide chain.

Multimerization Formation of multimers.

Multiple coagulation factor deficiency 2 Cofactor for lectin mannose binding protein 1.

Musculoaponeurotic A fibrous or membranous sheath that connects muscle to bone.

Mutagen A chemical substance known to induce DNA mutations.

Myeloablative therapy The process whereby the bone marrow is destroyed by high doses of chemotherapeutic agents in preparation for stem cell transplantation. Stem cell transplantation provides a potential cure for patients receiving this therapy.

Myeloblast The least mature, morphologically identifiable cell of the myeloid lineage.

Myelodysplasia Abnormal development of myeloid cells.

Myelofibrosis A bone marrow condition characterized by overgrowth of normal haemopoietic tissue by fibroblasts.

Myeloma A malignancy of white blood cells within the bone marrow, characterized by anaemia and, generally, a high erythrocyte sedimentation rate (ESR) and increased plasma viscosity.

Myelomagenesis The process that leads to myeloma.

Myeloperoxidase (MPO) A haem-containing enzyme important for the non-specific elimination of bacteria, viruses, and fungi.

Myelosuppression The failure of the myeloid component of the bone marrow to produce red cells, white cells, and platelets as a consequence of either a therapeutic agent or a malignant clone.

Myocardium Heart muscle.

Naive B cell Surface expression of both IgM and IgD, these cells are yet to encounter antigen. Also called mature B cells.

Naturally occurring inhibitors (non-pathological) Molecules that regulate secondary haemostasis by inhibiting activated clotting factors.

Necrotic Localized tissue death resulting from insufficient blood flow to the tissue due to injury, radiation, or toxic substances.

Negative predictive index The probability that a patient with a normal D-dimer level does not have a DVT or pulmonary embolism (PE).

Neoepitope An epitope is a localized area of a molecule that is recognized by an antibody; a neoepitope is a structure, recognizable by an antibody, that is only exposed upon formation of a complex.

Neoplasia An abnormal proliferation of cells that can be benign or progress to malignancy.

Nephelometry Measurement of the intensity of light scatter when transmitted through a reaction mixture containing particulate matter by detecting light at an angle (usually at right angles or about 75°).

Neutropenia A neutrophil count below the lower limit of the reference range.

Neutrophil A granulocyte characterized morphologically by small granules that generally do not obscure the multilobed nucleus.

Neutrophilia A neutrophil count above the upper limit of the reference range.

NFκB Is a transcription factor that is involved in cell development, growth, and apoptosis. Overexpression of NFκB is associated with a number of diseases, including cancer.

Nijmegen modification Altered version of Bethesda inhibitor assay that reduces pH drift.

Nilotinib A next-generation signal transduction inhibitor employed for the treatment of chronic myeloid leukaemia.

Nonsense mutation A mutation in a DNA sequence that generates a codon not encoding an amino acid.

Normoblast, or erythroblast Blast of the erythrocyte (red blood cell) pathway.

Normocellular The percentage of haemopoietic cells to bone marrow falls within the expected range based on a patient's age.

Normochromic Used to refer to a situation where the mean cell haemoglobin is in the reference range.

Normocyte A normal-sized red blood cell.

Normocytic Describes an anaemia where the MCV is within the reference range.

Normocytic normochromic anaemia A particular form of anaemia associated with red cells of a normal size and coloration. Although there can be a number of causes, in the context of a malignancy the most likely causes are anaemia of chronic disease or myelosuppression.

Nucleated red blood cell A stage in the development of the red blood cell when it still retains a nucleus and before it becomes a reticulocyte.

Nucleocytoplasmic ratio The ratio between the size of a cell's nucleus and its cytoplasmic volume.

Nucleolus (plural nucleoli) An area of the nucleus composed of genes that encode ribosomal RNA essential in the translation of transcribed proteins.

Ocular adnexa Accessory structures of the eye, including the eyelids, lacrimal glands, orbit, and paraorbital areas.

Oligomer A molecule containing a small number of repeating units; oligo means 'few' in Greek.

Oncogenic The process of producing cancer.

Oncology The area of medicine that deals with the development, diagnosis, treatment, and prevention of tumours.

One-stage coagulation assay Assay based (mainly) on prothrombin time or activated partial thromboplastin time that quantifies activity levels of individual coagulation factors.

Oocyst Spore phase.

Ookinete Motile malarial zygote.

Opportunistic infection An infection caused by organisms that do not normally cause disease in the presence of a competent immune system.

Opsonins Molecules coating the surface of cells which enhance the recognition and destruction of these coated cells by phagocytes.

Orthogonal regression line Generating a best-fit line through a series of points commonly involves ordinary regression, which assumes the measurements were made without error. Orthogonal regression allows for random error.

Osteolytic lesions Regions of bone degradation due to the overactivity of osteoclasts.

Osteoporosis A skeletal disease that leads to loss of bone mass and structure, resulting in an increased risk of fractures.

Oximetry A non-invasive method for determining the amount of oxygen within arterial blood.

Oxyhaemoglobin Haemoglobin that is carrying oxygen.

Palatal petechiae Small blood spots within the oral cavity (see also Petechiae).

Pallor A pale coloration to the skin.

Pancytopenia Low levels of all blood cells due to bone marrow invasion, suppression, or another pathology.

Pappenheimer bodies Iron-containing granules within siderocytes.

Paracaspase A caspase-related protein.

Paracentric inversion A region of a particular chromosomal arm rotated through 180°.

Paraprotein Circulating monoclonal immunoglobulin or light chain.

Parasitaemia Quantitation of the number of parasitized red blood cells (normally expressed as a percentage).

Parasitism A form of symbiosis where the parasitic organism benefits from the association to the detriment of the host organism.

Parasitophorous vacuole (PV) Vacuole that forms around a malarial parasite upon entering a red cell.

Parasitophorous vacuole membrane (PVM) Invagination of the red cell membrane that forms the membrane of the vacuole that surrounds a malarial parasite when invading a red cell.

Parenterally Not given orally through the alimentary canal, but administered by injection.

p-arm Located above the centromere; this is the shorter chromosomal arm.

Pelger-Hüet anomaly Hyposegmented neutrophils with a single or bilobed nucleus.

Peptidases Enzymes that cleave peptide bonds in proteins.

Peptide Short amino acid chain.

Pericentric inversion Rotation of chromosomal material through 180° and involving the centromere.

Peripheral blood The blood that is contained within the circulatory system.

Perls' stain The principal stain of blood, bone marrow, or tissues such as the liver for deposits of iron.

Pernicious anaemia The anaemia defined by autoimmune destruction of gastric parietal cells and/or intrinsic factor, which causes vitamin B_{12} deficiency.

PEST domains These are composed of proline, glutamic acid, serine, and threonine and provide proteolytic signals to enhance protein degradation.

Petechiae Red, pinpoint-sized haemorrhages of small capillaries in the skin or mucous membranes that form as a consequence of blood leaking from vessels.

Peyer's patches A group of lymph nodes in the wall of the ileum.

Phagocytosis The process of the ingestion and destruction of foreign and unwanted material, such as bacteria and effete red blood cells. Phagocytosis is performed by phagocytes, principally neutrophils and monocytes/macrophages.

Phagolysosome Membrane-enclosed vesicle formed from fusion of a lysosome (organelle containing digestive enzymes) and a phagosome (vacuole formed around a foreign body inside a phagocytic cell).

Phagosome Neutrophil membrane forming a vacuole around the membrane of an ingested bacterium.

Pharmacodynamics The action, effect, and breakdown of drugs in the body.

Pharmacokinetics The absorption, distribution, metabolism, and excretion of a drug.

Pharyngitis Inflammation of the pharynx caused by an infection.

Phenindione An oral anticoagulant and inhibitor of vitamin K metabolism.

Phenprocoumarin An oral anticoagulant and inhibitor of vitamin K metabolism.

Philadelphia chromosome This denotes the derived chromosome 22 der(22q−), not to be confused with t(9;22). The translocation process is abbreviated to t(9;22).

Phosphorylation The process of adding a phosphate group (PO_4) to an organic molecule.

Pitting The process of removing red cell inclusions, such as malarial parasites, which occurs in the spleen.

Placental insufficiency Placental failure to supply nutrients to the fetus and remove toxic waste.

Plasma The clear, straw-coloured fluid which remains when all blood cells have been removed from blood.

Plasmablast An intermediate stage of development between B cells and plasma cells.

Plasma cell Antibody-secreting terminal stage in B-cell maturation following exposure to antigen.

Plasma viscosity A global property of the 'thickness' or 'thinness' of the plasma.

Plasmin An enzyme that degrades fibrin.

Plasminogen The circulating zymogen of plasmin.

Plasminogen activator inhibitor type 1 (PAI-1) The prime inhibitor of tissue plasminogen activator.

Plasminogen activator inhibitor type 2 (PAI-2) An intracellular molecule involved in cell proliferation that is also produced by the placenta and exerts some activity against t-PA and u-PA.

Plasminogen activator inhibitor type 3 (PAI-3) An inhibitor of activated protein C.

Platelet A blood cell that helps prevent blood loss by combining with other platelets and certain proteins to form a clot, or thrombus.

Platelet-activating factor (PAF) A phospholipid released from platelets that contributes to maintenance of vessel endothelial cell junctions.

Platelet adhesion Process of adhering platelets to exposed subendothelium via ligands and receptors.

Platelet clumps Aggregates of platelets within a FBC, demonstrable on a blood film. Often artefactual, it can be facilitated by the presence of EDTA in blood tubes.

Platelet factor 4 (PF4) A small cytokine released from activated platelets that modulates the effects of heparin-like molecules, stimulates protein C activation, and is chemotactic for neutrophils and monocytes.

Platelet shape change Platelet shape alteration from discoid to a spiny sphere.

Pleomorphic Variations in the appearance of a particular type of cell.

Ploidy Describes the number of homologous chromosomes within a cell.

Pluripotent stem cell A stem cell that has the potential to give rise to many other different types of stem cells.

Poikilocytosis A variation in the shapes of the red cell population. Causes include abnormal erythropoiesis and myelofibrosis.

Point mutation The alteration of a single nucleotide in a gene sequence.

Point-of-care test (POCT) Tests analysed on portable analytical devices that can be used on wards, in clinics, GP surgeries, and even by patients themselves.

Polychromasia A finding on blood film that translates as 'many colours'. In practice, there will be red blood cells of a normal colour, but others (reticulocytes) with a blue tinge.

Polycythaemia A haemoproliferative disorder characterized by inappropriately increased numbers of red blood cells.

Polycythaemia vera (PV) Characterized by an increase in an individual's red cell mass >25% above the patient's mean predicted value.

Polymorphism Variations in the sequence of a particular gene between individuals.

Polymorphonuclear leucocyte (polymorph) A white blood cell with an irregular nucleus. Neutrophils, eosinophils, and basophils are all polymorphs.

Polyploidy The presence of more than one complete set of chromosomes.

Porphyria A disease caused by abnormalities in the production of haem characterized in the laboratory by anaemia and increased iron.

Post-thrombotic syndrome A late complication of DVT, where tissue damage occurs from diversion of blood into other veins to avoid blockage.

Pre-B cell An early B cell containing cytoplasmic IgM.

Pre-eclampsia Pregnancy-induced high blood pressure associated with protein in the urine. The only cure is delivery or abortion of the fetus.

Primary care Often the initial stages of health care where people seek help from their general practitioner.

Primary haemostasis The initial haemostatic response that results from interplay between vessel endothelium, von Willebrand factor, and platelets.

Primary lymphoid organs Includes the bone marrow and thymus as main sites of lymphopoiesis.

Procoagulant Physiological or pharmacological mechanism that promotes coagulation processes.

Proerythroblast A post-stem cell stage of differentiation of the red blood cell.

Professional body A learned organization that represents a particular profession.

Progenitor (Pro) B cell Progenitor (Pro) B cell is a precursor cell of B-lineage.

Pro-inflammatory cytokines Regulatory cellular-communication signalling molecules that favour and promote inflammation.

Prolymphocyte A stage in lymphoid development between lymphoblast and lymphocyte.

Promastigote The infective stage of *L. donovani* and *L. infantum*.

Promonocyte and monoblasts Immature cells that are bone marrow precursors of mature monocytes.

Propagation The activation of various haemostatic components by the trace thrombin formed at initiation of coagulation.

Prophylaxis A procedure to prevent a condition occurring rather than treat or cure it after it has occurred.

Prostacyclin A potent inhibitor of platelet aggregation generated through the arachidonic acid pathway within the vessel wall.

Prostaglandin A lipid-derived potent physiological mediator.

Protamine sulphate A strongly basic heparin neutralizer used therapeutically and analytically.

Protease nexin-2 A naturally occurring inhibitor of FXIa, FIXa, and FXa.

Protein C (PC) The zymogen of activated protein C, the naturally occurring inhibitor of the activated forms of factor V and factor VIII.

Protein S (PS) The non-enzymatic cofactor for activated protein C and tissue factor pathway inhibitor.

Proteolytic Enzymes that digest or lyse proteins into smaller sections or amino acids are said to be proteolytic.

Prothrombin time (PT) A coagulation screening test for global assessment of extrinsic and common pathways.

Prothrombin time (PT) ratio Derived from dividing the patient prothrombin time by that of a normal pooled control preparation or previously determined mean normal value.

Prothrombinase complex The enzyme/cofactor/substrate complex (FXa/FVa/FII + phospholipid and calcium ions) that generates activated factor II on the platelet surface.

Prothrombinase-induced clotting time (PiCT) A snake venom-based assay for monitoring direct thrombin inhibitors.

Prothrombotic The promotion of processes that form a blood clot.

Proto-oncogene The normal counterpart of a gene that, when mutated, can produce an abnormal protein which will lead to increased proliferation or cell survival.

Prozone A concentration of antibody or antigen is so high that the optimal concentration for maximal reaction with antigen is exceeded and binding is reduced, or does not occur.

Pulmonary angiography A technique performed to assess blood circulation to the lungs by adding a contrast dye through a catheter into a vein and taking X-rays of the lung.

Pulmonary embolism (PE) A fragment of a clot (the embolus) that travels through the circulation and lodges in the pulmonary artery or one of its branches causing partial or total blockage. It is potentially fatal.

Punched-out lesions Caused by the catabolic effects of osteoclast activating factors on bony structures.

Pure red cell aplasia (PRCA) A condition of suppression of the production of red blood cells which does not influence white blood cells or platelets.

Purpura Red or purple skin discoloration produced by bleeding small blood vessels near the surface.

Purpura fulminans A life-threatening disorder characterized by cutaneous haemorrhage and necrosis (tissue death), and disseminated intravascular coagulation.

Pyrexia A rise in the body's core temperature.

Pyruvate kinase A metabolic enzyme of the glycolytic pathway involved in the generation of ATP.

q-arm Located below the centromere, this is the longer of the chromosomal arms.

Qualitative Non-numerical data; for example, the ability of a platelet to function correctly.

Quantitative Types of data that deal with numerical values; for example, the number of platelets in one litre of blood.

Quantitative buffy coat (QBC) Laboratory technique for concentrating blood parasites and staining their DNA for microscopical detection.

Racemic mixture Equal amounts of two stereoisomers of an optically active substance which does not rotate plane-polarized light.

Reactive oxygen species (ROS) Toxic forms of oxygen that can attack and destroy many components of the red cell.

Receptor A molecule on the surface of (or in) a cell that acts as a recognition site for other molecules that bind in order to promote specific functions.

Recombinant technology The artificial production of a molecule produced by inserting foreign DNA into a host DNA which will produce the product.

Red blood cell count (RBCC) The number of red blood cells in a sample of blood.

Red blood cells (RBCs) Blood cells that carry oxygen.

Red cell distribution width (RDW) A **quantitative** method for describing anisocytosis. The greater the variation in red cell size, the greater the RDW. If the RBCs are of a uniform size, the RDW is low.

Refractory Does not respond to basic/standard treatment.

Refractory anaemia An anaemia that does not respond to B_{12}, folate, or iron therapy.

Release reaction Secretion of platelet granule contents via surface-connected open canalicular system.

Reptilase time Coagulation screening test employing a snake venom enzyme that mainly assesses fibrinogen activity but is not affected by the action of heparin.

Respiratory burst An increase in a phagocyte's oxygen consumption followed by the release of reactive oxygen species.

Restriction enzymes Enzymes that recognize specific nucleotide sequences and can cut DNA wherever the sequence occurs. This selective cutting allows fragments to be formed which can be indicative of mutations within a specified sequence through the loss or gain of these restriction sites.

Reticulocyte The final stage of the development of the red blood cell before it reaches maturity.

Reticuloendothelial system A collection of white blood cells (such as macrophages) present in organs such as the liver, spleen, and lymph nodes, with roles in immunology, phagocytosis, and in scavenging debris and dead cells.

Reversed ratio A lymphocyte count that is higher than the neutrophil count.

Rh-associated glycoprotein Component of the membrane of the red cell that provides recognition, transport, and anchorage sites.

Rheology/rheological The study of the physical nature of blood or plasma; principal measurements are ESR and viscosity.

Rheumatoid arthritis An inflammatory disease that mainly affects the joints.

Rhoptry Organelle found in malarial parasite merozoites.

Right-shifted Greater than 3% of the neutrophil population showing in excess of five nuclear lobes.

Ring sideroblasts Abnormal erythroblasts with iron granules arranged in a ring around the nucleus.

Risk assessment The determination and documentation of the **quantitative** or **qualitative** value of risk related to a specific situation/procedure and a recognized hazard.

Ristocetin cofactor assay A platelet aggregation assay to quantify the GPIb binding capacity of VWF.

Ristocetin-induced platelet aggregation (RIPA) test A platelet aggregation assay that subjects VWF to decreasing concentrations of ristocetin in order to identify type 2B VWD and pseudo-VWD.

Rivaroxaban Orally administered anticoagulant that directly inhibits activated factor X.

Romanowsky A type of stain specially developed to enable the examination of different blood cells.

Sarcoma A tumour that arises from connective tissue.

Schistocytes Distortions and fragments of red cells that are very common in different types of haemolytic anaemia. Consequently, they are not diagnostic of a particular condition, but may be useful in confirmation.

Schizogony Asexual reproductive process of multiple fission in malarial parasites.

Schizont The stage of the malarial parasite life cycle resulting from asexual reproduction in the liver of the human.

Schüffner's dots Multiple small brick-red dots inside red cells infected with *P. vivax* or *P. ovale*.

Scissile Easily split or cleaved.

Secondary care The stage in health care subsequent to primary care, where people are referred to the hospital.

Secondary haemostasis Regulated and localized formation of fibrin polymers via interplay of clotting-factor enzymes and cofactors.

Secondary leukaemia Leukaemia attributable to prior treatment for another disease.

Self-peptides Short protein sequences presented to developing lymphocytes in order to inactivate self-reactive lymphocytes.

Self-renewal A normal property of stem cells enabling the stem cell population to be regenerated without necessarily undergoing differentiation of daughter cells into specific cell lines.

Sequester To remove or separate.

Seropositive Identification of specific antibodies within a patient's serum.

Serotonin Also known as 5-hydroxytryptamine (5-HT), a hormone that acts as a neurotransmitter in the brain and contributes to haemostasis by initiating vasoconstriction.

Serum The fluid that remains after the blood has been allowed to clot.

Shear forces Forces produced when surfaces are pressed together or move over each other.

Sickle cell disease A **qualitative** haemoglobin disorder generally characterized in the laboratory by a chronic haemolytic microcytic anaemia, and in the clinic by infection and microvascular occlusion.

Side scatter In terms of a flow cytometric measurement, represents the internal complexity of a cell as may be due to granules.

Sideroblastic anaemia An anaemia that results from impaired inclusion of iron into haem, and thus into haemoglobin.

Sideroblast An erythroblast—therefore generally found in the bone marrow—containing granules of non-haem iron complexed to other molecules.

Siderocyte A red blood cell characterized by the presence of iron granules as a consequence of sideroblastic anaemia.

Sinus rhythm A natural heart (cardiac) rhythm generated through the sinus node (normal rate 60–100 beats/min.).

Smear cells Lymphocytes that have broken apart due to mechanical damage when making a blood film—commonly seen in leukaemia.

Sodium citrate An anticoagulant chemical added to blood to allow the measurement of coagulation proteins, prothrombin time, and the activated partial thromboplastin time.

Somatic hypermutation (SHM) Occurs within the germinal centre and incorporates point mutations within the variable region of immunoglobulin in order to improve antigenic binding.

Spherocyte An abnormal and small red blood cell characterized by the lack of central pallor.

Splenectomy The procedure of removing the spleen that generally increases the lifespan of cells and as such is a therapeutic option for individuals with certain diseases of the haemopoietic system.

Splenomegaly Enlarged spleen.

Spliceosomes These are complex macromolecular structures composed of small nucleoproteins, involved in the removal of introns from pre-mRNA.

Sporozoite The stage of the malarial parasite life cycle resulting from sexual reproduction in the midgut of the mosquito.

Standard operating procedures (SOP) Document that explicitly states how every aspect of a given procedure must be performed.

Stasis A condition in which the normal flow of blood through a vein is slowed or halted.

Stem cell The cell in the bone marrow that gives rise to all the mature peripheral blood cells.

Stoichiometric The **quantitative** relationship between reactants and products in a balanced chemical reaction.

Streptokinase Exogenous plasminogen activator produced by some strains of streptococcal bacteria.

Stroke A clinical condition where the brain's supply of oxygen is cut off by a blockage or interruption to the blood flow, which causes severe neurological damage.

Subcutaneous Administration of a drug in liquid form into the body by injection just under the skin.

Sulphaemoglobin A subtype of haemoglobin characterized by binding to sulphur-rich compounds.

Sumoylation This is a post-translational modification that results from the binding of SUMO. SUMO is a *S*mall *U*biquitin like *MO*difier that can attach to lysines on target proteins to alter their function or lead to their degradation.

Superoxide A toxic form of oxygen and powerful oxidizing agent that can attack and destroy many components of the red cell.

Superoxide dismutase, catalase Antioxidant enzymes that neutralize toxic reactive species such as superoxide and hydrogen peroxide.

Surface-connected open canalicular system (SCOCS) Intracellular network of channels/canals inside platelets that facilitate release of platelet granule constituents.

Symbiosis Close interaction between different species.

Systemic lupus erythematosus (SLE) A chronic autoimmune disease with multiple organ/system involvement that is often associated with antibodies directed against cell nuclei.

Talin Ubiquitous cytoskeletal protein linking integrins to the actin cytoskeleton.

Telangiectasia Visibly dilated (widened) blood vessels near the surface of the skin.

Terminal deletion The loss of the end of a chromosomal arm.

Ternary complex A complex containing three different molecules.

Tertiary care When a patient is passed from one hospital to a specialist referral centre, often part of a university teaching hospital.

Tetramer Complex comprised of four subunits.

Thalassaemia A **quantitative** haemoglobinopathy caused by a mutation in alpha, beta, or both genes that leads to a reduction, or even abolition, in the expression of alpha or beta globin.

Therapeutic anticoagulation The use of anticoagulants as a treatment for the increased risk of thrombosis.

Therapeutic range The range of concentrations at which a therapeutic agent is effective. For anticoagulants that are monitored with screening tests such as INR and APTT, the therapeutic range is referred to in units of the test, not concentration of the drug.

Thrombin The activated form of factor II (i.e. factor IIa) whose prime functions are coagulation factor activation, converting fibrinogen to fibrin, and activating protein C.

Thrombin activatable fibrinolysis inhibitor Inhibitor of fibrinolysis activated by thrombin bound to thrombomodulin.

Thrombin time (TT) Coagulation screening test that mainly assesses fibrinogen activity.

Thrombocytopenia Low numbers of platelets in the blood, often cited as less than 100×10^9/L.

Thrombocytosis When the platelet count exceeds the top end of the reference range.

Thrombogenic Causing blood to clot; causing thrombosis.

Thrombomodulin A transmembrane receptor that binds and modifies thrombin to facilitate activation of protein C and thrombin activatable fibrinolysis inhibitor.

Thrombophilia A defect in the haemostatic system predisposing an individual to increased risk of venous thromboembolism.

Thromboplastin A reagent used in the prothrombin time containing tissue factor and phospholipid in order to assess the extrinsic and common pathways. Calcium ions may be integral to the reagent or added separately.

Thrombopoiesis The development of platelets (also known as thrombocytes).

Thrombopoietin A growth factor that stimulates megakaryocyte maturation and platelet production.

Thrombosis Inappropriate or pathological formation of a blood clot in an artery or vein.

Thrombospondin Adhesive glycoprotein with roles in platelet aggregation, angiogenesis, and plasmin inhibition.

Thrombotic disease Disorders that lead to thrombosis, which is a partial or complete obstruction of a blood vessel by a blood clot.

Thrombotic thrombocytopenic purpura A thrombotic microangiopathy characterized by spontaneous formation of platelet thrombi in the microvessels leading to mechanical haemolysis and thrombocytopenia.

Thromboxane A_2 (TXA_2) A platelet agonist produced from membrane arachidonic acid as part of the cyclooxygenase pathway within platelets.

Thromboxane B_2 Metabolite of thromboxane A_2 which can be measured as a marker of cyclooxygenase activity and thus of the effect of aspirin.

Thrombus A blood clot that usually develops in a deep vein of the body, normally in the calf or higher up the leg in the thigh (the saphenous vein).

Thymocytes Immature T-cells entering the thymus.

Tissue factor (TF) The initiator of coagulation by acting as a cofactor to activated factor VII; often found at the surface of cells or vessel subendothelium, but also in plasma.

Tissue factor pathway inhibitor (TFPI) The inhibitory regulator of the initiation phase of coagulation.

Tissue plasminogen activator (t-PA) The prime activator of plasminogen.

Toxic granulation Occurs when neutrophils have been mobilized to fight an infection. The primary granules are more abundant in preparation for fighting infection and have a higher concentration of acid mucosubstances, which have prominent azurophilic-staining characteristics.

Trabeculae Bony processes which extend from the outer bony tables into the marrow cavity.

Transcobalamin The molecule that carries vitamin B_{12} in the blood and present in two forms: transcobalamin-1 and transcobalamin-2.

Transcription factor A protein which plays a regulatory role in the transcription of particular genes.

Transferrin Iron is actively carried through the blood as transferrin.

Transferrin receptor A molecule of the surface of red blood cell precursors that mediate the passage of transferrin into the cell. It may appear in the plasma as 'soluble transferrin receptor'.

Transient abnormal myelopoiesis (TAM) In children with trisomy 21, a brief period in which myeloid production within the bone marrow becomes grossly abnormal.

Trephine The process of obtaining a sample of bone that includes both bone tissue and bone marrow.

Trisomy The gain of a particular chromosome.

Trophoblasts Cells of the outer layer of the blastocyst ('pre-embryo') that attach a fertilized ovum to the uterine wall and develop into the placenta.

Trophozoite The feeding stage in the life cycle of malarial parasites.

Tropical pulmonary eosinophilia Nocturnal cough, wheezing, fever, and eosinophilia arising from marked sensitivity to microfilariae in the lungs.

Trypanosome Flagellated *Trypanosoma* trypomastigotes.

Trypanosomiasis Parasitic protozoan infection by *Trypanosoma* organisms.

Trypomastigotes *Trypanosoma* life cycle stages.

Tumour burden The size of an individual's tumour or the number of tumour cells involved in a particular malignant case.

Tumour lysis syndrome The release of a number of intracellular components following lysis of tumour cells.

Tumour suppressor genes Encode proteins that can protect the genome from DNA damage by, for example, inducing DNA repair or apoptosis.

Turbidimetry Measurement of the loss of intensity of light transmitted through a reaction mixture containing particulate matter; unscattered light is measured.

Two-stage factor assay Assays to measure activity of coagulation factors (usually FVIII) where the first stage involves generation of FXa, which then reacts with either a chromogenic substrate or exogenous clotting factors in the second stage.

Type I defect Reduced concentration of a normally functioning protein.

Type II defect Dysfunctional proteins which are often produced in normal amounts.

Ubiquitination The addition of ubiquitin monomers to a peptide which allows for degradation of the peptide via the proteosome.

Unfractionated heparin (UFH) Highly sulphated glycosaminoglycan with high negative charge used as an immediate acting therapeutic anticoagulant due to its potentiating effect on circulating antithrombin. Its use must be monitored by the APTT or anti-Xa assays.

United Kingdom Accreditation Service Accreditation body for pathology laboratories in the UK.

Upshaw–Schulman syndrome Hereditary thrombotic thrombocytopenic purpura due to ADAMTS-13 deficiency.

Uraemia An excess of end products of amino acid and protein metabolism in blood that would normally be excreted in urine, such as urea and creatinine.

Urinary plasminogen activator (u-PA) Plasminogen activator with minor role in blood clot lysis (sometimes called urokinase).

Variable region Generates antigen binding diversity within a specific class of antibodies.

Vascular Relating to the network of blood vessels in the body, namely arteries, veins, and capillaries and associated medical conditions.

Vasculature The network of blood vessels through the body or organs.

Vasoconstriction A narrowing of the diameter of a blood vessel.

Venepuncture The process of obtaining intravenous access to obtain a sample of blood via a needle.

Venography A radiology procedure where a contrasting agent is injected into the vein to enable a blockage or obstruction to be visualized by X-ray examination.

Venous thromboembolism A disorder that encompasses both DVT and PE.

Ventilated-perfusion (V/Q) scan A procedure using small amounts of inhaled or injected radioisotopes to measure flow of blood and air in the lungs. Images taken capture the flow patterns.

Vitamin B_{12}, and B_6, and folate Micronutrients essential for erythropoiesis: their absence results in anaemia.

Vitamin K Cofactor in the γ-carboxylation of glutamic acid side chains in factors II, VII, IX, and X and the inhibitor molecules protein C, protein S, and protein Z.

Vitronectin A multifunctional protein that stabilizes PAI-1, contributes to binding of platelets to vascular cell walls, and promotes adherence, migration, proliferation, and differentiation of many different cell types.

von Willebrand factor (VWF) A large protein that enables platelets to bind to exposed subendothelium and also stabilizes FVIII.

von Willebrands disease (VWD) Deficiency of von Willebrand factor.

Warfarin Most commonly used oral anticoagulant that works by antagonizing the action of vitamin K in the production of functional factors II, VII, IX, and X and the inhibitory proteins C, S, and Z.

Warfarin resistance Present in patients who require higher-than-average warfarin dosage to achieve a therapeutic international normalized ration (INR) and due to a genetic mutation in the vitamin K epoxide reductase enzyme, which reduces its inhibitory effect on vitamin K metabolism.

Warfarin sensitivity Present in patients who require less-than-average warfarin dosage to achieve a therapeutic INR; it is due to polymorphism variants in cytochrome P450 CYP2C9, which result in slower breakdown and clearance of the more potent enantiomer S-warfarin.

Weibel–Palade bodies Intracellular storage bodies for von Willebrand factor in endothelial cells.

Wells' score A clinical assessment method based on scoring the presence of specific symptoms in order to ascertain a probability of the presence of venous thromboembolism (VTE).

White blood cell differential A breakdown of the numbers of the different white blood cells in a sample of blood.

White blood cells Blood cells that function primarily as defence against infection.

Wild-type DNA Denotes germline DNA. Mutations found within tumour cells when compared to wild-type DNA are considered to be acquired.

Xenodiagnosis A method of diagnosing certain vector-borne diseases by allowing uninfected vectors to feed on the patient and then examining them for infections.

Zygote Derived from the Greek for 'joined', it is the unicellular product of joining male and female genetic material—the product of fertilization.

Zymogen Inactive precursor of an enzyme, also called proenzyme.

References

British Committee for Standards in Haematology (BCSH). *Guidelines for the Diagnosis, Investigation and Management of Polycythaemia/Erythrocytosis*. Blackwell Publishing, Oxford, 2005 (diagnostic algorithm updated 2007). Available at www.bcshguidelines.com.

Burkitt D. *A sarcoma involving the jaws in African children. British Journal of Surgery* 1958: 46(197): 218–23.

Delsol G, Lamant L, Mariamé B, Pulford K, Dastugue N, Brousset P, Rigal-Huguet F, al Saati T, Cerretti DP, Morris SW, and Mason DY. *A new subtype of large B-cell lymphoma expressing the ALK kinase and lacking the 2;5 translocation. Blood* 1997: 89(5): 1483–90.

Draper G, Vincent T, Kroll ME, and Swanson J. *Childhood cancer in relation to distance from high voltage power lines in England and Wales: A case-control study. British Medical Journal* 2005: 330(7503): 1290.

Köhler G and Milstein C. *Continuous cultures of fused cells secreting antibody of predefined specificity. Nature* 1975: 256: 495–7.

Nowell P and Hungerford D. *A minute chromosome in chronic granulocytic leukemia. Science* 1960: 132: 1497.

Pratt G, Harding S, Holder R, Fegan C, Pepper C, Oscier D, Gardiner A, Bradwell AR, and Mead G. *Abnormal serum free light chain ratios are associated with poor survival and may reflect biological subgroups in patients with chronic lymphocytic leukaemia. British Journal of Haematology* 2009: 144(2): 217–22.

Index

Note: *b*, *f* and *t* denote box, figure and table

A

B

C

H

N

O

Q

R

S

T

W

Z